Skills in Mathematics for

JEE Main & Advanced

Differential Calculus

With Sessionwise Theory & Exercises

Amit M. Agarwal

arihant

ARIHANT PRAKASHAN (Series), MEERUT

ARIHANT PRAKASHAN (Series), MEERUT

Administrative & Production Offices

Regd. Office

'Ramchhaya' 4577/15, Agarwal Road, Darya Ganj, New Delhi -110002
Tele: 011- 47630600, 43518550

Head Office

Kalindi, TP Nagar, Meerut (UP) - 250002 Tel: 0121-7156203, 7156204

Sales & Support Offices

Agra, Ahmedabad, Bengaluru, Bareilly, Chennai, Delhi, Guwahati, Hyderabad, Jaipur, Jhansi, Kolkata, Lucknow, Nagpur & Pune.

ISBN : 978-93-25298-65-1

PRICE : ₹495.00

PO No : TXT-XX-XXXXXXX-X-XX

Published by Arihant Publications (India) Ltd.

For further information about the books published by Arihant, log on to www.arihantbooks.com or e-mail at info@arihantbooks.com

Follow us on

Skills in Mathematics for

JEE Main & Advanced

Differential Calculus

With Sessionwise Theory & Exercises

PREFACE

"YOU CAN DO ANYTHING IF YOU SET YOUR MIND TO IT, I TEACH CALCULUS TO JEE ASPIRANTS BUT BELIEVE THE MOST IMPORTANT FORMULA IS COURAGE + DREAMS = SUCCESS"

Iit is a matter of great pride and honour for me to have received such an overwhelming response to the previous editions of this book from the readers. In a way, this has inspired me to revise this book thoroughly as per the changed pattern of JEE Main & Advanced. I have tried to make the contents more relevant as per the needs of students, many topics have been re-written, a lot of new problems of new types have been added in etcetc. All possible efforts are made to remove all the printing errors that had crept in previous editions. The book is now in such a shape that the students would feel at ease while going through the problems, which will in turn clear their concepts too.

A Summary of changes that have been made in Revised & Enlarged Edition

- Theory has been completely updated so as to accommodate all the changes made in JEE Syllabus & Pattern in recent years.
- The most important point about this new edition is, now the whole text matter of each chapter has been divided into small sessions with exercise in each session. In this way the reader will be able to go through the whole chapter in a systematic way.
- Just after completion of theory, Solved Examples of all JEE types have been given, providing the students a complete understanding of all the formats of JEE questions & the level of difficulty of questions generally asked in JEE.
- Along with exercises given with each session, a complete cumulative exercises have been given at the end of each chapter so as to give the students complete practice for JEE along with the assessment of knowledge that they have gained with the study of the chapter.
- Last 10 Years questions asked in JEE Main & Adv, IIT-JEE & AIEEE have been covered in all the chapters.

However I have made the best efforts and put my all calculus teaching experience in revising this book. Still I am looking forward to get the valuable suggestions and criticism from my own fraternity i.e. the fraternity of JEE teachers.

I would also like to motivate the students to send their suggestions or the changes that they want to be incorporated in this book. All the suggestions given by you all will be kept in prime focus at the time of next revision of the book.

Amit M. Agarwal

CONTENTS

1. ESSENTIAL MATHEMATICAL TOOLS 1-24

LEARNING PART

Session 1
- Basic Definitions

Session 2
- Intervals
- Modulus or Absolute Value Function

Session 3
- Number Line Rule
- Wavy Curve Method

Session 4
- Quadratic Expression
- Non-Negative Function

PRACTICE PART
- Chapter Exercises

2. DIFFERENTIATION 25-96

LEARNING PART

Session 1
- Geometrical Meaning of the Derivative
- Derivative of $f(x)$ from the First Principle or ab-Initio Method
- Rules of Differentiation

Session 2
- Chain Rule

Session 3
- Differentiation of Implicit Functions

Session 4
- Differentiation of Inverse Trigonometric Functions
- Graphical Approach for Differentiation of Inverse

Session 5
- Differentiation of a function in Parametric Form

Session 6
- Logarithmic Differentiation
- Differentiation of Infinite Series

Session 7
- Differentiation of a Function w.r.t. Another Function

Session 8
- Higher Derivatives of a Function

Session 9
- Differentiation of a function given in the form of a Determinant

Session 10
- Derivation form inverse function

PRACTICE PART
- JEE Type Examples
- Chapter Exercises

3. FUNCTIONS 97-200

LEARNING PART

Session 1
- Functions

Session 2
- Domain
- Algebraic Functions

Session 3
- Transcendental Functions

Session 4
- Piecewise Functions

Session 5
- Range

Session 6
- Odd and Even Functions

Session 7
- Periodic Functions

Session 8
- Mapping Functions

Session 9
- Identical Functions

Session 10
- Composite Functions

Session 11
- Inverse of a Function

Session 12
- Miscellaneous Problems of Functions

PRACTICE PART
- JEE Type Examples
- Chapter Exercises

4. GRAPHICAL TRANSFORMATIONS 201-244

LEARNING PART

Session 1
- When $f(x)$ Vertically transforms to $f(x) \pm a$

Session 2
- When $f(x)$ horizontally transforms to $f(x \pm a)$, where a is any positive constant
- When $f(x)$ transforms to a $f(x)$ or $\frac{1}{a} f(x)$
- When $f(x)$ transforms to $f(ax)$ or $f(x/a)$
- When $f(x)$ Transforms to $f(-x)$
- When $f(x)$ Transforms to $-f(x)$

Session 3
- To draw $y = |f(x)|$, when $y = f(x)$ is given
- To draw $y = f(|x|)$, when $f(x)$ is given
- $f(x)$ Transforms to ïf (ïxï), i.e. $f(x) \to |f(|x|)|$
- To draw the graph of $|y| = f(x)$ when $y = f(x)$ is given

Session 4
- To draw $y = f([x])$ when the graph of $y = f(x)$ is given
- When $f(x)$ transforms to $y = f(\{x\})$ where $\{x\}$ denotes fractional part of x, i.e. $\{x\} = x - [x]$ or $f(x) \to f(\{x\})$
- When $f(x)$ and $g(x)$ are two functions and are transformed to their sum
- When $f(x)$ transforms to $f(x) \to \frac{1}{f(x)} = h(x)$

- To draw the graph for $y=f(x) \times \sin x$ when graph of $y = f(x)$ is given
- When $f(x)$ and $g(x)$ are given, then find $h(x) = \max(f(x), g(x))$ or $h(x) = \min(f(x), g(x))$

PRACTICE PART
- JEE Type Examples
- Chapter Exercises

5. LIMITS 245-310

LEARNING PART

Session 1
- Definition of Limits
- Indeterminant Form
- L'Hospital's Rule
- Evaluation of Limit

Session 2
- Trigonometric Limits

Session 3
- Logarithmic Limits
- Exponential Limits

Session 4
- Miscellaneous Form

Session 5
- Left Hand and Right Hand Limit

Session 6
- Use of Standard Theorems/Results
- Use of Newton-Leibnitz's Formula in Evaluating the Limit

PRACTICE PART
- JEE Type Examples
- Chapter Exercises

6. CONTINUITY AND DIFFERENTIABILITY 311-396

LEARNING PART

Session 1
- Continuous Function
- Continuity of a Function at a Point

Session 2
- Continuity in an Interval or Continuity at End Points

Session 3
- Discontinuity of a Function

Session 4
- Theorems on Continuity
- Continuity of Composite Function

Session 5
- Intermediate Value Theorem

Session 6
- Differentiability : Meaning of Derivative

Session 7
- Differentiability in an Interval

PRACTICE PART
- JEE Type Examples
- Chapter Exercises

7. *dy/dx* AS A RATE MEASURER & TANGENTS, NORMALS 397-458

LEARNING PART

Session 1
- Derivative as the Rate of Change
- Velocity and Acceleration in Rectilinear Motion

Session 2
- Differential and Approximation
- Geometrical Meaning of Dx, Dy, dx and dt

Session 3
- Slope of Tangent and Normal
- Equation of Tangent
- Equation of Normal

Session 4
- Angle of Intersection of Two Curves
- Length of Tangent, Subtangent, Normal and Subnormal

Session 5
- Rolle's Theorem
- Lagrange's Mean Value Theorem

Session 6
- Application of Cubic Functions

PRACTICE PART
- JEE Type Examples
- Chapter Exercises

8. MONOTONICITY, MAXIMA AND MINIMA 459-550

LEARNING PART

Session 1
- Monotonicity

Session 2
- Critical Points

Session 3
- Comparison of Functions Using Calculus

Session 4
- Introduction to Maxima and Minima
- Methods of Finding Extrema of Continuous Functions
- Convexity/Concavity and Point of Inflection
- Concept of Global Maximum/Minimum

Session 5
- Maxima and Minima of Discontinuous Functions

PRACTICE PART
- JEE Type Examples
- Chapter Exercises

SYLLABUS

JEE MAIN

Limit, Continuity and Differentiability

Real valued functions, algebra of functions, polynomials, rational, trigonometric, logarithmic and exponential functions, inverse functions. Graphs of simple functions. Limits, continuity and differentiability.

Differentiation of the sum, difference, product and quotient of two functions. Differentiation of trigonometric, inverse trigonometric, logarithmic exponential, composite and implicit functions derivatives of order upto two. Rolle's and Lagrange's Mean Value Theorems. Applications of derivatives: Rate of change of quantities, monotonic - increasing and decreasing functions, Maxima and minima of functions of one variable, tangents and normals.

JEE ADVANCED

Differential Calculus

Real valued functions of a real variable, into, onto and one-to-one functions, sum, difference, product and quotient of two functions, composite functions, absolute value, polynomial, rational, trigonometric, exponential and logarithmic functions.

Limit and continuity of a function, limit and continuity of the sum, difference, product and quotient of two functions, l'Hospital rule of evaluation of limits of functions.

Even and odd functions, inverse of a function, continuity of composite functions, intermediate value property of continuous functions.

Derivative of a function, derivative of the sum, difference, product and quotient of two functions, chain rule, derivatives of polynomial, rational, trigonometric, inverse trigonometric, exponential and logarithmic functions.

Derivatives of implicit functions, derivatives up to order two, geometrical interpretation of the derivative, tangents and normals, increasing and decreasing functions, maximum and minimum values of a function, applications of Rolle's Theorem and Lagrange's Mean Value Theorem.

CHAPTER

01

Essential Mathematical Tools

Learning Part

Session 1

- Basic Definitions

Session 2

- Intervals
- Modulus or Absolute Value Function

Session 3

- Number Line Rule
- Wavy Curve Method

Session 4

- Quadratic Expression
- Non-negative Functions

Practice Part

- Chapter Exercises

Session 1

Basic Definitions

(i) Natural Numbers The set of numbers $\{1, 2, 3, 4, ...\}$ is called natural numbers and it is **denoted by N**.
i.e. $N = \{1, 2, 3, 4, ...\}$

(ii) Whole Numbers The set of numbers $\{0, 1, 2, 3, 4, ...\}$ is called whole numbers and it is **denoted by W**.
i.e. $W = \{0, 1, 2, 3, 4, ...\}$

(iii) Integers The set of numbers $\{......, -3, -2, -1, 0, 1, 2, 3, ...\}$ is called integers and it is **denoted by I or Z**,
i.e. I (or Z) $= \{....., -4, -3, -2, -1, 0, 1, 2, 3, 4, ...\}$
where we represent;

(a) **Positive integers** by $I^+ = \{1, 2, 3, 4, ...\}$ = natural numbers.

(b) **Negative integers** by $I^- = \{......., -4, -3, -2, -1\}$

(c) **Non-negative integers** $\{0, 1, 2, 3, 4, ...\}$ = whole numbers

(d) **Non-positive integers** $\{......., -3, -2, -1, 0\}$

(iv) Rational Numbers All the numbers of the form $\frac{a}{b}$, where a and b are integers and $b \neq 0$ are called rational numbers and their set is **denoted by Q**.

i.e. $Q = \left\{\frac{a}{b} : a, b \in I \text{ and } b \neq 0 \text{ and HCF of } a, b \text{ is } 1\right\}$.

Remarks

(a) Every integer is a rational number as it can be written as $\frac{a}{b}$ (where $b = 1$)

(b) All recurring decimals are rational numbers.
For example, $\frac{1}{3} = 0.3333.....$

(v) Irrational Numbers Those values which neither terminate nor could be expressed as recurring decimals are called irrational numbers, i.e. they can't be expressed as $\frac{a}{b}$ form and are **denoted by Q^C** (i.e. complement of Q).

e.g. $\sqrt{2}, 1+\sqrt{2}, \frac{1}{\sqrt{2}}, \frac{-3}{\sqrt{2}}, \frac{2}{\sqrt{2}}, \sqrt{3}, 1+\sqrt{3}, \pm\frac{1}{\sqrt{3}}, \pi, \ldots$ etc.

Remark

The set of rational and irrational numbers cannot be expressed in roster form.

(vi) Prime Numbers A counting number is called a prime number when it has exactly two factors, 1 and itself.

e.g. 2, 3, 5, 7, 11, 13, 17,... etc

Remarks

(a) 2 is the only even number which is prime.
(b) A prime is always greater than 1.

How to test a number is prime or not

Let given number be p, then

***Step* I** Find whole number x such that $x > \sqrt{p}$.

***Step* II** Take all prime numbers less than or equal to x.

***Step* III** If none of these divides with p exactly, then p is a prime otherwise p is non-prime.

e.g. Let $p = 193$; clearly, $14 > \sqrt{193}$
Prime numbers upto 14 are 2, 3, 5, 7, 11, 13.
No one of these divides 193 exactly.
Hence, 193 is a prime number.

(vii) Co-prime Numbers Two natural numbers are said to be co-primes, if their HCF is 1.

e.g. (7, 9), (15, 16) are called co-prime numbers.

Remark

Co-prime numbers may or may not be prime.

(viii) Twin Prime A prime number that is either 2 less or 2 more than another prime number, i.e. a twin prime is a prime that has a prime gap of two.

e.g. (3, 5), (5, 7), (11,13), (17, 19), (29, 31),... are called twin prime pairs.

(ix) Composite Numbers Composite numbers are non-prime natural numbers. They must have atleast one factor apart from 1 and itself.

e.g. 4, 6, 8, 9,... are called composite numbers.

Remarks

(a) Composite numbers can be both even or odd.
(b) 1 is neither prime nor composite number.

(x) Real Numbers The set which contains both rational and irrational numbers is called the real numbers and is **denoted by R**.

i.e. $\boldsymbol{R = Q \cup Q^C}$

$R = \{x : x \in Q \text{ or } x \in Q^C\}$

$$\therefore \quad R=\left\{\begin{array}{l} \ldots\ldots, -2,-1,0,1,2,3,\ldots\ldots,\frac{5}{6},\frac{3}{4},\frac{7}{9},\frac{1}{3},\frac{1}{7},\frac{1}{5}, \\ \qquad\qquad\qquad\qquad \ldots\ldots,\sqrt{2},\sqrt{3},\pi,\ldots\ldots\end{array}\right\}$$

Remarks

(i) As from above definitions;

$$N \subset W \subset I \subset Q \subset R,$$

It could be shown that real numbers can be expressed on number line with respect to origin as

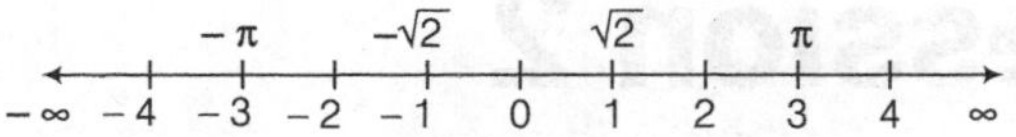

Figure 1.1

(ii) The sum, difference, product and quotient of two rational numbers is a rational number.

(iii) The sum, difference, product and quotient of two irrational numbers need not be irrational.

(iv) The sum, difference, product and quotient of a non-zero rational and an irrational number is always irrational.

Exercise for Session 1

■ **Directions** (Q. Nos. 1 to 6) *State whether following statements are True / False.*

1. All the rational numbers are irrational also.
2. All the integers are irrational also.
3. Irrational numbers are real numbers also.
4. Zero is a natural number.
5. Sum of two natural numbers is a rational number.
6. A positive integer is a natural number also.

■ **Directions** (Q. Nos. 7 to 10) *These questions have only one option correct.*

7. Sum of two rational numbers is
   (a) rational (b) irrational
   (c) Both (a) and (b) (d) None of these
8. Sum of two irrational numbers is
   (a) rational (b) irrational
   (c) real (d) None of these
9. Product of two rational numbers is
   (a) always rational (b) rational or irrational
   (c) always irrational (d) None of these
10. If *a* is an irrational number which is divisible by *b*, then the number *b*
   (a) must be rational (b) must be irrational
   (c) may be rational or irrational (d) None of these

Session 2

Intervals, Modulus or Absolute Value Function

Intervals

The set of numbers between any two real numbers is called intervals. The types of interval discussed below.

(i) **Open-Open Interval** If a and b are real numbers and $a < b$, then the set of all real numbers x such that $a < x < b$ is called open-open interval. It is denoted by (a, b) or $]a, b[$.

i.e. If $a < x < b \Rightarrow x \in]a, b[$ or $x \in (a, b)$

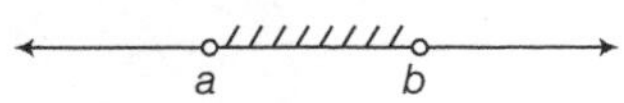

Figure 1.2

e.g. $1 < x < 2$

$\Rightarrow x \in (1, 2)$ or $x \in]1, 2[$ (using above definition)

Note At $\pm\infty$ brackets are always open.

(ii) **Open-Closed Interval** If a and b are real numbers and $a < b$, then the set of all real numbers x such that $a < x \le b$ is called open-closed interval. Here, a is excluded and b is included.

i.e. If $a < x \le b$

$\Rightarrow x \in]a, b]$ or $x \in (a, b]$

Figure 1.3

(iii) **Closed-Open Interval** If a and b are real numbers and $a < b$, then the set of all real 'x' such that $a \le x < b$ is called closed-open interval.

Here, a is included and b is excluded.

i.e. If $a \le x < b$

$\Rightarrow x \in [a, b[$

or $x \in [a, b)$

Figure 1.4

(iv) **Closed-Closed Interval** If a and b are real numbers and $a < b$, then the set of all real 'x' such that $a \le x \le b$ is called closed-closed interval. Here, a and b both are included.

i.e. If $a \le x \le b \Rightarrow x \in [a, b]$

Figure 1.5

Example 1 Solve $2x + 1 > 3$.

Sol. Here, $2x + 1 > 3$ or $2x > 2$ or $x > 1$, i.e. $x \in (1, \infty)$

This solution can be graphed on a real line as;

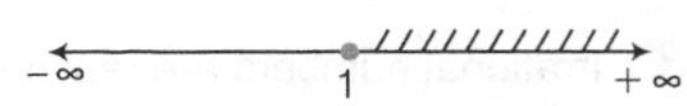

Example 2 Solve $-2 < 2x - 1 < 2$.

Sol. Here, $-2 < 2x - 1 < 2$

or $-1 < 2x < 3$

or $-\frac{1}{2} < x < \frac{3}{2}$, i.e. $x \in \left(-\frac{1}{2}, \frac{3}{2}\right)$

This solution can be graphed on a real line as;

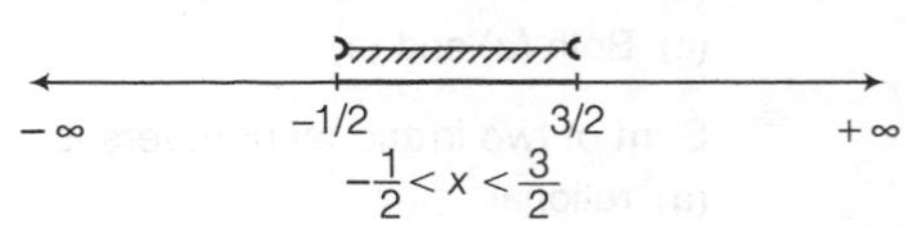

Example 3 Solve the following inequations.

(i) $\frac{3(x-2)}{5} \ge \frac{5(2-x)}{3}$ (ii) $\frac{2x-3}{4} + 9 \ge 3 + \frac{4x}{3}$

Sol. (i) We have, $\frac{3(x-2)}{5} \ge \frac{5(2-x)}{3}$

$\Rightarrow 3(3x - 6) \ge 5(10 - 5x)$

$\Rightarrow 9x - 18 \ge 50 - 25x$

$\Rightarrow 34x \ge 68 \Rightarrow x \ge 2$

Hence, the solution set is $x \in [2, \infty)$ and graphically it could be shown as

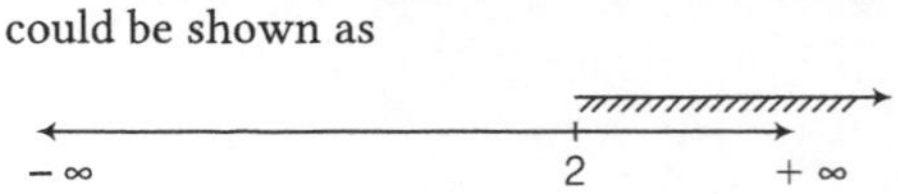

(ii) We have, $\frac{2x-3}{4} + 9 \ge 3 + \frac{4x}{3}$

$\Rightarrow \frac{2x-3}{4} - \frac{4x}{3} \ge 3 - 9$

$$\Rightarrow \quad \frac{3(2x-3)-16x}{12} \geq -6 \Rightarrow \frac{-10x-9}{12} \geq -6$$

$$\Rightarrow \quad -10x-9 \geq -72 \Rightarrow \quad -10x \geq -63$$

As we know the inequality sign changes, if multiplied by (−ve).

$$\therefore \quad 10x \leq 63 \text{ or } x \leq \frac{63}{10}$$

Hence, the solution set of the given inequation is $\left(-\infty, \frac{63}{10}\right]$. Graphically it could be shown as

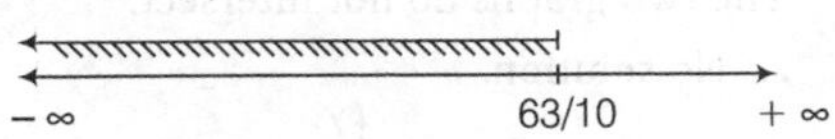

Example 4 Solve for x.

$$\frac{4}{x+1} \leq 3 \leq \frac{6}{x+1}, (x>0)$$

Sol. Consider the first inequality

$$\frac{4}{x+1} \leq 3 \Rightarrow 4 \leq 3(x+1)$$

$$\Rightarrow \quad 4 \leq 3x+3$$

$$\Rightarrow \quad 4-3 \leq 3x \Rightarrow 3x \geq 1$$

$$\Rightarrow \quad x \geq \frac{1}{3} \qquad \text{...(i)}$$

Again, $3 \leq \dfrac{6}{x+1} \Rightarrow 3(x+1) \leq 6$

$$\Rightarrow \quad 3x+3 \leq 6 \quad \Rightarrow 3x \leq 6-3$$

$$\Rightarrow \quad 3x \leq 3$$

$$\Rightarrow \quad x \leq 1 \qquad \text{...(ii)}$$

Combining the results (i) and (ii), we have

$$\frac{1}{3} \leq x \leq 1, \text{ i.e. } x \in \left[\frac{1}{3}, 1\right]$$

Hence, the solution set of the given in equations is $\left[\frac{1}{3}, 1\right]$.

Graphically, it could be shown as

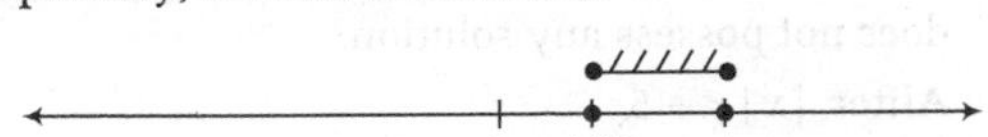

Modulus or Absolute Value Function

The function defined by $f(x) = |x| = \begin{cases} x, x \geq 0 \\ -x, x < 0 \end{cases}$ is called modulus function. This always gives a positive result.

As we know modulus means numerical value,

i.e. $\quad |3| = |-3| = 3,$

$$|-2| = 2, |-1.3291| = 1.3291 \text{..... etc.}$$

or it is the distance defined with respect to origin, as $|x| = 1$ means distance covered is one unit on right hand side or left hand side of origin shown as:

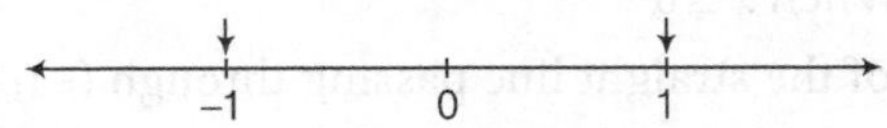

Figure 1.6

$$\therefore \quad |x| = 1 \Rightarrow x = \pm 1$$

Again, $|x| < 1$ means distance covered is less than one unit on right hand side or left hand side of origin shown as:

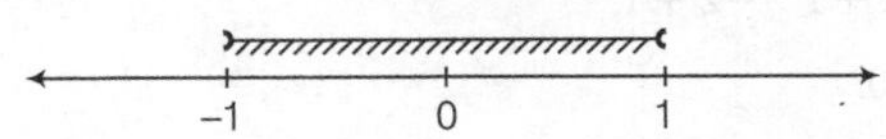

Figure 1.7

Similarly, $|x| > 1$ means distance covered is more than one unit on right hand side or left hand side of origin shown as:

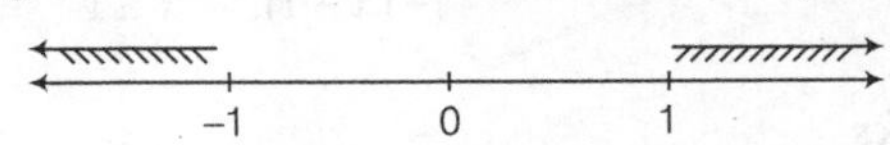

Figure 1.8

Graphical Representation of Modulus

As we define,

$$y = |x| = \begin{cases} x, & x \geq 0 \\ -x, & x < 0 \end{cases}$$

This behaviour is due to two straight lines represented by modulus.

For plotting the graph of the modulus function, we put

x	0	1	2	−1	−2
y	0	1	2	1	2

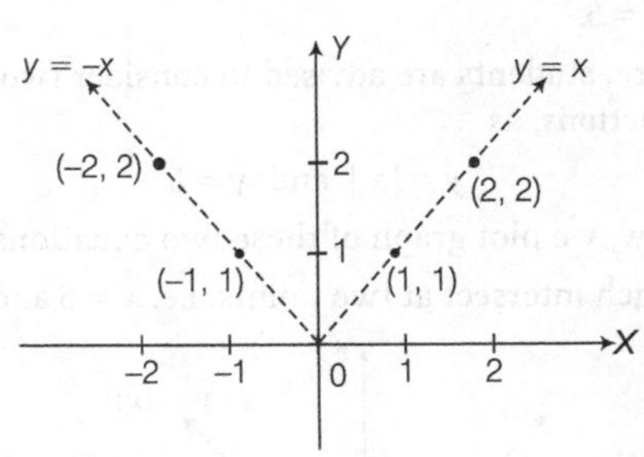

Figure 1.9

Now, we consider the following cases

***Case* I** When $x \geq 0$

Equation of the straight line, passing through (0, 0) and (1, 1) is

$$\frac{y-y_1}{x-x_1}=\frac{y_2-y_1}{x_2-x_1}, \text{ i.e. } \frac{y-0}{x-0}=\frac{1-0}{1-0}$$

$\Rightarrow$ $y = x$, when $x \geq 0$

***Case* II** When $x \leq 0$

Equation of the straight line passing through $(-1, 1)$ and $(-2, 2)$

$$\frac{y-y_1}{x-x_1}=\frac{y_2-y_1}{x_2-x_1}$$

i.e. $\dfrac{y-1}{x+1}=\dfrac{2-1}{-2+1} \Rightarrow -y+1=x+1$ or $y=-x$

when $x \leq 0$.

Thus for every modulus function it exhibits two values, which could be shown graphically.

Similarly, $y=|x-1|=\begin{cases}(x-1), & x\geq 1\\ -(x-1), & x\leq 1\end{cases}$

Remarks

(a) Every modulus function exhibits two values which are represented by +ve and –ve, signs but it only gives the positive outcomes. So, students shouldn't get confused by + ve or –ve, signs as these signs are in different intervals, but the outcomes are positive.

(b) Modulus function is never negative, thus $|x| \geq 0$ for any real x and $|x| \nless 0$.

Example 5 Explain the following :

(i) $|x|=5$ (ii) $|x|=-5$

(iii) $|x|<5$ (iv) $|x|<-5$

(v) $|x|>-5$ (vi) $|x|>5$

Sol. (i) If $|x|=5$

$\Rightarrow x=\pm 5$, which means, x is at a distance of 5 units from 0, which is certainly 5 and – 5.

Aliter

$|x|=5$.

Here, students are advised to consider two different functions, as

$$y=|x| \text{ and } y=5.$$

Now, we plot graph of these two equations.

which intersect at two points, i.e. $x=5$ and $x=-5$.

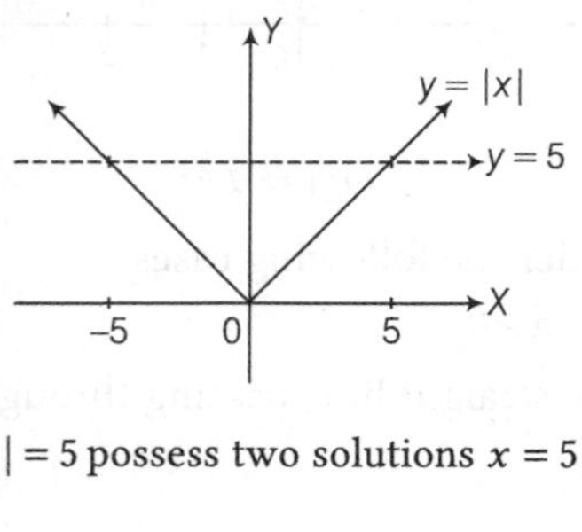

$\therefore$ $|x|=5$ possess two solutions $x=5$ and $x=-5$.

(ii) If $|x|=-5$

$\Rightarrow$ x has no solution.

As $|x|$ is always positive or zero, it can never be negative.

$\therefore$ RHS < LHS

or given relation has no solution.

Aliter

Same as in (i), plotting the graph of $y=|x|$ and $y=-5$.

The two graphs do not intersect.

$\therefore$ No solution.

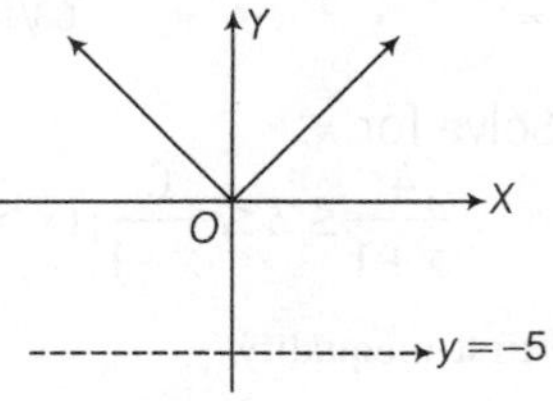

(iii) $|x|<5$

It means that x is the number, which is at distance less than 5 from 0.

Hence, $-5<x<5$

Aliter $|x|<5$. Plotting the graph of we have $y=|x|$ and $y=5$.

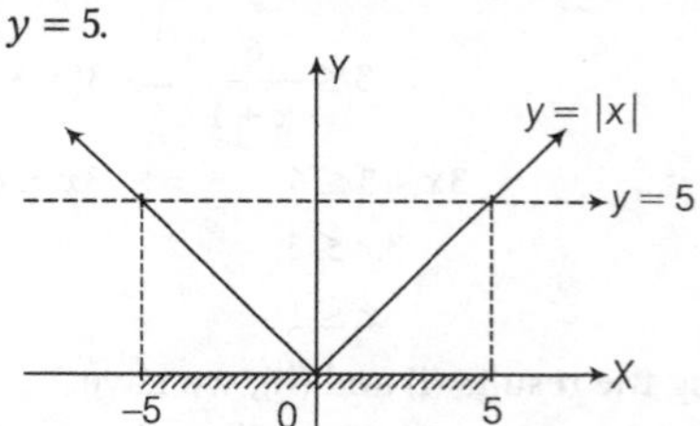

We see that, $|x|<5$, when $-5<x<5$

(iv) $|x|<-5$

which shows no solution.

As LHS is non-negative and RHS is negative or $|x|<-5$ does not possess any solution.

Aliter $|x|<-5$.

Plotting the graph of $y=|x|$ and $y=-5$.

We see that, $|x|<-5$ is not possible as $|x|>-5$, for all $x \in R$.

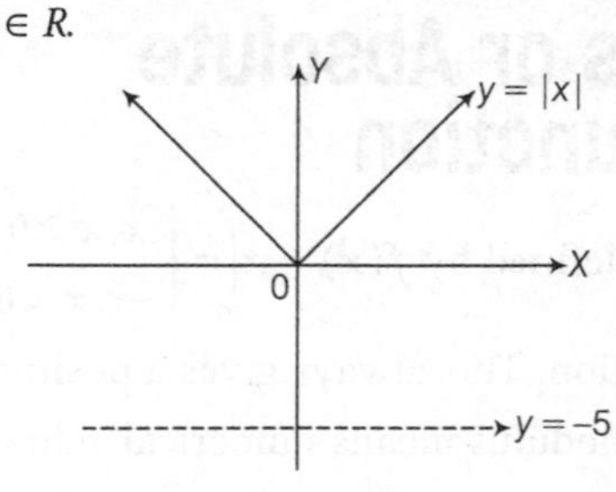

$\therefore$ $|x|<-5$ $\Rightarrow$ No solution.

(v) $|x| > -5$ We know, here LHS ≥ 0 and RHS < 0

So, LHS > RHS

i.e. above statement is true for all real x.

(as we know that non-negative number is always greater than negative).

Aliter

Here, $|x| > -5$

Plotting the graph of $y = |x|$ and $y = -5$.

We see that, $|x| > -5$, for all $x \in$ real number.

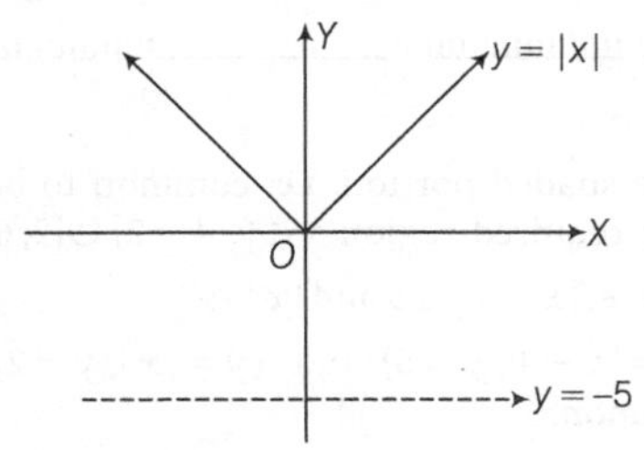

(vi) $|x| > 5$

It means that x is the number which is at distance greater than 5 from 0. Hence, $x < -5$ or $x > 5$

Aliter

$|x| > 5$

Plotting the graph of $y = |x|$ and $y = 5$. We have

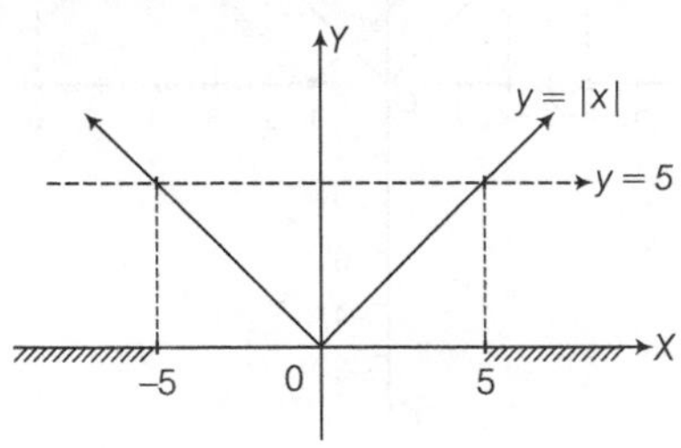

We see that, $x < -5$ or $x > 5$.

Generalized Results

(i) For any real number x, we have $x^2 = |x|^2$

(ii) For any real number x, we have $\sqrt{x^2} = |x|$

(iii) If $a < 0$, then

(a) $x^2 \leq a^2 \Leftrightarrow |x| \leq a \Leftrightarrow -a \leq x \leq a$

(b) $x^2 < a^2 \Leftrightarrow |x| < a \Leftrightarrow -a < x < a$

(c) $x^2 \geq a^2 \Leftrightarrow |x| \geq a \Leftrightarrow x \leq -a$ or $x \geq a$

(d) $x^2 > a^2 \Leftrightarrow |x| > a \Leftrightarrow x < -a$ or $x > a$

(e) $a^2 \leq x^2 \leq b^2 \Leftrightarrow a \leq |x| \leq b$
$\Leftrightarrow x \in [-b, -a] \cup [a, b]$

(f) $a^2 < x^2 < b^2 \Leftrightarrow a < |x| < b$
$\Leftrightarrow x \in (-b, -a) \cup (a, b)$

(iv) If $a < 0$, then

$|x| \leq a \Rightarrow$ No solution.

$|x| \geq a \Rightarrow$ All real numbers solutions.

(v) $|x + y| = |x| + |y| \Leftrightarrow (x \geq 0$ and $y \geq 0)$ or $(x \leq 0$ and $y \leq 0) \Leftrightarrow xy \geq 0$.

(vi) $|x - y| = |x| - |y| \Leftrightarrow (x \geq 0, y \geq 0$ and $|x| \geq |y|)$ or $(x \leq 0, y \leq 0$ and $|x| \geq |y|)$

(vii) $|x \pm y| \leq |x| + |y|$ (viii) $|x \pm y| \geq ||x| - |y||$

Example 6 Solve for x, where

(i) $f(x) = |x| \geq 0$ (ii) $f(x) = |x| > 0$

Sol. (i) $f(x) = |x| \geq 0$

As we know modulus is non-negative quantity.

(i.e. It is always greater than equal to zero)

$\therefore$ $x \in R$ is the required solution.

Aliter $f(x) = |x| \geq 0$.

Plotting two graphs $y = |x|$ and $y = 0$.

From graph $|x| \geq 0$, for all $x \in R$.

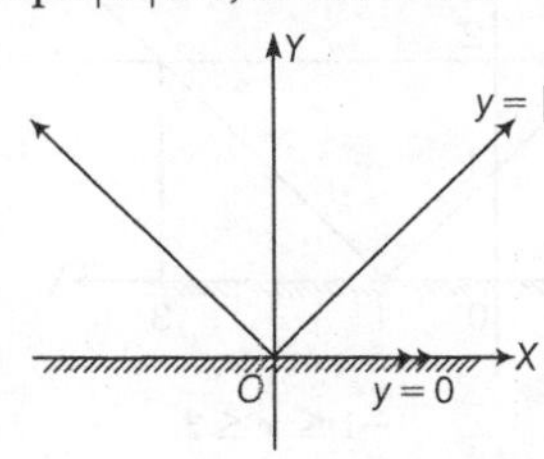

(ii) $f(x) = |x| > 0$

Here equal sign is absent, so we have to exclude those value of x for which $|x| = 0$.

$\therefore$ $x \in R$ except $x = 0$ or $x \in R - \{0\}$ is the required solution.

Aliter $f(x) = |x| > 0$.

Plotting two graphs $y = |x|$ and $y = 0$.

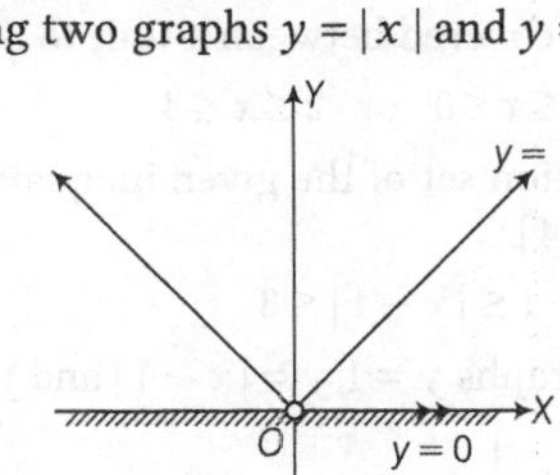

From graph $|x| > 0$, for all $x \in R$ except 0.

$\therefore$ $x \in R - \{0\}$.

Example 7 Solve $|x - 3| < 5$.

Sol. $|x - 3| < 5$

$\Rightarrow$ $-5 < x - 3 < 5$

$\Rightarrow$ $-5 + 3 < x < 5 + 3$

$\Rightarrow$ $-2 < x < 8$ or $x \in]-2, 8[$ or $x \in (-2, 8)$

Aliter For $|x-3|<5$.

Plotting two graphs $y=|x-3|$ and $y=5$

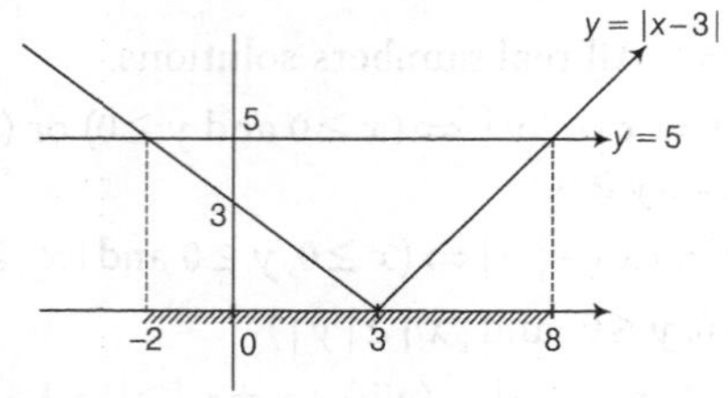

From graph, $-2<x<8$; $\therefore$ $x\in(-2,8)$.

Example 8 Solve $|x-1|\le 2$.

Sol. $|x-1|\le 2$

$\Rightarrow$ $-2\le x-1\le 2$ $\Rightarrow$ $-2+1\le x\le 2+1$

$\Rightarrow$ $-1\le x\le 3$ or $x\in[-1,3]$

Aliter For $|x-1|\le 2$.

Plotting two graphs $y=|x-1|$ and $y=2$

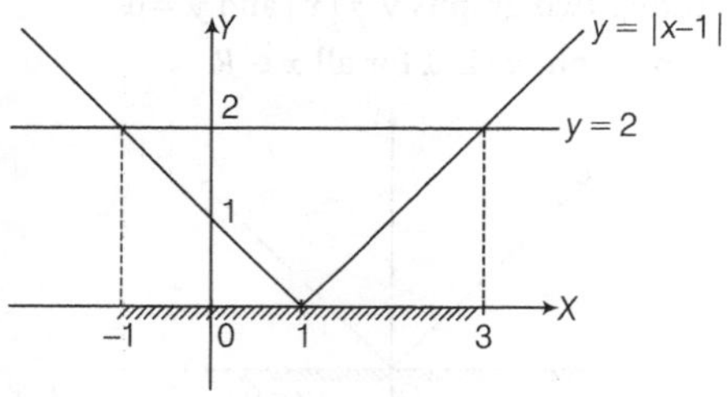

From graph, $-1\le x\le 3$

$\therefore$ $x\in[-1,3]$

Example 9 Solve $1\le|x-1|\le 3$.

Sol. Here, $1\le|x-1|\le 3$

$\Rightarrow$ $-3\le(x-1)\le -1$ or $1\le(x-1)\le 3$

$[\because a\le|x|\le b\Rightarrow x\in[-b,-a]\cup[b,a]]$

i.e. the distance covered between 1 unit to 3 units.

$\Rightarrow$ $-2\le x\le 0$ or $2\le x\le 4$

Hence, the solution set of the given inequation is $x\in[-2,0]\cup[2,4]$.

Aliter Here, $1\le|x-1|\le 3$

Plotting three graphs $y=1$, $y=|x-1|$ and $y=3$.

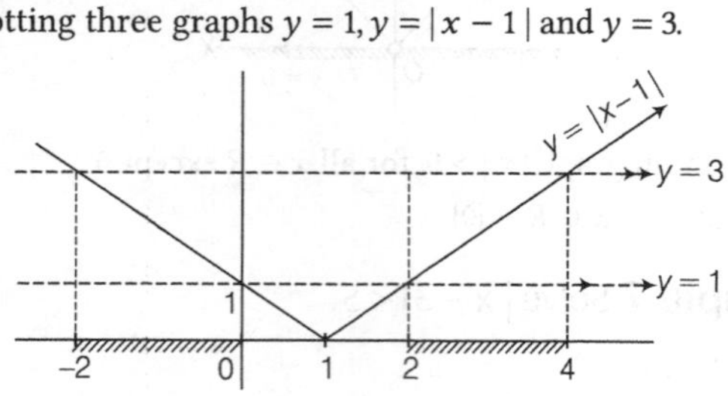

From graph, $-2\le x\le 0$ or $2\le x\le 4$.

$\therefore$ $x\in[-2,0]\cup[2,4]$

Example 10 Solve $|x-1|\le 5, |x|\ge 2$.

Sol. Here, $|x-1|\le 5$ and $|x|\ge 2$

i.e. $(-5\le x-1\le 5)$

$\Rightarrow$ $(-4\le x\le 6)$ ...(i)

Similarly, $(x\le -2$ or $x\ge 2)$

$\Rightarrow$ $(x\le -2$ or $x\ge 2)$ ...(ii)

From Eqs. (i) and (ii) could be graphically shown as

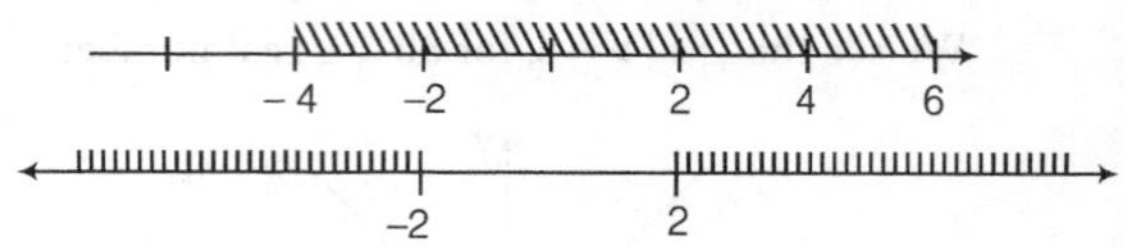

Thus, the shaded portion, i.e. common to both Eqs. (i) and (ii) is the required region $x\in[-4,-2]\cup[2,6]$.

Aliter Here, $|x-1|\le 5$ and $|x|\ge 2$.

Take $(y=|x-1|, y=5)$ and $(y=|x|, y=2)$ on graph and take common.

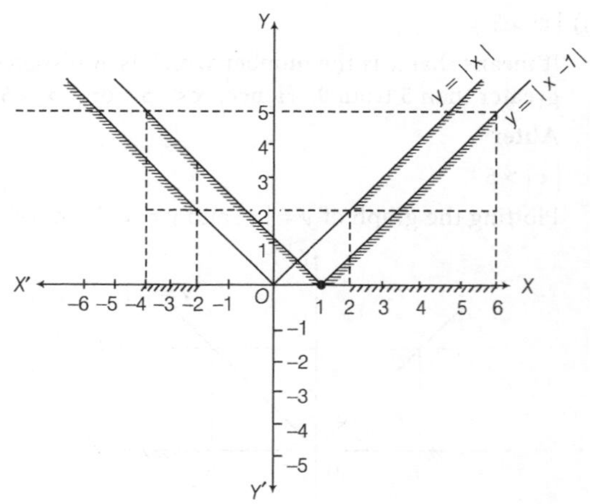

From the graph, $-4\le x\le -2$

or $2\le x\le 6$

$\therefore$ $x\in[-4,-2]\cup[2,6]$

Example 11 Solve $\left|\dfrac{2}{x-4}\right|>1, x\ne 4$.

Sol. We have, $\left|\dfrac{2}{x-4}\right|>1$, where $x\ne 4$ ...(i)

$\Rightarrow$ $\dfrac{2}{|x-4|}>1$ $\left[\because \left|\dfrac{a}{b}\right|=\dfrac{|a|}{|b|} \text{ and } |2|=2\right]$

$\Rightarrow$ $2>|x-4|$

$\Rightarrow$ $|x-4|<2$

$\Rightarrow$ $-2<x-4<2$

$\Rightarrow$ $2<x<6$

$\therefore$ $x\in(2,6)$, but $x\ne 4$ [from Eq. (i)]

Hence, the solution set of the given inequation is $x\in(2,4)\cup(4,6)$.

Aliter $\left|\dfrac{2}{x-4}\right| > 1, x \neq 4 \Rightarrow |x-4| < 2$

Plotting the graph of $y = |x-4|$ and $y = 2$

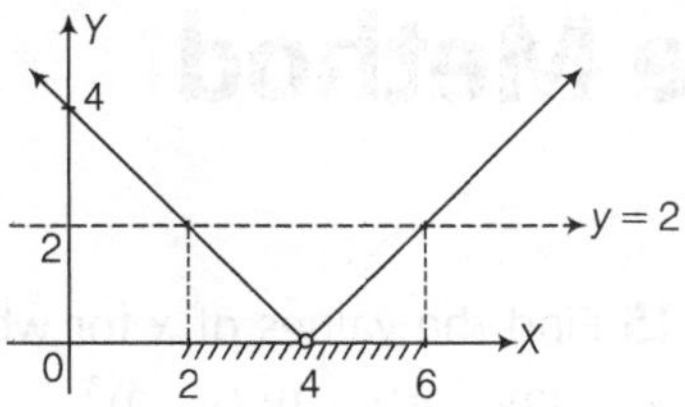

From graph, $2 < x < 6, x \neq 4$

i.e. $x \in (2, 6) - \{4\}$

Note

Students should always remember that they have to compare the solution set with the initial condition.

Example 12 Solve $|x-1|+|x-2| \geq 4$.

Sol. On the LHS of the given inequation, we have two modulus, so we should define each modulus i.e. by equating it to zero.

i.e. $|x-1| = \begin{cases} (x-1), & x \geq 1 \\ -(x-1), & x<1 \end{cases}$

and $|x-2| = \begin{cases} (x-2), & x \geq 2 \\ -(x-2), & x<2 \end{cases}$

Thus, it gives three cases :

***Case* I** When $-\infty < x < 1$

i.e. $|x-1|+|x-2| \geq 4$

$\Rightarrow -(x-1)-(x-2) \geq 4$

$\Rightarrow -2x+3 \geq 4 \Rightarrow -2x \geq 1$

$\Rightarrow x \leq -\dfrac{1}{2}$ ...(i)

But $-\infty < x < 1$

$\therefore$ Solution set is $x \in \left(-\infty, -\dfrac{1}{2}\right]$.

***Case* II** When $1 \leq x \leq 2$

i.e. $|x-1|+|x-2| \geq 4$

$\Rightarrow (x-1)-(x-2) \geq 4$

$\Rightarrow 1 \geq 4$, which is meaningless.

$\therefore$ No solution for $x \in [1, 2]$. ...(ii)

***Case* III** When $2 < x < \infty$

i.e. $|x-1|+|x-2| \geq 4$

$\Rightarrow (x-1)+(x-2) \geq 4$

$\Rightarrow 2x-3 \geq 4$

$\Rightarrow x \geq 7/2$

But $2 < x < \infty$

$\therefore$ Solution set is $\left[\dfrac{7}{2}, \infty\right)$. ...(iii)

From Eqs. (i), (ii) and (iii), we get

$$x \in \left(-\infty, -\frac{1}{2}\right] \cup \left[\frac{7}{2}, \infty\right)$$

Aliter

Plotting two graphs $y = |x-1|+|x-2|$ and $y = 4$ shown below.

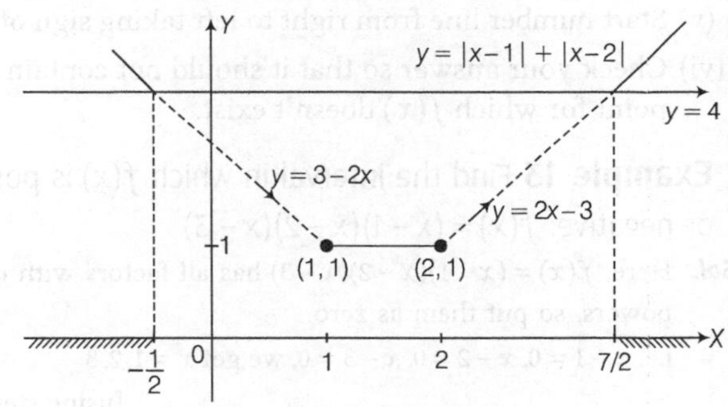

From graph, $x \leq -\dfrac{1}{2}$ or $x \geq \dfrac{7}{2}$

$\therefore$ $x \in \left(-\infty, -\dfrac{1}{2}\right] \cup \left[\dfrac{7}{2}, \infty\right)$

Exercise for Session 2

- **Directions** (Q. Nos. 1 to 5) *Solve the following inequalities for real values of x.*

1. $|x-1|<2$
2. $|x-3|>5$
3. $0<|x-1|<3$
4. $|x-1|+|2x-3|=|3x-4|$
5. $\left|\dfrac{x-3}{x^2-4}\right| \leq 1$

Session 3

Number Line Rule, Wavy Curve Method

Number Line Rule

It is used to solve algebraic inequalities using following steps:

(i) Factorize numerator as well as denominator.

(ii) Now, check the coefficients of x and make them positive.

(iii) Put only odd power factors in numerator and denominator and put them equal to zero separately and find the value of x.

(*As for polynomial function only numerator* $= 0$, *denominator* $\neq 0$).

(iv) Plot these points on number line in increasing order.

(v) Start number line from right to left taking sign of $f(x)$.

(vi) Check your answer so that it should not contain a point for which $f(x)$ doesn't exist.

Example 13 Find the interval in which $f(x)$ is positive or negative: $f(x) = (x-1)(x-2)(x-3)$.

Sol. Here, $f(x) = (x-1)(x-2)(x-3)$ has all factors with odd powers, so put them as zero.

i.e. $x - 1 = 0,\ x - 2 = 0,\ x - 3 = 0$, we get $x = 1, 2, 3$

[using step (iii)]

Using steps (iv) and (v), plotting on number line, we get

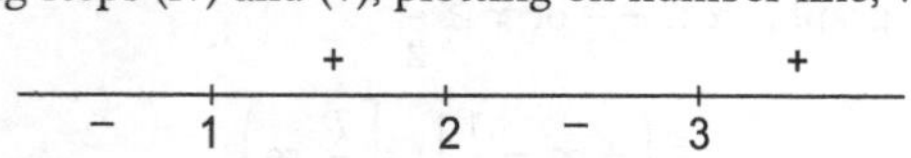

$f(x) > 0$ when $1 < x < 2$ and $x > 3$

$f(x) < 0$ when $x < 1$ and $2 < x < 3$

Example 14 Solve $f(x) = \dfrac{(x-1)(2-x)}{(x-3)} \geq 0$.

Sol. Here, $f(x) = \dfrac{(x-1)(2-x)}{(x-3)} \geq 0$

or $f(x) = -\dfrac{(x-1)(x-2)}{(x-3)}$, which gives

$x - 3 \neq 0$ or $x \neq 3$ ...(i)

Using number line rule as shown in the figure,

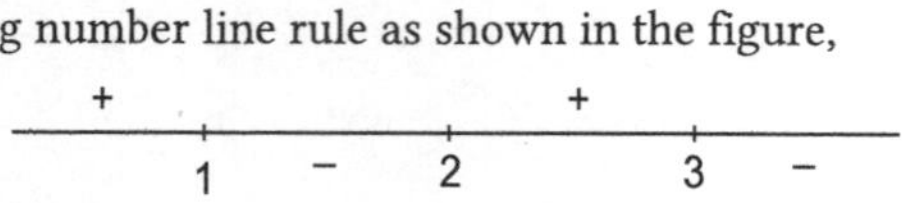

which shows $f(x) \geq 0$ when $x \leq 1$ or $2 \leq x < 3$

i.e. $x \in (-\infty, 1] \cup [2, 3)$. (as $x \neq 3$)

Example 15 Find the values of x for which

$$f(x) = \frac{(2x-1)(x-1)^2(x-2)^3}{(x-4)^4} > 0.$$

Sol. $f(x) = \dfrac{(2x-1)(x-1)^2(x-2)^3}{(x-4)^4}$, which gives $x \neq 4$ ...(i)

As denominator $\neq 0$ and $x \neq 1$, as at $x = 1$, $f(x)$ has even powers.

Putting zero to $(2x-1)$ and $(x-2)^3$ as they have odd powers and neglecting $(x-1)^2$ and $(x-4)^4$ on number line as shown in figure.

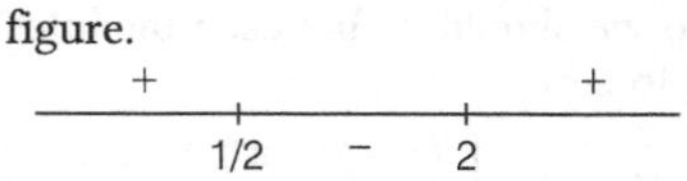

which shows $f(x) > 0$ when $x < 1/2$ or $x > 2$ but except for 4 and 1.

$\therefore$ $x \in (-\infty, 1/2) \cup (2, \infty) - \{4\}$

Example 16 Find the value of x for which

$$f(x) = \frac{(x-2)^2(1-x)(x-3)^3(x-4)^2}{(x+1)} \leq 0.$$

Sol. Here, $f(x) = \dfrac{(x-2)^2(1-x)(x-3)^3(x-4)^2}{(x+1)}$

or $f(x) = -\dfrac{(x-2)^2(x-1)(x-3)^3(x-4)^2}{(x+1)}, (x \neq -1)$ ...(i)

Putting zero to $(x-1), (x-3)^3, (x+1)$ as having odd powers and neglecting $(x-2)^2, (x-4)^2$, we get

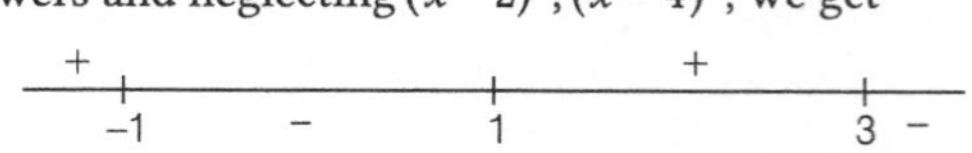

$f(x) \leq 0$ when $-1 < x \leq 1$ or $3 \leq x < \infty$ or $f(x) = 0$ at $x = 2$

or $x \in (-1, 1] \cup [3, \infty) \cup \{2\}$ [using Eq. (i) as $x \neq -1$]

Example 17 Solve $\dfrac{|x| - 1}{|x| - 2} \geq 0;\ x \in R,\ x \neq \pm 2$

Sol. We have, $\dfrac{|x|-1}{|x|-2} \geq 0 \Rightarrow \dfrac{y-1}{y-2} \geq 0$; where $y = |x|$

$\Rightarrow$ $y \leq 1$ or $y > 2$ using number line rule

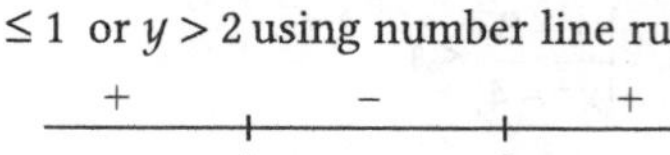

$\Rightarrow$ $|x| \leq 1$ or $|x| > 2$

$\Rightarrow$ $(-1 \leq x \leq 1)$ or $(x < -2$ or $x > 2)$

$\Rightarrow \quad x \in [-1, 1] \cup (-\infty, -2) \cup (2, \infty)$

Hence, the solution set is

$$x \in (-\infty, -2) \cup [-1, 1] \cup (2, \infty).$$

Example 18 Solve $\dfrac{-1}{|x|-2} \geq 1$, where $x \in R$, $x \neq \pm 2$.

Sol. We have, $\dfrac{-1}{|x|-2} \geq 1$

$$\Rightarrow \quad \frac{-1}{|x|-2} - 1 \geq 0 \Rightarrow \frac{-1-(|x|-2)}{|x|-2} \geq 0$$

$$\Rightarrow \quad \frac{1-|x|}{|x|-2} \geq 0 \Rightarrow \frac{-(|x|-1)}{(|x|-2)} \geq 0$$

Using number line rule,

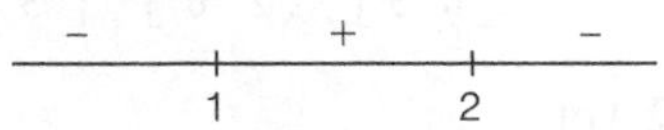

$$\Rightarrow \quad 1 \leq |x| < 2 \Rightarrow x \in (-2, -1] \cup [1, 2)$$

$$\{\because a \leq |x| < b \Leftrightarrow x \in (-b, -a] \cup [a, b)\}$$

Hence, the solution set is $(-2, -1] \cup [1, 2)$.

Example 19 Solve $\dfrac{|x+3|+x}{x+2} > 1$.

Sol. Here, $\dfrac{|x+3|+x}{x+2} - 1 > 0 \Rightarrow \dfrac{|x+3|+x-x-2}{x+2} > 0$

$$\Rightarrow \quad \frac{|x+3|-2}{x+2} > 0 \quad \text{...(i)}$$

Now, two cases arise :

***Case* I** When $x + 3 \geq 0$, i.e. $x \geq -3$ ...(ii)

$$\Rightarrow \quad \frac{x+3-2}{x+2} > 0 \Rightarrow \frac{x+1}{x+2} > 0$$

$\Rightarrow x \in (-\infty, -2) \cup (-1, \infty)$ using number line rule as shown in the figure.

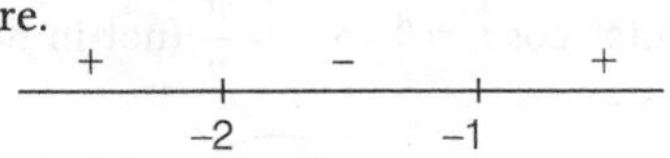

But $\quad x \geq -3 \quad$ [from Eq. (ii)]

$\Rightarrow \quad x \in [-3, -2) \cup (-1, \infty) \quad$ ...(a)

***Case* II** When $x + 3 < 0$, i.e. $x < -3$ ...(iii)

$$\Rightarrow \quad \frac{-(x+3)-2}{x+2} > 0 \Rightarrow \frac{-(x+5)}{(x+2)} > 0$$

$\Rightarrow x \in (-5, -2)$ using number line rule as shown in the figure.

But $\quad x < -3 \quad$ [from Eq. (iii)]

$\therefore \quad x \in (-5, -3) \quad$ ...(b)

Thus, from Eqs. (a) and (b), we have

$$x \in [-3, -2) \cup (-1, \infty) \cup (-5, -3)$$

$$\Rightarrow \quad x \in (-5, -2) \cup (-1, \infty)$$

Wavy Curve Method

The method of intervals (or wavy curve) is used for solving inequalities of the form :

$$f(x) = \frac{(x-a_1)^{n_1}(x-a_2)^{n_2} \ldots (x-a_k)^{n_k}}{(x-b_1)^{m_1}(x-b_2)^{m_2} \ldots (x-b_p)^{m_p}} > 0$$

$$(< 0, \leq 0 \text{ or } \geq 0)$$

where $n_1, n_2, \ldots\ldots, n_k$ and $m_1, m_2, \ldots\ldots, m_p$ are natural numbers.

$a_1, a_2, \ldots\ldots, a_k; b_1, b_2, \ldots\ldots, b_p$ are any real numbers such that $a_i \neq b_j$, where $i = 1, 2, 3, \ldots, k$ and $j = 1, 2, 3, \ldots, p$

It consists of the following statements :

1. All zeros of the function $f(x)$ (the values of x corresponding to numerator = 0) contained on the left hand side of the inequality should be marked on the number line with inked (black) circles.
2. All points of discontinuities of the function $f(x)$ (the values of x corresponding to denominator = 0) contained on the left hand side of the inequality should be marked on the number line with uninked (white) circles.
3. Check the value of $f(x)$ for any real number greater than the rightmost marked number on the number line.
4. From right to left, beginning above the number line (in case of value of $f(x)$ is positive in step (iii) otherwise from below the number line), a wavy curve should be drawn to pass through all the marked points so that when it passes through a simple point, the curve intersects the number line and when passing through a double point, the curve remains located on one side of the number line.
5. The appropriate intervals are chosen in accordance with the sign of inequality (the function $f(x)$ is positive whenever the curve is situated above the number line, it is negative if the curve is found below the number line). Their union just represents the solution of the inequality.

Remarks

(a) Points of discontinuity will never be include in the answers.

(b) If asked to find the intervals where $f(x)$ is non-negative or non-positive, then make the intervals closed, corresponding to the roots of the numerator and let it remain open corresponding to the roots of denominator.

Example 20 Let $f(x) = \dfrac{(x-1)^3(x+2)^4(x-3)^5(x+6)}{x^2(x-7)^3}$.

Solve the following inequalities :

(i) $f(x) > 0$ (ii) $f(x) \geq 0$

(iii) $f(x) < 0$ (iv) $f(x) \leq 0$

Sol. Given, $f(x) = \dfrac{(x-1)^3(x+2)^4(x-3)^5(x+6)}{x^2(x-7)^3}$

Put numerator = 0

$\Rightarrow \quad (x-1)^3(x+2)^4(x-3)^5(x+6) = 0$

$\Rightarrow \quad x = 1, -2, 3, -6$

Again, put denominator = 0

$\Rightarrow \quad x^2(x-7)^3 = 0$

$\Rightarrow \quad x = 0, 7$

We mark on the number line zeros of the function: 1, −2, 3 and −6 (with black circles) and the points of discontinuities 0 and 7 (with white circles). Isolate the double points: −2 and 0 and draw the curve of signs.

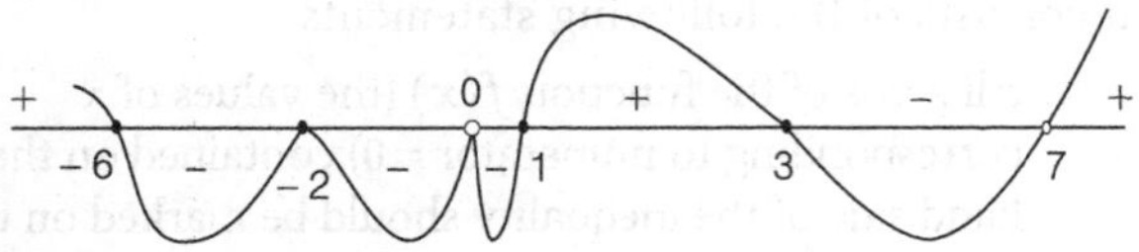

From the graph, we get

(i) if $f(x) > 0$, then $x \in (-\infty, -6) \cup (1, 3) \cup (7, \infty)$

(ii) if $f(x) \geq 0$, then $x \in (-\infty, -6] \cup \{-2\} \cup [1, 3] \cup (7, \infty)$

(iii) if $f(x) < 0$, then $x \in (-6, -2) \cup (-2, 0) \cup (0, 1) \cup (3, 7)$

(iv) if $f(x) \leq 0$, then $x \in [-6, 0) \cup (0, 1] \cup [3, 7)$

Example 21 Let

$$f(x) = \frac{\left(\sin x - \frac{1}{2}\right)(\ln x - 1)^2 (x-2)(\tan x - \sqrt{3})}{(e^x - e^2)(x-3)^2 \cdot \cos x}$$

Solve the following inequalities for $x \in [0, 2\pi]$:

(i) $f(x) > 0$ (ii) $f(x) \geq 0$

(iii) $f(x) < 0$ (iv) $f(x) \leq 0$

Sol. Clearly, $x \neq 2, 3, \dfrac{\pi}{2}, \dfrac{3\pi}{2}$ and $f(x) = 0$

for $x = \dfrac{\pi}{6}, \dfrac{\pi}{3}, \dfrac{5\pi}{6}, e, \dfrac{4\pi}{3}$

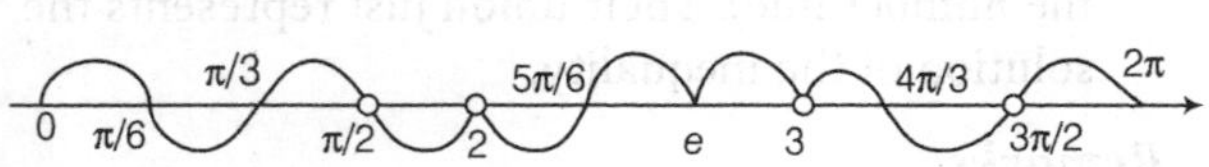

Now, sign of $f(x)$ will not change around $x = 2, e, 3$.

Then, for $f(x) > 0$

$\left(\sin x - \dfrac{1}{2}\right)(\tan x - \sqrt{3}) > 0$

$\Rightarrow \quad x \in \left(0, \dfrac{\pi}{6}\right) \cup \left(\dfrac{\pi}{3}, \dfrac{\pi}{2}\right) \cup \left(\dfrac{5\pi}{6}, \dfrac{4\pi}{3}\right) \cup \left(\dfrac{3\pi}{2}, 2\pi\right) - \{e, 3\}$

Hence, solution of

(i) $f(x) > 0$ is $x \in \left(0, \dfrac{\pi}{6}\right) \cup \left(\dfrac{\pi}{3}, \dfrac{\pi}{2}\right) \cup \left(\dfrac{5\pi}{6}, \dfrac{4\pi}{3}\right) \cup \left(\dfrac{3\pi}{2}, 2\pi\right) - \{e, 3\}$

(ii) $f(x) \geq 0$ is $x \in \left(0, \dfrac{\pi}{6}\right] \cup \left[\dfrac{\pi}{3}, \dfrac{\pi}{2}\right) \cup \left[\dfrac{5\pi}{6}, \dfrac{4\pi}{3}\right] \cup \left(\dfrac{3\pi}{2}, 2\pi\right) - \{3\}$

(iii) $f(x) < 0$ is $x \in \left(\dfrac{\pi}{6}, \dfrac{\pi}{3}\right) \cup \left(\dfrac{\pi}{2}, \dfrac{5\pi}{6}\right) \cup \left(\dfrac{4\pi}{3}, \dfrac{3\pi}{2}\right) - \{2\}$

(iv) $f(x) \leq 0$ is $x \in \left[\dfrac{\pi}{6}, \dfrac{\pi}{3}\right] \cup \left(\dfrac{\pi}{2}, \dfrac{5\pi}{6}\right] \cup \left[\dfrac{4\pi}{3}, \dfrac{3\pi}{2}\right) \cup \{e\} - \{2\}$

Example 22 Let

$$f(x) = \frac{(\cos x + |\cos x|)\left(\sin x - \frac{3}{2}\right)^3 (\tan x - 1)^5}{(\cos x - 2)^2 (\tan x - \sqrt{3})^3}$$

Find the interval of $x \in \left(\dfrac{-\pi}{2}, \dfrac{\pi}{2}\right)$ for which

(i) $f(x) > 0$ (ii) $f(x) < 0$

Sol. For $x \in \left(\dfrac{-\pi}{2}, \dfrac{\pi}{2}\right)$

$\cos x > 0$ for all $x \Rightarrow |\cos x| = \cos x$

$\Rightarrow \quad \cos x + |\cos x| = 2\cos x$

$\therefore \quad f(x) = \dfrac{2\cos x\left(\sin x - \frac{3}{2}\right)^3 (\tan x - 1)^5}{(\cos x - 2)^2 (\tan x - \sqrt{3})^3}$

Critical points $\cos x = 0 \Rightarrow x = \dfrac{\pi}{2}$ (not in domain)

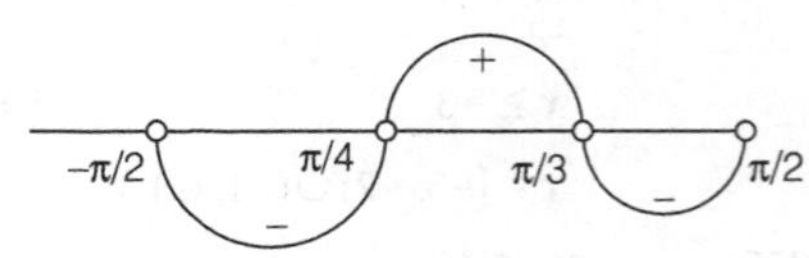

$\sin x = \dfrac{3}{2}$ (not possible)

$\tan x = 1 \Rightarrow x = \dfrac{\pi}{4} \Rightarrow \cos x = 2$ (not possible)

$\tan x = \sqrt{3} \Rightarrow x = \dfrac{\pi}{3}$.

Then, $\quad f(x) > 0 \Rightarrow x \in \left(\dfrac{\pi}{4}, \dfrac{\pi}{3}\right)$

and $\quad f(x) < 0 \Rightarrow x \in \left(\dfrac{-\pi}{2}, \dfrac{\pi}{4}\right) \cup \left(\dfrac{\pi}{3}, \dfrac{\pi}{2}\right)$

Exercise for Session 3

- **Directions** (Q. Nos. 1 to 5) *Solve the following inequalities.*

1. $\frac{x-2}{x+2} > \frac{2x-3}{4x-1}$

2. $\frac{2x-1}{2x^3+3x^2+x} > 0$

3. $\frac{4x}{x^2+3} \ge 1$

4. $\frac{(x-1)(x+1)(x+4)(x+6)}{7x^2+8x+4} > 0$

5. $\frac{x}{x^2-5x+9} \le 1$

6. Solution of inequality $|x-1| < 0$ is
   (a) $x = 0$
   (b) $x = 1$
   (c) $x \neq 1$
   (d) No solution

7. Solution of inequality $x^2 + x + |x| + 1 \le 0$ is
   (a) $(1, 2)$
   (b) $(0, 1)$
   (c) No solution
   (d) None of these

8. Solution of inequality $|x+3| > |2x-1|$ is
   (a) $\left(-\frac{2}{3}, 4\right)$
   (b) $(4, \infty)$
   (c) $\left(\frac{2}{3}, 1\right)$
   (d) None of these

9. Solution of inequality $\left|x + \frac{1}{x}\right| < 4$ is
   (a) $(2-\sqrt{3}, 2+\sqrt{3}) \cup (-2-\sqrt{3}, -2+\sqrt{3})$
   (b) $R - (2-\sqrt{3}, 2+\sqrt{3})$
   (c) $R - (-2-\sqrt{3}, 2+\sqrt{3})$
   (d) None of these

10. The solution of $|x^2+3x| + x^2 - 2 \ge 0$ is
    (a) $(-\infty, 1)$
    (b) $(0, 1)$
    (c) $\left(-\infty, -\frac{2}{3}\right] \cup \left[\frac{1}{2}, \infty\right)$
    (d) None of these

11. The solution of $||x| - 1| < |1 - x|, x \in R$ is
    (a) $(-1, 1)$
    (b) $(0, \infty)$
    (c) $(-1, \infty)$
    (d) None of these

12. The solution of $2^x + 2^{|x|} \ge 2\sqrt{2}$ is
    (a) $(-\infty, \log_2(\sqrt{2}+1))$
    (b) $(0, \infty)$
    (c) $\left(\frac{1}{2}, \log_2(\sqrt{2}-1)\right)$
    (d) $(-\infty, \log_2(\sqrt{2}-1)] \cup \left[\frac{1}{2}, \infty\right)$

Session 4

Quadratic Expression, Non-negative Functions

Quadratic Expression

The expression $ax^2 + bx + c$ is said to be a real quadratic expression in x, where a, b, c are real and $a \neq 0$. Let $f(x) = ax^2 + bx + c$, where $a, b, c \in R\,(a \neq 0)$. $f(x)$ can be rewritten as

$$f(x) = a\left\{\left(x + \frac{b}{2a}\right)^2 + \frac{4ac - b^2}{4a^2}\right\} = a\left[\left(x + \frac{b}{2a}\right)^2 - \frac{D}{4a^2}\right]$$

where $D = b^2 - 4ac$ is the **discriminant** of the quadratic expression.

Therefore, $y = f(x)$ represents a parabola whose axis is parallel to the Y-axis, with vertex at $A\left(-\frac{b}{2a}, \frac{-D}{4a}\right)$.

Graph for Quadratic Expressions

That if $a > 0$, the parabola will be concave upwards and if $a < 0$, the parabola will be concave downwards and it depends on the sign of $b^2 - 4ac$ that the parabola cuts the X-axis at two points $(b^2 - 4ac > 0)$, touches the X-axis $(b^2 - 4ac = 0)$ or never intersects with the X-axis $(b^2 - 4ac < 0)$.

This gives rise to the following cases:

(i) If $a > 0$ and $b^2 - 4ac < 0$.

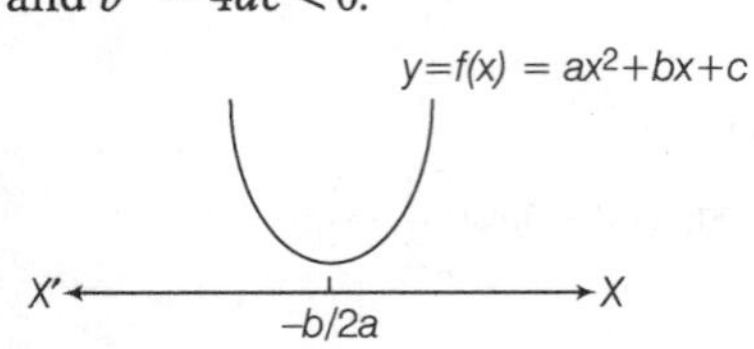

$\Leftrightarrow \quad f(x) > 0, \forall\, x \in R$

In this case the parabola always remains concave upwards and above the X-axis.

(ii) If $a > 0$ and $b^2 - 4ac = 0$.

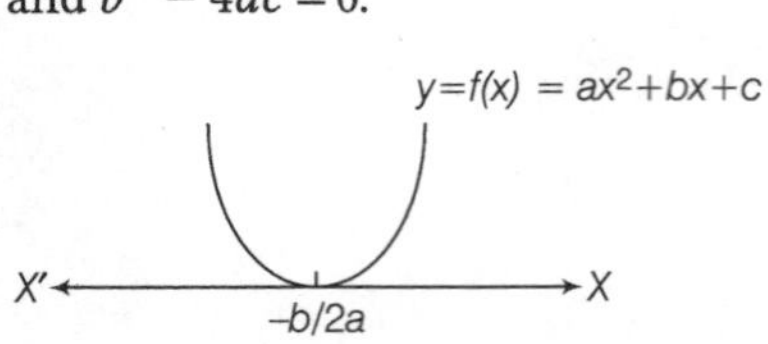

$\Leftrightarrow \quad f(x) \geq 0, \forall\, x \in R$

In this case the parabola touches the X-axis and remains concave upwards.

(iii) If $a > 0$ and $b^2 - 4ac > 0$.

Let $f(x) = 0$ has two real roots α and $\beta\,(\alpha < \beta)$.

Then, $f(x) > 0, \forall\, x \in (-\infty, \alpha) \cup (\beta, \infty)$,

$f(x) < 0, \forall\, x \in (\alpha, \beta)$ and $f(x) = 0$ for $x \in \{\alpha, \beta\}$.

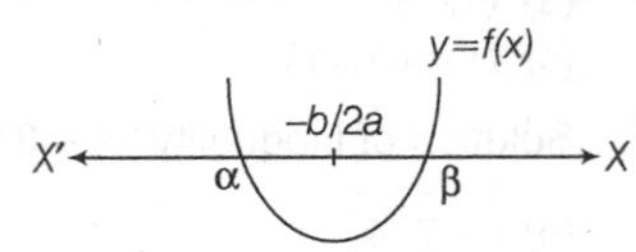

In this case the parabola cuts the X-axis at two points α and β and remains concave upwards.

(iv) If $a < 0$ and $b^2 - 4ac < 0 \Leftrightarrow f(x) < 0, \forall\, x \in R$

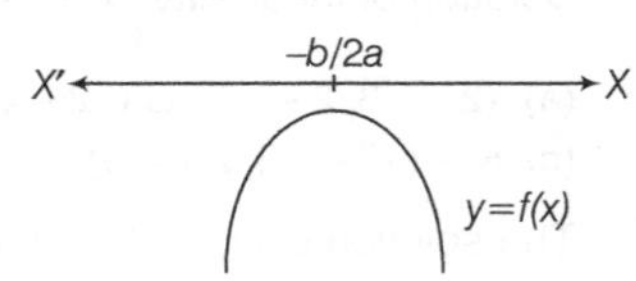

In this case the parabola remains concave downwards and always below the X-axis.

(v) If $a < 0$ and $b^2 - 4ac = 0$.

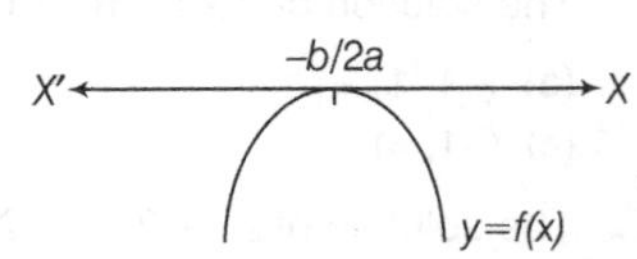

$\Leftrightarrow \quad f(x) \leq 0, \forall\, x \in R$

In this case the parabola touches the X-axis and remains concave downwards.

(vi) If $a < 0$ and $b^2 - 4ac > 0$.

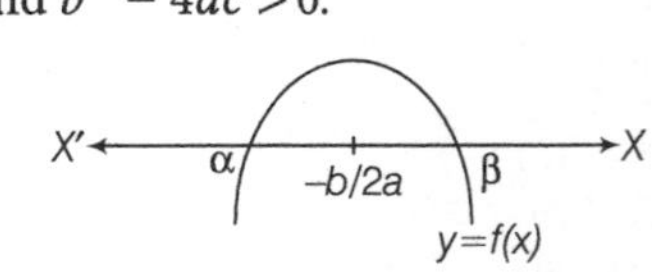

Let $f(x) = 0$ have two real roots α and $\beta\,(\alpha < \beta)$.

Then, $f(x) < 0, \forall\, x \in (-\infty, \alpha) \cup (\beta, \infty)$,

$f(x) > 0, \forall\, x \in (\alpha, \beta)$ and $f(x) = 0$ for $x \in \{\alpha, \beta\}$.

In this case the parabola cuts the X-axis at two points α and β and remains concave downwards.

Quick and Important Results for Characteristic Expression

1. The expression $ax^2 + bx + c$ will be at same sign for all real values of x, iff $D < 0$.
2. $ax^2 + bx + c$ will always be positive iff $D < 0$ and $a > 0$.
3. $ax^2 + bx + c$ will always be negative iff $D < 0$ and $a < 0$.
4. If $D > 0$, then sign of the expression between the roots will be opposite to that of a.
5. If $a > 0$, then minima of $f(x)$ occurs at $x = \dfrac{-b}{2a}$ and if $a < 0$, then maxima of $f(x)$ occurs at $x = \dfrac{-b}{2a}$ and maximum or minimum value of $f(x)$ will be given by $\left(\dfrac{-D}{4a}\right)$.
6. If $f(x) = 0$ has two distinct real roots, then $a \cdot f(d) < 0$ if and only if d lies between the roots and $a \cdot f(d) > 0$ if and only if d lies outside the roots.

Example 23 Find a for which $3x^2 + ax + 3 > 0, \forall\ x \in R$.

Sol. Here, $3x^2 + ax + 3 > 0, \forall\ x \in R$

$\Rightarrow \quad D < 0 \quad [\because \text{if } f(x) > 0 \text{ and } a > 0, \text{ then } D < 0]$

$\Rightarrow \quad (a)^2 - 4(3)(3) < 0 \quad \Rightarrow a^2 - 36 < 0$

$\Rightarrow (a-6)(a+6) < 0$, using number line rule as shown in figure,

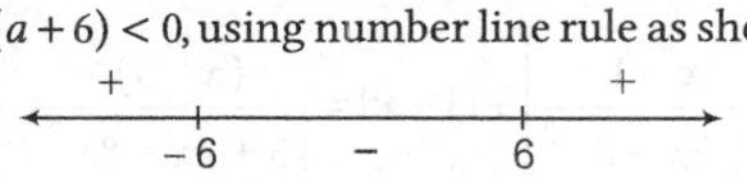

which shows $-6 < a < 6$ or $a \in (-6, 6)$

Example 24 Find a for which $ax^2 + x - 1 < 0, \forall\ x \in R$?

Sol. Here, $ax^2 + x - 1 < 0, \forall\ x \in R$

$\Rightarrow \quad a < 0$ and $D < 0 \quad [\because a < 0, D < 0, \text{ so } f(x) < 0]$

$\Rightarrow \quad a < 0$ and $1 + 4a < 0$

$\Rightarrow \quad a < 0$ and $a < -1/4$

$\therefore \quad a \in (-\infty, -1/4)$

Non-negative Functions

The sum of several non-negative terms is zero if and only if each term is zero.

i.e. $\quad a^2 + b^2 + c^2 = 0$

if $\quad a = b = c$

Example 25 Solve $(x+1)^2 + (x^2 + 3x + 2)^2 = 0$.

Sol. Here, $(x + 1)^2 + (x^2 + 3x + 2)^2 = 0$ if and only if each term is zero simultaneously,

$(x + 1) = 0$ and $(x^2 + 3x + 2) = 0 \Rightarrow (x+1)(x+2) = 0$

i.e. $\quad x = -1$ and $x = -1, -2$

$\therefore$ The common solution is $x = -1$.

Hence, solution of the above equation is $x = -1$.

Example 26 Solve $|x+1| + \sqrt{x-1} = 0$.

Sol. Here, $|x + 1| + \sqrt{x-1} = 0$, where each term is non-negative.

$\therefore \quad |x + 1| = 0$ and $\sqrt{x-1} = 0$

should be zero simultaneously.

i.e. $\quad x = -1$ and $x = 1$, which is not possible.

$\therefore$ There is no x for which each term is zero simultaneously. Hence, there is no solution.

Example 27 Solve

$$|x^2 - 1| + (x-1)^2 + \sqrt{x^2 - 3x + 2} = 0.$$

Sol. Here, each of the term is non-negative, thus each term must be zero simultaneously.

i.e. $(x^2 - 1) = 0, (x-1)^2 = 0$ and $x^2 - 3x + 2 = 0$

$\Rightarrow \quad x = \pm 1,\ x = 1$ and $x = 1, 2$

The common solution is $x = 1$.

Therefore, $x = 1$ is the solution of above equation.

Example 28 Let $f(x) = x$ and $g(x) = |x|$ be two real-valued functions, $\phi(x)$ be a function satisfying the condition;

$[\phi(x) - f(x)]^2 + [\phi(x) - g(x)]^2 = 0$. Then, find $\phi(x)$.

Sol. Here, $[\phi(x) - f(x)]^2 + [\phi(x) - g(x)]^2 = 0$ is only possible, if $\phi(x) - f(x) = 0$ and $\phi(x) - g(x) = 0$

$\Rightarrow \quad \phi(x) = f(x) = g(x)$ or $\phi(x) = x = |x|$,

which is only possible, if x is non-negative.

Therefore, $\phi(x) = x, \forall\ x \in [0, \infty)$

Exercise for Session 4

1. Find all values of 'm' for which $(2m - 3)x^2 + 2mx + 4 < 0$ for all real x.
2. If $ax^2 - bx + 5 = 0$ does not have two distinct real roots, then find the minimum value of $5a + b$.
3. If $a, b, c \in R, a \neq 0$ and the quadratic equation, $ax^2 + bx + c = 0$ has no real root, then show that $a(a + b + c) > 0$.
4. If $x, y \in [0, 10]$, then find the number of solutions (x, y) of the inequation $3^{\sec^2 x - 1} \cdot \sqrt{9y^2 - 6y + 2} \leq 1$.

Essential Mathematical Tools Exercise 1 : Subjective Questions

- **Directions** (Q. Nos. 1 to 11) *Solve each of the following system of equations :*

1. For $a<0$, determine all solutions of the equation $x^2-2a|x-a|-3a^2=0$.

2. Solve $|x^2+4x+3|+2x+5=0$.

3. Solve $|x^2-3x-4|=9-|x^2-1|$.

4. Solve $2^{|x+1|}-2^x=|2^x-1|+1$.

5. Find the set of all real 'a' such that $5a^2-3a-2$, a^2+a-2 and $2a^2+a-1$ are the lengths of the sides of a triangle?

6. Solve $(x+3)^5-(x-1)^5\geq 244$.

7. Solve $||x-2|-1|\geq 3$.

8. Solve $1\leq \dfrac{3x^2-7x+8}{x^2+1}\leq 2$.

9. Let $f(x)=\dfrac{2x}{2x^2+5x+2}$ and $g(x)=\dfrac{1}{x+1}$.
Find the set of real values of x for which $f(x)>g(x)$.

10. For $x\in R$, $||x||$ is defined as follows;

$$||x||=\begin{cases}x+1, & 0\leq x<2\\ |x-4|, & 2\leq x\end{cases}$$

Then, solve the equation, $||x||^2+x=||x||+x^2$.

11. Solve the inequality $|x-1|+|2-x|>3+x$.

12. Solve the equation

$$\sqrt{x^2+12y}+\sqrt{y^2+12x}=33,\ x+y=23.$$

13. Solve the equation

$$\sqrt{2x-1}+\sqrt{3x-2}=\sqrt{4x-3}+\sqrt{5x-4}.$$

14. If x, y and z are three real numbers such that $x+y+z=4$ and $x^2+y^2+z^2=6$, then show that each of x, y and z lies in the closed interval $\left[\frac{2}{3},2\right]$.

15. If $\{(\alpha+1)(\beta-1)+(\beta+1)(\alpha-1)\}a+(\alpha-1)(\beta-1)=0$ and $a(\alpha+1)(\beta+1)-(\alpha-1)(\beta-1)=0$

Also, let $A=\left\{\dfrac{\alpha+1}{\alpha-1},\dfrac{\beta+1}{\beta-1}\right\}$

and $B=\left\{\dfrac{2\alpha}{\alpha+1},\dfrac{2\beta}{\beta+1}\right\}$. If $A\cap B\neq\phi$, then find all the permissible values of the parameter 'a'.

16. Solve $\left|\dfrac{x-1}{3+2x-8x^2}\right|+|1-x|=\dfrac{(x-1)^2}{|3+2x-8x^2|}+1$

17. Let $f(x)=(x^2-2|x|)(2|x|-2)-9\dfrac{(2|x|-2)}{x^2-2|x|}$.

Solve the following inequalities

(i) $f(x)>0$ (ii) $f(x)\geq 0$

(iii) $f(x)<0$ (iv) $f(x)\leq 0$

18. Solve the inequality $\left|1-\dfrac{|x|}{1+|x|}\right|\geq\dfrac{1}{2}$.

Essential Mathematical Tools Exercise 2 : More Than One Correct Option Type Questions

19. If $\cos x-y^2-\sqrt{y-x^2-1}\geq 0$, then

(a) $y\geq 1$ (b) $x\in R$ (c) $y=1$ (d) $x=0$

20. If $(\sin\alpha)x^2-2x+b\geq 2$ for all real values of $x\leq 1$ and $\alpha\in\left(0,\frac{\pi}{2}\right)\cup\left(\frac{\pi}{2},\pi\right)$, then the possible real values of b is/are

(a) 2 (b) 3 (c) 4 (d) 5

21. If $|ax^2+bx+c|\leq 1$ for all x is [0, 1], then

(a) $|a|\leq 8$

(b) $|b|\leq 8$

(c) $|c|\leq 1$

(d) $|a|+|b|+|c|\leq 17$

Essential Mathematical Tools Exercise 3 : Passage Based Questions

Passage I

(Q. Nos. 22 to 24)

Let $f(x) = ax^2 + bx + c$; $a, b, c \in R$

It is given $|f(x)| \leq 1, \forall |x| \leq 1$.

Now, answer the following questions.

22. The possible value of $|a + c|$, if $\frac{8}{3}a^2 + 2b^2$ is maximum, is given by

(a) 1 (b) 0 (c) 2 (d) 3

23. The possible value of $|a + b|$, if $\frac{8}{3}a^2 + 2b^2$ is maximum, is given by

(a) 1 (b) 0 (c) 2 (d) 3

24. The possible maximum value of $\frac{8}{3}a^2 + 2b^2$ is given by

(a) 32 (b) $\frac{32}{3}$ (c) $\frac{2}{3}$ (d) $\frac{16}{3}$

Passage II

(Q. Nos. 25 to 27)

Consider the equation $|2x| - |x - 4| = x + 4$.

25. The least integer satisfying the equation, is

(a) -4 (b) 4 (c) 5 (d) -5

26. Total number of prime numbers less than 20 satisfying the equation, is

(a) 3 (b) 4 (c) 5 (d) 6

27. If P = greatest composite number less than 34 satisfying the given equation, then P^{2007} has the digit on its units place as

(a) 8 (b) 1
(c) 7 (d) 0

Passage III

(Q. Nos. 28 to 30)

Consider a number $N = 2\,1\,P\,5\,3\,Q\,4$.

28. The number of ordered pairs (P, Q) so that the number 'N' is divisible by 9, is

(a) 11 (b) 12
(c) 10 (d) 8

29. The number of values of Q so that the number 'N' is divisible by 8, is

(a) 4 (b) 3
(c) 2 (d) 6

30. The number of ordered pairs (P, Q) so that the number 'N' is divisible by 44, is

(a) 2 (b) 3 (c) 4 (d) 5

Passage IV

(Q. Nos. 31 to 35)

Consider the nine digit number $n = 7\,3\,\alpha\,4\,9\,6\,1\,\beta\,0$.

31. If p is the number of all possible distinct values of $(\alpha - \beta)$, then p is equal to

(a) 17 (b) 18 (c) 19 (d) 20

32. If q is the number of all possible values of β for which the given number is divisible by 8, then q is equal to

(a) 2 (b) 3 (c) 4 (d) 5

33. The number of ordered pairs (α, β) for which the given number is divisible by 88, is

(a) 1 (b) 2 (c) 3 (d) 4

34. The number of possible values of $(\alpha + \beta)$ for which the given number is divisible by 6, is

(a) 3 (b) 4 (c) 6 (d) 7

35. The number of possible values of β for which $i^N = 1$ (where $i = \sqrt{-1}$), is

(a) 2 (b) 3 (c) 4 (d) 5

Passage V

(Q. Nos. 36 to 38)

The set of integers can be classified into k classes, according to the remainder obtained when they are divided by k (where k is a fixed natural number). The classification enables is solving even some more difficult problems of number theory e.g.

(i) *even, odd classification is based on whether remainder is* 0 *or* 1 *when divided by* 2.

(ii) *when divided by* 3, *the remainder may be* 0, 1, 2. *Thus, there are three classes.*

36. The remainder obtained, when the square of an integer is divided by 3, is

(a) 0, 1 (b) 1, 2 (c) 0, 2 (d) 0, 1, 2

37. $n^2 + n + 1$ is never divisible by

(a) 2 (b) 3
(c) 111 (d) None of these

38. If n is odd, $n^5 - n$ is not divisible by

(a) 16 (b) 15 (c) 240 (d) 720

Essential Mathematical Tools Exercise 4 : Single Integer Answer Type Questions

39. The number of solutions of the equation $|x-1|-|2x-5|\geq 4$

40. The number of integral solution of the equation $|x^2-7|\leq 9$ are

41. The number of solutions of the system of equation $x+2y=6$ and $|x-3|=y$ is/are

Answers

Exercise for Session 1

1. False 2. False 3. True 4. False 5. True 6. True 7. (a) 8. (c) 9. (a) 10. (b)

Exercise for Session 2

1. $x\in(-1, 3)$
2. $x\in(-\infty,-2)\cup(8,\infty)$
3. $x\in(-2, 1)\cup(1, 4)$
4. $x\in(-\infty, 1]\cup\left[\frac{3}{2},\infty\right)$
5. $x\in\left(-\infty,\frac{-1-\sqrt{29}}{2}\right]\cup\left[\frac{1-\sqrt{5}}{2},\frac{1+\sqrt{5}}{2}\right]\cup\left[\frac{-1+\sqrt{29}}{2},\infty\right)$

Exercise for Session 3

1. $x\in(-\infty,-2)\cup\left(\frac{1}{4},1\right)\cup(4,\infty)$
2. $x\in(-\infty,-1)\cup\left(\frac{-1}{2},0\right)\cup\left(\frac{1}{2},\infty\right)$
3. $x\in[1, 3]$
4. $x\in[-\infty,-6)\cup(-4,-1)\cup(1,0)$ 5. $x\in R$
6. (d) 7. (c) 8. (a) 9. (a) 10. (c) 11. (d) 12. (d)

Exercise for Session 4

1. $m\in\phi$ 2. (−1) 4. (4)

Chapter Exercises

1. $\{(1-\sqrt{2})a, (-1+\sqrt{6})a\}$
2. $\{-4, -1-\sqrt{3}\}$
3. $\{-2, 2\}$
4. $\{-2\}\cup[0,\infty)$
5. $\left(\frac{3+\sqrt{57}}{8}<a<\frac{5+\sqrt{17}}{4}\right)$
6. $(-\infty,-2]\cup[0,\infty)$
7. $(-\infty,-2]\cup[6,\infty)$
8. $[1, 6]$
9. $x\in(-2,-1)\cup\left(-\frac{2}{3},-\frac{1}{2}\right)$
10. $x=\{0, 2\}$
11. $(-\infty, 0)\cup(6,\infty)$
12. (13, 10) and (10, 13)
13. $x=1$
14. $x, y, z\in\left[\frac{2}{3},2\right]$
15. $a\in\{-1, 1+i, 1-i\}$
16. $x\in\left\{0, 2, \frac{1\pm\sqrt{129}}{16}, \frac{3\pm\sqrt{73}}{16}\right\}$
17. (i) $x\in(-\infty,-3)\cup(-2,-1)\cup(1,2)\cup(3,\infty)$
 (ii) $x\in(-\infty,-3]\cup(-2,-1]\cup[1,2)\cup[3,\infty)$
 (iii) $x\in(-3,-2)\cup(-1, 0)\cup(0,1)\cup(2,3)$
 (iv) $x\in[-3,-2)\cup[-1, 0)\cup(0,1]\cup(2,3]$
18. $[-1, 1]$
19. (c,d)
20. (b,c,d)
21. (a,b,c,d) 22. (a) 23. (c) 24. (b)
25. (a) 26. (d) 27. (c) 28. (a) 29. (b) 30. (c)
31. (c) 32. (a) 33. (b) 34. (d) 35. (d) 36. (a)
37. (c) 38. (d) 39. (0) 40. (9) 41. (2)

Solutions

1. For $a < 0, |x - a| = \begin{cases} (x-a), & x \ge a \\ (a-x), & x \le a \end{cases}$.

***Case* I** $x \ge a$

$$x^2 - 2a(x-a) - 3a^2 = 0$$
$$\Rightarrow \quad x^2 - 2ax - a^2 = 0$$
$$\Rightarrow \quad x^2 - 2ax + a^2 = 2a^2$$
$$\Rightarrow \quad (x-a)^2 = 2a^2 \Rightarrow x = \pm\sqrt{2}a + a$$
$$\Rightarrow \quad x = a(1+\sqrt{2}), a(1-\sqrt{2})$$

Since, $x \ge a$ and $a < 0$

$\therefore$ Neglecting $\{a(1+\sqrt{2})\} \Rightarrow x = a(1-\sqrt{2})$

***Case* II** $x \le a$

$$\Rightarrow \quad x^2 + 2a(x-a) - 3a^2 = 0 \quad \Rightarrow x^2 + 2ax = 5a^2$$
$$\Rightarrow \quad (x+a)^2 = 6a^2 \quad \Rightarrow x = -a \pm \sqrt{6}a$$
$$\therefore \quad x = a(\sqrt{6}-1), -a(\sqrt{6}+1)$$

Since, $x \le a$ and $a < 0$

$\therefore$ Neglecting $\{-a(\sqrt{6}+1)\} \Rightarrow x = a(\sqrt{6}-1)$

Hence, $\quad x \in \{a(1-\sqrt{2}), a(\sqrt{6}-1)\}$.

2. $|x^2 + 4x + 3| + 2x + 5 = 0$

Take two cases, i.e. $|x^2 + 4x + 3| = \pm(x^2 + 4x + 3)$

***Case* I** $\quad x^2 + 4x + 3 \ge 0$

$$(x+3)(x+1) \ge 0$$

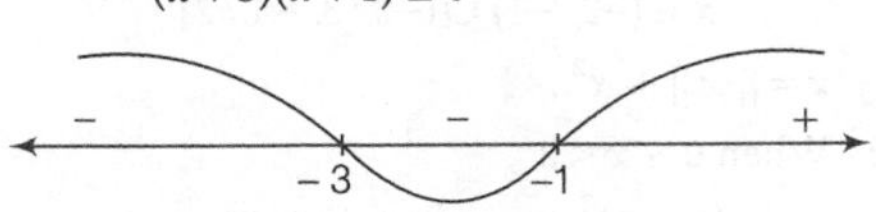

$$\Rightarrow \quad x \notin (-3, -1)$$
$$\text{or} \quad x \in (-\infty, -3] \cup [-1, \infty)$$
$$\Rightarrow \quad x^2 + 4x + 3 + 2x + 5 = 0$$
$$\Rightarrow \quad x^2 + 6x + 8 = 0$$
$$\Rightarrow \quad (x+4)(x+2) = 0$$
$$\Rightarrow \quad x = -2, -4.$$

But $x \notin (-3, -1)$

$\therefore$ Neglecting $\quad x = -2$

$\Rightarrow \quad x = -4 \qquad$ ...(i)

***Case* II** $x^2 + 4x + 3 \le 0$

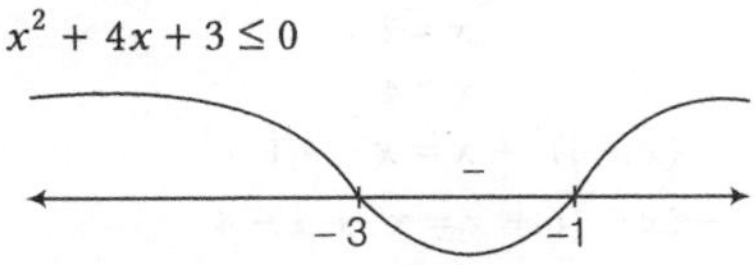

$$(x+3)(x+1) \le 0$$
$$\Rightarrow \quad x \in [-3, -1]$$
$$\Rightarrow \quad x^2 + 4x + 3 - 2x - 5 = 0$$
$$\Rightarrow \quad x^2 + 2x - 2 = 0$$
$$\Rightarrow \quad x = -1 \pm \sqrt{3}, \text{ as } x \in [-3, -1]$$

$\therefore$ Neglecting $x = -1 + \sqrt{3} \Rightarrow x = -1 - \sqrt{3} \qquad$ ...(ii)

$$\therefore \quad x \in \{-4, -1-\sqrt{3}\}.$$

3. $|x^2 - 3x - 4| + |x^2 - 1| = 9$

$\Rightarrow \quad |x+1|\{|x-4| + |x-1|\} = 9$

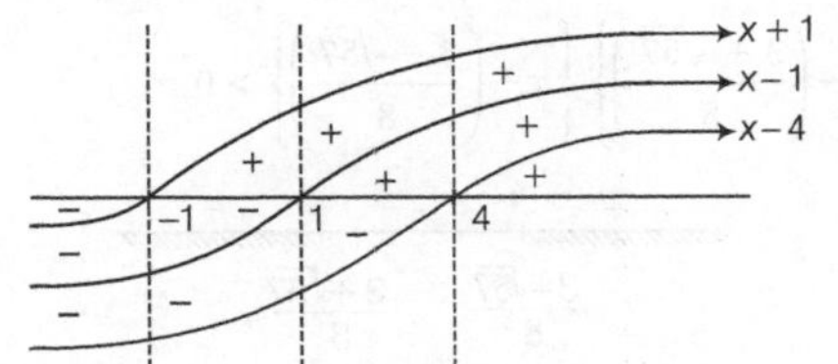

***Case* I** $x \ge 4$

$$(x+1)\{x - 4 + x - 1\} = 9$$
$$\Rightarrow \quad (x+1)(2x-5) = 9$$
$$\Rightarrow \quad 2x^2 - 3x - 14 = 0$$
$$\Rightarrow \quad x = \frac{7}{2}, -2$$

[neglecting both as $x \ge 4$]

***Case* II** $1 < x < 4$

$$(x+1)\{4 - x + x - 1\} = 9$$
$$\Rightarrow \quad (x+1)\cdot 3 = 9$$
$$\Rightarrow \quad x = 2 \in (1, 4) \qquad \text{...(i)}$$

***Case* III** $-1 < x < 1$

$$(x+1)\{4 - x + 1 - x\} = 9$$
$$\Rightarrow \quad (x+1)(5-2x) = 9$$

$\Rightarrow \quad 2x^2 - 3x + 4 = 0$, having imaginary roots.

$\therefore$ No solution.

***Case* IV** $x \le -1$

$$-(x+1)\{4 - x + 1 - x\} = 9$$
$$(x+1)(2x-5) = 9 \Rightarrow x = -2, \frac{7}{2}.$$

As $x \le -1$,

$\therefore \; x = -2$, neglecting $x = \frac{7}{2}$. $\qquad$ ...(ii)

Thus, from Eqs. (i) and (ii), we get $x \in \{-2, 2\}$.

4. $2^{|x+1|} - 2^x = |2^x - 1| + 1$

***Case* I** $\quad x \ge 0$

$$2^{x+1} - 2^x = 2^x - 1 + 1 \Rightarrow 2^x \cdot 2 = 2 \cdot 2^x$$

i.e. $\quad$ true for all $x \ge 0 \qquad$ ...(i)

***Case* II** $-1 \le x \le 0$

$$2^{+x+1} - 2^x = 1 - 2^x + 1$$
$$\Rightarrow \quad 2^{x+1} = 2$$
$$\Rightarrow \quad x = 0 \qquad \text{...(ii)}$$

***Case* III** $x \le -1$

$$\Rightarrow \quad 2^{-x-1} - 2^x = 1 - 2^x + 1$$
$$\Rightarrow \quad 2^{-x-1} = 2$$
$$\Rightarrow \quad x = -2 \qquad \text{...(iii)}$$

Thus, from Eqs. (i), (ii) and (iii), we get $x \in \{-2\} \cup [0, \infty)$.

5. If $5a^2 - 3a - 2, a^2 + a - 2, 2a^2 + a - 1$ are lengths of sides of triangle, then sum of any two sides must be greater than third side.

***Case* I** $(5a^2 - 3a - 2) + (a^2 + a - 2) > 2a^2 + a - 1$

$\Rightarrow \quad 6a^2 - 2a - 4 > 2a^2 + a - 1$

$\Rightarrow \quad 4a^2 - 3a - 3 > 0$

$\Rightarrow \left\{a - \left(\frac{3+\sqrt{57}}{8}\right)\right\}\left\{a - \left(\frac{3-\sqrt{57}}{8}\right)\right\} > 0$

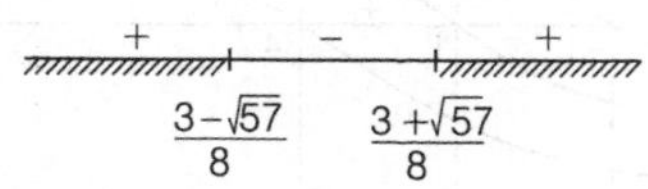

$\Rightarrow a \in \left(-\infty, \frac{3-\sqrt{57}}{8}\right) \cup \left(\frac{3+\sqrt{57}}{8}, \infty\right)$ ...(i)

***Case* II** $(a^2 + a - 2) + (2a^2 + a - 1) > 5a^2 - 3a - 2$

$\Rightarrow \quad 3a^2 + 2a - 3 > 5a^2 - 3a - 2$

$\Rightarrow \quad 2a^2 - 5a + 1 < 0$

$\Rightarrow \left\{a - \left(\frac{5+\sqrt{17}}{4}\right)\right\}\left\{a - \left(\frac{5-\sqrt{17}}{4}\right)\right\} < 0$

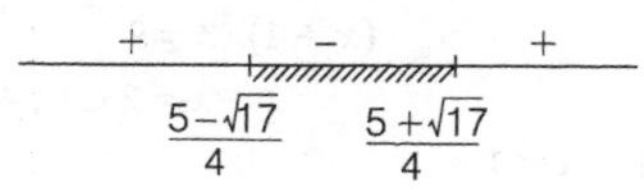

$\Rightarrow a \in \left(\frac{5-\sqrt{17}}{4}, \frac{5+\sqrt{17}}{4}\right)$ ...(ii)

***Case* III** $(5a^2 - 3a - 2) + (2a^2 + a - 1) > (a^2 + a - 2)$

$\Rightarrow \quad 7a^2 - 2a - 3 > a^2 + a - 2$

$\Rightarrow \quad 6a^2 - 3a - 1 > 0$

$\Rightarrow \left\{a - \left(\frac{3+\sqrt{33}}{12}\right)\right\}\left\{a - \left(\frac{3-\sqrt{33}}{12}\right)\right\} > 0$

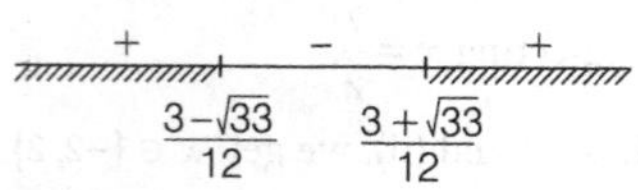

$\Rightarrow a \in \left(-\infty, \frac{3-\sqrt{33}}{12}\right) \cup \left(\frac{3+\sqrt{33}}{12}, \infty\right)$ ...(iii)

From Eqs. (i), (ii) and (iii), we get

$$a \in \left(\frac{3+\sqrt{57}}{8}, \frac{5+\sqrt{17}}{4}\right)$$

6. $(x+3)^5 - (x-1)^5 \ge 244$

Let $\quad y = \frac{(x+3)+(x-1)}{2} \Rightarrow y = x + 1$

$\therefore \quad (y+2)^5 - (y-2)^5 \ge 244$

$\Rightarrow \quad 2\{{}^5C_1 \cdot y^4 \cdot 2 + {}^5C_3 \cdot y^2 \cdot 2^3 + {}^5C_5 \cdot 2^5\} \ge 244$

$\Rightarrow \quad 2\{10y^4 + 80y^2 + 32\} \ge 244$

$\Rightarrow \quad 4\{5y^4 + 40y^2 + 16\} \ge 244$

$\Rightarrow \quad 5y^4 + 40y^2 + 16 \ge 61$

$\Rightarrow \quad y^4 + 8y^2 - 9 \ge 0$

$\Rightarrow \quad (y^2 + 9)(y^2 - 1) \ge 0$

$\Rightarrow \quad y^2 \ge 1$

i.e. $\quad y \le -1$ or $y \ge 1$

$\Rightarrow \quad x + 1 \le -1$ or $x + 1 \ge 1$

$\Rightarrow \quad x \in (-\infty, -2] \cup [0, \infty)$

7. $||x-2| - 1| \ge 3$

$\Rightarrow \quad |x-2| - 1 \le -3$ or $|x-2| - 1 \ge 3$

$\Rightarrow \quad |x-2| \le -2$ or $|x-2| \ge 4$

$\Rightarrow$ No solution or $x - 2 \le -4$ or $x - 2 \ge 4$

$\Rightarrow \quad x \le -2$ or $x \ge 6$

$\Rightarrow \quad x \in (-\infty, -2] \cup [6, \infty)$

8. $1 \le \frac{3x^2 - 7x + 8}{x^2 + 1} \le 2$

$\Rightarrow \quad 2x^2 - 7x + 7 \ge 0$ and $x^2 - 7x + 6 \le 0$

$\Rightarrow \quad x \in R$ and $x \in [1, 6]$

$\therefore \quad x \in [1, 6]$

9. $\quad f(x) > g(x) \Rightarrow \frac{2x}{2x^2 + 5x + 2} > \frac{1}{x+1}$

$\Rightarrow \quad \frac{2x}{(2x+1)(x+2)} - \frac{1}{(x+1)} > 0$

$\Rightarrow \quad \frac{2x^2 + 2x - (2x^2 + 5x + 2)}{(2x+1)(x+2)(x+1)} > 0$

$\Rightarrow \quad \frac{-(3x+2)}{(2x+1)(x+2)(x+1)} > 0$

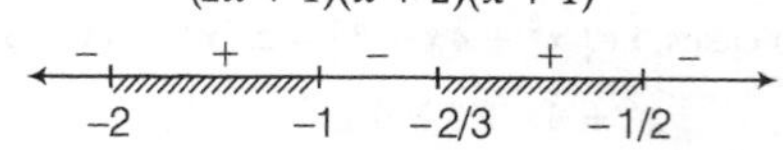

$\Rightarrow \quad x \in (-2, -1) \cup (-2/3, -1/2)$

10. $||x||^2 + x = ||x|| + x^2$.

***Case* I** When $0 \le x < 2$

$\Rightarrow \quad (x+1)^2 + x = (x+1) + x^2$

$\Rightarrow \quad x^2 + 2x + 1 + x = x^2 + x + 1$

$\Rightarrow \quad 2x = 0$

$\Rightarrow \quad x = 0$ ...(i)

***Case* II** When $x \ge 2$

$\Rightarrow \quad (x-4)^2 + x = |x-4| + x^2$

Now, if $2 \le x \le 4$

$\Rightarrow \quad (x^2 - 8x + 16) + x = 4 - x + x^2$

$\Rightarrow \quad 6x = 12$

$\Rightarrow \quad x = 2$ ...(ii)

Again, if $\quad x \ge 4$

$\Rightarrow \quad (x-4)^2 + x = x - 4 + x^2$

$\Rightarrow \quad x^2 - 8x + 16 + x = x^2 + x - 4$

$\Rightarrow \quad 8x = 20$

$\Rightarrow \quad x = \frac{5}{2}$,

but $\quad x \ge 4$.

$\therefore$ Only two solutions $x = 0$ and 2.

11. The points $x = 1$ and $x = 2$ divide the number axis into three intervals as follows :

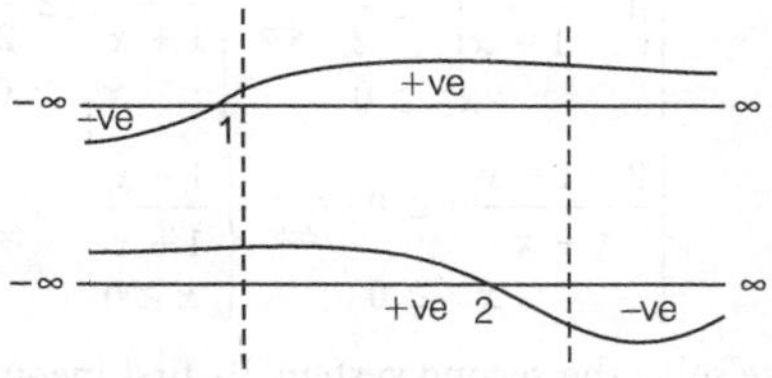

We solve the inequality on each intervals.

If $x < 1$, then $x - 1 < 0$ and $2 - x > 0$.

$\therefore \quad |x-1| + |2-x| > 3 - x$

$\Rightarrow \quad 1 - x + 2 - x > 3 + x$

$\Rightarrow \quad 3 - 2x > 3 + x$

$\Rightarrow \quad x < 0 \quad$ ...(i)

If $1 \le x \le 2$, then $x - 1 \ge 0$ and $2 - x \ge 0$, we have

$x - 1 + 2 - x > 3 + x$

$\Rightarrow \quad 1 > 3 + x \Rightarrow x < -2$

The system of inequalities obtained has no solution, for $1 \le x \le 2$. If $x > 2$, then $x - 1 > 0$ and $2 - x < 0$, we have

$x - 1 + x - 2 > 3 + x$

$\Rightarrow \quad 2x - 3 > 3 + x \Rightarrow x > 6$

Combining the solutions obtained on all parts of the domain of admissible values of the given inequality, we get the solution set $(-\infty, 0) \in (6, \infty)$.

12. Here, $\sqrt{x^2 + 12(23 - x)} + \sqrt{(23 - x)^2 + 12x} = 33 \quad (\because y = 23 - x)$

$\Rightarrow \quad \sqrt{x^2 - 12x + 276} + \sqrt{x^2 - 34x + 529} = 33$

Let $\quad a = \sqrt{x^2 - 12x + 276}$

and $\quad b = \sqrt{x^2 - 34x + 529} \quad$ ...(i)

$\Rightarrow \quad a + b = 33$ and $a^2 - b^2 = 11(2x - 23)$

$a - b = \dfrac{2x - 23}{3}$

$\therefore \quad a = \dfrac{x + 38}{3} \quad$ ...(ii)

$\Rightarrow \quad x^2 - 12x + 276 = \left(\dfrac{x + 38}{3}\right)^2 \quad$ [using Eqs. (i) and (ii)]

$\Rightarrow \quad 8x^2 - 184x + 1040 = 0 \quad \Rightarrow x = 13, 10$

$\Rightarrow \quad y = 10, 13$

$\therefore$ (13, 10) and (10, 13) are the required solutions.

13. Here, $u + v = p + q \quad$ ...(i)

Where $u = \sqrt{2x - 1}$, $v = \sqrt{3x - 2}$, $p = \sqrt{4x - 3}$,

$q = \sqrt{5x - 4}$

$\therefore \quad u^2 - v^2 = 1 - x$

$p^2 - q^2 = 1 - x \Rightarrow u^2 - v^2 = p^2 - q^2$

$\Rightarrow \quad u - v = p - q \quad$ ...(ii) {using Eq. (i), $u + v = p + q$}

$\therefore$ From Eqs. (i) and (ii), we get

$2u = 2p \quad \Rightarrow 2x - 1 = 4x - 3$

or $\quad x = 1$, which clearly satisfies

$\sqrt{2x - 1} + \sqrt{3x - 2} = \sqrt{4x - 3} + \sqrt{5x - 4}$

$\therefore \quad x = 1$ is the required solution.

14. Here, $x^2 + y^2 + (4 - x - y)^2 = 6 \quad (\because z = 4 - x - y)$

$x^2 + (y - 4)x + (y^2 + 5 - 4y) = 0$

Since, x is real.

$\Rightarrow \quad D \ge 0$

$\Rightarrow \quad (y - 4)^2 - 4(y^2 - 4y + 5) \ge 0$

$\Rightarrow \quad -3y^2 + 8y - 4 \ge 0$

$\Rightarrow \quad -(3y - 2)(y - 2) \ge 0$

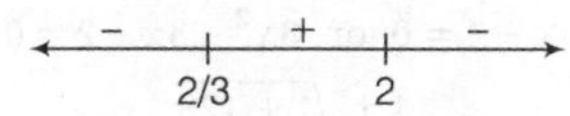

$\Rightarrow \quad \dfrac{2}{3} \le y \le 2$

Similarly, we can show that x, y and $z \in \left[\dfrac{2}{3}, 2\right]$.

15. Here, $\dfrac{\alpha + 1}{\alpha - 1} + \dfrac{\beta + 1}{\beta - 1} = -\dfrac{1}{a}$ and $\dfrac{(\alpha + 1)(\beta + 1)}{(\alpha - 1)(\beta - 1)} = \dfrac{1}{a}$

Now, the quadratic $x^2 - \left(-\dfrac{1}{a}\right)x + \dfrac{1}{a} = 0$ has roots $\left(\dfrac{\alpha + 1}{\alpha - 1}\right.$ and $\left.\dfrac{\beta + 1}{\beta - 1}\right)$

$\Rightarrow \quad ax^2 + x + 1 = 0 \quad$ ...(i)

Let $\quad x = \dfrac{2\alpha}{\alpha + 1} \Rightarrow \alpha = \dfrac{x}{2 - x} \Rightarrow \dfrac{\alpha + 1}{\alpha - 1} = \dfrac{1}{x - 1}$

Now, replacing x by $\dfrac{1}{x - 1}$ in Eq. (i), we get a quadratic whose roots are $\dfrac{2\alpha}{\alpha + 1}$ and $\dfrac{2\beta}{\beta + 1}$.

$\dfrac{a}{(x - 1)^2} + \dfrac{1}{(x - 1)} + 1 = 0$

$\Rightarrow \quad x^2 - x + a = 0 \quad$ ...(ii)

Hence, Eqs. (i) and (ii) are the two quadratics whose roots are elements of set 'A' and set 'B', respectively.

$\therefore$ They must have a root in common as $A \cap B \ne \phi$

***Case* I** Both roots are common.

$\Rightarrow \quad \dfrac{a}{1} = \dfrac{1}{-1} = \dfrac{1}{a}$

$\Rightarrow \quad a = -1$ and $a^2 = 1 \Rightarrow a = -1$

***Case* II** For a common root,

$\dfrac{x^2}{a + 1} = \dfrac{x}{1 - a^2} = \dfrac{1}{-(a + 1)}$

$\Rightarrow \quad a = 1 \pm i$

$\therefore \quad a \in \{-1, 1 + i, 1 - i\}$

16. Here, $\left|\dfrac{x - 1}{3 + 2x - 8x^2}\right| + |1 - x| = \dfrac{(x - 1)^2}{|3 + 2x - 8x^2|} + 1$

Let $a = \dfrac{x - 1}{3 + 2x - 8x^2}$; $b = (x - 1)$,

then $\quad |a| + |b| = |ab| + 1$

$\Rightarrow \quad |ab| - |a| - |b| + 1 = 0$

$\Rightarrow \quad (|a| - 1)(|b| - 1) = 0$

$\therefore \quad |a| = 1$ or $|b| = 1$

When $|a|=1$

$$\Rightarrow \left|\frac{x-1}{3+2x-8x^2}\right| = 1$$

$$\frac{x-1}{3+2x-8x^2}=1 \text{ or } \frac{x-1}{3+2x-8x^2}=-1$$

$$\Rightarrow \quad x-1 = 3+2x-8x^2$$

or $$x-1=-3-2x+8x^2$$

$$\Rightarrow \quad 8x^2-x-4=0 \text{ or } 8x^2-3x-2=0$$

$$\Rightarrow \quad x=\frac{1\pm\sqrt{1+128}}{16} \text{ or } x=\frac{3\pm\sqrt{9+64}}{16}$$

$$\Rightarrow \quad x=\frac{1\pm\sqrt{129}}{16} \text{ or } x=\frac{3\pm\sqrt{73}}{16} \quad \text{...(i)}$$

Again, when $|b|=1, |x-1|=1$

$$\Rightarrow \quad x-1=1 \text{ or } x-1=-1$$

$$\Rightarrow \quad x=0 \text{ or } x=2 \quad \text{...(ii)}$$

From Eqs. (i) and (ii), solution set is

$$x \in \left\{0, 2, \frac{1\pm\sqrt{129}}{16}, \frac{3\pm\sqrt{73}}{16}\right\}$$

17. We have, $f(x)=(x^2-2|x|)(2|x|-2)-9\cdot\dfrac{(2|x|-2)}{x^2-2|x|}$

$$=(2|x|-2)\left\{x^2-2|x|-\frac{9}{x^2-2|x|}\right\}$$

$$=(2|x|-2)\left\{\frac{(x^2-2|x|)^2-9}{x^2-2|x|}\right\}$$

$$=\frac{(2|x|-2)(x^2-2|x|+3)(x^2-2|x|-3)}{x^2-2|x|}$$

$$=\frac{(2|x|-2)\{(|x|-1)^2+2\}\{(|x|+1)(|x|-3)\}}{x^2-2|x|}$$

Taking, $N^r=0 \Rightarrow |x|=1, 3$

$\Rightarrow \quad x=\pm1, \pm3$ and $D^r=0$

$\Rightarrow \quad |x|(|x|-2)=0$

$\Rightarrow \quad 0, \pm2$

We mark these roots on a number line :

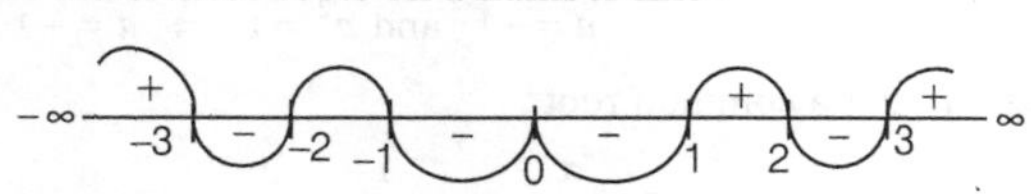

From the wavy curve method, we have

(i) $f(x)>0 \Rightarrow x\in(-\infty,-3)\cup(-2,-1)\cup(1,2)\cup(3,\infty)$

(ii) $f(x)\ge 0 \Rightarrow x\in(-\infty,-3]\cup(-2,-1]\cup[1,2)\cup[3,\infty)$

(iii) $f(x)<0 \Rightarrow x\in(-3,-2)\cup(-1,0)\cup(0,1)\cup(2,3)$

(iv) $f(x)\le 0 \Rightarrow x\in[-3,-2)\cup[-1,0)\cup(0,1]\cup(2,3]$

18. The domain of admissible values of this inequality consists of all the real numbers. The inequality is equivalent to the collection of two systems :

$$\begin{cases}\left|1-\dfrac{x}{1+x}\right| \ge \dfrac{1}{2} \\ x \ge 0\end{cases}, \quad \begin{cases}\left|1-\dfrac{-x}{1-x}\right| \ge \dfrac{1}{2} \\ x \le 0\end{cases}$$

We solve the first system :

$$\begin{cases}\left|1-\dfrac{x}{1+x}\right| \ge \dfrac{1}{2} \\ x \ge 0\end{cases} \Leftrightarrow \begin{cases}\dfrac{1}{1+x} \ge \dfrac{1}{2} \\ x \ge 0\end{cases}$$

$$\Leftrightarrow \begin{cases}\dfrac{2-1-x}{1+x} \ge 0 \\ x \ge 0\end{cases} \Leftrightarrow \begin{cases}\dfrac{1-x}{1+x} \\ x\ge 0\end{cases} \Leftrightarrow 0\le x\le 1$$

Again, we solve the second system. Its first inequality is equivalent to the inequality $\left|\dfrac{1}{1-x}\right| \ge \dfrac{1}{2}$.

If $x<0$, then $1-x>0$ and consequently the second system is equivalent to the system

$$\begin{cases}\dfrac{1}{1-x} \ge \dfrac{1}{2} \\ x \le 0\end{cases} \Leftrightarrow \begin{cases}\dfrac{2-1+x}{1-x} \ge 0 \\ x \le 0\end{cases}$$

$$\Leftrightarrow \begin{cases}\dfrac{1+x}{1-x} \ge 0 \\ x \le 0\end{cases} \Leftrightarrow -1\le x<0$$

Thus, the set of all solutions of the original inequality consists of the numbers belonging to the interval $[-1, 1]$.

19. Given, $\cos x - y^2 - \sqrt{y-x^2-1} \ge 0 \quad \text{...(i)}$

Clearly, $\sqrt{y-x^2-1}$ is defined when $y-x^2-1\ge 0$ or $y\ge x^2+1$. So, minimum value of y is 1.

From Eq. (i), we have $\cos x - y^2 \ge \sqrt{y-x^2-1}$

where $\cos x - y^2 \le 0$ [as when $\cos x$ is maximum $(=1)$ and y^2 is minimum $(=1)$. So, $\cos x - y^2$ is maximum.]

Also, $\sqrt{y-x^2-1}\ge 0$

Hence, $\cos x - y^2 = \sqrt{y-x^2-1} = 0$

$$\Rightarrow \quad y=1 \text{ and } \cos x = 1, y = x^2+1$$

$$\Rightarrow \quad x=0, y=1$$

20. Given, $(\sin\alpha)x^2-2x+b\ge 2$

Let $f(x)=(\sin\alpha)x^2-2x+b-2$

Abscissa of the vertex is given by $x=\dfrac{1}{\sin\alpha}>1$

The graph of $f(x)=(\sin\alpha)x^2-2x+b-2, \forall x\le 1$ is shown in the figure.

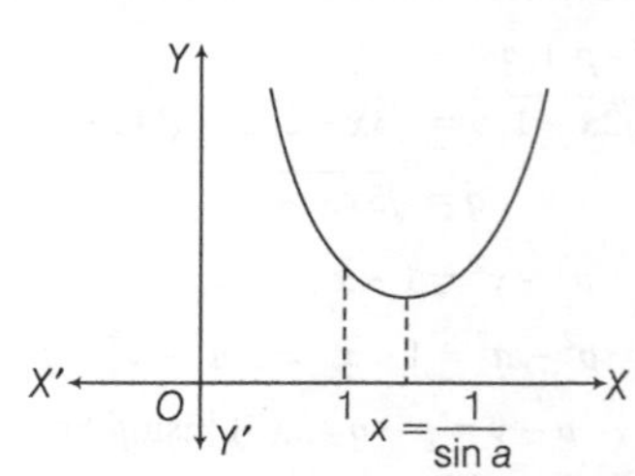

Therefore, minimum of $f(x)=(\sin\alpha)x^2-2x+b-2$ must be greater than zero but minimum is at $x=1$.

i.e. $\sin\alpha-2+b-2\ge 0,$

$$\Rightarrow \quad b\ge 4-\sin\alpha, \alpha\in(0,\pi)-\left\{\frac{\pi}{2}\right\}$$

21. Given, $|ax^2 + bx + c| \le 1, \forall x \in [0,1]$ ...(i)

Putting successively $x = 0$, 1 and $\frac{1}{2}$ in Eq. (i), we get

$$-1 \le c \le 1 \quad \text{...(ii)}$$
$$-1 \le a + b + c \le 1 \quad \text{...(iii)}$$
$$-4 \le a + 2b + 4c \le 4 \quad \text{...(iv)}$$

From Eqs. (ii), (iii) and (iv), we get

$$|b| \le 8 \text{ and } |a| \le 8$$
$$\Rightarrow \quad |a| + |b| + |c| \le 17$$

Sol. (Q. Nos. 22-24)

$|f(1) - f(0)| \le 2$ $\quad$ ($\because$ if $|u| \le 1, |v| \le 1$, then $|u - v| \le 2$)

$\Rightarrow \quad |a + b| \le 2$ $\quad$ $\left\{\because f(1) = a + b + c \text{ and } f(0) = c\right\}$

$\Rightarrow \quad (a + b)^2 \le 4$

Also, $|f(-1) - f(0)| \le 2$

$\Rightarrow \quad |a - b| \le 2$ $\quad$ $[\because f(-1) = a - b + c \text{ and } f(0) = c]$

$\Rightarrow \quad (a - b)^2 \le 4$

Now, $\quad 4a^2 + 3b^2 = 2(a + b)^2 + 2(a - b)^2 - b^2$

$$\le 2 \times 4 + 2 \times 4 - b^2 \le 16 - b^2 \le 16$$

$\Rightarrow$ Maximum value of $4a^2 + 3b^2 = 16$, when $b = 0$

$\Rightarrow \quad |a + b| + |a - b| = 4 \Rightarrow |a| = 2$

Also, $\quad |f(1) - f(0)| = |a + c - c| = |a| = 2$

$\Rightarrow \quad |a + c| = |c| = 1$

$\therefore$ The possible values of (a, b, c) are $(2, 0, -1)$ or $(-2, 0, 1)$

Also, $\frac{8}{3}a^2 + 2b^2 = \frac{2}{3}(4a^2 + 3b^2) \le \frac{2}{3}(16)$

22. $|a + c| = 1$, when $\frac{8a^2}{3} + 2b^2$ is maximum.

23. $|a + b| = 2$, when $\frac{8a^2}{3} + 2b^2$ is maximum.

24. The maximum value of $\frac{8a^2}{3} + 2b^2$ is $\frac{32}{3}$.

Sol. (Q. Nos. 25-27)

25. Given, $|2x| - |x - 4| = x + 4$

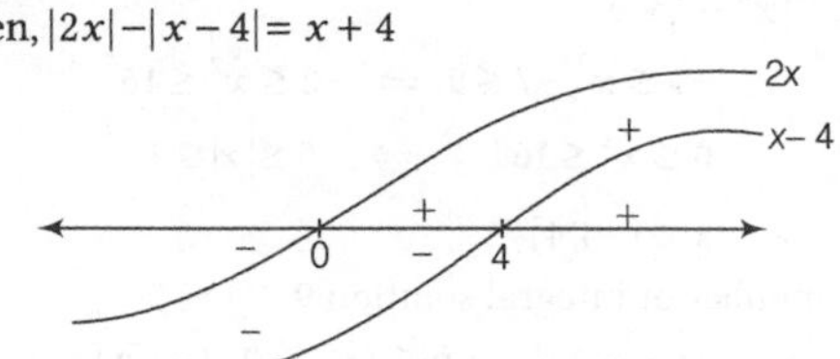

***Case* I** When, $-\infty < x \le 0$

$$-2x - 4 + x = x + 4$$
$$\Rightarrow \quad -2x = 8 \Rightarrow x = -4$$

***Case* II** When, $0 < x < 4$

$$2x - (4 - x) = x + 4$$
$$\Rightarrow \quad 2x = 8 \quad \Rightarrow x = 4$$

***Case* III** When, $4 \le x < \infty$

$$2x - x + 4 = x + 4$$
$$\Rightarrow \quad x + 4 = x + 4 \quad \Rightarrow x \in R$$

Solution of the given equation is $\{-4\} \cup [4, \infty)$.

$\therefore$ Least integer $= -4$.

26. Prime numbers less than 20 satisfying the equation are 5, 7, 11, 13, 17, 19 i.e. 6 prime number.

27. $P = 33$, $P^{2007} = (33)^{2007}$ = number has 7 at its units place.

Sol. (Q. Nos. 28-30)

28. Sum of digits $= P + Q + 15$

N is divisible by 9, if $P + Q + 15 = 18, 27$

$\Rightarrow \quad P + Q = 3$ ...(i)

or $\quad P + Q = 12$ ...(ii)

$$P = 0, 1, 2, 3, 4, 5, 6, 7, 8, 9$$

From Eq. (i), we get $\left.\begin{matrix} P = 0, Q = 3 \\ P = 1, Q = 2 \\ P = 2, Q = 1 \\ P = 3, Q = 0 \end{matrix}\right\}$ Number of ordered pairs is 4.

From Eq. (ii), we get

$\left.\begin{matrix} P = 3, Q = 9 \\ P = 4, Q = 8 \\ \dots \quad \dots \\ P = 8, Q = 4 \\ P = 9, Q = 3 \end{matrix}\right\}$ Number of ordered pairs is 7.

Total number of ordered pairs is 11.

29. N is divisible by 8, if $Q = 0, 4, 8$

Number of values of Q is 3.

30. $S_O = P + 9$

$S_E = Q + 6$

$$S_O - S_E = P - Q + 3$$

N is divisible is 11, if

$$P - Q + 3 = 0, 11$$
$$P - Q = -3 \quad \text{...(i)}$$

or $\quad P - Q = 8$ ...(ii)

N is divisible by 4, if

$$Q = 0, 2, 4, 6, 8$$

From Eq. (i), $\quad Q = 0, \quad P = -3$ $\quad$ (not possible)

$Q = 2, \quad P = -1$ $\quad$ (not possible)

$Q = 4, \quad P = 1$

$Q = 6, \quad P = 1$

$Q = 8, \quad P = 5$

$\therefore$ Number of ordered pairs is 3.

From Eq. (ii), $\quad Q = 0 \quad P = 8$

$Q = 2 \quad P = 10$ $\quad$ (not possible)

Similarly, $\quad Q \ne 4, 6, 8$

$\therefore$ Number of ordered pairs is 1.

$\therefore$ Total number of ordered pairs, so that the number N is divisible by 44 is 4.

Sol. (Q. Nos. 31-35)

31. If $\alpha = 0$, then the possible values of $\alpha - \beta$ are

$$\{0, -1, -2, -3, \dots, -9\}$$

If $\alpha = 1$, then the possible values of $\alpha - \beta$ are

$$\{1, 0, -1, -2, -3, \dots, -8\}$$

If $\alpha = 2$, then the possible values of $\alpha - \beta$ are

$$\{2, 1, 0, -1, -2, -3, \dots, -7\} \text{ and so on}$$

If $\alpha = 9$, then the possible values of $\alpha - \beta$ are

$$\{9, 8, 7, 6, 5, 4, 3, 2, 1, 0\}$$

Thus, in all the values of $\alpha - \beta$ are

$$\{-9, -8, -7, -6, -5, -4, -3, -2, -1, 0, 1, 2, 3, \ldots, 9\}$$

$\therefore$ Total number of values of $\alpha - \beta$ is 19.

$\therefore$ $p = 19$

32. Number N is divisible by 8, if $100 + 10\beta$ is divisible by 8.

$\therefore$ Possible values of β are 2, 6.

$\therefore$ $q = 2$

33. $N = 73\alpha 4961\beta 0$

N is divisible by 88, if N is divisible by 8 as well as 11.

N is divisible by 8, then $\beta = 2, 6$

For divisibility by 11,

$$S_O = 17 + \alpha,\ S_E = 13 + \beta$$

$$S_O - S_E = (17 + \alpha) - (13 + \beta) = \alpha - \beta + 4$$

$\alpha - \beta + 4$ be the integer lying between -5 and 13.

But, $\alpha - \beta + 4$ is divisible by 11, so $\alpha - \beta + 4 = 0,\ 11$

***Case* I** $\alpha - \beta + 4 = 0$

when $\beta = 2, \alpha = -2$ (rejected)

when $\beta = 6, \alpha = 2$

***Case* II** $\alpha - \beta + 4 = 11$ when $\alpha = 9, \beta = 2$

when $\beta = 6, \alpha = 13$ (rejected)

$\therefore$ $(\alpha, \beta) = (9, 2)\ (2, 6)$

$\therefore$ Number of ordered pairs are 2.

34. N is divisible by 6, if it is divisible is 2 and 3.

$N = 73\alpha 4961\beta 0$ is divisible by 2 for all values of α and β.

$$30 \le (30 + \alpha + \beta) \in I \le 48$$

But $30 + \alpha + \beta$ is divisible by 3, if

$$30 + \alpha + \beta = 30, 33, 36, 39, 42, 45, 48$$

$$\alpha + \beta = 0, 3, 6, 9, 12, 15, 18$$

Number of possible values of $(\alpha + \beta)$ is equal to 7.

35. $i^N = 1 \Rightarrow N$ must be divisible by 4.

$$\Rightarrow \beta = 0, 2, 4, 6, 8$$

Number of possible values of β is 5.

Sol. (Q. Nos. 36-38)

36. Let $n = 3k + 4, r = 0, 1, 2$

$$n^2 = 9k^2 + 6kr + r^2 = 3m + r^2$$

$$r^2 = 0, 1, 4$$

$\therefore$ Remainder = 0 or 1.

37. (a) $n^2 + n + 1$ is an odd number.

(b) when $n = 1$, $n^2 + n + 1$ is divisible by 3.

(c) $111 = 10^2 + 10 + 1$ $\therefore$ when $n = 10$

$n^2 + n + 1$ is divisible by 111.

38. n = odd

Let $n = 2k + 1$

$$n^5 - n = n(n^4 - 1) = n(n-1)(n+1)(n^2+1)$$
$$= (2k+1)(2k)(2k+2)(4k^2+4k+2)$$
$$= 8(k)(k+1)(2k+1)(2k^2+2k+1)$$

which is divisible by 15, 16 and 240 but not divisible by 720

When $n = 5 \Rightarrow n(n^4 - 1)$

$$= 5(5^4 - 1) = 5(625 - 1) = 5 \times 624 = 3120,$$

which is not divisible by 720.

39. Given, $|x-1| - |2x-5| \ge 4$

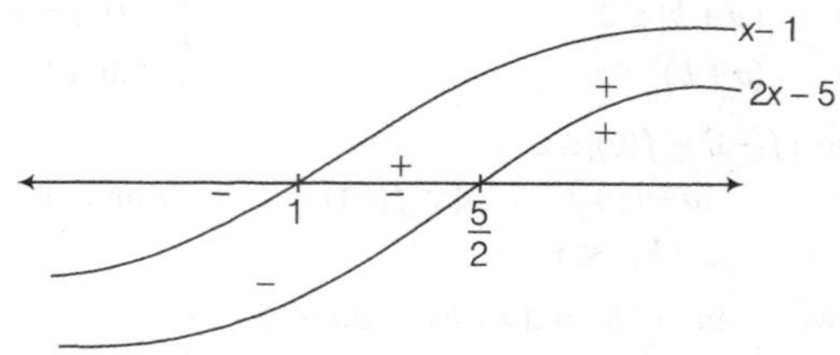

***Case* I** When, $x < 1$

$$1 - x - (5 - 2x) \ge 4 \Rightarrow x - 4 \ge 4 \Rightarrow x \ge 8$$

but $x < 1$

$\Rightarrow$ No solution

***Case* II** $1 \le x \le 5/2$

$$x - 1 + 2x - 5 \ge 4$$

$\Rightarrow$ $3x - 6 \ge 4 \Rightarrow 3x \ge 10$

$\Rightarrow$ $x \ge \dfrac{10}{3}$ but $1 \le x \le 5/2$

$\Rightarrow$ No such x exist.

***Case* III** $x \ge 5/2$

$\Rightarrow$ $x - 1 - (2x - 5) \ge 4 \Rightarrow x - 1 - 2x + 5 \ge 4$

$\Rightarrow$ $-x + 4 \ge 4 \Rightarrow x \le 0$

$\Rightarrow$ No such x exist.

$\therefore$ Number of solution is 0.

40. Given, $|x^2 - 7| \le 9$

$\Rightarrow$ $-9 \le x^2 - 7 \le 9 \Rightarrow -2 \le x^2 \le 16$

$\Rightarrow$ $0 \le x^2 \le 16 \Rightarrow 0 \le |x| \le 4$

$\Rightarrow$ $x \in [-4, 4]$

Hence, number of integral solution 9.

41. System of equations are $x + 2y = 6$ and $|x - 3| = y$

***Case* I** When $x \ge 3$, $x + 2y = 6$ and $x - 3 = y$

Solving these equations, we get $x = 4, y = 1$

***Case* II** When $x < 3$, $x + 2y = 6$ and $3 - x = y$

Solving these equations, we get $x = 0, y = 3$

Hence, two solutions are (0,3) and (4,1).

CHAPTER

02

Differentiation

Learning Part

Session 1

- Geometrical Meaning of the Derivative
- Derivative of $f(x)$ from the First Principle or *ab*-Initio Method
- Rules of Differentiation

Session 2

- Chain Rule

Session 3

- Differentiation of Implicit Functions

Session 4

- Differentiation of Inverse Trigonometric Functions
- Graphical Approach for Differentiation of Inverse Trigonometric Functions

Session 5

- Differentiation of a Function in Parametric Form

Session 6

- Logarithmic Differentiation
- Differentiation of Infinite Series

Session 7

- Differentiation of a Function w.r.t. Another Function

Session 8

- Higher Derivatives of a Function

Session 9

- Differentiation of a Function given in the form of a Determinant

Session 10

- Derivative of an Inverse Function

Practice Part

- JEE Type Examples
- Chapter Exercises

Arihant on Your Mobile !

Exercises with this symbol can be practised on your mobile. See title inside to activate for free.

Session 1

Geometrical Meaning of the Derivative, Derivative of *f*(*x*) from the First Principle or *ab*-Initio Method, Rules of Differentiation

Derivative is the instantaneous rate of change of a function with respect to dependent variable. Also, the derivative is the slope of tangent to a curve at a point.

Geometrical Meaning of the Derivative

Let us consider a function $y = f(x)$ in a rectangular coordinate system. We also consider a point $P(x, y)$ on the curve. If a point corresponding to an increased value of the argument $x + \Delta x$ is considered, its ordinate value is given by$y + \Delta y = f(x + \Delta x)$.

In figure, the point $(x + \Delta x, y + \Delta y)$ is represented by *A*.

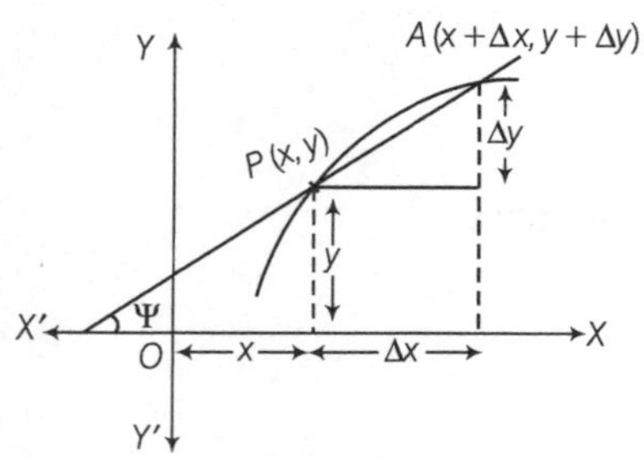

Figure 2.1

Here, *PA* is the secant to the curve.

Now, as $\Delta x \to 0$, $\Delta y \to 0 \Rightarrow PA \to 0$

(i.e. the distance *PA* tends to zero or to a single point *P*)

$\Rightarrow \lim_{\Delta x \to 0}$ (slope of chord *PA*) $\to$ (slope of tangent at *P*).

or $\tan \psi = \lim_{\Delta x \to 0} \tan \theta$

$$\Rightarrow \lim_{\Delta x \to 0} \frac{\Delta y}{\Delta x} = \frac{dy}{dx} \quad \text{or} \quad f'(x)$$

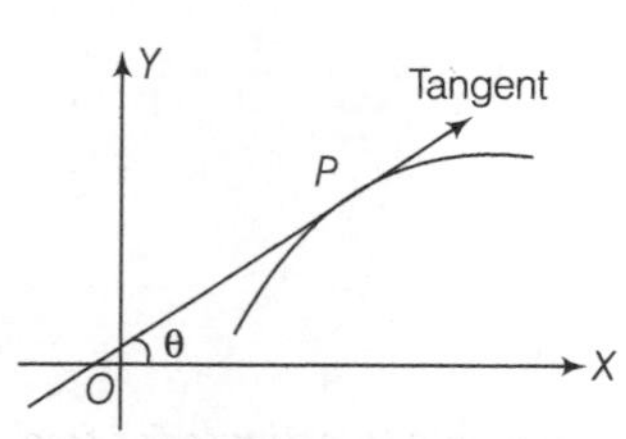

Figure 2.2

which means that **the value of the derivative $f'(x)$ for a given value of x is equal to the tangent of the angle formed by the tangent line to the graph of the function $y = f(x)$ at the point $P(x, y)$ with the positive X-axis.**

Relation between *dy*/*dx* and *dx*/*dy*

Let x and y be two variables connected by a relation of the form $f(x, y) = 0$.

Let Δx be a small change in x and Δy be a small change in y, then

$$\frac{dy}{dx} = \lim_{\Delta x \to 0} \frac{\Delta y}{\Delta x} \quad \text{and} \quad \frac{dx}{dy} = \lim_{\Delta y \to 0} \frac{\Delta x}{\Delta y}$$

Now, $$\frac{\Delta y}{\Delta x} \cdot \frac{\Delta x}{\Delta y} = 1$$

$$\Rightarrow \lim_{\Delta x \to 0} \left(\frac{\Delta y}{\Delta x} \cdot \frac{\Delta x}{\Delta y} \right) = 1$$

$$\Rightarrow \lim_{\Delta x \to 0} \frac{\Delta y}{\Delta x} \cdot \lim_{\Delta y \to 0} \frac{\Delta x}{\Delta y} = 1 \quad [\because \Delta x \to 0 \Leftrightarrow \Delta y \to 0]$$

$$\Rightarrow \frac{dy}{dx} \cdot \frac{dx}{dy} = 1$$

So, $$\frac{dy}{dx} = \frac{1}{dx/dy}$$

Derivatives of $f(x)$ from the First Principle

or ab-Initio Method or Delta Method or By Definition

If $f(x)$ is differentiable function, then

$$\lim_{\Delta x \to 0} \frac{\Delta y}{\Delta x} = \lim_{\Delta x \to 0} \frac{f(x + \Delta x) - f(x)}{\Delta x} = f'(x) = \frac{dy}{dx}$$

Simply, $$f'(x) = \lim_{h \to 0} \frac{f(x + h) - f(x)}{h}$$

Example 1 Differentiate the following functions w.r.t. x using first principle.

(i) $f(x)=\tan x$ (ii) $f(x)=e^{2x}$

(iii) e^{x^2} (iv) $\sqrt{\sin x}$

Sol. (i) Let $f(x)=\tan x$

Then, $f(x+h)=\tan(x+h)$

$$\therefore \frac{d}{dx}(\tan x)=\lim_{h\to 0}\frac{f(x+h)-f(x)}{h}$$

$$=\lim_{h\to 0}\frac{\tan(x+h)-\tan(x)}{h}$$

$$=\lim_{h\to 0}\frac{\tan(x+h-x)\cdot[1+\tan x\cdot\tan(x+h)]}{h}$$

$[\because \tan A-\tan B=\tan(A-B)\cdot(1+\tan A\tan B)]$

$$=\lim_{h\to 0}\frac{\tan h}{h}\cdot(1+\tan^2 x)\quad\left[\because \lim_{h\to 0}\frac{\tan h}{h}=1\right]$$

$$=\sec^2 x$$

$$\therefore \frac{d}{dx}(\tan x)=\sec^2 x$$

(ii) Let $f(x)=e^{2x}\Rightarrow f(x+h)=e^{2(x+h)}$

$$\therefore \frac{d}{dx}(f(x))=\lim_{h\to 0}\frac{f(x+h)-f(x)}{h}$$

$$\Rightarrow \frac{d}{dx}(e^{2x})=\lim_{h\to 0}\frac{e^{2(x+h)}-e^{2x}}{h}=\lim_{h\to 0}e^{2x}\cdot\frac{(e^{2h}-1)}{h}$$

$$=e^{2x}\cdot\left(\lim_{h\to 0}\frac{e^{2h}-1}{2h}\right)\times 2=2e^{2x}\quad\left[\because \lim_{x\to 0}\frac{e^x-1}{x}=1\right]$$

$$\therefore \frac{d}{dx}(e^{2x})=2e^{2x}$$

(iii) Let $f(x)=e^{x^2}\Rightarrow f(x+h)=e^{(x+h)^2}$

$$\therefore \frac{d}{dx}f(x)=\lim_{h\to 0}\frac{f(x+h)-f(x)}{h}=\lim_{h\to 0}\frac{e^{(x+h)^2}-e^{x^2}}{h}$$

$$=\lim_{h\to 0}\frac{e^{x^2+h^2+2hx}-e^{x^2}}{h}$$

$$=\lim_{h\to 0}\frac{e^{x^2}\cdot e^{h^2+2hx}-e^{x^2}}{h}=\lim_{h\to 0}e^{x^2}\left(\frac{e^{h^2+2hx}-1}{h}\right)$$

$$=e^{x^2}\lim_{h\to 0}\left(\frac{e^{h^2+2hx}-1}{h^2+2hx}\right)\times\left(\frac{h^2+2hx}{h}\right)$$

$$=e^{x^2}\lim_{t\to 0}\left(\frac{e^t-1}{t}\right)\lim_{h\to 0}(h+2x),\text{ where } t=h^2+2hx$$

$$=e^{x^2}\times 1\times\lim_{h\to 0}(h+2x)\quad\left[\because \lim_{x\to 0}\frac{e^x-1}{x}=1\right]$$

$$=e^{x^2}\times 2x=2xe^{x^2}$$

(iv) Let $f(x)=\sqrt{\sin x}$

$$\Rightarrow f(x+h)=\sqrt{\sin(x+h)}$$

$$\therefore \frac{d}{dx}f(x)=\lim_{h\to 0}\frac{f(x+h)-f(x)}{h}=\lim_{h\to 0}\frac{\sqrt{\sin(x+h)}-\sqrt{\sin x}}{h}$$

$$=\lim_{h\to 0}\frac{\sin(x+h)-\sin x}{h(\sqrt{\sin(x+h)})+\sqrt{\sin x}}$$

[rationalising the numerator]

$$=\lim_{h\to 0}\frac{2\sin\left(\frac{h}{2}\right)\cos\left(\frac{2x+h}{2}\right)}{h(\sqrt{\sin(x+h}+\sqrt{\sin x})}$$

$$\left[\because \sin C-\sin D=2\cos\left(\frac{C+D}{2}\right)\sin\left(\frac{C-D}{2}\right)\right]$$

$$=\lim_{h/2\to 0}\frac{\sin h/2}{h/2}\lim_{h\to 0}\frac{\cos(x+h/2)}{\sqrt{\sin(x+h)}+\sqrt{\sin x}}$$

[as $h\to 0\Rightarrow h/2\to 0$]

$$=1\cdot\frac{\cos x}{2\sqrt{\sin x}}=\frac{\cos x}{2\sqrt{\sin x}}$$

Differential Coefficients of Standard Functions

(i) $\frac{d}{dx}(\text{constant})=0$ (ii) $\frac{d}{dx}(x^n)=nx^{n-1}$

(iii) $\frac{d}{dx}(|x|)=\frac{x}{|x|}$ (iv) $\frac{d}{dx}(e^x)=e^x$

(v) $\frac{d}{dx}(a^x)=a^x\log a$ (vi) $\frac{d}{dx}(\log_e|x|)=\frac{1}{x}$

(vii) $\frac{d}{dx}(\log_a|x|)=\frac{1}{x\log_e a}$ (viii) $\frac{d}{dx}(\sin x)=\cos x$

(ix) $\frac{d}{dx}(\cos x)=-\sin x$ (x) $\frac{d}{dx}(\tan x)=\sec^2 x$

(xi) $\frac{d}{dx}(\cot x)=-\text{cosec}^2 x$ (xii) $\frac{d}{dx}(\sec x)=\sec x\tan x$

(xiii) $\frac{d}{dx}(\text{cosec } x)=-\text{cosec } x\cot x$

(xiv) $\frac{d}{dx}(\sin^{-1}x)=\frac{1}{\sqrt{1-x^2}},\ -1<x<1$ or $|x|<1$

(xv) $\frac{d}{dx}(\cos^{-1}x)=\frac{-1}{\sqrt{1-x^2}},\ -1<x<1$ or $|x|<1$

(xvi) $\frac{d}{dx}(\tan^{-1}x)=\frac{1}{1+x^2},\ -\infty<x<\infty$ or $x\in R$

(xvii) $\frac{d}{dx}(\sec^{-1}x)=\frac{1}{|x|\sqrt{x^2-1}},\ |x|>1$ or $x\in R-[-1,1]$

(xviii) $\frac{d}{dx}(\cot^{-1}x)=\frac{-1}{1+x^2},\ -\infty<x<\infty$ or $x\in R$

(xix) $\frac{d}{dx}(\text{cosec}^{-1}x)=\frac{-1}{|x|\sqrt{x^2-1}},\ |x|>1$ or $x\in R-[-1,1]$

Rules for Differentiation

Here, we will discuss some rules which are related to differentiation.

Differentiation of Sum and Difference of Functions

When a function is expressed as sum or difference of two or more functions then, to find the derivative of function, we need to differentiate each term seperately. i.e.

If $y = u(x) \pm v(x) \pm w(x) \pm \dots$, then

$$\frac{dy}{dx} = \frac{du(x)}{dx} \pm \frac{dv(x)}{dx} \pm \frac{dw(x)}{dx} \pm \dots$$

is known as term-by-term differentiation.

Example 2 If $y = \sin x + e^x$, then find $\frac{dy}{dx}$.

Sol. Here, $y = \sin x + e^x$

On differentiating both sides w.r.t. x, we get

$$\frac{dy}{dx} = \frac{d}{dx}(\sin x + e^x)$$

$$\Rightarrow \quad \frac{dy}{dx} = \frac{d}{dx}\sin x + \frac{d}{dx} e^x$$

$$\Rightarrow \quad \frac{dy}{dx} = \cos x + e^x$$

Example 3 If $y = x^2 + \sin^{-1} x + \log_e x$, then find $\frac{dy}{dx}$.

Sol. Here, $y = x^2 + \sin^{-1} x + \log_e x$

On differentiating, we get

$$\frac{dy}{dx} = \frac{d}{dx}(x^2) + \frac{d}{dx}(\sin^{-1} x) + \frac{d}{dx}(\log_e x)$$

$$\Rightarrow \quad \frac{dy}{dx} = 2(x)^{2-1} + \frac{1}{\sqrt{1-x^2}} + \frac{1}{x}$$

$$\therefore \quad \frac{dy}{dx} = 2x + \frac{1}{\sqrt{1-x^2}} + \frac{1}{x}$$

Example 4 If $y = \log_{10} x + \log_e 10 + \log_x x + \log_{10} 10$, then find $\frac{dy}{dx}$.

Sol. Here, $y = \log_{10} x + \log_e 10 + \log_x x + \log_{10} 10$

$$\Rightarrow \quad y = \log_{10} x + \log_e 10 + 1 + 1 \qquad [\because \log_a a = 1]$$

$$\Rightarrow \quad y = \log_{10} x + \log_e 10 + 2$$

On differentiating, we get

$$\frac{dy}{dx} = \frac{d}{dx}(\log_{10} x) + \frac{d}{dx}(\log_e 10) + \frac{d}{dx}(2)$$

$$= \frac{1}{x \log_e 10} + 0 + 0$$

$$= \frac{1}{x \log_e 10}$$

Example 5 If $y = x^{-1/2} + \log_5 x + \frac{\sin x}{\cos x} + 2^x$, then find $\frac{dy}{dx}$.

Sol. Here, $y = x^{-1/2} + \log_5 x + \tan x + 2^x$

On differentiating, we get

$$\frac{dy}{dx} = \frac{d}{dx}(x)^{-1/2} + \frac{d}{dx}(\log_5 x) + \frac{d}{dx}\tan x + \frac{d}{dx}(2^x)$$

$$= -\frac{1}{2}(x)^{-1/2-1} + \frac{1}{x \log_e 5} + \sec^2 x + 2^x \log 2$$

$$= -\frac{1}{2}x^{-3/2} + \frac{1}{x \log_e 5} + \sec^2 x + 2^x \log 2$$

Differentiation of a Function Multiplied with a Constant

If a function multiplied by a constant, then the derivative of constant times a function is the constant times the derivative of the function. i.e.

If $y = k\,f(x)$, then on differentiating, we get

$$\frac{dy}{dx} = k \cdot \frac{d}{dx} f(x)$$

Example 6 If $y = m^2 \sec^{-1} x$, then find $\frac{dy}{dx}$.

Sol. Here, $y = m^2 \sec^{-1} x$

On differentiating, we get

$$\frac{dy}{dx} = m^2 \frac{d}{dx}\sec^{-1} x$$

$$= m^2 \times \frac{1}{|x|\sqrt{x^2-1}}$$

$$= \frac{m^2}{|x|\sqrt{x^2-1}}$$

Example 7 If $y = \log x^3 + 3\sin^{-1} x + kx^2$, then find $\frac{dy}{dx}$.

Sol. Here, $y = \log x^3 + 3\sin^{-1} x + kx^2$

On differentiating, we get

$$\frac{dy}{dx} = \frac{d}{dx}[\log x^3] + \frac{d}{dx}[3\sin^{-1} x] + \frac{d}{dx}[kx^2]$$

$$= 3\frac{d}{dx}[\log x] + 3\frac{d}{dx}(\sin^{-1} x) + k\frac{d}{dx}(x^2) \quad [\because \log x^n = n\log x]$$

$$= 3\cdot\frac{1}{x} + 3\cdot\frac{1}{\sqrt{1-x^2}} + k(2x) = \frac{3}{x} + \frac{3}{\sqrt{1-x^2}} + 2kx$$

Product Rule

If $u(x)$ and $v(x)$ are two differentiable functions, then $u(x)\cdot v(x)$ is also differentiable.

If $y = u(x)\cdot v(x)$, then

$$\frac{dy}{dx} = \left\{\frac{d}{dx}u(x)\right\}\cdot v(x) + u(x)\cdot\left\{\frac{d}{dx}v(x)\right\}.$$

which is known as product rule.

i.e. **derivative of product of two functions**

= (first function) × (derivative of second function) + (second function) × (derivative of first function)

Example 8 If $y = e^x \sin x$, then find $\frac{dy}{dx}$.

Sol. Here, $y = e^x \sin x$

On differentiating, we get

$$\frac{dy}{dx} = \left\{\frac{d}{dx}(e^x)\right\}\cdot\sin x + e^x\cdot\left\{\frac{d}{dx}(\sin x)\right\}$$

$$= e^x\cdot\sin x + e^x\cdot\cos x = e^x(\sin x + \cos x)$$

Example 9 If $y = e^x \tan x + x\cdot\log_e x$, then find $\frac{dy}{dx}$.

Sol. Here, $y = e^x \tan x + x\cdot\log_e x$

On differentiating, we get

$$\frac{dy}{dx} = \frac{d}{dx}(e^x \tan x) + \frac{d}{dx}(x\log x)$$

$$= \left\{\frac{d}{dx}e^x\right\}\cdot\tan x + e^x\cdot\left\{\frac{d}{dx}\tan x\right\} + \left\{\frac{d}{dx}x\right\}\cdot\log x + x\cdot\left\{\frac{d}{dx}\log x\right\}$$

$$= e^x\cdot\tan x + e^x\cdot\sec^2 x + 1\cdot\log x + x\cdot\frac{1}{x}$$

Hence, $\frac{dy}{dx} = e^x(\tan x + \sec^2 x) + (\log x + 1)$

Differentiation of Product of More than Two Functions

(i) If 3 functions are involved, then remember

$$D\{f(x)\cdot g(x)\cdot h(x)\} = f(x)\cdot g(x)\cdot h'(x) + f(x)\cdot g'(x)\cdot h(x) + f'(x)\cdot g(x)\cdot h(x)$$

(ii) The result can be generalised to product of n terms as

$$D\{f_1(x) f_2(x) f_3(x)\dots f_n(x)\} = [f_1'(x)\cdot f_2(x)\cdot f_3(x)\dots f_n(x)] + [f_1(x)\cdot f_2'(x) f_3(x)\dots f_n(x)] + [f_1(x) f_2(x) f_3'(x)\dots f_n(x)] + \dots + [f_1(x) f_2(x) f_3(x)\dots f_n'(x)]$$

Example 10 Let f, g and h be differentiable functions. If $f(0)=1$, $g(0)=2$, $h(0)=3$ and the derivative of their pairwise product at $x=0$ are $(fg)'(0)=6$; $(gh)'(0)=4$ and $(hf)'(0)=5$, then compute the value of $(fgh)'(0)$.

Sol. We know that, $(fgh)' = \frac{(fg)'h + (gh)'f + (hf)'g}{2}$

$$\therefore \quad (fgh)'(0) = \frac{\left[\begin{array}{l}(fg)'(0)\cdot h(0) + (gh)'(0)\cdot f(0)\\ + (hf)'(0)\cdot g(0)\end{array}\right]}{2}$$

$$= \frac{(6)\cdot(3) + (4)\cdot(1) + (5)\cdot(2)}{2} = 16$$

Example 11 If $y = (1+x)(1+x^2)(1+x^4)\dots(1+x^{2n})$, then find $\frac{dy}{dx}$ at $x=0$.

Sol. Here, $y = (1+x)(1+x^2)(1+x^4)\dots(1+x^{2n})$

On differentiating, we get

$$\frac{dy}{dx} = (1)(1+x^2)(1+x^4)\dots(1+x^{2n}) + (1+x)(2x)(1+x^4)\dots(1+x^{2n}) + (1+x)(1+x^2)(4x^3)\dots(1+x^{2n}) + \dots + (1+x)(1+x^2)(1+x^4)(1+x^8)\dots(1+x^{2n-1})\times 2n\cdot x^{2n-1}$$

Now, $\left(\frac{dy}{dx}\right)_{x=0} = 1$

Quotient Rule

If $u(x)$ and $v(x)$ are two differentiable functions such that $v(x)\neq 0$, then $\frac{u(x)}{v(x)}$ is also differentiable.

If $y = \dfrac{u(x)}{v(x)}$, then

$$\frac{dy}{dx} = \frac{\left\{\frac{d}{dx}u(x)\right\}\cdot v(x) - \left\{\frac{d}{dx}v(x)\right\}\cdot u(x)}{\{v(x)\}^2}$$

is known as the quotient rule of differentiation.

Reader's can learn the given formula in following ways

Derivative of quotient of two functions

$$= \frac{\text{(denominator)} \times \text{(derivative of numerator)} - \text{(numerator)} \times \text{derivative of denominator}}{\text{(denominator)}^2}$$

Example 12 If $y = \dfrac{x}{x^2+1}$, then find $\dfrac{dy}{dx}$.

Sol. Here, $y = \dfrac{x}{x^2+1}$

On differentiating, we get

$$\frac{dy}{dx} = \frac{(x^2+1)\frac{d}{dx}(x) - x\frac{d}{dx}(x^2+1)}{(x^2+1)^2}$$

$$= \frac{(x^2+1)\cdot 1 - x\cdot(2x)}{(x^2+1)^2} = \frac{1-x^2}{(1+x^2)^2}$$

Example 13 If $y = \dfrac{x \sin x}{\log_e x}$, then find $\dfrac{dy}{dx}$.

Sol. Here, $y = \dfrac{x \sin x}{\log_e x}$

On differentiating, we get

$$\frac{dy}{dx} = \frac{\log_e x \frac{d}{dx}(x \sin x) - x \sin x \frac{d}{dx}(\log_e x)}{(\log_e x)^2}$$

$$= \frac{\log_e x(x \cos x + \sin x) - x \sin x \left(\frac{1}{x}\right)}{(\log_e x)^2}$$

$$= \frac{\log_e x\,(x \cos x + \sin x) - \sin x}{(\log_e x)^2}$$

Example 14 If $y = \dfrac{3a^2x - x^3}{a^3 - 3ax^2}$, then find $\dfrac{dy}{dx}$.

Sol. Here, $y = \dfrac{3a^2x - x^3}{a^3 - 3ax^2}$

On differentiating, we get

$$\frac{dy}{dx} = \frac{\left[(a^3 - 3ax^2)\frac{d}{dx}(3a^2x - x^3) - (3a^2x - x^3)\frac{d}{dx}(a^3 - 3ax^2)\right]}{(a^3 - 3ax^2)^2}$$

$$= \frac{(a^3 - 3ax^2)(3a^2 - 3x^2) - (3a^2x - x^3)(-6ax)}{(a^3 - 3ax^2)^2}$$

$$= \frac{3ax^4 + 6a^3x^2 + 3a^5}{(a^3 - 3ax^2)^2}$$

Example 15 If $y = \dfrac{e^x - \tan x}{x^n + \cot x}$, then find $\dfrac{dy}{dx}$.

Sol. Here, $y = \dfrac{e^x - \tan x}{x^n + \cot x}$

On differentiating, we get

$$\frac{dy}{dx} = \frac{\left[(x^n + \cot x)\frac{d}{dx}(e^x - \tan x) - (e^x - \tan x)\frac{d}{dx}(x^n + \cot x)\right]}{(x^n + \cot x)^2}$$

$$\therefore \quad \frac{dy}{dx} = \frac{\left[(e^x - \sec^2 x)(x^n + \cot x) - (e^x - \tan x)(nx^{n-1} - \operatorname{cosec}^2 x)\right]}{(x^n + \cot x)^2}$$

Example 16 If $y = \dfrac{\log x}{x} + e^x \sin x + \log_5 x$, then find $\dfrac{dy}{dx}$.

Sol. Here, $y = \dfrac{\log x}{x} + e^x \sin x + \log_5 x$

On differentiating, we get

$$\frac{dy}{dx} = \frac{d}{dx}\left(\frac{\log x}{x}\right) + \frac{d}{dx}(e^x \sin x) + \frac{d}{dx}(\log_5 x)$$

$$= \frac{\left\{\frac{d}{dx}(\log x)\right\}\cdot x - \log x\left\{\frac{d}{dx}x\right\}}{x^2} + \left\{\frac{d}{dx}e^x\right\}\cdot \sin x + e^x\cdot\left\{\frac{d}{dx}\sin x\right\} + \frac{1}{x \log_e 5}$$

$$= \frac{\frac{1}{x}\cdot x - \log x \cdot 1}{x^2} + e^x \sin x + e^x \cdot \cos x + \frac{1}{x \log_e 5}$$

Hence, $\dfrac{dy}{dx} = \left(\dfrac{1 - \log x}{x^2}\right) + e^x(\sin x + \cos x) + \dfrac{1}{x \log_e 5}$

Remark

While applying the quotient rule, think twice and check whether your function could be simplified prior to differentiation. (See Examples 17 and 18).

Example 17 If $y = \dfrac{x^4 + x^2 + 1}{x^2 + x + 1}$ and $\dfrac{dy}{dx} = ax + b$, then find a and b.

Sol. Here, $y = \dfrac{x^4 + x^2 + 1}{x^2 + x + 1}$

$$\Rightarrow \quad y = \frac{x^4 + 2x^2 + 1 - x^2}{x^2 + x + 1} = \frac{(x^2 + 1)^2 - x^2}{(x^2 + x + 1)}$$

$$\therefore \quad y = \frac{(x^2 + x + 1)(x^2 - x + 1)}{(x^2 + x + 1)}$$

$$[\because a^2 - b^2 = (a + b)(a - b)]$$

$$\Rightarrow \quad y = x^2 - x + 1 \Rightarrow \frac{dy}{dx} = 2x - 1$$

$$\Rightarrow \quad ax + b = 2x - 1 \qquad \left[\because \frac{dy}{dx} = ax + b\right]$$

$$\therefore \quad a = 2, b = -1$$

Example 18 If $y = \dfrac{\sec x + \tan x - 1}{(-\sec x) + \tan x + 1}$, then find $\dfrac{dy}{dx}$ at $x = \dfrac{\pi}{4}$.

Sol. Here, $y = \dfrac{\sec x + \tan x - 1}{-\sec x + \tan x + 1}$

$$\Rightarrow \quad y = \frac{(\sec x + \tan x) - (\sec^2 x - \tan^2 x)}{-\sec x + \tan x + 1}$$

$$(\because \sec^2 x - \tan^2 x = 1)$$

$$\Rightarrow \quad y = \frac{(\sec x + \tan x)\{1 - \sec x + \tan x\}}{\tan x - \sec x + 1}$$

$$\Rightarrow \quad y = \sec x + \tan x \Rightarrow \frac{dy}{dx} = \sec x \tan x + \sec^2 x$$

$$\therefore \quad \left(\frac{dy}{dx}\right)_{\text{at } x = \frac{\pi}{4}} = \sqrt{2} + 2$$

Example 19 If $y = \dfrac{\tan^{-1} x - \cot^{-1} x}{\tan^{-1} x + \cot^{-1} x}$, then find $\left(\dfrac{dy}{dx}\right)$ at $x = -1$.

Sol. Here, $y = \dfrac{\tan^{-1} x - \cot^{-1} x}{\tan^{-1} x + \cot^{-1} x}$, as $\tan^{-1} x + \cot^{-1} x = \dfrac{\pi}{2}$

$$\therefore \quad y = \frac{\tan^{-1} x - (\pi/2 - \tan^{-1} x)}{(\pi/2)}$$

$$\Rightarrow \quad y = \frac{2}{\pi}\left[2\tan^{-1} x - \frac{\pi}{2}\right] \Rightarrow \frac{dy}{dx} = \frac{2}{\pi}\left[\frac{2}{1 + x^2}\right]$$

$$\therefore \quad \left(\frac{dy}{dx}\right)_{x = -1} = \frac{2}{\pi}$$

Exercise for Session 1

■ **Directions** (Q. Nos. 1 to 9) *Differentiate the following functions w.r.t. x.*

1. $e^{x \log a} + e^{a \log x} + e^{a \log a}$

2. $\sqrt{\dfrac{1 - \cos 2x}{1 + \cos 2x}}$

3. $\log_3 x + 3 \log_e x + 2 \tan x$

4. $|x| + a_0 x^n + a_1 x^{n-1} + a_2 x^{n-2} + \ldots + a_{n-1}x + a_n$

5. $\sec^{-1}\left(\dfrac{x + 1}{x - 1}\right) + \sin^{-1}\left(\dfrac{x - 1}{x + 1}\right)$

6. $x^n \log_a x \, e^x$

7. $\dfrac{2^x \cot x}{\sqrt{x}}$

8. $\dfrac{\sin x - x \cos x}{x \sin x + \cos x}$

9. If $y = 1 + \dfrac{x}{1!} + \dfrac{x^2}{2!} + \dfrac{x^3}{3!} + \dfrac{x^4}{4!} + \ldots \infty$, then find the value of $\dfrac{dy}{dx}$.

10. Find the values of 'x' for which the rate of change of $\dfrac{x^4}{4} + \dfrac{x^3}{3} - x$ is more than the rate of change of $\dfrac{x^4}{4}$.

Session 2

Chain Rule (Derivative of Composite Function)

Chain Rule

(Derivative of Composite Function)

If $u(x)$ and $v(x)$ are differentiable functions, then $uov(x)$ or $u[v(x)]$ is also differentiable.

If $y[uov(x)] = [u\{v(x)\}]$, then

$$\frac{dy}{dx} = \frac{du\{v(x)\}}{d\{v(x)\}} \times \frac{d}{dx} v(x)$$

is known as chain rule. Or

If $y = f(u)$ and $u = g(x)$, then $\frac{dy}{dx} = \frac{dy}{du} \cdot \frac{du}{dx}$

The chain rule can be extended as follows

If $y[uovow(x)] = u[v\{w(x)\}]$, then

$$\frac{dy}{dx} = \frac{du[v\{w(x)\}]}{dv\{w(x)\}} \times \frac{dv\{w(x)\}}{dw(x)} \times \frac{dw(x)}{dx}$$

Example 20 If $y = \log(\sin x)$, then find $\frac{dy}{dx}$.

Sol. Here, $y = \log(\sin x)$

On differentiating, we get

$$\frac{dy}{dx} = \frac{d[\log(\sin x)]}{d(\sin x)} \times \frac{d}{dx}(\sin x) = \frac{1}{\sin x} \times \cos x$$

[by chain rule]

Hence, $\frac{dy}{dx} = \cot x$

Aliter

Here, $y = \log(\sin x)$

Put $\sin x = u$, we have, $y = \log u$ where $u = \sin x$

On differentiating, we get

$$\frac{dy}{dx} = \frac{dy}{du} \cdot \frac{du}{dx} \quad \text{[by chain rule]}$$

$$\frac{dy}{dx} = \frac{d}{dx}(\log u) \cdot \frac{d}{dx}(\sin x)$$

$$= \frac{1}{u} \times \cos x$$

$$\frac{dy}{dx} = \frac{1}{\sin x} \times \cos x = \cot x \quad [\because u = \sin x]$$

Example 21 If $y = e^{(\tan^{-1} x)^3}$, then find $\frac{dy}{dx}$.

Sol. Here, $y = e^{(\tan^{-1} x)^3}$

On differentiating, we get

$$\frac{dy}{dx} = \frac{d\{e^{(\tan^{-1} x)^3}\}}{d\{(\tan^{-1} x)^3\}} \cdot \frac{d\{(\tan^{-1} x)^3\}}{d(\tan^{-1} x)} \cdot \frac{d(\tan^{-1} x)}{dx}$$

[by chain rule]

$$= e^{(\tan^{-1} x)^3} \cdot 3(\tan^{-1} x)^2 \cdot \frac{1}{1+x^2}$$

Aliter

Here, $y = e^{(\tan^{-1} x)^3}$

Put $(\tan^{-1} x) = v$ and $(\tan^{-1} x)^3 = u$

$\Rightarrow$ $u = v^3$

$\therefore$ $y = e^u$, $u = v^3$ and $v = \tan^{-1} x$

On differentiating, we get

$$\frac{dy}{dx} = \frac{dy}{du} \cdot \frac{du}{dv} \cdot \frac{dv}{dx}$$

$$\frac{dy}{dx} = \frac{d}{du} e^u \times \frac{d(v)^3}{dv} \times \frac{d}{dx} \tan^{-1} x = e^u \cdot 3v^2 \cdot \frac{1}{1+x^2}$$

$$= e^{(\tan^{-1} x)^3} \cdot (3\tan^{-1} x)^2 \cdot \frac{1}{1+x^2}$$

Example 22 If $y = \log_e(\tan^{-1}\sqrt{1+x^2})$, then find $\frac{dy}{dx}$.

Sol. Here, $y = \log_e(\tan^{-1}\sqrt{1+x^2})$

On differentiating, we get

$$\frac{dy}{dx} = \frac{d[\log_e(\tan^{-1}\sqrt{1+x^2})]}{d(\tan^{-1}\sqrt{1+x^2})} \cdot \frac{d(\tan^{-1}\sqrt{1+x^2})}{d(\sqrt{1+x^2})} \cdot \frac{d(\sqrt{1+x^2})}{d(1+x^2)} \cdot \frac{d(1+x^2)}{dx}$$

$$= \frac{1}{\tan^{-1}(\sqrt{1+x^2})} \cdot \frac{1}{1+(\sqrt{1+x^2})^2} \cdot \frac{1}{2\sqrt{1+x^2}} \cdot 2x$$

$$= \frac{x}{\tan^{-1}(\sqrt{1+x^2})(2+x^2)\sqrt{1+x^2}}$$

Aliter

Here, $y = \log_e (\tan^{-1}\sqrt{1+x^2})$

Put $1 + x^2 = w, \sqrt{1+x^2} = v$

$\Rightarrow \quad v = \sqrt{w}$ and $\tan^{-1}\sqrt{1+x^2} = u$

$\Rightarrow \quad u = \tan^{-1} v$

$\therefore \quad y = \log u, u = \tan^{-1} v$

$v = \sqrt{w}$ and $w = 1 + x^2$

On differentiating, we get

$$\frac{dy}{dx} = \frac{dy}{du}\cdot\frac{du}{dv}\cdot\frac{dv}{dw}\cdot\frac{dw}{dx}$$

$$= \frac{d}{du}(\log u) \times \frac{d}{dv}\tan^{-1} v \cdot \frac{d}{dw}\sqrt{x}\cdot\frac{d}{dx}(1+x^2)$$

$$= \frac{1}{u} \times \frac{1}{1+v^2} \times \frac{1}{2\sqrt{w}} \times (2x)$$

$$= \frac{1}{\tan^{-1}\sqrt{1+x^2}} \times \frac{1}{2+x^2} \times \frac{1}{2\sqrt{1+x^2}} \times (2x)$$

$$= \frac{1}{\tan^{-1}\sqrt{1+x^2}} \times \frac{1}{2+x^2} \times \frac{x}{\sqrt{1+x^2}}$$

$$= \frac{x}{(\tan^{-1}\sqrt{1+x^2})(2+x^2)\sqrt{1+x^2}}$$

Example 23 If $y = e^{ax} \cdot \cos(bx + c)$, then find $\frac{dy}{dx}$.

Sol. Here, $y = e^{ax} \cdot \cos(bx + c)$

On differentiating, we get

$$\frac{dy}{dx} = \left\{\frac{d}{dx}(e^{ax})\right\}\cdot\cos(bx+c) + e^{ax}\cdot\left\{\frac{d}{dx}\{\cos(bx+c)\}\right\}$$

$$= \frac{d(e^{ax})}{d(ax)}\cdot\frac{d(ax)}{dx}\cdot\cos(bx+c) + e^{ax}\cdot\frac{d\cos(bx+c)}{d(bx+c)}\cdot\frac{d(bx+c)}{dx}$$

$$= e^{ax}\cdot a\cdot\cos(bx+c) + e^{ax}\cdot\{-\sin(bx+c)\}\cdot b$$

$$= ae^{ax}\cos(bx+c) - be^{ax}\sin(bx+c)$$

Remark

From now and onwards, we apply chain rule directly.

Example 24 Differentiate the following w.r.t. x.

(i) $\sqrt{\log\left\{\sin\left(\frac{x^2}{3}-1\right)\right\}}$ (ii) $\log_e\left(\frac{x+\sqrt{x^2-a^2}}{x-\sqrt{x^2-a^2}}\right)$

Sol. (i) Let $y = \sqrt{\log\left\{\sin\left(\frac{x^2}{3}-1\right)\right\}}$

On differentiating, we get

$$\frac{dy}{dx} = \frac{1}{2\sqrt{\log\left\{\sin\left(\frac{x^2}{3}-1\right)\right\}}} \times \frac{1}{\sin\left(\frac{x^2}{3}-1\right)} \times \cos\left(\frac{x^2}{3}-1\right) \times \frac{2x}{3}$$

$$= \frac{x\cot\left(\frac{x^2}{3}-1\right)}{3\sqrt{\log\left\{\sin\left(\frac{x^2}{3}-1\right)\right\}}}$$

(ii) Let $y = \log_e\left(\frac{x+\sqrt{x^2-a^2}}{x-\sqrt{x^2-a^2}}\right)$

$$\Rightarrow y = \log(x+\sqrt{x^2-a^2}) - \log(x-\sqrt{x^2-a^2})$$

$$\left[\because \log\frac{a}{b} = \log a - \log b\right]$$

On differentiating, we get

$$\frac{dy}{dx} = \left(\frac{1}{x+\sqrt{x^2-a^2}}\right)\left(1+\frac{2x}{2\sqrt{x^2-a^2}}\right) - \left[\left(\frac{1}{x-\sqrt{x^2-a^2}}\right)\left(1-\frac{2x}{2\sqrt{x^2-a^2}}\right)\right]$$

$$= \frac{1}{\sqrt{x^2-a^2}} + \frac{1}{\sqrt{x^2-a^2}} = \frac{2}{\sqrt{x^2-a^2}}$$

Exercise for Session 2

■ **Directions** (Q. Nos. 1 to 19) *Differentiate the following w.r.t. x.*

1. $(x^2+x+1)^4$
2. $\sqrt{x^2+x+1}$
3. $\sin^3 x$
4. $\frac{1}{\sqrt{a^2-x^2}}$
5. $e^{x\sin x}$
6. $\sin^{-1}\left(\frac{a+b\cos x}{b+a\cos x}\right), b > a > 1$

7. e^{e^x}

8. $\log(x + \sqrt{a^2 + x^2})$

9. $\log\left(\dfrac{a + b\sin x}{a - b\sin x}\right)$

10. $\log\sqrt{\dfrac{1+\sin x}{1-\sin x}}$

11. $\dfrac{e^x + \log x}{\sin 3x}$

12. $\sin(m\sin^{-1} x), |x| < 1$

13. $a^{(\sin^{-1} x)^2}, |x| < 1$

14. $e^{\cos^{-1}(\sqrt{1-x^2})}, |x| < 1$

15. $\dfrac{x\sin^{-1} x}{\sqrt{1-x^2}} + \log\sqrt{1-x^2}, |x| < 1$

16. $\log_{10} x + \log_x 10 + \log_x x + \log_{10} 10$

17. $5^{3-x^2} + (3-x^2)^5$

18. $\dfrac{\sqrt{a^2+x^2} + \sqrt{a^2-x^2}}{\sqrt{a^2+x^2} - \sqrt{a^2-x^2}}$

19. $\sqrt{4 + \sqrt{4 + \sqrt{4 + x^2}}}$

Multiple Choice Questions

20. The differential coefficient of $f(\log_e x)$ w.r.t. x, where $f(x) = \log_e x$, is

(a) $\dfrac{x}{\log_e x}$ (b) $\dfrac{1}{x}\log_e x$ (c) $\dfrac{1}{x\log_e x}$ (d) None of these

21. If $f(x) = \log_e |x|, x \neq 0$, then $f'(x)$ is equal to

(a) $\dfrac{1}{|x|}$ (b) $\dfrac{1}{x}$ (c) $-\dfrac{1}{x}$ (d) None of these

22. Let $f(x) = \sin x$, $g(x) = x^2$ and $h(x) = \log x$. If $F(x) = h(f(g(x)))$, then $F'(x)$ is

(a) $2x\cot x^2$ (b) $2\,\text{cosec}^3 x$ (c) $-2\,\text{cosec}^2 x$ (d) None of these

23. If $f(x) = \cos x \cdot \cos 2x \cdot \cos 4x \cdot \cos 8x \cdot \cos 16x$, then $f'\left(\dfrac{\pi}{4}\right)$ is

(a) $\sqrt{2}$ (b) $\dfrac{1}{\sqrt{2}}$ (c) 1 (d) None of these

24. If $y = f\left(\dfrac{3x+4}{5x+6}\right)$ and $f'(x) = \tan x^2$, then $\dfrac{dy}{dx}$ is equal to

(a) $-2\tan\left(\dfrac{3x+4}{5x+6}\right)^2 \cdot \dfrac{1}{(5x+6)^2}$

(b) $f\left(\dfrac{3\tan x^2 + 3}{5\tan x^2 + 6}\right)\tan x^2$

(c) $2x\tan\left(\dfrac{3x+4}{5x+6}\right)$

(d) $\tan x^2$

25. If $y = |\cos x| + |\sin x|$, then $\dfrac{dy}{dx}$ at $x = \dfrac{2\pi}{3}$ is

(a) $\dfrac{1}{2}(\sqrt{3}+1)$ (b) $2(\sqrt{3}-1)$ (c) $\dfrac{1}{2}(\sqrt{3}-1)$ (d) None of these

26. If $f'(x) = \sin x + \sin 4x \cdot \cos x$, then $f'\left(2x^2 + \dfrac{\pi}{2}\right)$ is

(a) $4x\{\cos(2x^2) - \sin 8x^2 \cdot \sin 2x^2\}$

(b) $4x\{\cos(2x^2) + \sin 8x^2 \cdot \cos 2x^2\}$

(c) $\{\cos(2x^2) - \sin 8x^2 \cdot \sin 2x^2\}$

(d) None of these

27. If $f'(x) = \sqrt{2x^2 - 1}$ and $y = f(x^2)$, then $\dfrac{dy}{dx}$ at $x = 1$, is

(a) 1 (b) -1 (c) -2 (d) 2

Session 3

Differentiation of Implicit Functions

Differentiation of Implicit Functions

In previous rules, we have studied functions in which y was only expressed in terms of x without complication called **explicit functions.**

But, if the relation between the variables x and y are given by an equation and this equation is not immediately solvable for y, then y is called an **implicit function** of x.

Implicit functions are given by $\phi(x, y) = 0$.

In this case, to find $\frac{dy}{dx}$, we differentiate both sides of the given relation with respect to x and collect the terms containing $\frac{dy}{dx}$ to the left hand side and transpose the other terms to the right hand side. Now, divide both sides by coefficient of $\frac{dy}{dx}$.

Shortcut for Differentiation of Implicit Functions

(Only to be Applied while Solving Objective MCQs)

For implicit function put; $\frac{d}{dx}\{f(x, y)\} = \frac{-\partial f / \partial x}{\partial f / \partial y}$,

where $\frac{\partial f}{\partial x}$ is a **partial differentiation** of given function w.r.t. x (i.e. differentiating f w.r.t. x keeping y as constant) and $\frac{\partial f}{\partial y}$ means partial differentiation of given function w.r.t. y (i.e. differentiating f w.r.t. y keeping x as constant). In simple words, we write this method as for implicit functions

$$\frac{dy}{dx} = \frac{\textbf{Differentiation of function w.r.t. } x \textbf{ keeping } y \textbf{ as constant}}{\textbf{Differentiation of function w.r.t. } y \textbf{ keeping } x \textbf{ as constant}}$$

Example 25 If $x^2 + y^2 + xy = 2$, then find $\frac{dy}{dx}$.

Sol. Here, $x^2 + y^2 + xy = 2$

Differentiating both sides, we get

$$\frac{d}{dx}(x^2) + \frac{d}{dx}(y^2) + \frac{d}{dx}(xy) = \frac{d}{dx}(2)$$

$$\Rightarrow 2x + 2y\frac{dy}{dx} + \left\{\frac{dx}{dx}\right\}y + x\left\{\frac{d}{dx}y\right\} = 0$$

$$\Rightarrow 2x + 2y\frac{dy}{dx} + 1 \cdot y + x \cdot \frac{dy}{dx} = 0$$

$$\Rightarrow (2y + x)\frac{dy}{dx} = -(2x + y)$$

$$\Rightarrow \frac{dy}{dx} = -\frac{(2x + y)}{(2y + x)}$$

Aliter

Put $f = x^2 + y^2 + xy - 2$

Now, $$\frac{dy}{dx} = \frac{-\frac{\partial f}{\partial x}}{\frac{\partial f}{\partial y}} \quad \text{...(i)}$$

where, $\frac{\partial f}{\partial x} = 2x + 0 + y - 0 = 2x + y$

and $\frac{\partial f}{\partial y} = 2y + x$

Substituting in Eq. (i), we get

$$\frac{dy}{dx} = -\frac{(2x + y)}{(2y + x)}$$

Example 26 If $y = x \cos y + y \cos x$, then find $\frac{dy}{dx}$.

Sol. Here, $y = x \cos y + y \cos x$

Differentiating both the sides, we get

$$\frac{dy}{dx} = \left\{\frac{d}{dx}(x)\right\}\cos y + x\left\{\frac{d}{dx}(\cos y)\right\} + y\left\{\frac{d}{dx}(\cos x)\right\} + \left\{\frac{d}{dx}(y)\right\} \cdot \cos x$$

$$\Rightarrow \frac{dy}{dx} = 1 \cdot \cos y + x(-\sin y)\frac{dy}{dx} + y(-\sin x) + \frac{dy}{dx}(\cos x)$$

$$\Rightarrow \frac{dy}{dx}(1 + x \sin y - \cos x) = \cos y - y \sin x$$

$$\therefore \frac{dy}{dx} = \frac{\cos y - y \sin x}{1 + x \sin y - \cos x}$$

Aliter

Let $f = x\cos y + y\cos x - y \Rightarrow \dfrac{\partial f}{\partial x} = \cos y - y\sin x$

and $\dfrac{\partial f}{\partial y} = -x\sin y + \cos x - 1$

$$\therefore \quad \frac{dy}{dx} = \frac{-\left(\dfrac{\partial f}{\partial x}\right)}{\left(\dfrac{\partial f}{\partial y}\right)} = \frac{-(\cos y - y\sin x)}{(-x\sin y + \cos x - 1)}$$

or $$\frac{dy}{dx} = \frac{\cos y - y\sin x}{1 + x\sin y - \cos x}$$

Example 27 If $y\sqrt{1-x^2} + x\sqrt{1-y^2} = 1$, when $|x| < 1$ and $|y| < 1$, then find $\dfrac{dy}{dx}$.

Sol. Here, $y\sqrt{1-x^2} + x\sqrt{1-y^2} = 1$, $|x| < 1$ and $|y| < 1$

Let $x = \sin\theta$, $y = \sin\phi$, then

$$\sin\phi\sqrt{1-\sin^2\theta} + \sin\theta\sqrt{1-\sin^2\phi} = 1$$

$$\Rightarrow \quad \sin\phi\cos\theta + \sin\theta\cos\phi = 1$$

$$\Rightarrow \quad \sin(\theta + \phi) = 1$$

$$\Rightarrow \quad \theta + \phi = \sin^{-1}(1)$$

$$\Rightarrow \quad \sin^{-1}x + \sin^{-1}y = \sin^{-1}(1)$$

$$\left[\begin{array}{l}\because \sin\theta = x \Rightarrow \theta = \sin^{-1}x \\ \text{and } \sin\phi = y \Rightarrow \phi = \sin^{-1}y\end{array}\right]$$

Differentiating both the sides, we get

$$\frac{1}{\sqrt{1-x^2}} + \frac{1}{\sqrt{1-y^2}}\cdot\frac{dy}{dx} = 0$$

Hence, $$\frac{dy}{dx} = -\sqrt{\frac{1-y^2}{1-x^2}}$$

Aliter

Let $f = y\sqrt{1-x^2} + x\sqrt{1-y^2} - 1$

$$\Rightarrow \frac{\partial f}{\partial x} = \frac{-xy}{\sqrt{1-x^2}} + \sqrt{1-y^2} \text{ and } \frac{\partial f}{\partial y} = \sqrt{1-x^2} - \frac{xy}{\sqrt{1-y^2}}$$

$$\therefore \quad \frac{dy}{dx} = \frac{-\left(\dfrac{\partial f}{\partial x}\right)}{\left(\dfrac{\partial f}{\partial y}\right)} = -\frac{[\sqrt{1-x^2}\sqrt{1-y^2} - xy]/\sqrt{1-x^2}}{[\sqrt{1-x^2}\sqrt{1-y^2} - xy]/\sqrt{1-y^2}}$$

$$\Rightarrow \frac{dy}{dx} = -\frac{\sqrt{1-y^2}}{\sqrt{1-x^2}}$$

Example 28 If $\sqrt{1-x^6} + \sqrt{1-y^6} = a^3(x^3 - y^3)$, then prove that $\dfrac{dy}{dx} = \dfrac{x^2}{y^2}\sqrt{\dfrac{1-y^6}{1-x^6}}$.

Sol. Here, $\sqrt{1-x^6} + \sqrt{1-y^6} = a^3(x^3 - y^3)$

Let $x^3 = \sin\theta$, $y^3 = \sin\phi$, then we get

$$\sqrt{1-\sin^2\theta} + \sqrt{1-\sin^2\phi} = a^3(\sin\theta - \sin\phi)$$

$$\Rightarrow \quad \cos\theta + \cos\phi = a^3(\sin\theta - \sin\phi)$$

$$\Rightarrow 2\cos\left(\frac{\theta+\phi}{2}\right)\cdot\cos\left(\frac{\theta-\phi}{2}\right) = a^3\left[2\cos\left(\frac{\theta+\phi}{2}\right)\sin\left(\frac{\theta-\phi}{2}\right)\right]$$

$$\Rightarrow \quad \cos\left(\frac{\theta-\phi}{2}\right) = a^3\sin\left(\frac{\theta-\phi}{2}\right) \Rightarrow \cot\left(\frac{\theta-\phi}{2}\right) = a^3$$

$$\Rightarrow \quad \theta - \phi = 2\cot^{-1}(a^3)$$

or $$\sin^{-1}x^3 - \sin^{-1}y^3 = 2\cot^{-1}(a^3)$$

$$\left[\begin{array}{l}\because \cos C + \cos D = 2\cos\left(\dfrac{C+D}{2}\right)\cos\left(\dfrac{C-D}{2}\right) \\ \text{and } \sin C - \sin D = 2\cos\left(\dfrac{C+D}{2}\right)\sin\left(\dfrac{C-D}{2}\right)\end{array}\right]$$

Differentiating both the sides, we get

$$\frac{1}{\sqrt{1-x^6}}\cdot 3x^2 - \frac{1}{\sqrt{1-y^6}}\cdot 3y^2\cdot\frac{dy}{dx} = 0$$

Hence, $$\frac{dy}{dx} = \frac{x^2\sqrt{1-y^6}}{y^2\sqrt{1-x^6}}$$

Example 29 If $y = \sqrt{\sin x + \sqrt{\sin x + \sqrt{\sin x + \ldots \infty}}}$, where $\sin x > 0$, then find $\dfrac{dy}{dx}$.

Sol. We have, $y = \sqrt{\sin x + \sqrt{\sin x + \sqrt{\sin x + \ldots \infty}}}$

$$\Rightarrow \quad y = \sqrt{(\sin x) + y}$$

$$\Rightarrow \quad y^2 = \sin x + y$$

Differentiating both the sides, we get

$$2y\frac{dy}{dx} = \cos x + \frac{dy}{dx}$$

$$\Rightarrow (2y-1)\frac{dy}{dx} = \cos x \Rightarrow \frac{dy}{dx} = \frac{\cos x}{2y-1}$$

Remark

In above examples, it is advised to use $\dfrac{dy}{dx} = -\left(\dfrac{\partial f/\partial x}{\partial f/\partial y}\right)$

Exercise for Session 3

1. If $\log(x^2 + y^2) = 2\tan^{-1}\left(\frac{y}{x}\right)$, then show that $\frac{dy}{dx} = \frac{x+y}{x-y}$.

2. If $x\sqrt{1+y} + y\sqrt{1+x} = 0$, then prove that $\frac{dy}{dx} = \frac{-1}{(x+1)^2}$.

3. If $\cos^{-1}\left(\frac{x^2 - y^2}{x^2 + y^2}\right) = \tan^{-1} a$, then prove that $\frac{dy}{dx} = \frac{y}{x}$.

4. If $\sin y = x \sin(a + y)$, then prove that $\frac{dy}{dx} = \frac{\sin^2(a+y)}{\sin a}$.

5. If $x^2 + y^2 = t - \frac{1}{t}$ and $x^4 + y^4 = t^2 + \frac{1}{t^2}$, then prove that $\frac{dy}{dx} = \frac{1}{x^3 y}$.

Multiple Choice Questions

6. If $\sin(xy) + \cos(xy) = 0$, then $\frac{dy}{dx}$ is

(a) $\frac{y}{x}$ (b) $-\frac{y}{x}$

(c) $-\frac{x}{y}$ (d) $\frac{x}{y}$

7. If $ax^2 + 2hxy + by^2 = 0$, then $\frac{dy}{dx}$ is

(a) $\frac{y}{x}$ (b) $\frac{x}{y}$

(c) $-\frac{x}{y}$ (d) None of these

8. If $x^2 \cdot e^y + 2xye^x + 13 = 0$, then $\frac{dy}{dx}$ is

(a) $\frac{-2xe^{y-x} - 2y(x+1)}{x(xe^{y-x} + 2)}$ (b) $\frac{2xe^{x-y} + 2y(x+1)}{x(xe^{y-x} + 2)}$

(c) $\frac{-2xe^{x-y} + 2y(x+1)}{x(xe^{y-x} + 2)}$ (d) None of these

9. If $\log(x + y) = 2xy$, then $y'(0)$ is

(a) 1 (b) −1 (c) 2 (d) 0

10. If $x \log_e y + y \log_e x = 5$, then $\frac{dy}{dx}$ is

(a) $-\frac{y}{x}\left(\frac{x\log y + y}{x + y\log x}\right)$ (b) $-\frac{x}{y}\left(\frac{x\log y + y}{x - y\log x}\right)$

(c) $-\frac{y}{x}\left(\frac{x\log y - y}{x + y\log x}\right)$ (d) None of these

Session 4

Differentiation of Inverse Trigonometric Functions

Differentiation of Inverse Trigonometric Functions

Inverse trigonometric functions are continuous functions, their domain and range are shown in the table given below.

S. No.	Function	Domain	Range	Principal value branch
1.	$\sin^{-1} x$	$[-1, 1]$	$\left[-\frac{\pi}{2}, \frac{\pi}{2}\right]$	$-\frac{\pi}{2} \leq y \leq \frac{\pi}{2}$, where $y = \sin^{-1} x$
2.	$\cos^{-1} x$	$[-1, 1]$	$[0, \pi]$	$0 \leq y \leq \pi$, where $y = \cos^{-1} x$
3.	$\tan^{-1} x$	R	$\left(-\frac{\pi}{2}, \frac{\pi}{2}\right)$	$-\frac{\pi}{2} < y < \frac{\pi}{2}$, where $y = \tan^{-1} x$
4.	$\text{cosec}^{-1} x$	$(-\infty, -1] \cup [1, \infty)$	$\left[-\frac{\pi}{2}, \frac{\pi}{2}\right] - \{0\}$	$-\frac{\pi}{2} \leq y \leq \frac{\pi}{2}$, where $y = \text{cosec}^{-1} x, y \neq 0$
5.	$\sec^{-1} x$	$(-\infty, -1] \cup [1, \infty)$	$[0, \pi] - \left\{\frac{\pi}{2}\right\}$	$0 \leq y \leq \pi$, where $y = \sec^{-1} x, y \neq \frac{\pi}{2}$
6.	$\cot^{-1} x$	R	$(0, \pi)$	$0 < y < \pi$, where $y = \cot^{-1} x$

To differentiate inverse trigonometric functions, we need to know some basic trigonometric identities and substitution, they can help you finding the differentiation of inverse trigonometric functions.

Useful Trigonometric Identities

(i) $\sin 2x = 2 \sin x \cos x$

(ii) $\cos 2x = 2 \cos^2 x - 1, \cos 2x = 1 - 2 \sin^2 x$

(iii) $\sin 2x = \dfrac{2 \tan x}{1 + \tan^2 x}$

(iv) $\cos 2x = \dfrac{1 - \tan^2 x}{1 + \tan^2 x}$

(v) $\tan 2x = \dfrac{2 \tan x}{1 - \tan^2 x}$, where $x \neq (2n + 1) \dfrac{\pi}{4}$

(vi) $\sin 3x = 3 \sin x - 4 \sin^3 x$

$= 4 \sin (60° - A) \cdot \sin A \cdot \sin (60° + A)$

(vii) $\cos 3x = 4 \cos^3 x - 3 \cos x$

$= 4 \cos (60° - A) \cdot \cos A \cdot \cos (60° + A)$

(viii) $\tan 3x = \dfrac{3 \tan x - \tan^3 x}{1 - 3 \tan^2 x}$

$= \tan (60° - A) \cdot \tan A \cdot \tan (60° + A)$

(ix) $\cos A \cdot \cos 2A \cdot \cos 2^2 A \ldots \cos 2^{n-1} A = \dfrac{\sin 2^n A}{2^n \sin A}$

Remark

If no branch of an inverse trigonometric function is mentioned, then it means that the principal value branch of the function have to be taken.

Substitutions for Inverse Trigonometric Functions

S.No.	Expression	Substitutions
1.	$\sqrt{a^2 - x^2}$	$x = a\sin\theta$ or $a\cos\theta$
2.	$\sqrt{a^2 + x^2}$	$x = a\tan\theta$ or $a\cot\theta$
3.	$\sqrt{x^2 - a^2}$	$x = a\sec\theta$ or $a\text{cosec}\,\theta$
4.	$\sqrt{\frac{a+x}{a-x}}$ or $\sqrt{\frac{a-x}{a+x}}$	$x = a\cos\theta$ or $a\cos 2\theta$
5.	$\sqrt{(a-x)(x-b)}$ or $\sqrt{\frac{a-x}{x-b}}$ or $\sqrt{\frac{x-b}{a-x}}$	$x = a\cos^2\theta - b\sin^2\theta$
6.	$\sqrt{(x-a)(x-b)}$ or $\sqrt{\frac{x-a}{x-b}}$ or $\sqrt{\frac{x-b}{a-x}}$	$x = a\sec^2\theta + b\tan^2\theta$
7.	$\sqrt{2ax - x^2}$	$x = a(1 - \cos\theta)$

Example 30 If $y=\sec^{-1}(\sqrt{1+x^2})$, when $-1<x<1$, then find $\frac{dy}{dx}$.

Sol. Here, $y=\sec^{-1}(\sqrt{1+x^2})$,

put $x=\tan\theta$

$$\Rightarrow \quad y=\sec^{-1}(\sqrt{\sec^2\theta})$$

$$=\sec^{-1}(\sec\theta)=\theta=\tan^{-1}x$$

$$\therefore \quad \frac{dy}{dx}=\frac{1}{1+x^2}$$

Example 31 If $y=\tan^{-1}\sqrt{\frac{1-x}{1+x}}$, then find $\frac{dy}{dx}$, when $-1<x<1$.

Sol. Here, $y=\tan^{-1}\sqrt{\frac{1-x}{1+x}}$, put $x=\cos\theta$

$$\Rightarrow \quad y=\tan^{-1}\sqrt{\frac{1-\cos\theta}{1+\cos\theta}}$$

$$\Rightarrow \quad y=\tan^{-1}\sqrt{\frac{2\sin^2\theta/2}{2\cos^2\theta/2}}$$

$$\Rightarrow \quad y=\tan^{-1}|\tan\theta/2| \quad \left[\begin{array}{l}\because -1<\cos\theta<1 \Rightarrow 0<\theta<\pi\\ \therefore 0<\frac{\theta}{2}<\frac{\pi}{2}\end{array}\right]$$

$$\therefore \quad y=\tan^{-1}(\tan\theta/2)$$

$$[\because \tan^{-1}(\tan x)=x, -\pi/2<x<\pi/2]$$

$$\Rightarrow \quad y=\frac{\theta}{2} \Rightarrow y=\frac{1}{2}\cos^{-1}x$$

$$\therefore \quad \frac{dy}{dx}=-\frac{1}{2\sqrt{1-x^2}}$$

Example 32 If $y=\tan^{-1}\sqrt{\frac{1-\cos x}{1+\cos x}}$, then find $\frac{dy}{dx}$, when $0<x<\frac{\pi}{2}$.

Sol. Here, $y=\tan^{-1}\sqrt{\frac{1-\cos x}{1+\cos x}}$

$$\Rightarrow \quad y=\tan^{-1}\sqrt{\frac{2\sin^2 x/2}{2\cos^2 x/2}}=\tan^{-1}\sqrt{\tan^2 x/2}$$

$$\Rightarrow \quad y=\tan^{-1}(\tan x/2)$$

$$\Rightarrow \quad y=\frac{x}{2} \quad \left[\because \tan^{-1}(\tan x)=x, -\frac{\pi}{2}<x<\frac{\pi}{2}\right]$$

Hence, $\frac{dy}{dx}=\frac{1}{2}$

Example 33 If $y=\tan^{-1}\left(\frac{\sqrt{1+x^2}-1}{x}\right)$, then find $\frac{dy}{dx}$ when $-1\le x\le 1$.

Sol. Here, $y=\tan^{-1}\left(\frac{\sqrt{1+x^2}-1}{x}\right)$, let $x=\tan\theta$

$$\Rightarrow \quad y=\tan^{-1}\left(\frac{\sqrt{1+\tan^2\theta}-1}{\tan\theta}\right)$$

$$\Rightarrow \quad y=\tan^{-1}\left(\frac{\sec\theta-1}{\tan\theta}\right)$$

$$\Rightarrow \quad y=\tan^{-1}\left(\frac{1-\cos\theta}{\sin\theta}\right)=\tan^{-1}\left(\frac{2\sin^2\frac{\theta}{2}}{2\sin\frac{\theta}{2}\cos\frac{\theta}{2}}\right)$$

$$\Rightarrow \quad y=\tan^{-1}\left(\tan\frac{\theta}{2}\right)=\frac{\theta}{2}$$

$$\Rightarrow \quad y=\frac{1}{2}\tan^{-1}x \Rightarrow \frac{dy}{dx}=\frac{1}{2(1+x^2)}$$

Example 34 Find $\frac{dy}{dx}$, if $y=\tan^{-1}\left(\frac{a\cos x-b\sin x}{b\cos x+a\sin x}\right)$, where $-\frac{\pi}{2}<x<\frac{\pi}{2}$ and $\frac{a}{b}\tan x>-1$.

Sol. Here, $y=\tan^{-1}\left(\frac{a\cos x-b\sin x}{b\cos x+a\sin x}\right)$

$$\Rightarrow \quad y=\tan^{-1}\left(\frac{\frac{a}{b}-\tan x}{1+\frac{a}{b}\tan x}\right)$$

$$=\tan^{-1}\left(\frac{a}{b}\right)-\tan^{-1}(\tan x)=\tan^{-1}\left(\frac{a}{b}\right)-x$$

$$\Rightarrow \quad \frac{dy}{dx}=0-1=-1$$

Example 35 Find $\frac{dy}{dx}$ for the function

$$y=\sin^{-1}\left(\frac{5x+12\sqrt{1-x^2}}{13}\right).$$

Sol. Here, $y=\sin^{-1}\left(\frac{5x+12\sqrt{1-x^2}}{13}\right)$

Put $x=\sin\theta$, $5=r\cos\phi$ and $12=r\sin\phi$

$$\Rightarrow \quad r=\sqrt{(r\cos\phi)^2+(r\sin\phi)^2}$$

$$=\sqrt{(5)^2+(12)^2}=\sqrt{25+144}=\sqrt{169}=13$$

$$\phi=\tan^{-1}\left(\frac{12}{5}\right)$$

$\therefore \quad y = \sin^{-1}\left[\dfrac{r\cos\phi\sin\theta + r\sin\phi\cos\theta}{13}\right]$

$\Rightarrow \quad y = \sin^{-1}(\sin(\theta+\phi)) = \theta + \phi$

$\Rightarrow \quad y = \sin^{-1}x + \tan^{-1}\left(\dfrac{12}{5}\right)$

$\left[\because \theta = \sin^{-1}x \text{ and } \phi = \tan^{-1}\left(\dfrac{12}{5}\right)\right]$

$\therefore \quad \dfrac{dy}{dx} = \dfrac{1}{\sqrt{1-x^2}}$

Example 36 Find $\dfrac{dy}{dx}$, for $y = \tan^{-1}\sqrt{\dfrac{a-x}{a+x}}$, where $-a < x < a$.

Sol. Here, $y = \tan^{-1}\sqrt{\dfrac{a-x}{a+x}}$, where $-a < x < a$

Put $x = a\cos\theta \Rightarrow \theta = \cos^{-1}\dfrac{x}{a}$

$\therefore \quad y = \tan^{-1}\left\{\sqrt{\dfrac{a - a\cos\theta}{a + a\cos\theta}}\right\}$

$\Rightarrow \quad y = \tan^{-1}\left\{\sqrt{\dfrac{1-\cos\theta}{1+\cos\theta}}\right\}$

$\Rightarrow \quad y = \tan^{-1}\left\{\sqrt{\tan^2\dfrac{\theta}{2}}\right\}$

$\Rightarrow \quad y = \tan^{-1}\left|\tan\dfrac{\theta}{2}\right|$

Also, $-a < x < a \Rightarrow -1 < \cos\theta < 1$

$\Rightarrow \quad \theta \in (0, \pi) \Rightarrow \dfrac{\theta}{2} \in \left(0, \dfrac{\pi}{2}\right)$

$\therefore \quad y = \tan^{-1}\left|\tan\dfrac{\theta}{2}\right| = \tan^{-1}\left(\tan\dfrac{\theta}{2}\right) = \dfrac{\theta}{2}$

$\Rightarrow \quad y = \dfrac{1}{2}\cos^{-1}\dfrac{x}{a}$

$\therefore \quad \dfrac{dy}{dx} = -\dfrac{1}{2} \times \dfrac{1}{\sqrt{1-\dfrac{x^2}{a^2}}}\dfrac{d}{dx}\left(\dfrac{x}{a}\right) = -\dfrac{1}{2\sqrt{a^2-x^2}}$

Graphical Approach for Differentiation of Inverse Trigonometric Functions

In this section, we are going to introduce, graphs of some important inverse trigonometric functions which are highly useful in differential calculus.

Example 37 Sketch the graph for $y = \sin^{-1}(\sin x)$ and hence find $\dfrac{dy}{dx}$.

Sol. As, $y = \sin^{-1}(\sin x)$ is periodic with period 2π.

$\therefore$ To draw this graph we should draw the graph for one interval of length 2π and repeat it for entire values of x.

As we know,

$$\sin^{-1}(\sin x) = \begin{cases} x, & -\dfrac{\pi}{2} \le x \le \dfrac{\pi}{2} \\ (\pi - x), & -\dfrac{\pi}{2} \le \pi - x < \dfrac{\pi}{2} \quad \left(\text{i.e.} \dfrac{\pi}{2} \le x \le \dfrac{3\pi}{2}\right) \end{cases}$$

or
$$\sin^{-1}(\sin x) = \begin{cases} x, & -\dfrac{\pi}{2} \le x \le \dfrac{\pi}{2} \\ \pi - x, & \dfrac{\pi}{2} \le x \le \dfrac{3\pi}{2} \end{cases}$$

which is defined for the interval of length 2π, plotted as;

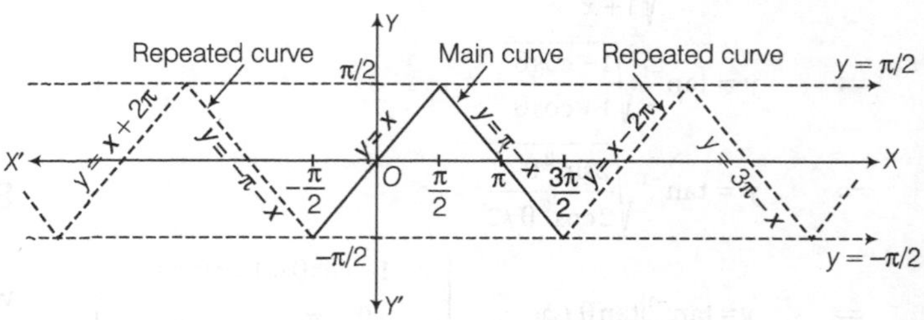

Thus, the graph for $y = \sin^{-1}(\sin x)$, is a straight line up and a straight line down with slopes 1 and -1 respectively lying between $\left[-\dfrac{\pi}{2}, \dfrac{\pi}{2}\right]$.

$$\text{Thus, } \frac{dy}{dx} = \begin{cases} 1, & \left(2n - \dfrac{1}{2}\right)\pi < x < \left(2n + \dfrac{1}{2}\right)\pi \\ -1, & \left(2n + \dfrac{1}{2}\right)\pi < x < \left(2n + \dfrac{3}{2}\right)\pi \\ \text{does not exist}, & x = \left\{(2n+1)\dfrac{\pi}{2}; n \in I\right\} \end{cases}$$

Remark

Students are advised to learn the definition of $\sin^{-1}(\sin x)$ as

$$y = \sin^{-1}(\sin x) = \begin{cases} x + 2\pi, & -\dfrac{5\pi}{2} \le x \le -\dfrac{3\pi}{2} \\ -\pi - x, & -\dfrac{3\pi}{2} \le x \le -\dfrac{\pi}{2} \\ x, & -\dfrac{\pi}{2} \le x \le \dfrac{\pi}{2} \\ \pi - x, & \dfrac{\pi}{2} \le x \le \dfrac{3\pi}{2} \\ x - 2\pi & \dfrac{3\pi}{2} \le x \le \dfrac{5\pi}{2} \text{... and so on} \end{cases}$$

Example 38 Sketch the graph for $y = \cos^{-1}(\cos x)$ and hence find $\dfrac{dy}{dx}$.

Sol. As, $y = \cos^{-1}(\cos x)$ is periodic with a period 2π.

$\therefore$ To draw this graph, we should draw the graph for one interval of length 2π and repeat it for entire values of x of length 2π.

As we know, $\cos^{-1}(\cos x) = \begin{cases} x, & 0 \le x \le \pi \\ 2\pi - x, & 0 \le 2\pi - x \le \pi \end{cases}$

or $\cos^{-1}(\cos x) = \begin{cases} x, & 0 \le x \le \pi \\ 2\pi - x, & \pi \le x \le 2\pi \end{cases}$

Thus, it has been defined for $0 < x < 2\pi$ that has length 2π. So, its graph could be plotted as

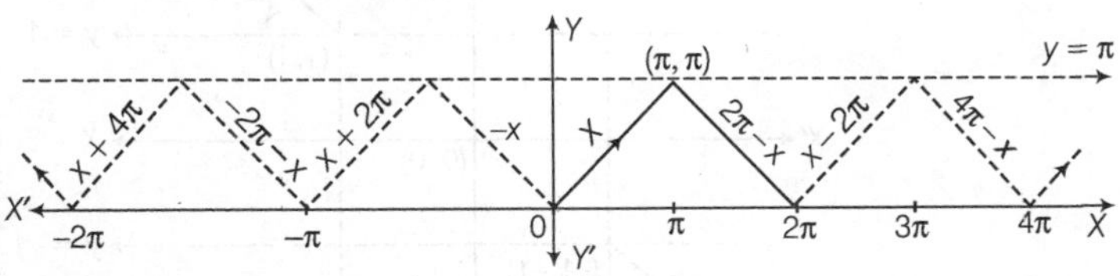

Thus, the curve $y = \cos^{-1}(\cos x)$ and hence

$$\frac{dy}{dx} = \begin{cases} \text{does not exist,} & x = \{n\pi, n \in I\} \\ 1, & 2n\pi < x < (2n+1)\pi \\ -1, & (2n+1)\pi < x < (2n+2)\pi \end{cases}$$

Example 39 Sketch the graph for $y = \tan^{-1}(\tan x)$ and hence find $\dfrac{dy}{dx}$.

Sol. As, $y = \tan^{-1}(\tan x)$ is periodic with period π.

$\therefore$ To draw this graph we should draw the graph for one interval of length π and repeat it for entire values of x.

As we know, $\tan^{-1}(\tan x) = \left\{x, -\dfrac{\pi}{2} < x < \dfrac{\pi}{2}\right\}$

Thus, it has been defined for $-\dfrac{\pi}{2} < x < \dfrac{\pi}{2}$ that has length π. So, its graph could be plotted as

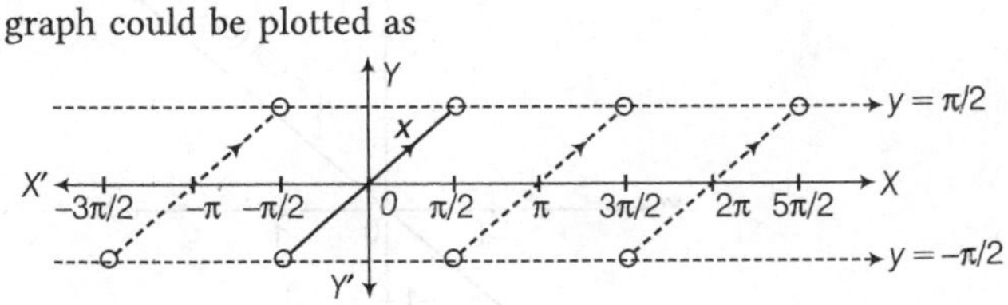

Thus, the curve for $y = \tan^{-1}(\tan x)$, where y is not defined for $x \in (2n+1)\dfrac{\pi}{2}$.

and hence $$\frac{dy}{dx} = \begin{cases} 1, & \left(2n - \dfrac{1}{2}\right)\pi < x < \left(2n + \dfrac{1}{2}\right)\pi \\ \text{does not exit,} & x = (2n+1)\dfrac{\pi}{2} \end{cases}$$

Example 40 Sketch the graphs for

(i) $y = \sin(\sin^{-1} x)$ (ii) $y = \cos(\cos^{-1} x)$

(iii) $y = \tan(\tan^{-1} x)$ (iv) $y = \text{cosec}(\text{cosec}^{-1} x)$

(v) $y = \sec(\sec^{-1} x)$

(vi) $y = \cot(\cot^{-1} x)$ and hence find $\dfrac{dy}{dx}$

Sol. As we know, all the above mentioned six curves are non-periodic, but have restricted domains and ranges.

So, we shall first define each curve for its domain and range and then sketch the curves.

(i) **Sketch for the curve $y = \sin(\sin^{-1} x)$**

We know that, domain $x \in [-1, 1]$, i.e. $-1 \le x \le 1$ and range $y = x \Rightarrow y \in [-1, 1]$.

Hence, we should sketch $y = \sin(\sin^{-1} x)$ only when $x \in [-1, 1]$ and $y = x$. So, its graph could be plotted as shown in the figure.

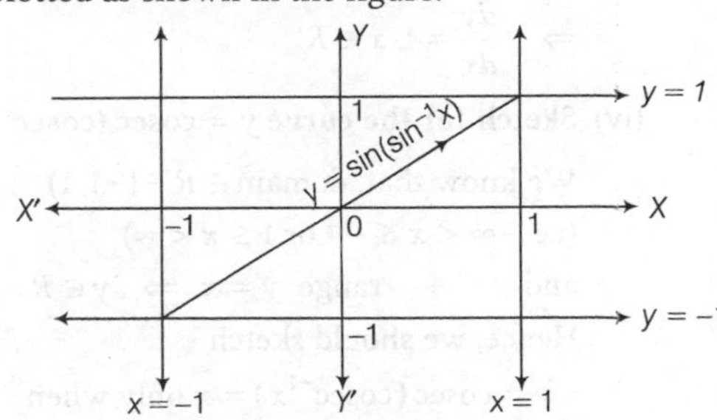

Thus, $y = \sin(\sin^{-1} x) = x, -1 \le x \le 1$.

$\Rightarrow \dfrac{dy}{dx} = 1, \ -1 < x < 1$.

(ii) **Sketch for the curve $y = \cos(\cos^{-1} x)$**

We know that, domain $x \in [-1, 1]$, i.e. $-1 \le x \le 1$ and range $y = x \Rightarrow y \in [-1, 1]$.

Hence, we should sketch $y = \cos(\cos^{-1} x) = x$ only when $x \in [-1, 1]$. So, its graph could be plotted as shown in the figure.

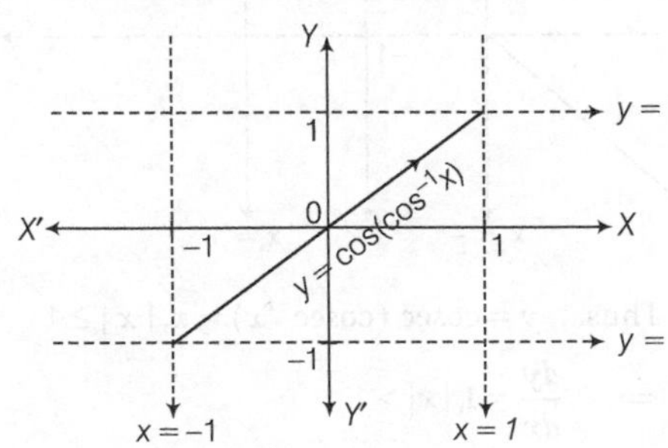

Thus, $y = \cos(\cos^{-1} x) = x, -1 \le x \le 1$

$\Rightarrow \dfrac{dy}{dx} = 1, \ -1 < x < 1$.

(iii) **Sketch for the curve $y = \tan(\tan^{-1} x)$**

We know that, domain $x \in R$, i.e. $-\infty < x < \infty$
and range $y = x \Rightarrow y \in R$.
Hence, we should sketch $y = \tan(\tan^{-1} x) = x$, $\forall\, x \in R$.
So, its graph could be plotted as shown in the figure.

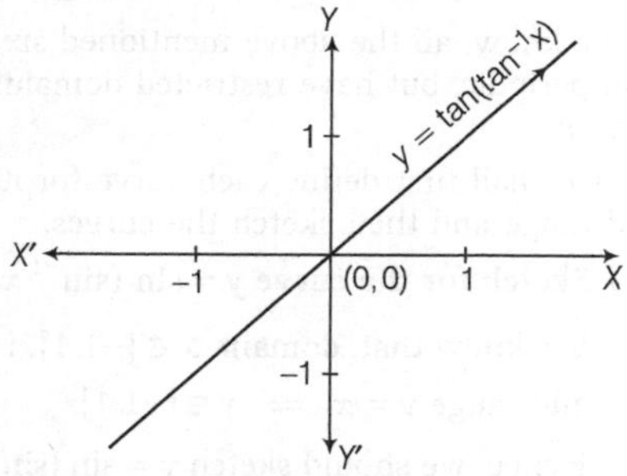

Thus, $y = \tan(\tan^{-1} x) = x, x \in R$

$$\Rightarrow \frac{dy}{dx} = 1, x \in R$$

(iv) **Sketch for the curve $y = \operatorname{cosec}(\operatorname{cosec}^{-1} x)$**

We know that, domain $\in R - (-1, 1)$
(i.e. $-\infty < x \le -1$ or $1 \le x < \infty$)
and range $y = x \Rightarrow y \in R - (-1, 1)$.
Hence, we should sketch
$y = \operatorname{cosec}(\operatorname{cosec}^{-1} x) = x$ only when
$x \in (-\infty, -1] \cup [1, \infty)$.
So, its graph could be plotted as shown in the figure.

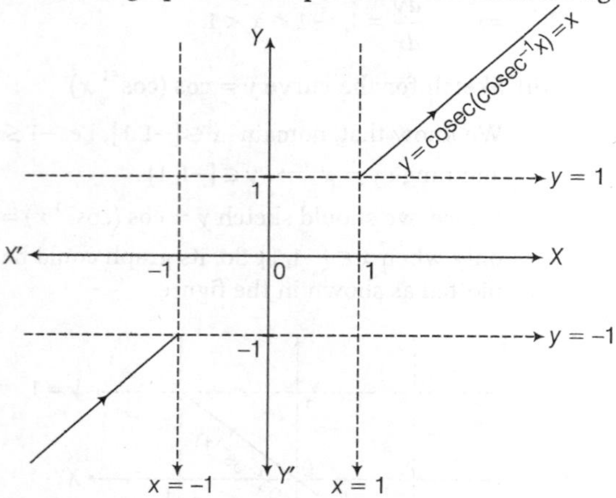

Thus, $y = \operatorname{cosec}(\operatorname{cosec}^{-1} x) = x, |x| \ge 1$

$$\Rightarrow \frac{dy}{dx} = 1, |x| > 1$$

(v) **Sketch for the curve $y = \sec(\sec^{-1} x)$**

We know that, domain $\in R - (-1, 1)$
(i.e. $-\infty < x \le -1$ or $1 \le x < \infty$)
and range $y = x$
$\Rightarrow y \in R - (-1, 1)$
Hence, we should sketch $y = \sec(\sec^{-1} x) = x$, only when

$$x \in (-\infty, -1] \cup [1, \infty)$$

So, its graph could be plotted as shown in the figure.

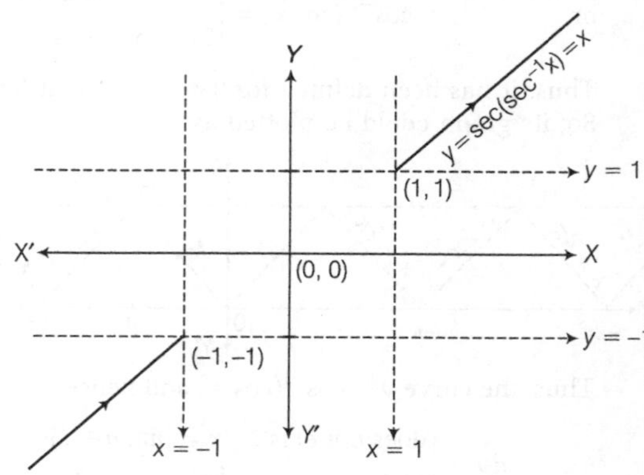

Thus, $y = \sec(\sec^{-1} x) = x, |x| \ge 1$

$$\Rightarrow \frac{dy}{dx} = 1, |x| > 1$$

(vi) **Sketch for the curve $y = \cot(\cot^{-1} x)$**

We know that, domain $\in R$ (i.e. $-\infty < x < \infty$)
and range $y = x$
$\Rightarrow y = R$.
Hence, we should sketch
$y = \cot(\cot^{-1} x) = x, \forall\, x \in R$.
So, its graph could be plotted as shown in the figure.

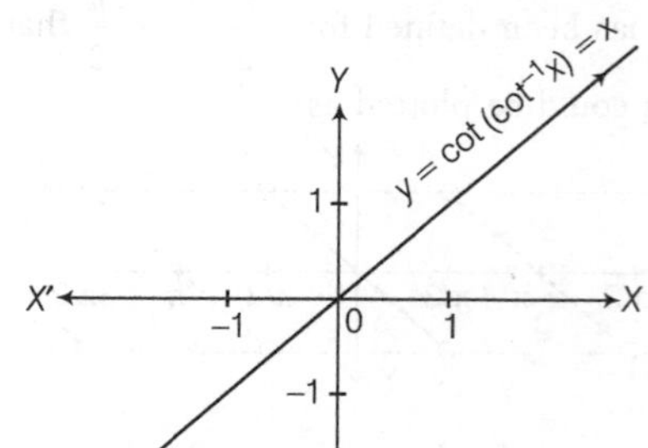

Thus, $y = \cos(\cot^{-1} x) = x, x \in R$

$$\Rightarrow \frac{dy}{dx} = 1, x \in R.$$

Exercise for Session 4

1. If $y = \tan^{-1}\left(\dfrac{1-\cos x}{\sin x}\right)$, then $\dfrac{dy}{dx}$ is

(a) $\begin{cases} \frac{1}{2}, & x \in R - \{n\pi\};\ n \in I \\ -\frac{1}{2}, & x = \{n\pi\};\ n \in I \end{cases}$

(b) $\begin{cases} \frac{1}{2}, & x \in R - \{n\pi\};\ n \in I \\ \text{does not exist}, & x = \{n\pi\};\ n \in I \end{cases}$

(c) $\begin{cases} -\frac{1}{2}, & x \in R - \{n\pi\};\ n \in I \\ \text{does not exist}, & x = \{n\pi\};\ n \in I \end{cases}$

(d) None of these

2. If $y = \cos^{-1}\left(\dfrac{x - x^{-1}}{x + x^{-1}}\right)$, then $\dfrac{dy}{dx}$ is

(a) $\begin{cases} \frac{2}{1+x^2}, & x > 0 \\ -\frac{2}{1+x^2}, & x < 0 \end{cases}$

(b) $\begin{cases} \frac{2}{1+x^2}, & x > 0 \\ \text{does not exist}, & x = 0 \\ \frac{-2}{1+x^2}, & x < 0 \end{cases}$

(c) $\begin{cases} \frac{2}{1+x^2}, & x < 0 \\ \text{does not exist}, & x = 0 \\ \frac{-2}{1+x^2}, & x > 0 \end{cases}$

(d) None of these

3. If $y = \tan^{-1}\left\{\dfrac{\sqrt{1+\sin x} + \sqrt{1-\sin x}}{\sqrt{1+\sin x} - \sqrt{1-\sin x}}\right\}$, then $\dfrac{dy}{dx}$ is

(a) $\begin{cases} \frac{1}{2}, & \cos\frac{x}{2} > \sin\frac{x}{2} \\ -\frac{1}{2}, & \cos\frac{x}{2} < \sin\frac{x}{2} \\ \text{does not exist}, & x = \{n\pi\}; n \in \text{integer} \end{cases}$

(b) $\begin{cases} -\frac{1}{2}, & \cos\frac{x}{2} > \sin\frac{x}{2} \\ \frac{1}{2}, & \cos\frac{x}{2} < \sin\frac{x}{2} \\ \text{does not exist}, & x = \{n\pi\}; n \in \text{integer} \end{cases}$

(c) $\begin{cases} -\frac{1}{2}, & \cos\frac{x}{2} \geq \sin\frac{x}{2} \\ \frac{1}{2}, & \cos\frac{x}{2} < \sin\frac{x}{2} \end{cases}$

(d) None of these

4. If $y = \cot^{-1}(\cot x)$, then $\dfrac{dy}{dx}$ is

(a) $1, x \in R$

(b) $1, x \in R - \{n\pi\}$

(c) $\begin{cases} 1, & x \in R - \{n\pi\} \\ \text{does not exist}, & x \in \{n\pi\}; n \in \text{integer} \end{cases}$

(d) None of these

5. Sketch the graph for

(i) $y = \sin^{-1}\left(\dfrac{2x}{1+x^2}\right)$

(ii) $y = \cos^{-1}\left(\dfrac{1-x^2}{1+x^2}\right)$

(iii) $y = \tan^{-1}\left(\dfrac{2x}{1-x^2}\right)$

(iv) $y = \tan^{-1}\left(\dfrac{3x-x^3}{1-3x^2}\right)$

(v) $y = \sin^{-1}(3x - 4x^3)$

(vi) $y = \cos^{-1}(4x^3 - 3x)$

Session 5

Differentiation of a Function in Parametric Form

Differentiation of a Function in Parametric Form

Until now we have differentiate the variable y w.r.t. x, but what can we do if the variable x and y are depend on third variable 't' (say) i.e. $x = g(t), y = f(t)$. In such case x and y are called parametric functions and t is called parameter.

In this case to find $\frac{dy}{dx}$, we first obtain the relationship between x and y by eliminating the parameter t, then we differentiate w.r.t. x. But always, it is not possible to eliminate t, then we use the following formula

$$\frac{dy}{dx} = \frac{\frac{dy}{dt}}{\frac{dx}{dt}} = \frac{f'(t)}{g'(t)}$$

Example 41 If $x = e^{-t^2}$ and $y = \tan^{-1}(2t+1)$, then find $\frac{dy}{dx}$.

Sol. Here, $x = e^{-t^2}$

On differentiating both sides, we get

$$\frac{dx}{dt} = e^{-t^2} \cdot (-2t) \text{ and } y = \tan^{-1}(2t+1)$$

On differentiating both sides, we get

$$\frac{dy}{dt} = \frac{1}{1+(2t+1)^2}(2)$$

$$\therefore \quad \frac{dy}{dx} = \frac{\frac{dy}{dt}}{\frac{dx}{dt}} = \frac{\frac{2}{1+(4t^2+4t+1)}}{-\frac{2t}{e^{t^2}}}$$

Hence, $$\frac{dy}{dx} = \frac{-e^{t^2}}{2t\,(2t^2+2t+1)}$$

Example 42 If $x = a(\theta - \sin\theta)$, $y = a(1-\cos\theta)$, then find $\frac{dy}{dx}$.

Sol. Here, $x = a(\theta - \sin\theta)$

On differentiating both the sides, we get

$$\frac{dx}{d\theta} = a(1-\cos\theta)$$

and $y = a(1-\cos\theta)$

On differentiating both sides, we get

$$\frac{dy}{d\theta} = a(\sin\theta)$$

So, $$\frac{dy}{dx} = \frac{\frac{dy}{d\theta}}{\frac{dx}{d\theta}} = \frac{a\sin\theta}{a(1-\cos\theta)} = \frac{2\sin\frac{\theta}{2}\cos\frac{\theta}{2}}{2\sin^2\frac{\theta}{2}} = \cot\frac{\theta}{2}$$

Example 43 If $x = \sqrt{a^{\sin^{-1}t}}$, $y = \sqrt{a^{\cos^{-1}t}}$, $a > 0$ and $-1 < t < 1$. Show that $\frac{dy}{dx} = -\frac{y}{x}$.

Sol. Given, $x = \sqrt{a^{\sin^{-1}t}}$ and $y = \sqrt{a^{\cos^{-1}t}}$

$$\Rightarrow \quad \frac{dx}{dt} = \frac{1}{2}(a^{\sin^{-1}t})^{-\frac{1}{2}} \frac{d}{dt}(a^{\sin^{-1}t})$$

$$\Rightarrow \quad \frac{dx}{dt} = \frac{1}{2}(a^{\sin^{-1}t})^{-\frac{1}{2}} \cdot a^{\sin^{-1}t}\log_e a \times \frac{1}{\sqrt{1-t^2}} = \frac{x\log_e a}{2\sqrt{1-t^2}}$$

Again, $$\frac{dy}{dt} = \frac{1}{2}(a^{\cos^{-1}t})^{-1/2}\frac{d}{dt}(a^{\cos^{-1}t})$$

$$= \frac{1}{2}(a^{\cos^{-1}t})^{-1/2} \cdot a^{\cos^{-1}t}\log_e a \times \left(\frac{1}{\sqrt{1-t^2}}\right) = \frac{-y\log_e a}{2\sqrt{1-t^2}}$$

So, $$\frac{dy}{dx} = \frac{\frac{dy}{dt}}{\frac{dx}{dt}} = -\frac{y}{x}$$

Aliter

We have, $xy = \sqrt{a^{\sin^{-1}t}} \cdot \sqrt{a^{\cos^{-1}t}} = \sqrt{a^{\sin^{-1}t + \cos^{-1}t}} = \sqrt{a^{\pi/2}}$

$$\Rightarrow \quad x^2y^2 = a^{\pi/2} \qquad \left[\because \sin^{-1}x + \cos^{-1}x = \frac{\pi}{2}\right]$$

On differentiating, we get

$$2xy^2 + 2x^2y\frac{dy}{dx} = 0 \Rightarrow \frac{dy}{dx} = -\frac{y}{x}$$

Example 44 If $x = a\left(\cos t + \frac{1}{2}\log\tan^2 t\right)$ and $y = a\sin t$, then find $\frac{dy}{dx}$ at $t = \frac{\pi}{4}$.

Sol. Here, $x = a\left(\cos t + \frac{1}{2}\log\tan^2 t\right)$ and $y = a\sin t$

On differentiating both equations w.r.t. t, we get

$$\frac{dx}{dt} = a\left(-\sin t + \frac{1}{\tan\frac{t}{2}}\sec^2\frac{t}{2}\times\frac{1}{2}\right)$$

$$\Rightarrow \frac{dx}{dt} = a\left(-\sin t + \frac{1}{2\sin\frac{t}{2}\cos\frac{t}{2}}\right)$$

$$\Rightarrow \frac{dx}{dt} = a\left(-\sin t + \frac{1}{\sin t}\right) = \frac{a\cos^2 t}{\sin t} \text{ and } \frac{dy}{dt} = a\cos t$$

So, $\frac{dy}{dx} = \frac{dy}{dt} / \frac{dx}{dt} = \frac{a\cos t}{\frac{a\cos^2 t}{\sin t}} = \tan t$

$$\Rightarrow \left(\frac{dy}{dx}\right)_{t=\frac{\pi}{4}} = \tan\frac{\pi}{4} = 1$$

Exercise for Session 5

1. If $x = 2\cos\theta - \cos 2\theta$ and $y = 2\sin\theta - \sin 2\theta$, prove that $\frac{dy}{dx} = \tan\left(\frac{3\theta}{2}\right)$.
2. If $x = e^{\cos 2t}$ and $y = e^{\sin 2t}$, then prove that $\frac{dy}{dx} = \frac{-y\log x}{x\log y}$.
3. If $x = \cos t$ and $y = \sin t$, then prove that $\frac{dy}{dx} = \frac{1}{\sqrt{3}}$, at $t = \frac{2\pi}{3}$.
4. If $x = a\left(t + \frac{1}{t}\right)$ and $y = a\left(t - \frac{1}{t}\right)$, then prove that $\frac{dy}{dx} = \frac{x}{y}$.
5. If $x = \sin^{-1}\left(\frac{2t}{1+t^2}\right)$ and $y = \tan^{-1}\left(\frac{2t}{1-t^2}\right)$, then prove that $\frac{dy}{dx} = 1$.
6. If $x = a\sec^3\theta$ and $y = a\tan^3\theta$, then find $\frac{dy}{dx}$ at $\theta = \frac{\pi}{3}$.
7. Let $y = x^3 - 8x + 7$ and $x = f(t)$. If $\frac{dy}{dt} = 2$ and $x = 3$ at $t = 0$, then find the value of $\frac{dx}{dt}$ at $t = 0$.

Session 6

Logarithmic Differentiation, Differentiation of Infinite Series

Logarithmic Differentiation

So far, we have discussed derivatives of the functions of the form $(f(x))^n$, $n^{f(x)}$ and n^n, where $f(x)$ is a function of x and n is a constant. In this section, we will mainly be discussing derivatives of the functions of the form $(f(x))^{g(x)}$, where $f(x)$ and $g(x)$ are functions of x or $f(x)s(x)h(x)\ldots$ or $\dfrac{f_1(x)f_2(x)\ldots}{g_1(x)g_2(x)\ldots}$

To find the derivatives of these types of functions, we proceed as follows

Let $\quad y=(f(x))^{g(x)}$

Taking logarithm on both sides, we have

$$\log y = g(x)\log\{f(x)\}$$

Differentiating w.r.t. x, we get

$$\frac{1}{y}\cdot\frac{dy}{dx} = g(x)\cdot\frac{1}{f(x)}\cdot\frac{d\,f(x)}{dx}+\log\{f(x)\}\cdot\frac{d\,g(x)}{dx}$$

$$\therefore\quad \frac{dy}{dx} = y\left[\frac{g(x)}{f(x)}\cdot\frac{d\,f(x)}{dx}+\log\{f(x)\}\cdot\frac{d\,g(x)}{dx}\right]$$

or $\quad \dfrac{dy}{dx} = (f(x))^{g(x)}\left\{\dfrac{g(x)}{f(x)}\cdot\dfrac{d\,f(x)}{dx}+\log\{f(x)\}\cdot\dfrac{d\,g(x)}{dx}\right\}$

Logarithmic Differentiation

(For Quickly Solving Objective MCQs)

If $y=\{f(x)\}^{\phi(x)}$, then

$\dfrac{dy}{dx}$ = **Differentiation of y taking $f(x)$ as constant + Differentiation of y taking $\phi(x)$ as constant.**

or when we have to differentiate the function of the form $y=(\text{Variable})^{\text{variable}}$, take log on both sides and then differentiate.

| Example 45 If $y = x^{\sin x}$, then find $\dfrac{dy}{dx}$.

Sol. $y = x^{\sin x}$, taking log on both sides, we get

$$\log y = \sin x \log x$$

On differentiating both sides, we get

$$\frac{1}{y}\frac{dy}{dx} = \sin x\cdot\left(\frac{1}{x}\right)+\log x\cdot(\cos x)$$

$$\Rightarrow\quad \frac{dy}{dx} = y\left[\frac{\sin x}{x}+(\cos x)\log x\right]$$

$$\therefore\quad \frac{dy}{dx} = x^{\sin x}\left[\frac{\sin x}{x}+(\cos x)\log x\right]$$

Aliter

Here, $y = x^{\sin x}$ could also be differentiated by using definition;

$$\frac{d}{dx}(\text{variable})^{\text{variable}} = \frac{d}{dx}(\text{variable})^{\text{constant}} + \frac{d}{dx}(\text{constant})^{\text{variable}}$$

i.e. $\quad \dfrac{dy}{dx} = \dfrac{d}{dx}(x)^{\sin x} + \dfrac{d}{dx}(x)^{\sin x}$

(variable $\;$ constant $\;$ constant $\;$ variable)

$$= \sin x\,(x)^{\sin x-1}+(x)^{\sin x}\cdot(\log x)\cdot\cos x$$

$$= (\sin x)\frac{(x)^{\sin x}}{x}+(x)^{\sin x}\cdot\log x\cdot\cos x$$

$$= x^{\sin x}\left\{\frac{\sin x}{x}+\cos x\cdot\log x\right\}$$

| Example 46 If $x^y\cdot y^x=1$, then find $\dfrac{dy}{dx}$.

Sol. Taking log on both the sides, $y\log x + x\log y = \log 1$

Differentiating both the sides, we get

$$y\cdot\frac{d}{dx}(\log x)+\left\{\frac{d}{dx}y\right\}\cdot\log x+\left\{\frac{d}{dx}x\right\}\cdot\log y + x\left\{\frac{d}{dx}\log y\right\}=0$$

$$\Rightarrow\quad y\cdot\frac{1}{x}+\log x\cdot\frac{dy}{dx}+1\cdot\log y + x\cdot\frac{1}{y}\cdot\frac{dy}{dx}=0$$

$$\Rightarrow\quad \left[\log x+\frac{x}{y}\right]\frac{dy}{dx} = -\left[\frac{y}{x}+\log y\right]$$

$$\Rightarrow\quad \frac{dy}{dx} = -\frac{(y+x\log y)}{(x+y\log x)}\cdot\frac{y}{x}$$

Example 47 If $y^{\cot x} + (\tan^{-1} x)^y = 1$, then find $\frac{dy}{dx}$.

Sol. Let $u = y^{\cot x}$ and $v = (\tan^{-1} x)^y$, we get

$$u + v = 1 \quad \text{...(i)}$$

Taking log on both sides for u and v, we get

$$\log u = \cot x \log y \text{ and } \log v = y \log (\tan^{-1} x)$$

On differentiating both sides, we get

$$\frac{1}{u} \cdot \frac{du}{dx} = \cot x \cdot \frac{1}{y} \frac{dy}{dx} + \log y \cdot (-\operatorname{cosec}^2 x)$$

$$\Rightarrow \frac{du}{dx} = y^{\cot x}\left[\frac{\cot x}{y} \cdot \frac{dy}{dx} - (\operatorname{cosec}^2 x) \log y\right] \quad \text{...(ii)}$$

Again, differentiating $\log v = y \log (\tan^{-1} x)$, we get

$$\frac{1}{v} \cdot \frac{dv}{dx} = y \cdot \frac{1}{\tan^{-1} x} \cdot \frac{1}{1+x^2} + \log (\tan^{-1} x) \cdot \frac{dy}{dx}$$

$$\Rightarrow \frac{dv}{dx} = (\tan^{-1} x)^y \left[\frac{y}{(1+x^2)\tan^{-1} x} + \log (\tan^{-1} x) \cdot \frac{dy}{dx}\right] \quad \text{...(iii)}$$

From Eq. (i), we get $\frac{du}{dx} + \frac{dv}{dx} = 0$

$$\Rightarrow y^{\cot x}\left\{\frac{\cot x}{y} \cdot \frac{dy}{dx} - (\operatorname{cosec}^2 x) \log y\right\}$$
$$+ (\tan^{-1} x)^y \left[\frac{y}{(1+x^2)\tan^{-1} x} + \log (\tan^{-1} x) \cdot \frac{dy}{dx}\right] = 0$$

$$\Rightarrow \left\{\frac{y^{\cot x} \cdot \cot x}{y} + (\tan^{-1} x)^y \cdot \log (\tan^{-1} x)\right\}\frac{dy}{dx}$$
$$= \left\{y^{\cot x} \cdot (\operatorname{cosec}^2 x) \log y - \frac{(\tan^{-1} x)^y \cdot y}{(1+x^2)\tan^{-1} x}\right\}$$

Hence, $$\frac{dy}{dx} = \frac{(y^{\cot x} \cdot \operatorname{cosec}^2 x \cdot \log y) - \left\{\frac{(\tan^{-1} x)^{y-1} \cdot y}{1+x^2}\right\}}{(y^{\cot x - 1} \cdot \cot x) + \{(\tan^{-1} x)^y \cdot \log (\tan^{-1} x)\}}$$

Derivative of the Product of Finite Number of Functions

Example 48 If $f(x) = (x+1)(x+2)(x+3)\ldots(x+n)$, then find $f'(0)$.

Sol. Here $f(x) = (x+1)(x+2)(x+3)\ldots(x+n)$

$$\log f(x) = \log \{(x+1)(x+2)\ldots(x+n)\}$$

Differentiating both sides w.r.t. x, we get

$$\frac{1}{f(x)} f'(x) = \frac{1}{x+1} + \frac{1}{x+2} + \frac{1}{x+3} + \ldots + \frac{1}{x+n}$$

$$\therefore \quad f'(0) = f(0)\left\{1 + \frac{1}{2} + \frac{1}{3} + \ldots + \frac{1}{n}\right\}$$

$$= (1 \cdot 2 \cdot 3 \cdot \ldots \cdot n)\left\{1 + \frac{1}{2} + \frac{1}{3} + \ldots + \frac{1}{n}\right\}$$

$$= (n)!\left\{1 + \frac{1}{2} + \frac{1}{3} + \ldots + \frac{1}{n}\right\}$$

Example 49 If $f(x) = \prod_{n=1}^{100} (x-n)^{n(101-n)}$, then find $\frac{f(101)}{f'(101)}$.

Sol. $f(x) = \prod_{n=1}^{100} (x-n)^{n(101-n)}$

$$\Rightarrow \quad \log f(x) = \left\{n(101-n) \log\left\{\prod_{n=1}^{100} (x-n)\right\}\right\}$$

{Here, Π changes to Σ when taken log}

$$\Rightarrow \quad \log f(x) = \sum_{n=1}^{100} n(101-n) \log (x-n)$$

Differentiating both sides, we get

$$\frac{f'(x)}{f(x)} = \sum_{n=1}^{100} n(101-n) \cdot \frac{1}{x-n}$$

$$\therefore \quad \frac{f'(101)}{f(101)} = \sum_{n=1}^{100} \frac{n(101-n)}{(101-n)} = \sum_{n=1}^{100} n = 5050$$

$$\Rightarrow \quad \frac{f(101)}{f'(101)} = \frac{1}{5050}$$

Differentiation of Infinite Series

When the value of y is given as an infinite series, then we use the fact that, if one term is deleted from an infinite series, it remains unaffected, to replace all terms except first term by y. Thus, we convert it into a finite series or function.

Example 50 If $y = \sqrt{\log x + \sqrt{\log x + \sqrt{\log x + \ldots \infty}}}$ then find $\frac{dy}{dx}$.

Sol. We have, $y = \sqrt{\log x + \sqrt{\log x + \sqrt{\log x + \ldots \infty}}}$

Then $\quad y = \sqrt{\log x + y}$

$$\Rightarrow \quad y^2 = \log x + y$$

Differentiating both sides, we get

$$2y \cdot \frac{dy}{dx} = \frac{1}{x} + \frac{dy}{dx}$$

$$\Rightarrow \quad \frac{dy}{dx}(2y - 1) = \frac{1}{x}$$

$$\Rightarrow \quad \frac{dy}{dx} = \frac{1}{x(2y-1)}$$

Exercise for Session 6

■ **Directions** (Q. Nos. 1 to 10) *Differentiate the following functions w.r.t. x.*

1. x^x
2. $x^{\sqrt{x}}$
3. x^{x^x}
4. $(x^x)^x$
5. $(x^x)\sqrt{x}$
6. $(\cos x)^x$
7. $(\sin x)^{\cos x}$
8. $(\sin x)^{\cos^{-1} x}$
9. $\cos(x^x)$
10. $\log(x^x + \text{cosec}^2 x)$
11. If $y = (\sin x)^{\tan x} + (\cos x)^{\sec x}$, then find $\dfrac{dy}{dx}$.
12. If $x^y = e^{x-y}$, then prove that $\dfrac{dy}{dx} = \dfrac{\log x}{(1+\log x)^2}$
13. If $x^y + y^x = 2$, then find $\dfrac{dy}{dx}$.
14. If $(\cos x)^y = (\sin y)^x$, then find $\dfrac{dy}{dx}$.
15. If $x \sin(a+y) + \sin a \cdot \cos(a+y) = 0$. Prove that $\dfrac{dy}{dx} = \dfrac{\sin^2(a+y)}{\sin a}$.
16. If $y = \sqrt{\cos x + \sqrt{\cos x + \sqrt{\cos x + \ldots \infty}}}$, then prove that $\dfrac{dy}{dx} = \dfrac{\sin x}{1-2y}$.
17. If $y = (\tan x)^{(\tan x)^{(\tan x)^{\cdots \infty}}}$, then prove that $\dfrac{dy}{dx} = 2$ at $x = \dfrac{\pi}{4}$.
18. If $y = e^{x^{e^x}} + x^{e^{e^x}} + e^{x^{x^e}}$, then prove that

$$\frac{dy}{dx} = e^{x^{e^x}} \cdot x^{e^x} \cdot \left\{\frac{e^x}{x} + e^x \cdot \log x\right\} + x^{e^{e^x}} \cdot e^{e^x} \cdot \left\{\frac{1}{x} + e^x \cdot \log x\right\} + e^{x^{x^e}} \cdot x^{x^e} \cdot x^{e-1}\{1 + e\log x\}.$$

19. If $y = \tan^{-1}\left(\dfrac{a_1 x - \alpha}{a_1\alpha + x}\right) + \tan^{-1}\left(\dfrac{a_2 - a_1}{1 + a_1 a_2}\right) + \tan^{-1}\left(\dfrac{a_3 - a_2}{1 + a_2 a_3}\right) + \ldots + \tan^{-1}\left(\dfrac{a_n - a_{n-1}}{1 + a_n a_{n-1}}\right) - \tan^{-1}(a_n)$, then find $\dfrac{dy}{dx}$.

Session 7

Differentiation of a Function w.r.t. Another Function

Differentiation of a Function w.r.t. Another Function

Suppose it is required to differentiate a function $f(x)$ with respect to another function $g(x)$. Let $y = f(x)$ and $z = g(x)$. Here, both functions are different but both are in same variable x.

Then, to find the derivative of $f(x)$ with respect to $g(x)$ or derivative of $g(x)$ with respect to $f(x)$, i.e. to find $\frac{dy}{dz}$ or $\frac{dz}{dy}$. We firstly differentiate both functions $f(x)$ and $g(x)$ with respect to x separately and then put these values in the following formulae:

$$\frac{dy}{dz} = \frac{dy/dx}{dz/dx} \text{ or } \frac{dz}{dy} = \frac{dz/dx}{dy/dx}$$

Example 51 Differentiate $\sin^{-1}\left(\frac{2x}{1+x^2}\right)$ w.r.t. $\tan^{-1} x$, $-1 < x < 1$.

Sol. Let $u = \sin^{-1}\left(\frac{2x}{1+x^2}\right)$ and $v = \tan^{-1} x$

Putting $x = \tan\theta \Rightarrow \theta = \tan^{-1} x$

$$\Rightarrow \quad u = \sin^{-1}\left(\frac{2\tan\theta}{1+\tan^2\theta}\right)$$

$$\Rightarrow \quad u = \sin^{-1}(\sin 2\theta) \quad \left[\because \sin 2\theta = \frac{2\tan\theta}{1+\tan^2\theta}\right]$$

$$\Rightarrow \quad u = 2\theta \Rightarrow u = 2\tan^{-1} x$$

$$\Rightarrow \quad \frac{du}{dx} = \frac{2}{1+x^2} \quad \text{...(i)}$$

Again, $v = \tan^{-1} x \Rightarrow \frac{dv}{dx} = \frac{1}{1+x^2}$ ...(ii)

Now, $$\frac{du}{dv} = \frac{du/dx}{dv/dx} = \frac{\frac{2}{1+x^2}}{\frac{1}{1+x^2}} = 2$$

Example 52 If $x \in \left(\frac{1}{\sqrt{2}}, 1\right)$. Differentiate $\tan^{-1}\left(\frac{\sqrt{1-x^2}}{x}\right)$ w.r.t. $\cos^{-1}(2x\sqrt{1-x^2})$.

Sol. Let $u = \tan^{-1}\left(\frac{\sqrt{1-x^2}}{x}\right)$

and $v = \cos^{-1}(2x\sqrt{1-x^2})$

Putting $x = \sin\theta$, then $x \in \left(\frac{1}{\sqrt{2}}, 1\right)$

$$\Rightarrow \quad \frac{1}{\sqrt{2}} < \sin\theta < 1 \Rightarrow \frac{\pi}{4} < \theta < \frac{\pi}{2}$$

Now, $$u = \tan^{-1}\left(\frac{\sqrt{1-x^2}}{x}\right) \Rightarrow u = \tan^{-1}\left(\frac{\sqrt{1-\sin^2\theta}}{\sin\theta}\right)$$

$$\Rightarrow \quad u = \tan^{-1}\left(\frac{\cos\theta}{\sin\theta}\right) \Rightarrow u = \tan^{-1}(\cot\theta)$$

$$\Rightarrow \quad u = \tan^{-1}\left\{\tan\left(\frac{\pi}{2} - \theta\right)\right\}$$

$$\left[\because \frac{\pi}{4} < \theta < \frac{\pi}{2} \Rightarrow 0 < \frac{\pi}{2} - \theta < \frac{\pi}{4}\right]$$

$$\Rightarrow \quad u = \frac{\pi}{2} - \theta \Rightarrow u = \frac{\pi}{2} - \sin^{-1} x$$

$$\Rightarrow \quad \frac{du}{dx} = -\frac{1}{\sqrt{1-x^2}} \quad \text{...(i)}$$

Again, $v = \cos^{-1}(2x\sqrt{1-x^2})$

$$\Rightarrow \quad v = \frac{\pi}{2} - \sin^{-1}(2x\sqrt{1-x^2})$$

$$\Rightarrow \quad v = \frac{\pi}{2} - \sin^{-1}(\sin 2\theta), \text{ where } x = \sin\theta$$

$$\Rightarrow \quad v = \frac{\pi}{2} - \sin^{-1}(\sin(\pi - 2\theta)) \Rightarrow v = \frac{\pi}{2} - (\pi - 2\theta)$$

$$\left[\because \frac{\pi}{4} < \theta < \frac{\pi}{2} \Rightarrow 0 < \pi - 2\theta > \frac{\pi}{2}\right]$$

$$\Rightarrow \quad v = -\frac{\pi}{2} + 2\theta \Rightarrow v = -\frac{\pi}{2} + 2\sin^{-1} x$$

$$\Rightarrow \quad \frac{dv}{dx} = \frac{2}{\sqrt{1-x^2}} \quad \text{...(ii)}$$

$$\therefore \quad \frac{du}{dv} = \frac{\frac{du}{dx}}{\frac{dv}{dx}} = \frac{-\frac{1}{\sqrt{1-x^2}}}{\frac{2}{\sqrt{1-x^2}}} = -\frac{1}{2}$$

Example 53 Find the derivative of $\sec^{-1}\left(\frac{1}{2x^2-1}\right)$ w.r.t. $\sqrt{1-x^2}$ at $x=\frac{1}{2}$.

Sol. Let $u=\sec^{-1}\left(\frac{1}{2x^2-1}\right)$ and $v=\sqrt{1-x^2}$

Put $x=\cos\theta$

$\therefore$ $u=\sec^{-1}(\sec 2\theta)=2\theta$

and $v=\sqrt{1-\cos^2\theta}=\sin\theta$

$\Rightarrow$ $\frac{du}{d\theta}=2$

and $\frac{dv}{d\theta}=\cos\theta$

$\Rightarrow$ $\frac{du}{dv}=\frac{du/d\theta}{dv/d\theta}=\frac{2}{\cos\theta}=\frac{2}{x}$

Now, $\left(\frac{du}{dv}\right)_{\text{at } x=\frac{1}{2}}=\frac{2}{\frac{1}{2}}=4$

Example 54 If $y=f(x^3)$, $z=g(x^5)$, $f'(x)=\tan x$ and $g'(x)=\sec x$, then find the value of $\lim_{x\to 0}\frac{dy/dz}{x}$.

Sol. Here, $y=f(x^3)\Rightarrow\frac{dy}{dx}=f'(x^3)\cdot 3x^2$

$\frac{dy}{dx}=3x^2\tan x^3$ $\quad[\because f'(x)=\tan x]$

and $z=g(x^5)\Rightarrow\frac{dz}{dx}=g'(x^5)\cdot 5x^4$

$\frac{dz}{dx}=5x^4\cdot\sec x^5$ $\quad[\because g'(x)=\sec x]$

Now, $\frac{dy}{dz}=\frac{3x^2\tan x^3}{5x^4\sec x^5}=\frac{3}{5x^2}\times\frac{\tan x^3}{\sec x^5}$

$\therefore$ $\lim_{x\to 0}\frac{\frac{dy}{dz}}{x}=\lim_{x\to 0}\frac{3\tan x^3}{5x^3\sec x^5}=\frac{3}{5}$

Example 55 Find the derivative of $f(\tan x)$ w.r.t. $g(\sec x)$ at $x=\frac{\pi}{4}$, where $f'(1)=2$ and $g'(\sqrt{2})=4$.

Sol. Let $u=f(\tan x)$ and $v=g(\sec x)$

$\Rightarrow$ $\frac{du}{dx}=f'(\tan x)\cdot\sec^2 x$

and $\frac{dv}{dx}=g'(\sec x)\cdot\sec x\cdot\tan x$

$\Rightarrow$ $\frac{du}{dv}=\frac{du/dx}{dv/dx}=\frac{f'(\tan x)\sec^2 x}{g'(\sec x)\sec x\tan x}$

$\Rightarrow$ $\left(\frac{du}{dv}\right)_{x=\frac{\pi}{4}}=\frac{f'\left(\tan\frac{\pi}{4}\right)}{g'\left(\sec\frac{\pi}{4}\right)\sin\frac{\pi}{4}}=\frac{f'(1)\sqrt{2}}{g'(\sqrt{2})}=\frac{2\sqrt{2}}{4}=\frac{1}{\sqrt{2}}$

Exercise for Session 7

1. Find the derivative of $\tan^{-1}\frac{2x}{1-x^2}$ w.r.t. $\sin^{-1}\frac{2x}{1+x^2}$, if
   (i) $x\in(-1,1)$ (ii) $x\in(1,\infty)$
   (iii) $x\in(-\infty,-1)$
2. Differentiate $x^{\sin^{-1}x}$ w.r.t. $\sin^{-1}x$.
3. Differentiate $\sin^{-1}(2ax\sqrt{1-a^2x^2})$ w.r.t. $\sqrt{1-a^2x^2}$.
4. Differentiate $\log\sin x$ w.r.t. $\sqrt{\cos x}$
5. Differentiate $\tan^{-1}\left\{\frac{\sqrt{1+x^2}-\sqrt{1-x^2}}{\sqrt{1+x^2}+\sqrt{1-x^2}}\right\}$ w.r.t. $\cos^{-1}x^2$.
6. Differentiate x^x w.r.t. $x\log x$.
7. Differentiate $\sin^{-1}(4x\sqrt{1-4x^2})$ w.r.t. $\sqrt{1-4x^2}$, if
   (i) $x\in\left(-\frac{1}{2\sqrt{2}},\frac{1}{2\sqrt{2}}\right)$ (ii) $x\in\left(\frac{1}{2\sqrt{2}},\frac{1}{2}\right)$
   (iii) $x\in\left(-\frac{1}{2},-\frac{1}{2\sqrt{2}}\right)$

Session 8

Higher Derivatives of a Function

Higher Derivatives of a Function

If $y = f(x)$, then the derivative of $\frac{dy}{dx}$ w.r.t. x is called the second derivative of y w.r.t. x and it is denoted by $\frac{d^2y}{dx^2}$.

Also, $\frac{d^2y}{dx^2} = \frac{d}{dx}\left(\frac{dy}{dx}\right)$; $\frac{d^3y}{dx^3} = \frac{d}{dx}\left(\frac{d^2y}{dx^2}\right)$ and so on.

The nth order derivative of y w.r.t. x is denoted by $\frac{d^ny}{dx^n}$. If $y = f(x)$, then other notations for $\frac{dy}{dx}, \frac{d^2y}{dx^2}, \frac{d^3y}{dx^3}, \ldots, \frac{d^ny}{dx^n}$ are $y_1, y_2, y_3, \ldots, y_n$ or $y', y'', y''', \ldots, y^n$ or $f'(x), f''(x), f'''(x), \ldots, f^n(x)$.

Remark

If y is a function of x, given parametrically by $x = \phi(t)$ $y = \psi(t)$, then

$$\frac{d^2y}{dx^2} = \frac{d}{dx}\left(\frac{dy}{dx}\right) = \frac{d}{dx}\left(\frac{\psi'(t)}{\phi'(t)}\right) = \frac{d}{dt}\cdot\left(\frac{\psi'(t)}{\phi'(t)}\right) \times \frac{dt}{dx}$$

Example 56 If $y = x^3 \log_e x$, then find y'' and y'''.

Sol. Here, $y = x^3 \log_e x \Rightarrow y' = x^3 \cdot \frac{1}{x} + 3x^2 \cdot \log_e x$

$$y' = x^2 + 3x^2 \cdot \log_e x$$

Again differentiating both sides, we get

$$y'' = 2x + 6x \cdot \log_e x + 3x^2 \cdot \frac{1}{x}$$

$$y'' = 5x + 6x \cdot \log_e x$$

Differentiating again,

$$y''' = 5 + 6 \cdot \log x + 6x \cdot \frac{1}{x}$$

$$y''' = 11 + 6 \log x$$

Example 57 If $y = \sin(\sin x)$, then prove that $y'' + (\tan x)y' + y\cos^2 x = 0$.

Sol. Here, $y = \sin(\sin x)$ ...(i)

$\Rightarrow$ $y' = \cos(\sin x) \cdot \cos x$

$\Rightarrow$ $(\sec x)\, y' = \cos(\sin x)$ ...(ii)

Again differentiating w.r.t. x, we get

$$(\sec x)y'' + (\sec x \tan x)y' = -\sin(\sin x) \cdot \cos x$$

$\Rightarrow$ $y'' + (\tan x)y' = -\sin(\sin x) \cdot \cos^2 x$

Using Eq. (i), we get $y = \sin(\sin x)$

$\Rightarrow y'' + (\tan x)y' + (\cos^2 x)y = 0$

Example 58 If $(a+bx)e^{y/x} = x$, show that $x^3y'' = (xy' - y)^2$.

Sol. Here, $(a+bx)e^{y/x} = x \Rightarrow e^{y/x} = \frac{x}{a+bx}$

Taking logarithm on both sides, we have

$$\frac{y}{x} = \log\left(\frac{x}{a+bx}\right) \quad \ldots(i)$$

$\Rightarrow$ $\frac{y}{x} = \log x - \log(a+bx)$

Differentiating both sides, we get

$$\frac{xy' - y \cdot 1}{x^2} = \frac{1}{x} - \frac{b}{a+bx}$$

$\Rightarrow$ $xy' - y = \frac{ax}{(a+bx)} = ae^{y/x}$ [from Eq. (i)]

Again taking logarithm on both sides, we get

$$\log(xy' - y) = \log a + \frac{y}{x} \cdot \log e$$

Again differentiating both sides w.r.t. x, we get

$$\frac{xy'' + y' - y'}{xy' - y} = 0 + \frac{(xy' - y \cdot 1)}{x^2}$$

$\Rightarrow$ $x^3y'' = (xy' - y)^2$.

Example 59 If $x = a(t + \sin t)$ and $y = a(1 - \cos t)$, then find $\frac{d^2y}{dx^2}$.

Sol. Here, $x = a(t + \sin t)$ and $y = a(1 - \cos t)$

Differentiating both sides w.r.t. t, we get

$$\frac{dx}{dt} = a(1 + \cos t) \text{ and } \frac{dy}{dt} = a(\sin t)$$

$$\therefore \quad \frac{dy}{dx} = \frac{a \sin t}{a(1 + \cos t)} = \frac{2\sin\frac{t}{2} \cdot \cos\frac{t}{2}}{2\cos^2\frac{t}{2}} = \tan\left(\frac{t}{2}\right)$$

Again, differentiating both sides, we get

$$\frac{d^2y}{dx^2} = \sec^2\left(\frac{t}{2}\right)\cdot\frac{1}{2}\cdot\frac{dt}{dx} = \frac{1}{2}\sec^2\left(\frac{t}{2}\right)\cdot\frac{1}{a\,(1+\cos t)}$$

$$= \frac{1}{2a}\cdot\frac{\sec^2\left(\frac{t}{2}\right)}{2\left(\cos^2\frac{t}{2}\right)}$$

Hence, $$\frac{d^2y}{dx^2} = \frac{1}{4a}\cdot\sec^4\left(\frac{t}{2}\right)$$

Example 60 Show that the function $y = f(x)$ defined by the parametric equations $x = e^t \cdot \sin t$, $y = e^t \cdot \cos t$, satisfies the relation $y''(x+y)^2 = 2(xy' - y)$.

Sol. Here, $x = e^t \cdot \sin t$

On differentiating w.r.t. t, we get

$$\frac{dx}{dt} = e^t \cdot \cos t + e^t \cdot \sin t = e^t\,(\sin t + \cos t)$$

and $y = e^t \cdot \cos t$

On differentiating w.r.t. t, we get

$$\frac{dy}{dt} = e^t \cdot (-\sin t) + e^t \cdot (\cos t) = e^t\,(\cos t - \sin t)$$

So, $$\frac{dy}{dx} = \frac{\frac{dy}{dt}}{\frac{dx}{dt}} = \frac{e^t\,(\cos t - \sin t)}{e^t\,(\cos t + \sin t)} = \frac{y-x}{y+x}$$

or $$\frac{dy}{dx} = \frac{y-x}{y+x} \qquad \text{...(i)}$$

On differentiating both sides w.r.t. x, we get

$$\frac{d^2y}{dx^2} = \frac{(y+x)\cdot\left\{\frac{dy}{dx} - 1\right\} - (y-x)\left\{\frac{dy}{dx} + 1\right\}}{(y+x)^2}$$

or $$y'' = \frac{(y+x)\cdot(y'-1) - (y-x)(y'+1)}{(y+x)^2}$$

Therefore, $(x+y)^2 \cdot y'' = 2(xy' - y)$

Exercise for Session 8

1. If $y = x^x$, then prove that $\frac{d^2y}{dx^2} - \frac{1}{y}\left(\frac{dy}{dx}\right)^2 - \frac{y}{x} = 0$.
2. If $y = A\cos(\log x) + B\sin(\log x)$, then prove that $x^2\frac{d^2y}{dx^2} + x\frac{dy}{dx} + y = 0$.
3. If $y = x\log\left(\frac{x}{a+bx}\right)$, then prove that $x^3\frac{d^2y}{dx^2} = \left(x\frac{dy}{dx} - y\right)^2$.
4. If $y = \log(x + \sqrt{x^2 + a^2})$, then prove that $(x^2 + a^2)\frac{d^2y}{dx^2} + x\frac{dy}{dx} = 0$.
5. If $y = (x + \sqrt{x^2 + 1})^m$, then prove that $(x^2 + 1)y_2 + xy_1 - m^2y = 0$.
6. If $x = at^2$, $y = 2at$, then find $\frac{d^2y}{dx^2}$.
7. If $x = a\cos^3\theta$, $y = a\sin^3\theta$, then find $\frac{d^2y}{dx^2}$.
8. If $x = \tan\left(\frac{1}{a}\log y\right)$, then show that $(1 + x^2)\frac{d^2y}{dx^2} = (a - 2x)\frac{dy}{dx}$.
9. If $x = a\cos\theta + b\sin\theta$ and $y = a\sin\theta - b\cos\theta$, then prove that $y^2\frac{d^2y}{dx^2} - x\frac{dy}{dx} + y = 0$.
10. If $y = \frac{ax+b}{cx+d}$, then prove that $2y_1y_3 = 3(y_2)^2$.
11. If $x = f(t)$ and $y = \phi(t)$, prove that $\frac{d^2y}{dx^2} = \frac{f_1\phi_2 - f_2\phi_1}{f_1^3}$, where suffixes denote differentiation w.r.t. 't'.
12. If $x = \sin t$, $y = \sin Kt$, then show that $(1 - x^2)y_2 - xy_1 + K^2y = 0$.
13. If $\frac{x+b}{2} = a\tan^{-1}(a\log_e y)$, $a > 0$, then prove that $yy'' - yy'\log y = (y')^2$.
14. Let $f(x)$ be a polynomial function of degree 2 and $f(x) > 0$ for all $x \in R$. If $g(x) = f(x) + f'(x) + f''(x)$, then for any x show that $g(x) > 0$.

Session 9

Differentiation of a Function Given in the Form of a Determinant

Differentiation of a Function Given in the Form of a Determinant

To find the derivative of a determinant, we differentiate the row (or columns) keeping other rows (or columns) unchanged, i.e.

If $y = \begin{vmatrix} u(x) & v(x) & w(x) \\ p(x) & q(x) & r(x) \\ \lambda(x) & \mu(x) & \gamma(x) \end{vmatrix}$, then

$$\frac{dy}{dx} = \begin{vmatrix} u'(x) & v'(x) & w'(x) \\ p(x) & q(x) & r(x) \\ \lambda(x) & \mu(x) & \gamma(x) \end{vmatrix} + \begin{vmatrix} u(x) & v(x) & w(x) \\ p'(x) & q'(x) & r'(x) \\ \lambda(x) & \mu(x) & \gamma(x) \end{vmatrix} + \begin{vmatrix} u(x) & v(x) & w(x) \\ p(x) & q(x) & r(x) \\ \lambda'(x) & \mu'(x) & \gamma'(x) \end{vmatrix}$$

The differentiation can also be done column-wise.

Example 61 If $f(x) = \begin{vmatrix} x & x^2 & x^3 \\ 1 & 2x & 3x^2 \\ 0 & 2 & 6x \end{vmatrix}$, then find $f'(x)$.

Sol. Here, $f(x) = \begin{vmatrix} x & x^2 & x^3 \\ 1 & 2x & 3x^2 \\ 0 & 2 & 6x \end{vmatrix}$

On differentiating, we get

$$f'(x) = \begin{vmatrix} \frac{d}{dx}(x) & \frac{d}{dx}(x^2) & \frac{d}{dx}(x^3) \\ 1 & 2x & 3x^2 \\ 0 & 2 & 6x \end{vmatrix} + \begin{vmatrix} x & x^2 & x^3 \\ \frac{d}{dx}(1) & \frac{d}{dx}(2x) & \frac{d}{dx}(3x^2) \\ 0 & 2 & 6x \end{vmatrix} + \begin{vmatrix} x & x^2 & x^3 \\ 1 & 2x & 3x^2 \\ \frac{d}{dx}(0) & \frac{d}{dx}(2) & \frac{d}{dx}(6x) \end{vmatrix}$$

or $f'(x) = \begin{vmatrix} 1 & 2x & 3x^2 \\ 1 & 2x & 3x^2 \\ 0 & 2 & 6x \end{vmatrix} + \begin{vmatrix} x & x^2 & x^3 \\ 0 & 2 & 6x \\ 0 & 2 & 6x \end{vmatrix} + \begin{vmatrix} x & x^2 & x^3 \\ 1 & 2x & 3x^2 \\ 0 & 0 & 6 \end{vmatrix}$

As we know, if any two rows or columns are equal, then value of the determinant is zero.

$\therefore \quad f'(x) = 0 + 0 + \begin{vmatrix} x & x^2 & x^3 \\ 1 & 2x & 3x^2 \\ 0 & 0 & 6 \end{vmatrix} = 6(2x^2 - x^2)$

Therefore, $\quad f'(x) = 6x^2$

Example 62 If $f(x) = \begin{vmatrix} x+a^2 & ab & ac \\ ab & x+b^2 & bc \\ ac & bc & x+c^2 \end{vmatrix}$, then find $f'(x)$.

Sol. We have, $f(x) = \begin{vmatrix} x+a^2 & ab & ac \\ ab & x+b^2 & bc \\ ac & bc & x+c^2 \end{vmatrix}$

$$\therefore \; f'(x) = \begin{vmatrix} 1 & 0 & 0 \\ ab & x+b^2 & bc \\ ac & bc & x+c^2 \end{vmatrix} + \begin{vmatrix} x+a^2 & ab & ac \\ 0 & 1 & 0 \\ ac & bc & x+c^2 \end{vmatrix} + \begin{vmatrix} x+a^2 & ab & ac \\ ab & x+b^2 & bc \\ 0 & 0 & 1 \end{vmatrix}$$

$= \{(x+b^2)(x+c^2) - b^2c^2\} + \{(x+a^2)(x+c^2) - a^2c^2\} + \{(x+a^2)(x+b^2) - a^2b^2\}$

$= 3x^2 + 2x(a^2 + b^2 + c^2)$

Example 63 Let $f(x) = \begin{vmatrix} x^3 & \sin x & \cos x \\ 6 & -1 & 0 \\ p & p^2 & p^3 \end{vmatrix}$, where p is constant. Then, find $\frac{d^3}{dx^3}[f(x)]$ at $x = 0$.

[IIT JEE 1997]

Sol. Given, $f(x)=\begin{vmatrix} x^3 & \sin x & \cos x \\ 6 & -1 & 0 \\ p & p^2 & p^3 \end{vmatrix}$

On differentiating, we get

$$\frac{d}{dx}[f(x)]=\begin{vmatrix} \frac{d}{dx}(x^3) & \frac{d}{dx}(\sin x) & \frac{d}{dx}(\cos x) \\ 6 & (-1) & 0 \\ p & p^2 & p^3 \end{vmatrix}$$

$$+\begin{vmatrix} x^3 & \sin x & \cos x \\ 0 & 0 & 0 \\ p & p^2 & p^3 \end{vmatrix}+\begin{vmatrix} x^3 & \sin x & \cos x \\ 6 & -1 & 0 \\ 0 & 0 & 0 \end{vmatrix}$$

So, $\frac{d}{dx}[f(x)]=\begin{vmatrix} 3x^2 & \cos x & -\sin x \\ 6 & -1 & 0 \\ p & p^2 & p^3 \end{vmatrix}$

Again differentiating, we get

$$\frac{d^2}{dx^2}[f(x)]=\begin{vmatrix} 6x & -\sin x & -\cos x \\ 6 & -1 & 0 \\ p & p^2 & p^3 \end{vmatrix} + \text{Remaining}$$

two determinants as zero.

Again differentiating, we get

$$\frac{d^3}{dx^3}[f(x)]=\begin{vmatrix} 6 & -\cos x & \sin x \\ 6 & -1 & 0 \\ p & p^2 & p^3 \end{vmatrix} + \text{Remaining two}$$

determinants as zero

At $x = 0$, $\frac{d^3}{dx^3}[f(x)]=\begin{vmatrix} 6 & -1 & 0 \\ 6 & -1 & 0 \\ p & p^2 & p^3 \end{vmatrix}=0$

$\therefore \left(\frac{d^3}{dx^3}[f(x)]\right)_{\text{at } x=0} = 0$ (i.e. independent of p)

Example 64 If $y=\cos ax$, prove that

$$\begin{vmatrix} y & y_1 & y_2 \\ y_3 & y_4 & y_5 \\ y_6 & y_7 & y_8 \end{vmatrix}=0, \text{ where } y_r=\frac{d^r}{dx^r}y.$$

Sol. Given, $y = \cos(ax)$

Then, $y_1 = -a\sin(ax) = a\cos\left(\frac{\pi}{2}+ax\right)$

$y_2 = -a^2\cos(ax) = a^2\cos\left(\frac{2\pi}{2}+ax\right)$

$y_3 = +a^3\sin(ax) = a^3\cos\left(\frac{3\pi}{2}+ax\right)$

....

....

$y_n = a^n\cos\left(\frac{n\pi}{2}+ax\right)$

$\therefore$ The determinant $\Delta(x)=\begin{vmatrix} y & y_1 & y_2 \\ y_3 & y_4 & y_5 \\ y_6 & y_7 & y_8 \end{vmatrix}$

$$\Delta(x)=\begin{vmatrix} \cos ax & a\cos\left(\frac{\pi}{2}+ax\right) & a^2\cos\left(\frac{2\pi}{2}+ax\right) \\ a^3\cos\left(\frac{3\pi}{2}+ax\right) & a^4\cos\left(\frac{4\pi}{2}+ax\right) & a^5\cos\left(\frac{5\pi}{2}+ax\right) \\ a^6\cos\left(\frac{6\pi}{2}+ax\right) & a^7\cos\left(\frac{7\pi}{2}+ax\right) & a^8\cos\left(\frac{8\pi}{2}+ax\right) \end{vmatrix}$$

$$\Delta(x)=a^3\times a^6\begin{vmatrix} \cos(ax) & -a\sin(ax) & -a^2\cos(ax) \\ \sin(ax) & a\cos(ax) & -a^2\sin(ax) \\ -\cos(ax) & a\sin(ax) & a^2\cos(ax) \end{vmatrix}$$

$$\Delta(x)=a^9\times(0)\{R_1\to -R_3\}$$

Hence, $\Delta(x)=0$

Example 65 If $f(x)=\begin{vmatrix} (1+x)^a & (1+2x)^b & 1 \\ 1 & (1+x)^a & (1+2x)^b \\ (1+2x)^b & 1 & (1+x)^a \end{vmatrix}$, then find

(i) constant term (ii) coefficient of x.

Sol. Here, $f(x)=\begin{vmatrix} (1+x)^a & (1+2x)^b & 1 \\ 1 & (1+x)^a & (1+2x)^b \\ (1+2x)^b & 1 & (1+x)^a \end{vmatrix}$

$= A + Bx + Cx^2 + \ldots$ ...(i)

Putting $x = 0$, we get

$$f(0)=\begin{vmatrix} 1 & 1 & 1 \\ 1 & 1 & 1 \\ 1 & 1 & 1 \end{vmatrix}=A+B(0)+C(0)^2+\ldots$$

$\Rightarrow A = 0$

Again, differentiating Eq. (i) w.r.t. x, we get

$$f'(x)=\begin{vmatrix} a(1+x)^{a-1} & 2b(1+2x)^{b-1} & 0 \\ 1 & (1+x)^a & (1+2x)^b \\ (1+2x)^b & 1 & (1+x)^a \end{vmatrix}$$

$$+\begin{vmatrix} (1+x)^a & (1+2x)^b & 1 \\ 0 & a(1+x)^{a-1} & 2b(1+2x)^{b-1} \\ (1+2x)^b & 1 & (1+x)^a \end{vmatrix}$$

$$+\begin{vmatrix} (1+x)^a & (1+2x)^b & 1 \\ 1 & (1+x)^a & (1+2x)^b \\ 2b(1+2x)^{b-1} & 0 & a(1+x)^{a-1} \end{vmatrix}=B+2Cx+\ldots$$

$$f'(0)=\begin{vmatrix} a & 2b & 0 \\ 1 & 1 & 1 \\ 1 & 1 & 1 \end{vmatrix}+\begin{vmatrix} 1 & 1 & 1 \\ 0 & a & 2b \\ 1 & 1 & 1 \end{vmatrix}+\begin{vmatrix} 1 & 1 & 1 \\ 1 & 1 & 1 \\ 2b & 0 & a \end{vmatrix}=B$$

$\Rightarrow \quad B=0$ ...(ii)

$\therefore$ Coefficient of constant term = coefficient of $x = 0$.

Example 66 If $a_i, b_i \in N$ for $i = 1, 2, 3$, then coefficient of x in the determinant;

$$\begin{vmatrix} (1+x)^{a_1b_1} & (1+x)^{a_1b_2} & (1+x)^{a_1b_3} \\ (1+x)^{a_2b_1} & (1+x)^{a_2b_2} & (1+x)^{a_2b_3} \\ (1+x)^{a_3b_1} & (1+x)^{a_3b_2} & (1+x)^{a_3b_3} \end{vmatrix}$$

Sol. Here,

$$\begin{vmatrix} (1+x)^{a_1b_1} & (1+x)^{a_1b_2} & (1+x)^{a_1b_3} \\ (1+x)^{a_2b_1} & (1+x)^{a_2b_2} & (1+x)^{a_2b_3} \\ (1+x)^{a_3b_1} & (1+x)^{a_3b_2} & (1+x)^{a_3b_3} \end{vmatrix}=A+Bx+Cx^2+\dots$$

Differentiating both the sides w.r.t. x, we get

$$\begin{vmatrix} a_1b_1(1+x)^{a_1b_1-1} & (1+x)^{a_1b_2} & (1+x)^{a_1b_3} \\ a_2b_1(1+x)^{a_2b_1-1} & (1+x)^{a_2b_2} & (1+x)^{a_2b_3} \\ a_3b_1(1+x)^{a_3b_1-1} & (1+x)^{a_3b_2} & (1+x)^{a_3b_3} \end{vmatrix}$$

$$+\begin{vmatrix} (1+x)^{a_1b_1} & a_1b_2(1+x)^{a_1b_2-1} & (1+x)^{a_1b_3} \\ (1+x)^{a_2b_1} & a_2b_2(1+x)^{a_2b_2-1} & (1+x)^{a_2b_3} \\ (1+x)^{a_3b_1} & a_3b_2(1+x)^{a_3b_2-1} & (1+x)^{a_3b_3} \end{vmatrix}$$

$$+\begin{vmatrix} (1+x)^{a_1b_1} & (1+x)^{a_1b_2} & a_1b_3(1+x)^{a_1b_3-1} \\ (1+x)^{a_2b_1} & (1+x)^{a_2b_2} & a_2b_3(1+x)^{a_2b_3-1} \\ (1+x)^{a_3b_1} & (1+x)^{a_3b_2} & a_3b_3(1+x)^{a_3b_3-1} \end{vmatrix}$$

$$= B + 2Cx + \dots$$

Putting $x = 0$,

$$B=\begin{vmatrix} a_1b_1 & 1 & 1 \\ a_2b_1 & 1 & 1 \\ a_3b_1 & 1 & 1 \end{vmatrix}+\begin{vmatrix} 1 & a_1b_2 & 1 \\ 1 & a_2b_2 & 1 \\ 1 & a_3b_2 & 1 \end{vmatrix}+\begin{vmatrix} 1 & 1 & a_1b_3 \\ 1 & 1 & a_2b_3 \\ 1 & 1 & a_3b_3 \end{vmatrix}$$

$\Rightarrow \quad B = 0$

Exercise for Session 9

1. If $f(x)$, $g(x)$ and $h(x)$ are three polynomials of degree 2, then prove that $\phi(x)=\begin{vmatrix} f(x) & g(x) & h(x) \\ f'(x) & g'(x) & h'(x) \\ f''(x) & g''(x) & h''(x) \end{vmatrix}$ is a constant polynomial.

2. If f, g and h are differentiable functions of x and $\Delta=\begin{vmatrix} f & g & h \\ (xf)' & (xg)' & (xh)' \\ (x^2f)'' & (x^2g)'' & (x^2h)'' \end{vmatrix}$.

   Prove that $\Delta'=\begin{vmatrix} f & g & h \\ f' & g' & h' \\ (x^3f'')' & (x^3g'')' & (x^3h'')' \end{vmatrix}$.

3. Find $\dfrac{dy}{dx}$ at $x=-1$, when $(\sin y)^{\sin\frac{\pi x}{2}}+\dfrac{\sqrt{3}}{2}\sec^{-1}(2x)+2^x\tan\{\log_e(x+2)\}=0$.

Session 10

Derivative of an Inverse Function

Derivative of an Inverse Function

Theorem If the inverse functions f and g are defined by $y = f(x)$ and $x = g(y)$ and if $f'(x)$ exists and $f'(x) \neq 0$, then $g'(y) = \dfrac{1}{f'(x)}$. This result can also be written as, if $\dfrac{dy}{dx}$ exists and $\dfrac{dy}{dx} \neq 0$, then

$$\frac{dx}{dy} = \frac{1}{\dfrac{dy}{dx}} \text{ or } \frac{dy}{dx} \cdot \frac{dx}{dy} = 1 \text{ or } \frac{dy}{dx} = \frac{1}{\dfrac{dx}{dy}} \quad \left[\because \frac{dx}{dy} \neq 0\right]$$

Example 67 If $y = f(x) = x^3 + x^5$ and g is the inverse of f, then find $g'(2)$ (means dx/dy when $y = 2$).

Sol. Here, $y = x^3 + x^5$

On differentiation, we get

$$\frac{dy}{dx} = 3x^2 + 5x^4 \Rightarrow g'(y) = \frac{dx}{dy} = \frac{1}{3x^2 + 5x^4}$$

when $y = 2$, then $2 = x^3 + x^5 \Rightarrow x = 1$

$$\therefore \quad g'(2) = \left.\frac{dx}{dy}\right]_{\substack{x=1 \\ y=2}} = \frac{1}{3+5} = \frac{1}{8}$$

Aliter

$$y = f(x),\ x = g(y)$$
$$(gof)(x) = x$$
$$g'[f(x)]\, f'(x) = 1$$

when $f(x) = 2$, then $x = 1$

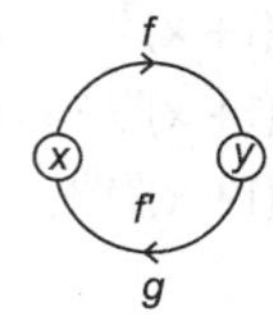

$$g'(2) \cdot f'(1) = 1 \qquad [\text{but } f'(1) = 8]$$
$$\therefore \quad g'(2) = 1/8$$

Example 68 Let $f(x) = \exp(x^3 + x^2 + x)$ for any real number x and let g be the inverse function for f. The value of $g'(e^3)$ is

(a) $\dfrac{1}{6e^3}$ (b) $\dfrac{1}{6}$ (c) $\dfrac{1}{34e^{39}}$ (d) 6

Sol. Let $y = e^{x^3 + x^2 + x}$,

On differentiating, $\dfrac{dy}{dx} = e^{x^3 + x^2 + x} \cdot (3x^2 + 2x + 1)$

$$\Rightarrow \quad g'(y) = \frac{dx}{dy} = \frac{1}{e^{x^3+x^2+x} \cdot [3x^2 + 2x + 1]}$$

$$y = e^3$$
$$\Rightarrow \quad e^3 = e^{x^3 + x^2 + x}$$
$$\Rightarrow \quad x^3 + x^2 + x - 3 = 0$$
$$\Rightarrow \quad (x-1)(x^2 + 2x + 3) = 0$$
$$\Rightarrow \quad x = 1 \qquad [\because x^2 + 2x + 3 > 0]$$
$$\therefore \quad g'(e^3) = \frac{1}{6e^3}$$

Hence, (a) is the correct option.

Exercise for Session 10

1. The function $f(x) = e^x + x$, being differentiable and one to one, has a differentiable increase $f^{-1}(x)$. Find $\dfrac{d}{dx}(f^{-1}(x))$ at point $f(\log 2)$.
2. Let $g(x)$ be the inverse of an invertible function $f(x)$ which is differentiable for all real x, then find $g''(f(x))$.
3. Let $g(x)$ be the inverse of an invertible function $f(x)$ which is differentiable at $x = c$, then $g'(f(c))$ equals
   (a) $f'(c)$ (b) $\dfrac{1}{f'(c)}$ (c) $f(c)$ (d) None of these
4. If $f(x) = x + \tan x$ and f is inverse of g, then $g'(x)$ equals
   (a) $\dfrac{1}{1 + [g(x) - x]^2}$ (b) $\dfrac{1}{2 - [g(x) - x]^2}$ (c) $\dfrac{1}{2 + [g(x) - x]^2}$ (d) None of these

JEE Type Solved Examples :

Single Option Correct Type Questions

Ex. 1 *Let f be a twice differentiable function such that $f''(x) = -f(x)$ and $f'(x) = g(x)$. If $h(x) = \{f(x)\}^2 + \{g(x)\}^2$, where $h(5) = 11$. Find $h(10)$.*

(a) 0 (b) 9

(c) 11 (d) None of these

Sol. (c) Given, $h(x) = \{f(x)\}^2 + \{g(x)\}^2$

On differentiating both the sides w.r.t. x, we get

$$h'(x) = 2f(x) \cdot f'(x) + 2g(x) \cdot g'(x) \quad \text{...(i)}$$

Now, $f'(x) = g(x)$.

Then, $f''(x) = g'(x)$

$\Rightarrow$ $-f(x) = g'(x)$ ...(ii)

$[\because f''(x) = -f(x), \text{ given}]$

From Eqs. (i) and (ii), we get

$$h'(x) = 2f(x) \cdot g(x) + 2g(x) \cdot \{-f(x)\}$$

[using $f'(x) = g(x)$ and $g'(x) = -f(x)$]

$\therefore$ $h'(x) = 0$

So, $h(x)$ must be constant. $\left[\because \dfrac{d}{dx}(\text{constant}) = 0\right]$

But $h(5) = 11$

$h(x) = 11$

So, $h(10) = 11$

Ex. 2 *Let $f(x)$ be a real valued differentiable function not identically zero such that $f(x + y^{2n+1}) = f(x) + \{f(y)\}^{2n+1}, n \in N$ and x, y are any real numbers and $f'(0) \geq 0$. Find the values of $f(5)$ and $f'(10)$.*

(a) 3, 2 (b) 0, 1

(c) 1, 5 (d) 5, 1

Sol. (d) Here, $f(x + y^{2n+1}) = f(x) + \{f(y)\}^{2n+1}$ ...(i)

Putting $x = 0$, $y = 0$, we get

$$f(0) = f(0) + \{f(0)\}^{2n+1} \Rightarrow f(0) = 0$$

$f'(0) \geq 0$ [given]

$$\Rightarrow \lim_{x \to 0} \frac{f(x) - f(0)}{x - 0} \geq 0 \Rightarrow \lim_{x \to 0} \frac{f(x)}{x} \geq 0$$

Now, if $x > 0 \Rightarrow f(x) \geq 0$ ...(ii)

Putting $x = 0$, $y = 1$ in Eq. (i), we get

$$f(1) = f(0) + \{f(1)\}^{2n+1} \text{ or } f(1)[1 - \{f(1)\}^{2n}] = 0$$

$\therefore$ $f(1) = 0$ or 1 [using Eq. (ii)]

Putting $y = 1$ in Eq. (i), for all real x.

$$f(x+1) = f(x) + \{f(1)\}^{2n+1} \quad \text{...(iii)}$$

Now, two cases arise either $f(1) = 0$ or 1

Case I If $f(1) = 0$

$\Rightarrow$ $f(x + 1) = f(x)$ [using Eq. (iii)]

$\Rightarrow$ $f(1) = f(2) = f(3) = \dots = 0$

$\therefore$ $f(x)$ is identically zero. (which is not possible)

Case II If $f(1) = 1$

$\Rightarrow$ $f(x + 1) = f(x) + 1$ [using Eq. (iii)]

$\therefore$ $f(2) = f(1) + 1 = 1 + 1 = 2$

$f(3) = f(2) + 1 = 2 + 1 = 3$

$f(4) = f(3) + 1 = 3 + 1 = 4$

$f(5) = f(4) + 1 = 4 + 1 = 5$

Proceeding in same way, we get

$f(x) = x$ and $f'(x) = 1 \Rightarrow f'(10) = 1$

So, $f(5) = 5$ and $f'(10) = 1$

Ex. 3 *Let $f\left(\dfrac{x+y}{2}\right) = \dfrac{f(x) + f(y)}{2}$ and $f'(0) = a$ and $f(0) = b$. Find $f''(x)$ (where y is independent of x), when $f(x)$ is differentiable.*

(a) 0 (b) 1

(c) a (d) None of these

Sol. (a) Here $f\left(\dfrac{x+y}{2}\right) = \dfrac{f(x) + f(y)}{2}$, this holds for any real x, y and y is independent of x. i.e. $\dfrac{dy}{dx} = 0$

On differentiating w.r.t. x, we get

$$f'\left(\frac{x+y}{2}\right) \cdot \frac{1}{2}\left(1 + \frac{dy}{dx}\right) = \frac{1}{2}\left\{f'(x) + f'(y) \cdot \frac{dy}{dx}\right\}$$

$$\therefore \quad \frac{1}{2} f'\left(\frac{x+y}{2}\right) = \frac{1}{2} f'(x) \quad \left[\text{as } \frac{dy}{dx} = 0\right]$$

$$\Rightarrow \quad f'\left(\frac{x+y}{2}\right) = f'(x) \quad \text{...(i)}$$

Taking $x = 0$ and $y = x$ in Eq. (i), we get

$$f'\left(\frac{0+x}{2}\right) = f'(0) \Rightarrow f'\left(\frac{x}{2}\right) = a \quad [\because f'(0) = a, \text{ given}]$$

which shows $f\left(\dfrac{x}{2}\right) = a\left(\dfrac{x}{2}\right) + c$ [using integration]

$\therefore$ $f(x) = ax + c$

Let $x = 0$

$\Rightarrow$ $f(0) = b = c$

$\therefore$ $f(x) = ax + b$

On differentiating,

$$f'(x) = a \text{ and } f''(x) = 0$$

Ex. 4 *If* $y = \dfrac{(a-x)\sqrt{a-x} - (b-x)\sqrt{x-b}}{\sqrt{a-x}+\sqrt{x-b}}$, *then find* $\dfrac{dy}{dx}$ *whenever defined.*

(a) $\dfrac{x-(a+b)}{\sqrt{(a-x)(x-b)}}$ (b) $\dfrac{2x+(a+b)}{\sqrt{a-x}(x-b)}$

(c) $\dfrac{2x-(a+b)}{2\sqrt{(a-x)(x-b)}}$ (d) None of these

Sol. (c) Here, $y = \dfrac{(a-x)^{3/2}+(x-b)^{3/2}}{(a-x)^{1/2}+(x-b)^{1/2}}$, use

$a^3+b^3=(a+b)(a^2-ab+b^2)$

$$\therefore \quad y = \frac{\left[\begin{array}{l}\{(a-x)^{1/2}+(x-b)^{1/2}\}\{(a-x)\\ \quad -\sqrt{(a-x)(x-b)}+(x-b)\}\end{array}\right]}{\{(a-x)^{1/2}+(x-b)^{1/2}\}}$$

$$\Rightarrow \quad y = (a-b) - \sqrt{(ax+bx-ab-x^2)}$$

$$\therefore \quad \frac{dy}{dx} = -\frac{(a+b-2x)}{2\sqrt{(a-x)(x-b)}}$$

$$\Rightarrow \quad \frac{dy}{dx} = \frac{2x-(a+b)}{2\sqrt{(a-x)(x-b)}}$$

Ex. 5 *If* $x^2+y^2=R^2$ $(R>0)$, *then* $K = \dfrac{y''}{\sqrt{(1+(y')^2)^3}}$.

Find K in terms of R.

(a) $\dfrac{1}{R}$ (b) $-\dfrac{1}{R}$

(c) R (d) $\dfrac{1}{2R}$

Sol. (b) Here, $x^2+y^2=R^2$

On differentiating w.r.t. x, we get

$$2x+2yy'=0 \Rightarrow x+yy'=0$$

$$\Rightarrow \quad y' = -\frac{x}{y} \quad \text{...(i)}$$

Again, differentiating both the sides,

$$1+yy''+(y')^2=0$$

$$\therefore \quad y'' = -\frac{(1+(y')^2)}{y} \quad \text{...(ii)}$$

Now, $K = \dfrac{y''}{\sqrt{(1+(y')^2)^3}} = -\dfrac{(1+(y')^2)}{y\cdot(1+(y')^2)^{3/2}}$

$$= -\frac{1}{y\sqrt{1+(y')^2}} = -\frac{1}{y\sqrt{1+\frac{x^2}{y^2}}} \quad \text{[from Eq. (i)]}$$

$$= -\frac{1}{\sqrt{x^2+y^2}} = -\frac{1}{R}$$

$$\therefore \quad K = -\frac{1}{R}$$

Ex. 6 *Let* $f(x)=x+\sin x$, *suppose g denotes the inverse function of f. Then, find the value of* $g'\left(\dfrac{\pi}{4}+\dfrac{1}{\sqrt{2}}\right)$.

(a) $2+\sqrt{2}$ (b) $\sqrt{2}-2$

(c) $2-\sqrt{2}$ (d) $2\sqrt{2}$

Sol. (c) Here, $f(x)=x+\sin x$

$$\therefore \quad \frac{dy}{dx} = 1+\cos x \Rightarrow g'(y) = \frac{dx}{dy} = \frac{1}{1+\cos x}$$

where, $y = \dfrac{\pi}{4}+\dfrac{1}{\sqrt{2}} = x+\sin x \Rightarrow x = \dfrac{\pi}{4}$

$$\therefore \quad g'\left(\frac{\pi}{4}+\frac{1}{\sqrt{2}}\right) = \frac{1}{1+\frac{1}{\sqrt{2}}} = \frac{\sqrt{2}}{\sqrt{2}+1} = 2-\sqrt{2}$$

Ex. 7 *Let* $e^{f(x)} = \log x$. *If* $g(x)$ *is the inverse of* $f(x)$, *then find* $g'(x)$.

(a) e^{e^x} (b) e^x (c) e^{e^x+x} (d) None of these

Sol. (c) Let $f(x)=y \Rightarrow x=f^{-1}(y)=g(y)$

$$\Rightarrow \quad x = e^{e^y} \Rightarrow \frac{dx}{dy} = e^{e^y}\cdot e^y = e^{e^y+y} = g'(y)$$

$$\therefore \quad g'(x) = e^{e^x+x}$$

Ex. 8 *If* $f(x) = 4x^3 - 6x^2\cos 2a + 3x\cdot\sin 2a\cdot\sin 6a + \sqrt{\log(2a-a^2)}$, *then*

(a) $f'\left(\dfrac{1}{2}\right)<0$ (b) $f'\left(\dfrac{1}{2}\right)\le 0$

(c) $f'\left(\dfrac{1}{2}\right)>0$ (d) None of these

Sol. (c) Here, $f(x) = 4x^3-6x^2\cos 2a+3x\sin 2a\cdot\sin 6a + \sqrt{\log(2a-a^2)}$ for $f(x)$ to exists.

$$\log(2a-a^2)\ge 0 \Rightarrow (2a-a^2)\ge e^0$$

i.e. $2a-a^2\ge 1$ or $a^2-2a+1\le 0 \Rightarrow (a-1)^2\le 0$

which is only possible, if $(a-1)^2=0$, i.e. $a=1$.

$$\therefore \quad f(x) = 4x^3-6x^2\cos 2+3x\sin 2\cdot\sin 6$$

$$\Rightarrow \quad f'(x) = 12x^2-12x\cos 2+3\sin 2\sin 6$$

$$\Rightarrow \quad f'\left(\frac{1}{2}\right) = 3-6\cos 2+3\sin 2\sin 6$$

$$= 3(1+\sin 2\sin 6)-6\cos 2 \Rightarrow f'\left(\frac{1}{2}\right)>0$$

Note

$\cos 2<0$ and $1+\sin 2\sin 6>0$

Ex. 9 *Suppose, $f(x)=e^{ax}+e^{bx}$, where $a\neq b$ and $f''(x)-2f'(x)-15f(x)=0$ for all $x\in R$. Then, find ab.*

(a) 15 (b) −15 (c) 10 (d) 16

Sol. (b) $f(x)=e^{ax}+e^{bx}$

$\Rightarrow\quad f'(x)=ae^{ax}+be^{bx},\ f''(x)=a^2e^{ax}+b^2e^{bx}$

$\therefore\quad f''(x)-2f'(x)-15f(x)=0$

$\Rightarrow\quad \{a^2e^{ax}+b^2e^{bx}\}-2\{ae^{ax}+be^{bx}\}-15\{e^{ax}+e^{bx}\}=0$, for all x.

$\Rightarrow\quad (a^2-2a-15)e^{ax}+(b^2-2b-15)e^{bx}=0$, for all x.

$\Rightarrow\quad a^2-2a-15=0$ and $b^2-2b-15=0$

$\Rightarrow\quad (a-5)(a+3)=0$

and $\quad (b-5)(b+3)=0$

$\Rightarrow\quad a=5$ or -3

and $\quad b=5$ or -3

But $\quad a\neq b$, hence, $a=5,\ b=-3$

or $\quad a=-3,\ b=5\Rightarrow ab=-15$

JEE Type Solved Examples : More than One Correct Option Type Questions

Ex. 10 *The functions $u=e^x\cdot\sin x$ and $v=e^x\cdot\cos x$ satisfy the equation*

(a) $v\dfrac{du}{dx}-u\dfrac{dv}{dx}=u^2+v^2$ (b) $v\dfrac{du}{dx}+u\dfrac{dv}{dx}=u^2+v^2$

(c) $\dfrac{du}{dx}+\dfrac{dv}{dx}=2v$ (d) $\dfrac{du}{dx}+\dfrac{dv}{dx}=2u$

Sol. (a, c) We have, $u=e^x\sin x,\ v=e^x\cos x$

Differentiating both the equations w.r.t. x, we get

$$\frac{du}{dx}=e^x(\sin x+\cos x) \quad \text{...(i)}$$

and
$$\frac{dv}{dx}=e^x(\cos x-\sin x) \quad \text{...(ii)}$$

Now, $\dfrac{du}{dx}+\dfrac{dv}{dx}=2e^x\cos x$ [using Eqs. (i) and (ii)]

$\Rightarrow\quad \dfrac{du}{dx}+\dfrac{dv}{dx}=2v$

Also, $\dfrac{du}{dx}=u+v$ and $\dfrac{dv}{dx}=v-u$

$\therefore\quad v\dfrac{du}{dx}=uv+v^2$ and $u\dfrac{dv}{dx}=uv-u^2$

$\Rightarrow\quad v\dfrac{du}{dx}-u\dfrac{dv}{dx}=u^2+v^2$

Ex. 11 *If $y+\log(1+x)=0$, then which of the following is true?*

(a) $e^y=xy'+1$ (b) $y'=-\dfrac{1}{(x+1)}$

(c) $y'+e^y=0$ (d) $y'=e^y$

Sol. (a, b, c) We have, $y+\log_e(1+x)=0$

$\Rightarrow\quad y=-\log(1+x)\Rightarrow y=\log(1+x)^{-1}$

$\Rightarrow\quad e^y=\dfrac{1}{1+x}$ ...(i)

$\Rightarrow\quad xe^y+e^y=1$

On differentiating w.r.t. x, we get

$$xe^yy'+e^y+e^yy'=0$$

$xy'+y'+1=0$ $[\because e^y\neq 0]$ ...(ii)

$\Rightarrow\quad y'=-\dfrac{1}{x+1}$

$\Rightarrow\quad y'=-e^y$ [from Eq. (i)]

$\Rightarrow\quad y'+e^y=0$

From Eq. (ii), we have

$$xy'+1=-y'\Rightarrow xy'+1=e^y$$

Ex. 12 *If $x^p\cdot y^q=(x+y)^{p+q}$, then $\dfrac{dy}{dx}$ is*

(a) independent of p

(b) independent of q

(c) dependent both p and q

(d) $\dfrac{y}{x}$

Sol. (a, b, d) We have, $x^p\cdot y^q=(x+y)^{p+q}$

On taking log both sides, we get

$$p\log x+q\log y=(p+q)\log(x+y)$$

On differentiating w.r.t. x, we get

$$\frac{p}{x}+\frac{q}{y}\frac{dy}{dx}=\frac{p+q}{x+y}\left(1+\frac{dy}{dx}\right)$$

$\Rightarrow\quad \dfrac{dy}{dx}\left(\dfrac{q}{y}-\dfrac{p+q}{x+y}\right)=\dfrac{p+q}{x+y}-\dfrac{p}{x}$

$\Rightarrow\quad \dfrac{dy}{dx}=\dfrac{y}{x}$

Ex. 13 *Two functions f and g have first and second derivatives at $x=0$ and satisfy the relations $f(0)=\dfrac{2}{g(0)}$, $f'(0)=2g'(0)=4g(0)$, $g''(0)=5f''(0)=6f(0)=3$ Then,*

(a) If $h(x)=\dfrac{f(x)}{g(x)}$, then $h'(0)=\dfrac{15}{4}$

(b) If $k(x)=f(x)\cdot g(x)\cdot \sin x$, then $k'(0)=2$

(c) $\lim\limits_{x\to 0}\dfrac{g'(x)}{f'(x)}=\dfrac{1}{2}$

(d) None of the above

Sol. (a, b, c) We have, $f(0)=\dfrac{2}{g(0)}$,

$$f'(0)=2g'(0)=4g(0)$$
$$g''(0)=5f''(0)=6f(0)=3$$

Now, on solving these equations, we get

$$f(0)=\frac{1}{2},\ g(0)=4,\ f'(0)=16,\ g'(0)=8$$
$$f''(0)=\frac{3}{5},\ g''(0)=3$$

(a) We have, $h(x)=\dfrac{f(x)}{g(x)}$,

$$h'(x)=\frac{g(x)f'(x)-g'(x)f'(x)}{[g(x)]^2}$$

$$\therefore \quad h'(0)=\frac{g(0)f'(0)-g'(0)f'(0)}{[g(0)]^2}$$

$$=\frac{4\times 16-8\times\frac{1}{2}}{(4)^2}=\frac{15}{4}$$

(b) $k(x)=f(x)\cdot g(x)\cdot \sin x$

$$k'(x)=f(x)g(x)\cos x+f(x)\sin x\cdot g'(x)+g(x)\sin x f'(x)$$
$$k'(0)=f(0)g(0)\cos 0+f(0)\sin 0 g'(0)+g(0)\sin 0 f'(0)$$
$$k'(0)=\frac{1}{2}\times 4\times 1+0+0=2$$

(c) $\lim\limits_{x\to 0}\dfrac{g'(x)}{f'(x)}=\dfrac{g'(0)}{f'(0)}=\dfrac{1}{2}$

● **Ex. 14** *If $f(x)=|x^2-3|x|+2|$, then which of the following is/are true?*

(a) $f'(x)=2x-3$ for $x\in(0,1)\cup(2,\infty)$

(b) $f'(x)=2x+3$ for $x\in(-\infty,-2)\cup(-1,0)$

(c) $f'(x)=-2x-3$ for $x\in(-2,-1)$

(d) None of the above

Sol. (a, b, c) We have,

$$f(x)=|x^2-3|x|+2|=\begin{cases}|x^2-3x+2|, & x\ge 0\\ |x^2+3x+2|, & x<0\end{cases}$$

$$\Rightarrow f(x)=\begin{cases}x^2-3x+2, & x^2-3x+2\ge 0,\ x\ge 0\\ -x^2+3x-2, & x^2-3x+2<0,\ x\ge 0\\ x^2+3x+2, & x^2+3x+2\ge 0,\ x<0\\ -x^2-3x-2, & x^2+3x+2<0,\ x<0\end{cases}$$

$$\Rightarrow f(x)=\begin{cases}x^2-3x+2, & x\in[0,1]\cup(2,\infty)\\ -x^2+3x+2, & x\in(1,2)\\ x^2+3x+2, & x\in(-\infty,-2]\cup[-1,0)\\ -x^2-3x+2, & x\in(-2,-1)\end{cases}$$

$$\Rightarrow f'(x)=\begin{cases}2x-3, & x\in(0,1)\cup(2,\infty)\\ -2x+3, & x\in(1,2)\\ 2x+3, & x\in(-\infty,-2)\cup(-1,0)\\ -2x-3, & x\in(-2,-1)\end{cases}$$

JEE Type Solved Examples : Statements I and II Type Questions

■ **Directions** (Ex. Nos. 15-17) *This section is based on Statement I and Statement II. Select the correct answer from the codes given below.*

(a) Both Statement I and Statement II are correct and Statement II is the correct explanation of Statement I

(b) Both Statement I and Statement II are correct but Statement II is not the correct explanation of Statement I

(c) Statement I is correct but Statement II is incorrect

(d) Statement II is correct but Statement I is incorrect

● **Ex. 15** *Consider $f(x)=\dfrac{x}{x^2-1}$ and $g(x)=f''(x)$*

Statement I *Graph of $g(x)$ is concave up for $x>1$.*

Statement II

$$\frac{d^n}{dx^n}f(x)=\frac{(-1)^n n!}{2}\left\{\frac{1}{(x+1)^{n+1}}+\frac{1}{(x-1)^{n+1}}\right\} n\in N$$

Sol. (b) We have, $f(x)=\dfrac{x}{x^2-1}\Rightarrow f(x)=\dfrac{x}{(x+1)(x-1)}$

$$\Rightarrow f(x)=\frac{1}{2}\left[\frac{1}{x+1}+\frac{1}{x-1}\right]$$

$$\Rightarrow f'(x)=\frac{1}{2}\left[-\frac{1}{(x+1)^2}-\frac{1}{(x-1)^2}\right]$$

$$\Rightarrow f''(x)=\frac{(-1)(-2)}{2}\left[\frac{1}{(x+1)^3}+\frac{1}{(x-1)^3}\right]$$

$$\vdots \qquad \vdots$$

$$\Rightarrow f^n(x)=\frac{(-1)^n n!}{2}\left[\frac{1}{(x+1)^{n+1}}+\frac{1}{(x-1)^{n+1}}\right]$$

Now,
$$g(x)=f''(x)$$
$$g'(x)=f'''(x)$$
$$\Rightarrow g''(x)=f''''(x)$$

$$\Rightarrow \quad g''(x) = \frac{(-1)^4 4!}{2}\left[\frac{1}{(x+1)^5} + \frac{1}{(x-1)^5}\right]$$

$$\Rightarrow \quad g''(x) = \frac{4!}{2}\left[\frac{1}{(x+1)^5} + \frac{1}{(x-1)^5}\right]$$

$$\therefore \quad g''(1) > 0$$

$\therefore$ Graph of $g(x)$ is concave up for $x > 1$.

Hence, both statements are correct but Statement II is not the correct explanation of Statement I.

Ex. 16 **Statement I** *If differentiable function $f(x)$ satisfies the relation $f(x) + f(x-2) = 0$, $\forall x \in R$, and if $\left(\frac{d}{dx} f(x)\right)_{x=a} = 6$, then $\left(\frac{d}{dx} f(x)\right)_{a+4000} = 6$*

Statement II *$f(x)$ is a periodic function with period 4.*

Sol. (*a*) We have, $\quad f(x) + f(x-2) = 0 \quad$ ...(i)

Replace x by $x - 2$ in Eq. (i), we have

$$f(x-2) + f(x-4) = 0 \quad \text{...(ii)}$$

From Eqs. (i) and (ii),

$$f(x) - f(x-4) = 0 \quad \text{...(iii)}$$

Replace x by $x + 4$ in Eq. (iii), we get

$$f(x+4) - f(x) = 0$$

$$\Rightarrow \quad f(x+4) = f(x)$$

$$\Rightarrow \quad f(x) = f(x+4) = f(x+8) \dots f(x+4000)$$

$$\Rightarrow \quad f'(x) = f'(x+4000)$$

Hence, both the statements are true and Statement II is the correct explanation of Statement I.

Ex. 17 **Statement I** *Let $f(x) = x[x]$ and $[\cdot]$ denotes greatest integer function, when x is not an integral, then rule for $f'(x)$ is given by $[x]$.*

Statement II *$f'(x)$ does not exist for any $x \in Z$.*

Sol. (*a*)

$$f(x) = x[x] = \begin{cases} -x, & -1 \le x < 0 \\ 0, & 0 \le x < 1 \\ x, & 1 \le x < 2 \\ 2x, & 2 \le x < 3 \\ \vdots & \vdots \end{cases}$$

$$\Rightarrow \quad f'(x) = \begin{cases} -1, & -1 < x < 0 \\ 0, & 0 < x < 1 \\ 1, & 1 < x < 2 \\ 2, & 2 < x < 3 \\ \vdots & \vdots \end{cases}$$

$$\Rightarrow \quad f'(x) = [x]$$

Hence, both statements are true and Statement II is the correct explanation of Statement I.

JEE Type Solved Examples : Passage Based Questions

Passage I

(Ex. Nos. 18 to 20)

A non-zero polynomial with real coefficients has the property that $f''(x) \cdot f'(x) = f(x)$. Then,

Ex. 18 *The leading coefficient of $f(x)$, is*

(a) 1/6 (b) 1/9
(c) 1/12 (d) 1/18

Ex. 19 *The degree of $f(x)$, is*

(a) 2 (b) 3
(c) 4 (d) 5

Ex. 20 *The value of $f'''(x)$, is*

(a) 1/18 (b) 1/3
(c) 1/9 (d) 1/6

Sol. (Ex. Nos. 18 to 20)

Let degree of $f(x) = n$

$\therefore$ Degree of $f'(x) = n - 1$, degree of $f''(x) = n - 2$

Since, $\quad f''(x) \cdot f'(x) = f(x)$

$\Rightarrow \quad (n-1) + (n-2) = n$

$\Rightarrow \quad 2n - 3 = n \quad$ or $\quad n = 3$

$\therefore$ Degree of $f(x) = 3$

Hence, $\quad f(x) = ax^3 + bx^2 + cx + d$

$$\Rightarrow \quad f'(x) = 3ax^2 + 2bx + c$$

$$\Rightarrow \quad f''(x) = 6ax + 2b$$

$$f'''(x) = 6a$$

As, $\quad f''(x) \cdot f'(x) = f(x)$

$$\Rightarrow \quad (6ax + 2b)(3ax^2 + 2bx + c) = (ax^3 + bx^2 + cx + d)$$

On comparing coefficients of x^3, we get

$$18a^2 = a \Rightarrow a = 0, \frac{1}{18}$$

$$\therefore \quad a = \frac{1}{18} \quad [\because a \ne 0]$$

The leading coefficient of $f(x) = \frac{1}{18}$ and $f'''(x) = 6a = \frac{1}{3}$.

18. (d) **19.** (b) **20.** (b)

Passage II

(Ex. Nos. 21 to 22)

The ends A and B of a rod of length $\sqrt{5}$ are sliding along the curve $y = 2x^2$. Let x_A and x_B be the x-coordinate of the ends. Then,

Ex. 21 *The moment when A is at $(0, 0)$ and B is at $(1, 2)$. The derivative $\frac{dx_B}{dx_A}$, is*

(a) $\frac{1}{9}$ (b) $\frac{1}{7}$

(c) $\frac{1}{5}$ (d) None of these

Ex. 22 *The moment when A is at $(1, 2)$ and B is at $(0, 0)$. The derivative $\frac{dx_B}{dx_A}$, is*

(a) 16 (b) 8 (c) 9 (d) 2

Sol. (Ex. Nos. 21 to 22)

We have, $y = 2x^2$

$$(AB)^2 = (x_B - x_A)^2 + (2x_B^2 - 2x_A^2)^2 = 5$$

$$\Rightarrow \quad (x_B - x_A)^2 + 4(x_B^2 - x_A^2)^2 = 5$$

Differentiating w.r.t. x_A and denoting $\frac{dx_B}{dx_A} = D$

$$2(x_B - x_A)(D - 1) + 8(x_B^2 - x_A^2) \cdot \{2x_B D - 2x_A\} = 0 \quad \text{...(i)}$$

Now, when $A(0, 0)$ and $B(1, 2) \Rightarrow x_A = 0, x_B = 1$

Then, Eq. (i) reduces to

$$2(1 - 0)(D - 1) + 8(1 - 0)\{2D - 0\} = 0$$

$$\Rightarrow \quad 2D - 2 + 16D = 0$$

$$\Rightarrow \quad D = \frac{1}{9} \Rightarrow \frac{dx_B}{dx_A} = \frac{1}{9}$$

Again, when $A(1, 2)$ and $B(0, 0) \Rightarrow x_A = 1, x_B = 0$

Eq. (i) reduces to

$$2(0 - 1)(D - 1) + 8(0 - 1)\{0 - 2\} = 0$$

$$\Rightarrow \quad -2D + 2 + 16 = 0 \Rightarrow D = 9$$

$$\therefore \quad \frac{dx_B}{dx_A} = 9$$

21. (a) 22. (c)

Passage III

(Ex. Nos. 23 to 24)

The figure of $y = P(x) = ax^5 + bx^4 + cx^3 + dx^2 + ex + f$ is given below

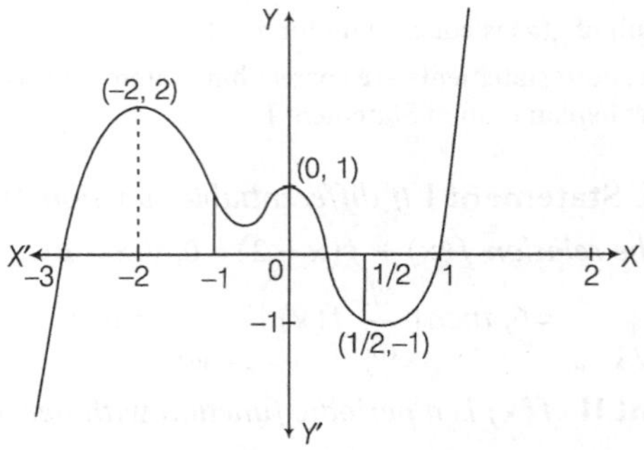

Ex. 23 *If $P''(x)$ has real roots α, β, γ. Then, $[\alpha] + [\beta] + [\gamma]$ is*

(a) -2 (b) -3

(c) -1 (d) 0

Sol. (b) Here, $P'(x) = 0$ at $x = -2, -1, 0, \frac{1}{2}$.

$\Rightarrow P''(x)$ will have real roots belongs to $(-2, -1), (-1, 0), \left(0, \frac{1}{2}\right)$.

[$\because$ between any two roots of a function, atleast one root of its derivative exist]

$$\therefore \quad [\alpha] + [\beta] + [\gamma] = -3$$

Ex. 24 *The minimum number of real roots of equation $(P''(x))^2 + P'(x) \cdot P'''(x) = 0$*

(a) 5

(b) 7

(c) 6

(d) 4

Sol. (c) Let $f(x) = (P''(x))^2 + P'(x) \cdot P'''(x)$

$$\Rightarrow \quad f(x) = \frac{d}{dx}(P'(x) \cdot P''(x))$$

Since, $P'(x)$ has 4 real roots.

So, $P''(x)$ has 3 real roots.

$\Rightarrow f(x)$ has 6 real roots.

JEE Type Solved Examples : Matching Type Questions

Ex. 25 *Match the Column I with Column II and mark the correct option from the codes given below.*

	Column I		Column II
(i)	If $f'(x)=\sqrt{3x^2+6}$ and $y=f(x^3)$ then at $x=1,\ \frac{dy}{dx}$ is	p.	-2
(ii)	If f is a differentiable function such that $f(xy)=f(x)+f(y): x, y \in R$, then $f(e)+f(1/e)$ is	q.	-1
(iii)	If f is a twice differentiable function such that $f''(x)=-f(x)$ and $f'(x)=g(x)$. If $h(x)=[f(x)]^2+[g(x)]^2$ and $h(5)=9$, then $h(10)$ is	r.	0
(iv)	$y=\tan^{-1}(\cot x)+\cot^{-1}(\tan x)$, $\frac{\pi}{2}<x<\pi$, then $\frac{dy}{dx}$ is	s.	9

Codes

	i	ii	iii	iv		i	ii	iii	iv
(a)	s	r	s	p	(b)	p	q	r	s
(c)	q	p	r	r	(d)	s	p	q	q

Sol. (a) (i) $\because \quad y=f(x^3)$

$\therefore \quad \frac{dy}{dx}=f'(x^3)\cdot 3x^2$

$$\left(\frac{dy}{dx}\right)_{x=1}=f'(1)\cdot 3=9$$

(ii) $\because \quad f(xy)=f(x)+f(y)$

Put $x=y=1$,

$f(1)=f(1)+f(1)$

$\therefore \quad f(1)=0$

Also, $\quad f(1)=f(e)+f(1/e)$

i.e. $\quad x=e,\ y=\frac{1}{e}$

$\therefore \quad f(e)+f(1/e)=0$

(iii) $\because \quad f''(x)=-f(x)$ and $g(x)=f'(x)$

$\therefore \quad g'(x)=f''(x)=-f(x) \quad \quad ...(i)$

Also, $\quad h(x)=[f(x)]^2+[g(x)]^2$

$\therefore \quad h'(x)=2f(x)f'(x)+2g(x)g'(x)$

$=2f(x)g(x)-2g(x)f(x)=0$

$\therefore \quad h(x)=c,\ \forall\ x \in R$

Here, $\quad h(5)=9 \Rightarrow h(10)=9$

(iv) $\because \quad y=\tan^{-1}(\cot x)+\cot^{-1}(\tan x),\ \frac{\pi}{2}<x<\pi$

$$\therefore \quad \frac{dy}{dx}=\frac{-\operatorname{cosec}^2 x}{1+\cot^2 x}+\frac{(-1)\sec^2 x}{1+\tan^2 x}=-1-1=-2$$

Hence, (i) → s; (ii) → r; (iii) → s; (iv) → p

Ex. 26 *Match the Column I with Column II and mark the correct option from the codes given below.*

	Column I		Column II
(i)	If $y=\cos^{-1}(\cos x)$, then y' at $x=5$ is equal to	p.	-1
(ii)	For the function $f(x)=\log_e \tan\left(\frac{\pi}{4}+\frac{x}{2}\right)$, if $\frac{dy}{dx}=\sec x+p$, then p is equal to	q.	0
(iii)	The derivative of $\tan^{-1}\left(\frac{1+x}{1-x}\right)$ at $x=-1$ is	r.	1/2
(iv)	The derivative of $\frac{\log\lvert x\rvert}{x}$ at $x=-1$ is	s.	1

Codes

	i	ii	iii	iv		i	ii	iii	iv
(a)	p	q	r	s	(b)	q	p	r	s
(c)	s	r	q	p	(d)	r	s	p	q

Sol. (a) (i) We have, $y=\cos^{-1}(\cos x)$

$$y=\begin{cases} x, & 0\le x\le \pi \\ 2\pi-x, & \pi\le x\le 2\pi \end{cases} \Rightarrow \frac{dy}{dx}=\begin{cases} 1, & 0\le x\le \pi \\ -1, & \pi\le x\le 2\pi \end{cases}$$

$\because \quad 5\in[\pi, 2\pi] \quad \therefore \left(\frac{dy}{dx}\right)_{x=5}=-1$

(ii) We have, $f(x)=\log\tan\left(\frac{\pi}{4}+\frac{x}{2}\right)$

$$f'(x)=\frac{1}{\tan\left(\frac{\pi}{4}+\frac{x}{2}\right)}\times\sec^2\left(\frac{\pi}{4}+\frac{x}{2}\right)\times\frac{1}{2}$$

$$f'(x)=\frac{1}{2\sin\left(\frac{\pi}{4}+\frac{x}{2}\right)\cos\left(\frac{\pi}{4}+\frac{x}{2}\right)}$$

$$f'(x)=\frac{1}{\sin\left(\frac{\pi}{2}+x\right)}=\frac{1}{\cos x}=\sec x$$

$\therefore \quad p=0$

(iii) Let $y=\tan^{-1}\left(\frac{1+x}{1-x}\right) \Rightarrow y=\tan^{-1}(1)+\tan^{-1}x$

$$\therefore \quad \frac{dy}{dx}=\frac{1}{1+x^2} \Rightarrow \left(\frac{dy}{dx}\right)_{x=-1}=\frac{1}{2}$$

(iv) Let $\quad y=\frac{\log\lvert x\rvert}{x} \Rightarrow \frac{dy}{dx}=\frac{-\log\lvert x\rvert}{x^2}+\frac{1}{x^2}$

$$\left(\frac{dy}{dx}\right)_{x=-1}=\frac{-\log 1}{1}+\frac{1}{1}=1$$

Hence, (i) → p; (ii) → q; (iii) → r; (iv) → s

JEE Type Solved Examples :
Single Integer Answer Type Questions

Ex. 27 *If* $y=\sqrt{(x-\sin x)+\sqrt{(x-\sin x)+\ldots}}$, *then* $\left|\left|\frac{dx}{dy}\right|^2_{x=\frac{\pi}{2}}-2\pi\right|=\ldots\ldots\ldots$

Sol. (3) Here, $y=\sqrt{(x-\sin x)+\sqrt{(x-\sin x)+\ldots\infty}}$

So, $y=\sqrt{(x-\sin x)+y}$

$\therefore \quad y^2=x-\sin x+y$

Differentiating, we get

$$2y\frac{dy}{dx}=1-\cos x+\frac{dy}{dx} \quad \ldots\text{(i)}$$

At $x=\frac{\pi}{2}, y^2-y=\frac{\pi}{2}-1 \Rightarrow y^2-y+\frac{1}{4}=\frac{\pi}{2}-\frac{3}{4}$

$$\Rightarrow \left(y-\frac{1}{2}\right)^2=\frac{2\pi-3}{4} \Rightarrow (2y-1)=\pm\sqrt{2\pi-3}$$

From Eq. (i), we get $(2y-1)\frac{dy}{dx}=1-\cos x$

$$\therefore \left|\frac{dy}{dx}\right|_{x=\frac{\pi}{2}}=\frac{1}{\sqrt{2\pi-3}} \Rightarrow \left|\left|\frac{dx}{dy}\right|^2_{x=\frac{\pi}{2}}-2\pi\right|=3$$

Ex. 28 *If* $f(x)=(x-a)(x-b)$ *for* $a,b\in R$. *Then, the minimum number of roots of equation* $\pi(f'(x))^2\cos(\pi f(x))+\sin(\pi f(x))f''(x)=0$ *in* (α,β), *where* $f(\alpha)=3=f(\beta)$, *is .... (where* $\alpha<a<b<\beta$*).*

Sol. (4) Let $g(x)=f'(x)\sin(\pi f(x))=2\left[x-\left(\frac{a+b}{2}\right)\right]\sin(\pi f(x))$

$$g'(x)=\pi(f'(x))^2\cos(\pi f(x))+\sin(\pi f(x))f''(x)=0$$

$$g(a)=0=g(b)=g\left(\frac{a+b}{2}\right)=g(\alpha)=g(\beta)$$

$\therefore$ $g'(x)$ must have atleast one root in the intervals (α, a), $\left(a,\frac{a+b}{2}\right),\left(\frac{a+b}{2},b\right)$ and (b,β). i.e. minimum 4 roots.

Ex. 29 *Let* $f(x)=\int_{-2}^{x}e^{(1+t)^2}dt$ *and* $g(x)=f(h(x))$, *where* $h(x)$ *is defined for all* $x\in R$. *If* $g'(2)=e^4$ *and* $h'(2)=1$, *then absolute value of sum of all possible values of* $h(2)$, *is ....*

Sol. (2) Here, $f(x)=\int_{-2}^{x}e^{(1+t)^2}dt$

$$g(x)=\int_{-2}^{h(x)}e^{(1+t)^2}dt$$

$\Rightarrow \quad g'(x)=h'(x)\cdot e^{(1+h(x))^2}$ [by Leibnitz Rule]

$\Rightarrow \quad g'(2)=h'(2)\cdot e^{(1+h(2))^2}$

$\Rightarrow \quad e^4=1\cdot e^{(1+h(2))^2}$ [given $g'(2)=e^4$ and $h'(2)=1$]

$\Rightarrow \quad (1+h(2))^2=4 \Rightarrow h(2)=-3,1$

$\therefore$ Absolute value of sum of all possible values of $h(2)=|-3+1|=2$.

Subjective Type Questions

Ex. 30 *If* $f(x)=\cos\left\{\frac{\pi}{2}[x]-x^3\right\}, 1<x<2$ *and* $[x]=$ *the greatest integer* $\leq x$, *then find* $f'\left(\sqrt[3]{\frac{\pi}{2}}\right)$.

Sol. As we know, $1<\sqrt[3]{\frac{\pi}{2}}<2$

$\therefore$ If $x=\sqrt[3]{\frac{\pi}{2}} \Rightarrow [x]=1$, so $f(x)=\cos\left\{\frac{\pi}{2}-x^3\right\}$

$$f(x)=\sin x^3 \quad \left[\text{at } x=\sqrt[3]{\frac{\pi}{2}}\in(1,2)\right]$$

$\Rightarrow \quad f'(x)=\cos x^3\cdot 3x^2$

$\therefore \quad f'\left(\sqrt[3]{\frac{\pi}{2}}\right)=3\left(\frac{\pi}{2}\right)^{2/3}\cdot\cos\frac{\pi}{2}=0 \Rightarrow f'\left(\sqrt[3]{\frac{\pi}{2}}\right)=0$

Ex. 31 *If* $u=f(x^3), v=g(x^2), f'(x)=\cos x$ *and* $g'(x)=\sin x$, *then find* $\frac{du}{dv}$.

Sol. Here, $u=f(x^3)$

$\Rightarrow \quad \frac{du}{dx}=f'(x^3)\cdot\frac{d}{dx}(x^3)=\{\cos(x^3)\}\cdot 3x^2=3x^2\cdot\cos x^3$

and $v=g(x^2)$

$\Rightarrow \quad \frac{dv}{dx}=g'(x^2)\cdot\frac{d}{dx}(x^2)=\{\sin x^2\}\cdot(2x)=2x\cdot\sin x^2$

$$\therefore \quad \frac{du}{dv} = \frac{\frac{du}{dx}}{\frac{dv}{dx}} = \frac{3x^2 \cdot \cos x^3}{2x \cdot \sin x^2}$$

$$\Rightarrow \quad \frac{du}{dv} = \frac{3}{2} x \cdot \cos x^3 \cdot \operatorname{cosec} x^2$$

Ex. 32 *Find a, b, c and d, where $f(x) = (ax + b) \cos x + (cx + d) \sin x$ and $f'(x) = x \cos x$ is identity in x.*

Sol. Here, $f'(x) = x \cos x$

$\Rightarrow a \cos x - (ax + b) \sin x + c \sin x + (cx + d) \cos x \equiv x \cos x$

or $(a + cx + d) \cos x + (-ax - b + c) \sin x \equiv x \cos x + 0 \cdot \sin x$

$\Rightarrow \quad a + d + cx = x$ and $-ax - b + c = 0$

which is again identity in x.

$\Rightarrow \quad a + d = 0, c = 1, -a = 0, -b + c = 0$

$\Rightarrow \quad a = 0, b = 1, c = 1, d = 0$

Ex. 33 *If $f(x) = x^3 + x^2 f'(1) + xf''(2) + f'''(3)$ for all $x \in R$. Then, find $f(x)$ independent of $f'(1)$, $f''(2)$ and $f'''(3)$.*

Sol. Here, $f(x) = x^3 + x^2 f'(1) + x f''(2) + f'''(3)$

Put $f'(1) = a, \; f''(2) = b, \; f'''(3) = c$ ...(i)

$\therefore \quad f(x) = x^3 + ax^2 + bx + c$

$\Rightarrow \quad f'(x) = 3x^2 + 2ax + b$ or $f'(1) = 3 + 2a + b$ ...(ii)

$\Rightarrow \quad f''(x) = 6x + 2a$ or $f''(2) = 12 + 2a$ ...(iii)

$\Rightarrow \quad f'''(x) = 6$ or $f'''(3) = 6$ ...(iv)

From Eqs. (i) and (iv), $c=6$

From Eqs. (i), (ii) and (iii), we have $a = -5, \quad b = 2$

$\therefore \quad f(x) = x^3 - 5x^2 + 2x + 6$

Ex. 34 *Let $f(x) = x^2 + xg'(1) + g''(2)$ and $g(x) = f(1) \cdot x^2 + xf'(x) + f''(x)$, then find $f(x)$ and $g(x)$.*

Sol. Here, put $g'(1) = a, \; g''(2) = b$ ...(i)

Then, $f(x) = x^2 + ax + b, \; f(1) = 1 + a + b$

$\Rightarrow \quad f'(x) = 2x + a, \quad f''(x) = 2$

$\therefore \quad g(x) = (1 + a + b)x^2 + (2x + a) \cdot x + 2$

$= x^2 (3 + a + b) + ax + 2$

$\Rightarrow \quad g'(x) = 2x (3 + a + b) + a$

Hence, $g'(1) = 2 (3 + a + b) + a$ ...(ii)

$g''(2) = 2 (3 + a + b)$ ...(iii)

From Eqs. (i), (ii) and (iii), we have

$a = 2 (3 + a + b) + a$ and $b = 2 (3 + a + b)$

i.e. $3 + a + b = 0$ and $b + 2a + 6 = 0$

Hence, $b = 0$ and $a = -3$

So, $f(x) = x^2 - 3x$ and $g(x) = -3x + 2$

Ex. 35 *If $y = \cfrac{\sin x}{1 + \cfrac{\cos x}{1 + \cfrac{\sin x}{1 + \cos x \ldots \infty}}}$, prove that*

$$\frac{dy}{dx} = \frac{(1 + y) \cos x + \sin x}{1 + 2y + \cos x - \sin x}.$$

Sol. Given function is $y = \dfrac{\sin x}{1 + \dfrac{\cos x}{1 + y}} = \dfrac{(1 + y) \sin x}{1 + y + \cos x}$

$\Rightarrow y + y^2 + y \cos x = (1 + y) \sin x$

On differentiating both the sides w.r.t. x, we get

$$\frac{dy}{dx} + 2y \frac{dy}{dx} + y(-\sin x) + \cos x \cdot \frac{dy}{dx} = (1 + y) \cos x + \frac{dy}{dx} \cdot \sin x$$

$$\Rightarrow \quad \frac{dy}{dx} \{1 + 2y + \cos x - \sin x\} = (1 + y) \cos x + y \sin x$$

$$\Rightarrow \quad \frac{dy}{dx} = \frac{(1 + y) \cos x + y \sin x}{(1 + 2y + \cos x - \sin x)}$$

Ex. 36 *If $y = \cfrac{x}{x + \cfrac{\sqrt[3]{x}}{x + \cfrac{\sqrt[3]{x}}{x + \sqrt[3]{x} + \ldots \infty}}}$, then find $\frac{dy}{dx}$.*

Sol.

$$y = \cfrac{x}{x + \cfrac{\sqrt[3]{x}}{x + \cfrac{\sqrt[3]{x}}{x + \sqrt[3]{x} + \ldots \infty}}}$$

$$\Rightarrow \quad y = \cfrac{x}{x + \cfrac{x}{x + \cfrac{\sqrt[3]{x}}{x + \sqrt[3]{x} + \ldots \infty}} \cdot x^{-2/3}}$$

$$\Rightarrow y = \frac{x}{x + y \cdot x^{-2/3}}$$

$\Rightarrow \quad y\{x^{5/3} + y\} = x^{5/3}$

or $\quad x^{5/3} y + y^2 = x^{5/3}$

Differentiating both the sides w.r.t. x, we get

$$x^{5/3} \cdot \frac{dy}{dx} + \frac{5}{3} \cdot x^{2/3} \cdot y + 2y \frac{dy}{dx} = \frac{5}{3} \cdot x^{2/3}$$

$$\Rightarrow \quad (x^{5/3} + 2y) \frac{dy}{dx} = \frac{5}{3} x^{2/3} - \frac{5}{3} x^{2/3} y$$

$$\text{or} \quad \frac{dy}{dx} = \frac{\frac{5}{3} x^{2/3} (1 - y)}{(x^{5/3} + 2y)}$$

Ex. 37 *If* $y=\tan^{-1}\dfrac{1}{x^2+x+1}+\tan^{-1}\dfrac{1}{x^2+3x+3}+\tan^{-1}\dfrac{1}{x^2+5x+7}+\ldots$ *to n terms, show that* $\dfrac{dy}{dx}=\dfrac{1}{(x+n)^2+1}-\dfrac{1}{x^2+1}$.

Sol. Given, $y=\tan^{-1}\dfrac{1}{x^2+x+1}+\tan^{-1}\dfrac{1}{x^2+3x+3}+\tan^{-1}\dfrac{1}{x^2+5x+7}+\ldots$ to n terms

$$=\tan^{-1}\left\{\frac{1}{1+x(x+1)}\right\}+\tan^{-1}\left\{\frac{1}{1+(x+1)(x+2)}\right\}+\tan^{-1}\left\{\frac{1}{1+(x+2)(x+3)}\right\}+\ldots+\tan^{-1}\left\{\frac{1}{1+\{x+(n-1)\}\{x+n\}}\right\}$$

$$=\tan^{-1}\left\{\frac{(x+1)-x}{1+(x+1)x}\right\}+\tan^{-1}\left\{\frac{(x+2)-(x+1)}{1+(x+2)(x+1)}\right\}+\tan^{-1}\left\{\frac{(x+3)-(x+2)}{1+(x+3)(x+2)}\right\}+\ldots+\tan^{-1}\left\{\frac{(x+n)-(x+n-1)}{1+(x+n)(x+n-1)}\right\}$$

$$\therefore\quad y=\{\tan^{-1}(x+1)-\tan^{-1}(x)\}+\{\tan^{-1}(x+2)-\tan^{-1}(x+1)\}+\{\tan^{-1}(x+3)-\tan^{-1}(x+2)\}+\ldots+\{\tan^{-1}(x+n)-\tan^{-1}(x+n-1)\}$$

$$y=\tan^{-1}(x+n)-\tan^{-1}(x)$$

On differentiating both the sides w.r.t. x, we get

$$\frac{dy}{dx}=\frac{1}{1+(x+n)^2}-\frac{1}{1+x^2}$$

Ex. 38 *If* $f(\theta)=\cos\theta_1\cdot\cos\theta_2\cdot\cos\theta_3\ldots\cos\theta_n$, *show that* $\{\tan\theta_1+\tan\theta_2+\tan\theta_3+\ldots+\tan\theta_n\}=-\left\{\dfrac{f'(\theta)}{f(\theta)}\right\}$, *where* $\dfrac{d\theta_1}{d\theta}=\dfrac{d\theta_2}{d\theta}=\ldots=\dfrac{d\theta_n}{d\theta}=1$.

Sol. $f(\theta)=\cos\theta_1\cdot\cos\theta_2\cdot\cos\theta_3\ldots\cos\theta_n$

Taking log on both the sides,

$$\log f(\theta)=\log(\cos\theta_1)+\log(\cos\theta_2)+\ldots+\log(\cos\theta_n)$$

On differentiating both the sides w.r.t. θ, we get

$$\frac{1}{f(\theta)}\cdot f'(\theta)=\frac{1}{\cos\theta_1}\cdot(-\sin\theta_1)+\frac{1}{\cos\theta_2}\cdot(-\sin\theta_2)+\ldots+\frac{1}{\cos\theta_n}\cdot(-\sin\theta_n)$$

Hence, $(\tan\theta_1+\tan\theta_2+\ldots+\tan\theta_n)=-\left\{\dfrac{f'(\theta)}{f(\theta)}\right\}$

Ex. 39 *Find the sum of the series* $\sin x+3\sin 3x+5\sin 5x+\ldots+(2k-1)\sin(2k-1)x$ *(using calculus).*

Sol. Let $S=\cos x+\cos 3x+\cos 5x+\ldots\ldots+\cos(2k-1)x$

Here, the angles are in AP whose first term is x and common difference is $2x$.

$$\therefore\quad S=\frac{\sin\left(\dfrac{k\cdot 2x}{2}\right)}{\sin\left(\dfrac{2x}{2}\right)}\cdot\cos\left\{\frac{x}{1}+\frac{(k-1)2x}{2}\right\}=\frac{\sin kx}{\sin x}\cdot\cos kx$$

$$\text{or }\{\cos x+\cos 3x+\cos 5x+\ldots+\cos(2k-1)x\}=\frac{\sin 2kx}{2\sin x}\quad\ldots\text{(i)}$$

On differentiating Eq. (i) w.r.t. x, we get

$$-\{\sin x+3\sin 3x+5\sin 5x+\ldots+(2k-1)\sin(2k-1)x\}=\frac{1}{2}\left\{\frac{2k(\cos 2kx)\cdot\sin x-(\sin 2kx)\cdot\cos x}{\sin^2 x}\right\}$$

$$\therefore\quad[\sin x+3\sin 3x+5\sin 5x+\ldots+(2k-1)\sin(2k-1)x]=-\frac{1}{2\sin^2 x}\left[k\{\sin(2k+1)x-\sin(2k-1)x\}-\frac{1}{2}\{\sin(2k+1)x+\sin(2k-1)x\}\right]$$

$$=\frac{1}{4\sin^2 x}[(2k+1)\sin(2k-1)x-(2k-1)\sin(2k+1)x]$$

Ex. 40 *Find the sum of series* $\sum_{r=1}^{n} r\,x^{r-1}$, *using calculus.*

Sol. Let $S=1+x+x^2+x^3+x^4+\ldots+x^n$

which is a geometric progression.

$$\therefore\quad S=1+x+x^2+x^3+\ldots+x^n=\frac{1(1-x^{n+1})}{1-x}$$

On differentiating both the sides, we get

$$0+1+2x+3x^2+4x^3+\ldots+nx^{n-1}=\frac{(1-x)\cdot[-(n+1)x^n]-(1-x^{n+1})\cdot(-1)}{(1-x)^2}$$

$$\therefore\quad\sum_{r=1}^{n} r x^{r-1}=\frac{1}{(1-x)^2}\cdot\{1-(n+1)x^n+n\cdot x^{n+1}\}$$

Ex. 41 *Use calculus to find the sum of*

$$\left[\frac{1}{x+1}+\frac{2}{x^2+1}+\frac{4}{x^4+1}+\ldots+\frac{2^n}{x^{2^n}+1}\right].$$

Sol. We know that, $(1-x)(1+x)=(1-x^2)$

$$(1-x)(1+x)(1+x^2)=(1-x^4)=(1-x^{2^2})$$

$$(1-x)(1+x)(1+x^2)(1+x^4)=(1-x^8)=(1-x^{2^3})$$

$$(1-x)(1+x)(1+x^2)(1+x^4)\ldots(1+x^{2^n})=(1-x^{2^{n+1}})\quad\ldots\text{(i)}$$

Taking log on both the sides of Eq. (i), we get

$$\log\{(1-x)(1+x)(1+x^2)(1+x^4)\dots(1+x^{2^n})\} = \log(1-x^{2^{n+1}})$$

or $\log(1-x)+\log(1+x)+\log(1+x^2)+\log(1+x^4)$

$$+\dots+\log(1+x^{2^n}) = \log(1-x^{2^{n+1}})$$

On differentiating the above equation, we get

$$\frac{1(-1)}{1-x}+\frac{1\cdot 1}{1+x}+\frac{1\,(2x)}{1+x^2}+\frac{1\cdot(4x^3)}{1+x^4}+\dots+\frac{1\cdot 2^n\cdot x^{2^n-1}}{1+x^{2^n}}$$

$$=\frac{-1\cdot 2^{n+1}\cdot x^{2^{n+1}-1}}{1-x^{2^{n+1}}}$$

or $$\frac{1}{1+x}+\frac{2x}{1+x^2}+\frac{4x^3}{1+x^4}+\dots+\frac{2^n\cdot x^{2^n-1}}{1+x^{2^n}}$$

$$=\frac{1}{1-x}-\frac{2^{n+1}\cdot x^{2^{n+1}-1}}{1-x^{2^{n+1}}}$$

Multiplying both the sides by x,

$$\left[\frac{x}{1+x}+\frac{2x^2}{1+x^2}+\frac{4x^4}{1+x^4}+\dots+\frac{2^n\cdot x^{2^n}}{1+x^{2^n}}\right]$$

$$=\frac{x}{1-x}-\frac{2^{n+1}\cdot x^{2^{n+1}}}{1-x^{2n+1}}$$

or $$\frac{(x+1)-1}{1+x}+\frac{2\,(x^2+1)-2}{1+x^2}+\frac{4\,(x^4+1)-4}{1+x^4}$$

$$+\dots+\frac{2^n\,(x^{2^n}+1)-2^n}{1+x^{2^n}}=\frac{x}{1-x}-2^{n+1}\cdot\frac{x^{2^{n+1}}}{1-x^{2^{n+1}}}$$

or $$[(1+2+2^2+\dots+2^n)-P]=\frac{x}{1-x}-\frac{2^{n+1}\cdot x^{2^{n+1}}}{1-x^{2^{n+1}}}$$

where, $P=\dfrac{1}{1+x}+\dfrac{2}{1+x^2}+\dfrac{4}{1+x^4}+\dots+\dfrac{2^n}{1+x^{2^n}}$

Then, $$P=\frac{2^{n+1}-1}{2-1}-\frac{x}{1-x}+\frac{2^{n+1}\cdot x^{2^{n+1}}}{1-x^{2^{n+1}}}$$

$\Rightarrow$ $$P=2^{n+1}-1-\frac{x}{1-x}+\frac{2^{n+1}\cdot x^{2^{n+1}}}{1-x^{2^{n+1}}}$$

$\Rightarrow$ $$P=2^{n+1}\left[1+\frac{x^{2^{n+1}}}{1-x^{2^{n+1}}}\right]-\left[1+\frac{x}{1-x}\right]$$

$\Rightarrow$ $$P=2^{n+1}\left[\frac{1}{1-x^{2^{n+1}}}\right]-\left[\frac{1}{1-x}\right]$$

Hence, $$\left\{\frac{1}{1+x}+\frac{2}{1+x^2}+\frac{4}{1+x^4}+\dots+\frac{2^n}{1+x^{2^n}}\right\}$$

$$=\left\{\frac{2^{n+1}}{1-x^{2^{n+1}}}\right\}-\left\{\frac{1}{1-x}\right\}$$

Ex. 42 *If $f_n(x)=e^{f_{n-1}(x)}$ for all $n\in N$ and $f_0(x)=x$, then find $\dfrac{d}{dx}\{f_n(x)\}$.*

Sol. Here, $$f_n(x)=e^{f_{n-1}(x)} \quad \text{...(i)}$$

$\Rightarrow$ $$\frac{d}{dx}\{f_n(x)\}=e^{f_{n-1}(x)}\cdot\frac{d}{dx}\{f_{n-1}(x)\}$$

or $$\frac{d}{dx}\{f_n(x)\}=f_n(x)\frac{d}{dx}\{f_{n-1}(x)\}$$

[using Eq. (i)] ...(ii)

Replacing n by $n-1$ in Eq. (ii), we have

$$\frac{d}{dx}\{f_{n-1}(x)\}=f_{n-1}(x)\cdot\frac{d}{dx}\{f_{n-2}(x)\} \quad \text{...(iii)}$$

$\therefore$ From Eqs. (ii) and (iii), we get

$$\frac{d}{dx}\{f_n(x)\}=f_n(x)\cdot f_{n-1}(x)\cdot\frac{d}{dx}\{f_{n-2}(x)\} \quad \text{...(iv)}$$

Similarly, $\dfrac{d}{dx}\{f_{n-2}(x)\}=f_{n-2}(x)\cdot\dfrac{d}{dx}\{f_{n-3}(x)\}$

$$\frac{d}{dx}\{f_1(x)\}=f_1(x)\frac{d}{dx}\{f_0(x)\} \quad \text{...(v)}$$

From Eqs. (iv) and (v), we get

$$\frac{d}{dx}\{f_n(x)\}=f_n(x)f_{n-1}(x)\cdot f_{n-2}(x)\dots f_2(x)\cdot f_1(x)\left\{\frac{d}{dx}f_0(x)\right\}$$

$\Rightarrow$ $$\frac{d}{dx}\{f_n(x)\}=f_n(x)f_{n-1}(x)\cdot f_{n-2}(x)\dots f_2(x)\cdot f_1(x)\cdot 1$$

$$\left[\because f_0(x)=x\Rightarrow\frac{d}{dx}\{f_0(x)\}=1\right]$$

Ex. 43 *If $y^3-y=2x$, then prove that $\dfrac{d^2y}{dx^2}=-\dfrac{24y}{(3y^2-1)^3}$.*

Hence, show that $\left(x^2-\dfrac{1}{27}\right)\dfrac{d^2y}{dx^2}+x\cdot\dfrac{dy}{dx}=\dfrac{1}{9}y$.

Sol. Given, $$y^3-y=2x \quad \text{...(i)}$$

On differentiating both the sides w.r.t. x, we get

$$(3y^2-1)\frac{dy}{dx}=2 \text{ or } \frac{dy}{dx}=\frac{2}{3y^2-1} \quad \text{...(ii)}$$

Again, differentiating w.r.t. x,

$$\frac{d^2y}{dx^2}=\frac{-2(6y)\cdot\frac{dy}{dx}}{(3y^2-1)^2}=\frac{-12y}{(3y^2-1)^2}\cdot\frac{2}{(3y^2-1)} \quad \text{[from Eq. (ii)]}$$

$\therefore$ $$\frac{d^2y}{dx^2}=-\frac{24y}{(3y^2-1)^3} \quad \text{...(iii)}$$

Now, LHS $=\left(x^2-\dfrac{1}{27}\right)\dfrac{d^2y}{dx^2}+x\cdot\dfrac{dy}{dx}$

$$=\left(\frac{y^2\,(y^2-1)^2}{4}-\frac{1}{27}\right)\left(\frac{-24y}{(3y^2-1)^3}\right)+\frac{y\,(y^2-1)}{2}\cdot\frac{2}{(3y^2-1)}$$

[from Eqs. (i), (ii) and (iii)]

$$= \left(\frac{27y^2(y^2-1)^2-4}{108}\right)\frac{-24y}{(3y^2-1)^3} + \frac{y(y^2-1)}{(3y^2-1)}$$

$$= \frac{y}{9}\left[\frac{-54y^2(y^2-1)^2+8}{(3y^2-1)^3} + \frac{9(y^2-1)}{(3y^2-1)}\right]$$

Let $(3y^2-1)=\alpha$

Then, $\text{LHS} = \frac{y}{9}\left[\frac{-2(1+\alpha)(\alpha-2)^2+8}{\alpha^3} + \frac{3(\alpha-1)}{\alpha}\right]$

$$= \frac{y}{9}\left[\frac{-2\alpha^3+6\alpha^2+3\alpha^3-6\alpha^2}{\alpha^3}\right]$$

$$= \frac{y}{9} = \text{RHS}$$

Hence, $\left(x^2-\frac{1}{27}\right)\frac{d^2y}{dx^2} + x\cdot\frac{dy}{dx} = \frac{y}{9}$

Ex. 44 *If $2x = y^{1/5} + y^{-1/5}$, then express y as an explicit function of x and prove that $(x^2-1)\frac{d^2y}{dx^2} + x\cdot\frac{dy}{dx} = 25y$.*

Sol. We have, $(y^{1/5}+y^{-1/5})^2 = (y^{1/5}-y^{-1/5})^2+4$

Then, $(y^{1/5}-y^{-1/5})^2 = 4x^2-4$ $\quad [\because 2x = y^{1/5}+y^{-1/5}]$

$\therefore \quad y^{1/5}-y^{-1/5} = 2\sqrt{x^2-1}$ ...(i)

and $\quad y^{1/5}+y^{-1/5} = 2x$ [given]...(ii)

Adding Eqs. (i) and (ii), we get

$$2y^{1/5} = 2x + 2\sqrt{x^2-1}$$

$\therefore \quad y^{1/5} = x+\sqrt{x^2-1}$

or $\quad y = (x+\sqrt{x^2-1})^5$ ...(iii)

On differentiating both the sides w.r.t. x, we get

$$\frac{dy}{dx} = 5(x+\sqrt{x^2-1})^4\cdot\left\{1+\frac{1\cdot 2x}{2\sqrt{x^2-1}}\right\}$$

$$\Rightarrow \quad \frac{dy}{dx} = \frac{5(x+\sqrt{x^2-1})^5}{\sqrt{x^2-1}}$$

$\Rightarrow \quad \frac{dy}{dx} = \frac{5y}{\sqrt{x^2-1}}$ [using Eq. (iii)]

$\Rightarrow \quad (x^2-1)\left(\frac{dy}{dx}\right)^2 = 25y^2$ ...(iv)

Again, differentiating both the sides w.r.t. x, we get

$$2x\left(\frac{dy}{dx}\right)^2 + (x^2-1)\cdot 2\frac{dy}{dx}\cdot\frac{d^2y}{dx^2} = 50y\cdot\frac{dy}{dx}$$

On dividing by $2\frac{dy}{dx}$ on both the sides,

$$x\cdot\frac{dy}{dx} + (x^2-1)\frac{d^2y}{dx^2} = 25y \quad \left[\because \frac{dy}{dx}\neq 0\right]$$

Ex. 45 *If $y = 1+\frac{a_1}{(x-a_1)}+\frac{a_2x}{(x-a_1)(x-a_2)} + \frac{a_3x^2}{(x-a_1)(x-a_2)(x-a_3)} + \ldots$ to $(n+1)$ terms, show that*

$$\frac{dy}{dx} = \frac{y}{x}\left[n-\frac{x}{x-a_1}-\frac{x}{x-a_2}-\frac{x}{x-a_3}-\ldots-\frac{x}{x-a_n}\right].$$

Sol. Here, $y = 1+\frac{a_1}{(x-a_1)}+\frac{a_2x}{(x-a_1)(x-a_2)}$

$$+\frac{a_3x^2}{(x-a_1)(x-a_2)(x-a_3)} + \ldots \text{ to } (n+1) \text{ terms}$$

$$\Rightarrow \quad y = \frac{x}{(x-a_1)}+\frac{a_2x}{(x-a_1)(x-a_2)} + \frac{a_3x^2}{(x-a_1)(x-a_2)(x-a_3)} + \ldots \text{to } (n) \text{ terms}$$

[adding first two terms]

$$\Rightarrow \quad y = \frac{x^2}{(x-a_1)(x-a_2)} + \frac{a_3x^2}{(x-a_1)(x-a_2)(x-a_3)} + \ldots \text{to } (n-1) \text{ terms}$$

[again adding first two terms]

Proceeding in the same way, we get

$$y = \frac{x^n}{(x-a_1)(x-a_2)(x-a_3)\ldots(x-a_n)} \quad \text{...(i)}$$

Taking logarithm on both the sides of Eq. (i), we have

$$\log y = n\log x - \log(x-a_1) - \log(x-a_2) - \log(x-a_3) - \ldots - \log(x-a_n)$$

On differentiating both the sides w.r.t. x, we get

$$\frac{1}{y}\cdot\frac{dy}{dx} = \frac{n}{x}-\frac{1}{(x-a_1)}-\frac{1}{(x-a_2)}-\ldots-\frac{1}{(x-a_n)}$$

or $\quad \frac{dy}{dx} = \frac{y}{x}\left[n-\frac{x}{(x-a_1)}-\frac{x}{(x-a_2)}-\ldots-\frac{x}{(x-a_n)}\right]$

Ex. 46 *If for all x, y the function f is defined by $f(x)+f(y)+f(x)\cdot f(y) = 1$ and $f(x) > 0$. Then, show $f'(x) = 0$, when $f(x)$ is differentiable.*

Sol. Here, $f(x)+f(y)+f(x)\cdot f(y) = 1$ ...(i)

Put $x = y = 0$, we get

$$2f(0) + \{f(0)\}^2 = 1$$

$$\Rightarrow \quad \{f(0)\}^2 + 2f(0) - 1 = 0$$

$$f(0) = \frac{-2\pm\sqrt{4+4}}{2} = -1-\sqrt{2} \text{ and } -1+\sqrt{2}$$

As $\quad f(0) > 0 \Rightarrow f(0) = \sqrt{2}-1$

[neglecting $f(0) = -1-\sqrt{2}$ as $f(0)$ is positive]

Again, putting $y = x$ in Eq. (i),

$$2f(x) + \{f(x)\}^2 = 1$$

On differentiating w.r.t. x, we have

$$2f'(x) + 2f(x)f'(x) = 0$$

$$2f'(x)\{1 + f(x)\} = 0$$

$$\Rightarrow \quad f'(x) = 0$$

Because $\quad f(x) > 0$

Therefore, $f'(x) = 0$ when $f(x) > 0$

Ex. 47 *If* $y = f\left(\frac{2x-1}{x^2+1}\right)$ *and* $f'(x) = \sin x^2$, *then find* $\frac{dy}{dx}$.

Sol. Here, $\frac{dy}{dx} = \frac{d}{dx} f\left(\frac{2x-1}{x^2+1}\right)$

$$= \frac{df\left(\frac{2x-1}{x^2+1}\right)}{d\left(\frac{2x-1}{x^2+1}\right)} \cdot \frac{d\left(\frac{2x-1}{x^2+1}\right)}{dx}$$

$$= f'\left(\frac{2x-1}{x^2+1}\right) \cdot \left\{\frac{(x^2+1)\cdot(2) - (2x-1)\cdot(2x)}{(x^2+1)^2}\right\}$$

$$= \sin\left(\frac{2x-1}{x^2+1}\right)^2 \cdot \frac{2\{x^2+1-2x^2+x\}}{(x^2+1)^2}$$

$$= \sin\left(\frac{2x-1}{x^2+1}\right)^2 \cdot \frac{2(1+x-x^2)}{(x^2+1)^2}$$

$$\Rightarrow \quad \frac{dy}{dx} = \frac{2(1+x-x^2)}{(x^2+1)^2} \cdot \sin\left(\frac{2x-1}{x^2+1}\right)^2$$

Ex 48 *Let* $f(x)$ *be a polynomial function of second degree. If* $f(1) = f(-1)$ *and* a_1, a_2, a_3 *are in AP, then show that* $f'(a_1), f'(a_2), f'(a_3)$ *are in AP.*

Sol. Let $f(x) = \lambda x^2 + \mu x + \nu$

Then, $\quad f'(x) = 2\lambda x + \mu$

Also, $\quad f(1) = f(-1)$

$$\Rightarrow \quad \lambda + \mu + \nu = \lambda - \mu + \nu$$

$$\Rightarrow \quad \mu = 0$$

$$\therefore \quad f'(x) = 2\lambda x$$

$$\Rightarrow \quad f'(a_1) = 2\lambda a_1, \; f'(a_2) = 2\lambda a_2$$

and $\quad f'(a_3) = 2\lambda a_3$

As, a_1, a_2, a_3 are in AP. Therefore, $f'(a_1), f'(a_2), f'(a_3)$ are in AP.

Ex. 49 *If* $5f(x) + 3f\left(\frac{1}{x}\right) = x + 2$ *and* $y = x\,f(x)$, *then find* $\frac{dy}{dx}$ *at* $x = 1$.

Sol. Here, $5f(x) + 3f\left(\frac{1}{x}\right) = x + 2 \quad$ ...(i)

Put $x = \frac{1}{x}$, we get $5f\left(\frac{1}{x}\right) + 3f(x) = \frac{1}{x} + 2 \quad$ ...(ii)

Solving Eqs. (i) and (ii), we get

$$16f(x) = 5x - \frac{3}{x} + 4 \quad \text{...(iii)}$$

$$\Rightarrow \quad y = x\,f(x)$$

$$\Rightarrow \quad y = x \cdot \frac{1}{16}\left\{5x - \frac{3}{x} + 4\right\}$$

$$\Rightarrow \quad y = \frac{1}{16}\{5x^2 - 3 + 4x\}$$

or $\quad \frac{dy}{dx} = \frac{1}{16}\{10x + 4\}$

Now, $\frac{dy}{dx}$ at $x = 1$,

$$\left(\frac{dy}{dx}\right)_{\text{at } x=1} = \frac{10+4}{16} = \frac{14}{16} = \frac{7}{8}$$

$$\therefore \quad \left(\frac{dy}{dx}\right)_{\text{at } x=1} = \frac{7}{8}$$

Differentiation Exercise 1 :
Single Option Correct Type Questions

1. If $y=\dfrac{\sec x-\tan x}{\sec x+\tan x}$, then $\dfrac{dy}{dx}$ equals

(a) $2\sec x(\sec x-\tan x)$ (b) $-2\sec x(\sec x-\tan x)^2$
(c) $2\sec x(\sec x+\tan x)^2$ (d) $-2\sec x(\sec x+\tan x)^2$

2. If $y=\dfrac{1+x^2+x^4}{1+x+x^2}$ and $\dfrac{dy}{dx}=ax+b$, then

(a) $a=2, b=1$ (b) $a=-2, b=1$
(c) $a=2, b=-1$ (d) $a=-2, b=-1$

3. Which of the following could be the sketch of graph of $y=\dfrac{d}{dx}(x\log x)$?

(a)
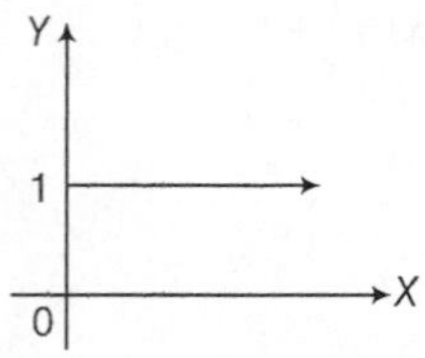

(b)
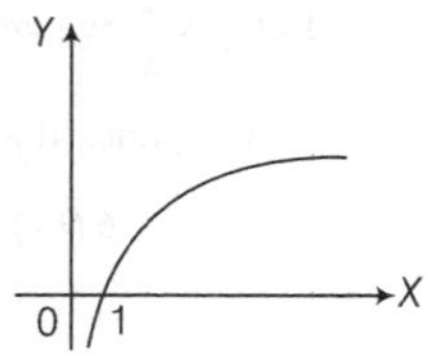

(c)
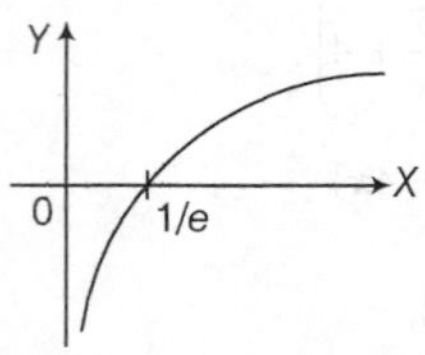

(d)
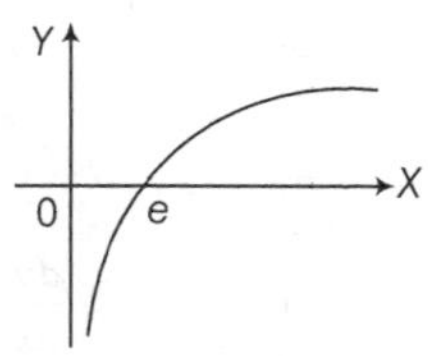

4. Let $f(x)=x+3\log(x-2)$ and $g(x)=x+5\log(x-1)$, then the set of x satisfying the inequality $f'(x)<g'(x)$ is

(a) $(2, 7/2)$ (b) $(1,2)\cup(7/2,\infty)$
(c) $(2,\infty)$ (d) $(7/2,\infty)$

5. If $\cos^{-1}\left(\dfrac{x^2-y^2}{x^2+y^2}\right)=\log a$, then $\dfrac{dy}{dx}$ is

(a) $-\dfrac{x}{y}$ (b) $-\dfrac{y}{x}$ (c) $\dfrac{y}{x}$ (d) $\dfrac{x}{y}$

6. If $f(x)=(|x|)^{|\sin x|}$, then $f'(-\pi/4)$ is equal to

(a) $\left(\dfrac{\pi}{4}\right)^{1/\sqrt{2}}\cdot\left(\dfrac{\sqrt{2}}{2}\log\dfrac{4}{\pi}-\dfrac{2\sqrt{2}}{\pi}\right)$

(b) $\left(\dfrac{\pi}{4}\right)^{1/\sqrt{2}}\cdot\left(\dfrac{\sqrt{2}}{2}\log\dfrac{4}{\pi}+\dfrac{2\sqrt{2}}{\pi}\right)$

(c) $\left(\dfrac{\pi}{4}\right)^{1/\sqrt{2}}\cdot\left(\dfrac{\sqrt{2}}{2}\log\dfrac{\pi}{4}-\dfrac{2\sqrt{2}}{\pi}\right)$

(d) $\left(\dfrac{\pi}{4}\right)^{1/\sqrt{2}}\cdot\left(\dfrac{\sqrt{2}}{2}\log\dfrac{\pi}{4}+\dfrac{2\sqrt{2}}{\pi}\right)$

7. If $y=\cfrac{x}{a+\cfrac{x}{b+\cfrac{x}{a+\cfrac{x}{b+\dots\infty}}}}$, then $\dfrac{dy}{dx}$ equals

(a) $\dfrac{a}{ab+2ay}$ (b) $\dfrac{b}{ab+2by}$
(c) $\dfrac{a}{ab+2by}$ (d) $\dfrac{b}{ab+2ay}$

8. If $y=x^{x^2}$, then $\dfrac{dy}{dx}$ equals

(a) $2\log x\cdot x^2$
(b) $(2\log x+1)\cdot x^{x^2}$
(c) $(\log x+1)\cdot x^{x^2+1}$
(d) $x^{x^2+1}\cdot(\log(ex^2))$

9. If $x\sqrt{1+y}+y\sqrt{1+x}=0$, then $\dfrac{dy}{dx}$ equals

(a) $\dfrac{1}{(1+x)^2}$ (b) $\dfrac{-1}{(1+x)^2}$
(c) $\dfrac{-1}{(1+x)}+\dfrac{1}{(1+x)^2}$ (d) None of these

10. If $x^2e^y+2xye^x+13=0$, then $\dfrac{dy}{dx}$ equals

(a) $\dfrac{-2xe^{y-x}+2y(x-1)}{x(xe^{y-x}+2)}$ (b) $-\left[\dfrac{2xe^{y-x}+2y(x+1)}{x(xe^{y-x}+2)}\right]$
(c) $\dfrac{2xe^{x-y}+2y(x-1)}{x(xe^{y-x}+2)}$ (d) None of these

11. If $x=e^{y+e^{y+\dots\infty}}$; $x>0$, then $\dfrac{dy}{dx}$ equals

(a) $\dfrac{x}{1+x}$ (b) $\dfrac{1+x}{x}$
(c) $\dfrac{1-x}{x}$ (d) $\dfrac{1}{x}$

12. Let g be the inverse function of f and $f'(x)=\dfrac{x^{10}}{1+x^2}$.

If $g(2)=a$, then $g'(2)$ is equal to

(a) $\dfrac{5}{2^{10}}$ (b) $\dfrac{1+a^2}{a^{10}}$
(c) $\dfrac{a^{10}}{1+a^2}$ (d) $\dfrac{1+a^{10}}{a^2}$

13. If f and g are the functions whose graphs are as shown, let $u(x) = f(g(x))$; $w(x) = g(g(x))$

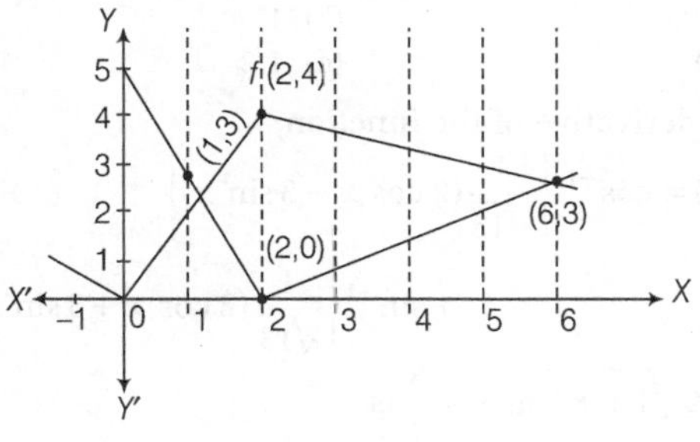

Then, the value of $u'(1) + w'(1)$ is

(a) $-\frac{1}{2}$ (b) $-\frac{3}{2}$ (c) $-\frac{5}{4}$ (d) does not exist

14. If $f'(x) = g(x)$ and $g'(x) = -f(x)$ for all real x and $f(5) = 2 = f'(5)$, then $f^2(10) + g^2(10)$ equals

(a) 2 (b) 4
(c) 8 (d) None of these

15. If $f(x) = x + \frac{x^2}{1!} + \frac{x^3}{2!} + \ldots + \frac{x^n}{(n-1)!}$, then $f(0) + f'(0) + f''(0) + \ldots + f^n(0)$ is equal to

(a) $\frac{n(n+1)}{2}$ (b) $\frac{(n^2+1)}{2}$

(c) $\left(\frac{n(n+1)}{2}\right)^2$ (d) $\frac{n(n+1)(2n+1)}{6}$

16. If $y = (fofof)(x)$ and $f(0) = 0, f'(0) = 2$, then $y'(0)$ is equal to

(a) 6 (b) 7 (c) 8 (d) 9

17. If $y^2 = p(x)$ is a polynomial of degree 3, then

$$2\frac{d}{dx}\left(y^3 \cdot \frac{d^2y}{dx^2}\right) \text{ is equal to}$$

(a) $p'''(x) \cdot p'(x)$ (b) $p''(x) \cdot p'''(x)$
(c) $p(x) \cdot p'''(x)$ (d) None of these

18. If $y = f(x)$ and $x = g(y)$ are inverse functions of each other, then

(a) $g''(y) = \frac{1}{f'(x)}$ (b) $g''(y) = -\frac{f''(x)}{(f'(x))^3}$

(c) $g''(y) = -\frac{f'(x)}{(f''(x))^3}$ (d) None of these

19. If y is a function of x, then $\frac{d^2y}{dx^2} + y \cdot \frac{dy}{dx} = 0$. If x is a function of y, then the equation becomes

(a) $\frac{d^2x}{dy^2} - x \cdot \frac{dx}{dy} = 0$ (b) $\frac{d^2x}{dy^2} + y\left(\frac{dx}{dy}\right)^3 = 0$

(c) $\frac{d^2x}{dy^2} - y\left(\frac{dx}{dy}\right)^2 = 0$ (d) $\frac{d^2x}{dy^2} - x\left(\frac{dx}{dy}\right)^2 = 0$

20. Let $g(x) = \log_e f(x)$, where $f(x)$ is twice differentiable positive function on $(0, \infty)$ such that $f(x+1) = x f(x)$, for $N = 1, 2, 3, \ldots$, then

$$g''\left(N + \frac{1}{2}\right) - g''\left(\frac{1}{2}\right) \text{ equals}$$

(a) $-4\left\{1 + \frac{1}{9} + \frac{1}{25} + \ldots + \frac{1}{(2N-1)^2}\right\}$

(b) $4\left\{1 + \frac{1}{9} + \frac{1}{25} + \ldots + \frac{1}{(2N-1)^2}\right\}$

(c) $-4\left\{1 + \frac{1}{9} + \frac{1}{25} + \ldots + \frac{1}{(2N+1)^2}\right\}$

(d) $4\left\{1 + \frac{1}{9} + \frac{1}{25} + \ldots + \frac{1}{(2N+1)^2}\right\}$

21. If the function $f(x) = x^3 + e^{x/2}$ and $g(x) = f^{-1}(x)$, then the value of $g'(1)$ is

(a) 1 (b) 2
(c) 3 (d) e

22. Let $f(\theta) = \sin\left(\tan^{-1}\left(\frac{\sin\theta}{\sqrt{\cos 2\theta}}\right)\right)$, where $-\frac{\pi}{4} < \theta < \frac{\pi}{4}$.

Then, the value of $\frac{d}{d(\tan\theta)} \cdot f(\theta)$, is

(a) 1 (b) 2
(c) 3 (d) 4

23. If $y = \log_{\sin x}(\tan x)$, then $\frac{dy}{dx}$ at $x = \frac{\pi}{4}$ is equal to

(a) $\frac{4}{\log 2}$ (b) $-4 \log 2$

(c) $\frac{-4}{\log 2}$ (d) None of these

24. If $y = \sum_{r=1}^{x} \tan^{-1}\left(\frac{1}{1 + r + r^2}\right)$, then $\frac{dy}{dx}$ is equal to

(a) $\frac{1}{1 + x^2}$ (b) $\frac{1}{1 + (1 + x)^2}$

(c) 0 (d) None of these

25. If $y = \sin^{-1}\left(\frac{\sin\alpha \sin x}{1 - \cos\alpha \sin x}\right)$, then $y'(0)$ is equal to

(a) 1 (b) $2\tan\alpha$

(c) $\left(\frac{1}{2}\right)\tan\alpha$ (d) $\sin\alpha$

26. If $f(x) = \cot^{-1}\left(\frac{x^x - x^{-x}}{2}\right)$, then $f'(1)$ equals

(a) -1 (b) 1
(c) $\log_e 2$ (d) $-\log_e 2$

27. The function $f(x) = e^x + x$, being differentiable and one-one, has a differentiable inverse $f^{-1}(x)$. The value of $\frac{d}{dx}(f^{-1})$ at the point $f(\log 2)$ is

(a) $\frac{1}{\ln 2}$ (b) $\frac{1}{3}$

(c) $\frac{1}{4}$ (d) None of these

28. If $f''(x) = -f(x)$, $g(x) = f'(x)$,

$$F(x) = \left(f\left(\frac{x}{2}\right)\right)^2 + \left(g\left(\frac{x}{2}\right)\right)^2 \text{ and } F(5) = 5, \text{ then } F(10) \text{ is}$$

equal to

(a) 5 (b) 10 (c) 0 (d) 15

29. If $x = \sec\theta - \cos\theta$ and $y = \sec^n\theta - \cos^n\theta$, then

$(x^2 + 4)\left(\frac{dy}{dx}\right)^2$ is equal to

(a) $n^2(y^2 - 4)$ (b) $n^2(4 - y^2)$

(c) $n^2(y^2 + 4)$ (d) None of these

30. If $x = f(t)\cos t - f'(t)\sin t$ and

$y = f(t)\sin t + f'(t)\cos t$, then $\left(\frac{dx}{dt}\right)^2 + \left(\frac{dy}{dt}\right)^2$ is

equal to

(a) $f(t) - f''(t)$ (b) $[f(t) - f''(t)]^2$

(c) $[f(t) + f''(t)]^2$ (d) None of these

31. If $y = at^2 + 2bt + c$ and $t = ax^2 + 2bx + c$, then $\frac{d^3y}{dx^3}$ is equal to

(a) $24a^2(at + b)$ (b) $24a(ax + b)^2$

(c) $24a(at + b)^2$ (d) $24a^2(ax + b)$

32. Differential coefficient of

$$\left(x^{\frac{l+m}{m-n}}\right)^{\frac{1}{n-l}} \cdot \left(x^{\frac{m+n}{n-l}}\right)^{\frac{1}{l-m}} \cdot \left(x^{\frac{n+l}{l-m}}\right)^{\frac{1}{m-n}} \text{ w.r.t. } x, \text{ is}$$

(a) 1 (b) 0 (c) -1 (d) x^{lmn}

33. If $y = (A + Bx)e^{mx} + (m-1)^{-2}e^x$, then

$\frac{d^2y}{dx^2} - 2m\frac{dy}{dx} + m^2y$ is equal to

(a) e^x (b) e^{mx} (c) e^{-mx} (d) $e^{(1-m)x}$

34. If $f(x) = -\frac{x^3}{3} + x^2\sin 1.5a - x\sin a \cdot \sin 2a - 5\sin^{-1}(a^2 - 8a + 17)$, then

(a) $f(x)$ is not defined at $x = \sin 8$

(b) $f'(\sin 8) > 0$

(c) $f'(x)$ is not defined at $x = \sin 8$

(d) $f'(\sin 8) < 0$

35. If f and g are differentiable functions such that $g'(a) = 2$ and $g(a) = b$ and if fog is an identity function, then $f'(b)$ has the value equal to

(a) 2/3 (b) 1

(c) 0 (d) 1/2

36. The derivative of the function,

$$f(x) = \cos^{-1}\left\{\frac{1}{\sqrt{13}}(2\cos x - 3\sin x)\right\} + \sin^{-1}\left\{\frac{1}{\sqrt{13}}(2\cos x + 3\sin x)\right\}$$

w.r.t. $\sqrt{1 + x^2}$ at $x = \frac{3}{4}$ is

(a) $\frac{3}{2}$ (b) $\frac{5}{2}$

(c) $\frac{10}{3}$ (d) 0

37. If $f(x) = \sqrt{x + 2\sqrt{2x - 4}} + \sqrt{x - 2\sqrt{2x - 4}}$, then the value of $10\,f'(102^+)$, is

(a) -1 (b) 0

(c) 1 (d) does not exist

38. Let $y = \ln(1 + \cos x)^2$, then the value of $\frac{d^2y}{dx^2} + \frac{2}{e^{y/2}}$ equals

(a) 0 (b) $\frac{2}{1 + \cos x}$

(c) $\frac{4}{(1 + \cos x)}$ (d) $\frac{-4}{(1 + \cos x)^2}$

39. If $f(x) = \frac{a + \sqrt{a^2 - x^2} + x}{\sqrt{a^2 - x^2} + a - x}$, where $a > 0$ and $x < a$, then $f'(0)$ has the value equal to

(a) $\sqrt{a}$ (b) a

(c) $\frac{1}{\sqrt{a}}$ (d) $\frac{1}{a}$

40. Let $u(x)$ and $v(x)$ be differentiable functions such that $\frac{u(x)}{v(x)} = 7$. If $\frac{u'(x)}{v'(x)} = p$ and $\left(\frac{u(x)}{v(x)}\right)' = q$, then $\frac{p+q}{p-q}$ has the value equal to

(a) 1 (b) 0

(c) 7 (d) -7

41. If $f(x) = |\log_e|x||$, then $f'(x)$ equals

(a) $\frac{1}{|x|}$, $x \neq 0$

(b) $\frac{1}{x}$ for $|x| > 1$ and $-\frac{1}{x}$ for $|x| < 1$

(c) $-\frac{1}{x}$ for $|x| > 1$ and $\frac{1}{x}$ for $|x| < 1$

(d) $\frac{1}{x}$ for $x > 0$ and $-\frac{1}{x}$ for $x < 0$

42. If $f(x)$ is given by, $f(x) = (\cos x + i \sin x)$ $(\cos 3x + i \sin 3x) \dots (\cos (2n-1)x + i \sin (2n-1)x)$, then $f''(x)$ is equal to

(a) $n^3 f(x)$ (b) $-n^4 f(x)$
(c) $-n^2 f(x)$ (d) $n^4 f(x)$

43. Let $f(x) = x^n$, n being a non-negative integer. The value of n for which the equality $f'(x+y) = f'(x) + f'(y)$ is valid for all $x, y > 0$, is

(a) 0, 1 (b) 1, 2
(c) 2, 4 (d) None of these

44. If $f(x) = \sin\left\{\frac{\pi}{3}[x] - x^2\right\}$ for $2 < x < 3$ and $[x]$ denotes the greatest integer less than or equal to x, then $f'(\sqrt{\pi/3})$ is equal to

(a) $\sqrt{\pi/3}$ (b) $-\sqrt{\pi/3}$
(c) $-\sqrt{\pi}$ (d) None of these

45. The functions $u = e^x \sin x$, $v = e^x \cos x$ satisfy the equation

(a) $v\frac{du}{dx} - u\frac{dv}{dx} = u^2 + v^2$ (b) $\frac{d^2u}{dx^2} = 2v$
(c) $\frac{d^2v}{dx^2} = -2u$ (d) All of these

46. If $f(x) = \log_x \{\ln(x)\}$, then $f'(x)$ at $x = e$, is

(a) e (b) $-e$
(c) e^2 (d) e^{-1}

47. Let f be a differentiable function satisfying

$$[f(x)]^n = f(nx) \text{ for all } x \in R$$

Then, $f'(x)f(nx)$ equals

(a) $f(x)$ (b) 0
(c) $f(x)f'(nx)$ (d) None of these

48. If $y = f(x)$ is an odd differentiable function defined on $(-\infty, \infty)$ such that $f'(3) = -2$, then $f'(-3)$ equals

(a) 4 (b) 2
(c) -2 (d) 0

49. If $y = \sqrt{x + \sqrt{y + \sqrt{x + \sqrt{y + \dots \infty}}}}$, then $\frac{dy}{dx}$ is equal to

(a) $\frac{y+x}{y^2 - 2x}$ (b) $\frac{y^3 - x}{2y^2 - 2xy - 1}$
(c) $\frac{y^3 + x}{2y^2 - x}$ (d) None of these

50. If $f(x) = |\cos x - \sin x|$, then $f'\left(\frac{\pi}{4}\right)$ is equal to

(a) $\sqrt{2}$ (b) $-\sqrt{2}$
(c) 0 (d) None of these

51. Let $f(x) = x^2 + xg'(1) + g''(2)$ and $g(x) = x^2 + xf'(2) + f''(3)$. Then,

(a) $f'(1) = 4 + f'(2)$
(b) $g'(2) = 8 + g'(1)$
(c) $g''(2) + f''(3) = 4$
(d) All of the above

52. If $f(x) = x^n$, then the value of

$$f(1) - \frac{f'(1)}{1!} + \frac{f''(1)}{2!} - \frac{f'''(1)}{3!} + \frac{f''''(1)}{4!} - \dots + \frac{(-1)^n f^n(1)}{n!}$$ is

(a) 1 (b) 2^n
(c) 2^{n-1} (d) 0

Differentiation Exercise 2 : More than One Option Correct Type Questions

53. If $y + \log(1+x) = 0$, which of the following is true?

(a) $e^y = xy' - 1$ (b) $y' = -\frac{1}{(x+1)}$
(c) $y' + e^y = 0$ (d) $y' = e^y$

54. If $y = 2^{3^x}$, then y' equals

(a) $3^x \cdot \log 3 \cdot \log 2$
(b) $y \cdot (\log_2 y) \cdot \log 3 \cdot \log 2$
(c) $2^{3^x} \cdot 3^x \cdot \log 6$
(d) $2^{3^x} \cdot 3^x \cdot \log 3 \cdot \log 2$

55. If g is the inverse of f and $f(x) = x^2 + 3x - 3, (x > 0)$, then $g'(1)$ equals

(a) $\frac{1}{2g(1)+3}$ (b) -1 (c) $\frac{1}{5}$ (d) $-\frac{f'(1)}{(f(1))^2}$

56. If $x^3 - 2x^2y^2 + 5x + y - 5 = 0$ and $y(1) = 1$, then

(a) $y'(1) = \frac{4}{3}$ (b) $y''(1) = -\frac{1}{3}$
(c) $y''(1) = -8\frac{22}{27}$ (d) $y'(1) = \frac{2}{3}$

57. Let $y=\sqrt{x+\sqrt{x+\sqrt{x+\ldots\infty}}}$, then dy/dx equals

(a) $\frac{1}{2y-1}$ (b) $\frac{x}{x+2y}$

(c) $\frac{1}{\sqrt{1-4x}}$ (d) $\frac{y}{2x+y}$

58. If $y=x^{(\ln x)^{\ln(\ln x)}}$, then $\frac{dy}{dx}$ is equal to

(a) $\frac{y}{x}(\ln x^{\ln x-1}+2\ln x\ln(\ln x))$

(b) $\frac{y}{x}(\ln x)^{\ln(\ln x)}(2\ln(\ln x)+1)$

(c) $\frac{y}{x\ln x}((\ln x)^2+2\ln x(\ln x))$

(d) $\frac{y}{x}\cdot\frac{\ln y}{\ln x}(2\ln(\ln x)+1)$

59. Which of the following functions are not derivable at $x=0$?

(a) $f(x)=\sin^{-1}2x\sqrt{1-x^2}$ (b) $g(x)=\sin^{-1}\left(\frac{2^{x+1}}{1+4^x}\right)$

(c) $h(x)=\sin^{-1}\left(\frac{1-x^2}{1+x^2}\right)$ (d) $k(x)=\sin^{-1}(\cos x)$

60. Let $f(x)=\frac{\sqrt{x-2\sqrt{x-1}}}{\sqrt{x-1}-1}\cdot x$, then

(a) $f'(10)=1$ (b) $f'(3/2)=-1$

(c) domain of $f(x)$ is $x\geq 1$ (d) None of these

61. If $2^x+2^y=2^{x+y}$, then $\frac{dy}{dx}$ is equal to

(a) $-\frac{2^y}{2^x}$ (b) $\frac{1}{1-2^x}$

(c) $1-2^y$ (d) $\frac{2^x(1-2^y)}{2^y(2^x-1)}$

62. For the function $y=f(x)=(x^2+bx+c)e^x$, which of the following holds?

(a) If $f(x)>0$ for all real $x\not\Rightarrow f'(x)>0$

(b) If $f(x)>0$ for all real $x\Rightarrow f'(x)>0$

(c) If $f'(x)>0$ for all real $x\Rightarrow f(x)>0$

(d) If $f'(x)>0$ for all real $x\not\Rightarrow f(x)>0$

63. If $\sqrt{y+x}+\sqrt{y-x}=c$, where $c\neq 0$, then $\frac{dy}{dx}$ has the value equal to

(a) $\frac{2x}{c^2}$ (b) $\frac{x}{y+\sqrt{y^2-x^2}}$

(c) $\frac{y-\sqrt{y^2-x^2}}{x}$ (d) $\frac{c^2}{2y}$

64. If $y=\tan x\tan 2x\tan 3x$, then $\frac{dy}{dx}$ has the value equal to

(a) $3\sec^2 3x\tan x\tan 2x+\sec^2 x\tan 2x\tan 3x$ $+2\sec^2 2x\tan 3x\tan x$

(b) $2y(\operatorname{cosec}2x+2\operatorname{cosec}4x+3\operatorname{cosec}6x)$

(c) $3\sec^2 3x-2\sec^2 2x-\sec^2 x$

(d) $\sec^2 x+2\sec^2 2x+3\sec^2 3x$

Differentiation Exercise 3 : Statements I and II Type Questions

- **Directions** (Q. Nos. 65 to 74) *This section is based on Statement* I *and Statement* II. *Select the correct answer from the codes given below.*

(a) Both Statement I and Statement II are correct and Statement II is the correct explanation of Statement I

(b) Both Statement I and Statement II are correct but Statement II is not the correct explanation of Statement I

(c) Statement I is correct but Statement II is incorrect

(d) Statement II is correct but Statement I is incorrect

65. Consider $f(x)=\frac{x}{x^2-1}$

Statement I Graph of $f(x)$ is concave up for $x>1$.

Statement II If $f(x)$ is concave up, then $f''(x)>0$.

66. If $f(x)=\sin^{-1}\left(\frac{2x}{1+x^2}\right)$, then

Statement I The value of $f'(2)=-\frac{2}{5}$.

Statement II $f(x)=\sin^{-1}\left(\frac{2x}{1+x^2}\right)=\frac{-2}{1+x^2}$, for $x>1$.

67. **Statement I** If $f(0)=a$, $f'(0)=b$, $g(0)=0$, $(fog)'(0)=c$, then $g'(0)=\frac{c}{b}$.

Statement II $(f(g(x))'=f'(g(x))\cdot g'(x)$, for all n.

68. Let f and g be real valued functions defined on interval $(-1,1)$ such that $g''(x)$ is continuous, $g(0)=0$, $g'(0)=0$, $g''(0)=0$ and $f(x)=g(x)\sin x$.

Statement I $\lim_{x\to 0}(g(x)\cot x-g(0)\operatorname{cosec}x)=f''(0)$.

Statement II $f'(0)=g'(0)$.

69. Statement I If $y = \sin^{-1}(3x - 4x^3)$, then $\dfrac{dy}{dx} = \dfrac{3}{\sqrt{1-x^2}}$ only when $\dfrac{-1}{2} \le x < \dfrac{1}{2}$.

Statement II $\sin^{-1}(3x - 4x^3)$

$$= \begin{cases} -\pi - 3\sin^{-1} x, & -1 \le x \le -\frac{1}{2} \\ 3\sin^{-1} x, & -\frac{1}{2} \le x \le \frac{1}{2} \\ \pi - 3\sin^{-1} x, & \frac{1}{2} \le x \le 1. \end{cases}$$

70. If $y = \cos^{-1}\left(\dfrac{1-x^2}{1+x^2}\right)$, then

Statement I $\dfrac{dy}{dx} = \dfrac{2}{1+x^2}$ for $x \in R$.

Statement II $\cos^{-1}\left(\dfrac{1-x^2}{1+x^2}\right) = \begin{cases} 2\tan^{-1} x, & x \ge 0 \\ -2\tan^{-1} x, & x < 0. \end{cases}$

71. If $y = x + [x]$, then

Statement I $\dfrac{dy}{dx} = 1$ for all $x \in R$.

Statement II $\dfrac{d([x])}{dx} = \begin{cases} 0, & x \notin \text{Integer} \\ \text{does not exist}, & x \in \text{Integer}. \end{cases}$

72. Statement I If $f(x)$ is a continuous function defined from R to Q and $f(5) = 3$, then differential coefficient of $f(x)$ w.r.t. x will be 0.

Statement II Differentiation of constant function is always zero.

73. Statement I Derivative of $\sin^{-1}\left(\dfrac{2x}{1+x^2}\right)$ w.r.t. $\cos^{-1}\left(\dfrac{1-x^2}{1+x^2}\right)$ is 1 for $0 < x < 1$.

Statement II $\sin^{-1}\left(\dfrac{2x}{1+x^2}\right) = \cos^{-1}\left(\dfrac{1-x^2}{1+x^2}\right)$ for $-1 \le x \le 1$.

74. Consider function $f(x)$ satisfies the relation $f(x + y^3) = f(x) + f(y^3)$, $\forall\ x, y \in R$ and differentiable for all x.

Statement I If $f'(2) = a$, then $f'(-2) = a$.

Statement II $f(x)$ is an odd function.

Differentiation Exercise 4 : Passage Based Questions

Passage I (Q. Nos. 75 to 77)

Let $\dfrac{f(x+y) - f(x)}{2} = \dfrac{f(y) - 1}{2} + xy$, *for all* $x, y \in R$, $f(x)$ *is differentiable and* $f'(0) = 1$. *Let* $g(x)$ *be a derivable function at* $x = 0$ *and follows the functional rule*

$$g\left(\frac{x+y}{k}\right) = \frac{g(x) + g(y)}{k};\ k \in R,\ k \ne 0, 2$$

and $g'(0) - \lambda g'(0) \ne 0$.

75. Domain of $\log(f(x))$, is

(a) R^+ (b) $R - \{0\}$
(c) R (d) R^-

76. Range of $y = \log_{3/4}(f(x))$ is

(a) $(-\infty, 1]$ (b) $[3/4, \infty)$
(c) $(-\infty, \infty)$ (d) R

77. If the graphs of $y = f(x)$ and $y = g(x)$ intersect in coincident points then λ can take values

(a) -3 (b) 1
(c) -1 (d) 4

Passage II (Q. Nos. 78 to 80)

Left hand derivative and right hand derivative of a function $f(x)$ *at a point* $x = a$ *are defined as*

$$f'(a^-) = \lim_{h \to 0^+} \frac{f(a) - f(a-h)}{h} = \lim_{h \to 0^+} \frac{f(a+h) - f(a)}{h}$$

$$\text{and } f'(a^+) = \lim_{h \to 0^+} \frac{f(a+h) - f(a)}{h} = \lim_{h \to 0^+} \frac{f(a) - f(a+h)}{h} = \lim_{h \to 0^+} \frac{f(a) - f(x)}{a - x} \text{ respectively.}$$

Let f be a twice differentiable function. We also know that derivative of an even function is odd function and derivative of an odd function is even function.

78. If f is odd, then which of the following is left hand derivative of f at $x = -a$?

(a) $\lim_{h \to 0^-} \dfrac{f(a-h) - f(a)}{-h}$

(b) $\lim_{h \to 0^-} \dfrac{f(h-a) - f(a)}{h}$

(c) $\lim_{h \to 0^+} \dfrac{f(a) + f(a-h)}{-h}$

(d) $\lim_{h \to 0^-} \dfrac{f(a) - f(a-h)}{-h}$

79. If f is even function, then which of the following is right hand derivative of f' at $x = a$?

(a) $\lim_{h \to 0^-} \dfrac{f'(a) + f'(-a+h)}{h}$

(b) $\lim_{h \to 0^+} \dfrac{f'(a) + f'(-a-h)}{h}$

(c) $\lim_{h \to 0^-} \dfrac{-f'(-a) + f'(-a-h)}{-h}$

(d) $\lim_{h \to 0^+} \dfrac{f'(a) + f'(-a+h)}{h}$

80. The statement $\lim_{h \to 0} \dfrac{f(-x) - f(-x-h)}{h} = \lim_{h \to 0} \dfrac{f(x) - f(x-h)}{-h}$ implies that for all $x \in R$,

(a) f is odd
(b) f is even
(c) f is neither even nor odd
(d) Nothing can be concluded

Passage III

(Q. Nos. 81 to 82)

If $f(x) = \sin^{-1}(3x - 4x^3)$. Then answer the following

81. The value of $f'(0)$ is

(a) -3 (b) 3
(c) $\sqrt{2}$ (d) $-\sqrt{2}$

82. The value of $f'\left(\dfrac{1}{\sqrt{2}}\right)$ is

(a) -3 (b) 3
(c) $-3\sqrt{2}$ (d) $3\sqrt{2}$

Passage IV (Q. Nos. 83 to 84)

Let the derivative of $f(x)$ be defined as

$$D^* f(x) = \lim_{h \to 0} \frac{f^2(x+h) - f^2(x)}{h}, \text{ where } f^2(x) = \{f(x)\}^2.$$

83. If $u = f(x)$, $v = g(x)$, then the value of $D^*(u \cdot v)$ is

(a) $(D^* u)v + (D^* v)u$ (b) $u^2(D^* v) + v^2(D^* u)$
(c) $D^* u + D^* v$ (d) $uv\, D^*(u + v)$

84. If $u = f(x)$, $v = g(x)$, then the value of $D^*\left(\dfrac{u}{v}\right)$ is

(a) $\dfrac{u^2(D^* v) - v^2(D^* u)}{v^4}$ (b) $\dfrac{u(D^* v) - v(D^* u)}{v^2}$

(c) $\dfrac{v^2(D^* u) - u^2(D^* v)}{v^4}$ (d) $\dfrac{v(D^* u) - u(D^* v)}{v^2}$

Passage V

(Q. Nos. 85 to 87)

A curve is represented parametrically by the equations $x = e^t \cos t$ and $y = e^t \sin t$, where t is a parameter. Then,

85. The relation between the parameter 't' and the angle α between the tangent to the given curve and the X-axis is given by, 't' equals

(a) $\dfrac{\pi}{2} - \alpha$ (b) $\dfrac{\pi}{4} + \alpha$

(c) $\alpha - \dfrac{\pi}{4}$ (d) $\dfrac{\pi}{4} - \alpha$

86. The value of $\dfrac{d^2y}{dx^2}$ at the point, where $t = 0$, is

(a) 1 (b) 2
(c) -2 (d) 3

87. If $F(t) = \int (x + y)\, dt$, then the value of $F\left(\dfrac{\pi}{2}\right) - F(0)$ is

(a) 1 (b) -1
(c) $e^{\pi/2}$ (d) 0

Passage VI

(Q. Nos. 88 to 90)

Equation $x^n - 1 = 0$, $n > 1$, $n \in N$ has roots $1, a_1, a_2, \ldots, a_n$.

88. The value of $(1 - a_1)(1 - a_2)\ldots(1 - a_n)$, is

(a) $\dfrac{n^2}{2}$ (b) n

(c) $(-1)^n n$ (d) None of these

89. The value of $\sum_{r=1}^{n} \dfrac{1}{2 - a_r}$, is

(a) $\dfrac{2^{n-1}(n-2) + 1}{2^n - 1}$ (b) $\dfrac{2^n(n-2) + 1}{2^n - 1}$

(c) $\dfrac{2^{n-1}(n-1) - 1}{2^n - 1}$ (d) None of these

90. The value of $\sum_{r=1}^{n} \dfrac{1}{1 - a_r}$, is

(a) $\dfrac{n}{4}$ (b) $\dfrac{n(n-1)}{2}$

(c) $\dfrac{n-1}{2}$ (d) None of these

Differentiation Exercise 5 : Matching Type Questions

91. Match the entries between following two columns.

	Column I	Column II
(A)	$y = f(x)$ be given by $x = t^5 - 5t^3 - 20t + 7$ and $y = 4t^3 - 3t^2 - 18t + 3$, then $-5 \times \frac{dy}{dx}$ at $t = 1$	(p) 0
(B)	$P(x)$ be a polynomial of degree 4 with $P(2) = -1, P'(2) = 0, P''(2) = 2, P'''(2) = -12$ and $P''''(2) = 24$, then $P''(3)$ is equal to	(q) −2
(C)	$y = \frac{1}{x}$, then $\dfrac{\frac{dy}{\sqrt{1+y^4}}}{\frac{dx}{\sqrt{1+x^4}}}$	(r) 2
(D)	$f\left(\frac{2x+3y}{5}\right) = \frac{2f(x)+3f(y)}{5}$ and $f'(0) = p$ and $f(0) = q$, then $f''(0)$	(s) −1

92. Match the following.

	Column I	Column II
(A)	$y = \sin^{-1}\left(\frac{2x}{1+x^2}\right)$, then $\frac{dy}{dx} = \frac{2}{1+x^2}$	(p) for $x < 0$
(B)	$y = \cos^{-1}\left(\frac{1}{\sqrt{1+x^2}}\right)$, then $\frac{dy}{dx} = -\frac{1}{1+x^2}$	(q) for $x > 1$
(C)	$y = \lvert e^{\lvert x \rvert} - e \rvert$, then $\frac{dy}{dx} > 0$	(r) for $x < -1$
(D)	$u = \log \lvert 2x \rvert, v = \lvert \tan^{-1} x \rvert$, then $\frac{du}{dv} > 2$	(s) for $-1 < x < 0$

Differentiation Exercise 6 : Single Integer Answer Type Questions

93. If $f(x) = x^3 + x^2 f'(1) + xf''(2) + f'''(3)$ for all $x \in R$. Then, $f(1)$ is ....

94. Let $f(x) = \sin^{-1}\left(\frac{2x+2}{\sqrt{4x^2+8x+13}}\right)$, then the value of $\frac{d(\tan f(x))}{d(\tan^{-1} x)}$, when $x = \frac{1}{2}$, is ....

95. Let $x, x_1, x_2, x_3, x_4, \ldots, x_8$ be 9 real zeroes, of the polynomial $P(x) = x^{10} + ax^2 + bx + c$, where $a, b, c \in R$. If the value of $Q(x_1) = \frac{p}{q}$, where p and q are coprime to each other. If $Q(x) = (x - x_2)(x - x_3)\ldots(x - x_8)$ and $x_1 = \frac{1}{2}$, then the value of $q - p$ is ....

96. If $f(x) = \begin{vmatrix} (x-a)^4 & (x-a)^3 & 1 \\ (x-b)^4 & (x-b)^3 & 1 \\ (x-c)^4 & (x-c)^3 & 1 \end{vmatrix}$ and

$f'(x) = \lambda \begin{vmatrix} (x-a)^4 & (x-a)^2 & 1 \\ (x-b)^4 & (x-b)^2 & 1 \\ (x-c)^4 & (x-c)^2 & 1 \end{vmatrix}$, then value of λ is ....

97. Let $P(x)$ be a polynomial of degree 4 such that $P(1) = P(3) = P(5) = P(7) = 0$. If the real number $x \neq 1, 3, 5$ is such that $P(x) = 0$ can be expressed as $x = \frac{p}{q}$, where p and q are relatively prime, then $(p - 8q)$ is ....

98. If $x^2 + y^2 = t - \frac{1}{t}$ and $x^4 + y^4 = t^2 + \frac{1}{t^2}$, then $\left(\frac{dy}{dx}\right)_{(1,1)}$ is ....

99. If $x^2 + y^2 + z^2 - 2xyz = 1$, then the value of $\dfrac{dx}{\sqrt{1-x^2}} + \dfrac{dy}{\sqrt{1-y^2}} + \dfrac{dz}{\sqrt{1-z^2}}$ is equal to ....

100. If y is twice differentiable function of x, then the expression $(1-x^2)\cdot\dfrac{d^2y}{dx^2} - x\dfrac{dy}{dx} + y$ by means of the transformation $x = \sin t$ in terms of t is $\dfrac{d^2y}{dt^2} + \lambda y$. Thus, λ is ....

101. The derivative of $f(x) = \cos^{-1}\left(\dfrac{1}{\sqrt{3}}(2\cos x - 3\sin x)\right) + \left\{\sin^{-1}\left(\dfrac{1}{\sqrt{3}}(2\cos x + 3\sin x)\right)\right\}$ w.r.t. $\sqrt{1+x^2}$ at $x = \dfrac{1}{\sqrt{3}}$ is ....

102. Suppose $f(x) = e^{ax} + e^{bx}$, where $a \neq b$ and $f''(x) - 2f'(x) - 15f(x) = 0$ for all x. Then, the value of $|a+b|$ is equal to ....

103. Suppose, $A = \dfrac{dy}{dx}$ of $x^2 + y^2 = 4$ at $(\sqrt{2}, \sqrt{2})$, $B = \dfrac{dy}{dx}$ of $\sin y + \sin x = \sin x \cdot \sin y$ at (π, π) and $C = \dfrac{dy}{dx}$ of $2e^{xy} + e^x e^y - e^x - e^y = e^{xy+1}$ at (1, 1), then $(A - B - C)$ has the value equal to ....

104. A function is represented parametrically by the equations $x = \dfrac{1+t}{t^3}$; $y = \dfrac{3}{2t^2} + \dfrac{2}{t}$, then $\dfrac{dy}{dx} - x\cdot\left(\dfrac{dy}{dx}\right)^3$ has the absolute value equal to ....

105. Suppose, the function $f(x) - f(2x)$ has the derivative 5 at $x = 1$ and derivative 7 at $x = 2$. The derivative of the function $f(x) - f(4x)$ at $x = 1$, has the value $10 + \lambda$, then the value of λ is equal to ....

106. If $x + y = 3e^2$, then $D(x^y)$ vanishes when x equals to λe^2, then the value of λ is equal to ....

107. Let $h(x)$ be differentiable for all x and let $f(x) = (kx + e^x)h(x)$, where k is some constant. If $h(0) = 5$, $h'(0) = -2$ and $f'(0) = 18$, then the value of k is equal to ....

Complex Number Exercise 7 :
Questions Asked in Previous 10 Years' Exams

(i) JEE Advanced & IIT-JEE

108. For $x \in R$, $f(x) = |\log 2 - \sin x|$ and $g(x) = f(f(x))$, then **[One Correct Option, 2016 Adv.]**

(a) g is not differentiable at $x = 0$
(b) $g'(0) = \cos(\log 2)$
(c) $g'(0) = -\cos(\log 2)$
(d) g is differentiable at $x = 0$ and $g'(0) = -\sin(\log 2)$

109. Let $f: R \to R$, $g: R \to R$ and $h: R \to R$ be differentiable functions such that $f(x) = x^3 + 3x + 2$, $g(f(x)) = x$ and $h(g(g(x))) = x$ for all $x \in R$. Then, **[More than One Correct Option, 2016 Adv.]**

(a) $g'(2) = \dfrac{1}{15}$
(b) $h'(1) = 666$
(c) $h(0) = 16$
(d) $h(g(3)) = 36$

(ii) JEE Main & AIEEE

110. If for $x \in \left(0, \dfrac{1}{4}\right)$, the derivative of $\tan^{-1}\left(\dfrac{6x\sqrt{x}}{1-9x^3}\right)$ is $\sqrt{x}\cdot g(x)$, then $g(x)$ equals **[2017 JEE Main]**

(a) $\dfrac{3}{1+9x^3}$
(b) $\dfrac{9}{1+9x^3}$
(c) $\dfrac{3x\sqrt{x}}{1-9x^3}$
(d) $\dfrac{3x}{1-9x^3}$

111. Let $g(x) = \log f(x)$, where $f(x)$ is a twice differentiable positive function on $(0, \infty)$ such that $f(x+1) = x f(x)$. Then, for $N = 1, 2, 3, \ldots$, $g''\left(N + \dfrac{1}{2}\right) - g''\left(\dfrac{1}{2}\right)$ is equal to **[2008 AIEEE]**

(a) $-4\left\{1 + \dfrac{1}{9} + \dfrac{1}{25} + \ldots + \dfrac{1}{(2N-1)^2}\right\}$
(b) $4\left\{1 + \dfrac{1}{9} + \dfrac{1}{25} + \ldots + \dfrac{1}{(2N-1)^2}\right\}$
(c) $-4\left\{1 + \dfrac{1}{9} + \dfrac{1}{25} + \ldots + \dfrac{1}{(2N+1)^2}\right\}$
(d) $4\left\{1 + \dfrac{1}{9} + \dfrac{1}{25} + \ldots + \dfrac{1}{(2N+1)^2}\right\}$

112. $\dfrac{d^2x}{dy^2}$ equals **[2007 AIEEE]**

(a) $\left(\dfrac{d^2y}{dx^2}\right)^{-1}$ (b) $-\left(\dfrac{d^2y}{dx^2}\right)^{-1}\left(\dfrac{dy}{dx}\right)^{-3}$

(c) $\left(\dfrac{d^2y}{dx^2}\right)\left(\dfrac{dy}{dx}\right)^{-2}$ (d) $-\left(\dfrac{d^2y}{dx^2}\right)\left(\dfrac{dy}{dx}\right)^{-3}$

113. If $f''(x) = -f(x)$, where $f(x)$ is a continuous double differentiable function and $g(x) = f'(x)$.

If $F(x) = \left\{f\left(\dfrac{x}{2}\right)\right\}^2 + \left\{g\left(\dfrac{x}{2}\right)\right\}^2$ and $F(5) = 5$, then $F(10)$ is **[2006 AIEEE]**

(a) 0 (b) 5 (c) 10 (d) 25

114. If y is a function of x and $\log(x+y) = 2xy$, then the value of $y'(0)$ is **[2004 AIEEE]**

(a) 1

(b) −1

(c) 2

(d) 0

115. If $x^2 + y^2 = 1$, then **[2000 AIEEE]**

(a) $yy'' - 2(y')^2 + 1 = 0$

(b) $yy'' + (y')^2 + 1 = 0$

(c) $yy'' + (y')^2 - 1 = 0$

(d) $yy'' + 2(y')^2 + 1 = 0$

Answers

Exercise for Session 1

1. $a^x \log a + ax^{a-1}$
2. $\dfrac{\tan x}{|\tan x|}\sec^2 x$
3. $\dfrac{1}{x\log_e 3} + \dfrac{3}{x} + 2\sec^2 x$
4. $\dfrac{x}{|x|} + na_0x^{n-1} + (n-1)a_1x^{n-2} + (n-2)a_2x^{n-3} + \ldots + a_{n-1}1$
5. 0
6. $e^x x^{n-1}\left\{n\log_a x + \dfrac{1}{\log_e a} + x\log_a x\right\}$
7. $\dfrac{2^x}{\sqrt{x}}\left\{\log 2\cot x - \operatorname{cosec}^2 x - \dfrac{\cot x}{2x}\right\}$
8. $\dfrac{x^2}{(x\sin x + \cos x)^2}$
9. y
10. $x \in (-\infty, -1) \cup (1, \infty)$
11. $\dfrac{(\sin 3x)\left(e^x + \dfrac{1}{x}\right) - 3(e^x + \log x)\cos 3x}{\sin^2 3x}$
12. $\dfrac{m}{\sqrt{1-x^2}}\cos(m\sin^{-1}x)$
13. $\dfrac{2\log a \cdot \sin^{-1}x}{\sqrt{1-x^2}} \cdot a^{(\sin^{-1}x)^2}$
14. $e^{\cos^{-1}\sqrt{1-x^2}} \cdot \dfrac{1}{\sqrt{1-x^2}}$
15. $\dfrac{\sin^{-1}x}{(1-x^2)^{3/2}}$
16. $\dfrac{1}{x\log_e 10} - \dfrac{\log_e 10}{x(\log_e x)^2}$
17. $-2x\{5^{3-x^2} \cdot \log_e 5 + 5(3-x^2)^4\}$
18. $\dfrac{-2a^2}{x^3}\left\{1 + \dfrac{a^2}{\sqrt{a^4 - x^4}}\right\}$
19. $\dfrac{x}{4\sqrt{4+\sqrt{4+\sqrt{4+x^2}}}\sqrt{4+\sqrt{4+x^2}}\sqrt{4+x^2}}$
20. (c) 21. (b) 22. (a) 23. (a)
24. (a) 25. (c) 26. (a) 27. (d)

Exercise for Session 2

1. $4(x^2+x+1)^3 \cdot (2x+1)$
2. $\dfrac{1}{2\sqrt{x^2+x+1}} \cdot (2x+1)$
3. $3\sin^2 x\cos x$
4. $\dfrac{x}{(a^2-x^2)^{3/2}}$
5. $e^{x\sin x}(x\cos x + \sin x)$
6. $-\dfrac{\sqrt{b^2-a^2}}{b + a\cos x}$
7. $e^{e^x} \cdot e^x$
8. $\dfrac{1}{\sqrt{a^2+x^2}}$
9. $\dfrac{2ab\cos x}{a^2 - b^2\sin^2 x}$
10. $\sec x$

Exercise for Session 3

6. (b) 7. (a) 8. (a) 9. (a) 10. (a)

Exercise for Session 4

1. (b) 2. (c) 3. (b) 4. (c)

5. (i) $y = \sin^{-1}\left(\dfrac{2x}{1+x^2}\right)$

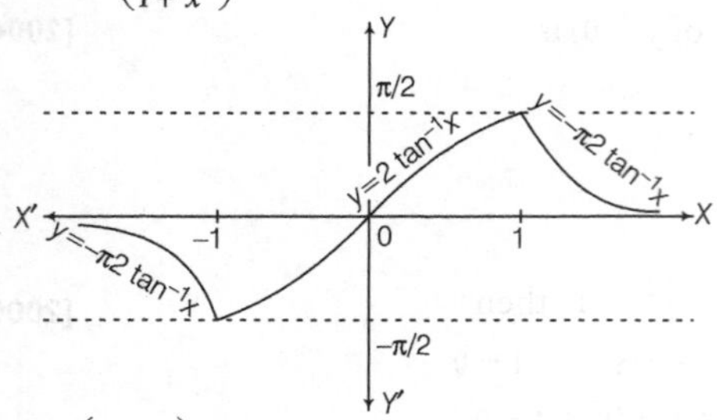

(ii) $y = \cos^{-1}\left(\dfrac{1-x^2}{1+x^2}\right)$

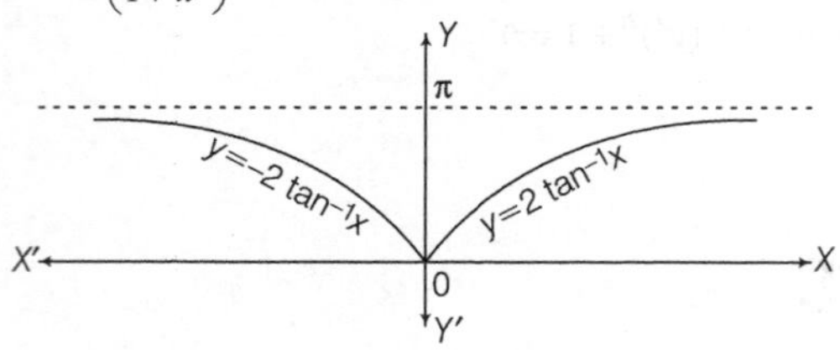

(iii) $y = \tan^{-1}\left(\dfrac{2x}{1-x^2}\right)$

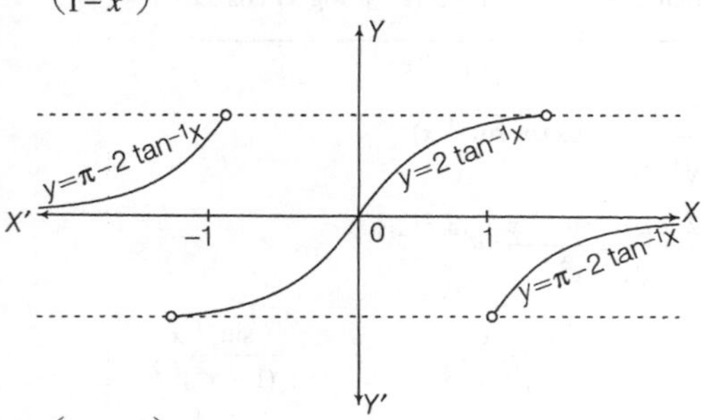

(iv) $y = \tan^{-1}\left(\dfrac{3x-x^3}{1-3x^2}\right)$

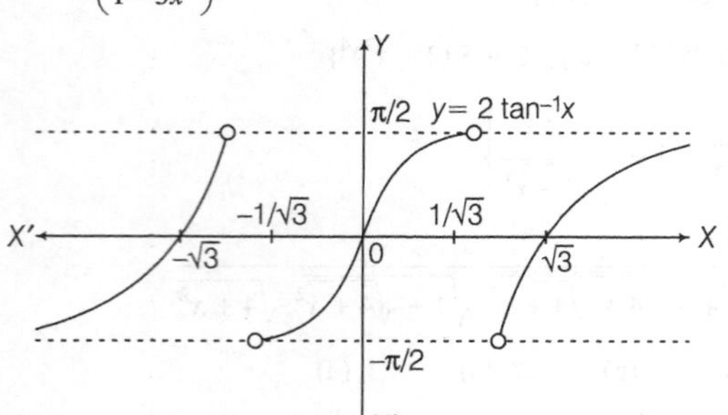

(v) $y = \sin^{-1}(3x-4x^3)$

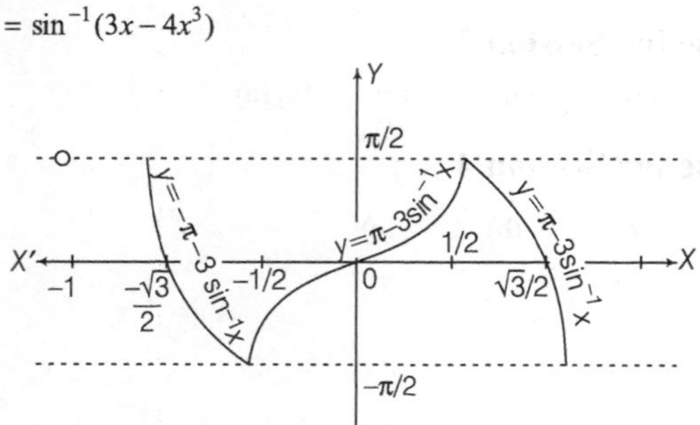

(vi) $y = \cos^{-1}(4x^3-3x)$

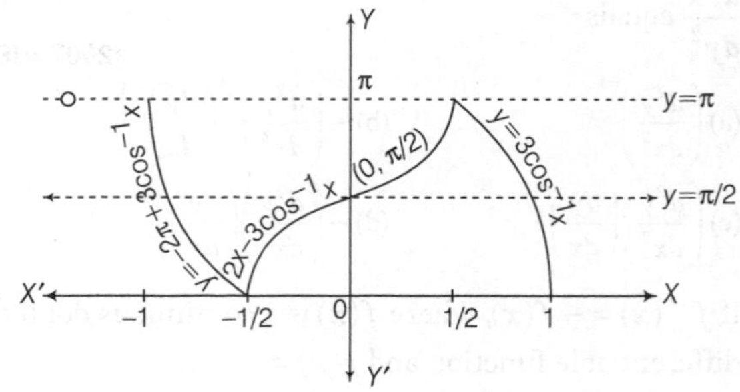

Exercise for Session 5

6. $\dfrac{\sqrt{3}}{2}$ 7. $\dfrac{2}{19}$

Exercise for Session 6

1. $x^x(1+\log x)$
2. $x^{\sqrt{x}}\left(\dfrac{\log x}{2\sqrt{x}}+\dfrac{1}{\sqrt{x}}\right)$
3. $x^{x^x}\cdot x^x\cdot\left\{(1+\log x)\cdot\log x+\dfrac{1}{x}\right\}$
4. $x\cdot x^{x^2}\cdot(2\log x+1)$
5. $x^{x+\frac{1}{2}}\left\{\left(\dfrac{2x+1}{2x}\right)+\log x\right\}$
6. $(\cos x)^x\,(\log\cos x - x\tan x)$
7. $(\sin x)^{\cos x}\left(-\sin x\log\sin x+\dfrac{\cos^2 x}{\sin x}\right)$
8. $(\sin x)^{\cos^{-1}x}\left(\cos^{-1}x\cot x-\dfrac{\log(\sin x)}{\sqrt{1-x^2}}\right)$
9. $-x^x\,(\sin x^x)\cdot\{1+\log x\}$
10. $\dfrac{1}{x^x+\operatorname{cosec}^2 x}\{x^x(1+\log x)-2\operatorname{cosec}^2 x\cot x\}$
11. $(\sin x)^{\tan x}\cdot\{\sec^2 x(\log\sin x)+1\}$
$+(\cos x)^{\sec x}\cdot\{\sec x\tan x\cdot\log(\cos x)-\sec x\tan x\}$
13. $-\left(\dfrac{yx^{y-1}+y^x\log y}{x^y\log x+xy^{x-1}}\right)$
14. $\dfrac{\log\sin y+y\tan x}{\log\cos x-x\cot y}$
19. $\dfrac{\alpha}{\alpha^2+x^2}$

Exercise for Session 7

1. (i) 1 (ii) – 1 (iii) – 1
2. $x^{\sin^{-1}x}\left(\log x+\dfrac{\sqrt{1-x^2}}{x}\sin^{-1}x\right)$
3. $-\dfrac{2}{ax}$ 4. $-2\sqrt{\cos x}\cdot\cot x\operatorname{cosec} x$
5. $-\dfrac{1}{2}$ 6. x^x 7. (i) $-\dfrac{1}{x}$ (ii) $\dfrac{1}{x}$ (iii) $\dfrac{1}{x}$

Exercise for Session 8

6. $-\dfrac{1}{2at^3}$ 7. $\dfrac{1}{3a}\sec^4\theta\,\text{cosec}\,\theta$

Exercise for Session 9

3. $\dfrac{3}{\pi\sqrt{\pi^2-3}}$

Exercise for Session 10

1. $\dfrac{1}{3}$ 2. $-\dfrac{f''(x)}{(f'(x))^3}$ 3. (b) 4. (c)

Chapter Exercises

1. (b) 2. (c) 3. (c) 4. (d) 5. (c)
6. (a) 7. (d) 8. (d) 9. (b)
10. (b) 11. (c) 12. (b) 13. (b) 14. (c)
15. (a) 16. (c) 17. (c) 18. (b) 19. (c)
20. (a) 21. (b) 22. (a) 23. (c) 24. (b)
25. (d) 26. (a) 27. (b) 28. (a)
29. (c) 30. (c) 31. (d) 32. (b)
33. (a) 34. (d) 35. (d) 36. (c) 37. (c)
38. (a) 39. (d) 40. (a) 41. (b) 42. (b)
43. (b) 44. (b) 45. (d) 46. (d) 47. (c)
48. (c) 49. (d) 50. (d) 51. (d) 52. (d)
53. (b, c) 54. (b, d) 55. (a, c) 56. (a, c) 57. (a, d)
58. (b, d) 59. (b, c, d) 60. (a, b) 61. (a, b, c, d)
62. (a, c) 63. (a, b) 64. (a, b, c) 65. (a)
66. (a) 67. (a) 68. (b) 69. (a)
70. (d) 71. (d) 72. (a) 73. (c)
74. (a) 75. (c) 76. (a) 77. (c)
78. (a) 79. (a) 80. (b) 81. (b)
82. (c) 83. (b) 84. (c) 85. (c)
86. (b) 87. (c) 88. (b) 89. (a)
90. (c)
91. (A) $\rightarrow$ (q), (B) $\rightarrow$ (r), (C) $\rightarrow$ (s), (D) $\rightarrow$ (p)
92. (A) $\rightarrow$ (q, r), (B) $\rightarrow$ (p), (C) $\rightarrow$ (q, s), (D) $\rightarrow$ (q, r)
93. (4) 94. (1) 95. (1) 96. (3) 97. (1) 98. (1)
99. (0) 100. (1) 101. (2) 102. (2) 103. (1) 104. (−1)
105. (9) 106. (1) 107. (3) 108. (b) 109. (b, c) 110. (c)
111. (a) 112. (d) 113. (b) 114. (a) 115. (b)

Solutions

1. Here, $y = \dfrac{\sec x - \tan x}{\sec x + \tan x} \cdot \dfrac{\sec x - \tan x}{\sec x - \tan x}$

$\therefore \quad y = (\sec x - \tan x)^2$

$\Rightarrow \quad \dfrac{dy}{dx} = 2(\sec x - \tan x)\cdot(\sec x \tan x - \sec^2 x)$

$= -2\sec x(\sec x - \tan x)^2$

2. Here, $y = \dfrac{1+x^2+x^4}{1+x+x^2} = \dfrac{(1+x^2)^2 - x^2}{(1+x+x^2)}$

$= \dfrac{(1+x+x^2)(1-x+x^2)}{(1+x+x^2)} = (1-x+x^2)$

$\therefore \quad \dfrac{dy}{dx} = 2x - 1 = ax + b$

$\Rightarrow \quad a = 2,\ b = -1$

3. $y = \dfrac{d}{dx}(x \log x) = x \cdot \dfrac{1}{x} + \log x \cdot 1$

$\Rightarrow \quad y = 1 + \log_e x$

$\therefore y = 0$, when $\log_e x = -1 \Rightarrow x = \dfrac{1}{e}$

Thus, $y = 1 + \log_e x$ can be shown as

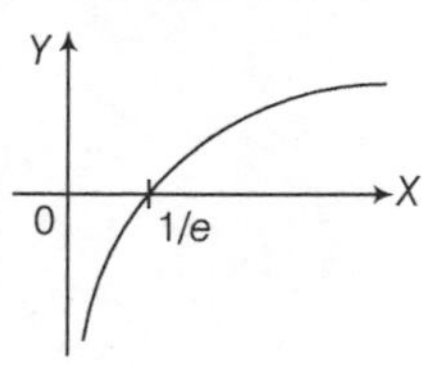

4. Here, $f'(x) = 1 + \dfrac{3}{x-2} = \dfrac{x+1}{x-2}$ and $g'(x) = 1 + \dfrac{5}{x-1} = \dfrac{x+4}{x-1}$

Since, $\quad f'(x) < g'(x)$

$\Rightarrow \quad \dfrac{x+1}{x-2} < \dfrac{x+4}{x-1}$

or $\quad \dfrac{x+1}{x-2} - \dfrac{x+4}{x-1} < 0$

$\Rightarrow \quad \dfrac{(x^2-1)-(x^2+2x-8)}{(x-1)(x-2)} < 0$

$\Rightarrow \quad \dfrac{-2x+7}{(x-1)(x-2)} < 0 \Rightarrow \dfrac{2x-7}{(x-1)(x-2)} > 0$

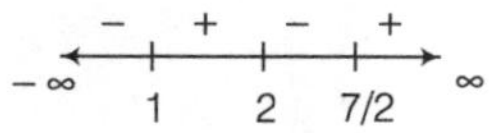

$\therefore \quad x \in (1,2) \cup (7/2, \infty)$, since $\log(x-2)$ exists, when $x > 2$.

$\Rightarrow \quad x \in \left(\dfrac{7}{2}, \infty\right)$

5. Here, $\cos^{-1}\left(\dfrac{x^2-y^2}{x^2+y^2}\right) = \log a$

$\Rightarrow \quad \cos^{-1}\left(\dfrac{1-(y/x)^2}{1+(y/x)}\right) = \log a$

Put $\dfrac{y}{x} = \tan\theta$

$\Rightarrow \quad \cos^{-1}\left(\dfrac{1-\tan^2\theta}{1+\tan^2\theta}\right) = \log a$

$\Rightarrow \quad \cos^{-1}(\cos 2\theta) = \log a$

$\Rightarrow \quad 2\theta = \log_e a$

$\Rightarrow \quad 2\tan^{-1}\left(\dfrac{y}{x}\right) = \log a$

$\Rightarrow \quad \dfrac{y}{x} = \tan\left(\dfrac{\log a}{2}\right)$

On differentiating both sides, we get

$$\frac{x\dfrac{dy}{dx} - y\cdot 1}{x^2} = 0$$

$\Rightarrow \quad x\dfrac{dy}{dx} - y = 0$

$\Rightarrow \quad \dfrac{dy}{dx} = \dfrac{y}{x}$

6. Here, $f(x) = (|x|)^{|\sin x|}$ as $x \to \dfrac{-\pi}{4}$

$\Rightarrow \quad f(x) = (-x)^{-\sin x}$

Taking logarithm on both sides, we get

$$\log(f(x)) = (-\sin x)\log(-x)$$

On differentiating both sides, we get

$$\frac{1}{f(x)}\cdot f'(x) = (-\sin x)\frac{1}{x} + \log(-x)\cdot(-\cos x)$$

At $x = -\dfrac{\pi}{4}$,

$$f'\left(-\frac{\pi}{4}\right) = f\left(-\frac{\pi}{4}\right)\left[\frac{-\frac{1}{\sqrt{2}}}{\frac{\pi}{4}} + \log\left(\frac{\pi}{4}\right)\cdot\left(-\frac{1}{\sqrt{2}}\right)\right]$$

$$= \left(\frac{\pi}{4}\right)^{\frac{1}{\sqrt{2}}}\left[\frac{\sqrt{2}}{2}\log\frac{4}{\pi} - \frac{2\sqrt{2}}{\pi}\right]$$

7. Here, $y = \dfrac{x}{a + \dfrac{x}{b+y}} = \dfrac{x(b+y)}{ab+ay+x}$

$\Rightarrow \quad aby + ay^2 + xy = bx + xy$

$\Rightarrow \quad aby + ay^2 = bx$

On differentiating w.r.t. x, we get

$$ab\frac{dy}{dx} + 2ay\frac{dy}{dx} = b$$

$\therefore \quad \dfrac{dy}{dx} = \dfrac{b}{a(b+2y)}$

8. Taking logarithm on both sides, we get
$$\log y = x^2 \log x$$
On differentiating both sides, we get
$$\frac{1}{y}\cdot\frac{dy}{dx} = x^2\cdot\frac{1}{x}+2x\cdot(\log x)$$
$$\Rightarrow \quad \frac{dy}{dx} = x^{x^2}\cdot(x+2x\cdot\log x)$$
$$= x^{x^2+1}\cdot(1+2\log x) = x^{x^2+1}\cdot(\log(e x^2))$$

9. Here, $x\sqrt{1+y} = -y\sqrt{1+x}$
On squaring both sides, we get
$$x^2(1+y) = y^2(1+x)$$
$$\Rightarrow \quad x^2+x^2y = y^2+y^2x$$
$$\Rightarrow \quad (x-y)(x+y)+xy(x-y) = 0$$
$$\Rightarrow \quad (x-y)(x+y+xy) = 0 \Rightarrow x+y+xy = 0$$
$$\Rightarrow \quad y = \frac{-x}{(1+x)}$$
$$\therefore \quad \frac{dy}{dx} = -\left[\frac{(1+x)\cdot 1 - x\cdot 1}{(1+x)^2}\right] \Rightarrow \frac{dy}{dx} = \frac{-1}{(1+x)^2}$$

10. Here, $x^2e^y + 2xye^x + 13 = 0$
On differentiating both sides, we get
$$x^2\cdot e^y\cdot\frac{dy}{dx}+2x\cdot e^y+2\left[xe^x\cdot\frac{dy}{dx}+ye^x+yx\cdot e^x\right]=0$$
$$\Rightarrow \quad \frac{dy}{dx}\cdot x(xe^y+2e^x)+2(xe^y+ye^x+xye^x)=0$$
$$\Rightarrow \quad \frac{dy}{dx} = -\frac{2[xe^y+ye^x(x+1)]}{x(xe^y+2e^x)}$$
$$\text{or} \quad \frac{dy}{dx} = -2\left[\frac{(xe^{y-x}+y(x+1))}{x(xe^{y-x}+2)}\right]$$

11. Here, $x = e^{y+x}$. Taking log on both sides, we get
$$\log x = (y+x)\log e$$
On differentiating w.r.t. x, we get
$$\frac{1}{x} = \frac{dy}{dx}+1 \Rightarrow \frac{dy}{dx} = \frac{1}{x}-1 = \frac{1-x}{x}$$

12. $g(f(x)) = x$, as g is the inverse of f.
$$\Rightarrow \quad g'(f(x))\cdot f'(x) = 1$$
$$\Rightarrow \quad g'(f(x)) = \frac{1+x^2}{x^{10}}, \text{ as } g(2) = a$$
Put $x = a$
$$\Rightarrow \quad g'(f(a)) = \frac{1+a^2}{a^{10}} \Rightarrow f(a)=2$$
$$\therefore \quad g'(2) = \frac{1+a^2}{a^{10}}$$

13. Here, $u'(x) = f'(g(x))\cdot g'(x)$
$$\Rightarrow \quad u'(1) = f'(g(1))\cdot g'(1) = f'(3)\cdot g'(1)$$
$$= \left(-\frac{1}{4}\right)\cdot(-3) = \frac{3}{4}$$
and $\quad w'(x) = g'(g(x))\cdot g'(x)$
$$\Rightarrow \quad w'(1) = g'(g(1))\cdot g'(1) = g'(3)\cdot g'(1)$$
$$= \frac{3}{4}\cdot(-3) = -\frac{9}{4}$$
$$\therefore \quad u'(1)+w'(1) = \frac{3}{4}-\frac{9}{4} = -\frac{6}{4} = -\frac{3}{2}$$

14. Let $h(x) = f^2(x)+g^2(x)$
On differentiating both sides, we get
$$h'(x) = 2f(x)\cdot f'(x)+2g(x)\cdot g'(x)$$
$$= -2g'(x)\cdot f'(x)+2f'(x)\cdot g'(x) = 0$$
$\therefore$ $h(x)$ is constant.
Thus, $\quad h(10) = f^2(10)+g^2(10) = h(5)$
$$\therefore \quad f^2(10)+g^2(10) = f^2(5)+g^2(5) = f^2(5)+(f'(5))^2 = 4+4 = 8$$

15. Here, $f(0)=0$
$$f'(0) = 1$$
$$f''(0) = 2$$
$$f'''(0) = 3$$
$$\vdots \quad \vdots$$
$$f^n(0) = n$$
$$\therefore \quad f(0)+f'(0)+f''(0)+f'''(0)+\ldots+f^n(0) = 1+2+3+\ldots+n$$
$$= \frac{n(n+1)}{2}$$

16. Here, $\quad \frac{dy}{dx} = f'(f(f(x)))\cdot f'(f(x))\cdot f'(x)$
$$\therefore \quad y'(0) = f'(f(f(0)))\cdot f'(f(0))\cdot f'(0)$$
$$= f'(f(0))\cdot f'(0)\cdot 2 = f'(0)\cdot f'(0)\cdot 2 = 2\times 2\times 2 = 8$$

17. Here, $y^2 = p(x)$
Differentiating both sides, we get $2y\cdot y' = p(x)$
Again, differentiating both sides, we get
$$2yy''+2(y')^2 = p''(x)$$
On multiplying by y^2, we get
$$2y^3\cdot y''+2(y\cdot y')^2 = y^2p''(x)$$
$$\because \quad y\cdot y' = \frac{p'(x)}{2} \qquad \left[\because y\cdot y' = \frac{p'(x)}{2}\right]$$
$$\Rightarrow \quad 2y^3\cdot y''+2\cdot\frac{(p'(x))^2}{4} = y^2p''(x)$$
$$\Rightarrow \quad 2y^3\cdot y''+\frac{1}{2}\cdot(p'(x))^2 = p''(x)\cdot y^2$$
Again, differentiating both sides, we get
$$2\cdot\frac{d}{dx}\left(y^3\cdot\frac{d^2y}{dx^2}\right)+\frac{1}{2}\cdot 2p'(x)\cdot p''(x) = p'''(x)y^2+2y\cdot y'p''(x)$$
$$\Rightarrow 2\cdot\frac{d}{dx}\left(y^3\cdot\frac{d^2y}{dx^2}\right) = p'''(x)\cdot p(x)+2y\cdot y'\cdot p''(x)-p'(x)\cdot p''(x)$$
$$[\because 2y\cdot y' = p'(x)]$$
$$= p'''(x)\cdot p(x)+p'(x)\cdot p''(x)-p'(x)\cdot p''(x)$$
$$= p'''(x)\cdot p(x)$$

18. As, $\frac{dy}{dx} = f'(x)$ and $\frac{dx}{dy} = g'(y)$
$$\Rightarrow \quad g'(y) = \frac{1}{f'(x)} \qquad \text{...(i)}$$

On differentiating w.r.t. y, we get

$$g''(y)=\frac{d}{dy}\left(\frac{1}{f'(x)}\right)=\frac{d}{dx}\left(\frac{1}{f'(x)}\right)\cdot\frac{dx}{dy}$$

$$=-\frac{f''(x)}{(f'(x))^2}\cdot\frac{dx}{dy}$$

$$=-\frac{f''(x)}{(f'(x))^2}\cdot\frac{1}{f'(x)}=-\frac{f''(x)}{(f'(x))^3}$$

$$\therefore \quad g''(y)=-\frac{f''(x)}{(f'(x))^3}$$

19. As discussed in Q.18,

$$\frac{d^2x}{dy^2}=-\frac{d^2y/dx^2}{\left(\frac{dy}{dx}\right)^3}$$

$$\Rightarrow \quad \frac{d^2y}{dx^2}=-\frac{d^2x}{dy^2}\cdot\left(\frac{dy}{dx}\right)^3 \quad \text{...(i)}$$

Now, $\quad \frac{d^2y}{dx^2}+y\cdot\frac{dy}{dx}=0$

$$\Rightarrow -\frac{d^2x}{dy^2}\cdot\left(\frac{dy}{dx}\right)^3+y\left(\frac{dy}{dx}\right)=0$$

$$\Rightarrow \quad \frac{d^2x}{dy^2}-y\left(\frac{dx}{dy}\right)^2=0$$

20. $g(x+1)=\log(f(x+1))=\log x+\log(f(x))$

$\Rightarrow g(x+1)-g(x)=\log x$

On differentiating both sides, we get

$$g'(x+1)-g'(x)=\frac{1}{x}$$

Again, differentiating both sides, we get

$$g''(x+1)-g''(x)=-\frac{1}{x^2} \quad \text{...(i)}$$

$$\therefore \quad g''\left(1+\frac{1}{2}\right)-g''\left(\frac{1}{2}\right)=-4$$

$$g''\left(2+\frac{1}{2}\right)-g''\left(1+\frac{1}{2}\right)=-\frac{4}{9}$$

$$g''\left(3+\frac{1}{2}\right)-g''\left(2+\frac{1}{2}\right)=\frac{-4}{25}$$

$$\vdots \qquad \vdots \qquad \vdots$$

$$g''\left(N+\frac{1}{2}\right)-g''\left(N-\frac{1}{2}\right)=-\frac{4}{(2N-1)^2}$$

On adding, we get

$$g''\left(N+\frac{1}{2}\right)-g''\left(\frac{1}{2}\right)=-4\left(1+\frac{1}{9}+\frac{1}{25}+\ldots+\frac{1}{(2N-1)^2}\right)$$

21. Here, $f(x)=x^3+e^{x/2}$ and $g(x)=f^{-1}(x)$

$\Rightarrow \quad g'(f(x))\cdot f'(x)=1$

Putting $f(x)=1\Rightarrow x^3+e^{x/2}=1\Rightarrow x=0$

$$\therefore \quad g'(1)\cdot f'(0)=1,\ f'(x)=3x^2+\frac{1}{2}e^{x/2}$$

$$\Rightarrow \quad g'(1)\cdot\frac{1}{2}=1\Rightarrow g'(1)=2$$

22. Let $f(\theta)=\sin\alpha$, where $\alpha=\tan^{-1}\left(\frac{\sin\theta}{\sqrt{\cos 2\theta}}\right)$

$$\Rightarrow \quad \tan\alpha=\frac{\sin\theta}{\sqrt{\cos 2\theta}}$$

$$\therefore \quad \sin\alpha=\frac{\sin\theta}{\cos\theta}=\tan\theta \quad [\because \theta\in(-\pi/4,\pi/4)]$$

$$\Rightarrow \quad f(\theta)=\tan\theta \quad \therefore \frac{d(f(\theta))}{d(\tan\theta)}=1$$

23. We have, $y=\frac{\log\tan x}{\log\sin x}$

$$\Rightarrow \quad \frac{dy}{dx}=\frac{(\log\sin x)\left(\frac{\sec^2 x}{\tan x}\right)-(\log\tan x)(\cot x)}{(\log\sin x)^2}$$

$$\therefore \quad \left[\frac{dy}{dx}\right]_{x=\pi/4}=\frac{-4}{\log 2} \quad \text{[on simplification]}$$

24. We have, $y=\sum_{r=1}^{x}\tan^{-1}\left(\frac{1}{1+r+r^2}\right)=\sum_{r=1}^{x}\tan^{-1}\left[\frac{(r+1)-r}{1+(r+1)r}\right]$

$$=\sum_{r=1}^{x}[\tan^{-1}(r+1)-\tan^{-1}r]$$

$$=[\tan^{-1}2-\tan^{-1}1+\tan^{-1}3-\tan^{-1}2+\ldots+\tan^{-1}x-\tan^{-1}(x-1)+\tan^{-1}(x+1)-\tan^{-1}x]$$

$$=[\tan^{-1}(x+1)-\tan^{-1}1]$$

$$\therefore \quad \frac{dy}{dx}=\frac{1}{1+(x+1)^2}$$

25. We have, $y=\sin^{-1}\left(\frac{\sin\alpha\sin x}{1-\cos\alpha\sin x}\right)$

$$\Rightarrow \quad \frac{dy}{dx}=\frac{1}{\sqrt{1-\frac{\sin^2\alpha\sin^2 x}{(1-\cos\alpha\sin x)^2}}}\cdot\frac{d}{dx}\left(\frac{\sin\alpha\sin x}{1-\cos\alpha\sin x}\right)$$

$$\Rightarrow \quad \frac{dy}{dx}=\frac{(1-\cos\alpha\sin x)}{\sqrt{1+\cos 2\alpha\sin^2 x-2\cos\alpha\sin x}}\cdot\frac{\sin\alpha}{\cos\alpha}$$

$$\Rightarrow \quad \frac{dy}{dx}=-\frac{\frac{d}{dx}\left(-\frac{1}{1-\cos\alpha\sin x}\right)\tan\alpha\,(1-\cos\alpha\sin x)}{\{\sqrt{1+\cos 2\alpha\sin^2 x-2\sin x\cos\alpha}\}}$$

$$\Rightarrow \quad \frac{dy}{dx}=\frac{\left[\frac{-\cos\alpha\cos x}{(1-\cos\alpha\sin x)^2}\right]\sin\alpha\cos x}{(1-\cos\alpha\sin x)\sqrt{[1+\cos 2\alpha\sin^2 x-2\cos\alpha\sin x]}}$$

$$\therefore \quad \left[\frac{dy}{dx}\right]_{x=0}=\sin\alpha \Rightarrow y'(0)=\sin\alpha$$

26. We have, $f(x)=\cot^{-1}\left(\frac{x^x-x^{-x}}{2}\right)$

$$\Rightarrow \quad f'(x)=-\frac{1}{1+\left(\frac{x^x-x^{-x}}{2}\right)^2}\cdot\frac{d}{dx}\left(\frac{x^x-x^{-x}}{2}\right)$$

$$\Rightarrow \quad f'(x) = \frac{-2}{4+(x^x - x^{-x})^2} \cdot \frac{d}{dx}(x^x - x^{-x})$$

$$\Rightarrow \quad f'(x) = \frac{-2}{(x^x + x^{-x})^2} \cdot \frac{d}{dx}(e^{x \log x} - e^{-x \log x})$$

$$\Rightarrow \quad f'(x) = \frac{-2}{(x^x + x^{-x})^2}$$

$$\left\{e^{x \log x} \cdot \frac{d}{dx}(x \log x) - e^{-x \log x} \cdot \frac{d}{dx}(-x \log x)\right\}$$

$$\Rightarrow \quad f'(x) = \frac{-2}{(x^x + x^{-x})^2}\{x^x(1+\log x) + x^{-x}(1+\log x)\}$$

$$\Rightarrow \quad f'(x) = \frac{-2(1+\log x)}{(x^x + x^{-x})^2} \cdot (x^x + x^{-x}) = \frac{-2(1+\log x)}{x^x + x^{-x}}$$

$$\therefore \quad f'(1) = \frac{-2}{(1+1)} = -1$$

27. $f(g(x)) = x \Rightarrow f'(g(x))\, g'(x) = 1$

$$\Rightarrow \quad (e^{g(x)} + 1)\, g'(x) = 1$$

$$\Rightarrow \quad (e^{g(f(\log 2))} + 1)\, g'(f(\log 2)) = 1$$

$$\Rightarrow \quad (e^{\log 2} + 1)\, g'(f(\log 2)) = 1$$

$$\Rightarrow \quad g'(f(\log 2)) = \frac{1}{3}$$

28. $F'(x) = \left[f\left(\frac{x}{2}\right) \cdot f'\left(\frac{x}{2}\right) + g\left(\frac{x}{2}\right) g'\left(\frac{x}{2}\right)\right]$

Here, $\quad g(x) = f'(x)$ and $g'(x) = f''(x) = -f(x)$

So, $\quad F'(x) = f\left(\frac{x}{2}\right) g\left(\frac{x}{2}\right) - f\left(\frac{x}{2}\right) g\left(\frac{x}{2}\right) = 0$

$\therefore$ $F(x)$ is a constant function. Hence, $F(10) = 5$

29. We have, $x = \sec\theta - \cos\theta$ and $y = \sec^n\theta - \cos^n\theta$

$$\Rightarrow \quad \frac{dx}{d\theta} = \sec\theta\tan\theta + \sin\theta = \tan\theta(\sec\theta + \cos\theta)$$

and $$\frac{dy}{d\theta} = n[\sec^n\theta\tan\theta + \cos^{n-1}\theta\sin\theta]$$

$$= n\tan\theta(\sec^n\theta + \cos^n\theta)$$

$$\therefore \quad \frac{dy}{dx} = \frac{\frac{dy}{d\theta}}{\frac{dx}{d\theta}} = \frac{n\tan\theta(\sec^n\theta + \cos^n\theta)}{\tan\theta(\sec\theta + \cos\theta)}$$

$$\Rightarrow \quad \left(\frac{dy}{dx}\right)^2 = n^2 \cdot \frac{(\sec^n\theta - \cos^n\theta)^2 + 4}{(\sec\theta - \cos\theta)^2 + 4} = \frac{n^2(y^2+4)}{(x^2+4)}$$

$$\Rightarrow \quad (x^2+4)\left(\frac{dy}{dx}\right)^2 = n^2(y^2+4)$$

30. We have, $x = f(t)\cos t - f'(t)\sin t$

and $\quad y = f(t)\sin t + f'(t)\cos t$

$$\Rightarrow \quad \frac{dx}{dt} = -f(t)\sin t + f'(t)\cos t - f'(t)\cos t - f''(t)\sin t$$

$$= -[f(t) + f''(t)]\sin t$$

and $$\frac{dy}{dt} = f(t)\cos t + f'(t)\sin t - f'(t)\sin t + f''(t)\cos t$$

$$= [f(t) + f''(t)]\cos t$$

$$\therefore \quad \left(\frac{dx}{dt}\right)^2 + \left(\frac{dy}{dt}\right)^2 = [f(t) + f''(t)]^2$$

31. Note that in y highest degree of x is 4 and therefore $\frac{d^3y}{dx^3}$ is a linear function of x, which is satisfied only in (d).

32. Exponent of $x = \frac{[l^2 - m^2 + m^2 - n^2 + n^2 - l^2]}{(l-m)(m-n)(n-l)}$

$$\Rightarrow \quad y = x^0 = 1 \Rightarrow y = 1 \Rightarrow y' = 0$$

33. $y = (A + Bx)e^{mx} + (m-1)^{-2} \cdot e^x$

$$\Rightarrow y \cdot e^{-mx} = (A + Bx) + (m-1)^{-2} \cdot e^{(1-m)x}$$

$$\Rightarrow e^{-mx} \cdot y_1 - mye^{-mx} = B - (m-1)^{-1} \cdot e^{-(m-1)x}$$

$$\Rightarrow e^{-mx} \cdot y_2 - y_1 e^{-mx} \cdot m - m[e^{-mx} \cdot y_1 - ye^{-mx} \cdot m] = e^{-(m-1)x}$$

$$\Rightarrow e^{-mx} \cdot y_2 - 2my_1 e^{-mx} + m^2 y \cdot e^{-mx} = e^{-(m-1)x}$$

$$\therefore \quad y_2 - 2my_1 + m^2 y = e^x$$

34. We have, $f(x) = -\frac{x^3}{3} + x^2 \sin 6 - x \sin 4 \cdot \sin 8$

$$- 5\sin^{-1}((a-4)^2 + 1)\ (a = 4)$$

$$\Rightarrow f'(x) = -x^2 + 2x\sin 6 - \sin 4 \sin 8$$

$$\Rightarrow f'(\sin 8) = -\sin^2 8 + 2\sin 6 \sin 8 - \sin 4 \sin 8$$

$$= \sin 8[-\sin 8 + 2\sin 6 - \sin 4]$$

$$= -\sin 8[\sin 8 + \sin 4 - 2\sin 6]$$

$$= -\sin 8[2\sin 6\cos 2 - 2\sin 6]$$

$$= 2\sin 8 \sin 6[1 - \cos 2] < 0$$

sin 8

sin 4 | sin 6

35. $f(g(x)) = x \Rightarrow f'(g(x)) \cdot g'(x) = 1;\ f'(g(a)) \cdot g'(a) = 1;\ f'(b) \cdot 2 = 1$

$$\Rightarrow \quad f'(b) = \frac{1}{2}$$

36. Put $\cos\phi = \frac{2}{\sqrt{13}}; \sin\phi = \frac{3}{\sqrt{13}}; \tan\phi = \frac{3}{2}$

$$y = \cos^{-1}\{\cos(x+\phi)\} + \sin^{-1}\{\cos(x-\phi)\}$$

$$= \cos^{-1}\{\cos(x+\phi)\} + \frac{\pi}{2} - \cos^{-1}\{\cos(\phi - x)\}$$

$$= x + \phi + \frac{\pi}{2} - \phi + x$$

$$y = 2x + \frac{\pi}{2}; z = \sqrt{1+x^2}$$

Now, $$\frac{dy}{dz} = \frac{\frac{dy}{dx}}{\frac{dz}{dx}} = \frac{2\sqrt{1+x^2}}{x}$$

$$\therefore \quad \left(\frac{dy}{dz}\right)_{x=\frac{3}{4}} = \frac{10}{3}$$

37. $f(x) = \sqrt{x + 2\sqrt{2x-4}} + \sqrt{x - 2\sqrt{2x-4}}$

$$\therefore \quad f(x) = \sqrt{(\sqrt{x-2} + \sqrt{2})^2} + \sqrt{(\sqrt{x-2} - \sqrt{2})^2}$$

$$= |\sqrt{x-2} + \sqrt{2}| + |\sqrt{x-2} - \sqrt{2}|$$

For $\sqrt{x-2}$ to exist, $x \ge 2$

Also, $\sqrt{x-2} + \sqrt{2} > 0$ [true]

But $\sqrt{x-2} - \sqrt{2} \ge 0$ only, if $x \ge 4 < 0$ only, if $x < 4$

Now, $f(x)$ becomes

$$f(x)=\begin{cases}\sqrt{x-2}+\sqrt{2}-\sqrt{x-2}+\sqrt{2}, & \text{for } 2\le x<4\\ \sqrt{x-2}+\sqrt{2}+\sqrt{x-2}-\sqrt{2}, & \text{for } x\ge 4\end{cases}$$

$$\therefore\quad f(x)=\begin{cases}2\sqrt{2}, & \text{for } 2\le x<4\\ 2\sqrt{x-2}, & \text{for } 4\le x<\infty\end{cases}$$

$\because$ f is continuous in $[2, 4)\cup[4, \infty)$.

$$\therefore\quad f'(x)=\begin{cases}0, & 2\le x<4\\ \dfrac{1}{\sqrt{x-2}}, & 4\le x<\infty\end{cases}$$

$$\therefore\quad f'(102^+)=\frac{1}{\sqrt{102-2}}=\frac{1}{10}$$

$$\therefore\quad 10f'(102^+)=1$$

38. Here, $y=2\ln(1+\cos x)\Rightarrow y_1=\dfrac{-2\sin x}{1+\cos x}$

$$y_2=-2\left[\frac{(1+\cos x)\cos x-\sin x(-\sin x)}{(1+\cos x)^2}\right]$$

$$=-2\left[\frac{\cos x+1}{(1+\cos x)^2}\right]=\frac{-2}{(1+\cos x)}$$

$$\therefore\ 2e^{-y/2}=2\cdot e^{\frac{-\ln(1+\cos x)^2}{2}}=\frac{2}{(1+\cos x)}$$

$$\therefore\quad y_2+\frac{2}{e^{y/2}}=0$$

39. $y=\dfrac{(a+x)+\sqrt{a-x}\cdot\sqrt{a+x}}{(a-x)+\sqrt{a-x}\cdot\sqrt{a+x}}$

$$y=\frac{\sqrt{(a+x)}\,(\sqrt{a+x}+\sqrt{a-x})}{\sqrt{a-x}\,(\sqrt{a+x}+\sqrt{a-x})}=\left(\frac{a+x}{a-x}\right)^{1/2}$$

$$\frac{dy}{dx}=\frac{1}{2}\sqrt{\frac{a-x}{a+x}}\left(\frac{(a-x)+(a+x)}{(a-x)^2}\right)=\frac{1}{2}\frac{\sqrt{a-x}}{\sqrt{a+x}}\times\frac{2a}{(a-x)^2}$$

$$\therefore\quad\left(\frac{dy}{dx}\right)_{x=0}=\frac{1}{a}$$

40. We have, $u(x)=7v(x)$

$$\Rightarrow\quad u'(x)=7v'(x)\Rightarrow p=7$$

Again, $\dfrac{u(x)}{v(x)}=7\Rightarrow\left(\dfrac{u(x)}{v(x)}\right)'=0$

$$\Rightarrow\quad q=0$$

Now, $\dfrac{p+q}{p-q}=\dfrac{7+0}{7-0}=1$

41. For $x>1$, we have $f(x)=|\log|x||=\log x$

$$\Rightarrow\quad f'(x)=\frac{1}{x}$$

For $x<-1$, we have

$$f(x)=|\log|x||=\log(-x)$$

$$\Rightarrow\quad f'(x)=\frac{1}{x}$$

For $0<x<1$, we have

$$f(x)=|\log|x||=-\log x$$

$$f'(x)=-\frac{1}{x}$$

For $-1<x<0$, we have

$$f(x)=-\log(-x)$$

$$f'(x)=-\frac{1}{x}$$

$$\text{Hence,}\quad f'(x)=\begin{cases}\dfrac{1}{x}, & |x|>1\\ -\dfrac{1}{x}, & |x|<1\end{cases}$$

42. We have, $f(x)=\cos\{x+3x+\ldots+(2n-1)x\}$
$+\,i\sin\{x+3x+5x+\ldots+(2n-1)x\}$

$$\Rightarrow\quad f(x)=\cos n^2x+i\sin n^2x$$

$$\Rightarrow\quad f'(x)=-n^2(\sin n^2x)+n^2(i\cos n^2x)$$

$$\Rightarrow\quad f''(x)=-n^4\cos n^2x-n^4i\sin n^2x$$

$$\Rightarrow\quad f''(x)=-n^4(\cos n^2x+i\sin n^2x)$$

$$\Rightarrow\quad f''(x)=-n^4f(x)$$

43. We have, $f(x)=x^n$

$$\Rightarrow\quad f(x+y)=(x+y)^n\Rightarrow f'(x+y)=n(x+y)^{n-1}$$

Also, $f'(x)=nx^{n-1}$ and $f'(y)=ny^{n-1}$

$$\therefore\quad f'(x+y)=f'(x)+f'(y)$$

$$\Rightarrow\ n(x+y)^{n-1}=n\cdot x^{n-1}+n\cdot y^{n-1}$$

$$\Rightarrow\ (x+y)^{n-1}=x^{n-1}+y^{n-1}\qquad\text{...(i)}$$

For $n-1>1$, we find that LHS of Eq. (i) is greater than the RHS. So, we must have $n-1\le 1$, i.e. $n-1=0$ or $n-1=1$.

Hence, $n=1$ or $n=2$

44. For $2<x<3$, we have $[x]=2$

$$\therefore\quad f(x)=\sin\left(\frac{2\pi}{3}-x^2\right)$$

$$\Rightarrow\quad f'(x)=-2x\cos\left(\frac{2\pi}{3}-x^2\right)$$

$$\Rightarrow\quad f'(\sqrt{\pi/3})=-2\sqrt{\pi/3}\cos\pi/3=-\sqrt{\pi/3}$$

45. We have, $u=e^x\sin x$

$$\Rightarrow\quad\frac{du}{dx}=e^x\sin x+e^x\cos x=u+v$$

and $v=e^x\cos x$

$$\Rightarrow\quad\frac{dv}{dx}=e^x\cos x+e^x\sin x=v-u$$

$$\therefore\quad v\frac{du}{dx}-u\frac{dv}{dx}=v(u+v)-u(v-u)=u^2+v^2$$

$$\frac{d^2u}{dx^2}=\frac{du}{dx}+\frac{dv}{dx}=u+v+v-u=2v$$

and $\dfrac{d^2v}{dx^2}=\dfrac{dv}{dx}-\dfrac{du}{dx}=(v-u)-(v+u)=-2u$

46. We have, $f(x) = \log_x \{\ln(x)\} = \dfrac{\ln\{\ln(x)\}}{\ln(x)}$

$$\therefore \quad f'(x) = \frac{\ln(x)\cdot\dfrac{1}{\ln(x)}\cdot\dfrac{1}{x} - \ln\{\ln(x)\}\dfrac{1}{x}}{\{\ln(x)\}^2} = \frac{1-\ln\{\ln(x)\}}{x\{\ln(x)\}^2}$$

$$\Rightarrow \quad f'(e) = \frac{1-\ln\{\ln(e)\}}{e\{\ln(e)\}^2} = \frac{1-\ln(1)}{e} = \frac{1}{e} \text{ or } e^{-1}$$

47. We have, $[f(x)]^n = f(nx)$ for all x

$$\Rightarrow \quad n[f(x)]^{n-1} f'(x) = nf'(nx)$$

$$\Rightarrow \quad n[f(x)]^n f'(x) = nf(x)f'(nx)$$

[multiplying both sides by $f(x)$]

$$\Rightarrow \quad nf(nx)f'(x) = nf(x)f'(nx) \qquad [\because [f(x)]^n = f(nx)]$$

$$\Rightarrow \quad f(nx)f'(x) = f(x)f'(nx)$$

48. Since, $f(x)$ is an odd differentiable function defined on R.

Therefore, $f(-x) = -f(x)$ for all $x \in R$

Differentiating both the sides w.r.t. x, we get

$$-f'(-x) = -f'(x) \text{ for all } x \in R$$

$$\Rightarrow \quad f'(-x) - f'(x) = 0 \text{ for all } x \in R$$

$$\Rightarrow \quad f'(-3) = f'(3) = -2$$

Aliter

We know that the derivative of a differentiable odd function is an even function. Therefore, $f'(x)$ is an even function.

Hence, $f'(-3) = f'(3) = -2$

49. We have, $y = \sqrt{x + \sqrt{y + \sqrt{x + \sqrt{y + \ldots \infty}}}}$

$$\Rightarrow \quad y^2 = x + \sqrt{y + \sqrt{x + \sqrt{y + \ldots \infty}}}$$

$$\Rightarrow \quad y^2 = x + \sqrt{y+y} \Rightarrow (y^2 - x)^2 = 2y$$

Differentiating both the sides w.r.t. x, we get

$$2(y^2 - x)\left(2y\frac{dy}{dx} - 1\right) = 2\frac{dy}{dx}$$

$$\Rightarrow \quad \frac{dy}{dx} = \frac{y^2 - x}{2y^3 - 2xy - 1}$$

50. We have, $f(x) = |\cos x - \sin x|$

$$\Rightarrow \quad f(x) = \begin{cases} \cos x - \sin x, & \text{for } 0 < x \le \dfrac{\pi}{x} \\ \sin x - \cos x, & \text{for } \dfrac{\pi}{4} < x < \dfrac{\pi}{2} \end{cases}$$

Clearly, $\left(\text{LHD at } x = \dfrac{\pi}{4}\right) = \left\{\dfrac{d}{dx}(\cos x - \sin x)\right\}_{\text{at } x = \frac{\pi}{4}}$

$$= (-\sin x - \cos x)_{x = \frac{\pi}{4}} = -\sqrt{2}$$

and $\left(\text{RHD at } x = \dfrac{\pi}{4}\right) = \left\{\dfrac{d}{dx}(\sin x - \cos x)\right\}_{\text{at } x = \frac{\pi}{4}}$

$$= (\cos x + \sin x)_{x = \frac{\pi}{4}} = \sqrt{2}$$

$$\therefore \quad \left(\text{LHD at } x = \frac{\pi}{4}\right) \ne \left(\text{RHD at } x = \frac{\pi}{4}\right)$$

Thus, $f'\left(\dfrac{\pi}{4}\right)$ doesn't exist.

51. We have, $f(x) = x^2 + xg'(1) + g''(2)$

and $g(x) = x^2 + xf'(2) + f''(3)$

$\Rightarrow \quad f'(x) = 2x + g'(1)$

and $g'(x) = 2x + f'(2)$ ...(i)

Putting $x = 1$ in Eq (i), we get

$f'(1) = 2 + g'(1)$

and $g'(1) = 2 + f'(2)$

$\Rightarrow \quad f'(1) = 4 + f'(2)$

Putting $x = 2$ in Eq. (i), we get

$f'(2) = 4 + g'(1)$

and $g'(2) = 4 + f'(2)$

$\Rightarrow \quad g'(2) = 4 + 4 + g'(1) = 8 + g'(1)$

Differentiating Eq. (i) w.r.t. x, we get

$f''(x) = 2$

and $g''(x) = 2$ for all x

$\Rightarrow \quad f''(3) = 2$

and $g''(2) = 2$

$\Rightarrow g''(2) + f''(3) = 2 + 2 = 4$

52. We have, $f(x) = x^n$

$$f^r(x) = n(n-1)(n-2)\ldots\{n-(r-1)\}x^{n-r}$$

$$\Rightarrow \quad f^r(x) = \frac{n!}{(n-r)!}x^{n-r}$$

$$\Rightarrow \quad f^r(1) = \frac{n!}{(n-r)!}$$

$$\therefore \; f(1) - \frac{f'(1)}{1!} + \frac{f''(1)}{2!} - \frac{f'''(1)}{3!} + \ldots + \frac{(-1)^n f^n(1)}{n!}$$

$$= \sum_{r=0}^{n}(-1)^r \frac{f^r(1)}{r!}, \text{ where } f^0(1) = f(1)$$

$$= \sum_{r=0}^{n}(-1)^r \frac{n!}{(n-r)!\,r!} = \sum_{r=0}^{n}(-1)^r\, {}^nC_r = 0$$

53. We have, $y + \log(1 + x) = 0$

$$\Rightarrow \quad \frac{dy}{dx} + \frac{1}{(1+x)} = 0$$

$$\Rightarrow \quad y' = -\frac{1}{(x+1)} \qquad \text{...(i)}$$

Also, $\log(1 + x) = -y$

$$\Rightarrow \quad 1 + x = e^{-y} \Rightarrow 1 = e^{-y}\cdot(-y')$$

$$\Rightarrow \quad e^y = -y'$$

or $\quad y' + e^y = 0$

54. As, $y = 2^{3^x}$

$$\Rightarrow \quad \frac{dy}{dx} = 2^{3^x} \cdot \log 2 \cdot \left(\frac{d}{dx}(3^x)\right)$$

$$= 2^{3^x} \cdot \log 2 \cdot 3^x \cdot \log 3$$

$$= 2^{3^x} \cdot 3^x \cdot \log 2 \cdot \log 3 \quad \text{...(i)}$$

Also, $y = 2^{3^x}$ and $\log_2 y = 3^x$

∴ Eq. (i) becomes,

$$\frac{dy}{dx} = y \cdot (\log_2 y) \cdot \log 3 \cdot \log 2$$

55. Here, $f(x) = x^2 + 3x - 3$

$$g(f(x)) = x$$

$$\Rightarrow \quad g'(f(x)) \cdot f'(x) = 1, f'(x) = 2x + 3$$

$$\Rightarrow \quad g'(f(x)) = \frac{1}{2x + 3}$$

$$\therefore \quad g'(1) = \frac{1}{2g(1) + 3} \quad \text{...(i)}$$

and $\quad g^{-1}(x) = f(x)$

When $f(x) = 1$,

then $g^{-1}(x) = 1 \Rightarrow x = g(1)$

Also, $\quad f(x) = 1$

$$\Rightarrow \quad 1 = x^2 + 3x - 3$$

$$\Rightarrow \quad x^2 + 3x - 4 = 0$$

$$\Rightarrow \quad (x + 4)(x - 1) = 0$$

$$\Rightarrow \quad x = 1 \quad [\text{as } x > 0]$$

$$\therefore \quad g'(1) = \frac{1}{2(1) + 3} = \frac{1}{5}$$

56. Here, $x^3 - 2x^2y^2 + 5x + y - 5 = 0$

On differentiating, we get

$3x^2 - 4xy^2 - 4x^2y\, dy/dx + 5 + dy/dx = 0$

$$\Rightarrow \quad y' = \frac{3x^2 - 4xy^2 + 5}{4x^2y - 1}$$

$$\therefore \quad y'(1) = \frac{3 - 4 + 5}{4 - 1} = \frac{4}{3}$$

Also, $y''(x)$

$$= \frac{[(6x - 4y^2 - 8xyy')](4x^2y - 1) - (8xy + 4x^2y')(3x^2 - 4xy^2 + 5)]}{(4x^2y - 1)^2}$$

$$\Rightarrow \quad y''(1) = \frac{\left(6 - 4 - 8 \cdot \frac{4}{3}\right) \cdot (4 - 1) - \left(8 + 4 \cdot \frac{4}{3}\right) \cdot (3 - 4 + 5)}{(4 - 1)^2}$$

$$= -8\frac{22}{27}$$

57. Here, $y = \sqrt{x + y}$

$$\Rightarrow \quad y^2 = x + y$$

On differentiating w.r.t. x, we get

$$2y \cdot \frac{dy}{dx} = 1 + \frac{dy}{dx}$$

$$\Rightarrow \quad \frac{dy}{dx} = \frac{1}{2y - 1}$$

Also, $\quad y = \frac{x}{y} + 1$

$$\Rightarrow \quad \frac{dy}{dx} = \frac{y \cdot 1 - x \cdot \frac{dy}{dx}}{y^2}$$

$$\Rightarrow \quad (y^2 + x)\frac{dy}{dx} = y$$

$$\Rightarrow \quad \frac{dy}{dx} = \frac{y}{y^2 + x} \quad [\because y^2 = x + y]$$

$$\Rightarrow \quad \frac{dy}{dx} = \frac{y}{(x + y) + x} = \frac{y}{2x + y}$$

58. We have, $y = x^{(\ln x)^{\ln (\ln x)}}$

$$\Rightarrow \quad \ln y = (\ln x)^{\ln (\ln x)} \ln x \quad \text{...(i)}$$

$$\Rightarrow \ln (\ln y) = \ln (\ln x) \cdot \ln (\ln x) + \ln (\ln x)$$

$$\Rightarrow \frac{1}{\ln y} \cdot \frac{1}{y} \cdot \frac{dy}{dx} = \frac{2 \ln (\ln x)}{\ln x} \cdot \frac{1}{x} + \frac{1}{x \ln x}$$

$$= \frac{2 \ln (\ln x) + 1}{x \ln x}$$

$$\therefore \quad \frac{dy}{dx} = \frac{y}{x} \cdot \frac{\ln y}{\ln x} (2 \ln (\ln x) + 1)$$

Substituting the value of $\ln y$ from Eq. (i), we get

$$\frac{dy}{dx} = \frac{y}{x} (\ln x)^{\ln (\ln x)} (2 \ln (\ln x) + 1)$$

59. (a) We have, $y = \sin^{-1} 2x\sqrt{1 - x^2}$

$$\Rightarrow \quad \frac{dy}{dx} = \frac{2 - 4x^2}{\sqrt{1 - 4x^2 + 4x^4}}$$

Clearly, it is not defined at $x = \frac{1}{\sqrt{2}}$.

Not derivable at $x = \frac{1}{\sqrt{2}}$, check with $x = \sin \theta$ or direct differentiation.

(b) $$g'(x) = \frac{1}{\sqrt{1 - \left(\frac{2 \cdot 2^x}{1 + 2^{2x}}\right)^2}} \cdot \frac{(1 + 4^x)(2^{x+1}) \ln 2 - 2^{x+1} \cdot 4^x \cdot \ln 4}{(1 + 4^x)^2}$$

$$= \frac{1 + 2^{2x}}{\sqrt{1 + 4^{2x} - 2 \cdot 2^{2x}}} \times \frac{2^{x+1} \ln 2\,(1 - 4^x)}{(1 + 4^x)^2}$$

$$= \frac{2^{x+1} \ln 2 (1 - 4^x)}{\left|1 - 2^{2x}\right| (1 + 4^x)}$$

Clearly, it is not defined, when $1 - 2^x = 0$

i.e. $\quad$ at $x = 0$

$\Rightarrow$ Not derivable at $x = 0$

(c) $h(x)=\sin^{-1}\left(\frac{1-x^2}{1+x^2}\right)=\frac{\pi}{2}-\cos^{-1}\left(\frac{1-x^2}{1+x^2}\right)$

$$=\begin{cases}\frac{\pi}{2}-2\tan^{-1}x, & x\ge 0\\ \frac{\pi}{2}+2\tan^{-1}x, & x<0\end{cases}$$

$$\therefore \quad h'(x)=\begin{cases}\frac{-2}{1+x^2}, & x>0\\ \text{not derivable}, & x=0\\ \frac{2}{1+x^2}, & x<0\end{cases}$$

$\therefore$ Not differentiable at $x=0$.

(d) $k(x)=\sin^{-1}(\cos x)=\frac{\pi}{2}-\cos^{-1}(\cos x)$

and graph for $\cos^{-1}(\cos x)$, is shown as

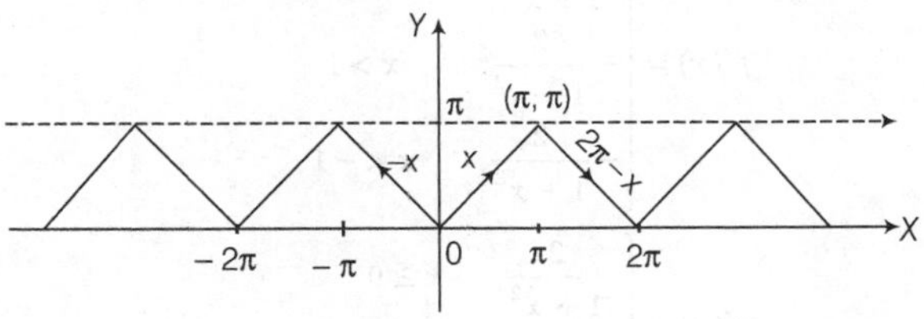

$\therefore k(x)$ is not differentiable at $x=0$.

60. $f(x)=\frac{\sqrt{(\sqrt{x-1})^2+1-2\sqrt{x-1}}}{\sqrt{x-1}-1}\cdot x$

$$=\frac{|\sqrt{x-1}-1|}{\sqrt{x-1}-1}\cdot x=\begin{cases}-x, & \text{if } x\in[1,2)\\ x, & \text{if } x\in(2,\infty)\end{cases}$$

$$\Rightarrow \quad f'(x)=\begin{cases}-1, & \text{if } x\in[1,2)\\ 1, & \text{if } x\in(2,\infty)\end{cases}$$

$\therefore f'(10)=1$ and $f'\left(\frac{3}{2}\right)=-1$

61. $2^x+2^y=2^{x+y}$

Differentiating both sides, we get

$$2^x\log 2+2^y\log 2\frac{dy}{dx}=2^{x+y}\cdot\log 2\left(1+\frac{dy}{dx}\right)$$

$$\Rightarrow \quad \log 2\cdot 2^y(2^x-1)\frac{dy}{dx}=2^x(1-2^y)\log 2$$

$$\Rightarrow \quad \frac{dy}{dx}=\frac{2^x(1-2^y)}{2^y(2^x-1)} \quad \text{...(i)}$$

Also, $\quad 2^x=2^y(2^x-1)$

$$\therefore \quad \frac{dy}{dx}=(1-2^y) \quad \text{...(ii)}$$

when $\quad 2^y=2^x(2^y-1)$ [from Eq. (i)]

$$\Rightarrow \quad \frac{dy}{dx}=\frac{-1}{2^x-1}=\frac{1}{1-2^x} \quad \text{...(iii)}$$

As, $\quad 2^x=2^y(2^x-1)$ and $2^y=2^x(2^y-1)$

$$\therefore \quad -\frac{2^y}{2^x}=\frac{2^x(1-2^y)}{2^y(2^x-1)}$$

Substituting this in Eq. (i), we get

$$\frac{dy}{dx}=-\frac{2^y}{2^x}$$

62. $f(x)=(x^2+bx+c)e^x$

$$\therefore \quad f'(x)=\{x^2+(b+2)x+(b+c)\}e^x$$

$f(x)>0$ iff $D=b^2-4c<0$

Now, $f'(x)>0$ iff $D'=(b+2)^2-4(b+c)$

$$=D+4<0$$

Thus, for $f'(x)>0$, $D+4<0$ holds.

$$\Rightarrow \quad D<0$$

$$\Rightarrow \quad f(x)>0$$

63. We have, $\sqrt{y+x}+\sqrt{y-x}=c$

On differentiating w.r.t. x, we get

$$\frac{dy}{dx}\left(\frac{1}{\sqrt{y+x}}+\frac{1}{\sqrt{y-x}}\right)=\frac{1}{\sqrt{y-x}}-\frac{1}{\sqrt{y+x}}$$

$$\Rightarrow \quad \frac{dy}{dx}=\frac{\sqrt{y+x}-\sqrt{y-x}}{(\sqrt{y+x}+\sqrt{y-x})}$$

$$\frac{dy}{dx}=\frac{(y+x)-(y-x)}{(\sqrt{y+x}+\sqrt{y-x})^2}=\frac{2x}{c^2}$$

$$\therefore \quad \frac{dy}{dx}=\frac{x}{y+\sqrt{y^2-x^2}}$$

Now, $\quad \frac{dy}{dx}=\frac{(\sqrt{y+x}-\sqrt{y-x})^2}{(y+x)-(y-x)}=\frac{y-\sqrt{y^2-x^2}}{x}$

64. $y=\tan x\cdot\tan 2x\cdot\tan 3x$ ...(i)

Differentiating both sides, we get

$$\frac{dy}{dx}=3\sec^2 3x\cdot\tan x\cdot\tan 2x+\sec^2 x\cdot\tan 2x\cdot\tan 3x$$
$$+2\sec^2 2x\cdot\tan x\cdot\tan 3x \quad \text{...(ii)}$$

Taking log on both sides of Eq. (i), we get

$$\log y=\log\tan x+\log\tan 2x+\log\tan 3x$$

Differentiating both sides, we get

$$\frac{1}{y}\cdot\frac{dy}{dx}=2\operatorname{cosec}2x+4\operatorname{cosec}4x+6\operatorname{cosec}6x$$

$$\therefore \quad \frac{dy}{dx}=2y(\operatorname{cosec}2x+2\operatorname{cosec}4x+3\operatorname{cosec}6x) \quad \text{...(iii)}$$

$$\tan(3x)=\tan(2x+x)=\frac{\tan 2x+\tan x}{1-\tan 2x\tan x}$$

$$\Rightarrow \tan 3x\cdot\tan 2x\cdot\tan x=\tan 3x-\tan 2x-\tan x$$

Differentiating both sides, we get

$$\frac{dy}{dx}=3\sec^2 3x-2\sec^2 2x-\sec^2 x$$

65. Here, $\quad f(x)=\frac{x}{(x-1)(x+1)}=\frac{1}{2}\left[\frac{(x+1)+(x-1)}{(x-1)(x+1)}\right]$

$$f(x)=\frac{1}{2}\left[\frac{1}{x-1}+\frac{1}{x+1}\right]$$

$$\Rightarrow \quad f'(x)=\frac{1}{2}\left[-\frac{1}{(x-1)^2}-\frac{1}{(x+1)^2}\right]$$

$$f''(x) = \frac{1}{2}\left[\frac{2}{(x-1)^3} + \frac{2}{(x+1)^3}\right] > 0, \text{ for all } x > 1$$

$\therefore f(x)$ is concave up.

66. Let $x = \tan\theta$, where $-\frac{\pi}{2} < \theta < \frac{\pi}{2}$.

$\Rightarrow \quad f(x) = \sin^{-1}(\sin 2\theta)$

$$\therefore f(x) = \begin{cases} \pi - 2\theta, & \frac{\pi}{2} < 2\theta < \pi \\ 2\theta, & -\frac{\pi}{2} \le 2\theta \le \frac{\pi}{2} \\ -\pi - 2\theta, & -\pi < 2\theta < -\frac{\pi}{2} \end{cases} = \begin{cases} \pi - 2\tan^{-1} x, & x > 1 \\ 2\tan^{-1} x, & -1 \le x \le 1 \\ -\pi - 2\tan^{-1} x, & x < -1 \end{cases}$$

$$\therefore f'(x) = \begin{cases} \frac{-2}{1+x^2}, & x > 1 \\ \frac{2}{1+x^2}, & -1 \le x \le 1 \\ \frac{-2}{1+x^2}, & x < -1 \end{cases} \Rightarrow f'(2) = \frac{-2}{1+4} = \frac{-2}{5}$$

67. As, $\quad (f(g(x))' = f'(g(x)) \cdot g'(x)$

$(f(g(0)))' = f'(g(0)) \cdot g'(0) = c$

$\Rightarrow \quad f'(0) \cdot g'(0) = c \Rightarrow g'(0) = \frac{c}{b}$

68. $\lim_{x \to 0} \frac{g(x)\cos x - g(0)}{\sin x} = \lim_{x \to 0} \frac{g'(x)\cos x - g(x)\sin x}{\cos x} = 0$

Now, $\quad f(x) = g(x)\sin x$

$f'(x) = g(x)\cos x + g'(x)\sin x$

$\therefore \quad f'(0) = 0$

On differentiating both sides, we get

$f''(x) = g'(x)\cos x - g(x)\sin x + g''(x)\sin x + g'(x)\cos x$

$= g''(x)\sin x + 2g'(x)\cos x - g(x)\sin x$

Also, $f'(0) = g'(0)$

$\therefore$ Both statements are correct but Statement II is not the correct explanation of Statement I.

69. Statement II is correct.

$\therefore \quad y = \sin^{-1}(3x - 4x^3) = 3\sin^{-1} x$, when $-\frac{1}{2} \le x \le \frac{1}{2}$

$\Rightarrow \quad \frac{dy}{dx} = \frac{3}{\sqrt{1-x^2}}$ only when $\frac{-1}{2} \le x \le \frac{1}{2}$

$\therefore$ Both statements are true and Statement II is the correct explanation of Statement I.

70. Clearly, Statement I is false.

$$\cos^{-1}\left(\frac{1-x^2}{1+x^2}\right) = \begin{cases} 2\tan^{-1} x, & x \ge 0 \\ -2\tan^{-1} x, & x < 0 \end{cases}$$

$\therefore$ Statement II is correct but Statement I is incorrect.

71. Clearly, Statement II is true.

$$\therefore \quad \frac{dy}{dx} = \begin{cases} 1, & x \notin \text{Integer} \\ \text{does not exist}, & x \in \text{Integer} \end{cases}$$

Thus, Statement I is incorrect but Statement II is correct.

72. We have, $f(x)$ is a continuous function $f : R \to Q$.

$f : R \to 0$, iff $f(x)$ is constant function.

$f(5) = 3$

$f(x) = 3$

$f'(x) = 0$

So, Statements I and II both are correct and Statement II is the correct explanation of Statement I.

73. Let $\quad f(x) = \sin^{-1}\frac{2x}{1+x^2}, \quad g(x) = \cos^{-1}\left(\frac{1-x^2}{1+x^2}\right)$

$$f(x) = \begin{cases} 2\tan^{-1} x, & -1 \le x \le 1 \\ \pi - 2\tan^{-1} x, & x > 1 \\ -\pi - 2\tan^{-1} x, & x < -1 \end{cases}$$

$$g(x) = \begin{cases} 2\tan^{-1} x, & x \ge 0 \\ -2\tan^{-1} x, & x < 0 \end{cases}$$

$$f'(x) = \begin{cases} \frac{2}{1+x^2}, & -1 \le x \le 1 \\ -\frac{2}{1+x^2}, & x > 1 \\ -\frac{2}{1+x^2}, & x < -1 \end{cases}$$

$$g'(x) = \begin{cases} \frac{2}{1+x^2}, & x \ge 0 \\ -\frac{2}{1+x^2}, & x < 0 \end{cases}$$

$\frac{f'(x)}{g'(x)} = 1, \ 0 < x < 1$

$\therefore \quad f'(x) \ne g'(x), -1 \le x \le 1$

Statement I is correct and Statement II is incorrect.

74. Given, $\quad f(x + y^3) = f(x) + f(y^3), \forall \ x, y \in R$

Put $x = y = 0$, we get

$f(0 + 0) = f(0) + f(0) \Rightarrow f(0) = 0$

Now, put $y = -x^{1/3}$, we get

$f(0) = f(x) + f(-x)$

$\Rightarrow \quad f(x) + f(-x) = 0$

$\Rightarrow f(x)$ is an odd function.

$\Rightarrow f'(x)$ is an even function.

$\Rightarrow \quad f'(2) = f'(-2) = a$

Sol. (Q. Nos. 75 to 77)

Consider, $f(x + y) - f(x) = f(y) - 1 + 2xy$

$\Rightarrow \quad f(0 + 0) - f(0) = f(0) - 1 + 0$

$\Rightarrow \quad f(0) = 1$ and $f'(0) = 1$ [given]

Also, $f'(x) = \lim_{h \to 0} \frac{f(x+h) - f(x)}{h} = \lim_{h \to 0}\left[\frac{f(h) - 1}{h} + \frac{2xh}{h}\right]$

$f'(x) = f'(0) + 2x$

$\therefore \quad f'(x) = 2x + 1$

Integrating both sides, $f(x) = x^2 + x + c$

As, $\quad f(0) = 1 \Rightarrow c = 1$

$\therefore \quad f(x) = x^2 + x + 1$

Thus, domain of $\log(f(x)) = \log(x^2 + x + 1)$ is $x \in R$

Range of $x^2 + x + 1 \geq 3/4$

$\therefore$ Range of $\log_{3/4}(x^2 + x + 1) \leq 1$

$\Rightarrow$ Range $\in (-\infty, 1]$

and $\quad g(0) = \dfrac{g(0) + g(0)}{k} \Rightarrow 2g(0) = kg(0)$

$\Rightarrow \quad g(0) = 0 \qquad [\because k \neq 2]$

$$g'(x) = \lim_{h \to 0} \frac{g\left(\dfrac{x+h}{1}\right) - g(x)}{h} = \lim_{h \to 0} \frac{g(x) + g(h) - g(x)}{h}$$

$$= \lim_{h \to 0} \frac{g(h) - g(0)}{h} = g'(0) = \lambda$$

$\therefore \quad g(x) = \lambda x + c$

As $\quad g(0) = 0 \Rightarrow c = 0$

$\Rightarrow \quad g(x) = \lambda x$

$\Rightarrow \quad \lambda x = x^2 + x + 1 \Rightarrow x^2 + x(1 - \lambda) + 1 = 0$

$\because \quad D = 0$

$\therefore \quad (1 - \lambda)^2 - 4 = 0 \Rightarrow \lambda = 3, -1$

75. (c) **76.** (a) **77.** (c)

Sol. (Q. Nos. 78 to 80)

$$\text{LHD} = \lim_{h \to 0^-} \frac{f(a-h) - f(a)}{h}$$

$$= \lim_{h \to 0^-} \frac{-f(a-h) + f(a)}{h} = \lim_{h \to 0^-} \frac{f(a-h) - f(a)}{-h}$$

$$\text{RHD} = \lim_{h \to 0^-} \frac{f'(a) + f'(-a+h)}{h} = \lim_{h \to 0^-} \frac{f'(a) + f'(h-a)}{h}$$

Also, $\displaystyle \lim_{h \to 0^-} \frac{f(-x) - f(-x-h)}{h} = \lim_{h \to 0^-} \frac{f(-x-h) - f(-x)}{-h}$

$$= f'(-x) \qquad \text{...(i)}$$

and $\displaystyle \lim_{h \to 0} \frac{f(x) - f(x-h)}{-h} = -f'(x) \qquad \text{...(ii)}$

From Eqs. (i) and (ii), $f'(x)$ is odd function and hence $f(x)$ is even function.

78. (a) **79.** (a) **80.** (b)

Sol. (Q. Nos. 81 to 82)

Since, $\quad f(x) = \sin^{-1}(3x - 4x^3)$

$$= \begin{cases} -\pi - 3\sin^{-1} x, & -1 \leq x \leq -\dfrac{1}{\sqrt{2}} \\ 3\sin^{-1} x, & -\dfrac{1}{2} \leq x \leq \dfrac{1}{2} \\ \pi - 3\sin^{-1} x, & \dfrac{1}{2} \leq x \leq 1 \end{cases}$$

$$\therefore \quad f'(x) = \begin{cases} \dfrac{3}{\sqrt{1-x^2}}, & -\dfrac{1}{2} \leq x \leq \dfrac{1}{2} \\ -\dfrac{3}{\sqrt{1-x^2}}, & \dfrac{1}{2} \leq |x| \leq 1 \end{cases}$$

$\Rightarrow \quad f'(0) = 3$ and $f'\left(\dfrac{1}{\sqrt{2}}\right) = -3\sqrt{2}$

81. (b) **82.** (c)

Sol. (Q. Nos. 83 to 84)

$$D^*(u \cdot v) = \lim_{h \to 0} \frac{f^2(x+h) \cdot g^2(x+h) - f^2(x) \cdot g^2(x)}{h}$$

$$= \lim_{h \to 0} \frac{f^2(x+h) \cdot g^2(x+h) - f^2(x+h)\, g^2(x) + f^2(x+h)\, g^2(x) - f^2(x)\, g^2(x)}{h}$$

$$= \lim_{h \to 0} f^2(x+h) \cdot \left(\frac{g^2(x+h) - g^2(x)}{h}\right) + g^2(x)\left(\lim_{h \to 0} \frac{f^2(x+h) - f^2(x)}{h}\right)$$

$$= f^2(x) \cdot (D^* v) + g^2(x) \cdot (D^* u)$$

$$= u^2(D^* v) + v^2(D^* u)$$

Also, $\displaystyle D^*\left(\frac{u}{v}\right) = \lim_{h \to 0} \frac{\dfrac{f^2(x+h)}{g^2(x+h)} - \dfrac{f^2(x)}{g^2(x)}}{h}$

$$= \lim_{h \to 0} \frac{f^2(x+h) \cdot g^2(x) - g^2(x+h) \cdot f^2(x)}{g^2(x+h) \cdot g^2(x) \cdot h}$$

$$= \lim_{h \to 0} \frac{\left[\begin{array}{l} f^2(x+h)\, g^2(x) - f^2(x)\, g^2(x) \\ + f^2(x)\, g^2(x) - g^2(x+h)\, f^2(x) \end{array}\right]}{g^2(x+h) \cdot g^2(x) \cdot h}$$

$$= \lim_{h \to 0} \frac{\left[\begin{array}{l} \left(\dfrac{f^2(x+h) - f^2(x)}{h}\right) \cdot g^2(x) \\ - f^2(x) \cdot \left(\displaystyle\lim_{h \to 0} \dfrac{g^2(x+h) - g^2(x)}{h}\right) \end{array}\right]}{g^2(x+h) \cdot g^2(x)}$$

$$= \frac{g^2(x) \cdot (D^* u) - f^2(x) \cdot (D^* v)}{v^2 \cdot v^2}$$

$$= \frac{v^2 \cdot (D^* u) - u^2(D^* v)}{v^4}$$

83. (b) **84.** (c)

85. $y = e^t \sin t \Rightarrow \dfrac{dy}{dt} = e^t\,[\cos t + \sin t]$

$x = e^t \cos t \Rightarrow \dfrac{dx}{dt} = e^t\,[\cos t - \sin t]$

$\therefore \quad \dfrac{dy}{dx} = \dfrac{\cos t + \sin t}{\cos t - \sin t} = \tan \alpha$

$\Rightarrow \quad \tan\left(\dfrac{\pi}{4} + t\right) = \tan \alpha$

$\Rightarrow \quad \left(\dfrac{\pi}{4} + t\right) = \alpha \Rightarrow t = \alpha - \dfrac{\pi}{4}$

86. $\dfrac{d^2y}{dx^2} = \dfrac{\sec^2\left(\dfrac{\pi}{4} + t\right)}{e^t(\cos t - \sin t)}$

$\therefore \quad \left(\dfrac{d^2y}{dx^2}\right)_{t=0} = 2$

87. $F(t)=\int e^t(\cos t+\sin t)\,dt=e^t\sin t+C$

$$F\left(\frac{\pi}{2}\right)-F(0)=(e^{\pi/2}+C)-0=e^{\pi/2}$$

Sol. (Q. Nos. 88 to 90)

88. Since, 1, $a_1, a_2, \ldots\ldots, a_n$ are roots of $x^n-1=0$.

$$\therefore \quad x^n-1=(x-1)(x-a_1)(x-a_2)\ldots(x-a_n) \quad \ldots(i)$$

$$\Rightarrow \quad \frac{x^n-1}{x-1}=(x-a_1)(x-a_2)\ldots(x-a_n)$$

$$\Rightarrow \quad \lim_{x\to 1}\frac{x^n-1}{x-1}=\lim_{x\to 1}[(x-a_1)(x-a_2)\ldots(x-a_n)]$$

$$\Rightarrow \quad (1-a_1)(1-a_2)\ldots(1-a_n)=n$$

89. From Eq. (i), $\log(x^n-1)=\log(x-1)+\log(x-a_1)+\ldots+\log(x-a_n)$

Differentiating w.r.t. x, we get

$$\frac{nx^{n-1}}{x^n-1}=\frac{1}{x-1}+\frac{1}{x-a_1}+\frac{1}{x-a_2}+\ldots+\frac{1}{x-a_n} \quad \ldots(ii)$$

Putting $x=2$ in Eq. (ii), we get

$$\frac{n2^{n-1}}{x^n-1}=1+\frac{1}{2-a_1}+\frac{1}{2-a_2}+\ldots+\frac{1}{2-a_n}$$

$$\Rightarrow \quad \frac{1}{2-a_1}+\frac{1}{2-a_2}+\ldots+\frac{1}{2-a_n}=\frac{n2^{n-1}}{2^n-1}-1$$

$$=\frac{n2^{n-1}-2^n+1}{2^n-1}=\frac{2^{n-1}(n-2)+1}{2^n-1}$$

90. From Eq. (ii), $\dfrac{nx^{n-1}}{x^n-1}-\dfrac{1}{x-1}=\dfrac{1}{x-a_1}+\dfrac{1}{x-a_2}+\ldots+\dfrac{1}{x-a_n}$

$$\Rightarrow \quad \frac{nx^{n-1}-1(1+x+x^2+\ldots+x^{n-1})}{x^n-1}=\frac{1}{x-a_1}+\frac{1}{x-a_2}+\ldots+\frac{1}{x-a_n}$$

$$\Rightarrow \quad \lim_{x\to 1}\frac{nx^{n-1}-1(1+x+x^2+\ldots+x^{n-1})}{x^n-1}=\lim_{x\to 1}\left(\frac{1}{x-a_1}+\frac{1}{x-a_2}+\ldots+\frac{1}{x-a_n}\right)$$

$$\Rightarrow \quad \lim_{x\to 1}\frac{n(n-1)x^{n-2}-\{1+2x+\ldots+(n-1)x^{n-2}\}}{nx^{n-1}}=\frac{1}{1-a_1}+\frac{1}{1-a_2}+\ldots+\frac{1}{1-a_n}$$

[applying L'Hospital's rule on LHS]

$$\Rightarrow \quad \frac{n(n-1)-\{1+2+\ldots+(n-1)\}}{n}=\frac{1}{1-a_1}+\frac{1}{1-a_2}+\ldots+\frac{1}{1-a_n}$$

$$\Rightarrow \quad \frac{1}{1-a_1}+\frac{1}{1-a_2}+\ldots+\frac{1}{1-a_n}=\frac{n-1}{2}$$

91. (A) $\dfrac{dy}{dx}=\dfrac{dy/dt}{dx/dt}=\dfrac{12t^2-6t-18}{5t^4-15t^2-20}$

$$\Rightarrow \quad \left[\frac{dy}{dx}\right]_{t=1}=\frac{12-6-18}{5-15-20}=\frac{2}{5}\Rightarrow\left[-5\frac{dy}{dx}\right]_{t=1}=-2$$

(B) Let us take

$$P(x)=a(x-2)^4+b(x-2)^3+c(x-2)^2+d(x-2)+e$$

$$-1=P(2)=e$$

$$0=P'(2)=d$$

$$2=P''(2)=2c$$

$$\Rightarrow \quad c=1$$

$$-12=P'''(2)=6b$$

$$\Rightarrow \quad b=-2$$

$$P^{iv}(2)=24\,a=24\Rightarrow a=1$$

Thus, $P''(x)=12(x-2)^2-12(x-2)+2$

$$\Rightarrow \quad P''(3)=12-12(1)+2=2$$

(C) Here, $\sqrt{(1+y^4)}=\sqrt{\left(1+\dfrac{1}{x^4}\right)}=\dfrac{\sqrt{1+x^4}}{x^2}$ $\quad\left[\because y=\dfrac{1}{x}\right]$

$$\Rightarrow \quad \frac{\sqrt{1+y^4}}{\sqrt{1+x^4}}=\frac{1}{x^2} \quad \ldots(i)$$

But $\quad y=\dfrac{1}{x}$

$$\therefore \quad \frac{dy}{dx}=-\frac{1}{x^2} \quad \ldots(ii)$$

From Eqs. (i) and (ii), we get

$$\frac{\sqrt{1+y^4}}{\sqrt{1+x^4}}=-\frac{dy}{dx}\Rightarrow\frac{dy}{\sqrt{(1+y^4)}}+\frac{dx}{\sqrt{(1+x^4)}}=0$$

$$\Rightarrow \quad \frac{\dfrac{dy}{\sqrt{1+y^4}}}{\dfrac{dx}{\sqrt{1+x^4}}}=-1$$

(D) Obviously, $f(x)$ is a linear function.

Also, from $f'(0)=p$ and $f(0)=q$,

$$f(x)=px+q\Rightarrow f''(0)=0$$

92. (A) We know that,

$$2\tan^{-1}x=\begin{cases}\sin^{-1}\left(\dfrac{2x}{1+x^2}\right), & \text{if } -1\le x\le 1\\ \pi-\sin^{-1}\left(\dfrac{2x}{1+x^2}\right), & \text{if } x>1\\ -\pi-\sin^{-1}\left(\dfrac{2x}{1+x^2}\right), & \text{if } x<-1\end{cases}$$

$$\Rightarrow \quad \frac{dy}{dx}=-\frac{2}{1+x^2}, \text{ if } x<-1 \text{ or } x>1$$

(B) $\cos^{-1}\left(\dfrac{1}{\sqrt{1+x^2}}\right)=\begin{cases}\tan^{-1}x, & x\ge 0\\ -\tan^{-1}x, & x<0\end{cases}$

$$\Rightarrow \quad \frac{dy}{dx}=-\frac{1}{1+x^2}, \text{ if } x<0$$

(C) $y = |e^{|x|} - e| = \begin{cases} |e^x - e|, & x \ge 0 \\ |e^{-x} - e|, & x < 0 \end{cases}$

$$= \begin{cases} e^x - e, & x \ge 1 \\ e - e^x, & 0 \le x < 1 \\ e - e^{-x}, & -1 \le x < 0 \\ e^{-x} - e, & x < -1 \end{cases}$$

$\Rightarrow \quad \dfrac{dy}{dx} > 0$, if $x > 1$ or $-1 < x < 0$

(D) $\dfrac{du}{dx} = \dfrac{1}{x}$ and $\dfrac{dv}{dx} = \begin{cases} \dfrac{1}{1+x^2}, & x > 0 \\ -\dfrac{1}{1+x^2}, & x < 0 \end{cases}$

$$\Rightarrow \quad \frac{du}{dv} = \begin{cases} \dfrac{1+x^2}{x}, & x > 0 \\ -\dfrac{1+x^2}{x}, & x < 0 \end{cases}$$

Now, we know that $\dfrac{1+x^2}{x} = x + \dfrac{1}{x} > 2$, if $x > 1$ and < -2, if $x < -1 \Rightarrow \dfrac{du}{dv} > 2$, if $x < -1$ or $x > 1$

93. Here, $\quad f(x) = x^3 + x^2 f'(1) + xf''(2) + f'''(3)$

Put $\quad f'(1) = a, f''(2) = b, f'''(3) = c$

$\Rightarrow \quad f(x) = x^3 + ax^2 + bx + c$

$\quad f'(x) = 3x^2 + 2ax + b$

$\Rightarrow \quad f'(1) = 3 + 2a + b$ or $a = 3 + 2a + b$

$\Rightarrow \quad a + b = -3 \quad$ ...(i)

$\quad f'' = 6x + 2a \Rightarrow f''(2) = 12 + 2a$

$\therefore \quad b = 12 + 2a \Rightarrow 2a - b = -12 \quad$ ...(ii)

$\quad f'''(x) = 6 \Rightarrow c = 6 \quad$ ...(iii)

From Eqs. (i) and (ii), we get

$$a = -5, \ b = 2$$

$\therefore \quad f(1) = 1 + a + b + c$

$\quad = 1 - 5 + 2 + 6 = 4$

94. Let $y = \tan\left(\sin^{-1}\left(\dfrac{2x+2}{\sqrt{4x^2+8x+13}}\right)\right)$ and $z = \tan^{-1} x$.

$$\Rightarrow \quad y = \tan\left(\sin^{-1}\left(\frac{2x+2}{\sqrt{(2x+2)^2+9}}\right)\right)$$

$$= \tan\left(\tan^{-1}\left(\frac{2x+3}{3}\right)\right) = \frac{2x+3}{3}$$

$$\Rightarrow \quad \frac{dy}{dx} = \frac{2}{3} \text{ and } \frac{dz}{dx} = \frac{1}{1+x^2}$$

$$\therefore \quad \frac{dy}{dx} = \frac{\frac{dy}{dx}}{\frac{dz}{dx}} = \frac{2}{3}(1+x^2)$$

$$\Rightarrow \quad \left(\frac{dy}{dz}\right)_{x=\frac{1}{\sqrt{2}}} = \frac{2}{3}\left(1+\frac{1}{2}\right) = \frac{2}{3} \times \frac{3}{2} = 1$$

95. $P(x) = (x - x_1)^3 (x - x_2)(x - x_3) \ldots (x - x_8)$

$\Rightarrow \quad P(x) = (x - x_1)^3 \cdot Q(x)$

$\therefore \quad Q(x) = \dfrac{P(x)}{(x - x_1)^3}$

As, $Q(x)$ is polynomial.

$\therefore$ $Q(x)$ must be continuous at $x = x_1$.

$\therefore \quad Q(x_1) = \lim\limits_{x \to x_1} Q(x)$

$$= \lim_{x \to x_1} \frac{x^{10} + ax^2 + bx + c}{(x - x_1)^3}$$

By L' Hospital's rule, we have

$$Q(x_1) = \frac{10 \times 9 \times 8 \times x_1^7}{3 \times 2 \times 1}$$

$$\therefore \quad Q\left(\frac{1}{2}\right) = \frac{10 \times 9 \times 8 \times \left(\frac{1}{2}\right)^7}{3 \times 2 \times 1} = \frac{15}{16} = \frac{p}{q}$$

$\Rightarrow \quad q - p = 1$

96. Here, $f'(x) = \begin{vmatrix} 4(x-a)^3 & (x-a)^3 & 1 \\ 4(x-b)^3 & (x-b)^3 & 1 \\ 4(x-c)^3 & (x-c)^3 & 1 \end{vmatrix}$

$$+ \begin{vmatrix} (x-a)^4 & 3(x-a)^2 & 1 \\ (x-b)^4 & 3(x-b)^2 & 1 \\ (x-c)^4 & 3(x-c)^2 & 1 \end{vmatrix} + \begin{vmatrix} (x-a)^4 & (x-a)^3 & 0 \\ (x-b)^4 & (x-b)^3 & 0 \\ (x-c)^4 & (x-c)^3 & 0 \end{vmatrix}$$

$$= 0 + 3\begin{vmatrix} (x-a)^4 & (x-a)^2 & 1 \\ (x-b)^4 & (x-b)^2 & 1 \\ (x-c)^4 & (x-c)^2 & 1 \end{vmatrix} + 0$$

$\Rightarrow \quad \lambda = 3$

97. Let $P(x) = a(x-1)(x-3)(x-5)(x-\lambda)$

$\Rightarrow \log P(x) = \log a + \log(x-1) + \log(x-3) + \log(x-5) + \log(x-7)$

Differentiating both sides, we get

$$\frac{1}{P(x)} \cdot P'(x) = \frac{1}{x-1} + \frac{1}{x-3} + \frac{1}{x-5} + \frac{1}{x-\lambda}$$

As, $\quad P'(7) = 0 \quad [\because P(x) = 0 \Rightarrow P'(x) = 0 \Rightarrow P'(7) = 0]$

$\Rightarrow \quad 0 = \dfrac{1}{6} + \dfrac{1}{4} + \dfrac{1}{2} + \dfrac{1}{7-\lambda}$

$\Rightarrow \quad \lambda = \dfrac{89}{11} = \dfrac{p}{q}$

$\therefore \quad p - 8q = 1$

98. We have, $\quad x^2 + y^2 = t - \dfrac{1}{t}$

$\Rightarrow \quad (x^2+y^2)^2 = \left(t - \dfrac{1}{t}\right)^2 = t^2 + \dfrac{1}{t^2} - 2$

$\Rightarrow \quad x^4 + y^4 + 2x^2y^2 = x^4 + y^4 - 2$

$\Rightarrow \quad 2x^2y^2 = -2$

$\Rightarrow \quad y^2 = -x^{-2}$

Differentiating w.r.t. x, we get

$$2y \cdot \frac{dy}{dx} = +2x^{-3}$$

$$\therefore \quad \frac{dy}{dx} = +\frac{1}{x^3 y}$$

$$\Rightarrow \quad \left(\frac{dy}{dx}\right)_{(1,1)} = 1$$

99. We have, $x^2 + y^2 + z^2 - 2xyz = 1$ ...(i)

$\Rightarrow \quad d(x^2 + y^2 + z^2 - 2xyz) = 0$

$\Rightarrow \quad 2x\,dx + 2y\,dy + 2z\,dz - 2(xy\,dz + yz\,dx + zx\,dy) = 0$

$\Rightarrow \quad (x - yz)\,dx + (y - zx)\,dy + (z - xy)dz = 0$

$$\Rightarrow \frac{dx}{(y - zx)(z - xy)} + \frac{dy}{(x - yz)(z - xy)} + \frac{dz}{(x - yz)(y - zx)} = 0$$

Now, $(y - zx)^2 (z - xy)^2 = (y^2 - 2xyz + z^2x^2)(z^2 - 2xyz + x^2y^2)$

$= (1 - x^2 - z^2 + z^2x^2)(1 - x^2 - y^2 + x^2y^2)$ [using Eq. (i)]

$= (1 - x^2)(1 - z^2)(1 - x^2)(1 - y^2)$

$= (1 - x^2)^2 (1 - y^2)(1 - z^2)$

$\therefore (y - zx)(z - xy) = (1 - x^2) \cdot \sqrt{(1 - y^2)(1 - z^2)}$

Similarly, $(x - yz)(z - xy) = (1 - y^2) \cdot \sqrt{(1 - x^2)(1 - z^2)}$

and $(x - yz)(y - zx) = (1 - z^2) \cdot \sqrt{(1 - x^2((1 - y^2)}$

Substituting these values in Eq. (i), we get

$$\frac{dx}{\sqrt{1 - x^2}} + \frac{dy}{\sqrt{1 - y^2}} + \frac{dz}{\sqrt{1 - z^2}} = 0$$

100. We have, $x = \sin t$ or $t = \sin^{-1} x$

Now, $$\frac{dy}{dx} = \frac{dy}{dt} \cdot \frac{dt}{dx} = \frac{dy}{dt} \cdot \frac{1}{\sqrt{1 - x^2}}$$

$$\Rightarrow \quad \sqrt{1 - x^2}\frac{dy}{dx} = \frac{dy}{dt}$$

Differentiating w.r.t. x, we get

$$\sqrt{1 - x^2} \cdot \frac{d^2y}{dx^2} + \frac{1(-2x)}{2\sqrt{1 - x^2}} \cdot \frac{dy}{dx} = \frac{d}{dx}\left(\frac{dy}{dt}\right)$$

$$\Rightarrow \quad \sqrt{1 - x^2} \cdot \frac{d^2y}{dx^2} - \frac{x}{\sqrt{1 - x^2}}\frac{dy}{dx} = \frac{d^2y}{dt^2} \cdot \frac{dt}{dx}$$

$$\Rightarrow \quad (1 - x^2)\frac{d^2y}{dx^2} - x\frac{dy}{dx} = \sqrt{1 - x^2} \cdot \frac{d^2y}{dt^2} \cdot \frac{1}{\sqrt{1 - x^2}}$$

$$\therefore \quad (1 - x^2)\frac{d^2y}{dx^2} - x\frac{dy}{dx} = \frac{d^2y}{dt^2}$$

Adding y on both sides, we get

$$(1 - x^{2)}\frac{d^2y}{dx^2} - x\frac{dx}{dy} + y = \frac{d^2y}{dx^2} + y$$

$$\therefore \quad \lambda = 1$$

101. Put $\cos\phi = \frac{2}{\sqrt{13}}$

and $\sin\phi = \frac{3}{\sqrt{13}}$

$\Rightarrow \quad y = \cos^{-1}\{\cos(x + \phi)\} + \sin^{-1}\{\cos(x - \phi)\}$

$= x + \phi + \frac{\pi}{2} - \phi + x$

$y = 2x + \frac{\pi}{2}$

and $z = \sqrt{1 + x^2}$

$$\therefore \quad \frac{dy}{dz} = \frac{2\sqrt{1 + x^2}}{x} = \left(\frac{2\sqrt{1 + x^2}}{x}\right)_{x = \frac{1}{\sqrt{3}}} = \frac{2\sqrt{1 + \frac{1}{3}}}{\frac{1}{\sqrt{3}}} = 4$$

102. Here, $f''(x) - 2f'(x) - 15f(x) = 0$

$\Rightarrow \quad (a^2 - 2a - 15)\,e^{ax} + (b^2 - 2b - 15)\,e^{bx} = 0$

$\Rightarrow \quad a^2 - 2a - 15 = 0$ and $b^2 - 2b - 15 = 0$

$\Rightarrow \quad a = (5 \text{ or } -3)$ and $(b = 5 \text{ or } -3)$

Since, $a \neq b$

$\therefore \quad a + b = 2$ or -2

$\Rightarrow \quad |a + b| = 2$

103. $A: 2x + 2yy' = 0 \Rightarrow y' = -\frac{x}{y}$

$\therefore \quad y'(\sqrt{2}) = -1$

$B: \cos y \cdot y' + \cos x = \sin x \cdot \cos y \cdot y' + \sin y \cdot \cos x$

when $x = y = \pi$

$\Rightarrow \quad -y' - 1 = 0 + 0$

$\Rightarrow \quad y'(\pi) = -1$

$C: 2e^{xy}(xy' + y) + e^x e^y y' + e^y e^x - e^x - e^y y'$

$= e \cdot e^{xy}(xy' + y)$

At $x = 1, y = 1$

$2e(y' + 1) + e^2y' + e^2 - e - ey' = e^2(y' + 1)$

$\Rightarrow \quad ey' + e = 0 \Rightarrow y' = -1$

Hence, $A - B - C = 1$

104. $$\frac{dx}{dt} = -\frac{3}{t^4} - \frac{2}{t^3} = -\left(\frac{3 + 2t}{t^4}\right)$$

$$\frac{dy}{dt} = -\left(\frac{3}{t^3} + \frac{2}{t^2}\right) = -\left(\frac{3 + 2t}{t^3}\right)$$

$$\therefore \quad \frac{dy}{dx} = \frac{dy/dt}{dx/dt} = t$$

$$= t - \left(\frac{1 + t}{t^3}\right) \cdot t^3 = -1$$

Now, $\frac{du}{dx} - x \cdot \left(\frac{dy}{dx}\right)^3$

105. We have, $y = f(x) - f(2x)$

$\Rightarrow \quad y' = f'(x) - 2f'(2x)$

$y'(1) = f'(1) - 2f'(2) = 5 \quad \text{...(i)}$

and $\quad y'(2) = f'(2) - 2f'(4) = 7 \quad \text{...(ii)}$

Now, let $\quad y = f(x) - f(4x)$

$\Rightarrow \quad y' = f'(x) - 4f'(4x)$

$\Rightarrow \quad y'(1) = f'(1) - 4f'(4) \quad \text{...(iii)}$

Substituting the value of $f'(2) = 7 + 2f'(4)$ in Eq. (i),

$f'(1) - 2[7 + 2f'(4)] = 5$

$\Rightarrow \quad f'(1) - 4f'(4) = 19$

Hence, $\quad 19 = 10 + \lambda$

$\Rightarrow \quad \lambda = 9$

106. $y = 3e^2 - x$

Let $\quad x^y = x^{3e^2 - x}$

$\Rightarrow \quad f(x) = x^{3e^2 - x}$

$\Rightarrow \quad \ln(f(x)) = (3e^2 - x)\ln x$

$\Rightarrow \quad \frac{1}{f(x)} \cdot f'(x) = \frac{3e^2 - x}{x} - \ln x$

$\therefore \quad f'(x) = 0$

$\Rightarrow \quad 3e^2 - x = x \ln x$

$\Rightarrow \quad 3e^2 = x(1 + \ln x)$

$\Rightarrow \quad x = e^2 \quad$ [by verification]

Hence, $\quad \lambda = 1$

107. $f'(x) = (kx + e^x)h'(x) + h(x)(k + e^x)$

$\therefore \quad f'(0) = h'(0) + h(0)(k + 1)$

$\Rightarrow \quad 18 = -2 + 5(k + 1)$

$\Rightarrow \quad k = 3$

108. We have, $f(x) = |\log 2 - \sin x|$

and $\quad g(x) = f(f(x)), x \in R$

Note that, for $x \to 0$, $\log 2 > \sin x$

$\therefore \quad f(x) = \log 2 - \sin x$

$\Rightarrow \quad g(x) = \log 2 - \sin(f(x))$

$= \log 2 - \sin(\log 2 - \sin x)$

Clearly, $g(x)$ is differentiable at $x = 0$ as $\sin x$ is differentiable.

Now, $\quad g'(x) = -\cos(\log 2 - \sin x)(-\cos x)$

$= \cos x \cdot \cos(\log 2 - \sin x)$

$\therefore \quad g'(0) = 1 \cdot \cos(\log 2)$

109. As, $g(f(x)) = x$

Thus, $g(x)$ is inverse of $f(x)$.

$\Rightarrow \quad g(f(x)) = x$

$\Rightarrow \quad g'(f(x)) \cdot f'(x) = 1$

$\therefore \quad g'(f(x)) = \frac{1}{f'(x)} \quad$ [where, $f'(x) = 3x^2 + 3$] ...(i)

When $f(x) = 2$, then

$x^3 + 3x + 2 = 2$

$\Rightarrow \quad x = 0$

i.e. when $x = 0$, then $f(x) = 2$

$\therefore \quad g'(f(x)) = \frac{1}{3x^2 + 3}$ at $(0, 2)$

$\Rightarrow \quad g'(2) = \frac{1}{3}$

$\therefore$ Option (a) is incorrect.

Now, $\quad h(g(g(x))) = x$

$\Rightarrow \quad h(g(g(f(x))) = f(x)$

$\Rightarrow \quad h(g(x)) = f(x) \quad \text{...(ii)}$

As $\quad g(f(x)) = x$

$\therefore \quad h(g(3)) = f(3)$

$= 3^3 + 3(3) + 2 = 38$

$\therefore$ Option (d) is incorrect.

From Eq. (ii),

$h(g(x)) = f(x)$

$\Rightarrow \quad h(g(f(x))) = f(f(x))$

$\Rightarrow \quad h(x) = f(f(x)) \quad \text{...(iii)}$

[using $g(f(x)) = x$]

$\Rightarrow \quad h'(x) = f'(f(x)) \cdot f'(x) \quad \text{...(iv)}$

Putting $x = 1$, we get

$h'(1) = f'(f(1)) \cdot f'(1)$

$= (3 \times 36 + 3) \times (6) = 111 \times 6 = 666$

$\therefore$ Option (b) is correct.

Putting $x = 0$ in Eq. (iii), we get

$h(0) = f(f(0))$

$= f(2) = 8 + 6 + 2 = 16$

$\therefore$ Option (c) is correct.

110. Let $y = \tan^{-1}\left(\frac{6x\sqrt{x}}{1 - 9x^3}\right)$, where $x \in \left(0, \frac{1}{4}\right)$

$= \tan^{-1}\left(\frac{2 . (3x^{3/2})}{1 - (3x^{3/2})^2}\right) = 2\tan^{-1}(3x^{3/2})$

As $3x^{3/2} \in \left(0, \frac{3}{8}\right)$

$\therefore \quad \frac{dy}{dx} = 2 \times \frac{1}{1 + 9x^3} \times 3 \times \frac{3}{2} \times x^{1/2}$

$= \frac{9}{1 + 9x^3}\sqrt{x}$

$\therefore \quad g(x) = \frac{9}{1 + 9x^3}$

111. Since, $f(x) = e^{g(x)}$

$\Rightarrow e^{g(x+1)} = f(x+1) = xf(x) = xe^{g(x)}$

and $\quad g(x+1) = \log x + g(x)$

i.e. $\quad g(x+1) - g(x) = \log x \quad \text{...(i)}$

Replacing x by $x-\frac{1}{2}$, we get

$$g\left(x+\frac{1}{2}\right)-g\left(x-\frac{1}{2}\right)=\log\left(x-\frac{1}{2}\right)$$

$$=\log(2x-1)-\log 2$$

$$\therefore\quad g''\left(x+\frac{1}{2}\right)-g''\left(x-\frac{1}{2}\right)=\frac{-4}{(2x-1)^2}\qquad\text{...(ii)}$$

On substituting $x=1,2,3,\ldots,N$ in Eq. (ii) and adding, we get

$$g''\left(N+\frac{1}{2}\right)-g''\left(\frac{1}{2}\right)=-4\left\{1+\frac{1}{9}+\frac{1}{25}+\ldots+\frac{1}{(2N-1)^2}\right\}.$$

112. Since, $\dfrac{dx}{dy}=\dfrac{1}{dy/dx}=\left(\dfrac{dy}{dx}\right)^{-1}$

$$\Rightarrow\quad \frac{d}{dy}\left(\frac{dx}{dy}\right)=\frac{d}{dx}\left(\frac{dy}{dx}\right)^{-1}\frac{dx}{dy}$$

$$\Rightarrow\quad \frac{d^2x}{dy^2}=-\left(\frac{d^2y}{dx^2}\right)\left(\frac{dy}{dx}\right)^{-2}\left(\frac{dx}{dy}\right)$$

$$=-\left(\frac{d^2y}{dx^2}\right)\left(\frac{dy}{dx}\right)^{-3}$$

113. Since, $f''(x)=-f(x)$

$$\Rightarrow\quad \frac{d}{dx}\{f'(x)\}=-f(x)$$

$$\Rightarrow\quad g'(x)=-f(x)\qquad [\because g(x)=f'(x),\text{ given}]\ \text{...(i)}$$

Also, $F(x)=\left\{f\left(\dfrac{x}{2}\right)\right\}^2+\left\{g\left(\dfrac{x}{2}\right)\right\}^2$

$$\Rightarrow\quad F'(x)=2\left(f\left(\frac{x}{2}\right)\right)\cdot f'\left(\frac{x}{2}\right)\cdot\frac{1}{2}$$

$$+2\left(g\left(\frac{x}{2}\right)\right)\cdot g'\left(\frac{x}{2}\right)\cdot\frac{1}{2}=0\ [\text{from Eq.(i)}]$$

$\therefore$ $F(x)$ is constant $\Rightarrow F(10)=F(5)=5$

114. Given that, $\log(x+y)=2xy$ ...(i)

$\therefore$ At $x=0$

$\Rightarrow$ $\log(y)=0$

$\Rightarrow$ $y=1$

$\therefore$ To find $\dfrac{dy}{dx}$ at $(0,1)$

On differentiating Eq. (i) w.r.t. x, we get

$$\frac{1}{x+y}\left(1+\frac{dy}{dx}\right)=2x\frac{dy}{dx}+2y\cdot 1$$

$$\Rightarrow\quad \frac{dy}{dx}=\frac{2y(x+y)-1}{1-2(x+y)x}$$

$$\Rightarrow\quad \left(\frac{dy}{dx}\right)_{(0,1)}=1$$

115. Given, $x^2+y^2=1$

On differentiating w.r.t. x, we get

$$2x+2yy'=0$$

$$\Rightarrow\quad x+yy'=0$$

Again, differentiating w.r.t. x, we get

$$1+y'y'+yy''=0$$

$$\therefore\quad 1+(y')^2+yy''=0$$

CHAPTER

03

Functions

Learning Part

Session 1
- Function

Session 2
- Domain
- Algebraic Functions

Session 3
- Transcendental Functions

Session 4
- Piecewise Functions

Session 5
- Range

Session 6
- Odd and Even Functions

Session 7
- Periodic Functions

Session 8
- Mapping of Functions

Session 9
- Identical Functions

Session 10
- Composite Functions

Session 11
- Inverse of a Function

Session 12
- Miscellaneous Problems on Functions

Practice Part

- JEE Type Examples
- Chapter Exercises

Session 1

Function

Function

Let A and B be two non-empty sets and f be a relation which associates each element of set A with unique element of set B, f is called a function from A to B.

Or

If for each element in a set A there is assigned a unique element of set B, we call such an assignment a function $f: A \rightarrow B$. "f is a function from A to B".

If $a \in A$, the element in B which is assigned to 'a' is called the image of 'a' and denoted by $f(a)$.

Here, set A is called the domain of f and B is called the codomain of f. The set of elements of B, which are the images of the elements of set A is called the range of f.

e.g. Let $A = \{a, b, c, d\}$, $B = \{1, 2, 3, 4, 5\}$. Here, $f(a) = 2$, $f(b) = 3$, $f(c) = 5$, $f(d) = 1$ given by

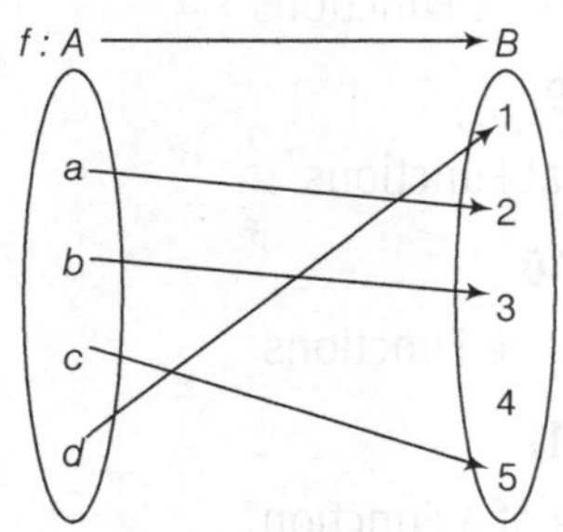

Figure 3.1

i.e. $A \rightarrow$ Domain of $f = \{a, b, c, d\}$

$B \rightarrow$ Codomain of $f = \{1, 2, 3, 4, 5\}$

Range of $f = \{1, 2, 3, 5\}$

An element of A (i.e. Domain) cannot associate with more than one element in B.

Example 1 In the given figure, find the domain, codomain and range.

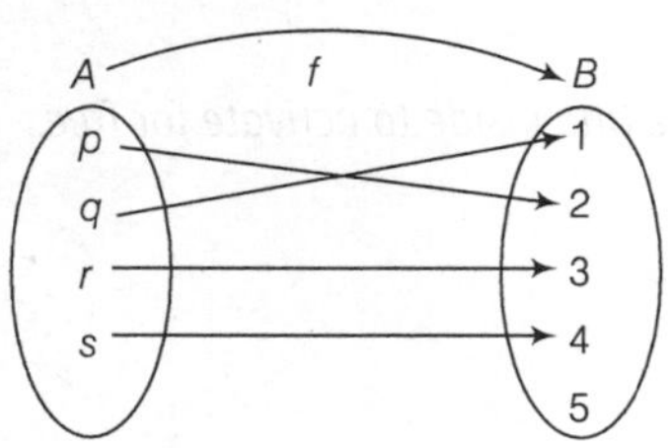

Sol. From the given figure, we conclude that

$A \rightarrow$ Domain of $f = \{p, q, r, s\}$

$B \rightarrow$ Codomain of $f = \{1, 2, 3, 4, 5\}$

Range of $f = \{1, 2, 3, 4\}$

Remark

We should always remember that "the range is a subset of codomain".

Number of Functions (or Mapping) from *A* to *B*

Let $A = \{x_1, x_2, x_3, \ldots, x_m\}$ [i.e. m elements]

and $B = \{y_1, y_2, y_3, \ldots, y_n\}$ [i.e. n elements]

Then, each element x_i's $(i = 1, 2, \ldots, m)$ of domain corresponds n images.

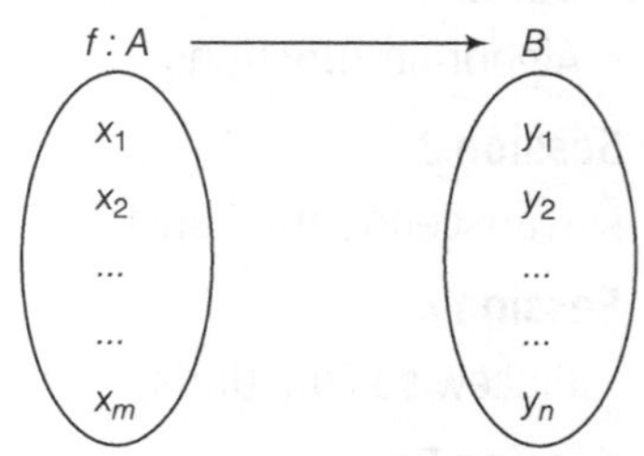

Figure 3.2

i.e. x_1 can take n images.

x_2 can take n images.

....

....

x_m can take n images.

Thus, **total number of functions from A to B**

$$= n \times n \times \ldots m \text{ times} = n^m$$

i.e. (Number of elements in codomain)$^{\text{number of elements in domain}}$.

Example 2 The number of functions $f: \{1, 2, 3, \ldots, n\} \rightarrow \{2016, 2017\}$, where $n \in N$, which satisfy the condition $f(1) + f(2) + \ldots + f(n)$ is an odd number are

(a) 2^n (b) $n \cdot 2^{n-1}$

(c) 2^{n-1} (d) $n!$

Sol.

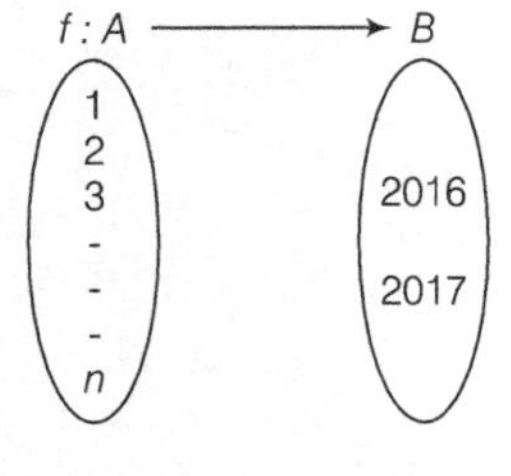

We can send 1, 2, 3, ..., $(n-1)$ any where in 2^{n-1} ways, the value $f(n)$ can uniquely determined according to sum of $f(1)+f(2)+...+f(n-1)$ is even or odd.

i.e. if $\{f(1)+f(2)+...+f(n-1)\} \in$ even, $f(n)=2017$

and if $\{f(1)+f(2)+...+f(n-1)\} \in$ odd, $f(n)=2016$

$\therefore$ Number of functions $=2^{n-1}$

Graphical Representation of Functions

The graph of $y=f(x)$, where f is a function, consists of points whose coordinates (x, y) satisfy $y=f(x)$. Also, it is such that lines drawn parallel to the Y-axis **do not intersect the curve at more than one point.**

Vertical line test for f to be a function If graph of a function is cut a line parallel to Y-axis at more than one point, then it does not form a function.

Example 3 Find whether $f(x)=x^3$ forms a mapping or not.

Sol. We have, $y=f(x)=x^3, \forall\, x \in R$.

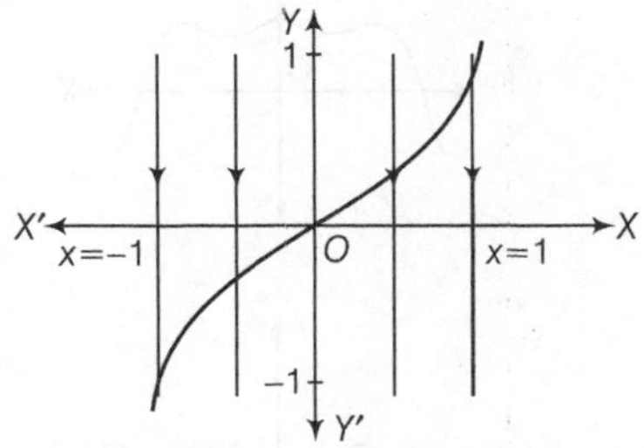

Here, all the straight lines drawn parallel to Y-axis cut $y=x^3$ only at one point. Thus, $y=f(x)$ forms a mapping.

Example 4 Find whether $\dfrac{x^2}{a^2}+\dfrac{y^2}{b^2}=1$, $a>b$, forms a mapping or not.

Sol. We have, $\dfrac{x^2}{a^2}+\dfrac{y^2}{b^2}=1$, i.e. $y=\pm\dfrac{b}{a}\sqrt{a^2-x^2}$, which represents a horizontal ellipse.

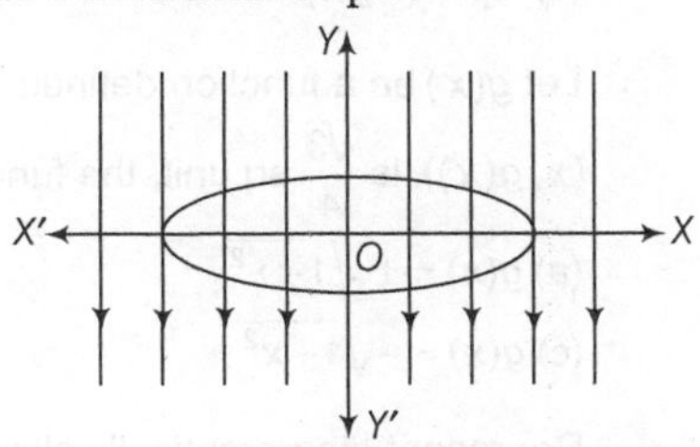

Here, straight lines drawn parallel to Y-axis meet the curve at more than one points. Thus, $f(x)=\pm\dfrac{b}{a}\sqrt{a^2-x^2}$ does not form a mapping.

Exercise for Session 1

1. Which of the following graphs are graphs of a function?

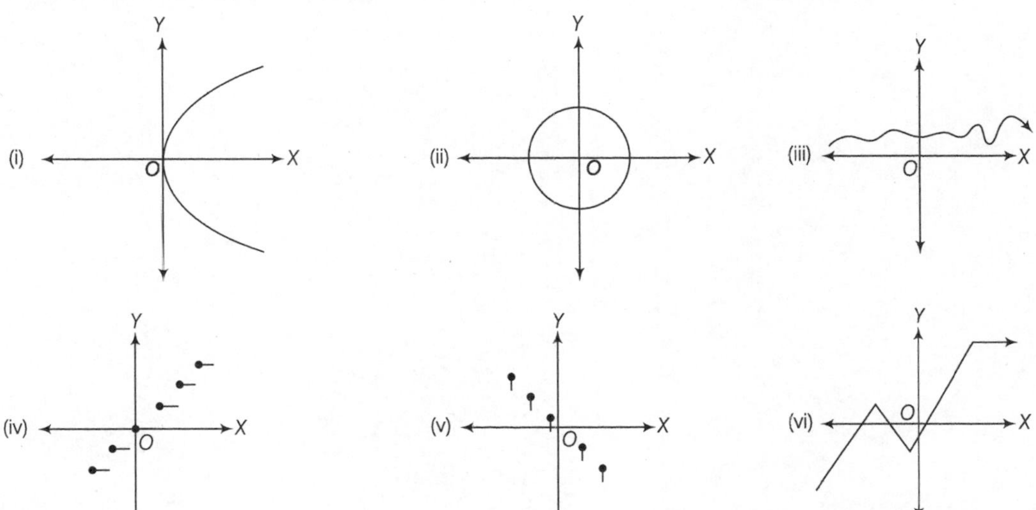

(vii) 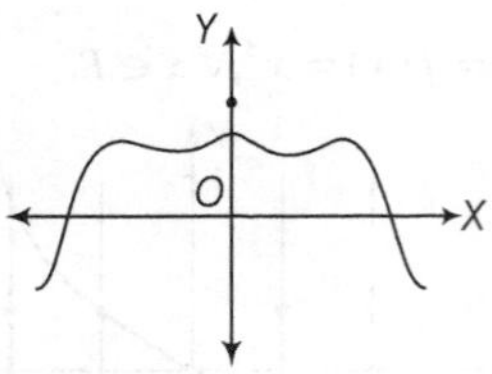

(viii) 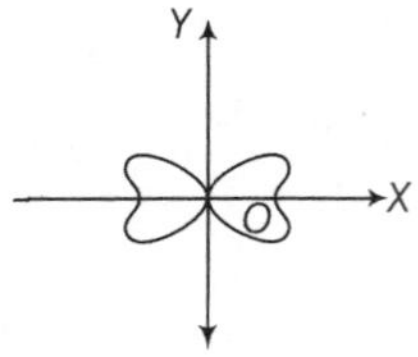

(ix) 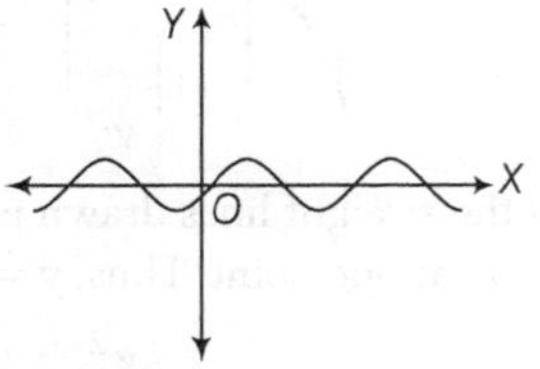

(x) 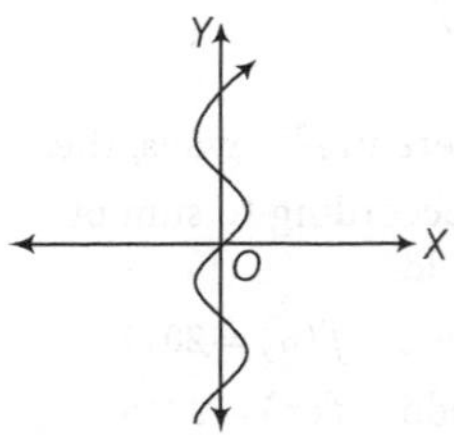

2. For which of the following, y can be a function of x, $(x \in R, y \in R)$?

(i) $(x-h)^2 + (y-k)^2 = r^2$ (ii) $y^2 = 4ax$

(iii) $x^4 = y^2$ (iv) $x^6 = y^3$

(v) $3y = (\log x)^2$

3. Let $g(x)$ be a function defined on $[-1, 1]$. If the area of equilateral triangle with two of its vertices at $(0, 0)$ and $(x, g(x))$, is $\frac{\sqrt{3}}{4}$ sq unit, the function $g(x)$ may be

(a) $g(x) = \pm\sqrt{1-x^2}$ (b) $g(x) = \sqrt{1-x^2}$

(c) $g(x) = -\sqrt{1-x^2}$ (d) $g(x) = \sqrt{1+x^2}$

4. Represent diagramatically all possible functions from $\{\alpha, \beta\}$ to $\{1, 2\}$.

5. The number of functions from $f: \{a_1, a_2, \ldots, a_{10}\} \to \{b_1, b_2, \ldots, b_5\}$ is

(a) 10^5 (b) 5^{10}

(c) $\frac{10!}{5!}$ (d) $5!$

Session 2

Domain, Algebraic Functions

Domain

The domain of $y = f(x)$ is the set of all real x for which $f(x)$ is defined (real).

Rules for Finding Domain

(i) If domain of $y = f(x)$ and $y = g(x)$ are D_1 and D_2 respectively, then the domain of $f(x) \pm g(x)$ or $f(x) \cdot g(x)$ is $D_1 \cap D_2$.

(ii) Domain of $\dfrac{f(x)}{g(x)}$ is $D_1 \cap D_2 - \{x : g(x) = 0\}$.

Example 5 Find the domain of the following functions.

(i) $y = \sqrt{5x - 3}$ (ii) $y = \sqrt[3]{5x - 3}$

(iii) $y = \dfrac{1}{(x-1)(x-2)}$ (iv) $y = \dfrac{1}{\sqrt[3]{(x-1)}}$

Sol. (i) Here, $y = \sqrt{5x - 3}$ is defined, if

$$5x - 3 \geq 0$$

$$\Rightarrow \quad x \geq \frac{3}{5} \quad \Rightarrow \quad x \in \left[\frac{3}{5}, \infty\right)$$

Thus, domain is $\left[\dfrac{3}{5}, \infty\right)$.

(ii) Here, $y = \sqrt[3]{5x - 3}$ is defined, if

$$5x - 3 \in R \Rightarrow x \in R$$

Thus, domain is R.

(iii) Here, $y = \dfrac{1}{(x-1)(x-2)}$ is defined, if

$(x-1)(x-2) \neq 0$

i.e. if $x \neq 1, 2$

Thus, domain is $R - \{1, 2\}$.

(iv) Here, $y = \dfrac{1}{\sqrt[3]{x-1}}$ is defined, if

$$x - 1 \neq 0 \Rightarrow x \neq 1$$

Thus, domain is $R - \{1\}$.

Example 6 Find the domain of $f(x) = \sqrt{\dfrac{1 - 5^x}{7^{-x} - 7}}$.

Sol. For $f(x)$ to be defined, $\dfrac{1 - 5^x}{7^{-x} - 7} \geq 0$ and $7^{-x} - 7 \neq 0$

$$\Rightarrow \quad \frac{5^x - 1}{7^{-x} - 7} \leq 0 \text{ and } 7^{-x} - 7 \neq 0$$

Now, $5^x - 1 = 0 \Rightarrow x = 0$ and $7^{-x} - 7 = 0 \Rightarrow x = -1$

$$\underset{}{-} \quad \underset{-1}{|} \quad + \quad \underset{0}{|} \quad -$$

The sign of $\dfrac{5^x - 1}{7^{-x} - 7}$

Hence, from the sign scheme $x \in (-\infty, -1) \cup [0, \infty)$.

$\therefore$ Domain of $f(x)$ is $(-\infty, -1) \cup [0, \infty)$.

Classification of Functions

Functions are classified by the type of mathematical equation which represent their relationship.

Functions

- 1. Algebraic
  - Monomial
  - Polynomial
    - Constant
    - Identity
  - Rational
  - Irrational
- 2. Transcendental
  - Exponential
  - Logarithmic
  - Trigonometric
  - Inverse trigonometric
- 3. Piecewise
  - Modulus
  - Signum
  - Greatest integer
  - Fractional part
  - Least integer

Algebraic Functions

A function f is called an algebraic function, if it can be constructed using algebraic operations such as addition, subtraction, multiplication, division and taking roots, starting with polynomials.

e.g. $f(x) = \sqrt{1 + x}$,

$$g(x) = \frac{x^3 - 16x}{x + \sqrt{x}} + (x - 2)\sqrt[3]{x - 1}$$

Different types of algebraic functions are described below

(i) Monomial Function

A function of the form $y = ax^n$, where a is constant and n is a non-negative integer, called monomial function.

e.g. $y = x^2, y = 2x, y = -x, \ldots\ldots$.

(ii) Polynomial Function

A function $y = f(x) = a_0 x^n + a_1 x^{n-1} + \ldots + a_n$, where $a_0, a_1, a_2, \ldots, a_n$ are real constants and n is a non-negative integer, then $f(x)$ is called a polynomial function. If $a_0 \neq 0$, then n is the degree of polynomial function. e.g.

Expression	Degree
$f(x) = x^{1920} + 5x^{1919} + 6x$	polynomial of degree 1920
$g(x) = x^2 + 3x + 3$	polynomial of degree 2
$h(x) = 7 = 7x^0$	polynomial of degree 0

A polynomial of degree one with no constant term is called an **odd linear function**, i.e. $f(x) = ax, a \neq 0$

In case $f(x) = 0$, it is constant function for which degree is not defined. A polynomial of odd degree has its range $(-\infty, \infty)$ but a polynomial of even degree has a range which is always subset of R.

Note: All polynomial functions are algebraic functions but not the converse.

Example 7 Draw the graph of polynomial functions and find their domain and range.

(i) $y = x + 1$ (ii) $y = x^2$

(iii) $y = x^3 + 1$ (iv) $y = x(x-1)(x-2)$

Sol. (i) Graph of $y = (x+1)$ is shown as

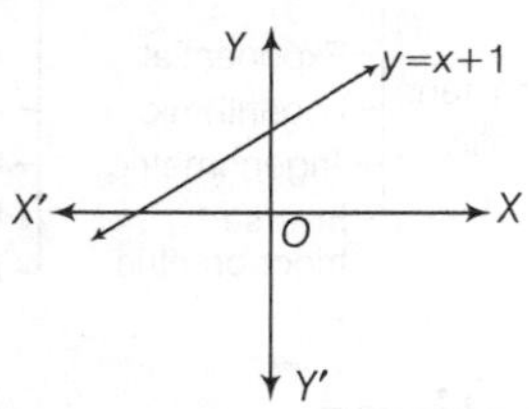

From graph domain is $x \in R$ and range is $y \in R$.

(ii) Graph of $y = x^2$ is shown as

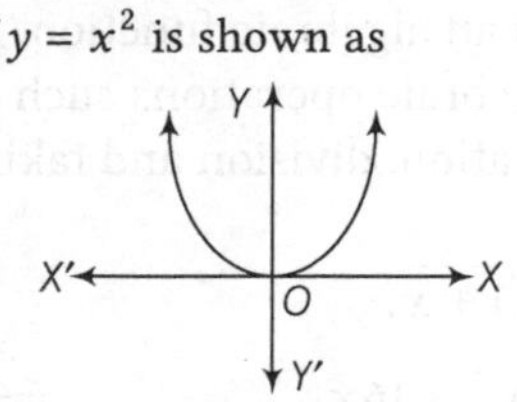

From graph domain is $x \in R$ and range is $y \in [0, \infty)$.

(iii) Graph of $y = x^3 + 1$ is shown as

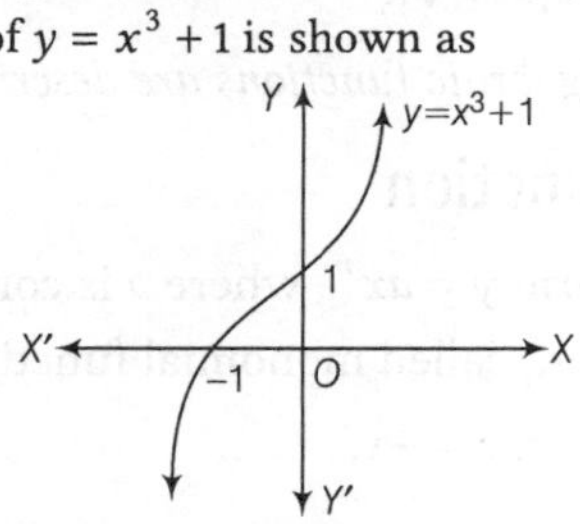

$\therefore$ Domain; $x \in R$ and Range; $y \in R$.

(iv) Graph of $y = x(x-1)(x-2)$ is shown as

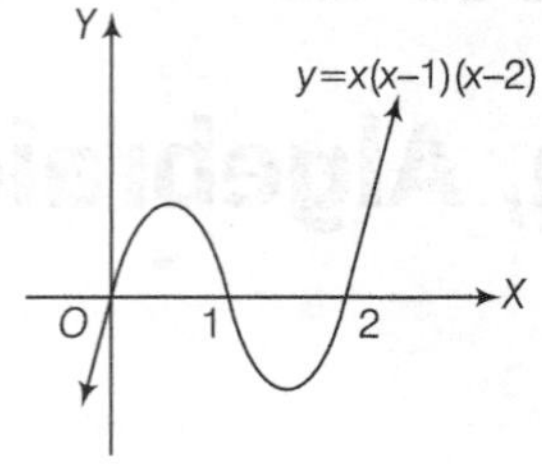

$\therefore$ Domain; $x \in R$ and Range; $y \in R$.

Constant Function

If the range of a function f consists of only one number, then f is called a constant function. The graph of a constant function is as shown below.

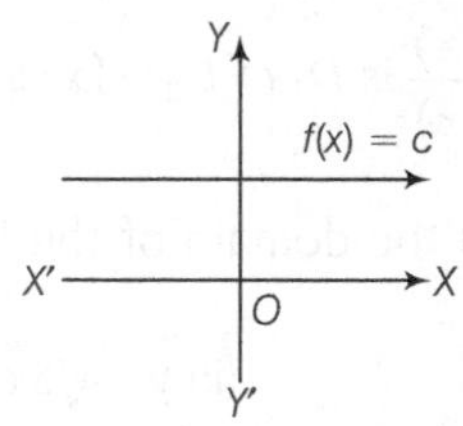

Figure 3.3

Thus, domain is R and range is $\{c\}$.

Identity Function

The function $y = f(x) = x$ for all $x \in R$ is called an identity function on R.

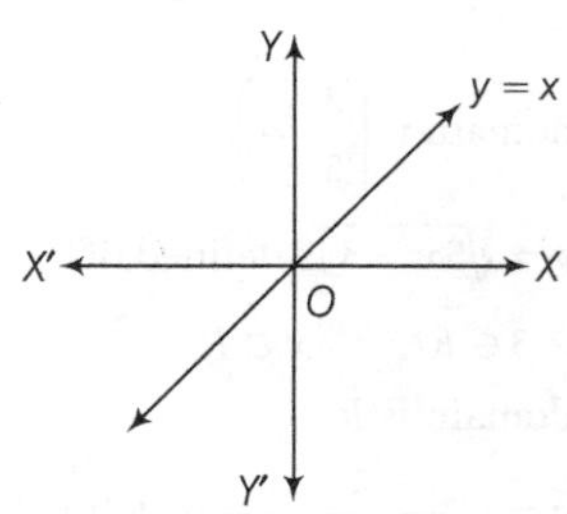

Figure 3.4

Here, domain of identity function is the set of real number. (As here for all values of x, $f(x)$ exists.)

Thus, domain is R and range is R.

(iii) Rational Function

It is defined as the ratio of two polynomials.

Let $P(x) = a_0 x^n + a_1 x^{n-1} + \ldots + a_n$

and $Q(x) = b_0 x^m + b_1 x^{m-1} + \ldots + b_m$.

Then, $f(x) = \dfrac{P(x)}{Q(x)}$ is a rational function provided $Q(x) \neq 0$.

Here, we can say that domain of $f(x)$ is all real numbers except the numbers where denominator is zero [i.e. Domain $(f) = R - \{x : Q(x) = 0\}$].

e.g. $f(x) = \dfrac{x^2 - 3x + 5}{(x-1)(x-2)}$; Domain is $R - \{1, 2\}$

e.g. $f(x) = \dfrac{1}{2 - \cos 3x}$; As $2 - \cos 3x \neq 0$. [$\because -1 \leq \cos 3x \leq 1$]

$\therefore$ Domain of $f(x)$ is R.

(iv) Irrational Function

An algebraic function having **non-integral** rational exponent is called irrational function.

e.g. $f(x) = x^{1/2}$,

e.g. $f(x) = \dfrac{x^2 + \sqrt{x}}{\sqrt{1 + 3x^2}}$, etc.

An Important Result

There are two polynomial functions, satisfying the relation;

$f(x) \cdot f\left(\dfrac{1}{x}\right) = f(x) + f\left(\dfrac{1}{x}\right)$.

They are

(i) $f(x) = 1 + x^n$ and (ii) $f(x) = 1 - x^n$,

where n is a positive integer.

For proof, refer Example 38 in Jee Type Solved Examples.

Exercise for Session 2

■ **Directions** (Q. Nos. 1 to 6) *Find the domain of the following.*

1. $f(x) = \sqrt{x^2 - 5x + 6}$
2. $f(x) = \sqrt{\dfrac{2x + 1}{x^3 - 3x^2 + 2x}}$
3. $f(x) = \sqrt{1 - \sqrt{1 - \sqrt{1 - x^2}}}$
4. $f(x) = \sqrt{x^{12} - x^9 + x^4 - x + 1}$
5. $f(x) = {}^{16 - x}C_{2x - 1} + {}^{20 - 3x}P_{4x - 5}$
6. $f(x) = \sqrt{{}^{x^2 + 4x}C_{2x^2 + 3}}$

Session 3

Transcendental Functions

Transcendental Functions

The functions which are not algebraic are called transcendental functions. Exponential, logarithmic, trigonometric and inverse trigonometric functions are transcendental functions.

(i) Exponential Functions

The function $f(x) = a^x, a > 0, a \neq 1, a \in$ constant is said to be an exponential function.

If $0 < a < 1$

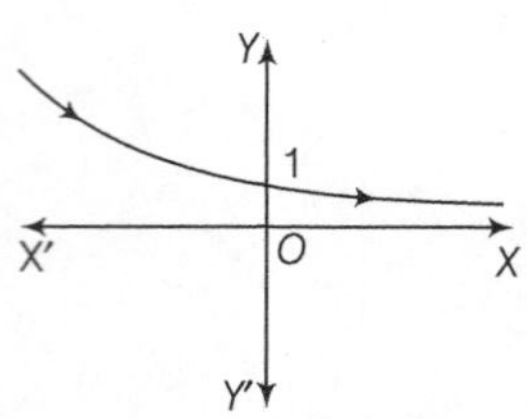

Figure 3.5

If $a > 1$

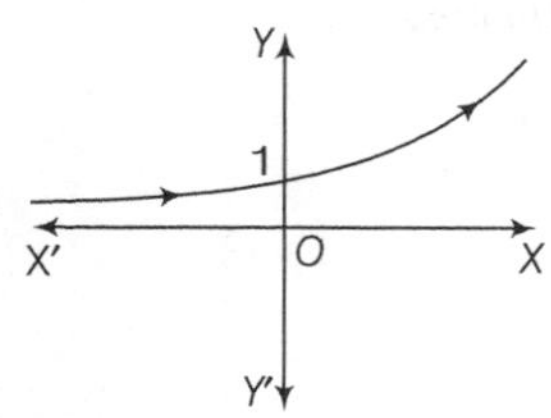

Figure 3.6

In the given graphs, none of them intersect the X-axis. Thus, we can say that they exist for all $x \in R$ and their corresponding values of y are greater than zero.

Thus, domain is R and range is $]\,0, \infty[$.

(ii) Logarithmic Functions

The function $f(x) = \log_a x$; $(x, a > 0)$ and $a \neq 1$ is a logarithmic function. Thus, domain of a logarithmic function is the set of all positive numbers and their range is the set R of real numbers.

If $0 < a < 1$

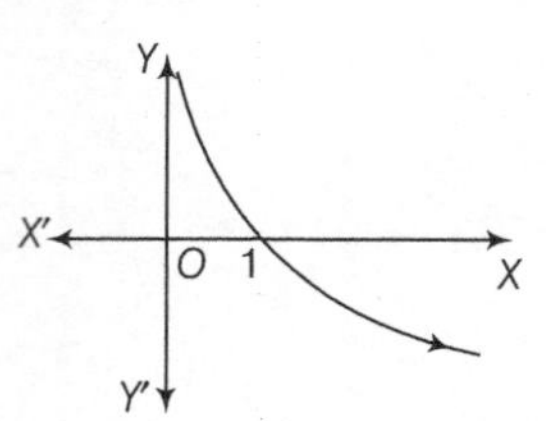

Figure 3.7

If $a > 1$

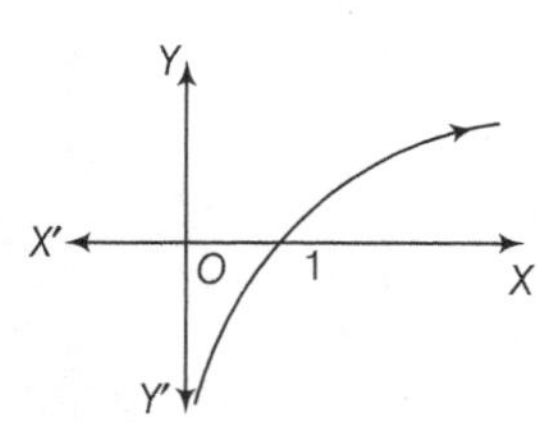

Figure 3.8

Properties of Logarithmic Functions

1. $\log_e(ab) = \log_e a + \log_e b$ $\quad [a, b > 0]$
2. $\log_e\left(\dfrac{a}{b}\right) = \log_e a - \log_e b$ $\quad [a, b > 0]$
3. $\log_e a^m = m \log_e a$ $\quad [a > 0, m \in R]$
4. $\log_a a = 1$ $\quad [a > 0, a \neq 1]$
5. $\log_{b^m} a = \dfrac{1}{m} \log_b a$ $\quad [a, b > 0, b \neq 1$ and $m \in R - \{0\}]$
6. $\log_b a = \dfrac{1}{\log_a b}$ $\quad [a, b > 0$ and $a, b \neq 1$
7. $\log_b a = \dfrac{\log_m a}{\log_m b}$ $\quad [a, b, m > 0$ and $b, m \neq 1]$
8. $a^{\log_a m} = m$ $\quad [a, m > 0$ and $a \neq 1]$
9. $a^{\log_c b} = b^{\log_c a}$ $\quad [a, b, c > 0$ and $c \neq 1]$
10. If $\log_m x > \log_m y \Rightarrow \begin{cases} x > y, & \text{if } m > 1 \\ x < y, & \text{if } 0 < m < 1 \end{cases}$ $\quad [m, x, y > 0, m \neq 1]$
11. $\log_m a = b \Rightarrow a = m^b$ $\quad [m, a > 0, m \neq 1, b \in R]$
12. $\log_m a > b \Rightarrow \begin{cases} a > m^b, & \text{if } m > 1 \\ a < m^b, & \text{if } 0 < m < 1 \end{cases}$
13. $\log_m a < b \Rightarrow \begin{cases} a < m^b, & \text{if } m > 1 \\ a > m^b, & \text{if } 0 < m < 1 \end{cases}$

Example 8 Find the domain of a single-valued function $y = f(x)$ given by the equation $10^x + 10^y = 10$.

Sol. Given, $10^x + 10^y = 10$

$\Rightarrow \quad 10^y = 10 - 10^x$

$\Rightarrow \quad y = \log_{10}(10 - 10^x)$ $\quad [\text{as } a^m = b \Rightarrow m = \log_a b]$

Now, $10^1 - 10^x > 0$

$\Rightarrow \quad 10^1 > 10^x \quad \Rightarrow 1 > x$

Therefore, domain of the single-valued function

$y = f(x)$ is $x < 1$ or $x \in (-\infty, 1)$

Example 9 Find the domain of

$$f(x)=\frac{1}{\sqrt{\log_{1/2}(x^2-7x+13)}}.$$

Sol. $f(x)=\dfrac{1}{\sqrt{\log_{1/2}(x^2-7x+13)}}$ exists, if

$$\log_{1/2}(x^2-7x+13)>0$$

$$\Rightarrow \quad (x^2-7x+13)<1 \quad ...(i)$$

$$\text{and} \quad x^2-7x+13>0 \quad ...(ii)$$

Considering Eq. (ii), $x^2-7x+13>0$, we get

$$\left(x^2-7x+\frac{49}{4}\right)+13-\frac{49}{4}>0$$

$$\Rightarrow \quad \left(x-\frac{7}{2}\right)^2+\frac{3}{4}>0$$

which is true for all $x \in R$. [As $\left(x-\dfrac{7}{2}\right)^2 \geq 0$ for all x.] ...(iii)

Now, considering Eq. (i), we get

$$x^2-7x+12<0$$

$$\Rightarrow \quad (x-3)(x-4)<0$$

$$\Rightarrow \quad 3<x<4 \quad ...(iv)$$

Combining Eqs. (iii) and (iv), we get

Domain of $f(x)$ is $]3, 4[$.

Example 10 Find the domain of function

$$f(x)=\frac{1}{\log_{10}(1-x)}+\sqrt{(x+2)}.$$

Sol. $f(x)=\dfrac{1}{\log_{10}(1-x)}+\sqrt{x+2}$

(As we know, $\log_a x$ is defined when $x, a>0$ and $a \neq 1$, also $\log_a 1=0$)

Thus, $\log_{10}(1-x)$ exists when

$$1-x>0 \quad ...(i)$$

Also, $\dfrac{1}{\log_{10}(1-x)}$ exists when

$$1-x>0 \text{ and } 1-x \neq 1 \quad ...(ii)$$

$$\Rightarrow \quad x<1 \text{ and } x \neq 0 \quad ...(iii)$$

Now, we have $\sqrt{x+2}$, which exists when $x+2 \geq 0$

$$\text{or} \quad x \geq -2 \quad ...(iv)$$

Thus, $f(x)=\dfrac{1}{\log_{10}(1-x)}+\sqrt{x+2}$ exists when Eqs. (iii) and (iv) both hold true.

$$\Rightarrow \quad -2 \leq x<1 \text{ and } x \neq 0$$

$$\Rightarrow \quad x \in [-2, 0) \cup (0, 1)$$

Thus, domain of $f(x)$ is $[-2, 0) \cup (0, 1)$.

Example 11 Find the domain of $f(x)=\log_{10}(1+x^3)$.

Sol. $f(x)=\log_{10}(1+x^3)$ exists,

$$\text{if} \quad 1+x^3>0$$

$$\Rightarrow \quad (1+x)(1-x+x^2)>0$$

[where $(1-x+x^2)$ is always positive as $D<0$ and $a>0$]

So, $1+x>0$ or $x>-1$ or $x \in]-1, \infty[$

Thus, domain of above function $f(x)$ is $]-1, \infty[$.

Example 12 Find the domain of

$$f(x)=\log_{10}\log_{10}(1+x^3).$$

Sol. $f(x)=\log_{10}\log_{10}(1+x^3)$ exists,

$$\text{if} \quad \log_{10}(1+x^3)>0 \text{ and } 1+x^3>0$$

$$\Rightarrow \quad 1+x^3>10^0 \text{ and } 1+x^3>0$$

$$\Rightarrow \quad 1+x^3>1 \text{ and } 1+x^3>0$$

$$\Rightarrow \quad 1+x^3>1 \Rightarrow x^3>0$$

$$\Rightarrow \quad x \in (0, \infty)$$

Thus, domain of above function $f(x)$ is $(0, \infty)$.

Example 13 Find the domain of

$$f(x)=\log_{10}\{\log_{10}\log_{10}\log_{10} x\}.$$

Sol. $f(x)$ exists, if $\{\log_{10}\log_{10}\log_{10} x\}>0$ and $x>0$

$$\Rightarrow \quad \log_{10}\log_{10} x>10^0 \text{ and } x>0$$

$$\Rightarrow \quad \log_{10}\log_{10} x>1 \text{ and } x>0$$

$$\Rightarrow \quad \log_{10} x>10^1 \text{ and } x>0$$

$$\Rightarrow \quad x>10^{10} \text{ and } x>0$$

Thus, $f(x)$ exists, if $x \in (10^{10}, \infty)$.

Hence, domain of above function $f(x)$ is $(10^{10}, \infty)$.

Example 14 Find the domain for

$$f(x)=\sqrt{\log_{0.4}\left(\frac{x-1}{x+5}\right)}.$$

Sol. $f(x)=\sqrt{\log_{0.4}\left(\dfrac{x-1}{x+5}\right)}$ exists, if

$$\log_{0.4}\left(\frac{x-1}{x+5}\right) \geq 0 \quad \text{and} \quad \left(\frac{x-1}{x+5}\right)>0$$

$$\Rightarrow \quad \frac{x-1}{x+5} \leq (0.4)^0 \quad \text{and} \quad \frac{x-1}{x+5}>0$$

$$\Rightarrow \quad \frac{x-1}{x+5} \leq 1 \quad \text{and} \quad \frac{x-1}{x+5}>0$$

$$\Rightarrow \quad \frac{x-1}{x+5}-1 \leq 0 \quad \text{and} \quad \frac{x-1}{x+5}>0$$

$$\Rightarrow \quad \frac{-6}{x+5} \leq 0 \quad \text{and} \quad \frac{x-1}{x+5}>0$$

$\Rightarrow \quad x+5>0 \quad \text{and} \quad \dfrac{x-1}{x+5}>0$

$\Rightarrow \quad x>-5 \quad \text{and} \quad x-1>0 \quad [\text{as } x+5>0]$

$\Rightarrow \quad x>-5 \quad \text{and} \quad x>1$

Thus, domain $f(x)$ is $(1, \infty)$.

Example 15 Find the domain of

$$f(x)=\log_{100x}\left(\frac{2\log_{10}x+1}{-x}\right)$$

Sol. $f(x)$ exists, if $100x>0$ ...(i)

$100x \neq 1$ ...(ii)

$\dfrac{2\log_{10}x+1}{-x}>0$ ...(iii)

$2\log_{10}x+1<0$ [as $x>0$ from Eq. (i)]

$\log_{10}x<-\dfrac{1}{2}$

$x<(10)^{-1/2}$ ...(iv)

Now, from Eqs. (i), (ii) and (iv), we get

$x\in(0,10^{-2})\cup(10^{-2},10^{-1/2})$.

Thus, domain of $f(x)$ is $(0,10^{-2})\cup(10^{-2},10^{-1/2})$.

Example 16 Find the domain of definition of $f(x)=\dfrac{\log_2(x+3)}{x^2+3x+2}$. *[IIT JEE 2001]*

Sol. Here, $f(x)=\dfrac{\log_2(x+3)}{x^2+3x+2}=\dfrac{\log_2(x+3)}{(x+1)(x+2)}$ exists, if

Numerator $x+3>0 \Rightarrow x>-3$ ...(i)

and denominator $(x+1)(x+2)\neq 0$

$\Rightarrow \quad x\neq -1,-2$ ...(ii)

Thus, from Eqs. (i) and (ii), we get domain of $f(x)$ is $(-3,\infty)-\{-1,-2\}$.

(iii) Trigonometric Functions

Sine Function $f(x)=\sin x$

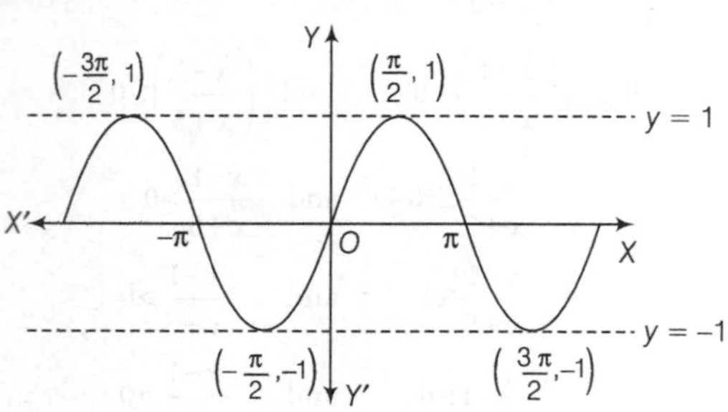

Figure 3.9

Here, domain is R and range is $[-1, 1]$.

Cosine Function $f(x)=\cos x$

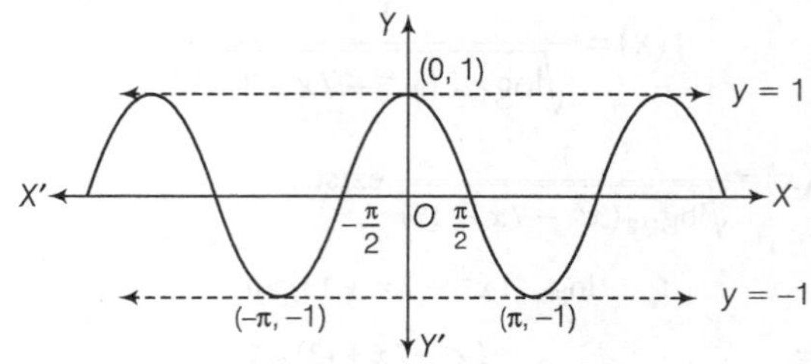

Figure 3.10

Here, domain is R and range is $[-1, 1]$.

Tangent Function $f(x)=\tan x$

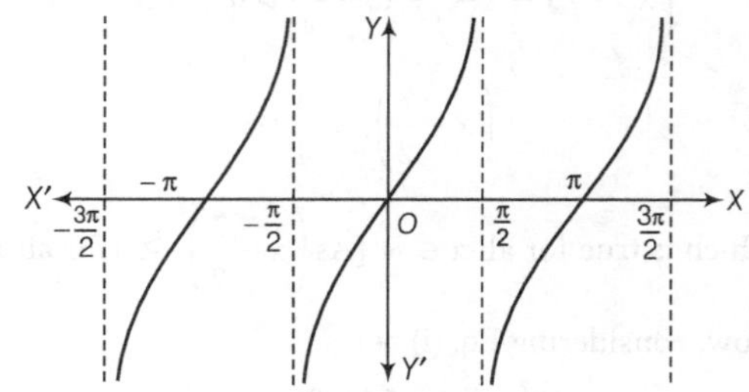

Figure 3.11

Here, domain is $R-\left\{\dfrac{(2n+1)\pi}{2}, n\in Z\right\}$ and range is R.

Cosecant Function $f(x)=\operatorname{cosec} x$

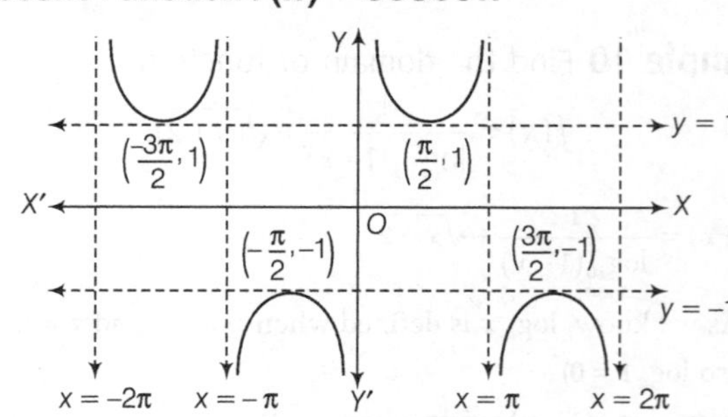

Figure 3.12

Here, domain is $R-\{n\pi, n\in Z\}$ and range is $R-(-1,1)$.

Secant Function $f(x)=\sec x$

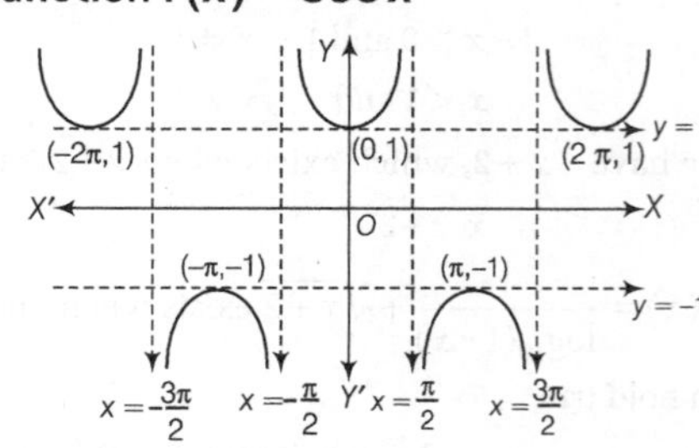

Figure 3.13

Here, domain is $R-\{(2n+1)\dfrac{\pi}{2}, n\in Z\}$ and range is $R-(-1,1)$.

Cotangent Function $f(x) = \cot x$

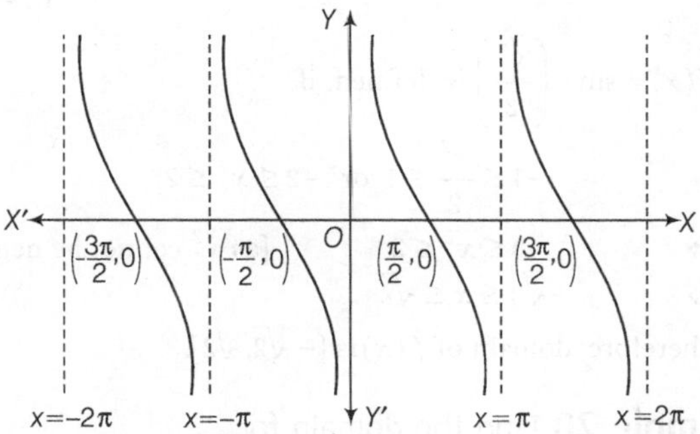

Figure 3.14

Here, domain is $R - \{n\pi, n \in Z\}$ and range is R.

Example 17 Find domain for $f(x) = \sqrt{\cos(\sin x)}$.

Sol. $f(x) = \sqrt{\cos(\sin x)}$ is defined, if

$$\cos(\sin x) \geq 0 \quad \text{...(i)}$$

As, we know $-1 \leq \sin x \leq 1$ for all x

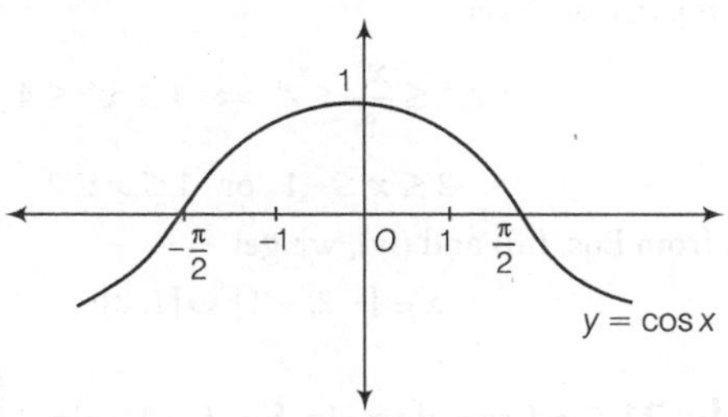

$\therefore$ $\cos(\sin x) \geq 0$, for all x (using graph of $y = \cos x$)

i.e. $x \in R$

Thus, domain $f(x)$ is R.

Example 18 The function $f(x)$ is defined in [0, 1]. Find the domain of $f(\tan x)$.

Sol. Here, $f(x)$ is defined in [0, 1].

$\Rightarrow x \in [0, 1]$, i.e. we can substitute only those values of x, which lie in [0, 1].

For $f(\tan x)$ to be defined, we must have

$$0 \leq \tan x \leq 1 \quad \text{[as } x \text{ is replaced by } \tan x]$$

i.e. $$n\pi \leq x \leq n\pi + \frac{\pi}{4} \quad \text{[in general]}$$

or $$0 \leq x \leq \frac{\pi}{4} \quad \text{[in particular]}$$

Thus, domain for $f(\tan x)$ is $\left[n\pi, n\pi + \frac{\pi}{4}\right]$.

Inverse Trigonometric Functions

We know that trigonometric functions are many-one in their actual domain. Hence, for inverse functions to get defined, the actual domain of trigonometric functions must be restricted to make the function one-one.

$y = \sin^{-1} x$

Since, $y = \sin x$ is not one-one. Therefore, it is not invertible. Let us consider the function $y = \sin x, x \in \left[\frac{-\pi}{2}, \frac{\pi}{2}\right]$ and $y \in [-1, 1]$ whose graph is the portion of the sin x curve in the interval $\left[\frac{-\pi}{2}, \frac{\pi}{2}\right]$, which is strictly increasing. Hence, one-one and its inverse is $y = \sin^{-1} x$.

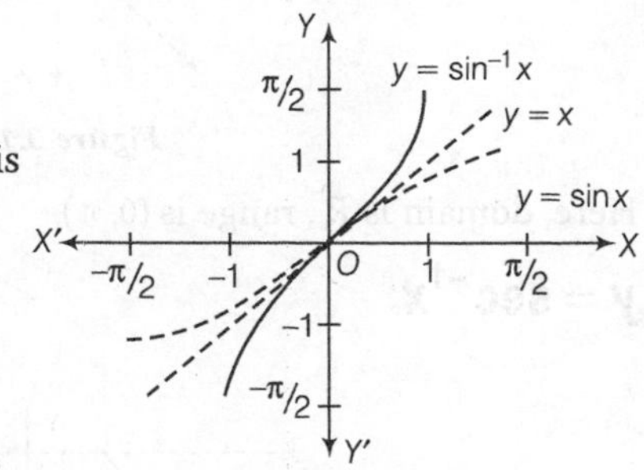

Figure 3.15

Here, domain is [− 1, 1]; range is $\left[-\frac{\pi}{2}, \frac{\pi}{2}\right]$.

$y = \cos^{-1} x$

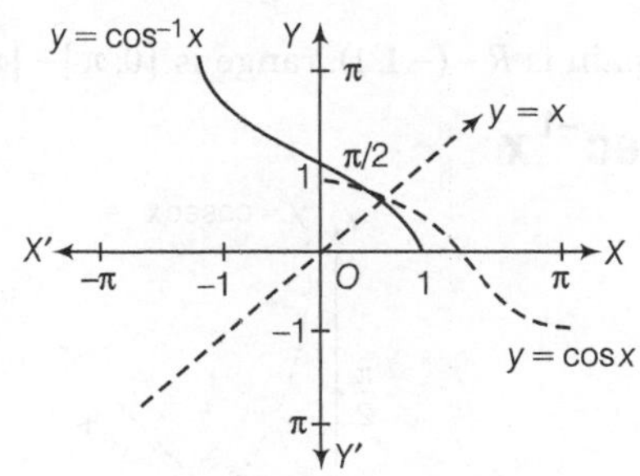

Figure 3.16

Here, domain is [− 1, 1]; range is [0, π].

$y = \tan^{-1} x$

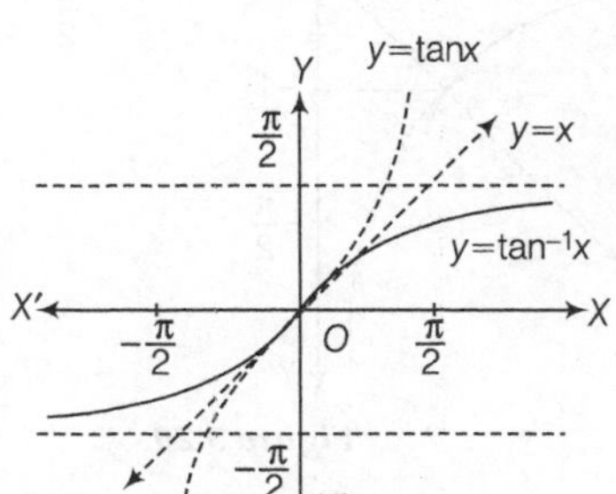

Figure 3.17

Here, domain is R; range is $\left(-\frac{\pi}{2}, \frac{\pi}{2}\right)$.

$y = \cot^{-1} x$

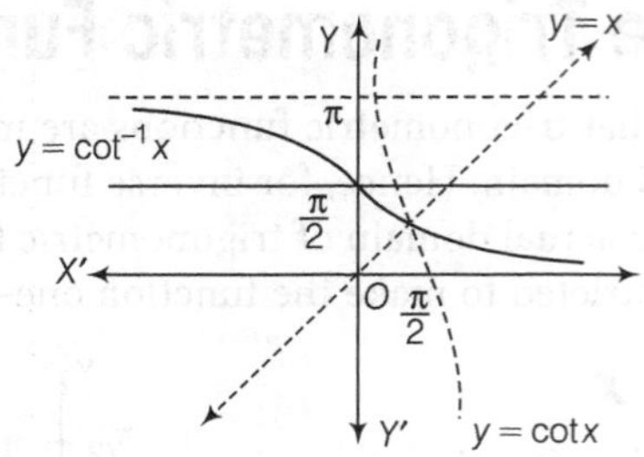

Figure 3.18

Here, domain is R ; range is $(0, \pi)$.

$y = \sec^{-1} x$

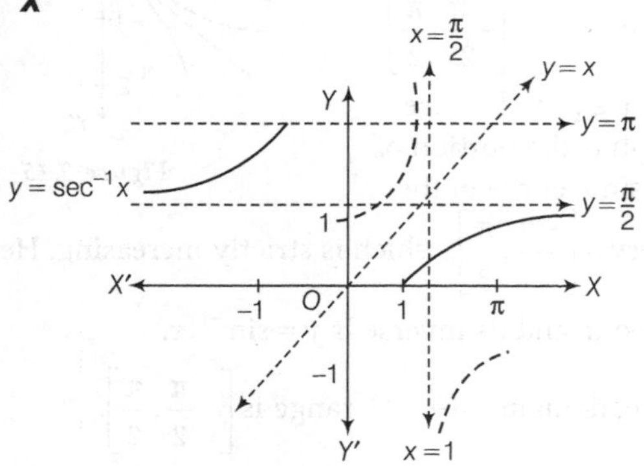

Figure 3.19

Here, domain is $R - (-1, 1)$; range is $[0, \pi] - \{\pi/2\}$.

$y = \operatorname{cosec}^{-1} x$

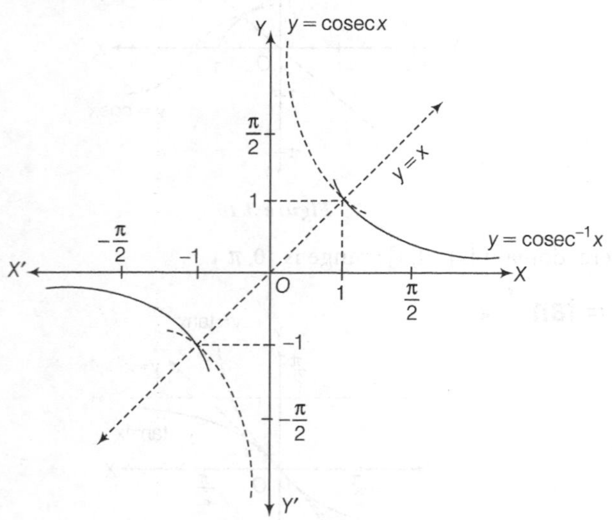

Figure 3.20

Here, domain is $R - (-1, 1)$; range is $\left[-\dfrac{\pi}{2}, \dfrac{\pi}{2}\right] - \{0\}$.

| Example 19 Find the domain for $f(x) = \sin^{-1}\left(\dfrac{x^2}{2}\right)$.

Sol. $f(x) = \sin^{-1}\left(\dfrac{x^2}{2}\right)$ is defined, if

$$-1 \le \frac{x^2}{2} \le 1 \text{ or } -2 \le x^2 \le 2$$

$$\Rightarrow \quad 0 \le x^2 \le 2 \quad [\text{as } x^2 \text{ cannot be negative}]$$

$$\Rightarrow \quad -\sqrt{2} \le x \le \sqrt{2}$$

Therefore, domain of $f(x)$ is $[-\sqrt{2}, \sqrt{2}]$.

| Example 20 Find the domain for

$$y = \sin^{-1}\left(\log_2\left(\frac{x^2}{2}\right)\right).$$

Sol. For y to be defined, $\dfrac{x^2}{2} > 0$ ...(i)

and $\quad -1 \le \log_2\left(\dfrac{x^2}{2}\right) \le 1$ ...(ii)

From Eq. (i), we have $x \in R - \{0\}$ ...(iii)

From Eq. (ii), we have

$$2^{-1} \le \frac{x^2}{2} \le 2^1 \Rightarrow 1 \le x^2 \le 4$$

$$\Rightarrow \quad -2 \le x \le -1 \text{ or } 1 \le x \le 2 \quad \text{...(iv)}$$

Thus, from Eqs. (iii) and (iv), we get

$$x \in [-2, -1] \cup [1, 2]$$

| Example 21 Find the domain for $f(x) = \sin^{-1}\left(\dfrac{1+x^2}{2x}\right)$.

Sol. $f(x) = \sin^{-1}\left(\dfrac{1+x^2}{2x}\right)$ is defined, if

$$-1 \le \frac{1+x^2}{2x} \le 1 \text{ or } \left|\frac{1+x^2}{2x}\right| \le 1$$

$$\Rightarrow \quad |1+x^2| \le |2x|$$

$$\Rightarrow \quad 1+x^2 \le |2x| \quad [\text{as } 1+x^2 > 0]$$

$$\Rightarrow \quad x^2 - 2|x| + 1 \le 0$$

$$\Rightarrow \quad |x|^2 - 2|x| + 1 \le 0 \quad [\text{as } x^2 = |x|^2]$$

$$\Rightarrow \quad (|x| - 1)^2 \le 0$$

But $(|x|-1)^2$ is either always positive or zero.

Thus, $\quad (|x|-1)^2 = 0$

$$\Rightarrow \quad |x| = 1 \Rightarrow x = \pm 1$$

Thus, domain for $f(x)$ is $\{-1, 1\}$.

Example 22 Find the domain for $y = \sqrt{\sin^{-1}(\log_2 x)}$.

Sol. y is defined, if $x > 0$, ...(i)

$$-1 \le \log_2 x \le 1$$

$$\Rightarrow \quad 2^{-1} \le x \le 2$$

$$\Rightarrow \quad \frac{1}{2} \le x \le 2 \quad \text{...(ii)}$$

and $\quad \sin^{-1}(\log_2 x) \ge 0$ ...(iii)

$$\Rightarrow \quad \log_2 x \ge 0$$

$$\Rightarrow \quad x \ge 2^0$$

$$\Rightarrow \quad x \ge 1 \quad \text{...(iv)}$$

From Eqs. (i), (ii) and (iv), we get

$$1 \le x \le 2$$

Hence, domain is $1 \le x \le 2$ or $x \in [1, 2]$.

Example 23 Find the domain of the function,

$$f(x) = \log\left\{\log_{|\sin x|}(x^2 - 8x + 23) - \frac{3}{\log_2|\sin x|}\right\}.$$

Sol. Here, $f(x)$ is defined, if

$$\log_{|\sin x|}(x^2 - 8x + 23) - \frac{3}{\log_2|\sin x|} > 0$$

$$\Rightarrow \quad \log_{|\sin x|}(x^2 - 8x + 23) - 3\log_{|\sin x|} 2 > 0$$

$$\Rightarrow \quad \log_{|\sin x|}\left(\frac{x^2 - 8x + 23}{8}\right) > 0$$

$$\Rightarrow \quad \frac{x^2 - 8x + 23}{8} < |\sin x|^0$$

and $\quad |\sin x| \ne 0, 1$

$$\Rightarrow \quad x^2 - 8x + 23 < 8 \text{ and } |\sin x| \ne 0, 1$$

$$\Rightarrow \quad x^2 - 8x + 15 < 0 \text{ and } \sin x \ne 0, \pm 1$$

$$\Rightarrow \quad (x-3)(x-5) < 0 \text{ and } x \ne n\pi, (2n+1)\frac{\pi}{2}$$

$$\Rightarrow \quad x \in (3, 5) \text{ and } x \ne \pi, \frac{3\pi}{2} \quad \text{[using number line rule]}$$

$$\Rightarrow \quad x \in (3, \pi) \cup \left(\pi, \frac{3\pi}{2}\right) \cup \left(\frac{3\pi}{2}, 5\right)$$

Thus, domain of $f(x)$ is $(3, \pi) \cup \left(\pi, \frac{3\pi}{2}\right) \cup \left(\frac{3\pi}{2}, 5\right)$.

Exercise for Session 3

- **Directions** (Q. Nos. 1 to 10) *Find the domain of the following.*

1. $f(x) = \log_{10}(\sqrt{x-4} + \sqrt{6-x})$
2. $f(x) = \sqrt{\log_{1/2}\left(\frac{5x - x^2}{4}\right)}$
3. $f(x) = \sqrt{\log_{0.3}\left(\frac{3x - x^2}{x - 1}\right)}$
4. $f(x) = \sqrt{\frac{\log_{0.3}(x-1)}{x^2 - 2x - 8}}$
5. $f(x) = \log_{10}(1 - \log_{10}(x^2 - 5x + 16))$
6. $f(x) = \sin|x| + \sin^{-1}(\tan x) + \sin(\sin^{-1} x)$
7. $f(x) = \sqrt{e^{\cos^{-1}(\log_4 x^2)}}$
8. $f(x) = \sin^{-1}\left(\frac{3 - 2x}{5}\right) + \sqrt{3 - x}$
9. $f(x) = \frac{\log_{2x} 3}{\cos^{-1}(2x - 1)}$
10. $f(x) = \log_{10}\log_2\log_{2/\pi}(\tan^{-1} x)^{-1}$

Session 4

Piecewise Functions

Piecewise Functions

(i) Absolute Value Function (or Modulus Function)

The modulus function is defined as

$$y = |x| = \begin{cases} x, & x \geq 0 \\ -x, & x < 0 \end{cases}.$$

It is the numerical value of x.

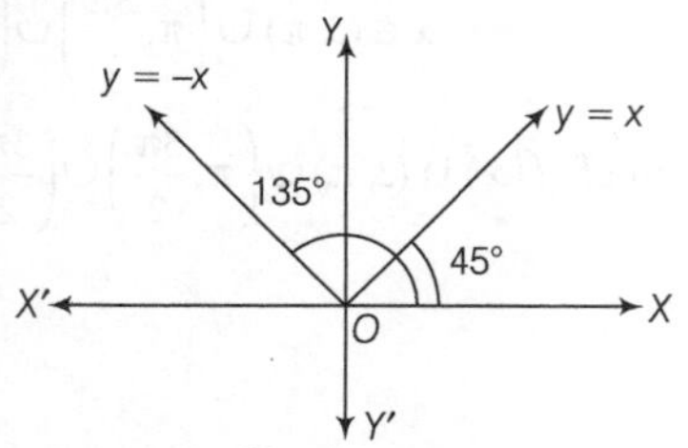

Figure 3.21

Geometrical Interpretation of Modulus of a Function

Geometrically, $|x|$ represents the distance of the point $P(x, 0)$ or $Q(-x, 0)$ from origin.

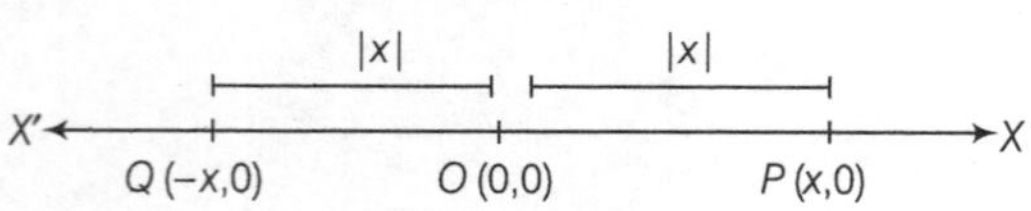

Figure 3.22

i.e. $\quad d(O, P) = \sqrt{(x-0)^2 + (0-0)^2} = \sqrt{x^2} = |x|$

$\quad d(O, Q) = \sqrt{(-x-0)^2 + (0-0)^2} = \sqrt{x^2} = |x|$

Properties of Modulus

(i) $|x|^2 = x^2$ (ii) $\sqrt{x^2} = |x|$

(iii) $||x|| = |-x| = |x|$ (iv) $|x| = \max\{-x, x\}$

(v) $-|x| = \min\{-x, x\}$ (vi) $\max(a, b) = \dfrac{a+b}{2} + \left|\dfrac{a-b}{2}\right|$

(vii) $\min(a, b) = \dfrac{a+b}{2} - \left|\dfrac{a-b}{2}\right|$

(viii) $|x+y| \leq |x| + |y|$

(ix) $|x+y| = |x| + |y|$, iff $xy \geq 0$

(x) $|x-y| = |x| + |y|$, iff $xy \leq 0$

(xi) $|x| \leq a$, (a is +ve) $\Rightarrow -a \leq x \leq a$ or $x \in [-a, a]$

(xii) $|x| \geq a$, (a is +ve) $\Rightarrow x \leq -a$ or $x \geq a$ or $x \in (-\infty, -a] \cup [a, \infty)$

(xiii) $|x| \leq a$, (a is −ve) $\Rightarrow$ No solution or $x \in \phi$

(xiv) $|x| \geq a$, (a is −ve) $\Rightarrow x \in$ Real number or $x \in (-\infty, \infty)$

(xv) $a \leq |x| \leq b$, where a and b are +ve.

$\Rightarrow -b \leq x \leq -a$ or $a \leq x \leq b$ or $x \in [-b, -a] \cup [a, b]$

Example 24 Solve for x.

(i) $(|x|-1)(|x|-2) \leq 0$ (ii) $\dfrac{|x|-1}{|x|-2} \geq 0$

(iii) $|x-3| + |4-x| = 1$ (iv) $|x| + |x+4| = 4$

Sol. (i) $(|x|-1)(|x|-2) \leq 0$

Using number line rule for $|x|$, we get

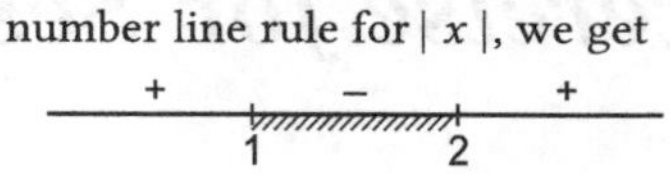

$\Rightarrow \quad 1 \leq |x| \leq 2$

i.e. $\quad x \in [-2, -1] \cup [1, 2]$

(ii) $\dfrac{|x|-1}{|x|-2} \geq 0$. Using number line rule for $|x|$, we get

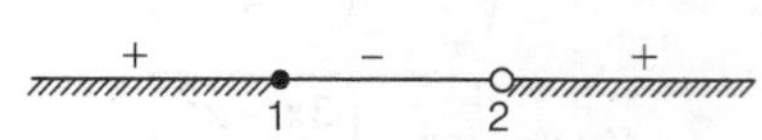

$\Rightarrow \quad |x| \leq 1$ or $|x| > 2$

i.e. $x \in [-1, 1] \cup (-\infty, -2) \cup (2, \infty)$

(iii) $|x-3| + |4-x| = 1$

As we know, $|x| + |y| = |x+y|$, iff $xy \geq 0$

$\therefore \quad (x-3)(4-x) \geq 0$ or $-(x-3)(x-4) \geq 0$

Using number line rule, we get

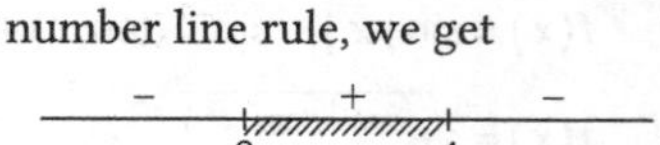

$\Rightarrow \quad x \in [3, 4]$

(iv) $|x| + |x+4| = 4$

As we know, $|x| + |y| = |x-y|$, iff $xy \leq 0$

$\therefore \quad x(x+4) \leq 0$

Using number line rule, we get

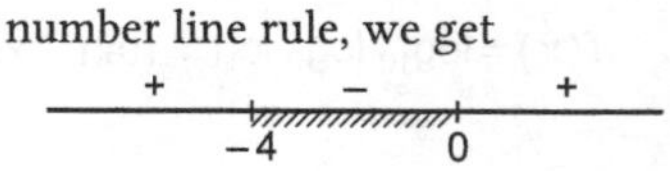

$\Rightarrow \quad x \in [-4, 0]$

Example 25 Solve $|x^2-1+\sin x| = |x^2-1| + |\sin x|$, where $x \in [-2\pi, 2\pi]$.

Sol. Let $f(x) = x^2 - 1$ and $g(x) = \sin x$

We know, $|f(x) + g(x)| = |f(x)| + |g(x)|$ iff $f(x) \cdot g(x) \geq 0$.

$\therefore \quad (x^2 - 1) \cdot \sin x \geq 0$ on $[-2\pi, 2\pi]$.

Using number line rule, we get

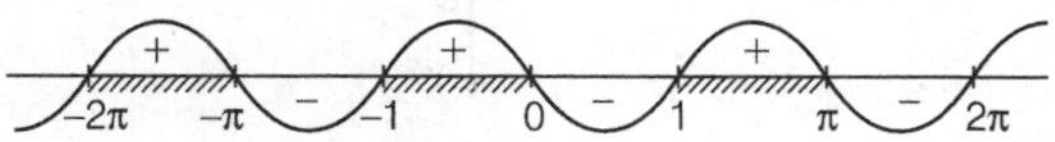

$\Rightarrow \quad x \in [-2\pi, -\pi] \cup [-1, 0] \cup [1, \pi]$

Example 26 Solve $\left|\dfrac{x}{x-1}\right| + |x| = \dfrac{x^2}{|x-1|}$.

Sol. Let $f(x) = \dfrac{x}{x-1}$ and $g(x) = x$

Then, $\quad f(x) + g(x) = \dfrac{x}{x-1} + x = \dfrac{x^2}{x-1}$

We know, $|f(x)| + |g(x)| = |f(x) + g(x)|$,

iff $\quad f(x) \cdot g(x) \geq 0$

$\therefore \quad \dfrac{x}{x-1} \cdot x \geq 0 \Rightarrow \dfrac{x^2}{x-1} \geq 0$

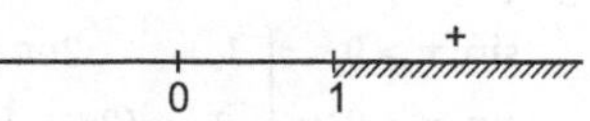

$\Rightarrow \quad x \in \{0\} \cup (1, \infty)$

Example 27 Find domain for $y = \dfrac{1}{\sqrt{|x| - x}}$.

Sol. y is defined, if $(|x| - x) > 0 \Rightarrow |x| > x$, which is true for negative x only.

Hence, domain $\in (-\infty, 0)$.

Example 28 Find domain for

$$y = \cos^{-1}\left(\frac{1-2|x|}{3}\right) + \log_{|x-1|} x.$$

Sol. Here, $\cos^{-1}\left(\dfrac{1-2|x|}{3}\right)$ exists, if $-1 \leq \dfrac{1-2|x|}{3} \leq 1$

$\Rightarrow \quad -3 \leq 1 - 2|x| \leq 3 \Rightarrow -4 \leq -2|x| \leq 2$

$\Rightarrow \quad 2 \geq |x| \geq -1 \quad \Rightarrow -2 \leq x \leq 2 \quad$ ...(i)

Also, $\log_{|x-1|} x$ exists, if

$x > 0$, $|x - 1| > 0$ and $|x - 1| \neq 1$

$x > 0$, $x \in R - \{1\}$ and $x \neq 0, 2 \quad$ ...(ii)

Hence, from Eqs. (i) and (ii), we get

$x \in (0, 1) \cup (1, 2)$

Example 29 Domain of the function

$$f(x) = \frac{1}{\sqrt{4x - |x^2 - 10x + 9|}} \text{ is}$$

(a) $(7 - \sqrt{40}, 7 + \sqrt{40})$ (b) $(0, 7 + \sqrt{40})$

(c) $(7 - \sqrt{40}, \infty)$ (d) None of these

Sol. Here, $f(x) = \dfrac{1}{\sqrt{4x - |x^2 - 10x + 9|}}$ would exists, if

$4x - |x^2 - 10x + 9| > 0$

i.e. $\quad |x^2 - 10x + 9| < 4x$,

where

$$|x^2 - 10x + 9| = \begin{cases} x^2 - 10x + 9, & x \leq 1 \text{ or } x \geq 9 \\ -(x^2 - 10x + 9), & 1 < x < 9 \end{cases}$$

$\Rightarrow$ Two cases

***Case* I** When $x \leq 1$ or $x \geq 9$

$\therefore \quad x^2 - 10x + 9 < 4x$

$\Rightarrow \quad x^2 - 14x + 9 < 0 \Rightarrow (x - 7)^2 < 40$

$\Rightarrow \quad x \in (7 - \sqrt{40}, 7 + \sqrt{40}) \quad$ [but $x \leq 1$ or $x \geq 9$]

$\Rightarrow \quad x \in (7 - \sqrt{40}, 1] \cup [9, 7 + \sqrt{40}) \quad$ ...(i)

***Case* II** When $1 < x < 9$

$-x^2 + 10x - 9 < 4x \Rightarrow x^2 - 6x + 9 > 0$

$\Rightarrow \quad (x - 3)^2 > 0$, which is always true except at $x = 3$

$\therefore \quad x \in (1, 9) - \{3\} \quad$ ...(ii)

From Eqs. (i) and (ii), domain of $f(x)$ is

$(7 - \sqrt{40}, 7 + \sqrt{40}) - \{3\}$

Hence, (d) is the correct answer.

Example 30 The domain of the function

$f(x) = \sqrt{|\sin^{-1}(\sin x)| - \cos^{-1}(\cos x)}$ in $[0, 2\pi]$ is

(a) $\left[0, \dfrac{\pi}{2}\right] \cup \left[\dfrac{3\pi}{2}, 2\pi\right]$ (b) $[\pi, 2\pi]$

(c) $[0, \pi] - \left\{\dfrac{\pi}{2}\right\}$ (d) $[0, 2\pi] - \left\{\dfrac{\pi}{2}, \dfrac{3\pi}{2}\right\}$

Sol. As, $|\sin^{-1}(\sin x)|$ could be sketched, as

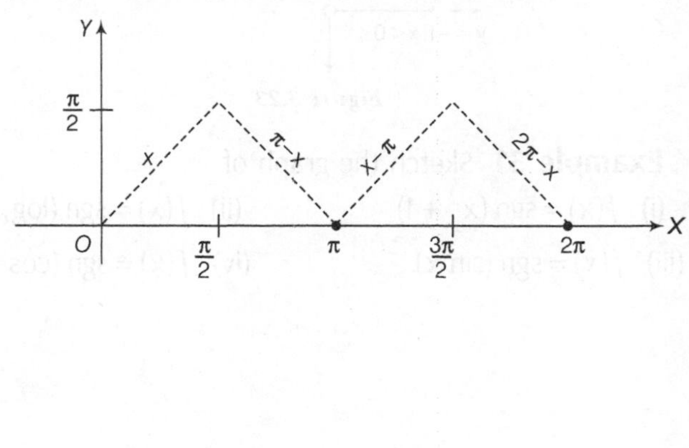

and $\cos^{-1}(\cos x)$ could be sketched as

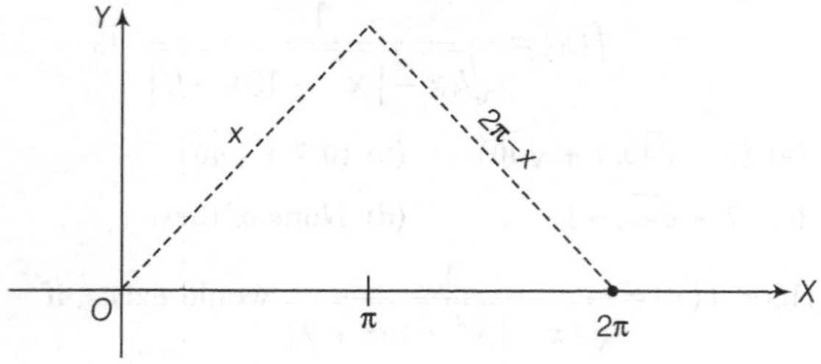

$\therefore |\sin^{-1}(\sin x)| > \cos^{-1}(\cos x)$ is not possible.

Only equality holds as

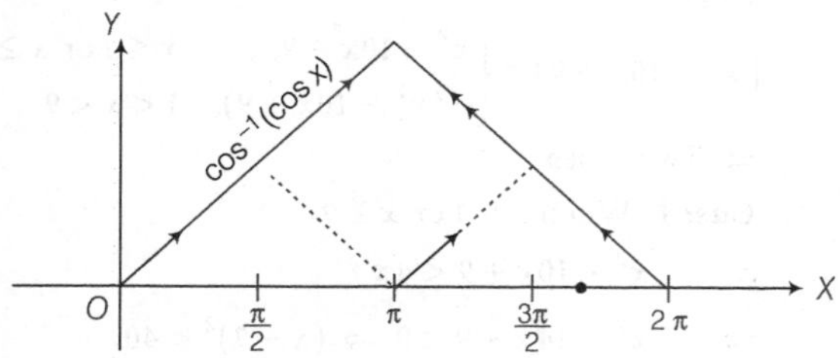

Thus, $|\sin^{-1}(\sin x)| = \cos^{-1}(\cos x)$

When $\quad x \in \left[0, \frac{\pi}{2}\right] \cup \left[\frac{3\pi}{2}, 2\pi\right]$

So, domain for $f(x)$ is $\left[0, \frac{\pi}{2}\right] \cup \left[\frac{3\pi}{2}, 2\pi\right]$.

Hence, (a) is the correct answer.

(ii) Signum Function

Signum function is denoted by sgn (x) and it is defined by

$$y = \text{sgn}(x) = \begin{cases} \dfrac{|x|}{x} \text{ or } \dfrac{x}{|x|}, & \text{if } x \neq 0 \\ 0, & \text{if } x = 0 \end{cases} = \begin{cases} 1, & \text{if } x > 0 \\ -1, & \text{if } x < 0 \\ 0, & \text{if } x = 0 \end{cases}$$

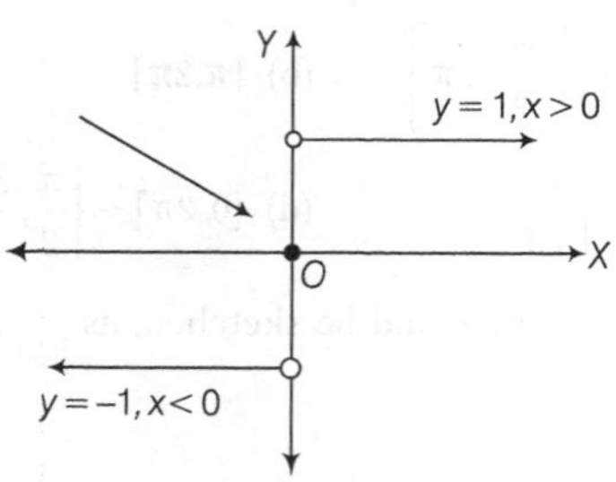

Figure 3.23

Example 31 Sketch the graph of

(i) $f(x) = \text{sgn}(x^2 + 1)$. (ii) $f(x) = \text{sgn}(\log_e x)$.

(iii) $f(x) = \text{sgn}(\sin x)$. (iv) $f(x) = \text{sgn}(\cos x)$.

Sol. (i) $f(x) = \text{sgn}(x^2 + 1) = \begin{cases} -1, & \text{if } x^2 + 1 < 0 \\ 0, & \text{if } x^2 + 1 = 0 \\ 1, & \text{if } x^2 + 1 > 0 \end{cases}$

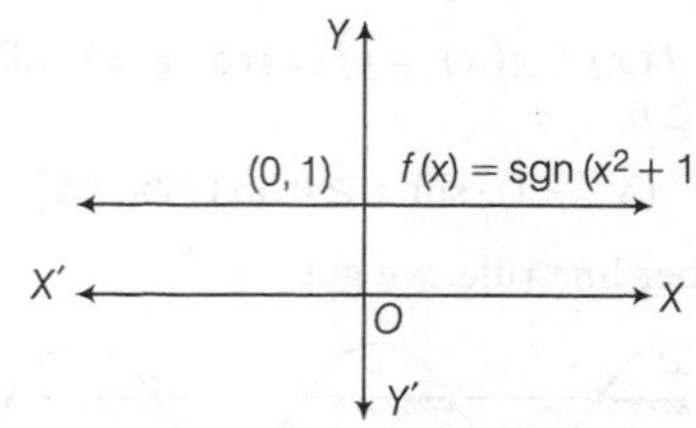

$$f(x) = 1, x \in R$$

(ii) $f(x) = \text{sgn}(\log_e x)$

$$= \begin{cases} 1, & \log_e x > 0 \\ -1, & \log_e x < 0 \\ 0, & \log_e x = 0 \end{cases} \quad \therefore f(x) = \begin{cases} 1, & x > 1 \\ -1, & 0 < x < 1 \\ 0, & x = 1 \end{cases}$$

Graph for, $f(x) = \text{sgn}(\log_e x)$ is

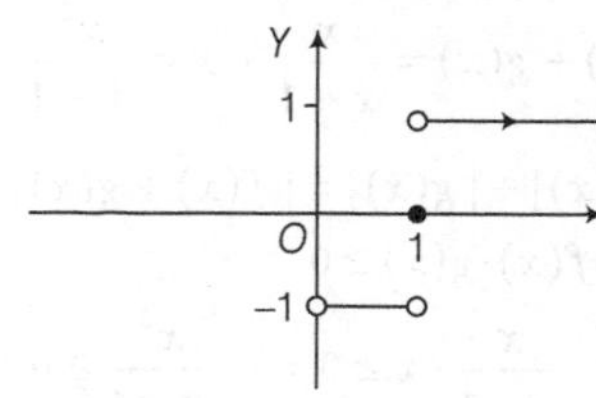

(iii) $f(x) = \text{sgn}(\sin x)$

$$f(x) = \begin{cases} 1, & \sin x > 0 \\ -1, & \sin x < 0 \\ 0, & \sin x = 0 \end{cases} = \begin{cases} 1, & 2n\pi < x < (2n+1)\pi \\ -1, & (2n+1)\pi < x < (2n+2)\pi \\ 0, & x = n\pi \end{cases}$$

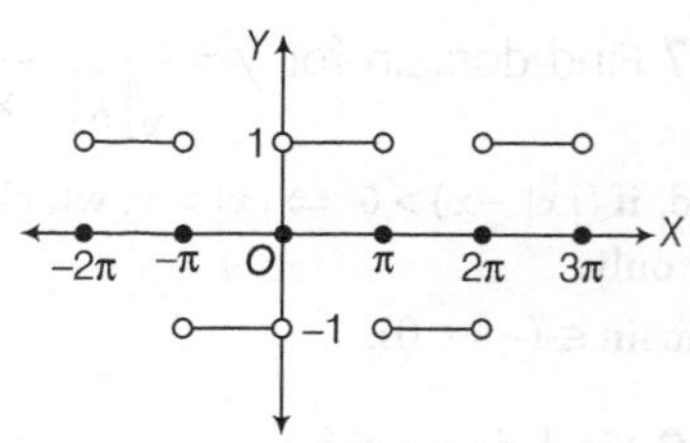

(iv) $f(x) = \text{sgn}(\cos x)$

$$f(x) = \begin{cases} 1, & \cos x > 0 \\ -1, & \cos x < 0 \\ 0, & \cos x = 0 \end{cases}$$

$$= \begin{cases} 1, & 2n\pi - \pi/2 < x < 2n\pi + \pi/2 \\ -1, & 2n\pi + \pi/2 < x < 2n\pi + 3\pi/2 \\ 0, & x = (2n+1)\pi/2 \end{cases}$$

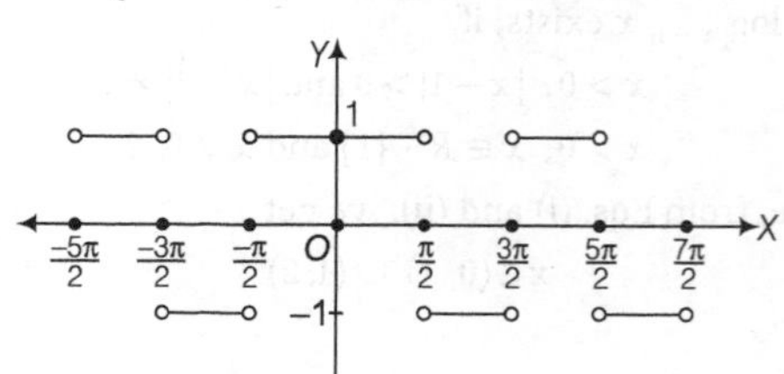

(iii) Greatest Integer Function or Step Function

$[x]$ indicates the integral part of x, which is nearest and smaller to x. It is also known as floor of x.

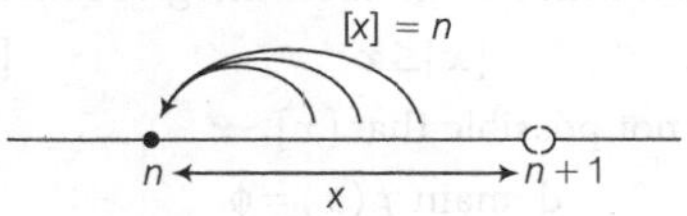

Figure 3.24

Thus, $[2.3202]=2,\ [0.23]=0,\ [5]=5,$
$[-8.0725]=-9,\ [-0.6]=-1$

In general, $n \le x < n+1$ $[n \in \text{Integer}]$

$\Rightarrow$ $[x]=n$

Here, $f(x)=[x]$ could be expressed graphically as

x	$[x]$
$0 \le x < 1$	0
$1 \le x < 2$	1
$2 \le x < 3$	2
$-1 \le x < 0$	-1
$-2 \le x < -1$	-2

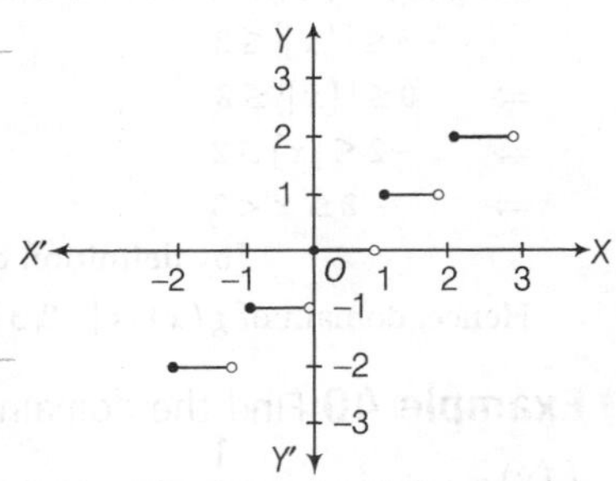

Figure 3.25

(•) darkened circle represents value is taken, (○) represents value is neglected.

Properties of the Greatest Integer Functions

(i) $[x] \le x < [x]+1$

(ii) $x-1 < [x] < x$

(iii) $I \le x < I+1 \Rightarrow [x]=I$, where $I \in$ Integer

(iv) $[[x]]=[x]$

(v) $[x]+[-x]=\begin{cases} 0, & \text{if } x \in \text{Integer} \\ -1, & \text{if } x \notin \text{Integer} \end{cases}$

i.e. $[-x]=\begin{cases} -x, & \text{if } x \in \text{Integer} \\ -1-[x], & \text{if } x \notin \text{Integer} \end{cases}$

(vi) $[x]-[-x]=\begin{cases} 2x, & \text{if } x \in \text{Integer} \\ 2[x]+1, & \text{if } x \notin \text{Integer} \end{cases}$

(vii) $[x \pm n]=[x] \pm n, n \in$ Integer

(viii) $[x] \ge n \Leftrightarrow x \ge n, n \in$ Integer

(ix) $[x] > n \Leftrightarrow x \ge n+1, n \in$ Integer

(x) $[x] \le n \Leftrightarrow x < n+1, n \in$ Integer

(xi) $[x] < n \Leftrightarrow x < n, n \in$ Integer

(xii) $[x]=\left[\dfrac{x}{2}\right]+\left[\dfrac{x+1}{2}\right]$

(xiii) $\left[\dfrac{n+1}{2}\right]+\left[\dfrac{n+2}{4}\right]+\left[\dfrac{n+4}{8}\right]+\ldots=n,$

$n \in$ Natural number

(xiv) $[x]+[y] \le [x+y] \le [x]+[y]+1$

(xv) $[x]+\left[x+\dfrac{1}{n}\right]+\left[x+\dfrac{2}{n}\right]+$

$\ldots+\left[x+\dfrac{n-1}{n}\right]=[nx], n \in N$

| Example 32 Find domain for, $f(x)=\cos^{-1}[x]$.

Sol. As, $y=\cos^{-1}x$ exists, when $-1 \le x \le 1$.

$\therefore$ $f(x)=\cos^{-1}[x]$ exists, when $-1 \le [x] \le 1 \Rightarrow -1 \le x < 2$

or $x \in [-1, 2)$

| Example 33 Find the value of

$$\left[\frac{3}{4}\right]+\left[\frac{3}{4}+\frac{1}{100}\right]+\left[\frac{3}{4}+\frac{2}{100}\right]+\ldots+\left[\frac{3}{4}+\frac{99}{100}\right].$$

Sol. $\left[\dfrac{3}{4}\right]+\left[\dfrac{3}{4}+\dfrac{1}{100}\right]+\ldots+\left[\dfrac{3}{4}+\dfrac{24}{100}\right]+\left[\dfrac{3}{4}+\dfrac{25}{100}\right]$

$+\ldots+\left[\dfrac{3}{4}+\dfrac{99}{100}\right]$

$\Rightarrow$ $\underbrace{[0.75]+\ldots+[0.99]}_{\text{25 terms are each equal to zero}}+\underbrace{[1.0]+\ldots+[1.74]}_{\text{75 terms are each equal to 1}} \Rightarrow 75$

Aliter Apply property (xv)

Here, $x=\dfrac{3}{4}$ and $n=100$

$\therefore$ $\left[\dfrac{3}{4}\right]+\left[\dfrac{3}{4}+\dfrac{1}{100}\right]+\ldots+\left[\dfrac{3}{4}+\dfrac{99}{100}\right]=\left[\dfrac{3}{4}\times 100\right]=75$

| Example 34 Given $y=2[x]+3$ and $y=3[x-2]+5$, find the value of $[x+y]$.

Sol. We have, $2[x]+3=3[x-2]+5$

$\Rightarrow$ $2[x]+3=3[x]-6+5$

$\Rightarrow$ $[x]=4 \Rightarrow 4 \le x < 5$

or $x=4+f$ $[f \rightarrow \text{fraction}]$

and $y=2[x]+3=11$

Hence, $[x+y]=[4+f+11]=[15+f]=15$

Example 35 Find domain for $f(x) = [\sin x] \cos\left(\dfrac{\pi}{[x-1]}\right)$.

Sol. $[\sin x]$ is always defined. $\cos\left(\dfrac{\pi}{[x-1]}\right)$ is also defined

except when $[x-1] = 0 \Rightarrow 0 \le x - 1 < 1$

$\Rightarrow \quad 1 \le x < 2$

Hence, domain of $f(x)$ is $R - [1,2)$.

Example 36 The domain of the function $f(x) = \dfrac{\log_4(5-[x-1]-[x]^2)}{x^2+x-2}$ is

(where $[x]$ denotes greatest integer function)

Sol. For domain of $f(x)$,

$$5-[x-1]-[x]^2 > 0$$

and $\quad x^2+x-2 \ne 0$

$\Rightarrow \quad (x+2)(x-1) \ne 0$

$\Rightarrow \quad x \ne 1, -2$

Now, $\quad 5-[x]+1-[x]^2 > 0$

$\Rightarrow \quad [x]^2+[x]-6 < 0$

$\Rightarrow \quad ([x]+3)([x]-2) < 0$

$\Rightarrow \quad -3 < [x] < 2$

$\Rightarrow \quad -2 \le x < 2$

$\because \quad x \ne 1, -2$

$\therefore$ Domain $\in (-2, 1) \cup (1, 2)$

Example 37 Let $[\sqrt{n^2+1}] = [\sqrt{n^2+\lambda}]$, where $n, \lambda \in N$. Show that λ can have $2n$ different values.

Sol. We have, $n^2+1 = (n+1)^2 - 2n < (n+1)^2; n \in N$

i.e. $\quad \sqrt{n^2+1} < n+1$

$\Rightarrow \quad n < \sqrt{n^2+1} < n+1$

$\Rightarrow \quad [\sqrt{n^2+1}] = n$

$\therefore \quad [\sqrt{n^2+\lambda}] = n$

$\Rightarrow \quad n < \sqrt{n^2+\lambda} < n+1$

or $\quad n^2 < (n^2+\lambda) < (n+1)^2$

$\Rightarrow \quad 0 < \lambda < 2n+1$

Thus, λ can take $2n$ different values.

Example 38 $f(x) = \dfrac{1}{\sqrt{[x]-x}}$, where $[\cdot]$ denotes the greatest integer function less than or equals to x. Then, find the domain of $f(x)$.

Sol. $f(x) = \dfrac{1}{\sqrt{[x]-x}}$ exists, if

$$[x]-x > 0$$

i.e. $\quad [x] > x$

But from definition of greatest integral function, we know

$[x] \le x \qquad$ [as $x = [x] + \{x\}$]

Thus, it is not possible that $[x] > x$.

Hence, $\quad$ domain $f(x) = \phi$

Example 39 If domain for $y = f(x)$ is $[-3, 2]$, find the domain of $g(x) = f\{|[x]|\}$.

Sol. Here, $f(x)$ is defined in $[-3, 2]$.

$\Rightarrow \quad x \in [-3, 2]$.

(i.e. we can substitute only those values of x, which lie between $[-3, 2]$).

For $g(x) = f\{|[x]|\}$ to be defined, we must have

$-3 \le |[x]| \le 2$

$\Rightarrow \quad 0 \le |[x]| \le 2 \qquad$ [as $|x| \ge 0$ for all x]

$\Rightarrow \quad -2 \le [x] \le 2 \qquad$ [as $|x| \le a \Rightarrow -a \le x \le a$]

$\Rightarrow \quad -2 \le x < 3$

[by definition of greatest integral function]

Hence, domain of $g(x)$ is $[-2, 3[$ or $[-2, 3)$.

Example 40 Find the domain of function $f(x) = \dfrac{1}{[|x-1|]+[|7-x|]-6}$, where $[\cdot]$ denotes the greatest integer function.

Sol. $f(x)$ is defined when

$$[|x-1|]+[|7-x|]-6 \ne 0$$

$$\Rightarrow \begin{cases} [1-x]+[7-x] \ne 6, & \text{when } x \le 1 & \text{...(i)} \\ [x-1]+[7-x] \ne 6, & \text{when } 1 \le x \le 7 & \text{...(ii)} \\ [x-1]+[x-7] \ne 6, & \text{when } x \ge 7 & \text{...(iii)} \end{cases}$$

$[1-x]+[7-x] \ne 6$

$1+[-x]+7+[-x] \ne 6 \Rightarrow 2[-x] \ne -2$

$\Rightarrow \quad [-x] \ne -1$

$\Rightarrow \quad x \notin (0, 1] \qquad$ ...(a)

From Eq. (ii), we have

$[x-1]+[7-x] \ne 6$

$\Rightarrow \quad [x]-1+7+[-x] \ne 6$

$\Rightarrow \quad [x]+[-x] \ne 0$

$\Rightarrow \quad x \notin$ Integer

$\Rightarrow \quad x \notin \{1, 2, 3, 4, 5, 6, 7\} \qquad$ ...(b)

From Eq. (iii), we have

$[x-1]+[x-7] \ne 6$

$\Rightarrow \quad [x]-1+[x]-7 \ne 6$

$\Rightarrow \quad 2[x] \ne 14 \Rightarrow [x] \ne 7$

$\Rightarrow \quad x \notin [7, 8) \qquad$ ...(c)

Hence, froms (a), (b) and (c), we get

Domain $f(x)$ is $R-\{(0,1)\cup\{1,2,3,4,5,6,7\}\cup(7,8)\}$.

Example 41 If the function $f(x)=[3.5+b\sin x]$ (where $[\cdot]$ denotes the greatest integer function) is an even function, the complete set of values of b is

(a) $(-0.5, 0.5)$ (b) $[-0.5, 0.5]$
(c) $(0, 1)$ (d) $[-1, 1]$

Sol. We have $f(x)=[3.5+b\sin x]$ For $f(x)$ to be an even funcion.

$$3<3.5+b\sin x<4,\ \forall x\in R$$

$$\Rightarrow \quad -0.5<b<0.5 \Rightarrow b\in(-0.5, 0.5)$$

Hence, (a) is the correct answer.

Example 42 The domain of the function $f(x)=\log_3\log_{1/3}(x^2+10x+25)+\dfrac{1}{[x]+5}$ (where $[\cdot]$ denotes the greatest integer function) is

(a) $(-4,-3)$ (b) $(-6,-5)$
(c) $(-6,-4)$ (d) None of these

Sol. Here, $\log_3\log_{1/3}(x^2+10x+25)$ is defined, when

I. $x^2+10x+25>0$, i.e. $(x+5)^2>0 \Rightarrow x\neq-5$ ...(i)

II. $\log_{1/3}(x^2+10x+25)>0$

$\Rightarrow \quad x^2+10x+25<1$

or $\quad x^2+10x+24<0$

or $\quad (x+6)(x+4)<0$

$$+\qquad -6 \qquad - \qquad -4 \qquad +$$

$\Rightarrow \quad x\in(-6,-4)$ ...(ii)

From Eqs. (i) and (ii), we get

$x\in(-6,-5)\cup(-5,-4)$ ...(iii)

III. $[x]+5\neq0$

$\Rightarrow \quad [x]\neq-5$

$\Rightarrow \quad x\notin[-5,-4)$ ...(iv)

From Eqs. (iii) and (iv), we get

Domain of $f(x)$ is $(-6,-5)$.

Hence, (b) is the correct answer.

Example 43 Let $[x]$ be the greatest integer less than or equal to x.
Then, the equation $\sin x=[1+\sin x]+[1-\cos x]$ has

(a) one solution in $\left[-\dfrac{\pi}{2},\dfrac{\pi}{2}\right]$ (b) one solution in $\left[\dfrac{\pi}{2},\pi\right]$

(c) one solution in R (d) no solution in R

Sol. Given that, $[1+\sin x]+[1-\cos x]=\sin x$

$\Rightarrow \quad 1+[\sin x]+1+[-\cos x]=\sin x$

$\Rightarrow \quad 2+[\sin x]+[-\cos x]=\sin x$

$\Rightarrow \quad 2+[-\cos x]=\{\sin x\}$

Here, LHS is 1, 2 or 3, but RHS $\in[0,1)$

$\therefore$ No solution. Hence, (d) is the correct answer.

(iv) Fractional Part Function

$y=\{x\}$, $\{x\}$ indicates fractional part of x.

Let $\quad x=I+f$, where $I=[x]$

and $\quad f=\{x\}$

Then, $\quad y=\{x\}$

$\quad = x-[x]$

Here, $y=\{x\}$ could be expressed graphically, as

x	$\{x\}$
$0\le x<1$	x
$1\le x<2$	$x-1$
$2\le x<3$	$x-2$
$-1\le x<0$	$x+1$
$-2\le x<-1$	$x+2$

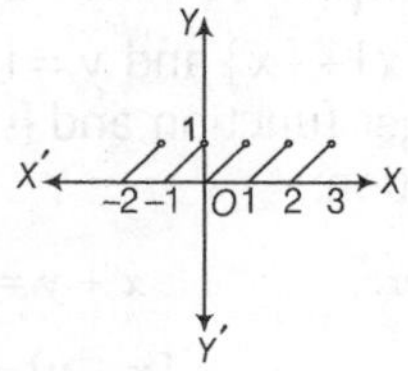

Figure 3.26

Properties of Fractional Part of x

(i) $\{x\}=x$, if $0\le x<1$.
(ii) $\{x\}=0$, if $x\in$ integer.
(iii) $\{-x\}=1-\{x\}$, if $x\notin$ integer.
(iv) $\{x\pm \text{integer}\}=\{x\}$.

Example 44 If $\{x\}$ and $[x]$ represent fractional and integral parts of x respectively, then find the value of $[x]+\sum_{r=1}^{2000}\dfrac{\{x+r\}}{2000}$

Sol. Clearly, $[x]+\sum_{r=1}^{2000}\dfrac{\{x+r\}}{2000}=[x]+\sum_{r=1}^{2000}\dfrac{\{x\}}{2000}$

[$\because$ we know that $\{x+r\}=\{x\}$, if $r\in$ Integer]

$$=[x]+\left[\frac{\{x\}}{2000}+\frac{\{x\}}{2000}+\ldots+\text{upto 2000 times}\right]$$

$$=[x]+\frac{2000\{x\}}{2000}$$

$$=[x]+\{x\}$$

$$=x$$

Thus, $\quad [x]+\sum_{r=1}^{2000}\dfrac{\{x+r\}}{2000}=x$

Example 45 Solve $4\{x\}=x+[x]$.

Sol. We know, $x=[x]+\{x\}$

$\therefore \quad 4\{x\}=[x]+\{x\}+[x] \Rightarrow \{x\}=\dfrac{2[x]}{3}$ ...(i)

But $0\le \{x\}<1$

So, $0\le \dfrac{2[x]}{3}<1$

$\Rightarrow \quad 0\le [x]<\dfrac{3}{2}$, therefore $[x]=0$ or 1

If $[x]=1$, then $\{x\}=\dfrac{2}{3}$ [from Eq. (i)]

Thus, $x=\dfrac{5}{3}$

If $[x]=0, \{x\}=0$ so, $x=0$

Thus, solutions of $4\{x\}=x+[x]$ are $x\in\left\{0,\dfrac{5}{3}\right\}$.

Example 46 Prove that $[x]+[y]\le[x+y]$, where $x=[x]+\{x\}$ and $y=[y]+\{y\}$, $[\cdot]$ represents greatest integer function and $\{\cdot\}$ represents fractional part function.

Sol. Here, $x+y=[x]+[y]+\{x\}+\{y\}$

$\therefore \quad [x+y]=[[x]+[y]+\{x\}+\{y\}]$

$=[x]+[y]+[\{x\}+\{y\}]$

[using $[x+I]=[x]+I, I\in$ Integer]

$\Rightarrow \quad [x+y]\ge[x]+[y]$

Remark

This could be generalised for n terms as

$$[x_1]+[x_2]+\ldots+[x_n]\le[x_1+x_2+\ldots+x_n].$$

Example 47 Find the number of solutions of $|[x]-2x|=4$, where $[x]$ is the greatest integer $\le x$.

Sol. Let $x=I+f, I\in$ integer, $f\in$ fractional part

[i.e. $0\le f<1$]

Then, $|[x]-2x|=4$

$\Rightarrow \quad |[I+f]-2(I+f)|=4$

$\Rightarrow \quad |I-2I-2f|=4 \Rightarrow |I+2f|=4,$

which is possible, only if $\left(f=\dfrac{1}{2}\text{ or }0\right)$

If $f=\dfrac{1}{2}$,

$|I+1|=4 \Rightarrow I+1=\pm 4$

So, $I=3,-5$ and $f=\dfrac{1}{2}$

If $f=0, |I|=4$

So, $I=\pm 4$ and $f=0$

Thus, number of solutions are $x=\left\{\pm 4, \dfrac{7}{2}, -\dfrac{9}{2}\right\}$,

i.e. 4 solutions.

(v) Least Integer Function

$y=(x)=[x],[x]$ or (x) indicates the integral part of x which is the nearest and greater to x.

It is known as **ceiling of x**.

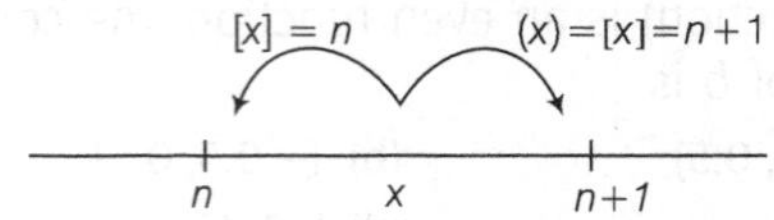

Figure 3.27

Thus, $[2.3203]=3$, $(0.23)=1$, $(-8.0725)=-8$, $(-0.6)=0$

In general, $n<x\le n+1$ ($n\in$ Integer)

$\Rightarrow \quad (x)=n+1=[x]$

Here, $f(x)=(x)=[x]$ can be expressed graphically as

x	$[x]=(x)$
$-1<x\le 0$	0
$0<x\le 1$	1
$1<x\le 2$	2
$2<x\le 3$	3
$-2<x\le -1$	-1

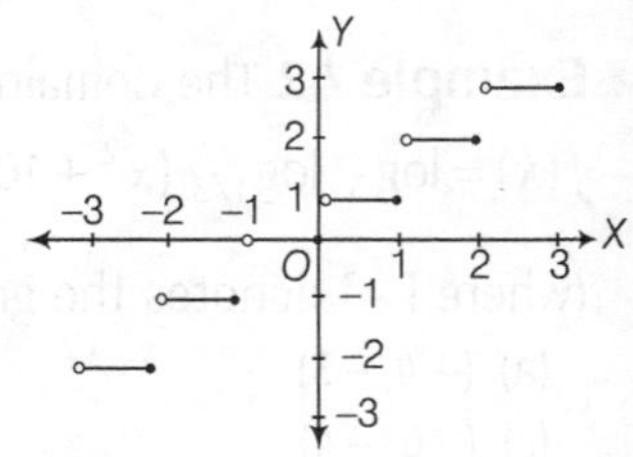

Figure 3.28

(•) represents value is taken.

(∘) represents value is neglected.

Properties of Least Integer Function

(i) $(x)=x=[x]$ holds, if x is Integer.

(ii) $(x+I)=[x+I]=(x)+I, I\in$ Integer.

(iii) Greatest integer function $[x]$ converts $x=I+f$ into I, while (x) converts to $I+1$.

(iv) $x=(x)+\{x\}-1$,

(v) $(-x)=\begin{cases}-(x), & \text{if } x\in I\\ -(x)+1, & \text{if } x\notin I\end{cases}$

(vi) $(x)-(-x)=\begin{cases}2(x), & \text{if } x\in I\\ 2(x)-1, & \text{if } x\notin I\end{cases}$

(vii) $(x)\ge n \Leftrightarrow x>n-1, n\in I$

(viii) $(x)>n \Leftrightarrow x>n, n\in I$

(ix) $(x)\le n \Leftrightarrow x\le n, n\in I$

(x) $(x)<n \Leftrightarrow x<n-1, n\in I$

(xi) $\left(\dfrac{(x)}{n}\right)=\left(\dfrac{x}{n}\right), n\in N$

(xii) $\left(\dfrac{n+1}{2}\right)+\left(\dfrac{n+2}{4}\right)+\left(\dfrac{n+4}{8}\right)+\ldots=2(n-1), n\in N$

(xiii) $(x)=\begin{cases}[x], & x\in I\\ [x]+1, & x\notin I\end{cases}$

(xiv) $(x)+\left(x+\dfrac{1}{n}\right)+\left(x+\dfrac{2}{n}\right)+\ldots+\left(x+\dfrac{n-1}{n}\right)=(nx)+n-1, n\in N$

Example 48 Find the solution set of $(x)^2 + (x+1)^2 = 25$, where (x) is the least integer greater than equals to x.

Sol. Let $x = I + f$, where I (integer) and f (fractional part) such that $0 < f < 1$.

Then, $(I + f)^2 + (I + f + 1)^2 = 25$

$\Rightarrow \{I + 1\}^2 + \{I + 2\}^2 = 25$

$\Rightarrow I^2 + 2I + 1 + I^2 + 4I + 4 = 25$

$\Rightarrow 2I^2 + 6I + 5 - 25 = 0$

$\Rightarrow 2I^2 + 6I - 20 = 0$

So, $I = 2, -5$

Thus, $x = 2 + f, -5 + f$

where $0 < f < 1$

$\Rightarrow 2 < 2 + f < 3, -5 < -5 + f < -4$ ...(i)

Again, let $x = I$

Then, $x^2 + (x+1)^2 = 25$

$\Rightarrow x = 3, -4$ ...(ii)

From Eqs. (i) and (ii), we get $x \in (-5, -4] \cup (2, 3]$

Example 49 If $[x]$ = the greatest integer less than or equal to x and (x) = the least integer greater than or equal to x and $[x]^2 + (x)^2 > 25$, then x belongs to ....

Sol. Let $x = I + f$, where $I \in$ integer, $f \in$ fractional part such that $0 \le f < 1$.

$\therefore [x]^2 + (x)^2 > 25$

$\Rightarrow [I + f]^2 + (I + f)^2 > 25$

$\Rightarrow I^2 + \{I + 1\}^2 > 25$

$\Rightarrow I^2 + I^2 + 2I + 1 > 25$

$\Rightarrow 2I^2 + 2I - 24 > 0$

$\Rightarrow I^2 + I - 12 > 0$

$\Rightarrow (I + 4)(I - 3) > 0$

$\therefore I < -4$

or $I > 3$

Here, $x = I + f$

So, $x < -4 + f$

or $x > 3 + f$ ...(i)

Now, let $x = I$, then $x^2 + x^2 > 25$

$\Rightarrow x^2 > 12.5$

$\Rightarrow x \le -4$ or $x \ge 4$

Form Eqs. (i) and (ii), we get $x \in (-\infty, -4] \cup [4, \infty)$

Exercise for Session 4

- **Directions** (Q. Nos. 1 to 18) *Find the domain of the following.*

1. $f(x) = \sqrt{x^2 - |x| - 2}$
2. $f(x) = \sqrt{2 - |x|} + \sqrt{1 + |x|}$
3. $f(x) = \log_e |\log_e x|$
4. $f(x) = \sin^{-1}\left(\dfrac{2 - 3[x]}{4}\right)$, which $[\cdot]$ denotes the greatest integer function.
5. $f(x) = \log(x - [x])$, where $[\cdot]$ denotes the greatest integer function.
6. $f(x) = \dfrac{1}{\sqrt{[x]^2 - [x] - 6}}$, where $[\cdot]$ denotes the greatest integer function.
7. $f(x) = \text{cosec}^{-1}[1 + \sin^2 x]$, where $[\cdot]$ denotes the greatest integer function.
8. $f(x) = \cos^{-1}\sqrt{\log_{[x]} \dfrac{|x|}{x}}$, where $[\cdot]$ denotes the greatest integer function.

9. $f(x)=\sqrt{\dfrac{x-1}{x-2\{x\}}}$, where $\{\cdot\}$ denotes the fractional part function.

10. $f(x)=\sin^{-1}\left(\dfrac{[x]}{\{x\}}\right)$, where $[\cdot]$ and $\{\cdot\}$ denotes the greatest integer and fractional part function.

11. $f(x)=\sin^{-1}[2x^2-3]$, where $[\cdot]$ denotes the greatest integer function.

12. $f(x)=\sin^{-1}\left[\log_2\left(\dfrac{x^2}{2}\right)\right]$, where $[\cdot]$ denotes the greatest integer function.

13. $f(x)=\sqrt{2\{x\}^2-3\{x\}+1}$, $x\in[-1,1]$, where $\{\cdot\}$ denotes the fractional part function.

14. $f(x)=\dfrac{1}{[|x-2|]+[|x-10|]-8}$, where $[\cdot]$ denotes the greatest integer function.

15. If a function is defined, as $g(x)=|\sin x|+\sin x$, $\phi(x)=\sin x+\cos x, 0\le x\le\pi$, then find the domain for $f(x)=\sqrt{\log_{\phi(x)} g(x)}$.

16. Solve the equations, $y=\dfrac{1}{3}[\sin x+[\sin x+[\sin x]]]$ and $[y+[y]]=2\cos x$, where $[\cdot]$ denotes the greatest integer function.

17. If $[x]=\left[\dfrac{x}{2}\right]+\left[\dfrac{x+1}{2}\right]$, where $[\cdot]$ denotes the greatest integer function and n be positive integer, then show that

$$\left[\frac{n+1}{2}\right]+\left[\frac{n+2}{4}\right]+\left[\frac{n+4}{8}\right]+\left[\frac{n+8}{16}\right]+\ldots=n.$$

18. Find the integral solutions to the equation $[x][y]=x+y$. Show that all the non-integral solutions lie on exactly two lines. Determine these lines.

Session 5

Range

Range

The range of a function is the set of all possible values that can be produced by that function.

e.g. Let $f(x)=x^2$

No matter what value we substitute for x, $f(x)$ will always be positive. Therefore, we would say that the range of this function is the set of all positive real numbers.

Rules for Finding the Range

First of all find the domain of given function $y = f(x)$

(i) If domain contains finite number of points, range is the set of corresponding $f(x)$ values.

(ii) If domain is R or $R-\{\text{some finite points}\}$,

express x in terms of y. From this, find y for x to be defined.

(i.e. Find the values of y for which x exists.)

(iii) If domain is *a* finite interval, then find the least and the greatest values for range, using monotonicity.

e.g. If $f(x)$ is defined on $[a,b]$, then

Step 1 Put, $f'(x)=0 \Rightarrow x=\alpha_1,\alpha_2,\ldots,\alpha_n \in (a,b)$

Step 2 Find, $\{f(a), f(\alpha_1), f(\alpha_2), \ldots, f(\alpha_n), f(b)\}$

The greatest and least values gives the range of $f(x)$.

| Example 50 Find the range for $y=\dfrac{x-[x]}{1-[x]+x}$.

Sol. Here, $y=\dfrac{x-[x]}{1+x-[x]}=\dfrac{\{x\}}{1+\{x\}}$

Thus, domain is R.

Now, from $y=\dfrac{\{x\}}{1+\{x\}}$,

we have, $y+y\{x\}=\{x\} \Rightarrow \{x\}=\dfrac{y}{1-y}$

Here, $0 \le \{x\} < 1$, so $0 \le \dfrac{y}{1-y} < 1 \Rightarrow 0 \le y < \dfrac{1}{2}$

Hence, range $=\left[0,\dfrac{1}{2}\right)$.

| Example 51 Find the range for $f(x)=\dfrac{e^x}{1+[x]}$ when $x \ge 0$.

Sol. Here, $f(x)$ is defined for all $x \ge 0$. Also, $f(x)$ is an increasing function in $[0,\infty)$.

Thus, range $=[f(0), f(\infty))$

Hence, range $=[1,\infty)$

| Example 52 Find the domain and range of the function $y=\log_e(3x^2-4x+5)$.

Sol. y is defined, if $3x^2-4x+5>0$

Where $D=16-4(3)(5)=-44<0$

and coefficient of $x^2=3>0$

Hence, $(3x^2-4x+5)>0,\ \forall x \in R$

Thus, domain $\in R$

Now, from $y=\log_e(3x^2-4x+5)$,

we have, $3x^2-4x+5=e^y$ or $3x^2-4x+(5-e^y)=0$

Since, x is real, therefore discriminant ≥ 0.

i.e. $(-4)^2-4(3)(5-e^y)\ge 0 \Rightarrow 12e^y \ge 44$

So, $e^y \ge \dfrac{11}{3}$. Thus, $y \ge \log\left(\dfrac{11}{3}\right)$

Hence, range is $\left[\log\left(\dfrac{11}{3}\right),\infty\right)$

Aliter Since, log is an increasing function and $3x^2-4x+5$ is minimum at $x=\dfrac{2}{3}$, i.e. $\log_e(3x^2-4x+5)$ is minimum at $x=\dfrac{2}{3}$ and minimum value of $y=\log_e\dfrac{11}{3}$, i.e. $\left(\log_e\dfrac{11}{3}\le y<\infty\right)$.

Thus, range is $\left[\log_e\left(\dfrac{11}{3}\right),\infty\right)$.

| Example 53 Find the range of $y=\sqrt{x-1}+\sqrt{5-x}$.

Sol. Here, domain $x \ge 1$ and $x \le 5$, i.e. $x \in [1,5]$.

Now, check monotonicity, $\dfrac{dy}{dx}=\dfrac{1}{2\sqrt{x-1}}-\dfrac{1}{2\sqrt{5-x}}$

When $\dfrac{dy}{dx}=0$, we get $x=3$

+ 1 min, 3 max, − 5 min

Thus, $y_{\min}$ at $x = 1$ or $x = 5$

$y_{\min}$ at $x = 1$ is 2;

$y_{\min}$ at $x = 5$ is 2.

Thus, minimum value of $y = 2$.

Also, $y_{\max}$ at $x = 3$ is $2\sqrt{2}$.

Hence, range $= [2, 2\sqrt{2}]$

Remarks

(i) When y is minimum at two or more points, the smallest value amongst them is taken.

(ii) When y is maximum at two or more points, the largest value amongst them is taken.

Example 54 Find the range of

$$\log_3\{\log_{1/2}(x^2+4x+4)\}.$$

Sol. Firstly, finding domain

$\log_3\{\log_{1/2}(x^2+4x+4)\}$ exists, if

$\log_{1/2}(x^2+4x+4) > 0$

$$\Rightarrow \quad x^2+4x+4 < \left(\frac{1}{2}\right)^0$$

[using $\log_a x > b \Rightarrow x < a^b$, if $0 < a < 1$]

$$\Rightarrow \quad x^2+4x+4 < 1$$

$$\Rightarrow \quad x^2+4x+3 < 0$$

$$\Rightarrow \quad (x+1)(x+3) < 0 \quad \Rightarrow -3 < x < -1 \quad \text{...(i)}$$

+ −3 − −1 +

and $\quad x^2+4x+4 > 0 \quad \Rightarrow (x+2)^2 > 0,$

which is always true except for $x = -2$. ...(ii)

From Eqs. (i) and (ii), we have

$$x \in (-3, -2) \cup (-2, -1)$$

Thus, domain is $(-3, -2) \cup (-2, -1)$

Now, we find out the range.

Since, $0 < \log_{1/2}(x^2+4x+4) < \infty, \forall\, x \in$ domain y

$\Rightarrow -\infty < \log_3\{\log_{1/2}(x^2+4x+4)\} < \infty$

Thus, range (y) is R.

Example 55 Range of the function $f(x) = (\cos^{-1}|1-x^2|)$ is

(a) $\left[0, \frac{\pi}{2}\right]$ (b) $\left[0, \frac{\pi}{3}\right]$ (c) $(0, \pi)$ (d) $\left(\frac{\pi}{2}, \pi\right)$

Sol. Here, $f(x) = \cos^{-1}|1-x^2|$ would exists, only when $|1-x^2| \le 1$

$$\Rightarrow \quad -1 \le 1-x^2 \le 1 \Rightarrow 2 \ge x^2 \ge 0$$

or $\quad x \in [-\sqrt{2}, \sqrt{2}]$

$\because \quad 0 \le |1-x^2| \le 1, \forall\, x \in [-\sqrt{2}, \sqrt{2}]$

$\therefore \quad 0 \le \cos^{-1}|1-x^2| \le \frac{\pi}{2}$

Hence, (a) is the correct answer.

Example 56 If x, y and z are real such that $x+y+z=4$, $x^2+y^2+z^2=6$, x belongs to

(a) $(-1, 1)$ (b) $[0, 2]$

(c) $[2, 3]$ (d) $\left[\frac{2}{3}, 2\right]$

Sol. Eliminating z,

$$x^2+y^2+(4-(x+y))^2 = 6$$

$$\Rightarrow \quad y^2+(x-4)y+x^2-4x+5 = 0$$

$$\Rightarrow \quad D \ge 0$$

$$\Rightarrow \quad (x-4)^2 - 4(x^2-4x+5) \ge 0$$

$$\Rightarrow \quad 3x^2-8x+4 \le 0$$

$$\Rightarrow \quad (x-2)(3x-2) \le 0$$

+ 2/3 − 2 +

$\therefore\ x \in [2/3, 2]$

Hence, (d) is the correct answer.

Example 57 The range of the function $f(x) = \dfrac{1}{|\sin x|} + \dfrac{1}{|\cos x|}$ is

(a) $[2\sqrt{2}, \infty)$ (b) $(\sqrt{2}, 2\sqrt{2})$ (c) $(0, 2\sqrt{2})$ (d) $(2\sqrt{2}, 4)$

Sol. $f(x) = \dfrac{1}{|\sin x|} + \dfrac{1}{|\cos x|}$

Using AM $\ge$ GM, we get

$$\frac{\dfrac{1}{|\sin x|} + \dfrac{1}{|\cos x|}}{2} \ge \left(\frac{1}{|\sin x||\cos x|}\right)^{1/2}$$

$$\Rightarrow \quad \frac{1}{|\sin x|} + \frac{1}{|\cos x|} \ge 2(2|\operatorname{cosec} 2x|)^{1/2}$$

[where, $|\operatorname{cosec} 2x| \ge 1$]

$$\Rightarrow \quad \frac{1}{|\sin x|} + \frac{1}{|\cos x|} \ge 2\sqrt{2}$$

$\therefore$ Range of $f(x)$ is $[2\sqrt{2}, \infty)$.

Hence, (a) is the correct answer.

Example 58 If $z = x + iy$ and $x^2 + y^2 = 16$, then the range of $||x| - |y||$ is

(a) $[0, 4]$ (b) $[0, 2]$

(c) $[2, 4]$ (d) None of these

Sol. Let $x = 4\cos\theta, y = 4\sin\theta$, then

$$\begin{aligned}|4|\cos\theta| - 4|\sin\theta|| &= 4||\cos\theta| - |\sin\theta||\\ &= 4\sqrt{1 - 2|\cos\theta||\sin\theta|}\\ &= 4\sqrt{1 - |\sin 2\theta|}\end{aligned}$$

$\therefore$ Range is [0, 4]. Hence, (a) is the correct answer.

Example 59 The range of

$f(x) = \frac{1}{\pi}(\sin^{-1} x + \tan^{-1} x) + \frac{x+1}{x^2 + 2x + 5}$ is

(a) $\left[-\frac{3}{4}, \frac{1}{5}\right]$ (b) $\left[-\frac{5}{4}, \frac{3}{4}\right]$

(c) $\left[-\frac{3}{4}, \frac{5}{4}\right]$ (d) $\left[-\frac{3}{4}, 1\right]$

Sol. Here, $f(x) = \frac{1}{\pi}(\sin^{-1} x + \tan^{-1} x) + \frac{1}{(x+1) + \frac{4}{(x+1)}}$

$= g(x) + h(x)$,

where domain of $g(x)$ is $[-1, 1]$

$\therefore$ Maximum value of $g(x) = g(1) = \frac{3}{4}$

and minimum value of $g(x) = g(-1) = -\frac{3}{4}$

Also, maximum value of $h(x)$ occurs, when $(x+1) + \frac{4}{(x+1)}$ is minimum at $x = 1$.

$\Rightarrow$ Range of $f(x)$ is $\left[-\frac{3}{4}, 1\right]$.

Hence, (d) is the correct answer.

Example 60 The range of the function $f(x) = \sin^2 x - 5\sin x - 6$ is

(a) $[-10, 0]$ (b) $[-1, 1]$ (c) $[0, \pi]$ (d) $\left[-\frac{49}{4}, 0\right]$

Sol. Here, $f(x) = \sin^2 x - 5\sin x - 6$

$$= \left(\sin^2 x - 5\sin x + \frac{25}{4}\right) - 6 - \frac{25}{4}$$

$$= \left(\sin x - \frac{5}{2}\right)^2 - \frac{49}{4} \quad \text{...(i)}$$

where $\frac{9}{4} \le \left(\sin x - \frac{5}{2}\right)^2 \le \frac{49}{4}$ ...(ii)

From Eqs. (i) and (ii), we get $-10 \le f(x) \le 0$

$\Rightarrow$ Range of $f(x)$ is $[-10, 0]$.

Hence, (a) is the correct answer.

Example 61 If $f(x) = [x^2] - [x]^2$, where [·] denotes the greatest integer function and $x \in [0, n], n \in N$, then the number of elements in the range of $f(x)$ is

(a) $(2n+1)$ (b) $4n - 3$ (c) $3n - 3$ (d) $2n - 1$

Sol. When $x = n - 1$, $f(x) = (n-1)^2 - (n-1)^2 = 0$

When $n - 1 < x < n$,

$[x] = n - 1$

and $(n-1)^2 \le [x^2] \le n^2 - 1$

$\therefore$ $0 \le [x^2] - [x]^2 \le n^2 - 1 - (n-1)^2$

$\Rightarrow$ $0 \le f(x) \le 2n - 2$, but $f(x)$ has to be an integer.

The set of values of $f(x)$ is $\{0, 1, 2, \ldots, 2n - 2\}$.

Hence, (d) is the correct answer.

Example 62 Range of the function

$f(x) = \sqrt{|\sin^{-1}|\sin x|| - \cos^{-1}|\cos x|}$, is

(a) {0} (b) $\left[0, \sqrt{\frac{\pi}{2}}\right]$ (c) $[0, \sqrt{\pi}]$ (d) None of these

Sol. We know that, $|\sin^{-1}|\sin x|| = \cos^{-1}|\cos x|, \forall x \in$ domain

$\therefore f(x) = \sqrt{|\sin^{-1}|\sin x|| - \cos^{-1}|\cos x|} = 0, \forall x \in$ domain

$\therefore$ Range of $f(x)$ is {0}.

Hence, (a) is the correct answer.

Example 63 The number of values of y in $[-2\pi, 2\pi]$ satisfying the equation $|\sin 2x| + |\cos 2x| = |\sin y|$ is

(a) 3 (b) 4 (c) 5 (d) 6

Sol. Here, $1 \le |\sin 2x| + |\cos 2x| \le \sqrt{2}$ and $|\sin y| \le 1$.

So, solution is possible only when $|\sin y| = 1$.

$\Rightarrow$ $\sin y = \pm 1 \Rightarrow y = \pm\frac{\pi}{2}, \pm\frac{3\pi}{2}$

$\therefore$ Number of values of y is 4.

Hence, (b) is the correct answer.

Example 64 Let $f(x) = \cot^{-1}(x^2 - 4x + 5)$, then range of $f(x)$ is equal to

(a) $\left(0, \frac{\pi}{2}\right)$ (b) $\left(0, \frac{\pi}{4}\right]$

(c) $\left[0, \frac{\pi}{4}\right)$ (d) None of these

Sol. Here, $x^2 - 4x + 5 \Rightarrow (x-2)^2 + 1 \ge 1$

$\therefore$ $1 \le x^2 - 4x + 5 < \infty$

$\Rightarrow$ $0 < \cot^{-1}(x^2 - 4x + 5) \le \frac{\pi}{4}$

[using, $x_1 < x_2 \Rightarrow \cot^{-1} x_1 > \cot^{-1} x_2$, since $\cot^{-1} x$ is decreasing]

$\Rightarrow$ Range of $f(x)$ is $\left(0, \frac{\pi}{4}\right]$.

Hence, (b) is the correct answer.

To Find Range for Rational Expressions

Let $f(x)=\dfrac{ax^2+bx+c}{px^2+qx+r}$

***Step* 1** Write the given function as a equation

$$y=\frac{ax^2+bx+c}{px^2+qx+r}$$

***Step* 2** Rewrite the above equation for x in standard form

$$x^2(py-a)+x(qy-b)+ry-c=0$$

***Step* 3** Find the discriminant to the above equation

$$D=(qy-b)^2-4(py-a)(ry-c)$$
$$D=q^2y^2-2qyb+b^2-4\ [pry^2-(pc+ar)y+ac]$$
$$D=y^2(q^2-4pr)+(4pc+4ar-2bq)y+b^2-4ac$$

***Step* 4** Find $y_1,y_2=\dfrac{-B\pm\sqrt{B^2-4AC}}{2A}$

where, $A=q^2-4pr,\ B=4pc+4ar-2bq, C=b^2-4ac.$

***Step* 5** Let $y_1<y_2$, then range of $f(x)$ is $[y_1,y_2]$ if $A<0$ and $A>0$, then range of $f(x)$ is $R-(y_1,y_2)$.

Example 65 Find the range of $f(x)=\dfrac{x^2+14x+9}{x^2+2x+3}$, where $x\in R$.

Sol. Here, $A=4-12=-8,\ B=12+36-56=-8,\ C=160$

Now, $\dfrac{-B\pm\sqrt{B^2-4AC}}{2A}=\dfrac{-1\pm\sqrt{1+80}}{-2}=-5,4$ and

here $A<0$

$\therefore$ Range is $[-5,4]$

Example 66 For what real values of a does the range of $f(x)=\dfrac{x+1}{a+x^2}$ contains the interval $[0,1]$?

Sol. Let $y=\dfrac{x+1}{a+x^2}$

$\Rightarrow\quad y(a+x^2)=x+1$

$\Rightarrow yx^2-x+(ay-1)=0$ has real roots for every $y\in[0,1]$.

$\therefore\quad 1-4y(ay-1)\geq 0$

$\Rightarrow\quad 1-4ay^2+4y\geq 0$ holds for $0\leq y\leq 1$ ...(i)

***Case* I** If $y=0$

Here, $y=0$ is assumed at $x=-1$ for any $a\neq-1$.

For $a=-1$, $y=\dfrac{x+1}{x^2-1}$ is undefined for $x=-1$.

***Case* II** If $0<y\leq 1$. Put $z=\dfrac{1}{y}$

$\Rightarrow\quad 1\leq z<\infty$

Now, from Eq. (i), we get

$z^2+4z-4a\geq 0$ holds for $1\leq z<\infty$

$\Rightarrow\quad (z+2)^2-4-4a\geq 0$ holds for $1\leq z<\infty$

Consider, $\quad (z+2)^2-4-4a\geq 0$

$\Rightarrow\quad (z+2)^2\geq 4+4a$

$\Rightarrow\quad z+2\geq\sqrt{4+4a};\ 4+4a\geq 0$

$a\geq-1$ ...(ii)

but $\quad z\geq 1$

$\therefore\quad 9\geq 4+4a$

$\Rightarrow\quad 5\geq 4a\quad$ [as $(u+2)^2\geq 9$ for $u=1$]

$\Rightarrow\quad a\leq\dfrac{5}{4}, a\neq-1$ ...(iii)

$\therefore$ From Eqs. (i), (ii) and (iii), we get $a\in\left(-1,\dfrac{5}{4}\right]$.

Example 67 Find the range of the function $f(x)=\dfrac{\sin^2x+\sin x-1}{\sin^2x-\sin x+2}$.

Sol. Let $\quad y=\dfrac{\sin^2x+\sin x-1}{\sin^2x-\sin x+2}$

Let $\quad t=\sin x\ \Rightarrow\ -1\leq t\leq 1$ and $y=\dfrac{t^2+t-1}{t^2-t+2}$

$\Rightarrow\quad (y-1)t^2-(y+1)t+(2y+1)=0$ ...(i)

Since t is real, $\quad\dfrac{3-2\sqrt{11}}{7}\leq y\leq\dfrac{3+2\sqrt{11}}{7}$

***Case* I** If both roots of Eq. (i) are greater than 1, i.e. $t_1>1$ and $t_2>1$.

$\Rightarrow\quad t_1+t_2>2$

and $\quad (t_1-1)(t_2-1)>0$

$\Rightarrow\quad \dfrac{y+1}{y-1}>2$

and $\quad \dfrac{2y+1}{y-1}-\dfrac{y+1}{y-1}+1>0$

$\Rightarrow\quad 1<y<3$

and $\quad y>1$ or $y<\dfrac{1}{2}$

$\therefore\quad y\in(1,3)$ ...(ii)

***Case* II** If $t_1<-1$ and $t_2<-1$

$\Rightarrow\quad t_1+t_2<-2$ and $(t_1+1)(t_2+1)>0$

$\Rightarrow\quad \dfrac{y+1}{y-1}+2<0$ and $\dfrac{2y+1}{y-1}+\dfrac{y+1}{y-1}+1>0$

$\Rightarrow \quad \frac{1}{3} < y < 1$ and $y > 1$ or $y < -\frac{1}{4}$

$\Rightarrow \quad y \in \phi$ ...(iii)

***Case* III** If $t_1 < -1$ and $t_2 > -1$, $t_1 < 1$ and $t_2 > 1$

$\Rightarrow \quad (t_1 + 1)(t_2 + 1) < 0$ and $(t_1 - 1)(t_2 - 1) < 0$

$\Rightarrow \quad \frac{2y+1}{y-1} + \frac{y+1}{y-1} + 1 < 0$ and $\frac{2y+1}{y-1} - \frac{y+1}{y-1} + 1 < 0$

$\Rightarrow \quad -\frac{1}{4} < y < 1$ and $\frac{1}{2} < y < 1 \Rightarrow \frac{1}{2} < y < 1$

$\therefore \quad R_f = \left[\frac{3 - 2\sqrt{11}}{7}, \frac{1}{2}\right] \cup \{1\}$

Exercise for Session 5

■ **Directions** (Q. Nos. 1 to 25) *Find the range of the following.*

1. $f(x) = \sqrt{9 - x^2}$
2. $f(x) = \frac{x}{1 + x^2}$
3. $f(x) = \sin x + \cos x + 3$
4. $f(x) = |x - 1| + |x - 2|, -1 \le x \le 3$
5. $f(x) = \log_3 (5 + 4x - x^2)$
6. $f(x) = \frac{x^2 - 2}{x^2 - 3}$
7. $f(x) = \frac{x^2 + 2x + 3}{x}$
8. $f(x) = |x - 1| + |x - 2| + |x - 3|$
9. $f(x) = \log_{[x-1]} \sin x$, where [·] denotes the greatest integer function.
10. $f(x) = \cos^{-1} \sqrt{\log_{[x]} \frac{|x|}{x}}$, where [·] denotes the greatest integer function.
11. $f(x) = \sqrt{[\sin 2x] - [\cos 2x]}$, where [·] denotes the greatest integer function.
12. $f(x) = \sin^{-1}\left[x^2 + \frac{1}{2}\right] + \cos^{-1}\left[x^2 - \frac{1}{2}\right]$, where [·] denotes the greatest integer function.
13. $f(x) = \sin^{-1}(\sqrt{x^2 + x + 1})$
14. $f(x) = \cos^{-1}\left(\frac{x^2}{\sqrt{1 + x^2}}\right)$
15. $f(x) = \sqrt{\log(\cos(\sin x))}$
16. $f(x) = \frac{x - 1}{x^2 - 2x + 3}$
17. $f(x) = \frac{\sin x}{\sqrt{1 + \tan^2 x}} - \frac{\cos x}{\sqrt{1 + \cot^2 x}}$
18. $f(x) = \frac{\tan(\pi [x^2 - x])}{1 + \sin(\cos x)}$
19. $f(x) = \frac{e^x}{[x + 1]}, x \ge 0$
20. $f(x) = [|\sin x| + |\cos x|]$, where [·] denotes the greatest integer function.
21. $f(x) = \sqrt{-x^2 + 4x - 3} + \sqrt{\sin \frac{\pi}{2}\left(\sin \frac{\pi}{2}(x - 1)\right)}$
22. Find the image of the following sets under the mapping $f(x) = x^4 - 8x^3 + 22x^2 - 24x + 10$ (i) $(-\infty, 1)$ (ii) $[1, 2]$.
23. Find the domain and range of $f(x) = \log\left[\cos|x| + \frac{1}{2}\right]$, where [·] denotes the greatest integer function.
24. Find the domain and range of $f(x) = \sin^{-1}(\log[x]) + \log(\sin^{-1}[x])$, where [·] denotes the greatest integer function.
25. Find the domain and range of $f(x) = [\log(\sin^{-1}\sqrt{x^2 + 3x + 2})]$, where [·] denotes the greatest integer function.

Session 6

Odd and Even Functions

Odd and Even Functions

Odd Functions

A function $f(x)$ is said to be an odd function, if $f(-x) = -f(x)$ for all x. Graph of an odd function is **symmetrical in opposite quadrants**, i.e. the curve in first quadrant is identical to the curve in the third quadrant and the curve in second quadrant is identical to the curve in fourth quadrant. Some graphs which are symmetrical in opposite quadrants (or about origin) are

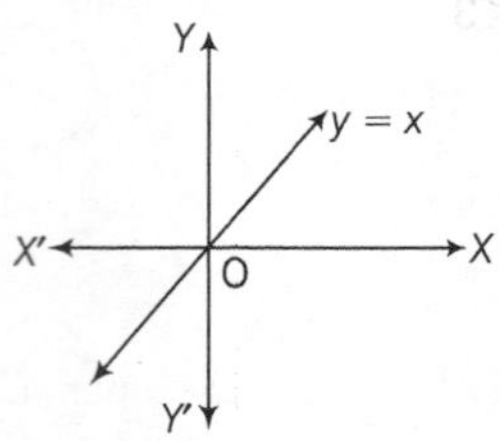

Figure 3.29

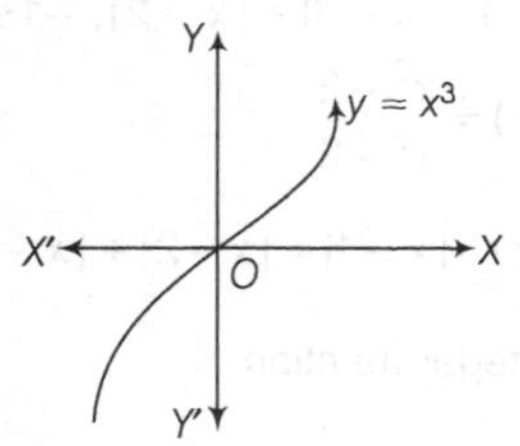

Figure 3.30

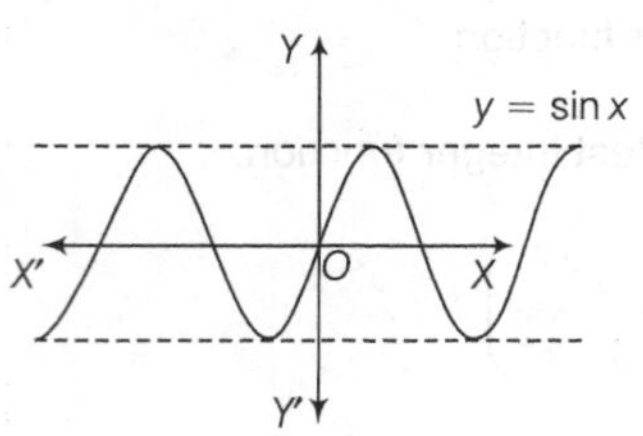

Figure 3.31

Even Functions

A function $f(x)$ is said to be an even, if $f(-x) = f(x)$ for all x. The graph is always **symmetrical about Y-axis**, i.e. the graph on left hand side of Y-axis is the mirror image of the curve on its right hand side.

Some graphs which are symmetrical about **Y-axis** are

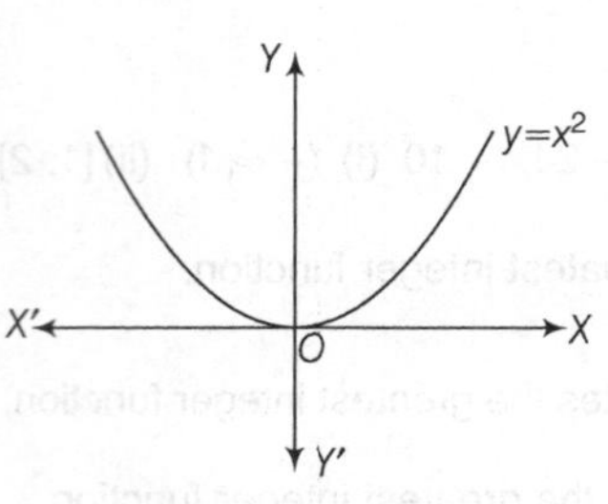

Figure 3.32

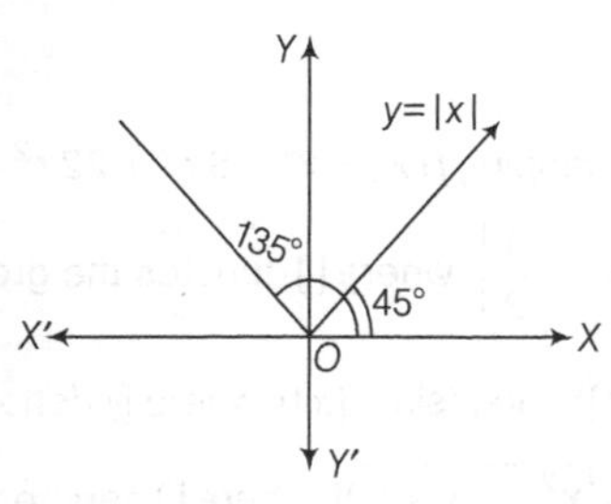

Figure 3.33

Properties of Odd and Even Functions

(i) Product of two odd functions or two even functions is an even function.

(ii) Product of odd and even function is an odd function.

(iii) Every function $y = f(x)$ can be expressed as the sum of an even and odd function.

(iv) The derivative of an odd function is an even function and derivative of an even function is an odd function.

(v) A function which is even or odd, when squared becomes an even function.

(vi) The only function which is both even and odd is $f(x) = 0$, i.e. zero function.

Example 68 If f is an even function, then find the real values of x satisfying the equation

$$f(x) = f\left(\frac{x+1}{x+2}\right).$$ ***[IIT JEE 1996, 2001]***

Sol. Since, $f(x)$ is even, so $f(-x) = f(x)$

Thus, $x = \dfrac{x+1}{x+2}$

or $-x = \dfrac{x+1}{x+2}$

$\Rightarrow$ $x^2 + 2x = x + 1$

or $-x^2 - 2x = x + 1$

$\Rightarrow$ $x^2 + x - 1 = 0$

or $-x^2 - 3x - 1 = 0$

$\Rightarrow$ $x = \dfrac{-1 \pm \sqrt{5}}{2}$

or $x = \dfrac{-3 \pm \sqrt{5}}{2}$

Thus, $x \in \left\{\dfrac{-1+\sqrt{5}}{2}, \dfrac{-1-\sqrt{5}}{2}, \dfrac{-3+\sqrt{5}}{2}, \dfrac{-3-\sqrt{5}}{2}\right\}$

Example 69 Find out whether the given function is even, odd or neither even nor odd,

$$\text{where } f(x) = \begin{cases} x|x| &, \quad x \le -1 \\ [1+x]+[1-x] &, \quad -1 < x < 1 \\ -x|x| &, \quad x \ge 1 \end{cases}$$

where | | and [ ] represent the modulus and greatest integer functions.

Sol. The given function can be written as

$$f(x)=\begin{cases}-x^2 & , \quad x\le -1\\ 2+[x]+[-x] & , \quad -1<x<1\\ -x^2 & , \quad x\ge 1\end{cases}$$

[using definition of modulus function and the properties of greatest integer functions]

$$\Rightarrow \quad f(x)=\begin{cases}-x^2 & , \quad x\le -1\\ 2-1+0 & , \quad -1<x<0\\ 2 & , \quad x=0\\ 2+0-1 & , \quad 0<x<1\\ -x^2 & , \quad x\ge 1\end{cases}$$

$$f(x)=\begin{cases}-x^2 & , \quad x\le -1\\ 1 & , \quad -1<x<0\\ 2 & , \quad x=0\\ 1 & , \quad 0<x<1\\ -x^2 & , \quad x\ge 1\end{cases}$$

which is clearly even as

if $\quad f(-x)=f(x)$.

Thus, $f(x)$ is even.

Example 70 Find whether the given function is even or odd function, where $f(x)=\dfrac{x(\sin x+\tan x)}{\left[\dfrac{x+\pi}{\pi}\right]-\dfrac{1}{2}}$, when $x\ne n\pi$, where $[\cdot]$ denotes the greatest integer function.

Sol. $f(x)=\dfrac{x(\sin x+\tan x)}{\left[\dfrac{x+\pi}{\pi}\right]-\dfrac{1}{2}}=\dfrac{x(\sin x+\tan x)}{\left[\dfrac{x}{\pi}\right]+1-\dfrac{1}{2}}$

$$f(x)=\frac{x(\sin x+\tan x)}{\left[\dfrac{x}{\pi}\right]+0.5}$$

Now, $f(-x)=\dfrac{-x(\sin(-x)+\tan(-x))}{\left[-\dfrac{x}{\pi}\right]+0.5}$

$$\Rightarrow \quad f(-x)=\begin{cases}\dfrac{x(\sin x+\tan x)}{-1-\left[\dfrac{x}{\pi}\right]+0.5} & , \quad x\ne n\pi\\ 0 & , \quad x=n\pi\end{cases}$$

So, $\quad f(-x)=-\dfrac{x(\sin x+\tan x)}{\left[\dfrac{x}{\pi}\right]+0.5}$ or $[\because x\ne n\pi]$

$\therefore f(-x)=-f(x)$. Hence, $f(x)$ is an odd function (if $x\ne n\pi$).

Exercise for Session 6

1. Determine whether the following functions are even or odd.

(i) $f(x)=\log(x+\sqrt{1+x^2})$ (ii) $f(x)=x\left(\dfrac{a^x+1}{a^x-1}\right)$

(iii) $f(x)=\sin x+\cos x$ (iv) $f(x)=x^2-|x|$

(v) $f(x)=\log\left(\dfrac{1-x}{1+x}\right)$ (vi) $f(x)=\{(\text{sgn } x)^{\text{sgn } x}\}^n$; n is an odd integer.

(vii) $f(x)=\text{sgn}(x)+x^2$ (viii) $f(x+y)+f(x-y)=2f(x)\cdot f(y)$; where $f(0)\ne 0$ and $x, y\in R$.

2. Determine whether function; $f(x)=(-1)^{[x]}$ is even, odd or neither of two (where [·] denotes the greatest integer function).

3. A function defined for all real numbers is defined for $x\ge 0$ as follows $f(x)=\begin{cases}x|x|, & 0\le x<1\\ 2x, & x\ge 1\end{cases}$.

How is f defined for $x\le 0$, if (i) f is even? (ii) f is odd?

4. Show that the function $f(x)=\dfrac{2x(\sin x+\tan x)}{2\left[\dfrac{x+21\pi}{\pi}\right]-41}$ is symmetric about origin.

5. If $f:[-20,20]\to R$ defined by $f(x)=\left[\dfrac{x^2}{a}\right]\sin x+\cos x$ is an even function, find the set of values of 'a' (where $[\cdot]$ denotes the greatest integer function).

Session 7

Periodic Functions

Periodic Functions

Definition A function $f(x)$ is said to be periodic function, if there exists a positive real number, T such that $f(x+T) = f(x)$, $\forall\ x \in \text{Dom}(f)$. Then, $f(x)$ is periodic with period T, where T is least positive value.

Graphically If the graph repeats at fixed interval, the function is said to be periodic and its period is the width of that interval.

Example 71 Prove that $\sin x$ is periodic and find its period.

Sol. Let $f(x) = \sin x$ and $T > 0$, then $f(x)$ is periodic, if $f(x+T) = f(x)$.

$\Rightarrow \quad \sin(x+T) = \sin(x), \forall\ x \in R$

$\Rightarrow \quad T = 2\pi, 4\pi, 6\pi, \ldots$

But period of $f(x)$ is smallest positive real number.

Thus, period of $f(x)$ is 2π.

Aliter $f(x) = \sin x$ could be expressed graphically as shown in figure.

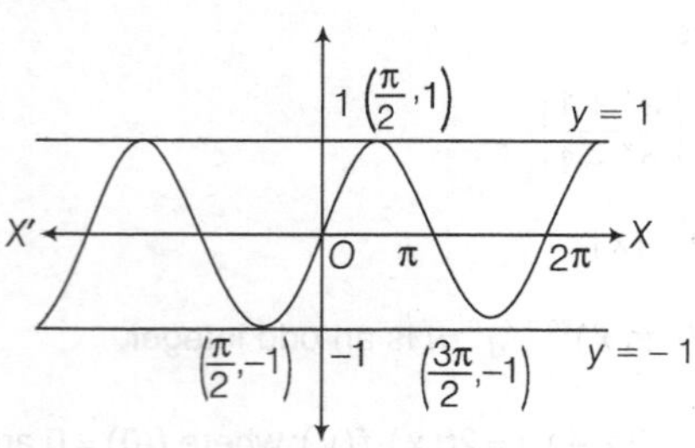

Here, graph repeats at an interval of 2π.

Thus, $f(x)$ is periodic with period 2π.

Example 72 Prove that $f(x) = x - [x]$ is periodic function. Also, find its period.

Sol. Let $T > 0$.

Then, $\quad f(x+T) = f(x), \forall\ x \in R$

$\Rightarrow \quad (x+T) - [x+T] = x - [x], \forall\ x \in R$

$\Rightarrow \quad [x+T] - [x] = T, \forall\ x \in R$

$\Rightarrow \quad T = 1, 2, 3, 4, \ldots$

[since, subtraction of two integers]

The smallest value of T satisfying $f(x+T) = f(x)$ is 1.

Thus, it is periodic with period 1.

Graphically $f(x) = x - [x] = \{x\}$

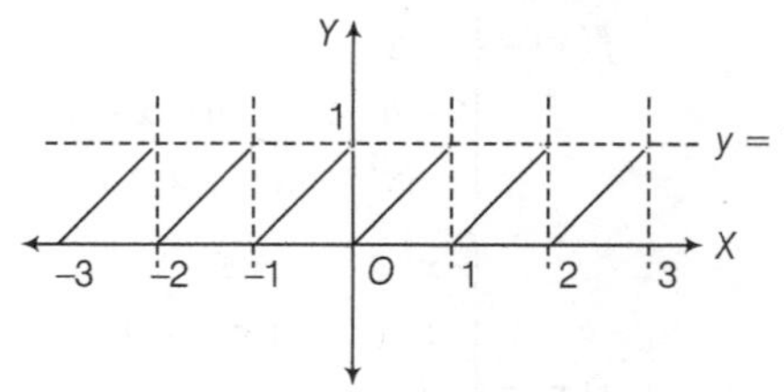

Clearly, from the given graph, the function repeats itself at an interval of 1 unit. Thus, period of $f(x) = 1$.

Remark

For those functions whose periods are not deducted by graphs, they can be judged by inspection method.

Example 73 Let $f(x)$ be periodic and k be a positive real number such that $f(x+k) + f(x) = 0$ for all $x \in R$. Prove that $f(x)$ is a periodic with period $2k$.

Sol. We have,

$f(x+k) + f(x) = 0, \forall\ x \in R$

$\Rightarrow \quad f(x+k) = -f(x), \forall\ x \in R$, put $x = x+k$

$\Rightarrow \quad f(x+2k) = -f(x+k), \forall\ x \in R$

[as $f(x+k) = -f(x)$]

$\Rightarrow \quad f(x+2k) = f(x), \forall\ x \in R$

which clearly shows that $f(x)$ is periodic with period $2k$.

Some Standard Results on Periodic Functions

Functions	Periods
(i) $\sin^n x$, $\cos^n x$ $\sec^n x$, $\text{cosec}^n x$	π, if n is even. 2π, if n is odd or fraction.
(ii) $\tan^n x$, $\cot^n x$	π, n is even or odd.
(iii) $\lvert\sin x\rvert, \lvert\cos x\rvert, \lvert\tan x\rvert$ $\lvert\cot x\rvert, \lvert\sec x\rvert, \lvert\text{cosec } x\rvert$	π
(iv) $x - [x]$	1
(v) Algebraic functions e.g. $\sqrt{x}, x^2, x^3 + 5, \ldots$ etc.	Period doesn't exist.
(vi) $f(x) =$ constant	Periodic with no fundamental period.

Example 74 Find periods for

(i) $\cos^4 x$. (ii) $\sin^3 x$. (iii) $\cos\sqrt{x}$. (iv) $\sqrt{\cos x}$.

Sol. (i) $\cos^4 x$ has a period π as n is even.

(ii) $\sin^3 x$ has a period 2π as n is odd.

(iii) $\cos\sqrt{x}$ is not periodic, as for no value of T,

$$f(x+T)=f(x) \Rightarrow \cos\sqrt{x+T}=\cos(\sqrt{x})$$

Thus, there exists no value of T for which $f(x+T)=f(x)$.

Hence, $\cos\sqrt{x}$ is not periodic.

(iv) $f(x)=\sqrt{\cos x}$ has the period 2π as n is in fraction.

Aliter $f(x+T)=f(x) \Rightarrow \sqrt{\cos(x+T)}=\sqrt{\cos(x)}$

$\Rightarrow \quad T=2\pi, 4\pi, \ldots$

But T is the least positive value, hence $f(x)$ is periodic with period 2π.

Properties of Periodic Functions

(i) If $f(x)$ is periodic with period T, then

(a) $c \cdot f(x)$ is periodic with period T

(b) $f(x+c)$ is periodic with period T

(c) $f(x) \pm c$ is periodic with period T, where c is any constant.

We know, $\sin x$ has period 2π.

Then, $f(x)=5(\sin x)+4$ is also periodic with period 2π.

i.e. "If constant is added, subtracted, multiplied or divided in a periodic function, its period remains the same."

(ii) If $f(x)$ is periodic with period T, then $kf(cx+d)$ has period $\dfrac{T}{|c|}$,

i.e. period is only affected by coefficient of x,

where, $k, c, d \in$ constant.

We know, $f(x)=\left\{7\sin\left(2x+\dfrac{\pi}{9}\right)\right\}-12$ has the period $\dfrac{2\pi}{|2|}=\pi$; as $\sin x$ is periodic with period 2π.

(iii) If $f_1(x)$, $f_2(x)$ are periodic functions with periods T_1, T_2 respectively, then $h(x)=f_1(x)+f_2(x)$ has period

$$=\begin{cases} \text{LCM of } \{T_1, T_2\}, \text{ if } h(x) \text{ is not an even function.} \\ \quad OR \\ \dfrac{1}{2}\text{ LCM of } \{T_1, T_2\}, \text{ if } f_1(x) \text{ and } f_2(x) \text{ are complementary} \\ \text{pair-wise comparable functions.} \end{cases}$$

While taking LCM we should always remember

(a) $\text{LCM of}\left(\dfrac{a}{b}, \dfrac{c}{d}, \dfrac{e}{f}\right)=\dfrac{\text{LCM of }(a,c,e)}{\text{HCF of }(b,d,f)}$

e.g. $\text{LCM of}=\dfrac{\text{LCM of }(2\pi,\pi,\pi)}{\text{HCF of }(3,6,12)}=\dfrac{2\pi}{3}$

$\therefore \quad \text{LCM of}\left(\dfrac{2\pi}{3}, \dfrac{\pi}{6}, \dfrac{\pi}{12}\right)=\dfrac{2\pi}{3}$

(b) LCM of rational with rational is possible.

LCM of irrational with irrational is possible.

But LCM of rational and irrational is not possible.

e.g. LCM of $(2\pi, 1, 6\pi)$ is not possible, as $2\pi, 6\pi \in$ irrational and $1 \in$ rational.

Periodicity of Constant Function

The LCM rule is not applicable, if function reduces to constant.

e.g. $\quad f(x)=\sin^2 x+\cos^2 x$.

Since, period of $\sin^2 x$ and $\cos^2 x$ are π.

$\therefore$ Period of $f(x)=\dfrac{1}{2}\text{ LCM}\{\pi,\pi\}=\dfrac{\pi}{2}$, which is not correct.

Whereas, $\sin^2 x+\cos^2 x=1$ is a constant function and period is undetermined.

Example 75 Find the period, if $f(x)=\sin x+\{x\}$, where $\{x\}$ is fractional part of x.

Sol. Here, $\sin x$ is periodic with period 2π and $\{x\}$ is periodic with period 1. Thus, LCM of 2π and $1 \Rightarrow$ Does not exist.

Hence, $f(x)$ is not periodic.

Example 76 Find period of $f(x)=\tan 3x+\sin\left(\dfrac{x}{3}\right)$.

Sol. Period for $\tan 3x$ is $\left|\dfrac{\pi}{3}\right|$.

Period for $\sin\dfrac{x}{3}$ is $\left|2\pi\times\dfrac{3}{1}\right|=|6\pi|$

Thus, LCM of $\dfrac{\pi}{3}$

and $\dfrac{6\pi}{1} \Rightarrow \dfrac{6\pi}{1}$

Hence, $f(x)$ is periodic with period 6π.

Example 77 Find the period of

$$f(x)=\sin x+\tan\frac{x}{2}+\sin\frac{x}{2^2}+\tan\frac{x}{2^3}+\ldots+\sin\frac{x}{2^{n-1}}+\tan\frac{x}{2^n}.$$

Sol. We have, Period of $\left(\sin x+\tan\dfrac{x}{2}\right)$ is 2π,

$\left(\sin\dfrac{x}{2^2}+\tan\dfrac{x}{2^3}\right)$ is $2^3\pi$,

....

....

$\left(\sin\dfrac{x}{2^{n-1}}+\tan\dfrac{x}{2^n}\right)$ is $2^n\pi$.

Thus, $\quad$ LCM of $\{2\pi, 2^3\pi, \ldots, 2^n\pi\}=2^n\pi$.

Hence, the period of $f(x)$ is $2^n\pi$.

Example 78 Find the period of $f(x)=|\sin x|+|\cos x|$.

Sol. $|\sin x|$ has period π and $|\cos x|$ has period π.

Here, $f(x)$ is an even function and $\sin x$, $\cos x$ are complementary.

Thus, period of $f(x)=\frac{1}{2}\{\text{LCM of } \pi \text{ and } \pi\}=\frac{\pi}{2}$

Thus, period for $f(x)$ is $\frac{\pi}{2}$.

Example 79 Find the period of $f(x)=\sin^4 x+\cos^4 x$.

Sol. $\sin^4 x$ and $\cos^4 x$ both has a period π. [as n is even]

But $f(x)$ is an even function and $\sin x$ and $\cos x$ are complementary.

Hence, $f(x)$ has period $=\frac{1}{2}\{\text{LCM of } \pi, \pi\}=\frac{\pi}{2}$

Thus, period of $f(x)$ is $\frac{\pi}{2}$.

Example 80 Find the period of $f(x)=\cos(\cos x)+\cos(\sin x)$.

Sol. Here, $\cos(\cos x)$ has period π; as it is even, also $\cos(\sin x)$ has period π; as it is even.

Thus, period of $f(x)=\frac{1}{2}\{\text{LCM of } \pi \text{ and } \pi\}$

Hence, period of $f(x)=\frac{\pi}{2}$

Example 81 Find the period of $f(x)=\cos^{-1}(\cos x)$.

Sol. Here, $f(x)=\cos^{-1}(\cos x)$; $f(x+T)=f(x)$

$\Rightarrow \cos^{-1}\{\cos(x+T)\}=\cos^{-1}\{\cos(x)\}$

$\Rightarrow T=2\pi, 4\pi, 6\pi, \ldots$

But, T is the least positive value. Hence, $T=2\pi$ or period is 2π.

Aliter $f(x)=\cos^{-1}(\cos x)$ has period 2π, since $\cos x$ has period 2π.

(i.e. In composition of function (fog) or (gof) are periodic, if $g(x)$ and $f(x)$ are periodic, respectively.)

Example 82 The period of $f(x)=\cos(|\sin x|-|\cos x|)$ is

(a) π (b) 2π

(c) $\frac{\pi}{2}$ (d) None of these

Sol. As, $\cos\theta$ is even and $|\sin x|-|\cos x|$ has the period π.

$\therefore$ $\cos(|\sin x|-|\cos x|)$ has period $\frac{\pi}{2}$.

(i.e. Half the period of $g(x)$, if $f(x)$ is even in fog).

Hence, (c) is the correct answer.

Example 83 Period of the function $f(x)=\sin(\sin(\pi x))+e^{\{3x\}}$, where {.} denotes the fractional part of x is

(a) 1 (b) 2

(c) 3 (d) None of these

Sol. As, $\sin(\pi x)$ has period $=\left|\frac{2\pi}{\pi}\right|=2$

$\therefore$ $\sin(\sin(\pi x))$ has period 2

and $e^{\{3x\}}$ has period $\frac{1}{3}$.

$\therefore$ Period of $f(x)=\sin(\sin(\pi x))+e^{\{3x\}}$ is LCM of $\left\{2, \frac{1}{3}\right\}=2$

Hence, (b) is the correct answer.

Example 84 $\sin ax+\cos ax$ and $|\cos x|+|\sin x|$ are periodic functions of same fundamental period, if 'a' equals

(a) 0 (b) 1 (c) 2 (d) 4

Sol. Fundamental period of $|\sin x|+|\cos x|$ is $\frac{\pi}{2}$.

Fundamental period of $(\sin ax+\cos ax)$ is $\frac{2\pi}{a}$.

$\therefore$ $a=4$

Hence, (d) is the correct answer.

Example 85 Let $f(x)=\sin x+\cos(\sqrt{4-a^2})x$. Then, the integral values of 'a' for which $f(x)$ is a periodic function, are given by

(a) $\{2,-2\}$ (b) $(-2, 2]$

(c) $[-2, 2]$ (d) None of these

Sol. $f(x)$ will be periodic, if $\sqrt{4-a^2}$ is a rational which is only possible when $(4-a^2)$ is a perfect square.

$\Rightarrow$ $a=0, 2, -2$ or $a\in\{-2, 0, 2\}$

Hence, (d) is the correct answer.

Example 86 Let $f(x)=\begin{cases}-1+\sin K_1\pi x, & x \text{ is rational.}\\ 1+\cos K_2\pi x, & x \text{ is irrational.}\end{cases}$

If $f(x)$ is a periodic function, then

(a) either $K_1, K_2 \in$ rational or $K_1, K_2 \in$ irrational

(b) $K_1, K_2 \in$ rational only

(c) $K_1, K_2 \in$ irrational only

(d) $K_1, K_2 \in$ irrational such that $\frac{K_1}{K_2}$ is rational

Sol. Range of $-1+\sin K_1\pi x$ is $[-2, 0]$ and range of $1+\cos K_2\pi x$ is $[0, 2]$.

$\Rightarrow$ If $g(x)=-1+\sin K_1\pi x$

$\Rightarrow g(x+T_1) = -1 + \sin K_1 \pi x$, where T_1 is a period of $1 + \sin K_1 \pi x$

$\Rightarrow T_1$ is rational

$\Rightarrow$ Period of $f(x)$ is rational.

$\Rightarrow K_1$ and K_2 are rational.

Hence, (b) is the correct answer.

Example 87 If $f(x) = \tan^2 \dfrac{\pi x}{n^2 - 5n + 8} + \cot (n+m)\pi x$; $(n \in N, m \in Q)$ is a periodic function with 2 as its fundamental period, then m can't belong to

(a) $(-\infty, -2) \cup (-1, \infty)$ (b) $(-\infty, -3) \cup (-2, \infty)$

(c) $(-2, -1) \cup (-3, -2)$ (d) $\left(-3, -\dfrac{5}{2}\right) \cup \left(-\dfrac{5}{2}, -2\right)$

Sol. Period is LCM of $n^2 - 5n + 8$ and $\dfrac{1}{n+m}$.

$\Rightarrow \quad n^2 - 5n + 8 = 1$ or $n^2 - 5n + 8 = 2$

$\Rightarrow \quad n^2 - 5n + 7 = 0$ or $n^2 - 5n + 6 = 0$

Since, $\quad n \in N \therefore n = 2, 3$

and $\quad \left(\dfrac{1}{n+m}\right) K_1 = 2, \ (K_1 \in I) \Rightarrow \dfrac{1}{n+m} \not> 1$

$\therefore \quad \dfrac{1}{2+m} \not> 1$ or $\dfrac{1}{3+m} \not> 1$

$\Rightarrow \quad m \notin (-2, -1) \cup (-3, -2)$

Hence, (c) is the correct answer.

Example 88 Let $f(x)$ be a periodic function with period 3 and $f\left(-\dfrac{2}{3}\right) = 7$ and $g(x) = \int_0^x f(t+n)\, dt$, where $n = 3K, K \in N$. Then, $g'\left(\dfrac{7}{3}\right)$ is equal to

(a) $-\dfrac{2}{3}$ (b) 7

(c) -7 (d) $\dfrac{7}{3}$

Sol. $g'(x) = f(x+n) = f(x + (n-3) + 3)$

$= f(x + n - 3)$

....

....

$= f(x+3) = f(x)$

$\Rightarrow \quad g'(x) = f(x)$

$\Rightarrow g'\left(\dfrac{7}{3}\right) = f\left(\dfrac{7}{3}\right) = f\left(-\dfrac{2}{3} + 3\right) = f\left(-\dfrac{2}{3}\right) = 7$

Hence, (b) is the correct answer.

Exercise for Session 7

1. Find the periods of following functions.

(i) $f(x) = [\sin 3x] + |\cos 6x|$

(ii) $f(x) = \dfrac{1}{2}\left\{\dfrac{|\sin x|}{\cos x} + \dfrac{|\cos x|}{\sin x}\right\}$

(iii) $f(x) = e^{\cos^4 \pi x + x - [x] + \cos^2 \pi x}$

(iv) $f(x) = 3 \sin \dfrac{\pi x}{3} + 4 \cos \dfrac{\pi x}{4}$

(v) $f(x) = \cos 3x + \sin \sqrt{3}\pi x$

(vi) $f(x) = \sin \dfrac{\pi x}{n!} - \cos \dfrac{\pi x}{(n+1)!}$

(vii) $f(x) = x - [x - b]$

(viii) $f(x) = e^{\ln (\sin x)} + \tan^3 x - \operatorname{cosec}(3x - 5)$

2. Find the period of the real-valued function satisfying $f(x) + f(x+4) = f(x+2) + f(x+6)$.

3. Check whether the function defined by $f(x + \lambda) = 1 + \sqrt{2f(x) - f^2(x)}, \forall x \in R$ is periodic. If yes, find its period.

4. Let $f(x+p) = 1 + \{2 - 3f(x) + 3(f(x))^2 - (f(x))^3\}^{1/3}, \forall x \in R$, where $p > 0$, prove that $f(x)$ is periodic.

5. Let $f(x)$ be a function such that ; $f(x-1) + f(x+1) = \sqrt{3} f(x), \forall x \in R$. If $f(5) = 100$, find $\sum_{r=0}^{99} f(5 + 12r)$.

Session 8

Mapping of Functions

Mapping of Functions

As discussed earlier, a function exists only if, "to every element in domain there exists unique image in the codomain".

i.e. To every element of A there exists one and only one element of B.

This is written as $f: A \to B$ and is read as f maps from A to B and this correspondence is denoted by $y = f(x)$.

From definition, it follows that there may exist some elements in B, which may not have any corresponding element in set A.

But there should not be any x left (element of A) for which there is no element in set B.

There are four types of mappings defined as

1. One-one Mapping or Injective or Monomorphic

A function $f: A \to B$ is said to be one-one mapping or injective, if different elements of A have different images in B.

Thus, no two elements of set A can have the same f image.

Verbal Description Let us consider set $A = \{1, 3, 5\}$ and $B = \{3, 7, 11, 15\}$, where $f: A \to B$ and $f(x) = 2x + 1$, then here every element in domain possess distinct images in codomain.

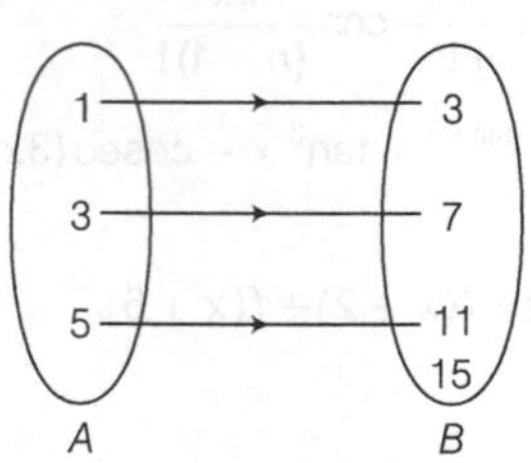

Figure 3.34

Thus, $f(x)$ is one-one or injective.

From above definition, following mappings are not one-one.

(i) $f: A \to B$ (ii) $f: A \to B$

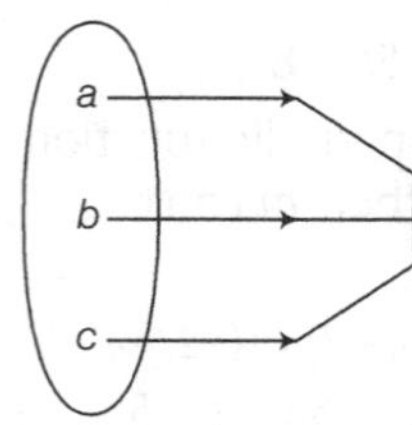

Figure 3.35

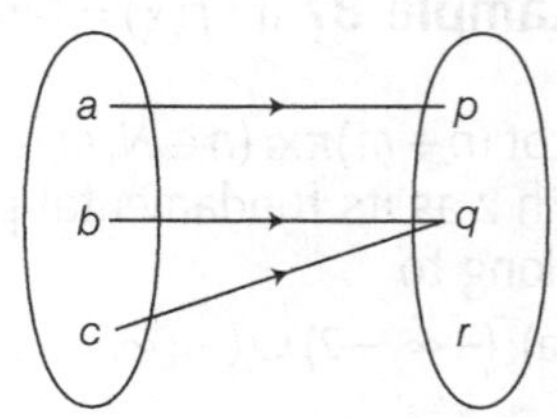

Figure 3.36

Method to Check One-one Mapping

***Method* 1. Theoretically** If $f(x) = f(y) \Rightarrow x = y$, then $f(x)$ is one-one.

***Method* 2. Graphically** A function is one-one iff no line parallel to X-axis meets the graph of function at more than one point.

Example 89 Let $f: [-\pi/2, \pi/2] \to [-1, 1]$ where $f(x) = \sin x$. Find whether $f(x)$ is one-one or not.

Sol. Here, $f: [-\pi/2, \pi/2] \to [-1, 1]$ indicates that domain $\in \left[-\frac{\pi}{2}, \frac{\pi}{2}\right]$ and codomain $\in [-1, 1]$.

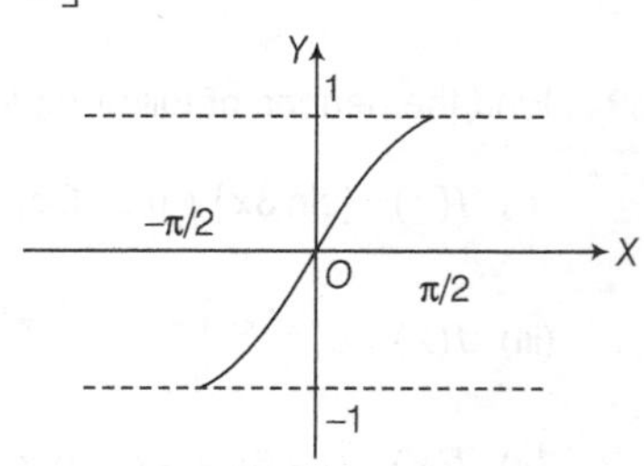

Thus, the graph of $f(x) = \sin x$ should be plotted in $[-\pi/2, \pi/2]$.

Which is clearly not intersected at more than one point by any straight line parallel to X-axis.

Thus, $f(x)$ is one-one.

***Method 3.* By Calculus** For checking whether $f(x)$ is one-one, find whether function is only increasing or only decreasing in its domain. If yes, then one-one.

i.e. if $f'(x) \geq 0, \ \forall \ x \in$ domain

or if $f'(x) \leq 0, \ \forall \ x \in$ domain, then one-one.

Remark

Students are advised to use the graphical or calculus method for finding one-one.

Number of One-one Mapping

If A and B are finite sets having m and n elements, then number of one-one function from A to B.

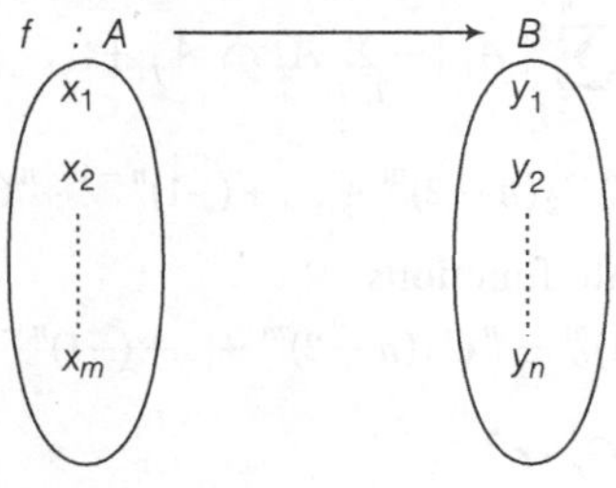

Figure 3.37

Here, x_1 can take n images,

x_2 can take $(n-1)$ images,

x_3 can take $(n-2)$ images,

....

....

x_m can take $(n-m+1)$ images.

Thus, number of mapping $\Rightarrow n(n-1)(n-2)\ldots(n-m+1)$

$$\Rightarrow \begin{cases} {}^nP_m, & \text{if } n \ge m \\ 0, & \text{if } n < m \end{cases}$$

Example 90 $f(x) = x^3 + 3x^2 + 4x + b \sin x + c \cos x$, $\forall x \in R$ is a one-one function, the value of $b^2 + c^2$ is

(a) ≥ 1 (b) ≥ 2 (c) ≤ 1 (d) None of these

Sol. Here, $f(x) = x^3 + 3x^2 + 4x + b \sin x + c \cos x$

$\Rightarrow f'(x) = 3x^2 + 6x + 4 + b \cos x - c \sin x$

Now, for $f(x)$ to be one-one, the only possibility is

$$f'(x) \ge 0, \forall x \in R$$

$\Rightarrow 3x^2 + 6x + 4 + b \cos x - c \sin x \ge 0, \forall x \in R$

$\Rightarrow 3x^2 + 6x + 4 \ge c \sin x - b \cos x, \forall x \in R$

$\Rightarrow 3x^2 + 6x + 4 \ge \sqrt{b^2 + c^2}, \forall x \in R$

$\Rightarrow \sqrt{b^2 + c^2} \le 3(x^2 + 2x + 1) + 1, \forall x \in R$

$\Rightarrow \sqrt{b^2 + c^2} \le 3(x+1)^2 + 1, \forall x \in R$

$\Rightarrow \sqrt{b^2 + c^2} \le 1, \forall x \in R \Rightarrow b^2 + c^2 \le 1, \forall x \in R$

Hence, (c) is the correct answer.

2. Many-one Mapping

A mapping $f : A \to B$ is said to be many-one function, if two or more elements of set A have the same image in B.

In other words; $f : A \to B$ is a many-one function, if it is not one-one function.

Verbal Description Let $f : A \to B$ and $g : X \to Y$ be two functions represented by

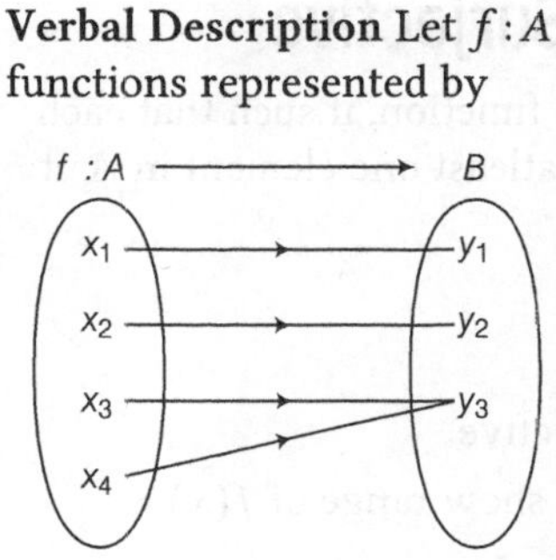

Figure 3.38

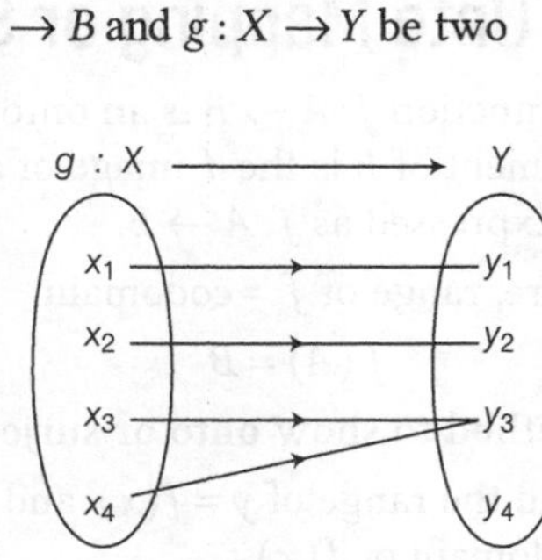

Figure 3.39

Clearly, f and g both are many-one as there are two elements x_3, x_4 which correspond to the same image.

i.e. $f(x_3) = f(x_4) = y_3$. Thus, many-one.

Method to Check Many-one

They are same as for one-one because, if mapping is not one-one it is many-one.

Example 91 Show $f : R \to R$ defined by $f(x) = x^2 + x$ for all $x \in R$ is many-one.

Sol. By graph

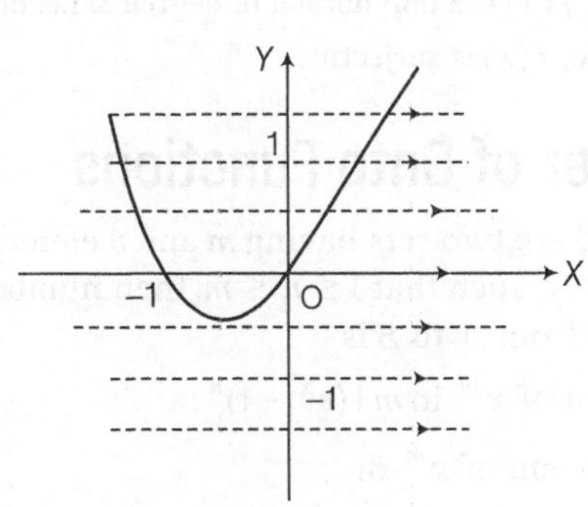

$f(x) = x^2 + x$ can be represented graphically, as shown in figure,

where the straight line parallel to X-axis meets the curve at two points (i.e. more than one point).

Thus, it is not one-one it is many-one.

Aliter By calculus,

$$f(x) = x^2 + x$$

$\Rightarrow f'(x) = 2x + 1$

where $f'(x) > 0$

if $x > -\frac{1}{2}$

and $f'(x) < 0$

if $x < -\frac{1}{2}$

which shows $f(x)$ is neither increasing nor decreasing i.e. not monotonic. Hence, many-one.

3. Onto Mapping or Surjective

A function $f: A \to B$ is an onto function, if such that each element of B is the f image of atleast one element in A. It is expressed as $f: A \to B$.

Here, range of f = codomain.

i.e. $f(A) = B$

Method to show onto or surjective

Find the range of $y = f(x)$ and show range of $f(x)$ = codomain of $f(x)$.

Remark

If range = codomain, then $f(x)$ is onto. Any polynomial of odd degree has range all real numbers and is onto for $f: R \to R$.

Example 92 Show $f: R \to R$ defined by $f(x) = (x-1)(x-2)(x-3)$ is surjective but not injective

Sol. We have,

$$f(x) = (x-1)(x-2)(x-3)$$
$$f(1) = f(2) = f(3) = 0$$

Hence, it is not injective, it is many-one function.

Now, $f(x)$ is a polynomial of degree 3, i.e. odd.

Hence, $f(x)$ is surjective.

Number of Onto Functions

If A and B are two sets having m and n elements respectively, such that $1 \le n \le m$, then number of onto functions from A to B is

Coefficient of x^m in $m!(e^x - 1)^n$

$\Rightarrow$ Coefficient of x^m in

$$m!\{{}^nC_0 e^{nx} - {}^nC_1 e^{(n-1)x} + \ldots + (-1)^r\, {}^nC_r e^{rx} + \ldots + (-1)^n\, {}^nC_n\}$$

$$= \sum_{r=1}^{n} (-1)^{n-r}\, {}^nC_r\, r^m$$

Or

Consider the set of all possible functions from A to B,

i.e. $\{f: A \to B\}$.

Now, let us define a subset Q such that

$$A_i = \{f \in Q \mid i \notin \text{Range of } f\}, i \in n$$

i.e. $\bigcup_{i=1}^{n} A_i = \{f \text{ is injective}\}$

To find number of onto functions

= Total number of functions – Number of injective functions

$= n^m$ – {Number of injective functions}

$$= n^m - \left|\bigcup_{i=1}^{n} A_i\right|, \quad \ldots(i)$$

where $\bigcup_{i=1}^{n} A_i = \sum_{i=1}^{n} |A_i| - \sum_{i,j} A_i \cap A_j + \ldots$

$$= n \cdot (n-1)^m - {}^nC_2 (n-2)^m + \ldots + (-1)^{n-1} \cdot {}^nC_{n-1}(1)^m \quad \ldots(ii)$$

$\therefore$ Number of onto functions

$$= n^m - \{n \cdot (n-1)^m - {}^nC_2(n-2)^m + \ldots + (-1)^{n-1} \cdot {}^nC_{n-1}(1)^m\}$$

$$= \sum_{r=1}^{n} (-1)^{n-r}\, {}^nC_r \cdot r^m$$

Example 93 If $f: R \to \left[\frac{\pi}{6}, \frac{\pi}{2}\right)$, $f(x) = \sin^{-1}\left(\frac{x^2 - a}{x^2 + 1}\right)$ is an onto function, the set of values of 'a' is

(a) $\left\{-\frac{1}{2}\right\}$ (b) $\left[-\frac{1}{2}, -1\right)$

(c) $(-1, \infty)$ (d) None of these

Sol. Here, $f(x)$ is onto.

$$\therefore \quad \frac{\pi}{6} \le \sin^{-1}\left(\frac{x^2 - a}{x^2 + 1}\right) < \frac{\pi}{2} \Rightarrow \frac{1}{2} \le \frac{x^2 - a}{x^2 + 1} < 1$$

$$\Rightarrow \frac{1}{2} \le 1 - \frac{(a+1)}{x^2 + 1} < 1, \forall x \in R \Rightarrow a + 1 > 0 \Rightarrow a \in (-1, \infty)$$

Hence, (c) is the correct answer.

4. Into Mapping

A function $f: A \to B$ is an into function, if there exists an element in B having no pre-image in A.

In other words, $f: A \to B$ is into function, if it is not onto function (mapping).

Example 94 Show $f: R \to R$ defined by $f(x) = x^2 + 4x + 5$ is into.

Sol. We have, $f(x) = x^2 + 4x + 5$

$$f(x) = (x+2)^2 + 1$$

Since, the codomain of f is R but the range of f is $[1, \infty)$. Hence, f is into.

One-one Onto Mapping or Bijective

A function is one-one onto or bijective, if it is both one-one and onto. A function is bijective if and only if every possible image is mapped to by exactly one argument.

The function $f: A \to B$ is bijective iff for all $y \in B$, there is a unique $x \in A$ such that $f(x) = y$.

Example 95 Let $A=\{x:-1\le x\le 1\}=B$ and a mapping $f:A\to B$. For each of the following functions from A to B, find whether it is surjective or bijective.

(i) $f(x)=|x|$ (ii) $f(x)=x|x|$ (iii) $f(x)=x^3$

(iv) $f(x)=[x]$ (v) $f(x)=\sin\frac{\pi x}{2}$

Sol. (i) $f(x)=|x|$

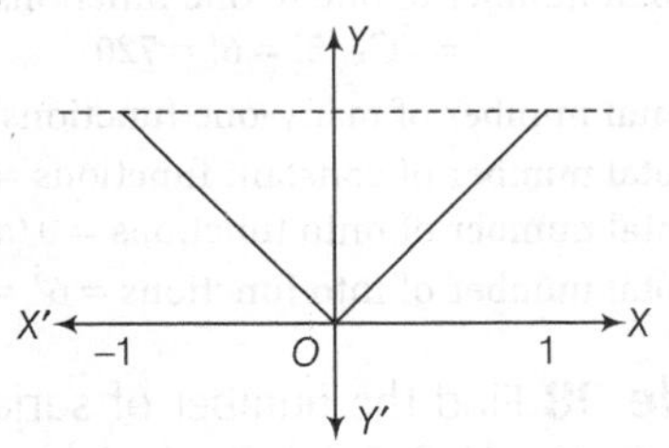

Which shows many-one, as the straight line is parallel to X-axis cuts at two points. Here, range for $f(x)\in[0,1]$. Which is clearly a subset of codomain.

i.e. $[0,1]\subseteq[-1,1]$

Thus, into. Hence, function is many-one-into.

$\therefore$ Neither injective nor surjective.

(ii) $f(x)=x|x|$

The graph shows that $f(x)$ is one-one, as the straight line parallel to X-axis cuts only at one point.

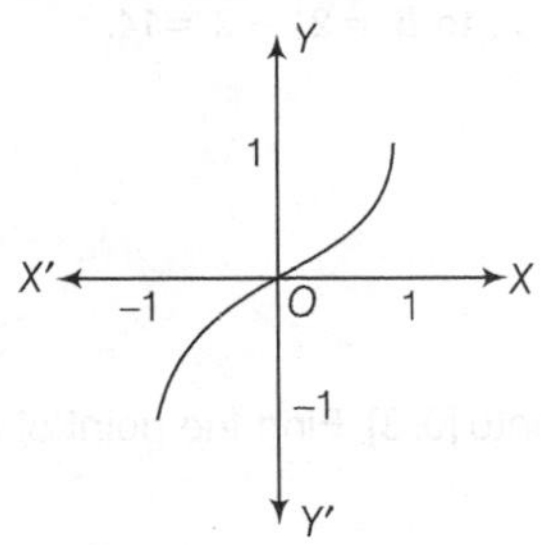

Here, range $f(x)\in[-1, 1]$

Thus, range = codomain

Hence, onto. Therefore, $f(x)$ is one-one onto or (bijective).

(iii) $f(x)=x^3$

Graph shows $f(x)$ is one-one onto (i.e. bijective). (as explained above)

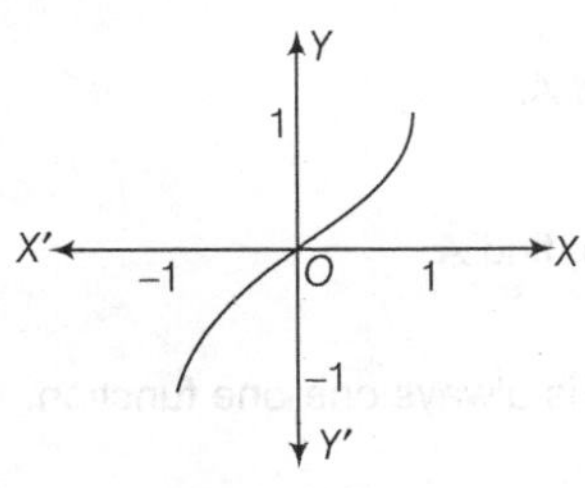

(iv) $f(x)=[x]$

Graph shows that $f(x)$ is many-one, as the straight line parallel to X-axis meets at more than one points.

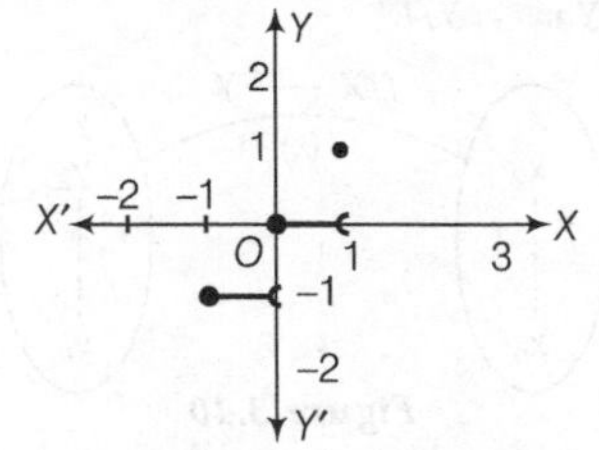

Here, range $f(x)\in\{-1, 0, 1\}$

which shows into as range $\subseteq$ codomain.

Hence, many-one-into.

(v) $f(x)=\sin\frac{\pi x}{2}$

Graph shows $f(x)$ is one-one and onto as range = codomain.

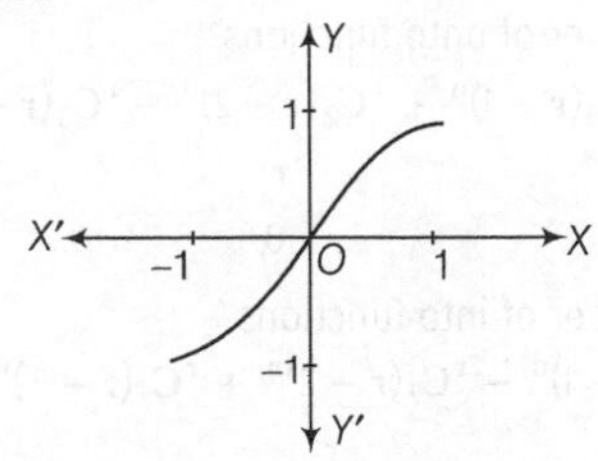

Therefore, $f(x)$ is bijective.

Example 96 The function $f:R\to R$ defined as $f(x)=\log_e(\sqrt{\sqrt{x^2+1}+x}+\sqrt{\sqrt{x^2+1}-x})$ is

(a) one-one and onto both

(b) one-one but not onto

(c) onto but not one-one

(d) Neither one-one nor onto

Sol. (d) Here, $f(x)=\frac{2}{2}\log\left(\sqrt{\sqrt{x^2+1}+x}+\sqrt{\sqrt{x^2+1}-x}\right)$

$$=\frac{1}{2}\log\left(\sqrt{\sqrt{x^2+1}+x}+\sqrt{\sqrt{x^2+1}-x}\right)^2$$

$$=\frac{1}{2}\log\left[\sqrt{x^2+1}+x+\sqrt{x^2+1}-x+2\sqrt{(\sqrt{x^2+1})^2-x^2}\right]$$

$$=\frac{1}{2}\log\left[2\sqrt{x^2+1}+2\sqrt{x^2+1-x^2}\right]$$

$$=\frac{1}{2}\log[2\sqrt{x^2+1}+2]$$

Now, $f(-x)=f(x)$

$\therefore$ $f(x)$ is even function.

Hence, neither one-one nor onto.

Quick Review

Let $X = \{x_1, x_2, x_3, \ldots, x_n\}$ [i.e. n elements]
and $Y = \{y_1, y_2, y_3, \ldots, y_r\}$ [i.e. r elements]

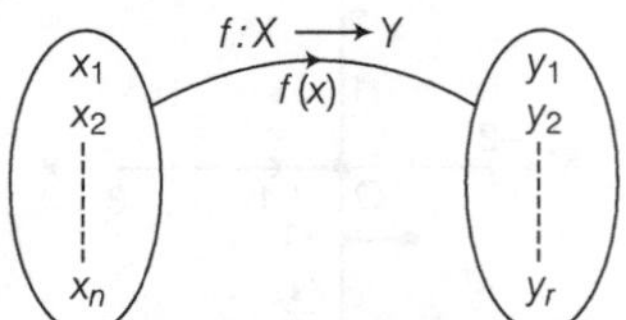

Figure 3.40

(a) Total number of functions $= r^n$
$=$ (Number of elements in codomain)$^{\text{number of elements in domain}}$

(b) Total number of one to one functions $= \begin{cases} {}^rC_n \cdot n!, & r \geq n \\ 0, & r < n \end{cases}$

(c) Total number of many-one functions $= \begin{cases} r^n - {}^rC_n \cdot n!, & r \geq n \\ r^n, & r < n \end{cases}$

(d) Total number of constant functions $= r$

(e) Total number of onto functions
$$= \begin{cases} r^n - {}^rC_1(r-1)^n + {}^rC_2(r-2)^n - {}^rC_3(r-3)^n + \ldots, & r < n \\ r!, & r = n \\ 0, & r > n \end{cases}$$

(f) Total number of into functions
$$= \begin{cases} {}^rC_1(r-1)^n - {}^rC_2(r-2)^n + {}^rC_3(r-3)^n - \ldots, & r \leq n \\ r^n, & r > n \end{cases}$$

Example 97 If $X = \{1, 2, 3, 4, 5\}$ and $Y = \{a, b, c, d, e, f\}$ and $f : X \to Y$, find the total number of

(i) functions (ii) one to one functions
(iii) many-one functions (iv) constant functions
(v) onto functions (vi) into functions

Sol. (i) Total number of functions $= 6^5 = 7776$

(ii) Total number of one to one functions
$= {}^6C_5 \cdot 5! = 6! = 720$

(iii) Total number of many-one functions $= 6^5 - 6! = 7056$

(iv) Total number of constant functions $= 6$

(v) Total number of onto functions $= 0$ (as $r > n$)

(vi) Total number of into functions $= 6^5 = 7776$

Example 98 Find the number of surjections from A to B, where $A = \{1, 2, 3, 4\}$, $B = \{a, b\}$. ***[IIT JEE 2000]***

Sol. Number of surjections from A to $B = \sum_{r=1}^{2} (-1)^{2-r}\, {}^2C_r (r)^4$

$$= (-1)^{2-1}\, {}^2C_1 (1)^4 + (-1)^{2-2}\, {}^2C_2 (2)^4 = -2 + 16 = 14$$

Therefore, number of onto mappings from A to $B = 14$.

Aliter Total number of mapping from A to B is 2^4 of which two functions $f(x) = a$ for all $x \in A$ and $g(x) = b$ for all $x \in A$ are not surjective.

Thus, total number of surjections from

A to $B = 2^4 - 2 = 14$.

Exercise for Session 8

1. There are exactly two distinct linear functions, which map $[-1, 1]$ onto $[0, 3]$. Find the point of intersection of the two functions.

2. Let f be one-one function with domain $\{x, y, z\}$ and range $\{1, 2, 3\}$. It is given that exactly one of the following statement is true and remaining two are false.
$f(x) = 1, f(y) \neq 1, f(z) \neq 2$. Determine $f^{-1}(1)$.

3. Let $A = R - \{3\}$, $B = R - \{1\}$ and $f : A \to B$ defined by $f(x) = \dfrac{x-2}{x-3}$. Is 'f' bijective? Give reasons.

4. Let $f : R \to R$ defined by $f(x) = \dfrac{x^2}{1+x^2}$. Prove that f is neither injective nor surjective.

5. If the function $f : R \to A$, given by $f(x) = \dfrac{x^2}{x^2+1}$ is surjection, find A.

6. If the function $f : R \to A$, given by $f(x) = \dfrac{e^x - e^{-|x|}}{e^x + e^{|x|}}$ is surjection, find A.

7. Let $f(x) = ax^3 + bx^2 + cx + d \sin x$. Find the condition that $f(x)$ is always one-one function.

8. Let $f : X \to Y$ be a function defined by $f(x) = a \sin\left(x + \dfrac{\pi}{4}\right) + b \cos x + c$. If f is both one-one and onto, find the sets X and Y.

Session 9

Identical (or Equal) Functions

Identical (or Equal) Functions

Two functions f and g are said to be identical (or equal) functions, if

(i) the domain of f = the domain of g,

(ii) the range of f = the range of g, and

(iii) $f(x) = g(x), \forall x \in$ domain.

Examples of Equal or Identical Functions

(i) $f(x) = \ln x^2, g(x) = 2 \ln x$ (NI)

$f(x) = \text{cosec } x, g(x) = \frac{1}{\sin x}$ (I)

(ii) $f(x) = \cot(\cot^{-1} x), g(x) = x$ (I)

$f(x) = \tan x, g(x) = \frac{1}{\cot x}$ (NI)

(iii) $f(x) = \sin^{-1}(3x - 4x^3), g(x) = 3\sin^{-1} x$ (NI)

(iv) $f(x) = \text{sgn}(x^2 + 1), g(x) = \sin^2 x + \cos^2 x$ (I)

(v) $f(x) = \tan^2 x \cdot \sin^2 x, g(x) = \tan^2 x - \sin^2 x$ (I)

(vi) $f(x) = \sec^2 x - \tan^2 x, g(x) = 1$ (NI)

(vii) $f(x) = \log_x e, g(x) = \frac{1}{\log_e x}$ (I)

(viii) $f(x) = \tan(\cot^{-1} x), g(x) = \cot(\tan^{-1} x)$ (I)

(ix) $f(x) = \sqrt{x^2 - 1}, g(x) = \sqrt{x-1} \cdot \sqrt{x+1}$ (NI)

(x) $f(x) = \tan x \cdot \cot x, g(x) = \sin x \cdot \text{cosec } x$ (NI)

(xi) $f(x) = e^{\ln e^x}, g(x) = e^x$ (I)

(xii) $f(x) = \sqrt{\frac{1 - \cos 2x}{2}}, g(x) = \sin x$ (NI)

(xiii) $f(x) = \sqrt{x^2}, \; g(x) = (\sqrt{x})^2$ (NI)

(xiv) $f(x) = \log(x+2) + \log(x-3)$,
$g(x) = \log(x^2 - x - 6)$ (NI)

(xv) $f(x) = x|x|, g(x) = x^2 \text{ sgn } x$ (I)

(xvi) $f(x) = \lim_{n \to \infty} \frac{x^{2n} - 1}{x^{2n} + 1}, g(x) = \text{sgn}(|x| - 1)$ (I)

(xvii) $f(x) = \sin(\sin^{-1} x), g(x) = \cos(\cos^{-1} x)$ (I)

(xviii) $f(x) = \frac{1}{1 + \frac{1}{x}}, g(x) = \frac{x}{1+x}$ (NI)

(xix) $f(x) = e^{\ln \sec^{-1} x}, g(x) = \sec^{-1} x$ (NI)

Identical, if $x \in (-\infty, -1] \cup [1, \infty)$

(xx) $F(x) = (fog)(x), G(x) = (gof)(x)$, where $f(x) = e^x$, $g(x) = \ln x$ (NI)

Example 99 If $f(x) = \log_{x^2} 25$ and $g(x) = \log_x 5$, then $f(x) = g(x)$ holds, now find the interval for x.

Sol. Domain of $f \in R - \{\pm 1, 0\}$

Domain of $g \in (0, \infty) - \{1\}$

For $\quad f(x) = g(x)$,

domain of f = domain of g.

i.e., $\quad f(x) = g(x)$, if $x \in (0, \infty) - \{1\}$

Example 100 Let $A = \{1, 2\}$, $B = \{3, 6\}$ and $f: A \to B$ given by $f(x) = x^2 + 2$ and $g: A \to B$ given by $g(x) = 3x$. Find whether they equal or not.

Sol.
$$f(1) = 3, f(2) = 6$$
$$g(1) = 3, g(2) = 6$$

which shows $f(x)$ and $g(x)$ have same domains and range, thus $f = g$.

Example 101 Which pair of functions is identical?

(a) $\sin^{-1}(\sin x)$ and $\sin(\sin^{-1} x)$

(b) $\log_e e^x, \; e^{\log_e x}$

(c) $\log_e x^2, 2\log_e x$

(d) None of the above

Sol. Here, (a) $\sin^{-1}(\sin x)$ is defined for $x \in \left[-\frac{\pi}{2}, \frac{\pi}{2}\right]$,

while $\sin(\sin^{-1} x)$ is defined only for $x \in [-1, 1]$.

(b) $\log_e e^x$ is defined for all x,

while $e^{\log_e x}$ is defined for $x > 0$.

(c) $\log_e x^2$ is defined for all $x \in R - \{0\}$, while $2\log_e x$ is defined for $x > 0$.

$\therefore$ None is identical.

Hence, (d) is the correct answer.

Exercise for Session 9

- **Directions** (Q. Nos. 1 to 10) *Which of the following are identical (or equal) functions?*

1. $f(x) = \ln e^{x^2}, g(x) = e^{\ln x}$
2. $f(x) = \sec x, g(x) = \dfrac{1}{\cos x}$
3. $f(x) = \sec^{-1} x + \operatorname{cosec}^{-1} x,\ g(x) = \dfrac{\pi}{2}$
4. $f(x) = \cot^2 x \cdot \cos^2 x,\ g(x) = \cot^2 x - \cos^2 x$
5. $f(x) = \operatorname{sgn}(\cot^{-1} x), g(x) = \operatorname{sgn}(x^2 - 4x + 5)$
6. $f(x) = \log_e x,\ g(x) = \dfrac{1}{\log_x e}$
7. $f(x) = \sqrt{1 - x^2}, g(x) = \sqrt{1 - x} \cdot \sqrt{1 + x}$
8. $f(x) = \dfrac{1}{|x|}, g(x) = \sqrt{x^{-2}}$
9. $f(x) = [\{x\}], g(x) = \{[x]\}$ [Note that $f(x)$ and $g(x)$ are constant functions]
10. $f(x) = e^{\ln \cot^{-1} x}, g(x) = \cot^{-1} x$

Session 10

Composite Functions

Composite Functions

Let us consider two functions, $f : X \to Y_1$ and $g : Y_1 \to Y$. We define function $h : X \to Y$, such that, $h(x) = g(f(x)) = (gof)(x)$.

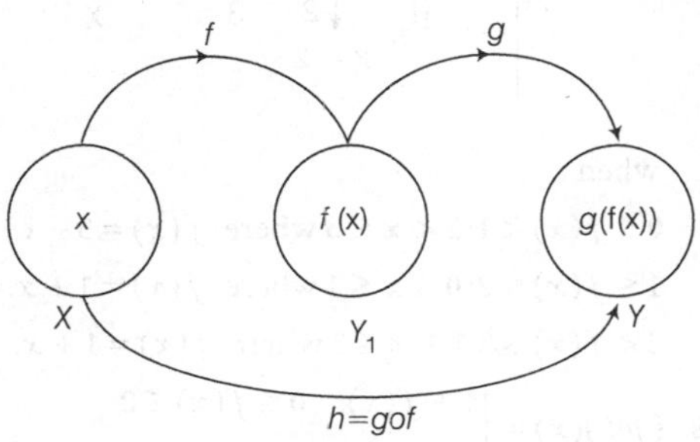

Figure 3.41

To obtain $h(x)$, we first take f-image of an element $x \in X$ so that $f(x) \in Y_1$, which is the domain of $g(x)$. Then, we take g-image of $f(x)$, i.e. $g(f(x))$ which would be an element of Y.

The diagram below shows the steps to be taken

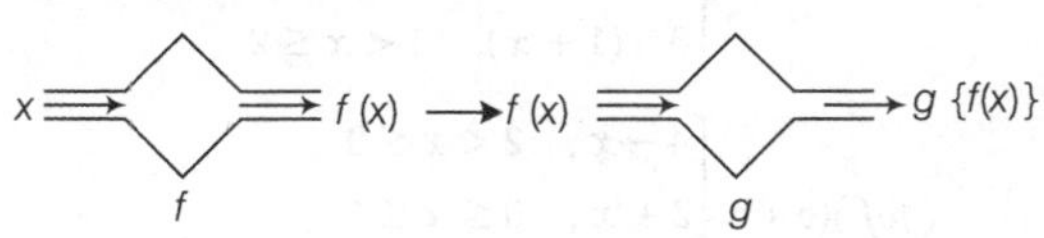

Figure 3.42

The function h defined in the diagram is called composition of f and g, and is denoted by (gof). Clearly, domain $(gof) = \{X : x \in \text{domain } (f), f(x) \in \text{domain } (g)\}$.

Similarly, we can write, $(fog)(x) = f\{g(x)\}$

and domain $(fog) = \{x : x \in \text{domain } (g), g(x) \in \text{domain } f\}$

In general, $fog \neq gof$.

Properties of Composite Functions

(i) It should be noted that gof exists, iff the range of $f \subseteq$ domain of g. Similarly, fog exists; iff the range of $g \subseteq$ domain of f.

(ii) Composite of functions is not commutative, i.e. $gof \neq fog$.

(iii) Composite of functions is associative, i.e. if f, g and h are three functions such that $fo(foh)$ and $(fog)oh$ are defined, then $fo\,(goh) = (fog)oh$.

e.g. Associativity $f : (N) \to I_0$ $\quad f(x) = 2x$

$g : I_0 \to Q \quad g(x) = \dfrac{1}{x}$

$h : Q \to R \quad h(x) = e^{1/x}$

$(hog)\,of = ho\,(gof) = e^{2x}$

(iv) The composite of two bijections is a bijection, i.e. if f and g are two bijections such that gof is defined, then gof is also a bijection.

Proof $f : A \to B$ and $g : B \to C$ be two bijections. Then, gof exists such that $gof : A \to C$.

We have to prove that gof is one-one and onto.

One-One Let $a_1, a_2 \in A$ such that $(gof)(a_1) = (gof)(a_2)$, then $(gof)(a_1) = (gof)\,(a_2) \Rightarrow g[f(a_1)] = g[f(a_2)]$

$\Rightarrow f(a_1) = f(a_2)$ $\quad$ [$\because$ g is one-one]

$\Rightarrow \quad a_1 = a_2$ $\quad$ [$\because$ f is one-one]

$\therefore$ gof is also one-one function.

Onto Let $c \in C$, then $c \in C$

$\Rightarrow \quad \exists\, b \in B$ such that $g(b) = c$ $\quad$ [$\because$ g is onto]

and $b \in B \Rightarrow \exists\, a \in A$ such that $f(a) = b$ $\quad$ [$\because$ f is onto]

Therefore, we see that

$c \in C \Rightarrow \exists\, a \in A$ such that $(gof)(a) = g[f(a)] = g(g) = c$

i.e. Every element of C is the gof image of some element of A. As such gof is an onto function. Hence, gof being one-one and onto is a bijection.

(v) f is even, g is even $\Rightarrow fog$ is even function.

(vi) f is odd, g is odd $\Rightarrow fog$ is odd function.

(vii) f is even, g is odd $\Rightarrow fog$ is even function.

(viii) f is odd, g is even $\Rightarrow fog$ is even function.

Example 102 Let $f : A \to B$ and $g : B \to C$ be functions and $gof : A \to C$. Which of the following statements is true?

(a) If gof is one-one, then f and g both are one-one

(b) If gof is one-one, then f is one-one

(c) If gof is a bijection, then f is one-one and g is onto

(d) If f and g are both one-one, then gof is one-one

Sol. (a) As shown gof is one-one, but g is many-one.

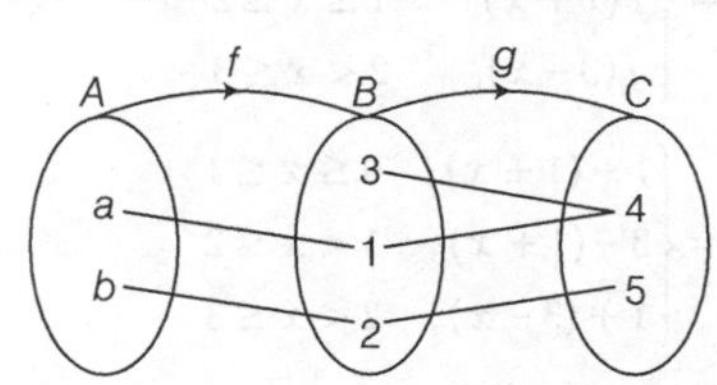

$\Rightarrow$ (a) is not correct.

(b) If gof is one-one, then f is also one-one,
if f is many-one, then gof cannot be one-one.

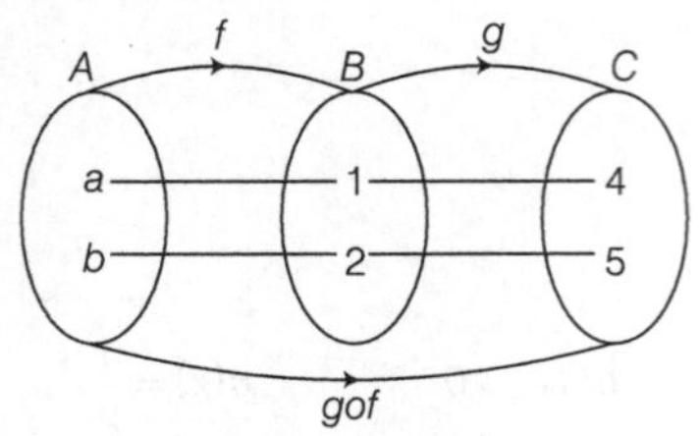

$\Rightarrow$ (c) and (d) are obviously true.

Hence, (b), (c) and (d) are correct answers.

Example 103 If $f: R \to R, f(x) = x^2$ and $g: R \to R$; $g(x) = 2x + 1$. Find fog and gof, also show $fog \neq gof$.

Sol. $\because$ $(gof)(x) = g\{f(x)\} = g\{x^2\}$

$(gof)(x) = 2x^2 + 1$ and

$(fog)(x) = f\{g(x)\} = f(2x + 1)$

$(fog)(x) = (2x + 1)^2$

where $(2x^2 + 1) \neq (2x + 1)^2$. Therefore, $(gof) \neq (fog)$.

Example 104 Let $g(x) = 1 + x - [x]$

and $$f(x) = \begin{cases} -1, & x < 0 \\ 0, & x = 0. \\ 1, & x > 0 \end{cases}$$

Then, for all x, find $f(g(x))$. *[IIT JEE 2001]*

Sol. Here, $g(x) = 1 + x - [x]$

$\Rightarrow$ $g(x) = 1 + \{x\}$ [as $x - [x] = \{x\}$]

i.e. $g(x) \geq 1$.

So, $f(g(x)) = 1$. Since, $f(x) = 1$ for all $x > 0$.

Thus, $f(g(x)) = 1$, for all $x \in R$.

Example 105 Let $f(x) = \begin{cases} 1 + x, & 0 \leq x \leq 2 \\ 3 - x, & 2 < x \leq 3 \end{cases}$, find $(fof)(x)$.

Sol. Clearly, $fof(x) = f\{f(x)\} = \begin{cases} f(1 + x), & 0 \leq x \leq 2 \\ f(3 - x), & 2 < x \leq 3 \end{cases}$

$$= \begin{cases} f(1 + x), & 0 \leq x \leq 1 \\ f(1 + x), & 1 \leq x \leq 2 \\ f(3 - x), & 2 < x \leq 3 \end{cases}$$

$$= \begin{cases} 1 + (1 + x), & 0 \leq x \leq 1 \\ 3 - (1 + x), & 1 < x \leq 2 \\ 1 + (3 - x), & 2 < x \leq 3 \end{cases}$$

$$= \begin{cases} 2 + x, & 0 \leq x \leq 1 \\ 2 - x, & 1 < x \leq 2 \\ 4 - x, & 2 < x \leq 3 \end{cases}$$

Aliter $f(x)$ can be expressed graphically as shown in figure below;

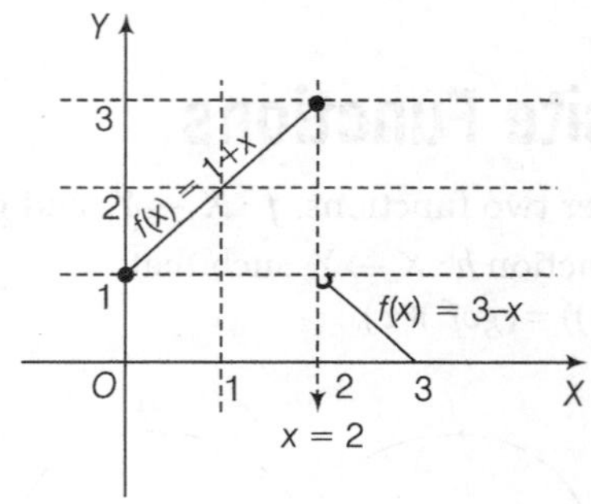

when

$0 \leq f(x) < 1; 2 < x \leq 3$ where $f(x) = 3 - x$.

$1 \leq f(x) \leq 2; 0 \leq x \leq 1$ where $f(x) = 1 + x$.

$2 < f(x) \leq 3; 1 \leq x \leq 2$ where $f(x) = 1 + x$.

Thus, $$(fof)(x) = \begin{cases} 1 + f(x), & 0 \leq f(x) \leq 2 \\ 3 - f(x), & 2 < f(x) \leq 3 \end{cases}$$

$$(fof)(x) = \begin{cases} 1 + f(x), & 0 \leq f(x) < 1 \\ 1 + f(x), & 1 \leq f(x) \leq 2 \\ 3 - f(x), & 2 < f(x) \leq 3 \end{cases}$$

$$(fof)(x) = \begin{cases} 1 + (3 - x), & 2 < x \leq 3 \\ 1 + (1 + x), & 0 \leq x \leq 1 \\ 3 - (1 + x), & 1 < x \leq 2 \end{cases}$$

$$(fof)(x) = \begin{cases} 4 - x, & 2 < x \leq 3 \\ 2 + x, & 0 \leq x \leq 1 \\ 2 - x, & 1 < x \leq 2 \end{cases}$$

$$(fof)(x) = \begin{cases} 2 + x, & 0 \leq x \leq 1 \\ 2 - x, & 1 < x \leq 2 \\ 4 - x, & 2 < x \leq 3 \end{cases}$$

Example 106 Let $f(x) = \begin{cases} x + 1, & x \leq 1 \\ 2x + 1, & 1 < x \leq 2 \end{cases}$

and $$g(x) = \begin{cases} x^2, & -1 \leq x < 2 \\ x + 2, & 2 \leq x \leq 3 \end{cases}.$$

Find (fog).

Sol. $f(g(x)) = \begin{cases} g(x) + 1, & g(x) \leq 1 \\ 2g(x) + 1, & 1 < g(x) \leq 2 \end{cases}$

Here, $g(x)$ becomes the variable that means we should draw the graph. It is clear that $g(x) \le 1;\ \forall\ x \in [-1, 1]$

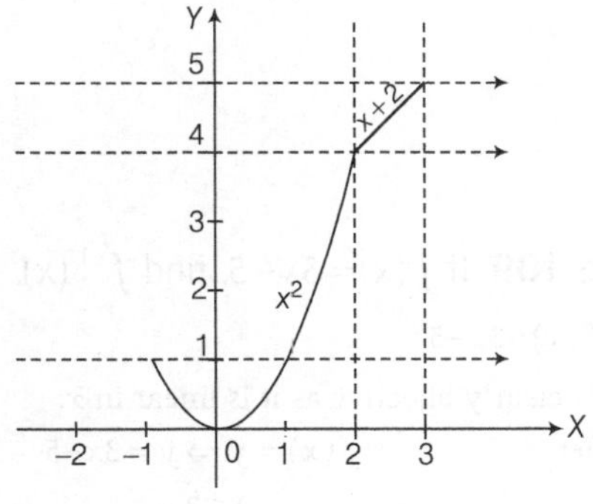

and $\qquad 1 < g(x) \le 2;\ \forall\ x \in (1, \sqrt{2}].$

$$\therefore \quad f(g(x)) = \begin{cases} x^2 + 1, & -1 \le x < 1 \\ 2x^2 + 1, & 1 < x \le \sqrt{2} \end{cases}$$

Example 107 If $f(x) = 2x + |x|$, $g(x) = \frac{1}{3}(2x - |x|)$ and $h(x) = f(g(x))$, domain of $\sin^{-1}(\underbrace{h(h(h(h\ldots h(x)\ldots))))}_{n \text{ times}}$ is

(a) $[-1, 1]$ (b) $\left[-1, -\frac{1}{2}\right] \cup \left[\frac{1}{2}, 1\right]$

(c) $\left[-1, -\frac{1}{2}\right]$ (d) $\left[\frac{1}{2}, 1\right]$

Sol. Since, $\quad f(x) = \begin{cases} 2x + x, & x \ge 0 \\ 2x - x, & x < 0 \end{cases} = \begin{cases} 3x, & x \ge 0 \\ x, & x < 0 \end{cases}$

and $\quad g(x) = \frac{1}{3}\begin{cases} 2x - x, & x \ge 0 \\ 2x + x, & x < 0 \end{cases} = \begin{cases} \frac{x}{3}, & x \ge 0 \\ x, & x < 0 \end{cases}$

$$\therefore \quad f(g(x)) = \begin{cases} 3\left(\frac{x}{3}\right), & x \ge 0 \\ x, & x < 0 \end{cases}$$

$\Rightarrow \quad f(g(x)) = x, \forall\ x \in R$

$\therefore \quad h(x) = x$

$\Rightarrow \sin^{-1}(h(h(h\ldots h(x)\ldots))) = \sin^{-1} x$

$\therefore$ Domain of $\sin^{-1}(h(h(h(h\ldots h(x)\ldots))))$ is $[-1, 1]$.

Hence, (a) is the correct answer.

Example 108 A function $f: R \to R$ satisfies

$$\sin x \cos y\,(f(2x + 2y) - f(2x - 2y)) = \cos x \sin y\,(f(2x + 2y) + f(2x - 2y)).$$

If $f'(0) = \frac{1}{2}$, then

(a) $f''(x) = f(x) = 0$ (b) $4f''(x) + f(x) = 0$

(c) $f''(x) + f(x) = 0$ (d) $4f''(x) - f(x) = 0$

Sol. We have, $\quad \frac{f(2x + 2y)}{f(2x - 2y)} = \frac{\sin(x + y)}{\sin(x - y)}$

$\Rightarrow \quad \frac{f(\alpha)}{\sin\frac{\alpha}{2}} = \frac{f(\beta)}{\sin\frac{\beta}{2}} = K$

$\Rightarrow \quad f(x) = K \sin\frac{x}{2}$

$\Rightarrow \quad f'(x) = \frac{K}{2}\cos\frac{x}{2}$

and $\quad f''(x) = \frac{-K}{4}\sin\frac{x}{2}$

$\Rightarrow \quad 4f''(x) + f(x) = 0$

Hence, (b) is the correct answer.

Exercise for Session 10

1. Consider the real-valued function satisfying $2f(\sin x) + f(\cos x) = x$. Find the domain and range of $f(x)$.
2. If $f(x)$ is defined in $[-3, 2]$, find the domain of definition of $f([|x|])$ and $f([2x + 3])$.
3. $f(x) = \begin{cases} x - 1, & -1 \le x \le 0 \\ x^2, & 0 < x \le 1 \end{cases}$ and $g(x) = \sin x$. Find $h(x) = f(|g(x)|) + |f(g(x))|$.
4. Let $f(x)$ be defined on $[-2, 2]$ and is given by $f(x) = \begin{cases} -1, & -2 \le x \le 0 \\ x - 1, & 0 \le x \le 2 \end{cases}$ and $g(x) = f(|x|) + |f(x)|$, find $g(x)$.
5. Let two functions are defined as $g(x) = \begin{cases} x^2, & -1 \le x < 2 \\ x + 2, & 2 \le x \le 3 \end{cases}$ and $f(x) = \begin{cases} x + 1, & x \le 1 \\ 2x + 1, & 1 < x \le 2 \end{cases}$, find gof.

Session 11

Inverse of a Function

Inverse of a Function

Let $f: A \rightarrow B$ be a one-one and onto function, then there exists a unique function, $g: B \rightarrow A$ such that $f(x) = y \Leftrightarrow g(y) = x, \; \forall \; x \in A$ and $y \in B$.

Then, g is said to be inverse of f.

Thus, $g = f^{-1}: B \rightarrow A = \{(f(x), x) | (x, f(x)) \in f\}$

Let us consider a one-one function with domain A and range B.

Where, $A = \{1, 2, 3, 4\}$ and $B = \{2, 4, 6, 8\}$ and $f: A \rightarrow B$ is given by $f(x) = 2x$, then write f and f^{-1} as a set of ordered pairs.

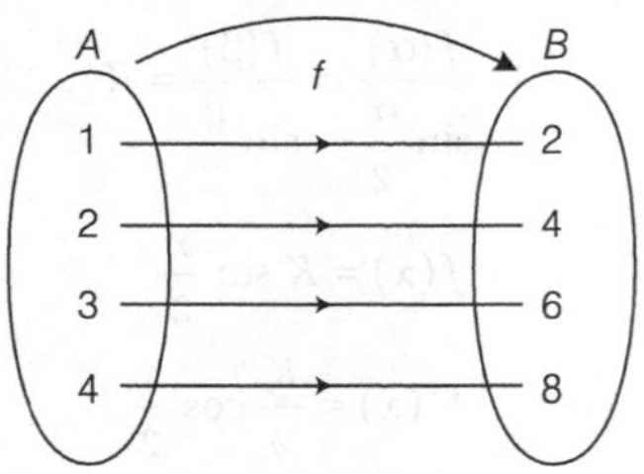

Figure 3.43

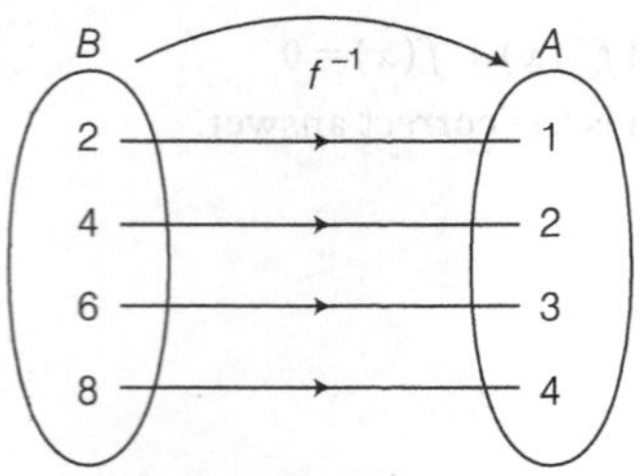

Figure 3.44

Here, member $y \in B$ arises from one and only one member $x \in A$.

So, $$f = \{(1, 2)\,(2, 4)\,(3, 6)\,(4, 8)\}$$

and $$f^{-1} = \{(2, 1)\,(4, 2)\,(6, 3)\,(8, 4)\}$$

In above function,

Domain of $f = \{1, 2, 3, 4\}$ = Range of f^{-1}.

Range of $f = \{2, 4, 6, 8\}$ = Domain of f^{-1}.

which represents for a function to have its inverse, it must be **one-one onto** or **bijective**.

Example 109 If $f(x) = 3x - 5$, find $f^{-1}(x)$.

Sol. Here, $f(x) = 3x - 5$

which is clearly bijective as it is linear in x.

Now, let $$f(x) = y \Rightarrow y = 3x - 5$$

$$\Rightarrow \quad x = \frac{y+5}{3}$$

$$\Rightarrow \quad f^{-1}(y) = \frac{y+5}{3} \qquad [\text{as } f(x) = y \Rightarrow x = f^{-1}(y)]$$

Therefore, $$f^{-1}(x) = \frac{x+5}{3}$$

Example 110 If $f: [1, \infty) \rightarrow [2, \infty)$ is given by $f(x) = x + \frac{1}{x}$, find $f^{-1}(x)$, (assume bijection).

[IIT JEE 2001, 2002]

Sol. Let $y = f(x)$

$$\therefore \quad y = \frac{x^2+1}{x} \quad \Rightarrow \quad x^2 - xy + 1 = 0$$

$$\Rightarrow \quad x = \frac{y \pm \sqrt{y^2 - 4}}{2}$$

$$\Rightarrow \quad f^{-1}(y) = \frac{y \pm \sqrt{y^2-4}}{2} \qquad [\text{as } f(x) = y \Rightarrow x = f^{-1}(y)]$$

$$\Rightarrow \quad f^{-1}(x) = \frac{x \pm \sqrt{x^2-4}}{2}$$

Since, range of inverse function is $[1, \infty)$, therefore neglecting the negative sign, we have

$$f^{-1}(x) = \frac{x + \sqrt{x^2-4}}{2}$$

Example 111 Let $f(x) = x^3 + 3$ be bijective, then find its inverse.

Sol. Let $$y = x^3 + 3 \qquad [\text{i.e. } y = f(x)]$$

$$\Rightarrow \quad x^3 = y - 3$$

$$\Rightarrow \quad x = (y-3)^{1/3}$$

$$\Rightarrow \quad f^{-1}(y) = (y-3)^{1/3} \quad [y = f(x) \Rightarrow f^{-1}(y) = x]$$

$$\Rightarrow \quad f^{-1}(x) = (x-3)^{1/3}$$

Thus, $$f^{-1}(x) = (x-3)^{1/3}$$

when $f(x) = x^3 + 3$ is bijective.

Example 112 Find the inverse of the function, (assuming onto).

$$y=\log_a(x+\sqrt{x^2+1}),\ (a>1).$$

Sol. We have, $y=\log_a(x+\sqrt{x^2+1})$

Since, $\sqrt{x^2+1}>|x|$

$\therefore$ It is defined for all x.

Now, $y=\log_a(x+\sqrt{x^2+1})$,

which is strictly increasing when $a>1$.

Thus, one-one. Also, given that $f(x)$ is onto. [where $y=f(x)$]

Hence, the given function is invertible.

Now, $y=\log_a(x+\sqrt{x^2+1})$

$\Rightarrow a^y=x+\sqrt{x^2+1}$ and $a^{-y}=\sqrt{x^2+1}-x$

$\Rightarrow x=\frac{1}{2}(a^y-a^{-y})$

Hence, the inverse in the form $y=f^{-1}(x)$ is,

$$y=\frac{1}{2}(a^x-a^{-x})$$

Graphical Representation of Invertible Functions

Let (h,k) be a point on the graph of the function f. Then, (k,h) is the corresponding point on the graph of inverse of f, i.e. g.

The line segment joining the points (h,k) and (k,h) is bisected at right angle by the line $y=x$.

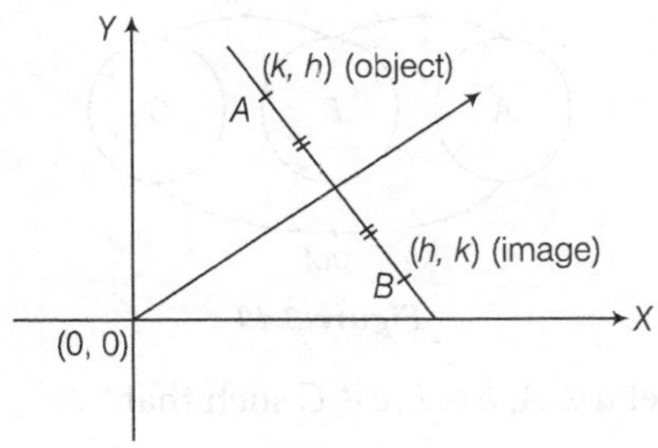

Figure 3.45

So, that the two points play object-image role in the line $y=x$ as plane mirror.

It follows that the graph of $y=f(x)$ and its inverse written in form $y=g(x)$ are symmetrical about the line $y=x$.

The graphs $y=f(x)$ and $y=f^{-1}(x)$, if they intersect then they meet on the line $y=x$ only. Hence, the solutions of $f(x)=f^{-1}(x)$ are also the solutions of $f(x)=x$.

Example 113 Let $f: R\to R$ be defined by $f(x)=\frac{e^x-e^{-x}}{2}$. Is $f(x)$ invertible? If so, find its inverse.

Sol. Let us check for invertibility of $f(x)$.

(a) **One-one** Here, $f'(x)=\frac{e^x+e^{-x}}{2}$

$\Rightarrow f'(x)=\frac{e^{2x}+1}{2e^x}$, which is strictly increasing as $e^{2x}>0$ for all x. Thus, it is one-one.

(b) **Onto** Let $y=f(x)$

$\Rightarrow y=\frac{e^x-e^{-x}}{2}$, where y is strictly monotonic.

Hence, range of $f(x)=(f(-\infty),f(\infty))$

$\Rightarrow$ Range of $f(x)=(-\infty,\infty)$

So, range of $f(x)=$ codomain

Hence, $f(x)$ is one-one and onto.

(c) **To find f^{-1}** $y=\frac{e^{2x}-1}{2e^x}$

$\Rightarrow e^{2x}-2e^xy-1=0\Rightarrow e^x=\frac{2y\pm\sqrt{4y^2+4}}{2}$

$\Rightarrow x=\log(y\pm\sqrt{y^2+1})$

$\Rightarrow f^{-1}(y)=\log(y\pm\sqrt{y^2+1})$

[as $f(x)=y\Rightarrow x=f^{-1}(y)$]

Since, $e^{f^{-1}(x)}$ is always positive.

So, neglecting the negative sign.

Hence, $f^{-1}(x)=\log(x+\sqrt{x^2+1})$

Example 114 Let $f:[1/2,\infty)\to[3/4,\infty)$, where $f(x)=x^2-x+1$. Find the inverse of $f(x)$. Hence, solve the equation $x^2-x+1=\frac{1}{2}+\sqrt{x-\frac{3}{4}}$.

Sol. (a) $f(x)=x^2-x+1$

$\Rightarrow f(x)=\left(x-\frac{1}{2}\right)^2+\frac{3}{4}$, which is clearly one-one and onto in given domain and codomain.

Thus, its inverse can be obtained.

Let $y=\left(x-\frac{1}{2}\right)^2+\frac{3}{4}$

$\Rightarrow x-\frac{1}{2}=\pm\sqrt{y-\frac{3}{4}}\Rightarrow x=\frac{1}{2}\pm\sqrt{y-\frac{3}{4}}$

$\Rightarrow f^{-1}(y)=\frac{1}{2}+\sqrt{y-\frac{3}{4}}$ [$f(x)=y\Rightarrow x=f^{-1}(y)$]

[neglecting – ve sign as x is always +ve]

$\Rightarrow \ f^{-1}(x) = \dfrac{1}{2} + \sqrt{x - \dfrac{3}{4}}$

(b) To solve $x^2 - x + 1 = \dfrac{1}{2} + \sqrt{x - \dfrac{3}{4}}$, since solution of $f(x) = f^{-1}(x)$ are solutions of $f(x) = x$.

i.e. $\quad f(x) = x \Rightarrow x^2 - x + 1 = x$

$\Rightarrow \quad x^2 - 2x + 1 = 0 \Rightarrow (x-1)^2 = 0$

$\therefore$ $x = 1$ is the required solution.

Properties of Inverse Functions

(i) *The inverse of a bijection is unique.*

Proof Let $f : A \to B$ be a bijection. If possible let $g : B \to A$ and $h : B \to A$ be two inverse functions of f. Also, let $a_1, a_2 \in A$ and $b \in B$ such that $g(b) = a_1$ and $h(b) = a_2$, then $g(b) = a_1 \Rightarrow f(a_1) = b$

$$h(b) = a_2 \Rightarrow f(a_2) = b$$

But, since f is one-one, so $f(a_1) = f(a_2) \Rightarrow a_1 = a_2$

$\Rightarrow \quad g(b) = h(b), \forall\, b \in B$

(ii) *If $f : A \to B$ is a bijection and $g : B \to A$ is the inverse of f, then $fog = I_B$ and $gof = I_A$, where I_A and I_B are identity functions on the sets A and B, respectively.*

Remarks

(a) The graphs of f and g are the mirror images of each other in the line $y = x$. As shown, in the figure given below a point (x', y') corresponding to $y = x^2 (x \geq 0)$ changes to (y', x') corresponding to $y = +\sqrt{x}$, the changed form of $x = \sqrt{y}$.

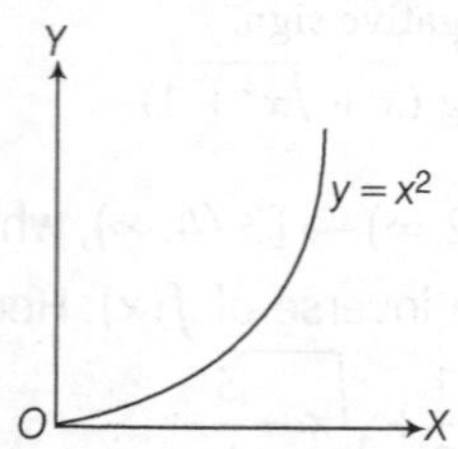

Figure 3.46

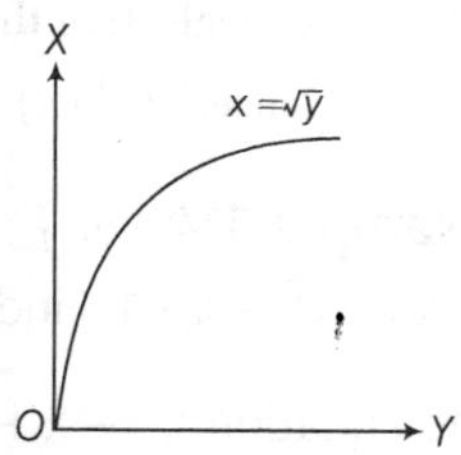

Figure 3.47

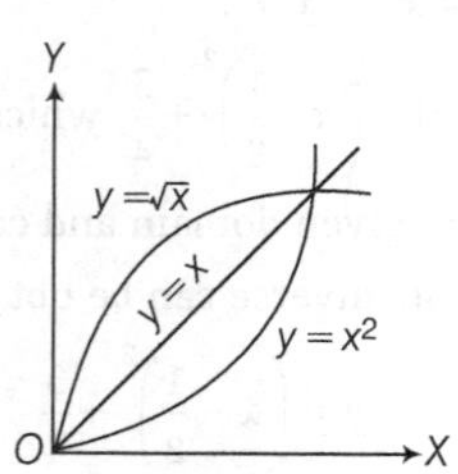

Figure 3.48

(b) If $f(x)$ has its own inverse as in $f(x) = \dfrac{1}{x}$, then $f(x) = f^{-1}(x)$ will have infinite solutions but $f(x) = f^{-1}(x) = x$ will have only one solution.

(iii) *The inverse of a bijection is also a bijection.*

Proof Let $f : A \to B$ be a bijection and $g : B \to A$ be its inverse. We have to show that g is one-one and onto.

One-one Let $g(b_1) = a_1$ and

$g(b_2) = a_2 : a_1, a_2 \in A$ and $b_1, b_2 \in B$

Then, $\quad g(b_1) = g(b_2)$

$\Rightarrow \quad a_1 = a_2$

$\Rightarrow \quad f(a_1) = f(a_2)$

$\Rightarrow \quad b_1 = b_2 \quad$ [$\because f$ is a bijection]

[$\because g(b_1) = a_1 \Rightarrow b_1 = f(a_1), g(b_2) = a_2 \Rightarrow b_2 = f(a_2)$]

which proves that g is one-one.

Onto Again, if $a \in A$, then

$a \in A \Rightarrow \exists\, b \in B$ such that $f(a) = b$

[by definition of f]

$\Rightarrow \exists\, b \in B$ such that $g(b) = a$ [$\because f(a) = b \Rightarrow a = g(b)$]

which proves that g is onto.

Hence, g is also a bijection.

(iv) If f *and* g are two bijections $f : A \to B, g : B \to C$, then the inverse of gof exists and $(gof)^{-1} = f^{-1}og^{-1}$.

Proof Since, $f : A \to B$ and $g : B \to C$ are two bijections.

$\therefore \quad gof : A \to C$ is also a bijection.

[by theorem, the composite of two bijections is a bijection]

As such gof has an inverse function $(gof)^{-1} : C \to A$. We have to show that $(gof)^{-1} = f^{-1}og^{-1}$.

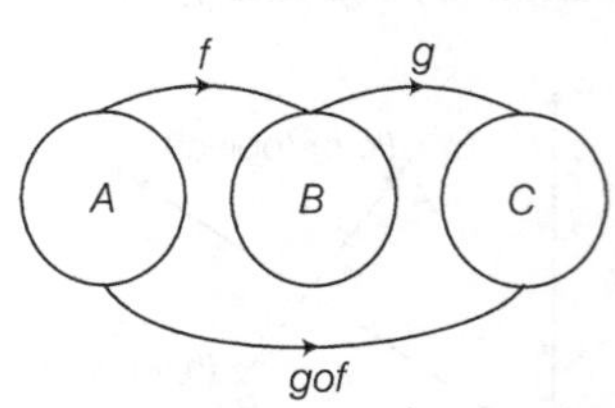

Figure 3.49

Now, let $a \in A, b \in B, c \in C$ such that

$$f(a) = b \text{ and } g(b) = c$$

So, $\quad (gof)(a) = g[f(a)] = g(b) = c$

Now, $\quad f(a) = b \Rightarrow \ a = f^{-1}(b) \quad$ ...(i)

$g(b) = c \Rightarrow \ b = g^{-1}(c) \quad$ ...(ii)

$(gof)(a) = c \Rightarrow \ a = (gof)^{-1}(c) \quad$ ...(iii)

Also, $(f^{-1}og^{-1})(c) = f^{-1}[g^{-1}(c)] \quad$ [by definition]

$$= f^{-1}(b) \quad \text{[by Eq. (ii)]}$$
$$= a \quad \text{[by Eq. (i)]}$$
$$= (gof)^{-1}(c) \quad \text{[by Eq. (iii)]}$$
$$\therefore \quad (gof)^{-1} = f^{-1}og^{-1},$$

which proves the theorem.

Example 115 Let $g(x)$ be the inverse of $f(x)$ and $f'(x) = \frac{1}{1+x^3}$. Find $g'(x)$ in terms of $g(x)$.

Sol. We know, if $g(x)$ is inverse of $f(x)$.

$$\Rightarrow g\{f(x)\} = x \Rightarrow g'\{f(x)\} \cdot f'(x) = 1$$
$$\Rightarrow g'\{f(x)\} = \frac{1}{f'(x)} = 1 + x^3 \Rightarrow g'\{f(g(x))\} = 1 + (g(x))^3$$
$$\Rightarrow \quad g'(x) = 1 + (g(x))^3 \quad [\because f(g(x)) = x]$$

Example 116 If $f: R \to R$ is defined by $f(x) = x^2 + 1$, find the value of $f^{-1}(17)$ and $f^{-1}(-3)$.

Sol. $f(x) = x^2 + 1$; $f^{-1}(17) \Rightarrow f(x) = 17 \Rightarrow x^2 + 1 = 17$

$$\Rightarrow \quad x = \pm 4 \quad \text{and} \quad f^{-1}(-3)$$
$$\Rightarrow \quad f(x) = -3 \Rightarrow x^2 + 1 = -3$$
$$\Rightarrow \quad x^2 = -4 \quad \text{[which is not possible]}$$

Hence, $f^{-1}(17) = \pm 4$ and $f^{-1}(-3) = \phi$

Example 117 If the function f and g are defined as $f(x) = e^x$ and $g(x) = 3x - 2$, where $f: R \to R$ and $g: R \to R$, find the function fog and gof. Also, find the domain of $(fog)^{-1}$ and $(gof)^{-1}$.

Sol. $(fog)(x) = f\{g(x)\}$

$$\Rightarrow \quad f\{g(x)\} = f(3x - 2)$$
$$\Rightarrow \quad f\{g(x)\} = e^{3x-2} \quad \text{...(i)}$$
and
$$(gof)(x) = g\{f(x)\}$$
$$\Rightarrow \quad g\{f(x)\} = g\{e^x\}$$
$$\Rightarrow \quad g\{f(x)\} = 3e^x - 2 \quad \text{...(ii)}$$

For finding $(fog)^{-1}$ and $(gof)^{-1}$.

Let $(fog)(x) = y = e^{3x-2}$

$$\Rightarrow \quad 3x - 2 = \log y \quad \Rightarrow \quad x = \frac{\log y + 2}{3}$$
$$\Rightarrow \quad (fog)^{-1} y = \frac{\log y + 2}{3} \text{ and } (fog)^{-1} x = \frac{\log x + 2}{3}$$

and domain of $(fog)^{-1}$ is $x > 0$, i.e. $x \in (0, \infty)$.

Again, let $(gof)x = y = 3e^x - 2 \Rightarrow e^x = \frac{y+2}{3}$

$$\Rightarrow \quad x = \log\left(\frac{y+2}{3}\right)$$
$$\Rightarrow \quad (gof)^{-1} y = \log\left(\frac{y+2}{3}\right)$$
$$\Rightarrow \quad (gof)^{-1} x = \log\left(\frac{x+2}{3}\right)$$

and domain of $(gof)^{-1}$ is $\frac{x+2}{3} > 0$.

Hence, domain of $(gof)^{-1}$ is $x > -2$, i.e. $x \in (-2, \infty)$.

Example 118 If $f(x) = ax + b$ and the equation $f(x) = f^{-1}(x)$ be satisfied by every real value of x, then

(a) $a = 2, b = -1$ (b) $a = -1, b \in R$

(c) $a = 1, b \in R$ (d) $a = 1, b = -1$

Sol. If $f(x) = ax + b$

$$\Rightarrow \quad f^{-1}(x) = \frac{x}{a} - \frac{b}{a}$$

Since, $f(x) = f^{-1}(x), \forall x \in R$

$$\Rightarrow \quad \frac{1}{a} = a \text{ and } b = -\frac{b}{a} \Rightarrow a = -1 \text{ and } b \in R$$

Hence, (b) is the correct answer.

Example 119 If $g(x)$ is the inverse of $f(x)$ and $f'(x) = \sin x$, then $g'(x)$ is equal to

(a) $\sin(g(x))$ (b) $\text{cosec}(g(x))$

(c) $\tan(g(x))$ (d) None of these

Sol. Given, $g(x) = f^{-1}(x)$

So, $x = f(g(x))$

On differentiating w.r.t. 'x', we get $1 = f'(g(x)) \cdot g'(x)$

Therefore, $g'(x) = \frac{1}{f'(g(x))} = \frac{1}{\sin(g(x))}$

$$\therefore \quad g'(x) = \text{cosec}(g(x))$$

Hence, (b) is the correct answer.

Example 120 If A and B are the points of intersection of $y = f(x)$ and $y = f^{-1}(x)$, then

(a) A and B necessarity lie on the line $y = x$

(b) A and B must be coincident

(c) slope of line AB may be -1

(d) None of the above

Sol. If solution of $f(x) = f^{-1}(x)$ doesn't lie on $y = x$, then they must be of the form (α, β) and (β, α).

$\therefore$ Slope of line AB may be -1. Hence, (c) is the correct answer.

General Results

If x, y are independent variables, then

(i) $f(xy) = f(x) + f(y) \Rightarrow f(x) = k \ln x$ or $f(x) = 0$.

(ii) $f(xy) = f(x) \cdot f(y) \Rightarrow f(x) = x^n, n \in R$.

(iii) $f(x + y) = f(x) \cdot f(y) \Rightarrow f(x) = a^{kx}$.

(iv) $f(x + y) = f(x) = f(y) \in f(x) = k$, where k is constant.

(v) $f(x)$ takes rational values for all $x \Rightarrow f(x)$ is a constant function.

Exercise for Session 11

1. Find the inverse of the following functions

   (i) $f(x) = \sin^{-1}\left(\frac{x}{3}\right), x \in [-3, 3]$

   (ii) $f(x) = \log_e (x^2 + 3x + 1), x \in [1, 3]$

   (iii) $f(x) = 5^{\log_e x}, x > 0$

   (iv) $f(x) = \log_e (x + \sqrt{x^2 + 1})$

   (v) $f(x) = \begin{cases} x, & x < 1 \\ x^2, & 1 \le x \le 4 \\ 8\sqrt{x}, & x > 4 \end{cases}$

2. If the function $f : [1, \infty) \to [1, \infty)$ is defined by $f(x) = 2^{x(x-1)}$, then find $f^{-1}(x)$.

Session 12

Miscellaneous Problems on Functions

Miscellaneous Problems On Functions

We should consider certain examples to make the concept clear.

Example 121 For $x \in R$, the function $f(x)$ satisfies $2f(x)+f(1-x)=x^2$. The value of $f(4)$ is equal to

(a) $\frac{13}{3}$ (b) $\frac{43}{3}$ (c) $\frac{23}{3}$ (d) None of these

Sol. We have, $2f(x)+f(1-x)=x^2$ ...(i)

In Eq. (i) x is replaced by $(1-x)$, we get

$$2f(1-x)+f(x)=(1-x)^2 \quad ...(ii)$$

On solving Eqs. (i) and (ii), we get

$$\therefore \quad 2f(x)+f(1-x)=x^2 \} \times 2$$

$$\Rightarrow \quad \underline{\underset{-}{2}f(1-x)\underset{-}{+}f(x)=\underset{-}{(}1-x)^2}$$

$$\Rightarrow \quad 3f(x)=2x^2-(1-x)^2$$

$$\Rightarrow \quad f(x)=\frac{1}{3}\{x^2+2x-1\}$$

$$\therefore \quad f(4)=\frac{1}{3}\{16+8-1\}=\frac{23}{3}$$

Example 122 Let $f(x)=ax^7+bx^3+cx-5$, where a, b and c are constants. If $f(-7)=7$, find $f(7)$.

Sol. As, $f(x)=ax^7+bx^3+cx-5$

$$\therefore \quad f(-x)=-ax^7-bx^3-cx-5$$

$$\Rightarrow \quad f(x)+f(-x)=-10$$

Put $x=7$, $f(7)+f(-7)=-10 \Rightarrow f(7)=-17$

Example 123 If $f(x)+f\left(1-\frac{1}{x}\right)=1+x$ for $x \in R-\{0,1\}$. The value of $4f(2)$ is equal to

Sol. Here, $f(x)+f\left(1-\frac{1}{x}\right)=1+x$ ...(i)

$$\Rightarrow \quad f(y)+f\left(1-\frac{1}{y}\right)=1+y, \text{ let } y=1-\frac{1}{x}$$

$$\Rightarrow \quad f\left(1-\frac{1}{x}\right)+f\left(1-\frac{x}{x-1}\right)=1+\left(1-\frac{1}{x}\right)$$

$$\Rightarrow \quad f\left(1-\frac{1}{x}\right)+f\left(\frac{-1}{x-1}\right)=2-\frac{1}{x}$$

$$\text{or} \quad f\left(1-\frac{1}{x}\right)+f\left(\frac{1}{1-x}\right)=2-\frac{1}{x} \quad ...(ii)$$

Also, $f(z)+f\left(1-\frac{1}{z}\right)=1+z$, put $z=\frac{1}{1-x}$

$$f\left(\frac{1}{1-x}\right)+f(1-(1-x)=1+\frac{1}{1-x}$$

$$\Rightarrow \quad f\left(\frac{1}{1-x}\right)+f(x)=1+\frac{1}{1-x} \quad ...(iii)$$

On subtracting Eq. (ii) from Eq. (iii), we get

$$f(x)-f\left(1-\frac{1}{x}\right)=\frac{1}{1-x}+\frac{1}{x}-1 \quad ...(iv)$$

From Eqs. (i) and (iv), we get

$$2f(x)=\frac{1}{1-x}+\frac{1}{x}+x$$

$$\therefore \quad 2f(2)=\frac{1}{-1}+\frac{1}{2}+2=\frac{3}{2}$$

$$\Rightarrow \quad f(2)=\frac{3}{4} \Rightarrow 4f(2)=3$$

Example 124 Let $f(x)=\max\{x, x^2\}$. Then, find equivalent definition of $f(x)$.

Sol. Note *These type of questions, where $f(x)$ are either maximum or minimum should be solved graphically for better representation.*

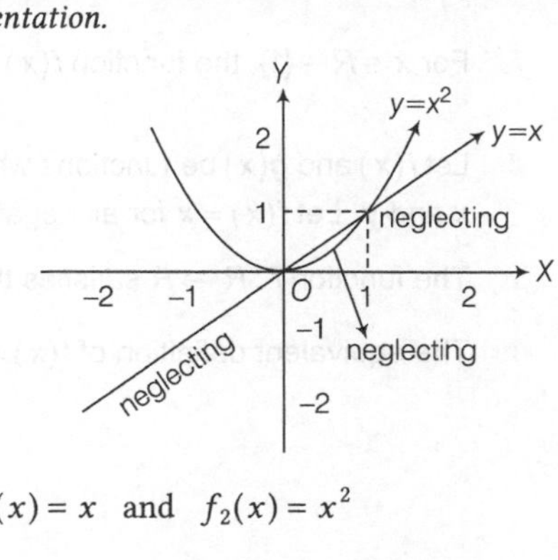

Let $f_1(x)=x$ and $f_2(x)=x^2$

Now, draw graph for $f_1(x) = x$ and $f_2(x) = x^2$.

Here, neglecting the graph, below point of intersection. Since, we want to find the maximum of two functions $f_1(x)$ and $f_2(x)$.

$$\therefore \quad f(x) = \begin{cases} x^2, & x \le 0 \text{ or } x \ge 1 \\ x, & 0 \le x \le 1 \end{cases}$$

Example 125 Let $f(x) = \max\{1+\sin x, 1-\cos x, 1\}, \forall\, x \in [0, 2\pi]$ and $g(x) = \max\{1, |x-1|\}, \forall\, x \in R$. Determine $f\{g(x)\}$ and $g\{f(x)\}$ in terms of x.

Sol. Here, $f(x) = \max\{1+\sin x, 1-\cos x, 1\}$.
Graphically it can be shown as

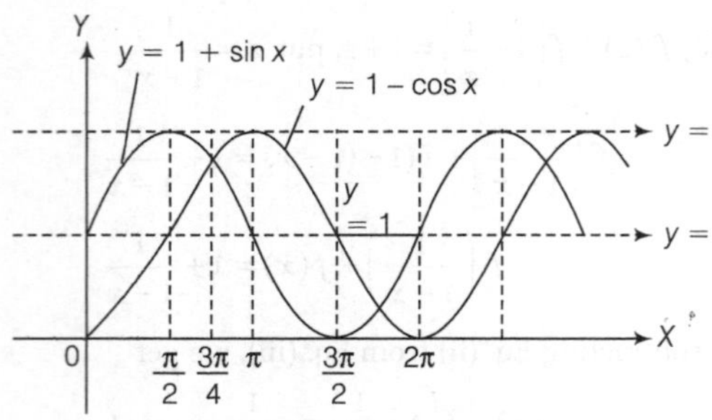

$$f(x) = \begin{cases} 1+\sin x, & 0 \le x \le \dfrac{3\pi}{4} \\ 1-\cos x, & \dfrac{3\pi}{4} < x \le \dfrac{3\pi}{2} \\ 1, & \dfrac{3\pi}{2} < x \le 2\pi \end{cases} \quad \text{[using above graph]}$$

Again, $g(x) = \max\{1, |x-1|\}$, graphically it can be shown as

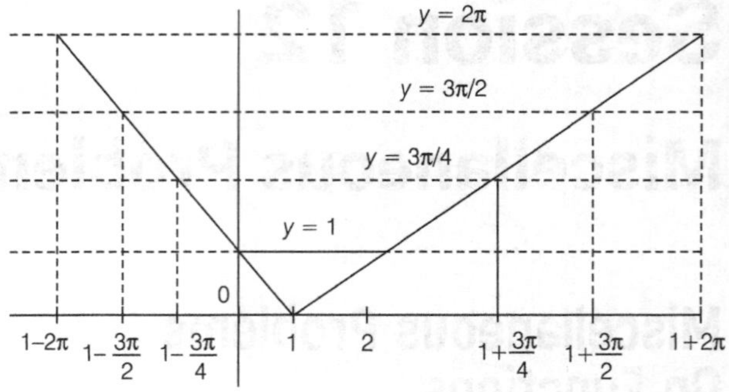

$$\therefore \quad g(x) = \begin{cases} 1-x, & x \le 0 \\ 1, & 0 < x \le 2 \\ x-1, & x > 2 \end{cases}$$

$$\text{Now,} \quad g\{f(x)\} = \begin{cases} 1-f(x), & f(x) \le 0 \\ 1, & 0 < f(x) \le 2 \\ f(x)-1, & f(x) > 2 \end{cases}$$

Since, $f(x) \in [1, 2], \forall\, x \in [0, 2\pi]$; $g(f(x)) = 1, \forall\, x \in [0, 2\pi]$

$$\text{Also,} \quad f\{g(x)\} = \begin{cases} 1+\sin\{g(x)\}, & 0 \le g(x) \le \dfrac{3\pi}{4} \\ 1-\cos\{g(x)\}, & \dfrac{3\pi}{4} < g(x) \le \dfrac{3\pi}{2} \\ 1, & \dfrac{3\pi}{2} < g(x) \le 2\pi \end{cases}$$

$$\Rightarrow f(g(x)) = \begin{cases} 1, & 1-2\pi \le x < 1-3\pi/2 \\ 1-\cos(1-x), & 1-3\pi/2 \le x < 1-3\pi/4 \\ 1+\sin(1-x), & 1-3\pi/4 \le x \le 0 \\ 1+\sin 1, & 0 < x \le 2 \\ 1+\sin(x-1), & 2 < x \le 1+3\pi/4 \\ 1-\cos(x-1), & 1+3\pi/4 < x \le 1+3\pi/2 \\ 1, & 1+3\pi/2 < x \le 2\pi+1 \end{cases}$$

Exercise for Session 12

1. For $x \in R - \{1\}$, the function $f(x)$ satisfies $f(x) + 2f\left(\dfrac{1}{1-x}\right) = x$. Find $f(2)$.
2. Let $f(x)$ and $g(x)$ be functions which take integers as arguments. Let $f(x+y) = f(x) + g(y) + 8$ for all integers x and y. Let $f(x) = x$ for all negative integers x and let $g(8) = 17$. Find $f(0)$.
3. The function $f: R \to R$ satisfies the condition $mf(x-1) + nf(-x) = 2|x| + 1$. If $f(-2) = 5$ and $f(1) = 1$, find $(m+n)$.
4. The equivalent definition of $f(x) = \max\{x^2, (1-x)^2, 2x(1-x)\}$, where $0 \le x \le 1$.

JEE Type Solved Examples :
Single Option Correct Type Questions

Ex. 1 *Let* $f(x) = \dfrac{a_{2k} x^{2k} + a_{2k-1} x^{2k-1} + \ldots + a_1 x + a_0}{b_{2k} x^{2k} + b_{2k-1} x^{2k-1} + \ldots + b_1 x + b_0}$, *where k is a positive integer,* $a_i, b_i \in R$ *and* $a_{2k} \neq 0$, $b_{2k} \neq 0$ *such that* $b_{2k} x^{2k} + b_{2k-1} x^{2k-1} + \ldots + b_1 x + b_0 = 0$ *has no real roots, then*

(a) $f(x)$ must be one to one
(b) $a_{2k} x^{2k} + a_{2k-1} x^{2k-1} + \ldots + a_1 x + a_0 = 0$ must have real roots
(c) $f(x)$ must be many to one
(d) Nothing can be said about the above options

Sol. (c) $f(x)$ is continuous, $\forall\ x \in R$.

$$\lim_{x \to -\infty} f(x) = \lim_{x \to \infty} f(x) = \frac{a_{2k}}{b_{2k}}$$

$\therefore f(x)$ is many to one.

Ex. 2 *If* $\log_{10}\left(\sin\left(x + \frac{\pi}{4}\right)\right) = \frac{\log_{10} 6 - 1}{2}$, *the value of* $\log_{10}(\sin x) + \log_{10}(\cos x)$ *is*

(a) -1 (b) -2 (c) 2 (d) 1

Sol. (a) $$2\log_{10}\left(\frac{\sin x + \cos x}{\sqrt{2}}\right) = \log_{10}\left(\frac{6}{10}\right)$$

$\Rightarrow$ $$\log_{10}\left(\frac{1 + 2\sin x \cos x}{2}\right) = \log_{10}\left(\frac{6}{10}\right)$$

$\Rightarrow$ $$\frac{1}{2} + \sin x \cos x = \frac{3}{5}$$

$\Rightarrow$ $$\sin x \cos x = \frac{1}{10}$$

$\therefore$ $$\log_{10}(\sin x) + \log_{10}(\cos x) = -1$$

Ex. 3 *The diagram shows the dimensions of the floor of an L-shaped room. (All the angles are right angles). The area of the largest circle that can be drawn on the floor of this room is*

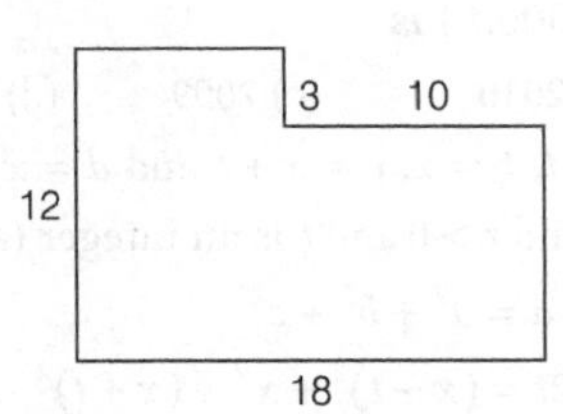

(a) 16π (b) 25π
(c) $\frac{81\pi}{4}$ (d) $\frac{145\pi}{4}$

Sol. (b)

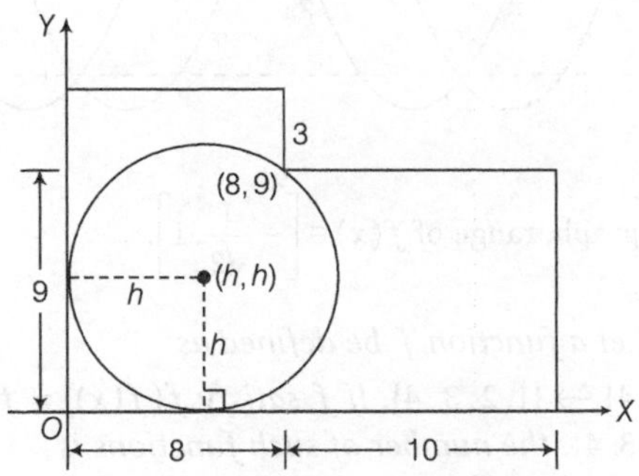

Here, $(x - h)^2 + (y - h)^2 = h^2$ passes through (8, 9).

$$(8 - h)^2 + (9 - h)^2 = h^2$$

$\Rightarrow$ $$h^2 - 34h + 145 = 0$$

$\Rightarrow$ $$(h - 5)(h - 29) = 0$$

$\Rightarrow$ $$h = 5, \text{ neglecting } h = 29$$

$\therefore$ $$r = 5$$

Area of the largest circle $= \pi(5)^2 = 25\pi$.

Ex. 4 *Suppose that the temperature T at every point* (x, y) *in the cartesian plane is given by the formula* $T = 1 - x^2 + 2y^2$. *The correct statement about the maximum and minimum temperature along the line* $x + y = 1$ *is*

(a) Minimum is -1. There is no maximum
(b) Maximum is -1. There is no minimum
(c) Maximum is 0. Minimum is -1
(d) There is neither a maximum nor a minimum along the line

Sol. (a) $T = 1 - x^2 + 2y^2$, where $x + y = 1$

$$T = 1 - x^2 + 2(1 - x)^2$$
$$= 1 - x^2 + 2(1 + x^2 - 2x)$$
$$T = x^2 - 4x + 3$$
$$= (x - 2)^2 - 1$$

$\therefore$ $T_{\max}$ = doesn't exist

and $T_{\min} = -1$

Ex. 5 *The domain of the function* $f(x) = \max\{\sin x, \cos x\}$ *is* $(-\infty, \infty)$. *The range of* $f(x)$ *is*

(a) $\left[-\frac{1}{\sqrt{2}}, 1\right]$ (b) $\left[-\frac{1}{\sqrt{2}}, \frac{1}{\sqrt{2}}\right]$
(c) $[0, 1]$ (d) $[-1, 1]$

Sol. (*a*) Here, $f(x) = \max\{\sin x, \cos x\}$

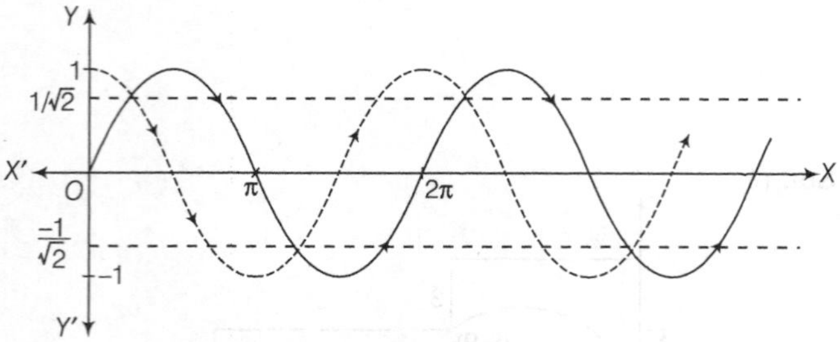

From graph, range of $f(x) \in \left[-\frac{1}{\sqrt{2}}, 1\right]$.

Ex. 6 *Let a function f be defined as* $f:\{1,2,3,4\} \to \{1,2,3,4\}$. *If f satisfy* $f(f(x)) = f(x)$, $\forall\, x \in \{1, 2, 3, 4\}$, *the number of such functions is*

(a) 10 (b) 40 (c) 41 (d) 31

Sol. (*c*) Here, $f(f(x)) = f(x)$, let $f(x) = y$

$\Rightarrow \qquad f(y) = y$

***Case* I** Range contains exactly one element.

It can be done in ${}^4C_1$ ways.

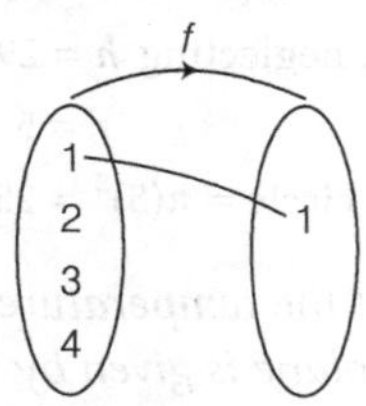

Say, $f(1) = 1$ remaining 3 elements 2, 3, 4 can be mapped only in one way $\Rightarrow$ Total $= {}^4C_1 \cdot 1 = 4$ ...(i)

***Case* II** Range contains two elements, this can be done in ${}^4C_2$ ways.

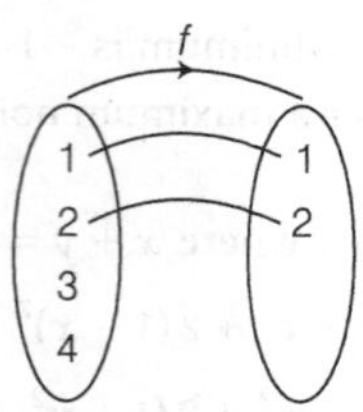

Say, $\qquad f(1) = 1, f(2) = 2$

Remaining 2 elements, i.e. 3 and 4 each can be mapped in 2 ways.

$\therefore$ Total $= {}^4C_2 \times 2 \times 2 = 24$ ...(ii)

***Case* III** Range contains exactly 3 elements which can be done in ${}^4C_3 = 4$ ways. Say, $f(1) = 1$, $f(2) = 2$, $f(3) = 3$.

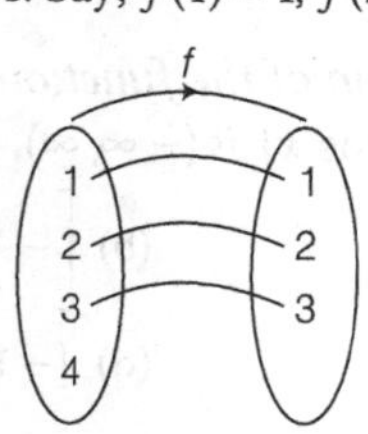

Now, remaining element can be mapped only in 3 ways.

$\therefore \qquad$ Total $= {}^4C_3 \cdot 3 = 12$ ways ...(iii)

***Case* IV** Range contains all 4 elements $f(1) = 1$, $f(2) = 2$, $f(3) = 3$, $f(4) = 4$.

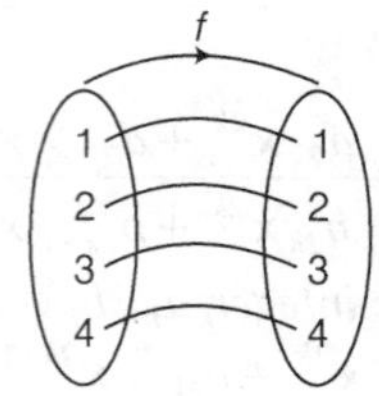

$\therefore$ Only 1 way.

$\therefore$ Total ways $= 4 + 24 + 12 + 1 = 41$

Ex. 7 *Area bounded by the relation* $[2x] + [y] = 5$, $x, y > 0$ *is (where* $[\cdot]$ *represent greatest integer function)*

(a) 2 (b) 3 (c) 4 (d) 5

Sol. (*b*) Here, $\qquad [2x] + [y] = 5$

Let us consider $[2x] = 0 \Rightarrow 0 \le 2x < 1$ and $[y] = 5$

i.e. $\qquad x \in [0, 1/2)$ and $y \in [5, 6)$.

Similarly, we can consider $[2x] = 1, 2, 3, 4$ and 5.

when, $[2x] = 1 \Rightarrow 1/2 \le x < 1$ and $y \in [4, 5)$

$[2x] = 2 \Rightarrow 1 \le x < 3/2$ and $y \in [3, 4)$

$[2x] = 3 \Rightarrow 3/2 \le x < 2$ and $y \in [2, 3)$

$[2x] = 4 \Rightarrow 2 \le x < \frac{5}{2}$ and $y \in [1, 2)$

$[2x] = 5 \Rightarrow 5/2 \le x < 3$ and $y \in [0, 1)$

which can be shown as

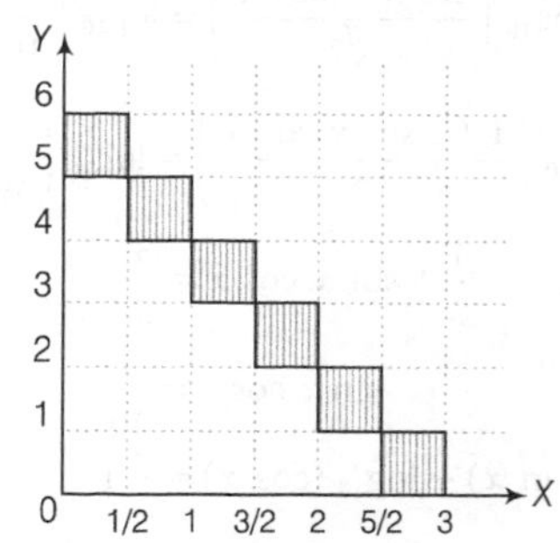

$\therefore$ From the graph, area is 3 sq units.

Ex. 8 *If the integers a, b, c, d are in arithmetic progression and* $a < b < c < d$ *and* $d = a^2 + b^2 + c^2$, *the value of* $(a + 10b + 100c + 1000d)$ *is*

(a) 2008 (b) 2010 (c) 2099 (d) 2016

Sol. (*c*) Let $a = x - t$, $b = x$, $c = x + t$ and $d = x + 2t$.

Where, $x \in I$ and $t > 0$ and t is an integer (as $a < b < c < d$).

Now, $\qquad d = a^2 + b^2 + c^2$

$\Rightarrow \qquad x + 2t = (x - t)^2 + x^2 + (x + t)^2$

$\Rightarrow \qquad x + 2t = 3x^2 + 2t^2$

$\Rightarrow \qquad x(1 - 3x) = 2t(t - 1)$ ...(i)

as $t > 0$ and $t \in I$, hence $t \ge 1 \Rightarrow$ RHS ≥ 0

$\therefore \quad LHS = x(1-3x) \geq 0 \Rightarrow x \in [0, 1/3]$

The only integer is zero $\Rightarrow x = 0$

From Eq. (i), $t = 0$ or $t = 1$

But $\quad t > 0 \Rightarrow t = 1$

$\therefore \quad a = -1, b = 0, c = 1, d = 2$

$\Rightarrow \quad a + 10b + 100c + 1000d = 2099$

Ex. 9 *Let $f(n)$ denotes the square of the sum of the digits of natural number n, where $f^2(n)$ denotes $f(f(n))$, $f^3(n)$ denotes $f(f(f(n)))$ and so on. Then, the value of $\dfrac{f^{2017}(2011) - f^{2016}(2011)}{f^{2017}(2011) - f^{2018}(2011)}$, is*

(a) 1 (b) 3 (c) 5 (d) 7

Sol. (a) Here, $f(n) = (\text{sum of digits of natural number } n)^2$

$$\therefore \quad f(2011) = (2+0+1+1)^2 = 16$$

$$f^2(2011) = f(f(2011)) = f(16) = (1+6)^2 = 49$$

$$f^3(2011) = f(49) = (4+9)^2 = 169$$

$$f^4(2011) = f(169) = (1+6+9)^2 = 256$$

$$f^5(2011) = f(256) = (2+5+6)^2 = 169$$

$$f^6(2011) = f(169) = (1+6+9)^2 = 256$$

Thus, $f^{2n}(2011) = 256$ and $f^{2n+1}(2011) = 169\ n \geq 2$

$$\Rightarrow \quad \frac{f^{2017}(2011) - f^{2016}(2011)}{f^{2017}(2011) - f^{2018}(2011)} = \frac{169-256}{169-256} = 1$$

JEE Type Solved Examples :
More Than One Correct Option Type Questions

Ex. 10 *If $\sum_{i=1}^{4} a_i^2 x^2 - 2\sum_{i=1}^{4} a_i a_{i+1} x + \sum_{i=1}^{4} a_{i+1}^2 \leq 0$, where $a_i > 0$ and all are distinct. Then,*

(a) $a_1 + a_5 > 2a_3$ (b) $\sqrt{a_1 a_5} = a_3$

(c) $\dfrac{2}{\sqrt{a_1 a_4}} > \dfrac{1}{a_1} + \dfrac{1}{a_4}$ (d) $\prod_{i=1}^{5} a_i = a_3^5$

Sol. (a, b, d) $(a_1^2 + a_2^2 + a_3^2 + a_4^2)x^2 - 2(a_1a_2 + a_2a_3 + a_3a_4 + a_4a_5)x + (a_2^2 + a_3^2 + a_4^2 + a_5^2) \leq 0$

$\Rightarrow (a_1^2 x^2 - 2a_1a_2 x + a_2^2) + (a_2^2 x^2 - 2a_2a_3 x + a_3^2) + \ldots + (a_4^2 x^2 - 2a_4a_5 x + a_5^2) \leq 0$

$\Rightarrow (a_1x - a_2)^2 + (a_2x - a_3)^2 + \ldots + (a_4x - a_5)^2 \leq 0$

$\therefore (a_1x - a_2)^2 + (a_2x - a_3)^2 + \ldots + (a_4x - a_5)^2 = 0$

$$\therefore \quad x = \frac{a_2}{a_1} = \frac{a_3}{a_2} = \frac{a_4}{a_3} = \frac{a_5}{a_4}$$

$\Rightarrow a_1, a_2, a_3, a_4, a_5$ are in GP.

$\therefore \quad \dfrac{a_1 + a_5}{2} > \sqrt{a_1 a_5} \Rightarrow a_1 + a_5 > 2a_3. \quad (\because \sqrt{a_1 a_5} = a_3)$

HM of a_1 and a_4 < GM of a_1 and $a_4 \Rightarrow \dfrac{2}{\sqrt{a_1 a_4}} < \dfrac{1}{a_1} + \dfrac{1}{a_4}$

$a_1 \cdot a_2 \cdot a_3 \cdot a_4 \cdot a_5 = a_3^5. \quad [\because a_1a_5 = a_3^2, a_2 \cdot a_4 = a_3^2]$

Ex. 11 *If $f(x-y)$, $f(x) \cdot f(y)$, $f(x+y)$ are in AP for all $x, y \in R$ and $f(0) \neq 0$, then*

(a) $f'(x)$ is an even function

(b) $f'(1) + f'(-1) = 0$

(c) $f'(2) - f'(-2) = 0$

(d) $f'(3) + f'(-3) = 0$

Sol. (b, d) Given, $2f(x) \cdot f(y) = f(x+y) + f(x-y)$

At $x = y = 0$,

$$2f(0) \cdot f(0) = 2f(0)$$

$$\Rightarrow \quad 2f(0) \cdot (f(0) - 1) = 0$$

$$\Rightarrow \quad f(0) = 1 \quad [\text{as } f(0) \neq 0] \ldots \text{(i)}$$

At $x = 0$,

$$2f(0) \cdot f(y) = f(y) + f(-y) \quad [\text{using Eq. (i)}, f(0) = 1]$$

$$\Rightarrow \quad 2f(y) = f(y) + f(-y)$$

$\therefore \quad f(y) = f(-y) \Rightarrow f(x)$ is even function.

We know, if $f(x)$ is even, then $f'(x)$ is odd.

$$\Rightarrow \quad f'(-x) = -f'(x)$$

$\therefore \quad f'(1) = -f'(-1)$ and $f'(3) = -f'(-3)$

Ex. 12 *If the equation $x^2 + 4 + 3\cos(ax + b) = 2x$ has atleast one solution where $a, b \in [0, 5]$, the value of $(a + b)$ equal to*

(a) 5π (b) 3π (c) 2π (d) π

Sol. (b, d) $x^2 - 2x + 4 = -3\cos(ax + b)$

$$(x-1)^2 + 3 = -3\cos(ax + b)$$

Here, LHS ≥ 3 and RHS ≤ 3. Thus, the above equation have solution if LHS = RHS = 3

$\Rightarrow \quad x = 1$ and $\cos(ax + b) = -1$

Hence, $\cos(a + b) = -1 \therefore a + b = \pi, 3\pi, 5\pi$

But, $0 \leq a + b \leq 10 \quad \Rightarrow a + b = \pi$ or 3π

Hence, (b) and (d) are the correct answers.

Ex. 13 *Which of the following functions have the same graph?*

(a) $f(x)=\log_e e^x$ (b) $g(x)=|x|\text{ sgn } x$

(c) $h(x)=\cot^{-1}(\cot x)$ (d) $k(x)=\lim_{n\to\infty}\dfrac{2|x|}{\pi}\tan^{-1}(nx)$

Sol. (a,b,d) (a) $f(x)=\log_e e^x = x\cdot\log_e e = x$ shown as,

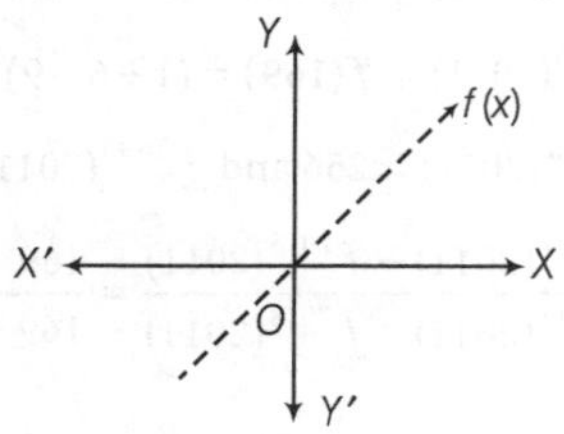

(b) $g(x)=|x|\text{ sgn } x=\begin{cases}|x|\cdot\dfrac{x}{|x|}, & x\neq 0\\ 0, & x=0\end{cases}=\begin{cases}x, & x\neq 0\\ 0, & x=0\end{cases}$

$\therefore\quad g(x)=x$, which is same as $f(x)$.

(c) $h(x)=\cot^{-1}(\cot x)$ is shown as,

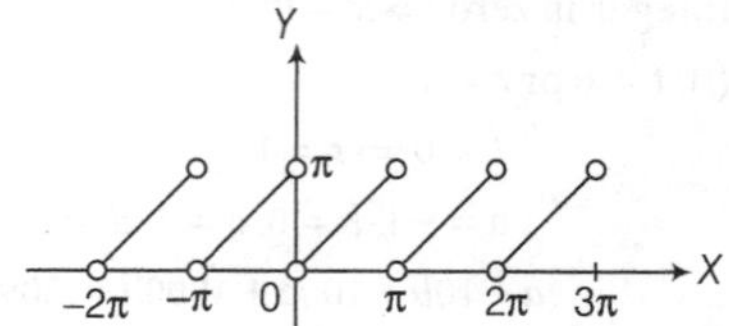

which is not same as $f(x)$ and $g(x)$.

(d) $k(x)=\lim_{n\to\infty}\dfrac{2|x|}{\pi}\cdot\tan^{-1}(nx)$

$$=\begin{cases}\dfrac{2x}{\pi}\cdot\dfrac{\pi}{2}, & x>0\\ -\dfrac{2x}{\pi}\cdot-\dfrac{\pi}{2}, & x<0\\ 0, & x=0\end{cases}=\begin{cases}x, & x>0\\ x, & x<0\\ 0, & x=0\end{cases}$$

$\therefore k(x)=x,\forall x.$

Hence, (a), (b) and (d) are the correct answers.

JEE Type Solved Examples : Statements I and II Type Questions

Directions (Q. Nos. 14 to 15) *For the following questions, choose the correct answers from the codes* (a), (b), (c) *and* (d) *defined as follows*

(a) Statement I is true, Statement II is also true; Statement II is the correct explanation of Statement I

(b) Statement I is true, Statement II is also true; Statement II is not the correct explanation of Statement I

(c) Statement I is true, Statement II is false

(d) Statement I is false, Statement II is true

Ex. 14 *Let* $f(\theta)=\dfrac{\sin^2\theta\cos\theta}{(\sin\theta+\cos\theta)}-\dfrac{1}{4}\tan\left(\dfrac{\pi}{4}-\theta\right)$, $\forall\theta\in R-\left\{n\pi-\dfrac{\pi}{4}\right\}, n\in I.$

Statement I *The largest and smallest value of* $f(\theta)$ *differ by* $\dfrac{1}{\sqrt{2}}$.

Statement II $a\sin x+b\cos x+c\in[c-\sqrt{a^2+b^2},\ c+\sqrt{a^2+b^2}],\forall x\in R$, *where* $a,b,c\in R$.

Sol. (a) We have, $f(\theta)=\dfrac{4\sin^2\theta\cos\theta-\cos\theta+\sin\theta}{4(\cos\theta+\sin\theta)}$

$$=\frac{2\sin\theta((\sin\theta+\cos\theta)^2-1)-(\cos\theta+\sin\theta)+2\sin\theta}{4(\cos\theta+\sin\theta)}$$

$$=\frac{1}{2}\sin\theta(\sin\theta+\cos\theta)-\frac{1}{4}$$

$$=\frac{1}{4}(\sin 2\theta-\cos 2\theta)\in\left[-\frac{\sqrt{2}}{4},\frac{\sqrt{2}}{4}\right]$$

Ex. 15 *Consider two functions* $f(x)=1+e^{\cot^2 x}$ *and* $g(x)=\sqrt{2|\sin x|-1}+\dfrac{1-\cos 2x}{1+\sin^4 x}$.

Statement I *The solutions of the equation* $f(x)=g(x)$ *is given by* $x=(2n+1)\dfrac{\pi}{2}$, $\forall n\in I$.

Statement II *If* $f(x)\geq k$ *and* $g(x)\leq k$ *(where* $k\in R$*), then solutions of the equation* $f(x)=g(x)$ *is the solution corresponding to the equation* $f(x)=k$.

Sol. (c) LHS $=1+e^{\cot^2 x}\geq 2$

As, $\sqrt{2|\sin x|-1}\leq 1$

and $\quad \dfrac{1-\cos 2x}{1+\sin^4 x}=\dfrac{2\sin^2 x}{1+\sin^4 x}=\dfrac{2}{\dfrac{1}{\sin^2 x}+\sin^2 x}\le 1$

$\therefore$ RHS $=\sqrt{2|\sin x|-1}+\dfrac{1-\cos 2x}{1+\sin^4 x}\le 2$

Equation will satisfy, if LHS = RHS = 2 which is possible when $\cot^2 x=0$ and $|\sin x|=1$.

$\Rightarrow \quad x=(2n+1)\dfrac{\pi}{2}, n\in I$

$\therefore$ Statement I is correct.

Statement II is not always correct because solution of the equation $f(x)=g(x)$ will be solutions corresponding to $f(x)=g(x)=k$ in the domain of $f(x)$ and $g(x)$ both. Hence, (c) is the correct answer.

JEE Type Solved Examples : Passage Based Questions

Passage I
(Ex. Nos. 16 to 18)

Let a_m $(m=1,2,\ldots,p)$ be the possible integral values of a for which the graphs of

$$f(x)=ax^2+2bx+b$$

and $\quad g(x)=5x^2-3bx-a$

meet at some points for all real values of b.

Let $\quad t_r=\prod_{m=1}^{p}(r-a_m)$ *and* $S_n=\sum_{r=1}^{n}t_r, n\in N.$

● **Ex. 16** *The minimum possible value of a is*

(a) $\frac{1}{5}$ (b) $\frac{5}{26}$ (c) $\frac{3}{28}$ (d) $\frac{2}{43}$

● **Ex.17** *The sum of values of n for which S_n vanishes is*

(a) 8 (b) 9
(c) 10 (d) 15

● **Ex. 18** *The value of $\sum_{r=5}^{\infty}\frac{1}{t_r}$ is equal to*

(a) $\frac{1}{3}$ (b) $\frac{1}{6}$ (c) $\frac{1}{15}$ (d) $\frac{1}{18}$

■ **Sol.** (Q. Nos. 16 to 18)

16. (a) 17. (c) 18. (d)

$$ax^2+2bx+b=5x^2-3bx-a$$

$\Rightarrow \quad (a-5)x^2+5bx+(b+a)=0$

If $a\ne 5$, then since $x\in R$,

$$D=25b^2-4(b+a)(a-5)\ge 0, \forall\, b\in R$$

$\Rightarrow \quad 25b^2-4(a-5)b-4a(a-5)\ge 0, \forall\, b\in R$

$\therefore \quad 16(a-5)^2+16(25)a(a-5)\le 0$

$\Rightarrow \quad 16(a-5)(a-5+25a)\le 0$

$\Rightarrow \quad (a-5)(26a-5)\le 0 \;\therefore\; a\in\left[\dfrac{5}{26},5\right)$

If $a=5$, $5bx+(b+5)=0$ is not satisfied for $b=0$.

$\therefore \quad a_m\in\{1,2,3,4\};\ t_r=(r-1)(r-2)(r-3)(r-4)$

$$S_n=\frac{1}{5}\sum_{r=1}^{n}(r-4)(r-3)(r-2)(r-1)(r-(r-5))$$

$$=\frac{1}{5}\sum_{r=1}^{n}[(r-4)(r-3)(r-2)(r-1)r-(r-5)(r-4)(r-3)(r-2)(r-1)]$$

$$=\frac{1}{5}n(n-1)(n-2)(n-3)(n-4)$$

$S_n=0$

$\Rightarrow \quad n=1,2,3,4$

$\therefore$ Sum of values of n for which S_n vanishes is $1+2+3+4=10$ $\quad$ [$n=0$ rejected]

$$\sum_{r=5}^{\infty}\frac{1}{t_r}=\frac{1}{3}\lim_{n\to\infty}\sum_{r=5}^{n}\frac{(r-1)-(r-4)}{(r-4)(r-3)(r-2)(r-1)}$$

$$=\frac{1}{3}\lim_{n\to\infty}\sum_{r=5}^{n}\left[\frac{1}{(r-4)(r-3)(r-2)}-\frac{1}{(r-3)(r-2)(r-1)}\right]$$

$$=\lim_{n\to\infty}\frac{1}{3}\left[\frac{1}{6}-\frac{1}{(n-3)(n-2)(n-1)}\right]=\frac{1}{18}$$

Passage II
(Ex. Nos. 19 to 21)

Let w be non-real fifth root of 3 and $x=w^3+w^4$. If $x^5=f(x)$, where $f(x)$ is real quadratic polynomial, with roots α and β, $(\alpha,\beta\in C)$, then determine $f(x)$ and answer the following questions.

● **Ex. 19** *Every term of the sequence $\{f(n)\}, n\in N$ is divisible by*

(a) 12 (b) 18
(c) 24 (d) 27

● **Ex. 20** *Which of the following is not true?*

(a) $\alpha+\beta=-3$ (b) $\alpha\beta=12/5$
(c) $|\alpha-\beta|=\sqrt{3/5}$ (d) $|\alpha|+|\beta|=2\sqrt{3/5}$

● **Ex. 21** *If α and β are represented by points A and B in argand plane, then circumradius of ΔOAB, where O is origin, is*

(a) 4 / 5 (b) 8 / 5 (c) 16 / 5 (d) 32 / 5

■ ***Sol.*** (Q. Nos. 19 to 21)

19. (b) 20. (d) 21. (a)

Here, $x=w^3+w^4$

$$\Rightarrow \quad x^5=(w^2)^5(w+w^2)^5$$
$$=(w^5)^2\ [w^5+5w^4\cdot w^2+10w^3\cdot w^4+10w^2\cdot w^6+5w\cdot w^8+w^{10}]$$

Put $w^5=3$ and $w^{10}=9$

Now, $x^5=9\ [3+5\cdot(3w)+10\cdot(3w^2)+10\cdot 3w^3+5\cdot(3w^4)+3^2]$

$$=9[12+15(w+w^2)+15(w^2+w^3)+15(w^3+w^4)]$$
$$=9\left[12+15\left(\frac{x}{w^2}\right)+15\cdot\left(\frac{x}{w}\right)+15x\right]$$
$$=9\left[12+15x\left(\frac{1}{w^2}+\frac{1}{w}\right)+15x\right]$$
$$=9\left[12+15x\cdot\frac{w+w^2}{w^3}\cdot\frac{w^2}{w^2}+15x\right]$$
$$=9\left[12+15x\cdot\frac{(w^3+w^4)}{w^5}+15x\right]$$
$$=9\left[12+15x\cdot\frac{x}{3}+15x\right]$$
$$=9[12+15x+5x^2]=f(x)$$

Now, $f(x)=9(5x^2+15x+12)$

$\therefore\ f(x)=0$

$$\Rightarrow \quad x=\frac{-15\pm\sqrt{-15}}{10}=\frac{-15\pm i\sqrt{15}}{10}=\alpha,\beta$$

(i) For sequence $\{f(n)\}$,

$$f(n)=9(5n^2+15n+12)=9[5n(n+1)+10n+12]$$
$$=9\times\text{an even number}$$

$\therefore$ Every term of $\{f(n)\}$ is always divisible by 18.

(ii) Clearly, $\alpha+\beta=-3, \alpha\beta=\frac{12}{5}$, $|\alpha-\beta|=\left|\frac{2i\sqrt{15}}{10}\right|=\sqrt{\frac{3}{5}}$

$$\text{But } |\alpha|+|\beta|=2\sqrt{\left(\frac{15}{10}\right)^2+\left(\frac{\sqrt{15}}{10}\right)^2}=4\sqrt{\frac{3}{5}}$$

(iii) In argand plane,

$$OA=OB=\sqrt{\frac{12}{5}} \text{ and } AB=2\frac{\sqrt{15}}{10}=\sqrt{\frac{3}{5}}$$

and area of ΔOAB,

$$\Delta=\frac{1}{2}AB(OM)=\frac{1}{2}\cdot\sqrt{\frac{3}{5}}\cdot\left(\frac{15}{10}\right)$$

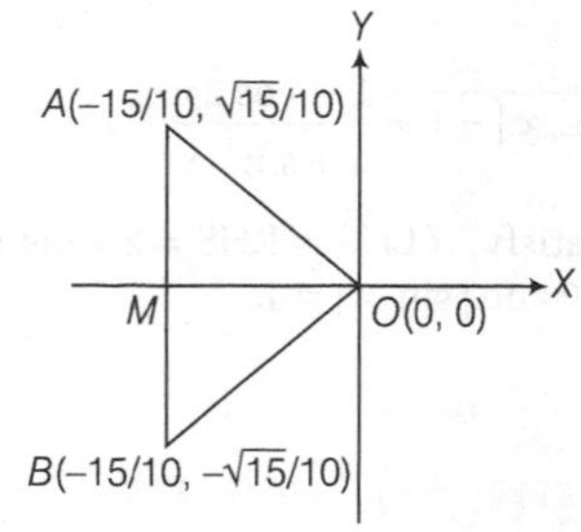

$$\therefore \text{ Circumradius, } R=\frac{OA\cdot AB\cdot OB}{4\Delta}=\frac{\sqrt{\frac{12}{5}}\sqrt{\frac{3}{5}}\sqrt{\frac{12}{5}}}{4\cdot\frac{1}{2}\cdot\sqrt{\frac{3}{5}}\left(\frac{15}{10}\right)}=\frac{4}{5}$$

Passage III
(Ex. Nos. 22 to 24)

Let $A=\{1,2,3,4,5\}$ and $B=\{-2,-1,0,1,2,3,4,5\}$.

● **Ex. 22** *Increasing function from A to B is*

(a) 120 (b) 72 (c) 60 (d) 56

● **Ex. 23** *Non-decreasing functions from A to B is*

(a) 216 (b) 540 (c) 792 (d) 840

● **Ex. 24** *Onto functions from A to A such that $f(i)\neq i$ for all i, is*

(a) 44 (b) 120 (c) 56 (d) 76

■ ***Sol.*** (Q. Nos. 22 to 24)

22. (*d*) The desired number is ${}^8C_5=56$

23. (*c*) Here, for non-decreasing functions from *A* to *B*, is

$${}^8C_1\cdot 1+{}^8C_2\cdot 4+{}^8C_3\cdot 6+{}^8C_4\cdot 4+{}^8C_5\cdot 1=792$$

Explanation for case ${}^8C_2$, say two elements of set *B* are selected in ${}^8C_2$ is $\{-1,0\}$.

Now, $x_1+x_2=5$ $[x_1\geq 1, x_2\geq 1]$

where, x_1 denotes number of elements of set *A* maps to −1 and x_2 denotes number of elements of set *A* maps to 0.

$\therefore$ Total number of solutions is ${}^4C_1$.

Similarly, explanation for case ${}^8C_3$ say three elements selected in ${}^8C_3$ is $\{-1,0,1\}$.

Now, $x_1+x_2+x_3=5$ $(x_1\geq 1, x_2\geq 1, x_3\geq 1)$

$\therefore$ Total number of solutions is ${}^4C_2$ etc.

24. (*a*) Onto functions from *A* to *A* such that $f(i)\neq i$ for all *i*.

$$\Rightarrow \quad 5!\left(1-\frac{1}{1!}+\frac{1}{2!}-\frac{1}{3!}+\frac{1}{4!}-\frac{1}{5!}\right)=44$$

JEE Type Solved Examples :
Matching Type Questions

Ex. 25 *Match the statements of Column I with values of Column II.*

Column I	Column II
(A) Let $f(x)$ be a function on $(-\infty, \infty)$ and $f(x+2) = f(x-2)$. If $f(x) = 0$ has only three real roots in $[0, 4]$ and one of them is 4, then the number of real roots of $f(x) = 0$ in $(-8, 10]$ is	(p) 4
(B) Let $r_1, r_2, r_3, \ldots, r_n$ be n positive integers, not necessarily distinct, such that $(x+r_1)(x+r_2)\ldots(x+r_n) = x^n + 56x^{n-1} + \ldots + 2009$. The possible value of n is	(q) 5
(C) If x and y are positive integers and $2xy = 2009 - 3y$, then the number of ordered pairs of (x, y) is	(r) 8
(D) If $x, y \in R$, satisfying the equation $\frac{(x-4)^2}{4} + \frac{y^2}{9} = 1$, then the difference between the largest and smallest value of the expression $\frac{x^2}{4} + \frac{y^2}{9}$ is	(s) 9

Sol. (A) $\to$ (s), (B) $\to$ (p), (C) $\to$ (q), (D) $\to$ (r)

(A) Given, $f(x+2) = f(x-2)$, domain is R $x \to x+2$

$$f(x+4) = f(x)$$

$\Rightarrow f$ is periodic with period 4.

Put $x = 0$, $f(4) = f(0)$

But $f(4) = 0$

$\Rightarrow f(0) = 0$

$\Rightarrow x = 0$, is also the root.

Since, in $[0, 4]$, f has only 3 roots.

So, $f(x)$ is the form of $\sin\left(\frac{\pi x}{2}\right)$ fulfilling all conditions given in the question, hence the 3rd root in $[0, 4]$ is 2 as $f(2) = 0$.

$\therefore$ The number of real roots in $(-8, 10]$ is 9 and the roots are $(-6, -4, -2, 0, 2, 4, 6, 8, 10)$.

(B) $(r_1 + r_2 + \ldots + r_n) = 56$

and $r_1 r_2 \ldots r_n = 2009 = 7^2 \cdot 41$

[using theory of equation]

As $2009 = 7^2 \cdot 41$

Therefore, no value of r_i can be 49 or larger factor of 2009 otherwise their sum is larger than 56.

As $56 = 41 + 7 + 7 + 1$

Therefore, $r_1 = 41; r_2 = r_3 = 7$

and $r_4 = 1$

Hence, $n = 4$

(C) $2xy = 2009 - 3y$ yields $2xy + 3y = 2009$

$$\Rightarrow y = \frac{2009}{2x+3}$$

So, $2x + 3$ must be the factors of 2009.

Since, $2009 = 7^2 \cdot 41$

Thus, $2x + 3$ can be one of the factors 7, 49, 41, 287 and 2009.

or $2x + 3 = 7; 2x + 3 = 49; 2x + 3 = 41;$ $2x + 3 = 287$

and $2x + 3 = 2009$.

Hence, $x = 2, 23, 19, 142$ and 1003

$\Rightarrow$ Ordered pairs are (2, 287), (23, 41), (19, 49), (142, 7), (1003, 1).

(D) Let $x - 4 = 2\cos\theta$

$\Rightarrow x = 2\cos\theta + 4$

and $y = 3\sin\theta$

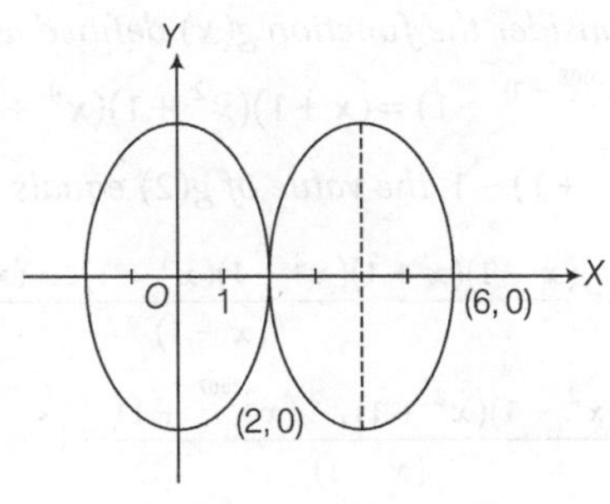

Now, $z = \frac{x^2}{4} + \frac{y^2}{9} = \frac{(2\cos\theta + 4)^2}{4} + \sin^2\theta$

$$= \frac{4\cos^2\theta + 16 + 16\cos\theta + 4\sin^2\theta}{4} = \frac{20 + 16\cos\theta}{4}$$

$z = 5 + 4\cos\theta$

Hence, $z_{max} - z_{min} = (9 - 1) = 8$

JEE Type Solved Examples : Single Integer Answer Type Questions

Ex. 26 *Let $f(x) = \sin^{23} x - \cos^{22} x$ and*

$$g(x) = 1 + \frac{1}{2}\tan^{-1}|x|.$$

The number of values of x in interval $[-10\pi, 20\pi]$ satisfying the equation $f(x) = \text{sgn}(g(x))$, is 5a. Then, a is equal to ....

Sol. (3) $g(x) = \frac{1}{2}\tan^{-1}|x| + 1$

$\Rightarrow \quad \text{sgn}(g(x)) = 1$

$\Rightarrow \quad \sin^{23} x - \cos^{22} x = 1$

$\Rightarrow \quad \sin^{23} x = 1 + \cos^{22} x$

which is possible, if $\sin x = 1$ and $\cos x = 0$

$$x = 2n\pi + \frac{\pi}{2}$$

Hence, $\quad -10\pi \le 2n\pi + \frac{\pi}{2} \le 20\pi$

$\Rightarrow \quad -\frac{21}{2} \le 2n \le \frac{39}{2}$

$\Rightarrow \quad -\frac{21}{4} \le n \le \frac{39}{4}$

$\Rightarrow \quad -5 \le n \le 9$

Hence, number of values of $x = 15 \Rightarrow a = 3$.

Ex. 27 *Consider the function g(x) defined as*

$$g(x)\cdot(x^{(2^{2008}-1)} - 1) = (x+1)(x^2+1)(x^4+1)\ldots(x^{2^{2007}}+1) - 1$$ *the value of g(2) equals ....*

Sol. (2) RHS $= \dfrac{(x-1)(x+1)(x^2+1)(x^4+1)\ldots(x^{2^{2007}}+1)}{(x-1)} - 1$

$$= \frac{(x^2-1)(x^2+1)\ldots(x^{2^{2007}}+1)}{(x-1)} - 1$$

$$= \frac{(x^{2^2}-1)(x^{2^2}+1)\ldots(x^{2^{2007}}+1)}{(x-1)} - 1$$

Hence, $g(x)\cdot(x^{(2^{2008}-1)} - 1) = \dfrac{(x^{2^{2008}}-1)-(x-1)}{(x-1)}$

$$= \frac{x(x^{2^{2008}-1}-1)}{(x-1)}$$

$\Rightarrow \quad g(x) = \dfrac{x}{x-1}$

$\therefore \quad g(2) = 2$

Ex. 28 *If $f(x) = \sqrt[3]{\dfrac{9}{\log_2(3-2x)}} - 1$, the value of 'a' which satisfies $f^{-1}(2a-4) = \frac{1}{2}$, is*

Sol. (3) Given, $f^{-1}(2a-4) = \frac{1}{2}$

$\therefore \quad f\left(\frac{1}{2}\right) = 2a - 4$

Put $x = \frac{1}{2}$ in $f(x) = \sqrt[3]{\dfrac{9}{\log_2(3-2x)}} - 1$

$\therefore \quad 2a - 4 = \sqrt[3]{\dfrac{9}{\log_2 2}} - 1$

$\Rightarrow \quad 2a - 4 = \sqrt[3]{9} - 1$

$\Rightarrow \quad 2a - 4 = 2$

$\Rightarrow \quad 2a = 6$

$\therefore \quad a = 3$

Ex. 29 *Let f be defined on the natural numbers as follow:*

$f(1) = 1$ and for $n > 1$, $f(n) = f[f(n-1)] + f[n - f(n-1)]$, the value of $\frac{1}{30}\sum_{r=1}^{20} f(r)$ is

Sol. (7) Here, $\quad f(1) = 1$

$f(2) = f[f(1)] + f[2 - f(1)]$, using $f(1) = 1$

$f(2) = f(1) + f(1) = 2.$

$f(3) = f[f(z)] + f[3 - f(2)]$

$= f(2) + f(1) = 2 + 1 = 3$

Thus, $\quad f(n) = n$

$\therefore \quad \frac{1}{30}\sum_{r=1}^{20} f(r) = \frac{1}{30}[1+2+3+\ldots+20]$

$$= \frac{1}{30} \times \frac{20(20+1)}{2} = 7$$

Ex. 30 *If a, b, c are real roots of the cubic equation $f(x) = 0$ such that $(x-1)^2$ is a factor of $f(x) + 2$ and $(x+1)^2$ is a factor of $f(x) - 2$, then $|ab + bc + ca|$ is equal to*

Sol. (3) Let $f(x) = Ax^3 + Bx^2 + Cx + D$

Since, $(x-1)^2$ is a factor of $f(x) + 2$.

$\Rightarrow \quad f(1) + 2 = 0$

$\Rightarrow \quad A+B+C+D+2=0 \quad$ ...(i)

Also, $(x+1)^2$ is a factor of $f(x)-2$.

$\Rightarrow \quad f(-1)-2=0$

$\Rightarrow \quad -A+B-C+D-2=0 \quad$ ...(ii)

From Eq. (i), $\quad D=-A-B-C-2$

$\therefore \quad f(x)=Ax^3+Bx^2+Cx-A-B-C-2$

$\Rightarrow \quad f(x)+2=A(x^3-1)+B(x^2-1)+C(x-1)$

Since, $(x-1)^2$ is factor of $f(x)+2$.

$\Rightarrow \quad f'(1)=0 \Rightarrow 3A+2B+C=0 \quad$ ...(iii)

Similarly, $(x+1)^2$ is factor of $f(x)-2$.

$\Rightarrow \quad f'(-1)=0 \Rightarrow 3A-2B+C=0 \quad$ ...(iv)

From Eqs. (i), (ii), (iii) and (iv), $A=1, C=-3, B=D=0$.

$\therefore \quad f(x)=x^3-3x$

Hence, $\quad |ab+bc+ca|=3$

Ex. 31 *The minimum value of $k (k \in I)$ for which the equation $e^x = kx^2$ has exactly three real solutions, is*

Sol. (2) Here, $\quad e^x = kx^2 \Rightarrow \dfrac{e^x}{x^2}=k \quad$ ...(i)

Let $\quad f(x)=\dfrac{e^x}{x^2}$

$$f'(x)=\frac{(x-2)e^x}{x^3}$$

$\therefore f(x)$ is increasing on $(-\infty,0)\cup[2,\infty)$ and decreasing on $(0,2)$.

$\therefore f(x)$ can be plotted as

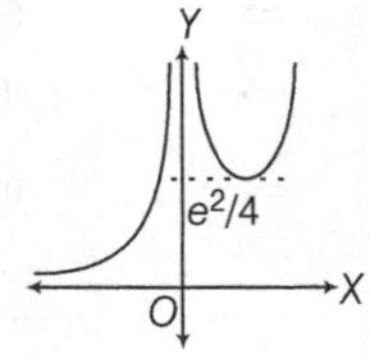

So, clearly from graph, $k>\dfrac{e^2}{4}, k\in I$

Hence, $\quad k_{\min}=2$

Ex. 32 *The value of expression*

$$\sqrt{1+2\sqrt{1+3\sqrt{1+4\sqrt{1+5\sqrt{\ldots\infty}}}}} \text{ is}$$

Sol. (3) As, we can write

$$n=\sqrt{n^2}=\sqrt{1+(n^2-1)}=\sqrt{1+(n-1)(n+1)}$$

$$=\sqrt{1+(n-1)\cdot\sqrt{(n+1)^2}}.$$

$$=\sqrt{1+(n-1)\cdot\sqrt{1+(n+1)^2-1}}$$

$$=\sqrt{1+(n-1)\cdot\sqrt{1+(n)\cdot(n+2)}}.$$

$$=\sqrt{1+(n-1)\cdot\sqrt{1+n\cdot\sqrt{(n+2)^2}}}$$

$$=\sqrt{1+(n-1)\cdot\sqrt{1+n\cdot\sqrt{1+(n+2)^2-1}}}$$

$$=\sqrt{1+(n-1)\cdot\sqrt{1+n\cdot\sqrt{1+(n+1)(n+3)}}}$$

$$=\sqrt{1+(n-1)\cdot\sqrt{1+n\cdot\sqrt{1+(n+1)\cdot\sqrt{(n+3)^2}}}} \ldots \text{ and}$$

so on.

$$\therefore \quad \sqrt{1+2\cdot\sqrt{1+3\cdot\sqrt{1+4\cdot\sqrt{1+5\sqrt{\ldots\ldots\infty}}}}}=3.$$

Subjective Type Questions

Ex. 33 *Let a sequence $x_1, x_2, x_3, \ldots$ of complex numbers be defined by $x_1=0, x_{n+1}=x_n^2-i$ for all $n>1$, where $i^2=-1$. Find the distance of x_{2000} from x_{1997} in the complex plane.*

Sol. Given, $\quad x_1=0$

Then, $\quad x_2=0^2-i=-i$

$x_3=(-i)^2-i=-1-i=-(1+i)$

$x_4=[-(1+i)]^2-i=2i-i=i$

$x_5=(i)^2-i=-1-i=x_3$

$\therefore \quad x_6=x_4$ and hence $x_7=x_5$ and so on.

$\therefore \quad x_{2n}=i,\ x_{2n+1}=-(1+i)$

$\therefore \quad x_{2000}=i=(0,1)$ and

$x_{1997}=-1-i=(-1,-1)$ in the complex plane.

So, the distance between x_{2000} and x_{1997} is $\sqrt{1+4}=\sqrt{5}$.

Ex. 34 *If a, b, c, d, e are positive real numbers, such that $a+b+c+d+e=8$ and $a^2+b^2+c^2+d^2+e^2=16$, find the range of e.*

Sol. As we know,

$$\left(\frac{a+b+c+d}{4}\right)^2 \le \frac{a^2+b^2+c^2+d^2}{4} \quad \text{...(i)}$$

[using Tchebycheff's Inequality]

where $\quad a+b+c+d+e=8$

and $\quad a^2+b^2+c^2+d^2+e^2=16$

From Eq (i), reduces to

$$\left(\frac{8-e}{4}\right)^2 \le \frac{16-e^2}{4}$$

$\Rightarrow \quad 64+e^2-16e \le 4(16-e^2)$

$\Rightarrow \quad 5e^2 - 16e \le 0$

$\Rightarrow \quad e(5e-16) \le 0$

$\Rightarrow \quad 0 \le e \le \dfrac{16}{5}$ [using number line rule]

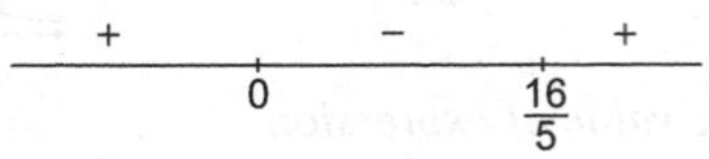

Thus, $e \in \left[0, \dfrac{16}{5}\right]$.

Ex. 35 *Find the set of all solutions of the equation* $2^{|y|} - \left|2^{y-1} - 1\right| = 2^{y-1} + 1.$ *[IIT JEE 1997]*

Sol. Here, $2^{|y|} - \left|2^{y-1}-1\right| = 2^{y-1}+1$

We know, how to define modulus, we have three cases as

Case I $y < 0$

$$\Rightarrow \quad 2^{-y} + (2^{y-1}-1) = 2^{y-1}+1; \quad \left\{\begin{array}{l}\text{as when } y<0\\ |y| = -y\\ \text{and } \left|2^{y-1}-1\right| = -(2^{y-1}-1)\end{array}\right.$$

$$\Rightarrow \quad 2^{-y} = 2^1$$

Hence, $y = -1$, which is true when $y < 0$. ...(i)

Case II $0 \le y < 1$

$$\Rightarrow \quad 2^{y} + (2^{y-1}-1) = 2^{y-1}+1; \quad \left\{\begin{array}{l}\text{as when } 0 \le y<1\\ |y| = y\\ \text{and } \left|2^{y-1}-1\right| = -(2^{y-1}-1)\end{array}\right.$$

$$\Rightarrow \quad 2^{y} = 2$$

$\Rightarrow \quad y = 1$, which shows no solution as,

$0 \le y < 1$ ...(ii)

Case III $\quad y \ge 1$

$$\Rightarrow \quad 2^{y} - (2^{y-1}-1) = 2^{y-1}+1$$

$$\Rightarrow \quad 2^{y} = 2^{y-1} + 2^{y-1}; \quad \left\{\begin{array}{l}\text{as when } y \ge 0\\ |y| = y\\ \text{and } \left|2^{y-1}-1\right| = (2^{y-1}-1)\end{array}\right.$$

$$\Rightarrow \quad 2^{y} = 2 \cdot 2^{y-1}$$

$\Rightarrow 2^y = 2^y$, which is an identity, therefore it is true, $\forall\ y \ge 1$. ...(iii)

Hence, from Eqs. (i), (ii) and (iii) the solution set is $\{y : y \ge 1 \cup y = -1\}$.

Ex. 36 *Solve the equation* $[x]\{x\} = x$, *where* [ ] *and* { } *denote the greatest integer function and fractional part, respectively.*

Sol. We know that, $\quad x = [x] + \{x\}$ ...(i)

Thus, we have $\quad [x]\{x\} = [x] + \{x\}$

$\Rightarrow \quad \{x\} = \dfrac{[x]}{[x]-1}$ ...(ii)

Here, in Eq. (ii), $[x] \ne 1$

But, if $[x] = 1$, then given equation

$\{x\} = x$ ...(iii)

which is true only when $\quad x \in [0, 1)$

and $\quad [x] = 1 \Rightarrow x \in [1, 2)$ ...(iv)

From Eqs. (iii) and (iv) no value of x,

when $[x] = 1$ ...(v)

Now, from Eq. (ii), $\quad \{x\} = \dfrac{[x]}{[x]-1}$,

when $\quad [x] \ne 1$

Again, as we know $\quad \{x\} \in [0, 1)$

$\therefore \quad 0 \le \dfrac{[x]}{[x]-1} < 1$

i.e. $\quad \dfrac{[x]}{[x]-1} < 1$ and $\dfrac{[x]}{[x]-1} \ge 0$

i.e. $\quad \dfrac{1}{[x]-1} < 0$ and $\{[x] \le 0$ or $[x] > 1\}$

i.e. $\quad [x] < 1$ and $\{[x] \le 0$ or $[x] > 1\}$ [using number line rule]

i.e. $\quad [x] \le 0$

$\therefore \quad x = [x] + \{x\} \Rightarrow x = [x] + \dfrac{[x]}{[x]-1}$

$\Rightarrow \quad x = \dfrac{[x]^2}{[x]-1}$, [where x takes values less than 1]

$\therefore \quad x = \dfrac{[x]^2}{[x]-1}$, [where $x < 1$]

$\Rightarrow \quad x = \dfrac{[x]^2}{[x]-1}$, [where $[x]$ is any non-positive integer]

Ex. 37 *Find all possible values of x satisfying*

$$\frac{[x]}{[x-2]} - \frac{[x-2]}{[x]} = \frac{8\{x\}+12}{[x-2][x]}$$

(where [ ] *denotes the greatest integer function and* { } *is the fractional part).*

Sol. Here, $\quad \dfrac{[x]}{[x-2]} - \dfrac{[x-2]}{[x]} = \dfrac{8\{x\}+12}{[x-2][x]}$

$$\Rightarrow \quad \frac{[x]^2 - [x-2]^2}{[x-2][x]} = \frac{8\{x\}+12}{[x][x-2]}$$

$$\Rightarrow \quad \frac{([x]-[x-2])([x]+[x-2])}{[x-2][x]} = \frac{8\{x\}+12}{[x][x-2]}$$

$\Rightarrow ([x]-[x-2])([x]+[x-2]) = 8\{x\}+12$

$\Rightarrow ([x]-[x]+2)([x]+[x]-2) = 8\{x\}+12$

$[\because [x+I] = [x]+I]$

$\Rightarrow \quad 4([x]-1) = 8\{x\}+12$

$\Rightarrow \quad [x]-4 = 2\{x\}$ ...(i)

Now, as we know $\quad 0 \le \{x\} < 1 \Rightarrow 0 \le 2\{x\} < 2$

$$\Rightarrow \quad 0 \le [x] - 4 < 2 \Rightarrow 4 \le [x] < 6$$

$$\Rightarrow \quad [x] = 4, 5$$

$$\left.\begin{array}{l}\text{if } [x] = 4 \Rightarrow 2\{x\} = [x] - 4 \\ \text{becomes } \{x\} = 0\end{array}\right\} \quad \text{...(ii)}$$

$$\left.\begin{array}{l}\text{and if } [x] = 5 \Rightarrow 2\{x\} = [x] - 4 \\ \text{becomes } \{x\} = \dfrac{1}{2}\end{array}\right\} \quad \text{...(iii)}$$

Thus, from Eqs. (ii) and (iii), we have

$$x = [x] + \{x\}$$

i.e. $x = 4 + 0 = 4$ [using Eq. (ii)]

and $x = 5 + \frac{1}{2} = \frac{11}{2}$ [using Eq. (iii)]

$$\Rightarrow \quad x \in \left\{4, \frac{11}{2}\right\}$$

Ex. 38 *If $f(x)$ is a polynomial function satisfying $f(x) \cdot f\left(\frac{1}{x}\right) = f(x) + f\left(\frac{1}{x}\right)$ and $f(4) = 65$. Find $f(6)$.*

Sol. Here, $f(x) \cdot f\left(\frac{1}{x}\right) = f(x) + f\left(\frac{1}{x}\right)$

$$\Rightarrow \quad f(x) \cdot f\left(\frac{1}{x}\right) - f(x) = f\left(\frac{1}{x}\right)$$

$$\Rightarrow \quad f(x) = \frac{f(1/x)}{f(1/x) - 1} \quad \text{...(i)}$$

Also, $f(x) \cdot f\left(\frac{1}{x}\right) = f(x) + f\left(\frac{1}{x}\right)$

$$\Rightarrow \quad f(x) \cdot f\left(\frac{1}{x}\right) - f\left(\frac{1}{x}\right) = f(x)$$

$$\Rightarrow \quad f\left(\frac{1}{x}\right) = \frac{f(x)}{f(x) - 1} \quad \text{...(ii)}$$

On multiplying Eqs. (i) and (ii), we get

$$f(x) \cdot f\left(\frac{1}{x}\right) = \frac{f(1/x) \cdot f(x)}{\{f(1/x) - 1\}\{f(x) - 1\}}$$

$$\Rightarrow \quad \left[f\left(\frac{1}{x}\right) - 1\right][f(x) - 1] = 1 \quad \text{...(iii)}$$

Since, $f(x)$ is a polynomial function, so $\{f(x) - 1\}$ and $\left\{f\left(\frac{1}{x}\right) - 1\right\}$ are reciprocals of each other. Also, x and $\frac{1}{x}$ are reciprocals of each other.

Thus, Eq. (iii) can hold only when

$$f(x) - 1 = \pm x^n, \quad \text{where } n \in R$$

$$\therefore \quad f(x) = \pm x^n + 1, \quad \text{but } f(4) = 65$$

$$\Rightarrow \quad \pm 4^n + 1 = 65 \Rightarrow 4^n = 64$$

$$\Rightarrow \quad 4^n = 4^3 \Rightarrow n = 3 \quad [\because 4^n > 0]$$

So, $f(x) = x^3 + 1$

Hence, $f(6) = 6^3 + 1 = 217$

Aliter Let $f(x) = a_0 x^n + a_1 x^{n-1} + a_2 x^{n-2} + \dots + a_{n-1} x + a_n$

Then, $f(x) \cdot f\left(\frac{1}{x}\right) = f(x) + f\left(\frac{1}{x}\right)$

$$\Rightarrow (a_0 x^n + a_1 x^{n-1} + \dots + a_n)\left(\frac{a_0}{x^n} + \frac{a_1}{x^{n-1}} + \dots + a_n\right)$$

$$= (a_0 x^n + a_1 x^{n-1} + \dots + a_n) + \left(\frac{a_0}{x^n} + \frac{a_1}{x^{n-1}} + \dots + a_n\right)$$

On comparing the coefficient of x^n, we have

$$a_0 a_n = a_0 \Rightarrow a_n = 1 \quad [\text{as } a_0 \ne 0]$$

Comparing the coefficient of x^{n-1}, we have

$$a_0 a_{n-1} + a_n a_1 = a_1$$

$$\Rightarrow \quad a_0 a_{n-1} + a_1 = a_1 \quad [\text{as } a_n = 1]$$

$$\Rightarrow \quad a_0 a_{n-1} = 0$$

$$\Rightarrow \quad a_{n-1} = 0 \quad [\text{as } a_0 \ne 0]$$

Similarly, $a_{n-1} = a_{n-2} = \dots\dots = a_1 = 0$

and $a_0 = \pm 1$

$$\therefore \quad f(x) = \pm x^n + 1$$

$$\Rightarrow \quad f(4) = \pm 4^n + 1$$

$$\Rightarrow \quad 4^n + 1 = 65 \Rightarrow 4^n = 64$$

$$\Rightarrow \quad n = 3$$

So, $f(x) = x^3 + 1$

Hence, $f(6) = 6^3 + 1 = 217$

Ex. 39 *If $f(x)$ satisfies the relation, $f(x + y) = f(x) + f(y)$ for all $x, y \in R$ and $f(1) = 5$, then find $\sum_{n=1}^{m} f(n)$. Also, prove that $f(x)$ is odd.*

Sol. Here, $f(x + y) = f(x) + f(y)$, put $x = r - 1, y = 1$

$$f(r) = f(r - 1) + f(1) \quad \text{...(i)}$$

$$\therefore \quad f(r) = f(r - 1) + 5$$

$$\Rightarrow \quad f(r) = \{f(r - 2) + 5\} + 5 \quad [\text{using definition}]$$

$$\Rightarrow \quad f(r) = f(r - 2) + 2 \cdot (5)$$

$$\Rightarrow \quad f(r) = f(r - 3) + 3 \cdot (5)$$

....

....

$$\Rightarrow \quad f(r) = f\{r - (r - 1)\} + (r - 1) \cdot 5$$

$$\Rightarrow \quad f(r) = f(1) + (r - 1) \cdot 5$$

$$\Rightarrow \quad f(r) = 5 + (r - 1) \cdot 5$$

$$\Rightarrow \quad f(r) = 5r$$

$$\therefore \quad \sum_{n=1}^{m} f(n) = \sum_{n=1}^{m} (5n) = 5[1 + 2 + 3 + \dots + m]$$

$$= \frac{5m(m + 1)}{2}$$

Hence, $$\sum_{n=1}^{m} f(n) = \frac{5m(m+1)}{2} \qquad \text{...(ii)}$$

Now, putting $x = 0$, $y = 0$ in the given function, we have

$$f(0+0) = f(0) + f(0)$$

$$\therefore \quad f(0) = 0$$

Also, putting $(-x)$ for (y) in the given function,

$$f(x-x) = f(x) + f(-x)$$

$$\Rightarrow \quad f(0) = f(x) + f(-x)$$

$$\Rightarrow \quad 0 = f(x) + f(-x)$$

$$\Rightarrow \quad f(-x) = -f(x) \qquad \text{...(iii)}$$

Thus, $\sum_{n=1}^{m} f(n) = \frac{5m(m+1)}{2}$ and $f(x)$ is odd.

Ex. 40 *Let $f(x) = \frac{9^x}{9^x + 3}$. Show that $f(x) + f(1-x) = 1$ and hence evaluate*

$$f\left(\frac{1}{1996}\right) + f\left(\frac{2}{1996}\right) + f\left(\frac{3}{1996}\right) + \ldots + f\left(\frac{1995}{1996}\right).$$

Sol. Given, $$f(x) = \frac{9^x}{9^x + 3} \qquad \text{...(i)}$$

$$\Rightarrow \quad f(1-x) = \frac{9^{1-x}}{9^{1-x} + 3}$$

$$\Rightarrow \quad f(1-x) = \frac{\frac{9}{9^x}}{\frac{9}{9^x} + 3} = \frac{9}{9 + 3\cdot 9^x}$$

$$f(1-x) = \frac{9}{3(3+9^x)} \qquad \text{...(ii)}$$

On adding Eqs. (i) and (ii), we get

$$f(x) + f(1-x) = \frac{9^x}{9^x + 3} + \frac{9}{3(3+9^x)}$$

$$= \frac{3\cdot 9^x + 9}{3(9^x + 3)} = \frac{3(9^x + 3)}{3(9^x + 3)}$$

$$\therefore \quad f(x) + f(1-x) = 1 \qquad \text{...(iii)}$$

Now, putting $x = \frac{1}{1996}, \frac{2}{1996}, \frac{3}{1996}, \ldots, \frac{998}{1996}$ in Eq. (iii), we get

$$f\left(\frac{1}{1996}\right) + f\left(\frac{1995}{1996}\right) = 1$$

$$\Rightarrow \quad f\left(\frac{2}{1996}\right) + f\left(\frac{1994}{1996}\right) = 1$$

$$\Rightarrow \quad f\left(\frac{3}{1996}\right) + f\left(\frac{1993}{1996}\right) = 1$$

....

....

$$\Rightarrow \quad f\left(\frac{997}{1996}\right) + f\left(\frac{999}{1996}\right) = 1$$

$$\Rightarrow \quad f\left(\frac{998}{1996}\right) + f\left(\frac{998}{1996}\right) = 1$$

or $$f\left(\frac{998}{1996}\right) = \frac{1}{2}$$

On adding all the above expressions, we get

$$f\left(\frac{1}{1996}\right) + f\left(\frac{2}{1996}\right) + \ldots + f\left(\frac{1995}{1996}\right)$$

$$= (1 + 1 + 1 + \ldots + \text{to } 997 \text{ times}) + \frac{1}{2}$$

$$= 997 + \frac{1}{2}$$

$$= 997.5$$

Ex. 41 *ABCD is a square of side a. A line parallel to the diagonal BD at a distance x from the vertex A cuts the two adjacent sides. Express the area of the segment of the square with A at a vertex, as a function of x.*

Sol. There are two different situations,

(i) when $x = AP \le OA$,

i.e. $$x \le \frac{a}{\sqrt{2}}$$

(ii) when $x = AP > OA$,

i.e. $x > \frac{a}{\sqrt{2}}$ but $x \le \sqrt{2}a$

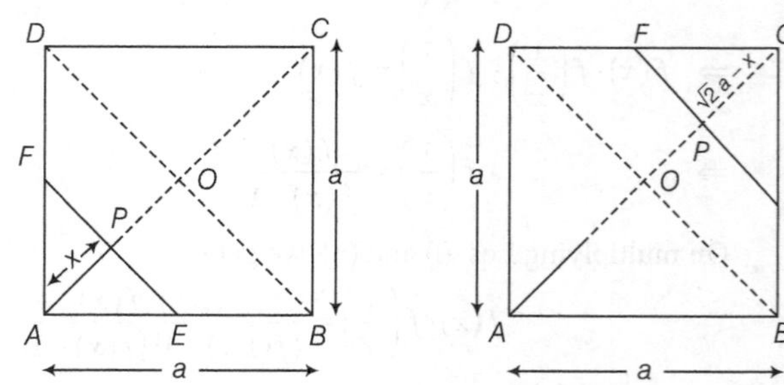

***Case* I** ar $(\Delta AEF) = \frac{1}{2} x \cdot 2x = x^2$

***Case* II** ar $(ABEFDA)$ = ar $(ABCD)$ − ar (ΔCEF)

$$= a^2 - \frac{1}{2}(\sqrt{2}a - x)2(\sqrt{2}a - x)$$

$$\text{ar }(ABEFDA) = 2\sqrt{2}\,ax - x^2 - a^2$$

$$\therefore \quad f(x) = \begin{cases} x^2, & 0 \le x \le \frac{a}{\sqrt{2}} \\ 2\sqrt{2}\,ax - x^2 - a^2, & \frac{a}{\sqrt{2}} < x \le \sqrt{2}a \end{cases}$$

Ex. 42 *A function $R \to R$ is defined by $f(x) = \dfrac{\alpha x^2 + 6x - 8}{\alpha + 6x - 8x^2}$. Find the interval of values of α for which f is onto. Is the function one-one for $\alpha = 3$? Justify your answer.* **[IIT JEE 1996]**

Sol. Since, $f : R \longrightarrow R$ is an onto mapping.

$\therefore$ Range of $f = R$

$$\Rightarrow \frac{\alpha x^2 + 6x - 8}{\alpha + 6x - 8x^2} \text{ assumes all real values of } x.$$

Let $y = \dfrac{\alpha x^2 + 6x - 8}{\alpha + 6x - 8x^2}$

Then, y assumes all real values for real values of x.

$\Rightarrow \alpha y + 6xy - 8x^2 y = \alpha x^2 + 6x - 8, \ \forall\, y \in R$

$\Rightarrow x^2(\alpha + 8y) + 6x(1 - y) - (8 + \alpha y) = 0, \ \forall\, y \in R$

We know, above equation assumes all real values.

So, $D \geq 0$

$$\Rightarrow 36(1 - y)^2 + 4(\alpha + 8y)(8 + \alpha y) \geq 0$$

$$\Rightarrow 4[9(1 - 2y + y^2) + (8\alpha + \alpha^2 y + 64y + 8\alpha y^2)] \geq 0$$

$$\Rightarrow [9 - 18y + 9y^2 + 8\alpha + \alpha^2 y + 64y + 8\alpha y^2] \geq 0$$

$$\Rightarrow [y^2(8\alpha + 9) + y(\alpha^2 + 46) + (8\alpha + 9)] \geq 0$$

We know, if $ax^2 + bx + c \geq 0, \forall\, x$ and $a > 0 \Rightarrow D \leq 0$

So, $(\alpha^2 + 46)^2 - 4(8\alpha + 9)(8\alpha + 9) \leq 0$ and $(8\alpha + 9) > 0$

$\Rightarrow (\alpha^2 + 46)^2 - [2(8\alpha + 9)]^2 \leq 0$ and $\alpha > -9/8$

$\Rightarrow (\alpha^2 + 46 - 16\alpha - 18)(\alpha^2 + 46 + 16\alpha + 18) \leq 0$

and $\alpha > -9/8$

$(\alpha^2 - 16\alpha + 28)(\alpha^2 + 16\alpha + 64) \leq 0$ and $\alpha > -9/8$

$\Rightarrow (\alpha - 14)(\alpha - 2)(\alpha + 8)^2 \leq 0$ and $\alpha > -9/8$

+ 2 − 14 +

$\alpha \in [2, 14] \cup \{-8\}$ and $\alpha > -9/8$

Thus, $\alpha \in [2, 14]$

Hence, f is onto when $\alpha \in [2, 14]$.

Again, if $\alpha = 3$, we have, $f(x) = \dfrac{3x^2 + 6x - 8}{3 + 6x - 8x^2}$.

Clearly, $f(1) = f(-1)$, which shows $f(x)$ is not one-one.

Ex. 43 *Let $f(x, y)$ be a periodic function satisfying the condition $f(x, y) = f((2x + 2y), (2y - 2x)), \forall\, x, y \in R$. Now, define a function g by $g(x, 0) = f(2^x, 0)$. Then, prove that $g(x)$ is a periodic function and find its period.*

Sol. We have $f(x, y) = f(2x + 2y, 2y - 2x)$ ...(i)

$\Rightarrow f(2x + 2y, 2y - 2x) = f\{2(2x + 2y) + 2(2y - 2x), 2(2y - 2x) - 2(2x + 2y)\}$ [using Eq. (i)]

$\Rightarrow f(x, y) = f(2x + 2y, 2y - 2x) = f(8y, -8x)$

$\Rightarrow f(x, y) = f(8y, -8x)$ ...(ii)

$\Rightarrow f(8y, -8x) = f\{8(-8x), -8(8y)\}$ [using Eq. (ii)]

$\Rightarrow f(x, y) = f(2x + 2y, 2y - 2x) = f(8y, -8x) = f(-64x, -64y)$

$\Rightarrow f(x, y) = f(-64x, -64y)$ ...(iii)

$\Rightarrow f(-64x, -64y) = f(64 \times 64x, 64 \times 64y) = f(2^{12}x, 2^{12}y)$

$\Rightarrow f(x, y) = f(2^{12}x, 2^{12}y)$ [using Eq. (iii)]

$\Rightarrow f(2^x, 0) = f(2^{12} \cdot 2^x, 0) = f(2^{12 + x}, 0)$ ...(iv)

Given, $g(x, 0) = f(2^x, 0)$

$\Rightarrow g(x, 0) = f(2^x, 0) = f(2^{12 + x}, 0)$ [using Eq. (iv)]

$\Rightarrow g(x, 0) = g(x + 12, 0)$

Hence, $g(x)$ is periodic with period 12.

Ex. 44 *Solve the equation* $10^{(x+1)(3x+4)} - 2 \cdot 10^{(x+1)(x+2)} = 10^{1 - x - x^2}$.

Sol. Given equation is

$$10^{3x^2 + 7x + 4} - 2 \cdot 10^{x^2 + 3x + 2} = \frac{10}{10^{x + x^2}}$$

$$\Rightarrow 10^{4x^2 + 8x + 4} - 2 \cdot 10^{2x^2 + 4x + 2} = 10$$

$$\Rightarrow 10^{4(x^2 + 2x + 1)} - 2 \cdot 10^{2(x^2 + 2x + 1)} = 10$$

$$\Rightarrow 10^{4(x+1)^2} - 2 \cdot 10^{2(x+1)^2} = 10$$

$$\Rightarrow \{10^{2(x+1)^2}\}^2 - 2 \cdot \{10^{2(x+1)^2}\} = 10 \quad \text{...(i)}$$

Let $10^{2(x+1)^2} = y$ ...(ii)

$\Rightarrow y^2 - 2y = 10 \Rightarrow y^2 - 2y - 10 = 0$

$$\Rightarrow y = \frac{2 \pm \sqrt{4 + 40}}{2}$$

$\Rightarrow y = 1 \pm \sqrt{11}$

$\Rightarrow y = 1 + \sqrt{11}$ [neglecting –ve sign as $y > 0$]

$\Rightarrow 10^{2(x+1)^2} = 1 + \sqrt{11}$

$\Rightarrow 2(x + 1)^2 = \log_{10}(1 + \sqrt{11})$

$\Rightarrow (x + 1)^2 = \dfrac{1}{2}\log_{10}(1 + \sqrt{11})$

$\Rightarrow x + 1 = \pm\sqrt{\dfrac{1}{2}\log_{10}(1 + \sqrt{11})}$

Hence, $x = -1 \pm \sqrt{\dfrac{1}{2}\log_{10}(1 + \sqrt{11})}$

Ex. 45 *Find the real solution of $[x] + [5x] + [10x] + [20x] = 36K + 35, K \in$ integer, where $[\cdot]$ denotes the greatest integral function.*

Sol. Let $x = a + r$, where $a \in$ integer and $r \in [0, 1)$. Then, there are four cases

(i) $0 \le r < \frac{1}{20}$ (ii) $\frac{1}{20} \le r < \frac{1}{10}$

(iii) $\frac{1}{10} \le r < \frac{1}{5}$ (iv) $\frac{1}{5} \le r < 1$

Case I If $0 \le r < \frac{1}{20}$.

$\therefore$ $[x]+[5x]+[10x]+[20x]=36K+35$

$\Rightarrow [a+r]+[5a+5r]+[10a+10r]+[20a+20r]=36K+35$

$\Rightarrow a+5a+10a+20a=36K+35$

[by definition of the greatest integer function]

$\Rightarrow$ $36a=36K+35$,

which is not possible for any values of a and K as they belongs to integers. ...(i)

Case II If $\frac{1}{20} \le r < \frac{1}{10}$.

$\therefore$ $[x]+[5x]+[10x]+[20x]=36K+35$

$\Rightarrow$ $[a+r]+[5a+5r]+[10a+10r]+[20a+20r]$ $=36K+35$

$\Rightarrow$ $a+5a+10a+20a+1=36K+35$

$\Rightarrow$ $36a+1=36K+35$

$\Rightarrow$ $36a=36K+34$,

which is again not possible, $\forall$ $a, K \in$ integer. ...(ii)

Case III If $\frac{1}{10} \le r < \frac{1}{5}$.

$\therefore$ $[x]+[5x]+[10x]+[20x]=36K+35$

$\Rightarrow [a+r]+[5a+5r]+[10a+10r]+[20a+20r]=36K+35$

$\Rightarrow$ $a+5a+10a+1+20a+2=36K+35$

$\Rightarrow$ $36a+3=36K+35$

$\Rightarrow$ $36a=36K+32$,

which shows no value of a and K can be given. ...(iii)

Case IV If $\frac{1}{5} \le r < 1$.

$\therefore$ $[x]+[5x]+[10x]+[20x]=36K+35$

$\Rightarrow$ $[a+r]+[5a+5r]+[10a+10r]+[20a+20r]=36K+35$

$\Rightarrow$ $a+5a+1+10a+2+20a+4=36K+35$

$\Rightarrow$ $36a+7=36K+35$

$\Rightarrow$ $36a=36K+28$

Again, no solution is possible. ...(iv)

Hence, from Eqs. (i), (ii), (iii), (iv) there exists no real value which possesses solution.

Ex. 46 *Let $f: N \to N$ be a function such that*

(i) $x - f(x) = 19\left[\frac{x}{19}\right] - 90\left[\frac{f(x)}{90}\right], \forall x \in N$, where $[\cdot]$ denotes the greatest integer function.

(ii) $1900 < f(1990) < 2000$.

Find all the possible values of $f(1990)$.

Sol. Since, $1900 < f(1990) < 2000$

$\Rightarrow$ $\frac{1900}{90} < \frac{f(1990)}{90} < \frac{2000}{90} \Rightarrow 21.1 < \frac{f(1990)}{90} < 22.2$

$\therefore$ $\left[\frac{f(1990)}{90}\right] = 21, 22$ ...(i)

Given, $x - f(x) = 19\left[\frac{x}{19}\right] - 90\left[\frac{f(x)}{90}\right]$.

Now, there are two cases arise.

Case I $\left[\frac{f(1990)}{90}\right] = 21$

We have, $1990 - f(1990) = 19\left[\frac{1990}{19}\right] - 90\left[\frac{f(1990)}{90}\right]$

$1990 - f(1990) = 19 \cdot (104) - 90 \cdot (21)$

$\Rightarrow$ $f(1990) = 1904$ ... (ii)

Case II $\left[\frac{f(1990)}{90}\right] = 22$

We have, $1990 - f(1990) = 19\left[\frac{1990}{19}\right] - 90\left[\frac{f(1990)}{90}\right]$

$\Rightarrow$ $f(1990) = 1994$ ... (iii)

From Eqs. (ii) and (iii), we have $f(1990) = 1904$ or 1994.

Ex. 47 *Solve the system of equations;*

$$|x^2 - 2x| + y = 1, x^2 + |y| = 1.$$

Sol. Here, $|x^2 - 2x| + y = 1$ and $x^2 + |y| = 1$ gives four cases as

Case I $x^2 - 2x \ge 0$ and $y \ge 0$ ...(i)

Equations are $x^2 - 2x + y = 1$ and $x^2 + y = 1$

$\Rightarrow$ $1 - 2x = 1 \Rightarrow x = 0$ and $y = 1$ ...(ii)

From Eq. (i), $x^2 - 2x \ge 0 \Rightarrow x \le 0$ $x \ge 2$ and $y \ge 0$

Thus, $x = 0$ and $y = 1$ is the only solution when

$x^2 - 2x \ge 0$ and $y \ge 0$

Case II $x^2 - 2x \ge 0$ and $y < 0$

$\Rightarrow$ $(x \le 0$ or $x \ge 2)$ and $(y < 0)$ ...(iii)

$\therefore$ Equations are $x^2 - 2x + y = 1$ and $x^2 - y = 1$.

Adding these equations, we get

$2x^2 - 2x = 2$

or $x = \frac{1 \pm \sqrt{5}}{2}$ and $y = \frac{1 \pm \sqrt{5}}{2}$ ...(iv)

From Eqs. (iii) and (iv), $x = \frac{1-\sqrt{5}}{2}$ and $y = \frac{1-\sqrt{5}}{2}$

Case III $x^2 - 2x \le 0$ and $y \ge 0$.

$\Rightarrow$ $(0 \le x \le 2)$ and $(y \ge 0)$ ...(v)

Equations are $-x^2 + 2x + y = 1$ and $x^2 + y = 1$, subtracting, we get

$-2x^2 + 2x = 0 \Rightarrow x = 0, 1$

i.e. $x = 0, y = 1$ or $x = 1, y = 0$ ...(vi)

which satisfy Eq. (v).

Case IV $x^2 - 2x \le 0$ and $y \le 0$.

$\Rightarrow \quad (0 \le x \le 2)$ and $(y \le 0)$ ...(vii)

$\therefore$ Equations are $-x^2 + 2x + y = 1$ and $x^2 - y = 1$,

adding, we get $2x = 2 \Rightarrow x = 1$

and $y = 0$ which satisfies Eq. (vii).

Thus, solutions are

$$(x = 0, y = 1)(x = 1, y = 0) \text{ and } \left(x = y = \frac{1-\sqrt{5}}{2}\right).$$

Ex. 48 *Let f and g be real-valued functions such that $f(x+y) + f(x-y) = 2f(x) \cdot g(y), \forall\, x, y \in R$. Prove that, if $f(x)$ is not identically zero and $|f(x)| \le 1, \forall\, x \in R$, then $|g(y)| \le 1, \forall\, y \in R$.*

Sol. Let maximum value of $f(x)$ be M.

$\Rightarrow \quad \max |f(x)| = M$, where $0 < M \le 1$ ...(i)

[since, f is not identically zero and $|f(x)| \le 1, \forall\, x \in R$]

Now, $f(x+y) + f(x-y) = 2f(x) \cdot g(y)$

$\Rightarrow \quad |2f(x)| \cdot |g(y)| = |f(x+y) + f(x-y)|$

$\Rightarrow \quad 2|f(x)| \cdot |g(y)| \le |f(x+y)| + |f(x-y)|$

[as $|a+b| \le |a| + |b|$]

$\Rightarrow \quad 2|f(x)| \cdot |g(y)| \le M + M$

[using Eq. (i), i.e. $\max |f(x)| = M$]

$\Rightarrow \quad |g(y)| \le 1, \forall\, y \in R$

Thus, if $f(x)$ is not identically zero and $|f(x)| \le 1, \forall\, x \in R$, then $|g(y)| \le 1, \forall\, y \in R$.

Ex. 49 *If p and q are positive integers, f is a function defined for positive numbers and attains only positive values such that $f(x\, f(y)) = x^p y^q$, then prove that $q = p^2$.*

Sol. For $x = \dfrac{1}{f(y)}$, we have

$$f\left(x \cdot \frac{1}{x}\right) = \frac{1}{f(y)^p} \cdot y^q$$

$$\Rightarrow \quad f(1) = \frac{y^q}{\{f(y)\}^p}$$

$$\Rightarrow \quad f(y) = \frac{y^{q/p}}{\{f(1)\}^{1/p}}$$

For $y = 1$, we have $f(1) = 1$

$$\therefore \quad f(y) = y^{q/p}$$

$$\text{or} \quad f(x) = x^{q/p} \quad \text{...(i)}$$

$$\text{Hence,} \quad f(x \cdot y^{q/p}) = x^p \cdot y^q$$

$$\text{Let} \quad y^{q/p} = z$$

$$\Rightarrow \quad y = z^{p/q}$$

$$\Rightarrow \quad f(x \cdot z) = x^p \cdot z^p$$

$$\text{or} \quad f(x) = x^p \quad \text{...(ii)}$$

From Eqs. (i) and (ii), we have $x^{q/p} = x^p$

$$\Rightarrow \quad \frac{q}{p} = p \text{ or } q = p^2$$

Functions Exercise 1 :
Single Option Correct Type Questions

1. $f_1(x) = \begin{cases} x, \text{ for } 0 \le x \le 1 \\ 1, \text{ for } x > 1 \\ 0, \quad \text{otherwise} \end{cases}$ $\quad f_2(x) = f_1(-x)$, for all x and $f_3(x) = -f_2(x)$, for all x. $f_4(x) = f_3(-x)$, for all x

Which of the following is necessarily true?

(a) $f_4(x) = f_1(x)$, for all x (b) $f_1(x) = -f_3(-x)$, for all x
(c) $f_2(-x) = f_4(x)$, for all x (d) $f_1(x) + f_3(x) = 0$, for all x

2. Which one of the following is an odd function?

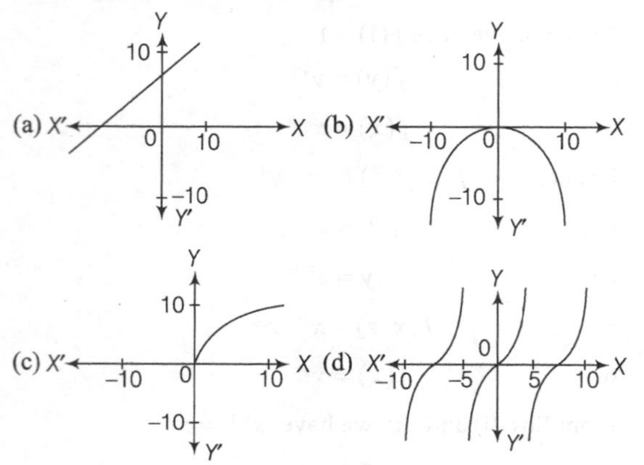

3. If $f(x) = \sqrt{\dfrac{8}{1-x} + \dfrac{8}{1+x}}$ and $g(x) = \dfrac{4}{f(\sin x)} + \dfrac{4}{f(\cos x)}$, the $g(x)$ is periodic with period

(a) $\dfrac{\pi}{2}$ (b) π (c) $\dfrac{3\pi}{2}$ (d) 2π

4. Let f be a function defined by $f(xy) = \dfrac{f(x)}{y}$ for all positive real numbers x and y. If $f(30) = 20$, the value of $f(40)$ is

(a) 15 (b) 20 (c) 40 (d) 60

5. Let $f(x) = e^{\{e^{|x|} \text{ sgn } x\}}$ and $g(x) = e^{[e^{|x|} \text{ sgn } x]}$, $x \in R$, where $\{x\}$ and $[x]$ denote fractional part and greatest integer function, respectively. Also, $h(x) = \log(f(x)) + \log(g(x))$, then for all real x, $h(x)$ is

(a) an odd function
(b) an even function
(c) neither odd nor even function
(d) both odd as well as even function

6. Which of the following function is surjective but not injective?

(a) $f : R \to R$, $f(x) = x^4 + 2x^3 - x^2 + 1$
(b) $f : R \to R$, $f(x) = x^3 + x + 1$
(c) $f : R \to R^+$, $f(x) = \sqrt{1 + x^2}$
(d) $f : R \to R$, $f(x) = x^3 + 2x^2 - x + 1$

7. If $f(x) = 2x^3 + 7x - 9$, then $f^{-1}(4)$ is

(a) 1 (b) 2 (c) 1/3 (d) non-existent

8. The range of the function

$$f(x) = \frac{e^x \cdot \log x \cdot 5^{x^2+2} \cdot (x^2 - 7x + 10)}{2x^2 - 11x + 12} \text{ is}$$

(a) $(-\infty, \infty)$ (b) $[0, \infty)$ (c) $\left(\dfrac{3}{2}, \infty\right)$ (d) $\left(\dfrac{3}{2}, 4\right)$

9. If $x = \cos^{-1}(\cos 4)$ and $y = \sin^{-1}(\sin 3)$, then which of the following holds?

(a) $x - y = 1$ (b) $x + y + 1 = 0$
(c) $x + 2y = 2$ (d) $x + y = 3\pi - 7$

10. Let $f(x) = \left(\dfrac{2\sin x + \sin 2x}{2\cos x + \sin 2x} \cdot \dfrac{1 - \cos x}{1 - \sin x}\right)^{2/3}$; $x \in R$.

Consider the following statements.

I. Domain of f is R.
II. Range of f is R.
III. Domain of f is $R - (4n - 1)\dfrac{\pi}{2}, n \in I$.
IV. Domain of f is $R - (4n + 1)\dfrac{\pi}{2}, n \in I$.

Which one of the following is correct?

(a) I and II (b) II and III
(c) III and IV (d) II, III and IV

11. If $f(x) = e^{\sin(x - [x]) \cos \pi x}$, where $[x]$ denotes the greatest integer function, then $f(x)$ is

(a) non-periodic
(b) periodic with no fundamental period
(c) periodic with period 2
(d) periodic with period π

12. The range of the function,
$f(x) = \cot^{-1}(\log_{0.5}(x^4 - 2x^2 + 3))$ is

(a) $(0, \pi)$ (b) $\left(0, \dfrac{3\pi}{4}\right]$ (c) $\left[\dfrac{3\pi}{4}, \pi\right)$ (d) $\left[\dfrac{\pi}{2}, \dfrac{3\pi}{4}\right]$

13. Range of $f(x) = \left[\dfrac{1}{\log(x^2 + e)}\right] + \dfrac{1}{\sqrt{1 + x^2}}$, where $[\cdot]$ denotes greatest integer function, is

(a) $\left(0, \dfrac{e+1}{e}\right) \cup \{2\}$ (b) $(0, 1)$
(c) $(0, 1] \cup \{2\}$ (d) $[0, 1) \cup \{2\}$

14. The period of the function $f(x) = \sin(x + 3 - [x + 3])$, where $[\cdot]$ denotes greatest integer function, is

(a) $2\pi + 3$ (b) 2π (c) 1 (d) 4

15. Which one of the following functions best represent the graph as shown below?

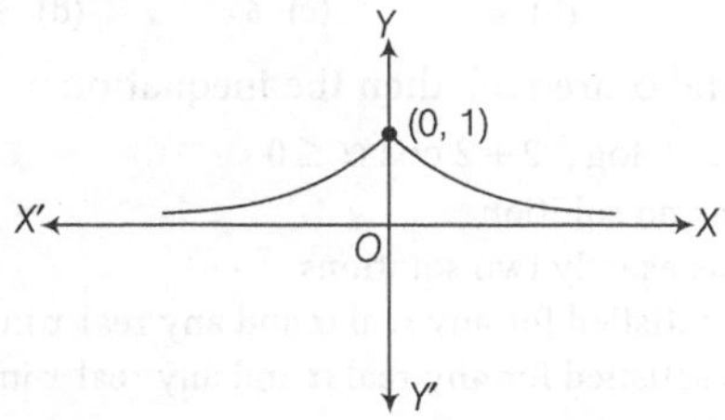

(a) $f(x)=\dfrac{1}{1+x^2}$ (b) $f(x)=\dfrac{1}{\sqrt{1+|x|}}$

(c) $f(x)=e^{-|x|}$ (d) $f(x)=a^{|x|}, a>1$

16. The solution set for $[x]\{x\}=1$, where $\{x\}$ and $[x]$ denote fractional part and greatest integer functions, is

(a) $R^+-(0,1)$ (b) $R^+-\{1\}$

(c) $\left\{m+\dfrac{1}{m}; m\in I-\{0\}\right\}$ (d) $\left\{m+\dfrac{1}{m}; m\in N-\{1\}\right\}$

17. The domain of definition of function

$f(x)=\log(\sqrt{x^2-5x-24}-x-2)$, is

(a) $(-\infty,-3]$ (b) $(-\infty-3]\cup[8,\infty)$

(c) $\left(-\infty,\dfrac{-28}{9}\right]$ (d) None of these

18. If $f(x)$ is a function $f:R\to R$, we say $f(x)$ has property

I. If $f(f(x))=x$ for all real numbers x.

II. If $f(-f(x))=-x$ for all real numbers x.

How many linear functions, have both property I and II?

(a) 0 (b) 2

(c) 3 (d) Infinite

19. Let $f(x)=\dfrac{x}{1+x}$ and $g(x)=\dfrac{rx}{1-x}$. Let S be the set of all real numbers r, such that $f(g(x))=g(f(x))$ for infinitely many real numbers x. The number of elements in set S is

(a) 1 (b) 2

(c) 3 (d) 5

20. Let $f(x)$ be linear functions with the properties that $f(1)\le f(2)$, $f(3)\ge f(4)$ and $f(5)=5$. Which one of the following statements is true?

(a) $f(0)<0$ (b) $f(0)=0$

(c) $f(1)<f(0)<f(-1)$ (d) $f(0)=5$

21. Suppose R is relation whose graph is symmetric to both X-axis and Y-axis and that the point (1, 2) is on the graph of R. Which one of the following is not necessarily on the graph of R?

(a) $(-1,2)$ (b) $(1,-2)$

(c) $(-1,-2)$ (d) $(2,1)$

22. The area between the curve $2\{y\}=[x]+1, 0\le y<1$, where $\{\cdot\}$ and $[\cdot]$ are the fractional part and greatest integer functions, respectively and the X-axis, is

(a) $\dfrac{1}{2}$ (b) 1 (c) 0 (d) $\dfrac{3}{2}$

23. If $f(x)=\sin^{-1}x$ and $g(x)=[\sin(\cos x)]+[\cos(\sin x)]$, then range of $f(g(x))$ is (where, $[\cdot]$ denotes greatest integer function)

(a) $\left\{\dfrac{-\pi}{2},\dfrac{\pi}{2}\right\}$ (b) $\left\{\dfrac{-\pi}{2},0\right\}$

(c) $\left\{0,\dfrac{\pi}{2}\right\}$ (d) $\left\{-\dfrac{\pi}{2},0,\dfrac{\pi}{2}\right\}$

24. The number of solutions of the equation $e^{2x}+e^x-2=[\{x^2+10x+11\}]$ is (where, $\{x\}$ denotes fractional part of x and $[x]$ denotes greatest integer function)

(a) 0 (b) 1 (c) 2 (d) 3

25. Total number of values of x, of the form $\dfrac{1}{n}, n\in N$ in the interval $x\in\left[\dfrac{1}{25},\dfrac{1}{10}\right]$, which satisfy the equation $\{x\}+\{2x\}+\ldots+\{12x\}=78x$ (where, $\{\cdot\}$ represents fractional part function), is

(a) 12 (b) 13

(c) 14 (d) 15

26. The sum of the maximum and minimum values of the function $f(x)=\dfrac{1}{1+(2\cos x-4\sin x)^2}$ is

(a) $\dfrac{22}{21}$ (b) $\dfrac{21}{20}$ (c) $\dfrac{22}{20}$ (d) $\dfrac{21}{11}$

27. Let $f:X\to Y, f(x)=\sin x+\cos x+2\sqrt{2}$ be invertible, then $X\to Y$ is/are

(a) $\left[\dfrac{\pi}{4},\dfrac{5\pi}{4}\right]\to[\sqrt{2},3\sqrt{2}]$

(b) $\left[-\dfrac{\pi}{4},\dfrac{3\pi}{4}\right]\to[\sqrt{2},3\sqrt{2}]$

(c) $\left[-\dfrac{3\pi}{4},\dfrac{3\pi}{4}\right]\to[\sqrt{2},-3\sqrt{2}]$

(d) $\left[-\dfrac{3\pi}{4},-\dfrac{\pi}{4}\right]\to[\sqrt{2},3\sqrt{2}]$

28. The range of values of a, so that all the roots of the equation $2x^3-3x^2-12x+a=0$ are real and distinct, belongs to

(a) (7, 20) (b) (−7, 20)

(c) (−20, 7) (d) (−7, 7)

29. If $f(x)$ is continuous such that $|f(x)| \le 1, \forall x \in R$ and $g(x) = \dfrac{e^{f(x)} - e^{|f(x)|}}{e^{f(x)} + e^{|f(x)|}}$, then range of $g(x)$ is

(a) $[0, 1]$ (b) $\left[0, \dfrac{e^2+1}{e^2-1}\right]$

(c) $\left[0, \dfrac{e^2-1}{e^2+1}\right]$ (d) $\left[\dfrac{1-e^2}{1+e^2}, 0\right]$

30. Let $f(x) = \sqrt{|x| - \{x\}}$, where {} denotes the fractional part of x and X, Y and its domain and range respectively, then

(a) $f: X \to Y : y = f(x)$ is one-one function

(b) $X \in \left(-\infty, -\dfrac{1}{2}\right] \cup [0, \infty)$ and $Y \in \left[\dfrac{1}{2}, \infty\right)$

(c) $X \in \left(-\infty, -\dfrac{1}{2}\right] \cup [0, \infty)$ and $Y \in [0, \infty)$

(d) None of the above

31. If the graphs of the functions $y = \ln x$ and $y = ax$ intersect at exactly two points, then a must be

(a) $(0, e)$ (b) $\left(\dfrac{1}{e}, 0\right)$

(c) $\left(0, \dfrac{1}{e}\right)$ (d) None of these

32. A quadratic polynomial maps from $[-2, 3]$ onto $[0, 3]$ and touches X-axis at $x = 3$, then the polynomial is

(a) $\dfrac{3}{16}(x^2 - 6x + 16)$ (b) $\dfrac{3}{25}(x^2 - 6x + 9)$

(c) $\dfrac{3}{25}(x^2 - 6x + 16)$ (d) $\dfrac{3}{16}(x^2 - 6x + 9)$

33. The range of the function $y = \sqrt{2\{x\} - \{x\}^2 - \dfrac{3}{4}}$ (where, $\{\cdot\}$ denotes the fractional part) is

(a) $\left[-\dfrac{1}{4}, \dfrac{1}{4}\right]$ (b) $\left[0, \dfrac{1}{2}\right)$

(c) $\left[0, \dfrac{1}{4}\right]$ (d) $\left[\dfrac{1}{4}, \dfrac{1}{2}\right]$

34. Let $f(x)$ be a fourth differentiable function such that $f(2x^2 - 1) = 2xf(x), \forall x \in R$, then $f^{iv}(0)$ is equal to (where, $f^{iv}(0)$ represents fourth derivative of $f(x)$ at $x = 0$)

(a) 0 (b) 1

(c) -1 (d) Data insufficient

35. Number of solutions of the equation $[y + [y]] = 2 \cos x$ is (where, $y = \dfrac{1}{3}[\sin x + [\sin x + [\sin x]]]$ and $[\cdot]$ denotes the greatest integer function)

(a) 1 (b) 2

(c) 3 (d) None of these

36. If a function satisfies $f(x+1) + f(x-1) = \sqrt{2} f(x)$, then period of $f(x)$ can be

(a) 2 (b) 4 (c) 6 (d) 8

37. If x and α are real, then the inequation $\log_2 x + \log_x 2 + 2 \cos \alpha \le 0$

(a) has no solution

(b) has exactly two solutions

(c) is satisfied for any real α and any real x in $(0, 1)$

(d) is satisfied for any real α and any real x in $(1, \infty)$

38. The range of values of 'a' such that $\left(\dfrac{1}{2}\right)^{|x|} = x^2 - a$ is satisfied for maximum number of values of 'x'

(a) $(-\infty, -1)$ (b) $(-\infty, \infty)$ (c) $(-1, 1)$ (d) $(-1, \infty)$

39. Let $f: R \to R$ be a function defined by $f(x) = \{|\cos x|\}$, where $\{x\}$ represents fractional part of x. Let S be the set containing all real values x lying in the interval $[0, 2\pi]$ for which $f(x) \ne |\cos x|$. The number of elements in the set S is

(a) 0 (b) 1 (c) 3 (d) infinite

40. The domain of the function $f(x) = \sqrt{\log_{\sin x + \cos x}(|\cos x| + \cos x)}, 0 \le x \le \pi$ is

(a) $(0, \pi)$ (b) $\left(0, \dfrac{\pi}{2}\right)$

(c) $\left(0, \dfrac{\pi}{3}\right)$ (d) None of these

41. If $f(x) = (x^2 + 2\alpha x + \alpha^2 - 1)^{1/4}$ has its domain and range such that their union is set of real numbers, then α satisfies

(a) $-1 < \alpha < 1$ (b) $\alpha \le -1$

(c) $\alpha \ge 1$ (d) $\alpha \le 1$

42. Let $f: (e, \infty) \to R$ be a function defined by $f(x) = \log(\log(\log x))$, the base of the logarithm being e. Then,

(a) f is one-one and onto

(b) f is one-one but not onto

(c) f is onto but not one-one

(d) the range of f is equal to its domain

43. The expression $x^2 - 4px + q^2 > 0$ for all real x and also $r^2 + p^2 < qr$, the range of $f(x) = \dfrac{x + r}{x^2 + qx + p^2}$ is

(a) $\left[\dfrac{p}{2r}, \dfrac{q}{2r}\right]$ (b) $(0, \infty)$

(c) $(-\infty, 0)$ (d) $(-\infty, \infty)$

44. Let $f(x) = \dfrac{x^4 - \lambda x^3 - 3x^2 + 3\lambda x}{x - \lambda}$. If range of $f(x)$ is the set of entire real numbers, the true set in which λ lies is

(a) $[-2, 2]$ (b) $[0, 4]$

(c) $(1, 3)$ (d) None of these

45. Let $a = 3^{1/224} + 1$ and for all $n \geq 3$,
let $f(n) = {}^nC_0\, a^{n-1} - {}^nC_1 a^{n-2} + {}^nC_2 a^{n-3} + \ldots + (-1)^{n-1} \cdot {}^nC_{n-1} \cdot a^0$.
If the value of $f(2016) + f(2017) = 3^K$, the value of K is
(a) 6 (b) 8
(c) 9 (d) 10

46. The area bounded by $f(x) = \sin^{-1}(\sin x)$ and
$g(x) = \frac{\pi}{2} - \sqrt{\frac{\pi^2}{2} - \left(x - \frac{\pi}{2}\right)^2}$ is
(a) $\frac{\pi^3}{8}$ sq units (b) $\frac{\pi^2}{8}$ sq units
(c) $\frac{\pi^3}{2}$ sq units (d) $\frac{\pi^2}{2}$ sq units

47. If $f : R \to R$, $f(x) = \frac{x^2 + bx + 1}{x^2 + 2x + b}$, $(b > 1)$ and $f(x)$, $\frac{1}{f(x)}$ have the same bounded set as their range, the value of b is
(a) $2\sqrt{3} - 2$ (b) $2\sqrt{3} + 2$
(c) $2\sqrt{2} - 2$ (d) $2\sqrt{2} + 2$

48. The period of $\sin \frac{\pi[x]}{12} + \cos \frac{\pi[x]}{4} + \tan \frac{\pi[x]}{3}$, where $[x]$ represents the greatest integer less than or equal to x is
(a) 12 (b) 4
(c) 3 (d) 24

49. If $f(2x + 3y, 2x - 7y) = 20x$, then $f(x, y)$ equals
(a) $7x - 3y$ (b) $7x + 3y$
(c) $3x - 7y$ (d) $x - y$

50. The range of the function $f(x) = \sqrt{x - 1} + 2\sqrt{3 - x}$ is
(a) $[\sqrt{2}, 2\sqrt{2}]$ (b) $[\sqrt{2}, \sqrt{10}]$
(c) $[2\sqrt{2}, \sqrt{10}]$ (d) $[1, 3]$

51. The domain of the function
$f(x) = \cos^{-1}(\sec(\cos^{-1} x)) + \sin^{-1}(\operatorname{cosec}(\sin^{-1} x))$ is
(a) $x \in R$ (b) $x = 1, -1$
(c) $-1 \leq x \leq 1$ (d) $x \in \phi$

52. Let $f(x)$ be a polynomial one-one function such that
$f(x)f(y) + 2 = f(x) + f(y) + f(xy)$, $\forall\ x, y \in R - \{0\}$,
$f(1) \neq 1, f'(1) = 3$.
Let $g(x) = \frac{x}{4}(f(x) + 3) - \int_0^x f(x)\, dx$, then
(a) $g(x) = 0$ has exactly one root for $x \in (0, 1)$
(b) $g(x) = 0$ has exactly two roots for $x \in (0, 1)$
(c) $g(x) \neq 0$, $\forall\ x \in R - \{0\}$
(d) $g(x) = 0$, $\forall\ x \in R - \{0\}$

53. Let $f(x)$ be a polynomial with real coefficients such that $f(x) = f'(x) \times f'''(x)$. If $f(x) = 0$ is satisfied $x = 1, 2, 3$ only, then the value of $f'(1)\, f'(2)\, f'(3)$ is
(a) positive (b) negative
(c) 0 (d) Inadequate data

54. Let $A = \{1, 2, 3, 4, 5\}$ and $f : A \to A$ be an into function such that $f(i) \neq i, \forall\ i \in A$, then number of such functions f are
(a) 1024 (b) 904
(c) 980 (d) None of these

55. If functions $f : \{1, 2, \ldots, n\} \to \{1995, 1996\}$ satisfying $f(1) + f(2) + \ldots + f(1996) =$ odd integer are formed, the number of such functions can be
(a) 2^n (b) $2^{n/2}$ (c) n^2 (d) 2^{n-1}

56. The range of $y = \sin^3 x - 6\sin^2 x + 11\sin x - 6$ is
(a) $[-24, 2]$ (b) $[-24, 0]$
(c) $[0, 24]$ (d) None of these

57. Let $f(x) = x^2 - 2x$ and $g(x) = f(f(x) - 1) + f(5 - f(x))$, then
(a) $g(x) < 0, \forall x \in R$
(b) $g(x) < 0$, for some $x \in R$
(c) $g(x) \geq 0$, for some $x \in R$
(d) $g(x) \geq 0$, $\forall x \in R$

58. If $f(x)$ and $g(x)$ are non-periodic functions, then $h(x) = f(g(x))$ is
(a) non-periodic
(b) periodic
(c) may be periodic
(d) always periodic, if domain of $h(x)$ is a proper subset of real numbers

59. If $f(x)$ is a real-valued function discontinuous at all integral points lying in $[0, n]$ and if $(f(x))^2 = 1$, $\forall\ x \in [0, n]$, then number of functions $f(x)$ are
(a) 2^{n+1} (b) 6×3^n (c) $2 \times 3^{n-1}$ (d) 3^{n+1}

60. A function f from integers to integers is defined as
$$f(x) = \begin{cases} n + 3, & n \in \text{odd} \\ n/2, & n \in \text{even} \end{cases}$$
Suppose $k \in$ odd and $f(f(f(k))) = 27$, then the sum of digits of k is
(a) 3 (b) 6 (c) 9 (d) 12

61. If $f : R \to R$ and $f(x) = \frac{\sin(\pi\{x\})}{x^4 + 3x^2 + 7}$, where $\{\}$ is a fractional part of x, then
(a) f is injective
(b) f is not one-one and non-constant
(c) f is a surjective
(d) f is a zero function

62. Let $f: R \rightarrow R$ and $g: R \rightarrow R$ be two one-one and onto functions, such that they are the mirror images of each other about the line $y = a$. If $h(x) = f(x) + g(x)$, then $h(x)$ is
(a) one-one and onto
(b) only one-one and not onto
(c) only onto but not one-one
(d) None of the above

63. Domain of the function $f(x)$, if $3^x + 3^{f(x)} =$ minimum of $\phi(t)$, where $\phi(t) = \min\{2t^3 - 15t^2 + 36t - 25, 2 + |\sin t|\}$ is
(a) $(-\infty, 1)$
(b) $(-\infty, \log_3 e)$
(c) $(0, \log_3 2)$
(d) $(-\infty, \log_3 2)$

64. Let x be the elements of the set $A = \{1,2,3,4,5,6,8,10,12,15,20,24,30,40,60,120\}$ and x_1, x_2, x_3 be positive integers and d be the number of integral solutions of $x_1 x_2 x_3 = x$, then d is
(a) 100 (b) 150
(c) 320 (d) 250

65. If $A > 0, c, d, u, v$ are non-zero constants and the graph of $f(x) = |Ax + c| + d$ and $g(x) = -|Ax + u| + v$ intersect exactly at two points $(1, 4)$ and $(3, 1)$, then the value of $\frac{u+c}{A}$ equals
(a) 4 (b) -4
(c) 2 (d) -2

66. If $f(x) = x^3 + 3x^2 + 4x + a\sin x + b\cos x, \forall x \in R$ is a one-one function, then the greatest value of $(a^2 + b^2)$ is
(a) 1 (b) 2
(c) $\sqrt{2}$ (d) None of these

67. If two roots of the equation $(p-1)(x^2 + x + 1)^2 - (p+1)(x^4 + x^2 + 1) = 0$ are real and distinct and $f(x) = \frac{1-x}{1+x}$, then $f(f(x)) + f\left(f\left(\frac{1}{x}\right)\right)$ is equal to
(a) p (b) $-p$
(c) $2p$ (d) $-2p$

68. Let $f(x) = x^{13} + 2x^{12} + 3x^{11} + \ldots + 13x + 14$ and $\alpha = \cos\frac{2\pi}{15} + i\sin\frac{2\pi}{15}$. If $N = f(\alpha)f(\alpha^2)\ldots f(\alpha^{14})$, then
(a) number of divisors of N is 144
(b) number of divisors of N is 196
(c) number of divisors of N which are perfect squares of 49
(d) number of divisors of N which are perfect square of 12

69. The sum of the maximum and minimum values of function $f(x) = \sin^{-1} 2x + \cos^{-1} 2x + \sec^{-1} 2x$ is
(a) π (b) $\frac{\pi}{2}$
(c) 2π (d) $\frac{3\pi}{2}$

70. The complete set of values of 'a' for which the function $f(x) = \tan^{-1}(x^2 - 18x + a) > 0, \forall x \in R$, is
(a) $(81, \infty)$ (b) $[81, \infty)$
(c) $(-\infty, 81)$ (d) $(-\infty, 81]$

71. The domain of the function $f(x) = \sin^{-1}\frac{1}{|x^2 - 1|} + \frac{1}{\sqrt{\sin^2 x + \sin x + 1}}$ is
(a) $(-\infty, \infty)$
(b) $(-\infty, -\sqrt{2}] \cup [\sqrt{2}, \infty)$
(c) $(-\infty, -\sqrt{2}] \cup [\sqrt{2}, \infty) \cup \{0\}$
(d) None of the above

72. The domain of $f(x) = \frac{\log(\sin^{-1}\sqrt{x^2 + x + 1})}{\log(x^2 - x + 1)}$ is
(a) $(-1, 1)$ (b) $(-1, 0) \cup (0, 1)$
(c) $(-1, 0) \cup \{1\}$ (d) None of these

73. The domain of $f(x) = \sqrt{\sin^{-1}(3x - 4x^3)} + \sqrt{\cos^{-1} x}$ is equal to
(a) $\left[-1, -\frac{\sqrt{3}}{2}\right] \cup \left[0, \frac{\sqrt{3}}{2}\right]$ (b) $\left[-1, -\frac{1}{2}\right] \cup \left[0, \frac{1}{2}\right]$
(c) $\left[0, \frac{1}{2}\right]$ (d) None of these

74. The domain of the function $f(x) = \sqrt[6]{4^x + 8^{2/3(x-2)} - 52 - 2^{2(x-1)}}$ is
(a) $(0, 1)$ (b) $[3, \infty)$
(c) $[1, 0)$ (d) None of these

75. The domain of derivative of the function $f(x) = |\sin^{-1}(2x^2 - 1)|$ is
(a) $(-1, 1)$ (b) $(-1, 1) \sim \left\{0, \pm\frac{1}{\sqrt{2}}\right\}$
(c) $(-1, 1) \sim \{0\}$ (d) $(-1, 1) \sim \left\{\pm\frac{1}{\sqrt{2}}\right\}$

76. The range of a function $f(x) = \tan^{-1}\{\log_{5/4}(5x^2 - 8x + 4)\}$ is
(a) $\left(\frac{-\pi}{4}, \frac{\pi}{2}\right)$ (b) $\left[\frac{-\pi}{4}, \frac{\pi}{2}\right)$
(c) $\left(\frac{-\pi}{4}, \frac{\pi}{2}\right]$ (d) $\left[\frac{-\pi}{4}, \frac{\pi}{2}\right]$

Functions Exercise 2 :
More Than One Option Correct Type Questions

77. Which of the following function(s) is/are transcendal?

(a) $f(x)=5\sin(\sqrt{x})$ (b) $f(x)=\dfrac{2\sin 3x}{x^2+2x-1}$

(c) $f(x)=\sqrt{x^2+2x+1}$ (d) $f(x)=(x^2+3)\cdot 2^x$

78. Let $f(x)=\dfrac{\sqrt{x-2\sqrt{x-1}}}{\sqrt{x-1}-1}\cdot x$, then

(a) domain of $f(x)$ is $x\geq 1$ (b) domain of $f(x)$ is $[1,\infty)-\{2\}$

(c) $f'(10)=1$ (d) $f'\left(\dfrac{3}{2}\right)=-1$

79. $f(x)=\cos^2 x+\cos^2\left(\dfrac{\pi}{3}+x\right)-\cos x\cdot\cos\left(x+\dfrac{\pi}{3}\right)$ is

(a) an odd function (b) an even function

(c) a periodic function (d) $f(0)=f(1)$

80. If the following functions are defined from $[-1,1]$ to $[-1,1]$, identify these which are into.

(a) $\sin(\sin^{-1}x)$ (b) $\dfrac{2}{\pi}\cdot\sin^{-1}(\sin x)$

(c) $\operatorname{sgn}(x)\cdot\log(e^x)$ (d) $x^3\operatorname{sgn}(x)$

81. Let $f(x)=\begin{cases}x^2-4x+3, & x<3\\ x-4, & x\geq 3\end{cases}$

and $g(x)=\begin{cases}x-3, & x<4\\ x^2+2x+2, & x\geq 4\end{cases}$, which one of the following is/are true?

(a) $(f+g)(3.5)=0$ (b) $f(g(3))=3$

(c) $f(g(2))=1$ (d) $(f-g)(4)=0$

82. If $f(x)=x^2-2ax+a(a+1)$, $f:[a,\infty)\to[a,\infty)$. If one of the solutions of the equation $f(x)=f^{-1}(x)$ is 5049, the other may be

(a) 5051 (b) 5048 (c) 5052 (d) 5050

83. The function g defined by $g(x)=\sin\alpha+\cos\alpha-1$; $\alpha=\sin^{-1}\sqrt{\{x\}}$, where {} denotes fractional part function, is

(a) an even function (b) periodic function

(c) odd function (d) neither even nor odd

84. The graph of $f:R\to R$ defined by $y=f(x)$ is symmetric with respect to $x=a$ and $x=b$. Which of the following is true?

(a) $f(2a-x)=f(x)$ (b) $f(2a+x)=f(-x)$

(c) $f(2b+x)=f(-x)$ (d) f is periodic

85. Let f be the continuous and differentiable function such that $f(x)=f(2-x)$, $\forall\, x\in R$ and $g(x)=f(1+x)$, then

(a) $g(x)$ is an odd function

(b) $g(x)$ is an even function

(c) $f(x)$ is symmetric about $x=1$

(d) None of the above

86. Let $f(x)=|x-1|+|x-2|+|x-3|+|x-4|$, then

(a) least value of $f(x)$ is 4

(b) least value is not attained at unique point

(c) the number of integral solution of $f(x)=4$ is 2

(d) the value of $\dfrac{f(\pi-1)+f(e)}{2f\left(\frac{12}{5}\right)}$ is 1

87. Let $A=\{1,2,3,4,5\}$, $B=\{1,2,3,4\}$ and $f:A\to B$ is a function, the

(a) number of onto functions, if $n(f(A))=4$ is 240

(b) number of onto functions, if $n(f(A))=3$ is 600

(c) number of onto functions, if $n(f(A))=2$ is 180

(d) number of onto functions, if $n(f(A))=1$ is 4

88. If $2f(x)+x\,f\left(\dfrac{1}{x}\right)-2f\left(\left|\sqrt{2}\sin\pi\left(x+\dfrac{1}{4}\right)\right|\right)$

$=4\cos^2\left(\dfrac{\pi x}{2}\right)+x\cos\left(\dfrac{\pi}{x}\right)$, $\forall\, x\in R-\{0\}$, which of the following statement(s) is/are true?

(a) $f(2)+f\left(\dfrac{1}{2}\right)=1$ (b) $f(2)+f(1)=0$

(c) $f(2)+f(1)=f\left(\dfrac{1}{2}\right)$ (d) $f(1)\cdot f\left(\dfrac{1}{2}\right)\cdot f(2)=1$

89. If $f(x)$ is a differentiable function satisfying the condition $f(100x)=x+f(100x-100)$, $\forall\, x\in R$ and $f(100)=1$, then $f(10^4)$ is

(a) 5049 (b) $\sum\limits_{r=1}^{100} r$ (c) $\sum\limits_{r=2}^{100} r$ (d) 5050

90. If $[x]$ denotes the greatest integer function then the extreme values of the function $f(x)=[1+\sin x]+[1+\sin 2x]+\ldots+[1+\sin nx]$, $n\in I^+$, $x\in(0,\pi)$ are

(a) $(n-1)$ (b) n (c) $(n+1)$ (d) $(n+2)$

91. Which of the following is/are periodic?

(a) $f(x)=\begin{cases}1, & \text{if } x \text{ is rational}\\ 0, & \text{if } x \text{ is irrational}\end{cases}$

(b) $f(x)=\begin{cases}x-[x], & 2n\leq x<2n+1\\ \dfrac{1}{2}, & 2n+1\leq x<2n+2\end{cases}$, where [·] denotes the greatest integer function

(c) $f(x)=(-1)^{\left[\frac{2x}{\pi}\right]}$, where [·] denotes the greatest integer function

(d) $f(x)=ax-[ax+a]+\tan\left(\dfrac{\pi x}{2}\right)$, where [·] denotes the greatest integer function

92. If $f(x)$ is a polynomial of degree n, such that $f(0)=0$, $f(1)=1/2, \dots, f(n)=\dfrac{n}{n+1}$, then the value of $f(n+1)$ is

(a) 1, when n is even (b) $\dfrac{n}{n+2}$, when n is odd

(c) 1, when n is odd (d) $\dfrac{n}{n+2}$, when n is even

93. Let $f: R \to R$ be a function defined by $f(x+1)=\dfrac{f(x)-5}{f(x)-3}, \forall x \in R$. Then, which of the following statement(s) is/are true?

(a) $f(2008)=f(2004)$ (b) $f(2006)=f(2010)$

(c) $f(2006)=f(2002)$ (d) $f(2006)=f(2018)$

94. Let $f(x)=1-x-x^3$. Then, the real values of x satisfying the inequality, $1-f(x)-f^3(x)>f(1-5x)$, are

(a) $(-2, 0)$ (b) $(0, 2)$

(c) $(2, \infty)$ (d) $(-\infty, -2)$

95. If a function satisfies $(x-y)f(x+y)-(x+y)f(x-y)=2(x^2y-y^3)$, $\forall x, y \in R$ and $f(1)=2$, then

(a) $f(x)$ must be polynomial function

(b) $f(3)=12$

(c) $f(0)=0$

(d) $f(x)$ may not be differentiable

96. If the fundamental period of function $f(x)=\sin x+\cos(\sqrt{4-a^2})x$ is 4π, then the value of a is/are

(a) $\dfrac{\sqrt{15}}{2}$ (b) $-\dfrac{\sqrt{15}}{2}$ (c) $\dfrac{\sqrt{7}}{2}$ (d) $-\dfrac{\sqrt{7}}{2}$

97. Let $f(x)$ be a real valued function such that $f(0)=\dfrac{1}{2}$ and $f(x+y)=f(x)f(a-y)+f(y)f(a-x)$, $\forall x, y \in R$, then for some real a,

(a) $f(x)$ is a periodic function

(b) $f(x)$ is a constant function

(c) $f(x)=\dfrac{1}{2}$ (d) $f(x)=\dfrac{\cos x}{2}$

98. If $f(g(x))$ is one-one function, then

(a) $g(x)$ must be one-one (b) $f(x)$ must be one-one

(c) $f(x)$ may not be one-one (d) $g(x)$ may not be one-one

99. Which of the following functions have their range equal to R (the set of real numbers)?

(a) $x \sin x$

(b) $\dfrac{[x]}{\tan 2x} \cdot x \in \left(-\dfrac{\pi}{4} \cdot \dfrac{\pi}{4}\right)-\{0\}$, where $[\cdot]$ denotes the greatest integer function

(c) $\dfrac{x}{\sin x}$

(d) $[x]+\sqrt{\{x\}}$, where $[\cdot]$ and $\{\cdot\}$, respectively denote the greatest integer and fractional part functions

100. Which of the following pairs of function are identical?

(a) $f(x)=e^{\ln \sec^{-1} x}$ and $g(x)=\sec^{-1} x$

(b) $f(x)=\tan(\tan^{-1} x)$ and $g(x)=\cot(\cot^{-1} x)$

(c) $f(x)=\text{sgn}(x)$ and $g(x)=\text{sgn}(\text{sgn}(x))$

(d) $f(x)=\cot^2 x \cdot \cos^2 x$ and $g(x)=\cot^2 x-\cos^2 x$

101. Let $f: R \to R$ defined by $f(x)=\cos^{-1}(-\{-x\})$, where $\{x\}$ denotes fractional part of x. Then, which of the following is/are correct?

(a) f is many one but not even function

(b) Range of f contains two prime numbers

(c) f is non-periodic

(d) Graph of f does not lie below X-axis

Functions Exercise 3 : Statements I and II Type Questions

■ Directions (Q. Nos. 102 to 112) For the following questions, choose the correct answers from the codes (a), (b), (c) and (d) defined as follows :

(a) Statement I is true, Statement II is also true; Statement II is the correct explanation of Statement I

(b) Statement I is true, Statement II is also true; Statement II is not the correct explanation of Statement I

(c) Statement I is true, Statement II is false

(d) Statement I is false, Statement II is true

102. Statement I The function $f(x)=x \sin x$ and $f'(x)=x\cos x+\sin x$ are both non-periodic.

Statement II The derivative of differentiable function (non-periodic) is non-periodic function.

103. Statement I The maximum value of $\sin\sqrt{2}\,x+\sin ax$ cannot be 2 (where a is positive rational number).

Statement II $\dfrac{\sqrt{2}}{a}$ is irrational.

104. Let $f: R \to R$ be a function such that $f(x)=\dfrac{e^{|x|}-e^{-x}}{e^x+e^{-x}}$.

Statement I $f(x)$ is into function.

Statement II $f(x)$ is many-one function and the many-one function is not onto.

105. Statement I The range of

$f(x)=\sin\left(\frac{\pi}{5}+x\right)-\sin\left(\frac{\pi}{5}-x\right)-\sin\left(\frac{2\pi}{5}+x\right)+\sin\left(\frac{2\pi}{5}-x\right)$ is $[-1, 1]$.

Statement II $\cos\frac{\pi}{5}-\cos\frac{2\pi}{5}=\frac{1}{2}$.

106. Statement I The period of $f(x)=2\cos\frac{1}{3}(x-\pi)+4\sin\frac{1}{3}(x-\pi)$ is 3π.

Statement II If T is the period of $f(x)$, then the period of $f(ax+b)$ is $\frac{T}{|a|}$.

107. f is a function defined on the interval $[-1, 1]$ such that $f(\sin 2x)=\sin x+\cos x$.

Statement I If $x\in\left[-\frac{\pi}{4},\frac{\pi}{4}\right]$, then $f(\tan^2 x)=\sec x$

Statement II $f(x)=\sqrt{1+x}, \forall\ x\in[-1, 1]$

108. Statement I The equation $f(x)=4x^5+20x-9=0$ has only one real root.

Statement II $f'(x)=20x^4+20=0$ has no real root.

109. Statement I The range of $\log\left(\frac{1}{1+x^2}\right)$ is $(-\infty,\infty)$.

Statement II When $0<x\le 1$, $\log x\in(-\infty, 0]$.

110. Let $f: X\to Y$ be a function defined by $f(x)=2\sin\left(x+\frac{\pi}{4}\right)-\sqrt{2}\cos x+c$.

Statement I For set X, $x\in\left[0,\frac{\pi}{2}\right]\cup\left[\pi,\frac{3\pi}{2}\right]$, $f(x)$ is one-one function.

Statement II $f'(x)\ge 0, x\in\left[0,\frac{\pi}{2}\right]$

111. Let $f(x)=\sin x$

Statement I f is not a polynomial function.

Statement II nth derivative of $f(x)$, w.r.t. x, is not a zero function for any positive integer n.

112. Statement I The function $f: R\to R$, given $f(x)=\log_a(x+\sqrt{x^2+1}), a>0, a\ne 1$ is invertible.

Statement II f is many-one and into.

Functions Exercise 4 : Passage Based Questions

Passage I (Q. Nos. 113 to 115)

Let $f: R\to R$ be a continuous function such that $f(x)-2f\left(\frac{x}{2}\right)+f\left(\frac{x}{4}\right)=x^2$.

113. $f(3)$ is equal to

(a) $f(0)$ (b) $4+f(0)$ (c) $9+f(0)$ (d) $16+f(0)$

114. The equation $f(x)-x-f(0)=0$ have exactly

(a) no solution (b) one solution
(c) two solutions (d) infinite solutions

115. $f'(0)$ is equal to

(a) 0 (b) 1 (c) $f(0)$ (d) $-f(0)$

Passage II (Q. Nos. 116 to 117)

Consider the equation $x+y-[x][y]=0$, where $[\cdot]$ is the greatest integer function.

116. The number of integral solutions to the equation is

(a) 0 (b) 1
(c) 2 (d) None of these

117. Equation of one of the lines on which the non-integral solution of given equation lies, is

(a) $x+y=-1$ (b) $x+y=0$
(c) $x+y=1$ (d) $x+y=5$

Passage III (Q. Nos. 118 to 120)

Let $f(x)=\frac{1}{2}\left[f(xy)+f\left(\frac{x}{y}\right)\right]$ for $x, y\in R^+$ such that $f(1)=0$; $f'(1)=2$.

118. $f(x)-f(y)$ is equal to

(a) $f\left(\frac{y}{x}\right)$ (b) $f\left(\frac{x}{y}\right)$ (c) $f(2x)$ (d) $f(2y)$

119. $f'(3)$ is equal to

(a) $\frac{1}{3}$ (b) $\frac{2}{3}$ (c) $\frac{1}{2}$ (d) $\frac{1}{4}$

120. $f(e)$ is equal to

(a) 2 (b) 1 (c) 3 (d) 4

Passage IV (Q. Nos. 121 to 123)

If $f: R\to R$ and $f(x)=g(x)+h(x)$, where $g(x)$ is a polynomial and $h(x)$ is a continuous and differentiable bounded function on both sides, then $f(x)$ is one-one, we need to differentiate $f(x)$. If $f'(x)$ changes sign in domain of f, then f, if many-one else one-one.

121. If $f: R\to R$ and $f(x)=a_1x+a_3x^3+a_5x^5+\ldots+a_{2n+1}x^{2n+1}-\cot^{-1}x$ where $0<a_1<a_3<\ldots<a_{2n+1}$, then the function $f(x)$ is

(a) one-one into (b) many-one onto
(c) one-one onto (d) many-one into

122. If $f: R \rightarrow R$ and $f(x)=\dfrac{x(x^4+1)(x+1)+x^4+2}{x^2+x+1}$, then $f(x)$ is
(a) one-one into (b) many-one onto
(c) one-one onto (d) many-one into

123. If $f: R \rightarrow R$ and $f(x)=2ax+\sin 2x$, then the set of values of a for which $f(x)$ is one-one and onto is
(a) $a \in \left(-\frac{1}{2}, \frac{1}{2}\right)$ (b) $a \in (-1, 1)$
(c) $a \in R-\left(-\frac{1}{2}, \frac{1}{2}\right)$ (d) $a \in R-(-1, 1)$

Passage V (Q. Nos. 124 to 126)

Let $g(x)=a_0+a_1x+a_2x^2+a_3x^3$ and $f(x)=\sqrt{g(x)}$, $f(x)$ has its non-zero local minimum and maximum values at -3 and 3, respectively. If $a_3 \in$ the domain of the function $h(x)=\sin^{-1}\left(\dfrac{1+x^2}{2x}\right)$.

124. The value of a_1+a_2 is
(a) 30 (b) -30 (c) 27 (d) -27

125. The value of a_0 is
(a) equal to 50 (b) greater than 54
(c) less than 54 (d) less than 50

126. $f(10)$ is defined for
(a) $a_0>830$ (b) $a_0<830$
(c) $a_0=830$ (d) None of these

Passage VI (Q. Nos. 127 to 129)

Let $f:[2, \infty) \rightarrow [1, \infty)$ defined by $f(x)=2^{x^4-4x^2}$ and $g:\left[\frac{\pi}{2}, \pi\right] \rightarrow A$ defined by $g(x)=\dfrac{\sin x+4}{\sin x-2}$ be two invertible functions.

127. $f^{-1}(x)$ is equal to
(a) $\sqrt{2+\sqrt{4-\log_2 x}}$ (b) $\sqrt{2+\sqrt{4+\log_2 x}}$
(c) $\sqrt{2-\sqrt{4+\log_2 x}}$ (d) None of these

128. The set A is equal to
(a) $[-5,-2]$ (b) $[2,5]$ (c) $[-5,2]$ (d) $[-3,-2]$

129. The domain of $f^{-1}g^{-1}(x)$ is
(a) $[-5, \sin 1]$ (b) $\left[-5, \dfrac{\sin 1}{2-\sin 1}\right]$
(c) $\left[-5, -\dfrac{(4+\sin 1)}{2-\sin 1}\right]$ (d) $\left[-\dfrac{(4+\sin 1)}{2-\sin 1}, -2\right]$

Passage VII (Q. Nos. 130 to 132)

Let $P(x)$ be polynomial of degree atmost 5 which leaves remainders -1 and 1 upon division by $(x-1)^3$ and $(x+1)^3$, respectively.

130. Numbers of real roots of $P(x)=0$ is
(a) 1 (b) 3 (c) 5 (d) 2

131. The maximum value of $y=P''(x)$ can be obtained at x is equal to
(a) $-\frac{1}{\sqrt{3}}$ (b) 0 (c) $\frac{1}{\sqrt{3}}$ (d) 1

132. The sum of pairwise product of all roots (real and complex) of $P(x)=0$ is
(a) $-\frac{5}{3}$ (b) $-\frac{10}{3}$ (c) 2 (d) -5

Passage VIII (Q. Nos. 133 to 135)

Consider $\alpha>1$ and $f:\left[\frac{1}{\alpha}, \alpha\right] \rightarrow \left[\frac{1}{\alpha}, \alpha\right]$ be bijective function. Suppose that $f^{-1}(x)=\dfrac{1}{f(x)}$; for all $x \in \left[\frac{1}{\alpha}, \alpha\right]$.

133. $f(1)$ is equal to
(a) 1 (b) 0
(c) -1 (d) does'nt attain a unique value

134. Which of the following statements can be concluded about $(f(x))$?
(a) $f(x)$ is discontinuous in $\left[\frac{1}{\alpha}, \alpha\right]$
(b) $f(x)$ is increasing in $\left[\frac{1}{\alpha}, \alpha\right]$
(c) $f(x)$ is decreasing in $\left[\frac{1}{\alpha}, \alpha\right]$
(d) None of the above

135. Which of the following statements can be concluded about $f(f(x))$?
(a) $f(f(x))$ is continuous in $\left[\frac{1}{\alpha}, \alpha\right]$
(b) $f(f(x))$ is increasing in $\left[\frac{1}{\alpha}, \alpha\right]$
(c) $f(f(x))$ is decreasing in $\left[\frac{1}{\alpha}, \alpha\right]$
(d) None of the above

Passage IX (Q. Nos. 136 to 137)

Let f be a real valued function from N to N satisfying. The relation $f(m+n)=f(m)+f(n)$ for all $m, n \in N$.

136. The range of f contains all the even numbers, the value of $f(1)$ is
(a) 1 (b) 2 (c) 1 or 2 (d) 4

137. If domain of f is first $3m$ natural numbers and if the number of elements common in domain and range is m, then the value of $f(1)$ is
(a) 2 (b) 3
(c) 6 (d) Can't say

Functions Exercise 5 : Matching Type Questions

138. Match the statements of Column I with values of Column II.

Column I	Column II
(A) $\sqrt{\sin(\cos x)}$ has domain	(p) $x \in R$
(B) $(\sqrt{\cos(\sin x)})^{-1}$ has domain	(q) $R - \left\{n\pi \pm \frac{\pi}{6}\right\}$
(C) $\tan(\pi \sin x)$ has domain	(r) $x \in \left(n\pi, n\pi + \frac{\pi}{2}\right)$
(D) $\ln(\tan x)$ has domain	(s) $x \in \left[2n\pi - \frac{\pi}{2}, 2n\pi + \frac{\pi}{2}\right]$

139. Match the statements of Column I with values of Column II.

Column I	Column II
(A) $\lvert 4\sin x - 1 \rvert < \sqrt{5}, x \in [0, \pi]$, the domain is	(p) $\left[0, \frac{\pi}{4}\right] \cup \left[\frac{3\pi}{4}, \pi\right]$
(B) $4\sin^2 x - 8 \sin x + 3 \le 0, [0, 2\pi]$, the domain is	(q) $\left[\frac{3\pi}{2}, 2\pi\right] \cup \{0\}$
(C) $\lvert \tan x \rvert \le 1$ and $x \in [0, \pi]$, the domain is	(r) $\left[0, \frac{3\pi}{10}\right)$
(D) $\cos x - \sin x \ge 1$ and $[0, 2\pi]$, the domain is	(s) $\left[\frac{\pi}{6}, \frac{5\pi}{6}\right]$

140. Match the statements of Column I with values of Column II.

Column I	Column II
(A) If $f(x) = \begin{cases} x+1, & \text{when } x < 0 \\ x^2 - 1, & \text{when } x \ge 0 \end{cases}$, the $fof(x)$ for $-1 \le x < 0$ is	(p) $\frac{x-3}{2}$
(B) If $f\left(\frac{2\tan x}{1+\tan^2 x}\right) = \frac{(\cos 2x + 1)(\sec^2 x + 2\tan x)}{2}$, then $f(x)$ is	(q) $x^2 + 2x$
(C) If $f(x+y+1) = (\sqrt{f(x)} + \sqrt{f(y)})^2$ for all $x, y \in R$ and $f(0) = 1$, then $f(x)$ is	(r) $1 + x$
(D) If $4 < x < 5$ and $f(x) = \left[\frac{x}{4}\right] + 2x + 2$, where $[y]$ is the greatest integer $\le y$, then $f^{-1}(x)$ is	(s) $(x+1)^2$

141. Match the statements of Column I with values of Column II.

Column I	Column II
(A) $f(x) = \sin^2 x + \sin^2\left(x + \frac{\pi}{3}\right) + \cos x \cos\left(x + \frac{\pi}{3}\right)$ then $f(x)$ is	(p) Defined for all real 'x'
(B) $g(x) = \tan(e^{\{x\}}) + [x + \alpha] - 5 - x$, where $[\cdot]$ denotes the greatest integer less than or equal to x, then $g(x)$ is	(q) Even function
(C) $h(x) = \frac{x}{5^x - 1} + \frac{x}{2} + 5$, then $h(x)$ is	(r) Odd function
(D) $k(x) = 2\sin^2 x - \cos 2\alpha + 4\sin\alpha \cdot \sin x \cos(x+\alpha) + \cos 2(x+\alpha), \alpha \in R$, then $k(x)$ is	(s) Periodic function

Functions Exercise 6 : Single Integer Answer Type Questions

142. A function $f : R \to R$ is defined by $f(x+y) - k\,xy = f(x) + 2y^2, \forall\, x, y \in R$ and $f(1) = 2$; $f(2) = 8$, where k is some real constant, then $f(x+y) \cdot f\left(\frac{1}{x+y}\right)$ is ....

143. If $f : R \to R$ satisfying $f(x - f(y)) = f(f(y)) + x\,f(y) + f(x) - 1$, for all $x, y \in R$, then $\frac{-f(10)}{7}$ is ....

144. Let $f : N \to R$ be such that $f(1) = 1$ and $f(1) + 2f(2) + 3f(3) + \ldots + nf(n) = n(n+1)f(n)$, for $n \ge 2$, then $\frac{1}{2010\, f(2010)}$ is ....

145. If $f(x) = \frac{2010x + 163}{165x - 2010}$, $x > 0$ and $x \ne \frac{2010}{165}$, the least value of $f(f(x)) + f\left(f\left(\frac{4}{x}\right)\right)$ is ....

146. If $\alpha, \beta, \gamma \in R$; $\alpha+\beta+\gamma=4$ and $\alpha^2+\beta^2+\gamma^2=6$, the number of integers lie in the exhaustive range of α is ....

147. The number of linear functions satisfying $f(x+f(x))=x+f(x), \forall\ x$ is ....

148. If $A=\{1, 2, 3\}$, $B=\{1, 3, 5, 7, 9\}$, the ratio of number of one-one functions to the number of strictly monotonic functions is ....

149. If $n(A)=4$, $n(B)=5$ and number of functions from A to B such that range contains exactly 3 elements is k, $\frac{k}{60}$ is ....

150. If a and b are constants, such that $f(x)=a\sin x+bx\cos x+2x^2$ and $f(2)=15$, $f(-2)$ is ....

151. If the functions $f(x)=x^5+e^{x/3}$ and $g(x)=f^{-1}(x)$, the value of $g'(1)$ is ....

152. If $f(x)=x^3-12x+p$; $p\in\{1, 2, 3, \ldots, 15\}$ and for each 'p', the number of real roots of equation $f(x)=0$ is denoted by θ, the $\frac{1}{5}\Sigma\theta$ is equal to ....

153. Let $f(x)$ denotes the number of zeroes in $f'(x)$. If $f(m)-f(n)=3$, the value of $\frac{(m-n)_{\max}-(m-n)_{\min}}{2}$ is ....

154. If $x^2+y^2=4$, the maximum value of $\left(\frac{x^3+y^3}{x+y}\right)$ is

155. Let $f(n)$ denotes the square of the sum of the digits of natural number n, where $f^2(n)$ denotes $f(f(n))$, $f^3(n)$ denotes $f(f(f(n)))$ and so on. The value of $\frac{f^{2011}(2011)-f^{2010}(2011)}{f^{2013}(2011)-f^{2012}(2011)}$ is ....

156. If $[\sin x]+\left[\frac{x}{2\pi}\right]+\left[\frac{2x}{5\pi}\right]=\frac{9x}{10\pi}$, where $[\cdot]$ denotes the greatest integer function, the number of solutions in the interval $(30, 40)$ is ....

157. The number of integral solutions of $\frac{1}{x}+\frac{1}{y}=\frac{1}{6}$ with $x\le y$ is 'α'. The value of '$\alpha-6$' is ....

158. If $f(x)$ is a polynomial of degree 4 with leading coefficient '1' satisfying $f(1)=10$, $f(2)=20$ and $f(3)=30$, then $\left(\frac{f(12)+f(-8)}{19840}\right)$ is ....

159. If $a+b=3-\cos 4\theta$ and $a-b=4\sin 2\theta$, then ab is always less than or equal to ....

160. Let 'n' be the number of elements in the domain set of the function $f(x)=\left|\ln\sqrt{{}^{x^2+4x}C_{2x^2+3}}\right|$ and 'Y' be the global maximum value of $f(x)$, then $[n+[Y]]$ is .... (where $[\cdot]$ = greatest integer function).

161. If $f(x)$ is a function such that $f(x-1)+f(x+1)=\sqrt{3}f(x)$ and $f(5)=10$, then the sum of digits of the value of $\sum_{r=0}^{19} f(5+12r)$ is ....

162. If $2f(x)=f(xy)+f\left(\frac{x}{y}\right)$ for all positive values of x and y, $f(1)=0$ and $f'(1)=1$, then $f(e)$ is ....

163. Let f be a function from the set of positive integers to the set of real number such that $f(1)=1$ and $\sum_{r=1}^{n} rf(r)=n(n+1)f(n)$, $\forall\ n\ge 2$, the value of 2126 $f(1063)$ is ....

164. If $f(x)=\frac{x^4+x^2+1}{x^2-x+1}$, the value of $f(\omega^n)$ (where 'ω' is the non-real root of the equation $z^3=1$ and 'n' is a multiple of 3), is ....

165. If $f^2(x)\cdot f\left(\frac{1-x}{1+x}\right)=x^3$, $[x\ne -1, 1$ and $f(x)\ne 0]$, the value of $|[f(-2)]|$ (where $[\cdot]$ is the greatest integer function), is ....

166. An odd function is symmetric about the vertical line $x=a\,(a>0)$ and if $\sum_{r=0}^{\infty}[f(1+4r)]^r=8$, find the numerical value of $8f(1)$.

167. Let $\frac{e^x-e^{-x}}{e^x+e^{-x}}=\ln\sqrt{\frac{1+x}{1-x}}$, then find x.

168. If the maximum value of $f(x)=\frac{3x^2+9x+17}{3x^2+9x+7}$ is $5k+1$, the value of k is ....

169. The period of the function $f(x)$ which satisfies the relation $f(x)+f(x+4)=f(x+2)+f(x+6)$ is ....

170. If a non-zero function $f(x)$ is symmetrical about $y=x$, then the value of p (constant) such that $f^2(x)=(f^{-1}(x))^2-px\cdot f(x)\cdot f^{-1}(x)+2x^2f(x)$ for all $x\in R^+$ is ....

171. Let $f:R\to R$ and $f(x)=\dfrac{3x^2+mx+n}{x^2+1}$. If the range of this function is $[-4,3]$, then the value of $\dfrac{m^2+n^2}{4}$ is....

172. Let $f(x)$ be a monotonic polynomial of $(2m-1)$ degree where $m\in N$, then the equation $f(x)+f(3x)+f(5x)+...+f(2m-1)x=(2m-1)$ has .... roots.

Functions Exercise 7 : Subjective Type Questions

173. Let x be a real number. $[x]$ denotes the greatest integer function, $\{x\}$ denotes the fractional part and (x) denotes the least integer function, then solve the following :

(i) $(x)^2=[x]^2+2x$

(ii) $[2x]-2x=[x+1]$

(iii) $[x^2]+2[x]=3x, 0\le x\le 2$

(iv) $y=4-[x]^2$ and $[y]+y=6$

(v) $[x]+|x-2|\le 0$ and $-1\le x\le 3$

174. Let n be a positive integer and define $f(n)=1!+2!+3!+...+n!$. Find polynomials $P(x)$ and $Q(x)$ such that $f(n+2)=Q(n)f(n)+P(n)f(n+1)$ for all $n\ge 1$.

175. If $f(x)=\dfrac{a^x}{a^x+\sqrt{a}}\,(a>0)$, evaluate $\sum_{r=1}^{2n-1}2f\left(\dfrac{r}{2n}\right)$.

176. Find the domain of the function
$$f(x)=\log\left\{\log_{|\sin x|}(x^2-8x+23)-\frac{3}{\log_2|\sin x|}\right\}$$

177. Let $S(n)$ denotes the number of ordered pairs (x,y) satisfying $\dfrac{1}{x}+\dfrac{1}{y}=\dfrac{1}{n}$, where $n>1$ and $x,y,n\in N$.

(i) Find the value of $S(6)$.

(ii) Show that, if n is prime, then $S(n)=3$, always.

178. Solve $\dfrac{1}{[x]}+\dfrac{1}{[2x]}=\{x\}+\dfrac{1}{3}$, where $[\cdot]$ denotes the greatest integer function and $\{\}$ denotes fractional part of x.

179. Let $f(x)=x^2+3x-3, x\ge 0$. n points $x_1, x_2, \ldots, x_n$ are so chosen on the X-axis that

(i) $\dfrac{1}{n}\sum_{i=1}^{n}f^{-1}(x_i)=f\left(\dfrac{1}{n}\sum_{i=1}^{n}x_i\right)$

(ii) $\sum_{i=1}^{n}f^{-1}(x_i)=\sum_{i=1}^{n}x_i$, where f^{-1} denotes the inverse of f. Find the AM of x_i's.

180. Let $f(x)=x^2-2x, x\in R$ and $g(x)=f(f(x)-1)+f(5-f(x))$, show that $g(x)\ge 0$, $\forall\, x\in R$.

181. If f is a polynomial function satisfying $2+f(x)\cdot f(y)=f(x)+f(y)+f(xy), \forall\; x,y\in R$ and if $f(2)=5$, find $f(f(2))$.

182. If $a+b+c=abc$, a, b and $c\in R^+$, prove that $a+b+c\ge 3\sqrt{3}$.

183. Consider the function $f(x)=\begin{cases}x-[x]-\dfrac{1}{2}, & \text{if } x\notin I\\ 0, & \text{if } x\in I\end{cases}$, where $[\cdot]$ denotes the greatest integral function and I is the set of integers. Find $g(x)=\max\{x^2, f(x), |x|\}$; $-2\le x\le 2$.

184. If $f(x)$ is continuous function in $[0,2\pi]$ and $f(0)=f(2\pi)$, then prove that there exists a point $c\in(0,\pi)$ such that $f(c)=f(c+\pi)$.

185. Let $g(t)=|t-1|-|t|+|t+1|, \forall\, t\in R$. Find $f(x)=\max\{g(t): -\dfrac{3}{2}\le t\le x\}, \forall\; x\in\left(\dfrac{-3}{2},\infty\right)$

186. Find the integral solution for $n_1n_2=2n_1-n_2$, where $n_1, n_2\in$ integer.

Functions Exercise 8 : Questions Asked in Previous 10 Year's Exams

(i) JEE Advanced & IIT-JEE

187. Let $f_1 : R \to R, f_2 : [0, \infty) \to R, f_3 : R \to R$ and $f_4 : R \to [0, \infty)$ be defined by

[Match Type Question, 2014 Adv.]

$$f_1(x) = \begin{cases} |x|, & \text{if } x < 0 \\ e^x, & \text{if } x \geq 0 \end{cases}; f_2(x) = x^2; f_3(x) = \begin{cases} \sin x, & \text{if } x < 0 \\ x, & \text{if } x \geq 0 \end{cases}$$

and $f_4(x) = \begin{cases} f_2[f_1(x)], & \text{if } x < 0 \\ f_2[f_1(x)] - 1, & \text{if } x \geq 0 \end{cases}$

Column I	Column II
A. f_4 is	p. onto but not one-one
B. f_3 is	q. neither continuous nor one-one
C. $f_2 o f_1$, is	r. differentiable but not one-one
D. f_2 is	s. continuous and one-one

Codes

	A	B	C	D		A	B	C	D
(a)	r	p	s	q	(b)	p	r	s	q
(c)	r	p	q	s	(d)	p	r	q	s

188. If function $f(x) = x^2 + e^{x/2}$ and $g(x) = f^{-1}(x)$, then the value of $g'(1)$ is **[Integer Type Question, 2009]**

189. For the following questions, choose the correct answer from the codes (a), (b), (c) and (d) defined as follows.

(a) Statement I is true, Statement II is also true; Statement II is the correct explanation of Statement I.

(b) Statement I is true, Statement II is also true; Statement II is not the correct explanation of Statement I.

(c) Statement I is true; Statement II is false.

(d) Statement I is false; Statement II is true.

Let $F(x)$ be an indefinite integral of $\sin^2 x$.

Statement I The function $F(x)$ satisfies $F(x + \pi) = F(x)$ for all real x. Because

Statement II $\sin^2(x + \pi) = \sin^2 x$, for all real x.

[Assertion and Reason Type Question, 2007]

190. Match the conditons/expressions in Column I with statement in Column II.

Let $f(x) = \dfrac{x^2 - 6x + 5}{x^2 - 5x + 6}$. **[Match Type Question, 2007]**

Column I	Column II
A. If $-1 < x < 1$, then $f(x)$ satisfies	p. $0 < f(x) < 1$
B. If $1 < x < 2$, then $f(x)$ satisfies	q. $f(x) < 0$
C. If $3 < x < 5$, then $f(x)$ satisfies	r. $f(x) > 0$
D. If $x > 5$, then $f(x)$ satisfies	s. $f(x) < 1$

191. Find the range of values of t for which

$$2 \sin t = \frac{1 - 2x + 5x^2}{3x^2 - 2x - 1}, \; t \in \left[-\frac{\pi}{2}, \frac{\pi}{2}\right].$$

[Subjective Type Question, 2005]

(ii) JEE Main & AIEEE

192. If $f_k(x) = \frac{1}{k}(\sin^k x + \cos^k x)$, where $x \in R$ and $k \geq 1$, then $f_4(x) - f_6(x)$ is equal to **[2014 JEE Main]**

(a) 1/6 (b) 1/3 (c) 1/4 (d) 1/12

193. The function $f : [0, 3] \to [1, 29]$, defined by

$$f(x) = 2x^3 - 15x^2 + 36x + 1, \text{ is}$$ **[2012 AIEEE]**

(a) one-one and onto

(b) onto but not one-one

(c) one-one but not onto

(d) neither one-one nor onto

194. Let $f(x) = x^2$ and $g(x) = \sin x$ for all $x \in R$. Then, the set of all x satisfying $(fogogof)(x) = (gogof)(x)$, where $(fog)(x) = f(g(x))$, is **[2011 AIEEE]**

(a) $\pm \sqrt{n\pi}, n \in \{0, 1, 2, \ldots\}$

(b) $\pm \sqrt{n\pi}, n \in \{1, 2, \ldots\}$

(c) $\pi/2 + 2n\pi, n \in \{\ldots, -2, -1, 0, 1, 2, \ldots\}$

(d) $2n\pi, n \in \{\ldots, -2, -1, 0, 1, 2, \ldots\}$

195. Let $f : (0, 1) \to R$ be defined by $f(x) = \dfrac{b - x}{1 - bx}$, where b is a constant such that $0 < b < 1$. Then, **[2011 AIEEE]**

(a) f is not invertible on (0, 1)

(b) $f \neq f^{-1}$ on (0, 1) and $f'(b) = \dfrac{1}{f'(0)}$

(c) $f = f^{-1}$ on (0, 1) and $f'(b) = \dfrac{1}{f'(0)}$

(d) f^{-1} is differentiable on (0, 1)

196. Let f be a real-valued function defined on the interval $(-1, 1)$ such that $e^{-x} f(x) = 2 + \int_0^x \sqrt{t^4 + 1} \cdot dt, \forall x \in (-1, 1)$ and let f^{-1} be the inverse function of f. Then, $[f^{-1}(2)]'$ is equal to **[2010 AIEEE]**

(a) 1 (b) 1/3 (c) 1/2 (d) $1/e$

197. If X and Y are two non-empty sets, where $f : X \to Y$, is function defined such that

$f(C) = \{f(x) : x \in C\}$ for $C \subseteq X$ and

$f^{-1}(D) = \{x : f(x) \in D\}$ for $D \subseteq Y$,

for any $A \subseteq Y$ and $B \subseteq Y$, then **[2005 AIEEE]**

(a) $f^{-1}\{f(A)\} = A$

(b) $f^{-1}\{f(A)\} = A$, only if $f(X) = Y$

(c) $f\{f^{-1}(B)\} = B$, only if $B \subseteq f(x)$

(d) $f\{f^{-1}(B)\} = B$

198. $f(x) = \begin{cases} x, & \text{if } x \text{ is rational} \\ 0, & \text{if } x \text{ is irrational} \end{cases}, g(x) = \begin{cases} 0, & \text{if } x \text{ is rational} \\ x, & \text{if } x \text{ is irrational.} \end{cases}$

Then, $f - g$ is **[2005 AIEEE]**

(a) one-one and into (b) neither one-one nor onto

(c) many one and onto (d) one-one and onto

199. If $f(x) = \sin x + \cos x$, $g(x) = x^2 - 1$, then $g\{f(x)\}$ is invertible in the domain **[2004 AIEEE]**

(a) $\left[0, \frac{\pi}{2}\right]$ (b) $\left[-\frac{\pi}{4}, \frac{\pi}{4}\right]$ (c) $\left[-\frac{\pi}{2}, \frac{\pi}{2}\right]$ (d) $[0, \pi]$

200. Domain of definition of the function

$f(x) = \sqrt{\sin^{-1}(2x) + \frac{\pi}{6}}$ for real valued x, is **[2003 AIEEE]**

(a) $\left[-\frac{1}{4}, \frac{1}{2}\right]$ (b) $\left[-\frac{1}{2}, \frac{1}{2}\right]$ (c) $\left(-\frac{1}{2}, \frac{1}{9}\right)$ (d) $\left[-\frac{1}{4}, \frac{1}{4}\right]$

201. Range of the function $f(x) = \frac{x^2 + x + 2}{x^2 + x + 1}; x \in R$ is **[2003 AIEEE]**

(a) $(1, \infty)$ (b) $(1, 11/7)$

(c) $(1, 7/3]$ (d) $(1, 7/5)$

202. If $f : [0, \infty) \to [0, \infty)$ and $f(x) = \frac{x}{1 + x}$, then f is **[2003 AIEEE]**

(a) one-one and onto (b) one-one but not onto

(c) onto but not one-one (d) neither one-one nor onto

203. Let function $f : R \to R$ be defined by $f(x) = 2x + \sin x$ for $x \in R$. Then, f is **[2002 AIEEE]**

(a) one-to-one and onto

(b) one-to-one but not onto

(c) onto but not one-to-one

(d) neither one-to-one nor onto

204. Let $E = \{1, 2, 3, 4\}$ and $F = \{1, 2\}$. Then, the number of onto functions from E to F is **[2001 AIEEE]**

(a) 14 (b) 16 (c) 12 (d) 8

205. Suppose $f(x) = (x + 1)^2$ for $x \geq -1$. If $g(x)$ is the function whose graph is reflection of the graph of $f(x)$ with respect to the line $y = x$, then $g(x)$ equals **[2002 AIEEE]**

(a) $-\sqrt{x} - 1, x \geq 0$ (b) $\frac{1}{(x + 1)^2}, x > -1$

(c) $\sqrt{x + 1}, x \geq -1$ (d) $\sqrt{x} - 1, x \geq 0$

206. If $f : [1, \infty) \to [2, \infty)$ is given by $f(x) = x + \frac{1}{x}$, then $f^{-1}(x)$ equals **[2001 AIEEE]**

(a) $\frac{x + \sqrt{x^2 - 4}}{2}$ (b) $\frac{x}{1 + x^2}$

(c) $\frac{x - \sqrt{x^2 - 4}}{2}$ (d) $1 + \sqrt{x^2 - 4}$

207. Let $f(x) = (1 + b^2) x^2 + 2bx + 1$ and let $m(b)$ be the minimum value of $f(x)$. As b varies, the range of $m(b)$ is **[2001 AIEEE]**

(a) $[0, 1]$ (b) $\left[0, \frac{1}{2}\right]$

(c) $\left[\frac{1}{2}, 1\right]$ (d) $(0, 1]$

208. The domain of definition of $f(x) = \frac{\log_2(x + 3)}{(x^2 + 3x + 2)}$ is **[2001 AIEEE]**

(a) $R / \{-1, -2\}$ (b) $(-2, \infty)$

(c) $R / \{-1, -2, -3\}$ (d) $(-3, \infty) / \{-1, -2\}$

209. Let $f(x) = \frac{\alpha x}{x + 1}, x \neq -1$. Then, for what value of α is $f[f(x)] = x$? **[2001 AIEEE]**

(a) $\sqrt{2}$ (b) $-\sqrt{2}$

(c) 1 (d) -1

210. Let $g(x) = 1 + x - [x]$ and $f(x) = \begin{cases} -1, & x < 0 \\ 0, & x = 0 \\ 1, & x > 0 \end{cases}$, then for all x, $f[g(x)]$ is equal to **[2001 AIEEE]**

(a) x (b) 1

(c) $f(x)$ (d) $g(x)$

211. The domain of definition of the function $y(x)$ is given by the equation $2^x + 2^y = 2$, is **[2000 AIEEE]**

(a) $0 < x \leq 1$

(b) $0 \leq x \leq 1$

(c) $-\infty < x \leq 0$

(d) $-\infty < x < 1$

212. Let $f(\theta) = \sin\theta (\sin\theta + \sin 3\theta)$. Then, $f(\theta)$ **[2000 AIEEE]**

(a) ≥ 0, only when $\theta \geq 0$ (b) ≤ 0, for all real θ

(c) ≥ 0, for all real θ (d) ≤ 0, only when $\theta \leq 0$

Answers

Exercise for Session 1

1. (i) not a function (ii) not a function (iii) function
   (iv) function (v) not a function (vi) function
   (vii) function (viii) not a function (ix) function
   (x) not a function.
2. (i) not a function (ii) not a function (iii) not a function
   (iv) is a function (v) not a function
3. (a)
4. (a) $f: A \to B$

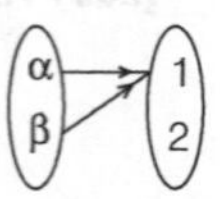

(b) $f: A \to B$

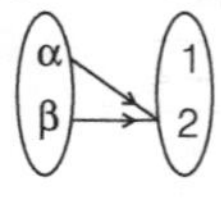

(c) $f: A \to B$

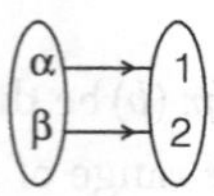

(d) $f: A \to B$

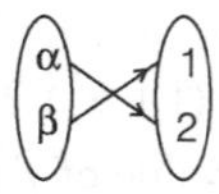

5. (b)

Exercise for Session 2

1. $(-\infty, 2] \cup [3, \infty)$
2. $(-\infty, -1/2] \cup (0, 1) \cup (2, \infty)$
3. $[-1, 1]$
4. $(-\infty, \infty)$
5. $\{2, 3\}$
6. $\{1, 2, 3\}$

Exercise for Session 3

1. $[4, 6]$
2. $(0, 1] \cup [4, 5)$
3. $[1-\sqrt{2}, 0) \cup [1+\sqrt{2}, 3)$
4. $x \in [2, 4]$
5. $(2, 3)$
6. $[-\pi/4, \pi/4]$
7. $[-2, -1/2] \cup [1/2, 2]$
8. $[-1, 3]$
9. $\left(0, \frac{1}{2}\right) \cup \left(\frac{1}{2}, 1\right)$
10. ϕ

Exercise for Session 4

1. $R - (-2, 2)$
2. $[-2, 2]$
3. $(0, 1) \cup (1, \infty)$
4. $[0, 3)$
5. $R - I$
6. $(-\infty, -2) \cup (4, \infty)$
7. R
8. $(2, \infty)$
9. $(-\infty, 0) \cup (0, 1] \cup [2, \infty)$
10. $(0, 1)$
11. $\left(-\sqrt{\frac{5}{2}}, -1\right] \cup \left[1, \sqrt{\frac{5}{2}}\right)$
12. $(-\sqrt{8}, -1] \cup [1, \sqrt{8})$
13. $\left[-1, -\frac{1}{2}\right] \cup \left[0, \frac{1}{2}\right] \cup \{1\}$
14. $R - (1, 2) \cup \{2, 3, 4, 5, 6, 7, 8, 9, 10\} \cup (10, 11)$
15. $\left[\frac{\pi}{6}, \frac{\pi}{2}\right)$
16. No solution
18. Integral solutions are (2, 2) and (0, 0), all non-integral solutions lie on exactly two lines $x + y = 0$ and $x + y = 6$.

Exercise for Session 5

1. $[0, 3]$
2. $[-1/2, 1/2]$
3. $[3 - \sqrt{2}, 3 + \sqrt{2}]$
4. $[1, 5]$
5. $(-\infty, \log_3 9)$
6. $\left(-\infty, \frac{2}{3}\right] \cup (1, \infty)$
7. $R - (2 - 2\sqrt{3}, 2 + 2\sqrt{3})$
8. $[2, \infty)$
9. $(-\infty, 0]$
10. $\left\{\frac{\pi}{2}\right\}$
11. $\{0, 1\}$
12. $\{\pi\}$
13. $\left[\frac{\pi}{3}, \frac{\pi}{2}\right]$
14. $\left(0, \frac{\pi}{2}\right]$
15. $\{0\}$
16. $\left[\frac{-1}{2\sqrt{2}}, \frac{1}{2\sqrt{2}}\right]$
17. $[-1, 1]$
18. $\{0\}$
19. $[1, \infty)$
20. $\{1\}$
21. $[0, 2]$
22. Image of $(-\infty, 1)$ under f is $(1, \infty)$, Image of $[1, 2]$ under f is $[1, 2]$
23. Range of $f(x)$ is $\{0\}$
    and domain is $\cup \left[\left(2n\pi - \frac{\pi}{3}\right), \left(2n\pi + \frac{\pi}{3}\right)\right]$
24. Range of $f(x)$ is $\left\{\log \frac{\pi}{2}\right\}$ and domain is $[1, 2)$
25. Range of $f(x)$ is the set of all non-positive integers and domain is $\left[\frac{-3-\sqrt{5}}{2}, -2\right) \cup \left(-1, \frac{-3+\sqrt{5}}{2}\right]$.

Exercise for Session 6

1. (i) odd (ii) even (iii) neither even nor odd (iv) even
   (v) odd (vi) odd (vii) neither even nor odd (viii) even
2. f is an even function when $x \in$ integer f is an odd function when $x \notin$ integer.
3. (i) $f(x) = \begin{cases} -2x, & x \le -1 \\ -x|x|, & -1 < x \le 0 \end{cases}$ (ii) $f(x) = \begin{cases} 2x, & x \le -1 \\ x|x|, & -1 < x \le 0 \end{cases}$
5. $a \in (400, \infty)$

Exercise for Session 7

1. (i) $\pi/3$ (ii) 2π (iii) 1 (iv) 24 (v) Does not exist
   (vi) $2(n+1)!$ (vii) 1 (viii) 2π
2. Period is 8.
3. $f(x)$ is periodic with period 2λ.
4. $f(x)$ is periodic with period $2p$.
5. $f(x)$ is periodic with period 12 and $\sum_{r=0}^{99} f(5 + 12r) = 10000$

Exercise for Session 8

1. $x = 0$ and $y = \frac{3}{2}$
2. $f^{-1}(1) = y$
3. As monotonic and range = Codomain $\Rightarrow$ Bijective
5. $A \in [0, 1)$
6. $A \in \left[0, \frac{1}{2}\right)$
7. $b^2 < 3a(c - |d|)$
8. $X \in \left[-\frac{\pi}{2} - \alpha, \frac{\pi}{2} - \alpha\right]$ and $Y \in [c - r, c + r]$, where
   $\alpha = \tan^{-1}\left(\frac{a + b\sqrt{2}}{a}\right)$ and $r = \sqrt{a^2 + \sqrt{2}ab + b^2}$

Exercise for Session 9

1. Not Identical
2. Identical
3. Not Identical
4. Identical
5. Identical
6. Not Identical
7. Identical
8. Identical
9. Identical
10. Identical

Exercise for Session 10

1. Domain $\in [-1, 1]$ and range $\in \left[\frac{-2\pi}{3}, \frac{\pi}{3}\right]$

2. Domain for $f([|x|]) \in (-3, 3)$
   Domain for $f([2x+3]) \in [-3, 0)$

3. $h(x) = \begin{cases} \sin^2 x - \sin x + 1, & -1 < x < 0 \\ 2\sin^2 x, & 0 < x \le 1 \end{cases}$

4. $g(x) = \begin{cases} -x, & -2 \le x \le 0 \\ 0, & 0 \le x \le 1 \\ 2(x-1), & 1 \le x \le 2 \end{cases}$ 5. $gof = \{(x+1)^2, -2 \le x \le 1\}$

Exercise for Session 11

1. (i) $f^{-1}(x) = 3\sin x$ (ii) $f^{-1}(x) = \dfrac{-3 + \sqrt{5 + 4e^x}}{2}$
   (iii) $f^{-1}(x) = x^{\log_5 e}, x > 0$ (iv) $f^{-1}(x) = \frac{1}{2}(e^x - e^{-x})$
   (v) $f^{-1}(x) = \begin{cases} x, & x < 1 \\ \sqrt{x}, & 1 \le x \le 16 \\ \dfrac{x^2}{64}, & x > 16 \end{cases}$

2. $f^{-1}(x) = \dfrac{1 + \sqrt{1 + 4\log_2 x}}{2}, x > 0$

Exercise for Session 12

1. $\frac{2}{3}$ 2. 17 3. $\frac{4}{3}$ 4. $f(x) = \begin{cases} (1-x)^2, & 0 \le x \le \frac{1}{3} \\ 2x(1-x), & \frac{1}{3} \le x \le \frac{2}{3} \\ x^2, & \frac{2}{3} \le x \le 1 \end{cases}$

Chapter Exercises

1. (b) 2. (d) 3. (a) 4. (a) 5. (a) 6. (d) 7. (a) 8. (a) 9. (d) 10. (c)
11. (c) 12. (c) 13. (d) 14. (c) 15. (c) 16. (d) 17. (a) 18. (b) 19. (b) 20. (d)
21. (d) 22. (a) 23. (d) 24. (b) 25. (b) 26. (a) 27. (a) 28. (b) 29. (d) 30. (c)
31. (c) 32. (b) 33. (c) 34. (a) 35. (d) 36. (d) 37. (c) 38. (d) 39. (c) 40. (d)
41. (b) 42. (a) 43. (d) 44. (a) 45. (c) 46. (a) 47. (a) 48. (d) 49. (b) 50. (b)
51. (b) 52. (d) 53. (c) 54. (c) 55. (d) 56. (b) 57. (d) 58. (c) 59. (c) 60. (b)
61. (b) 62. (d) 63. (d) 64. (c) 65. (b) 66. (a) 67. (a) 68. (b) 69. (c) 70. (a)
71. (c) 72. (d) 73. (a) 74. (b) 75. (b) 76. (b) 77. (a,b,d) 78. (b,c,d) 79. (b,c,d) 80. (b,c,d)
81. (a, b) 82. (b, d) 83. (a,b) 84. (a,b, c, d) 85. (b, c) 86. (a, b, c, d)
87. (a, b, c, d) 88. (a,b,c) 89. (b, d) 90. (b, c) 91. (a, b, c, d) 92. (c, d) 93. (a,b, c, d)
94. (a, c) 95. (a, b, c) 96. (a, b, c, d) 97. (a, b, c) 98. (a, c)
99. (a, d) 100. (b, c, d) 101. (a,b,d) 102. (c) 103. (b) 104. (c) 105. (a)
106. (d) 107. (a) 108. (a) 109. (d) 110. (d) 111. (a) 112. (c) 113. (d) 114. (c) 115. (a)
116. (c) 117. (b) 118. (b) 119. (b) 120. (a) 121. (c) 122. (d) 123. (d) 124. (c) 125. (b)
126. (d) 127. (b) 128. (a) 129. (c) 130. (a) 131. (c) 132. (b) 133. (a) 134. (b) 135. (b)
136. (c) 137. (a)
138. (A) → (s), (B) → (p), (C) → (q), (D) → (r) 139. (A) → (r), (B) → (s), (C) → (p), (D) → (q)
140. (A) → (q), (B) → (r), (C) → (s), (D) → (p) 141. (A) → (p,q,s), (B) → (s), (C) → (q), (D) → (p,q,r,s)
142. (4) 143. (7) 144. (2) 145. (4) 146. (2) 147. (2) 148. (3) 149. (6) 150. (1) 151. (3)
152. (9) 153. (9) 154. (6) 155. (1) 156. (1) 157. (4) 158. (1) 159. (1) 160. (5) 161. (2)
162. (1) 163. (2) 164. (3) 165. (2) 166. (7) 167. (0) 168. (8) 169. (8) 170. (2) 171. (4)
172. (1)

173. (i) 0, $n + \frac{1}{2}$, where $n \in Z$
   (ii) $\left\{-1, -\frac{1}{2}\right\}$ (iii) $\left\{0, 1, \frac{4}{3}, \frac{5}{3}\right\}$
   (iv) $\{1, -1, \pm 1 + k$, where k is any positive proper fraction$\}$
   (v) no solution

174. $P(x) = x + 3$ and $Q(x) = -x - 2$

175. $(2n - 1)$

176. $x \in (3, \pi) \cup \left(\pi, \frac{3\pi}{2}\right) \cup \left(\frac{3\pi}{2}, 5\right)$

177. (i) $S(6) = 9$

178. Possible solutions are $\frac{29}{12}, \frac{19}{6}, \frac{97}{24}$.

179. $\frac{1}{n}\sum_{i=1}^{n} x_i = 1$ 181. $f(f(2)) = 26$

183. $g(x) = \begin{cases} x^2, & -2 \le x \le -1 \\ -x, & -1 \le x \le -1/4 \\ x + \frac{1}{2}, & -\frac{1}{4} \le x \le 0 \\ x, & 0 \le x \le 1 \\ x^2, & 1 \le x \le 2 \end{cases}$ 185. $f(x) = \begin{cases} 3/2, & -3/2 < x \le -1/2 \\ 2 + x, & -1/2 \le x < 0 \\ 2, & 0 \le x \le 2 \\ x, & 2 \le x \end{cases}$

186. (− 3, 3), (− 2, 4), (0, 0) and (1, 1) 187. (d)

188. (2) 189. (d) 190. A → p, B → q, C → q, D → p

191. $\left[-\frac{\pi}{2}, -\frac{\pi}{10}\right] \cup \left[\frac{3\pi}{10}, \frac{\pi}{2}\right]$ 192. (d) 193. (d)

194. (a) 195. (a) 196. (b) 197. (c) 198. (d) 199. (b)
200. (a) 201. (c) 202. (b) 203. (a) 204. (a) 205. (d)
206. (a) 207. (d) 208. (d) 209. (d)
210. (b) 211. (d) 212. (c)

Solutions

1. Here,

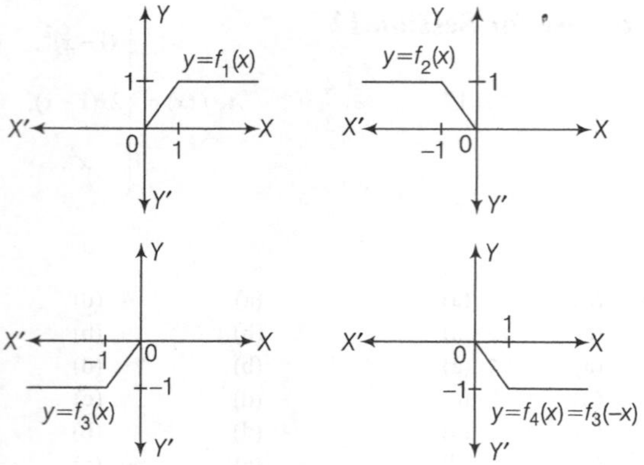

$\Rightarrow \qquad f_1(x) = -f_3(-x)$

2. Only in option (d), the graph has a symmetry w.r.t. origin.

3. Here, $\qquad f(x) = \dfrac{4}{\sqrt{1-x^2}}$

$\Rightarrow \qquad f(\sin x) = \dfrac{4}{|\cos x|}$

and $\qquad f(\cos x) = \dfrac{4}{|\sin x|}$

$\therefore \qquad g(x) = |\sin x| + |\cos x|$

$\therefore \qquad$ Period of $g(x) = \dfrac{\pi}{2}$

4. If $x = 1$, we see $f(y) = \dfrac{f(1)}{y}$ for all y

Put $\qquad y = 30$

$\Rightarrow \qquad f(1) = 30 \cdot f(30) = 30(20)$

$= 600 \qquad$ ...(i)

Now, $\qquad$ let $y = 40$

$\Rightarrow \qquad f(40) = \dfrac{f(1)}{40} = \dfrac{600}{40} = 15$

5. $h(x) = \log(f(x) \cdot g(x)) = \log e^{\{e^{|x|} \text{ sgn } x\}} \cdot e^{[e^{|x|} \text{ sgn } x]}$

$= \log e^{e^{|x|} \text{ sgn } x} = e^{|x|} \text{ sgn } x$

$\therefore \qquad h(x) = e^{|x|} \text{ sgn } x$

$$= \begin{cases} e^x, & x > 0 \\ 0, & x = 0 \\ -e^{-x}, & x < 0 \end{cases}$$

$\Rightarrow \qquad h(x) + h(-x) = 0$ for all x

$\therefore \quad h(x)$ is odd.

6. (a) $f(x) = x^4 + 2x^3 - x^2 + 1$

A polynomial of degree even will always be into.

(b) $f(x) = x^3 + x + 1$

$\Rightarrow f'(x) = 3x^2 + 1$, i.e. injective as well as surjective.

(c) $f(x) = \sqrt{1 + x^2}$, neither injective nor surjective as range $\in [1, \infty)$.

(d) $f(x) = x^3 + 2x^2 - x + 1$

$\Rightarrow \quad f'(x) = 3x^2 + 4x - 1$

$\Rightarrow \qquad D > 0$

$\therefore \quad f(x)$ is surjective but not injective.

7. Here, $f(x)$ is bijective, hence $f^{-1}(4)$ exists when $y = 4$.

$\therefore \qquad 2x^3 + 7x - 9 = 0$

$\Rightarrow \qquad (2x^2 + 2x + 9)(x - 1) = 0$

$\Rightarrow x = 1$ only, as $2x^2 + 2x + 9 = 0$ has no other root.

8. Here, $f(x) = \dfrac{e^x \cdot \log x \cdot 5^{x^2+2} \cdot (x^2 - 7x + 10)}{2x^2 - 11x + 12}$

$= \dfrac{e^x \cdot \log x \cdot 5^{x^2+2} \cdot (x-2)(x-5)}{(2x-3)(x-4)}$

Note that at $x = \dfrac{3}{2}$ and $x = 4$, function is not defined and in open interval $\left(\dfrac{3}{2}, 4\right)$, function is continuous.

$\therefore \qquad \lim\limits_{x \to \frac{3}{2}^+} \dfrac{e^x \cdot \log x \cdot 5^{x^2+2} \cdot (x-2)(x-5)}{(2x-3)(x-4)}$

$= \dfrac{(+\text{ ve})(+\text{ve})(-\text{ ve})(-\text{ ve})}{(+\text{ ve})(-\text{ ve})} = -\infty$

and $\quad \lim\limits_{x \to 4^-} \dfrac{e^x \cdot \log x \cdot 5^{x^2+2} \cdot (x-2)(x-5)}{(2x-3)(x-4)}$

$= \dfrac{(+\text{ve})(+\text{ve})(-\text{ve})}{(+\text{ve})(-\text{ve})} = \infty$

In the open interval $\left(\dfrac{3}{2}, 4\right)$, the function is continuous and takes up all real values from $(-\infty, \infty)$.

Hence, range of the function is $(-\infty, \infty)$.

9. As, $x = \cos^{-1}(\cos 4) = \cos^{-1}(\cos(2\pi - 4)) = 2\pi - 4$

and $\qquad y = \sin^{-1}(\sin 3)$

$= \sin^{-1}(\sin(\pi - 3)) = \pi - 3$

$\therefore \qquad x + y = 3\pi - 7$

10. $f(x) = \left(\dfrac{2\sin x + \sin 2x}{2\cos x + \sin 2x} \cdot \dfrac{1 - \cos x}{1 - \sin x}\right)^{2/3}$

$= \left(\dfrac{2\sin x \cdot (1 + \cos x)}{2\cos x \cdot (1 + \sin x)} \cdot \dfrac{(1 - \cos x)}{(1 - \sin x)}\right)^{2/3}$

For domain, $\sin x \neq \pm 1$

$\Rightarrow \qquad x \in R - \left\{(4n-1)\dfrac{\pi}{2}, (4n+1)\dfrac{\pi}{2}\right\}$

$\therefore \quad f(x) = \left(\tan x \cdot \frac{(1-\cos^2 x)}{(1-\sin^2 x)}\right)^{2/3} = \tan^2 x$

$\Rightarrow \quad$ Range $\in [0, \infty)$

11. As, $f(x+2) = e^{\sin\{x+2\} \cdot \cos \pi(x+2)} = e^{\sin\{x\} \cdot \cos \pi x} = f(x)$

$\therefore$ Periodic with period 2.

12. Here, $y = x^4 - 2x^2 + 3 = (x^2-1)^2 + 2 \Rightarrow y_{\min} = 2$

$\Rightarrow \quad \log_{0.5}(x^4 - 2x^2 + 3) \le \log_{1/2} 2 = -1$

$\therefore \quad f(x) = \cot^{-1}(\log_{0.5}(x^4 - 2x^2 + 3)) \ge \cot^{-1}(-1)$

$\Rightarrow$ Range $\in \left[\frac{3\pi}{4}, \pi\right)$

13. Here, $\log_e(x^2 + e) \ge 1$, for all $x \in R$

$\therefore \quad 0 < \frac{1}{\log(x^2+e)} \le 1$

$\Rightarrow \quad \left[\frac{1}{\log(x^2+e)}\right] = \begin{cases} 0, & x \ne 0 \\ 1, & x = 0 \end{cases}$

$\Rightarrow \quad f(x) = \begin{cases} \frac{1}{\sqrt{1+x^2}}, & x \ne 0 \\ 2, & x = 0 \end{cases}$

Which is shown in figure as

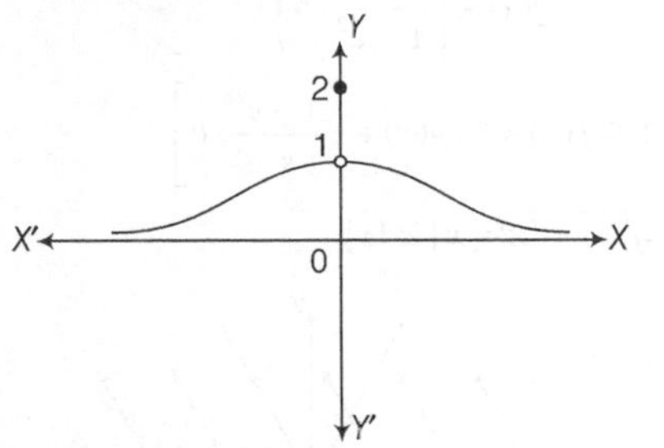

$\therefore$ Range of $f(x) \in [0, 1) \cup \{2\}$.

14. Let $x + 3 = t$

$\therefore \quad f(t) = \sin(\{t\}) \Rightarrow$ Period $= 1$

15. As, $e^{-|x|}$ can be shown by graphical transformation.

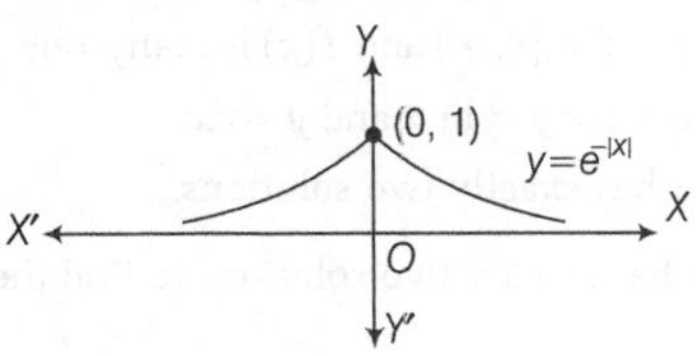

16. Here, $[x]\{x\} = 1$

$\Rightarrow \quad \{x\} = \frac{1}{[x]}$ or $x - [x] = \frac{1}{[x]} \Rightarrow x = [x] + \frac{1}{[x]}$

Obviously, $x > 2$

$\therefore \quad x = m + \frac{1}{m}, m \in N - \{1\}$

17. Here, $\sqrt{x^2 - 5x - 24} > x + 2$ is equivalent to the collection of two system of inequations.

***Case* I** $x^2 - 5x - 24 \ge 0$ and $x + 2 < 0$

$\Rightarrow \quad (x-8)(x+3) \ge 0$ and $x < -2$

$\Rightarrow \quad x \le -3 \quad$ ...(i)

***Case* II** $x^2 - 5x - 24 \ge 0, x + 2 \ge 0$

and $\quad x^2 - 5x - 24 > (x+2)^2$

$\Rightarrow \quad x \in (-\infty, -3] \cup [8, \infty), x \in [-2, \infty)$

and $\quad x^2 - 5x - 24 > x^2 + 4x + 4$

$\Rightarrow \quad 9x < -28$

$\Rightarrow \quad x < \frac{-28}{9}$

$\therefore$ No solution, i.e. $x \in \phi$. $\quad$ ...(ii)

From Eqs. (i) and (ii), we get

$x \in (-\infty, -3]$

18. Here, $\quad f(f(x)) = x$

$\Rightarrow \quad m(mx + b) + b = x$

$\Rightarrow \quad m^2x + b(m+1) = x$

$\Rightarrow \quad m = \pm 1, b = 0$

If $f(-f(x)) = -x \Rightarrow -m(mx+b) + b = -x$

$\Rightarrow \quad -m^2x + b(-m+1) = -x$

$\Rightarrow \quad m = \pm 1$ and $b = 0$

$\therefore$ Only 2 straight lines, i.e. $y = \pm x$.

19. Here, $f(g(x)) = \frac{rx}{1 + (r-1)x}$ and $g(f(x)) = rx$

If $f(g(x)) = g(f(x))$, then $\frac{rx}{1+(r-1)x} = rx$

$\Rightarrow \quad rx = rx(1 + (r-1)x)$

$\Rightarrow \quad r(r-1)x^2 = 0$

If this is to be true for infinitely many x, then

$r(r-1) = 0 \Rightarrow r = 0, 1$

20. Since, f is a linear function, so it has the form $f(x) = mx + b$ because $f(1) \le f(2)$, we have $m \ge 0$. Similarly, $f(3) \le f(4)$ $\Rightarrow m \le 0$.

Hence, $m = 0$ and f is constant function. Thus, $f(0) = f(5) = 5$

21. Suppose R is just a rectangle whose four vertices are $(1, 2), (1, -2), (-1, -2), (-1, 2)$.

The X-axis and Y-axis symmetries in the problem are satisfied, but the point $(2, 1)$ is not contained in R.

22. Here, $2\{y\} = [x] + 1$

Since, $0 \le y < 1$

$\therefore \quad \{y\} = y$

and $\quad 0 \le [x] + 1 < 2$, i.e. $-1 \le [x] < 1$

When $-1 \le x < 0 \Rightarrow y = 0$

When $\quad 0 \le x < 1 \Rightarrow y = \frac{1}{2}$

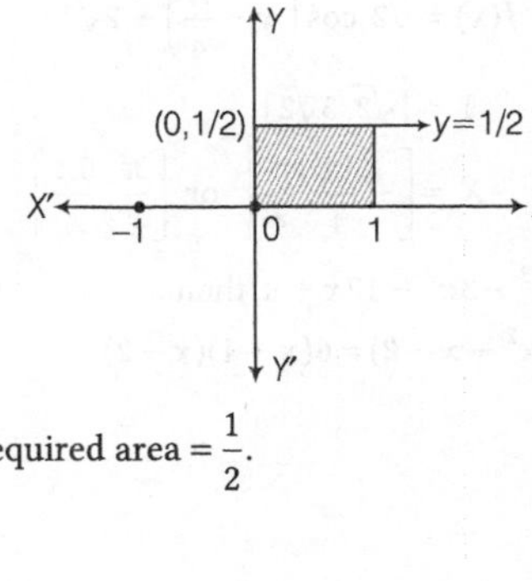

$\therefore \quad$ The required area $= \frac{1}{2}$.

23. Here, $-1 \le \cos x \le 1$

$\Rightarrow \quad -\sin 1 \le \sin(\cos x) \le \sin 1$

$\Rightarrow \quad [\sin(\cos x)] = 0, -1$

Also, $\quad -1 \le \sin x \le 1$

$\Rightarrow \quad \cos 1 \le \cos(\sin x) \le 1$

$\therefore \quad [\cos(\sin x)] = 0, 1$

$\therefore \quad f(g(x))$ has range $\left\{\frac{-\pi}{2}, 0, \frac{\pi}{2}\right\}$.

24. As we know, $\quad \{f(x)\} \in [0,1)$

$\Rightarrow \quad [\{f(x)\}] = 0$

$\therefore \quad e^{2x} + e^x - 2 = [\{x^2 + 10x + 11\}]$

$\Rightarrow \quad e^{2x} + e^x - 2 = 0$

$\Rightarrow \quad (e^x + 2)(e^x - 1) = 0$

$\therefore \quad e^x = 1$ is the only solution.

$\Rightarrow \quad x = 0$

$\Rightarrow$ Number of solutions is 1.

25. We know that, $\{x\} = x - [x]$.

$\therefore \quad \{x\} + \{2x\} + \{3x\} + \ldots + \{12x\} = 78x.$

$\Rightarrow \quad x - [x] + 2x - [2x] + 3x - [3x] + \ldots + 12x - [12x] = 78x$

$\Rightarrow \quad (x + 2x + 3x + \ldots + 12x) - ([x] + [2x] + \ldots + [12x]) = 78x$

$\Rightarrow \quad [x] + [2x] + \ldots + [12x] = 0$

$\therefore \quad 0 \le 12x < 1$

$\Rightarrow \quad 0 \le x < \frac{1}{12}, \ \forall \frac{1}{25} \le x \le \frac{1}{10}$

$\therefore$ Common values of $x \in \left[\frac{1}{25}, \frac{1}{12}\right)$.

Since, x is of the form $\frac{1}{n}$.

$\therefore \quad x = \frac{1}{25}, \frac{1}{24}, \frac{1}{23}, \ldots, \frac{1}{14}, \frac{1}{13}$

i.e. 13 solutions.

26. $-\sqrt{20} \le 2\cos x - 4\sin x \le \sqrt{20}$

$\Rightarrow \quad 0 \le (2\cos x - 4\sin x)^2 \le 20$

$$\min = \frac{1}{1+20} = \frac{1}{21}; \max = 1$$

$\Rightarrow \quad M + m = \frac{22}{21}$

27. $f(x) = \sqrt{2}\sin\left(x + \frac{\pi}{4}\right) + 2\sqrt{2}$

or $\quad f(x) = \sqrt{2}\cos\left(x - \frac{\pi}{4}\right) + 2\sqrt{2}$

$\Rightarrow \quad Y = [\sqrt{2}, 3\sqrt{2}]$

and $\quad X = \left[-\frac{3\pi}{4}, \frac{\pi}{4}\right]$ or $\left[\frac{\pi}{4}, \frac{5\pi}{4}\right]$

28. Let $f(x) = 2x^3 - 3x^2 - 12x + a$, then

$f'(x) = 6(x^2 - x - 2) = 6(x+1)(x-2)$

So, the roots of $f'(x) = 0$ are $x = -1, 2$.

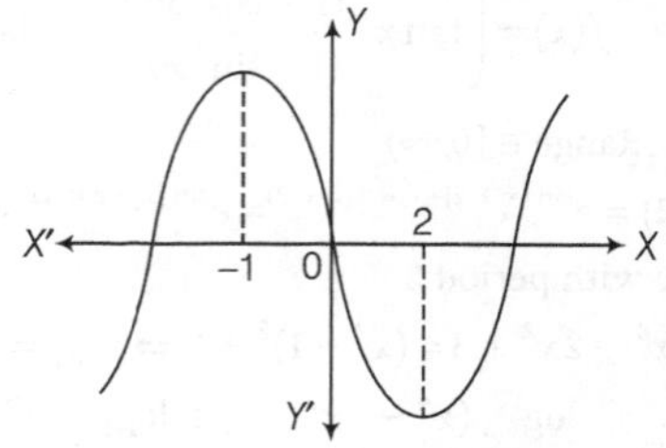

Now, $f(x) = 0$ will have all real roots, if $f(-1) > 0$ and $f(2) < 0$.

$\Rightarrow \quad -2 - 3 + 12 + a > 0$ and $16 - 12 - 24 + a < 0$

$\Rightarrow \quad -7 < a < 20$

29. $g(x) = \frac{e^{f(x)} - e^{|f(x)|}}{e^{f(x)} + e^{|f(x)|}}, -1 \le f(x) \le 1$

For $\quad 0 \le f(x) \le 1$,

$g(x) = 0$

$-1 \le f(x) < 0$

$$g(x) = \frac{e^{f(x)} - e^{-f(x)}}{e^{f(x)} + e^{-f(x)}} = \frac{e^{2f(x)} - 1}{e^{2f(x)} + 1} = 1 - \frac{2}{e^{2f(x)} + 1}$$

For $\quad -1 \le f(x) < 0$,

$$g(x) \in \left[\frac{1-e^2}{1+e^2}, 0\right)$$

For $\quad -1 \le f(x) < 1$, $g(x) \in \left[\frac{1-e^2}{1+e^2}, 0\right]$

30. $f(x) = \sqrt{|x| - \{x\}}; |x| \ge \{x\}$

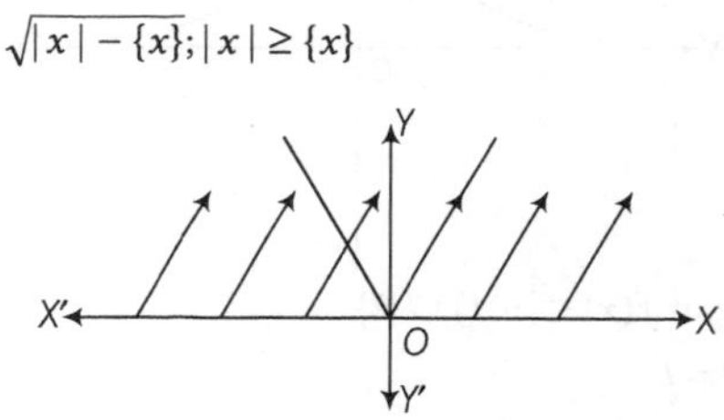

$\Rightarrow \quad X \in \left(-\infty, -\frac{1}{2}\right] \cup [0, \infty)$

$\Rightarrow \quad Y \in [0, \infty)$ and $f(x)$ is many-one.

31. Given, curves are $y = \ln x$ and $y = ax$.

$\Rightarrow \ln x = ax$ has exactly two solutions.

$\Rightarrow \frac{\ln x}{x} = a$ has exactly two solutions to find the range of $\frac{\ln x}{x}$.

Let $y = \frac{\ln x}{x}, x > 0$; $\frac{dy}{dx} = \frac{x \cdot \frac{1}{x} - \ln x}{x^2} = \frac{1 - \ln x}{x^2}$

y is increasing, if $1 - \ln x > 0$ or $\ln x < 1 \Rightarrow 0 < x < e$

Range of $y \in \left(-\infty, \frac{1}{e}\right]$ graph of $y = \frac{\ln x}{x}$

For exactly two solutions of $\frac{\ln x}{x} = a$

$\Rightarrow \quad a \in \left(0, \frac{1}{e}\right)$

32. Let $f(x) = ax^2 + bx + c$ as it touches X-axis at $x = 3$

$\Rightarrow \quad \frac{-b}{2a} = 3$

$\Rightarrow \quad b = -6a \quad$ ...(i)

Also, $\quad 9a + 3b + c = 0 \quad$ ...(ii)

$4a - 2b + c = 3 \quad$ ...(iii)

$\Rightarrow \quad a = \frac{3}{25}, b = -\frac{18}{25}, c = \frac{27}{25}$

$\Rightarrow \quad f(x) = \frac{3}{25}(x^2 - 6x + 9)$

33. $y = \sqrt{2\{x\} - \{x\}^2 - \frac{3}{4}}$

$\Rightarrow \quad 2\{x\} - \{x\}^2 - \frac{3}{4} \ge 0 \Rightarrow \frac{1}{2} \le \{x\} \le \frac{3}{2}$

$\Rightarrow \quad \frac{1}{2} \le \{x\} < 1 \quad (\because 0 \le \{x\} < 1)$

$\therefore \quad 2\{x\} - \{x\}^2 - \frac{3}{4}$ is increasing for $\frac{1}{2} \le \{x\} < 1$.

$\Rightarrow$ Range $= \left[0, \frac{1}{4}\right]$

34. Replace x by $-x$

$\Rightarrow \quad x[f(x) + f(-x)] = 0$

$\Rightarrow \quad f(x)$ is an odd function.

$\Rightarrow \quad f^{iv}(x)$ is also odd $\Rightarrow f^{iv}(0) = 0$

35. $[y + [y]] = 2\cos x \quad \Rightarrow \quad [y] = \cos x$

$\Rightarrow \quad y = \frac{1}{3}[\sin x + [\sin x + [\sin x]]] = [\sin x]$

$\Rightarrow [\sin x] = \cos x$

Number of solutions in $[0, 2\pi]$ is 0.

Hence, total solution is 0.

$\therefore$ Both are periodic with period 2π.

36. By replacing $x = x + 1$ and $x = x - 1$, we get

$f(x + 2) + f(x) = \sqrt{2} f(x + 1) \quad$ ...(i)

$f(x) + f(x - 2) = \sqrt{2} f(x - 1) \quad$ ...(ii)

From Eqs. (i) and (ii), gives

$f(x + 2) + f(x - 2) + 2f(x)$

$= \sqrt{2}\,[f(x + 1) + f(x - 1)]$

$= \sqrt{2}\sqrt{2} f(x)$

$\therefore \quad f(x + 2) + f(x - 2) = 0$

On replacing x by $x + 2$, we get

$f(x + 4) + f(x) = 0$

$f(x + 8) = -f(x + 4) = f(x), \forall\, x$

$\therefore \quad f(x)$ is periodic with period 8.

37. The equation has meaning, if $x > 0, x \ne 1$.

$\therefore \quad$ Domain $= (0, 1) \cup (1, \infty)$

If $x \in (0, 1)$, then $\log_2 x < 0$

and $\quad \log_2 x + \log_x 2 = \frac{\log x}{\log 2} + \frac{\log 2}{\log x}$

$=$ sum of a negative number ≤ -2

In this case any α will satisfy, since $2\cos\alpha$ can never be more than 2.

Thus, the inequation is satisfied for any x in $(0, 1)$ and for any α.

If $x \in (1, \infty)$, then $\log_2 x > 0 \Rightarrow \frac{\log x}{\log 2} + \frac{\log 2}{\log x} > 0$

The inequation cannot be satisfied unless

$\cos\alpha = -1$ and $x = 2$

i.e. $\quad \log_2 x = 1$

Option (d) is wrong, since in the last case there are infinite solution.

38. If we draw the graph of $\left(\frac{1}{2}\right)^{|x|}$ and $x^2 - a$, then the range of value of a will be $(-1, \infty)$.

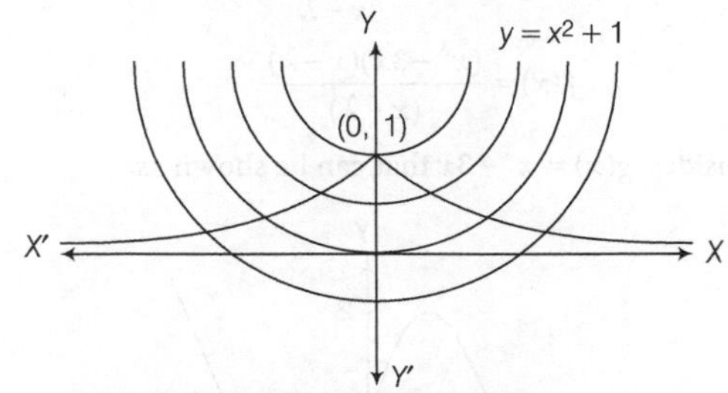

Maximum possible solution for 'x' is '2'.

39. $f(x) \ne |\cos x|$ is true only when

$|\cos x| = 1 \Rightarrow x = 0, \pi, 2\pi$

40. $|\cos x| + \cos x > 0 \to x \in \left[0, \frac{\pi}{2}\right)$

For $\quad x \in \left[0, \frac{\pi}{2}\right), 1 \le \sin x + \cos x \le \sqrt{2}$

But $\quad \sin x + \cos x \ne 1 \Rightarrow x \in \left(0, \frac{\pi}{2}\right)$

Now, for $x \in \left(0, \frac{\pi}{2}\right)$,

$\log_{\sin x + \cos x}(|\cos x| + \cos x) \ge 0$

$\Rightarrow \quad \cos x \ge \frac{1}{2} \Rightarrow x \in \left(0, \frac{\pi}{3}\right]$

41. $y = ((x + \alpha)^2 - 1)^{1/4} = [(x + \alpha - 1)(x + \alpha + 1)]^{1/4}$

$(x + \alpha - 1)(x + \alpha + 1) \ge 0$

$x \ge 1 - \alpha$ and $x \le -1 - \alpha$ for $\alpha > 0$

For $\quad \alpha < 0, x \le 1 - \alpha, x \ge -1 - \alpha$

$(x + \alpha) \le 1$ and $(x + \alpha) \ge -1$

For $\alpha \le -1, x \le 0$ and range is $[0, \infty)$.

42. $f(x) = \log(\log(\log x))$

$\log x > 1$ when $x \in (e, \infty)$

$\therefore \log(\log x) > 0$ and hence, $\log(\log(\log x))$ is well defined and uniquely.

It is evidently one-one. Since, the range of $\log x = R$, $f(x)$ is one-one and onto.

43. $\quad x^2 - 4px + q^2 > 0, \forall\, x \in R$

$\Rightarrow \quad 4p^2 - q^2 < 0 \quad$ ...(i)

$r^2 + p^2 < qr \quad$ ...(ii)

Let $\qquad y = \dfrac{x + r}{x^2 + qx + p^2}$

$\Rightarrow \qquad x^2y + x(qy - 1) + p^2y - r = 0 \qquad \ldots\text{(iii)}$

x is real,

$\Rightarrow \qquad (q^2 - 4p^2)y^2 + y(-2q + 4r) + 1 > 0$

From Eq. (i) $\Rightarrow$ Coefficient of y^2 is a positive discriminant.

$$= (4r - 2q)^2 - 4(q^2 - 4p^2)$$
$$= 16(r^2 + p^2 - qr) < 0 \qquad \text{[from Eq. (ii)]}$$

Hence, Eq. (iii) is true for all real y or $y \in (-\infty, \infty)$.

44. Here,

$$f(x) = \frac{x^4 - \lambda x^3 - 3x^2 + 3\lambda x}{x - \lambda}$$

$$\therefore \qquad f(x) = \frac{(x^3 - 3x)(x - \lambda)}{(x - \lambda)}$$

Consider, $g(x) = x^3 - 3x$ that can be shown as

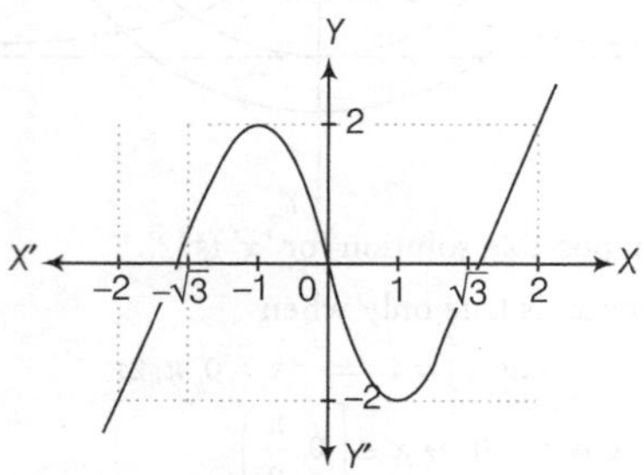

Now, the range of $f(x)$ is the set of entire real numbers, if $\lambda \in [-2, 2]$.

Because, if $\lambda > 2$ or $\lambda < -2$, then range of $f(x) \notin R$.

45. We know that,

$$(a - 1)^n = {}^nC_0 a^n - {}^nC_1 a^{n-1} + {}^nC_2 a^{n-2} - \ldots + (-1)^{n-1} \cdot {}^nC_{n-1} \cdot a + (-1)^n \cdot {}^nC_n$$

$$\therefore \qquad \frac{(a-1)^n}{a} = {}^nC_0 a^{n-1} - {}^nC_1 \cdot a^{n-2} + {}^nC_2 \cdot a^{n-3} + \ldots + (-1)^{n-1} \cdot {}^nC_{n-1} + (-1)^n \cdot \frac{{}^nC_n}{a}$$

Hence, $f(n) = \dfrac{(a-1)^n - (-1)^n}{a}$

Now, $f(2016) + f(2017) = \dfrac{(a-1)^{2016} - 1}{a} + \dfrac{(a-1)^{2017} + 1}{a}$

$$= \frac{(a-1)^{2016}[1 + a - 1]}{a} = (a-1)^{2016}, \text{ where } a = 3^{\frac{1}{224}} + 1$$

$$\therefore \qquad f(2016) + f(2017) = (3^{\frac{1}{224}})^{2016} = 3^9 = 3^K$$

$\Rightarrow \qquad K = 9$

46. Here,

$$g(x) = \frac{\pi}{2} - \sqrt{\frac{\pi^2}{2} - \left(x - \frac{\pi}{2}\right)^2}$$

or $\qquad \left(y - \dfrac{\pi}{2}\right)^2 = \dfrac{\pi^2}{2} - \left(x - \dfrac{\pi}{2}\right)^2$

$\Rightarrow \qquad \left(x - \dfrac{\pi}{2}\right)^2 + \left(y - \dfrac{\pi}{2}\right)^2 = \dfrac{\pi^2}{2}$, i.e. circle

and $f(x) = \sin^{-1}(\sin x)$ is shown as

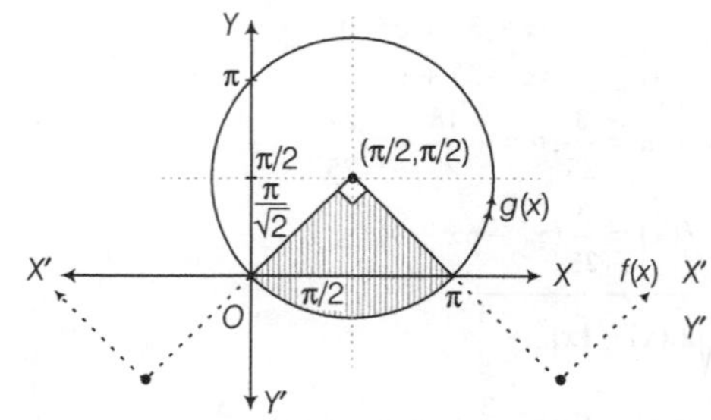

$\therefore \quad$ Area of shaded part $= \dfrac{\theta}{360^\circ} \times \pi r^2$

$$= \frac{90^\circ}{360^\circ} \times \pi \times \frac{\pi^2}{2} = \frac{\pi^3}{8} \text{ sq units}$$

47. Let $\qquad y = \dfrac{x^2 + bx + 1}{x^2 + 2x + b} \; (b > 1)$

$\Rightarrow (y - 1)x^2 + (2y - b)x + (by - 1) = 0$

$\Rightarrow D \geq 0 \Rightarrow (4 - 4b)y^2 + 4y + (b^2 - 4) \geq 0 \qquad \ldots\text{(i)}$

Since, $f(x)$ and $\dfrac{1}{f(x)}$ have the same bounded set as their range.

Thus, $(4 - 4b)y^2 + 4y + (b^2 - 4) \geq 0$ have roots α and $\dfrac{1}{\alpha}$.

$\therefore \quad$ Product of roots $= 1$

$\Rightarrow \qquad \dfrac{b^2 - 4}{4(1 - b)} = 1 \Rightarrow b^2 - 4 = 4 - 4b$

or $\qquad b^2 + 4b - 8 = 0$

$$b = \frac{-4 \pm \sqrt{16 + 32}}{2} = \frac{-4 \pm 4\sqrt{3}}{2} = 2\sqrt{3} - 2$$

48. Since, $\sin \dfrac{\pi[x + 24]}{12} = \sin \dfrac{\pi}{12}(24 + [x])$

$$= \sin\left(2\pi + \frac{\pi[x]}{2}\right) = \sin \frac{\pi[x]}{12}$$

The period of $\sin \dfrac{\pi[x]}{12}$ is 24.

Similarly, period of $\cos \dfrac{\pi[x]}{4}$ is 8 and period of $\tan \dfrac{\pi[x]}{3} = 3$.

Hence, the period of the given function = LCM of 24, 8, 3 = 24.

49. Let $\quad f(x, y) = ax + by$

Then, $f(2x + 3y, 2x - 7y)$

$$= a(2x + 3y) + b(2x - 7y) = 20x \qquad \text{[given]}$$

$\therefore \qquad 2a + 2b = 20$ and $3a - 7b = 0$

$\therefore \qquad a = 7$ and $b = 3$

$\therefore \qquad f(x, y) = 7x + 3y$

50. Domain of $f(x)$ is [1, 3] and the function is continuous.

$$f'(x) = \frac{1}{2\sqrt{x-1}} - \frac{1}{\sqrt{3-x}} = 0$$

$\Rightarrow \qquad \sqrt{3 - x} = 2\sqrt{x - 1}$

$$\Rightarrow \quad 3 - x = 4x - 4 \Rightarrow x = \frac{7}{5}$$

$\therefore$ Critical points in [1, 3] are 1, $\frac{7}{5}$ and 3.

$$f(1) = 2\sqrt{2}, f(3) = \sqrt{2}$$

and $f\left(\frac{7}{5}\right) = \sqrt{\frac{2}{5}} + 2\sqrt{\frac{8}{5}} = \sqrt{10}, 2\sqrt{2}$ being $< \sqrt{10}$, the range $= [\sqrt{2}, \sqrt{10}]$.

51. $f(x) = \cos^{-1}(\sec(\cos^{-1} x)) + \sin^{-1}(\text{cosec}(\sin^{-1} x))$

$\Rightarrow \quad -1 \le \sec(\cos^{-1} x) \le 1$

and $\quad -1 \le \text{cosec}(\sin^{-1} x) \le 1$

$\Rightarrow \quad \sec(\cos^{-1} x) = \pm 1$

and $\quad \text{cosec}(\sin^{-1} x) = \pm 1$

$\Rightarrow \cos^{-1} x = 0, \pi$ and $\sin^{-1} x = \frac{\pi}{2}, -\frac{\pi}{2}$

$\Rightarrow \quad x = \pm 1$ and $x = \pm 1$

$\therefore$ Domain is $x = \pm 1$.

52. Put $x = y = 1 \Rightarrow f(1) = 2$

Put $\quad y = \frac{1}{x} \Rightarrow f(x) + f\left(\frac{1}{x}\right) = f(x) f\left(\frac{1}{x}\right)$

$\Rightarrow \quad f(x) = x^3 + 1$

$\Rightarrow \quad g(x) = 0, \forall x \in R - \{0\}$

53. $f(x) = f'(x) \times f''(x)$ is satisfied by only the polynomial of degree 4.

Since, $f(x) = 0$ satisfies $x = 1, 2, 3$ only. It is clear one of the roots is twice repeated.

$\therefore \quad f'(1) f'(2) f'(3) = 0$

54. Total number of functions for which $f(i) \ne i = 4^5$ and number of onto functions in which $f(i) \ne i = 44$.

$\therefore$ Required number of functions = 980

55. We can send 1, 2, ..., $n - 1$ anywhere and the value of $f(n)$ will then be uniquely determined.

56. Put $\sin x = t$,

$$y = t^3 - 6t^2 + 11t - 6, -1 \le t \le 1$$

$$f(-1) = -24, f(1) = 0$$

57. $g(x) = f(x^2 - 2x - 1) + f(5 - x^2 + 2x)$

$= 2x^4 - 8x^3 - 4x^2 + 24x + 18$

$g'(x) = 8x^3 - 24x^2 - 8x + 24$

$g'(x) = 0 \Rightarrow x = -1, 1, 3$

We observe that,

$g(x) \ge \min\{g(-1), g(1), g(3)\} = 0$

$\therefore \quad g(x) \ge 0, \forall x \in R$

58. Let $f(x) = [x], g(x) = \frac{e^{-|x|}}{2}$

$$\Rightarrow \quad h(x) = \left[\frac{e^{-|x|}}{2}\right] = 0, \forall x \in R$$

59. There are four possible functions defined in $0 \le x < 1$, of them 2 are continuous and two are discontinuous, now for each of the points (1, 2, ..., $n - 1$), keep functions fixed from left of the point, so there are 4 possible functions defined in between next two consecutive integral points of them only one is continuous and at last for $x = n$, there is only one possibility of discontinuity of the function. So, total number of functions

$$= 2 \times 3^{n-1} \times 1$$

60. $\because \quad k \in$ odd

$f(x) = k + 3 \quad$ [even]

$f(f(k)) = \frac{k+3}{2}$

If $\frac{k+3}{2} \in$ odd $\Rightarrow 27 = \frac{k+3}{2} + 3$

$\Rightarrow \quad k = 45$ not possible.

Now, let $\frac{k+3}{2} \in$ even

$\therefore \quad 27 = f(f(f(k))) = f\left(\frac{k+3}{2}\right) = \frac{k+3}{4}$

$\therefore \quad k = 105$

Verifying $f(f(f(105))) = f(f(108)) = f(54) = 27$

$\therefore \quad k = 105$

Hence, sum of digits of $k = 1 + 0 + 5 = 6$

61. $$f(x) = \frac{\sin(\pi\{x\})}{x^4 + 3x^2 + 7}$$

Here, $\quad f(1/2) = f(-1/2)$

Clearly, $f(x)$ is not one-one and also it is dependent on x.

62. Since, $f(x)$ and $g(x)$ are one-one and onto and are also the mirror images of each other which respect to the line $y = a$. It clearly indicates that $h(x) = f(x) + g(x)$ will be a constant function and will always be equal to $2a$.

63. Let $g(t) = 2t^3 - 15t^2 + 36t - 25$

$g'(t) = 6t^2 - 30t + 36 = 6(t^2 - 5t + 6)$

$= 6(t-2)(t-3) = 0 \Rightarrow t = 2, 3$

For $\quad 2 \le t \le 3$,

$g(t)_{min} = g(3) = 2 \times 27 - 15 \times 9 + 36 \times 3 - 25 = 2$

Also, $\quad 2 + |\sin t| \ge 2$

Hence, minimum $\phi(t) = 2$

$\therefore \quad 3^x + 3^{f(x)} = 2$

$\Rightarrow \quad 3^{f(x)} = 2 - 3^x \quad \because 3^{f(x)} > 0$

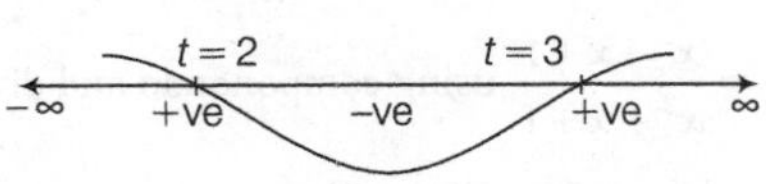

$\Rightarrow 2 - 3^x > 0 \Rightarrow 3^x < 2 \Rightarrow x < \log_3 2$

$\therefore \quad x \in (-\infty, \log_3 2)$

64. Here, $x_1 x_2 x_3 = x$ and x be the element of A.

$\therefore \quad x_1 x_2 x_3 = \frac{120}{x_4}$

$\Rightarrow x_1 x_2 x_3 x_4 = 120 = 2^3 \times 3^1 \times 5^1$.

Thus, to find number of positive integer solutions of

$$x_1 x_2 x_3 x_4 = 2^3 \times 3^1 \times 5^1.$$

i.e. $\alpha + \beta + \gamma + \delta = 3, \alpha + \beta + \gamma + \delta = 1, \alpha + \beta + \gamma + \delta = 1$

$$\Rightarrow \; ^{3+4-1}C_{4-1} \cdot ^{1+4-1}C_{4-1} \cdot ^{1+4-1}C_{4-1} = {}^6C_3 \cdot {}^4C_3 \cdot {}^4C_3$$

$$= 20 \times 4 \times 4 = 320 \quad \therefore \; d = 320$$

65.

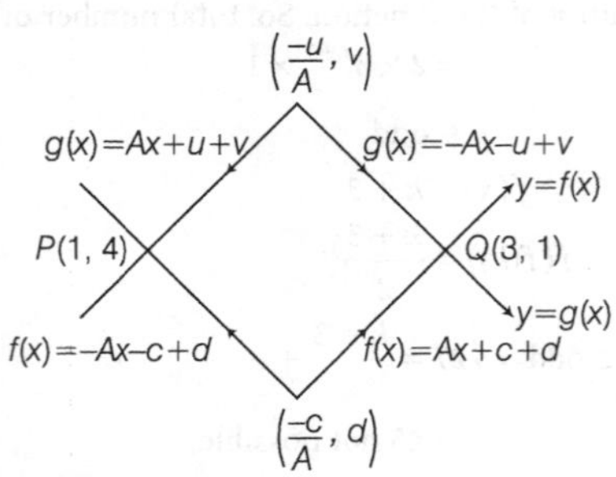

From above figure,

$$4 = A + u + v, \quad 1 = -3A - u + v. \qquad \text{...(i)}$$

and $1 = 3A + c + d, 4 = -A - c + d$ ...(ii)

From Eq. (i), $3 = 4A + 2u$ ...(iii)

From Eq. (ii), $-3 = 4A + 2c$ ...(iv)

On adding Eqs. (iii) and (iv), we get

$$8A + 2u + 2c = 0$$

$$\therefore \quad \frac{u+c}{A} = -4$$

66. As, $f(x)$ is one-one, if $f'(x) \geq 0, \forall x \in R$.

$\therefore \quad 3x^2 + 2x + 4 + a\cos x - b\sin x \geq 0, \forall x \in R.$

$\Rightarrow 3x^2 + 2x + 4 \geq b\sin x - a\cos x, \forall x \in R$

$\Rightarrow 3x^2 + 6x + 4 \geq \sqrt{a^2 + b^2}, \forall x \in R$

$$[\text{as}, |b\sin x - a\cos x| \leq \sqrt{a^2 + b^2}]$$

$\Rightarrow 3x^2 + 6x + 3 + 1 \geq \sqrt{a^2 + b^2}, \forall x \in R$

$\Rightarrow \quad 3(x+1)^2 + 1 \geq \sqrt{a^2 + b^2}, \forall x \in R$

$\therefore \quad \sqrt{a^2 + b^2} \leq 1 + 3(x+1)^2$, since $1 + 3(x+1)^2 \geq 1$

$\Rightarrow \quad \sqrt{a^2 + b^2} \leq 1$

$\therefore$ Greatest value of $(a^2 + b^2) = 1$

67. Here, $\dfrac{p-1}{p+1} = \dfrac{x^4 + x^2 + 1}{(x^2 + x + 1)^2} = \dfrac{x^4 + 2x^2 + 1 - x^2}{(x^2 + x + 1)^2} = \dfrac{(x^2 + 1)^2 - x^2}{(x^2 + x + 1)^2}$

$$= \frac{(x^2 + x + 1)(x^2 - x + 1)}{(x^2 + x + 1)^2}$$

$\therefore \quad \dfrac{p-1}{p+1} = \dfrac{x^2 - x + 1}{x^2 + x + 1}$, using componendo and dividendo

$$\Rightarrow \quad \frac{2p}{2} = \frac{2(x^2 + 1)}{2x}$$

$$\Rightarrow \quad p = x + \frac{1}{x} \qquad \text{...(i)}$$

As, $f(x) = \dfrac{1-x}{1+x} \Rightarrow f(f(x)) + f\left(f\left(\dfrac{1}{x}\right)\right) = x + \dfrac{1}{x}$ ...(ii)

From Eqs. (i) and (ii), we get $f(f(x)) + f\left(f\left(\dfrac{1}{x}\right)\right) = p$

68. As, α^{15} and α is the root of $x^{14} + x^{13} + \ldots + x + 1 = 0$

Now, $f(x) = x^{13} + 2x^{12} + 3x^{11} + \ldots + 13x + 14$

$$\therefore \quad \frac{f(x)}{x} = x^{12} + 2x^{11} + 3x^{10} + \ldots + \frac{14}{x}$$

On subtracting, we get

$$\left(1 - \frac{1}{x}\right) f(x) = x^{13} + x^{12} + x^{11} + \ldots + x + 1 - \frac{14}{x}$$

Put $x = \alpha$, $\left(\dfrac{\alpha - 1}{\alpha}\right) f(\alpha) = -\alpha^{14} - \dfrac{14}{\alpha} = \dfrac{-\alpha^{15} - 14}{\alpha} = -\dfrac{15}{\alpha}$

$$\therefore \quad f(\alpha) = \frac{15}{1-\alpha}.$$

Hence, $N = \dfrac{15^{14}}{(1-\alpha)(1-\alpha^2)\ldots(1-\alpha^{14})}$

and we know that

$$x^{15} - 1 = (x-1)(x-\alpha)(x-\alpha^2)(x-\alpha^3)\ldots(x-\alpha^{14})$$

$$\therefore \quad (x-\alpha)(x-\alpha^2)(x-\alpha^3)\ldots(x-\alpha^{14}) = \frac{x^{15}-1}{x-1}$$

As, $x \to 1$.

$$(1-\alpha)(1-\alpha^2)(1-\alpha^3)\ldots(1-\alpha^{14}) = \lim_{x \to 1} \frac{x^{15}-1}{x-1} = 15$$

$$\therefore \quad N = 15^{3} = 3^{13} \cdot 5^{13}$$

Thus, number of divisors $= (13+1)(13+1) = 14^2 = 196.$

69. Here, domain of $f(x) \Rightarrow x \in \left\{-\dfrac{1}{2}, \dfrac{1}{2}\right\}$ only

$\therefore$ $f(x)$ is minimum when $x = \dfrac{1}{2}$, i.e. $f_{\min}\left(\dfrac{1}{2}\right) = \dfrac{\pi}{2}$

and $f(x)$ is maximum when $x = -\dfrac{1}{2}$, i.e. $f_{\max}\left(-\dfrac{1}{2}\right) = \dfrac{3\pi}{2}$

$\therefore$ Sum of maximum and minimum value of function is 2π.

70. Here, $\tan^{-1}(x^2 - 18x + a) > 0, \forall x \in R$

$$\Rightarrow \quad x^2 - 18x + a > 0, \forall x \in R$$

$$\Rightarrow \quad (18)^2 - 4a < 0$$

$$\Rightarrow \quad a > 81$$

$$\Rightarrow \quad a \in (81, \infty)$$

71. As, $\sin^2 x + \sin x + 1 > 0, \forall x \in R.$

$\therefore \quad \dfrac{1}{\sqrt{\sin^2 x + \sin x + 1}}$ is always exists.

For $\sin^{-1}\left(\dfrac{1}{|x^2 - 1|}\right)$ to exists,

$$0 < \frac{1}{|x^2 - 1|} \leq 1$$

$$\Rightarrow \quad |x^2 - 1| \geq 1$$

$$\Rightarrow \quad x^2 - 1 \leq -1 \quad \text{or} \quad x^2 - 1 \geq 1$$

$$\Rightarrow \quad x^2 \leq 0 \quad \text{or} \quad x^2 \geq 2$$

$$\Rightarrow \quad x = 0 \quad \text{or} \quad (x \leq -\sqrt{2} \text{ or } x \geq \sqrt{2})$$

$\therefore \quad x \in (-\infty, -\sqrt{2}] \cup [\sqrt{2}, \infty) \cup \{0\}.$

72. Here, $\log(x^2 - x + 1)$ is defined, when

$$x^2 - x + 1 > 0 \text{ and } x^2 - x + 1 \neq 1$$

$$\Rightarrow \quad x \in R \text{ and } x \neq 0, 1 \Rightarrow x \in R - \{0, 1\} \quad \text{...(i)}$$

Again, $\log(\sin^{-1}\sqrt{x^2 + x + 1})$ exists, when

$$0 < x^2 + x + 1 \leq 1$$

$$\Rightarrow \quad x^2 + x \leq 0 \Rightarrow x \in [-1, 0] \quad \text{...(ii)}$$

$$x \in [-1, 0)$$

73. Here, $\sqrt{\cos^{-1} x}$ is defined for $x \in [-1, 1]$ for $\sqrt{\sin^{-1}(3x - 4x^3)}$ for x defined $0 \leq 3x - 4x^3 \leq 1$

$$\therefore \quad 3x - 4x^3 \geq 0 \text{ and } 3x - 4x^3 \leq 1$$

$$\Rightarrow x \in \left(-\infty, -\frac{\sqrt{3}}{2}\right] \cup \left[0, \frac{\sqrt{3}}{2}\right] \text{ and } x \in [-1, \infty)$$

$$\Rightarrow x \in \left[-1, -\frac{\sqrt{3}}{2}\right] \cup \left[0, \frac{\sqrt{3}}{2}\right]$$

74. Here, $f(x)$ exists only, if

$$4^x + 8^{\frac{2}{3}(x-2)} - 52 - 2^{2(x-1)} \geq 0.$$

$$\Rightarrow \quad 2^{2x} + 2^{2(x-2)} - 2^{2(x-1)} \geq 52$$

$$\Rightarrow \quad 2^{2x} \geq 64 \Rightarrow x \geq 3$$

75. $y = |\sin^{-1}(2x^2 - 1)|$

$$\therefore \quad \frac{dy}{dx} = \frac{|\sin^{-1}(2x^2 - 1)|}{\sin^{-1}(2x^2 - 1)} \cdot \frac{4x}{|2x|\sqrt{1 - x^2}}$$

which would exist, if

$$|2x| \neq 0, \ \sin^{-1}(2x^2 - 1) \neq 0 \text{ and } 1 - x^2 > 0$$

$$\Rightarrow \quad x \neq 0, \ 2x^2 - 1 \neq 0$$

and $\quad |x| < 1 \Rightarrow x \neq 0, \pm\frac{1}{\sqrt{2}}$ and $|x| < 1$

$$\Rightarrow \quad x \in (-1, 1) \sim \left\{0, \pm\frac{1}{\sqrt{2}}\right\}$$

76. The given function is defined for $5x^2 - 8x + 4 > 0$ which is true, $\forall\ x \in R$.

Since, coefficient of $x^2 = 5 > 0$ and $D = 64 - 80 = -16 < 0$

Let $\quad g(x) = 5x^2 - 8x + 4$

Here, $\quad a = 5 > 0$

$$\therefore \text{Range of } g(x) = \left[-\frac{D}{4a}, \infty\right) = \left[-\frac{-16}{4 \times 5}, \infty\right) = \left[\frac{4}{5}, \infty\right)$$

As, $\frac{4}{5} \leq 5x^2 - 8x + 4 < \infty$

$$\therefore \quad \log_{5/4}\left(\frac{4}{5}\right) \leq \log_{5/4}(5x^2 - 8x + 4) < \log_{5/4}\infty$$

$$\Rightarrow \quad -1 \leq \log_{5/4}(5x^2 - 8x + 4) < \infty$$

$$\Rightarrow \tan^{-1}(-1) \leq \tan^{-1}\{\log_{5/4}(5x^2 - 8x + 4)\} < \tan^{-1}(\infty)$$

$$\Rightarrow \quad -\frac{\pi}{4} \leq f(x) < \frac{\pi}{2}$$

$$\therefore \quad R_f = \left[\frac{-\pi}{4}, \frac{\pi}{2}\right)$$

77. Functions which are not algebraic, are known as transcendal functions.

78. $f(x) = \dfrac{\sqrt{(\sqrt{x-1})^2 + 1 - 2\sqrt{x-1}}}{\sqrt{x-1} - 1} \cdot x$

$$= \frac{|\sqrt{x-1} - 1|}{\sqrt{x-1} - 1} \cdot x = \begin{cases} x, & \text{when } x \in [2, \infty) \\ -x, & \text{when } x \in [1, 2) \end{cases}$$

$$f'(x) = \begin{cases} 1, & x \in [2, \infty) \\ -1, & x \in [1, 2) \end{cases}$$

$$\therefore f'(10) = 1, f'\left(\frac{3}{2}\right) = -1$$

79. $f(x) = \dfrac{1 + \cos 2x}{2} + \dfrac{1 + \cos\left(\frac{2\pi}{3} + 2x\right)}{2}$

$$- \frac{\left[\cos\left(2x + \frac{\pi}{3}\right) + \cos\left(\frac{\pi}{3}\right)\right]}{2}$$

$$= \frac{1}{2}\left[2 + \cos 2x + \cos\left(\frac{2\pi}{3} + 2x\right) - \cos\left(2x + \frac{\pi}{3}\right) - \frac{1}{2}\right]$$

$$= \frac{1}{2}\left[\frac{3}{2} + 2\cos\left(2x + \frac{\pi}{3}\right) \cdot \cos\frac{\pi}{3} - \cos\left(2x + \frac{\pi}{3}\right)\right] = \frac{3}{4}$$

$\therefore$ $f(x)$ is even function, periodic function and

$$f(0) = f(1) = \frac{3}{4}.$$

80. (a) Let $f(x) = \sin(\sin^{-1} x) = x, \forall\ x \in [-1, 1]$

So, $f(x)$ is one-one and onto.

(b) Let $f(x) = \dfrac{2}{\pi}\sin^{-1}(\sin x) = \dfrac{2}{\pi}x$

The range is $\left[-\frac{2}{\pi}, \frac{2}{\pi}\right]$.

So, $f(x)$ is one-one and into.

(c) Let $f(x) = \text{sgn}(x) \cdot \log e^x = \begin{cases} x, & x > 0 \\ -x, & x < 0 \\ 0, & x = 0 \end{cases}$

$\therefore$ Range is $[0, 1]$ for $x \in [-1, 1]$

So, $f(x)$ is many-one and into.

(d) Let $f(x) = x^3 \cdot \text{sgn}(x) = \begin{cases} x^3, & x > 0 \\ -x^3, & x < 0 \\ 0, & x = 0 \end{cases}$

So, $f(x)$ is many-one and into.

81. (a) $(f + g)(3.5) = f(3.5) + g(3.5) = (-0.5) + (0.5) = 0$

(b) $f(g(3)) = f(0) = 3$

(c) $f(g(2)) = f(-1) = 8$

(d) $(f - g)(4) = f(4) - g(4) = 0 - 26 = -26$

82. Here, $f(x) = x^2 - 2ax + a(a + 1)$

$$\Rightarrow \quad f(x) = (x - a)^2 + a, \ x \in [a, \infty)$$

Let $\quad y = f(x) = (x - a)^2 + a$

Clearly, $y \geq a$

$\Rightarrow \quad (x-a)^2 = (y-a) \Rightarrow x = a + \sqrt{y-a}$

$\therefore \quad f^{-1}(x) = a + \sqrt{x-a}$

Now, $\quad f(x) = f^{-1}(x)$

$\Rightarrow \quad (x-a)^2 + a = a + \sqrt{x-a}$

$\Rightarrow \quad (x-a)^2 = \sqrt{x-a}$

or $\quad (x-a)((x-a)^3 - 1) = 0$

$\Rightarrow \quad x = a \text{ or } a+1$

If $a = 5049$, then $a + 1 = 5050$

If $a + 1 = 5049$, then $a = 5048$

83. Here, $g(x) = \sin(\sin^{-1}\sqrt{\{x\}}) + \cos(\sin^{-1}\sqrt{\{x\}}) - 1$

$= \sqrt{\{x\}} + \cos(\cos^{-1}\sqrt{1-\{x\}}) - 1$

$= \sqrt{x} + \sqrt{1-\{x\}} - 1$

If $x \in I$, then $\{x\} = 0 \Rightarrow g(x) = 0$

Also, $g(-x) = g(x) \Rightarrow g(x)$ is even.

If $x \notin I$, then $\{-x\} = 1 - \{x\}$

$\Rightarrow \quad g(-x) = \sqrt{1-\{x\}} + \sqrt{\{x^2\}} - 1 = g(x)$

$\Rightarrow g(x)$ is even function.

$\therefore g(x) = 0, x \in I$ and $g(x) = g(-x), x \notin I$

$\Rightarrow g(x)$ is periodic function.

84. Given, $f(a+x) = f(a-x)$

(a) $f(2a-x) = f(a+(a-x)) = f(a-(a-x)) = f(x)$

$\therefore \quad f(2a-x) = f(x) \quad$ ...(i)

(b) $f(2a+x) = f(a+(a+x)) = f(a-(a+x)) = f(-x)$

$\therefore \quad f(2a+x) = f(-x) \quad$ ...(ii)

(c) $f(2b+x) = f(-x) \quad$ [from Eq. (ii)] ...(iii)

(d) From Eqs. (ii) and (iii), we get

$$f(2a+x) = f(2b+x)$$

$\therefore$ Period is $(2b-2a)$.

85. Here, $f(1+x) = f(1-x)$

$\Rightarrow f(x)$ is symmetric about $x = 1$.

$\therefore \quad g(x) = f(1+x) \Rightarrow g(-x) = f(1-x) = f(1+x) = g(x)$

$\Rightarrow g(x)$ is an even function.

86. $f(x) = \begin{cases} 10-4x, & \text{if } -\infty < x < 1 \\ 8-2x, & \text{if } 1 < x \leq 2 \\ 4, & \text{if } 2 < x \leq 3 \\ 2x-2, & \text{if } 3 < x \leq 4 \\ 4x-10, & \text{if } 4 < x < \infty \end{cases}$

Could be shown as

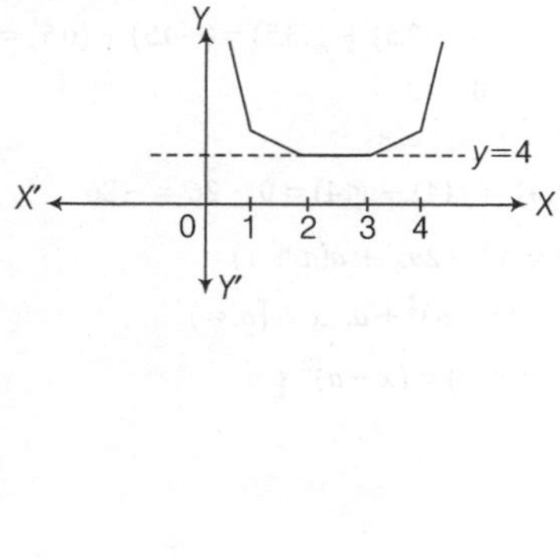

Clearly, the least value of $f(x)$ is 4.

The number of integral solutions of $f(x) = 4$ are two, i.e. $\{2, 3\}$.

Also, $\quad \pi - 1, e, \frac{12}{5} \in \{2, 3\}$

$\therefore \quad \dfrac{f(\pi-1) + f(e)}{2f\left(\frac{12}{5}\right)} = 1$

87. (a) Number of onto functions

$= 4^5 - 4 \cdot 3^5 + 6 \cdot 2^5 - 4 = 240$

(b) Number of onto functions, whose range is 3 elements

$= {}^4C_3(3^5 - 3 \cdot 2^5 + 3) = 4 \times 150 = 600$

(c) Number of onto functions, whose range is 2 elements

$= {}^4C_2(2^5 - 2) = 6 \times 30 = 180$

(d) Number of onto functions, whose range is 1 element

$= {}^4C_1 = 4$

88. Replacing x by 2,

$\Rightarrow \quad 2f(2) + 2f\left(\frac{1}{2}\right) - 2f(1) = 4$

$\Rightarrow \quad f(2) + f\left(\frac{1}{2}\right) = 2 + f(1) \quad$ ...(i)

Replacing x by 1,

$f(1) = -1 \quad$ ...(ii)

Replacing x by $\frac{1}{2}$, $2f\left(\frac{1}{2}\right) + \frac{1}{2}f(2) + 2 = \frac{5}{2}$

$\therefore \quad 2f(2) + \frac{1}{2}f\left(\frac{1}{2}\right) = \frac{1}{2} \quad$ ...(iii)

From Eqs. (i) and (iii), we get

$$f(2) = 1, f\left(\frac{1}{2}\right) = 0$$

89. Let $g(x) = f(100x)$, $g(1) = 1$

$\Rightarrow \quad g(x+1) - g(x) = x + 1$

Putting $x = 1, 2, 3, \ldots, 99$ and adding, we get

$$g(100) = 5050 = \sum_{r=1}^{100} r = \frac{100}{2}(100+1) = 5050$$

90. We have, $f(x) = [1 + \sin x] + [1 + \sin 2x] + \ldots + [1 + \sin nx]$

$= n + [\sin x] + [\sin 2x] + \ldots + [\sin nx]$

$\because \quad x \in (0, \pi)$

$\Rightarrow \quad 0 < \sin x \leq 1$

$\Rightarrow \quad f(x) = \begin{cases} n, & \text{if } n \text{ is even} \\ n+1, & \text{if } n \text{ is odd} \end{cases}$

91. (a) $f(x) = \begin{cases} 1, & \text{if } x \text{ is rational} \\ 0, & \text{if } x \text{ is irrational} \end{cases}$

$\Rightarrow f(x+k) = \begin{cases} 1, & \text{if } x \text{ is rational} \\ 0, & \text{if } x \text{ is irrational} \end{cases}$

$\Rightarrow f(x)$ is periodic but period cannot be determined.

(b) $f(x) = \begin{cases} x - [x], & 2n \le x < 2n+1 \\ \dfrac{1}{2}, & 2n+1 \le x < 2n+2 \end{cases}$

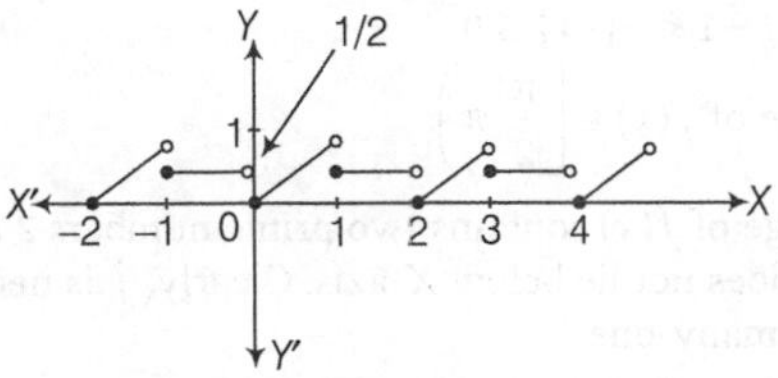

$\therefore$ $f(x)$ is periodic with period 2.

(c) $$f(x) = (-1)^{\left[\frac{2x}{\pi}\right]}$$

$$\Rightarrow f(x+\pi) = (-1)^{\left[\frac{2(x+\pi)}{\pi}\right]} = (-1)^{\left[\frac{2x}{\pi}\right]}$$

$\therefore$ $f(x)$ is periodic with period π.

(d) $$f(x) = (ax+a) - [ax+a] + \tan\left(\frac{\pi x}{2}\right)$$

$$= \{ax+a\} + \tan\left(\frac{\pi x}{2}\right)$$

$\therefore$ Period of $f(x)$ is LCM of $\left\{\frac{1}{a}, \frac{2}{1}\right\} = 2$

92. Here, $(x+1)\,f(x) = x$

$\Rightarrow$ $(x+1)\,f(x) - x = 0$ is $(n+1)$th degree

$\Rightarrow$ $(x+1)f(x) - x = (x-0)\,(x-1)(x-2)\ldots(x-n)\cdot k$

Put $x = n+1 \Rightarrow (n+2)f(n+1) - (n+1) = (n+1)!\,(k)$ ...(i)

Put $x = -1 \Rightarrow 1 = (-1)^{n+1}\cdot(n+1)!(k)$ ...(ii)

From Eqs. (i) and (ii), we get

$$f(n+1) = \frac{(n+1)+(-1)^{n+1}}{(n+2)} = \begin{cases} 1, & \text{when } n \text{ is odd} \\ \dfrac{n}{n+2}, & \text{when } n \text{ is even} \end{cases}$$

93. We have, $f(x+1) = \dfrac{f(x)-5}{f(x)-3}$ ...(i)

$$f(x)\cdot f(x+1) - 3f(x+1) = f(x) - 5$$

$$\Rightarrow \qquad f(x) = \frac{3f(x+1)-5}{f(x+1)-1}$$

On replacing x by $(x-1)$, we have $f(x-1) = \dfrac{3f(x)-5}{f(x)-1}$ ...(ii)

Using Eq. (i), we get $f(x+2) = \dfrac{f(x+1)-5}{f(x+1)-3}$

$$= \frac{\dfrac{f(x)-5}{f(x)-3} - 5}{\dfrac{f(x)-5}{f(x)-3} - 3} = \frac{2f(x)-5}{f(x)-2} \quad \text{...(iii)}$$

Using Eq. (ii), we get $f(x-2) = \dfrac{3f(x-1)-5}{f(x-1)-1}$

$$= \frac{3\left(\dfrac{3f(x)-5}{f(x)-1}\right) - 5}{\dfrac{3f(x)-5}{f(x)-1}} = \frac{2f(x)-5}{f(x)-2} \quad \text{...(iv)}$$

Using Eqs. (iii) and (iv), we have $f(x+2) = f(x-2)$

$$\Rightarrow \qquad f(x+4) = f(x)$$

$\therefore$ $f(x)$ is periodic with period 4.

94. $f(x) = 1 - x - x^3$

Replacing x by $f(x)$, $f(f(x)) = 1 - f(x) - f^3(x)$

Hence, the given equation is

$$f(f(x)) > f(1-5x),\ f(x) < 1-5x$$

$$f(x) = 1 - x - x^3$$

$$1 - x - x^3 < 1 - 5x$$

$$x^3 - 4x > 0$$

$$x(x-2)(x+2) > 0$$

So, $\qquad x \in (-2, 0) \cup (2, \infty)$

95. $(x-y)f(x+y) - (x+y)f(x-y) = 2y\,((x-y)(x+y))$

Let $\quad x - y = u,\ x + y = v$

$$uf(v) - vf(u) = uv\,(v-u)$$

$$\frac{f(v)}{v} - \frac{f(u)}{u} = v - u$$

$$\Rightarrow \qquad \left(\frac{f(v)}{v} - v\right) = \left(\frac{f(u)}{u} - u\right) = \text{constant}$$

Let $\quad \dfrac{f(x)}{x} - x = \lambda \Rightarrow f(x) = (\lambda x + x^2)$

$$f(1) = 2$$

$$\lambda + 1 = 2 \quad \Rightarrow \lambda = 1$$

$$f(x) = x^2 + x$$

96. Period of $\sin x = 2\pi$

and period of $\cos(\sqrt{4-a^2}\,x) = \dfrac{2\pi}{\sqrt{4-a^2}}$.

$$\Rightarrow \quad \text{LCM}\left(2\pi, \frac{2\pi}{\sqrt{4-a^2}}\right) = 4\pi \qquad \text{[given]}$$

i.e. $\qquad \sqrt{4-a^2} = \dfrac{p}{2}$, where $p = 1, 3$

Hence, $a^2 = \dfrac{15}{4}, \dfrac{7}{4}$; $a = \pm\dfrac{\sqrt{15}}{2}, \pm\dfrac{\sqrt{7}}{2}$

97. $f(x+y) = f(x)f(a-y) + f(y)\,f(a-x)$ ...(i)

Put $x = y = 0$, we get $f(a) = \dfrac{1}{2}$

Let $y = 0$

$$\Rightarrow \qquad f(x) = f(x)f(a) + f(0)\cdot f(a-x)$$

$$\Rightarrow \qquad f(x) = \frac{1}{2}f(x) + \frac{1}{2}f(a-x)$$

$$\Rightarrow \qquad f(x) = f(a-x)$$

Put $y = a - x$ in Eq. (i),

$$f(a) = (f(x))^2 + (f(a-x))^2$$

$$\Rightarrow \qquad (f(x))^2 = \frac{1}{4}$$

$$\Rightarrow \qquad f(x) = \pm\frac{1}{2} \qquad \left[\because f(x) \ne -\frac{1}{2}\right]$$

Hence, $\qquad f(x) = \dfrac{1}{2}$

98. Since, $f(g(x))$ is one-one function.

$\Rightarrow \quad f(g(x_1)) = f(g(x_2))$ whenever $x_1 \neq x_2$

$\Rightarrow \quad g(x_1) = g(x_2)$ whenever $x_1 \neq x_2$

$\Rightarrow$ $g(x)$ must be one-one.

Let $\quad f(x) = y$ is satisfied by $x = x_1$ and x_2.

If $g(x)$ is such that its range has only one of x_1 and x_2, then $f(g(x))$ can be one-one even, if $f(x)$ is many-one.

99. $x \sin x$ is a continuous function value of $|x \sin x|$ can be made as large as we like for sufficiently large values of x.

Therefore, range of $x \sin x = R \Rightarrow \dfrac{[x]}{\tan 2x} = 0$

For $x \in \left(0, \dfrac{\pi}{4}\right), [x] = 0$

For $x \in \left(-\dfrac{\pi}{4}, 0\right), \dfrac{[x]}{\tan 2x} > 0.$

Therefore, values of $\dfrac{[x]}{\tan 2x}$ are never negative.

Thus, range of $\dfrac{[x]}{\tan 2x} \neq R.$

$$\left|\frac{x}{\sin x}\right| > 1, \text{ whenever defined.}$$

Thus, range of $\dfrac{x}{\sin x}$ is not R.

$|x| + \sqrt{\{x\}}$ is a continuous function and

$$\lim_{x \to \infty} ([x] + \sqrt{\{x\}}) = \infty,$$

$$\lim_{x \to -\infty} ([x] - \sqrt{\{x\}}) = -\infty$$

Thus, range of $[x] + \sqrt{\{x\}} = R.$

100. (a) $f(x) = e^{\ln \sec^{-1} x}$. $g(1) = 0$ but $f(1)$ is not defined.

Thus, f and g are not identical.

(b) $f(x) = \tan(\tan^{-1} x) = x, \forall\ x \in R$

and $g(x) = \cot(\cot^{-1} x) = x, \forall\ x \in R$

Thus, f and g are identical.

(c) $f(x) = \begin{cases} 1, & x > 0 \\ 0, & x = 0 \\ -1, & x < 0 \end{cases}$

$$g(x) = \text{sgn}(\text{sgn } x) = \text{sgn}\begin{cases} 1, & x > 0 \\ 0, & x = 0 \\ -1, & x < 0 \end{cases}$$

$$g(x) = \begin{cases} 1, & x > 0 \\ 0, & x = 0 \\ -1, & x < 0 \end{cases}$$

Thus, f and g are identical.

(d) $g(x) = \cot^2 x - \cos^2 x = \dfrac{\cos^2 x}{\sin^2 x} - \cos^2 x$

$= \cot^2 x \cos^2 x = f(x), \forall\ x \neq \pi$

Thus, f and g are identical.

101. We have, $f(x) = \cos^{-1}(-\{-x\})$

Domain of $f(x) \in R.$

As, $0 \le \{-x\} < 1$ for all $x \in R$

$\Rightarrow \quad -1 < -\{-x\} \le 0$

So, range of $f(x) \in \left[\dfrac{\pi}{2}, \pi\right).$

$\therefore$ Range of $f(x)$ contains two prime numbers 2 and 3. Graph of $f(x)$ does not lie below X-axis. Clearly, f is neither even nor odd but many-one.

$\therefore f(x+1) = f(x) \Rightarrow f(x)$ is periodic.

102. If we take example of $x + \sin x$, it is non-periodic whereas its derivative $1 + \cos x$ is periodic.

103. The value of $\sin \sqrt{2}x + \sin ax$ can be equal to 2, if $\sin \sqrt{2}x$ and $\sin ax$ both are equal to one but both are not equal one for any common value of x.

104. Clearly, $f(x)$ is many-one and into function.

105. $f(x) = 2 \cos \dfrac{\pi}{5} \sin x - 2 \cos \dfrac{2\pi}{5} \sin x$

$$= 2 \sin x \left[\cos \frac{\pi}{5} - \cos \frac{2\pi}{5}\right]$$

$$= 2 \sin x [\cos 36^\circ - \cos 72^\circ]$$

$$= 2 \sin x [\cos 36^\circ - \sin 18^\circ]$$

$$= 2 \sin x \left[\frac{\sqrt{5}+1}{4} - \left(\frac{\sqrt{5}-1}{4}\right)\right] = \sin x$$

$\therefore$ Range is $[-1, 1]$.

106. Period of $2 \cos \dfrac{1}{3}(x - \pi)$ and $4 \sin \dfrac{1}{3}(x - \pi)$ are $\dfrac{2\pi}{\frac{1}{3}}, \dfrac{2\pi}{\frac{1}{3}}$

or $6\pi, 6\pi$.

$\therefore$ Period of their sums $= 6\pi$

107. $(f(\sin 2x))^2 = \sin^2 x + \cos^2 x + 2 \sin x \cos x$

$$= 1 + \sin 2x$$

$\Rightarrow \quad f(x) = \sqrt{1 + x}, \forall\ x \in [-1, 1]$

$\Rightarrow \quad$ If $x \in \left[-\dfrac{\pi}{4}, \dfrac{\pi}{4}\right]$, then

$$f(\tan^2 x) = \sqrt{1 + \tan^2 x} = \sec x$$

108. Fifth degree equation must have atleast one real root. If it had two real roots, $f'(x) = 0$ must have one real root.

109. Range of $\dfrac{1}{1 + x^2} \in (0, 1]$

For domain R, $\log\left(\dfrac{1}{1 + x^2}\right) \in (-\infty, 0]$

110. $f(x) = \sqrt{2} \sin x + \sqrt{2} \cos x - \sqrt{2} \cos x + c = \sqrt{2} \sin x + c$

$\Rightarrow \quad f(0) = f(\pi) = c$

Hence, many-one function.

111. If $f(x) = a_0 + a_1x + a_2x^2 + \ldots + a_nx^n, a_n \neq 0.$

Then, $f^{(n+1)}(x) = 0$, for all x while any derivative of $\sin x$ is never a zero function.

Hence, $\sin x$ is not polynomial function.

112. f is injective, since $x \neq y\ (x, y \in R)$.

$$\Rightarrow \log_a \{x + \sqrt{x^2+1}\} \neq \log_a \{y + \sqrt{y^2+1}\}$$

$$\Rightarrow \quad f(x) \neq f(y)$$

f is onto because $\log_a (x + \sqrt{x^2+1}) = y$

$$\Rightarrow \quad x = \frac{a^y - a^{-y}}{2}$$

113. $f(x) - 2f\left(\frac{x}{2}\right) + f\left(\frac{x}{4}\right) = x^2$

$$\therefore \quad f\left(\frac{x}{2}\right) - 2f\left(\frac{x}{4}\right) + f\left(\frac{x}{8}\right) = \left(\frac{x}{2}\right)^2$$

$$f\left(\frac{x}{4}\right) - 2f\left(\frac{x}{8}\right) + f\left(\frac{x}{16}\right) = \left(\frac{x}{4}\right)^2$$

$$\vdots \qquad \vdots \qquad \vdots$$

$$f\left(\frac{x}{2^n}\right) - 2f\left(\frac{x}{2^{n+1}}\right) + f\left(\frac{x}{2^{n+2}}\right) = \left(\frac{x}{2^n}\right)^2$$

On adding, we get

$$f(x) - f\left(\frac{x}{2}\right) - f\left(\frac{x}{2^{n+1}}\right) + f\left(\frac{x}{2^{n+2}}\right) = x^2\left(1 + \frac{1}{2^2} + \ldots + \frac{1}{2^{2n}}\right)$$

As $n \to \infty$, we get $f(x) - f\left(\frac{x}{2}\right) = \frac{4x^2}{3}$

Repeating the same procedure again, we get

$$f(x) - f(0) = \frac{16x^2}{9}$$

$$\therefore \quad f(3) = 16 + f(0)$$

114. $f(x) - f(0) - x = 0$

$$\Rightarrow \quad \frac{16x^2}{9} - x = 0 \Rightarrow x = 0, \frac{9}{16}$$

$\therefore$ Two solutions.

115. $f'(x) = \frac{32x}{9} \quad \therefore f'(0) = 0$

116. For integral solution,

$$x + y - [x][y] = 0 \Rightarrow x + y - xy = 0$$

$$\Rightarrow \quad -1 + x + y - xy = -1 \Rightarrow (x-1)(y-1) = 1$$

i.e. only possible, if

$$(x - 1 = 1, y - 1 = 1) \text{ or } (x - 1 = -1, y - 1 = -1)$$

$$\Rightarrow \quad x = 2, y = 2 \text{ or } x = 0, y = 0$$

$\therefore$ Solutions are (0, 0) or (2, 2).

117. For non-integral solution,

Let $\quad x = [x] + f_1$

and $\quad y = [y] + f_2$

$$\therefore \quad [x][y] = [x] + f_1 + [y] + f_2$$

$$([x] - 1)([y] - 1) = f_1 + f_2 + 1$$

Now, $\quad 0 \leq f_1 + f_2 < 2$

$$\therefore \quad f_1 + f_2 = 1$$

$$\Rightarrow ([x] - 1)([y] - 1) = 2$$

Which is possible for

$$\therefore \quad [x] = 3 \text{ and } [y] = 2$$

$$\Rightarrow \quad [x] = 2 \text{ and } [y] = 3$$

$$\Rightarrow \quad [x] = -1 \text{ and } [y] = 0$$

$$\Rightarrow \quad [x] = 0 \text{ and } [y] = -1$$

$\therefore$ The $x + y = [x][y]$ becomes $x + y = 6 \Rightarrow x + y = 0$

$\therefore$ Non-integer solution lies on $x + y = 6$

$$\Rightarrow \quad x + y = 0$$

118. We have, $f(x) = \frac{1}{2}\left[f(xy) + f\left(\frac{x}{y}\right)\right]$ ...(i)

Interchanging x and y, we get

$$f(y) = \frac{1}{2}\left[f(xy) + f\left(\frac{y}{x}\right)\right] \quad \text{...(ii)}$$

On subtracting, we get

$$f(x) - f(y) = \frac{1}{2}\left\{f\left(\frac{x}{y}\right) - f\left(\frac{y}{x}\right)\right\} \quad \text{...(iii)}$$

From Eq. (i), put $x = 1$, we get

$$f(1) = \frac{1}{2}\left[f(y) + f\left(\frac{1}{y}\right)\right] \quad \text{...(iv)}$$

$$f(1) = 0 \Rightarrow f(y) = -f\left(\frac{1}{y}\right)$$

Hence, $\quad f\left(\frac{x}{y}\right) = -f\left(\frac{y}{x}\right)$

$\therefore$ From Eq. (iii) becomes $f(x) - f(y) = f\left(\frac{x}{y}\right)$

119. $f'(x) = \lim_{h \to 0} = \frac{f(x+h) - f(x)}{h}$

Using the result in Q. No. 118.

$$\lim_{h \to 0} \frac{f\left(\frac{x+h}{x}\right)}{h} = \lim_{h \to 0} \frac{f\left(1 + \frac{h}{x}\right)}{h}$$

$$= \lim_{h \to 0} \frac{f\left(1 + \frac{h}{x}\right)}{\frac{h}{x} \cdot x} = \frac{1}{x} f'(1) = \frac{1}{x} \cdot 2 = \frac{2}{x}$$

$$\therefore \quad f'(3) = \frac{2}{3}$$

120. From Q. No. 119, $f'(x) = \frac{2}{x}$

$$f(x) = 2 \log x + C$$

$$\Rightarrow \quad f(1) = 0 \Rightarrow C = 0$$

$$\therefore \quad f(x) = 2 \log x$$

$$\therefore \quad f(e) = 2 \log e = 2$$

121. $f(x)$ = odd degree polynomial + bounded function

$$\cot^{-1} x \in (0, \pi)$$

Also, $\quad f'(x) > 0$

$\therefore$ $f(x)$ is one-one and range of $f(x) \in R$

$\therefore$ $f(x)$ is onto.

$\Rightarrow$ $f(x)$ is one-one onto.

122. We have, $f(x) = x^4 + 1 + \dfrac{1}{x^2 + x + 1}$

= even degree polynomial + bounded function

$$\frac{1}{x^2 + x + 1} \in \left(0, \frac{4}{3}\right)$$

$$f'(x) = \frac{4x^3(x^2 + x + 1)^2 - 2x - 1}{(x^2 + x + 1)^2}$$

$\Rightarrow f'(x) = 0$ has at least one root which is repeated odd number of times or it has one root which is not repeated, since numerator of $f'(x)$ is a polynomial of degree 7.

$\Rightarrow$ $f(x) = 0$ has a point of extreme.

$\therefore$ $f(x)$ is many-one into.

123. $f(x)$ = odd degree polynomial + bounded function $\sin 2x \Rightarrow$ $f(x)$ is onto.

$f(x)$ is one-one, if $f'(x) \ge 0$ or $f'(x) \le 0, \forall x$

$\Rightarrow \quad a \ge 1 \cup a \le -1 \Rightarrow a \in R - (-1, 1)$

124. $D_h = \{-1, 1\}$, as minimum occurs before maxima for $f(x)$.

$\therefore \quad a_3 = -1$

Now, $\quad g(x) = a_0 + a_1 x + a_2 x^2 - x^3$

$$g'(x) = a_1 + 2a_2 x - 3x^2 = -3(x-3)(x+3) = -3x^2 + 27$$

$\therefore \quad a_1 = 27, a_2 = 0$

$\therefore \quad a_1 + a_2 = 27$

125. Also, $\quad g(-3) > 0$ and $g(3) > 0$

$\Rightarrow \quad a_0 > 54$ and $a_0 < -54$

$\therefore \quad a_0 > 54$

126. Now, $\quad g(x) = a_0 + 27x - x^3$

$$f(x) = \sqrt{a_0 + 27x - x^3}$$

$$f(10) = \sqrt{a_0 + 270 - 1000}$$

Clearly, $f(10)$ is defined for $a_0 > 730$.

127. $\because ff^{-1}(x) = x$

$$2^{(f^{-1}(x))^4 - 4(f^{-1}(x))^2} = x$$

$\Rightarrow (f^{-1}(x))^4 - 4(f^{-1}(x))^2 - \log_2 x = 0$

$\therefore \quad (f^{-1}(x))^2 = 2 + \sqrt{4 + \log_2 x}$

$\therefore$ Range of $f^{-1}(x)$ is $[2, \infty)$.

$\therefore \quad f^{-1}(x) = \sqrt{2 + \sqrt{4 + \log_2 x}}$

$\therefore \quad f^{-1}(x) > 0$

128. $$g(x) = \frac{\sin x + 4}{\sin x - 2}$$

$\Rightarrow \quad g'(x) = \dfrac{-6\cos x}{(\sin x - 2)^2} \ge 0 \quad \left[\because x \in \left[\dfrac{\pi}{2}, \pi\right]\right]$

$\Rightarrow g(x)$ is an increasing function, hence one-one function.

$\therefore$ Range is $\left[g\left(\dfrac{\pi}{2}\right), g(\pi)\right]$ and lies in $[-5, -2]$.

129. $x : x$ in domain of $g^{-1}(x)$,

$g^{-1}(x)$ in domain of $f^{-1}(x)$

$g^{-1}(x) = \sin^{-1}\dfrac{2(x+2)}{x-1}, \forall x \in [-5, -2]$ ...(i)

$\Rightarrow \quad \sin^{-1}\dfrac{2(x+2)}{x-1} \ge 1$

i.e. $\quad 1 \le \sin^{-1}\dfrac{2(x+2)}{x-1} \le \dfrac{\pi}{2}$

$\Rightarrow \quad \sin 1 \le \dfrac{2(x+2)}{x-1} \le 1$

Solving this, we get

$$x \le -\frac{(4 - \sin 1)}{2 - \sin 1} \text{ or } x > 1 \quad \text{...(ii)}$$

From Eqs. (i) and (ii), we get

$$x \in \left[-5, -\frac{(4 + \sin 1)}{2 - \sin 1}\right]$$

Sol. (Q. Nos. 130 to 132)

$P(x) + 1 = 0$ has a thrice repeated root at $x = 1$.

$P'(x) = 0$ has a twice repeated root at $x = 1$.

Similarly, $P'(x) = 0$ has a twice repeated root at $x = -1$.

$\Rightarrow P'(x)$ is divisible by $(x-1)^2(x+1)^2$

$\because \quad P'(x)$ is degree at most 4.

$\Rightarrow \quad P'(x) = \alpha(x-1)^2(x+1)^2$

$\therefore \quad P(x) = \alpha\left(\dfrac{x^5}{5} - \dfrac{2}{3}x^3 + x\right) + c$

Now, $\quad P(1) = -1$ and $P(-1) = 1 \Rightarrow \alpha = -\dfrac{15}{8}$ and $c = 0$

$\therefore \quad P(x) = -\dfrac{15}{8}\left(\dfrac{x^5}{5} - \dfrac{2}{3}x^3 + x\right) = -\dfrac{15}{8}x\left(\dfrac{x^4}{5} - \dfrac{2x^2}{3} + 1\right)$

has only one real root $x = 0$, as $\dfrac{x^4}{5} - \dfrac{2x^2}{3} + 1$ has imaginary roots.

Also, sum of pairwise product of all roots $= -\dfrac{10}{3}$.

Also, $\quad P''(x) = -\dfrac{15}{2}(x^3 - x)$

Let $\quad f(x) = -\dfrac{15}{2}(x^3 - x)$

$$f'(x) = -\frac{15}{2}(3x^2 - 1) = -\frac{15}{2}(\sqrt{3}x - 1)(\sqrt{3}x + 1)$$

− + −

$\Rightarrow$ Maximum at $x = \dfrac{1}{\sqrt{3}}$.

130. (a) **131.** (c) **132.** (b)

Sol. (Q. Nos. 133 to 135)

Given, $\quad f^{-1}(x) = \dfrac{1}{f(x)}$ ...(i)

Replace x by $f(x)$.

$\Rightarrow \quad f(f(x)) = \dfrac{1}{x} \Rightarrow f^{-1}\left(\dfrac{1}{x}\right) = f(x)$

$\Rightarrow \quad f^{-1}(x) = f\left(\dfrac{1}{x}\right)$ ...(ii)

From Eqs. (i) and (ii), we get,

$$\frac{1}{f(x)} = f\left(\frac{1}{x}\right) \Rightarrow f(x)\cdot f\left(\frac{1}{x}\right) = 1 \quad ...(iii)$$

Putting $x = 1$, we get

$$(f(1))^2 = 1$$

133. $f(1) = 1$ [as $f(1) > 0$]

134. If $f(x)$ is continuous, then being bijective, it will be monotonic.

135. $f(f(x))$ would be increasing but $f(f(x)) = \frac{1}{x}$, i.e. decreasing.

Thus, contradicts the facts of $f(x)$ being continuous.

Sol. (Q. Nos. 136 to 137)

$$f(m+n) = f(m) + f(n)$$
$$n = N-1$$
$$\therefore \quad f(N) = f(1) + f(N-1)$$
$$f(N-1) = f(1) + f(N-2)$$
$$f(N-2) = f(1) + f(N-3)$$
$$\vdots \qquad \vdots \qquad \vdots$$
$$f(2) = f(1) + f(1)$$

On adding all, we get

$$f(N) = Nf(1)$$

Now, if $f(1) = 1$

$\Rightarrow f(N) = N$, which contains even numbers

If $f(1) = 2 \Rightarrow f(N) = 2N$, which contains even numbers

$\therefore \quad f(1)$ is 1 or 2.

136. (c) **137.** (a)

138. (A) $\sqrt{\sin(\cos x)}$, $\sin(\cos x) \ge 0$

$$2n\pi \le \cos x \le (2n+1)\pi$$
$$0 \le \cos x \le 1$$
$$x \in \left[2n\pi - \frac{\pi}{2}, 2n\pi + \frac{\pi}{2}\right]$$

(B) $(\sqrt{\cos(\sin x)})^{-1}$

$$\cos(\sin x) > 0$$
$$-\frac{\pi}{2} < \sin x < \frac{\pi}{2} \Rightarrow x \in R$$

(C) $\tan(\pi \sin x)$

$$\pi \sin x \ne \pm\frac{\pi}{2}$$
$$\sin x \ne \pm\frac{1}{2}$$
$$\therefore \quad x \ne n\pi \pm \frac{\pi}{6} \Rightarrow x \in R - \left\{n\pi \pm \frac{\pi}{6}\right\}$$

(D) $\ln(\tan x)$

$$\tan x > 0 \Rightarrow 0 < x < \frac{\pi}{2}$$
$$n\pi < x < n\pi + \frac{\pi}{2} \Rightarrow x \in \left(n\pi, n\pi + \frac{\pi}{2}\right)$$

139. (A) $|4\sin x - 1| < \sqrt{5} \Rightarrow -\sqrt{5} < 4\sin x - 1 < \sqrt{5}$

$$\Rightarrow 1-\sqrt{5} < 4\sin x < 1+\sqrt{5}$$
$$\Rightarrow \frac{1-\sqrt{5}}{4} < \sin x < \frac{1+\sqrt{5}}{4}$$
$$\Rightarrow -\frac{\pi}{10} < x < \frac{3\pi}{10}$$

But $x \in [0, \pi]$ domain $x \in [0, 3\pi/10)$.

(B) $4\sin^2 x - 8\sin x + 3 \le 0, \quad 0 \le x \le 2\pi$

$$\Rightarrow (2\sin x - 1)(2\sin x - 3) \le 0 \Rightarrow 2\sin x - 1 \ge 0$$
$$\Rightarrow \sin x \ge \frac{1}{2} \Rightarrow x \in \left[\frac{\pi}{6}, \frac{5\pi}{6}\right]$$

(C) $|\tan x| \le 1; x \in [-\pi, \pi], \quad -1 \le \tan x \le 1$

$$n\pi - \frac{\pi}{4} \le x \le n\pi + \frac{\pi}{4}$$

Put $n = 0$, $\quad -\frac{\pi}{4} \le x \le \frac{\pi}{4}$

Put $n = 1$, $\quad \pi - \frac{\pi}{4} \le x \le \pi + \frac{\pi}{4}$

$$\frac{3\pi}{4} \le x \le \frac{5\pi}{4}$$

But, from domain $\frac{3\pi}{4} \le x \le \pi$

Put $x = -1$, $-\pi - \frac{\pi}{4} \le x \le -\pi + \frac{\pi}{4} \Rightarrow -\frac{5\pi}{4} \le x \le -\frac{3\pi}{4}$

But from domain $-\pi \le x \le -\frac{3\pi}{4}$

Finally, $\left[-\pi, -\frac{3\pi}{4}\right] \cup \left[-\frac{\pi}{4}, \frac{\pi}{4}\right] \cup \left[\frac{3\pi}{4}, \pi\right]$

But, $x \in [0, \pi]$ $\therefore x \in \left[0, \frac{\pi}{4}\right] \cup \left[\frac{3\pi}{4}, \pi\right]$

(D) $\cos x - \sin x \ge 1, 0 \le x \le 2\pi \Rightarrow \sqrt{2}\left(\frac{1}{\sqrt{2}}\cos x - \frac{\sin x}{\sqrt{2}}\right) \ge 1$

$$\Rightarrow \left(\cos x \cdot \cos\frac{\pi}{4} - \sin x \cdot \sin\frac{\pi}{4}\right) \ge \frac{1}{\sqrt{2}}$$
$$\Rightarrow \cos\left(x + \frac{\pi}{4}\right) \ge \frac{1}{\sqrt{2}}$$
$$\Rightarrow 2n\pi - \frac{\pi}{4} \le x + \frac{\pi}{4} \le 2n\pi + \frac{\pi}{4}$$
$$\Rightarrow 2n\pi - \frac{\pi}{2} \le x \le 2n\pi$$

On substituting suitable values of n, according to domain

$$x \in \left[\frac{3\pi}{2}, 2\pi\right] \cup \{0\}$$

140. (A) When $-1 \le x < 0$,

$$f(x) \in [0, 1) \text{ and } f(x) = x + 1$$
$$\therefore \quad f(f(x)) = (x+1)^2 - 1 = x^2 + 2x$$

(B) $\frac{(\cos 2x + 1)(\sec^2 x + 2\tan x)}{2}$

$$= \left(\frac{1-\tan^2 x}{1+\tan^2 x} + 1\right)\left(\frac{1+\tan^2 x + 2\tan x}{2}\right)$$
$$= \frac{1+\tan^2 x + 2\tan x}{1+\tan^2 x}$$

i.e. $f\left(\frac{2\tan x}{1+\tan^2 x}\right) = 1 + \frac{2\tan x}{1+\tan^2 x}$

$\therefore f(x) = 1 + x$

(C) Put $x = 0, y = 0, f(1) = (1+1)^2 = 2^2$

Put $x = 0, y = 1, f(2) = (1+2)^2 = 3^2$

By induction, $f(x) = (x+1)^2$

(D) When $4 < x < 5, \left[\frac{x}{4}\right] = 1$

$$\Rightarrow \quad f(x) = 2x + 3$$

$$\therefore \quad y = 2x + 3 \Rightarrow x = \frac{y-3}{2}$$

$$\therefore \quad f^{-1}(x) = \frac{x-3}{2}$$

141. (A) Simplifying the expression, we get $f(x) = \frac{5}{4}$.

(B) Period of $\tan(e^{\{x\}}) = 1$, the period of $[x+\alpha] - (x+\alpha) + \alpha - 5$ is also 1, hence 1 is the period.

(C) $h(x) - h(-x) = 0$, hence even function.

(D) Simplifying the expression, $k(x) = 0$.

142. Here, $f(x+y) - kxy = f(x) + 2y^2$

Put $x = 0 \Rightarrow f(y) = f(0) + 2y^2$

$$\Rightarrow \quad f(x) = 2x^2 + f(0)$$

Now, $f(1) = 2 + f(0) \Rightarrow f(0) = 0$

$$\therefore \quad f(x) = 2x^2$$

Hence, $f(x+y) \cdot f\left(\frac{1}{x+y}\right) = 4$

143. Given, $f(x - f(y)) = f(f(y)) + xf(y) + f(x) - 1$ ...(i)

On putting $x = 0$ and $f(y) = 0$, we get

$$f(0) = f(0) + f(0) - 1$$

$$\Rightarrow \quad f(0) = 1 \quad \text{...(ii)}$$

On putting $x = a$ and $f(y) = a$, we get

$$f(0) = f(a) + a^2 + f(a) - 1$$

$$\Rightarrow \quad f(a) = 1 - \frac{a^2}{2} \therefore f(x) = 1 - \frac{x^2}{2}$$

$$\Rightarrow \quad -f(10) = -1 + \frac{(10)^2}{2} = 49$$

Hence, $\frac{-f(10)}{7} = 7$

144. Given,

$$f(1) + 2f(2) + 3f(3) + \dots + nf(n) = n(n+1)f(n) \quad \text{...(i)}$$

On putting $(n+1)$ in place of n, we get

$$f(1) + 2f(2) + \dots + nf(n) + (n+1)f(n+1) = (n+1)(n+2)f(n+1) \quad \text{...(ii)}$$

On subtracting Eq. (i) from Eq. (ii), we get

$$(n+1)f(n+1) = (n+1)(n+2)f(n+1) - n(n+1)f(n)$$

$$\Rightarrow (n+1)f(n+1) - nf(n) = 0$$

$$\Rightarrow \quad nf(n) = (n+1)f(n+1)$$

$$\Rightarrow \quad 2f(2) = 3f(3) = \dots = nf(n)$$

From Eq. (i),

$$f(1) + [nf(n) + nf(n) + \dots \text{ to } (n-1) \text{ terms}] = n(n+1)f(n)$$

$$\Rightarrow \quad f(1) + (n-1)nf(n) = n(n+1)f(n)$$

$$\Rightarrow \quad f(1) = 2nf(n) \Rightarrow 1 = 2nf(n)$$

$$\Rightarrow \quad \frac{1}{f(n)} = 2n$$

$$\Rightarrow \quad \frac{1}{f(2010)} = 4020 \Rightarrow \frac{1}{(2010)\,f(2010)} = 2$$

145. If $f(x) = \frac{ax+b}{cx-a}, x \neq \frac{a}{c}$

Then, $f(f(x)) = x$

$$\therefore \quad f(f(x)) = x \text{ and } f\left(f\left(\frac{4}{x}\right)\right) = \frac{4}{x}$$

$$\therefore \quad f(f(x)) + f\left(f\left(\frac{4}{x}\right)\right) = x + \frac{4}{x} \geq 4$$

146. Here, $\gamma = 4 - (\alpha + \beta)$

$$\Rightarrow \quad \alpha^2 + \beta^2 + \gamma^2 = \alpha^2 + \beta^2 + (4 - (\alpha+\beta))^2 = 6$$

$$\Rightarrow \quad 2\beta^2 + 2\beta(\alpha - 4) + (2\alpha^2 - 8\alpha + 10) = 0$$

But, $\beta \in R \Rightarrow D \geq 0$

$$\Rightarrow \quad 3\alpha^2 - 8\alpha + 4 \geq 0$$

$$\Rightarrow \quad \alpha \in [2/3, 2]$$

$\therefore$ Integer values of α are $\{1, 2\}$.

147. Here, $f(x) = ax + b$

$$\Rightarrow \quad a(x + f(x)) + b = x + ax + b$$

$$\Rightarrow \quad a\,f(x) = x$$

$$\Rightarrow \quad f(x) = \frac{x}{a}$$

$$\Rightarrow \quad \frac{x}{a} = ax + b$$

$$\Rightarrow \quad b = 0 \Rightarrow a^2 = 1$$

$$\Rightarrow \quad a = \pm 1$$

$$\Rightarrow \quad f(x) = \pm x, \text{ i.e. 2 functions.}$$

148. Number of one-one functions $= {}^5P_3 = 60$

Number of strictly monotonic functions $= 10 + 10 = 20$

$$\therefore \quad \frac{\text{Number of one - one functions}}{\text{Number of strictly monotonic functions}} = \frac{60}{20} = 3$$

149. The number of different sets contains exactly 3 elements of $B = {}^5C_3 = 10$

The number of onto functions from A to the set contains 3 elements $= 10[3^4 - 3(2)^4 + 3] = 360$

$$\therefore \quad \frac{k}{60} = 6$$

150. Here, $f(2) = a\sin(2) + 2b\cos(2) + 8$

and $f(-2) = -a\sin(2) - 2b\cos(2) + 8$

On adding, we get

$$f(2) + f(-2) = 16$$

$$\therefore \quad f(-2) = 1$$

151. Let $y = f(x) = x^5 + e^{x/3}$

Then, $y(0) = 1$

Also, $g'(y) = \frac{1}{f'(x)}$

$$\therefore \quad g'(1) = \frac{1}{f'(0)} = 3$$

152. $f(x) = x^3 - 12x + p$

$\Rightarrow \quad f'(x) = 3x^2 - 12 = 0$

$\Rightarrow \quad x = \pm 2$

$\therefore \quad f(-2) = p + 16$ and $f(2) = p - 16$ for all $p \in \{1, 2, ..., 15\}$

$f(-2) > 0$, $f(2) < 0$. It has 3 roots in each case.

$\therefore \quad \Sigma\theta = 3 \times 15 = 45 \Rightarrow \frac{1}{5} \cdot \Sigma\theta = 9$

153. Given, $f(m) - f(n) = 3$

If $m = 125, n = 124 \Rightarrow (m - n)_{\min} = 1$

$m = 24, n = 5$

$\Rightarrow \quad f(m) - f(n) = 3 \Rightarrow (m-n)_{\max} = 19$

$\therefore \quad \frac{(m-n)_{\max} - (m-n)_{\min}}{2} = \frac{19-1}{2} = 9$

154. As, $x^2 + y^2 = 4$

$\Rightarrow \quad x = 2\cos\theta, y = 2\sin\theta$

$\therefore \quad \frac{x^3 + y^3}{x + y} = x^2 + y^2 - xy = 4 - 2\sin 2\theta$

$\Rightarrow \quad \left(\frac{x^3 + y^3}{x + y}\right)_{\max} = (4 - 2(\sin 2\theta))_{\max} = 4 + 2 = 6$

155. $f(2011) = (2 + 0 + 1 + 1)^2 = 16$

$f^2(2011) = (1 + 6)^2 = 49$

$f^3(2011) = (4 + 9)^2 = 169$

$f^4(2011) = (1 + 6 + 9)^2 = 256$

$f^5(2011) = (2 + 5 + 6) = 169$

$\vdots \quad \vdots$

$f^{2n}(2011) = 256$

$\therefore \quad f^{2n+1}(2011) = 169$

$\Rightarrow \quad \frac{f^{2011}(2011) - f^{2010}(2011)}{f^{2013}(2011) - f^{2012}(2011)} = \frac{169 - 256}{169 - 256} = 1$

156. As, $[\sin x] = \frac{x}{2\pi} - \left[\frac{x}{2\pi}\right] + \frac{2x}{5\pi} - \left[\frac{2x}{5\pi}\right]$

$[\sin x] = \left\{\frac{x}{2\pi}\right\} + \left\{\frac{2x}{5\pi}\right\}$

Thus, only one solution, i.e. $x = 10\pi$.

157. As, $xy - 6(x + y) = 0 \Rightarrow (x - 6)(y - 6) = 36$

$\therefore (x - 6, y - 6) = (1, 36), (2, 18), (3, 12), (4, 9), (6, 6),$
$(-36, -1), (-18, -2), (-12, -3), (-4, -9), (-6, -6),$

$\because \quad x \le y \Rightarrow \alpha = 10$

$\therefore \quad (\alpha - 6) = 4$

158. Here, $f(x) - 10x = (x - 1)(x - 2)(x - 3)(x - \alpha)$

$\therefore \quad f(12) + f(-8) = 19840$

$\Rightarrow \quad \frac{f(12) + f(-8)}{19840} = 1$

159. On adding, $a = \frac{3 - \cos 4\theta + 4\sin 2\theta}{2} = (1 + \sin 2\theta)^2$

On subtracting, $b = (1 - \sin 2\theta)^2$

$\Rightarrow \quad ab = \cos^4 2\theta \le 1$

160. Clearly, $x^2 + 4x \ge 0$

$2x^2 + 3 \ge 0 \Rightarrow x^2 + 4x \ge 2x^2 + 3$

and x is an integer.

$\therefore \quad x \in \{1, 2, 3\}$

$\therefore \quad n = 3$

Now, maximum value ${}^{x^2+4x}C_{2x^2+3} = 12$

$\therefore \quad Y = |\ln 12|$

$\therefore \quad [Y] = 2 \; (\ln e^2 < \ln 12 < \ln e^3)$

$\therefore \quad [n + [Y]] = [3 + 2] = 5$

161. $f(x - 1) + f(x + 1) = \sqrt{3} f(x)$

$\Rightarrow \quad f(x) + f(x + 2) = \sqrt{3} f(x + 1)$

Putting $x = x + 2$,

$f(x + 1) + f(x + 3) = \sqrt{3} f(x + 2)$

$f(x - 1) + 2f(x + 2) + f(x + 3) = \sqrt{3}[\sqrt{3} f(x + 1)]$

$f(x - 1) + f(x + 3) = f(x + 1)$

Again, putting $x = x + 2$,

$f(x + 1) + f(x + 5) = f(x + 3)$

$f(x - 1) + f(x + 5) = 0$

$f(x + 5) = -f(x - 1)$

$f(x) = -f(x + 6)$

$f(x + 12) = f(x)$

$\Rightarrow \quad \sum_{r=0}^{19} f(5 + 12r) = 20 f(5) = 20 \times 10 = 200$

Hence, the sum of digits $= 2 + 0 + 0 = 2$

162. Put $x = 1$,

$$2f(x) = f(xy) + f\left(\frac{x}{y}\right) \quad ...(i)$$

$$2f(1) = f(y) + f\left(\frac{1}{y}\right) \quad ...(ii)$$

$\Rightarrow \quad f(y) = -f\left(\frac{1}{y}\right)$

Replacing x by y and y by x in Eq. (i), we get

$$2f(y) = f(yx) + f\left(\frac{y}{x}\right) \quad ...(iii)$$

From Eqs. (i) and (iii), we get

$$2\{f(x) - f(y)\} = f\left(\frac{x}{y}\right) - \left\{-f\left(\frac{x}{y}\right)\right\} = 2f\left(\frac{x}{y}\right) \quad ...(iv)$$

$$f(x) - f(y) = f\left(\frac{x}{y}\right) \quad ...(v)$$

$\lim_{h \to 0} \frac{f(1 + h) - f(1)}{h} = f'(1) = 1$

$\lim_{h \to 0} \frac{f(1 + h)}{h} = 1$, as $f(1) = 0$

$f'(x) = \lim_{h \to 0} \frac{f(x + h) - f(x)}{h} = \lim_{h \to 0} \frac{f\left(1 + \frac{h}{x}\right)}{h} = \frac{1}{x}$

$f(x) = \log|x| + c$

$f(1) = 0 \Rightarrow c = 0$

$f(e) = 1$

163. $f(1)+2f(2)+3f(3)+\ldots+n(f(n))=n(n+1)f(n)$

$\Rightarrow f(1)+2f(2)+3f(3)+\ldots+(n+1)f(n+1)$
$=(n+1)(n+2)f(n+1)$

$\Rightarrow n(n+1)f(n)=(n+1)^2f(n+1)$

$\Rightarrow \quad nf(n)=(n+1)f(n+1)$

i.e. $\quad 2f(2)=3f(3)=\ldots=n(f(n))$

i.e. $\quad f(n)=\frac{1}{n}$

$\therefore \quad 2126\ f(1063)=2126\times\frac{1}{1063}=2$

164. Now, $f(x)=\frac{x^4+x^2+1}{x^2-x+1}\Rightarrow f(x)=x^2+x+1$

Now, $f(\omega^n)=\omega^{2n}+\omega^n+1=3$

$\because \quad \omega^n=1$, when n is a multiple of 3.

165. $f^2(x)\cdot f\left(\frac{1-x}{1+x}\right)=x^3$ ...(i)

Replacing x by $\frac{1-x}{1+x}$, we get

$$f^2\left(\frac{1-x}{1+x}\right)f(x)=\left(\frac{1-x}{1+x}\right)^3 \quad ...(ii)$$

By using Eqs. (i) and (ii), we get $f^3(x)=x^6\left(\frac{1+x}{1-x}\right)^3$

$\Rightarrow \quad f(x)=x^3\left(\frac{1+x}{1-x}\right)\Rightarrow f(-2)=\frac{8}{3}$

$\Rightarrow \quad [f(-2)]=2\Rightarrow |[f(-2)]|=2$

166. $f(2a-x)=f(x)$

$\Rightarrow f(2a-x)=-f(x)$

$\because$ f is odd $\Rightarrow f(x+4a)=f(x)$

$\Rightarrow$ f is periodic with period $4a$.

$\Rightarrow f(1+4r)=f(1)$

Now, $\sum_{r=0}^{\infty}[f(1)]^r=8 \Rightarrow \frac{1}{1-f(1)}=8$

$\Rightarrow \quad f(1)=\frac{7}{8}\Rightarrow 8f(1)=7$

167. Consider, $f(x)=\frac{e^x-e^{-x}}{e^x+e^{-x}}=\frac{e^{2x}-1}{e^{2x}+1}=1-\frac{2}{e^{2x}+1}$

$\Rightarrow \quad f'(x)=\frac{4e^{2x}}{(e^{2x}+1)^2}>0,\ \forall\ x\in R$

$\Rightarrow f(x)$ is an increasing function.

$\Rightarrow$ Domain : R, Range : $(-1,1)$

For $f:R\to(-1,1)$,

$$f(x)=\frac{e^x-e^{-x}}{e^x+e^{-x}},\ f^{-x}:(-1,1)\to R$$

$\Rightarrow \quad x=\frac{e^y-e^{-y}}{e^y+e^{-y}}$

$\Rightarrow \quad e^y=\sqrt{\frac{1+x}{1-x}}\Rightarrow f^{-1}(x)=\ln\sqrt{\frac{1+x}{1-x}}$

Hence, given equation is equivalent to $f(x)=f^{-1}(x)$.

$\Leftrightarrow \quad f(x)=x$ [as f is an increasing function]

$\Rightarrow \quad \ln\sqrt{\frac{1+x}{1-x}}=x\Rightarrow\frac{1+x}{1-x}=e^{2x}$

Now, draw the graph of $y=\frac{1+x}{1-x}$ and $y=e^{2x}$. They intersect each other at $x=0$.

168. Let $y=\frac{3x^2+9x+17}{3x^2+9x+7}=1+\frac{10}{3x^2+9x-7}$

Now, $\quad 3x^2+9x+7=3(x^2+3x)+7$

$=3\left(x+\frac{3}{2}\right)^2+\frac{1}{4}\geq\frac{1}{4}$ for all $x\in R$

Maximum value of $\frac{10}{3x^2+9x+7}$ is 40.

Maximum value of y is $1+40=41$

$\therefore \quad 5k+1=41\Rightarrow k=8$

169. The period of the function is found as follows

Given, $f(x)+f(x+4)=f(x+2)+f(x+6)$ ...(i)

$\therefore$ Replacing x by $x+2$, we get

$f(x+2)+f(x+6)=f(x+4)+f(x+8)$ ...(ii)

From Eqs. (i) and (ii), we get

$f(x)+f(x+4)=f(x+4)+f(x+8)$

$\Rightarrow \quad f(x)=f(x+8)\Rightarrow$ Thus, 8 is the period.

170. Since, $f(x)$ is symmetric about $y=x$.

$\Rightarrow f^{-1}(x)=f(x)$

$\therefore \quad f^2(x)=(f^{-1}(x))^2-px\cdot f(x)f^{-1}(x)+2x^2f(x)$

$\Rightarrow \quad f^2(x)=f^2(x)-px\cdot f(x)\cdot f(x)+2x^2f(x)$

$\Rightarrow \quad f(x)\cdot\{2x^2-px\cdot f(x)\}=0$

$\Rightarrow \quad f(x)=\frac{2x^2}{px}$ [as $f(x)\neq 0$]

$\Rightarrow \quad f(x)=\frac{2x}{p}\Rightarrow p=2$

171. Here, $f(x)=\frac{3x^2+mx+n}{x^2+1}=3+\frac{mx+n-3}{x^2+1}$

$\therefore \quad y=3+\frac{mx+n-3}{1+x^2}$

For y to lie in $[-4,3)$.

$mx+n-3<0,\forall x\in R.$

This is possible only if $m=0$.

When $m=0$, then $y=3+\frac{n-3}{1+x^2}$.

If $\quad x\to\infty\Rightarrow y_{\max}\to 3^-$

Now, $y_{\min}$ occurs at $x=0$. [as $1+x^2$ is maximum]

$y_{\min}=3+\frac{n-3}{1}=n$ ...(i)

Since, $\quad -4\leq y<3$

$\therefore \quad y_{\min}=-4$ ...(ii)

From Eqs. (i) and (ii), $n=-4$

$\therefore \quad \frac{m^2+n^2}{4}=\frac{16}{4}=4$

172. Since, $f(x)$ is monotonic.

$\Rightarrow f'(x) < 0$ or $f'(x) > 0, \forall x \in R.$

$\Rightarrow f'(px) < 0$ or $f'(px) > 0, \forall x \in R.$

$\Rightarrow f(px)$ is monotonic

$\Rightarrow f(x) + f(3x) + f(5x) + \ldots + f((2m-1)x)$

is a monotonic polynomial of odd degree $(2m-1)$, so it will attains all real values only once.

$\therefore \quad f(x) + f(3x) + f(5x) + \ldots + f((2m-1)x) = (2m-1)$ has only one root.

173. (i) We have,

$$(x)^2 = [x]^2 + 2x \quad \ldots(i)$$

$$\Rightarrow (i+f)^2 = [i+f]^2 = [i+f]^2 + 2i + 2f$$

[$\because x = i + f$, where i is integer and f is fraction]

$$\Rightarrow \{i+f\}^2 = (i)^2 + 2i + 2f$$

$$\Rightarrow i^2 + 2i + 1 = i^2 + 2i + 2f$$

$$\Rightarrow f = \frac{1}{2} \Rightarrow [x] = 0, (x) = 1$$

Put the value of (x) and $[x]$ in Eq. (i), we get

$$x = 0,\ n + \frac{1}{2}, n \in Z$$

(ii) We have,

$$[2x] - 2x = [x+1]$$

$$\Rightarrow -\{2x\} = [x] + 1 \Rightarrow [x] + \{2x\} = -1$$

$$\therefore x = -1, -\frac{1}{2}$$

(iii) We have, $[x]^2 + 2(x) = 0,\ [x]^2 = 3x, 0 \le x \le 2$

For $0 \le x < 1, [x] = 0, [x^2] = 0 \Rightarrow x = 0$

$1 \le x < \sqrt{2}, [x] = 1, [x^2] = 1 \Rightarrow x = 1$

$\sqrt{2} \le x < \sqrt{2}, [x] = 1,\ [x^2] = 2 \Rightarrow x = \frac{4}{3}$

$\sqrt{3} \le x < 2, [x] = 1,\ [x^2] = 3 \Rightarrow x = \frac{5}{3}$

$x = 2,\ [x] = 2, [x^2] = 4 \Rightarrow x = \frac{8}{3}$

$\therefore x \in \left\{0, 1, \frac{4}{3}, \frac{5}{3}\right\}$ $\left[\because \frac{8}{3} \notin [0, 2]\right]$

(iv) We have, $y = 4 - [x]^2$ and $[y] + y = 6$

Now, $[y] + y = 6, \Rightarrow y = 3$

On substituting the value of y in $y = 4 - [x]^2$

$\Rightarrow [x]^2 = 1,\ x = \pm 1, \pm 1 + k$, where k is fraction.

(v) We have,

$[x] + |x-2| \le 0, -1 \le x \le 3$

For $-1 \le x < 0, [x] = - \Rightarrow -1 - x + 2 \le 0 \quad x \ge 1$, Not possible.

$0 \le x \le 3$, $[x]$ and $|x-2|$ are positive.

Hence, no solution.

174. Given, $f(n) = 1! + 2! + 3! + \ldots + n!$

$\therefore \quad f(n+2) = Q(n)\, f(n) + P(n) f(n+1)$

$\Rightarrow \{1! + 2! + \ldots + (n+2)!\} = Q(n)\{1! + 2! + \ldots + n!\}$
$+ P(n)\{1! + 2! + \ldots + (n+1)!\}$

$\Rightarrow (1! + 2! + \ldots + n!) + \{(n+1)! + (n+2)!\} = \{Q(n) + P(n)\}$
$\{1! + 2! + 3! + \ldots + n!\} + P(n).(n+1)!$

Equating coefficients of $\{1! + 2! + \ldots + n!\}$ and $(n+1)!$ on both sides, we get

$$Q(n) + P(n) = 1 \text{ and } P(n) = (n+3)$$

So, $\quad P(n) = n + 3$

or $\quad P(x) = x + 3$ and $Q(x) = 1 - P(x)$

Hence, $\quad P(x) = x + 3$ and $Q(x) = -x - 2.$

175. Given, $f(x) = \dfrac{a^x}{a^x + \sqrt{a}}$ ...(i)

Now, $f(1-x) = \dfrac{a^{1-x}}{a^{1-x} + \sqrt{a}} = \dfrac{\sqrt{a}}{\sqrt{a} + a^x}$ ...(ii)

From Eqs. (i) and (ii), we have $f(x) + f(1-x) = 1$ ...(iii)

$$\Rightarrow f\left(\frac{r}{2n}\right) + f\left(\frac{2n-r}{2n}\right) = 1$$

$$\Rightarrow \sum_{r=1}^{2n-1} f\left(\frac{r}{2n}\right) + \sum_{r=1}^{2n-1} f\left(\frac{2n-r}{2n}\right) = 2n - 1$$

$$\Rightarrow \sum_{r=1}^{2n-1} f\left(\frac{r}{2n}\right) + \sum_{t=1}^{2n-1} f\left(\frac{t}{2n}\right) = 2n - 1 \quad [\text{putting } 2n - r = t]$$

Hence, $$\sum_{r=1}^{2n-1} 2f\left(\frac{r}{2n}\right) = 2n - 1$$

176. $f(x)$ is defined if $\left(\log_{|\sin x|}(x^2 - 8x + 23) - \dfrac{3}{\log_2 |\sin x|}\right) > 0$

$$\Rightarrow \log_{|\sin x|}\left(\frac{x^2 - 8x + 23}{8}\right) > 0$$

$$\left[\text{as } \frac{3}{\log_2 |\sin x|} = \frac{\log_2 8}{\log_2 |\sin x|} = \log_{|\sin x|} 8\right]$$

This is true, if

$$|\sin x| \ne 0, 1 \quad \text{and} \quad \frac{x^2 - 8x + 23}{8} < 1$$

[as $|\sin x| < 1 \Rightarrow \log_{|\sin x|} a > 0 \Rightarrow a < 1$]

Now, $$\frac{x^2 - 8x + 23}{8} < 1$$

$$\Rightarrow x^2 - 8x + 15 < 0$$

$$\Rightarrow x \in (3, 5) - \left\{\pi, \frac{3\pi}{2}\right\}$$

Hence, domain $= (3, \pi) \cup \left(\pi, \dfrac{3\pi}{2}\right) \cup \left(\dfrac{3\pi}{2}, 5\right).$

177. Given, $\quad nx + ny = xy$

$\Rightarrow xy - nx - ny + n^2 = n^2 \Rightarrow (x-n)(y-n) = n^2$

$\Rightarrow (x-n)$ and $(y-n)$ are two integral factors of n^2.

[as $x, y, n \in N$]

Obviously, if d is one divisor of n^2, then for each such divisor, there will be an ordered pair (x, y).

So, $S(n)$ = number of divisors of n^2.

(i) For $n = 6$, we have $d = 1, 2, 3, 6, 9, 12, 18, 36.$

Thus, $\quad S(6) = 9$

(ii) If n is prime, then $d = 1, n$ and n^2.

Hence, $S(n) = 3$, whenever n is prime.

178. We have, $\frac{1}{[x]} + \frac{1}{[2x]} = \{x\} + \frac{1}{3}$

Case I $x \in I$, $[x] = x$, $[2x] = 2x$ $\{x\} = 0$

$$\therefore \quad \frac{1}{x} + \frac{1}{2x} = \frac{1}{3} \Rightarrow \frac{3}{2x} = \frac{1}{3} \Rightarrow x = \frac{9}{2}$$

$\frac{9}{2} \notin I$, Not possible.

Case II, $x \in I + f$, where I is an integer and f is fraction

$$[x] = I, [2x] = 2I \;\; \{x\} = f, 0 \le f < \frac{1}{2}$$

$$\Rightarrow \quad \frac{1}{I} + \frac{1}{2I} = f + \frac{1}{3} \Rightarrow f = \frac{3}{2I} - \frac{1}{3}$$

$$\Rightarrow \quad 0 \le \frac{3}{2I} = -\frac{1}{3} < \frac{1}{2} \Rightarrow \frac{1}{3} \le \frac{3}{2I} < \frac{5}{6}$$

$$\Rightarrow \quad \frac{9}{5} < I \le \frac{9}{2} \Rightarrow I = 2, 3, 4$$

When $I = 2$, $f = \frac{5}{12}$; $I = 3$, $f = \frac{1}{6}$; $I = 4$, $f = \frac{1}{24}$

$$\therefore \quad x \in \left\{\frac{29}{12}, \frac{19}{6}, \frac{97}{24}\right\}.$$

179. The condition, $\sum_{i=1}^{n} f^{-1}(x_i) = \sum_{i=1}^{n} x_i$ can be written as;

$$\frac{1}{n}\sum_{i=1}^{n} f^{-1}(x_i) = \frac{1}{n}\sum_{i=1}^{n} x_i$$

i.e. AM of y-coordinates of f^{-1} = AM of x-coordinates of f

The given two conditions hold if and only if

$$x^2 + 3x - 3 = x$$

$$\Rightarrow \quad x^2 + 2x - 3 = 0$$

So, $\quad x = -3, +1$

But $\quad x \ge 0 \Rightarrow x = 1$ [neglecting $x = -3$]

Hence, we can write $\frac{1}{n}\sum_{i=1}^{n} x_i = 1$, which is the required result.

180. We have, $f(x) = x^2 - 2x$ ...(i)

and $g(x) = f(f(x) - 1) + f(5 - f(x))$ ...(ii)

$$\therefore \quad g(x) = f(x^2 - 2x - 1) + f(5 - x^2 + 2x)$$

$$= [\{x^2 - 2x - 1\}^2 - 2\{x^2 - 2x - 1\}] + [(5 - x^2 + 2x)^2 - 2\{5 - x^2 + 2x\}]$$

$$\Rightarrow \quad g(x) = 2x^4 - 8x^3 - 4x^2 + 24x + 18$$

To show $g(x) \ge 0$, we find its range.

i.e. $g'(x) = 8x^3 - 24x^2 - 8x + 24$ let $g'(x) = 0$

$$\Rightarrow \quad x = -1, 1 \text{ and } 3$$

$$\Rightarrow \quad g(x) \ge \min\{g(-1), g(1), g(3)\} = 0$$

Hence, $g(x) \ge 0, \forall x \in R$.

181. Given, $2 + f(x)f(y) = f(x) + f(y) + f(xy)$

or $1 - f(x) - f(y) + f(x)f(y) = f(xy) - 1$

or $(1 - f(x))(1 - f(y)) = f(xy) - 1$

The above result holds if and only if,

$$f(x) = 1 + x^n$$

If $\quad f(x) = a_n x^n + a_{n-1} x^{n-1} + \ldots + a_0$

Then, consider $(1 + f(x))(1 - f(y)) = f(xy) - 1$

Compare constant term on either side, we have

$$1 - a_0 = a_0 - 1 \Rightarrow a_0 = 1$$

Comparing coefficient of $x^n y^n$, we get

$a_n^2 = a_n \Rightarrow a_n = 1$ or otherwise polynomial would not be of n degree.

Comparing coefficient of $x, x^1, \ldots, x^{n-1}$ on either sides, we have

$$a_1 = a_2 = \ldots = a_{n-1} = 0$$

$$\Rightarrow \quad a_n = 1 \text{ and } f(x) = x^n + 1$$

Given, $\quad f(2) = 5$, i.e. $2^n + 1 = 5$

$$\Rightarrow \quad n = 2$$

Thus, $\quad f(x) = x^2 + 1$

$$\Rightarrow \quad f(f(2)) = f(5) = 5^2 + 1 = 26$$

182. Using AM $\ge$ GM, we have $\frac{a + b + c}{3} \ge (abc)^{1/3}$

$$\Rightarrow \quad \frac{a + b + c}{3} \ge (a + b + c)^{1/3} \quad [\because a + b + c = abc]$$

$$\Rightarrow \quad (a + b + c)^{2/3} \ge 3$$

$$\Rightarrow \quad (a + b + c) \ge 3\sqrt{3}$$

183. Clearly, $g(x) =$

$$\begin{cases} x^2, -2 \le x \le -1 \\ -x, -1 \le x \le -\frac{1}{4} \\ x + \frac{1}{2}, -\frac{1}{4} \le x < 0 \text{ as graphically if,} \\ x, 0 \le x \le 1 \\ x^2, 1 \le x \le 2 \text{ can be expressed as shown in the following figure} \end{cases}$$

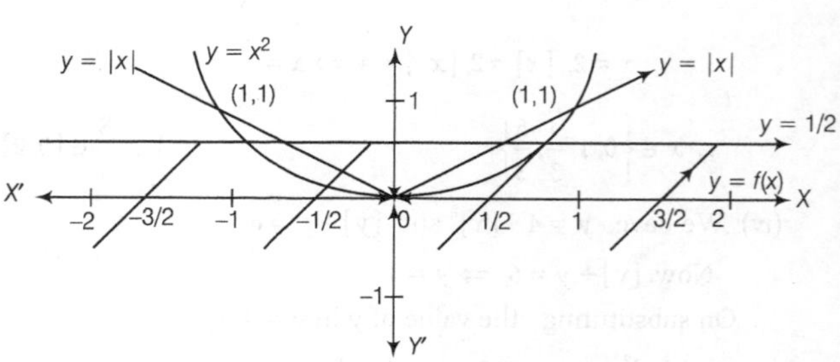

184. Let $\quad g(x) = f(x) - f(x + \pi)$ ...(i)

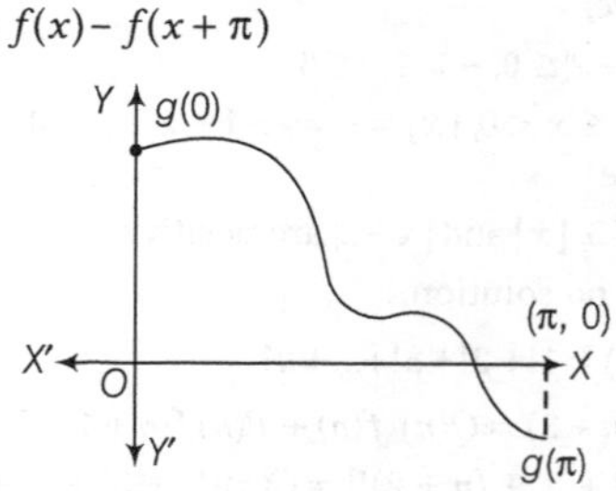

At $x = \pi$, $g(\pi) = f(\pi) - f(2\pi)$ ...(ii)

At $x = 0$, $g(0) = f(0) - f(\pi)$ ...(iii)

Adding Eqs. (ii) and (iii), we have

$$g(0) + g(\pi) = f(0) - f(2\pi) = 0$$

$$\Rightarrow \quad g(0) = -g(\pi)$$

$\Rightarrow g(0)$ and $g(\pi)$ are of opposite sign.

$\Rightarrow$ There is a point c between 0 and π such that

$$g(c) = 0 \quad \text{...(iv)}$$

From Eq. (i) putting $x = c$, we have

$$g(c) = f(c) - f(\pi + c) \quad \text{...(v)}$$

From Eqs. (iv) and (v), we have

$$f(c) - f(\pi + c) = 0$$

Hence, $\quad f(c) = f(\pi + c)$

185. We have, $g(t) = |t-1| - |t| + |t+1|$

$$\Rightarrow \quad g(t) = \begin{cases} -t, & t < -1 \\ t+2, & -1 \le t < 0 \\ 2-t, & 0 \le t < 1 \\ t, & t \ge 1 \end{cases}$$

Graph of $g(t)$ is

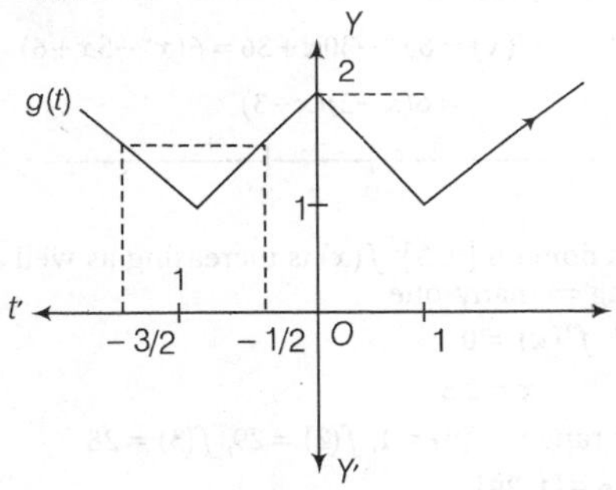

$$f(x) = \max\left\{g(t) : \frac{-3}{2} \le t \le x\right\} \quad x \in \left(\frac{-3}{2}, \infty\right)$$

Clearly from the graph,

$$f(x) = \begin{cases} 3/2, & \frac{-3}{2} < x \le \frac{-1}{2} \\ x+2, & \frac{-1}{2} \le x < 0 \\ 2, & 0 \le x \le 2 \\ x, & x \ge 2 \end{cases}$$

186. $n_1 n_2 = 2n_1 - n_2$ [given]

$\Rightarrow n_2(n_1 + 1) = 2n_1 \Rightarrow n_2 = \dfrac{2n_1}{n_1 + 1}$

$\Rightarrow \quad n_2 = 2 - \dfrac{2}{n_1 + 1}$

Since, n_1, n_2 are integers.

$\therefore \quad \dfrac{2}{n_1 + 1} \in$ Integer

$\Rightarrow \quad n_1 + 1 = -2, -1, 1, 2$

$\Rightarrow \quad n_1 = -3, -2, 0, 1$

$\Rightarrow \quad n_2 = 3, 4, 0, 1$

$\Rightarrow$ Integral solutions of the form (n_1, n_2) are $(-3, 3), (-2, 4), (0, 0), (1, 1)$.

187. PLAN

(i) For such questions, we need to properly define the functions and then we draw their graphs.

(ii) From the graphs, we can examine the function for continuity, differentiability, one-one and onto.

$$f_1(x) = \begin{cases} -x, & x < 0 \\ e^x, & x \ge 0 \end{cases} \Rightarrow f_2(x) = x^2, x \ge 0$$

$$f_3(x) = \begin{cases} \sin x, & x < 0 \\ x, & x \ge 0 \end{cases} \Rightarrow f_4(x) = \begin{cases} f_2(f_1(x)), & x < 0 \\ f_2(f_1(x)) - 1, & x \ge 0 \end{cases}$$

Now, $f_2(f_1(x)) = \begin{cases} x^2, & x < 0 \\ e^{2x}, & x \ge 0 \end{cases} \Rightarrow f_4 = \begin{cases} x^2, & x < 0 \\ e^{2x} - 1, & x \ge 0 \end{cases}$

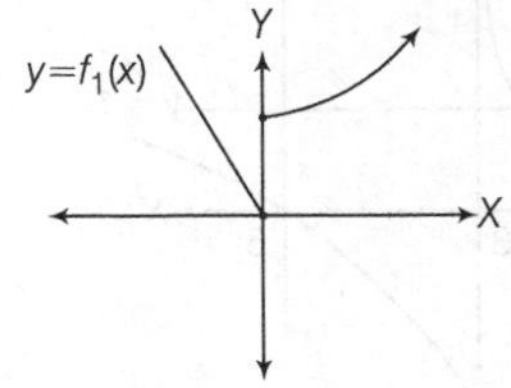

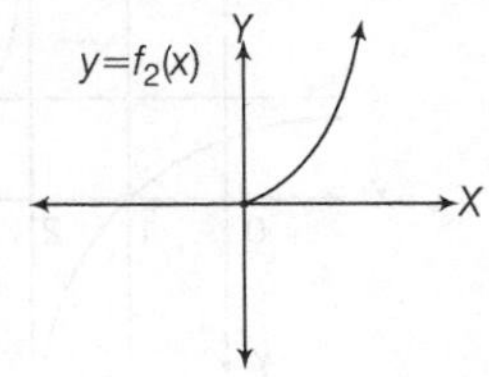

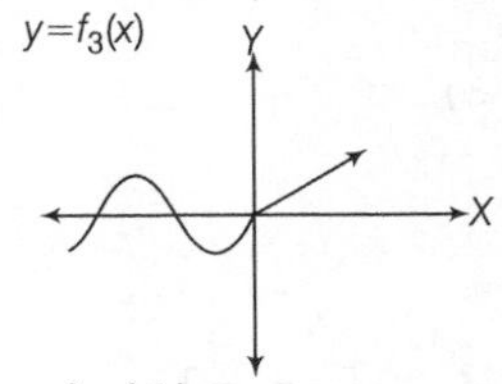

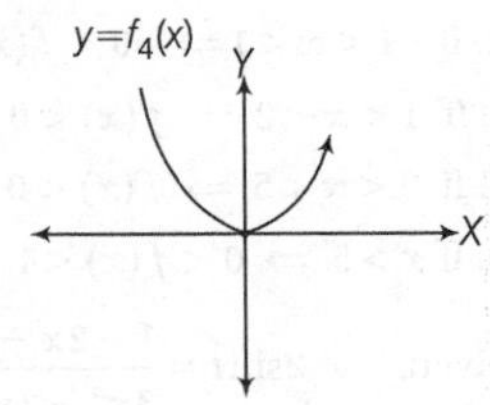

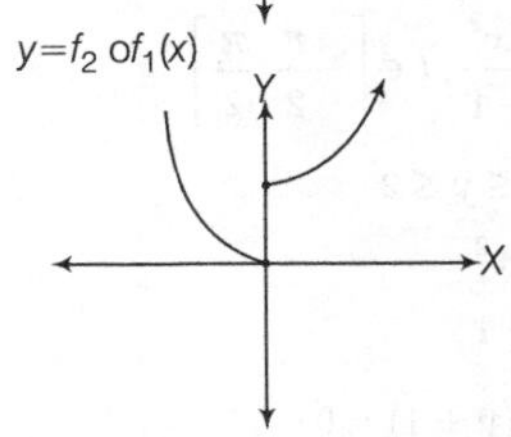

As, $f_4(x)$ is continuous, $f'_4(x) = \begin{cases} 2x, & x < 0 \\ 2e^{2x}, & x > 0 \end{cases}$

$f_4'(0)$ is not defined. Its range is $[0, \infty)$, thus f_4 is onto.

Also, horizontal line (drawn parallel to X-axis) meets the curve more than once, thus function is not one-one.

188. Given, $g\{f(x)\} = x \Rightarrow g'\{f(x)\} f'(x) = 1$

If $\quad f(x) = 1 \Rightarrow x = 0, f(0) = 1$

Substitute $x = 0$ in Eq. (i), we get

$$g'(1) = \frac{1}{f'(0)}$$

$\Rightarrow \quad g'(1) = 2 \quad \left[\because f'(x) = 3x^2 + \frac{1}{2}e^{x/2} \Rightarrow f'(0) = \frac{1}{2}\right]$

Alternate Solution

Given, $\quad f(x) = x^2 + e^{x/2}$

$\Rightarrow \quad f'(x) = 3x^2 + \dfrac{1}{2}e^{x/2}$

For $x = 0$, $f(0) = 1$, $f'(0) = \dfrac{1}{2}$ and $g(x) = f^{-1}(x)$

Replacing x by $f(x)$, we have

$$g(f(x)) = x \Rightarrow g'(f(x)) \cdot f'(x) = 1$$

Put $x = 0$, we get $g'(1) = \dfrac{1}{f'(0)} = 2$

189. Given, $F(x) = \int \sin^2 x\, dx = \int \frac{1-\cos 2x}{2} dx$

$$F(x) = \frac{1}{4}(2x - \sin 2x) + C$$

Since, $F(x+\pi) \neq F(x)$

Hence, Statement I is false.

But Statement II is true, as $\sin^2 x$ is periodic with period π.

190. Given, $f(x) = \frac{(x-1)(x-5)}{(x-2)(x-3)}$

The graph of $f(x)$ is shown as

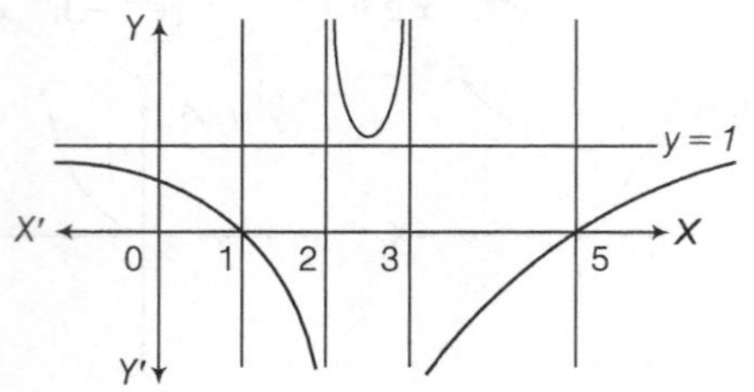

A. If $-1 < x < 1 \Rightarrow 0 < f(x) < 1$

B. If $1 < x < 2 \Rightarrow f(x) < 0$

C. If $3 < x < 5 \Rightarrow f(x) < 0$

D. If $x > 5 \Rightarrow 0 < f(x) < 1$

191. Given, $2\sin t = \frac{1-2x+5x^2}{3x^2-2x-1}, t \in \left[-\frac{\pi}{2}, \frac{\pi}{2}\right]$

Put $2\sin t = y \Rightarrow -2 \le y \le 2$

$\therefore \quad y = \frac{1-2x+5x^2}{3x^2-2x-1}$

$\Rightarrow (3y-5)x^2 - 2x(y-1) - (y+1) = 0$

Since, $x \in R - \{1, -1/3\}$

$[\text{as}, 3x^2 - 2x - 1 \neq 0 \Rightarrow (x-1)(x+1/3) \neq 0]$

$\therefore \quad D \ge 0$

$\Rightarrow 4(y-1)^2 + 4(3y-5)(y+1) \ge 0$

$\Rightarrow y^2 - y - 1 \ge 0$

$\Rightarrow \left(y - \frac{1}{2}\right)^2 - \frac{5}{4} \ge 0$

$\Rightarrow \left(y - \frac{1}{2} - \frac{\sqrt{5}}{2}\right)\left(y - \frac{1}{2} + \frac{\sqrt{5}}{2}\right) \ge 0$

$\Rightarrow y \le \frac{1-\sqrt{5}}{2}$ or $y \ge \frac{1+\sqrt{5}}{2}$

$\Rightarrow 2\sin t \le \frac{1-\sqrt{5}}{2}$ or $2\sin t \ge \frac{1+\sqrt{5}}{2}$

$\Rightarrow \sin t \le \sin\left(-\frac{\pi}{10}\right)$ or $\sin t \ge \sin\left(\frac{3\pi}{10}\right)$

$\Rightarrow t \le -\frac{\pi}{10}$ or $t \ge \frac{3\pi}{10}$

Hence, range of t is $\left[-\frac{\pi}{2}, -\frac{\pi}{10}\right] \cup \left[\frac{3\pi}{10}, \frac{\pi}{2}\right]$.

192. Given, $f_k(x) = \frac{1}{k}(\sin^k x + \cos^k x)$,

where $x \in R$ and $k > 1$

$f_4(x) - f_6(x) = \frac{1}{4}(\sin^4 x + \cos^4 x) - \frac{1}{6}(\sin^6 x + \cos^6 x)$

$= \frac{1}{4}(1 - 2\sin^2 x \cdot \cos^2 x) - \frac{1}{6}(1 - 3\sin^2 x \cdot \cos^2 x)$

$= \frac{1}{4} - \frac{1}{6} = \frac{1}{12}$

193. Plan To check nature of function.

(i) **One-one** To check one-one, we must check whether $f'(x) > 0$ or $f'(x) < 0$ in given domain.

(ii) **Onto** To check onto, we must check

Range = Codomain

Description of Situation To find range in given domain $[a, b]$, put $f'(x) = 0$ and find $x = \alpha_1, \alpha_2, \ldots, \alpha_n \in [a, b]$

Now, find $\{f(a), f(\alpha_1), f(\alpha_2), \ldots, f(\alpha_n), f(b)\}$ its greatest and least values gives you range.

Now, $f: [0,3] \to [1, 29]$

$$f(x) = 2x^3 - 15x^2 + 36x + 1$$

$\therefore \quad f'(x) = 6x^2 - 30x + 36 = 6(x^2 - 5x + 6)$

$= 6(x-2)(x-3)$

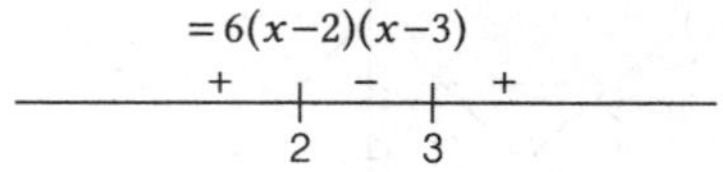

For given domain [0, 3], $f(x)$ is increasing as well as decreasing $\Rightarrow$ many-one

Now, put $f'(x) = 0$

$\Rightarrow \quad x = 2, 3$

Thus, for range $f(0) = 1, f(2) = 29, f(3) = 28$

$\Rightarrow$ Range $\in [1, 29]$

$\therefore$ Hence, the function is into.

194. $f(x) = x^2, g(x) = \sin x$

$(gof)(x) = \sin x^2$

$go(gof)(x) = \sin(\sin x^2)$

$(fogogof)(x) = (\sin(\sin x^2))^2$

Given, $(fogogof)(x) = (gogof)(x)$

$\Rightarrow (\sin(\sin x^2))^2 = \sin(\sin x^2)$

$\Rightarrow \sin(\sin x^2)\{\sin(\sin x^2) - 1\} = 0$

$\Rightarrow \sin(\sin x^2) = 0$ or $\sin(\sin x^2) = 1$

$\Rightarrow \sin x^2 = 0$ or $\sin x^2 = \frac{\pi}{2}$

$\therefore \quad x^2 = n\pi$ [i.e. not possible as $-1 \le \sin\theta \le 1$]

$x = \pm\sqrt{n\pi}$

195. Here, $f(x) = \frac{b-x}{1-bx}$, where $0 < b < 1, 0 < x < 1$

For function to be invertible, it should be one-one onto.

Check Range

Let $f(x) = y \Rightarrow y = \frac{b-x}{1-bx}$

$\Rightarrow y - bxy = b - x \Rightarrow x(1-by) = b - y$

$\Rightarrow \quad x = \dfrac{b-y}{1-by}$, where $0 < x < 1$

$\therefore \quad 0 < \dfrac{b-y}{1-by} < 1$

$\Rightarrow \quad \dfrac{b-y}{1-by} > 0$ and $\dfrac{b-y}{1-by} < 1$

$\Rightarrow \quad y < b$ or $y > \dfrac{1}{b}$ ...(i)

and $\quad \dfrac{(b-1)(y+1)}{1-by} < 0 \Rightarrow -1 < y < \dfrac{1}{b}$ ...(ii)

From Eqs. (i) and (ii), we get

$y \in (-1, b) \subset$ codomain

196. We have, $e^{-x} f(x) = 2 + \int_0^x \sqrt{t^4+1}\, dt, \forall x \in (-1, 1)$

On differentiating w.r.t. x, we get

$e^{-x} [f'(x) - f(x)] = \sqrt{x^4+1}$

$\Rightarrow \quad f'(x) = f(x) + \sqrt{x^4+1} \cdot e^x$

$\because f^{-1}$ is the inverse of f.

$\therefore \quad f^{-1}\{f(x)\} = x$

$\Rightarrow \quad [f^{-1}\{f(x)\}]' f'(x) = 1$

$\Rightarrow \quad [f^{-1}\{f(x)\}]' = \dfrac{1}{f'(x)}$

$\Rightarrow \quad [f^{-1}\{f(x)\}]' = \dfrac{1}{f(x) + \sqrt{x^4+1} \cdot e^x}$

At $x = 0$, $f(x) = 2$

$\Rightarrow \quad \{f^{-1}(2)\}' = \dfrac{1}{2+1} = \dfrac{1}{3}$

197. Since, only (c) satisfy given definition,

i.e. $\quad f\{f^{-1}(B)\} = B$

Only, if $\quad B \subseteq f(x)$

198. Let $\phi(x) = f(x) - g(x) = \begin{cases} x, & x \in Q \\ -x, & x \notin Q \end{cases}$

Now, to check one-one.

Take any straight line parallel to X-axis which will intersect $\phi(x)$ only at one point.

$\Rightarrow \phi(x)$ is one-one.

To check onto

As, $f(x) = \begin{cases} x, & x \in Q \\ -x, & x \notin Q \end{cases}$, which shows

$y = x$ and $y = -x$ for rational and irrational values

$\Rightarrow y \in$ real numbers.

$\therefore \quad$ Range = Codomain $\Rightarrow$ onto

Thus, $f - g$ is one-one and onto.

199. By definition of composition of function,

$g(f(x)) = (\sin x + \cos x)^2 - 1$, is invertible (i.e. bijective)

$\Rightarrow \quad g\{f(x)\} = \sin 2x$ is bijective.

We know, $\sin x$ is bijective, only when $x \in \left[-\dfrac{\pi}{2}, \dfrac{\pi}{2}\right]$

Thus, $g\{f(x)\}$ is bijective, if $-\dfrac{\pi}{2} \le 2x \le \dfrac{\pi}{2}$

$\Rightarrow \quad -\dfrac{\pi}{4} \le x \le \dfrac{\pi}{4}$

200. Here, $f(x) = \sqrt{\sin^{-1}(2x) + \dfrac{\pi}{6}}$, to find domain we must have,

$\sin^{-1}(2x) + \dfrac{\pi}{6} \ge 0 \quad \left[\text{but } -\dfrac{\pi}{2} \le \sin^{-1}\theta \le \dfrac{\pi}{2}\right]$

$-\dfrac{\pi}{6} \le \sin^{-1}(2x) \le \dfrac{\pi}{2} \Rightarrow \sin\left(-\dfrac{\pi}{6}\right) \le 2x \le \sin\dfrac{\pi}{2}$

$\dfrac{-1}{2} \le 2x \le 1$

$\Rightarrow \quad \dfrac{-1}{4} \le x \le \dfrac{1}{2}$

$\because \quad x \in \left[\dfrac{-1}{4}, \dfrac{1}{2}\right]$

201. Let $y = f(x) = \dfrac{x^2+x+2}{x^2+x+1}, x \in R$

$\therefore \quad y = \dfrac{x^2+x+2}{x^2+x+1}$

$y = 1 + \dfrac{1}{x^2+x+1}$ [i.e. $y > 1$]...(i)

$\Rightarrow \quad yx^2 + yx + y = x^2 + x + 2$

$\Rightarrow \quad x^2(y-1) + x(y-1) + (y-2) = 0, \forall x \in R$

Since, x is real, $D \ge 0$

$\Rightarrow \quad (y-1)^2 - 4(y-1)(y-2) \ge 0$

$\Rightarrow \quad (y-1)\{(y-1) - 4(y-2)\} \ge 0$

$\Rightarrow \quad (y-1)(-3y+7) \ge 0$

$\Rightarrow \quad 1 \le y \le \dfrac{7}{3}$ ...(ii)

From Eqs. (i) and (ii), Range $\in \left(1, \dfrac{7}{3}\right]$

202. Given, $f: [0, \infty) \to [0, \infty)$

Here, domain is $[0, \infty)$ and codomain is $[0, \infty)$. Thus, to check one-one

Since, $\quad f(x) = \dfrac{x}{1+x}$

$\Rightarrow \quad f'(x) = \dfrac{1}{(1+x)^2} > 0, \forall x \in [0, \infty)$

$\therefore f(x)$ is increasing in its domain. Thus, $f(x)$ is one-one in its domain. To check onto (we find range)

Again, $\quad y = f(x) = \dfrac{x}{1+x}$

$\Rightarrow \quad y + yx = x$

$\Rightarrow \quad x = \dfrac{y}{1-y} \Rightarrow \dfrac{y}{1-y} \ge 0$

Since, $x \ge 0$, therefore $0 \le y < 1$

i.e. Range $\ne$ Codomain

$\therefore f(x)$ is one-one but not onto.

203. Given, $f(x) = 2x + \sin x$

$\Rightarrow \quad f'(x) = 2 + \cos x$

$\Rightarrow \quad f'(x) > 0, \forall x \in R$

which shows $f(x)$ is one-one, as $f(x)$ is strictly increasing.

Since, $f(x)$ is increasing for every $x \in R$.

$\therefore$ $f(x)$ takes all intermediate values between $(-\infty, \infty)$.

Range of $f(x) \in R$.

Hence, $f(x)$ is one-to-one and onto.

204. The number of onto functions from

$E = \{1, 2, 3, 4\}$ to $F = \{1, 2\}$

= Total number of functions which map E to F

– Number of functions for which map $f(x) = 1$ and

$f(x) = 2$ for all $x \in E = 2^4 - 2 = 14$

205. It is only to find the inverse.

Let $y = f(x) = (x+1)^2$ for $x \geq -1$

$\Rightarrow \pm\sqrt{y} = x + 1, x \geq -1$

$\Rightarrow \sqrt{y} = x + 1 \Rightarrow y \geq 0, x + 1 \geq 0$

$\Rightarrow \quad x = \sqrt{y} - 1$

$\Rightarrow \quad f^{-1}(y) = \sqrt{y} - 1$

$\Rightarrow \quad f^{-1}(x) = \sqrt{x} - 1$

$\Rightarrow \quad x \geq 0$

206. Let $y = x + \dfrac{1}{x} \Rightarrow y = \dfrac{x^2+1}{x}$

$\Rightarrow \quad xy = x^2 + 1$

$\Rightarrow \quad x^2 - xy + 1 = 0$

$\Rightarrow \quad x = \dfrac{y \pm \sqrt{y^2 - 4}}{2}$

$\Rightarrow \quad f^{-1}(y) = \dfrac{y \pm \sqrt{y^2 - 4}}{2}$

$\Rightarrow \quad f^{-1}(x) = \dfrac{x \pm \sqrt{x^2 - 4}}{2}$

Since, the range of the inverse function is $[1, \infty)$, then

we take $\quad f^{-1}(x) = \dfrac{x + \sqrt{x^2 - 4}}{2}$

If we consider $f^{-1}(x) = \dfrac{x - \sqrt{x^2 - 4}}{2}$, then $f^{-1}(x) > 1$

This is possible only if $(x-2)^2 > x^2 - 4$

$\Rightarrow \quad x^2 + 4 - 4x > x^2 - 4$

$\Rightarrow \quad 8 > 4x$

$\Rightarrow \quad x < 2$, where $x > 2$

Therefore, (a) is the correct answer.

207. Given, $f(x) = (1 + b^2)x^2 + 2bx + 1$

$$= (1 + b^2)\left(x + \frac{b}{1 + b^2}\right)^2 + 1 - \frac{b^2}{1 + b^2}$$

$m(b)$ = minimum value of $f(x) = \dfrac{1}{1 + b^2}$ is positive and $m(b)$ varies from 1 to 0, so range = $(0,1]$

208. Given, $f(x) = \dfrac{\log_2(x+3)}{(x^2 + 3x + 2)} = \dfrac{\log_2(x+3)}{(x+1)(x+2)}$

For numerator, $x + 3 > 0$

$\Rightarrow \quad x > -3 \quad$ ...(i)

and for denominator, $(x+1)(x+2) \neq 0$

$\Rightarrow \quad x \neq -1, -2 \quad$ ...(ii)

From Eqs. (i) and (ii), we get

Domain is $(-3, \infty) / \{-1, -2\}$

209. Given, $f(x) = \dfrac{\alpha x}{x+1}$

$$f[f(x)] = f\left(\frac{\alpha x}{x+1}\right) = \frac{\alpha\left(\dfrac{\alpha x}{x+1}\right)}{\dfrac{\alpha x}{x+1} + 1}$$

$$= \frac{\dfrac{\alpha^2 x}{x+1}}{\dfrac{\alpha x + (x+1)}{x+1}} = \frac{\alpha^2 x}{(\alpha + 1)x + 1} = x \text{ [given]} \quad \text{...(i)}$$

$\Rightarrow \quad \alpha^2 x = (\alpha + 1)x^2 + x$

$\Rightarrow \quad x[\alpha^2 - (\alpha + 1)x - 1] = 0$

$\Rightarrow \quad x(\alpha + 1)(\alpha - 1 - x) = 0$

$\Rightarrow \quad \alpha - 1 = 0$ and $\alpha + 1 = 0 \Rightarrow \alpha = -1$

But $\alpha = 1$ does not satisfy the Eq. (i).

210. $g(x) = 1 + x - [x] \geq 1$

since $x - [x] \geq 0$

$f[g(x)] = 1$, since $f(x) = 1$ for all $x > 0$

211. Given, $2^x + 2^y = 2, \forall x, y \in R$

But $2^x, 2^y > 0, \forall x, y \in R$

Therefore, $2^x = 2 - 2^y < 2 \Rightarrow 0 < 2^x < 2$

Taking log on both sides with base 2, we get

$$\log_2 0 < \log_2 2^x < \log_2 2 \Rightarrow -\infty < x < 1$$

212. It is given,

$f(\theta) = \sin\theta(\sin\theta + \sin 3\theta)$

$= (\sin\theta + 3\sin\theta - 4\sin^3\theta)\sin\theta$

$= (4\sin\theta - 4\sin^3\theta)\sin\theta = \sin^2\theta(4 - 4\sin^2\theta)$

$= 4\sin^2\theta\cos^2\theta = (2\sin\theta\cos\theta)^2 = (\sin 2\theta)^2 \geq 0$

which is true for all θ.

CHAPTER

04

Graphical Transformations

Learning Part

Session 1

- When $f(x)$ vertically transforms to $f(x) \pm a$

Session 2

- When $f(x)$ horizontally transforms to $f(x \pm a)$, where a is any positive constant
- When $f(x)$ transforms to $a\,f(x)$ or $\dfrac{1}{a}f(x)$
- When $f(x)$ transforms to $f(ax)$ or $f(x/a)$
- When $f(x)$ transforms to $f(-x)$
- When $f(x)$ transforms to $-f(x)$

Session 3

- To draw $y = |f(x)|$, when $y = f(x)$ is given
- To draw $y = f(|x|)$, when $f(x)$ is given
- $f(x)$ transforms to $|f(|x|)|$, i.e. $f(x) \to |f(|x|)|$
- To draw the graph of $|y| = f(x)$ when $y = f(x)$ is given

Session 4

- To draw $y = [f(x)]$ when the graph of $y = f(x)$ is given and $[\cdot]$ denotes greatest integer or integral function
- To draw $y = f([x])$ when the graph of $y = f(x)$ is given
- When $f(x)$ transforms to $y = f(\{x\})$, where $\{\cdot\}$ denotes fractional part of x, i.e. $\{x\} = x - [x]$ or $f(x) \to f(\{x\})$
- When $f(x)$ and $g(x)$ are two functions and are transformed to their sum
- When $f(x)$ transforms to $f(x) \to \dfrac{1}{f(x)} = h(x)$
- To draw the graph for $y = f(x) \cdot \sin x$ when graph of $y = f(x)$ is given
- When $f(x)$ and $g(x)$ are given, then find $h(x) = \max(f(x), g(x))$ or $h(x) = \min(f(x), g(x))$

Practice Part

- JEE Type Solved Examples
- Chapter Exercises

Arihant on Your Mobile !

Exercises with this symbol can be practised on your mobile. See title inside to activate for free.

Graph and diagram provide a simple and powerful approach to a variety of problems, from which we can solve them in a very short time. In this chapter, we will discuss how the graph of a function may be transformed either by shifting, stretching or compressing or reflection.

Graphical Transformations

A graphical transformation takes whatever its basic function $f(x)$. The graph of new function is the graph of the original function shifted vertically or horizontally.

Session 1

When $f(x)$ Vertically Transforms to $f(x) \pm a$ (Where a is Any Positive Constant)

For $f(x) \to f(x)+a$,

First, draw the graph of $f(x)$, then shift the graph of $f(x)$ upwards through 'a' units such that the distance between graphs of $f(x)$ and $f(x)+a$ at every point is same and is equal to 'a' units.

For $f(x) \to f(x)-a$,

First, draw the graph of $f(x)$, then shift the graph of $f(x)$ downwards through 'a' units such that the distance between the graphs of $f(x)$ and $f(x)-a$ at every point is same and is equal to 'a' units.

Graphically, it could be stated as shown in figure.

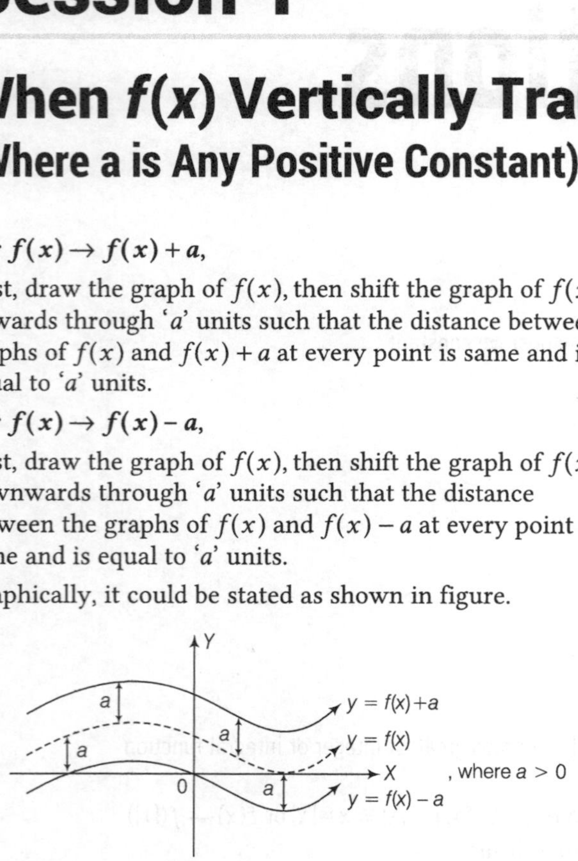

Figure 4.1

Example 1 Plot $y=|x|$ and $y=|x|+2$.

Sol. We know, $y=|x|$ (modulus function) is define as

$$y=\begin{cases} x, & x \ge 0 \\ -x, & x<0 \end{cases}$$

So, we have to plot two straight lines $y=x$ and $y=-x$.

For $y=x,\ x\ge 0$

x	1	2	3	4
y	1	2	3	4

$\therefore$ Locate the points (1, 1), (2, 2), (3, 3), (4, 4) and join them.

For $y=-x,\ x<0$

x	–1	–2	–3
y	1	2	3

$\therefore$ Locate the points (–1, 1), (–2, 2), (–3, 3) and join them.

So, we obtained graph of $y=|x|$.

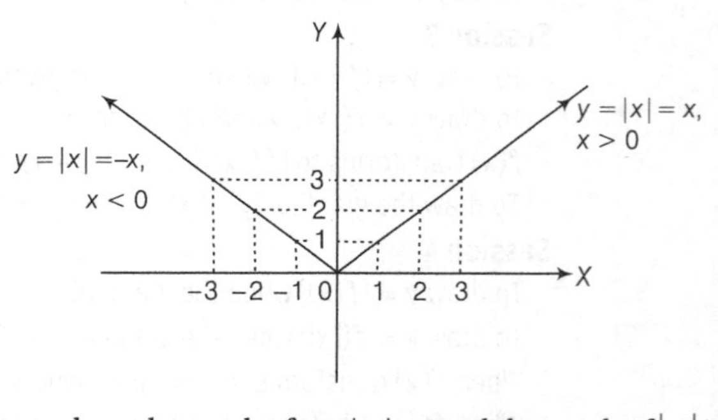

Now, to draw the graph of $y=|x|+2$ and the graph of $|x|$ is shifted upwards by 2 units as shown below.

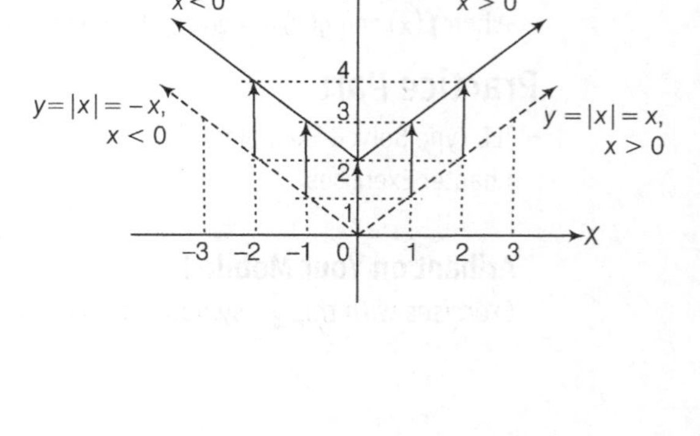

Example 2 Plot $y = |x|$ and $y = |x| - 2$.

Sol. First, we plot the graph of $y = |x|$ as explained in Example 1, then shift the graph of $|x|$ through 2 units downwards as shown below.

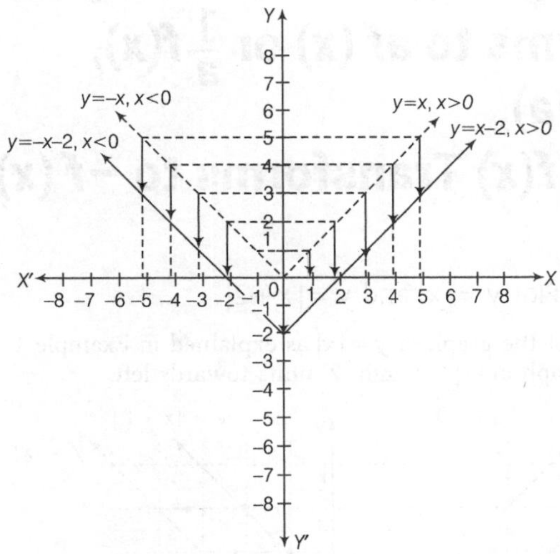

Example 3 Plot $y = e^x$, $y = e^x + 1$ and $y = e^x - 1$.

Sol. We know, $y = e^x$ (exponential function) could be plotted as shown below which cuts the Y-axis at (0,1).

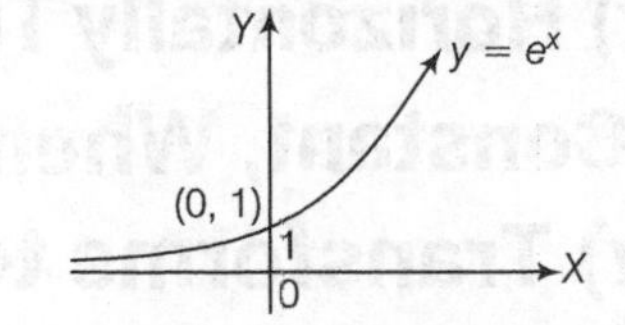

$\Rightarrow y = e^x + 1$ is shifted upward by 1 unit such that it intersects Y-axis at (0, 2) and $y = e^x - 1$ shifted downwards by 1 unit. Such that it intersects Y-axis at (0,0).

The graph of the given functions is shown below.

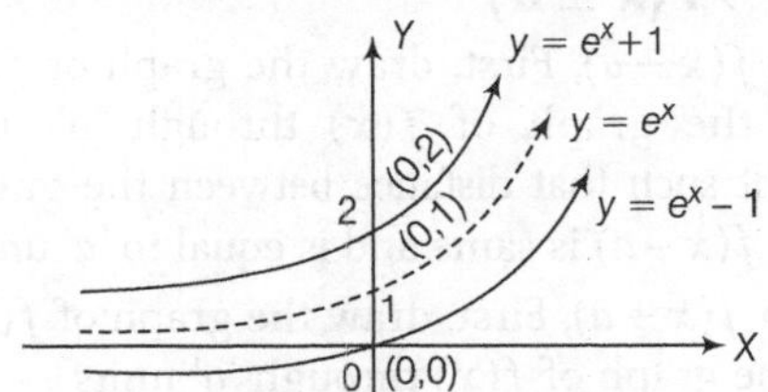

Exercise for Session 1

- **Directions** (Q. Nos. 1 to 8) *Plot the following functions.*

1. $y = x^2 + 1$
2. $y = x^2 - 1$
3. $y = x^3 + 1$
4. $y = x^3 - 1$
5. $y = \sin x + 1$
6. $y = \sin x - 1$
7. $y = (\log_e x) + 1$
8. $y = (\log_e x) - 1$

Session 2

When $f(x)$ Horizontally Transforms to $f(x \pm a)$, where a is any Positive Constant, When $f(x)$ Transforms to $af(x)$ or $\frac{1}{a}f(x)$, When $f(x)$ Transforms to $f(ax)$ or $f(x/a)$, When $f(x)$ Transforms to $f(-x)$, When $f(x)$ Transforms to $-f(x)$

For $f(x) \to f(x \pm a)$

For $f(x) \to f(x-a)$, First, draw the graph of $f(x)$, then shift the graph of $f(x)$ through 'a' units towards right such that distance between the graphs of $f(x)$ and $f(x-a)$ is same and is equal to 'a' units.

For $f(x) \to f(x+a)$, First, draw the graph of $f(x)$, then shift the graph of $f(x)$ through 'a' units towards left such that the distance between the graphs of $f(x)$ and $f(x+a)$ is same and is equal to 'a' units.

Graphically, it could be stated as shown in figure.

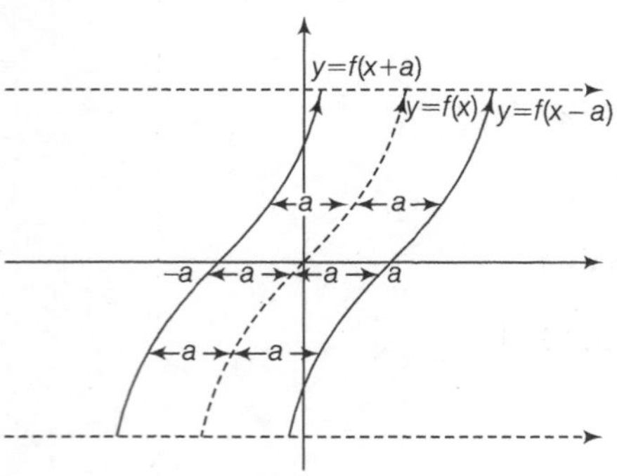

Figure 4.2

Example 4 Plot $y=|x|$ and $y=|x-2|$.

Sol. First we plot the graph of $y=|x|$ as explained in Example 1, then shift the graph of $|x|$ through '2' units towards right as shown below.

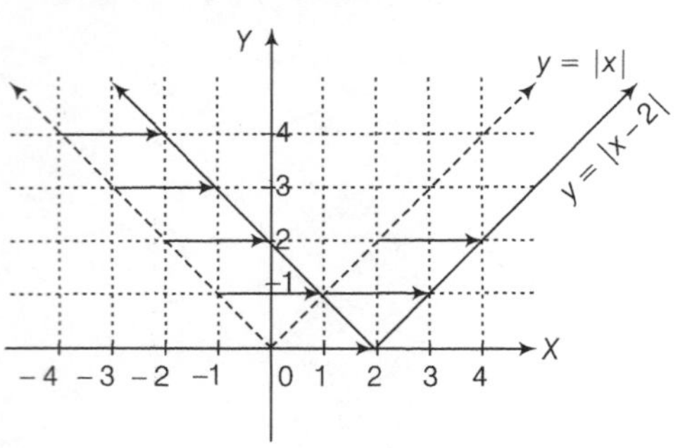

Example 5 Plot $y=|x|$ and $y=|x+2|$.

Sol. First we plot the graph of $y=|x|$ as explained in Example 1, then shift the graph of $|x|$ through '2' units towards left.

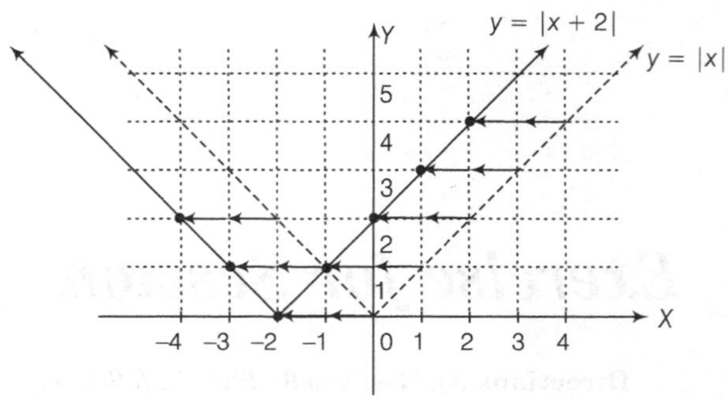

Example 6 Plot $y=\sin\left(x+\frac{\pi}{4}\right)$ and $y=\sin\left(x-\frac{\pi}{4}\right)$.

Sol. Let us consider $y=\sin x$, domain of sine functions is R and range is $[-1, 1]$. $\sin x$ increases strictly from -1 to 1 as x increases from $-\frac{\pi}{2}$ to $\frac{\pi}{2}$, decreases strictly from 1 to -1 as x increases from $\frac{\pi}{2}$ to $\frac{3\pi}{2}$. Also, for $y=\sin x$.

x	-2π	$-\frac{3\pi}{2}$	$-\pi$	$-\frac{\pi}{2}$	0	$\frac{\pi}{2}$	π	$\frac{3\pi}{2}$	2π
y	0	1	0	-1	0	1	0	-1	0

Locate the points $(-2\pi,0)$, $\left(-\frac{3\pi}{2},1\right)$, $(-\pi,0)$, $\left(-\frac{\pi}{2},-1\right)$, $(0,0)$, $\left(\frac{\pi}{2},1\right)$, $(\pi,0)$, $\left(\frac{3\pi}{2},-1\right)$ and $(2\pi,0)$ on the graph paper, then we get the graph of $\sin x$ as shown below.

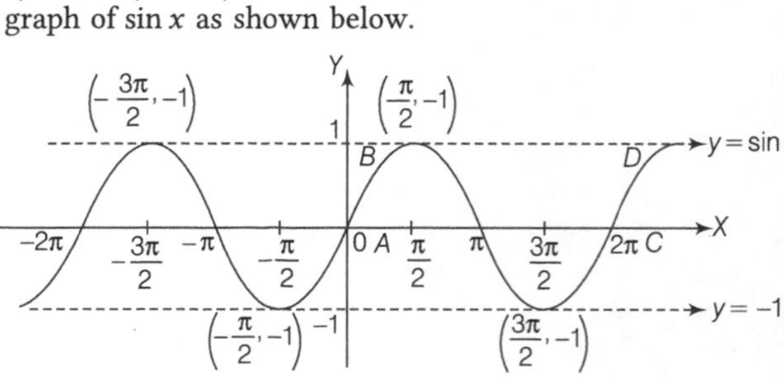

Now, for the graph of $y = \sin\left(x + \frac{\pi}{4}\right)$ shift the graph of $y = \sin x$ through $\frac{\pi}{4}$ towards left as shown below.

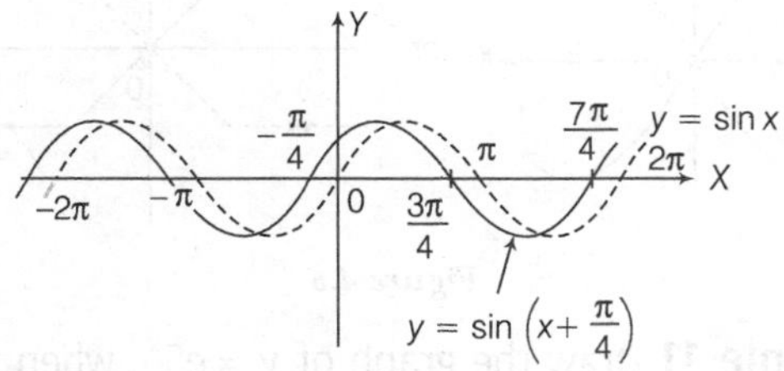

And for the graph of $y = \sin\left(x - \frac{\pi}{4}\right)$, shift the graph of $y = \sin x$ through $\frac{\pi}{4}$ towards right as shown below.

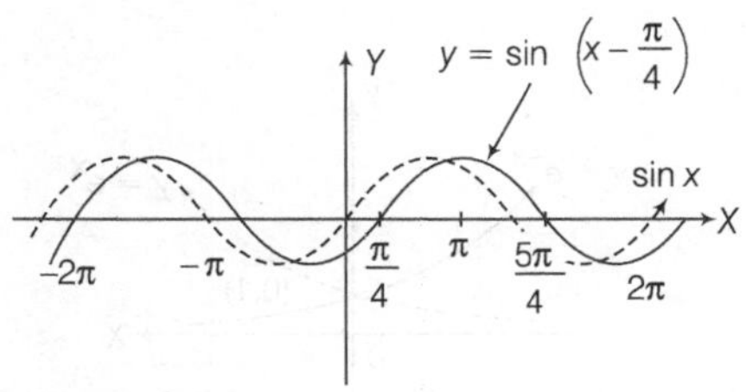

$f(x) \rightarrow \{af(x)\}, a > 1$

First draw the graph of $f(x)$, then stretch the graph of $f(x)$, 'a' times along Y-axis.

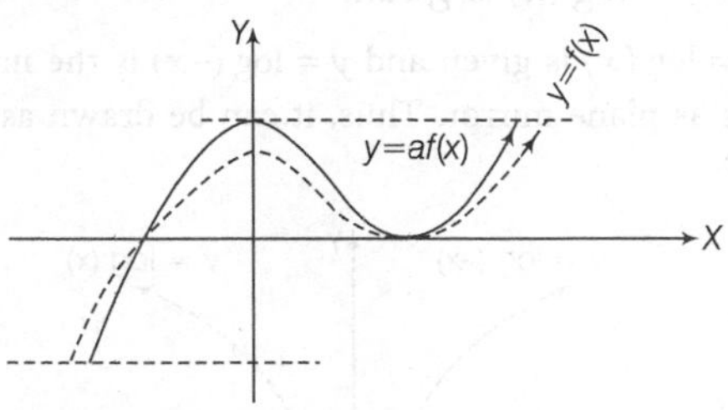

Example 7 Plot $y = \sin x$ and $y = 2\sin x$.

Sol. First we draw the graph of $\sin x$ as explained in Example 6, then stretch the graph of $\sin x$ by 2 times along the Y-axis, we get the graph of $2\sin x$ as shown below.

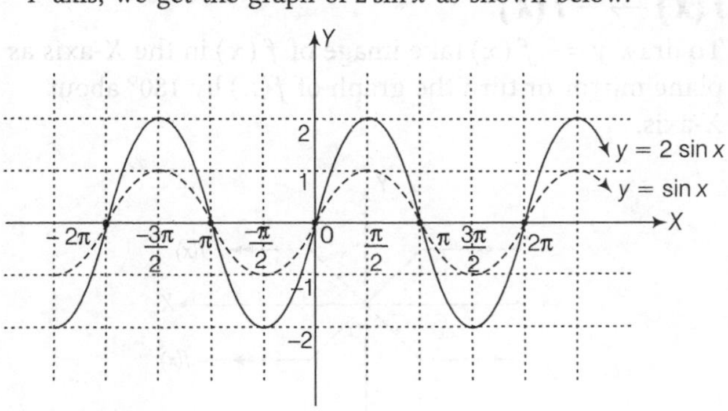

$f(x) \rightarrow \frac{1}{a}f(x), a > 1$

First, draw the graph of $f(x)$, then shrink the graph of $f(x)$, 'a' times along Y-axis. As shown in the figure.

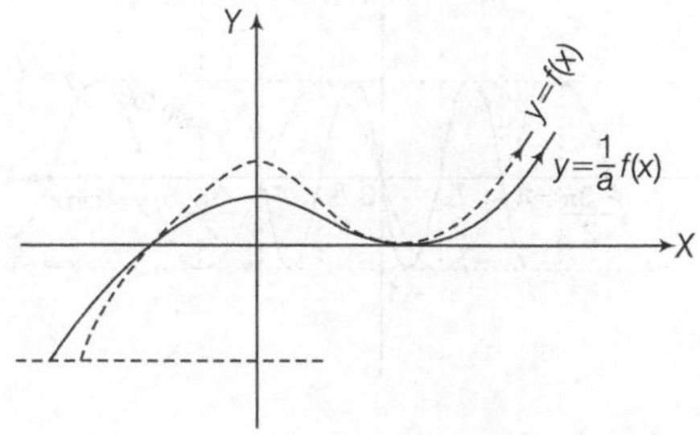

Example 8 Plot $y = \sin x$ and $y = \frac{1}{2}\sin x$.

Sol. First we draw the graph of $y = \sin x$ as explained in Example 6, then shrink the graph of $\sin x$ by 2 times along Y-axis, we get the graph of $\frac{1}{2}\sin x$ as shown below.

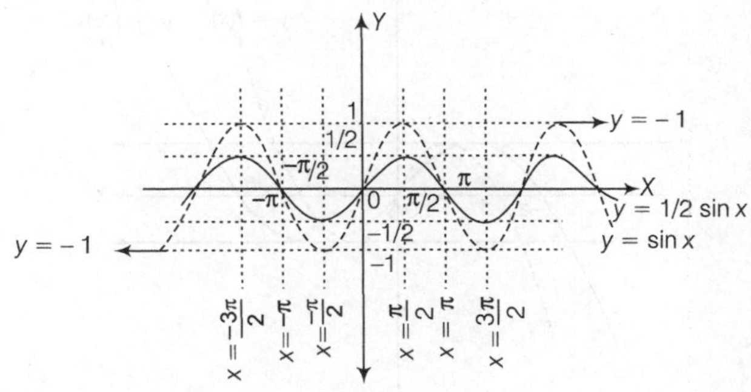

$f(x) \rightarrow f(ax), a > 1$

First draw the graph of $f(x)$, then shrink the graph of $f(x)$, 'a' times along X-axis as shown in figure.

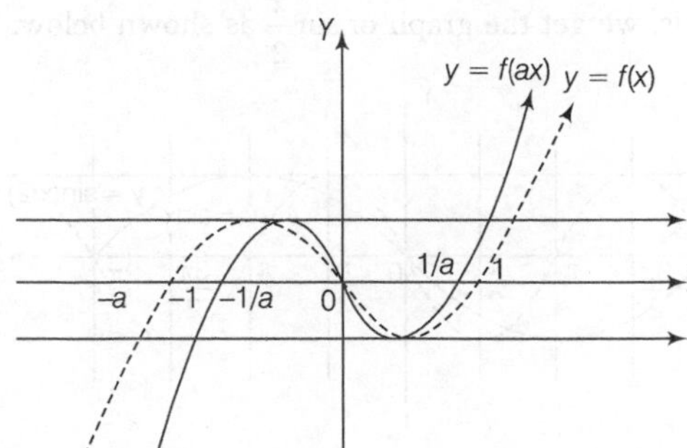

Figure 4.4

Example 9 Plot $y = \sin x$ and $y = \sin 2x$.

Sol. First draw the graph of $\sin x$ as explained in Example 6, then shrink the graph of $\sin x$, 2 times along X-axis, we get the graph of $\sin 2x$ as shown below.

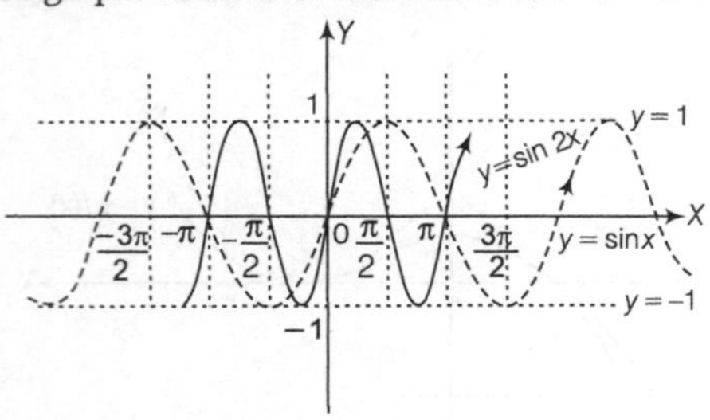

$f(x) \to f\left(\dfrac{x}{a}\right), a > 1$

First draw the graph of $f(x)$, then stretch the graph of $f(x)$, 'a' times along X-axis, as shown in figure.

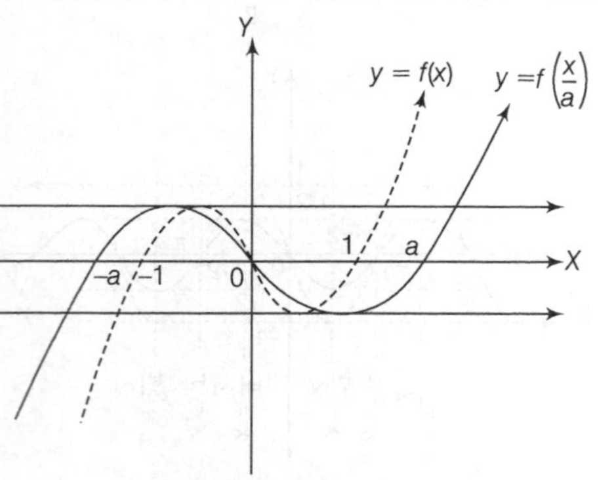

Figure 4.5

Example 10 Plot $y = \sin x$ and $y = \sin \dfrac{x}{2}$.

Sol. First we draw the graph of $\sin x$ as explained in Example 6, then stretch the graph of $\sin x$, 2 times along X-axis, we get the graph of $\sin \dfrac{x}{2}$ as shown below.

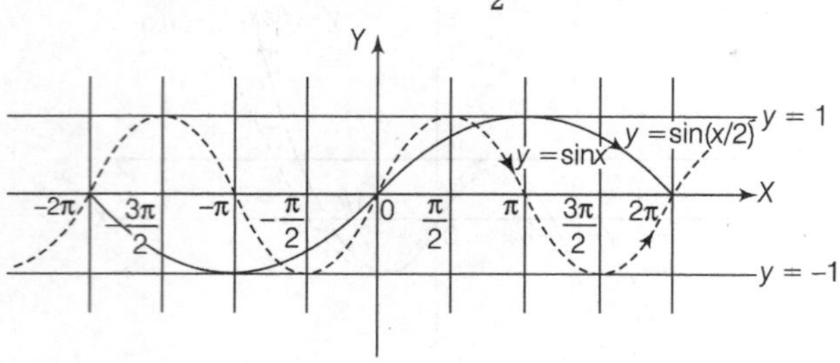

$f(x) \to f(-x)$

To draw $y = f(-x)$, take the image of the curve $y = f(x)$ in the Y-axis as plane mirror or turn the graph of $f(x)$ by 180° about Y-axis.

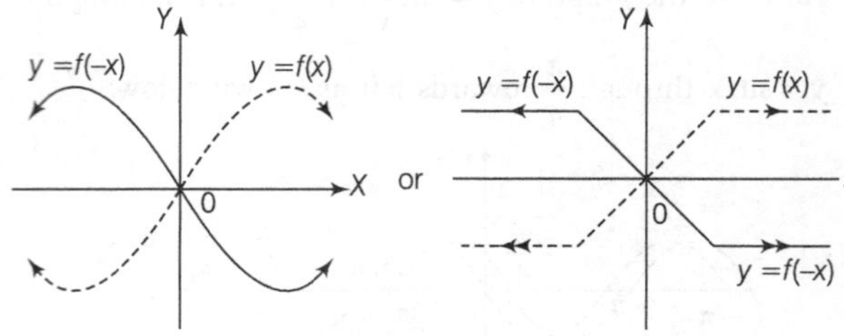

Figure 4.6

Example 11 Draw the graph of $y = e^{-x}$, when the graph of $y = e^{x}$ is known.

Sol. As $y = e^{x}$ is given, then $y = e^{-x}$ is the image in Y-axis as plane mirror. Thus, it can be drawn as shown in the figure.

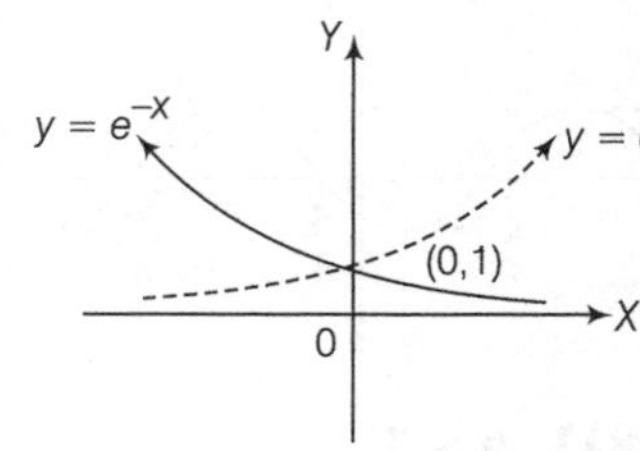

Example 12 Draw graph of $y = \log(-x)$, when the graph of $y = \log(x)$ is given.

Sol. As $y = \log(x)$ is given and $y = \log(-x)$ is the image in Y-axis as plane mirror. Thus, it can be drawn as shown in figure.

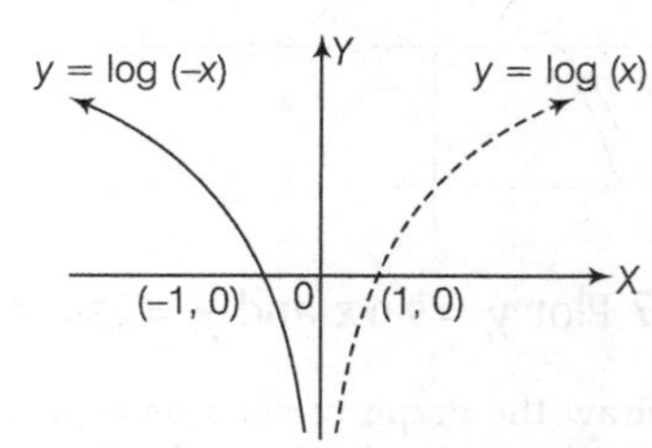

$f(x) \to -f(x)$

To draw $y = -f(x)$ take image of $f(x)$ in the X-axis as plane mirror or turn the graph of $f(x)$ by 180° about X-axis.

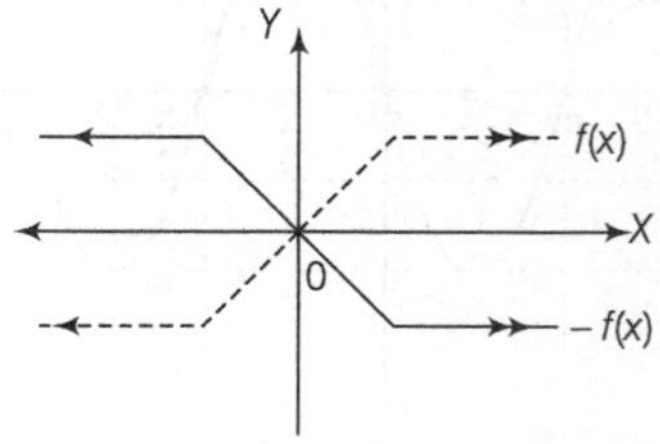

Figure 4.7

Example 13 Draw the graph of $y = -e^x$, when the graph of $y = e^x$ is known.

Sol. As $y = e^x$ is given, then $y = -e^x$ is the image of e^x in the X-axis as plane mirror.

Thus, it can be plotted as shown in figure

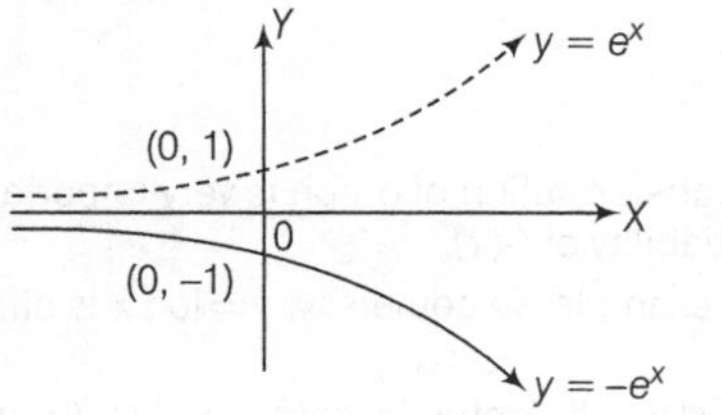

Example 14 Draw the graph of $y = -\log(x)$, when the graph of $y = \log x$ is known.

Sol. As $y = \log x$ is given, then $y = -\log x$ is the image of $\log x$ in the X-axis as plane mirror. Thus, it can be drawn as shown in figure.

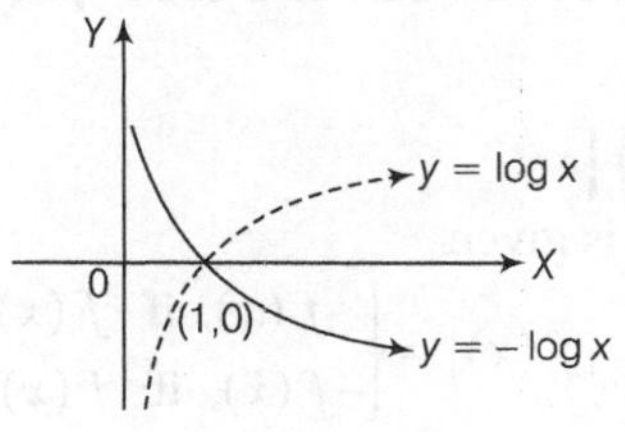

Exercise for Session 2

1. Consider the following function f, whose graph is given below.

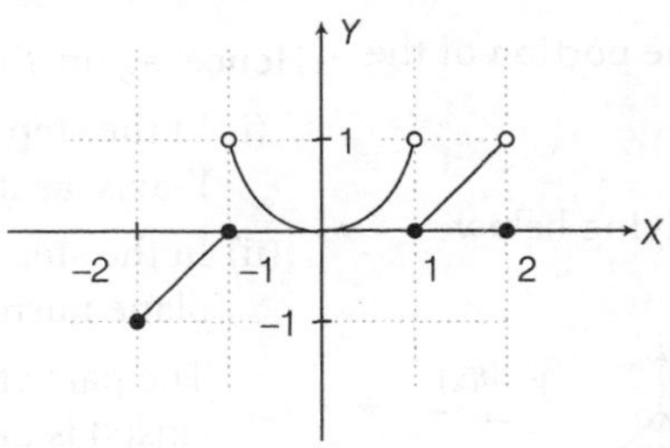

Draw the graph of the following functions.

(i) $f(x+1)$ (ii) $f(x-1)$ (iii) $-f(x)$ (iv) $f(-x)$

(v) $2f(x)$ (vi) $\frac{1}{2}f(x)$ (vii) $f(2x)$ (viii) $f\left(\frac{x}{2}\right)$

Session 3

To Draw $y = f(|x|)$, When $f(x)$ is Given, $f(x)$ Transforms to Draw $|f(|x|)|$, i.e. $f(x) \rightarrow |f(|x|)|$ To Draw $|y| = f(x)$

$f(x) \rightarrow |f(x)|$

When $y = f(x)$ is given

We know that $|f(x)| = \begin{cases} f(x), & \text{if } f(x) \geq 0 \\ -f(x), & \text{if } f(x) \leq 0 \end{cases}$

Hence, $y = |f(x)|$ is drawn in two steps

(i) In the step I, leave the positive part of $f(x)$, (i.e. the part of $f(x)$ above X-axis) as it is.

(ii) In the step II, take the image of negative part of $f(x)$, (i.e. the part of $f(x)$ below X-axis) in the X-axis as plane mirror.

Or

Take the mirror image (in X-axis) of the portion of the graph of $f(x)$ which lies below X-axis.

Or

Turn the portion of the graph of $f(x)$ lying below X-axis by 180° about X-axis.

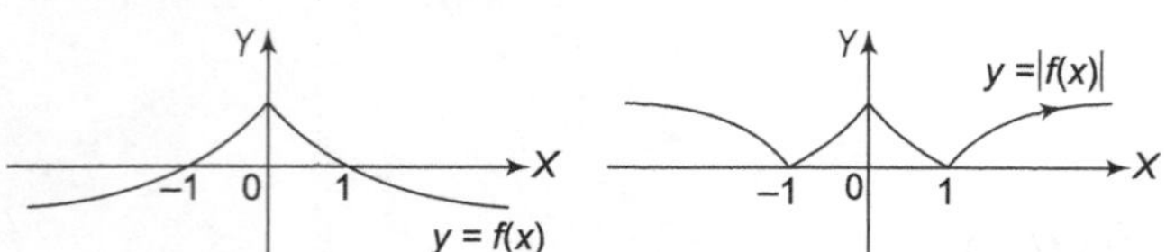

Figure 4.8

Example 15 Draw the graph of $y = |\log x|$, when the graph of $y = \log(x)$ is known.

Sol. To draw the graph of $y = |\log x|$, we have to proceed in two steps.

(i) In the first step, leave the positive part of $y = \log x$, i.e. the part of $y = \log x$ above X-axis as it is.

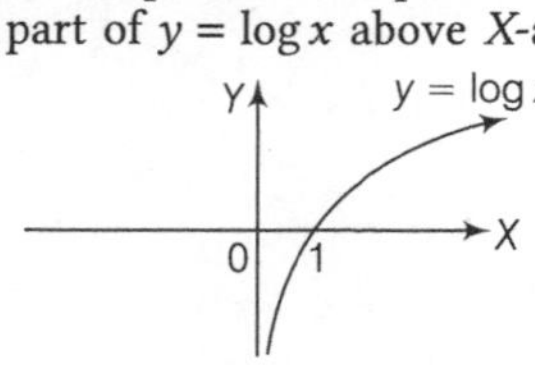

(ii) In the second step, take the image of negative part of $y = \log x$, i.e. the part below X-axis in the X-axis as plane mirror. Thus, the graph can be

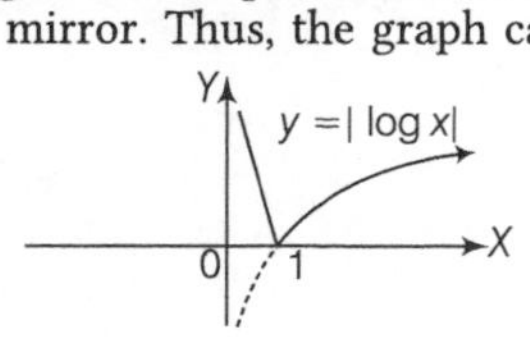

Remark

The above transformation of graph is very important in explaining the differentiability of $f(x)$.

From above example we could say, $y = \log x$ is differentiable for $(0, \infty)$.

But $y = |\log x|$ is differentiable for $(0, \infty) - \{1\}$. As, we have a sharp edge at $x = 1$ (which indicates that it is not differentiable at $x = 1$).

$f(x) \rightarrow f(|x|)$

When $f(x)$ is given

We know that, $y = f|x| = \begin{cases} f(x), & \text{if } x \geq 0 \\ f(-x), & \text{if } x \leq 0 \end{cases}$.

Hence, again $f(|x|)$ could be drawn in two steps

(i) In the step I, leave the graph lying right side of the Y-axis, as it is.

(ii) In the step II, take the image of $f(x)$ in the Y-axis as plane mirror.

The part of $f(x)$ lying left side of the Y-axis (if it exists) is omitted.

e.g. Let us assume the graph of $f(x)$ as shown

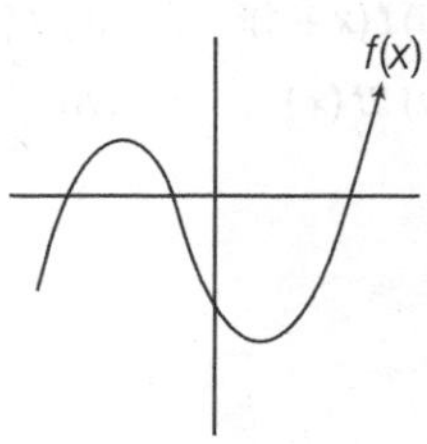

Figure 4.9

Now, **On the right of Y-axis** Plot the graph of $f(x)$ as such

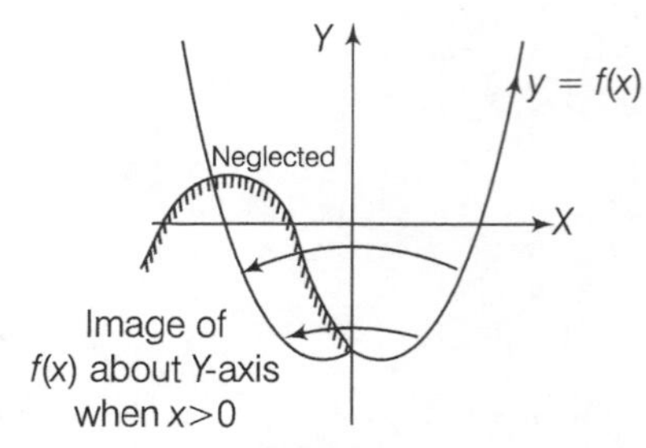

Figure 4.10

On the left of *Y*-axis Take the mirror image of the graph of $f(x)$ (of the portion lying on right of Y-axis).

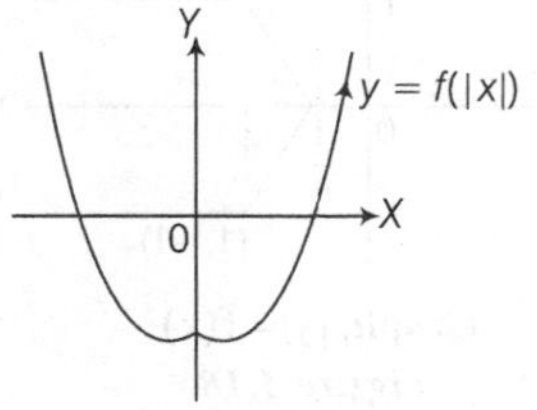

Figure 4.11

Example 16 Draw the graph of $y = \log|x|$, when graph of $y = \log(x)$ is given.

Sol. We know, $y = \log|x|$ could be drawn in two steps

(i) In step I, leave the graph lying right side of Y-axis as it is.

(ii) In step II, take the image of $f(x)$ in the Y-axis as plane mirror.

Thus, it can be plotted as shown in figure.

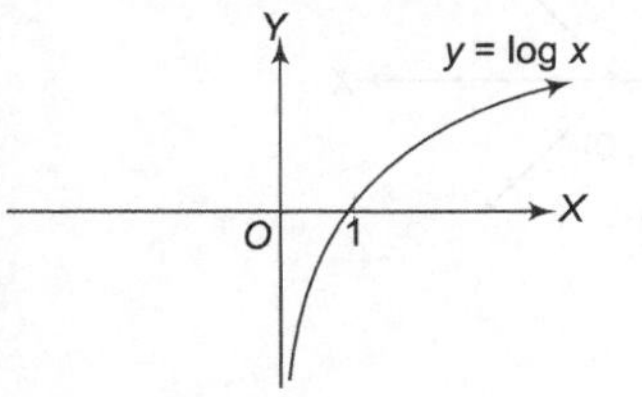

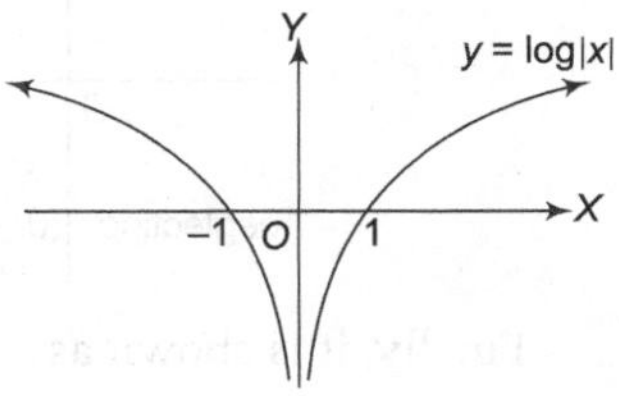

Note $f(|x|)$ is always an even function.

$f(x) \to |f(|x|)|$

Here, we plot $y = |f(|x|)|$ when graph of $f(|x|)$ is given, in two steps

(i) In step I, $f(x)$ transforms to $|f(x)|$.

(ii) In step II, $|f(x)|$ transforms to $|f(|x|)|$.

or

(i) $f(x) \to |f(x)|$ (ii) $f(x) \to |f(|x|)|$

e.g. Let graph for $f(x)$

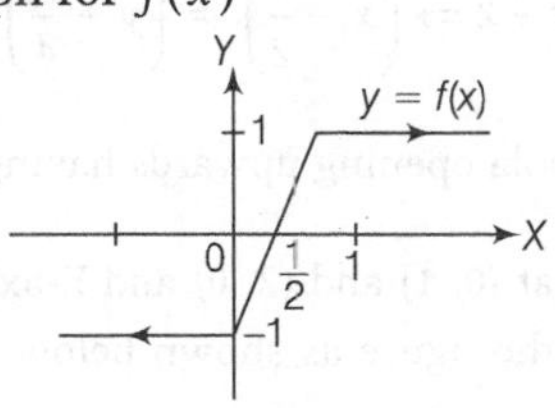

Figure 4.12

Now, $f(x) \to |f(x)|$ can be shown as,

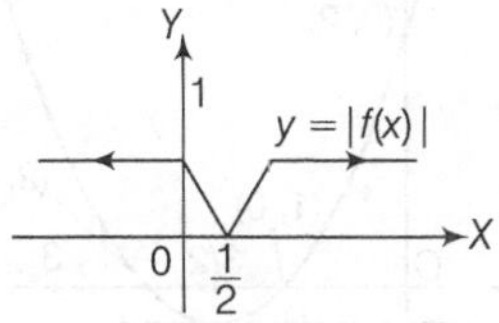

Figure 4.13

Again, $|f(x)| \to |f(|x|)|$ can be shown as,

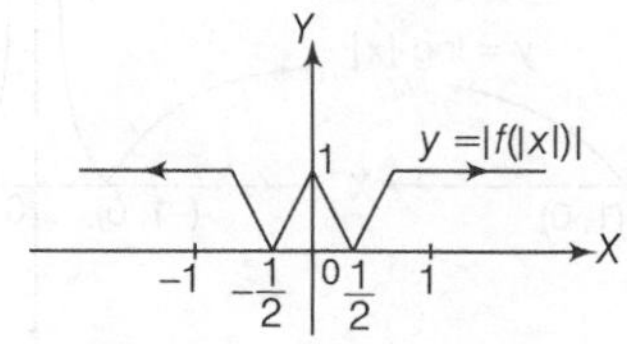

Figure 4.14

Example 17 Draw the graph for $y = \left||x|^2 - 2|x| - 3\right|$, if the graph for $y = x^2 - 2x - 3$ is given.

Sol. First we draw the graph for $y = x^2 - 2x - 3$

$$\Rightarrow \quad (y + 4) = (x - 1)^2$$

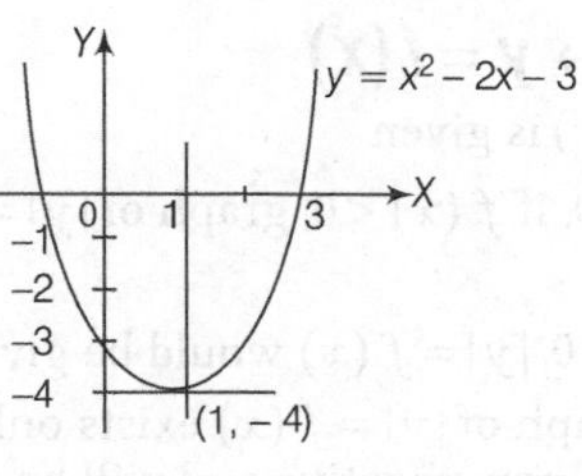

Now, we form $f(x) \to f(|x|)$,

i.e. $y = x^2 - 2x - 3 \Rightarrow y = |x|^2 - 2|x| - 3$ as shown below

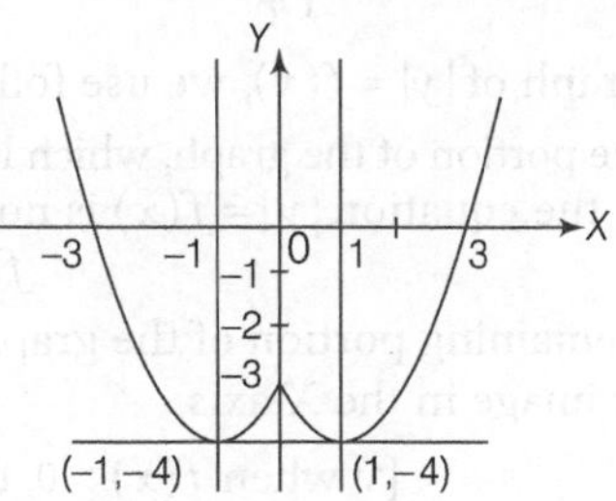

In next step, we transform $f(x) = |f(|x|)|$,

i.e. $y = |x|^2 - 2|x| - 3 \to ||x|^2 - 2|x| - 3|$ as shown below

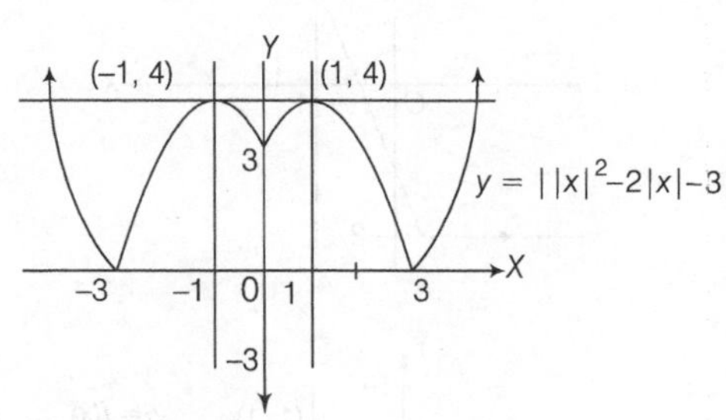

Remarks

Above mentioned transformations are very useful in explaining continuity and differentiability of functions as, if $y = \log(x)$, then

(i) $y = \log|x|$ can be plotted as shown in figure (4.15).

(ii) $y = |\log|x||$ can be plotted as shown in figure (4.16).

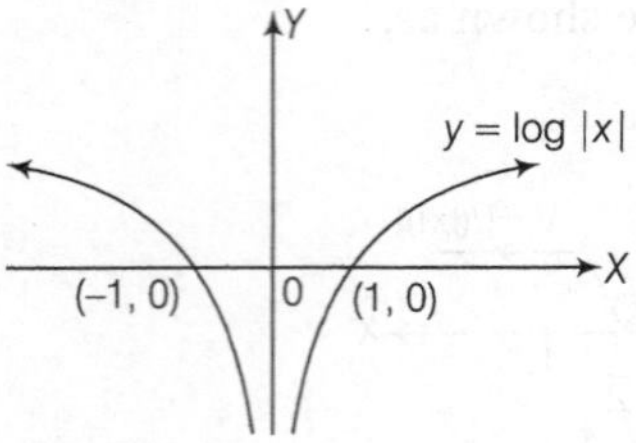

Figure 4.15

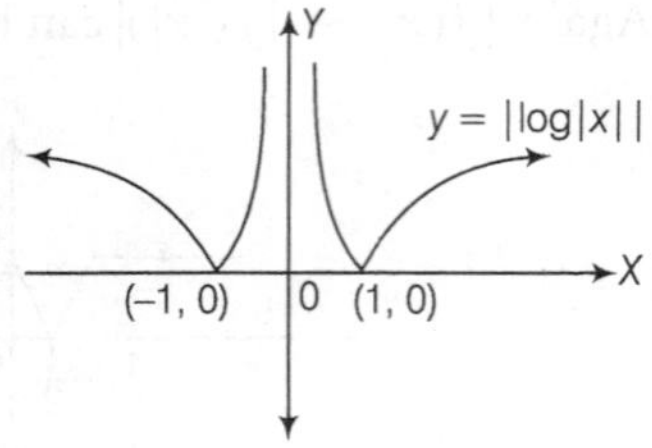

Figure 4.16

While discussing the case (i), we could say that it is continuous for all $x \in R - \{0\}$ and differentiable for all $x \in R - \{0\}$. Again, while discussing the case (ii), we could say that it is continuous for all $x \in R - \{0\}$ and differentiable at all $x \in R - \{-1, 0, 1\}$. (As now it has sharp edges at $\{-1, 1\}$ and is broken at $\{0\}$).

$|y| = f(x) \rightarrow y = f(x)$

When $y = f(x)$ is given

Clearly, $|y| > 0$, if $f(x) < 0$, graph of $|y| = f(x)$ would not exist.

And if $f(x) \geq 0$, $|y| = f(x)$ would be given as $y = \pm f(x)$. Hence, the graph of $|y| = f(x)$ exists only in the regions, where $f(x)$ is non-negative and will be reflected about X-axis only when $f(x) \geq 0$. Region where $f(x) < 0$ will be neglected.

Or

To draw the graph of $|y| = f(x)$, we use following steps

(i) Remove the portion of the graph, which lies below X-axis. [$\because$ the equation $|y| = f(x)$ is not satisfied when $f(x)$ is negative.]

(ii) Plot the remaining portion of the graph and also take its mirror image in the X-axis. [$\because$ when $f(x) > 0$, then $y = \pm f(x)$]

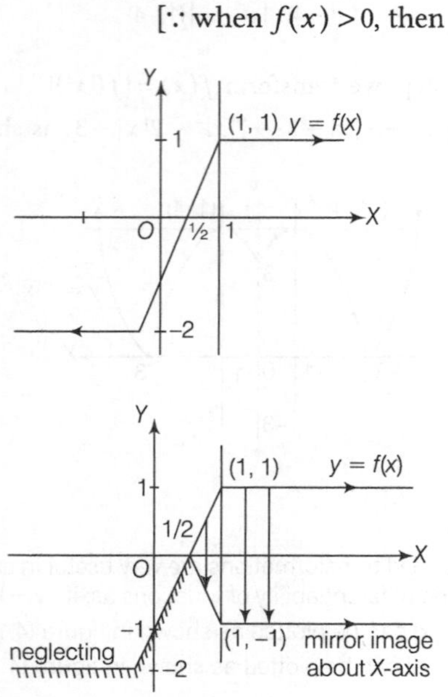

Figure 4.17

Finally, it is as shown

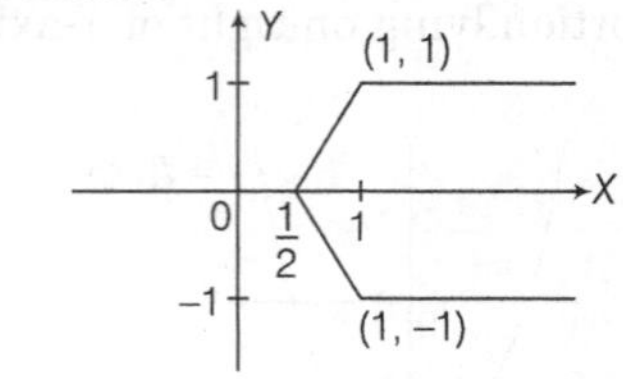

Graph, $|y| = f(x)$

Figure 4.18

Example 18 Draw the graph for $|y| = (x - 1)$.

Sol. We know, here $|y| = (x - 1)$ exists only when $(x - 1)$ is positive. Clearly $y = x - 1$ represent straight line the passing through (0, 1) and (0, –1).

$\therefore$ Neglecting negative values and will be reflected about X-axis. Thus, it can be drawn as shown.

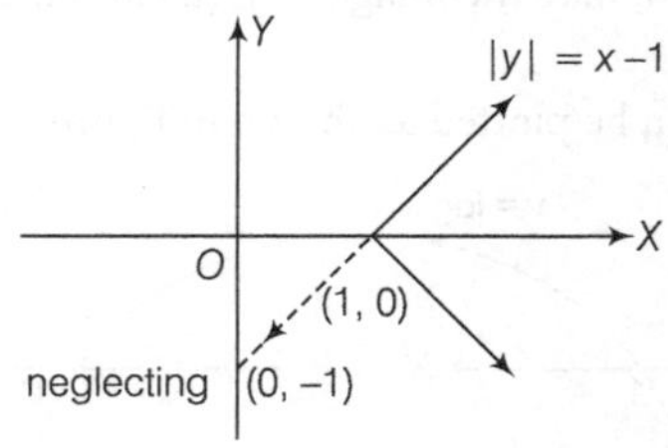

Finally, it is shown as

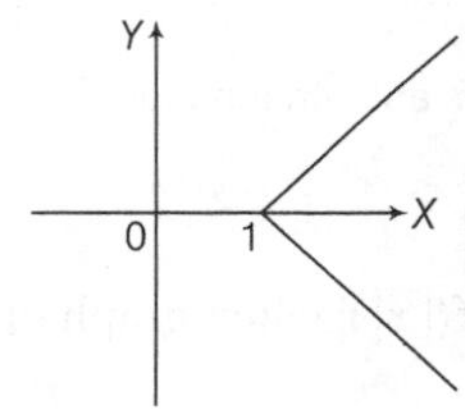

Graph for $|y| = x - 1$

Example 19 Draw the graph for $|y| = (x - 1)(x - 2)$.

Sol. First we plot $y = (x - 1)(x - 2)$

$$\Rightarrow \quad y = x^2 - 3x + 2 \Rightarrow \left(x - \frac{3}{2}\right)^2 = \left(y + \frac{1}{4}\right)$$

It represent a parabola opening upwards having vertex $\left(\frac{3}{2}, -\frac{1}{4}\right)$.

It intersect X-axis at (0, 1) and (2, 0) and Y-axis at (0, 2). Thus, we can plot the figure as shown below.

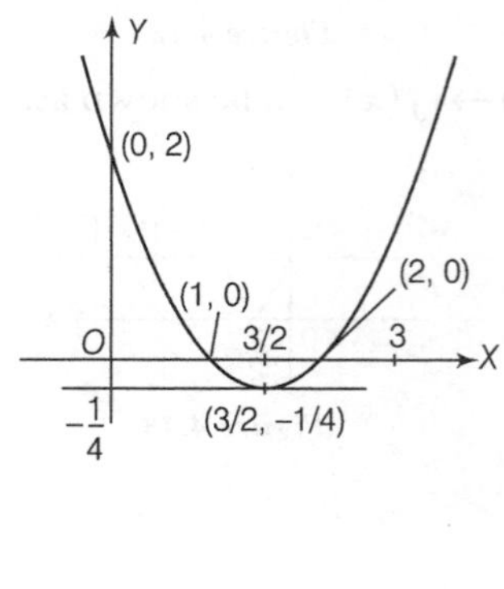

We know, $|y| = (x-1)(x-2)$ exists only, when $(x-1)(x-2) \geq 0$ and neglecting, if $(x-1)(x-2) < 0$.

Now, $f(x)$ will be reflected about X-axis, when $(x-1)(x-2) > 0$.

Thus, it can be plotted as shown.

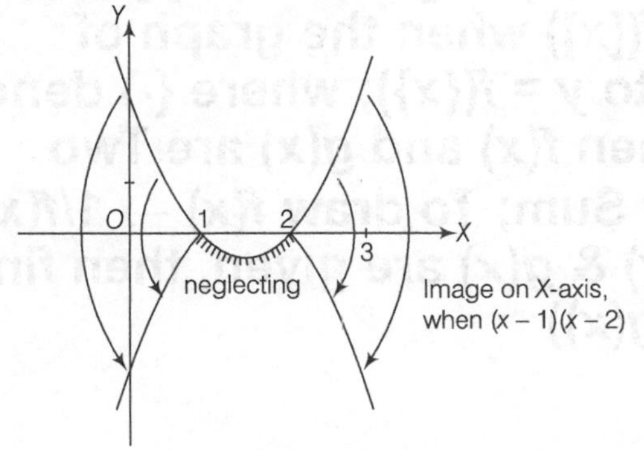

Finally, $|y| = (x-1)(x-2)$ can be shown as

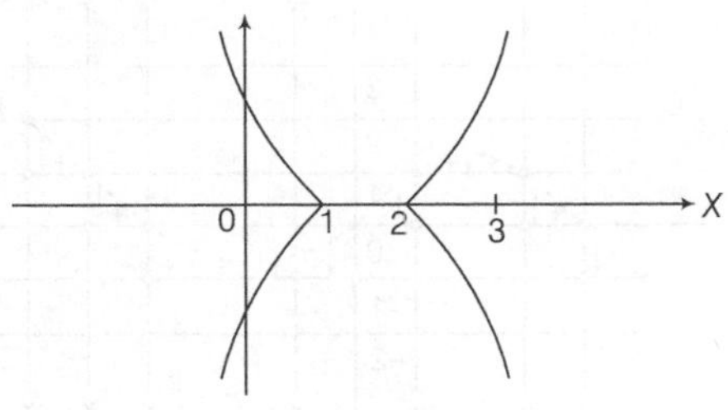

Summary

- $y = f(x)$ to $y = -f(x)$ — *Flip about the X-axis.*
- $y = f(x)$ to $y = f(-x)$ — *Flip about the Y-axis.*
- $y = f(x)$ to $y = |f(x)|$ — *Reflect the parts of the graph that lie in the lower half (negative parts) into the upper half of the axes.*
- $y = f(x)$ to $y = f(|x|)$ — *Discard the left part of the graph (for $x < 0$) and take a reflection of the right part of the graph into the left half of the axes.*
- $y = f(x)$ to $|y| = f(x)$ — *Discard the lower part of the graph ($f(x) < 0$) and take a reflection of the upper part of the graph into the lower half of the axes.*
- $y = f(x)$ to $y = f(x) + k$ — *Shift the graph $|k|$ units upwards or downwards depending on whether k is positive or negative respectively.*
- $y = f(x)$ to $y = f(x+k)$ — *Advance (shift left) or delay (shift right) the graph by $|k|$ units depending on whether k is positive or negative, respectively.*
- $y = f(x)$ to $y = kf(x)$ — *Stretch or compress the graph along the Y-axis depending on whether $|k| > 1$ or $|k| < 1$, respectively. Also, flip it about the X-axis, if k is negative (this latter statement follows from part 1).*
- $y = f(x)$ to $y = f(kx)$ — *Stretch or compress the graph along the X-axis depending on whether $|k| < 1$ or $|k| > 1$, respectively. Also, flip it about the Y-axis, if k is negative.*

Exercise for Session 3

1. Consider the following function f, whose graph is given below.

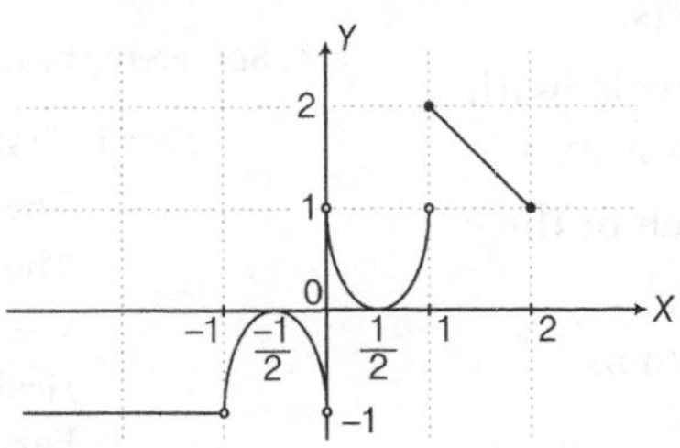

Draw the graph of the following functions.

(i) $|f(x)|$ (ii) $f(|x|)$ (iii) $|f(|x|)-1|$

2. Plot the following functions.

(i) $y = |x^2 - 2x - 3|$ (ii) $y = x^2 - 2|x| - 3$ (iii) $y = |\log_2 x|$ (iv) $y = |\log_2 |x||$

(v) $y = \log_2 |1-x|$ (vi) $y = \log_2 (2-x)^2$ (vii) $y = |\cos |x||$ (viii) $y = |2 - 2^x|$

(ix) $y = \sin(|x|)$ (x) $y = \cos(|x|)$

3. Plot the following functions.

(i) $|f(x)| = \log_2 x$ (ii) $|f(x)| = \log_2 (-x)$

4. Find the number of solutions of $\sin \pi x = |\log_e |x||$.

5. Find the number of solutions of

(i) $2^{|x|} = \sin x^2$ (ii) $\sin x = x^2 + x + 1$

Session 4

To draw $y = [f(x)]$, when the graph of $y = f(x)$ is given and $[\cdot]$ denotes greatest integer fuctions; To draw $y = f([x])$ when the graph of $y = f(x)$ is given; When $f(x)$ transforms to $y = f(\{x\})$; where $\{\cdot\}$ denotes fractional part of x, i.e. $\{x\} = x - [x]$; When $f(x)$ and $g(x)$ are Two Functions and are transformed to their Sum; To draw $f(x) \to 1/f(x) = h(x)$, To draw $y = f(x) \cdot \sin x$, When $f(x)$ & $g(x)$ are given, then find $h(x) = \max(f(x), g(x))$ or $h(x) = \min(f(x), g(x))$

To Draw $y = [f(x)]$

Where, $[\cdot]$ denotes greatest integral function {i.e. now y cannot be fraction}.

Here, in order to draw $y = [f(x)]$ mark the integer on Y-axis. Draw the horizontal lines through integers till they intersect the graph. Draw vertical dotted lines from these intersection points.

Finally, draw horizontal lines parallel to X-axis from any intersection point to the nearest vertical dotted line with blank dot at right end in case $f(x)$ increases.

Or

To plot $y = [f(x)]$, we use the following steps

***Step* I** Plot $f(x)$.

***Step* II** Mark the intervals of unit length with integers as end points on Y-axis.

***Step* III** Mark the corresponding intervals {with the help of graph of $f(x)$} on X-axis.

***Step* IV** Plot the value of $[f(x)]$ for each of the marked intervals.

Graphically, it could be shown as

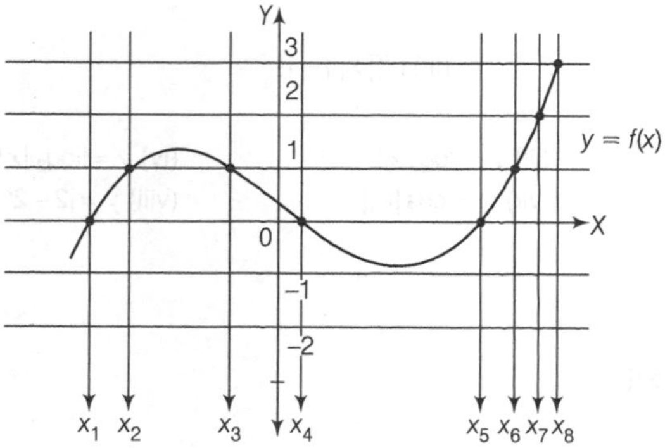

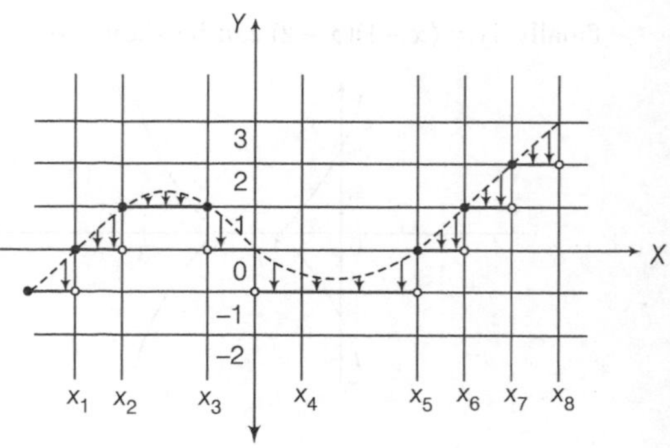

Figure 4.19

Example 20 Draw the graph of $y = [x^3]$, when $-2^{1/3} \le x \le 2^{1/3}$.

Sol. Here, in order to draw the graph of $y = [x^3]$.

***Step* I** First we draw the graph at $y = x^3$.
The function $f(x) = x^3$ is called the cube function. The domain and range of cube function is R. Since, $y = x^3$ is an odd function, so its graph is symmetrical in opposite quardrant.
For $y = x^3$, we have following table.

x	0	1	$2^{1/3}$	–1	$-2^{1/3}$
y	0	1	2	–1	–2

Now, plot the points (0,0) (1,1), $(2^{1/3},2)$, $(-1,-1)$ and $(-2^{1/3},2)$ and join them by free hand, we get the graph of $y = x^3$.

***Step* II** Draw horizontal lines through integers located on Y-axis till they intersect the graph.

Step III Draw vertical dotted lines from these intersection points to intersect X-axis.

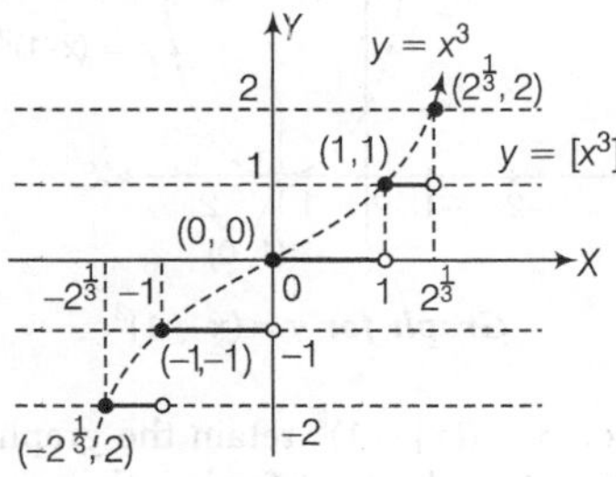

Step IV Finally, draw horizontal lines parallel to X-axis from any intersection point to the nearest vertical dotted line with blank dot at right end.

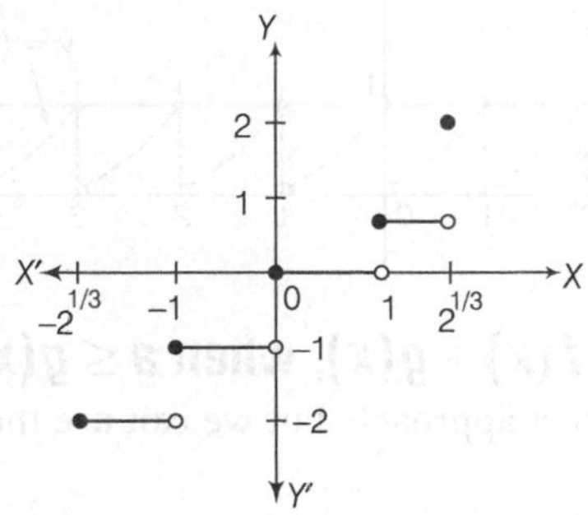

Graph for $y = [x^3]$, $x \in [-2^{1/3}, 2^{1/3}]$

Example 21 Draw the graph of $y = [\sin x]$.

Sol. Step I First we draw the graph of $y = \sin x$ as (explained in Example 6) shown below.

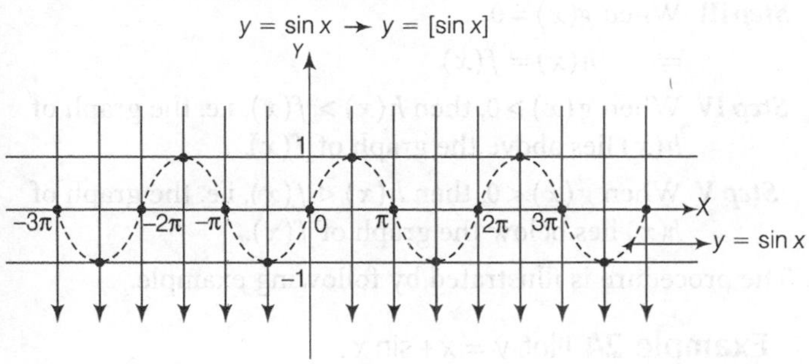

Step II Now, draw horizontal lines through integers located as Y-axis till they intersect the graph.

Step III Draw vertical lines from the intersection points to intersect X-axis.

Step IV Finally, draw horizontal lines parallel to X-axis from any intersection point to the nearest vertical dotted line with blank dot at right end as shown below.

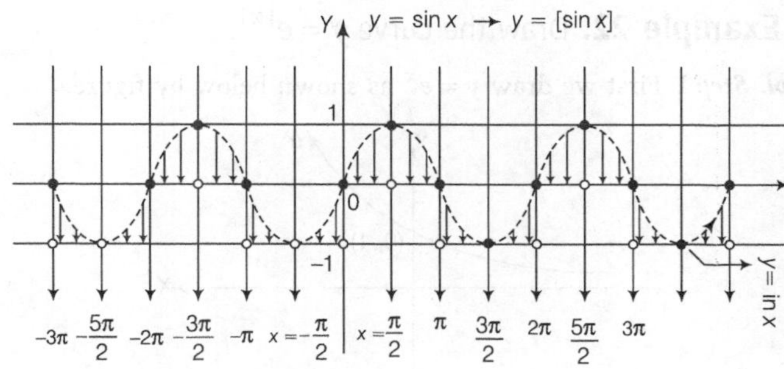

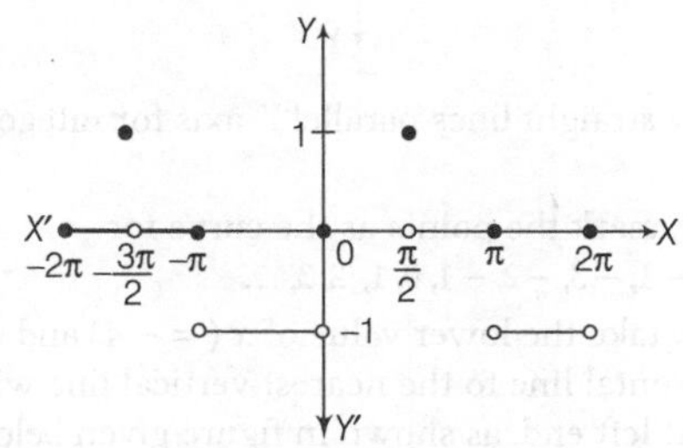

Graph for $y = [\sin x]$

To draw $y = f[x]$

when the graph of $y = f(x)$ is given, following steps are used.

Step I Plot the straight lines parallel to Y-axis for integral value of x, i.e. $n = \ldots -3, -2, -1, 0, 1, 2, 3$.

Step II Now, mark the points of which $x = -3, x = -2, x = -1, x = 0, x = 1, \ldots$ on the curve.

Step III Take the lower marked point for x (say) if $n < x < n + 1$, then take the point at $x = n$ and draw a horizontal line to the nearest vertical line formed by $x = n + 1$, proceeding in this way, we get required curve.

Mark the integers on the X-axis. Draw vertical lines till they intersect the graph of $f(x)$. From these intersection points, draw horizontal lines (parallel to X-axis) to meet the nearest right vertical line with a blank dot on each nearest right vertical line which can be shown as in the figure.

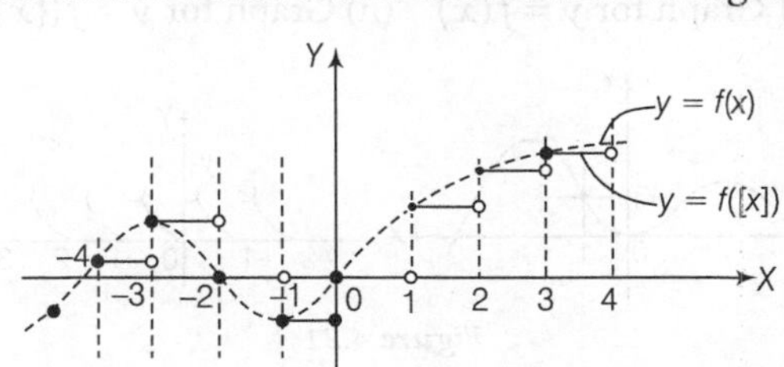

Figure 4.20

Example 22. Draw the curve $y = e^{[x]}$.

Sol. ***Step*** I. First we draw $y = e^x$ as shown below by figure.

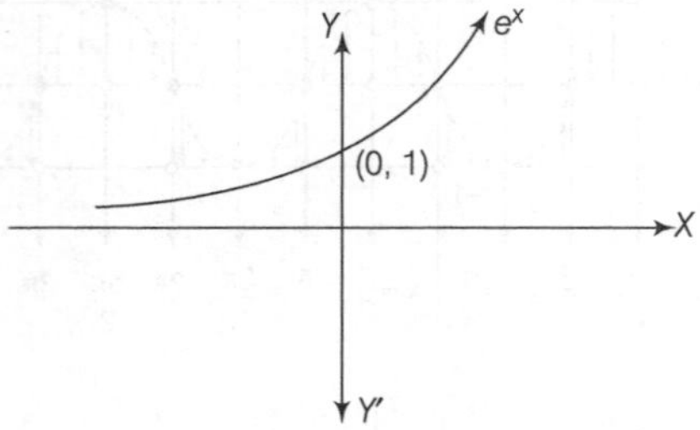

Step II. Draw straight lines parallel Y-axis for integer values of x.

Step III. Now, mark the points as the curve for $n = -4, -3, -2, -1, 0, 1, 2, 3, \ldots$

Step IV. Now, take the lower value of $x\ (= -4)$ and draw a horizontal line to the nearest vertical line with a blank dot at left end, as shown in figure given below.

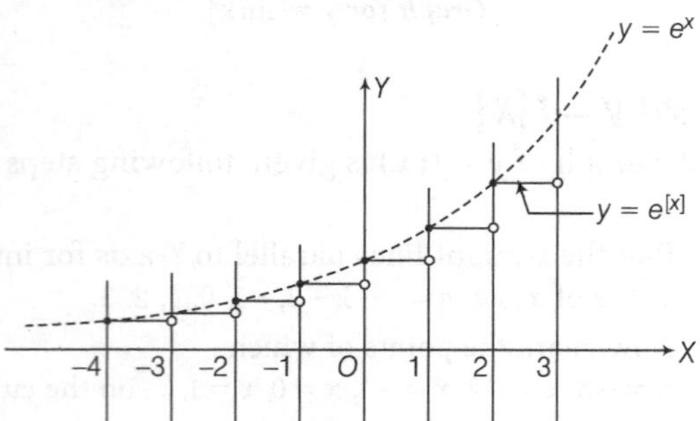

To Plot $y = f(\{x\})$

Graph of $f(\{x\})$ can be obtained from the graph of $f(x)$ by following rule. "Retain the graph of $f(x)$ for values of x lying between interval $[0, 1)$. Now, it can be repeated for rest of the points. (taking periodicity 1).

Now, obtained function is graph for $y = f(\{x\})$.

Graphically, it could be stated as

(i) Graph for $y = f(x)$ (ii) Graph for $y = f(\{x\})$

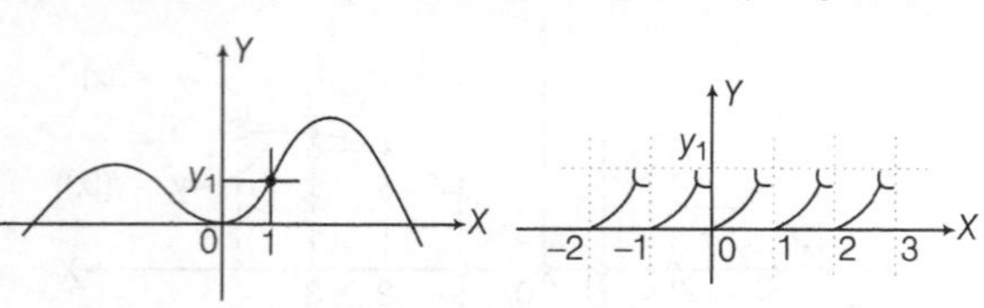

Figure 4.21

Example 23. Draw the graph for $y = (\{x\} - 1)^2$.

Sol. First we will draw the graph for $y = (x - 1)^2$, clearly it represent a parabola opening upward and having vertex (1, 0) which could be ploted as shown below.

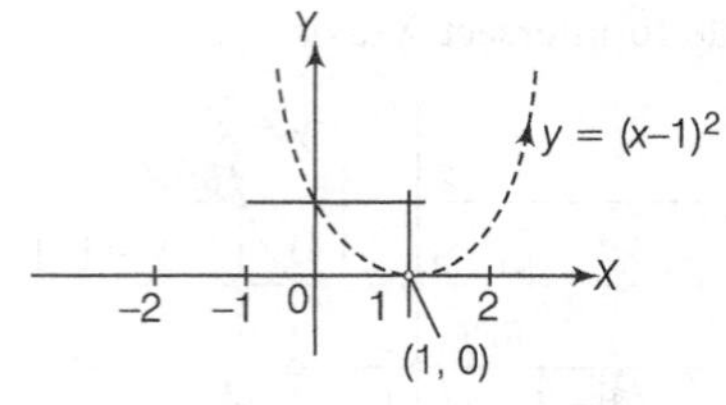

Graph for $y = (x - 1)^2$

Now, to plot $y = (\{x\} - 1)^2$ retain the graph for the interval $x \in [0, 1)$ and repeat for length one as shown below.

Graph for $y = (\{x\} - 1)^2$

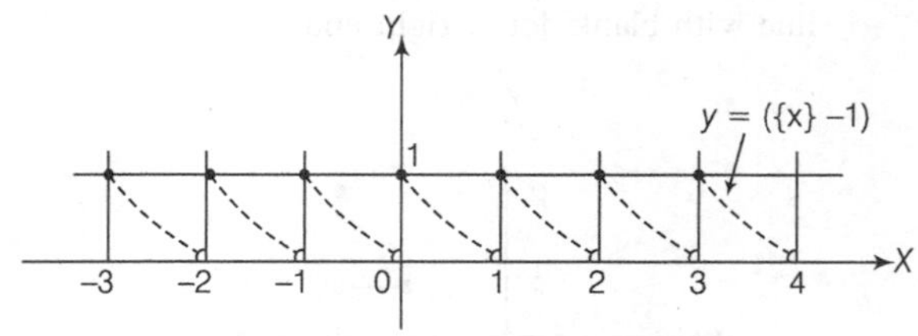

To Plot $y = f(x) + g(x)$, when $a \le g(x) \le b$

There is no direct approach, but we can use the following steps :

If maximum or minimum value of any one is known.

Step I Find the maximum and minimum value of $g(x)$ say $a \le g(x) \le b$.

Step II Plot the curve $f(x)$ between $f(x) + a$ to $f(x) + b$, i.e. $f(x) + a \le h(x) \le f(x) + b$.

Step III When $g(x) = 0$

$\Rightarrow \quad h(x) = f(x)$

Step IV When $g(x) > 0$, then $h(x) > f(x)$, i.e. the graph of $h(x)$ lies above the graph of $f(x)$.

Step V When $g(x) < 0$, then $h(x) < f(x)$, i.e. the graph of $h(x)$ lies below the graph of $f(x)$.

The procedure is illustrated by following example.

Example 24 Plot $y = x + \sin x$.

Sol. Here, $y = f(x) + g(x) = x + \sin x$

where $f(x) = x$ and $g(x) = \sin x$.

As we know, $\quad g(x) = \sin x \in [-1, 1]$

$\therefore \quad x - 1 \le y \le x + 1 \qquad$ ...(i)

To sketch the curve between two parallel line $y = x + 1$ and $y = x - 1$.

Here, $g(x) = 0 \Rightarrow y = x \qquad$ ...(ii)

Also, when $g(x) > 0$

$\Rightarrow \quad x + \sin x > x \qquad$ ...(iii)

and when $g(x) < 0$

$\Rightarrow \quad x + \sin x < x \qquad$ ...(iv)

Using Eqs. (i), (ii), (iii) and (iv), we have the following graph.

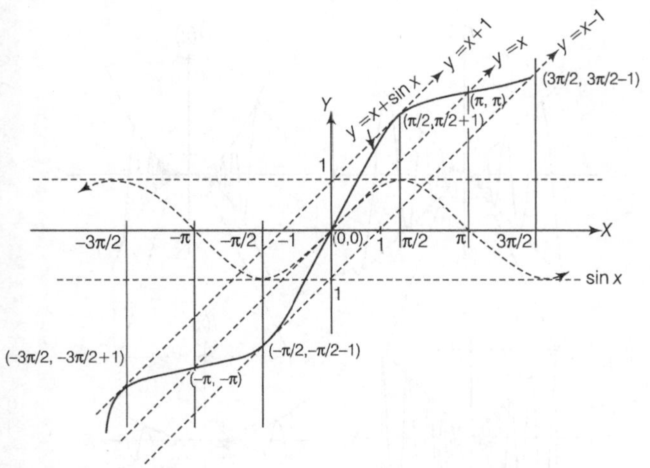

When $f(x)$ Transforms to $\frac{1}{f(x)} = h(x)$

(i) when $f(x)$ increases, then $h(x)$ decreases.

(ii) when $f(x)$ decreases, then $h(x)$ increases.

(iii) as $f(x) \to 0, h(x) \to +\infty$.

(iv) as $f(x) \to \infty, h(x) \to 0$.

(v) when $f(x) = \pm 1$, then $h(x)$ is also equal to ± 1.

Example 25 Plot $y = |x| - 2$ and hence $f(x) = \frac{1}{|x| - 2}$.

Sol. First we plot $y = |x| - 2$ as explained in Example 2. The graph is shown below.

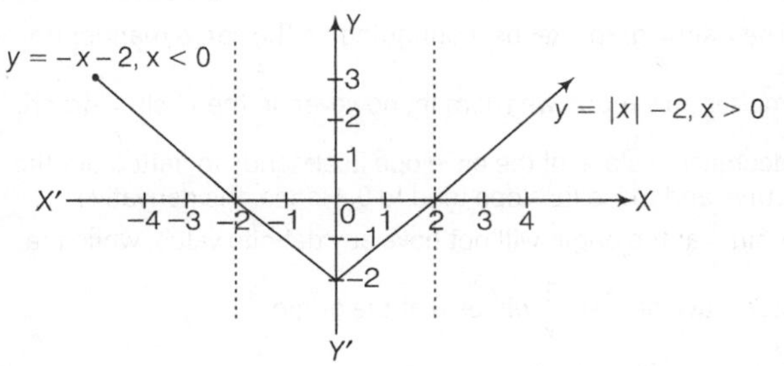

Here, (i) $y = |x| - 2$ increases when $x > 0$

$\Rightarrow \quad f(x) = \frac{1}{|x| - 2}$ decreases when $x > 0$

and (ii) $y = |x| - 2$ decreases when $x < 0$

$\Rightarrow \quad f(x) = \frac{1}{|x| - 2}$ increases when $x < 0$

(iii) As, $y \to 0 \Rightarrow x = \pm 2 \Rightarrow f(x) \to \infty$ as $x = \pm 2$

(iv) As, $f(x) \to 0$, i.e. $\frac{1}{|x| - 2} \to 0$ as $x \to \infty$

and when $x \to \infty \Rightarrow y = |x| - 2 \to \infty$

(v) $y = \pm 1 \Rightarrow f(x) = \pm 1$

$\therefore$ $f(x)$ could be plotted as,

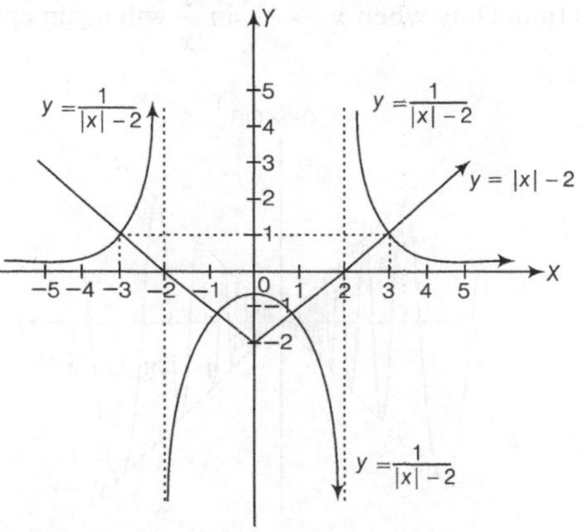

To Plot the Graph of $y = f(x) \cdot \sin x$, when Graph of $y = f(x)$ is given

Clearly, $-f(x) \le f(x) \cdot \sin x \le f(x)$ [as $-1 \le \sin x \le 1$]

Hence, graph for $y = f(x) \cdot \sin x$ would be lying between the graph of $y = f(x)$ and $y = -f(x)$. It amounts to just drawing graph of $\sin x$ in between the graphs of $y = \pm f(x)$.

The procedure is illustrated by following example.

Example 26 Let $f(x) = \begin{cases} x \sin \frac{1}{x}, & x \ne 0 \\ 0, & x = 0 \end{cases}$

and $g(x) = \begin{cases} x^2 \sin \frac{1}{x}, & x \ne 0 \\ 0, & x = 0 \end{cases}$.

Discuss the graph for $f(x)$ and $g(x)$, and evaluate the continuity and differentiability of $f(x)$ and $g(x)$.

Sol. We will first try to graphically understand the behaviour of these two functions and then verify our results analytically. Notice that no matter what the argument of the sine function is, its magnitude will always remain between − 1 and 1.

Therefore, $\left| x \sin \frac{1}{x} \right| \le |x|$ and $\left| x^2 \sin \frac{1}{x} \right| \le |x|^2$

This means that the graph of $x \sin \frac{1}{x}$ will always lie between the lines $y = \pm x$ and the graph of $x^2 \sin \frac{1}{x}$ will always lie between the two curves $y = \pm x^2$. Also, notice that as $|x|$ increases, $\frac{1}{x}$ decreases in a progressively slower manner while when $|x|$ is close to 0, the increase in $\frac{1}{x}$ is very fast (as $|x|$ decreases visualise the graph of $y = 1/x$). This means that near the origin, the variation in the graph of $\sin \frac{1}{x}$ will be extremely rapid because the successive zeroes of the graph will become closer and closer. As we keep on increasing x, the variation will become slower and slower and the graph will 'spread out'.

For example, for $x > \frac{1}{\pi}$ there will not be finite root of the function. Only when $x \to \infty$, $\sin \frac{1}{x}$ will again approach 0.

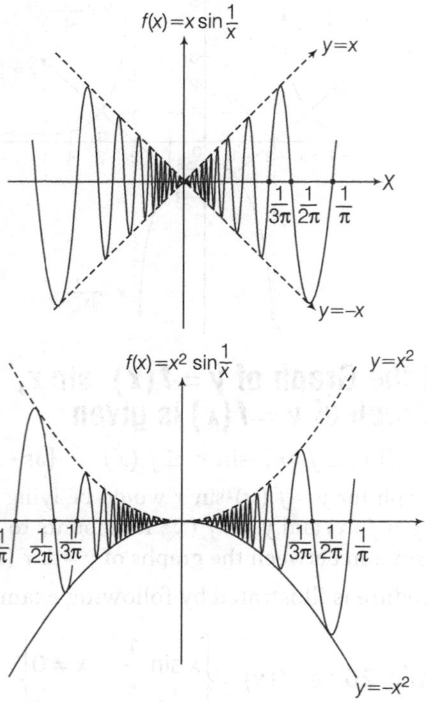

Figure 4.22

Remarks

Notice how the lines $y = \pm x$ 'envelope' the graph of the function in the first case and the curves $y = \pm x^2$ 'envelope' the graph of the function in the second case.

The envelopes shrink to zero vertical width at the origin in both cases. Therefore, we must have

$$\lim_{x \to 0} x \sin\frac{1}{x} = 0 \text{ and } \lim_{x \to 0} x^2 \sin\frac{1}{x} = 0.$$

(This is also analytically obvious; $\sin\frac{1}{x}$ is a finite number between -1 and 1; when it gets multiplied by x (where $x \to 0$) the whole product gets infinitesimally small).

Now lets try to get a 'feel' on what will happen to the derivatives of these two functions at the origin.

For $f(x) = x \sin\frac{1}{x}$, the slope of the envelope is constant ($\pm$ 1). Thus, the sinusoidal function inside the envelope will keep on oscillating as we approach the origin, while shrinking in width due to the shrinking envelope. The slope of the curve also keeps on changing and does not approach a fixed value.

However, for $g(x) = x^2 \sin\frac{1}{x}$, the slope of the envelope is itself decreasing as we approach the origin, apart from shrinking in width. This envelope will 'compress' or 'harmmer out' or 'flatten' the sin oscillations near the origin. What should therefore happen to the derivative? It should become 0 at the origin!

Let us 'zoom in' on the graphs of both the functions around the origin, to see what is happening.

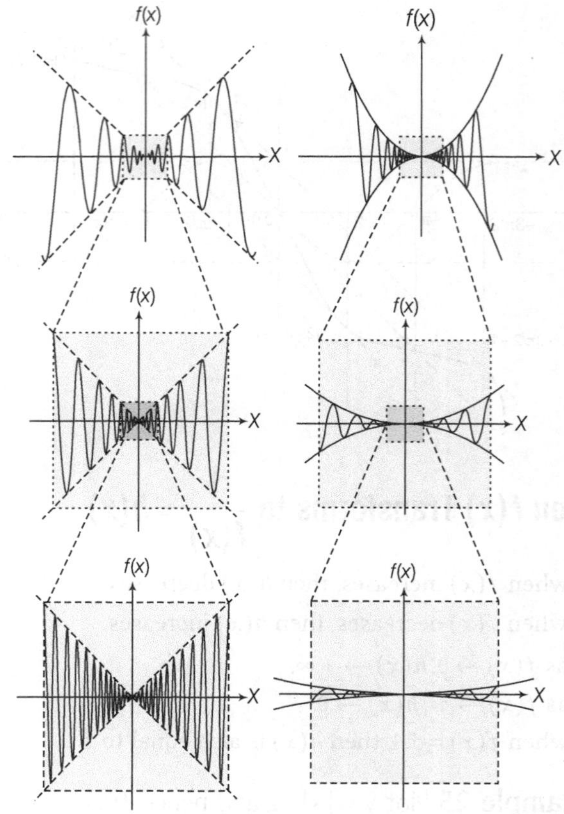

Figure 4.23

These graphs are not very accurate and are only of an approximate nature; but they do give us some feeling on the behaviour of these two functions near the origin.

The $x \sin\frac{1}{x}$ graph keeps 'continuing' in the same manner no matter how much we zoom in; however, in the $x^2 \sin\frac{1}{x}$ graph, the decreasing slope of the envelope itself tends to flatten out the curve and make its slope tend to 0. Hence, the derivative of $x \sin\frac{1}{x}$ at the origin will not have any definite value, while the derivative of $x^2 \sin\frac{1}{x}$ will be 0 at the origin.

Lets verify this analytically

(i) $f(x) = x \sin\frac{1}{x}$ at $x = 0$.

$$\text{LHD} = \text{RHD} = \lim_{h \to 0} \frac{f(h) - f(0)}{h} = \lim_{h \to 0} \frac{h \sin\frac{1}{h}}{h} = \lim_{h \to 0}\left(\sin\frac{1}{h}\right)$$

This limit, as we know does not exist; hence, the derivative for $f(x)$ does not exist at $x = 0$

(ii) $f(x) = x^2 \sin\frac{1}{x}$ at $x = 0$

$$\text{LHD} = \text{RHD} = \lim_{h \to 0} \frac{h^2 \sin\frac{1}{h} - 0}{h} = \lim_{h \to 0} h \sin\frac{1}{h} = 0.$$

To Find $h(x)$ max $(f(x), g(x))$ or $h(x) = \min(f(x), g(x))$ when $f(x)$ and $g(x)$ are given

(i) $h(x) = \max\{f(x), g(x)\}$

$$\because \quad h(x) = \begin{cases} f(x), & \text{when } f(x) > g(x) \\ g(x), & \text{when } f(x) < g(x) \end{cases}$$

To draw the graph of $y = \max\{f(x), g(x)\}$, first we draw the graphs of both the functions $f(x)$ and $g(x)$ and find their point of intersection.

Then, we find any two consecutive points of intersection. In between these points $f(x) > g(x)$, then in order to draw the graph of max $\{f(x), g(x)\}$, we take those segments of $f(x)$ for which $f(x) > g(x)$ between any two consecutive points of intersection of $f(x)$ and $g(x)$.

(ii) $h(x) = \min\{f(x), g(x)\}$

$$\because \quad h(x) = \begin{cases} f(x), & \text{when } f(x) < g(x) \\ g(x), & \text{when } g(x) < f(x) \end{cases}$$

Similarly, in order to draw the graph of min $\{f(x), g(x)\}$ we take those segments of $f(x)$ for which $f(x) < g(x)$, between any two consecutive points of intersection of $f(x)$ and $g(x)$.

Example 27 Draw graph for $y = \max\{2x, x^2\}$ and discuss the continuity and differentiability.

Sol. Here, to draw, $y = \max\{2x, x^2\}$.

First, plot $y = 2x$ and $y = x^2$ on graph and put $2x = x^2$

$\Rightarrow x = 0, 2$ (i.e. their point of intersection).

Clearly, $y = x^2$ represent a parabola open upward with vertex (0, 0) and $y = 2x$ represent a straight line pointing through (0, 0).

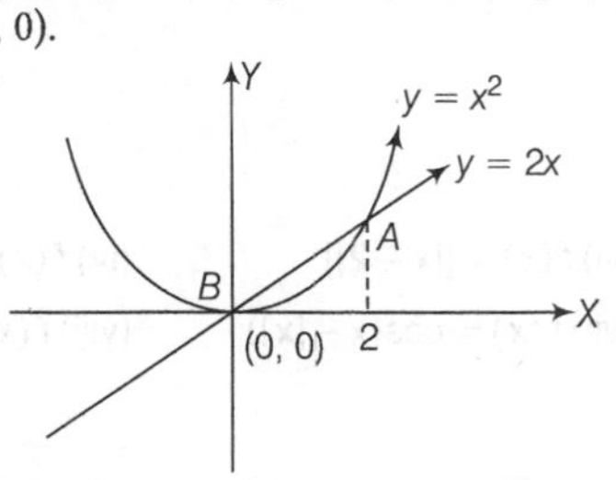

Now, since $y = \max\{2x, x^2\}$ we have to neglect the curve below the point of intersections thus, the required graph is, as shown below.

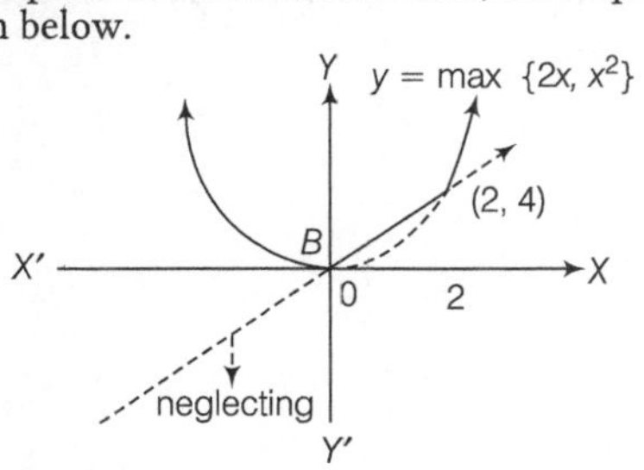

Thus, from the given graph $y = \max\{2x, x^2\}$, we can say $y = \max\{2x, x^2\}$ is continuous for all $x \in R$.

But $y = \max\{2x, x^2\}$ is differentiable for all $x \in R - \{0, 2\}$. As at $x = 0, 2$ curve has corner point.

Remarks

One must remember the formula, we can write

$$\max\{f(x), g(x)\} = \frac{f(x) + g(x)}{2} + \left|\frac{f(x) - g(x)}{2}\right|$$

$$\text{and } \min\{f(x), g(x)\} = \frac{f(x) + g(x)}{2} - \left|\frac{f(x) - g(x)}{2}\right|.$$

Example 28 Draw the graph for $y = |2 - |x - 1||$.

Sol. First we plot $y = x - 1$, it is a straight line intersecting X and Y-axes at (1, 0) and (0, –1), respectively. The graph is shown below.

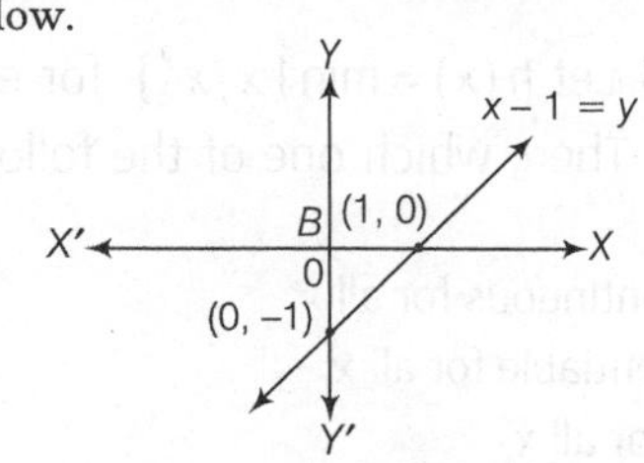

$\Rightarrow$ Now, we plot $y = |x - 1|$, For this, the portion of the graph remain same which lines above the X-axis and reflect the negative portion of the graph about X-axis.

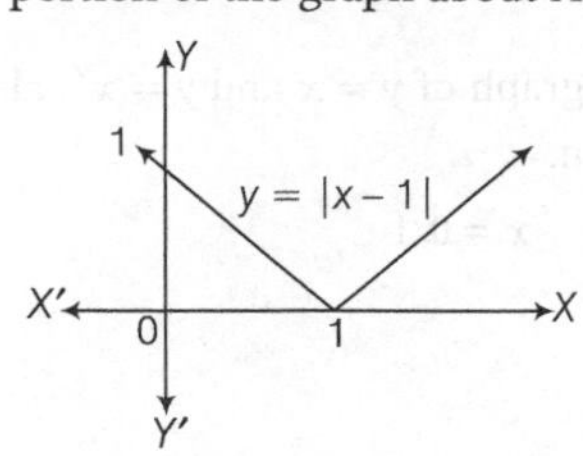

$\Rightarrow$ Now, to plot $y = -|x - 1|$ take the mirror image of the graph of $|x - 1|$ in the X-axis as plane mirror.

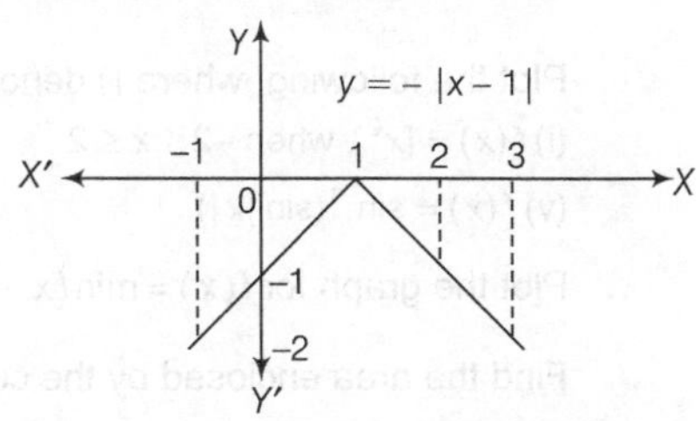

To find the graph of $2 - |x - 1|$. Shift the graph of $y = -|x - 1|$ upward through 2 units.

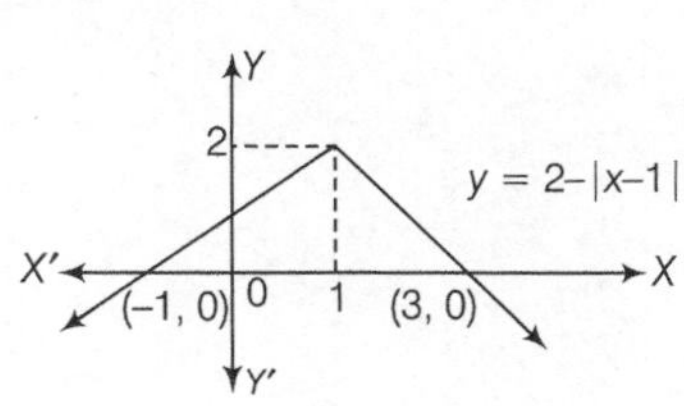

To find the graph of $|2-|x-1||$, reflect the negative portion of the graph of $2-|x-1|$ in X-axis.

Thus, $y=|2-|x-1||$ can be plotted as shown below.

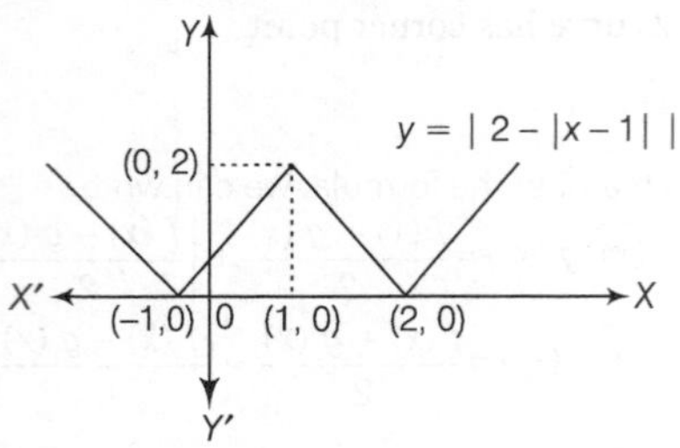

Remark

From above figure we could say $y=|2-|x-1||$ is not differentiable at $x=\{-1, 1, 3\}$ as sharp edges at $x=-1, 1, 3$.

Example 29 Let $h(x)=\min\{x; x^2\}$ for every real number of x. Then, which one of the following is true?

[IIT JEE 1998]

(a) h is not continuous for all x

(b) h is differentiable for all x

(c) $h'(x)=1$, for all x

(d) h is not differentiable at two values of x

Sol. Here, $h(x)=\min\{x; x^2\}$ can be drawn on graph in two steps.

(a) Draw the graph of $y=x$ and $y=x^2$, also find their point of intersection.

i.e. $x=x^2 \Rightarrow \quad x=0, 1$

Thus, points of intersection are (0, 0) and (1,1).

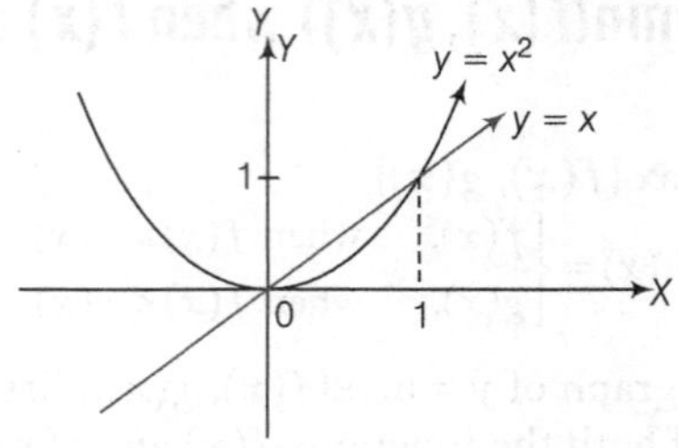

(b) To find $h(x)=\min\{x; x^2\}$ neglecting the graph above the point of intersection, we get

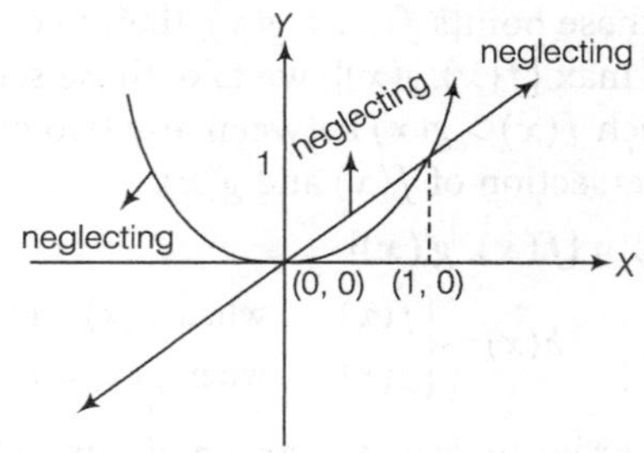

Thus, from the given graph,

$$h(x)=\begin{cases} x, & x \le 0 \text{ or } x \ge 1 \\ x^2, & 0 \le x \le 1 \end{cases}$$

which shows $h(x)$ is continuous for all x. But not differentiable at $x=\{0, 1\}$.

Thus, $h(x)$ is not differentiable at two values of x.

Hence, (d) is the correct answer.

Exercise for Session 4

1. Plot the following, where [·] denotes greatest integer function.

(i) $f(x)=[x^2]$, when $-2 \le x \le 2$ (ii) $f(x)=[|x|]$ (iii) $f(x)=[|x-2|]$ (iv) $f(x)=[|x|-2]$

(v) $f(x)=\sin^{-1}(\sin|x|)$ (vi) $f(x)=[\cos^{-1}x]$ (vii) $f(x)=\cos(x-[x])$ (viii) $f(x)=[\sin^{-1}(\sin x)]$

2. Plot the graph for $f(x)=\min(x-[x], -x-[-x])$.

3. Find the area enclosed by the curves

(i) $\max(|x|, |y|)=1$ (ii) $\max(2|x|, 2|y|)=1$ (iii) $\max(|x+y|, |x-y|)=1$

JEE Type Solved Examples :
Single Option Correct Type Questions

Ex. 1 *The number of real solutions of the equation* $e^x + x = 0$ *is*

(a) 0 (b) 1
(c) 2 (d) None of these

Sol. (*b*) It is evident from figure that the curves $y = e^x$ and $y = -x$ intersect exactly at one point. So, the equation $e^x = -x$ or $e^x + x = 0$ has one real solution.

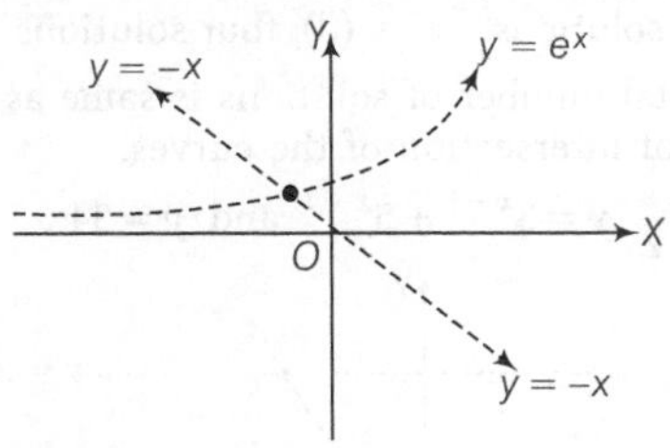

Ex. 2 *The number of real solutions of the equation* $\log_a x = |x|, 0 < a < 1$, *is*

(a) 0 (b) 1
(c) 2 (d) None of these

Sol. (*b*) The number of real solutions of the equation $\log_a x = |x|$ is equal to the number of points of intersection of the curves.

$$y = \log_a x \text{ and } y = |x| \quad (0 < a < 1)$$

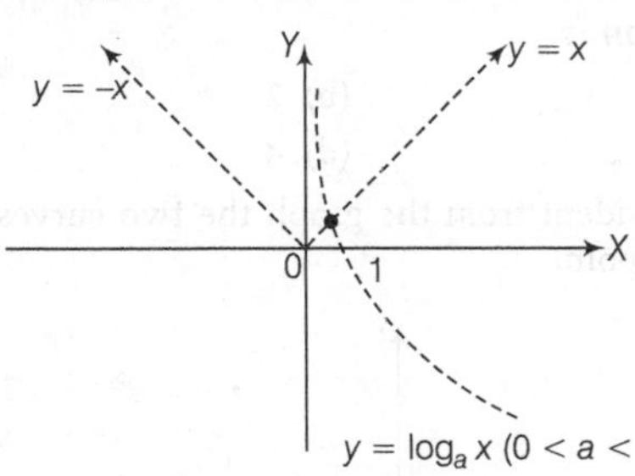

It is evident from the graph that the two curves intersect at one point only.

∴ One real solution lying in (0, 1).

Ex. 3 *The number of solutions of the equation* $|x| = \cos x$ *is*

(a) 0 (b) 1
(c) 2 (d) 3

Sol. (*c*) The number of real solutions of the equation $|x| = \cos x$ is equal to the number of point of intersection of the curves.

$y = |x|$ and $y = \cos x$ shown as;

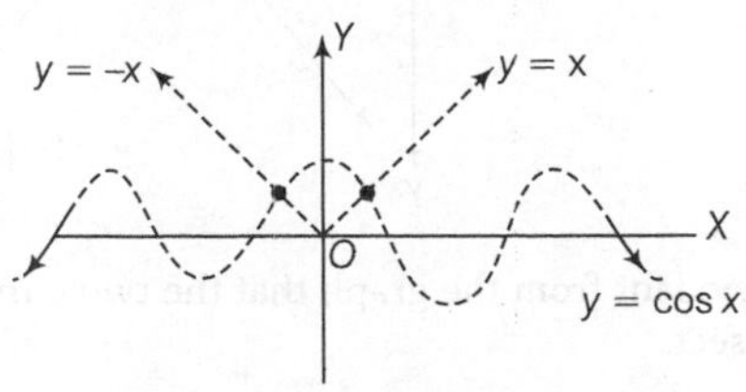

It is evident from the graph that the two curves intersect at two points only.

∴ Two real solutions lying in $\left(-\frac{\pi}{2}, \frac{\pi}{2}\right)$.

Ex. 4 *How many roots does the following equation possess* $3^{|x|}\{|2 - |x||\} = 1$?

(a) 1 (b) 2 (c) 3 (d) 4

Sol. (*b*) Here, $3^{|x|}\{|2 - |x||\} = 1$

$$\Rightarrow \quad 2 - |x| = 3^{-|x|}$$

In order to determine the number of roots, it is sufficient to find the points of intersection of the curves $y = 2 - |x|$ and $y = 3^{-|x|}$, shown as;

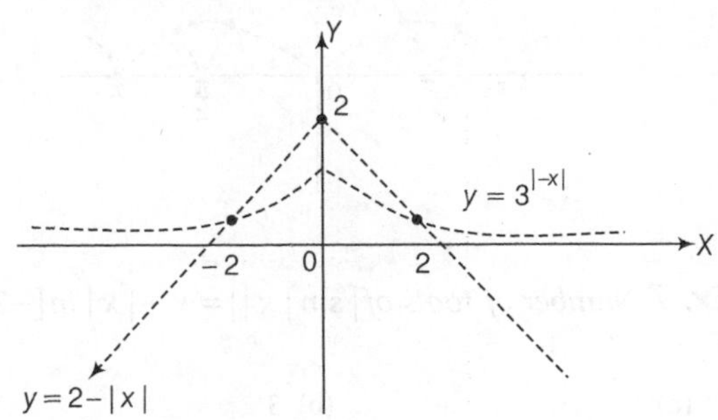

We observe the two curves intersect at two points.
∴ Two real solutions ∈ (−2, 2).

Ex. 5 *The number of real solutions of the equation* $x^2 = 1 - |x - 5|$ *is*

(a) 1 (b) 2
(c) 4 (d) None of these

Sol. (*d*) The number of real solutions of the equation

$$x^2 = 1 - |x - 5|$$

is equal to the number of points of intersection of the curves.

$y = x^2$ and $y = 1 - |x - 5|$, shown as;

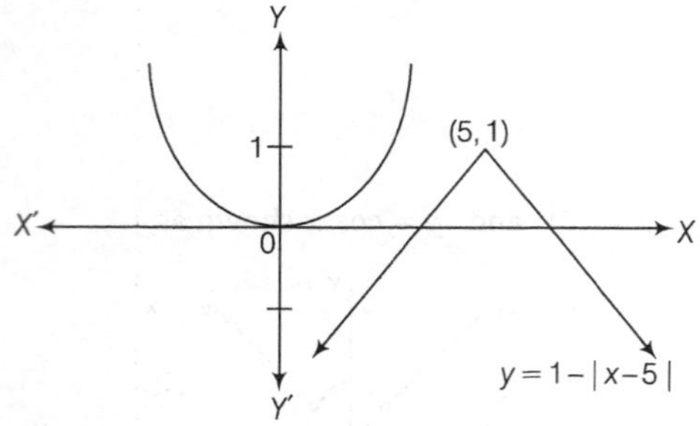

It is evident from the graph that the two curves do not intersect.

∴ No solution.

Ex. 6 *Number of solutions for* $2^{\sin|x|} = 4^{|\cos x|}$ *in* $[-\pi, \pi]$ *is equal to*

(a) 2 (b) 4
(c) 6 (d) 8

Sol. (*b*) The total number of solutions for the given equation is equal to the number of points of intersection of curves $y = 4^{|\cos x|}$ and $y = 2^{\sin|x|}$.

Clearly, the two curves intersect at four points. So, there are four solutions of the given equation.

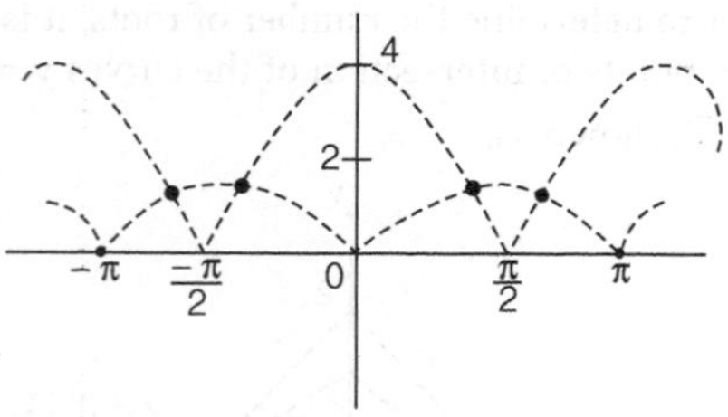

Ex. 7 *Number of roots of* $|\sin|x|| = x + |x|$ *in* $[-2\pi, 2\pi]$ *is*

(a) 2 (b) 3
(c) 4 (d) 6

Sol. (*b*) The total number of solutions for the given equation is equal to the number of points of intersection of the curves.

$$y = |\sin|x||$$

and $$y = x + |x| = \begin{cases} 2x, & x \geq 0 \\ 0, & x < 0 \end{cases}$$

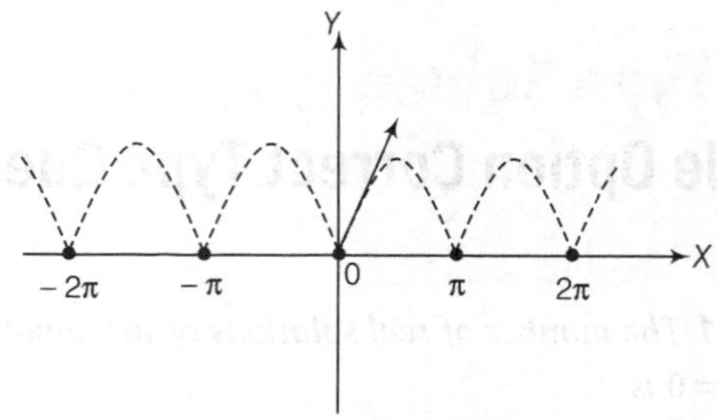

Clearly, the two curves intersect at three points.

∴ There are three solutions.

Ex. 8 *The equation* $3^{x-1} + 5^{x-1} = 34$ *has*

(a) one solution (b) two solutions
(c) three solutions (d) four solutions

Sol. (*a*) The total number of solutions is same as the number of points of intersection of the curves.

$$y = 3^{x-1} + 5^{x-1} \text{ and } y = 34$$

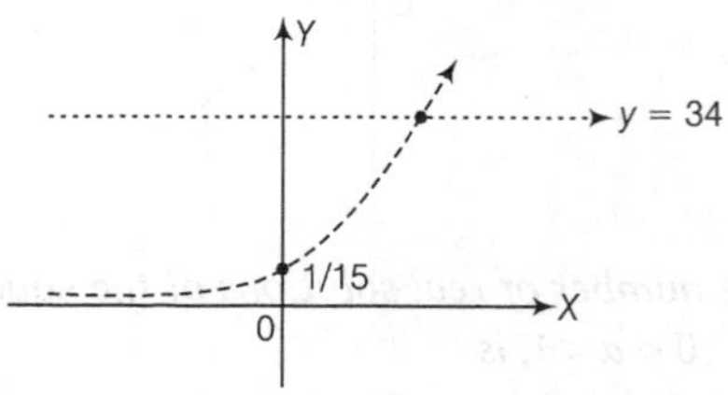

It is evident that these two curves intersect at exactly one point.

∴ Exactly one solution.

Ex. 9 *The number of solutions of the equation* $\cos[x] = e^{2x-1}$, $x \in [0, 2\pi]$, *where* [.] *denotes the greatest integer function is*

(a) 1 (b) 2
(c) 3 (d) 4

Sol. (*a*) It is evident from the graph the two curves intersect at only one point.

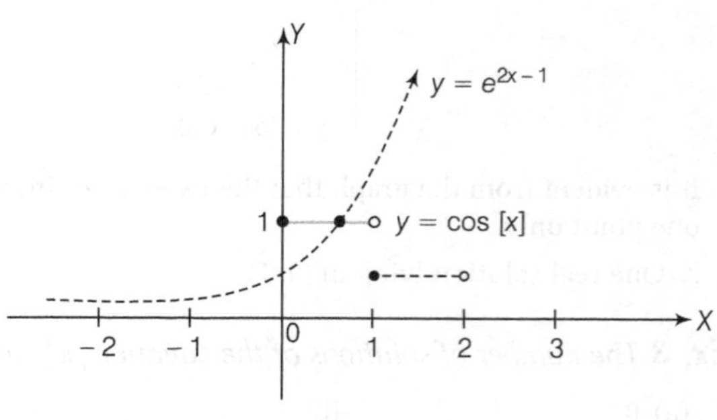

JEE Type Solved Examples : More than One Correct Option Type Questions

Ex. 10 *Let* $g(x) = \sqrt{x - 2k}, \forall\ 2k \le x < 2(k+1)$ *where,* $k \in I$*, then*

(a) $g(x) = \sqrt{x+2}, -2 \le x < 0$

(b) $g(x) = \sqrt{x-2}, 2 \le x < 4$

(c) $g(x) = \sqrt{x}, 0 \le x < 2$

(d) period of $g(x)$ is 2

Sol. (a,b,c,d) Given, $g(x) = \sqrt{x - 2k}, \forall\ 2k \le x < 2(k+1), \forall\ k \in I$

$$= \begin{cases} \dots \\ \dots \\ \sqrt{x+2}, & -2 \le x < 0 \\ \sqrt{x}, & 0 \le x < 2 \\ \sqrt{x-2}, & 2 \le x < 4 \\ \sqrt{x-4}, & 4 \le x < 6 \\ \dots \\ \dots \end{cases}$$

The graph of $g(x)$ is shown below

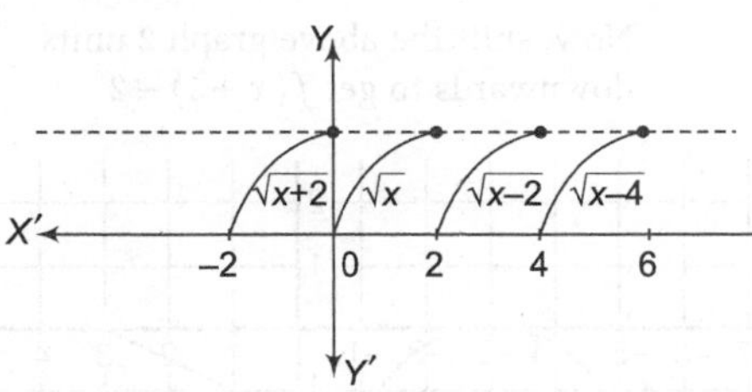

Clearly, $g(x)$ is periodic with period 2.

Ex. 11 *The graph of* $f(x)$ *is given below.*

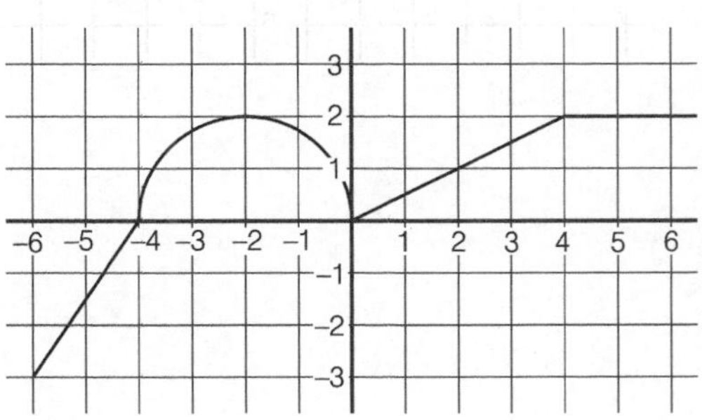

Then, (a) Graph of $-f(x)+2$ is

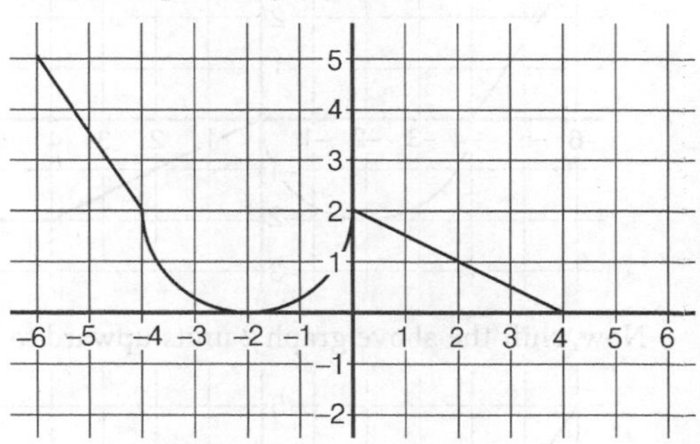

(b) Graph of $-f(x-1)$ is

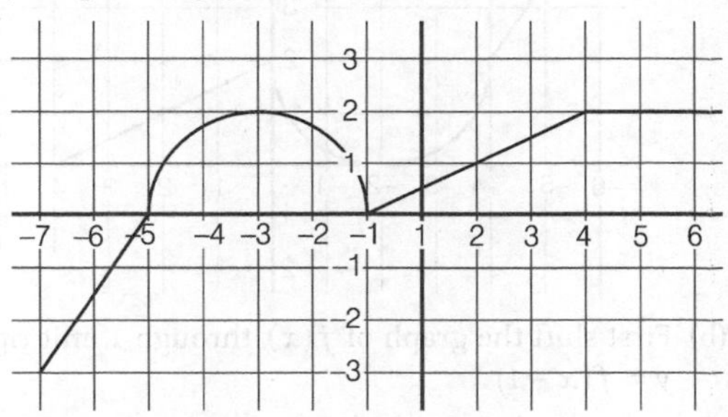

(c) Graph of $f(|x|)$ is

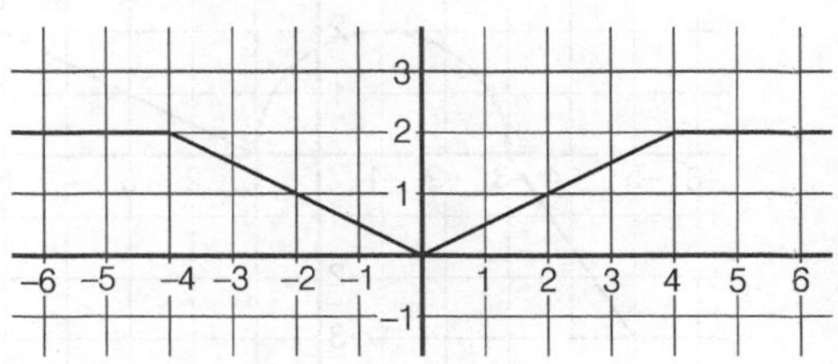

(d) Graph of $f(x+1)-2$ is

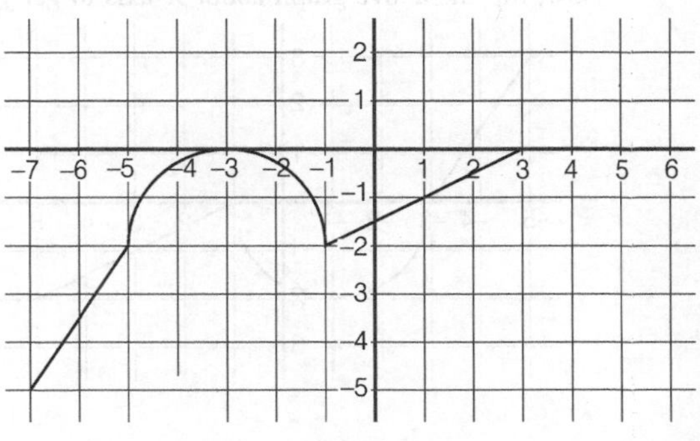

Sol. (*a,c,d*)

(a) For the graph of $-f(x)+2$, first flip the graph of $y=f(x)$ about, X-axis to get $y=-f(x)$

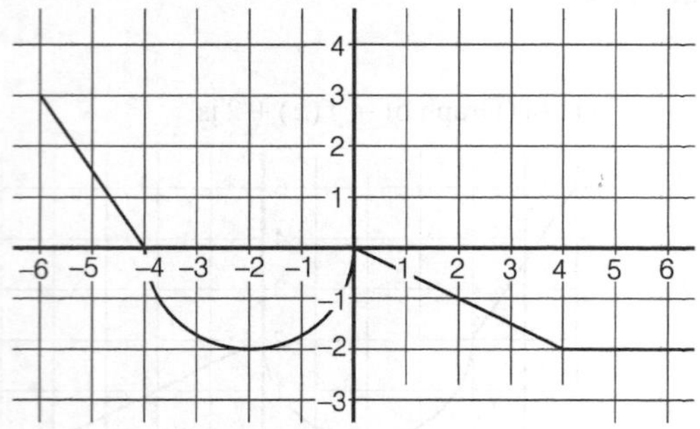

Now, shift the above graph 2 units upward to get $y=2-f(x)$

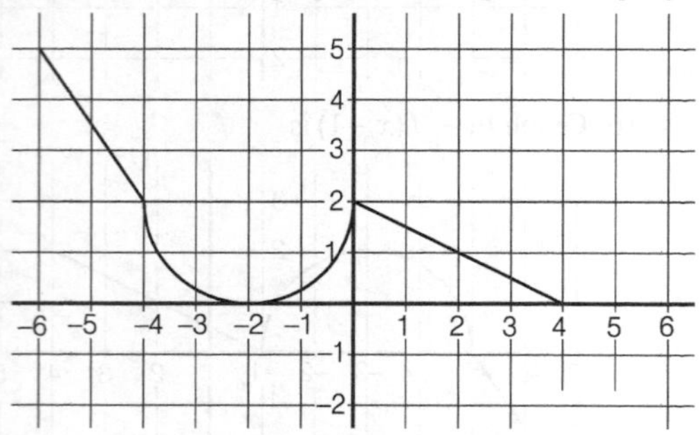

(b) First shift the graph of $f(x)$, through 1 unit right to get $y=f(x-1)$

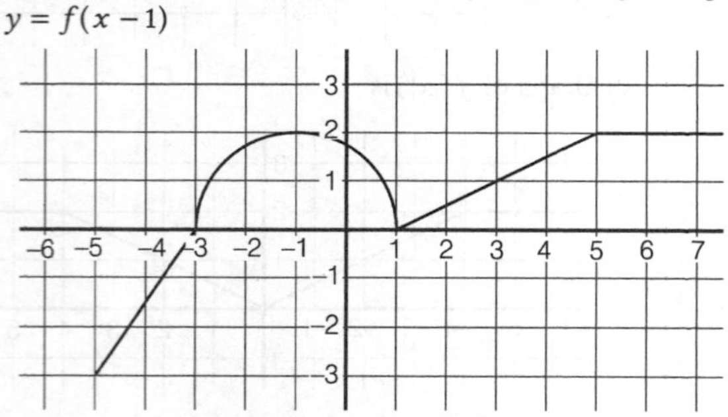

Now, flip the above graph about X-axis to get $y=-f(x-1)$

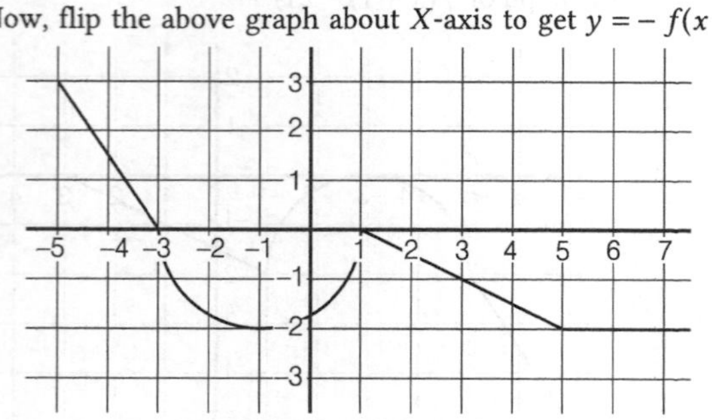

(c) For $y=f(|x|)$, neglect the graph of $y=f(x)$ for $x<0$, and take the mirror image of $y=f(x)$ for $x>0$ about Y-axis, keeping $y=f(x)$ for $x>0$

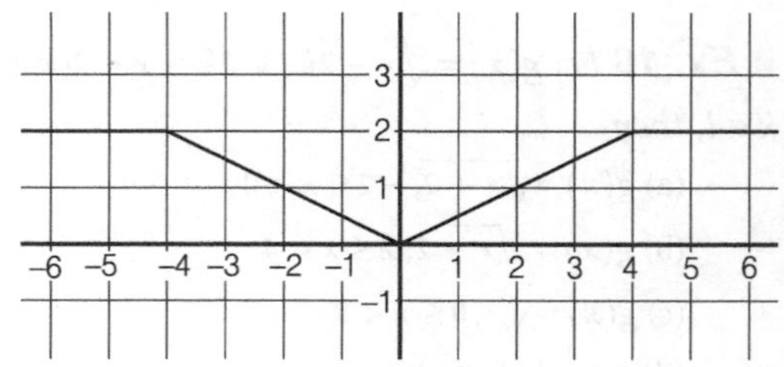

(d) For $y=f(x+1)-2$, shift the graph of $y=f(x)$, 1 unit left to get $y=f(x+1)$

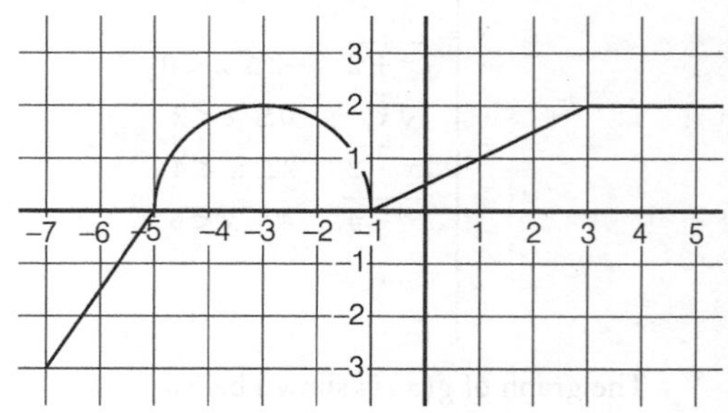

Now, shift the above graph 2 units downwards to get $f(x+1)-2$

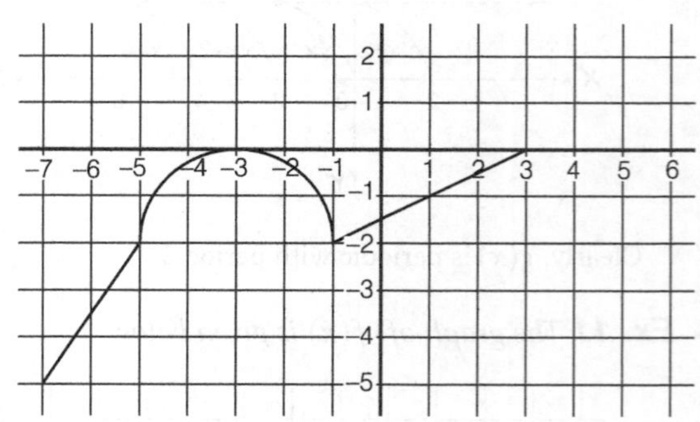

JEE Type Solved Examples : Single Integer Answer Type Questions

- ***Ex. 12*** *The number of solutions of the equation* $[y+[y]] = 2\cos x$, *where* $y = \frac{1}{3}[\sin x + [\sin x + [\sin x]]]$, *where* $[\cdot]$ *denotes the greatest integer function, is*

Sol. Here, $[y + [y]] = 2\cos x$

or $[y] + [y] = 2\cos x$

$[\because [x+h] = [x] + h, \forall\, x \in I]$

or $2[y] = 2\cos x$

or $[y] = \cos x$ ...(i)

Also, $y = \frac{1}{3}[\sin x + [\sin x + [\sin x]]]$

$y = \frac{1}{3}(3[\sin x])$

$y = [\sin x]$ ...(ii)

From Eqs. (i) and (ii), we have

$[[\sin x]] = \cos x$

or $[\sin x] = \cos x$

The graph of $y = [\sin x]$ and $y = \cos x$ is shown below.

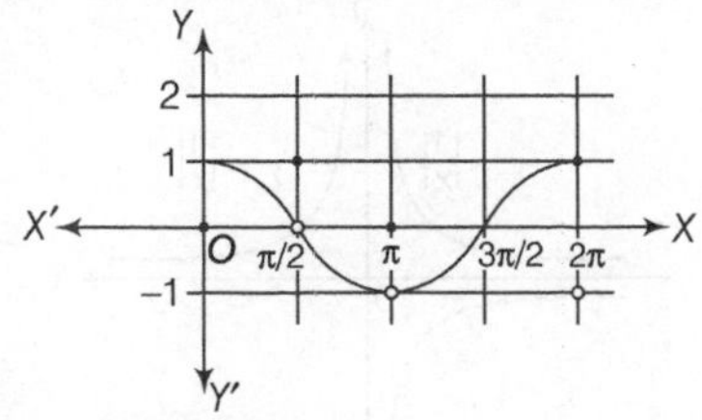

$\therefore$ The number of solution zero.

- ***Ex. 13*** *The sum of the roots of the equation* $\cos^{-1}(\cos x) = [x]$, *where* $[x]$ *denotes greatest integer function, is*

Sol. The graph of $\cos^{-1}(\cos x) = [x]$ is shown below.

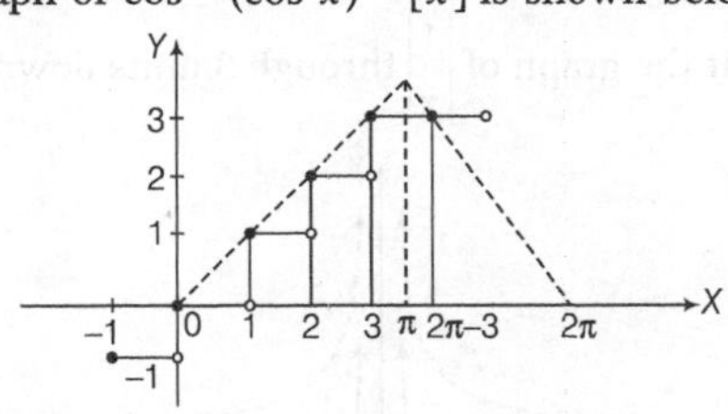

Solutions are clearly 0, 1, 2, 3 and $2\pi - 3$.
Hence, there are 5 solutions.

Subjective Type Questions

- ***Ex. 14*** *Sketch the graph of* $y = \log_{0.5}|x|$.

Sol. As we know, $y = \log_{0.5} x$ is a decreasing graph given as;

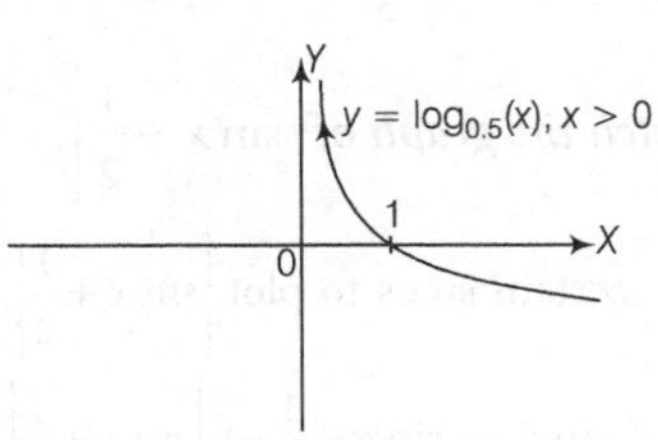

Since, $f(x) \to f(|x|)$, then taking the images about Y-axis for $x > 0$

$\therefore$ $y = \log_{0.5}|x|$ could be plotted as;

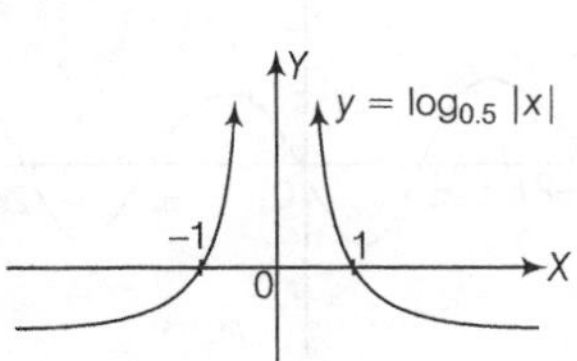

- ***Ex. 15*** *Sketch the graph for* $y = \left|\left|\frac{1}{x}\right| - 3\right|$.

Sol. Here, we follow certain steps to plot

$y = \left|\left|\frac{1}{x}\right| - 3\right|$, as first we plot $\frac{1}{x}$

and then successively $\left|\frac{1}{x}\right|, \left|\frac{1}{x}\right| - 3, \left|\left|\frac{1}{x}\right| - 3\right|$

(i) The graph of $\frac{1}{x}$ is shown below

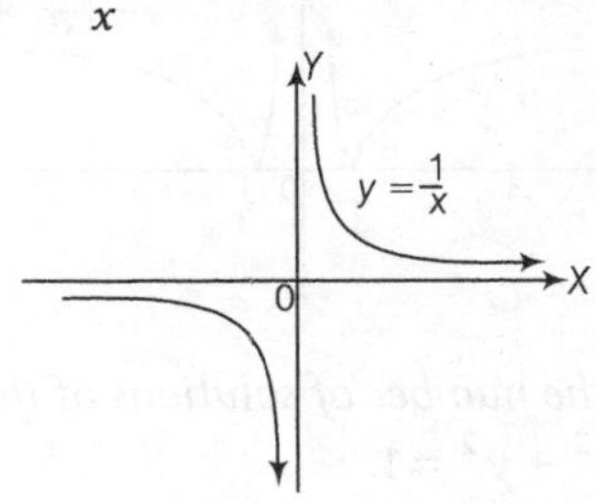

(ii) $\frac{1}{x} \to \left|\frac{1}{x}\right|$ To draw the graph of $\left|\frac{1}{x}\right|$ taking mirror images about X-axis for negative values of $\frac{1}{x}$ in the graph of $\frac{1}{x}$.

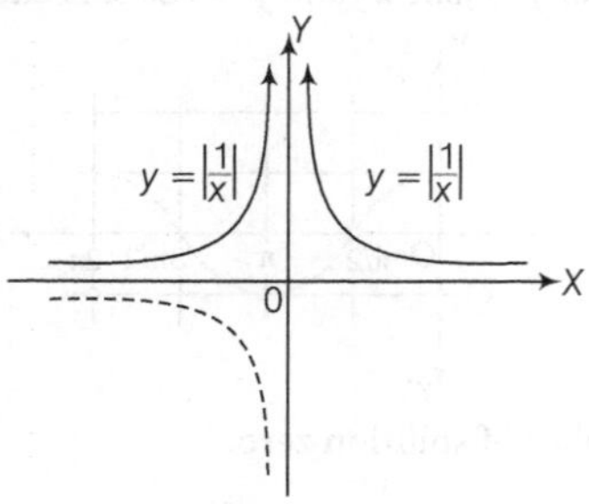

(iii) $\left|\frac{1}{x}\right| \to \left|\frac{1}{x}\right| - 3$

Shift the graph of $\left|\frac{1}{x}\right|$ through 3 units downward.

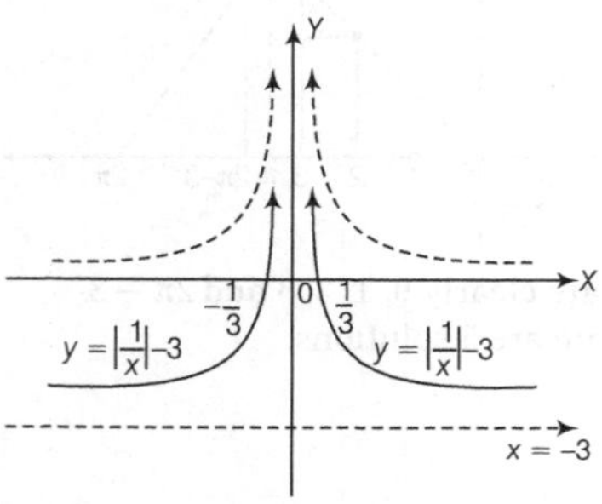

(iv) $\left|\frac{1}{x}\right| - 3 \to \left|\left|\frac{1}{x}\right| - 3\right|$

To draw the graph of $\left|\left|\frac{1}{x}\right| - 3\right|$, taking mirror images about X-axis for negative value of $\left|\frac{1}{x}\right| - 3$ in the graph of $\left|\frac{1}{x}\right| - 3$.

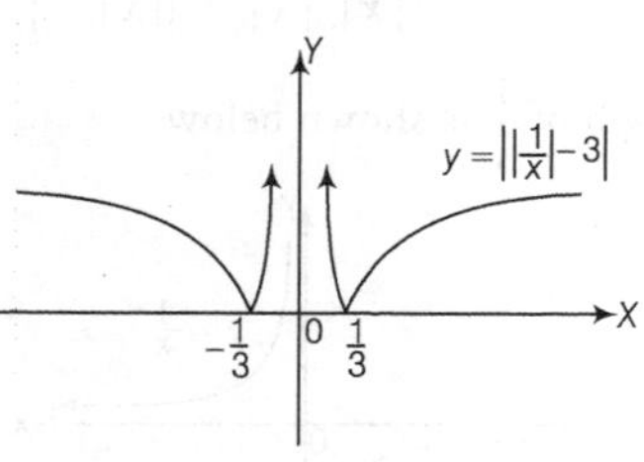

Ex. 16 *Find the number of solutions of the equations $y = |\sin x|$ and $x^2 + y^2 = 1$.*

Sol. To find the number of solutions of two curves we should find the point of intersection of two curves.

As we know, $x^2 + y^2 = 1$ is a circle and $y = |\sin x|$ is the image of negative values of $y = \sin x$ about X-axis. Thus, we can plot them as; which shows the two curves intersects at two points.

$\therefore$ Number of solutions is 2.

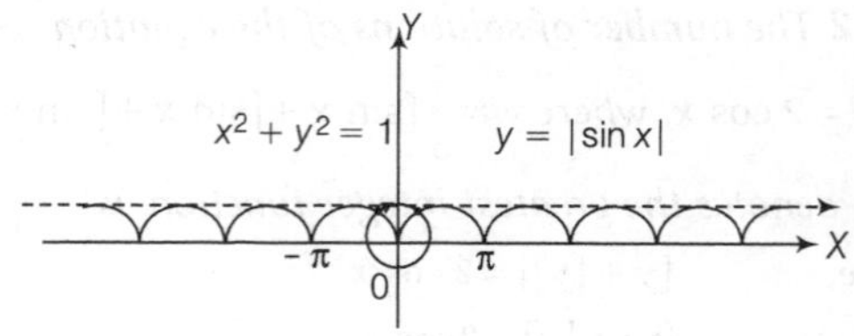

Ex. 17 *Find the number of solutions of $4\{x\} = x + [x]$.*

Sol. Here, $4\{x\} = x + [x]$

$\Rightarrow \quad 4(x - [x]) = x + [x] \quad [\because \{x\} = x - [x]]$

$\Rightarrow \quad 3x = 5[x]$

$\Rightarrow \quad [x] = \frac{3}{5}x$

To find their solution we plot the graph of both $y = [x]$ and $y = \frac{3}{5}x$,

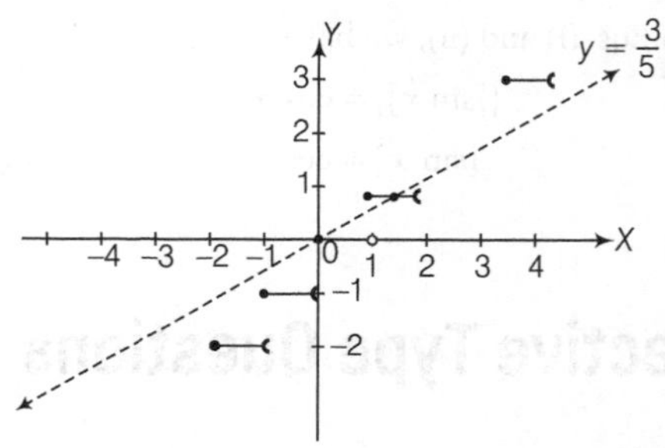

i.e. The two graphs intersects when $[x] = 0$ and 1

$\Rightarrow \quad x = 0$ and $x = \frac{5}{3}$

Ex. 18 *Sketch the graph of $\left|\sin x + \frac{1}{2}\right|$.*

Sol. We follow certain steps to plot $\left|\sin x + \frac{1}{2}\right|$,

i.e. $\quad \sin x \to \sin x + \frac{1}{2} \to \left|\sin x + \frac{1}{2}\right|$

(i) (ii) (iii)

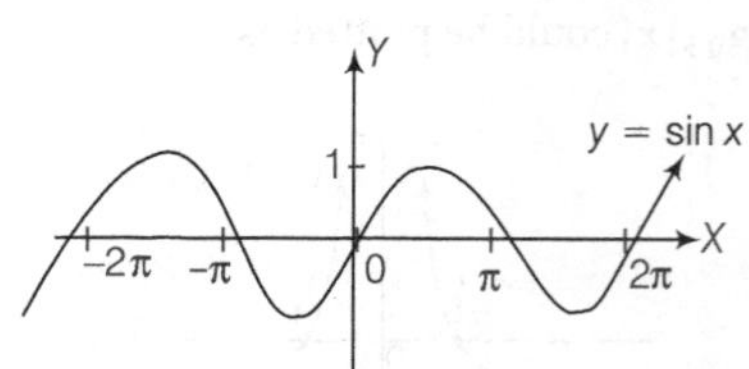

(i)

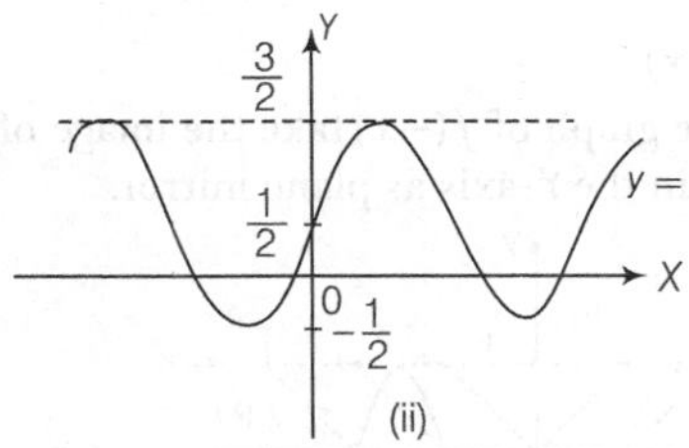

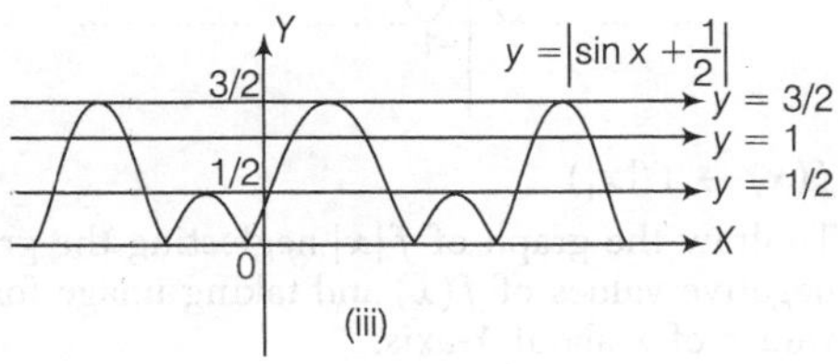

Remarks

Plotting graph of $f(x - [x])$:

Graph of $f(x - [x])$ can be obtained from the graph of $f(x)$ by the following rule.

"Retain the graph of $f(x)$ for values of x lying between interval $[0, 1)$. Now, it can be repeated for rest of points. Now, obtained function is periodic with period 1."

Ex. 19 *Sketch the graph of* $y = \dfrac{2^x}{2^{[x]}}$.

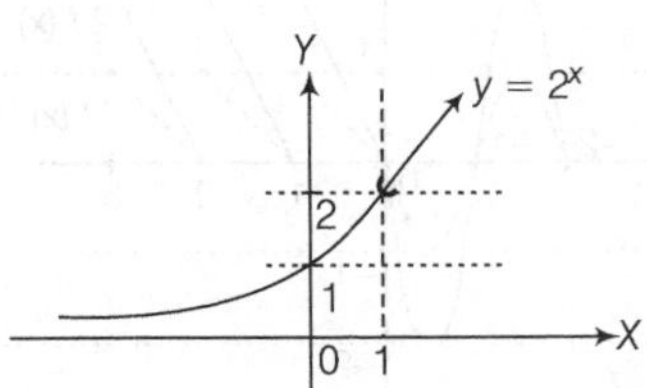

Sol. As we know, 2^x is exponential function and we want to transform it to $2^{x-[x]}$, it retains, the graph for $x \in [0, 1)$ and repeat for rest points.

Here, to retain graph between $x \in [0, 1)$, so we get

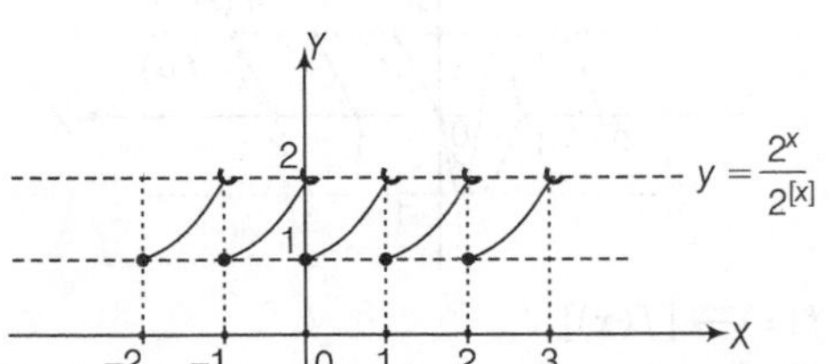

Ex. 20 *Sketch the region for* $y = \sin(x - [x])$.

Sol. We know, $y = \sin x$ is the periodic function. So, to plot the graph between $x \in [0, 1)$ and repeat for all values of x.

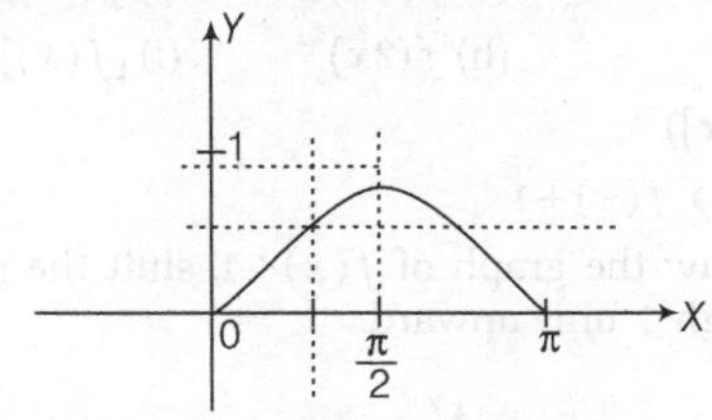

i.e. To retain the graph between $x \in [0, 1)$.

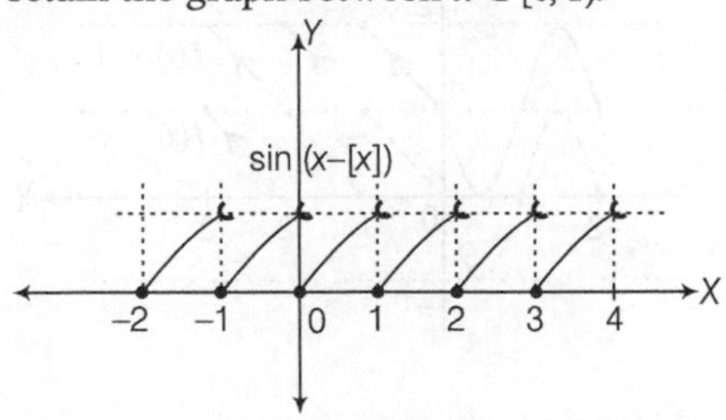

Ex. 21 *Sketch the region for* $|y| = \sin x$.

Sol. We know, $y = \sin x$ is a periodic function. So, to plot $|y| = \sin x$ neglect all the points for which y is negative and take the image for positive values of y about X-axis shown as;

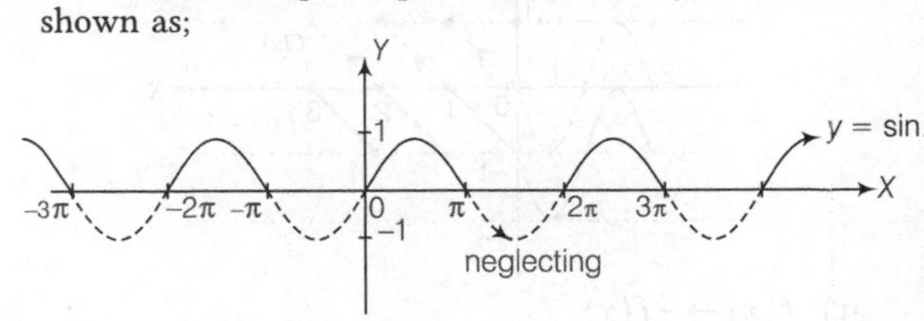

Now, plotting, $|y| = \sin x$.

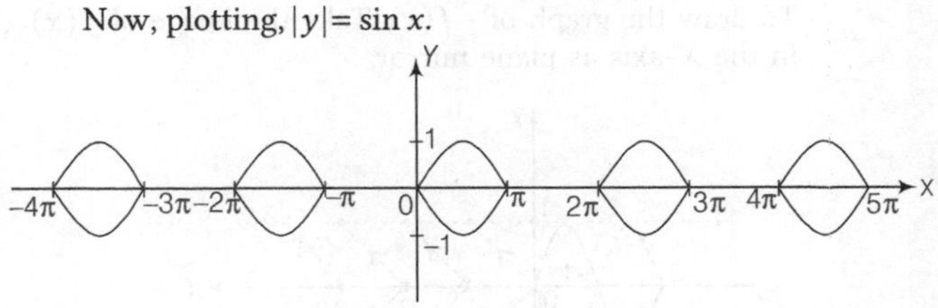

Ex. 22 *Consider the following function f whose graph is given below.*

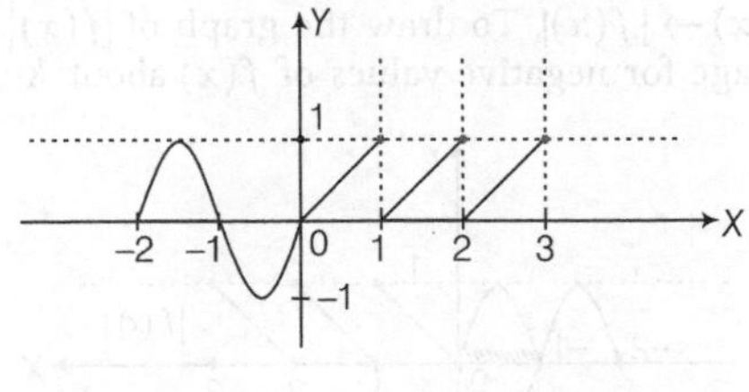

Draw the graph of following functions.

(a) $f(x)+1$ (b) $f(x)-1$ (c) $-f(x)$

(d) $|f(x)|$ (e) $f(-x)$ (f) $f(|x|)$

(g) $2f(x)$ (h) $f(2x)$ (i) $[f(x)]$

(j) $f(x-[x])$

Sol. (a) $f(x)\to f(x)+1$

To draw the graph of $f(x)+1$, shift the graph of $f(x)$ through 1 unit upward.

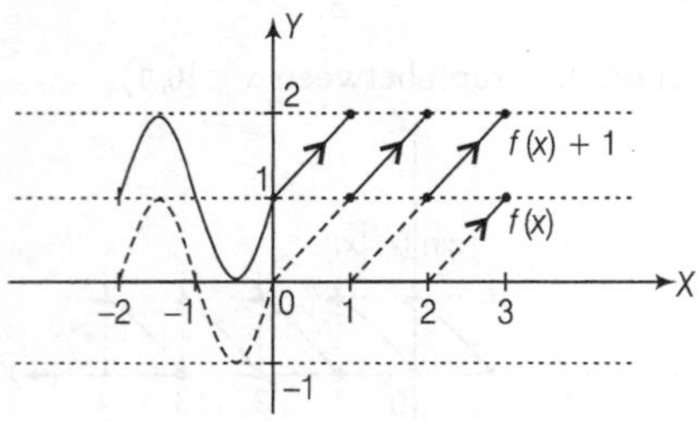

(b) $f(x)\to f(x)-1$

To draw the graph of $f(x)-1$, shift the graph of $f(x)$ through 1 unit downward.

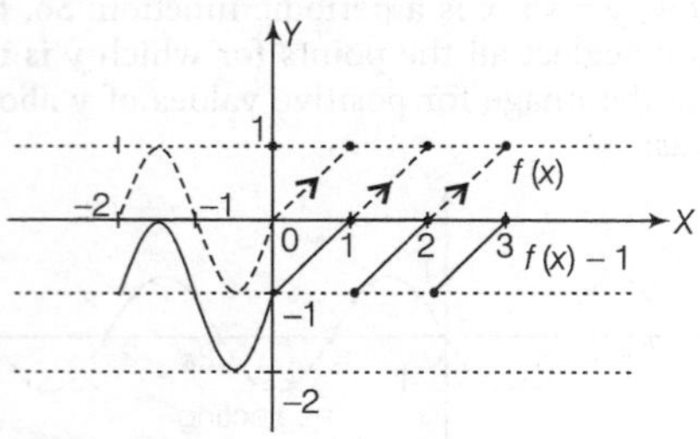

(c) $f(x)\to -f(x)$

To draw the graph of $-f(x)$. Take the image of $f(x)$ in the X-axis as plane mirror.

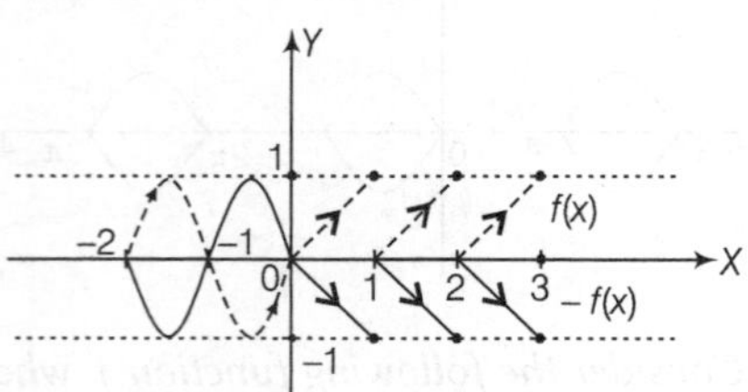

(d) $f(x)\to|f(x)|$, To draw the graph of $|f(x)|$ taking image for negative values of $f(x)$ about X-axis.

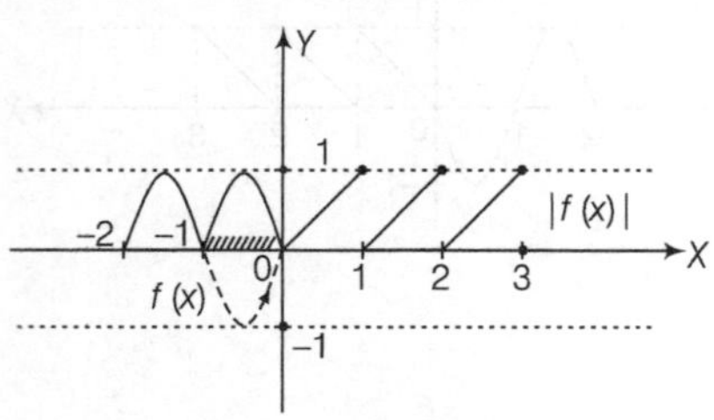

(e) $f(x)\to f(-x)$

To draw the graph of $f(-x)$ take the image of the curve $f(x)$ in the Y-axis as plane mirror.

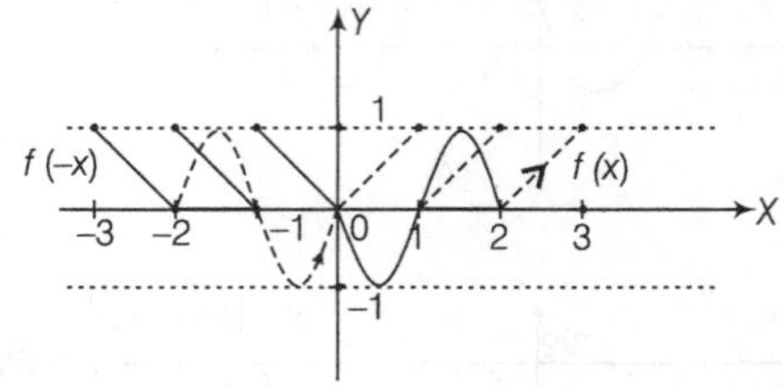

(f) $f(x)\to f(|x|)$

To draw the graph of $f|x|$ neglecting the graph for negative values of $f(x)$ and taking image for positive values of x about Y-axis.

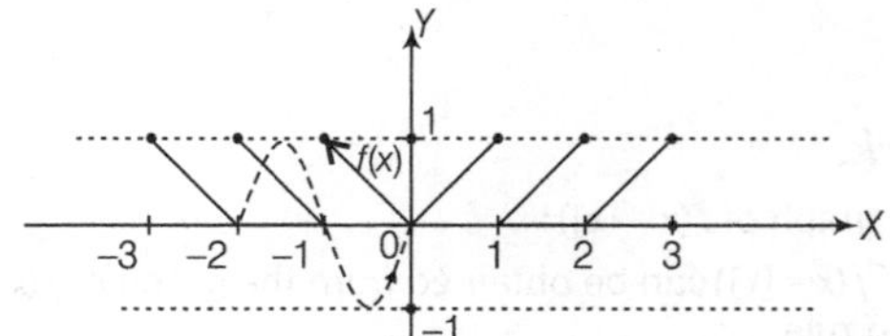

(g) $f(x)\to 2f(x)$

To draw the graph of $2f(x)$ stretch the graph of $f(x)$, 2 times along Y-axis.

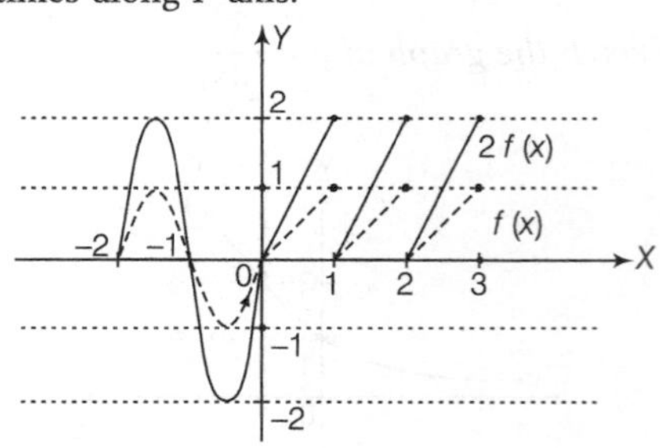

(h) $f(x)\to f(2x)$

To draw the graph of $f(2x)$, stretch the graph of $f(x)$, 2 times along X-axis.

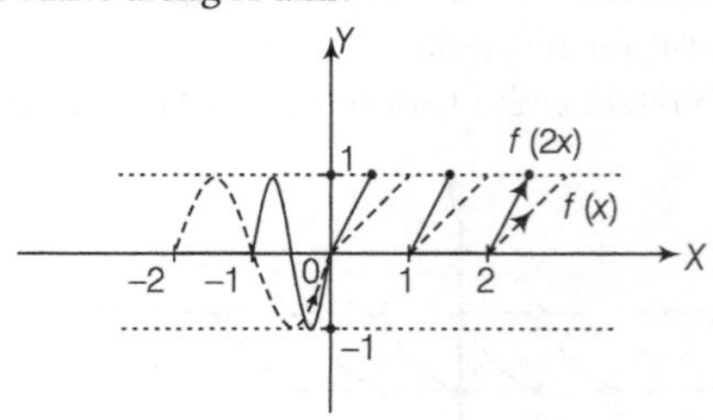

(i) $f(x)\to[f(x)]$

To draw $[f(x)]$ mark the integer on Y-axis. Draw the horizontal lines till they intersect the graph, now draw vertical dotted lines from these intersection point. Finally, draw horizontal lines parallel to X-axis from any intersection point to the nearest vertical dotted line with blank dot at right end.

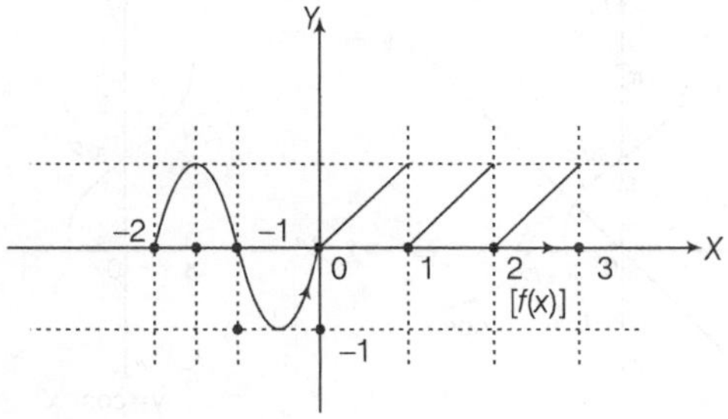

(j) $f(x) \to f(x-[x])$

Retain the graph of $f(x)$ for values of x lying between interval $[0,1)$. Now, it can be repeated for rest of the point taking periodicity '1'.

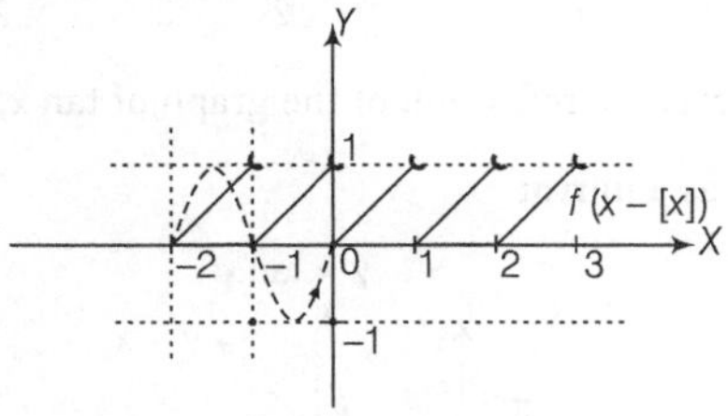

Ex. 23 *Sketch the graph of $y = \max(\sin x, \cos x)$, $\forall x \in \left(-\pi, \frac{3\pi}{2}\right)$.*

Sol. First plot both $y = \sin x$ and $y = \cos x$ by a dotted curve as can be seen from the graph in the interval $\left(-\pi, \frac{3\pi}{2}\right)$ and then darken those dotted line for which $f(x) > g(x)$ or $g(x) > f(x)$.

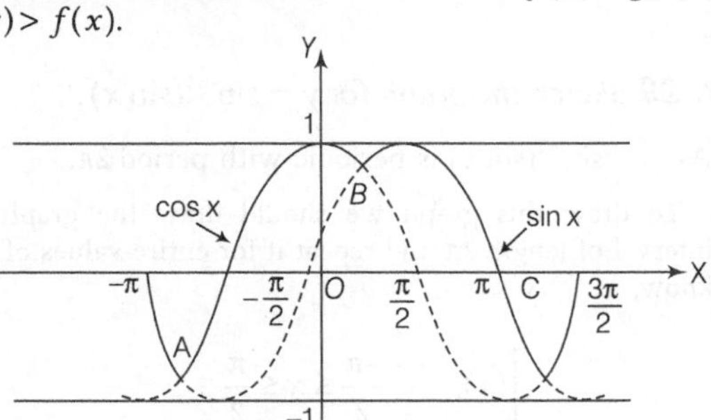

∴ **Graph of max {sin x, cos x}**

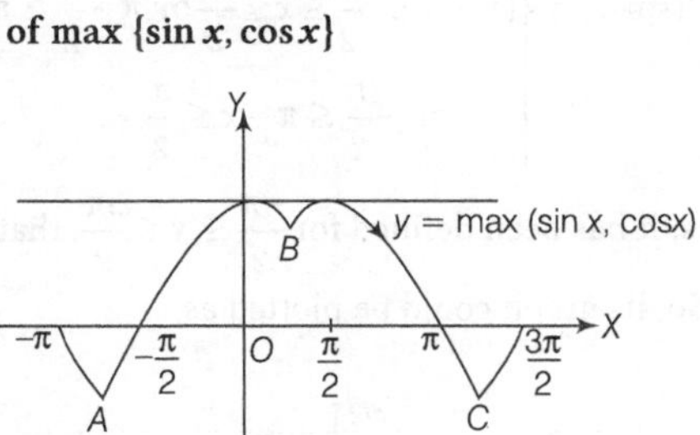

As from the above graph,

$$\max(\sin x, \cos x) = \begin{cases} \sin x, & -\pi < x \le -\frac{3\pi}{4} \\ \cos x, & -\frac{3\pi}{4} \le x \le \frac{\pi}{4} \\ \sin x, & \frac{\pi}{4} \le x \le \frac{5\pi}{4} \\ \cos x, & \frac{5\pi}{4} \le x \le \frac{3\pi}{2} \end{cases}$$

Ex. 24 *Sketch the graph for $y = \min\{\tan x, \cot x\}$.*

Sol. First plot both $f(x) = \tan x$ and $g(x) = \cot x$ by a dotted curve as can be seen from the graph and then darken those dotted lines for which $f(x) < g(x)$ and $g(x) < f(x)$.

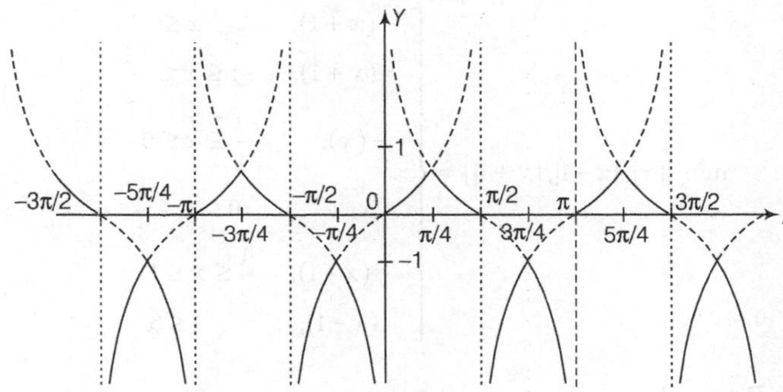

∴ **Graph of min {tan x, cot x}**

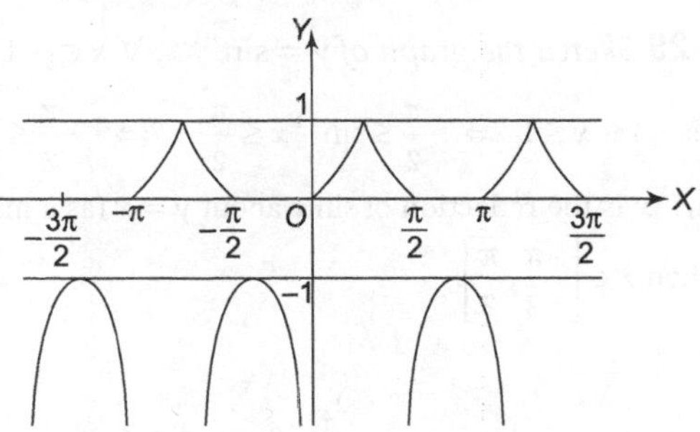

As from the graph, we have

$$\min\{\tan x, \cot x\} = \begin{cases} \ldots\ldots\ldots\ldots \\ \ldots\ldots\ldots\ldots \\ \tan x, & -\frac{\pi}{2} < x \le -\frac{\pi}{4} \\ \cot x, & -\frac{\pi}{4} \le x < 0 \\ \tan x, & 0 \le x \le \frac{\pi}{4} \\ \cot x, & \frac{\pi}{4} \le x \le \frac{\pi}{2} \\ \ldots\ldots\ldots\ldots \\ \ldots\ldots\ldots\ldots \end{cases}$$

Ex. 25 *Sketch the graph of* $y = \min\{|x|, |x-1|, |x+1|\}$.

Sol. First plot the graph for, $y = |x|$, $y = |x-1|$ and $y = |x+1|$ by a dotted curve as can be seen from the graph and then darken those dotted lines for $|x| < \{|x-1|, |x+1|\}$, $|x-1| < \{|x|, |x+1|\}$ and $|x+1| < \{|x|, |x-1|\}$.

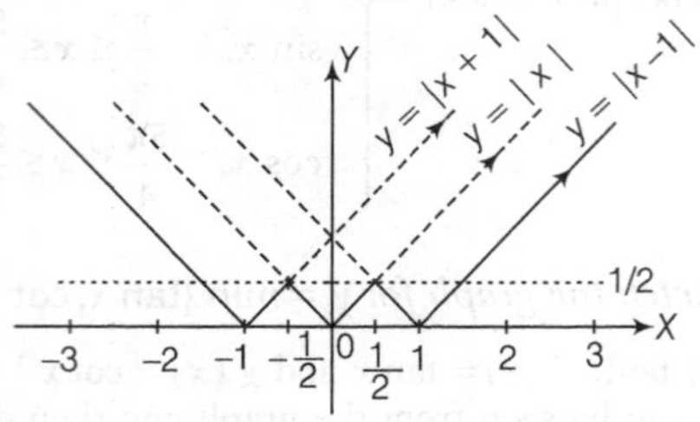

As from the above graph,

$$\min\{|x|, |x-1|, |x+1|\} = \begin{cases} -(x+1), & x \le -1 \\ (x+1), & -1 \le x \le -\frac{1}{2} \\ -(x), & -\frac{1}{2} \le x \le 0 \\ x, & 0 \le x \le \frac{1}{2} \\ -(x-1), & \frac{1}{2} \le x \le 1 \\ (x-1), & 1 \le x \end{cases}$$

Remark

To plot the graph of $f^{-1}(x)$, take reflection of $f(x)$ in $y = x$ line as a mirror.

Ex. 26 *Sketch the graph of* $y = \sin^{-1} x$, $\forall\ x \in [-1, 1]$.

Sol. As, $-1 \le x \le 1 \Rightarrow -\frac{\pi}{2} \le \sin^{-1} x \le \frac{\pi}{2} \Rightarrow -\frac{\pi}{2} \le y \le \frac{\pi}{2}$

$\sin^{-1} x$ is the reflection of $\sin x$ about $y = x$ (as a mirror) when $x \in \left[-\frac{\pi}{2}, \frac{\pi}{2}\right]$.

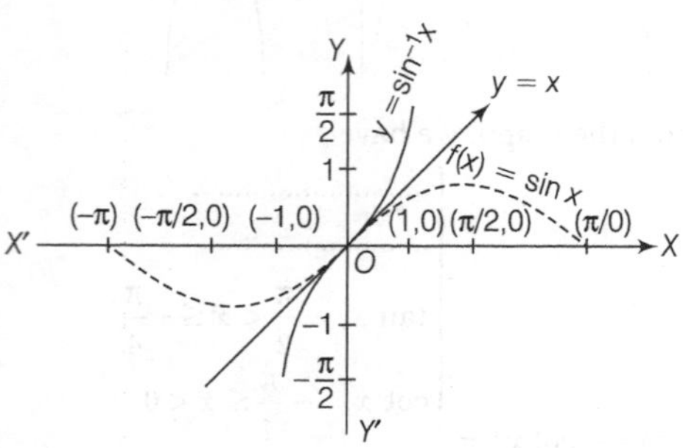

Ex. 27 *Sketch the graph for* $y = \cos^{-1} x$, $\forall\ x \in [-1, 1]$.

Sol. As, $-1 \le x \le 1$

$\Rightarrow \quad 0 \le \cos^{-1} x \le \pi \quad \Rightarrow \quad 0 \le y \le \pi$

$\therefore$ $\cos^{-1} x$ is reflection of the graph of $\cos x, 0 \le x \le \pi$ in $y = x$ line as a mirror.

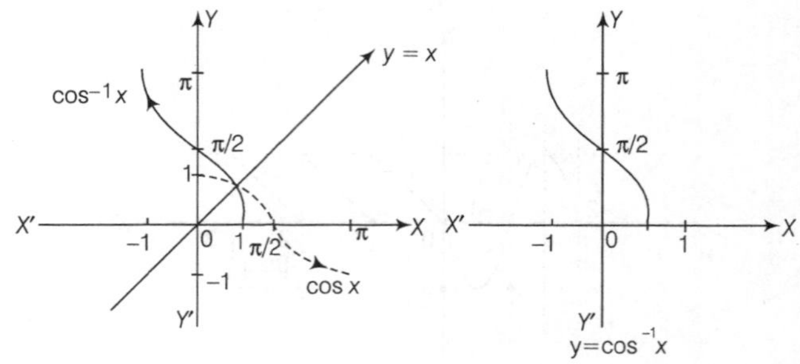

Ex. 28 *Sketch the graph for* $y = \tan^{-1} x$, $\forall\ x \in R$.

Sol. As, $x \in R \Rightarrow \frac{-\pi}{2} < \tan^{-1} x < \frac{\pi}{2} \Rightarrow \frac{-\pi}{2} < y < \frac{\pi}{2}$

$\therefore$ $\tan^{-1} x$ is the reflection of the graph of $\tan x$, $\frac{-\pi}{2} < x < \frac{\pi}{2}$ in $y = x$ as a mirror.

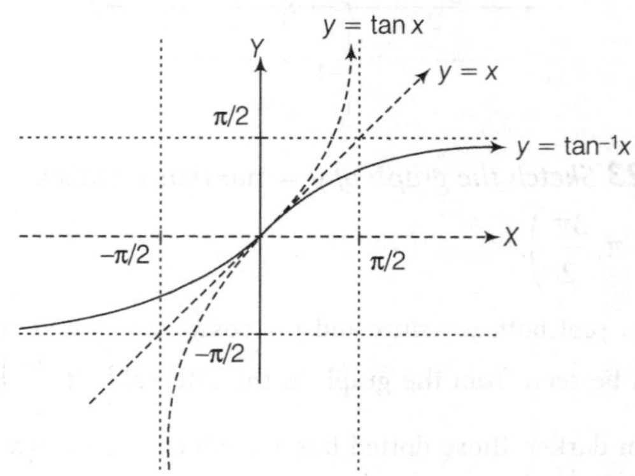

Ex. 29 *Sketch the graph for* $y = \sin^{-1}(\sin x)$.

Sol. As, $y = \sin^{-1}(\sin x)$ is periodic with period 2π.

$\therefore$ To draw this graph we should draw the graph for one interval of length 2π and repeat it for entire values of x. As we know,

$$\sin^{-1}(\sin x) = \begin{cases} (x), & \frac{-\pi}{2} \le x \le \frac{\pi}{2} \\ (\pi - x), & \frac{\pi}{2} \le x \le \frac{3\pi}{2} \text{ or } \pi - \frac{\pi}{2} \ge \pi - x \ge \pi - \frac{3\pi}{2} \\ & \text{or } -\frac{\pi}{2} \le \pi - x \le \frac{\pi}{2} \end{cases}$$

Thus, it has been defined for $\frac{-\pi}{2} \le x \le \frac{3\pi}{2}$, that has length 2π. So, its graph could be plotted as;

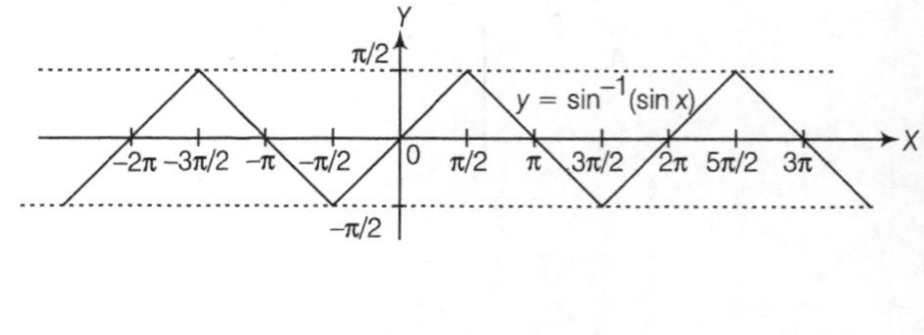

Ex. 30 *Sketch the graph for* $y = \cos^{-1}(\cos x)$.

Sol. As, we have that $y = \cos^{-1}(\cos x)$ is periodic with period 2π.

$\therefore$ To draw this graph we should draw the graph for one interval of length 2π and repeat it for entire values of x of length 2π.

As we know,

$$\cos^{-1}(\cos x) = \begin{cases} x, & 0 \le x \le \pi \\ (2\pi - x), & \pi < x \le 2\pi \text{ or } 0 < 2\pi - x < \pi \end{cases}$$

Thus, it has been defined for $0 < x < 2\pi$, that has length 2π. So, its graph could be plotted as;

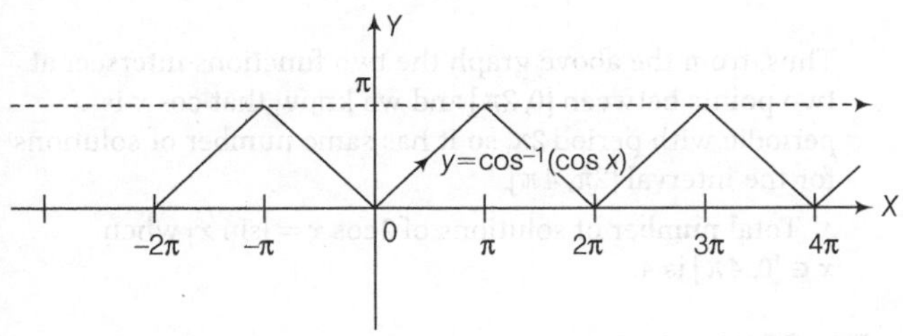

Ex. 31 *Sketch the graph for* $y = \tan^{-1}(\tan x)$.

Sol. As, we have that $y = \tan^{-1}(\tan x)$ is periodic with period π.

$\therefore$ To draw this graph we should draw the graph for one interval of length π and repeat for entire values of x.

As we know, $\tan^{-1}(\tan x) = \left\{ x, \quad \frac{-\pi}{2} < x < \frac{\pi}{2} \right.$

Thus, it has been defined for $\frac{-\pi}{2} < x < \frac{\pi}{2}$ that has length π.

So, its graph could be plotted as;

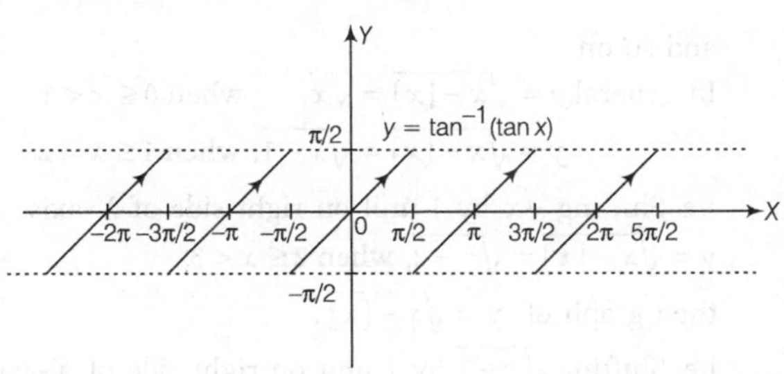

Ex. 32 *Find the values of x graphically which satisfy* $\left|\frac{x^2}{x-1}\right| \le 1$.

Sol. We have, $\qquad \left|\frac{x^2}{x-1}\right| \le 1$

$\Rightarrow \qquad |x^2| \le |x-1|, \forall\ x \in R - \{1\}$

or $\qquad x^2 \le |x-1|, \forall\ x \in R - \{1\}$

Thus, to find the points for which $f(x) = x^2$ is less than or equal to $g(x) = |x-1|$.

Where the two functions $f(x)$ and $g(x)$ could be plotted as;

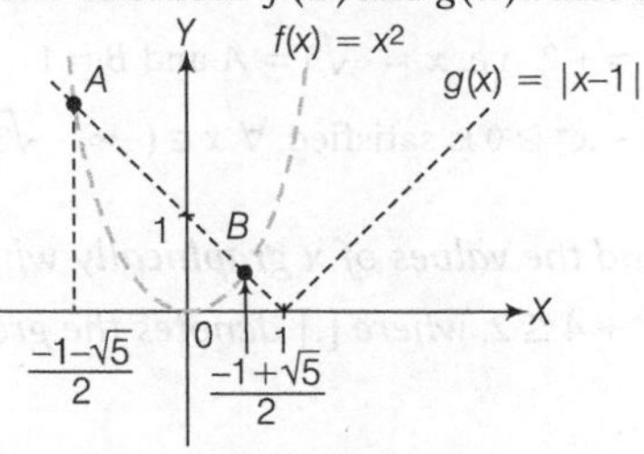

Thus, from the graph $f(x) \le g(x)$, when $x \in [A, B]$. So that, A and B are point of intersection of x^2 and $1 - x$.

$\therefore$ Solving $x^2 = 1 - x$, we get $x = \frac{-1-\sqrt{5}}{2} = A$

and $\qquad x = \frac{-1+\sqrt{5}}{2} = B.$

$\therefore \qquad \left|\frac{x^2}{x-1}\right| \le 1$ is satisfied, $\forall\ x \in \left[\frac{-1-\sqrt{5}}{2}, \frac{-1+\sqrt{5}}{2}\right]$.

Ex. 33 *Find the values of x graphically satisfying* $[x] - 1 + x^2 \ge 0$; *where [·] denotes the greatest integer function.*

Sol. We have, $\quad [x] - 1 + x^2 \ge 0 \Rightarrow x^2 - 1 \ge -[x]$

Thus, to find the points for which $f(x) = x^2 - 1$ is greater than or equal to $g(x) = -[x]$.

Where the two functions $f(x)$ and $g(x)$ could be plotted as;

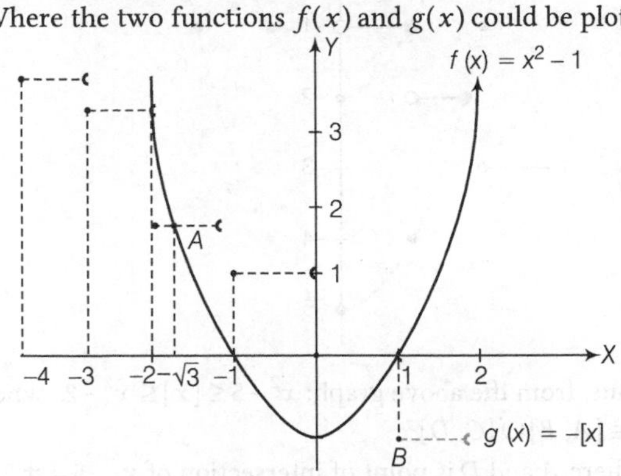

Finally.

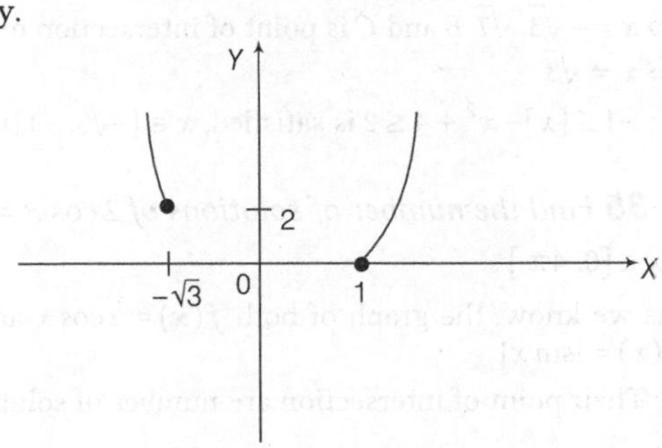

Thus, from the above graph $f(x) \geq g(x)$ when $x \in (-\infty, A] \cup [B, \infty)$, where A is point of intersection of $x^2 - 1$ and $-[x]$ when $-[x] = +2$.

$\therefore\ x^2 - 1 = +2$, i.e. $x = -\sqrt{3} = A$ and $B = 1$

$\therefore\ [x] - 1 + x^2 \geq 0$ is satisfied, $\forall\ x \in (-\infty, -\sqrt{3}] \cup [1, \infty)$.

Ex. 34 *Find the values of x graphically which satisfy; $-1 \leq [x] - x^2 + 4 \leq 2$, where [.] denotes the greatest integer function.*

Sol. We have, $-1 \leq [x] - x^2 + 4 \leq 2 \Rightarrow x^2 - 5 \leq [x] \leq x^2 - 2$

Thus, to find the points for which $f(x) = x^2 - 5$ is less than or equal to $g(x) = [x]$ and $g(x) = [x]$ is less than or equal to $h(x) = x^2 - 2$.

Where the three functions $f(x)$, $g(x)$ and $h(x)$ could be plotted as;

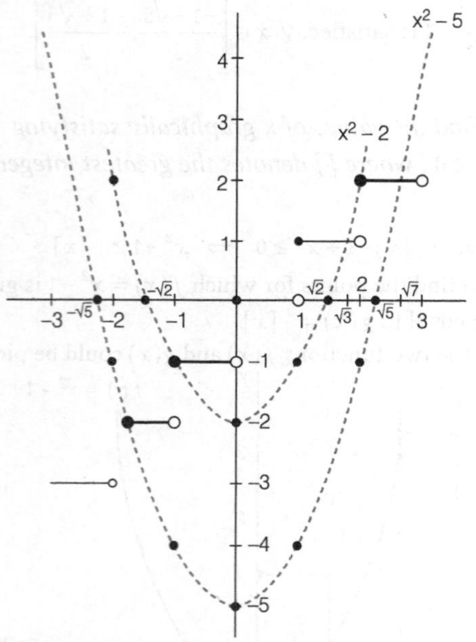

Thus, from the above graph; $x^2 - 5 \leq [x] \leq x^2 - 2$, when $x \in [A, B] \cup [C, D]$.

Where A and D is point of intersection of $x^2 - 5 = \pm 2$

$\Rightarrow x = -\sqrt{3}, \sqrt{7}$ B and C is point of intersection of $x^2 - 2 = \pm 1$

$\Rightarrow x = \sqrt{3}$

$\therefore\ -1 \leq [x] - x^2 + 4 \leq 2$ is satisfied, $x \in [-\sqrt{3}, -1] \cup [\sqrt{3}, \sqrt{7}]$.

Ex. 35 *Find the number of solutions of $2\cos x = |\sin x|$ when $x \in [0, 4\pi]$.*

Sol. As we know, the graph of both $f(x) = 2\cos x$ and $g(x) = |\sin x|$

$\therefore$ Their point of intersection are number of solutions.

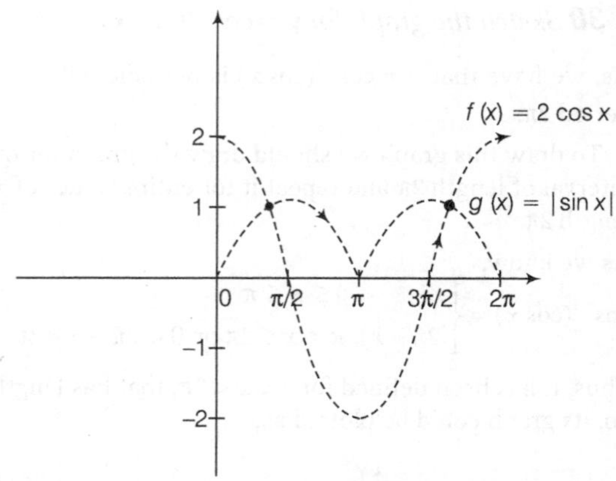

Thus, from the above graph the two functions intersect at two points between $[0, 2\pi]$ and we know that $\cos x$ is periodic with period 2π, so it has same number of solutions for the interval $[2\pi, 4\pi]$.

$\therefore$ Total number of solutions of $2\cos x = |\sin x|$ when $x \in [0, 4\pi]$ is 4.

Ex. 36 *Sketch the curves*

(i) $y = \sqrt{x - [x]}$ (ii) $y = [x] + \sqrt{x - [x]}$

(iii) $y = |[x] + \sqrt{x - [x]}|$

(where $[\cdot]$ denotes the greatest integer function).

Sol. (i) Here, $0 \leq x - [x] < 1, \forall\ x \in R$ and $x^2 \leq x \leq \sqrt{x}, \forall x \in [0, 1)$

$\therefore\ x - [x] \leq \sqrt{x - [x]}$

$$\Rightarrow\ y = \sqrt{x - [x]} = \begin{cases} \sqrt{x - 1}, & 1 \leq x < 2 \\ \sqrt{x}, & 0 \leq x < 1 \\ \sqrt{x + 1}, & -1 \leq x < 0 \end{cases}$$

and so on.

In general, $y = \sqrt{x - [x]} = \sqrt{x}$, when $0 \leq x < 1$

$y = \sqrt{x - [x]} = \sqrt{x - 1}$, when $1 \leq x < 2$

i.e. Shifting $\sqrt{x}$ by 1 unit on right side of X-axis

$y = \sqrt{x - [x]} = \sqrt{x - 2}$, when $2 \leq x < 3$,

then graph of $y = \sqrt{x - [x]}$.

i.e. Shifting $\sqrt{x - 1}$ by 1 unit on right side of X-axis and so on. Thus, the graph for $y = \sqrt{x - [x]}$ is shown as in figure, which is periodic with period 1.

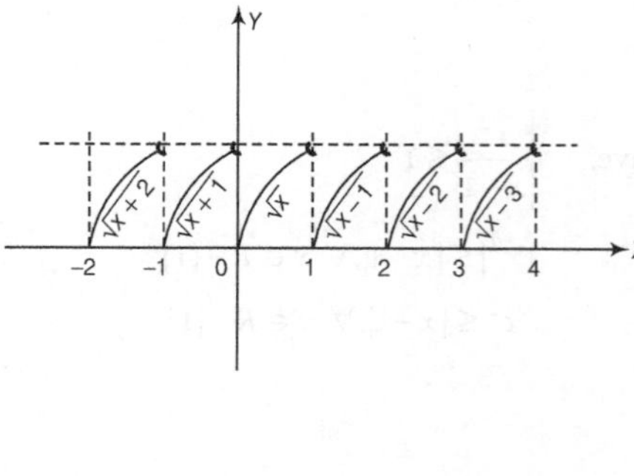

(ii) Again, $y = [x] + \sqrt{x - [x]} \Rightarrow y = k + \sqrt{x - k}$,

$k \le x < k + 1$; $k \in$ integer $-1 + \sqrt{x + 1}, -1 \le x < 0$

$$\Rightarrow \quad y = \begin{cases} \sqrt{x}, & 0 \le x < 1 \\ 1 + \sqrt{x - 1}, & 1 \le x < 2 \\ 2 + \sqrt{x - 2}, & 2 \le x < 3 \text{ and so on.} \end{cases}$$

Thus, the graph for $y = [x] + \sqrt{x - [x]}$ is obtained by the graph of $y = \sqrt{x - [x]}$ by translating it by $[x]$ units in upward or downward directions according as $[x] > 0$ or $[x] < 0$. The graph is shown as

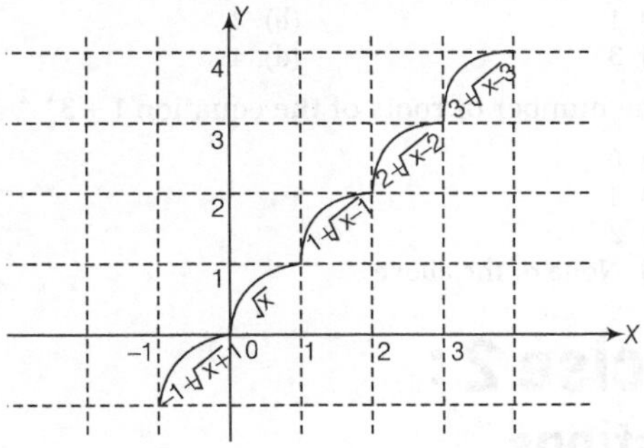

(iii) Graph for $y = |[x] + \sqrt{x - [x]}|$ is obtained by reflecting the portion lying below X-axis of the graph of $y = [x] + \sqrt{x - [x]}$. About X-axis and keeping the portion lying above the X-axis (as it is).

Thus, the graph for $y = \left|[x] + \sqrt{x - [x]}\right|$ is shown below.

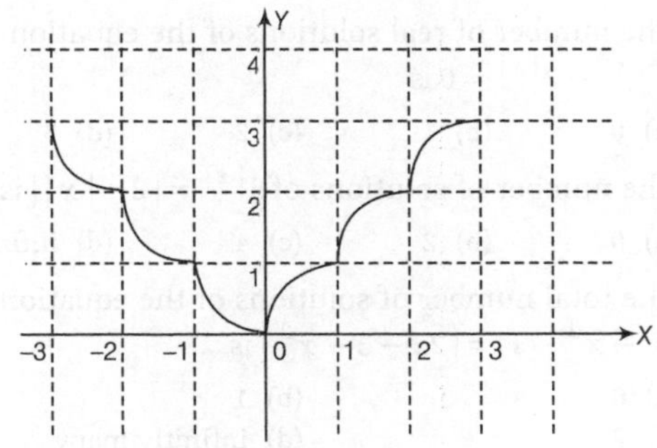

Graphical Transformations Exercise 1 : Single Option Correct Type Questions

1. The number of real solutions of the equation $e^{|x|} - |x| = 0$ is

(a) 0 (b) 1 (c) 2 (d) None of these

2. The number of real solutions of the equation $3^{-|x|} - 2^{|x|} = 0$ is

(a) 0 (b) 1 (c) 2 (d) 3

3. The number of solutions of $3^{|x|} = |2 - |x||$ is

(a) 0 (b) 2 (c) 4 (d) infinite

4. The total number of solutions of the equation $|x - x^2 - 1| = |2x - 3 - x^2|$ is

(a) 0 (b) 1 (c) 2 (d) infinitly many

5. The equation $e^x = m(m+1)$, $m < -1$ has

(a) no real root
(b) exactly one real root
(c) two real roots
(d) None of the above

6. The number of real solutions of the equation $1 - x = [\cos x]$ is

(a) 1 (b) 2
(c) 3 (d) 4

7. The number of roots of the equation $1 + 3^{x/2} = 2^x$ is

(a) 0
(b) 1
(c) 2
(d) None of the above

Graphical Transformations Exercise 2 : More than One Option Correct Type Questions

8. The equation $x^2 - 2 = [\sin x]$, where $[\cdot]$ denotes the greatest integer function, has

(a) infinity many roots
(b) exactly one integer root
(c) exactly one irrational root
(d) exactly two roots

9. Consider the function $f(x) = \begin{cases} x - [x] - \frac{1}{2}, & \text{if } x \notin I \\ 0, & \text{if } x \in I \end{cases}$,

where $[\cdot]$ denotes greatest integer function and I is the set of integers, then $g(x) = \max\{x^2, f(x), |x|\}$, $-2 \le x \le 2$ is defined as

(a) $x^2, -2 \le x \le -1$
(b) $1 - x, -1 < x \le -\frac{1}{4}$
(c) $\frac{1}{2} + x, -\frac{1}{4} < x < 0$
(d) $1 + x, 0 \le x < 1$

10. If $f(x)$ is defined on $[-2, 2]$ and is given by

$f(x) = \begin{cases} -1, & -2 \le x < 0 \\ x - 1, & 0 < x \le 2 \end{cases}$ and $g(x) = f|x| + |f(x)|$, then $g(x)$ is defined as

(a) $-x, -2 \le x \le 0$ (b) $x, -2 \le x \le 0$
(c) $0, 0 < x \le 1$ (d) $2(x - 1), 1 < x \le 2$

Graphical Transformations Exercise 3 : Statements I and II Type Questions

- **Directions** (Q. Nos. 11 to 12) *This section is based on Statement I and Statement II. Select the correct answer from the code given below.*

(a) Both Statement I and Statement II are correct and Statement II is the correct explanation of Statement I
(b) Both Statement I and Statement II are correct but Statement II is not the correct explanation of Statement I
(c) Statement I is correct but Statement II is incorrect
(d) Statement II is correct but Statement I is incorrect

11. Statement I The graph of $y = \sec^2 x$ is symmetrical about the Y-axis.

Statement II The graph of $y = \tan x$ is symmetrical about the origin.

12. Statement I The equation $|(x - 2) + a| = 4$ can have four distinct real solutions for x if a belongs to the interval $(-\infty, -4)$.

Statement II The number of point of intersection of the curve represent the solution of the equation.

Graphical Transformations Exercise 4 : Passage Based Questions

Passage I (Q. Nos. 13 to 14)

Let $f(x) = f_1(x) - 2f_2(x)$, *where*

$$f_1(x) = \begin{cases} \min\{x^2, |x|\}, & |x| \le 1 \\ \max\{x^2, |x|\}, & |x| > 1 \end{cases}$$

$$f_2(x) = \begin{cases} \min\{x^2, |x|\}, & |x| > 1 \\ \max\{x^2, |x|\}, & |x| \le 1 \end{cases},$$

and let $g(x) = \begin{cases} \min\{f(t), & -3 \le t \le x, -3 \le x < 0\} \\ \max\{f(t), & 0 \le t < x, 0 \le x \le 3\}. \end{cases}$

13. For $x \in (-1, 0)$, $f(x) + g(x)$ is

(a) $x^2 - 2x + 1$ (b) $x^2 + 2x - 1$

(c) $x^2 + 2x + 1$ (d) $x^2 - 2x - 1$

14. The graph of $y = g(x)$ in its domain is broken at

(a) 1 point (b) 2 points (c) 3 points (d) None of these

Passage II (Q. Nos. 15 to 16)

Consider the functions $f(x) = \begin{cases} x - 1, & -1 \le x \le 0 \\ x^2, & 0 \le x \le 1 \end{cases}$

and $g(x) = \sin x$.

If $h_1(x) = f(|g(x)|)$

and $h_2(x) = |f(g(x))|$.

15. Which of the following is not true about $h_1(x)$?

(a) It is a periodic function with period π

(b) The range is $[0, 1]$

(c) Domain R

(d) None of the above

16. Which of the following is not true about $h_2(x)$?

(a) The domain is R (b) It is periodic with period 2π

(c) The range is $[0, 1]$ (d) None of these

Graphical Transformations Exercise 5 : Matching Type Questions

■ **Directions** (Q.No. 17) *Choices for the correct combination of elements from Column I and Column II are given as option (a), (b), (c) and (d) out of which are correct.*

17.

	Column I (Equation)		Column II (Number of Roots)
A.	$x^2 \tan x = 1$, $x \in [0, 2\pi]$	p.	5
B.	$2^{\cos x} = \|\sin x\|$, $x \in [0, 2\pi]$	q.	2
C.	If $f(x)$ is a polynomial of degree 5 with real coefficient such that $f(x) = 0$ has 8 real roots, then the number of roots of $f(x) = 0$	r.	3
D.	$7^{\|x\|}(5 - \|x\|) = 1$	s.	4

Codes

	A	B	C	D		A	B	C	D
(a)	p	q	r	s	(b)	q	s	p	s
(c)	q	p	s	r	(d)	s	p	q	r

Graphical Transformations Exercise 6 : Single Integer Answer Type Questions

18. Let $f(x) = x + 2|x+1| + 2|x-1|$. If $f(x) = k$ has exactly one real solution, then the value of k is

19. The number of roots of the equation $x \sin x = 1$, $x \in [-2\pi, 0] \cup (0, 2\pi)$ is

20. The number of solutions of $\tan x - mx = 0$, $m > 1$ in $\left(-\frac{\pi}{2}, \frac{\pi}{2}\right)$ is

Graphical Transformations Exercise 7 : Subjective Type Questions

21. Find the number of solutions of the equation $\frac{x^2}{1-|x-2|} = 1$, graphically.

22. Find the number of solutions for $\tan 4x = \cos x$, when $x \in (0, \pi)$.

23. Find number of solutions for equation $[\sin^{-1} x] = x - [x]$, where $[\cdot]$ denotes the greatest integer function.

24. If x and y satisfy the equations $\max(|x+y|, |x-y|) = 1$ and $|y| = x - [x]$, the number of ordered pairs (x, y).

25. Find the area enclosed by $|x+y-1| + |2x+y+1| = 1$.

26. Find $f(x)$ when it is given by

$$f(x) = \max\left\{x^3, x^2, \frac{1}{64}\right\}, \forall x \in [0, \infty).$$

27. Find a formula for the function f graphed as

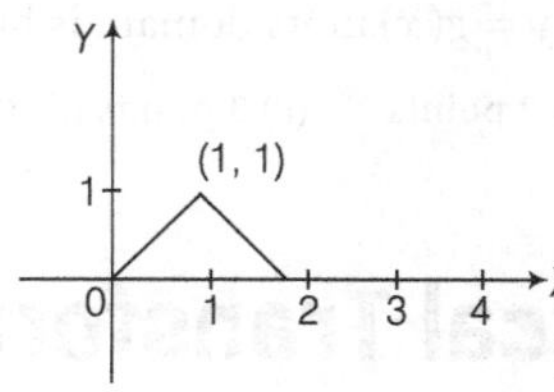

28. Find the domain for $f(x) = \frac{1}{[|x-1|] + [|5-x|] - 4}$, graphically.

29. Draw the graph for $y = \sqrt{\{x\}}$ and $|y| = \sqrt{\{x\}}$.

30. Draw the graph for $y = -[x] + \sqrt{\{x\}}$.

Answers

Exercise for Session 1

1.

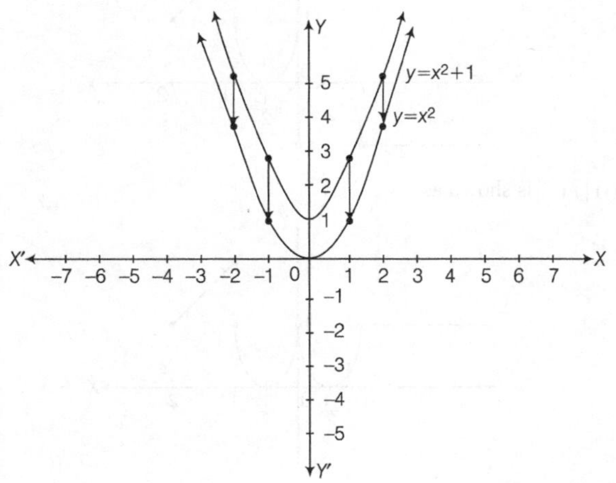

2.

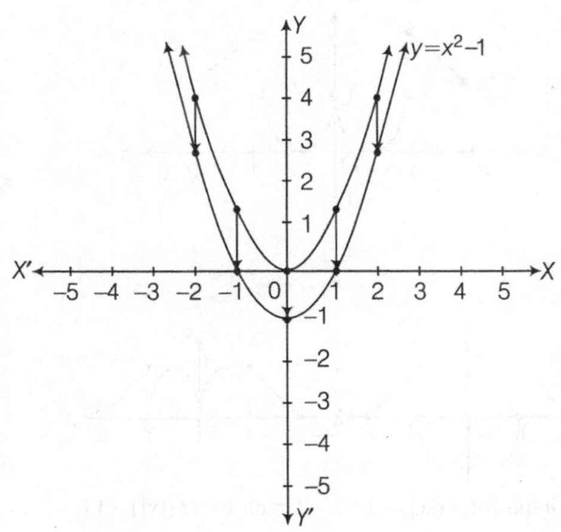

3.

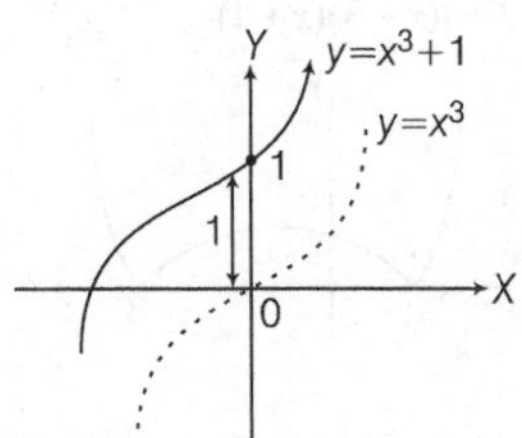

4.

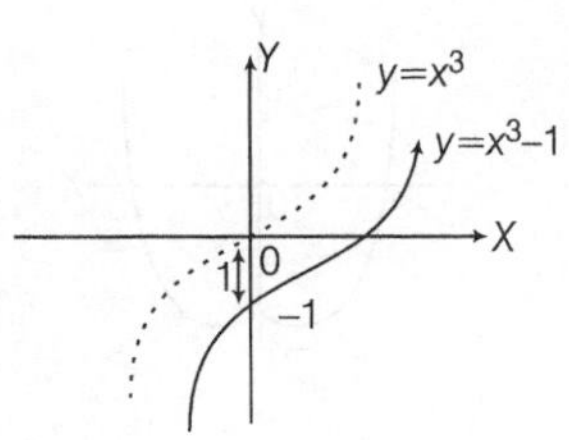

5.

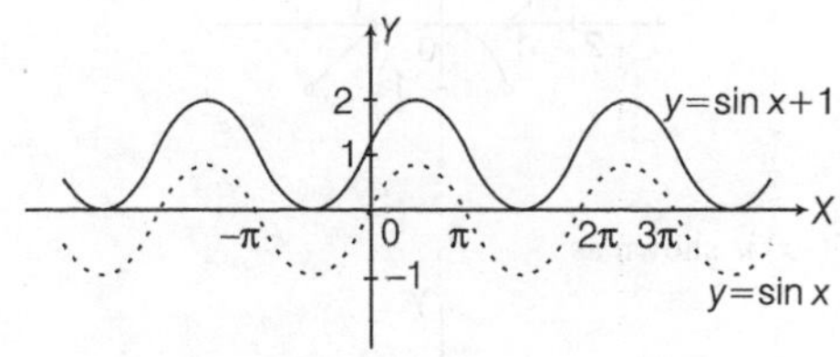

6.

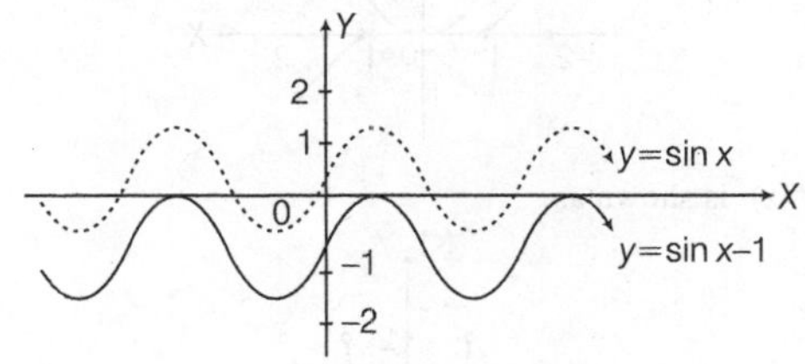

7.

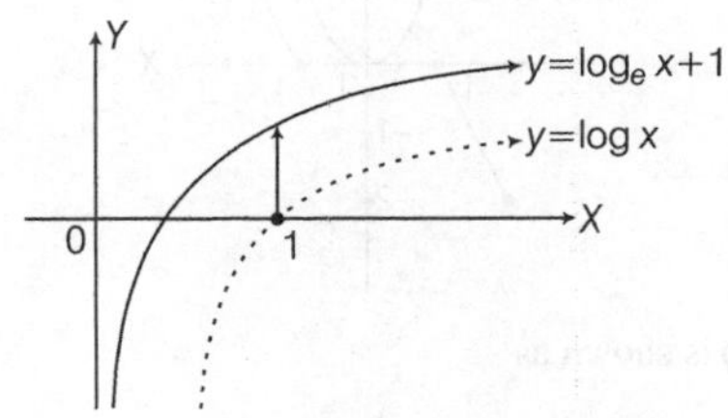

8.

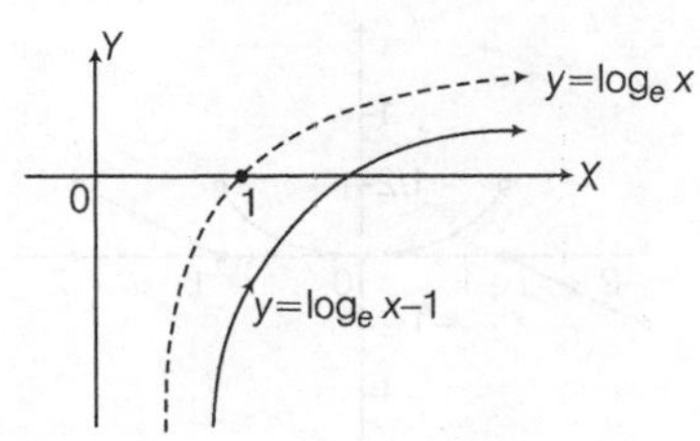

Exercise for Session 2

1. (i) $f(x+1)$ is shown as

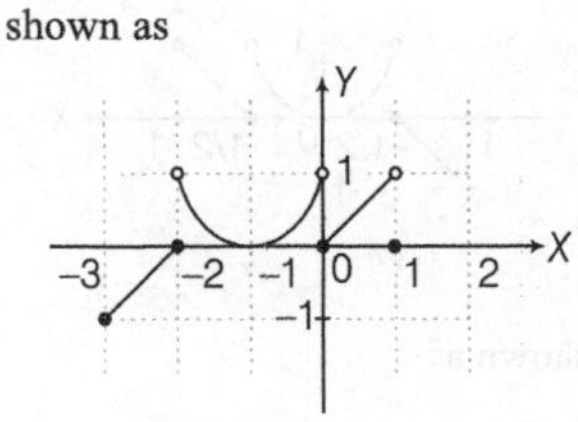

(ii) $f(x-1)$ is shown as

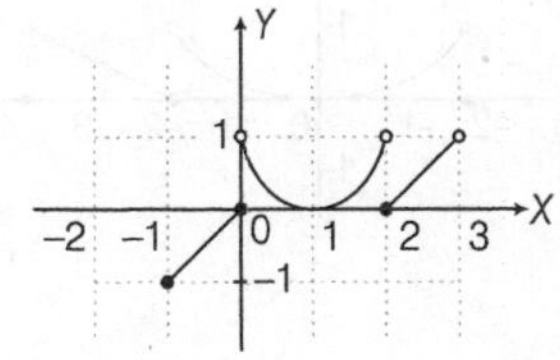

(iii) $-f(x)$ is shown as

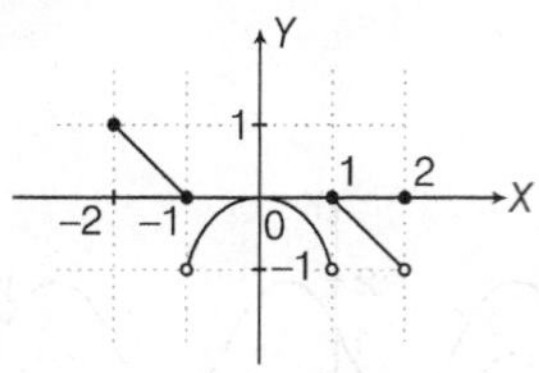

(iv) $f(-x)$ is shown as

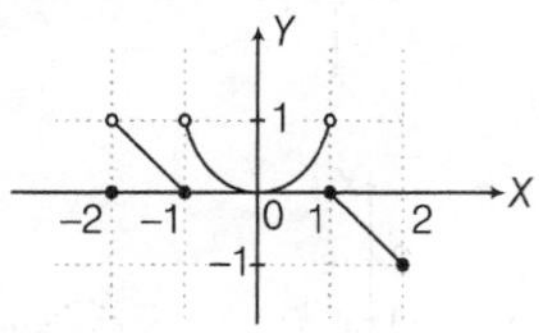

(v) $2f(x)$ is shown as

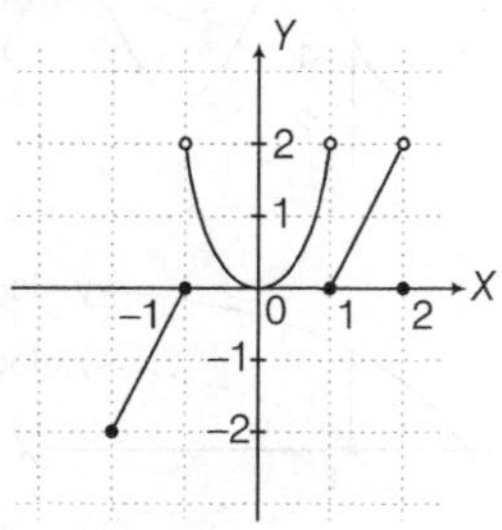

(vi) $\frac{1}{2}f(x)$ is shown as

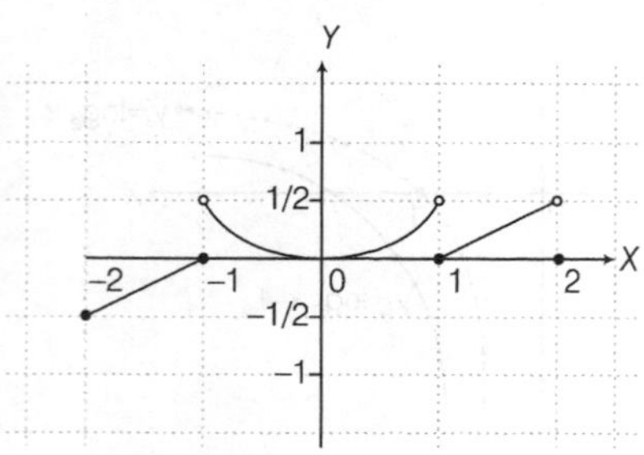

(vii) $f(2x)$ is shown as

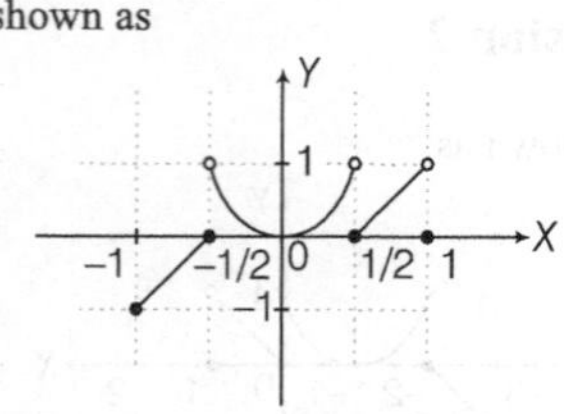

(viii) $f\left(\frac{x}{2}\right)$ is shown as

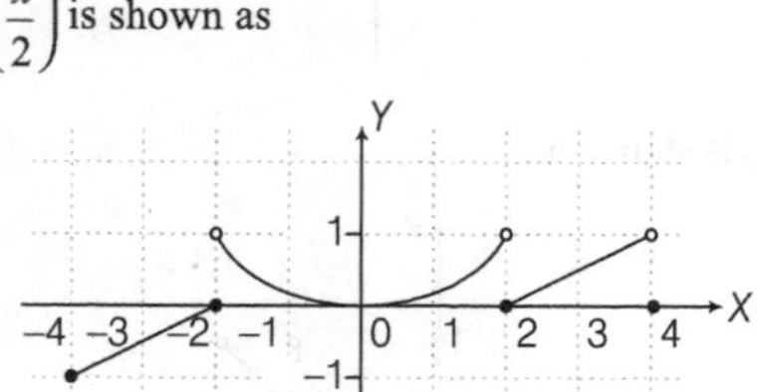

Exercise for Session 3

1. Graph for $f(x)$

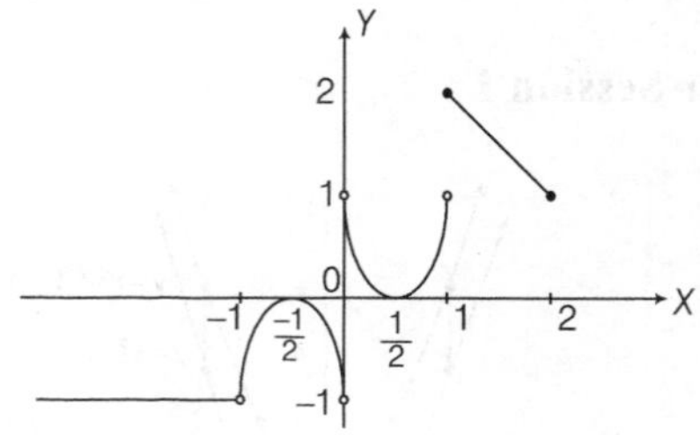

(i) $|f(x)|$ is shown as

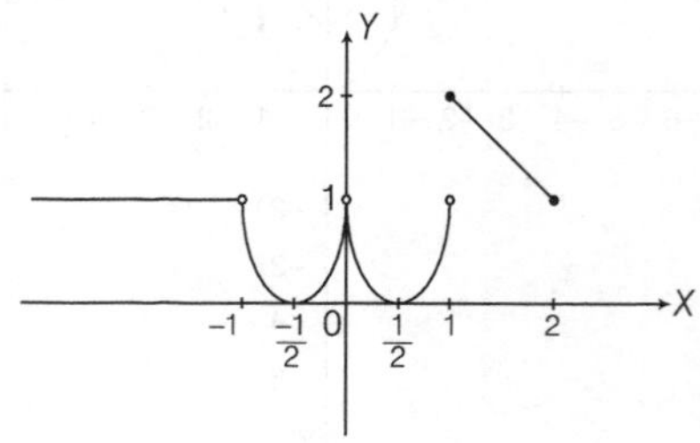

(ii) $f(|x|)$ is shown as

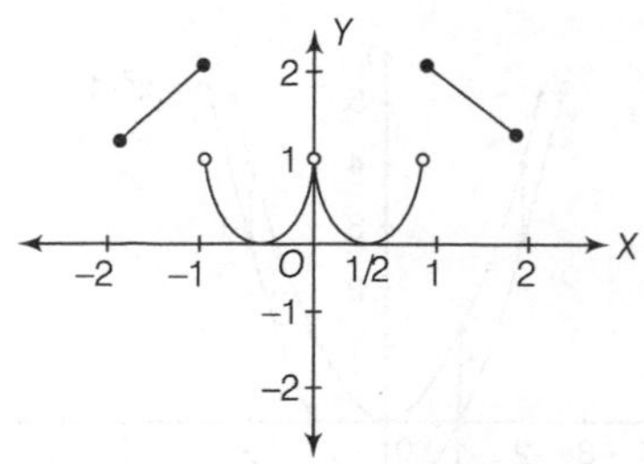

(iii)

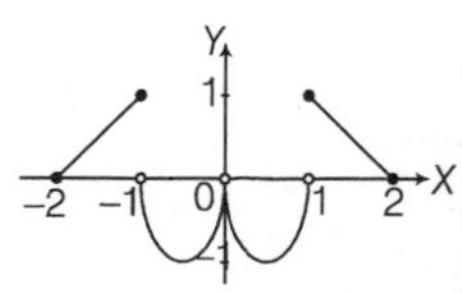

Graph for $f(|x|)-1$

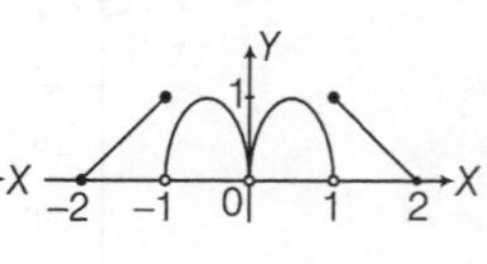

Graph for $|f(|x|)-1|$

2. (i) $y=|x^2-2x-3|=|(x-3)(x+1)|$

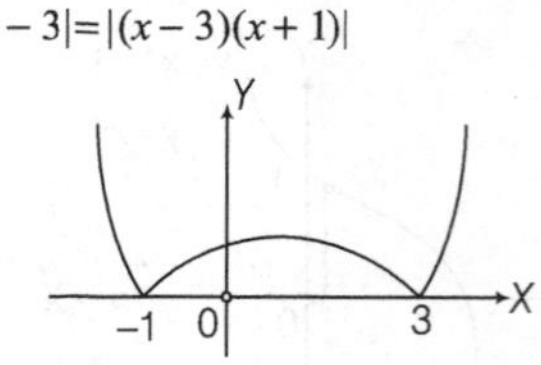

(ii) $y=(x+1)(x-3)=x^2-2x-3,\ y=x^2-2|x|-3$

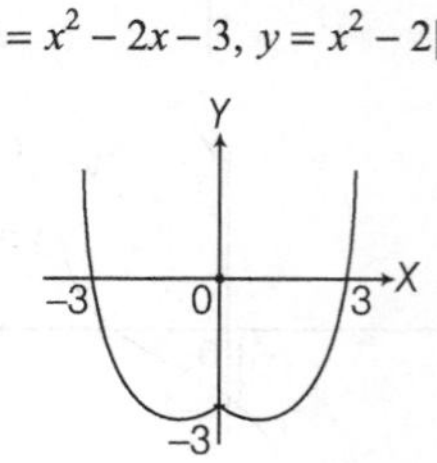

(iii) $y = |\log_2 x|$

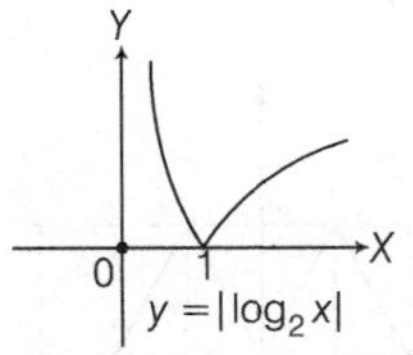

(iv) $y = |\log_2 |x||$

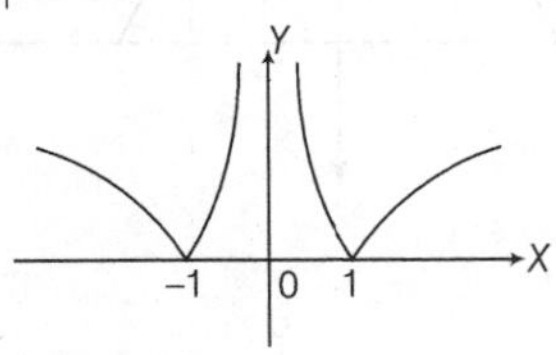

(v) $y = \log_2(1-x) \Rightarrow y = \log_2|1-x|$

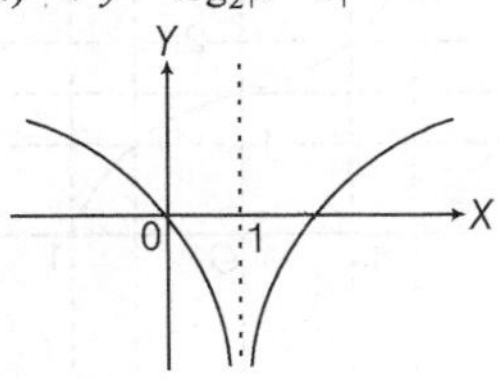

(vi) $y = \log_2(2-x)^2 = 2\log_2|2-x|$

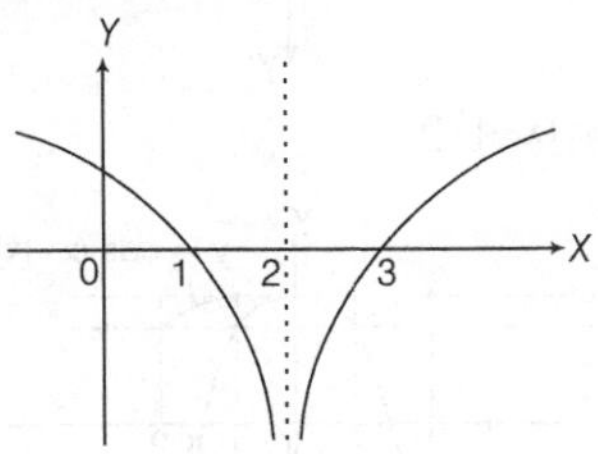

(vii) $y = |\cos|x||$

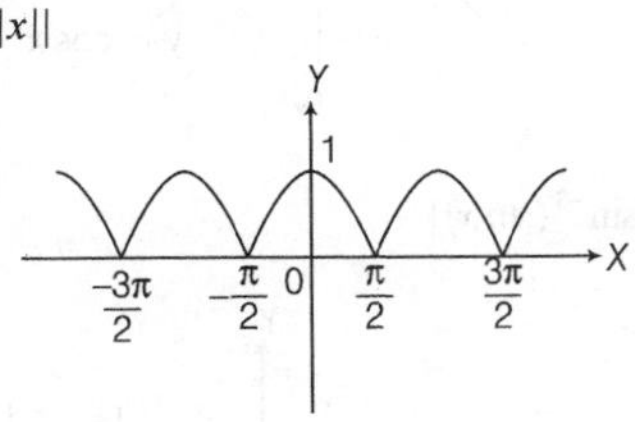

(viii) $y = 2 - 2^x \Rightarrow y = |2 - 2^x|$

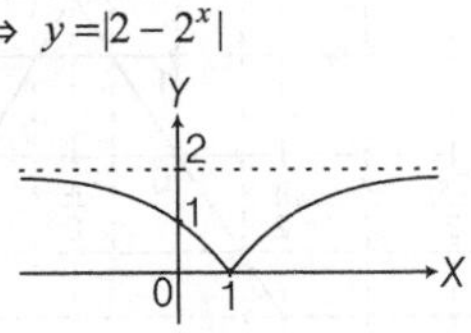

(ix) $y = \sin|x|$

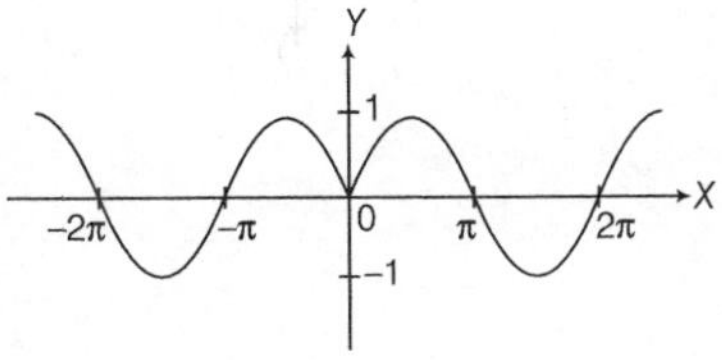

(x) $y = \cos|x|$

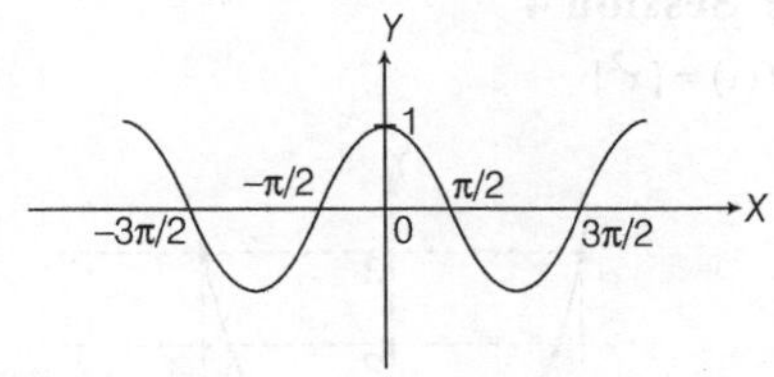

3. (i) $|f(x)| = \log_2 x$

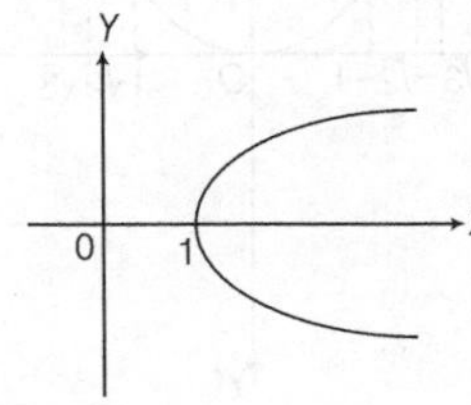

(ii) $|f(x)| = \log_2(-x)$

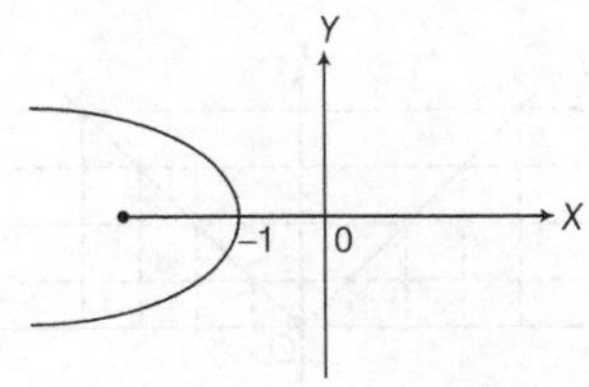

4. $\sin \pi x = |\log_e |x||$

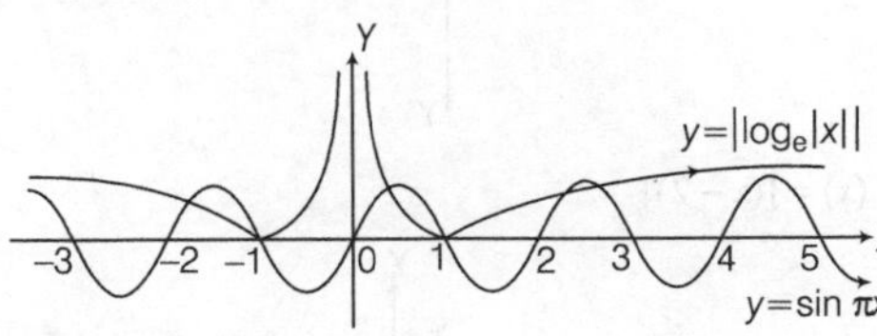

$\therefore$ Number of solutions is 6.

5. (i) $2^{|x|} = \sin x^2$

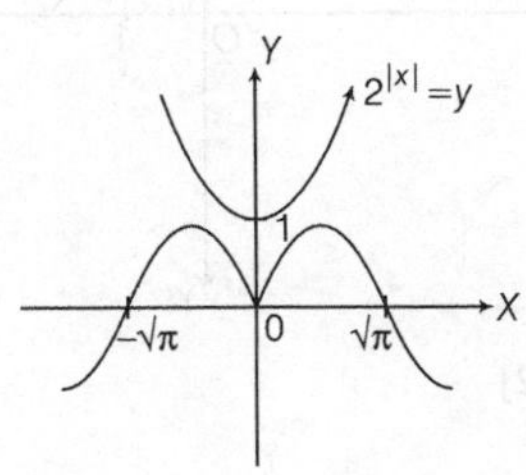

$\therefore$ No solution

(ii) $\sin x = x^2 + x + 1$

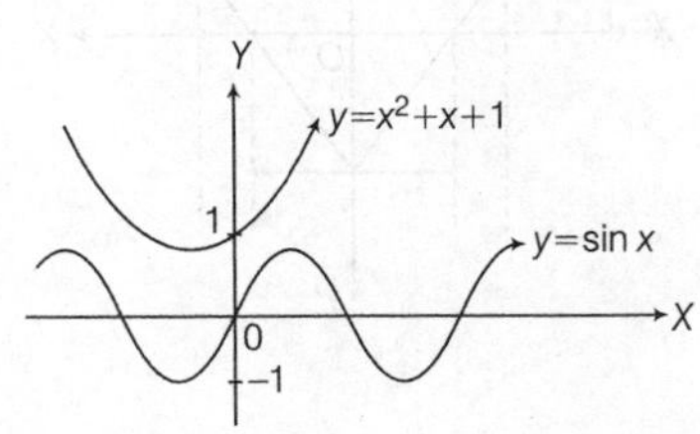

$\therefore$ No solution

Exercise for Session 4

1. (i) Here, $f(x) = [x^2]$

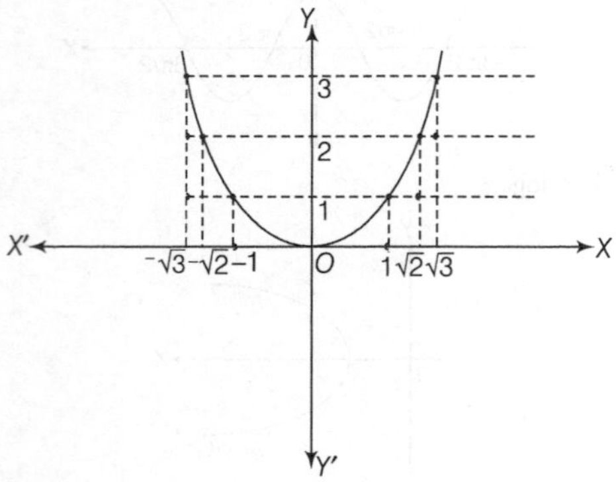

(ii) Here, $f(x) = [|x|]$

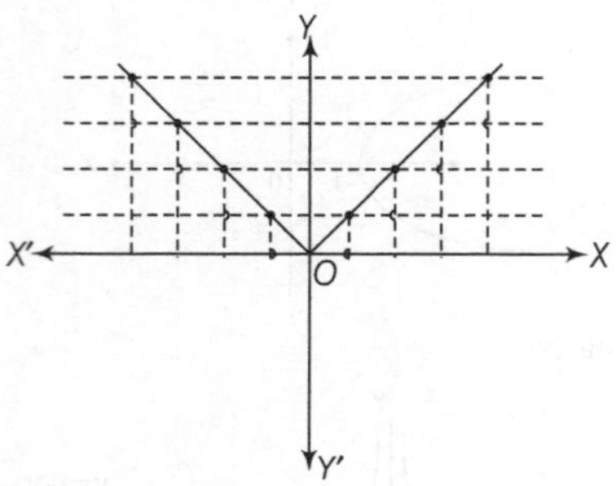

(iii) $f(x) = [(x-2)]$

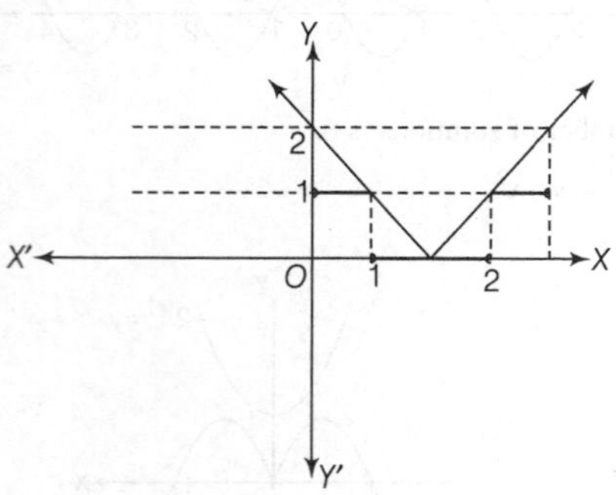

(iv) $f(x) = [|x|-2]$

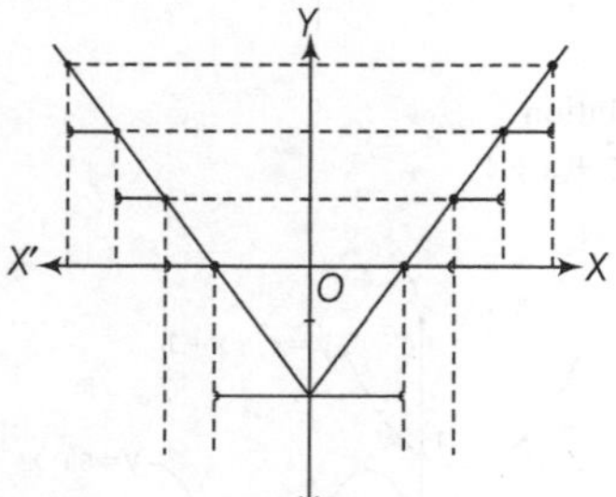

(v) $y = \sin^{-1}(\sin |x|)$

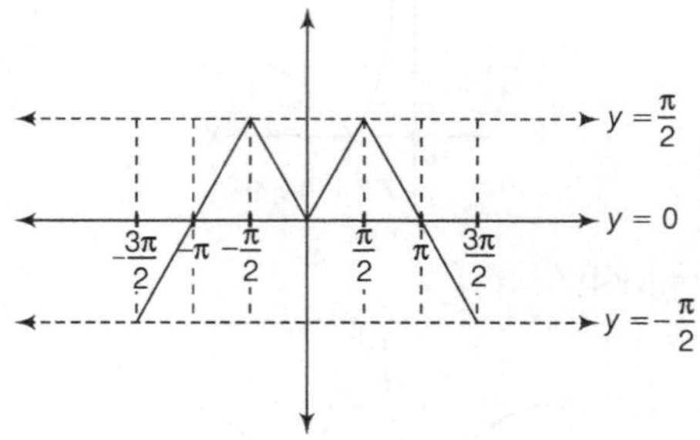

(vi) $y = [\cos^{-1} x]$

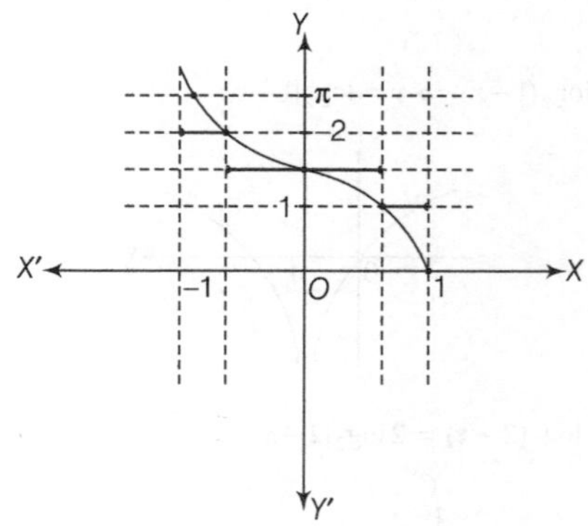

(vii) $f(x) = \cos(x - [x])$

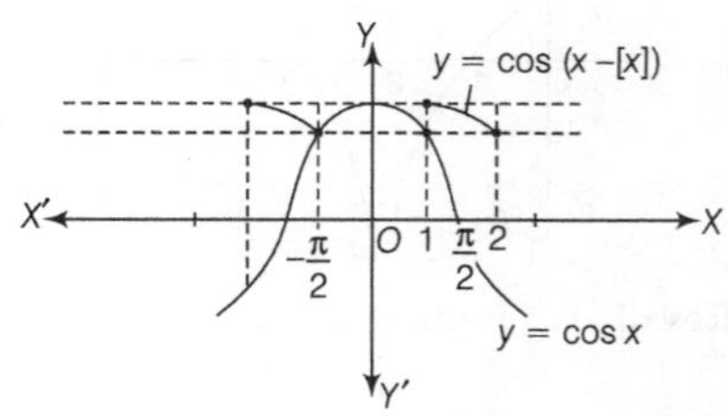

(viii) $f(x) = [\sin^{-1}(\sin x)]$

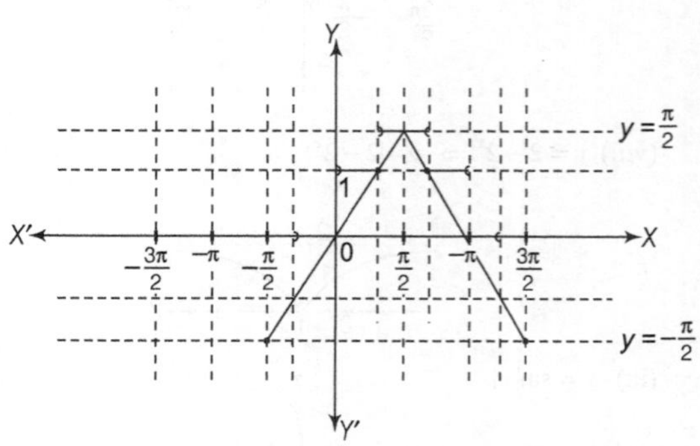

2. Here, $f(x) = \min(x - [x], -x - [-x])$

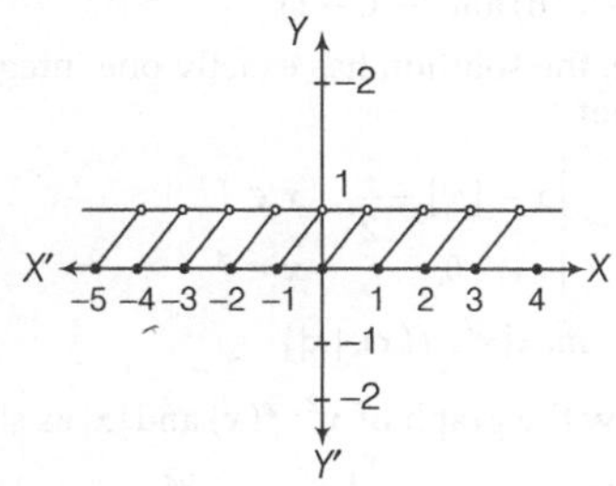

3. (i) 4 sq units (ii) 1 sq unit (iii) 2 sq units

Chapter Exercises

1. (a) 2. (b) 3. (b) 4. (b) 5. (b) 6. (b) 7. (b) 8. (b, c, d) 9. (a, b, c, d) 10. (a, c, d) 11. (a) 12. (a) 13. (b) 14. (a) 15. (d) 16. (c) 17. (b) 18. (3) 19. (4) 20. (3) 21. 0 22. 5 23. 1 24. 0 25. 4 sq units

26. $f(x) = \begin{cases} \dfrac{1}{64}, & 0 < x \le \dfrac{1}{8} \\ x^2, & \dfrac{1}{8} < x \le 1 \\ x^3, & x > 1 \end{cases}$

27. $f(x) = \begin{cases} x, & 0 \le x \le 1 \\ 2 - x, & 1 < x \le 2 \\ 0, & x > 2 \end{cases}$

28. $D = R / \{(0,1], 2, 3, 4, [5, 6\}$.

Solutions

1. Given, $e^{|x|} - |x| = 0 \Rightarrow e^{|x|} = |x|$

Now, the number of solution will be the point of intersection of $y = e^{|x|}$ and $y = |x|$.

Both of the curves shown below

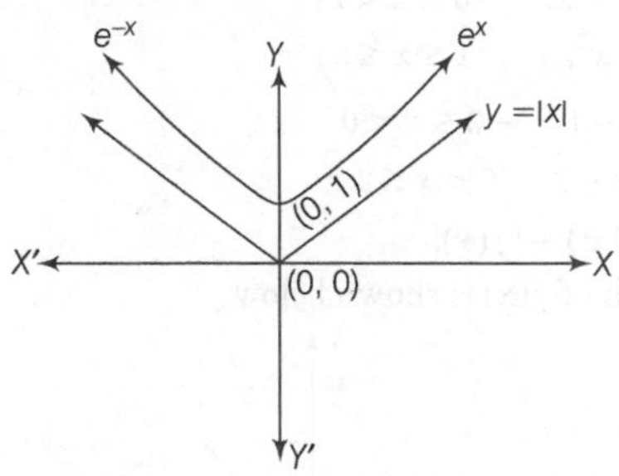

Clearly, from the graph we see that the both curves do not intersect.

$\therefore$ No solution.

2. Given, $3^{-|x|} - 2^{|x|} = 0 \Rightarrow 3^{-|x|} = 2^{|x|}$

Now, we plot the graph $y = 3^{-|x|}$ and $y = 2^{|x|}$ as shown below

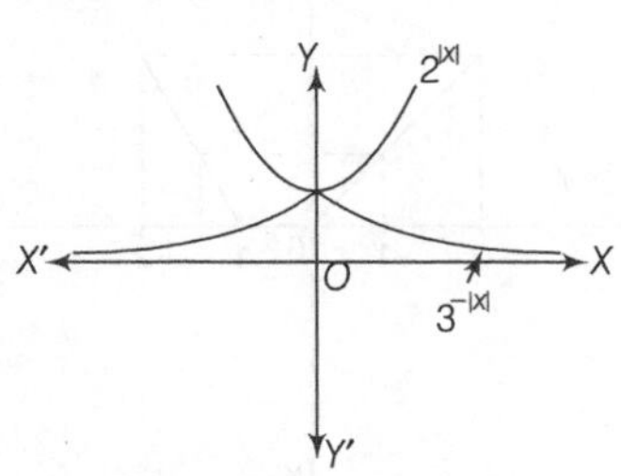

Clearly, the two curves intersect at one points.

Hence, the number of solutions is 1.

3. Given, $3^{|x|} = |2 - |x||$

The graph of $y = 3^{|x|}$ and $y = |2 - |x||$ is shown below

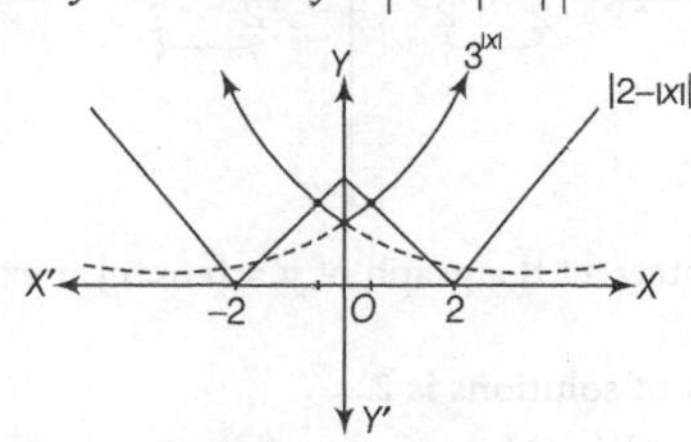

Clearly, both the curve intersect of two points.

$\therefore$ Number of solutions is 2.

4. Given, $|x - x^2 - 1| = |2x - 3 - x^2|$

Let $f(x) = x - x^2 - 1$ and $g(x) = 2x - 3 - x^2$

First we draw the graph of $f(x)$ and $g(x)$ then, transforms it into $|f(x)|$ and $|g(x)|$.

Here, $f(x) = x - x^2 - 1 = y \Rightarrow \left(x - \dfrac{1}{2}\right)^2 = -\left(y + \dfrac{3}{4}\right)$

Clearly, it represents a parabola having vertex $\left(-\dfrac{1}{2}, -\dfrac{3}{4}\right)$.

Also, $g(x) = 2x - 3 - x^2 = y \Rightarrow (x - 1)^2 = -(y + 2)$

It also represent a parabola having vertex $(1, -2)$.

The graph of both curve is given below

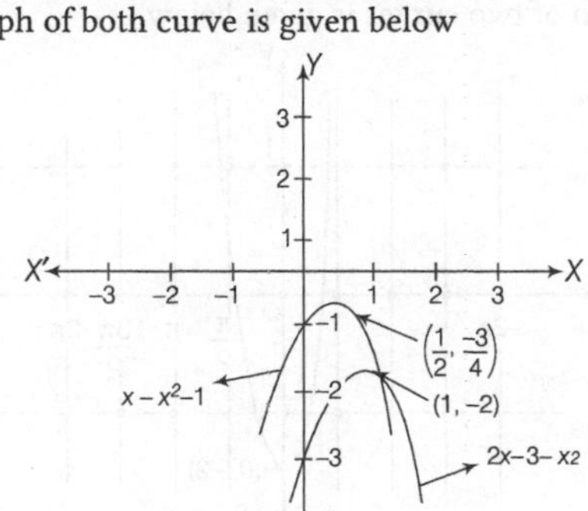

Clearly, both the curves intersect at only one point.

$\therefore$ Number of solutions is 1.

5. Given, $e^x = m(m+1)$

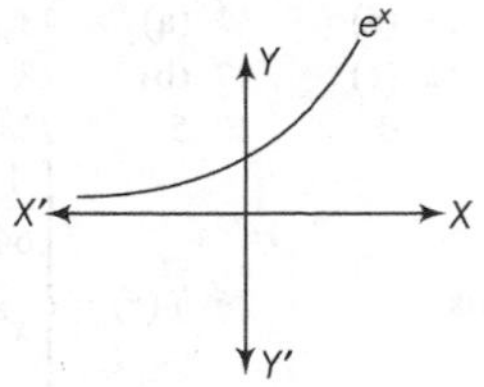

Clearly, for $m < -1$, it will represent a straight line parallel to X-axis which intersect the e^x at only one point.
Hence, exactly one real root.

6. Given, $1 - x = [\cos x]$

The graph of both the curves is given below

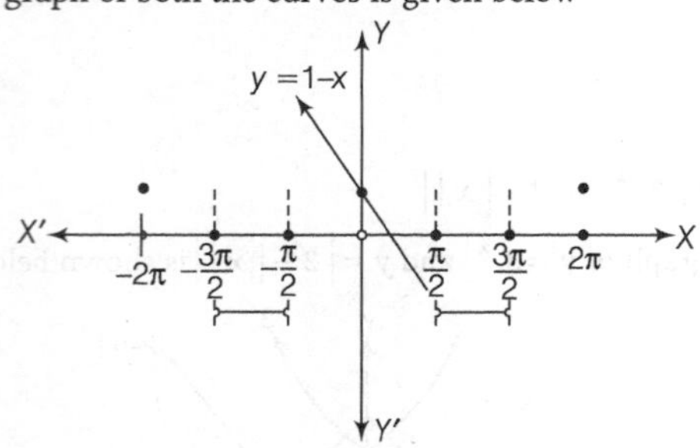

$y = 1 - x$ intersect the graph of $y = [\cos x]$ at exactly two points.

$\therefore$ Number of solutions is 2.

7. Given, $1 + 3^{x/2} = 2^x$

The graph is shown below

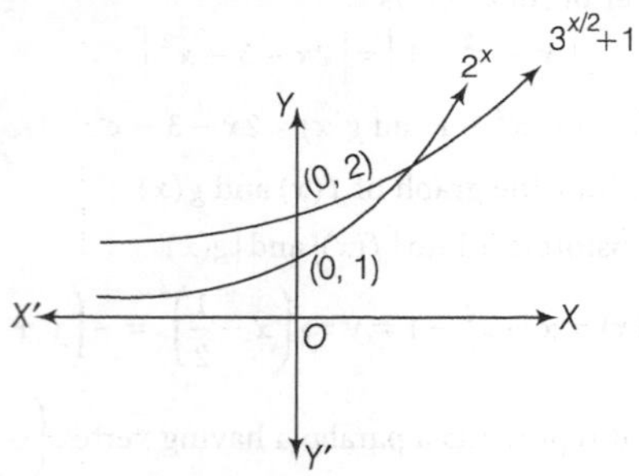

Clearly, both the curves intersect at one point.

8. Given, $x^2 - 2 = [\sin x]$

The graph of two curves is given below

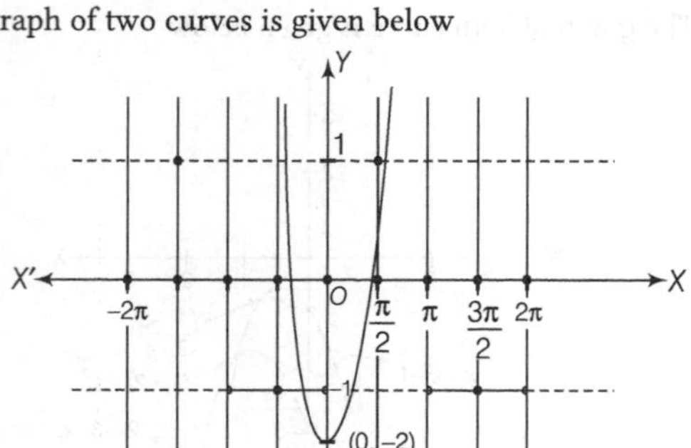

Clearly, the curve $y = x^2 - 2$ intersect the curves $y = [\sin x]$ at two points $(\sqrt{2}, 0)$ and $(-1, -1)$.

$\therefore$ Therefore, the solution has exactly one integer root and one irrational root.

9. Given, $f(x) = \begin{cases} x - [x] - \frac{1}{2}, & x \notin I \\ 0, & x \in I \end{cases}$

and $g(x) = \max\{x^2, f(x), |x|\}$

First we draw the graph of x^2, $f(x)$ and $|x|$ as shown below

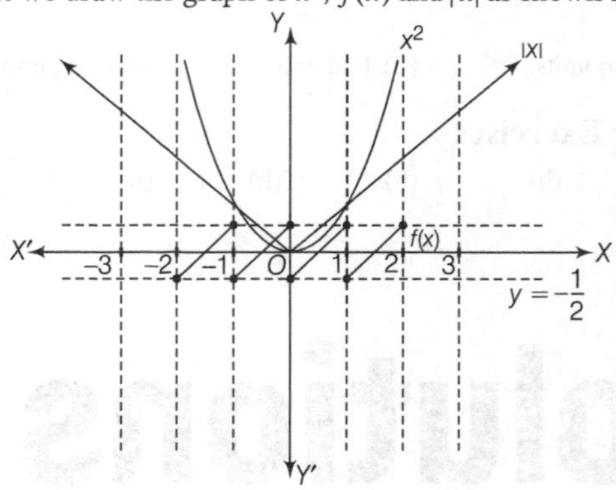

Clearly, from the graph

$$g(x) = \begin{cases} x^2, & -2 \le x \le -1 \\ 1 - x, & -1 < x \le -\frac{1}{4} \\ \frac{1}{2} + x, & -\frac{1}{4} < x < 0 \\ 1 + x, & 0 \le x < 1 \\ x^2, & 1 \le x \le 2 \end{cases}$$

10. $f(x) = \begin{cases} -1 & -2 \le x < 0 \\ x - 1, & 0 < x \le 2 \end{cases}$

$g(x) = f(|x|) + |f(x)|$

The graph of $f(x)$ is shown below

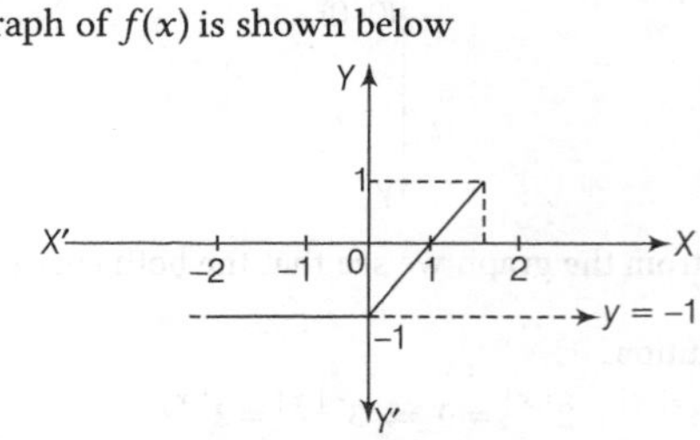

$\therefore$ $g(x) = |f(x)| + f(|x|)$ is shown below

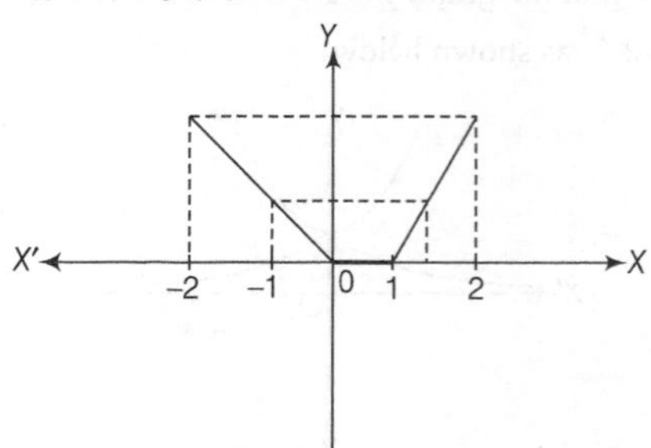

Clearly, from the graph

$$g(x) = \begin{cases} -x, & -2 \le x \le 0 \\ 0, & 0 < x < 1 \\ 2(x-1) & 1 < x \le 2 \end{cases}$$

11. We have the graph of $y = \tan x$ as shown below

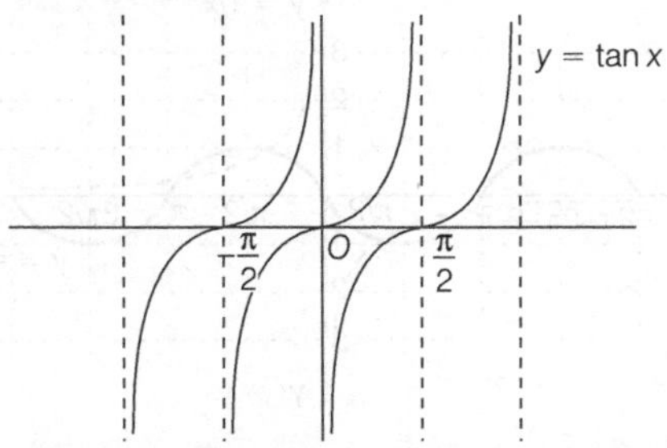

Clearly, from the graph it is symmetrical about the origin. Also, derivative of an odd function is an even function and $\sec^2 x$ is derivative of $\tan x$.

Hence, both the statements are true and Statement II is a correct explanation of Statement I.

12. We know that, the number of point of intersection of two curve is the real solutions of the equation involving these functions.

Also, $||x-2| + a| = 4$

$\Rightarrow$ $|x-2| = \pm 4 - a$

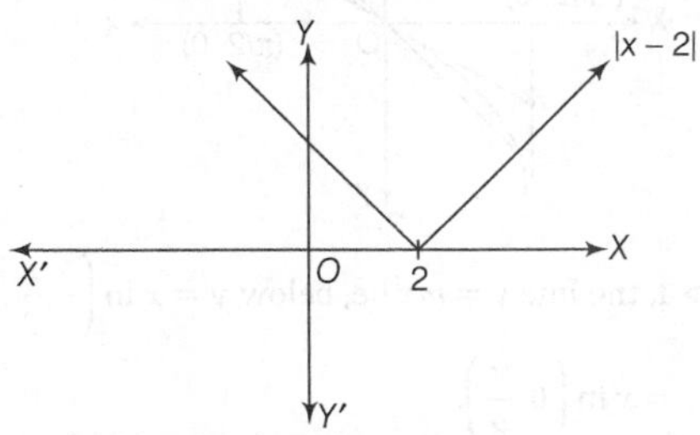

For four real roots $4 - a > 0$ and $-4 - a > 0$

$\Rightarrow$ $a \in (-\infty, -4)$

Both Statement I and Statement II are correct and Statement II is the correct explanation of Statement I

Sol. (Q.Nos. 13 to 14)

Here, $f_1(x) = x^2$

$f_2(x) = |x|$

or $f(x) = f_1(x) - 2f_2(x)$

$= x^2 - 2|x|$

Graph for $f(x)$

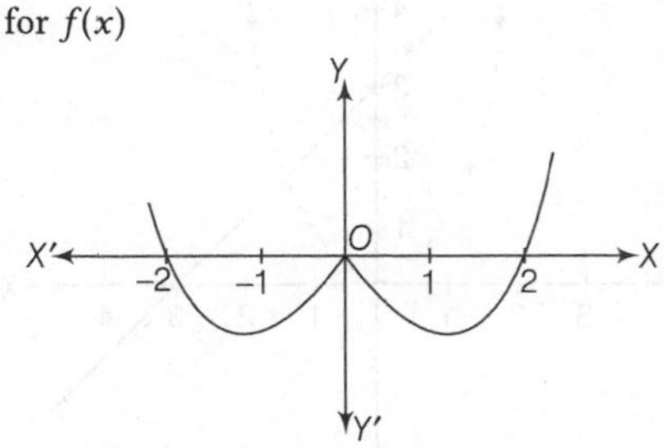

$$\text{Now, } g(x) = \begin{cases} f(x), & -3 \le x < -1 \\ -1, & -1 \le x < 0 \\ 0, & 0 \le x \le 2 \\ f(x), & 2 < x \le 3 \end{cases}$$

The graph of $g(x)$ is shown below

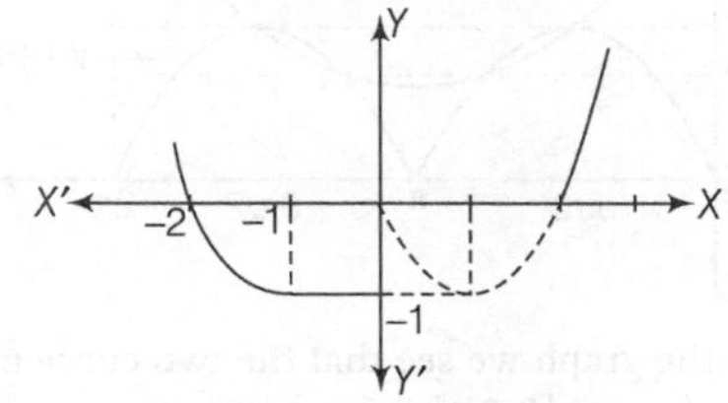

$$g(x) = \begin{cases} x^2 + 2x, & -3 \le x < -1 \\ -1, & -1 \le x < 0 \\ 0, & 0 \le x \le 2 \\ x^2 - 2x, & 2 < x < 3 \end{cases}$$

13. For $x \in (-1, 0)$, $f(x) + g(x) = x^2 + 2x - 1$

14. Obviously, the graph is broken at $x = 0$, i.e. in one point.

15. Clearly, $h_1(x) = f(|g(x)|) = \sin^2 x$ has period π, range $[0, 1]$ and domain R.

16. $h_2(x) = |f(g(x))|$

The graph for $h_2(x)$ is

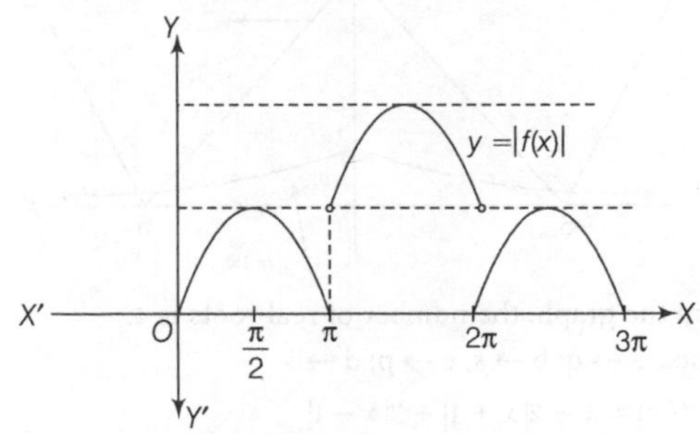

It is clear from the graph that the function is periodic with period 2π and has range $[0, 2]$.

17. Let $y = \tan x = \dfrac{1}{x^2}$

We plot the graph for $\tan x$ and $\dfrac{1}{x^2}$ as shown below

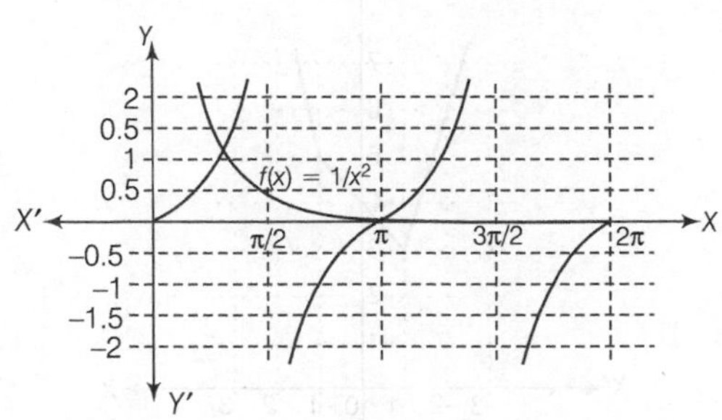

From the graph, it is clear that, it will have two real roots.

(b) $2^{\cos x} = |\sin x|, x \in [0, 2\pi]$

Plotting the graph of $y = 2^{\cos x}$ and $y = |\sin x|$, as shown below

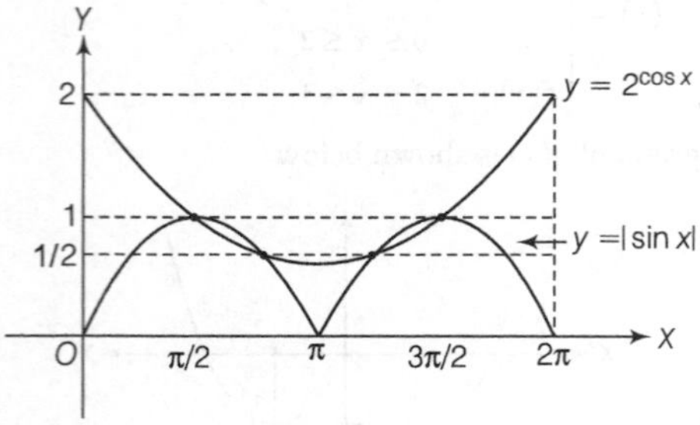

From the graph we see that the two curve meet at four points for $x \in [0, 2\pi]$.

So, the equation $2^{\cos x} = |\sin x|$ has four solutions.

(c) Given, $f(|x|) = 0$ has a real root or $f(x) = 0$ has four positive roots. Since, $f(x)$ is a polynomial of degree 5, $f(x)$ cannot have even number of real roots. Hence, $f(x)$ has all the five roots real and one root is negative.

(d) Here, $7^{|x|}(|5 - |x||) = 1$ or $|5 - |x|| = 7^{-|x|}$

Draw the graph of $y = 7^{-|x|}$ and $y = |5 - |x||$

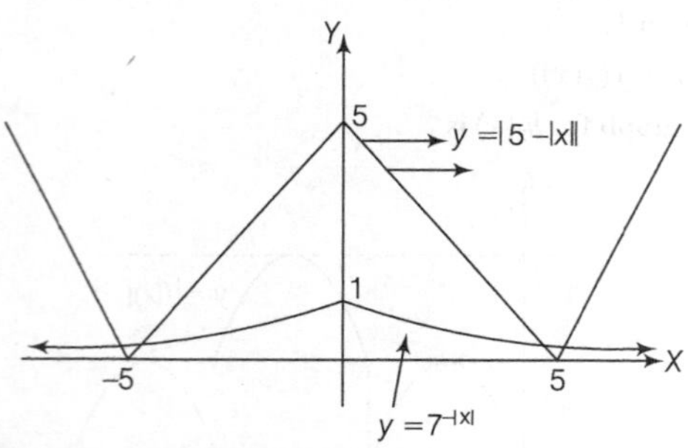

From the graph, the number of real roots is 4.

Hence, a → q; b → s; c → p; d → s.

18. Let $f(x) = x + 2|x + 1| + 2|x - 1|$

$$= \begin{cases} x - 2(x+1) - 2(x-1), & x < -1 \\ x + 2(x+1) - 2(x-1), & -1 \le x \le 1 \\ x + 2(x+1) + 2(x-1), & x > 1 \end{cases} = \begin{cases} -3x, & x < -1 \\ x + 4, & -1 \le x \le 1 \\ 5x, & x > 1 \end{cases}$$

The graph of $f(x)$ is shown below

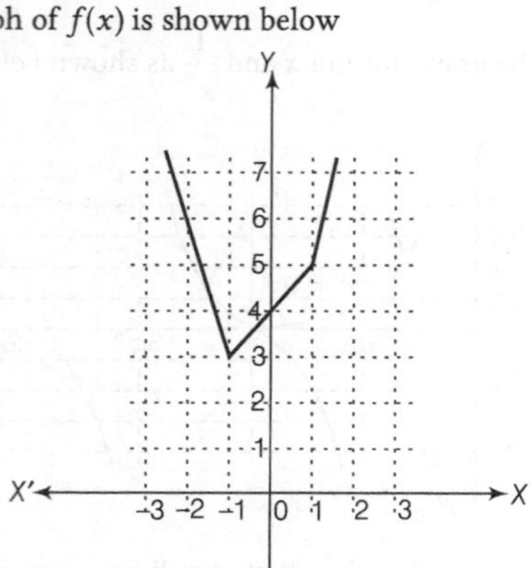

Clearly, from the graph, $y = k$ can intersect $y = f(x)$ at exactly one point only if $k = 3$.

19. Here, $x \sin x = 1$ or $y = \sin x = \dfrac{1}{x}$

The roots of Eq. (i) will be given by the points of intersection of the graphs $y = \sin x$ and $y = \dfrac{1}{x}$, which is shown below.

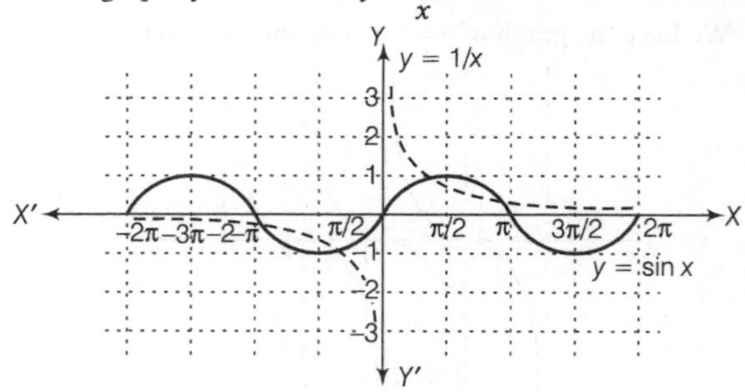

Graphically, it is clear that the given equation has four roots.

20. In $\left(-\dfrac{\pi}{2}, 0\right)$, the graph of $y = \tan x$ lies below the line $y = x$, which is the tangent at $x = 0$ and in $\left(0, \dfrac{\pi}{2}\right)$, it lies above the line $y = x$.

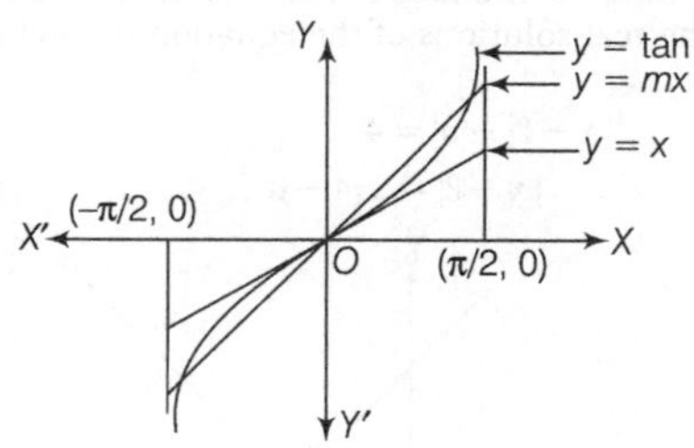

For $m > 1$, the line $y = mx$ lie, below $y = x$ in $\left(-\dfrac{\pi}{2}, 0\right)$ and above $y = x$ in $\left(0, \dfrac{\pi}{2}\right)$.

Thus, graph of $y = \tan x$ and $y = mx$, $m > 1$, meet at three points including $x = 0$ in $\left(-\dfrac{\pi}{2}, \dfrac{\pi}{2}\right)$ independent of m.

21. We have, $\dfrac{x^2}{1 - |x - 2|} = 1$

(i) When $x \ge 2$, $\dfrac{x^2}{1 - (x - 2)} = 1 \Rightarrow \dfrac{x^2}{3 - x} = 1$

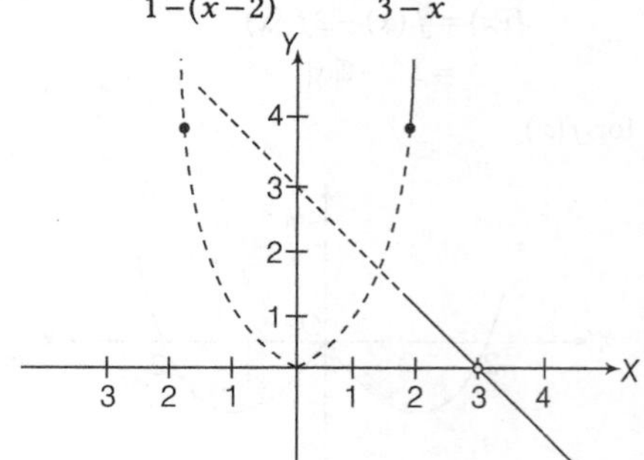

$\Rightarrow \qquad x^2 = 3 - x, x \ne 3$

No point of intersection.

(ii) When $x < 2, \dfrac{x^2}{1-(2-x)} = 1 \Rightarrow \dfrac{x^2}{-1+x} = 1$

$\Rightarrow \qquad x^2 = -1 + x, x \neq 1$

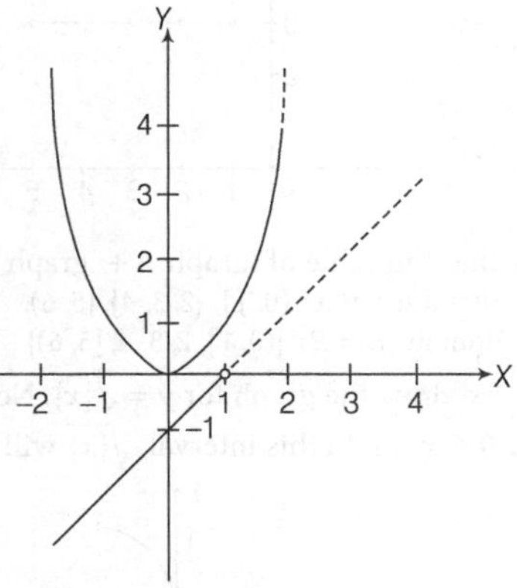

No point of intersection.

Here, the number of solution is zero.

22. Given, $\tan 4x = \cos x$

The number of solutions will be the point of intersection of curves $y = \tan 4x$ and $\cos x$

The graph of $y = \tan 4x$ and $y = \cos x$ is given below.

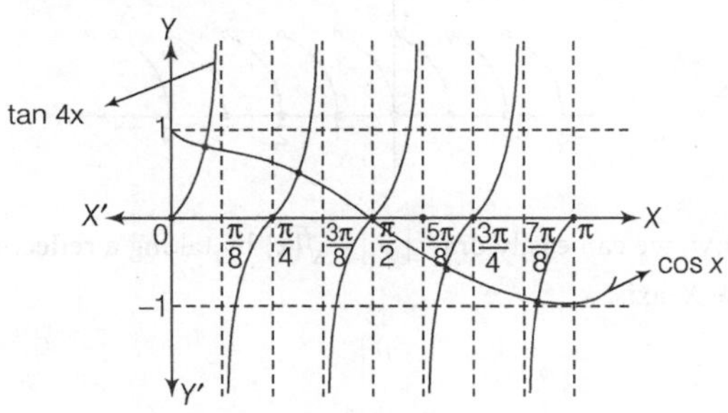

From the graph, it is clear that both curve intersect at five points.

$\therefore$ Number of solutions is 5.

23. Given, $[\sin^{-1} x] = x - [x]$

The number of solution will be the point of intersection of $y = [\sin^{-1} x]$ and $y = x - [x] = \{x\}$

The graph is shown below

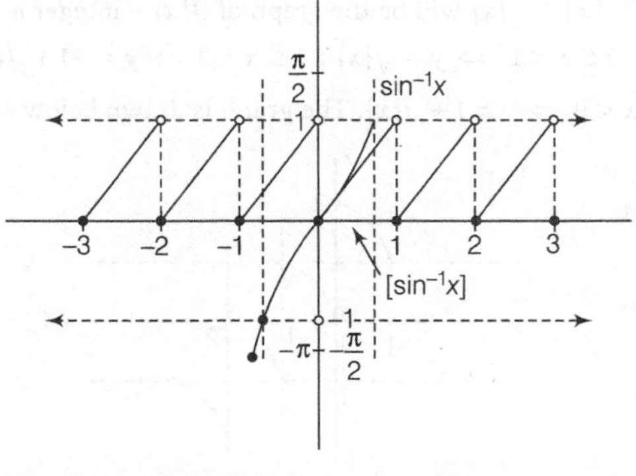

Clearly, the two graphs intersect at only one point.

$\therefore$ Number of solutions is 1.

24. Now, $|x + y| = 1$

$\Rightarrow \qquad x + y = \pm 1 \qquad \text{...(i)}$

and $\qquad |x - y| = 1$

$\Rightarrow \qquad x - y = \pm 1 \qquad \text{...(ii)}$

Also, $\qquad |y| = x - [x]$

$\Rightarrow \qquad \pm y = x - [x]$

Here, we see that there is no point of intersection between the two curves.

Here, no ordered pair of (x, y) satisfy the given equations.

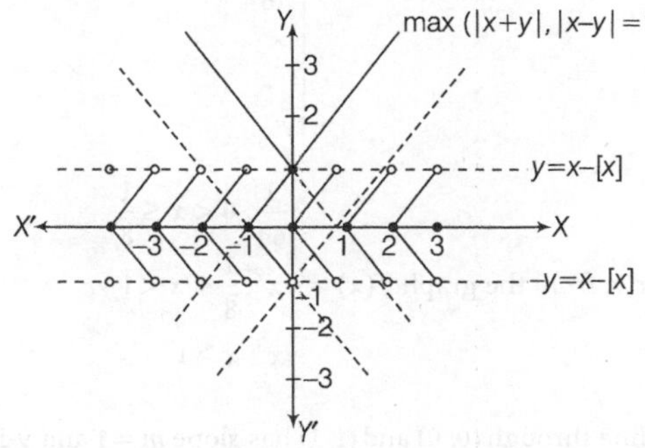

25. We have, $\quad |x + y - 1| + |2x + y + 1| = 1$

$\Rightarrow \pm (x + y - 1) \pm (2x + y + 1) = 1$

(i) $(x + y - 1) + (2x + y + 1) = 1$

$\Rightarrow \qquad 3x + 2y = 1 \qquad \text{...(i)}$

(ii) $+(x + y - 1) - (2x + y + 1) = 1$

$\Rightarrow \qquad -x = 3 \qquad \text{...(ii)}$

(iii) $-(x + y - 1) + (2x + y + 1) = 1$

$\Rightarrow \qquad x = -1 \qquad \text{...(iii)}$

(iv) $-(x + y - 1) - (+2x + y + 1) = 1$

$\Rightarrow \qquad -3x - 2y = 1 \qquad \text{...(iv)}$

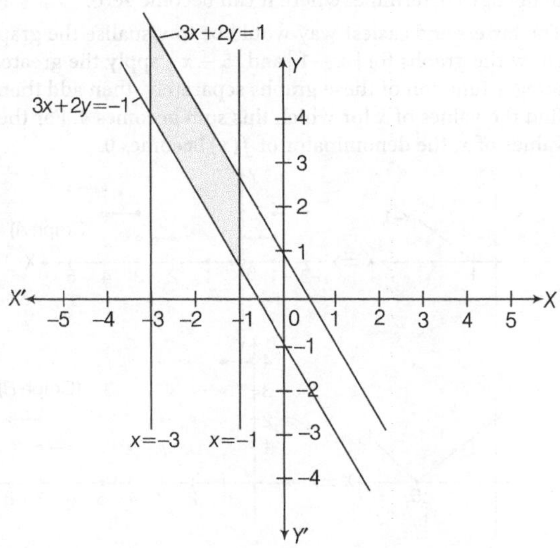

$\therefore$ Area of shaded region $= \int_{-1}^{-3} [(3x + 2y - 1) - (3x + 2y + 1)]dx$

$= \int_{-1}^{-3} -2\,dx = -2[x]_{-1}^{-3} = -2[-3 + 1] = 4$ sq units

26. Given, $f(x) = \max\left\{x^3, x^2, \frac{1}{64}\right\}, \forall\, x \in (0, \infty)$

Now, we draw the graph for the curves $y = x^3$, $y = x^2$ and $y = \frac{1}{64}$

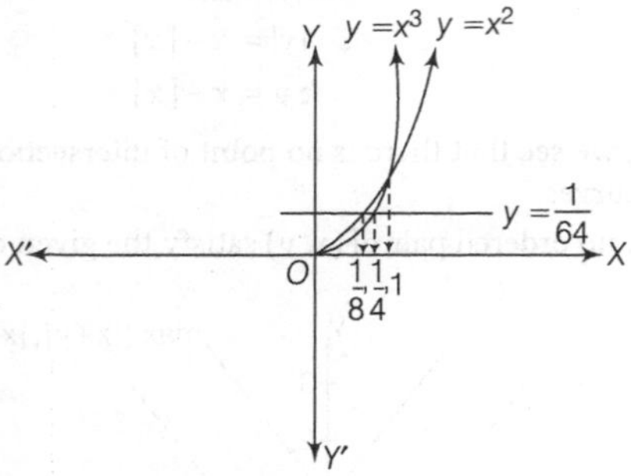

Clearly, from the graph $f(x) = \begin{cases} \frac{1}{64}, & 0 < x \le \frac{1}{8} \\ x^2, & \frac{1}{8} < x < 1 \\ x^3, & x > 1 \end{cases}$

27. The line through (0, 0) and (1, 1) has slope $m = 1$ and y-intercept $b = 0$, so its equation is $y = x$. Thus, for the part of the graph f that joins (0, 0) to (1,1), we have $f(x) = x$, if $0 \le x \le 1$

The line through (1, 1) and (2, 0) has slope $m = -1$, so its point slope form is $y - 0 = (-1)(x - 2)$ or $y = 2 - x$.

So, we have $f(x) = 2 - x$, if $1 < x \le 2$

We also see that the graph of f coincides with the X-axis for $x > 2$. Putting this information together, we have the following three-piece formula for $f(x) = \begin{cases} x, & \text{if } 0 \le x \le 1 \\ 2 - x, & \text{if } 1 < x \le 2 \\ 0, & \text{if } x > 2 \end{cases}$

28. The denominator is a bit complicated and we need to analyse it in detail to determine, where it can become zero.

The fastest and easiest way would be to visualise the graph. Draw the graphs for $|x - 1|$ and $|5 - x|$, apply the greatest integer function of these graphs separately, then add them and find the values of x for which this sum becomes 4. For these values of x, the denominator of $f(x)$ becomes 0.

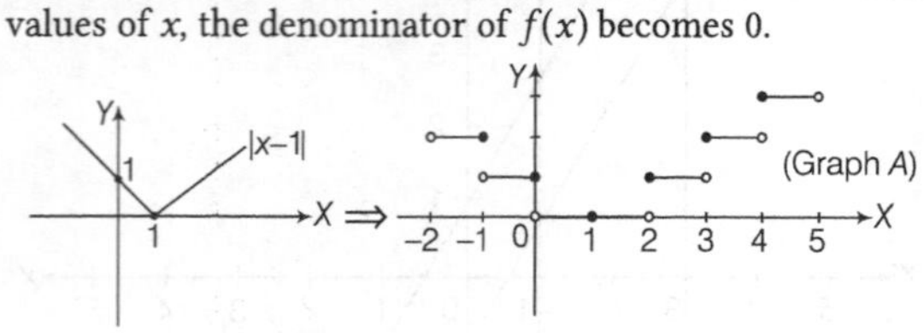

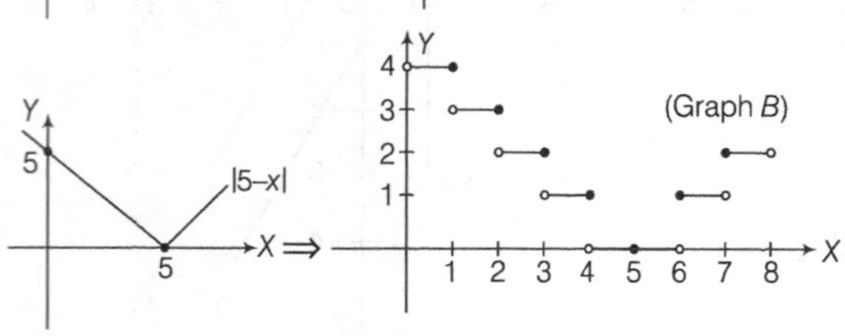

Now, add the graphs of A and B point by point

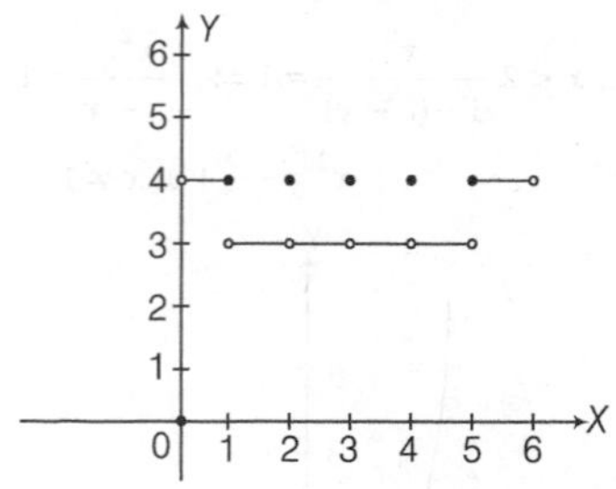

We see that the value of (graph $A +$ graph B) is 4 for the following values of x : (0, 1], {2, 3, 4}, [5, 6).

Hence, domain $D = R / \{(0, 1], 2, 3, 4, [5, 6)\}$.

29. Let us first draw the graph for $y = \sqrt{\{x\}}$. Now, $\{x\}$ is the same as x for $0 \le x < 1$. In this interval, $\sqrt{\{x\}}$ will be the same as $\sqrt{x}$.

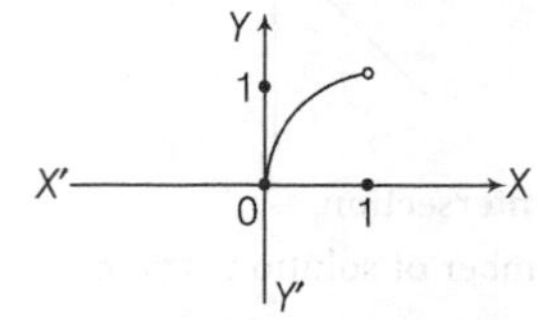

We can easily see that this same curve will be repeated in every previous and subsequent unit interval, since $\{x\}$ is the same in all such intervals. Hence, we obtain the graph of $y = \sqrt{\{x\}}$.

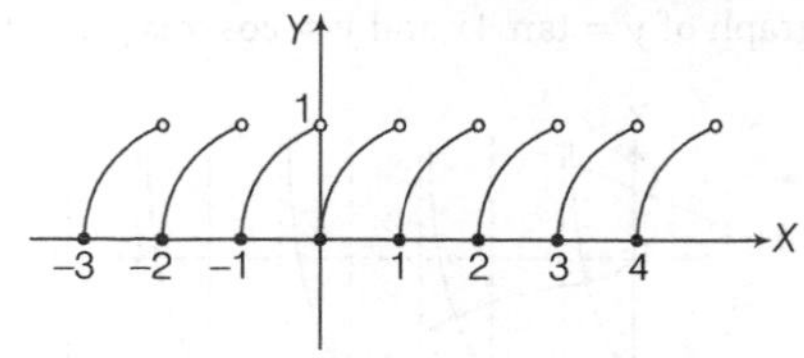

Now, we can easily draw $|\,y\,| = \sqrt{\{x\}}$ by taking a reflection in the X-axis.

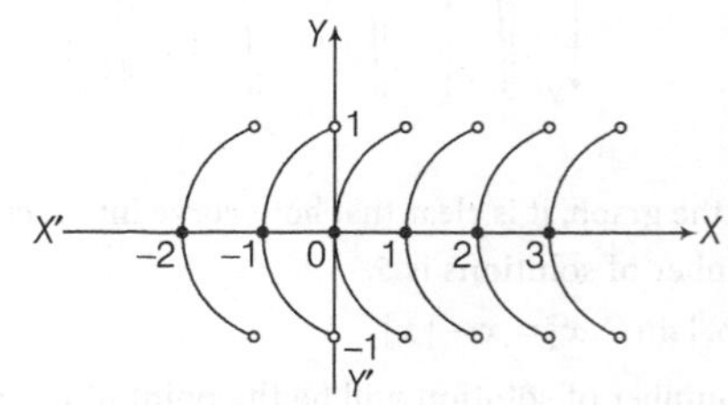

30. In any interval $n \le x < n + 1$, (where n is an integer) $[x]$ has the value n. In any interval, therefore the graph of $y = -[x] + \sqrt{\{x\}}$ will be the graph of $\sqrt{\{x\}}$ – integer n.

e.g. $0 \le x < 1 \Rightarrow y = \sqrt{\{x\}}$; $1 \le x < 2 \Rightarrow y = -1 + \sqrt{\{x\}}$; $-1 \le x < 0 \Rightarrow y = 1 + \sqrt{\{x\}}$. The graph is drawn below

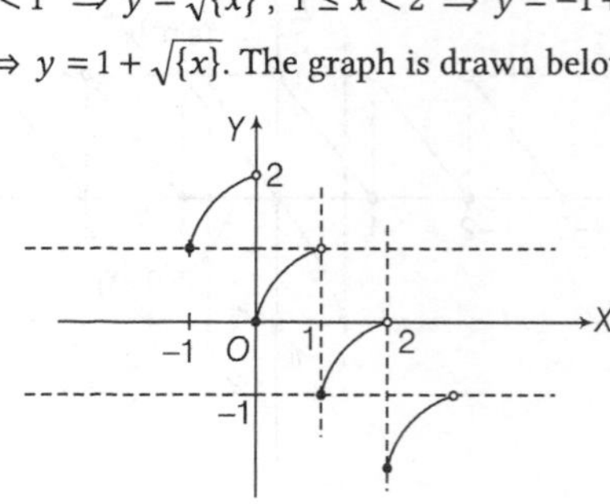

CHAPTER

05

Limits

Learning Part

Session 1

- Definition of Limits
- Indeterminate Forms
- L'Hospital's Rule
- Evaluation of Limits

Session 2

- Trigonometric Limits

Session 3

- Logarithmic Limits
- Exponential Limits

Session 4

- Miscellaneous Forms

Session 5

- Left Hand and Right Hand Limits

Session 6

- Use of Standard Theorems/Results
- Use of Newton-Leibnitz's Formula in Evaluating the Limits
- Summation of Series Using Definite Integral as the Limit

Practice Part

- JEE Type Examples
- Chapter Exercises

Arihant on Your Mobile !

Exercises with this symbol can be practised on your mobile. See title inside to activate for free.

Session 1

Definition of Limits, Indeterminate Forms, L'Hospital's Rule, Evaluation of Limits

Definition of Limits

Let $\lim_{x \to a} f(x) = l$. It would mean that when we approach the point $x = a$ from the values which are just greater than or smaller than $x = a$, $f(x)$ would have a tendency to move closer to the value 'l'.

This is same as saying 'difference between $f(x)$ and l can be made as small as we feel like by suitably choosing x in the neighbourhood of $x = a$'.

Mathematically, we write this, as $\lim_{x \to a} f(x) = l$, which is equivalent of saying that, $|f(x) - l| < \varepsilon, \forall x$ **whenever** $0 < |x - a| < \delta$ **and** ε **and** δ **sufficiently small +ve numbers.**

It is clear from the above discussion that, if we are interested in finding the limit of $f(x)$ at $x = a$, the first thing we have to make sure that $f(x)$ is well defined in the neighbourhood of $x = a$ and not necessarily at $x = a$ (that means $x = a$ may or may not be in the domain of $f(x)$), because we have to examine its behaviour or tendency in the neighbourhood of $x = a$.

Following possibilities may arise :

(a) **Left tendency is same as its right tendency** As shown in figure 5.1, when we approach $x = a$ from the values which are just less than a, $f(x)$ has a tendency to move towards the value l (**left tendency**).

Similarly, when we approach $x = a$ from the values which are just greater than a, $f(x)$ has a tendency to move towards the value l (**right tendency**).

In this case, we say $f(x)$ has limit l at $x = a$,
i.e. $\lim_{x \to a} f(x) = l$.

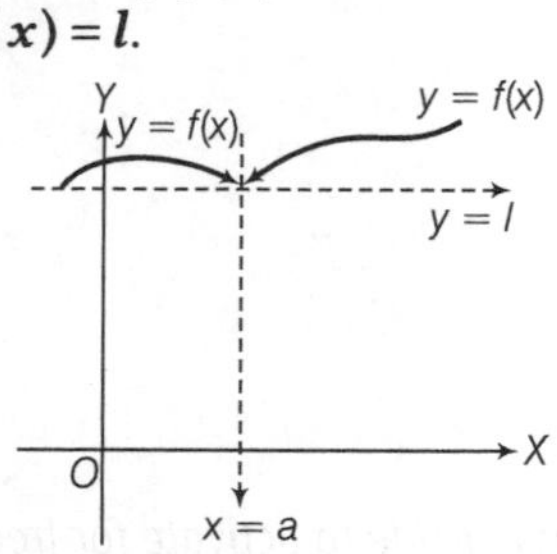

Figure 5.1

(b) **When the left tendency is not the same as its right tendency** Here, left tendency is l_1 and right tendency is l_2, clearly left tendency (l_1) is not same as right tendency (l_2). In this case, we say that the limit of $f(x)$ at $x = a$ will not exist.

i.e. $\lim_{x \to a} f(x) =$ **Doesn't exist.**

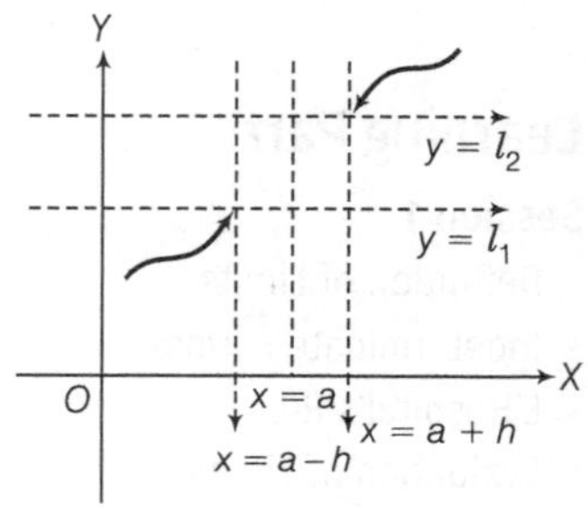

Figure 5.2

(c) **When the left tendency and/or right tendency is not fixed** As shown in the figure 5.3, it is clear that in this case, the function has **erratic** behaviour in the neighbourhood of $x = a$ and it will not be possible to talk about the left and right tendencies of the function in the neighbourhood of $x = a$.

In this case, we conclude that the limit of $f(x)$ at $x = a$ will not exist.

i.e. $\lim_{x \to a} f(x) =$ **Doesn't exist.**

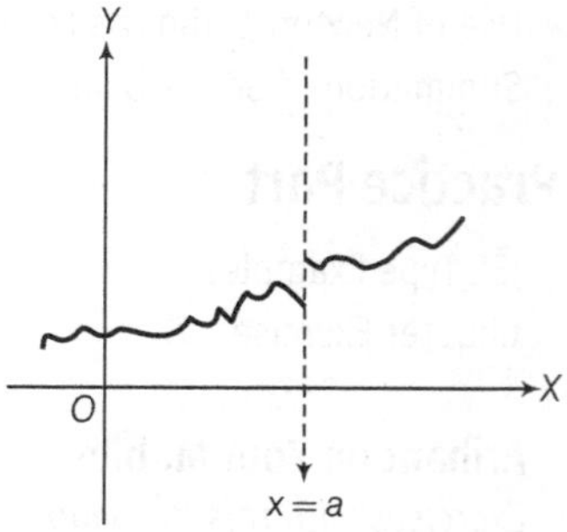

Figure 5.3

Remarks

1. Normally, students have the perception that limit should be a finite number. But, it is not always so. It is quite possible that $f(x)$ has infinite limit at $x = a$.
   If $\lim_{x \to a} f(x) = \infty$, it would simply mean that function has tendency to assume very large positive values in neighbourhood of $x = a$ (as shown in figure 5.4).

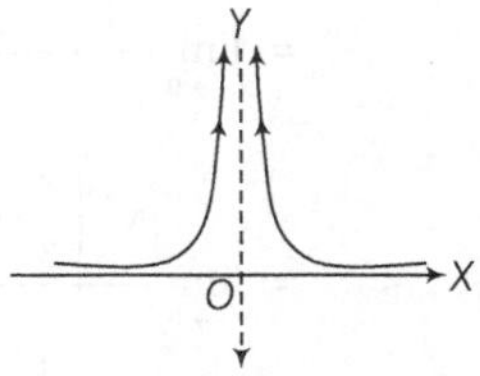

Figure 5.4

   For example $\lim_{x \to 0} \frac{1}{|x|} = \infty$

   which indicates the left tendency as well as right tendency are the same.

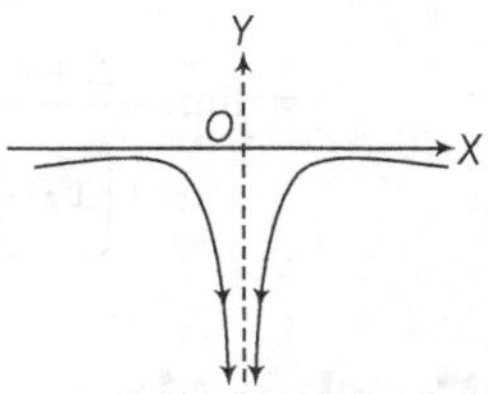

Figure 5.5

   Again, if $\lim_{x \to a} f(x) = -\infty$, it would simply mean that the function has tendency to assume very large negative values in the neighbourhood of $x = a$ as shown in figure 5.5.

   For example (as shown in figure 5.5) $\lim_{x \to 0} \frac{-1}{|x|} = -\infty$

   At the end we discuss the case when left tendency is $(-\infty)$ and right tendency is $(+\infty)$ (i.e. $f(x)$ does not have unique tendency).

   Thus, in this case limit does not exist.

   For example $\lim_{x \to 0} \frac{1}{x}$ does not exist, since left tendency is $(-\infty)$ and right tendency is $(+\infty)$ as shown in figure 5.6.

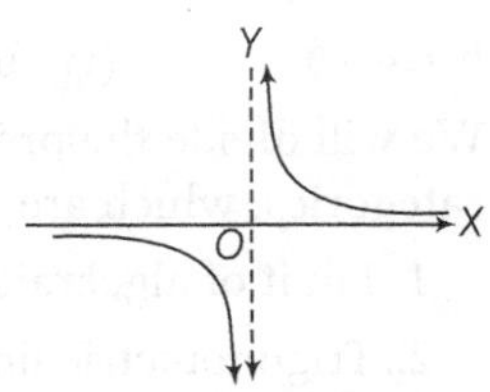

Figure 5.6

2. If $f(x)$ is well defined at $x = a$, it doesn't imply that $\lim_{x \to a} f(x) = f(a)$. Because, it is quite possible that $f(x)$ is well defined at $x = a$ but not in the neighbourhood of $x = a$ or $f(x)$ is well defined in the neighbourhood of $x = a$, but doesn't have a unique tendency.

Indeterminate Forms

If direct substitution of $x = a$ while evaluating $\lim_{x \to a} (x)$ leads to one of the following forms

$\frac{0}{0}, \frac{\infty}{\infty}, \infty - \infty, 1^{\infty}, 0^0, \infty^0, \infty \times 0$ then it is called

indeterminate form.

e.g. $\lim_{x \to 1} \frac{x^2 - 1}{x - 1} = \frac{0}{0}$ indeterminate form.

$\lim_{x \to a} \frac{x^n - a^n}{x - a} = \frac{0}{0}$ indeterminate form.

$\lim_{x \to 0} \frac{\sin x}{x} = \frac{0}{0}$ indeterminate form.

Infinity (∞) is a symbol and not a number. It is a symbol for the behaviour of a variable which continuously increases and passes through all limits. Thus, the statement $x = \infty$ is meaningless, we should write $x \to \infty$.

Similarly, $-\infty$ is a symbol for the behaviour of a variable which continuously decreases and passes through all limits. Thus, the statement $x = -\infty$ is meaningless, we should write $x \to -\infty$.

Also, $\frac{1}{x} \to 0$, if $x \to +\infty$ and $\frac{1}{x} \to 0$, if $x \to -\infty$.

We cannot plot ∞ on paper. Infinity does not obey laws of elementary algebra.

(i) $\infty + \infty = \infty$ is indeterminate

(ii) $\infty - \infty$ is indeterminate.

L'Hospital's Rule

This rule states that, if $\lim_{x \to a} \frac{f(x)}{g(x)}$, reduces to $\frac{0}{0}$ or $\frac{\infty}{\infty}$.

Then, differentiate numerator and denominator till this form is removed.

i.e. $\lim_{x \to a} \frac{f(x)}{g(x)} = \lim_{x \to a} \frac{f'(x)}{g'(x)}$, provided the later limit exists.

But, if it again take form $\left(\frac{0}{0} \text{ or } \frac{\infty}{\infty}\right)$,

then $\lim_{x \to a} \frac{f(x)}{g(x)} = \lim_{x \to a} \frac{f'(x)}{g'(x)} = \lim_{x \to a} \frac{f''(x)}{g''(x)}$

and this process is continued till $\left(\frac{0}{0} \text{ or } \frac{\infty}{\infty}\right)$ form is removed.

Remark

L'Hospital's rule is applicable to only two indeterminate forms $\left(\frac{0}{0} \text{ or } \frac{\infty}{\infty}\right)$.

e.g. Evaluate $\lim_{x \to 2} \frac{x^6 - 24x - 16}{x^3 + 2x - 12}$.

Sol. We have, $\lim_{x \to 2} \frac{x^6 - 24x - 16}{x^3 + 2x - 12}$ $\left[\frac{0}{0} \text{ form}\right]$

$= \lim_{x \to 2} \frac{6x^5 - 24}{3x^2 + 2}$ [by L'Hospital's rule]

$= \frac{6(2)^5 - 24}{3(2)^2 + 2} = \frac{168}{14} = 12$

Frequently Used Series Expansions

1. $e^x = 1 + \frac{x}{1!} + \frac{x^2}{2!} + \frac{x^3}{3!} + \dots$

2. $a^x = 1 + \frac{x \cdot \log a}{1!} + \frac{(\log a)^2 x^2}{2!} + \dots$ [where, $a \in R^+$]

3. $(1+x)^n = 1 + \frac{nx}{1!} + \frac{n(n-1)x^2}{2!} + \frac{n(n-1)(n-2)x^3}{3!} + \dots n \in R$ and $|x| < 1$

4. $\log(1+x) = x - \frac{x^2}{2} + \frac{x^3}{3} - \frac{x^4}{4} + \dots$ [where $-1 \le x \le 1$]

5. $\frac{x^n - a^n}{x - a} = x^{n-1} + x^{n-2}a + x^{n-3}a^2 + \dots + a^{n-1}$

6. $(1+x)^{1/x} = e\left(1 - \frac{x}{2} + \frac{11x^2}{24} + \dots\right)$

7. $\sin x = x - \frac{x^3}{3!} + \frac{x^5}{5!} - \frac{x^7}{7!} + \dots$

8. $\cos x = 1 - \frac{x^2}{2!} + \frac{x^4}{4!} - \frac{x^6}{6!} + \dots$

9. $\tan x = x + \frac{x^3}{3} + \frac{2}{15}x^5 + \frac{17}{315}x^7 + \dots$

10. $\sin^{-1} x = x + \frac{1^2}{3!}x^3 + \frac{1^2 \cdot 3^2}{5!}x^5 + \frac{1^2 \cdot 3^2 \cdot 5^2}{7!}x^7 + \dots$

11. $\tan^{-1} x = x - \frac{x^3}{3} + \frac{x^5}{5} + \dots$

12. $\sec^{-1} x = 1 + \frac{x^2}{2!} + \frac{5x^4}{4!} + \frac{61x^6}{6!} + \dots$

13. $(\sin^{-1} x)^2 = \frac{2}{2!}x^2 + \frac{2 \cdot 2^2}{4!}x^4 + \frac{2 \cdot 2^2 \cdot 4^2}{6!}x^6 + \dots$

14. $x \cot x = 1 - \frac{x^3}{3} + \frac{x^4}{45} - \frac{2x^6}{945} + \dots$

15. $\sec x = 1 + \frac{x^2}{2} + \frac{5x^4}{24} + \frac{61x^6}{720} + \dots$

16. $x \operatorname{cosec} x = 1 + \frac{x^2}{6} + \frac{7x^4}{360} + \frac{31x^6}{15120} + \dots$

| Example 1 Evaluate $\lim_{x \to 0} \frac{x \tan 2x - 2x \tan x}{(1 - \cos 2x)^2}$.

[IIT JEE 1999]

Sol. We have, $\lim_{x \to 0} \frac{x \tan 2x - 2x \tan x}{(1 - \cos 2x)^2}$

$$= \lim_{x \to 0} \frac{x\left[2x + \frac{2^3 x^3}{3} + 2\frac{2^5 x^5}{15} + \dots\right] - 2x\left[x + \frac{x^3}{3} + 2\frac{x^5}{15} + \dots\right]}{(2\sin^2 x)^2}$$

$$= \lim_{x \to 0} \frac{x^4\left(\frac{8}{3} - \frac{2}{3}\right) + x^6\left(\frac{64}{15} - \frac{4}{15}\right) + \dots}{4\left(x - \frac{x^3}{3!} + \frac{x^5}{5!} - \dots\right)^4}$$

$$= \lim_{x \to 0} \frac{2 + 4x^2 + \dots}{4\left(1 - \frac{x^2}{3!} + \dots\right)^4} = \frac{2}{4} = \frac{1}{2}$$

Evaluation of Limits

Now, according to our plan, first of all we shall learn the evaluation of limits of different forms and then learn the existence of limits.

There are eight indeterminate or meaningless forms, which are

(i) $\frac{0}{0}$ (ii) $\frac{\infty}{\infty}$ (iii) $\infty - \infty$ (iv) $\infty \times \infty$

(v) $\infty \cdot 0$ (vi) 0^0 (vii) ∞^0 (viii) 1^∞

We will divide the problems of evaluation of limits in five categories, which are

1. Limit of algebraic functions
2. Trigonometric limits
3. Logarithmic limits
4. Exponential limits
5. Miscellaneous forms

Now, we discuss one by one in details.

Limit of Algebraic Functions

In this section, we evaluate limit of algebraic functions when variable tends to a finite or infinite value. While evaluating algebraic limits the form $\frac{0}{0}, \frac{\infty}{\infty}$ and $\infty - \infty$ arise, which we will discussed here.

(i) $\frac{0}{0}$ Form

This form can be resolved by factorisation method, rationalisation method or by using the formula $\lim_{x \to a} \frac{x^n - a^n}{x - a} = na^{n-1}$, which are discussed below.

(a) Factorisation Method

In this method, numerators and denominators are factorised. The common factors are cancelled and the rest output is the result.

Example 2 Evaluate $\lim_{x\to 1}\dfrac{x^2-3x+2}{x-1}$.

Sol. **Method I** We have, $\lim_{x\to 1}\dfrac{x^2-3x+2}{x-1}$ $\left[\dfrac{0}{0}\text{ form}\right]$

$$=\lim_{x\to 1}\frac{(x-1)(x-2)}{(x-1)}$$

[as $x^2-3x+2=(x-1)(x-2)$]

$$=\lim_{x\to 1}(x-2) \quad [\text{as } x-1\neq 0]$$

$$=1-2=-1$$

Method II We have, $L=\lim_{x\to 1}\dfrac{x^2-3x+2}{x-1}$ $\left[\dfrac{0}{0}\text{ form}\right]$

So, applying L'Hospital's rule,

$$L=\lim_{x\to 1}\frac{2x-3}{1}=\frac{2-3}{1}=-1$$

[i.e. differentiating numerator and denominator separately]

Example 3 Evaluate $\lim_{x\to 1}\dfrac{x^3-x^2\log x+\log x-1}{x^2-1}$.

Sol. We have, $\lim_{x\to 1}\dfrac{x^3-x^2\log x+\log x-1}{x^2-1}$

$$=\lim_{x\to 1}\frac{(x^3-1)-(x^2-1)\log x}{(x^2-1)} \quad \left[\frac{0}{0}\text{ form}\right]$$

$$=\lim_{x\to 1}\frac{(x-1)\{x^2+x+1-(x+1)\log x\}}{(x-1)(x+1)}$$

$\left[\because x^3-1=(x-1)(x^2+x+1),\ x^2-1=(x-1)(x+1)\right]$

$$=\lim_{x\to 1}\frac{x^2+x+1-(x+1)\log x}{(x+1)}$$

$$=\frac{1^2+1+1-(1+1)\log 1}{1+1}=\frac{3}{2} \quad [\text{as } \log 1=0]$$

(b) Rationalisation Method

Rationalisation is followed when we have fractional powers (like $\frac{1}{2},\frac{1}{3}$ etc.) on expressions in numerator or denominator or in both. After rationalisation the terms are factorised which on cancellation gives the result.

Example 4 Evaluate $\lim_{h\to 0}\dfrac{\sqrt{x+h}-\sqrt{x}}{h}$.

Sol. **Method I** We have, $\lim_{h\to 0}\dfrac{\sqrt{x+h}-\sqrt{x}}{h}$

$$=\lim_{h\to 0}\frac{\sqrt{x+h}-\sqrt{x}}{h}\times\frac{\sqrt{x+h}+\sqrt{x}}{\sqrt{x+h}+\sqrt{x}} \quad \left[\frac{0}{0}\text{ form}\right]$$

$$=\lim_{h\to 0}\frac{(x+h)-(x)}{h(\sqrt{x+h}+\sqrt{x})}=\lim_{h\to 0}\frac{h}{h(\sqrt{x+h}+\sqrt{x})}$$

$$=\lim_{h\to 0}\frac{1}{\sqrt{x+h}+\sqrt{x}}=\frac{1}{2\sqrt{x}}$$

Method II (L'Hospital's rule) We have,

$$L=\lim_{h\to 0}\frac{\sqrt{x+h}-\sqrt{x}}{h} \quad \left[\frac{0}{0}\text{ form}\right]$$

$\therefore$ Applying L'Hospital's rule,

$$L=\lim_{h\to 0}\frac{\frac{1}{2\sqrt{x+h}}-0}{1}$$

[differentiating numerator and denominator w.r.t. h]

$$=\frac{1}{2\sqrt{x}}$$

Example 5 Evaluate $\lim_{x\to 2a}\dfrac{\sqrt{x-2a}+\sqrt{x}-\sqrt{2a}}{\sqrt{x^2-4a^2}}$.

Sol. We have, $\lim_{x\to 2a}\dfrac{\sqrt{x-2a}+\sqrt{x}-\sqrt{2a}}{\sqrt{x^2-4a^2}}$ $\left[\dfrac{0}{0}\text{ form}\right]$

$$=\lim_{x\to 2a}\frac{\sqrt{x-2a}+\dfrac{x-2a}{\sqrt{x}+\sqrt{2a}}}{\sqrt{x-2a}\sqrt{x+2a}}$$

$$=\lim_{x\to 2a}\frac{1}{\sqrt{x+2a}}+\frac{\sqrt{x-2a}}{\sqrt{x+2a}(\sqrt{x}+\sqrt{2a})}$$

$$=\frac{1}{\sqrt{4a}}=\frac{1}{2\sqrt{a}}$$

(c) Based on Standard Formula

$$\lim_{x\to a}\frac{x^n-a^n}{x-a}=na^{n-1}, \text{ where } n \text{ is a rational number.}$$

Proof Let $f(x)=\dfrac{x^n-a^n}{x-a}$

$$=x^{n-1}+ax^{n-2}+a^2x^{n-3}+\ldots+a^{n-1}$$

$$\therefore \lim_{x\to a}f(x)=\lim_{x\to a}(x^{n-1}+ax^{n-2}+a^2x^{n-3}+\ldots+a^{n-1})$$

$$=\underbrace{a^{n-1}+a\cdot a^{n-2}+a^2\cdot a^{n-3}+\ldots+a^{n-1}}_{n\text{ terms}}$$

$$=a^{n-1}+a^{n-1}+a^{n-1}+\ldots\text{ upto } n \text{ terms}=n\cdot a^{n-1}$$

Example 6 Evaluate $\lim_{x \to 2} \frac{x^3 - 2^3}{x - 2}$.

Sol. We have, $\lim_{x \to 2} \frac{x^3 - 2^3}{x - 2}$ $\left[\frac{0}{0} \text{ form}\right]$

$$= 3(2)^{3-1} \quad \left[\because \lim_{x \to a} \frac{x^n - a^n}{x - a} = n \cdot a^{n-1}\right]$$

$$= 3(2)^2 = 12$$

Example 7 Evaluate $\lim_{x \to 1} \frac{x + x^2 + \ldots + x^n - n}{x - 1}$.

Sol. We have, $\lim_{x \to 1} \frac{x + x^2 + x^3 + \ldots + x^n - n}{x - 1}$

$$= \lim_{x \to 1} \frac{(x-1) + (x^2 - 1^2) + (x^3 - 1^3) + \ldots + (x^n - 1^n)}{(x - 1)}$$

$$= \lim_{x \to 1} \left\{\frac{x - 1}{x - 1} + \frac{x^2 - 1^2}{x - 1} + \frac{x^3 - 1^3}{x - 1} + \ldots + \frac{x^n - 1^n}{x - 1}\right\}$$

$$= 1 + 2(1)^{2-1} + 3(1)^{3-1} + \ldots + n(1)^{n-1}$$

$$= 1 + 2 + 3 + \ldots + n$$

$$= \frac{n(n+1)}{2}$$

Example 8 The value of $\lim_{\substack{x \to 1 \\ y \to 0}} \frac{y^3}{x^3 - y^2 - 1}$ as $(x, y) \to (1, 0)$ along the line $y = x - 1$ is

(a) 1

(b) –1

(c) 0

(d) Doesn't exist

Sol. As, $y \to x - 1$ or $x \to y + 1$

$$\therefore \quad \lim_{y \to 0} \frac{(y)^3}{(y+1)^3 - y^2 - 1}$$

Using L, Hospital's rule,

$$\lim_{y \to 0} \frac{3y^2}{3(y+1)^2 - 2y} = \lim_{y \to 0} \frac{6y}{6(y+1) - 2}$$

$$= \frac{0}{6} = 0$$

Hence, (c) is the correct answer.

(ii) Algebraic Function of ∞ Type

(a) $\frac{\infty}{\infty}$ Form

First we should know the limiting values of a^x $(a > 0)$ as $x \to \infty$. *See the graphs of this function.*

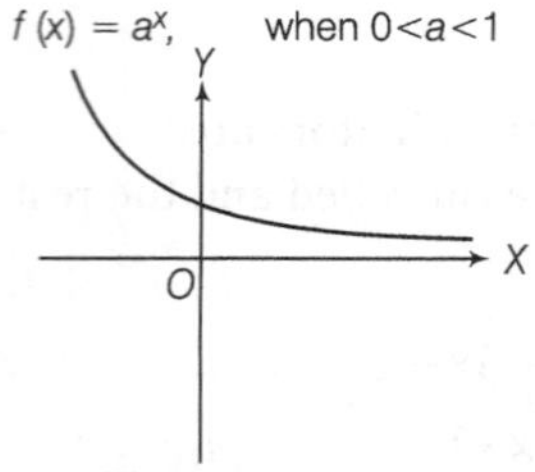

Figure 5.7

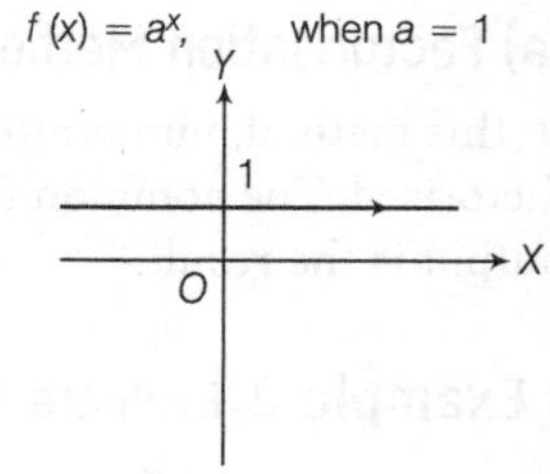

Figure 5.8

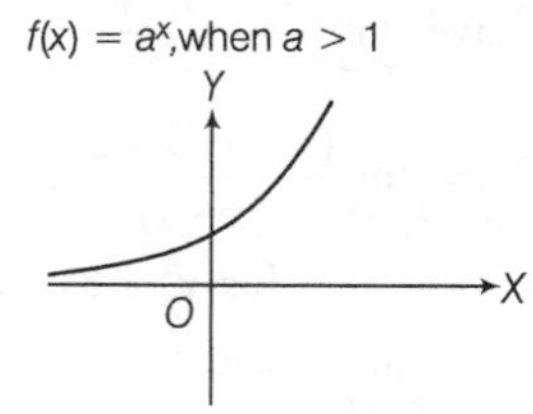

Figure 5.9

Now, see the graph for a^x, when $a > 1$. This graph appears to touch X-axis in the negative side of X-axis and thereafter it increases rapidly. This is why because $\lim_{x \to -\infty} a^x \to 0$, again you will also find the result,

$$\lim_{x \to \infty} a^x \to \infty$$

Thus, we have $\lim_{x \to \infty} a^x = \begin{cases} \infty, & \text{if } a > 1 \\ 1, & \text{if } a = 1 \\ 0, & \text{if } 0 \le a < 1 \end{cases}$

This type of problems are solved by taking the highest power of the terms tending to infinity as common from numerator and denominator. That is after they are cancelled and the rest output is the result (or apply L'Hospital's rule).

Example 9 Evaluate $\lim_{x \to \infty} \frac{x^2 + 5}{x^2 + 4x + 3}$.

Sol. Let $L = \lim_{x \to \infty} \frac{x^2 + 5}{x^2 + 4x + 3}$

Dividing numerator and denominator by x^2, we get

$$L = \lim_{x \to \infty} \frac{1 + \frac{5}{x^2}}{1 + \frac{4}{x} + \frac{3}{x^2}} = \frac{1 + 0}{1 + 0 + 0} = 1$$

[because $\frac{K}{x} \to 0$, when $x \to \infty$, where K is any constant]

Aliter We have, $L = \lim_{x \to \infty} \frac{x^2 + 5}{x^2 + 4x + 3}$ $\left[\frac{\infty}{\infty} \text{ form}\right]$

Applying L'Hospital's rule, $L = \lim_{x \to \infty} \frac{2x}{2x + 4}$ $\left[\frac{\infty}{\infty} \text{ form}\right]$

Again, applying L'Hospital's rule, $L = \lim_{x \to \infty} \frac{2}{2} = 1$

Example 10 Evaluate $\lim_{n\to\infty} \frac{(n+2)!+(n+1)!}{(n+2)!-(n+1)!}$.

Sol. We have, $\lim_{n\to\infty} \frac{(n+2)(n+1)!+(n+1)!}{(n+2)(n+1)!-(n+1)!}$

$$= \lim_{n\to\infty} \frac{(n+1)![n+2+1]}{(n+1)![n+2-1]} = \lim_{n\to\infty} \frac{(n+3)}{(n+1)} \quad \left[\frac{\infty}{\infty} \text{ form}\right]$$

$$= \lim_{n\to\infty} \frac{1+3/n}{1+1/n} = \frac{1+0}{1+0} = 1 \quad \left[\because \frac{1}{n} \to 0, \text{ as } n \to \infty\right]$$

(b) $\infty - \infty$ Form

Such problems are simplified (generally rationalised) first, thereafter they generally acquire $\left(\frac{\infty}{\infty}\right)$ form.

Example 11 Evaluate $\lim_{x\to\infty} (x-\sqrt{x^2+x})$.

Sol. We have, $\lim_{x\to\infty} (x-\sqrt{x^2+x})$

$$= \lim_{x\to\infty} \frac{x-\sqrt{x^2+x}}{1} \times \frac{x+\sqrt{x^2+x}}{x+\sqrt{x^2+x}} \quad [\infty-\infty \text{ form}]$$

$$= \lim_{x\to\infty} \frac{x^2-(x^2+x)}{x+\sqrt{x^2+x}} = \lim_{x\to\infty} \frac{-x}{x+\sqrt{x^2+x}}$$

$$= \lim_{x\to\infty} \frac{-x}{x\left\{1+\sqrt{1+\frac{1}{x}}\right\}}$$

$$= \lim_{x\to\infty} \frac{-1}{1+\sqrt{1+\frac{1}{x}}} = -\frac{1}{2} \quad \left[\because \frac{1}{x} \to 0, \text{ as } x \to \infty\right]$$

Example 12 Evaluate $\lim_{x\to\infty} (\sqrt{x^2+x+1}-\sqrt{x^2+1})$.

Sol. We have, $\lim_{x\to\infty} (\sqrt{x^2+x+1}-\sqrt{x^2+1})$ $[\infty-\infty \text{ form}]$

$$= \lim_{x\to\infty} \frac{(\sqrt{x^2+x+1}-\sqrt{x^2+1})}{1} \times \frac{(\sqrt{x^2+x+1}+\sqrt{x^2+1})}{(\sqrt{x^2+x+1}+\sqrt{x^2+1})}$$

$$= \lim_{x\to\infty} \frac{(x^2+x+1)-(x^2+1)}{\sqrt{x^2+x+1}+\sqrt{x^2+1}}$$

$$= \lim_{x\to\infty} \frac{x}{\sqrt{x^2+x+1}+\sqrt{x^2+1}}$$

$$= \lim_{x\to\infty} \frac{1}{\sqrt{1+\frac{1}{x}+\frac{1}{x^2}}+\sqrt{1+\frac{1}{x^2}}} = \frac{1}{1+1} = \frac{1}{2}$$

$$\left[\because \frac{1}{x} \to 0, \text{ as } x \to \infty\right]$$

An Important Result

If m, n are positive integers and $a_0, b_0 \neq 0$ and non-zero real numbers, then

$$\lim_{x\to\infty} \frac{a_0x^m + a_1x^{m-1} + \ldots + a_{m-1}x + a_m}{b_0x^n + b_1x^{n-1} + \ldots + b_{n-1}x + b_n} = \begin{cases} 0, & m < n \\ \frac{a_0}{b_0}, & m = n \\ \infty, & m > n \text{ when } a_0b_0 > 0 \\ -\infty, & m > n \text{ when } a_0b_0 < 0 \end{cases}$$

Example 13 Evaluate $\lim_{x\to\infty} \frac{ax^2+b}{x+1}$, when $a \geq 0$.

Sol. Here, if $a \neq 0$; $\lim_{x\to\infty} \frac{ax^2+b}{x+1}$

[as degree of numerator > degree of denominator]

$$= \lim_{x\to\infty} \frac{ax^2+b}{x+1} = \infty \quad [\text{as } a > 0]$$

Again, if $a = 0$; $\lim_{x\to\infty} \frac{0\cdot x^2+b}{x+1} = \lim_{x\to\infty} \frac{b}{x+1} = 0$

[as degree of numerator < degree of denominator]

$$\therefore \quad \lim_{x\to\infty} \frac{ax^2+b}{x+1} = \begin{cases} \infty, & a > 0 \\ 0, & a = 0 \end{cases}$$

Example 14 If $\lim_{x\to\infty} \left(\frac{x^2+1}{x+1} - ax - b\right) = 0$, find the values of a and b.

Sol. Given, $\lim_{x\to\infty} \left(\frac{x^2+1}{x+1} - ax - b\right) = 0$

$$\Rightarrow \lim_{x\to\infty} \frac{x^2+1-ax^2-ax-bx-b}{x+1} = 0$$

$$\Rightarrow \lim_{x\to\infty} \frac{x^2(1-a)-x(a+b)+(1-b)}{x+1} = 0$$

Since, the limit of above expression is zero.

$\therefore$ Degree of numerator < Degree of denominator.

So, numerator must be a constant, i.e. a zero degree polynomial.

$\therefore \quad 1-a=0$ and $a+b=0$

Hence, $\quad a=1$ and $b=-1$

Example 15 If $\lim_{x\to\infty} \left(\frac{x^2+1}{x+1} - ax - b\right) = 2$, find the values of a and b.

Sol. We have, $\lim_{x\to\infty} \left(\frac{x^2+1}{x+1} - ax - b\right)$

$$= \lim_{x\to\infty} \frac{x^2(1-a)-x(a+b)+(1-b)}{(x+1)} = 2$$

Since, limit of above expression is a finite non-zero number.

$\therefore$ Degree of numerator = Degree of denominator

$$\Rightarrow \quad 1-a=0 \Rightarrow a=1$$

Putting $a=1$ in above limit, we get

$$\lim_{x\to\infty} \frac{-x\,(1+b)+(1+b)}{x+1}=2$$

$$\Rightarrow \quad -(1+b)=2 \Rightarrow b=-3$$

Hence, $a=1$ and $b=-3$.

| Example 16 If $\lim_{x\to\infty}\left(\frac{x^2+1}{x+1}-ax-b\right)=\infty$, find a and b.

Sol. Given, $\lim_{x\to\infty}\left(\frac{x^2+1}{x+1}-ax-b\right)=\infty$

$$\Rightarrow \quad \lim_{x\to\infty}\frac{x^2(1-a)-x(a+b)+(1-b)}{x+1}=\infty$$

The limit of above expression is infinity.

$\therefore$ Degree of numerator > Degree of denominator

$$\Rightarrow \quad 1-a>0 \Rightarrow a\neq 1$$

Hence, $a<1$ and b can assume any real value.

| Example 17 Let $S_n=1+2+3+\cdots+n$

and $P_n=\frac{S_2}{S_2-1}\cdot\frac{S_3}{S_3-1}\cdot\frac{S_4}{S_4-1}\cdots\frac{S_n}{S_n-1}$,

where $n\in N\,(n\geq 2)$. Find $\lim_{n\to\infty}P_n$.

Sol. As, $S_n=\frac{n(n+1)}{2}$

$$\Rightarrow \quad S_n-1=\frac{n(n+1)}{2}-1=\frac{n^2+n-2}{2}=\frac{(n+2)(n-1)}{2}$$

$$\therefore \quad \frac{S_n}{S_n-1}=\left(\frac{n}{n-1}\right)\left(\frac{n+1}{n+2}\right)$$

$$\Rightarrow \quad P_n=\left(\frac{2}{1}\cdot\frac{3}{2}\cdot\frac{4}{3}\cdots\frac{n}{n-1}\right)\left(\frac{3}{4}\cdot\frac{4}{5}\cdot\frac{5}{6}\cdots\frac{n+1}{n+2}\right)$$

$$=\left(\frac{n}{1}\right)\left(\frac{3}{n+2}\right)$$

$$\therefore \quad \lim_{n\to\infty}P_n=\lim_{n\to\infty}\frac{3n}{n+2}=3$$

| Example 18 If $\lim_{n\to\infty}\frac{1}{(\sin^{-1}x)^n+1}=1$, x lies in the interval

(a) $(-\sin 1, \sin 1)$ (b) $(-1, 1)$

(c) $(0, 1)$ (d) $(-1, 0)$

Sol. Here, $\lim_{n\to\infty}\frac{1}{(\sin^{-1}x)^n+1}=1$ is possible only, if

$$-1<\sin^{-1}x<1$$

$$\Rightarrow \quad x\in(-\sin 1, \sin 1)$$

Hence, (a) is the correct answer.

Exercise for Session 1

1. If $f(a)=2, f'(a)=1, g(a)=-1, g'(a)=-2$, then $\lim_{x\to a}\frac{g(x)f(a)-g(a)f(x)}{x-a}$ is

(a) -5 (b) 3 (c) -3 (d) 5

2. The value of $\lim_{x\to 0}\frac{x\cos x-\log(1+x)}{x^2}$ is

(a) 1 (b) $\frac{1}{4}$ (c) $\frac{1}{2}$ (d) None of these

3. The value of $\lim_{x\to 0}\frac{e^{x^2}-\cos x}{x^2}$ is

(a) $\frac{3}{2}$ (b) $-\frac{3}{2}$ (c) $\frac{1}{2}$ (d) $-\frac{1}{2}$

4. The value of $\lim_{x\to a}\frac{\cos x-\cos a}{\cot x-\cot a}$ is

(a) $-\sin^3 a$ (b) $\cos^3 a$ (c) $\sin^3 a$ (d) $\cot a$

5. The value of $\lim_{x \to 0} \left(\frac{1}{x^2} - \cot x\right)$ is

(a) 0 (b) 1

(c) $\frac{1}{4}$ (d) None of these

6. The value of $\lim_{x \to \infty} (\sqrt{a^2x^2 + ax + 1} - \sqrt{a^2x^2 + 1}), (a > 0)$ is

(a) $\frac{1}{2}$ (b) $-\frac{1}{2}$

(c) Doesn't exist (d) None of these

7. The value of $\lim_{n \to \infty} \frac{1^3 + 2^3 + 3^3 + \ldots + n^3}{(n^2+1)^2}$ is

(a) $\frac{1}{4}$ (b) $\frac{1}{2}$

(c) $\frac{1}{2\sqrt{2}}$ (d) None of these

8. The value of $\lim_{n \to \infty} \frac{1 \cdot n + 2 \cdot (n-1) + 3 \cdot (n-2) + \ldots + n \cdot 1}{1^2 + 2^2 + \ldots + n^2}$ is

(a) 1 (b) -1

(c) $\frac{1}{\sqrt{2}}$ (d) $\frac{1}{2}$

9. The value of $\lim_{n \to \infty} \frac{a^n + b^n}{a^n - b^n}$, (where $a > b > 1$) is

(a) 1 (b) -1

(c) $\frac{1}{2}$ (d) $\frac{1}{\sqrt{2}}$

Session 2

Trigonometric Limits

Trigonometric Limits

To evaluate trigonometric limits the following results are given below

(i) $\lim_{x\to 0} \frac{\sin x}{x} = 1$ (ii) $\lim_{x\to 0} \frac{\tan x}{x} = 1$

(iii) $\lim_{x\to 0} \frac{\sin^{-1} x}{x} = 1$ (iv) $\lim_{x\to 0} \frac{\tan^{-1} x}{x} = 1$

(v) $\lim_{x\to 0} \frac{\sin x^0}{x} = \frac{\pi}{180°}$ (vi) $\lim_{x\to 0} \cos x = 1$

(vii) $\lim_{x\to a} \frac{\sin(x-a)}{x-a} = 1$ (viii) $\lim_{x\to a} \frac{\tan(x-a)}{x-a} = 1$

Example 19 Evaluate $\lim_{x\to 0} \frac{1-\cos x}{x^2}$.

Sol. We have, $\lim_{x\to 0} \frac{1-\cos x}{x^2}$

$$= \lim_{x\to 0} \frac{2\sin^2 x/2}{x^2} = \lim_{x\to 0} \frac{2}{4} \frac{\sin^2 x/2}{x^2/4} \quad \left[\frac{0}{0} \text{ form}\right]$$

$$= \lim_{x\to 0} \frac{1}{2}\left(\frac{\sin x/2}{x/2}\right)^2 = \frac{1}{2}(1)^2 = \frac{1}{2}$$

Example 20 Solve $\lim_{x\to 0} \frac{1-\cos(1-\cos x)}{\sin^4 x}$

Sol. $\lim_{x\to 0} \frac{1-\cos(1-\cos x)}{\sin^4 x} = \lim_{x\to 0} \frac{1-\cos\left(2\sin^2\frac{x}{2}\right)}{x^4} \Bigg/ \frac{\sin^4 x}{x^4}$

$$= \lim_{x\to 0} \frac{1-\cos\left(2\sin^2\frac{x}{2}\right)}{x^4} \Bigg/ \lim_{x\to 0}\left(\frac{\sin x}{x}\right)^4$$

$$= \lim_{x\to 0} \frac{2\sin^2\left(\sin^2\frac{x}{2}\right)}{x^4} \Bigg/ 1$$

$$= \lim_{x\to 0} 2\cdot\left(\frac{\sin\left(\sin^2\frac{x}{2}\right)}{\sin^2\frac{x}{2}} \cdot \frac{\sin^2\frac{x}{2}}{4\cdot\frac{x^2}{4}}\right)^2$$

$$= 2\times\frac{1}{4^2} = \frac{1}{8}$$

Example 21 Evaluate $\lim_{x\to\infty} 2^{-x}\sin(2^x)$.

Sol. Since, $2^{-x} = \frac{1}{2^x}$. We know that, as $x\to\infty$, $2^x\to\infty$

$\therefore$ The given limit $= 0 \times$ [A finite number between -1 and $+1$] $=0$

Hence, $\lim_{x\to\infty} \frac{\sin(2^x)}{(2^x)} = 0$

Example 22 Evaluate $\lim_{x\to\infty} e^x \sin(d/e^x)$.

Sol. When $x\to\infty$, $e^x\to\infty$

But, angle of sine $= \frac{d}{e^x} = \frac{\text{finite}}{\infty} = 0$

$\therefore$ The given limit $= \lim_{x\to\infty} \frac{\sin(d/e^x)}{1/e^x} = \lim_{\frac{d}{e^x}\to 0} \frac{\sin d/e^x}{d/e^x}\times d$

$$= 1\times d = d$$

Example 23 Evaluate $\lim_{x\to\infty} \sqrt{\frac{x-\sin x}{x+\cos^2 x}}$.

Sol. We have, $\lim_{x\to\infty} \sqrt{\frac{x-\sin x}{x+\cos^2 x}} = \lim_{x\to\infty} \sqrt{\frac{x\left(1-\frac{\sin x}{x}\right)}{x\left(1+\frac{\cos^2 x}{x}\right)}}$

$$= \lim_{x\to\infty} \sqrt{\frac{1-\frac{\sin x}{x}}{1+\frac{\cos^2 x}{x}}} = \sqrt{\frac{1-0}{1+0}} = 1$$

Example 24 Evaluate $\lim_{x\to y} \frac{\sin^2 x - \sin^2 y}{x^2 - y^2}$.

Sol. $\lim_{x\to y} \frac{\sin^2 x - \sin^2 y}{x^2-y^2} = \lim_{x\to y} \frac{\sin(x+y)\sin(x-y)}{(x+y)(x-y)}$

$$= \lim_{x\to y} \frac{\sin(x+y)}{(x+y)} \times \lim_{x\to y} \frac{\sin(x-y)}{(x-y)} = \frac{\sin(2y)}{2y}\times 1$$

[as $x\to y \Rightarrow (x-y)\to 0$, but $x+y\to 2y$]

$$= \frac{\sin 2y}{2y}$$

Example 25

$\lim_{x \to \infty} [(x+5)\tan^{-1}(x+5) - (x+1)\tan^{-1}(x+1)]$ is equal to

(a) π (b) 2π

(c) $\frac{\pi}{2}$ (d) None of these

Sol. Here, $\lim_{x \to \infty} ((x+5)\tan^{-1}(x+5) - (x+1)\tan^{-1}(x+1))$

$$= \lim_{x \to \infty} (x+5)\left\{\frac{\pi}{2} - \cot^{-1}(x+5)\right\} - (x+1)\left\{\frac{\pi}{2} - \cot^{-1}(x+1)\right\}$$

$$= \lim_{x \to \infty} 2\pi - \frac{\tan^{-1}\left(\frac{1}{x+5}\right)}{\left(\frac{1}{x+5}\right)} + \frac{\tan^{-1}\left(\frac{1}{x+1}\right)}{\left(\frac{1}{x+1}\right)} = 2\pi$$

Hence, (b) is the correct answer.

Example 26 Evaluate $\lim_{x \to 0} \frac{\sin(\pi \cos^2 x)}{x^2}$.

[IIT JEE 2001]

Sol. $\lim_{x \to 0} \frac{\sin(\pi \cos^2 x)}{x^2} = \lim_{x \to 0} \frac{\sin\{\pi(1-\sin^2 x)\}}{x^2}$

$$= \lim_{x \to 0} \frac{\sin(\pi - \pi\sin^2 x)}{x^2}$$

$$= \lim_{x \to 0} \frac{\sin(\pi\sin^2 x)}{x^2} = \lim_{x \to 0}\left\{\frac{\sin(\pi\sin^2 x)}{\pi\sin^2 x} \times \frac{\pi}{1} \times \frac{\sin^2 x}{x^2}\right\}$$

$$= \lim_{x \to 0} \frac{\sin(\pi\sin^2 x)}{\pi\sin^2 x} \times \pi \times \lim_{x \to 0} \frac{\sin^2 x}{x^2}$$

$$= 1 \times \pi \times 1 = \pi$$

Example 27 Let $a = \min\{x^2 + 2x + 3, x \in R\}$ and $b = \lim_{\theta \to 0} \frac{1-\cos\theta}{\theta^2}$. The value of $\sum_{r=0}^{n} a^r \cdot b^{n-r}$ is

Sol. Here, $a = \min\{x^2 + 2x + 3, x \in R\}$

i.e. $x^2 + 2x + 3 = x^2 + 2x + 1 + 2$

$$= (x+1)^2 + 2$$

$\therefore$ a is minimum, when $x = -1$, i.e. $a = 2$

Again, $b = \lim_{\theta \to 0} \frac{2\sin^2\theta/2}{\theta^2} = \lim_{\theta \to 0} \frac{2}{4} \cdot \frac{\sin^2\theta/2}{\theta^2/4}$

$$= \lim_{\theta \to 0} \frac{2}{4} \cdot \frac{\sin^2\theta/2}{(\theta/2)^2} = \frac{1}{2}$$

Hence, $\sum_{r=0}^{n} a^r b^{n-r} = \sum_{r=0}^{n} 2^r \cdot \left(\frac{1}{2}\right)^{n-r}$

$$\Rightarrow \quad \frac{1}{2^n}\{2^0 + 2^2 + 2^4 + \dots + 2^{2n}\}$$

$$\Rightarrow \quad \frac{1}{2^n}\left\{\frac{1(4^{n+1}-1)}{4-1}\right\} = \frac{4^{n+1}-1}{2^n \times 3}$$

[i.e. sum of $(n+1)$ terms of GP]

$$\therefore \quad \sum_{r=0}^{n} a^r b^{n-r} = \frac{4^{n+1}-1}{2^n \times 3}$$

Example 28 Evaluate

$$\lim_{n \to \infty} \cos\left(\frac{x}{2}\right)\cos\left(\frac{x}{4}\right)\cos\left(\frac{x}{8}\right)\dots\cos\left(\frac{x}{2^n}\right).$$

Sol. Here, $\lim_{n \to \infty} \cos\left(\frac{x}{2^n}\right)\cos\left(\frac{x}{2^{n-1}}\right)\dots\cos\left(\frac{x}{8}\right)\cos\left(\frac{x}{4}\right)\cos\left(\frac{x}{2}\right)$

We know, $\cos A \cos 2A \cos 2^2 A \dots \cos 2^{n-1} A = \frac{\sin 2^n A}{2^n \sin A}$

Thus, $\lim_{n \to \infty} \cos\left(\frac{x}{2^n}\right)\cos\left(\frac{x}{2^{n-1}}\right)\dots\cos\left(\frac{x}{8}\right)\cos\left(\frac{x}{4}\right)\cos\left(\frac{x}{2}\right)$

$$= \lim_{n \to \infty} \frac{\sin 2^n\left(\frac{x}{2^n}\right)}{2^n \sin\left(\frac{x}{2^n}\right)} = \lim_{n \to \infty} \frac{\sin x}{2^n \sin\left(\frac{x}{2^n}\right)}$$

$$= \sin x \cdot \lim_{n \to \infty} \frac{1}{2^n \sin\left(\frac{x}{2^n}\right)} \cdot \frac{x}{x}$$

$$= \frac{\sin x}{x} \cdot \lim_{n \to \infty} \frac{1}{2^n \sin\left(\frac{x}{2^n}\right)} \cdot x$$

$$= \frac{\sin x}{x} \cdot 1 = \frac{\sin x}{x} \quad \left[n \to \infty, \frac{x}{2^n} \to 0 \text{ and } \lim_{\theta \to 0} \frac{\theta}{\sin\theta} = 1\right]$$

Exercise for Session 2

1. If $\lim\limits_{x \to 0} (x^{-3} \sin 3x + ax^{-2} + b)$ exists and is equal to zero, then

(a) $a = -3,\ b = \dfrac{9}{2}$ (b) $a = 3, b = \dfrac{9}{2}$

(c) $a = -3, b = \dfrac{-9}{2}$ (d) None of these

2. The value of $\lim\limits_{x \to a} \dfrac{x \sin a - a \sin x}{x - a}$ is

(a) $a \sin a - \cos a$ (b) $\sin a - a \cos a$

(c) $\cos a + a \sin a$ (d) $\sin a + a \cos a$

3. The value of $\lim\limits_{x \to \pi} \dfrac{\sqrt{2 + \cos x} - 1}{(\pi - x)^2}$ is

(a) $\dfrac{1}{4}$ (b) $\dfrac{1}{2}$

(c) 2 (d) Doesn't exist

4. The value of $\lim\limits_{\theta \to \pi/4} \dfrac{(\sqrt{2} - \cos\theta - \sin\theta)}{(4\theta - \pi)^2}$ is

(a) $\dfrac{1}{16\sqrt{2}}$ (b) $\dfrac{1}{16}$

(c) $\dfrac{1}{8\sqrt{2}}$ (d) $\dfrac{1}{2\sqrt{2}}$

5. The value of $\lim\limits_{x \to \pi/4} \dfrac{(\cos x + \sin x)^3 - 2\sqrt{2}}{1 - \sin 2x}$ is

(a) $\dfrac{\sqrt{3}}{2}$ (b) $-\dfrac{3}{\sqrt{2}}$

(c) $\dfrac{1}{2}$ (d) $-\dfrac{1}{\sqrt{2}}$

Session 3

Logarithmic Limits, Exponential Limits

Logarithmic Limits

In this section, we will deal with the problems based on expansion of logarithmic series, which is given below

$$\log(1+x) = x - \frac{x^2}{2} + \frac{x^3}{3} + \ldots \infty$$

where, $-1 \le x \le 1$ and it should be noted that the expansion is true only if the base is e. To evaluate the logarithmic limit, we use $\lim_{x \to 0} \frac{\log(1+x)}{x} = 1$.

Example 29 Evaluate $\lim_{x \to a} \frac{\log\{1+(x-a)\}}{(x-a)}$.

Sol. We have, $\lim_{x \to a} \frac{\log\{1+(x-a)\}}{(x-a)}$

Let $x - a = y$, when $x \to a$; $y \to 0$

$\therefore$ The given limit $= \lim_{y \to 0} \frac{\log\{1+y\}}{y} = 1$

Example 30 Evaluate $\lim_{h \to 0} \frac{\log_{10}(1+h)}{h}$.

Sol. We have, $\lim_{h \to 0} \frac{\log_{10}(1+h)}{h} = \lim_{h \to 0} \frac{\log_e(1+h) \times \log_{10} e}{h}$

$$= \lim_{h \to 0} \frac{\log_e(1+h)}{h} \times \log_{10} e$$

$$= \log_{10} e \times 1 = \log_{10} e \qquad \left[\because \lim_{x \to 0} \frac{\log(1+x)}{x} = 1\right]$$

Example 31 Evaluate $\lim_{x \to 0} \frac{\log(5+x) - \log(5-x)}{x}$.

Sol. We have, $\lim_{x \to 0} \frac{\log\left\{5\left(1+\frac{x}{5}\right)\right\} - \log\left\{5\left(1-\frac{x}{5}\right)\right\}}{x}$

$$= \lim_{x \to 0} \frac{\log 5 + \log\left(1+\frac{x}{5}\right) - \log 5 - \log\left(1-\frac{x}{5}\right)}{x}$$

$$= \lim_{x \to 0} \frac{\log\left(1+\frac{x}{5}\right)}{5\left(\frac{x}{5}\right)} - \frac{\log\left(1-\frac{x}{5}\right)}{-5\left(-\frac{x}{5}\right)} = \frac{1}{5} + \frac{1}{5} = \frac{2}{5}$$

$$\left[\because \lim_{x \to 0} \frac{\log(1+x)}{x} = 1\right]$$

Example 32 Evaluate $\lim_{h \to 0} \frac{\log_e(1+2h) - 2\log_e(1+h)}{h^2}$.

[IIT JEE 1999]

Sol. We have, $\lim_{h \to 0} \frac{\log_e(1+2h) - 2\log_e(1+h)}{h^2}$

$$= \lim_{h \to 0} \frac{\left[(2h) - \frac{(2h)^2}{2} + \frac{(2h)^3}{3} - \ldots \infty\right] - 2\left(h - \frac{h^2}{2} + \frac{h^3}{3} - \ldots\right)}{h^2}$$

$$= \lim_{h \to 0} \frac{-h^2 + 2h^3 - \ldots}{h^2}$$

$$= \lim_{h \to 0} \frac{h^2\{-1 + 2h - \ldots\}}{h^2}$$

$$= \lim_{h \to 0} \{-1 + 2h - \ldots\} = -1$$

Example 33 Solve $\lim_{x \to \infty} \left\{x - x^2 \cdot \log\left(1 + \frac{1}{x}\right)\right\}$.

Sol. Here, put $x = \frac{1}{y}$ in $\lim_{x \to \infty} \left\{x - x^2 \cdot \log\left(1 + \frac{1}{x}\right)\right\}$

$$= \lim_{y \to 0} \left\{\frac{1}{y} - \frac{\log(1+y)}{y^2}\right\}$$

$$= \lim_{y \to 0} \frac{y - \log(1+y)}{y^2}$$

$$= \lim_{y \to 0} \frac{y - \left\{y - \frac{y^2}{2} + \frac{y^3}{3} - \frac{y^4}{4} + \ldots\right\}}{y^2}$$

$$= \lim_{y \to 0} \frac{y^2\left\{\frac{1}{2} - \frac{y}{3} + \ldots\right\}}{y^2}$$

$$= \lim_{y \to 0} \left\{\frac{1}{2} - \frac{y}{3} + \ldots\right\} = \frac{1}{2}$$

Remark

If we solve above question as

$$\lim_{y \to 0} \left\{\frac{1}{y} - \frac{\log(1+y)}{y^2}\right\} = \lim_{y \to 0} \left\{\frac{1}{y} - \frac{\log(1+y)}{y} \cdot \frac{1}{y}\right\}$$

$$\left[\text{as } \lim_{y \to 0} \frac{\log(1+y)}{y} = 1\right]$$

$= \lim_{y \to 0} \left(\frac{1}{y} - \frac{1}{y}\right) = 0$, then it is not correct.

Exponential Limits

There are two types of exponential limits discussed below

(i) Based on Series Expansion

$$e^x = 1 + x + \frac{x^2}{2!} + \frac{x^3}{3!} + \dots \infty$$

To evaluate the exponential limit, we use the following results.

(a) $\lim_{x \to 0} \frac{e^x - 1}{x} = 1$ (b) $\lim_{x \to 0} \frac{a^x - 1}{x} = \log_e a$

Example 34 Evaluate $\lim_{x \to 0} \frac{a^x - b^x}{x}$.

Sol. We have, $\lim_{x \to 0} \frac{a^x - b^x}{x} = \lim_{x \to 0} \frac{(a^x - 1) - (b^x - 1)}{x}$

$$= \lim_{x \to 0} \frac{a^x - 1}{x} - \lim_{x \to 0} \frac{b^x - 1}{x} = \log a - \log b = \log (a/b)$$

Example 35 Evaluate $\lim_{x \to 0} \frac{(ab)^x - a^x - b^x + 1}{x^2}$.

Sol. We have, $\lim_{x \to 0} \frac{(ab)^x - a^x - b^x + 1}{x^2} = \lim_{x \to 0} \frac{a^x b^x - a^x - b^x + 1}{x^2}$

$$= \lim_{x \to 0} \frac{a^x (b^x - 1) - (b^x - 1)}{x^2}$$

$$= \lim_{x \to 0} \frac{(a^x - 1)}{x} \times \lim_{x \to 0} \frac{(b^x - 1)}{x} = \log a \times \log b$$

$$\left[\because \lim_{x \to 0} \frac{a^x - 1}{x} = \log_e a \right]$$

Example 36 Evaluate $\lim_{x \to 0} \frac{e^{\tan x} - e^x}{\tan x - x}$.

Sol. We have, $\lim_{x \to 0} \frac{e^{\tan x} - e^x}{\tan x - x} = \lim_{x \to 0} \frac{e^x \times e^{(\tan x - x)} - e^x}{(\tan x - x)}$

$$= \lim_{x \to 0} \frac{e^x \{e^{\tan x - x} - 1\}}{(\tan x - x)}$$

$$= e^0 \times 1 \text{ [as } x \to 0, \tan x - x \to 0]$$

$$= 1 \times 1 = 1$$

Example 37 Evaluate $\lim_{x \to 0} \frac{ae^x - b}{x} = 2$. Find a and b.

Sol. Given, $\lim_{x \to 0} \frac{ae^x - b}{x} = \lim_{x \to 0} \frac{a\left(1 + \frac{x}{1!} + \frac{x^2}{2!} + \dots \infty\right) - b}{x} = 2$

$$\Rightarrow \quad \lim_{x \to 0} \frac{(a - b) + xa + \frac{ax^2}{2!} + \dots \infty}{x} = 2$$

Since, limit is finite, so $(a - b) = 0 \Rightarrow b = a$

$$\therefore \quad \lim_{x \to 0} \frac{xa + \frac{ax^2}{2!} + \dots \infty}{x} = 2$$

$$\Rightarrow \quad \lim_{x \to 0} a + \frac{ax}{2!} + \dots \infty = 2 \Rightarrow a = 2$$

$$\therefore \quad b = 2$$

Example 38 Solve $\lim_{x \to 0} \frac{a \sin x - bx + cx^2 + x^3}{2x^2 \log (1 + x) - 2x^3 + x^4}$, if it exists and is finite, also find a, b and c.

Sol. Let $L = \lim_{x \to 0} \frac{a \sin x - bx + cx^2 + x^3}{2x^2 \log (1 + x) - 2x^3 + x^4}$

$$= \lim_{x \to 0} \frac{a\left(x - \frac{x^3}{3!} + \frac{x^5}{5!} - \dots\right) - bx + cx^2 + x^3}{2x^2\left(x - \frac{x^2}{2} + \frac{x^3}{3} - \frac{x^4}{4} + \dots\right) - 2x^3 + x^4}$$

$$= \lim_{x \to 0} \frac{(a - b)x + cx^2 + \left(1 - \frac{a}{3!}\right)x^3 + \frac{a}{5!}x^5 + \dots}{\frac{2}{3}x^5 - \frac{1}{2}x^6 + \dots}$$

For finite limit, $a - b = 0$, $c = 0$, $1 - \frac{a}{3!} = 0$, i.e. $a = b = 6$, $c = 0$]

$$\therefore \quad L = \lim_{x \to 0} \frac{\frac{a}{5!} + \text{higher powers of } x}{\frac{2}{3} - \frac{1}{2}x + \dots}$$

$$= \frac{a}{5!} \cdot \frac{3}{2} = \frac{a}{80} = \frac{6}{80} = \frac{3}{40} \quad [\because a = 6]$$

So, $\lim_{x \to 0} \frac{a \sin x - bx + cx^2 + x^3}{2x^2 \log (1 + x) - 2x^3 + x^4} = \frac{3}{40}$

Where, $a = 6 = b, c = 0$

Example 39 Find the values of a, b and c such that

$$\lim_{x \to 0} \frac{axe^x - b \log (1 + x) + cxe^{-x}}{x^2 \sin x} = 2.$$

Sol. We have, $\lim_{x \to 0} \frac{axe^x - b\log (1 + x) + c\, xe^{-x}}{x^2 \sin x} = 2$

Using the expansion, we have

$$\lim_{x \to 0} \frac{\left[ax\left(1 + x + \frac{x^2}{2!} + \dots\right) - b\left(x - \frac{x^2}{2} + \frac{x^3}{3} - \dots\right) + cx\left(1 - \frac{x}{1!} + \frac{x^2}{2!} - \frac{x^3}{3!} + \dots\right) \right]}{x^2\left(x - \frac{x^3}{3!} + \frac{x^5}{5!} - \dots\right)} = 2$$

$$\lim_{x\to 0}\frac{x(a-b+c)-x^2\left(a+\frac{b}{2}-c\right)+x^3\left(\frac{a}{2}-\frac{b}{3}+\frac{c}{2}\right)+\ldots}{x^2\left(x-\frac{x^3}{3!}+\frac{x^5}{5!}-\ldots\right)}=2$$

Now, above limit would exist, if least power in numerator is greater than or equal to least power in denominator, i.e. Coefficient of x and x^2 must be zero and coefficient of x^3 should be 2.

i.e. $a-b+c=0,\ a+\frac{b}{2}-c=0,\quad \frac{a}{2}-\frac{b}{3}+\frac{c}{2}=2$

On solving, we get $a=3$, $b=12$, $c=9$.

(ii) Evaluation of Exponential Limits of the Form 1^∞

To evaluate the exponential limits of the form 1^∞, we use the following results.

If $\lim_{x\to a} f(x)=\lim_{x\to a} g(x)=0$,

then $\lim_{x\to a}\{1+f(x)\}^{1/g(x)}=e^{\lim_{x\to a}\frac{f(x)}{g(x)}}$

or when $\lim_{x\to a} f(x)=1$ and $\lim_{x\to a} g(x)=\infty$

Then, $\lim_{x\to a}\{f(x)\}^{g(x)}=\lim_{x\to a}[1+f(x)-1]^{g(x)}$

$=e^{\lim_{x\to a}(f(x)-1)\,g(x)}$

Particular Cases

(i) $\lim_{x\to 0}(1+x)^{1/x}=e$ (ii) $\lim_{x\to\infty}\left(1+\frac{1}{x}\right)^x=e$

(iii) $\lim_{x\to 0}(1+\lambda x)^{1/x}=e^\lambda$ (iv) $\lim_{x\to\infty}\left(1+\frac{\lambda}{x}\right)^x=e^\lambda$

Example 40 Evaluate $\lim_{x\to\infty}\left(1+\frac{2}{x}\right)^x$.

Sol. We have, $\lim_{x\to\infty}\left(1+\frac{2}{x}\right)^x=e^{\lim_{x\to\infty}\frac{2}{x}\cdot x}=e^2$ $[1^\infty$ form$]$

Example 41 Evaluate $\lim_{x\to 1}(\log_3 3x)^{\log_x 3}$.

Sol. We have, $\lim_{x\to 1}(\log_3 3x)^{\log_x 3}=\lim_{x\to 1}(\log_3 3+\log_3 x)^{\log_x 3}$

$=\lim_{x\to 1}(1+\log_3 x)^{1/\log_3 x}$ $\left[\text{as } \log_b a=\frac{1}{\log_a b}\right]$

$=e^{\lim_{x\to 1}\log_3 x\times\frac{1}{\log_3 x}}$

$=e^1$

Example 42 Evaluate $\lim_{x\to a}\left(2-\frac{a}{x}\right)^{\tan\frac{\pi x}{2a}}$.

Sol. We have, $\lim_{x\to a}\left(2-\frac{a}{x}\right)^{\tan\frac{\pi x}{2a}}$

$=\lim_{x\to a}\left\{1+\left(1-\frac{a}{x}\right)\right\}^{\tan\frac{\pi x}{2a}}=e^{\lim_{x\to a}\left(1-\frac{a}{x}\right)\cdot\tan\frac{\pi x}{2a}}$ $[1^\infty$ form$]$

$=e^{\lim_{x\to a}\left(\frac{x-a}{x}\right)\cdot\tan\frac{\pi x}{2a}}=e^{\lim_{x\to a}f(x)}$ ...(i)

Let $x-a=h$, we get

$e^{\lim_{h\to 0}\left(\frac{h}{a+h}\right)\cdot\tan\frac{\pi}{2a}(a+h)}=e^{\lim_{h\to 0}\frac{h}{a+h}\cdot\tan\left(\frac{\pi}{2}+\frac{\pi h}{2a}\right)}$

$=e^{\lim_{h\to 0}\frac{h}{a+h}\left[-\cot\left(\frac{\pi h}{2a}\right)\right]}=e^{\lim_{h\to 0}\frac{-h}{(a+h)\tan(\pi h/2a)}\cdot\frac{\pi}{2a}\cdot\frac{1}{\pi/2a}}$

$=e^{\lim_{h\to 0}\frac{-2a}{\pi(a+h)}\cdot\frac{\pi h/2a}{\tan(\pi h/2a)}}=e^{\frac{-2a}{\pi(a)}}=e^{-2/\pi}$ $\left[\text{as }\lim_{x\to 0}\frac{x}{\tan x}=1\right]$

Example 43 Evaluate $\lim_{x\to\infty}\left(\frac{x+6}{x+1}\right)^{x+4}$.

Sol. We have, $\lim_{x\to\infty}\left(\frac{x+6}{x+1}\right)^{x+4}$

As $x\to\infty$, $\lim_{x\to\infty}\frac{x+6}{x+1}=1$ and $(x+4)\to\infty$ $[1^\infty$ form$]$

$\therefore \lim_{x\to\infty}\left(\frac{x+6}{x+1}\right)^{x+4}=\lim_{x\to\infty}\left[1+\left(\frac{x+6}{x+1}-1\right)\right]^{x+4}$

$=\lim_{x\to\infty}\left[1+\frac{5}{x+1}\right]^{x+4}=e^{\lim_{x\to\infty}\left(\frac{5}{x+1}\right)\cdot(x+4)}$

$=e^{\lim_{x\to\infty}5\cdot\frac{x+4}{x+1}}=e^{5\cdot(1)}=e^5$ $\left[\text{as } x\to\infty;\ \lim_{x\to\infty}\frac{x+4}{x+1}=1\right]$

Example 44 Evaluate $\lim_{x\to 0}(1+\tan^2\sqrt{x})^{\frac{1}{2x}}$.

Sol. We have, $\lim_{x\to 0}(1+\tan^2\sqrt{x})^{\frac{1}{2x}}=e^{\lim_{x\to 0}\tan^2\sqrt{x}\cdot\frac{1}{2x}}$

$=e^{\frac{1}{2}\lim_{x\to 0}\left(\frac{\tan\sqrt{x}}{\sqrt{x}}\right)^2}=e^{1/2}$ $\left[\text{as }\lim_{x\to 0}\frac{\tan x}{x}=1\right]$

Example 45 Evaluate $\lim_{x\to 0}\left\{\tan\left(\frac{\pi}{4}+x\right)\right\}^{\frac{1}{x}}$.

[IIT JEE 1993]

Sol. We have, $\lim_{x\to 0}\left\{\tan\left(\frac{\pi}{4}+x\right)\right\}^{\frac{1}{x}}=\lim_{x\to 0}\left\{\frac{\tan\frac{\pi}{4}+\tan x}{1-\tan\frac{\pi}{4}\tan x}\right\}^{\frac{1}{x}}$

$$= \lim_{x \to 0} \left\{\frac{1+\tan x}{1-\tan x}\right\}^{\frac{1}{x}} = \lim_{x \to 0} \left\{1 + \frac{2\tan x}{1-\tan x}\right\}^{\frac{1}{x}}$$

$$= e^{\lim_{x\to 0}\left(\frac{2\tan x}{1-\tan x}\right)\cdot\frac{1}{x}} = e^{\lim_{x\to 0} 2\cdot\frac{\tan x}{x}\cdot\frac{1}{1-\tan x}} = e^2$$

Example 46 Evaluate $\lim_{x\to 0}\left(\frac{1+5x^2}{1+3x^2}\right)^{1/x^2}$. *[IIT JEE 1996]*

Sol. We have, $\lim_{x\to 0}\left(\frac{1+5x^2}{1+3x^2}\right)^{1/x^2} = \lim_{x\to 0}\left(1+\frac{2x^2}{1+3x^2}\right)^{1/x^2}$ [1^∞ form]

$$= e^{\lim_{x\to 0}\left(\frac{2x^2}{1+3x^2}\right)\cdot\frac{1}{x^2}} = e^{\frac{2}{1+0}} = e^2$$

Example 47 Evaluate $\lim_{x\to\infty}\left(\frac{x-3}{x+2}\right)^x$. *[IIT JEE 2000]*

Sol. We have, $\lim_{x\to\infty}\left(\frac{x-3}{x+2}\right)^x = \lim_{x\to\infty}\left(1+\frac{(-5)}{x+2}\right)^x = e^{\lim_{x\to\infty}\left(\frac{-5}{x+2}\right)\cdot x}$

$$= e^{\lim_{x\to\infty}\frac{-5}{1+\frac{2}{x}}} = e^{-5}$$

Example 48 The value of

$$\lim_{x\to 0}\left\{\sin^2\left(\frac{\pi}{2-ax}\right)\right\}^{\sec^2\frac{\pi}{2-bx}}$$ is equal to

(a) $e^{-a/b}$ (b) e^{-a^2/b^2} (c) $a^{2a/b}$ (d) $e^{4a/b}$

Sol. Here, $\lim_{x\to 0}\left(\sin^2\left(\frac{\pi}{2-ax}\right)\right)^{\sec^2\left(\frac{\pi}{2-bx}\right)}$

$$= \lim_{x\to 0}\left\{1-\cos^2\left(\frac{\pi}{2-ax}\right)\right\}^{\sec^2\left(\frac{\pi}{2-bx}\right)}$$ [1^∞ form]

$$= e^{\lim_{x\to 0} -\left\{\cos^2\left(\frac{\pi}{2-ax}\right)\cdot\frac{1}{\cos^2\left(\frac{\pi}{2-bx}\right)}\right\}}$$

$$= e^{\lim_{x\to 0} -\left\{\frac{2\sin\left(\frac{\pi}{2-ax}\right)\cos\left(\frac{\pi}{2-ax}\right)\times\frac{\pi a}{(2-ax)^2}}{2\sin\left(\frac{\pi}{2-bx}\right)\cos\left(\frac{\pi}{2-bx}\right)\times\frac{\pi b}{(2-bx)^2}}\right\}}$$

[using L'Hospital's rule]

$$= e^{-\lim_{x\to 0}\frac{\sin\left(\frac{2\pi}{2-ax}\right)}{\sin\left(\frac{2\pi}{2-bx}\right)}\cdot\frac{a}{b}\cdot\frac{(2-bx)^2}{(2-ax)^2}} = e^{-\lim_{x\to 0}\frac{a}{b}\cdot\frac{(2-bx)^3}{(2-ax)^3}} = e^{-\frac{a}{b}}$$

Hence, (a) is the correct answer.

Example 49 The value of $\lim_{x\to 7/2}(2x^2-9x+8)^{\cot(2x-7)}$ is equal to

(a) $e^{5/2}$ (b) $e^{-5/2}$
(c) $e^{7/2}$ (d) $e^{3/2}$

Sol. Here, $\lim_{x\to 7/2}(2x^2-9x+8)^{\cot(2x-7)}$ [1^∞ form]

$$= \lim_{x\to 7/2}\{1+(2x^2-9x+7)\}^{\cot(2x-7)}$$

$$= e^{\lim_{x\to 7/2}(2x^2-9x+7)\cdot\cot(2x-7)}$$

$$= e^{\lim_{x\to 7/2}\frac{4x-9}{\sec^2(2x-7)\cdot 2}} = e^{5/2}$$

Hence, (a) is the correct answer.

Example 50 The value of $\lim_{x\to 1}\left(\tan\left(\frac{\pi}{4}+\log x\right)\right)^{\frac{1}{\log x}}$ is equal to

(a) e (b) e^{-1}
(c) e^2 (d) e^{-2}

Sol. Here, $\lim_{x\to 1}\left(\tan\left(\frac{\pi}{4}+\log x\right)\right)^{\frac{1}{\log x}}$ [1^∞ form]

$$= \lim_{x\to 1}\left(1+\frac{2\tan(\log x)}{1-\tan(\log x)}\right)^{\frac{1}{\log x}} = e^{\lim_{x\to 1}\frac{2\tan(\log x)}{\{1-\tan(\log x)\}}\cdot\frac{1}{\log x}}$$

$$= e^{2\lim_{x\to 1}\frac{\tan(\log x)}{\log x}\cdot\frac{1}{1-\tan(\log x)}} = e^{2\cdot(1)} = e^2$$

[as $x\to 1$, $\log x\to 0$]

Hence, (c) is the correct answer.

Example 51 The value of $\lim_{x\to 0}\left(\sin\frac{x}{m}+\cos\frac{3x}{m}\right)^{\frac{2m}{x}}$ is equal to

(a) e (b) 1 (c) e^{-1} (d) e^2

Sol. Here, $\lim_{x\to 0}\left(\sin\frac{x}{m}+\cos\frac{3x}{m}\right)^{\frac{2m}{x}}$

$$= \lim_{x\to 0}\left\{1+\left(\sin\frac{x}{m}+\cos\frac{3x}{m}-1\right)\right\}^{\frac{2m}{x}}$$ [1^∞ form]

$$= e^{\lim_{x\to 0}\frac{2m}{x}\left(\sin\frac{x}{m}+\cos\frac{3x}{m}-1\right)}$$

$$= e^{\lim_{x\to 0}\frac{2m\left(\frac{1}{m}\cos\left(\frac{x}{m}\right)-\frac{3}{m}\sin\frac{3x}{m}\right)}{1}} = e^{2m\left(\frac{1}{m}\right)} = e^2$$

Hence, (d) is the correct answer.

Example 52 The value of $\lim\limits_{n \to \infty} \left(\dfrac{a - 1 + \sqrt[n]{b}}{a} \right)^n$ $(a > 0, b > 0)$ is equal to

(a) $\sqrt[a]{b}$ (b) $\sqrt[b]{a}$ (c) $\sqrt{b}$ (d) $\sqrt{a}$

Sol. Here, $\lim\limits_{n \to \infty} \left(\dfrac{a - 1 + \sqrt[n]{b}}{a} \right)^n = \lim\limits_{n \to \infty} \left(1 + \dfrac{\sqrt[n]{b} - 1}{a} \right)^n$

$$= e^{\lim\limits_{n \to \infty} \left(\frac{\sqrt[n]{b} - 1}{a} \right) \cdot n} \qquad [1^\infty \text{ form}]$$

$$= e^{\lim\limits_{n \to \infty} \frac{1}{a} \cdot \frac{b^{1/n} - 1}{1/n}} = e^{\lim\limits_{n \to \infty} \frac{1}{a} \cdot \frac{b^{1/n} \log b \left(\frac{-1}{n^2} \right)}{-1/n^2}}$$

$$= e^{\frac{1}{a} \log_e b} = e^{\log_e b^{1/a}} = b^{1/a}$$

Hence, (a) is the correct answer.

Exercise for Session 3

1. The value of $\lim\limits_{x \to 0} \dfrac{e^x - e^{x \cos x}}{x + \sin x}$ is

(a) 0 (b) 1 (c) -1 (d) None of these

2. The value of $\lim\limits_{x \to y} \dfrac{x^y - y^x}{x^x - y^y}$ is

(a) $\dfrac{1 - \log x}{1 + \log x}$ (b) $\dfrac{1 - \log y}{1 + \log y}$ (c) $\dfrac{\log x - \log y}{\log x + \log y}$ (d) None of these

3. The value of $\lim\limits_{x \to 0} \dfrac{p^x - q^x}{r^x - s^x}$ is

(a) $\dfrac{1 - \log p}{1 + \log p}$ (b) $\dfrac{\log p - \log q}{\log r - \log s}$ (c) $\dfrac{\log p \cdot \log q}{\log r \cdot \log s}$ (d) None of these

4. The value of $\lim\limits_{x \to \infty} (x + 2) \tan^{-1}(x + 2) - (x \tan^{-1} x)$ is

(a) $\dfrac{\pi}{2}$ (b) Doesn't exist (c) $\dfrac{\pi}{4}$ (d) None of these

5. The value of $\lim\limits_{x \to 4} \dfrac{(\cos \alpha)^x - (\sin \alpha)^x - \cos 2\alpha}{x - 4}$, $\alpha \in \left(0, \dfrac{\pi}{2}\right)$ is

(a) $\log(\cos \alpha) + (\sin \alpha)^4 \log(\sin \alpha)$ (b) $(\cos^4 \alpha) \log(\cos \alpha) - (\sin \alpha)^4 \log(\sin \alpha)$

(c) $(\cos^4 \alpha) \log(\cos \alpha)$ (d) None of these

6. The value of $\lim\limits_{x \to 0} \left(\dfrac{1^x + 2^x + 3^x + \ldots + n^x}{n} \right)^{a/x}$ is

(a) $(n!)^{a/n}$ (b) $n!$ (c) $a^{n!}$ (d) Doesn't exist

7. If $\lim\limits_{x \to 0} (1 + ax + bx^2)^{2/x} = e^3$, the values of a and b are

(a) $a = \dfrac{3}{2}, b \in R$ (b) $a = \dfrac{1}{2}, b \in R$ (c) $a \in R, b \in R$ (d) None of these

8. If α and β are roots of $ax^2 + bx + c = 0$, the value of $\lim\limits_{x \to \alpha} (1 + ax^2 + bx + c)^{2/x - \alpha}$ is

(a) $e^{2a(\alpha - \beta)}$ (b) $e^{a(\alpha - \beta)}$ (c) $e^{\frac{2a}{3}(\alpha - \beta)}$ (d) None of these

9. The value of $\lim\limits_{n \to \infty} ((1.5)^n + [(1 + 0.0001)^{10000}]^n)^{1/n}$, where [.] denotes the greatest integer function, is

(a) 1 (b) $\dfrac{1}{2}$ (c) Doesn't exist (d) 2

10. The value of $\lim\limits_{x \to 0} |x|^{[\cos x]}$, where [·] denotes the greatest integer function, is

(a) 0 (b) Doesn't exist (c) 1 (d) None of these

Session 4

Miscellaneous Forms

Miscellaneous Forms

(i) 0^0 Form

When $\lim_{x \to a} f(x) \neq 1$ but $f(x)$ is positive in the neighbourhood of $x = a$.

In this case, we write $\{f(x)\}^{g(x)} = e^{\log_e \{f(x)\}^{g(x)}}$

$$\Rightarrow \quad \lim_{x \to a} [f(x)]^{g(x)} = e^{\lim_{x \to a} g(x) \log_e f(x)}$$

Example 53 Evaluate $\lim_{x \to 0} |x|^{\sin x}$.

Sol. $\lim_{x \to 0} |x|^{\sin x} = \lim_{x \to 0} e^{\sin x \log_e |x|} = e^{\lim_{x \to 0} \frac{\log_e |x|}{\operatorname{cosec} x}}$ [0^0 form]

$$= e^{\lim_{x \to 0} \frac{1/x}{-\operatorname{cosec} x \cot x}} \quad \text{[by L' Hospital's rule]}$$

$$= e^{\lim_{x \to 0} -\frac{\sin^2 x}{x \cos x}} = e^{\lim_{x \to 0} -\left(\frac{\sin x}{x}\right)^2 \cdot \left(\frac{x}{\cos x}\right)}$$

$$= e^{-(1)^2 \cdot (0)} = e^0 = 1$$

Example 54 Evaluate $\lim_{n \to \infty} (\pi n)^{2/n}$.

Sol. Let $A = \lim_{n \to \infty} (\pi n)^{2/n}$ [∞^0 form]

$$\therefore \quad \log A = \lim_{n \to \infty} \frac{2 \log (\pi n)}{n} \quad \left[\frac{\infty}{\infty} \text{ form}\right]$$

$$= \lim_{n \to \infty} \frac{2 \cdot \frac{1}{\pi n} \cdot \pi}{1} \quad \text{[by L'Hospital's rule]}$$

$$= \lim_{n \to \infty} \frac{2}{n} = 0$$

$$\therefore \quad \log_e A = 0 \Rightarrow A = 1$$

Example 55 Evaluate $\lim_{n \to \infty} \left(\frac{e^n}{\pi}\right)^{1/n}$.

Sol. Let $A = \lim_{n \to \infty} \left(\frac{e^n}{\pi}\right)^{1/n}$ [∞^0 form]

$$\therefore \quad \log A = \lim_{n \to \infty} \frac{1}{n} \log \left(\frac{e^n}{\pi}\right)$$

$$= \lim_{n \to \infty} \frac{n \log e - \log \pi}{n} \quad \left[\frac{\infty}{\infty} \text{ form}\right]$$

$$= \lim_{n \to \infty} \frac{\log e - 0}{1} \quad \text{[by L'Hospital's rule]}$$

$$\log A = 1$$

$$\Rightarrow \quad A = e^1 \text{ or } \lim_{n \to \infty} \left(\frac{e^n}{\pi}\right)^{1/n} = e$$

Example 56 Evaluate $\lim_{x \to 0} (\operatorname{cosec} x)^x$.

Sol. Let $A = \lim_{x \to 0} (\operatorname{cosec} x)^x$ [∞^0 form]

$$\Rightarrow \quad \log A = \lim_{x \to 0} x \log (\operatorname{cosec} x)$$

$$= \lim_{x \to 0} \frac{\log (\operatorname{cosec} x)}{\frac{1}{x}} \quad \left[\frac{\infty}{\infty} \text{ form}\right]$$

$$= \lim_{x \to 0} \frac{\frac{1}{\operatorname{cosec} x} \cdot (-\operatorname{cosec} x \cot x)}{-\frac{1}{x^2}} \quad \text{[by L'Hospital's rule]}$$

$$= \lim_{x \to 0} \frac{x^2}{\tan x} = 0$$

$$\therefore \quad \log A = 0 \text{ or } A = 1 \Rightarrow \lim_{x \to 0} (\operatorname{cosec} x)^x = 1$$

(ii) 1^0 Form

Example 57 Solve $\lim_{x \to 0} \frac{(1+x)^{1/x} - e}{x}$.

Sol. Let $L = \lim_{x \to 0} \frac{(1+x)^{1/x} - e}{x}$

$$= \lim_{x \to 0} \frac{e^{\log (1+x)^{1/x}} - e}{x} = \lim_{x \to 0} \frac{e \left\{e^{\frac{\log (1+x)}{x} - 1}\right\}}{x}$$

$$= \lim_{x \to 0} \frac{e \{e^M - 1\}}{M} \cdot \lim_{x \to 0} \frac{\log(1+x) - x}{x^2},$$

where $M = \frac{\log(1+x)}{x} - 1$

$$\therefore \quad L = \lim_{M \to 0} \frac{e \cdot \{e^M - 1\}}{M} \cdot \lim_{x \to 0} \frac{\frac{1}{1+x} - 1}{2x}$$

$$= e \times 1 \times \lim_{x \to 0} \frac{-x}{2x(1+x)} = -\frac{e}{2} \quad [\text{as } x \to 0 \Rightarrow M \to 0]$$

Exercise for Session 4

1. Evaluate $\lim_{x \to 0^+} (\sin x)^x$.

2. Evaluate $\lim_{x \to 0^+} (\sin x)^{\tan x}$.

3. Evaluate $\lim_{n \to \infty} \left(\frac{e^n}{\pi}\right)^{1/n}$.

4. The value of $\lim_{x \to 0} \frac{e - (1+x)^{1/x}}{\tan x}$ is

(a) e (b) $\frac{11e}{24}$ (c) $\frac{e}{2}$ (d) None of these

Session 5

Left Hand and Right Hand Limits

Left Hand and Right Hand Limits

Let $y = f(x)$ be a given function and $x = a$ is the point under consideration. Left tendency of $f(x)$ at $x = a$ is called its **left hand limit** and right tendency is called its **right hand limit.**

Left tendency (left hand limit) is denoted by $f(a-0)$ or $f(a-)$ and right tendency (right hand limit) is denoted by $f(a+0)$ or $f(a+)$ and are written as

$$\left.\begin{aligned} f(a-0) &= \lim_{h \to 0} f(a-h) \\ f(a+0) &= \lim_{h \to 0} f(a+h) \end{aligned}\right\}$$

where, h is a small positive number.

Thus, for the existence of the limit of $f(x)$ at $x = a$, it is necessary and sufficient that

$f(a-0) = f(a+0)$, if these are finite or $f(a-0)$ and $f(a+0)$ both should be either $+\infty$ or $-\infty$.

Example 58 Evaluate the right hand limit and left hand limit of the function

$$f(x) = \begin{cases} \dfrac{|x-4|}{x-4}, & x \neq 4 \\ 0, & x = 4 \end{cases}.$$

Sol. Given, $f(x) = \begin{cases} \dfrac{|x-4|}{x-4}, & x \neq 4 \\ 0, & x = 4 \end{cases}$

Now, RHL of $f(x)$ at $x = 4$

$$= \lim_{x \to 4^+} f(x) = \lim_{h \to 0} f(4+h)$$

$$= \lim_{h \to 0} \frac{|4+h-4|}{4+h-4} = \lim_{h \to 0} \frac{|h|}{h} = 1$$

LHL of $f(x)$ at $x = 4$

$$= \lim_{x \to 4^-} f(x) = \lim_{h \to 0} f(4-h)$$

$$= \lim_{h \to 0} \frac{|4-h-4|}{4-h-4} = \lim_{h \to 0} \frac{|h|}{-h} = \lim_{h \to 0} \frac{h}{-h} = -1$$

Thus, RHL $\neq$ LHL. So, $\lim_{x \to 4} f(x)$ does not exist.

Example 59 If $f(x) = \begin{cases} 5x-4, & 0 < x \le 1 \\ 4x^3 - 3x, & 1 < x < 2 \end{cases}$, show that $\lim_{x \to 1} f(x)$ exists.

Sol. We have, $f(x) = \begin{cases} 5x-4, & 0 < x \le 1 \\ 4x^3 - 3x, & 1 < x < 2 \end{cases}$

LHL of $f(x)$ at $x = 1 = \lim_{x \to 1^-} f(x) = \lim_{h \to 0} f(1-h)$

$$= \lim_{h \to 0} 5(1-h) - 4 = \lim_{h \to 0} 1 - 5h = 1$$

RHL of $f(x)$ at $x = 1 = \lim_{x \to 1^+} f(x) = \lim_{h \to 0} f(1+h)$

$$= \lim_{h \to 0} 4(1+h)^3 - 3(1+h) = 4(1)^3 - 3(1) = 1$$

Thus, RHL = LHL = 1. So, $\lim_{x \to 1} f(x)$ exists and is equal to 1.

Example 60 Show that $\lim_{x\to 0} \frac{e^{1/x}-1}{e^{1/x}+1}$ does not exist.

Sol. Let $f(x)=\frac{e^{1/x}-1}{e^{1/x}+1}$.

Then, LHL $= \lim_{x\to 0^-} f(x) = \lim_{h\to 0} f(0-h)$

$$= \lim_{h\to 0} \frac{e^{-1/h}-1}{e^{-1/h}+1}$$

$$= \lim_{h\to 0} \frac{(1/e^{1/h}-1)}{(1/e^{1/h}+1)} = \frac{0-1}{0+1} = -1$$

$$\left[\text{as } h\to 0 \Rightarrow \frac{1}{h}\to\infty \Rightarrow e^{1/h}\to\infty \Rightarrow 1/e^{1/h}\to 0\right] \quad ...(i)$$

RHL $= \lim_{x\to 0^+} f(x) = \lim_{h\to 0} f(0+h)$

$$= \lim_{h\to 0} \frac{e^{1/h}-1}{e^{1/h}+1} = \lim_{h\to 0} \frac{(1-1/e^{1/h})}{(1+1/e^{1/h})}$$

[dividing numerator and denominator both by $e^{1/h}$]

$$= \frac{1-0}{1+0} = 1 \quad \text{[using Eq. (i)]}$$

Clearly, $\lim_{x\to 0^-} f(x) \neq \lim_{x\to 0^+} f(x)$

Hence, $\lim_{x\to 0} f(x)$ doesn't exist.

Example 61 Evaluate $\lim_{x\to 1} \frac{\sqrt{1-\cos 2(x-1)}}{(x-1)}$.

[IIT JEE 1998]

Sol. We have, $\lim_{x\to 1} \frac{\sqrt{1-\cos 2(x-1)}}{(x-1)}$

$$= \lim_{x\to 1} \frac{\sqrt{2\sin^2 (x-1)}}{(x-1)} = \lim_{x\to 1} \frac{\sqrt{2}|\sin (x-1)|}{(x-1)}$$

$\therefore$ LHL $= \lim_{x\to 1^-} \frac{\sqrt{2}|\sin (x-1)|}{(x-1)}$

$$= \lim_{h\to 0} \frac{\sqrt{2}|\sin (-h)|}{(-h)} = \lim_{h\to 0} \frac{\sqrt{2}\sin h}{-h} = -\sqrt{2}$$

Again, RHL $= \lim_{x\to 1^+} \frac{\sqrt{2}|\sin (x-1)|}{(x-1)}$

$$= \lim_{h\to 0} \frac{\sqrt{2}|\sin (h)|}{h}$$

$$= \lim_{h\to 0} \frac{\sqrt{2}\sin h}{h} = \sqrt{2}$$

Clearly, $\lim_{x\to 1^+} f(x) \neq \lim_{x\to 1^-} f(x)$

Hence, $\lim_{x\to 1} f(x)$ doesn't exist.

Example 62 Solve (i) $\lim_{x\to 1} [\sin^{-1} x]$ (ii) $\lim_{x\to 0^+} \left[\frac{\sin x}{x}\right]$

(iii) $\lim_{x\to 0^-} \left[\frac{\sin x}{x}\right]$

(where $[\cdot]$ denotes greatest integer function.)

Sol. (i) Here, $L = \lim_{x\to 1} [\sin^{-1} x]$

Put $\sin^{-1} x = t$

$\therefore$ $x = \sin t$ and $t \to \frac{\pi}{2}$ as $x\to 1$

$\Rightarrow$ $L = \lim_{t\to \pi/2} [t] = \left[\frac{\pi}{2}\right] = 1$

$\therefore$ $\lim_{x\to 1} [\sin^{-1} x] = 1$

(ii) Here, $L = \lim_{x\to 0^+} \left[\frac{\sin x}{x}\right]$

Put $\frac{\sin x}{x} = t \Rightarrow t\to 1^-$ as $x\to 0^+$ and $[t]=0$

$\Rightarrow$ $L = \lim_{t\to 1^-} [t] = [1-h] = 0$

[as $x\to 0+h \Rightarrow t\to 1-h$]

(iii) Here, $L = \lim_{x\to 0^-} \left[\frac{\sin x}{x}\right]$

Put $\frac{\sin x}{x} = t$

$\Rightarrow$ $t\to 1^-$ as $x\to 0^-$

$\therefore$ $L = \lim_{t\to 1^-} [t] = [1-h] = 0$

[as $x\to 0-h \Rightarrow t\to 1-h$ $\therefore$ $[t]=0$]

Example 63 Solve (i) $\lim_{x\to\infty} [\tan^{-1} x]$

(ii) $\lim_{x\to -\infty} [\tan^{-1} x]$

(where $[\cdot]$ denotes greatest integer function.)

Sol. (i) Here, $\lim_{x\to\infty} [\tan^{-1} x]$

Put $\tan^{-1} x = t \Rightarrow t\to \frac{\pi}{2}$ as $x\to\infty$

$\therefore$ $\lim_{t\to \pi/2} [t] = \left[\frac{\pi}{2}\right] = 1$

(ii) Here, $\lim_{x\to -\infty} [\tan^{-1} x]$

Put $\tan^{-1} x = t \Rightarrow t\to -\frac{\pi}{2}$ as $x\to -\infty$

$\therefore$ $\lim_{t\to -\pi/2} [t] = \left[-\frac{\pi}{2}\right] = [-1.57] = -2$

Example 64 Solve (i) $\lim_{x \to 0^+} \left[\frac{\tan x}{x}\right]$

(ii) $\lim_{x \to 0^-} \left[\frac{\tan x}{x}\right]$

(where $[\cdot]$ denotes greatest integer function.)

Sol. (i) Here, $\lim_{x \to 0^+} \left[\frac{\tan x}{x}\right]$

Put $\frac{\tan x}{x} = t \Rightarrow t \to 1^+$ as $x \to 0^+$

$\therefore \lim_{t \to 1^+} [t] = [1 + h] = 1$ [as $x \to 0 + h \Rightarrow t \to 1 + h$]

(ii) Here, $\lim_{x \to 0^-} \left[\frac{\tan x}{x}\right]$

Put $\frac{\tan x}{x} = t \Rightarrow t \to 1^+$ as $x \to 0^-$

$\therefore \lim_{t \to 1^+} [t] = [1 + h] = 1$ [as $x \to 0 - h \Rightarrow t \to 1 + h$]

Example 65 Solve

(i) $\lim_{x \to 1^-} [\sin (\sin^{-1} x)]$ (ii) $\lim_{x \to \pi/2} [\sin^{-1}(\sin x)]$

(where $[\cdot]$ denotes greatest integer function.)

Sol. (i) Here, $\lim_{x \to 1^-} [\sin (\sin^{-1} x)]$

$= \lim_{x \to 1^-} [x]$ [as $\sin (\sin^{-1} x) = x$, if $-1 \le x \le 1$]

$= \lim_{h \to 0} [1 - h]$ [as $x \to 1^- \Rightarrow x = 1 - h$]

$= 0$

$\therefore \lim_{x \to 1} [\sin (\sin^{-1} x)] = \lim_{x \to 1^-} [\sin (\sin^{-1} x)]$

and no need to check for $\lim_{x \to 1^+} [\sin (\sin^{-1} x)]$

(ii) $\lim_{x \to \frac{\pi}{2}} [\sin^{-1}(\sin x)] = \lim_{x \to \frac{\pi^-}{2}} [\sin^{-1}(\sin x)]$

$= \lim_{x \to \frac{\pi^-}{2}} [x] = \lim_{h \to 0} \left[\frac{\pi}{2} - h\right] = 1$

Remark

If $\lim_{x \to 1} [\sin (\sin^{-1} x)]$, it means you have to calculate only left hand limit and not right hand limit as for $x > 1$, $\sin (\sin^{-1} x)$ is not defined.

Example 66 Solve (i) $\lim_{x \to 0} [\cot x]$

(ii) $\lim_{x \to +\infty} [\cot^{-1} x]$

(where $[\cdot]$ denotes greatest integer function.)

Sol. (i) Here, $\lim_{x \to 0} [\cot x]$

Put $\cot x = t$, now as $x \to 0$; $\cot x$ exhibits two values for $x \to 0^+$ and $x \to 0^-$. i.e. $\cot x \to +\infty$ and $\cot x \to -\infty$, respectively.

$\therefore$ We should apply right hand and left hand limit;

i.e. $\lim_{x \to 0^+} [\cot x] = \lim_{t \to +\infty} [t] = \infty$

[$\because \cot x = t \Rightarrow t \to +\infty$ as $x \to 0^+$]

and $\lim_{x \to 0^-} [\cot x] = \lim_{t \to -\infty} [t] = -\infty$

[$\because \cot x = t \Rightarrow t \to -\infty$ as $x \to 0^-$]

$\therefore$ Limit doesn't exist.

(ii) Here, $\lim_{x \to +\infty} [\cot^{-1} x] = \lim_{t \to 0^+} [t]$

[$\because \cot^{-1} x = t \Rightarrow t \to 0^+$ as $x \to +\infty$]

$= \lim_{h \to 0} [0 + h] = \lim_{h \to 0} 0 = 0$

Example 67 Solve $\lim_{x \to 0} \left[\sin \frac{|x|}{x}\right]$, where $[\cdot]$ denotes greatest integer function.

Sol. Here, $\lim_{x \to 0} \left[\sin \frac{|x|}{x}\right]$, since we have greatest integer function, we must define function.

Now, RHL (put $x = 0 + h$) $= \lim_{h \to 0} \left[\frac{\sin |0 + h|}{0 + h}\right]$

We know that, $\frac{\sin h}{h} \to 1$ as $h \to 0$ but less than 1.

$\therefore$ RHL $= \lim_{h \to 0} 0 = 0$ $\left[\because \left(\frac{\sin h}{h}\right) = 0 \text{ as } h \to 0\right]$

Again, LHL (put $x = 0 - h$) $= \lim_{h \to 0} \left[\sin \frac{|0 - h|}{0 - h}\right]$,

we know $\frac{\sin h}{-h} \to -1$ as $h \to 0$ but greater than -1.

$\therefore$ LHL $= \lim_{h \to 0} -1 = -1$ $\left[\because \left(\frac{\sin h}{h}\right) = -1 \text{ as } h \to 0\right]$

Thus, Limit doesn't exist, as RHL = 0 and LHL = -1.

Example 68 Solve $\lim_{x \to 0} \left[\frac{\sin |x|}{|x|}\right]$, where $[\cdot]$ denotes greatest integer function.

Sol. Here, $\lim_{x \to 0} \left[\frac{\sin |x|}{|x|}\right]$ ...(i)

We know that, $\frac{\sin x}{x} \to 1$ as $x \to 0$ or $\frac{\sin |x|}{|x|} \to 1$ as $x \to 0$ from right or left,

i.e. at $x = 0 + h$ or $x = 0 - h$, $\frac{\sin |x|}{|x|} \to 1^-$

$\therefore \quad \left[\dfrac{\sin|x|}{|x|}\right] = 0$ as $\dfrac{\sin|x|}{|x|} < 1$

From Eq. (i) whether we find RHL or LHL

$$\lim_{x\to 0}\left[\frac{\sin|x|}{|x|}\right] = \lim_{x\to 0} 0 = 0 \Rightarrow \lim_{x\to 0}\left[\frac{\sin|x|}{|x|}\right],$$

exists and is 0.

Example 69 $\lim_{x\to 0}\left[\dfrac{-2x}{\tan x}\right]$, where $[\cdot]$ denotes greatest integer function is

(a) −1 (b) 4
(c) 5 (d) None of these

Sol. We know, when $x \to 0$

$$\Rightarrow \quad \frac{x}{\tan x} < 1 \Rightarrow \frac{-x}{\tan x} > -1$$

$$\Rightarrow \quad \frac{-2x}{\tan x} > -2$$

So, $\lim_{x\to 0}\left[\dfrac{-2x}{\tan x}\right] = -2$

Hence, (d) is the correct answer.

Exercise for Session 5

1. The value of $\lim_{x\to 1}\{1 - x + [x-1] + [1-x]\}$ (where $[\cdot]$ denotes the greatest integral function) is

(a) −1 (b) Doesn't exist
(c) 1 (d) None of these

2. The value of $\lim_{x\to 0}\dfrac{\sin[x]}{[x]}$ (where $[\cdot]$ denotes the greatest integer function) is

(a) 1 (b) sin 1
(c) Doesn't exist (d) None of these

3. The value of $\lim_{x\to 0}\sin^{-1}\{x\}$ (where $\{\cdot\}$ denotes fractional part of x) is

(a) 0 (b) $\dfrac{\pi}{2}$
(c) Doesn't exist (d) None of these

4. The value of $\lim_{x\to 0}\left[\dfrac{x^2}{\sin x \tan x}\right]$ (where $[\cdot]$ denotes the greatest integer function) is

(a) 0 (b) 1
(c) Doesn't exist (d) None of these

Session 6

Use of Standard Theorems/Results, Use of Newton-Leibnitz's Formula in Evaluating the Limits, Summation of Series Using Definite Integral as the Limit

Use of Standard Theorems/Results

Theorem 1 : Sandwich/Squeeze Play Theorem

General The Squeeze principle is used on limit problems where the usual algebraic methods, factorisation or algebraic manipulation etc., are not effective. However, it requires to "squeeze" our problem in between two other simpler function, whose limits can be easily computed and equal. Use of Squeeze principle requires accurate analysis, indepth algebra skills and careful use of inequalities.

Statement If f, g and h are three functions such that $f(x) \le g(x) \le h(x)$ for all x in some interval containing the point $x = c$ and if

$$\lim_{x \to c} f(x) = \lim_{x \to c} h(x) = L$$

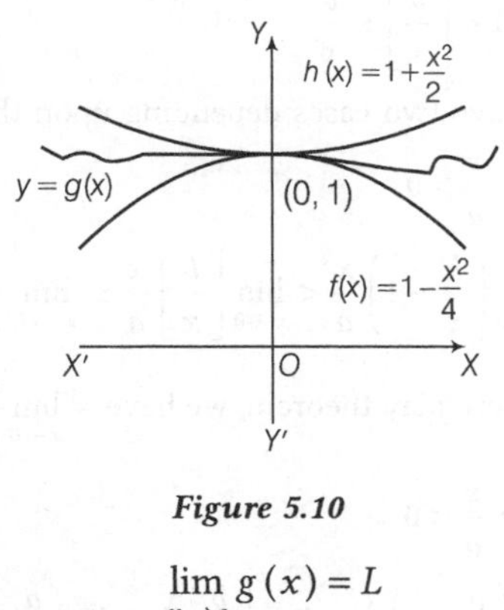

Figure 5.10

Then, $$\lim_{x \to c} g(x) = L$$

From the figure, note that $\lim_{x \to 0} g(x) = 1$.

Remark

The quantity c may be a finite number, $+\infty$ or $-\infty$. Similarly, L may also be finite number, $+\infty$ or $-\infty$.

Theorem 2 : Limits of Trigonometric Functions

If x is small and is measured in radians, then

$$\lim_{x \to 0} \frac{\sin x}{x} = 1 = \lim_{x \to 0} \frac{x}{\sin x}$$

$$= \lim_{x \to 0} x \operatorname{cosec} x = \lim_{x \to 0} \frac{\sin^{-1} x}{x} = \lim_{x \to 0} \frac{x}{\sin^{-1} x}$$

Proof Consider a circle with unit radius.

Area of $\Delta OAP <$ Area of sector $\widehat{OAP} <$ Area of ΔOAT

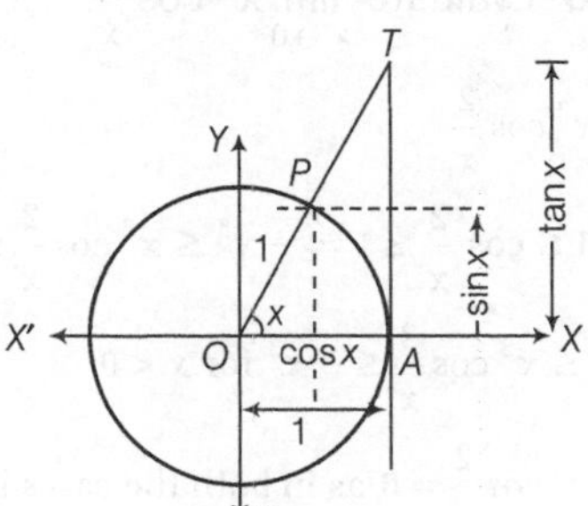

Figure 5.11

$$\frac{\sin x}{2} < \frac{x}{2} < \frac{\tan x}{2} \Rightarrow 1 < \frac{x}{\sin x} < \frac{1}{\cos x} \quad [\because 0 < x < \pi/2]$$

$$\Rightarrow \qquad \cos x < \frac{\sin x}{x} < 1$$

Now, using Sandwich theorem,

$$\lim_{x \to 0} \cos x < \lim_{x \to 0^+} \frac{\sin x}{x} < 1$$

Obviously, we have $$\lim_{x \to 0^+} \frac{\sin x}{x} = 1$$

Put $x = -y$, $$\lim_{y \to 0^-} \frac{\sin y}{y} = 1$$

Hence, $$\lim_{x \to 0} \frac{\sin x}{x} = 1$$

Similarly, $\lim_{x\to 0} \frac{\sin x}{x} = 1 = \lim_{x\to 0} \frac{x}{\sin x} = \lim_{x\to 0} x \operatorname{cosec} x$

$$= \lim_{x\to 0} \frac{\sin^{-1} x}{x} = \lim_{x\to 0} \frac{x}{\sin^{-1} x} \quad \text{...(i)}$$

Using Eq. (i), we can deduce

$$\lim_{x\to 0} \frac{\tan x}{x} = 1 = \lim_{x\to 0} \frac{x}{\tan x} = \lim_{x\to 0} x \cot x$$

$$= \lim_{x\to 0} \frac{\tan^{-1} x}{x} = \lim_{x\to 0} \frac{x}{\tan^{-1} x}$$

Important Results

The $\lim_{x\to 0} \frac{\sin x}{x}$ always approaches 1 from its left hand,

i.e. 0.9999....

$\Rightarrow \lim_{x\to 0} \left[\frac{\sin x}{x}\right] = 0$, where $[\cdot]$ denotes step up function.

$$\left[\text{Note that } \left(\lim_{x\to 0} \frac{\sin x}{x}\right) = 1\right]$$

Note that the $\lim_{x\to 0} \frac{\tan x}{x}$ approaches 1 from RHS.

$\Rightarrow \lim_{x\to 0} \left[\frac{\tan x}{x}\right] = 1$, where $[\cdot]$ denotes step up function.

Example 70 Evaluate $\lim_{x\to 0} x^3 \cos \frac{2}{x}$.

Sol. Here, $\lim_{x\to 0} x^3 \cos \frac{2}{x}$

As, $\quad -1 \le \cos \frac{2}{x} \le 1 \Rightarrow -x^3 \le x^3 \cos \frac{2}{x} \le x^3$ for $x > 0$

and $\quad x^3 \le x^3 \cos \frac{2}{x} \le -x^3$ for $x < 0$,

Thus, $\lim_{x\to 0} x^3 \cos \frac{2}{x} = 0$, as in both the cases limit is zero.

Example 71 Evaluate $\lim_{x\to\infty} \frac{x^2(2+\sin^2 x)}{x+100}$.

Sol. We have, $\quad \lim_{x\to\infty} \frac{x^2(2+\sin^2 x)}{x+100}$

Now, $\frac{2x^2}{x+100} \le \frac{x^2(2+\sin^2 x)}{x+100} \le \frac{3x^2}{x+100}$, as $0 \le \sin^2 x \le 1$

$$\therefore \quad \lim_{x\to\infty} \frac{2x^2}{x+100} \le \lim_{x\to 0} \frac{x^2(2+\sin^2 x)}{x+100} \le \lim_{x\to\infty} \frac{3x^2}{x+100}$$

$$\Rightarrow \quad \lim_{x\to\infty} \frac{x^2(2+\sin^2 x)}{x+100} = \infty$$

$$\left[\because \lim_{x\to\infty} \frac{2x^2}{x+100} = \lim_{x\to\infty} \frac{3x^2}{x+100} = \infty\right]$$

Example 72 Evaluate

$$\lim_{n\to\infty} \left(\frac{n}{n^2+1} + \frac{n}{n^2+2} + \frac{n}{n^2+3} + \ldots + \frac{n}{n^2+n}\right)$$

Sol. Let $f(n) = \frac{n}{n^2+1} + \frac{n}{n^2+2} + \frac{n}{n^2+3} + \ldots + \frac{n}{n^2+n}$.

Note that $f(n)$ has n terms which are decreasing.

Suppose

$$h(n) = \left(\frac{n}{n^2+1} + \frac{n}{n^2+1} + \frac{n}{n^2+1} + \ldots + \frac{n}{n^2+1}\right), n \text{ terms}$$

$$= \frac{n^2}{n^2+1} \quad [\text{obviously } f(n) < h(n)]$$

and

$$g(n) = \left(\frac{n}{n^2+n} + \frac{n}{n^2+n} + \frac{n}{n^2+n} + \ldots + \frac{n}{n^2+n}\right), n \text{ terms}$$

$$= \frac{n^2}{n^2+n} \quad [\text{obviously } g(n) < f(n)]$$

Hence, $\quad \lim_{n\to\infty} g(n) < \lim_{n\to\infty} f(n) < \lim_{n\to\infty} h(n)$

Since, $\quad \lim_{n\to\infty} g(n) = 1 = \lim_{n\to\infty} h(n)$

Hence, using Sandwich theorem, $\lim_{n\to\infty} f(n) = 1$.

Example 73 The value of the $\lim_{x\to 0} \frac{x}{a}\left[\frac{b}{x}\right]$ $(a \ne 0)$ (where $[\cdot]$ denotes the greatest integer function) is

(a) a (b) b

(c) $\frac{b}{a}$ (d) $1-\frac{b}{a}$

Sol. Since, $\frac{b}{x} - 1 < \left[\frac{b}{x}\right] \le \frac{b}{x}$

Now, we have two cases depending upon the value of $\frac{x}{a}$.

***Case* I** For $\frac{x}{a} > 0$

$$\Rightarrow \quad \lim_{x\to 0}\left(\frac{b}{x} - 1\right)\frac{x}{a} < \lim_{x\to 0}\left[\frac{b}{x}\right]\frac{x}{a} \le \lim_{x\to 0} \frac{b}{x}\cdot\frac{x}{a}$$

Using Squeeze play theorem, we have $= \lim_{x\to 0} \frac{x}{a}\left[\frac{b}{x}\right] = \frac{b}{a}$

***Case* II** For $\frac{x}{a} < 0$

$$\Rightarrow \quad \lim_{x\to 0}\left(\frac{b}{x} - 1\right)\frac{x}{a} > \lim_{x\to 0}\left[\frac{b}{x}\right]\frac{x}{a} \ge \lim_{x\to 0} \frac{b}{x}\cdot\frac{x}{a}$$

Using Squeeze play theorem, we have

$$\lim_{x\to 0} \frac{x}{a}\left[\frac{b}{x}\right] = \frac{b}{a}$$

Hence, (c) is the correct answer.

Example 74 Evaluate

$$\lim_{n \to \infty} \frac{[x]+[2x]+[3x]+...+[nx]}{n^2},$$

where $[\cdot]$ denotes the greatest integer function.

Sol. We know that, $x-1 < [x] \le x$

$$\Rightarrow \quad 2x-1 < [2x] \le 2x$$

$$\Rightarrow \quad 3x-1 < [3x] \le 3x$$

$$\text{.....................}$$

$$\Rightarrow \quad nx-1 < [nx] \le nx$$

$$\therefore \ (x+2x+3x+...+nx) - n < [x]+[2x]+...+[nx] \le (x+2x+...+nx)$$

$$\Rightarrow \quad \frac{xn(n+1)}{2} - n < \sum_{r=1}^{n} [rx] \le \frac{x \cdot n(n+1)}{2}$$

Thus, $\lim_{n \to \infty} \frac{[x]+[2x]+...+[nx]}{n^2}$

$$\Rightarrow \lim_{n \to \infty} \frac{x}{2}\left(1+\frac{1}{n}\right) - \frac{1}{n} < \lim_{n \to \infty} \frac{[x]+[2x]+...+[nx]}{n^2} \le \lim_{n \to \infty} \frac{x}{2}\left(1+\frac{1}{n}\right)$$

$$\Rightarrow \quad \frac{x}{2} < \lim_{n \to \infty} \frac{[x]+[2x]+...+[nx]}{n^2} \le \frac{x}{2}$$

$$\therefore \quad \lim_{n \to \infty} \frac{[x]+[2x]+...+[nx]}{n^2} = \frac{x}{2}$$

Aliter We know that, $[x] = x - \{x\}$

$$\sum_{r=1}^{n} rx = [x]+[2x]+...+[nx]$$

$$= x - \{x\} + 2x - \{2x\} + ... + nx - \{nx\}$$

$$= (x+2x+3x+...+nx) - (\{x\}+\{2x\}+...+\{nx\})$$

$$= \frac{xn(n+1)}{2} - (\{x\}+\{2x\}+...+\{nx\})$$

$$\therefore \frac{1}{n^2}\sum_{r=1}^{n}[rn] = \frac{x}{2}\left(1+\frac{1}{n}\right) - \frac{\{x\}+\{2x\}+...+\{nx\}}{n^2}$$

Since, $0 \le \{rx\} < 1$

$$\therefore \quad 0 \le \sum_{r=1}^{n} \{rx\} < n$$

$$\Rightarrow \quad \lim_{n \to \infty} \frac{\sum_{r=1}^{n} \{rx\}}{n^2} = 0$$

$$\therefore \quad \lim_{n \to \infty} \frac{\sum_{r=1}^{n} [rx]}{n^2} = \lim_{n \to \infty} \frac{x}{2}\left(1+\frac{1}{n}\right) - \lim_{n \to \infty} \frac{\sum_{r=1}^{n} \{rx\}}{n^2}$$

$$\Rightarrow \quad \lim_{n \to \infty} \frac{\sum_{r=1}^{n} [rx]}{n^2} = \frac{x}{2}$$

Use of Newton-Leibnitz's Formula in Evaluating the Limits

Let us consider the definite integral,

$$I(x) = \int_{\phi(x)}^{\psi(x)} f(t)\, dt$$

Newton-Leibnitz's formula states that,

$$\frac{d}{dx}\{I(x)\} = f\{\psi(x)\} \cdot \left\{\frac{d}{dx}\psi(x)\right\} - f\{\phi(x)\}\left\{\frac{d}{dx}\phi(x)\right\}$$

Example 75 Evaluate

$$\lim_{x \to 0} \left(\frac{1}{x^5}\int_0^x e^{-t^2} dt - \frac{1}{x^4} + \frac{1}{3x^2}\right).$$

Sol We have, $\lim_{x \to 0} \left(\frac{1}{x^5}\int_0^x e^{-t^2} dt - \frac{1}{x^4} + \frac{1}{3x^2}\right)$

$$= \lim_{x \to 0} \frac{3\int_0^x e^{-t^2} dt - 3x + x^3}{3x^5} \qquad \left[\frac{0}{0} \text{ form}\right]$$

$$= \lim_{x \to 0} \frac{3\frac{d}{dx}\int_0^x e^{-t^2} dt - 3 + 3x^2}{15x^4} \qquad \text{[by L'Hospital's rule]}$$

Applying Newton-Leibnitz's formula,

$$\frac{d}{dx}\int_0^x e^{-t^2} dt = e^{-x^2} \cdot \frac{d}{dx}(x) - e^{-0}\frac{d}{dx}(0) = e^{-x^2}$$

$$\therefore \lim_{x \to 0} \frac{3\frac{d}{dx}\int_0^x e^{-t^2} dt - 3 + 3x^2}{15x^4} = \lim_{x \to 0} \frac{3e^{-x^2} - 3 + 3x^2}{15x^4} \qquad \left[\frac{0}{0} \text{ form}\right]$$

$$= \lim_{x \to 0} \frac{-3(2x)e^{-x^2} + 6x}{60x^3} \qquad \text{[again, apply L'Hospital's rule]}$$

$$= \lim_{x \to 0} \frac{-6x(e^{-x^2} - 1)}{60x^3} = \lim_{x \to 0} \frac{-(e^{-x^2} - 1)}{10x^2} = \frac{1}{10}\lim_{x \to 0}\left(\frac{e^{-x^2} - 1}{-x^2}\right)$$

$$= \frac{1}{10} \times 1 = \frac{1}{10} \qquad \left[\because \lim_{x \to 0}\left(\frac{e^x - 1}{x}\right) = 1\right]$$

Example 76 Evaluate $\lim_{x \to 0} \frac{x - \int_0^x \cos t^2\, dt}{x^3 - 6x}$.

Sol. Let $L = \lim_{x \to 0} \frac{x - \int_0^x \cos t^2\, dt}{x^3 - 6x} \qquad \left[\frac{0}{0} \text{ form}\right]$

Applying L'Hospital's rule, we get

$$L = \lim_{x \to 0} \frac{1 - \frac{d}{dx}\int_0^x \cos t^2\, dt}{3x^2 - 6}$$

Applying Newton-Leibnitz's rule,

$$\frac{d}{dx}\int_0^x (\cos t^2)\, dt = \cos(x^2)\cdot 1 - 0 = \cos(x^2)$$

$$\therefore \quad L = \lim_{x\to 0} \frac{1 - \frac{d}{dx}\int_0^x \cos t^2\, dt}{3x^2 - 6} = \lim_{x\to 0} \frac{1-\cos(x^2)}{3(x^2-2)}$$

$$= \frac{1-\cos 0}{3(0-2)} = \frac{1-1}{3(-2)} = \frac{0}{-6} = 0$$

Example 77 Evaluate $\lim_{x\to 0} \frac{\int_0^{x^2} \cos t^2\, dt}{x \sin x}$.

[IIT JEE 1997]

Sol. Applying Newton-Leibnitz's rule, followed by L' Hospital's rule, we get

$$\lim_{x\to 0} \frac{\cos(x^2)^2 \cdot \{2x\} - 0}{x\cos x + \sin x} = \lim_{x\to 0} \frac{2\cos x^4}{\cos x + \left(\frac{\sin x}{x}\right)}$$

$$= \frac{2\cos 0}{\cos 0 + 1} = \frac{2}{1+1} = 1$$

Summation of Series Using Definite Integral as the Limit

The expression of the form,

$$\lim_{n\to\infty} \frac{1}{n} \sum_{r=\phi(x)}^{\psi(x)} f\left(\frac{r}{n}\right) = \int_a^b f(x)\, dx$$

where, (i) Σ is replaced by $\int$,

(ii) $\frac{r}{n}$ is replaced by x,

(iii) $\frac{1}{n}$ is replaced by dx,

(iv) To obtain, $a = \lim_{n\to\infty} \frac{\phi(x)}{n}$ and $b = \lim_{n\to\infty} \frac{\psi(x)}{n}$

The value so obtained is the required sum of the given series.

Example 78 Evaluate $\lim_{n\to\infty} \frac{1}{n} \sum_{r=1}^{2n} \frac{r}{\sqrt{n^2+r^2}}$.

[IIT JEE 1997]

Sol. Let $L = \lim_{n\to\infty} \frac{1}{n} \sum_{r=1}^{2n} \frac{r}{\sqrt{n^2+r^2}}$ dividing numerator and denominator both by n, we get

$$L = \lim_{n\to\infty} \frac{1}{n} \sum_{r=1}^{2n} \frac{r/n}{\sqrt{1+(r/n)^2}}$$

Put $\frac{r}{n} = x$; $\frac{1}{n} = dx$; $\lim_{n\to\infty} \sum_{r=1}^{2n} = \int_a^b$

where, $a = \lim_{n\to\infty} \frac{1}{n} = 0$ and $b = \lim_{n\to\infty} \frac{2n}{n} = 2$

$$\therefore \quad L = \lim_{n\to\infty} \frac{1}{n} \sum_{r=1}^{2n} \frac{r/n}{\sqrt{1+(r/n)^2}} = \int_0^2 \frac{x}{\sqrt{1+x^2}}\, dx$$

$$= \left(\sqrt{1+x^2}\right)_0^2 = \sqrt{5} - 1$$

Example 79 Evaluate

$$\lim_{n\to\infty}\left(\frac{n}{n^2+1^2} + \frac{n}{n^2+2^2} + \ldots + \frac{n}{n^2+n^2}\right).$$

Sol. Let $S = \lim_{n\to\infty}\left(\frac{n}{n^2+1^2} + \frac{n}{n^2+2^2} + \ldots + \frac{n}{n^2+n^2}\right)$

$$= \lim_{n\to\infty} \sum_{r=1}^{n} \frac{n}{n^2+r^2} = \lim_{n\to\infty} \frac{1}{n} \sum_{r=1}^{n} \frac{n^2}{n^2+r^2}$$

[dividing numerator and denominator both by n]

$$= \lim_{n\to\infty} \frac{1}{n} \sum_{r=1}^{n} \frac{1}{1+(r/n)^2}$$

Replace $\frac{r}{n} = x$; $\frac{1}{n} = dx$; $\lim_{n\to\infty} \sum_{r=1}^{n} = \int_a^b$

where, $a = \lim_{n\to\infty} \frac{1}{n} = 0$ and $b = \lim_{x\to\infty} \frac{n}{n} = 1$

we get, $S = \lim_{n\to\infty} \frac{1}{n} \sum_{r=1}^{n} \frac{1}{1+(r/n)^2}$

$$= \int_0^1 \frac{1}{1+x^2}\, dx = [\tan^{-1} x]_0^1 = \tan^{-1}(1) - \tan^{-1}(0) = \frac{\pi}{4}$$

Example 80 The value of $\lim_{n\to\infty}\left(\frac{n!}{n^n}\right)^{1/n}$ is equal to

(a) $\frac{1}{e}$ (b) e (c) e^2 (d) $\frac{1}{e^2}$

Sol. Let $A = \lim_{n\to\infty}\left(\frac{n!}{n^n}\right)^{1/n}$

$$\therefore \log A = \lim_{n\to\infty} \frac{1}{n} \log\left(\frac{n}{n}\cdot\frac{(n-1)}{n}\cdot\frac{(n-2)}{n}\ldots\frac{3}{n}\cdot\frac{2}{n}\cdot\frac{1}{n}\right)$$

$$= \lim_{n\to\infty} \frac{1}{n} \sum_{r=0}^{n-1} \log\left(\frac{n-r}{n}\right)$$

$$= \int_0^1 1\cdot\log(1-x)\, dx = -1$$

[replace $\frac{r}{n}$ by n and $\frac{1}{n}$ by dx $\lim_{n\to\infty} \sum_{r=0}^{n-1} = \int_a^b$, where $\frac{0}{n} = 0$,

$a = \lim_{n\to\infty}$, $b = \lim_{n\to\infty} \frac{n-1}{n} = 1$]

$$\Rightarrow \quad \log_e A = -1 \Rightarrow A = e^{-1}$$

Hence, (a) is the correct answer.

Exercise for Session 6

1. The value of $\lim_{n \to \infty} \frac{1}{n^3}([1^2 x + 1^2] + [2^2 x + 2^2] + \ldots + [n^2 x + n^2])$ (where $[\cdot]$ denotes the greatest integer function) is

(a) $\frac{x}{3}$ (b) $x + \frac{1}{3}$

(c) $\frac{x}{3} + \frac{1}{3}$ (d) None of these

2. The value of $\lim_{x \to \infty} \left\{ \lim_{n \to \infty} \frac{[1^2 (\sin x)^x] + [2^2 (\sin x)^x] + \ldots + [n^2 (\sin x)^x]}{n^3} \right\}$ (where $[\cdot]$ denotes the greatest integer function) is

(a) $\frac{x}{3} + \frac{\sin x}{3}$ (b) $\frac{x}{3} + (\sin x)^x$

(c) $\frac{1}{3}$ (d) 0

3. The value of $\lim_{x \to 1^+} \frac{\int_1^x |t - 1| \, dt}{\sin (x - 1)}$ is

(a) 0 (b) 1

(c) Doesn't exist (d) None of these

4. The value of $\lim_{n \to \infty} \sum_{K=1}^{n} \log \left(1 + \frac{K}{n}\right)^{1/n}$ is

(a) $\log_e \left(\frac{e}{4}\right)$ (b) $\log_e \left(\frac{4}{e}\right)$

(c) $\log_e 4$ (d) None of these

5. The value of $\lim_{n \to \infty} \left(\frac{1}{na} + \frac{1}{na + 1} + \frac{1}{na + 2} + \ldots + \frac{1}{nb} \right)$ is

(a) $\log \left(\frac{a}{b}\right)$ (b) $\log \left(\frac{b}{a}\right)$

(c) $\log (ab)$ (d) None of these

JEE Type Solved Examples :
Single Option Correct Type Questions

Ex. 1 *The value of* $\lim_{x \to 0} \frac{x \sin(\sin x) - \sin^2 x}{x^6}$ *equals*

(a) $\frac{1}{6}$ (b) $\frac{1}{12}$

(c) $\frac{1}{18}$ (d) $\frac{1}{24}$

Sol. (c) Here, $\lim_{x \to 0} \frac{x \sin(\sin x) - \sin^2 x}{x^6}$

$$= \lim_{t \to 0} \frac{\sin^{-1}(t) \cdot \sin t - t^2}{(\sin^{-1} t)^6} \qquad [\text{put } \sin x = t]$$

$$= \lim_{t \to 0} \frac{\sin^{-1}(t) \cdot \sin t - t^2}{t^6} \cdot \frac{t^6}{(\sin^{-1} t)^6}$$

$$= \lim_{t \to 0} \frac{\left\{t + \frac{t^3}{6} + \frac{9t^5}{120} + \frac{5t^7}{112} + \dots\right\} \cdot \left\{t - \frac{t^3}{3!} + \frac{t^5}{5!} - \dots\right\} - t^2}{t^6} \times 1$$

$$= \lim_{t \to 0} \frac{t^6\left(\frac{1}{5!} - \frac{1}{6 \cdot (3!)} + \frac{9}{120}\right) + \text{higher powers to } t}{t^6}$$

$$= \frac{1}{120} + \frac{9}{120} - \frac{1}{36} = \frac{1}{18}$$

Ex. 2 *If* $\lim_{x \to 0} \frac{\log_e \cot\left(\frac{\pi}{4} - K_1 x\right)}{\tan K_2 x} = 1$, *then*

(a) $K_1 = K_2$ (b) $2K_1 = K_2$

(c) $K_1 = 2K_2$ (d) $K_1 = 4K_2$

Sol. (b) Here, $\lim_{x \to 0} \frac{\log_e \left[\cot\left(\frac{\pi}{4} - K_1 x\right)\right]}{\tan K_2 x} = 1$

$$\Rightarrow \lim_{x \to 0} \frac{\log\left[\cot\left(\frac{\pi}{4} - K_1 x\right) - 1 + 1\right]}{\tan K_2 x} = 1$$

$$\Rightarrow \lim_{x \to 0} \frac{\log\left(1 + \frac{2 \tan K_1 x}{1 - \tan K_1 x}\right)}{\tan K_2 x} = 1$$

$$\Rightarrow \lim_{x \to 0} \frac{\log\left(1 + \frac{2 \tan K_1 x}{1 - \tan K_1 x}\right)}{\frac{2 \tan K_1 x}{1 - \tan K_1 x}} \cdot \frac{\frac{2 \tan K_1 x}{1 - \tan K_1 x}}{\tan K_2 x} = 1$$

$$\Rightarrow \lim_{x \to 0} \frac{2 \tan K_1 x}{\tan K_2 x} = 1 \Rightarrow \lim_{x \to 0} \frac{\frac{2 \tan K_1 x}{K_1 x} \cdot K_1 x}{\frac{\tan K_2 x}{K_2 x} \cdot K_2 x} = 1$$

$$\Rightarrow \frac{2K_1}{K_2} = 1 \Rightarrow 2K_1 = K_2$$

Ex. 3 *Let* $f(\theta) = \frac{1}{\tan^2 \theta} \{(1 + \tan\theta)^3 + (2 + \tan\theta)^3 + \dots + (10 + \tan\theta)^3\} - 10 \tan \theta$.

Then, $\lim_{\theta \to \frac{\pi}{2}^-} f(\theta)$ *is equal to*

(a) 170 (b) 166 (c) 165 (d) None of these

Sol. (c) Here,

$$\lim_{\theta \to \frac{\pi}{2}^-} \frac{(1 + \tan\theta)^3}{\tan^2\theta} = \lim_{\theta \to \frac{\pi}{2}^-} \frac{1 + 3\tan\theta + 3\tan^2\theta + \tan^3\theta}{\tan^2\theta}$$

$$[(a+b)^3 = a^3 + b^3 + 3a^2b + 3ab^2]$$

$$= \lim_{\theta \to \frac{\pi}{2}^-} 3 + \tan\theta + \left(\frac{3\tan\theta + 1}{\tan^2\theta}\right) = \lim_{\theta \to \frac{\pi}{2}^-} (3 + \tan\theta)$$

$$\left[\text{as } \lim_{\theta \to \frac{\pi}{2}^-} \left(\frac{3 \tan\theta + 1}{\tan^2\theta}\right) = 0\right]$$

$$\Rightarrow \lim_{\theta \to \frac{\pi}{2}^-} f(\theta) = \lim_{\theta \to \frac{\pi}{2}^-} [(3 + \tan\theta) + (3(2) + \tan\theta) + \dots + (3(10) + \tan\theta)] - 10 \lim_{\theta \to \frac{\pi}{2}^-} \tan\theta$$

$$= \lim_{\theta \to \frac{\pi}{2}^-} 3(1 + 2 + 3 + \dots + 10) + 10 \cdot \lim_{\theta \to \frac{\pi}{2}^-} (\tan\theta) - 10 \cdot \lim_{\theta \to \frac{\pi}{2}^-} (\tan\theta)$$

$$= \frac{3 \times 10 \times 11}{2} = 165$$

Ex. 4 *For positive integers* $K = 1, 2, 3, \dots, n$. *Let* S_k *denotes the area of* ΔOAB_K *(where 'O' is the origin) such that* $\angle AOB_K = \frac{K\pi}{2n}$, $OA = 1$ *and* $OB_K = K$. *The value of the* $\lim_{n \to \infty} \frac{1}{n^2} \cdot \sum_{K=1}^{n} S_K$ *is*

(a) $\frac{2}{\pi^2}$ (b) $\frac{4}{\pi^2}$ (c) $\frac{8}{\pi^2}$ (d) $\frac{1}{2\pi^2}$

Sol. (a) Here, $\quad OB_K = K$ and $\angle AOB_K = \dfrac{K\pi}{2n}$

$$\therefore \quad S_K = \frac{1}{2}\cdot(1)(K)\sin\left(\frac{K\pi}{2n}\right) \quad \left[\text{using, } \Delta = \frac{1}{2}ab\sin\theta\right]$$

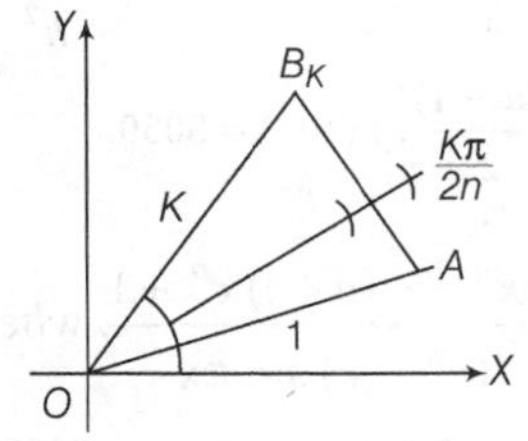

$$\text{Then, } L = \lim_{n\to\infty} \frac{1}{n^2}\cdot\frac{K}{2}\cdot\sin\left(\frac{K\pi}{2n}\right)$$

$$= \frac{1}{2n}\cdot \lim_{n\to\infty}\frac{K}{n}\cdot\sin\left(\frac{\pi}{2}\cdot\frac{K}{n}\right) = \frac{1}{2}\cdot\int_0^1 x\cdot\sin\left(\frac{\pi}{2}x\right)dx$$

$$= \frac{1}{2}\left[\underbrace{\left(\frac{-2}{\pi}\cdot x\cdot\cos\frac{\pi x}{2}\right)_0^1}_{0} + \frac{2}{\pi}\int_0^1 \cos\frac{\pi x}{2}\cdot dx\right]$$

$$= \frac{1}{2}\left[\frac{2}{\pi}\cdot\frac{2}{\pi}\left(\sin\frac{\pi x}{2}\right)_0^1\right] = \frac{2}{\pi^2}$$

Ex. 5 *If*

$$S_n = \left(1-\tan^4\frac{\pi}{2^3}\right)\left(1-\tan^4\frac{\pi}{2^4}\right)\ldots\left(1-\tan^4\frac{\pi}{2^n}\right).$$

The value of $\lim_{n\to\infty} S_n$, *is*

(a) $\dfrac{\pi^3}{4}$ (b) $\dfrac{\pi^3}{16}$

(c) $\dfrac{\pi^3}{32}$ (d) $\dfrac{\pi^3}{256}$

Sol. (c) Here,

$$S_n = \left(1-\tan^4\frac{\pi}{2^3}\right)\left(1-\tan^4\frac{\pi}{2^4}\right)\ldots\left(1-\tan^4\frac{\pi}{2^n}\right)$$

$$= \frac{\left(\cos^2\frac{\pi}{2^3}-\sin^2\frac{\pi}{2^3}\right)\left(\cos^2\frac{\pi}{2^4}-\sin^2\frac{\pi}{2^4}\right)\ldots\left(\cos^2\frac{\pi}{2^n}-\sin^2\frac{\pi}{2^n}\right)}{\left(\cos^4\frac{\pi}{2^3}\cdot\cos^4\frac{\pi}{2^4}\ldots\cos^4\frac{\pi}{2^n}\right)}$$

$$= \frac{\left(\cos\frac{\pi}{2^2}\right)\cdot\left(\cos\frac{\pi}{2^3}\right)\ldots\left(\cos\frac{\pi}{2^{n-1}}\right)}{\left(\cos^4\frac{\pi}{2^3}\cdot\cos^4\frac{\pi}{2^4}\ldots\cos^4\frac{\pi}{2^n}\right)}$$

$$= \frac{\frac{1}{\sqrt{2}}}{\left(\cos\frac{\pi}{2^3}\cdot\cos\frac{\pi}{2^3}\ldots\cos\frac{\pi}{2^{n-1}}\right)^3}\cdot\frac{1}{\cos^4\frac{\pi}{2^n}} \quad \ldots\text{(i)}$$

Let $\quad M = \cos\dfrac{\pi}{2^3}\cdot\cos\dfrac{\pi}{2^4}\ldots\cos\dfrac{\pi}{2^{n-1}}$

$$= \frac{\sin\left(2^{n-3}\cdot\frac{\pi}{2^{n-1}}\right)}{2^{n-3}\cdot\sin\left(\frac{\pi}{2^{n-1}}\right)} = \frac{\sin\left(\frac{\pi}{2^2}\right)}{2^{n-3}\cdot\sin\left(\frac{\pi}{2^{n-1}}\right)}$$

$$\therefore \quad S_n = \lim_{n\to\infty}\frac{\frac{1}{\sqrt{2}}}{\left(\frac{1}{\sqrt{2}}\right)^3}\times\left\{2^{n-3}\cdot\sin\frac{\pi}{2^{n-1}}\right\}^3\times\frac{1}{\cos^n\left(\frac{\pi}{2^n}\right)}$$

$$= \lim_{n\to\infty} 2\cdot\left(\frac{\sin\frac{\pi}{2^{n-1}}}{\frac{\pi}{2^{n-1}}}\right)^3\cdot\left(\frac{\pi}{4}\right)^3\times\frac{1}{\cos^n\left(\frac{\pi}{2^n}\right)}$$

$$= 2\left(\frac{\pi}{4}\right)^3\cdot\frac{1}{1} = \frac{\pi^3}{32}$$

Ex. 6 *The value of* $\lim_{x\to 2}\dfrac{\sqrt{x+7}-3\sqrt{2x-3}}{\sqrt[3]{x+6}-2\sqrt[3]{3x-5}}$ *is*

(a) $\dfrac{33}{23}$ (b) $\dfrac{34}{23}$ (c) $\dfrac{54}{25}$ (d) None of these

Sol. (b) We have, $L = \lim_{x\to 2}\dfrac{\sqrt{x+7}-3\sqrt{2x-3}}{\sqrt[3]{x+6}-2\sqrt[3]{3x-5}}$ $\quad\left[\dfrac{0}{0}\text{ form}\right]$

Let $\quad x-2=t$, such that $x\to 2 \Rightarrow t\to 0$

$$\therefore \quad L = \lim_{t\to 0}\frac{(t+9)^{1/2}-3(2t+1)^{1/2}}{(t+8)^{1/3}-2(3t+1)^{1/3}}$$

$$= \frac{3}{2}\cdot\lim_{t\to 0}\frac{(1+t/9)^{1/2}-(2t+1)^{1/2}}{\left(1+\frac{t}{8}\right)^{1/3}-(1+3t)^{1/3}} \quad \left[\frac{0}{0}\text{ form}\right]$$

$$= \frac{3}{2}\cdot\lim_{t\to 0}\frac{\left(1+\frac{1}{2}\cdot\frac{t}{9}\right)-\left(1+\frac{1}{2}\cdot(2t)\right)}{\left(1+\frac{1}{3}\cdot\frac{t}{8}\right)-\left(1+\frac{1}{3}\cdot(3t)\right)};$$

$$\left[\text{using } (1+x)^n = 1+nx, \text{ as } x\to 0\right]$$

$$= \frac{3}{2}\cdot\lim_{t\to 0}\frac{\frac{t}{18}-t}{\frac{t}{24}-t} = \frac{3}{2}\frac{\left(\frac{1}{18}-1\right)}{\left(\frac{1}{24}-1\right)} = \frac{34}{23}$$

Ex. 7 *Let* $\Delta(x) = \begin{vmatrix} f(x+\alpha) & f(x+2\alpha) & f(x+3\alpha) \\ f(\alpha) & f(2\alpha) & f(3\alpha) \\ f'(\alpha) & f'(2\alpha) & f'(3\alpha)\end{vmatrix}$ *for some real values differential function* f *and constant* α, $\lim_{x\to 0}\dfrac{\Delta(x)}{x}$ *is equal to*

(a) 0 (b) 1 (c) 2 (d) 3

Sol. (a) Here, $\Delta(0)=\begin{vmatrix} f(\alpha) & f(2\alpha) & f(3\alpha)\\ f(\alpha) & f(2\alpha) & f(3\alpha)\\ f'(\alpha) & f'(2\alpha) & f'(3\alpha)\end{vmatrix}$

Thus, $\lim_{x\to 0}\dfrac{\Delta(x)}{x}$ $\left[\dfrac{0}{0}\text{ form}\right]$

$=\lim_{x\to 0}\dfrac{\Delta'(x)}{1}=\Delta'(0)$ [applying L'Hospital's rule]

$\therefore \lim_{x\to 0}\dfrac{\Delta(x)}{x}=\Delta'(0)$ ...(i)

Given, $\Delta(x)=\begin{vmatrix} f(x+\alpha) & f(x+2\alpha) & f(x+3\alpha)\\ f(\alpha) & f(2\alpha) & f(3\alpha)\\ f'(\alpha) & f'(2\alpha) & f'(3\alpha)\end{vmatrix}$

Using definition of differentiation of determinant,

$$\Delta'(x)=\begin{vmatrix} f'(x+\alpha) & f'(x+2\alpha) & f'(x+3\alpha)\\ f(\alpha) & f(2\alpha) & f(3\alpha)\\ f'(\alpha) & f'(2\alpha) & f'(3\alpha)\end{vmatrix}+\begin{vmatrix} f(x+\alpha) & f(x+2\alpha) & f(x+3\alpha)\\ 0 & 0 & 0\\ f'(\alpha) & f'(2\alpha) & f'(3\alpha)\end{vmatrix}+\begin{vmatrix} f(x+\alpha) & f(x+2\alpha) & f(x+3\alpha)\\ f(\alpha) & f(2\alpha) & f(3\alpha)\\ 0 & 0 & 0\end{vmatrix}$$

[as α is constant $\Rightarrow \dfrac{d}{dx}(\alpha)=0$]

$$\therefore \Delta'(x)=\begin{vmatrix} f'(x+\alpha) & f'(x+2\alpha) & f'(x+3\alpha)\\ f(\alpha) & f(2\alpha) & f(3\alpha)\\ f'(\alpha) & f'(2\alpha) & f'(3\alpha)\end{vmatrix}$$

or $\Delta'(0)=\begin{vmatrix} f'(\alpha) & f'(2\alpha) & f'(3\alpha)\\ f(\alpha) & f(2\alpha) & f(3\alpha)\\ f'(\alpha) & f'(2\alpha) & f'(3\alpha)\end{vmatrix}=0$

[$\because R_1$ and R_3 are identical]

Hence, $\lim_{x\to 0}\dfrac{\Delta(x)}{x}=\Delta'(0)=0$

Ex. 8 *Let* $\lim_{x\to 1}\dfrac{x^a-ax+a-1}{(x-1)^2}=f(a)$*. The value of* $f(101)$ *equals*

(a) 5050
(b) 5151
(c) 4950
(d) 101

Sol. (a) We have, $\lim_{x\to 1}\dfrac{x^a-ax+a-1}{(x-1)^2}$

Put $x=1+h$, we have

$$\lim_{h\to 0}\frac{(1+h)^a-a(1+h)+a-1}{h^2}$$

$$=\lim_{h\to 0}\frac{\left(1+ah+\dfrac{a(a-1)}{2!}h^2+\dots\right)-a-ah+a-1}{h^2}$$

$\therefore f(a)=\dfrac{a(a-1)}{2}$; $f(101)=5050$

Ex. 9 $\lim_{x\to 1}\dfrac{nx^{n+1}-(n+1)x^n+1}{(e^x-e)\sin\pi x}$*, where* $n=100$ *is equal to*

(a) $\dfrac{5050}{\pi e}$ (b) $\dfrac{100}{\pi e}$ (c) $-\dfrac{5050}{\pi e}$ (d) $-\dfrac{4950}{\pi e}$

Sol. (c) Let $l=\lim_{x\to 1}\dfrac{nx^n(x-1)-(x^n-1)}{(e^x-e)\sin\pi x}$

Put $x=1+h$, so that as $x\to 1$, $h\to 0$

$\therefore\quad l=-\lim_{h\to 0}\dfrac{h\cdot n(1+h)^n-\{(1+h)^n-1\}}{e(e^h-1)\sin\pi h}$

$$\Rightarrow\quad l=-\lim_{x\to 1}\frac{n\cdot h(1+{}^nC_1h+{}^nC_2h^2+{}^nC_3h^3+\dots)-(1+{}^nC_1h+{}^nC_2h^2+{}^nC_3h^3+\dots-1)}{\pi e(h^2)\left(\dfrac{e^h-1}{h}\right)\left(\dfrac{\sin\pi h}{\pi h}\right)}$$

$$=-\frac{n^2-{}^nC_2}{\pi e}=-\left[\frac{2n^2-n(n-1)}{2\pi e}\right]=-\frac{n^2+n}{2(\pi e)}=-\frac{n(n+1)}{2(\pi e)}$$

If $n=100\Rightarrow l=-\left(\dfrac{5050}{\pi e}\right)$

Ex. 10 $\lim_{n\to\infty}\dfrac{1^2n+2^2(n-1)+3^2(n-2)+\dots+n^2\cdot 1}{1^3+2^3+3^3+\dots+n^3}$ *is equal to*

(a) $\dfrac{1}{3}$ (b) $\dfrac{2}{3}$ (c) $\dfrac{1}{2}$ (d) $\dfrac{1}{6}$

Sol. (a) We have,

$$\lim_{n\to\infty}\frac{1^2n+2^2(n-1)+3^2(n-2)+\dots+n^2\{n-(n-1)\}}{\Sigma n^3}$$

Numerator $=n(1^2+2^2+\dots+n^2)-\{1\cdot 2^2+2\cdot 3^2+3\cdot 4^2+\dots+(n-1)n^2\}$

$=n\Sigma n^2-\sum_{r=2}^{n}(r-1)\cdot r^2$

$=n\Sigma n^2-\sum_{r=1}^{n}(r^3-r^2)$

$=n\Sigma n^2-[\Sigma n^3-\Sigma n^2]=(n+1)\Sigma n^2-\Sigma n^3$

$$\therefore\ l = \lim_{n\to\infty} \frac{(n+1)\,\Sigma n^2 - \Sigma n^3}{\Sigma n^3}$$

$$= \lim_{n\to\infty} \frac{4(n+1)\,n(n+1)\,(2n+1)}{6n\,(n+1)\,n\,(n+1)} - 1 = \frac{4}{3} - 1 = \frac{1}{3}$$

Aliter $l = \dfrac{1^2 n + 2^2\,(n-1) + \ldots + n^2 \cdot 1}{\underbrace{1^3 + 2^3 + 3^3 + \ldots + n^3}} + 1 - 1$

$$l = \frac{1^2(n+1) + 2^2\,(n+1) + \ldots + n^2\,(n+1)}{\Sigma n^3} - 1$$

$$l = \frac{(n+1)\,\Sigma n^2}{\Sigma n^3} - 1$$

$$\Rightarrow \lim_{n\to\infty} \frac{(n+1)\cdot n(n+1)\,(2n+1)}{6\cdot \dfrac{n^2\,(n+1)^2}{4}} - 1 = \frac{4}{3} - 1 = \frac{1}{3}$$

● ***Ex. 11*** *The value of* $\lim_{x\to\infty} \dfrac{\cot^{-1}(x^{-a}\log_a x)}{\sec^{-1}(a^x \log_x a)}$ $(a > 1)$ *is equal to*

(a) 1 (b) 0 (c) $\dfrac{\pi}{2}$ (d) Doesn't exist

Sol. *(a)* We have, $\lim_{x\to\infty} \dfrac{\cot^{-1}\left(\dfrac{\log_a x}{x^a}\right)}{\sec^{-1}\left(\dfrac{a^x}{\log_a x}\right)}$;

as $\lim_{x\to\infty} \left(\dfrac{\log_a x}{x^a}\right) \to 0$ and $\left(\dfrac{a^x}{\log_a x}\right) \to \infty$

[using L' Hospital's rule]

$$\therefore\quad l = \frac{\frac{\pi}{2}}{\frac{\pi}{2}} = 1$$

● ***Ex. 12*** *Suppose that a and b $(b \ne a)$ are real positive numbers, thene the value of* $\lim_{t\to 0}\left(\dfrac{b^{t+1} - a^{t+1}}{b-a}\right)^{1/t}$ *is equal to*

(a) $\dfrac{a\ln b - b\ln a}{b-a}$ (b) $\dfrac{b\ln b - a\ln a}{b-a}$

(c) $b\ln b - a\ln a$ (d) $\left(\dfrac{b^b}{a^a}\right)^{\frac{1}{b-a}}$

Sol. *(d)* Here, $\lim_{t\to 0}\left(\dfrac{b^{t+1} - a^{t+1}}{b-a}\right)^{\frac{1}{t}}$.

Obviously, limit is of the form 1^∞.

Hence, $l = e^{\lim_{t\to 0}\frac{1}{t}\left[\frac{b^{t+1} - a^{t+1}}{b-a} - 1\right]} = e^{\lim_{t\to 0}\left(\frac{b^{t+1} - a^{t+1} - b + a}{t(b-a)}\right)}$

$$= e^{\lim_{t\to 0}\left(\frac{b(b^t - 1) - a(a^t - 1)}{t(b-a)}\right)} = e^{\frac{b\ln b - a\ln a}{b-a}}$$

$$= e^{\frac{\ln b^b - \ln a^a}{b-a}} = e^{\frac{\ln \frac{b^b}{a^a}}{b-a}} = e^{\ln\left(\frac{b^b}{a^a}\right)^{\frac{1}{b-a}}} = \left(\frac{b^b}{a^a}\right)^{\frac{1}{b-a}}$$

● ***Ex. 13*** $\lim_{x\to\infty} \dfrac{\cot^{-1}(\sqrt{x+1} - \sqrt{x})}{\sec^{-1}\left\{\left(\dfrac{2x+1}{x-1}\right)^x\right\}}$ *is equal to*

(a) 1 (b) 0 (c) $\dfrac{\pi}{2}$ (d) None of these

Sol. *(a)* As, $\lim_{x\to\infty} \sqrt{x+1} - \sqrt{x} = 0 \Rightarrow \cot^{-1}(0) = \dfrac{\pi}{2}$

$$\lim_{x\to\infty}\left(\frac{2x+1}{x-1}\right)^x \to \infty \quad \Rightarrow \sec^{-1}(\infty) = \frac{\pi}{2}$$

$$\therefore\quad l = 1$$

● ***Ex. 14*** $\lim_{n\to\infty} \cos(\pi\sqrt{n^2 + n})$ *(when n is an integer) is equal to*

(a) 1 (b) − 1 (c) 0 (d) Doesn't exist

Sol. *(c)* Let $l = \lim_{n\to\infty} \pm\cos(n\pi - \pi\sqrt{n^2 + n})$

$$= \lim_{n\to\infty} \pm\cos(\pi(n - \sqrt{n^2 + n}))$$

$$= \lim_{n\to\infty} \pm\cos\left(\frac{(\pi(+n))}{n + \sqrt{n^2 + n}}\right)$$

$$= \lim_{n\to\infty} \pm\cos\left(\frac{n\pi}{n + n\sqrt{1 + \frac{1}{n}}}\right)$$

$$= \lim_{n\to\infty} \pm\cos\left(\frac{\pi}{1 + \sqrt{1 + \frac{1}{n}}}\right) = \cos\frac{\pi}{2} \to 0$$

Aliter

We have, $\pi\sqrt{n^2 + n}$

$$= \pi n\left(1 + \frac{1}{n}\right)^{1/2} = n\pi\left[1 + \frac{1}{2n} + \frac{1}{2}\left(\frac{1}{2} - 1\right)\frac{1}{2!}\frac{1}{n^2} + \ldots\right]$$

$$= \pi\left[n + \frac{1}{2} + \frac{1}{2}\left(\frac{1}{2} - 1\right)\frac{1}{2!}\frac{1}{n} + \ldots\right]$$

As $n \to \infty$; $\dfrac{\pi}{2}\cdot\left[2n + 1 + \left(\dfrac{1}{2} - 1\right)\dfrac{1}{2!}\dfrac{1}{n} + \ldots\right] = (2n+1)\dfrac{\pi}{2}$

$$\therefore\quad \lim_{n\to\infty} \cos\left((2n+1)\frac{\pi}{2}\right) = 0$$

Ex. 15 *The value of* $\lim_{x\to 0}\frac{(\tan(\{x\}-1))\sin\{x\}}{\{x\}(\{x\}-1)}$, *where* $\{x\}$ *denotes the fractional part function is*

(a) 1 (b) tan 1 (c) sin 1 (d) Doesn't exist

Sol. (d) Let $f(x)=\lim_{x\to 0}\frac{(\tan(\{x\}-1))\sin\{x\}}{\{x\}(\{x\}-1)}$

$$\lim_{x\to 0^+} f(x)=\lim_{h\to 0^+}\frac{\tan(h-1)\cdot\sin h}{h(h-1)}=\frac{\tan(-1)}{-1}=\tan 1$$

$$\text{LHL}=\lim_{h\to 0}\frac{\tan((1-h)-1)\sin(1-h)}{(1-h)(1-h-1)}=\frac{\sin 1}{1}=\sin 1$$

Hence, $\lim_{x\to 0} f(x)$ doesn't exist.

Ex. 16 $\lim_{x\to 0^-}(-\ln(\{x\}+|[x]|))^{\{x\}}$ *is equal to*

(a) 0 (b) 1 (c) ln 2 (d) $\ln\frac{1}{2}$

Sol. (d) We have, $\lim_{x\to 0^-}=(-\ln(\{x\}+|[x]|))^{\{x\}}$

$$=\lim_{x\to 0^-}(-\ln(\{0-h\}+|[-h]|))^{\{-h\}}$$

$$=\lim_{x\to 0^-}(-\ln(1-h+1))^{1-h}=-\ln 2=\ln\left(\frac{1}{2}\right)$$

Ex. 17 $\lim_{x\to\infty}\frac{2+2x+\sin 2x}{(2x+\sin 2x)\,e^{\sin x}}$ *is equal to*

(a) 0 (b) 1 (c) − 1 (d) Non-existent

Sol. (d) We have, $\lim_{x\to\infty}\frac{2+2x+\sin 2x}{(2x+\sin 2x)e^{\sin x}}$

$$=\lim_{x\to\infty}\frac{\frac{2}{x}+2+\frac{\sin 2x}{x}}{\left(2+\frac{\sin 2x}{x}\right)e^{\sin x}},\text{ as } x\to\infty$$

$\Rightarrow l=\lim_{x\to\infty}\frac{2}{2\cdot e^{\sin x}}=$ oscillatory between $\frac{1}{e}$ and $\frac{1}{e^{-1}}\Rightarrow$ Non-existent

Ex. 18 *The value of* $\lim_{x\to 0}(\cos ax)^{\text{cosec}^2 bx}$ *is*

(a) $e^{\left(-\frac{8b^2}{a^2}\right)}$ (b) $e^{\left(-\frac{8a^2}{b^2}\right)}$

(c) $e^{\left(-\frac{a^2}{2b^2}\right)}$ (d) $e^{\left(-\frac{b^2}{2a^2}\right)}$

Sol. (c) Let $l=\lim_{x\to 0}(\cos ax)^{\text{cosec}^2 bx}$

$$\Rightarrow\quad l=e^{\lim_{x\to\infty}}\text{cosec}^2\,bx\,(\cos ax-1)$$

Now, $-\lim_{x\to 0}\frac{1-\cos ax}{\sin^2 bx}=-\lim_{x\to 0}\frac{\sin^2 ax}{\sin^2 bx}\cdot\frac{1}{1+\cos ax}$

$$=-\frac{a^2}{2b^2},\quad l=e^{-\frac{a^2}{2b^2}}$$

Ex. 19 $\lim_{n\to\infty}\sum_{r=1}^{n}\frac{r}{n^2+n+r}$ *equals*

(a) 0 (b) 1/3 (c) 1/2 (d) 1

Sol. (c) Let $f(n)=\frac{1}{n^2+n+1}+\frac{2}{n^2+n+2}+\ldots+\frac{n}{n^2+n+n}$

Consider $g(n)=\frac{1}{n^2+n+n}+\frac{2}{n^2+n+n}+\ldots+\frac{n}{n^2+n+n}$

$$=\frac{1+2+3+\ldots+n}{n^2+2n}=\frac{n(n+1)}{2(n^2+2n)}$$

$$g(n)<f(n)\qquad\ldots(\text{i})$$

Similarly, $h(n)=\frac{1}{n^2+n+1}+\frac{2}{n^2+n+1}+\ldots+\frac{n}{n^2+n+1}$

$$=\frac{n(n+1)}{2(n^2+n+1)}$$

$$\therefore\quad f(n)<h(n)\qquad\ldots(\text{ii})$$

From Eqs. (i) and (ii), $g(n)<f(n)<h(n)$

But $\lim_{n\to\infty}g(n)=\lim_{n\to\infty}h(n)=\frac{1}{2}$.

Hence, using Sandwich theorem, $\lim_{n\to\infty}f(n)=\frac{1}{2}$.

Ex. 20 *The value of* $\lim_{n\to\infty}(\sqrt{n^2+n+1}-[\sqrt{n^2+n+1}])$, *where* $[\cdot]$ *denotes the greatest integer function is*

(a) 0 (b) 1/2 (c) 2/3 (d) 1/4

Sol. (b) We know that, $n<\sqrt{n^2+n+1}<n+1$.

Hence, $[\sqrt{n^2+n+1}]=n$

$$\therefore\ l=\lim_{n\to\infty}(\sqrt{n^2+n+1}-n)=\lim_{n\to\infty}\frac{n+1}{\sqrt{n^2+n+1}+n}=\frac{1}{2}$$

Ex. 21 $\lim_{x\to 1}\frac{\sin^2(x^3+x^2+x-3)}{1-\cos(x^2-4x+3)}$ *has the value equal to*

(a) 18 (b) 9/2 (c) 9 (d) None of these

Sol. (a) We have, $\lim_{x\to 1}\frac{\sin^2(x^3+x^2+x-3)}{1-\cos(x^2-4x+3)}$

$$=\lim_{x\to 1}\frac{\sin^2(x^3+x^2+x-3)}{(x^3+x^2+x-3)^2}\cdot\frac{(x^3+x^2+x-3)^2}{1-\cos(x^2-4x+3)}$$

$$=(1)\lim_{x\to 1}\frac{(x^2-4x+3)^2}{1-\cos(x^2-4x+3)}\cdot\frac{(x^3+x^2+x-3)^2}{(x^2-4x+3)^2}$$

$$= (1)(2) \lim_{x \to 1} \left(\frac{x^3 + x^2 + x - 3}{x^2 - 4x + 3} \right)^2 = 2l^2$$

where, $l = \lim_{x \to 1} \frac{3x^2 + 2x + 1}{2x - 4}$ [using L' Hospital's rule]

$$= \frac{6}{-2} = -3 \Rightarrow l = 2(-3)^2 = 18$$

● **Ex. 22** *The graph of function $y = f(x)$ has a unique tangent at $(e^a, 0)$ through which the graph passes, then $\lim_{x \to e^a} \frac{\log(1 + 7f(x)) - \sin(f(x))}{3f(x)}$ equals*

(a) 1 (b) 2
(c) 7 (d) None of these

Sol. *(b)* Here, $\lim_{x \to e^a} \frac{\log(1 + 7f(x)) - \sin(f(x))}{3f(x)}$ $\left[\frac{0}{0} \text{ form}\right]$

$$= \lim_{x \to e^a} \frac{7f'(x) - \{\cos(f(x)) \cdot f'(x)\}\{1 + 7f(x)\}}{3f'(x) \cdot \{1 + 7f(x)\}}$$

[using L'Hospital's rule]

$$= \lim_{x \to e^a} \frac{7 - \cos(f(x))\{1 + 7f(x)\}}{3\{1 + 7f(x)\}} = \frac{7 - 1}{3} = 2$$

JEE Type Solved Examples : More than One Correct Option Type Questions

● **Ex. 23** *If $f(x) = \frac{3x^2 + ax + a + 1}{x^2 + x - 2}$, which of the following can be correct?*

(a) $\lim_{x \to 1} f(x)$ exists $\Rightarrow a = -2$

(b) $\lim_{x \to -2} f(x)$ exists $\Rightarrow a = 13$

(c) $\lim_{x \to 1} f(x) = 4/3$

(d) $\lim_{x \to -2} f(x) = -1/3$

Sol. *(a, b, c, d)* Given, $f(x) = \frac{3x^2 + ax + a + 1}{(x + 2)(x - 1)}$

As $x \to 1$, Dr. $\to 0$, hence as $x \to 1$, Nr. $\to 0$

$\therefore \quad 3 + 2a + 1 = 0 \Rightarrow a = -2$

As $x \to -2$, Dr. $\to 0$, hence as $x \to -2$, Nr. $\to 0$

$\therefore \quad 12 - 2a + a + 1 = 0 \Rightarrow a = 13$

Now, $\lim_{x \to 1} f(x) = \lim_{x \to 1} \frac{3x^2 - 2x - 1}{(x + 2)(x - 1)}$

$$= \lim_{x \to 1} \frac{(3x + 1)(x - 1)}{(x + 2)(x - 1)} = \frac{4}{3}$$

and $\lim_{x \to -2} \frac{3x^2 + 13x + 14}{(x + 2)(x - 1)} = \lim_{x \to -2} \frac{(3x + 7)(x + 2)}{(x + 2)(x - 1)} = -\frac{1}{3}$

● **Ex. 24** *Let $f(x) = \begin{cases} \frac{\tan^2\{x\}}{x^2 - [x]^2}, & \text{for } x > 0 \\ 1, & \text{for } x = 0 \\ \sqrt{\{x\} \cot \{x\}}, & \text{for } x < 0 \end{cases}$, where $[x]$ is the step up function and $\{x\}$ is the fractional part function of x, then*

(a) $\lim_{x \to 0^+} f(x) = 1$ (b) $\lim_{x \to 0^-} f(x) = 1$

(c) $\cot^{-1}\left(\lim_{x \to 0^-} f(x)\right)^2 = 1$ (d) None of these

Sol. *(a, c)* We have, $f(x) = \begin{cases} \frac{\tan^2\{x\}}{x^2 - [x]^2}, & x > 0 \\ 1, & x = 0 \\ \sqrt{\{x\} \cot \{x\}}, & x < 0 \end{cases}$

$$\text{RHL} = \lim_{x \to 0^+} f(x) = \lim_{h \to 0} \frac{\tan^2\{h\}}{h^2 - [h]^2} = \lim_{h \to 0} \frac{\tan^2 h}{h^2} = 1$$

$$\text{LHL} = \lim_{x \to 0^-} f(x) = \lim_{h \to 0} \sqrt{\{-h\} \cot \{-h\}}$$

$$= \lim_{h \to 0} \sqrt{(1 - h)\cot(1 - h)} = \sqrt{\cot 1}$$

$$\therefore \quad \cot^{-1}\left(\lim_{x \to 0^-} f(x)\right)^2 = \cot^{-1}(\sqrt{\cot 1})^2 = 1$$

● **Ex. 25** *Given that the derivative $f'(a)$ exists. Indicate which of the following statement(s) is/are always true?*

(a) $f'(a) = \lim_{h \to a} \frac{f(h) - f(a)}{h - a}$

(b) $f'(a) = \lim_{h \to 0} \frac{f(a) - f(a - h)}{h}$

(c) $f'(a) = \lim_{t \to 0} \frac{f(a + 2t) - f(a)}{t}$

(d) $f'(a) = \lim_{t \to 0} \frac{f(a + 2t) - f(a + t)}{2t}$

Sol. *(a, b)* Here, options (a) and (b) are true by definition.

Option (c) is false, as

$$\lim_{t \to 0} \frac{f(a+2t)-f(a)}{t} = 2f'(a)$$

and $\lim_{t \to 0} \frac{f(a+2t)-f(a+t)}{2t} = \frac{1}{2}[2f'(a)-f'(a)] = \frac{1}{2}f'(a)$

Hence, option (d) is false.

Ex. 26 *Let* $f(x) = \lim_{n \to \infty} \sum_{r=0}^{n} \frac{x}{(rx+1)\{(r+1)x+1\}}$. *Then,*

(a) $f(0) = 0$ (b) $f(0) = x$

(c) $f(0^+) = 1$ (d) $f(0^-) = 1$

Sol. *(a, c, d)* Given, $f(x) = \lim_{n \to \infty} \sum_{r=0}^{n} \frac{x}{(rx+1)\{(r+1)x+1\}}$

$$= \lim_{n \to \infty} \sum_{r=0}^{n} \frac{[(r+1)x+1)]-(rx+1)}{(rx+1)[(r+1)x+1]}$$

$$= \lim_{n \to \infty} \sum_{r=0}^{n} \left(\frac{1}{rx+1} - \frac{1}{(r+1)x+1}\right)$$

$$= \lim_{n \to \infty} \left[\left(\frac{1}{1} - \frac{1}{x+1}\right) + \left(\frac{1}{x+1} - \frac{1}{2x+1}\right) + \ldots + \left(\frac{1}{nx+1} - \frac{1}{(n+1)x+1}\right)\right]$$

$$= \lim_{n \to \infty} \left[1 - \frac{1}{(n+1)x+1}\right] = \lim_{n \to \infty} \frac{(n+1)x}{(n+1)x+1}$$

$$\therefore \quad f(x) = \begin{cases} 0, & x = 0 \\ 1, & x \neq 0 \end{cases} \Rightarrow f(0) = 0, f(0^+) = f(0^-) = 1.$$

JEE Type Solved Examples : Passage Based Questions

Passage I

(Q. Nos. 27 to 29)

Let $f(x) = \begin{cases} e^{\{x^2\}} - 1, & x > 0 \\ \dfrac{\sin x - \tan x + \cos x - 1}{2x^2 + \ln(2+x) + \tan x}, & x < 0, \\ 0, & x = 0 \end{cases}$

where { } represents fractional part function. Suppose lines L_1 *and* L_2 *represent tangent and normal to curve* $y = f(x)$ *at* $x = 0$. *Consider the family of circles touching both the lines* L_1 *and* L_2.

Ex. 27 *Ratio of radii of two circles belonging to this family cutting each other orthogonally is*

(a) $2+\sqrt{3}$ (b) $\sqrt{3}$

(c) $2+\sqrt{2}$ (d) $2-\sqrt{2}$

Ex. 28 *A circle having radius unity is inscribed in the triangle formed by* L_1 *and* L_2 *and a tangent to it. Then, the minimum area of the triangle possible is*

(a) $3+\sqrt{2}$ (b) $2+\sqrt{3}$

(c) $3+2\sqrt{2}$ (d) $3-2\sqrt{2}$

Ex. 29 *If centres of circles belonging to family having equal radii r are joined, the area of figure formed is*

(a) $2r^2$ (b) $4r^2$

(c) $8r^2$ (d) r^2

Sol. (Q. Nos. 27 to 29)

$$\text{LHD} = \lim_{h \to 0} \frac{\dfrac{-\sin h + \tan h + \cos h - 1}{2h^2 + \ln(2-h) - \tan h} - 0}{-h}$$

$$= \lim_{h \to 0} \frac{\dfrac{\sin h}{h} - \dfrac{\tan h}{h} + \dfrac{1-\cos h}{h^2} \times h}{2h^2 + \ln(2-h) - \tan h} = 0$$

$$f'(0^+) = \text{RHD} = \lim_{h \to 0} \frac{e^{h^2} - 1 - 0}{h} = h \times \frac{e^{h^2} - 1}{h^2} = 0$$

$L_1 \equiv y = 0$ and $L_2 \equiv x = 0$

27. *(a)* $(x-r)^2 + (y-r)^2 = r^2$ [family of circle]

$x^2 + y^2 - 2rx - 2ry + r^2 = 0$

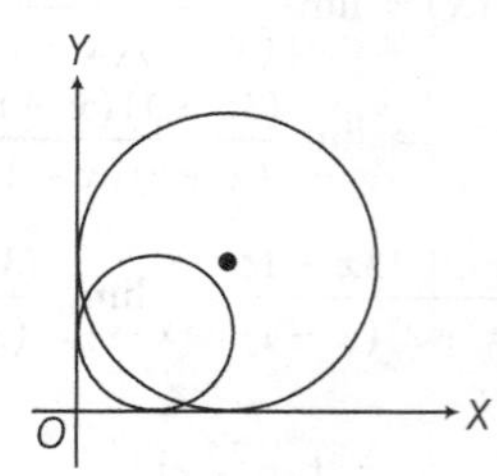

$\because \quad 2(r_1r_2 + r_1r_2) = r_1^2 + r_2^2$

or $\quad 4r_1r_2 = r_1^2 + r_2^2$

$$\left(\frac{r_2}{r_1}\right)^2 - 4\left(\frac{r_2}{r_1}\right) + 1 = 0$$

$$\Rightarrow \quad \frac{r_2}{r_1} = \frac{4 \pm \sqrt{12}}{2} = 2 \pm \sqrt{3}$$

28. (c) $2[\Delta_1+\Delta_2+\Delta_3]$

$$\Delta = 2\times\frac{1}{2}\left(\cot\left(\frac{\pi}{4}-\frac{\theta}{2}\right)+\cot\frac{\theta}{2}+1\right)\quad\left[\text{using, }\frac{1}{2}ab\right]$$

$$\Delta = \frac{\cos\left(\frac{\pi}{4}-\frac{\theta}{2}\right)}{\sin\left(\frac{\pi}{4}-\frac{\theta}{2}\right)}+\frac{\cos\frac{\theta}{2}}{\sin\frac{\theta}{2}}+1$$

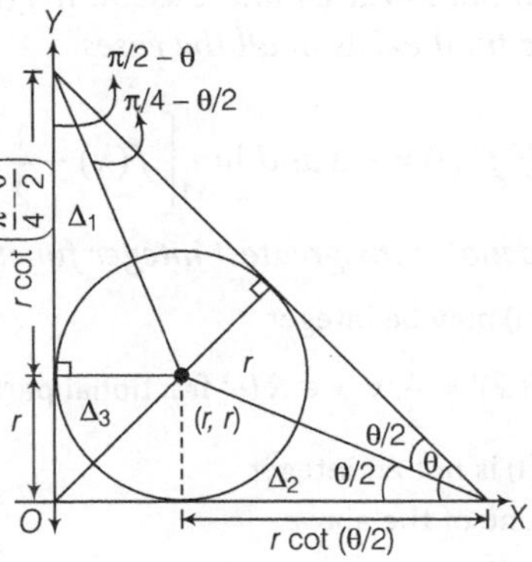

$$\Rightarrow\quad \Delta = 1+\frac{2\sin\frac{\pi}{4}}{2\sin\frac{\theta}{2}\cdot\sin\left(\frac{\pi}{4}-\frac{\theta}{2}\right)}$$

$$\Delta = 1+\frac{\sqrt{2}}{\cos\left(\theta-\frac{\pi}{4}\right)-\cos\left(\frac{\pi}{4}\right)}$$

Δ is minimum, if denominators is maximum when $\theta=\frac{\pi}{4}$,

$$\Delta_{\min} = 1+\frac{\sqrt{2}}{1-\frac{1}{\sqrt{2}}} = 1+\frac{2}{\sqrt{2}-1} = 1+2(\sqrt{2}+1) = 3+2\sqrt{2}$$

29. (b) Area $=(2r)^2=4r^2$

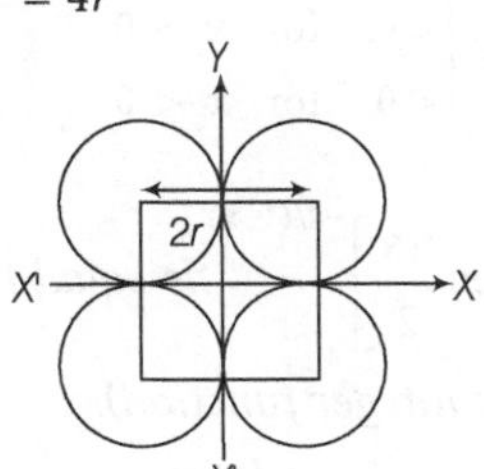

Passage II

(Q. Nos. 30 to 32)

Let A be $n\times n$ matrix given by

$$A=\begin{bmatrix} a_{11} & a_{12} & a_{13}\dots a_{1n}\\ a_{21} & a_{22} & a_{23}\dots a_{2n}\\ \vdots & \vdots & \vdots\\ a_{n1} & a_{n2} & a_{n3}\dots a_{nn}\end{bmatrix}$$

Such that each horizontal row is arithmetic progression and each vertical column is a geometrical progression. It is known that each column in geometric progression have the same common ratio. Given that $a_{24}=1$, $a_{42}=\frac{1}{8}$ and $a_{43}=\frac{3}{16}$.

● **Ex.30** *Let $S_n=\sum_{j=1}^{n}a_{4j}$, $\lim_{n\to\infty}\frac{S_n}{n^2}$ is equal to*

(a) $\frac{1}{4}$ (b) $\frac{1}{8}$ (c) $\frac{1}{16}$ (d) $\frac{1}{32}$

Sol. (d) Here, $a_{43}=a_{42}+d$

$$\Rightarrow\quad \frac{3}{16}=\frac{1}{8}+d \Rightarrow d=\frac{1}{16}$$

[common difference of 4th row]

$$\therefore\quad a_{41}=a_{42}-d=\frac{1}{8}-\frac{1}{16}=\frac{1}{16}$$

$$\therefore\quad a_{41}=\frac{1}{16}, a_{42}=\frac{2}{16}, a_{43}=\frac{3}{16},\dots,a_{4n}=\frac{n}{16}$$

Now, all elements of 4th row are known

$$S_n=\sum_{j=1}^{n}a_{4j}=\frac{n(n+1)}{2(16)};\ \lim_{n\to\infty}\frac{S_n}{n^2}=\frac{1}{32}$$

● **Ex.31** *Let d_i be the common difference of the elements in ith row, then $\sum_{i=1}^{n}d_i$ is*

(a) n (b) $\frac{1}{2}-\frac{1}{2^{n+1}}$ (c) $1-\frac{1}{2^n}$ (d) $\frac{n+1}{2^n}$

Sol. (c) Also, $a_{24}r^2=a_{44}=\frac{4}{16}$

$$\Rightarrow\quad r^2=\frac{4}{16}\Rightarrow r=\frac{1}{2}\quad\left[\text{common ratio of all GP is }\frac{1}{2}\right]$$

$$\therefore\quad A=\begin{bmatrix}\frac{1}{2} & \frac{2}{2} & \frac{3}{2} & \dots & \frac{n}{2}\\ \frac{1}{2^2} & \frac{2}{2^2} & \frac{3}{2^2} & \dots & \frac{n}{2^2}\\ \frac{1}{2^3} & \frac{2}{2^3} & \frac{3}{2^3} & \dots & \frac{n}{2^3}\\ \vdots & \vdots & \vdots & \vdots & \vdots\\ \frac{1}{2^n} & \frac{2}{2^n} & \frac{3}{2^n} & \dots & \frac{n}{2^n}\end{bmatrix}_{n\times n}$$

Now, $d_i=$ common difference in ith row $=\frac{1}{2^i}$

$$\therefore\quad \sum_{i=1}^{n}d_i=\sum_{i=1}^{n}\frac{1}{2^i}=\frac{\frac{1}{2}\left(1-\frac{1}{2^n}\right)}{1-\frac{1}{2}}=1-\frac{1}{2^n}$$

Ex.32 *The value of* $\lim_{n\to\infty} \sum_{i=1}^{n} a_{ii}$ *is equal to*

(a) $\frac{1}{4}$ (b) $\frac{1}{2}$ (c) 1 (d) 2

Sol. (d) Given, $\sum_{i=1}^{n} a_{ii} = S = \frac{1}{2} + \frac{2}{2^2} + \frac{3}{2^3} + \ldots + \frac{n}{2^n}$ ...(i)

$$\Rightarrow \frac{1}{2} S_n = \frac{1}{2^2} + \frac{2}{2^3} + \frac{3}{2^4} + \ldots + \frac{n-1}{2^n} + \frac{n}{2^{n+1}} \quad \ldots\text{(ii)}$$

On subtracting Eq. (ii) from Eq. (i), we get

$$\frac{1}{2} S_n = \left(\frac{1}{2} + \frac{1}{2^2} + \frac{1}{2^3} + \ldots + \frac{1}{2^n}\right) - \frac{n}{2^{n+1}}$$

$$\Rightarrow \frac{1}{2} S_n = \frac{\frac{1}{2}\left(1 - \frac{1}{2^n}\right)}{1 - \frac{1}{2}} - \frac{n}{2^{n+1}} \Rightarrow S_n = 2\left(1 - \frac{1}{2^n}\right) - \frac{n}{2^n}$$

$$\therefore \lim_{n\to\infty} \sum_{i=1}^{n} a_{ii} = \lim_{n\to\infty} 2 - \frac{2}{2^n} - \frac{2n}{2^n} = 2 - 0 - 0 = 2$$

$$\Rightarrow \lim_{n\to\infty} \sum_{i=1}^{n} a_{ii} = 2$$

Passage III

(Q. Nos. 33 to 35)

To evaluate $\lim_{x\to a} [f(x)]$, *we must analyse the* $f(x)$ *in right hand neighbourhood as well as in left hand neighbourhood of* $x = a$. *e.g. In case of continuous function, we may come across following cases.*

Case I

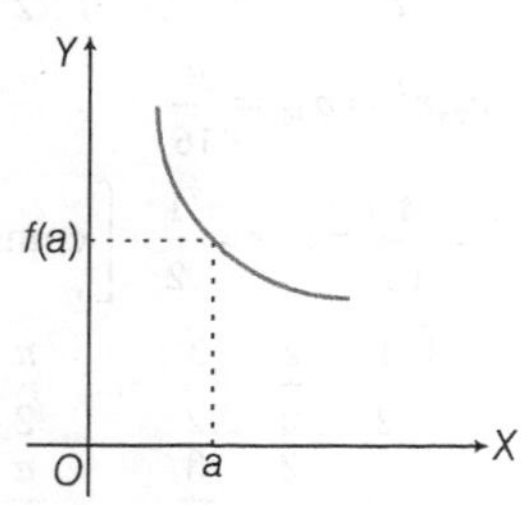

Case II

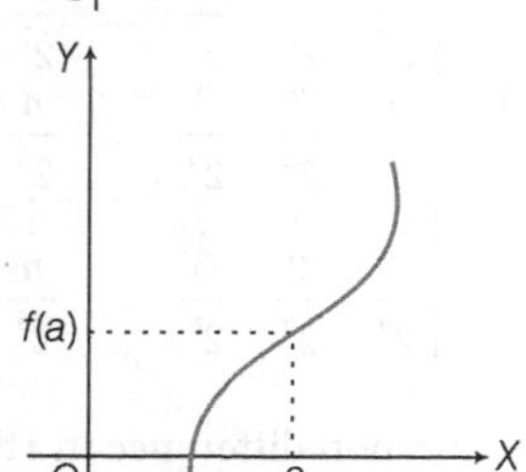

Case III

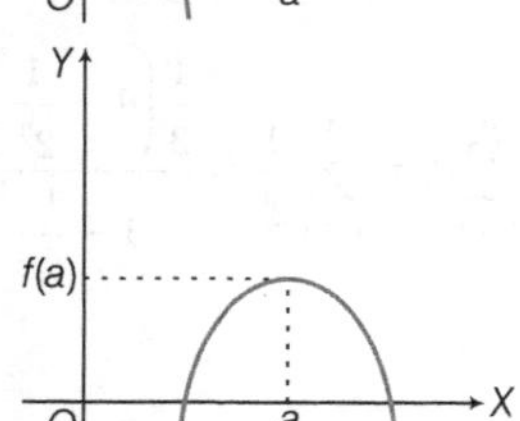

Case IV

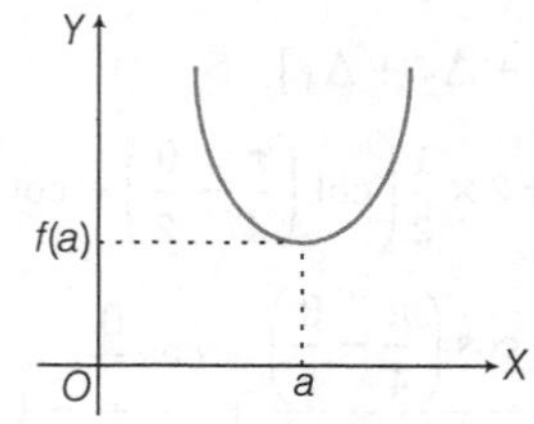

If $f(a)$ *is an integer, the limit will exist in Case III and Case IV but not in Case I and Case II. If* $f(a)$ *is not an integer, the limit exists in all the cases.*

Ex.33 *If* $f'(1) = -3$ *and* $\lim_{x\to 1}\left[f(x) - \frac{1}{2}\right]$ *does not exist, (where* $[\cdot]$ *denotes the greatest integer function), then*

(a) $f(1)$ may be integer

(b) $\{f(x)\} = \frac{1}{2}, \forall x \in R$ ({·} fractional part of x)

(c) $f(1)$ is not an integer

(d) None of the above

Sol. (c) $\lim_{x\to 1}\left[f(x) - \frac{1}{2}\right]$ does not exist, when $f(1)$ is not an integer, as $f'(1) = -3$, i.e. decreasing in the neighbourhood at $x = 1$.

Ex.34 $\lim_{x\to 0}\left[(1 - e^x) \cdot \frac{\sin x}{|x|}\right]$ *(where* $[\cdot]$ *denotes the greatest integer function), equals*

(a) 0 (b) 1 (c) −1 (d) Doesn't exist

Sol. (c) Let $f(x) = \begin{cases} (1 - e^x) \cdot \frac{\sin x}{x}, & x > 0 \\ (e^x - 1) \cdot \frac{\sin x}{x}, & x < 0 \end{cases}$.

$$\therefore \quad f(x) = \begin{cases} < 0, & \text{for } x > 0 \\ < 0, & \text{for } x < 0 \end{cases} \Rightarrow \lim_{x\to 0} [f(x)] = -1$$

Ex.35 $\lim_{x\to 1}\left[\operatorname{cosec}\frac{\pi x}{2}\right]^{-1/(1-x)}$ *is equal to (where* $[\cdot]$ *denotes the greatest integer function).*

(a) 0 (b) 1 (c) ∞ (d) Doesn't exist

Sol. (b) Let $f(x) = \operatorname{cosec}\left(\frac{\pi x}{2}\right)$

$$\Rightarrow \quad f(1^+) = \lim_{h\to 0} \operatorname{cosec}\left(\frac{\pi}{2}(1+h)\right) > 1$$

$$\text{and} \quad f(1^-) = \lim_{h\to 0} \operatorname{cosec}\left(\frac{\pi}{2}(1-h)\right) > 1$$

$$\therefore \lim_{x\to 1}\left[\operatorname{cosec}\frac{\pi x}{2}\right] = 1 \Rightarrow \lim_{x\to 1}\left[\operatorname{cosec}\frac{\pi x}{2}\right]^{\frac{-1}{1-x}} = 1$$

JEE Type Solved Examples : Statements I and II Type Questions

- **Directions** (Ex. Nos. 36 to 38) This section is based on Statements I and II. Select the correct answer from the codes given below.
  (a) Statement I is true, Statement II is true; Statement II is correct explanation for Statement I
  (b) Statement I is true, Statement II is true; Statement II is not the correct explanation for Statement I
  (c) Statement I is true, Statement II is false
  (d) Statement I is false, Statement II is true

- ***Ex. 36***

Statement I $\lim_{x \to \pi/2} \dfrac{\sin(\cot^2 x)}{(\pi - 2x)^2} = \dfrac{1}{2}$.

Statement II $\lim_{\theta \to 0} \dfrac{\sin\theta}{\theta} = 1$ *and* $\lim_{\theta \to 0} \dfrac{\tan\theta}{\theta} = 1$, *where* θ *is measured in radians.*

Sol. *(d)* $\lim_{x \to \pi/2} \dfrac{\sin(\cot^2 x)}{\cot^2 x} \cdot \dfrac{\cot^2 x}{(\pi - 2x)^2}$; put $x = \dfrac{\pi}{2} - h$

$\Rightarrow \lim_{h \to 0} \dfrac{\tan^2 h}{4h^2} = \dfrac{1}{4}$

$\therefore$ Statement I is false and Statement II is true.

- ***Ex. 37*** **Statement I** $\lim_{m \to \infty} \lim_{n \to \infty} \{\sin^{2m} n!\pi n\} = 0 \; m, n \in N$, when x is rational.

Statement II When $n \to \infty$ and x is rational $n!x$, is integer.

Sol. *(a)* When $n \to \infty$ and x is rational $x = \dfrac{b}{q}$.

Where p and q are integers and $q \neq 0$ $n!x = n! \times \dfrac{p}{q}$ is integer as $n!$ has factor q, when $n \to \infty$

Also, when $n!\,x$ is integer, $\sin(n!\pi x) = 0$.

Therefore, the given limit is zero.

- ***Ex. 38*** **Statement I** *If* $\lim_{x \to 0} \left\{ f(x) + \dfrac{\sin x}{x} \right\}$ *does not exist, then* $\lim_{x \to 0} f(x)$ *does not exist.*

Statement II $\lim_{x \to 0} \left(\dfrac{e^{1/x} - 1}{e^{1/x} + 1} \right)$ does not exist.

Sol. *(b)* If $\lim_{x \to 0} f(x)$ exist, then $\lim_{x \to 0} \left(f(x) + \dfrac{\sin x}{x} \right)$ always exist as $\lim_{x \to 0} \dfrac{\sin x}{x}$ exists finitely.

Hence, $\lim_{x \to 0} f(x)$ must not exist.

JEE Type Solved Examples : Matching Type Questions

- ***Ex. 39*** *Match the following.*

	Column I	Column II	
(A)	$\lim_{x \to \infty} \left(\dfrac{x}{1 + x} \right)^x$ equals	(p)	e^2
(B)	$\lim_{x \to \infty} \left(\sin\dfrac{1}{x} + \cos\dfrac{1}{x} \right)^x$ equals	(q)	$e^{-1/2}$
(C)	$\lim_{x \to 0} (\cos x)^{\cot^2 x}$ equals	(r)	e
(D)	$\lim_{x \to 0} \left[\tan\left(\dfrac{\pi}{4} + x \right) \right]^{1/x}$ equals	(s)	e^{-1}

Sol. (A) $\to$ (s); (B) $\to$ (r); (C) $\to$ (q); (D) $\to$ (p)

(A) Put $x = \dfrac{1}{y}$, $\lim_{y \to 0} \left(\dfrac{1}{1 + y} \right)^{1/y} = e^{\lim_{y \to 0} \frac{1 - y - 1}{y(1 + y)}} = e^{-1}$

(B) $\lim_{y \to 0} (\sin y + \cos y)^{1/y} = e^{\lim_{y \to 0} \frac{\sin y + \cos y - 1}{y}} = e$

(C) $e^{\lim_{x \to 0} \frac{\cos x - 1}{\frac{\tan^2 x}{x^2} \cdot x^2}} = e^{-\frac{1}{2}}$

(D) $e^{\lim_{x \to 0} \frac{\tan((\pi/4) + x) - \tan(\pi/4)}{x}}$

$= e^{\lim_{x \to 0} \frac{\tan x [1 + \tan((\pi/4) + x) \cdot 1]}{x}} = e^2$

Ex. 40 *Match the following.*

	Column I	Column II	
(A)	$\lim_{x \to \infty} (\sqrt{x+\sqrt{x}} - \sqrt{x-\sqrt{x}})$ equals	(p)	-2
(B)	The value of the $\lim_{x \to 0} \frac{\sin 2x - 2\tan x}{\ln(1+x^3)}$ is	(q)	-1
(C)	$\lim_{x \to 0^+} (\ln \sin^3 x - \ln(x^4 + ex^3))$ equals	(r)	0
(D)	Let $\tan(2\pi \mid \sin\theta \mid) = \cot(2\pi \mid \cos\theta \mid)$, where $\theta \in R$ and $f(x) = (\mid\sin\theta\mid + \mid\cos\theta\mid)^x$. The value of $\lim_{x \to \infty} \left[\frac{2}{f(x)}\right]$ equals (here, [·] represents greatest integer function)	(s)	1

Sol. (A) → (s); (B) → (p); (C) → (q); (D) → (r)

(A) $\lim_{x \to \infty} \frac{(x+\sqrt{x})-(x-\sqrt{x})}{\sqrt{x+\sqrt{x}}+\sqrt{x-\sqrt{x}}}$

$$= \lim_{x \to \infty} \frac{2\sqrt{x}}{\sqrt{x+\sqrt{x}}+\sqrt{x-\sqrt{x}}} = 1$$

(B) $\lim_{x \to 0} \frac{\sin 2x - 2\tan x}{x^3} \lim_{x \to 0} \frac{x^3}{\log(1+x^3)}$

$$= \lim_{x \to 0} \frac{2\sin x(\cos^2 x - 1)}{\cos x \cdot x^3} \cdot 1 = \lim_{x \to 0} -\frac{2\tan^3 x}{x^3} = -2$$

(C) $\lim_{x \to 0^+} [\ln \sin^3 x - \ln(x^4 + ex^3)] = \lim_{x \to 0} \ln\left(\frac{\sin^3 x}{x^3(x+e)}\right)$

$$= \ln\left(\frac{1}{e}\right) = -1$$

(D) $\tan(2\pi \mid \sin\theta \mid) = \tan\left(\frac{\pi}{2} - \mid\cos\theta\mid 2\pi\right)$

$$\Rightarrow \quad 2\pi \mid\sin\theta\mid = n\pi + \frac{\pi}{2} - \mid\cos\theta\mid 2\pi$$

$$\Rightarrow 2\pi(\mid\sin\theta\mid + \mid\cos\theta\mid) = n\pi + \frac{\pi}{2}$$

$$\Rightarrow \quad \mid\sin\theta\mid + \mid\cos\theta\mid = \frac{n}{2} + \frac{1}{4} \quad \text{...(i)}$$

Since, $1 \le \mid\sin\theta\mid + \mid\cos\theta\mid \le \sqrt{2}; 1 \le \frac{n}{2} + \frac{1}{4} \le \sqrt{2}$

$$4 \le 2n + 1 \le 4\sqrt{2}$$

$$\Rightarrow \frac{3}{2} \le n \le \frac{4\sqrt{2}-1}{2}$$

Thus, $n = 2$ is only possible value. Putting in Eq. (i),

$$\mid\sin\theta\mid + \mid\cos\theta\mid = \frac{5}{4},$$

$$g(x) = \lim_{x \to \infty} \left[2\left(\frac{4}{5}\right)^x\right] = 0$$

JEE Type Solved Examples : Single Integer Answer Type Questions

Ex. 41 *If $\lim_{x \to 0} (x^{-3}\sin 3x + ax^{-2} + b)$ exists and is equal to zero, the value of $a + 2b$ is ....*

Sol. (6) We have, $\lim_{x \to 0} \frac{\sin 3x}{x^3} + \frac{a}{x^2} + b = \lim_{x \to 0} \frac{\sin 3x + ax + bx^3}{x^3}$

$$= \lim_{x \to 0} \frac{3\frac{\sin 3x}{3x} + a + bx^2}{x^2} \text{ for}$$

existence of limit $3 + a = 0 \Rightarrow a = -3$

$$\therefore \quad l = \lim_{x \to 0} \frac{\sin 3x - 3x + bx^3}{x^3}$$

$$= 27 \cdot \frac{\sin t - t}{t^3} + b = 0 \quad [\because 3x = t]$$

$$= -\frac{27}{6} + b = 0 \Rightarrow b = \frac{9}{2}$$

Hence, $\quad a + 2b = -3 + 2 \times \frac{9}{2} = 6$

[using L' Hospital's rule]

Ex. 42 *For a certain value of c, $\lim_{x \to -\infty} [(x^5 + 7x^4 + 2)^c - x] = \lambda$, is finite and non-zero. The value of $3c + \lambda$ is ....*

Sol. (2) We have, $\lim_{x \to -\infty} [(x^5 + 7x^4 + 2)^c - x]$

$$= \lim_{x \to -\infty} \left(x^{5c}\left(1 + \frac{7}{x} + \frac{2}{x^5}\right)^c - x\right)$$

$$= \lim_{x \to -\infty} x\left(x^{5c-1}\left(1 + \frac{7}{x} + \frac{2}{x^5}\right)^c - 1\right)$$

This will be of the form $\infty \times 0$ only, if

$5c - 1 = 0 \Rightarrow c = \frac{1}{5}$ substituting $c = \frac{1}{5}$, λ becomes

$$\lambda = \lim_{x \to -\infty} x[(1+x)^{1/5} - 1], \text{ where } x = \frac{7}{x} + \frac{2}{x^5}$$

$$= \lim_{x \to -\infty} x\left[1 + \frac{x}{5} + \ldots - 1\right] = \lim_{x \to -\infty} x\left(\frac{7}{x} + \frac{2}{x^5}\right) \cdot \frac{1}{5} = \frac{7}{5}$$

Hence, $c = \frac{1}{5}$ and $\lambda = \frac{7}{5} \Rightarrow 3c + \lambda = 2$

Ex. 43 *Consider the curve $y = \tan^{-1} x$ and a point $A\left(1, \frac{\pi}{4}\right)$ on it. If the variable point $P_i(x_i, y_i)$ moves on the curve for $i = 1, 2, 3, \ldots, n\ (n \in N)$ such that $y_r = \sum_{m=1}^{r} \tan^{-1}\left(\frac{1}{2m^2}\right)$ and $B(x, y)$ be the limiting position of variable point P_n as $n \to \infty$, the value of reciprocal of the slope of AB will be ....*

Sol. (2) Here, $y = \lim_{n \to \infty} y_n = \lim_{n \to \infty} \sum_{m=1}^{n} \tan^{-1}\left(\frac{1}{2m^2}\right)$

$$= \lim_{n \to \infty} \sum_{m=1}^{n} \tan^{-1}\left(\frac{2}{1 + (2m+1)(2m-1)}\right)$$

$$= \lim_{n \to \infty} \sum_{m=1}^{n} \tan^{-1}\left[\frac{(2m+1) - (2m-1)}{1 + (2m+1)(2m-1)}\right]$$

$$= \lim_{n \to \infty} \sum_{m=1}^{n} (\tan^{-1}(2m+1) - \tan^{-1}(2m-1))$$

$$= \lim_{n \to \infty} (\tan^{-1}3 - \tan^{-1}1) + (\tan^{-1}5 - \tan^{-1}3) + \ldots + (\tan^{-1}(2n+1) - \tan^{-1}(2n-1))$$

$$= \lim_{n \to \infty} (\tan^{-1}(2n+1) - \tan^{-1}(1))$$

$$= \frac{\pi}{2} - \frac{\pi}{4} = \frac{\pi}{4}$$

$\because B \to \left[1, \frac{\pi}{4}\right]$, i.e. coordinates of B approaches towards those of A.

Chord AB approaches to the tangent to $y = f(x)$ at A.

$\therefore$ Slope of $AB = \left(\frac{d}{dx}\tan^{-1} x\right)_{\text{at } x = 1}$

$$= \left(\frac{1}{1 + x^2}\right)_{\text{at } x = 1} = \frac{1}{2}$$

$\Rightarrow$ (Slope of $AB)^{-1} = 2$

Ex. 44 *If $\lim_{x \to 0} \int_0^x \frac{2t\,dt}{(e^x - 1 - x)\sqrt{\frac{2a}{3} - \frac{t}{2} + 104}} = \frac{1}{19}$, then $\frac{a}{2010}$ equals ....*

Sol. (1) Here, $\lim_{x \to 0} \frac{\int_0^x \frac{2t}{\sqrt{\frac{2a}{3} - \frac{t}{2} + 104}}\,dt}{e^x - 1 - x} = \frac{1}{19}$

[using L'Hospital's rule]

$$\Rightarrow \lim_{x \to 0} \frac{2x}{\sqrt{\frac{2a}{3} - \frac{x}{2} + 104}\,(e^x - 1)} = \frac{1}{19}$$

$$\Rightarrow 2\lim_{x \to 0} \frac{1}{\sqrt{\frac{2a}{3} - \frac{x}{2} + 104}} \cdot \lim_{x \to 0} \frac{x}{e^x - 1} = \frac{1}{19}$$

$$\Rightarrow 2 \cdot \frac{1}{\sqrt{\frac{2a}{3} + 104}} \cdot 1 = \frac{1}{19}$$

$$\Rightarrow \frac{2a}{3} + 104 = 1444 \Rightarrow 2a = 4020 \Rightarrow \frac{a}{2010} = 1$$

Ex. 45 *Evaluate $\lim_{x \to 0^+} \log_{\sin x} \sin 2x$.*

Sol. (1) Here, $\lim_{x \to 0^+} \log_{\sin x} \sin 2x = \lim_{x \to 0^+} \frac{\log \sin 2x}{\log \sin x}$ $\left[\frac{-\infty}{-\infty}\text{ form}\right]$

$$= \lim_{x \to 0^+} \frac{\frac{1}{\sin 2x} \cdot 2\cos 2x}{\frac{1}{\sin x} \cdot \cos x}$$ [using L'Hospital's rule]

$$= \lim_{x \to 0^+} \frac{\left[\frac{(2x)}{\sin(2x)}\right]\cos 2x}{\left(\frac{x}{\sin x}\right)\cos x} = \lim_{x \to 0^+} \frac{\cos 2x}{\cos x} = 1$$

$\left[\because \lim_{x \to 0} \frac{\sin x}{x} = 1\right]$

Subjective Type Questions

Ex. 46 *Evaluate a, b, c and d, if*

$$\lim_{x \to \infty} \left(\sqrt{x^4 + ax^3 + 3x^2 + bx + 2} - \sqrt{x^4 + 2x^3 - cx^2 + 3x - d}\right) = 4.$$

Sol. Here, $\lim_{x \to \infty} \left(\sqrt{x^4 + ax^3 + 3x^2 + bx + 2} - \sqrt{x^4 + 2x^3 - cx^2 + 3x - d}\right) = 4$ $[\infty - \infty \text{ form}]$

Rationalising

$$\lim_{x \to \infty} \frac{(a-2)x^3 + (3+c)x^2 + (b-3)x + (2+d)}{\left[\sqrt{x^4 + ax^3 + 3x^2 + bx + 2} + \sqrt{x^4 + 2x^3 - cx^2 + 3x - d}\right]} = 4$$

Since, limit is finite, the degree of the numerator must be 2. So, $a - 2 = 0$, i.e. $a = 2$.

$$\Rightarrow \lim_{x \to \infty} \frac{(3+c)x^2 + (b-3)x + (2+d)}{\left[\sqrt{x^4 + ax^3 + 3x^2 + bx + 2} + \sqrt{x^4 + 2x^3 - cx^2 + 3x - d}\right]} = 4$$

On dividing numerator and denominator by x^2, we get

$$\lim_{x \to \infty} \frac{(3+c) + (b-3)/x + (2+d)/x^2}{\sqrt{1 + \frac{a}{x} + \frac{3}{x^2} + \frac{b}{x^3} + \frac{2}{x^4}} + \sqrt{1 + \frac{2}{x} - \frac{c}{x^2} + \frac{3}{x^3} - \frac{d}{x^4}}} = 4 \Rightarrow \frac{3+c}{2} = 4$$

$\Rightarrow \quad c = 5$

$\therefore \quad c = 5,\ a = 2$

Hence, $a = 2$, $c = 5$ and b, d are any real numbers.

Ex. 47 *If x is a real number in [0, 1]. Find the value of*

$$\lim_{m \to \infty} \lim_{n \to \infty} [1 + \cos^{2m}(n!\pi x)].$$

Sol. If x is a real number in [0, 1], then we have two cases either.

Case I $x \in Q$ [rational number]

As $x \in Q$, we have $x = 0, \frac{1}{2}, \frac{1}{3}, \frac{1}{4}, \ldots, 1$

which shows $(n!\pi x)$ is integral multiple of π for large values of n.

$\therefore \quad \cos(n!\pi x) = \pm 1$

Thus, $\lim_{m \to \infty} \lim_{n \to \infty} [1 + \cos^{2m}(n!\pi x)] = \lim_{m \to \infty} \lim_{n \to \infty} (1+1) = 2$

Case II $x \notin Q$ [irrational number]

As $x \notin Q$, we have $x = \frac{1}{\sqrt{2}}, \frac{1}{\sqrt{3}}, \frac{1}{\sqrt{5}}, \ldots$

which shows $(n!\pi x)$ is not an integral multiple of π and so, $\cos(n!\pi x)$ will lie between -1 and $+1$.

Thus, $\lim_{m \to \infty} \lim_{n \to \infty} [1 + \cos^{2m}(n!\pi x)]$

$= \lim_{m \to \infty} \lim_{n \to \infty} [1 + (\text{a value between } -1 \text{ and } +1)^{2m}]$

$= \lim_{n \to \infty} [1+0] = 1$ $\quad$ [as $0 < x < 1 \Rightarrow x^\infty = 0$]

Thus, $\lim_{m \to \infty} \lim_{n \to \infty} [1 + \cos^{2m}(n!\pi x)] = \begin{cases} 2, & x \in Q \\ 1, & x \notin Q \end{cases}$

Ex. 48 *If $a_1 = 1$ and $a_{n+1} = \dfrac{4 + 3a_n}{3 + 2a_n}$, $n \geq 1$, show that $a_{n+2} > a_{n+1}$ and if a_n has a limit l as $n \to \infty$, then evaluate $\lim_{n \to \infty} a_n$.*

Sol Here, $a_1 = 1$

$\therefore \quad a_2 = \frac{4+3}{3+2} = \frac{7}{5} > 1$

$\therefore \quad a_2 > a_1$

Assuming $a_{n+1} > a_n$

$$\therefore a_{n+2} - a_{n+1} = \frac{4 + 3a_{n+1}}{3 + 2a_{n+1}} - \frac{4 + 3a_n}{3 + 2a_n}$$

$$= \frac{(4 + 3a_{n+1})(3 + 2a_n) - (4 + 3a_n)(3 + 2a_{n+1})}{(3 + 2a_{n+1})(3 + 2a_n)}$$

$$= \frac{a_{n+1} - a_n}{(3 + 2a_{n+1})(3 + 2a_n)} > 0 \quad [\because a_{n+1} > a_n]$$

$\therefore \quad a_{n+2} - a_{n+1} > 0$

$\Rightarrow \quad a_{n+2} > a_{n+1}$

whenever $a_{n+1} > a_n$.

$\therefore$ The sequence of values a_n is increasing and since $a_1 = 1$, $a_n > 0$, for all n.

Now, let $\quad l = \lim_{n \to \infty} a_n = \lim_{n \to \infty} a_{n+1}$

$\therefore \quad l = \lim_{n \to \infty} \frac{4 + 3a_n}{3 + 2a_n}$

$\Rightarrow \quad l = \frac{4 + 3l}{3 + 2l}$ $\quad \left[\because \lim_{n \to \infty} a_n = l\right]$

$\Rightarrow \quad 3l + 2l^2 = 4 + 3l$

$\Rightarrow \quad l^2 = 2$

$\therefore \quad l = \sqrt{2}$ $\quad$ [neglecting $l = -\sqrt{2}$, as $l > 0$]

Ex. 49 *Let $a_1, a_2, \ldots, a_n$ be sequence of real numbers with $a_{n+1} = a_n + \sqrt{1 + a_n^2}$ and $a_0 = 0$. Prove that* $\lim_{n \to \infty} \left(\frac{a_n}{2^{n-1}} \right) = \frac{4}{\pi}$.

Sol. Here, $a_{n+1} = a_n + \sqrt{1 + a_n^2}$, let $a_n = \cot(\alpha_n)$

$$\Rightarrow \quad a_{n+1} = \cot(\alpha_n) + \operatorname{cosec}(\alpha_n)$$

$$\Rightarrow \quad a_{n+1} = \frac{\cos(\alpha_n) + 1}{\sin(\alpha_n)}$$

$$= \frac{2\cos^2(\alpha_n/2)}{2\sin(\alpha_n/2)\cos(\alpha_n/2)} = \cot\left(\frac{\alpha_n}{2}\right)$$

Putting $n = 1, a_1 = \cot(\alpha_1)$

and $a_1 = a_0 + \sqrt{1 + a_0^2} = 1$

$$\Rightarrow \quad \cot(\alpha_1) = 1 \text{ or } \alpha_1 = \frac{\pi}{4}$$

Again, $a_2 = \cot\left(\frac{\alpha_1}{2}\right) = \cot\left(\frac{\pi}{8}\right)$

$$a_3 = \cot\left(\frac{\alpha_2}{2}\right) = \cot\left(\frac{\pi}{4 \cdot 2^2}\right)$$

$$a_4 = \cot\left(\frac{\alpha_3}{2}\right) = \cot\left(\frac{\pi}{4 \cdot 2^3}\right)$$

....

....

$$a_n = \cot\left(\frac{\pi}{4 \cdot 2^{n-1}}\right);$$

Hence,

$$\lim_{n \to \infty} \left(\frac{a_n}{2^{n-1}} \right) = \lim_{n \to \infty} \frac{\cot\left(\frac{\pi}{4 \cdot 2^{n-1}}\right)}{2^{n-1}} = \lim_{x \to 0} \frac{\frac{1}{\tan\left(\frac{\pi}{4} x\right)}}{x}$$

$$\left[\text{put } \frac{1}{2^{n-1}} = x\right]$$

$$\therefore \quad \lim_{n \to \infty} \left(\frac{a_n}{2^{n-1}} \right) = \frac{4}{\pi}$$

Ex. 50 *A square is inscribed in a circle of radius R, a circle is inscribed in this square then a square in this circle and so on, n times. Find the limit of the sum of areas of all the squares as $n \to \infty$.*

Sol. Let the side of a square, S_1 be 'a' units.

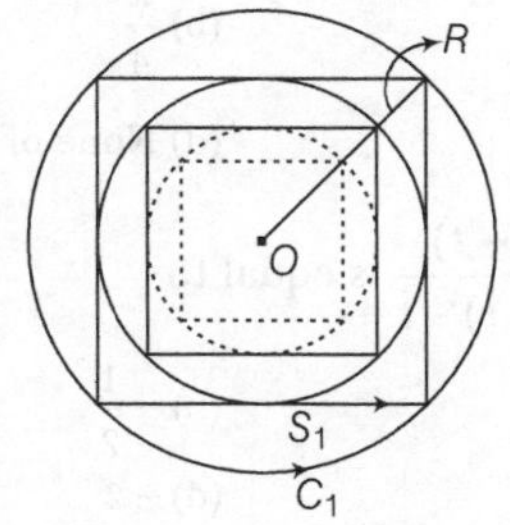

Then, $a\sqrt{2} = 2R \Rightarrow R = \frac{a}{\sqrt{2}}$ is radius of circle C_1.

If a_1 be the side of another square, then

$$a_1\sqrt{2} = a \quad \Rightarrow \quad a_1 = \frac{a}{\sqrt{2}}$$

$$a_2\sqrt{2} = a_1 \Rightarrow a_2 = \frac{a_1}{\sqrt{2}} = \frac{a}{2}$$

....

....

So, sum of areas of all the squares,

$$S_n = a^2 + a_1^2 + a_2^2 + \ldots \text{ upto } n \text{ terms}$$

$$= a^2 + \frac{a^2}{2} + \frac{a^2}{4} + \ldots \text{ upto } n \text{ terms}$$

$$= a^2 \left(1 + \frac{1}{2} + \frac{1}{4} + \frac{1}{8} + \ldots \text{ upto } n \text{ terms}\right) = a^2 \left\{\frac{1 - \frac{1}{2^n}}{1 - \frac{1}{2}}\right\}$$

$$= 2a^2 \left(1 - \frac{1}{2^n}\right)$$

$$\therefore \quad \lim_{n \to \infty} S_n = \lim_{n \to \infty} 2a^2 \left(1 - \frac{1}{2^n}\right) = 2a^2 = 4R^2$$

$$\left[\text{as } n \to \infty \Rightarrow \frac{1}{2^n} \to 0\right]$$

Limits Exercise 1 : Single Option Correct Type Questions

1. $\lim_{x \to 0} \frac{\sin(\pi \cos^2(\tan(\sin x)))}{x^2}$ is equal to

(a) π (b) $\frac{\pi}{4}$

(c) $\frac{\pi}{2}$ (d) None of these

2. $\lim_{t \to 0} \frac{1-(1+t)^t}{\ln(1+t)-t}$ is equal to

(a) $\frac{1}{2}$ (b) $-\frac{1}{2}$

(c) 2 (d) -2

3. If $I_1 = \lim_{x \to \infty} (\tan^{-1} \pi x - \tan^{-1} x) \cos x$ and $I_2 = \lim_{x \to 0} (\tan^{-1} \pi x - \tan^{-1} x) \cos x$, then (I_1, I_2) is

(a) (0, 0)
(b) (0, 1)
(c) (1, 0)
(d) None of the above

4. If $f(x) = 0$ is a quadratic equation such that $f(-\pi) = f(\pi) = 0$ and $f\left(\frac{\pi}{2}\right) = -\frac{3\pi^2}{4}$,

then $\lim_{x \to -\pi} \frac{f(x)}{\sin(\sin x)}$ is equal to

(a) 0 (b) π

(c) 2π (d) None of these

5. $\lim_{x \to 0} \frac{\sqrt[3]{1+\sin^2 x} - \sqrt[4]{1-2\tan x}}{\sin x + \tan^2 x}$ is equal to

(a) -1 (b) 1

(c) $\frac{1}{2}$ (d) $-\frac{1}{2}$

6. If $x_{n+1} = \sqrt{\frac{1+x_n}{2}}$ and $|x_0| < 1, n \geq 0$, then

$\lim_{n \to \infty} \left(\frac{\sqrt{1-x_0^2}}{x_1 x_2 x_3 \ldots x_{n+1}} \right)$ is equal to

(a) -1
(b) 1
(c) $\cos^{-1}(x_0 + 1)$
(d) $\cos^{-1}(x_0)$

7. For $n \in N$, let $f_n(x) = \tan \frac{x}{2}(1+\sec x)(1+\sec 2x)$ $(1+\sec 4x)\ldots(1+\sec 2^n x)$. Then, $\lim_{x \to 0} \frac{f_n(x)}{2x}$ is

(a) 0 (b) 2^n

(c) 2^{n-1} (d) 2^{n+1}

8. Let $f(x)$ be a real valued function defined for all $x \geq 1$, satisfying $f(1) = 1$ and $f'(x) = \frac{1}{x^2 + (f(x))}$; then $\lim_{x \to \infty} f(x)$

(a) doesn't exist
(b) exists and less than $\frac{\pi}{4}$
(c) exists and less than $1 + \frac{\pi}{4}$
(d) exists and equal to 0

9. The quadratic equation whose roots are the minimum value of $\sin^2 \theta - \sin \theta + \frac{1}{2}$ and $\lim_{x \to \infty} \sqrt{(x+1)(x+2)} - x$ is

(a) $3x^2 - 7x + 3 = 0$ (b) $8x^2 - 14x + 3 = 0$

(c) $x^2 - 7x + 3 = 0$ (d) $2x^2 - 7x + 3 = 0$

10. If $x_1 = \sqrt{3}$ and $x_{n+1} = \frac{x_n}{1+\sqrt{1+x_n^2}}, \forall n \in N$, then $\lim_{x \to \infty} 2^n x_n$ equal to

(a) $\frac{3}{2\pi}$ (b) $\frac{2}{3\pi}$ (c) $\frac{2\pi}{3}$ (d) $\frac{3\pi}{2}$

11. $\lim_{x \to a^-} \frac{\sqrt{x-b} - \sqrt{a-b}}{(x^2 - a^2)}, (a > b)$ is

(a) $\frac{1}{4a}$ (b) $\frac{1}{a\sqrt{a-b}}$

(c) $\frac{1}{2a\sqrt{a-b}}$ (d) $\frac{1}{4a\sqrt{a-b}}$

12. $\lim_{n \to \infty} (\sin^n 1 + \cos^n 1)^n$ is equal to

(a) cot 1 (b) tan 1 (c) cos 1 (d) sin 1

13. The value of $\lim_{x \to 0} \left(\frac{2}{x^3} (\sin^{-1} x - \tan^{-1} x) \right)^{2/x^2}$ equals

(a) e (b) $\sqrt{e}$ (c) $\frac{1}{e}$ (d) $\frac{1}{\sqrt{e}}$

14. The value of $\lim_{n \to \infty} \sum_{k=1}^{n} \frac{n-K}{n^2} \cos\left(\frac{4K}{n}\right)$ equals

(a) $\frac{1}{4}\sin 4 + \frac{1}{16}\cos 4 - \frac{1}{16}$
(b) $\frac{1}{4}\sin 4 - \frac{1}{16}\cos 4 + \frac{1}{16}$
(c) $\frac{1}{16}(1 - \sin 4)$
(d) $\frac{1}{16}(1 - \cos 4)$

15. If $\lim_{x\to 0} \dfrac{1-\sqrt{\cos 2x}\cdot\sqrt[3]{\cos 3x}\cdot\sqrt[4]{\cos 4x}\ldots\sqrt[n]{\cos nx}}{x^2}$ has the value equal to 10, the value of n is

(a) 6 (b) 7 (c) 8 (d) 9

16. $\lim_{z\to\infty} \dfrac{\int_{1/2}^{z}[\cot^{-1}x]dx}{\int_{1/2}^{z}\left[1+\dfrac{1}{x}\right]dx}$, where [·] denotes the greatest integer function, equals

(a) 0 (b) 1

(c) cot 1 (d) not defined

17. If α and β are roots of $x^2-(\sqrt{1-\cos 2\theta})x+\theta=0$, where $0<\theta<\dfrac{\pi}{2}$. Then, $\lim_{\theta\to 0^+}\left(\dfrac{1}{\alpha}+\dfrac{1}{\beta}\right)$ is

(a) $\dfrac{1}{\sqrt{2}}$ (b) $-\sqrt{2}$

(c) $\sqrt{2}$ (d) None of these

18. If $f(x)=\dfrac{1}{3}\left(f(x+1)+\dfrac{5}{f(x+2)}\right)$ and $f(x)>0, \forall x\in R$, then $\lim_{x\to\infty} f(x)$ is

(a) 0 (b) $\sqrt{\dfrac{2}{5}}$ (c) $\sqrt{\dfrac{5}{2}}$ (d) ∞

19. Let $f:(1,2)\to R$ satisfies the inequality $\dfrac{\cos(2x+4)-33}{2}<f(x)<\dfrac{x^2|4x-8|}{x-2}, \forall x\in(1,2)$. Then, $\lim_{x\to 2^-} f(x)$ is

(a) 16 (b) −16

(c) 8 (d) doesn't exist

20. Let $f(x)$ be polynomial of degree 4 with roots 1, 2, 3, 4 and leading coefficient 1 and $g(x)$ be the polynomial of degree 4 with roots $1, \dfrac{1}{2}, \dfrac{1}{3}$ and $\dfrac{1}{4}$ with leading coefficient 1. Then, $\lim_{x\to 1}\dfrac{f(x)}{g(x)}$ equals

(a) $\dfrac{1}{24}$ (b) −24 (c) $\dfrac{1}{12}$ (d) $-\dfrac{1}{12}$

21. The value of $\lim_{x\to\frac{\pi}{4}}\dfrac{4\sqrt{2}-(\cos x+\sin x)^5}{1-\sin 2x}$ is

(a) $\sqrt{2}$ (b) $3\sqrt{5}$

(c) $5\sqrt{2}$ (d) $-5\sqrt{2}$

22. If $\lim_{n\to\infty}\dfrac{n\cdot 3^n}{n(x-2)^n+n\cdot 3^{n+1}-3^n}=\dfrac{1}{3}$, where $n\in N$, the number of integer(s) in the range of 'x' is

(a) 3 (b) 4

(c) 5 (d) infinite

23. If $f(x)=\lim_{n\to\infty}\dfrac{1}{\left(\dfrac{3}{\pi}\tan^{-1}2x\right)^{2n}+5}$, then the set of values of x for which $f(x)=0$, is

(a) $|2x|>\sqrt{3}$ (b) $|2x|<\sqrt{3}$

(c) $|2x|\ge\sqrt{3}$ (d) $|2x|\le\sqrt{3}$

24. The integer 'n' for which the

$\lim_{x\to 0}\dfrac{\cos^2 x-\cos x-e^x\cos x+e^x-\dfrac{x^3}{2}}{x^n}$ is a finite non-zero number, is

(a) 2 (b) 3

(c) 4 (d) None of these

25. If $I_1=\lim_{x\to 0}\sqrt{\dfrac{\tan^{-1}x}{x}-\dfrac{\sin^{-1}x}{x}}$

and $I_2=\lim_{x\to 0}\sqrt{\dfrac{\sin^{-1}x}{x}-\dfrac{\tan^{-1}x}{x}}$,

where $|x|<1$, which of the following statement is true?

(a) Neither I_1 nor I_2 exist

(b) I_1 exists and I_2 doesn't exist

(c) I_1 doesn't exist and I_2 exists

(d) None of the above

26. The value of $\lim_{x\to\pi/2}\dfrac{\left[\dfrac{x}{2}\right]}{\log(\sin x)}$ is equal to

(a) 0 (b) 1

(c) −1 (d) doesn't exist

27. The value of $\lim_{n\to\infty}\left(1+\dfrac{1}{a_1}\right)\left(1+\dfrac{1}{a_2}\right)\ldots\left(1+\dfrac{1}{a_n}\right)$, where $a_1=1$ and $a_n=n(1+a_{n-1}), \forall n\ge 2$, is

(a) e (b) $\dfrac{e}{2}$

(c) $2e$ (d) $3e$

28. If $f(x+y)=f(x)+f(y)$ for all $x, y\in R$ and $f(1)=1$, then $\lim_{x\to 0}\dfrac{2^{f(\tan x)}-2^{f(\sin x)}}{x^2 f(\sin x)}$ equals

(a) log 2 (b) log 4 (c) $\log\sqrt{2}$ (d) log 8

29. The value of $\lim_{n\to\infty} n^{-n^2}\left[(n+1)\left(n+\dfrac{1}{2}\right)\left(n+\dfrac{1}{2^2}\right)\ldots\left(n+\dfrac{1}{2^{n-1}}\right)\right]^n$ is

(a) e

(b) e^2

(c) e^3

(d) e^4

30. If $f(x)=\begin{cases}\dfrac{\sin[x]}{[x]}, & \text{for } [x]\neq 0\\ 0, & \text{for } [x]=0\end{cases}$, where $[x]$ denotes the greatest integer less than or equal to x, then $\lim_{x\to 0} f(x)$ equals

(a) 1 (b) 0
(c) −1 (d) doesn't exist

31. Let $f(x)=\dfrac{|x^3-6x^2+11x-6|}{x^3-6x^2+11x-6}$, then the number of solutions of 'a', where $\lim_{x\to a} f(x)$ doesn't exist is

(a) 3 (b) 2 (c) 1 (d) 4

32. Let the rth term, t_r of a series is given by $t_r=\dfrac{r}{1+r^2+r^4}$. The value of $\lim_{n\to\infty}\sum_{r=1}^{n} t_r$ is

(a) 2 (b) $\frac{1}{2}$ (c) 1 (d) $\frac{1}{4}$

33. The value of $\lim_{n\to\infty}\sum_{r=1}^{n}\cot^{-1}\left(\dfrac{r^3-r+\frac{1}{r}}{2}\right)$ is

(a) π (b) $\frac{\pi}{4}$
(c) $\frac{\pi}{2}$ (d) π

34. Let $(\tan\alpha)\,x+(\sin\alpha)\,y=\alpha$ and $(\alpha\operatorname{cosec}\alpha)\,x+(\cos\alpha)\,y=1$ be two variable straight lines, α being the parameter. Let P be the point of intersection of the lines. In the limiting position, when $\alpha\to 0$, the point of intersection of straight lines is

(a) (2, −1) (b) (2, 1)
(c) (−2, 1) (d) (−2, −1)

35. The polynomial of least degree, such that $\lim_{x\to 0}\left(1+\dfrac{x^2+f(x)}{x^2}\right)^{1/x}=e^2$ is

(a) x^2 (b) x^3+2x^2
(c) $-x^2+2x^3$ (d) None of these

36. Let [ ] represents the greatest integer function less than or equal to x. The value of $\lim_{x\to 0}\left(\left[\dfrac{n\sin x}{x}\right]+\left[\dfrac{n\tan x}{x}\right]\right)$ is

(a) $n+1$ (b) $2n$
(c) $n-1$ (d) $2n-1$

37. The value of $\lim_{x\to a}[\sqrt{2-x}+\sqrt{1+x}]$, where $a\in\left[0,\frac{1}{2}\right]$ and $[\cdot]$ denotes the greatest integer function is

(a) 1 (b) 2
(c) 3 (d) 4

38. The value of

$$\lim_{x\to 0^+}\frac{-1+\sqrt{(\tan x-\sin x)+\sqrt{(\tan x-\sin x)+\sqrt{(\tan x-\sin x)+\ldots\infty}}}}{-1+\sqrt{x^3+\sqrt{x^3+\sqrt{x^3+\ldots\infty}}}}$$ is

(a) $\frac{1}{2}$ (b) $\frac{1}{4}$
(c) $\frac{1}{8}$ (d) 1

39. The value of

$$\lim_{\theta\to 0}\frac{\cos^2(1-\cos^2(1-\cos^2(1\ldots\cos^2\theta))\ldots)}{\sin\left(\dfrac{\pi(\sqrt{\theta+4}-2)}{\theta}\right)}$$ is

(a) 2 (b) $\sqrt{2}$
(c) $\frac{1}{2}$ (d) $\frac{1}{\sqrt{2}}$

40. The value of $\lim_{n\to\infty} a_n$ when $a_{n+1}=\sqrt{2+a_n}$, $n=1,2,3,\ldots$ is

(a) 1 (b) 2
(c) 3 (d) 4

Limits Exercise 2 : More than One Option Correct Type Questions

41. If $\lim_{x\to\infty} 4x\left(\dfrac{\pi}{4}-\tan^{-1}\dfrac{x+1}{x+2}\right)=y^2+4y+5$, then y can be equal to

(a) 1 (b) −1
(c) −4 (d) −3

42. $\lim_{x\to 0}\dfrac{1-\cos(x^2)}{x^3(4^x-1)}$ is equal to

(a) $\frac{1}{2}\ln 2$ (b) $\ln 2$
(c) $\ln 4$ (d) $1-\frac{1}{2}\ln\left(\frac{e^2}{4}\right)$

43. If $f(x)=e^{[\cot x]}$, where $[y]$ represents the greatest integer less than or equal to y, then

(a) $\lim\limits_{x\to\frac{\pi^+}{2}} f(x)=1$ (b) $\lim\limits_{x\to\frac{\pi^+}{2}} f(x)=\frac{1}{e}$

(c) $\lim\limits_{x\to\frac{\pi^-}{2}} f(x)=\frac{1}{e}$ (d) $\lim\limits_{x\to\frac{\pi^-}{2}} f(x)=1$

44. $\lim\limits_{x\to 0}\left[m\dfrac{\sin x}{x}\right]$ is equal to (where $m\in I$ and $[\cdot]$ denotes greatest integer function)

(a) m, if $m<0$

(b) $m-1$, if $m>0$

(c) $m-1$, if $m<0$

(d) m, if $m>0$

45. If $\lim\limits_{x\to 0}(1+ax+bx^2)^{2/x}=e^3$, then

(a) $a=3, b=0$ (b) $a=\frac{3}{2}, b=1$

(c) $a=\frac{3}{2}, b=4$ (d) $a=2, b=3$

46. The graph of the function $y=f(x)$ is shown in the adjacent figure, then correct statement is

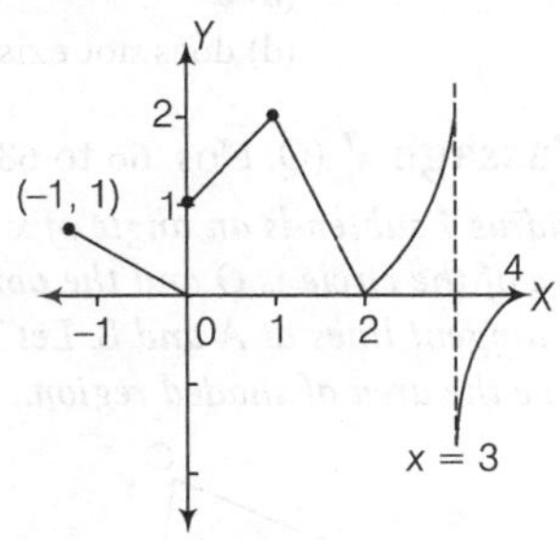

(a) $\lim\limits_{x\to 0^+} f(x)=1$ (b) $\lim\limits_{x\to 1} f(x)=2$

(c) $\lim\limits_{x\to 3} f(x)=$ does not exist (d) $\lim\limits_{x\to 4} f(x)=4$

47. For $\lim\limits_{x\to 0}\dfrac{\cot^{-1}\left(\frac{1}{x}\right)}{x}$

(a) RHL exists (b) LHL does not exist

(c) limit does not exist as RHL is 1 and LHL is -1

(d) limit does not exist as RHL and LHL both are non-existent

48. If $l=\lim\limits_{x\to\infty}\left(\dfrac{x+1}{x-1}\right)^x$, the value of $\{l\}$ and $[l]$ are

(a) 7 (b) $7-e^2$ (c) -7 (d) e^2-7

49. If $f(x)=\begin{cases}\dfrac{\sin[x]}{[x]}, & [x]\neq 0\\ 0, & [x]=0\end{cases}$,

where $[\cdot]$ denotes the greatest integer less than or equal to x, then

(a) $\lim\limits_{x\to 0^-} f(x)=\sin 1$

(b) $\lim\limits_{x\to 0^+} f(x)=0$

(c) limit does not exist at $x=0$

(d) limit exists at $x=0$

50. $\lim\limits_{x\to c} f(x)$ does not exist when

(a) $f(x)=[[x]]-[2x-1]$, $c=3$ (b) $f(x)=[x]-x$, $c=1$

(c) $f(x)=\{x\}^2-\{-x\}^2$, $c=0$ (d) $f(x)=\dfrac{\tan(\text{sgn}\,x)}{(\text{sgn}\,x)}$, $c=0$

(where $[\cdot]$ and $\{\cdot\}$ denotes greatest integer and fractional part of x)

51. Identify the correct statement.

(a) $\lim\limits_{n\to\infty}\left[\sum\limits_{r=1}^{n}\dfrac{1}{2^r}\right]=1$

(b) If $f(x)=(x-1)\{x\}$, where $[\cdot]$ and $\{\cdot\}$ denotes greatest integer function and fractional part of x respectively, the limit of $f(x)$ does not exist at $x=1$

(c) $\lim\limits_{x\to 0^+}\left[\dfrac{\tan x}{x}\right]=1$

(d) $\left[\lim\limits_{x\to 0^+}\dfrac{\tan x}{x}\right]=1$

52. For $a>0$, let $l=\lim\limits_{x\to\frac{\pi}{2}}\dfrac{a^{\cot x}-a^{\cos x}}{\cot x-\cos x}$ and

$m=\lim\limits_{x\to-\infty}(\sqrt{x^2+ax}-\sqrt{x^2-ax})$, then

(a) $l>m$, for all $a>0$ (b) $l>m$, only when $a\geq 1$

(c) $l>m$, for all $a>e^{-a}$ (d) $e^l+m=0$

53. Consider the function $f(x)=\left(\dfrac{ax+1}{bx+2}\right)^x$, where $a, b>0$, the $\lim\limits_{x\to\infty} f(x)$ is

(a) exists for all values of a and b

(b) zero for $a<b$

(c) non-existent for $a>b$

(d) $e^{-(1/a)}$ or $e^{-(1/b)}$, if $a=b$

54. If $f(x)=\dfrac{x\cdot 2^x-x}{1-\cos x}$ and $g(x)=2^x\cdot\sin\left(\dfrac{\log 2}{2^x}\right)$, then

(a) $\lim\limits_{x\to 0} f(x)=\log 2$ (b) $\lim\limits_{x\to\infty} g(x)=\log 4$

(c) $\lim\limits_{x\to 0} f(x)=\log 4$ (d) $\lim\limits_{x\to\infty} g(x)=\log 2$

55. If $\lim\limits_{x\to 3}\dfrac{x^3+cx^2+5x+12}{x^2-7x+12}=l$ (finite real number), then

(a) $l=4$ (b) $c=-6$

(c) $c=4$ (d) $x\in R$

Limits Exercise 3 : Passage Based Questions

Passage I (Q. Nos. 56 to 58)

If $\lim_{x \to a} f(x) = 1$ and $\lim_{x \to a} g(x) = \infty$, *then*

$$\lim_{x \to a} \{f(x)\}^{g(x)} = e^{\lim_{x \to a} (f(x)-1) \times g(x)}.$$

56. $\lim_{x \to 0} \left(\frac{\sin x}{x}\right)^{\frac{\sin x}{x - \sin x}}$ is equal to

(a) $1/e$ (b) $-1/e$
(c) e (d) $-e$

57. $\lim_{x \to 0} \left(\frac{x - 1 + \cos x}{x}\right)^{\frac{1}{x}}$ is equal to

(a) $e^{1/2}$ (b) $e^{-1/2}$ (c) e^{1} (d) $\frac{1}{e}$

58. $\lim_{x \to 0} \left(\frac{a^x + b^x + c^x}{3}\right)^{\frac{2}{x}}$ is equal to

(a) $a^{2/3} + b^{2/3} + c^{2/3}$ (b) abc
(c) $(abc)^{2/3}$ (d) 1

Passage II (Q. Nos. 59 to 61)

Let $f(x) = \lim_{n \to \infty} \left(\cos \sqrt{\frac{x}{n}}\right)^n$, $g(x) = \lim_{n \to \infty} (1 - x + x\sqrt[n]{e})^n$.

Now, consider the function $y = h(x)$, *where* $h(x) = \tan^{-1}(g^{-1}f^{-1}(x))$.

59. $\lim_{x \to 0} \frac{\ln (f(x))}{\ln (g(x))}$ is equal to

(a) $\frac{1}{2}$ (b) $-\frac{1}{2}$
(c) 0 (d) 1

60. Domain of the function $y = h(x)$ is

(a) $(0, \infty)$ (b) R
(c) $(0, 1)$ (d) $[0, 1]$

61. Range of the function $y = h(x)$ is

(a) $\left(0, \frac{\pi}{2}\right)$ (b) $\left(-\frac{\pi}{2}, 0\right)$
(c) R (d) $\left(-\frac{\pi}{2}, \frac{\pi}{2}\right)$

Passage III (Q. Nos. 62 to 63)

Let K_1 = *Total number of ways of selecting of ball from a bag which contains n balls of first colour, (n + 1) balls of second colour, (n + 2) balls of third colour, ..., (2n − 1) balls of n colour.*

K_2 = *number of n-digit numbers using the digits 1, 2, 3, ..., n and* K_3 = *number of ways of arranging (n + 1) objects on a circle.*

62. The value of $\lim_{n \to \infty} \left(\frac{K_1}{K_2}\right)^{1/n}$ is

(a) e^2 (b) $\log 4 - 1$ (c) $\frac{4}{e}$ (d) $\frac{1}{e}$

63. The value of $\lim_{n \to \infty} \left(\frac{K_1}{K_2 + K_3}\right)$ is

(a) e (b) 1
(c) 0 (d) does not exist

Passage IV (Q. Nos. 64 to 65)

Let $f: N \to R$ *and* $g: N \to R$ *be two functions and* $f(1) = 0.8$, $g(1) = 0.6$; $f(n+1) = f(n)\cos(g(n)) - g(n)\sin(g(n))$ *and* $g(n+1) = f(n)\sin(g(n)) + g(n)\cos(g(n))$ *for* $n \geq 1$.

64. $\lim_{n \to \infty} f(n)$ is equal to

(a) -1 (b) 0
(c) 1 (d) does not exist

65. $\lim_{n \to \infty} g(n)$ is equal to

(a) -1 (b) 0
(c) 1 (d) does not exist

Passage V (Q. Nos. 66 to 68)

A circular arc of radius 1 subtends an angle of x radian as shown in figure. The centre of the circle is O and the point C is the intersection of two tangent lines at A and B. Let T(x) be the area of ΔABC *and S(x) be the area of shaded region.*

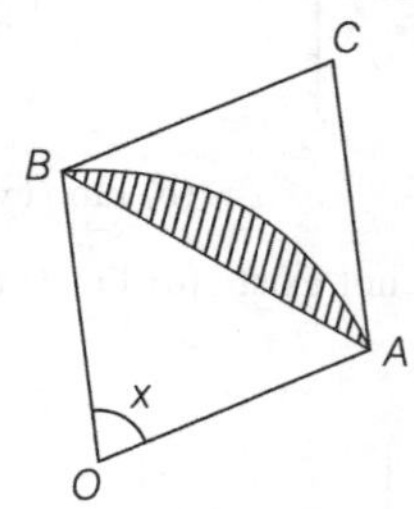

66. $\lim_{x \to 0} \frac{T(x)}{x^3}$ is

(a) $\frac{1}{2}$ (b) $\frac{1}{3}$ (c) $\frac{1}{4}$ (d) $\frac{1}{8}$

67. $\lim_{x \to 0} \frac{S(x)}{x}$ is

(a) 0 (b) $\frac{1}{2}$ (c) 1 (d) None of these

68. $\lim_{x \to 0} \frac{T(x)}{S(x)}$ is

(a) $\frac{1}{4}$ (b) $\frac{3}{4}$ (c) $\frac{3}{2}$ (d) 0

Limits Exercise 4 : Matching Type Questions

69. Match the statements of Column I with values of Column II.

	Column I		Column II
(A)	$\lim\limits_{x \to \frac{\pi^+}{2}} \tan^{-1}(\tan x)$	(p)	0
(B)	$\lim\limits_{n \to \infty} \left[\sum\limits_{r=1}^{n} \frac{1}{2^r}\right]$ ([·] denotes the greatest integer function)	(q)	Doesn't exist
(C)	$\lim\limits_{x \to \infty} \sec^{-1}\left(\frac{x}{x+1}\right)$	(r)	$-\frac{\pi}{2}$
(D)	$\lim\limits_{x \to \frac{\pi}{2}} \frac{\cos x}{(1-\sin x)^{2/3}}$	(s)	$\frac{\pi}{2}$

70. A right angled triangle has legs 1 and x. The hypotenuse is y and the angle opposite to the side x is θ. Shown as

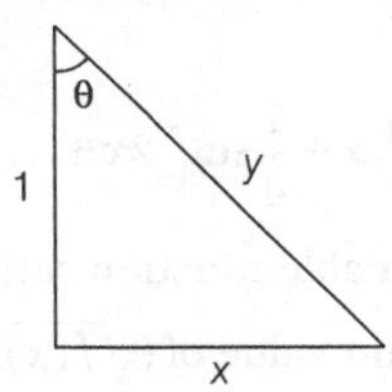

Match the column.

	Column I		Column II
(A)	$\lim\limits_{\theta \to \frac{\pi}{2}} (\sqrt{y} - \sqrt{x})$	(p)	0
(B)	$\lim\limits_{\theta \to \frac{\pi}{2}} (y - x)$	(q)	$\frac{1}{2}$
(C)	$\lim\limits_{\theta \to \frac{\pi}{2}} (y^2 - x^2)$	(r)	1
(D)	$\lim\limits_{\theta \to \frac{\pi}{2}} (y^3 - x^3)$	(s)	∞

71. Match the column.

	Column I		Column II
(A)	$\lim\limits_{x \to 0} \frac{1-\cos 2x}{e^{x^2} - e^x + x}$ equals	(p)	1
(B)	If the value of $\lim\limits_{x \to 0^+} \left(\frac{(3/x)+1}{(3/x)-1}\right)^{1/x}$ can be expressed in the form of $e^{p/q}$, where p and q are relative prime, then $(p+q)$ is equal to	(q)	2
(C)	$\lim\limits_{x \to 0} \frac{\tan^3 x - \tan x^3}{x^5}$ equals	(r)	4
(D)	$\lim\limits_{x \to 0} \frac{x + 2\sin x}{\sqrt{x^2 + 2\sin x + 1} - \sqrt{\sin^2 x - x + 1}}$ equals	(s)	5

72. Match the column.

	Column I		Column II
(A)	$\lim\limits_{n \to \infty} \cos^2(\pi(\sqrt[3]{n^3 + n^2 + 2n} - n))$, where n is an integer, equals	(p)	$\frac{1}{2}$
(B)	$\lim\limits_{n \to \infty} n \sin(2\pi\sqrt{1+n^2})\ (n \in N)$ equals	(q)	$\frac{1}{4}$
(C)	$\lim\limits_{n \to \infty} (-1)^n \sin(\pi\sqrt{n^2 + 0.5n + 1})\left(\sin\frac{(n+1)\pi}{4n}\right)$ is (where $n \in N$)	(r)	π
(D)	If $\lim\limits_{x \to \infty} \left(\frac{x+a}{x-a}\right)^x = e$, where a is some real constant, the value of a is equal to	(s)	non -existent

Limits Exercise 5 : Single Integer Answer Type Questions

73. Let $L = \lim\limits_{x \to \infty} \left(x \log x + 2x \cdot \log \sin\left(\frac{1}{\sqrt{x}}\right)\right)$, then the value of $\left(-\frac{2}{L}\right)$ is ....

74. For $n \in N$, let x_n be defined as $\left(1 + \frac{1}{n}\right)^{(n+x_n)} = e$, then $2\lim\limits_{n \to \infty}(x_n)$ equals ....

75. Let $f(n) = \left[\sqrt{n} + \frac{1}{2}\right]$, where [·] denotes greatest integer function, $\forall\ n \in N$.

Then, $\sum\limits_{n=1}^{\infty} \frac{2^{f(n)} + 2^{-f(n)}}{2^n}$ is equal to ....

76. If the arithmetic mean of the product of all distinct pairs of positive integers whose sum is n, is S_n, then $\lim\limits_{n\to\infty}\dfrac{n^2}{S_n}$ must equal to ....

77. If $k=\sum\limits_{n=1}^{\infty}\dfrac{6^n}{(3^n-2^n)(3^{n+1}-2^{n+1})}$, the value of k is ....

78. The value of $\lim\limits_{x\to\frac{\pi}{2}}\sqrt{\dfrac{\tan x-\sin\{\tan^{-1}(\tan x)\}}{\tan x+\cos^2(\tan x)}}$ is ....

79. The figure shows two regions in the first quadrant. $A(t)$ is the area under the curve $y=\sin x^2$ from 0 to t and $B(t)$ is the area of the triangle with vertices 0, P and $M(t,0)$. If $\lim\limits_{t\to 0}\dfrac{A(t)}{B(t)}=\dfrac{m}{n}$, then $(m+n)$ is ....

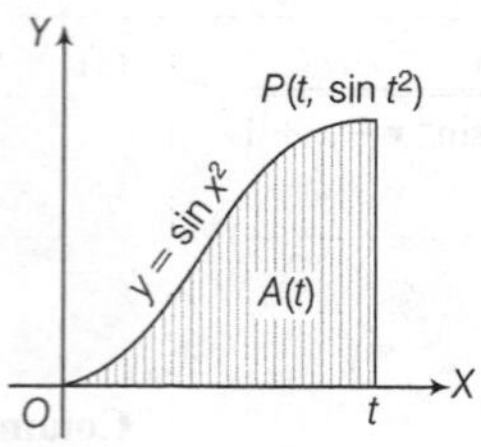

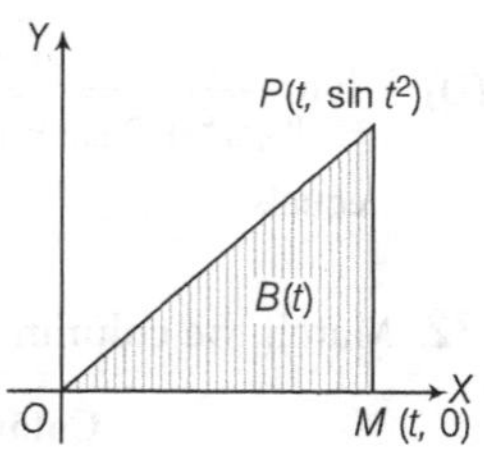

80. If the two lines $AB:\left(\int_0^{2t}\left(\dfrac{\sin x}{x}+1\right)dx\right)x+y=3t$ and $AC:2tx+y=0$ intersect at a point A, the x-coordinate of a point A as $t\to 0$, is equal to $\dfrac{p}{q}$ (p and q are in their lowest form), the $(p+q)$ is ....

81. Consider a parabola $y=\dfrac{x^2}{4}$ and the point $F(0,1)$. Let $A_1(x_1,y_1), A_2(x_2,y_2), A_3(x_3,y_3),\ldots, A_n(x_n,y_n)$ are 'n' points on the parabola such that $x_k>0$ and $\angle OFA_k=\dfrac{k\pi}{2n}(k=1,2,\ldots,n)$. If the value of $\lim\limits_{n\to\infty}\dfrac{1}{n}\cdot\sum\limits_{k=1}^{n}FA_k=\dfrac{m}{\pi}$, then m is ....

82. If $L=\lim\limits_{x\to\frac{\pi}{2}^+}\dfrac{\cos(\tan^{-1}(\tan x))}{x-\pi/2}$, then $\cos(2\pi L)$ is ....

83. Number of solutions of the equation $\operatorname{cosec}\theta=k$ in $[0,\pi]$, where $k=\lim\limits_{n\to\infty}\prod\limits_{r=2}^{n}\left(\dfrac{r^3-1}{r^3+1}\right)$, is ....

84. If C satisfies the equation

$\lim\limits_{x\to\infty}\left(\dfrac{x+c}{x-c}\right)^x=4$, then $\left|\dfrac{e^c}{2}\right|$ is ....

85. If $\lim\limits_{x\to-\infty}\dfrac{(3x^4+2x^2)\cdot\sin\left(\dfrac{1}{x}\right)+|x|^3+5}{|x|^3+|x|^2+|x|+1}=k$, then $\left|\dfrac{k}{2}\right|$ is ....

86. If $f(x)=\lim\limits_{t\to 0}\left[\dfrac{2x}{\pi}\cdot\tan^{-1}\left(\dfrac{x}{t^2}\right)\right]$, then $f(1)$ is ....

87. The value of $\lim\limits_{n\to\infty}\left(\dfrac{1}{\sqrt{n^2}}+\dfrac{1}{\sqrt{n^2+1}}+\ldots+\dfrac{1}{\sqrt{n^2+2n}}\right)$ is

88. If $l=\lim\limits_{x\to 1^+}2^{-2^{\frac{1}{1-x}}}$ and $m=\lim\limits_{x\to 1^+}\dfrac{x\sin(x-[x])}{x-1}$ (where $[\cdot]$ denotes greatest integer function). Then, $(l+m)$ is ....

89. The value of $\lim\limits_{x\to 0}\left[\dfrac{\sin x\cdot\tan x}{x^2}\right]$ is ....

(where $[\cdot]$ denotes greatest integer function).

90. If $\lim\limits_{n\to\infty}\sum\limits_{r=1}^{n}\dfrac{kr}{1\times 3\times 5\times\ldots\times(2r-1)\times(2r+1)}=1$, then k^2 is ....

91. If $f(x)=\lim\limits_{n\to\infty}\sin^4 x+\dfrac{1}{4}\sin^4 2x+\ldots+\dfrac{1}{4^n}\cdot\sin^4(2^n x)$ and $g(x)$ is a differentiable function satisfying $g(x)+f(x)=1$, then the maximum value of $(\sqrt{f(x)}+\sqrt{g(x)})^4$ is ....

92. If $f(x+y+z)=f(x)+f(y)+f(z)$ with $f(1)=1$ and $f(2)=2$ and $x,y,z\in R$, the value of $\lim\limits_{n\to\infty}\sum\limits_{r=1}^{n}\dfrac{(4r)f(3r)}{n^3}$ is ....

93. If $\lim\limits_{n\to\infty}\dfrac{n\cdot 3^n}{n\cdot(x-1)^n+n\cdot 3^{n+1}-3^n}=\dfrac{1}{3}$, the number of the integral values of x is ....

94. The value of

$\lim\limits_{x\to\infty}(((x-1)(x-2)(x+3)(x+10)(x+15))^{1/5}-x)$ is ....

95. If $\lim\limits_{x\to\infty}([f(x)]+x^2)\{f(x)\}=k$, where $f(x)=\dfrac{\tan x}{x}$ and $[\cdot]$, $\{\cdot\}$ denotes greatest integer and fractional part of x respectively, the value of $[k/e]$ is ....

96. The value of $\lim\limits_{n\to\infty}\{(\sqrt{3}+1)^{2n}\}$ is .... (where $\{\cdot\}$ denotes fractional part of x).

97. If $f(x)$ is a polynomial of degree 4, having extremum at $x=1,2$ and $\lim\limits_{x\to 0}\left(1+\dfrac{f(x)}{x^2}\right)=2$.

Then, $f(2)$ is ....

98. If α is the number of solutions of $|x| = \log(x - [x])$, (where $[\cdot]$ denotes greatest integer function) and $\lim_{x \to \alpha} \frac{x e^{ax} - b \sin x}{x^3}$ is finite, the value of $(a + b)$ is

99. Suppose $x_1 = \tan^{-1} 2 > x_2 > x_3 > \ldots$ are the real numbers satisfying $\sin(x_{n+1} - x_n) + 2^{-(n+1)} \cdot \sin x_n \cdot \sin x_{n+1} = 0$ for all $n > 1$ and $l = \lim_{n \to \infty} x_n$, the value of $[4l]$ is .... (where $[\cdot]$ denotes greatest integer function).

Limits Exercise 6 : Questions Asked in Previous 10 Years' Exams

(i) JEE Advanced & IIT-JEE

100. Let $\alpha, \beta \in R$ be such that $\lim_{x \to 0} \frac{x^2 \sin(\beta x)}{\alpha x - \sin x} = 1$. Then, $6(\alpha + \beta)$ equals **[Integer Answer Type, 2016 Adv]**

101. Let m and n be two positive integers greater than 1. If $\lim_{\alpha \to 0} \left(\frac{e^{\cos(\alpha^n)} - e}{\alpha^m} \right) = -\left(\frac{e}{2}\right)$, then the value of $\frac{m}{n}$ is **[Integer Answer Type, 2015 Adv]**

102. Let $f : R \to R$ be a continuous odd function, which vanishes exactly at one point and $f(1) = \frac{1}{2}$. Suppose that

$$F(x) = \int_{-1}^{x} f(t)dt \text{ for all } x \in [-1, 2]$$

and $G(x) = \int_{-1}^{x} t|f\{f(t)\}|\, dt$ for all $x \in [-1, 2]$.

If $\lim_{x \to 1} \frac{F(x)}{G(x)} = \frac{1}{14}$, then the value of $f\left(\frac{1}{2}\right)$ is **[Integer Answer Type, 2015 Adv]**

103. The largest value of the non-negative integer a for which $\lim_{x \to 1} \left\{ \frac{-ax + \sin(x-1) + a}{x + \sin(x-1) - 1} \right\}^{\frac{1-x}{1-\sqrt{x}}} = \frac{1}{4}$, is **[Integer Answer Type , 2014 Adv]**

104. Let $L = \lim_{x \to 0} \frac{a - \sqrt{a^2 - x^2} - \frac{x^2}{4}}{x^4}$, $a > 0$. If L is finite, then **[More than One Correct Option, 2009]**

(a) $a = 2$ (b) $a = 1$

(c) $L = \frac{1}{64}$ (d) $L = \frac{1}{32}$

(ii) JEE Main & AIEEE

105. The $\lim_{x \to \frac{\pi}{2}} \frac{\cot x - \cos x}{(\pi - 2x)^3}$ equals **[2017 JEE Main]**

(a) $\frac{1}{4}$ (b) $\frac{1}{24}$ (c) $\frac{1}{16}$ (d) $\frac{1}{8}$

106. Let $p = \lim_{x \to 0^+} (1 + \tan^2 \sqrt{x})^{1/2x}$, then $\log p$ is equal to **[2016 JEE Main]**

(a) 2 (b) 1 (c) $\frac{1}{2}$ (d) $\frac{1}{4}$

107. $\lim_{n \to \infty} \left[\frac{(n+1)(n+2)\ldots 3n}{n^{2n}} \right]^{1/n}$ is equal to **[2016 JEE Main]**

(a) $\frac{18}{e^4}$ (b) $\frac{27}{e^2}$ (c) $\frac{9}{e^2}$ (d) $3 \log 3 - 2$

108. $\lim_{x \to 0} \frac{\sin(\pi \cos^2 x)}{x^2}$ is equal to **[2014 JEE Main]**

(a) $\frac{\pi}{2}$ (b) 1 (c) $-\pi$ (d) π

109. $\lim_{x \to 0} \frac{(1 - \cos 2x)(3 + \cos x)}{x \tan 4x}$ is equal to **[2013 JEE Main]**

(a) 4 (b) 3 (c) 2 (d) $\frac{1}{2}$

110. If $\lim_{x \to \infty} \left(\frac{x^2 + x + 1}{x + 1} - ax - b \right) = 4$, then **[2012 AIEEE]**

(a) $a = 1, b = 4$ (b) $a = 1, b = -4$

(c) $a = 2, b = -3$ (d) $a = 2, b = 3$

111. Let $\alpha(a)$ and $\beta(a)$ be the roots of the equation $(\sqrt[3]{1+a} - 1)x^2 - (\sqrt{1+a} - 1)x + (\sqrt[6]{1+a} - 1) = 0$, where $a > -1$. Then, $\lim_{a \to 0^+} \alpha(a)$ and $\lim_{a \to 0^+} \beta(a)$ are **[2012 AIEEE]**

(a) $-\frac{5}{2}$ and 1 (b) $-\frac{1}{2}$ and -1

(c) $-\frac{7}{2}$ and 2 (d) $-\frac{9}{2}$ and 3

112. If $\lim_{x \to 0} [1 + x \log(1 + b^2)]^{\frac{1}{x}} = 2b \sin^2 \theta, b > 0$ and $\theta \in (-\pi, \pi]$, then the value of θ is **[2011 AIEEE]**

(a) $\pm \frac{\pi}{4}$ (b) $\pm \frac{\pi}{3}$

(c) $\pm \frac{\pi}{6}$ (d) $\pm \frac{\pi}{2}$

113. For $x > 0$, $\lim_{x \to 0} \left[(\sin x)^{1/x} + \left(\frac{1}{x}\right)^{\sin x} \right]$ is **[2006 AIEEE]**

(a) 0 (b) -1 (c) 1 (d) 2

114. $\lim_{h \to 0} \frac{f(2h+2+h^2) - f(2)}{f(h-h^2+1) - f(1)}$, given that $f'(2) = 6$ and $f'(1) = 4$, **[2003 AIEEE]**

(a) does not exist (b) is equal to $-3/2$
(c) is equal to $3/2$ (d) is equal to 3

115. If $\lim_{x \to 0} \frac{\{(a-n)nx - \tan x\} \sin nx}{x^2} = 0$, where n is non-zero real number, then a is equal to **[2003 AIEEE]**

(a) 0 (b) $\frac{n+1}{n}$ (c) n (d) $n + \frac{1}{n}$

116. The integer n for which $\lim_{x \to 0} \frac{(\cos x - 1)(\cos x - e^x)}{x^n}$ is a finite non-zero number, is **[2002 AIEEE]**

(a) 1 (b) 2 (c) 3 (d) 4

117. Let $f : R \to R$ be such that $f(1) = 3$ and $f'(1) = 6$. Then, $\lim_{x \to 0} \left[\frac{f(1+x)}{f(1)} \right]^{1/x}$ equals **[2002 AIEEE]**

(a) 1 (b) $e^{\frac{1}{2}}$ (c) e^2 (d) e^3

118. For $x \in R$, $\lim_{x \to \infty} \left(\frac{x-3}{x+2} \right)^x$ is equal to **[2000 AIEEE]**

(a) e (b) e^{-1}
(c) e^{-5} (d) e^5

Answers

Exercise for Session 1

1. (c) 2. (c) 3. (a) 4. (c) 5. (d) 6. (a)
7. (a) 8. (d) 9. (a)

Exercise for Session 2

1. (a) 2. (b) 3. (a) 4. (a) 5. (b)

Exercise for Session 3

1. (a) 2. (b) 3. (b) 4. (d) 5. (b) 6. (a)
7. (a) 8. (a) 9. (d) 10. (c)

Exercise for Session 4

1. (1) 2. (1) 3. (e) 4. (c)

Exercise for Session 5

1. (a) 2. (c) 3. (c) 4. (a)

Exercise for Session 6

1. (c) 2. (c) 3. (a) 4. (b) 5. (b)

Chapter Exercises

1. (a) 2. (c) 3. (a) 4. (c) 5. (c) 6. (d) 7. (c) 8. (c) 9. (b) 10. (c)
11. (d) 12. (d) 13. (d) 14. (d) 15. (a) 16. (a) 17. (c) 18. (c) 19. (b) 20. (b)
21. (c) 22. (c) 23. (a) 24. (c) 25. (c) 26. (a) 27. (a) 28. (c) 29. (b) 30. (d)
31. (a) 32. (b) 33. (c) 34. (a) 35. (c) 36. (d) 37. (b) 38. (a) 39. (b) 40. (b)
41. (b, d) 42. (b, d) 43. (b, d) 44. (a, b) 45. (b, c) 46. (a, b, c) 47. (a, b) 48. (a, d)
49. (a,b) 50. (b, c) 51. (c, d) 52. (b, c, d) 53. (b, c,d) 54. (c,d) 55. (a, b) 56. (a)
57. (b) 58. (c) 59. (b) 60. (c) 61. (d) 62. (c) 63. (b) 64. (a) 65. (b) 66. (d)
67. (a) 68. (c)
69. (A) $\to$ (r); (B) $\to$ (p); (C) $\to$ (q); (D) $\to$ (q)
70. (A) $\to$ (p); (B) $\to$ (p); (C) $\to$ (r); (D) $\to$ (s)
71. (A) $\to$ (r); (B) $\to$ (s); (C) $\to$ (p); (D) $\to$ (q)
72. (A) $\to$ (q); (B) $\to$ (r); (C) $\to$ (p); (D) $\to$ (p)
73. (6) 74. (1) 75. (3) 76. (6) 77. (2) 78. (1) 79. (5) 80. (5) 81. (4) 82. (1)
83. (0) 84. (1) 85. (1) 86. (1) 87. (2) 88. (2) 89. (1) 90. (4) 91. (4) 92. (4)
93. (3) 94. (5) 95. (7) 96. (1) 97. (0) 98. (1) 99. (3) 100. (7) 101. (2) 102. (7)
103. (2) 104. (c) 105. (c) 106. (b) 107. (d) 108. (c) 109. (b) 110. (d) 111. (b) 112. (d)
113. (c) 114. (d) 115. (d) 116. (c) 117. (c) 118. (c)

Solutions

1. $\lim\limits_{x \to 0} \dfrac{\sin(\pi - \pi \sin^2 \tan(\sin x))}{x^2}$

$$= \lim_{x \to 0} \frac{\sin(\pi \sin^2 \tan(\sin x))}{\pi \sin^2 \tan(\sin x)} \left(\frac{\pi \sin^2 \tan(\sin x)}{\tan^2(\sin x)}\right) \left(\frac{\tan^2(\sin x)}{\sin^2 x}\right)\left(\frac{\sin^2 x}{x^2}\right) = \pi$$

2. $\lim\limits_{t \to 0} \dfrac{1-(1+t)^t}{\ln(1+t)-t} = \lim\limits_{t \to 0} \dfrac{\dfrac{1-(1+t)^t}{t^2}}{\dfrac{\ln(1+t)-t}{t^2}}$

$$= \lim_{t \to 0} \frac{\dfrac{1-e^{t(\ln(1+t))}}{t^2}}{-1/2} = 2$$

3. $I_1 = \lim\limits_{x \to \infty} (\tan^{-1} \pi x - \tan^{-1} x)$

$$= \lim_{x \to \infty} \tan^{-1}\left(\frac{(\pi - 1)x}{1 + \pi x^2}\right) = 0 \Rightarrow I_1 = 0 \text{ and } I_2 = 0$$

4. From given $f(x) = x^2 - \pi^2$,

$$\lim_{x \to -\pi} \frac{x^2 - \pi^2}{\sin(\sin x)} = \lim_{x \to -\pi} \frac{(-\pi + h)^2 - \pi^2}{\sin \sin(-\pi + h)}$$

$$= \lim_{h \to 0} \frac{-2h\pi + h^2}{-\sin(\sin h)} = \lim_{h \to 0} \frac{h - 2\pi}{\dfrac{-\sin(\sin h)}{\sin h} \times \dfrac{\sin h}{h}} = 2\pi$$

5. Using approximations

$$L = \lim_{x \to 0} \frac{\dfrac{\sin^2 x}{3} + \dfrac{2\tan x}{4}}{\sin x + \tan^2 x} = \lim_{x \to 0} \frac{\dfrac{x^2}{3} + \dfrac{2x}{4}}{x + x^2}$$

$$= \lim_{x \to 0} \frac{x\left(\dfrac{x}{3} + \dfrac{1}{2}\right)}{x(1+x)} = \frac{1}{2}$$

6. Let $x_0 = \cos\theta$, $x_1 = \cos\dfrac{\theta}{2}$, $x_2 = \cos\dfrac{\theta}{2^2}$

$$= \lim_{n \to \infty} \left(\frac{\sqrt{1 - x_0{}^2}}{x_1 x_2 x_3 \ldots x_{n+1}}\right) = \lim_{n \to \infty} \left(\frac{\sin\theta}{\cos\dfrac{\theta}{2}\cos\dfrac{\theta}{2^2}\ldots\cos\dfrac{\theta}{2^{n+1}}}\right)$$

$$= \lim_{n \to \infty} \left(\frac{2^n \cdot \sin\dfrac{\theta}{2^n}}{\cos\dfrac{\theta}{2^{n+1}}}\right) = \lim_{n \to \infty} \left(\frac{\sin\dfrac{\theta}{2^n}}{\dfrac{\theta}{2^n}}\right)\left(\frac{\theta}{\cos\dfrac{\theta}{2^{n+1}}}\right)$$

$$= \theta = \cos^{-1}(x_0)$$

7. $f_n(x) = \tan\dfrac{x}{2}\left(\dfrac{1 + \cos x}{\cos x}\right)(1 + \sec 2x)(1 + \sec 4x) \ldots (1 + \sec 2^n x)$

$$= \tan x \left(\frac{1 + \cos 2x}{\cos 2x}\right)(1 + \sec 4x) \ldots (1 + \sec 2^n x)$$

$$= \tan 2x\,(1 + \sec 2^2 x) \ldots (1 + \sec 2^n x)$$

$$= \tan 2^{n-1} x\,(1 + \sec 2^n x) = \tan 2^n x$$

Now, $\lim\limits_{x \to 0} \dfrac{f_n(x)}{2x} = \lim\limits_{x \to 0} \dfrac{\tan 2^n x}{2^n x} \cdot 2^{n-1} = 2^{n-1}$

8. As, $f'(x) > 0 \Rightarrow f(x)$ is increasing.

So, for $t > 1$; $f(t) > 1$

Now, $\quad f'(t) = \dfrac{1}{t^2 + f(t)} < \dfrac{1}{1 + t^2}$

$\therefore \quad f(x) = 1 + \displaystyle\int_1^x f'(t)\,dt < 1 + \int_1^x \frac{dt}{1 + t^2}$

$\Rightarrow \quad \lim\limits_{x \to \infty} f(x) < 1 + \displaystyle\int_1^\infty \frac{dt}{1 + t^2} \Rightarrow \lim\limits_{x \to \infty} f(x) < 1 + \frac{\pi}{4}$

9. $E = \sin^2\theta - \sin\theta + \dfrac{1}{2} = \left(\sin\theta - \dfrac{1}{2}\right)^2 + \dfrac{1}{4}$

$\Rightarrow$ Minimum value is $\dfrac{1}{4}$.

Let $K = \sqrt{(x+1)(x+2)}$, then $K - x = \dfrac{K^2 - x^2}{K + x} = \dfrac{3x + 2}{K + x}$

$$\lim_{x \to \infty} (K - x) = \lim_{x \to \infty} \frac{3 + \dfrac{2}{x}}{\dfrac{K}{x} + 1} \Rightarrow \lim_{x \to \infty} \frac{K}{x} = 1$$

$\therefore \quad \lim\limits_{x \to \infty} (K - x) = \dfrac{3 + 0}{1 + 1} = \dfrac{3}{2} \qquad \left[\lim\limits_{x \to \infty} \dfrac{K}{x} = 1\right]$

The required equation is $x^2 - \dfrac{7}{4}x + \dfrac{3}{8} = 0$,

i.e. $\quad 8x^2 - 14x + 3 = 0.$

10. Let $x_n = \tan\theta_n$

Now, $\quad \tan\theta_{n+1} = x_{n+1} = \dfrac{x_n}{1 + \sqrt{1 + x_n^2}} = \dfrac{\tan\theta}{1 + \sqrt{1 + \tan^2\theta_n}}$

$\Rightarrow \quad \tan\theta_{n+1} = \dfrac{\tan\theta_n}{1 + \sec\theta_n} = \dfrac{\sin\theta_n}{1 + \cos\theta_n} = \tan\dfrac{\theta_n}{2}$

$\Rightarrow \quad \theta_{n+1} = \dfrac{\theta_n}{2} \quad \Rightarrow \theta_n = \dfrac{\theta_{n-1}}{2}$

Now, $\theta_1 = \dfrac{\pi}{3} \Rightarrow \theta_n = \dfrac{\pi}{3 \cdot 2^{n-1}} \Rightarrow x_n = \tan\left(\dfrac{2\pi}{3 \cdot 2^n}\right)$

$\Rightarrow \quad \lim\limits_{n \to \infty} 2^n x_n = \lim\limits_{n \to \infty} \dfrac{\tan\left(\dfrac{2\pi}{3 \cdot 2^n}\right)}{\left(\dfrac{1}{2^n}\right)} = \dfrac{2\pi}{3}$

11. $L = \lim\limits_{x \to a^-} = \dfrac{(x - b) - (a - b)}{\sqrt{x - b} + \sqrt{a - b}} \times \dfrac{1}{(x^2 - a^2)}$

$$= \lim_{x \to a^-} \frac{1}{(x + a)\{\sqrt{x - b} + \sqrt{a - b}\}} = \frac{1}{4a\sqrt{a - b}}$$

12. $\lim\limits_{n \to \infty} (\sin^n 1 + \cos^n 1)^n = \sin 1 \lim\limits_{n \to \infty} (1 + \cot^n 1)^n$

$$= \sin 1 \cdot e^{\lim\limits_{n \to \infty} \frac{n}{\tan^n 1}} = \sin 1 \cdot e^{\lim\limits_{n \to \infty} \frac{1}{\tan^n 1 \cdot \ln(\tan 1)}} = \sin 1$$

13. As, $\sin^{-1}x = x + \dfrac{x^3}{6} + \dfrac{3x^5}{40} + \dfrac{5}{112}x^7 + \cdots$

and $\tan^{-1}x = x - \dfrac{x^3}{3} + \dfrac{x^5}{5} - \dfrac{x^7}{7} + \cdots$

$$\therefore \lim_{x\to 0}\frac{2}{x^3}(\sin^{-1}x - \tan^{-1}x)$$

$$= \lim_{x\to 0}\frac{2}{x^3}\left[\left(x + \frac{x^3}{6} + \frac{3x^5}{40} + \cdots\right) - \left(x - \frac{x^3}{3} + \frac{x^5}{5} + \cdots\right)\right]$$

$$= \lim_{x\to 0}\frac{2}{x^3}\left[x^3\left(\frac{1}{2}\right) + x^5\left(\frac{3}{40} - \frac{1}{5}\right) + \cdots\right] = 1$$

$$\therefore \lim_{x\to 0}\left[\frac{2}{x^3}(\sin^{-1}x - \tan^{-1}x)\right]^{2/x^2} \qquad [1^\infty \text{ form}]$$

$$= e^{\lim_{x\to 0}\left[\frac{2}{x^3}(\sin^{-1}x - \tan^{-1}x) - 1\right]\cdot\frac{2}{x^2}}$$

$$= e^{\lim_{x\to 0}\frac{2}{x^2}\,\frac{2\left[\left(x + \frac{x^3}{6} + \frac{3x^5}{5} + \cdots\right) - \left(x - \frac{x^3}{3} + \frac{x^5}{5} - \cdots\right)\right] - x^3}{x^3}}$$

$$= e^{\lim_{x\to 0}\frac{2}{x^5}\left[-\frac{1}{4}x^5 + \text{higher powers of } x\right]} = e^{-1/2} = \frac{1}{\sqrt{e}}$$

14. We have, $S = \dfrac{1}{n}\cdot\displaystyle\sum_{k=1}^{n}\left(1 - \frac{k}{n}\right)\cdot\cos 4\left(\frac{k}{n}\right)$

$$S = \int_0^1 (1-x)\cos 4x \cdot dx = \left((1-x)\cdot\frac{\sin 4x}{4}\right)_0^1 + \frac{1}{4}\int_0^1 \sin 4x\, dx$$

$$= 0 + \frac{1}{4}\cdot\frac{1}{(-4)}(\cos 4x)_0^1 = -\frac{1}{16}(\cos 4 - \cos 0) = \frac{1}{16}(1 - \cos 4)$$

15. Let $y = \sqrt{\cos 2x}\cdot\sqrt[3]{\cos 3x}\cdot\sqrt[4]{\cos 4x}\ \ldots\ \sqrt[n]{\cos nx}$

$$\log_e y = \frac{1}{2}\log\cos 2x + \frac{1}{3}\log\cos 3x + \frac{1}{4}\log\cos 4x + \ldots + \frac{1}{n}\log\cos nx$$

On differentiating both sides, we get

$$\frac{1}{y}\cdot dy/dx = \frac{1}{2}\cdot\frac{(-2\sin 2x)}{\cos 2x} + \frac{1}{3}\cdot\left(\frac{-3\sin 3x}{\cos 3x}\right) + \ldots + \frac{1}{n}\left[\frac{(-n\sin nx)}{\cos nx}\right]$$

$$\therefore \frac{dy}{dx} = \prod_{r=2}^{n}(\cos rx)^{1/r}\cdot\sum_{r=2}^{n}[-\tan(rx)] \qquad \text{...(i)}$$

$$\therefore L = \lim_{x\to 0}\frac{1 - \prod_{r=2}^{n}[\cos(rx)]^{1/r}}{x^2}$$

$$= \lim_{x\to 0}\frac{-\prod_{r=2}^{n}(\cos rx)^{1/2}\sum_{r=2}^{n}[-\tan(rx)]}{2x} \qquad \text{[using L' Hospital's rule]}$$

$$= \lim_{x\to 0}\frac{1}{2}\cdot\left[\frac{\tan 2x + \tan 3x + \cdots + \tan nx}{x}\right]$$

$$= \frac{1}{2}[2 + 3 + \cdots + n] = \frac{n^2 + n - 2}{4} = 10 \qquad \text{[given]}$$

$$\Rightarrow (n+7)(n-6) = 0 \Rightarrow n = 6$$

16. Here, $\displaystyle\lim_{z\to\infty}\frac{\int_{1/2}^{z}[\cot^{-1}x]\,dx}{\int_{1/2}^{z}\left[1 + \frac{1}{x}\right]dx} = \frac{\int_{1/2}^{\cot 1} 1\cdot dx + \int_{\cot 1}^{z} 0\cdot dx}{\int_{1/2}^{1} 2\cdot dx + \int_{1}^{z} 1\cdot dx}$

As,

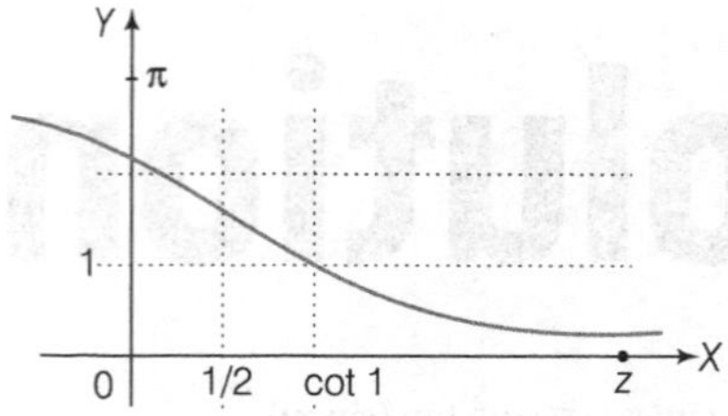

$\because$ $[\cot^{-1}x] = 1$, when $1/2 < x < \cot 1$

and $[\cos^{-1}x] = 0$, when $\cot 1 < x < z$

$$\Rightarrow \lim_{z\to\infty}\frac{\int_{1/2}^{\cot 1} 1\cdot dx}{\int_{1/2}^{1} 2\cdot dx + \int_{1}^{z} 1\cdot dx} = \lim_{z\to\infty}\frac{(\cot 1 - 1/2)}{2(1 - 1/2) + (z - 1)} = 0$$

17. $\displaystyle\lim_{\theta\to 0^+}\left(\frac{1}{\alpha} + \frac{1}{\beta}\right) = \lim_{\theta\to 0^+}\frac{\alpha + \beta}{\alpha\beta} = \lim_{\theta\to 0^+}\frac{\sqrt{2}|\sin\theta|}{\theta} = \sqrt{2}$

18. Here, $\displaystyle\lim_{x\to\infty} f(x) = \lim_{x\to\infty}\frac{1}{3}\left[f(x+1) + \frac{5}{f(x+2)}\right]$

$$\Rightarrow \qquad l = \frac{1}{3}\cdot\left(l + \frac{5}{l}\right) \Rightarrow l^2 = \frac{1}{3}(l^2 + 5)$$

$$\Rightarrow \qquad 2l^2 = 5 \Rightarrow l = \sqrt{\frac{5}{2}}$$

19. $\displaystyle\lim_{x\to 2^-}\frac{\cos(2x+4) - 33}{2} = -16$ and $\displaystyle\lim_{x\to 2^-}\frac{x^2|4x - 8|}{2} = -4(4) = -16$

By Sandwich theorem, $\displaystyle\lim_{x\to 2^-} f(x) = -16$

20. As, $f(x)$ being polynomial having roots 1, 2, 3, 4 and leading coefficient 1.

$\therefore$ $\qquad f(x) = (x-1)(x-2)(x-3)(x-4)$

Similarly, $\qquad g(x) = (x-1)\left(x - \dfrac{1}{2}\right)\left(x - \dfrac{1}{3}\right)\left(x - \dfrac{1}{4}\right)$

$$\therefore \quad \lim_{x\to 1}\frac{f(x)}{g(x)} = \lim_{x\to 1}\frac{(x-1)(x-2)(x-3)(x-4)}{(x-1)\left(x - \frac{1}{2}\right)\left(x - \frac{1}{3}\right)\left(x - \frac{1}{4}\right)}$$

$$= \frac{(-1)(-2)(-3)}{\left(\frac{1}{2}\right)\left(\frac{2}{3}\right)\left(\frac{3}{4}\right)} = -24$$

21. We have, $\displaystyle\lim_{x\to\frac{\pi}{4}}\frac{4\sqrt{2} - (\cos x + \sin x)^5}{1 - \sin 2x}$

$$= \lim_{x\to\frac{\pi}{4}}\frac{[(\cos x + \sin x)^2]^{5/2} - (2)^{5/2}}{(1 + \sin 2x) - 2} = \lim_{x\to\frac{\pi}{4}}\frac{(1 + \sin 2x)^{5/2} - (2)^{5/2}}{(1 + \sin 2x) - 2}$$

Put $y = 1 + \sin 2x$

$$= \lim_{x\to\frac{\pi}{4}}\frac{y^{5/2} - 2^{5/2}}{y - 2} = \frac{5}{2}(2)^{\frac{5}{2}-1} = 5\sqrt{2}\left[\text{as } x \to \frac{\pi}{4} \Rightarrow y \to 2\right]$$

22. We have, $\displaystyle\lim_{n\to\infty}\frac{n\cdot 3^n}{n(x-2)^n + n\cdot 3^{n+1} - 3^n} = \frac{1}{3}$

$$\Rightarrow \lim_{n\to\infty}\frac{1}{\left(\frac{x-2}{3}\right)^n + 3 - \frac{1}{n}} = \frac{1}{3}$$

For, limit to exists and is equal to $\frac{1}{3}$.

We must have, $-1 < \frac{x-2}{3} < 1$.

$\Rightarrow -3 < x-2 < 3 \Rightarrow -1 < x < 5$, as $x \in$ integer

$\therefore \quad x \in \{0, 1, 2, 3, 4\}$

$\Rightarrow$ Number of integer solutions = 5.

23. Given, $f(x) = \lim_{n \to \infty} \frac{1}{\left(\frac{3}{\pi}\tan^{-1}2x\right)^{2n} + 5}$

Now, $f(x) = 0$, if and only if $\left(\frac{3}{\pi}\tan^{-1}2x\right)^2 > 1$

$\Rightarrow \tan^{-1}2x > \frac{\pi}{3}$ or $\tan^{-1}2x < -\frac{\pi}{3}$

$\Rightarrow \quad 2x > \sqrt{3}$ or $2x < -\sqrt{3} \Rightarrow |2x| > \sqrt{3}$

24. We have, $\lim_{x \to 0} \frac{\cos^2 x - \cos x - e^x \cos x + e^x - \frac{x^3}{2}}{x^n}$

$$= \lim_{x \to 0} \frac{(\cos x - 1)(\cos x - e^x) - \frac{x^3}{2}}{x^n}$$

$$= \lim_{x \to 0} \frac{\left(1 - \frac{x^2}{2!} + \frac{x^4}{4!} - \frac{x^6}{6!} + \ldots - 1\right)\left[\left(1 - \frac{x^2}{2!} + \frac{x^4}{4!} - \ldots\right) - \left(1 + \frac{x}{1!} + \frac{x^2}{2!} + \ldots\right)\right] - \frac{x^3}{2}}{x^n}$$

$$= \lim_{x \to 0} \frac{\left(-\frac{x^2}{2!} + \frac{x^4}{4!} - \frac{x^6}{6!} + \ldots\right)\left(-x - x^2 - \frac{x^3}{3!} - \frac{2x^5}{5!} - \ldots\right) - \frac{x^3}{2}}{x^n}$$

$$= \lim_{x \to 0} \frac{\left(\frac{x^3}{2} + \frac{x^4}{2} + \frac{x^5}{12} - \ldots\right) - \frac{x^3}{2}}{x^n} = \lim_{x \to 0} \frac{\frac{x^4}{2} + \frac{x^5}{12} - \frac{x^5}{24}}{x^n}$$

= a non-zero finit number, if $n = 4$.

25. We know that,

$\frac{\tan^{-1} x}{x} < 1$ and $\frac{\sin^{-1} x}{x} > 1, \forall x \in R$

$\therefore \quad \frac{\tan^{-1} x}{x} - \frac{\sin^{-1} x}{x} < 0$ and $\frac{\sin^{-1} x}{x} - \frac{\tan^{-1} x}{x} > 0$

$\Rightarrow I_1$ doesn't exist and I_2 exists.

26. Here, $\lim_{x \to \frac{\pi}{2}} \frac{[x/2]}{\log(\sin x)}$

As, $x \to \frac{\pi}{2} \Rightarrow \frac{x}{2} \to \frac{\pi}{4}$, where $\frac{x}{2} \to \frac{\pi}{4} < 1$

for $\quad x \to \frac{\pi^+}{2}$ or $x \to \frac{\pi^-}{2}$

$\therefore \quad \left[\frac{x}{2}\right] = 0$, since $0 < \left(\frac{x}{2} \to \frac{\pi}{4}\right) < 1$

$$\Rightarrow \quad \lim_{x \to \pi/2} \frac{\left[\frac{x}{2}\right]}{\log \sin x} = \lim_{x \to \pi/2} 0 = 0$$

27. We have, $\lim_{n \to \infty} \left(\frac{a_1 + 1}{a_1}\right)\left(\frac{a_2 + 1}{a_2}\right) \ldots \left(\frac{a_n + 1}{a_n}\right)$

and $\quad a_{n-1} + 1 = \frac{a_n}{n} \quad$ ...(i)

$$\therefore \lim_{n \to \infty} \left(\frac{a_2}{2}\right)\left(\frac{a_3}{3}\right)\left(\frac{a_4}{4}\right) \ldots \left(\frac{a_{n+1}}{n+1}\right) \cdot \frac{1}{a_1 \cdot a_2 \ldots a_n}$$

$$= \lim_{n \to \infty} \frac{a_{n+1}}{(n+1)!} = \lim_{n \to \infty} \frac{1 + a_n}{n!} \quad \text{[using Eq. (i)]}$$

$$= \lim_{n \to \infty} \left(\frac{1}{n!} + \frac{a_n}{n!}\right) = \lim_{n \to \infty} \left(\frac{1}{n!} + \frac{1}{(n-1)!} + \frac{a_{n-1}}{(n-1)!}\right) \quad \text{[using Eq. (i)]}$$

$$= \lim_{n \to \infty} \left(\frac{1}{n!} + \frac{1}{(n-1)!} + \frac{1}{(n-2)!} + \ldots + \frac{1}{(2)!} + \frac{1}{1!} + \frac{a_1}{1!}\right)$$

$[a_1 = 1$, given$]$

$$= \lim_{n \to \infty} \left(\frac{1}{n!} + \frac{1}{(n-1)!} + \frac{1}{(n-2)!} + \ldots + \frac{1}{(2)!} + \frac{1}{1!} + \frac{1}{1}\right)$$

$$= e \quad \left[\text{as } e = 1 + \frac{1}{1!} + \frac{1}{2!} + \frac{1}{3!} + \ldots \infty\right]$$

28. Here, $\quad f(x+y) = f(x) + f(y)$

$f(2) = f(1) + f(1) = 2$

$f(3) = f(1+2) = f(1) + f(2) = 3$

$f(4) = f(1+3) = f(1) + f(3) = 4$

: :

$f(x) = x$, for all $x \in R$

$\Rightarrow \quad f(\tan x) = \tan x, \; f(\sin x) = \sin x$

$$\therefore \quad \lim_{x \to 0} \frac{2^{f(\tan x)} - 2^{f(\sin x)}}{x^2 f(\sin x)} = \lim_{x \to 0} \frac{2^{\tan x} - 2^{\sin x}}{x^2 \sin x}$$

$$= \lim_{x \to 0} \frac{2^{\sin x} \cdot \{2^{\tan x - \sin x} - 1\}}{\sin x \cdot x^2} \times \frac{\tan x - \sin x}{\tan x - \sin x}$$

$$= \lim_{x \to 0} \frac{\{2^{\tan x - \sin x} - 1\}}{\tan x - \sin x} \times \left\{\frac{\tan x - \sin x}{x^2 \sin x}\right\} \times 2^{\sin x}$$

$$= \lim_{x \to 0} \frac{2^{\tan x - \sin x} - 1}{\tan x - \sin x} \times \frac{1 - \cos x}{x^2 \cos x} \times 2^{\sin x}$$

$$= \lim_{x \to 0} \frac{2^{\tan x - \sin x} - 1}{\tan x - \sin x} \times \frac{2\sin^2 x/2}{4(x/2)^2} \times 2^{\sin x} \times \frac{1}{\cos x}$$

$$= \log 2 \times \frac{1}{2} \times 1 = \frac{1}{2} \log 2$$

29. We have, $\lim_{n \to \infty} n^{-n^2} \left[(n+1)\left(n + \frac{1}{2}\right) \ldots \left(n + \frac{1}{2^{n-1}}\right)\right]^n$

$$= \lim_{n \to \infty} \left[\frac{(n+1)\left(n+\frac{1}{2}\right) \ldots \left(n + \frac{1}{2^{n-1}}\right)}{n^n}\right]^n$$

$$= \lim_{n \to \infty} \left(\frac{n+1}{n}\right)^n \cdot \left(\frac{n + \frac{1}{2}}{n}\right)^n \ldots \left(\frac{n + \frac{1}{2^{n-1}}}{n}\right)^n$$

$$= \lim_{n\to\infty}\left(1+\frac{1}{n}\right)^n\cdot\left(1+\frac{1}{2n}\right)^n\ldots\left(1+\frac{1}{2^{n-1}n}\right)^n \qquad [1^\infty \text{ form}]$$

$$= \lim_{n\to\infty}\left(1+\frac{1}{n}\right)^n\cdot\left(1+\frac{1}{2n}\right)^{\frac{2n}{2}}\ldots\left(1+\frac{1}{2^{n-1}n}\right)^{\frac{2^{n-1}\cdot n}{2^{n-1}}}$$

$$= e^1\cdot e^{1/2}\cdot e^{1/4}\ldots e^{1/2^{n-1}}\ldots\infty \qquad \left[\text{using } \lim_{n\to\infty}\left(1+\frac{1}{n}\right)^{an}=e^a\right]$$

$$= e^{(1+1/2+1/4+\ldots)} = e^{\frac{1}{1-\frac{1}{2}}} = e^2$$

30. Here, limit can be calculated only after removing greatest integral function (i.e. $[x]$).

$\therefore$ LHL at $x=0$

$$= \lim_{x\to 0^-} f(x) = \lim_{x\to 0^-}\frac{\sin[x]}{[x]} = \lim_{h\to 0}\frac{\sin[0-h]}{[0-h]}$$

$$= \lim_{h\to 0}\frac{\sin(-1)}{(-1)} = \sin 1$$

$$[\text{as } -1\le[0-h]<0 \;\therefore\; [0-h]=-1, \text{ i.e. } [x]\ne 0]$$

Again, RHL at $x=0$

$$= \lim_{x\to 0^+} f(x) = \lim_{x\to 0^+} 0 = 0 \qquad \left[\text{as } f(x)=\begin{cases}\dfrac{\sin[x]}{[x]}, & [x]\ne 0\\ 0, & [x]=0\end{cases}\right]$$

$$\therefore \qquad \lim_{x\to 0^+} f(x) = 0$$

Here, $0\le[0+h]<1$

$\therefore \qquad [x]=0 \Rightarrow f(x)=0$

So, $\lim_{x\to 0^-} f(x) \ne \lim_{x\to 0^+} f(x)$. Thus, limit doesn't exist.

31. We write, $f(x) = \dfrac{|(x-1)(x-2)(x-3)|}{(x-1)(x-2)(x-3)}$

$$f(x)=\begin{cases}-1, & x<1\\ 1, & 1<x<2\\ -1, & 2<x<3\\ 1, & x>3\end{cases}$$

[leaving $x=1, 2, 3$ as denominator $\ne 0$]

Using Wavy-curve method, as shown in figure

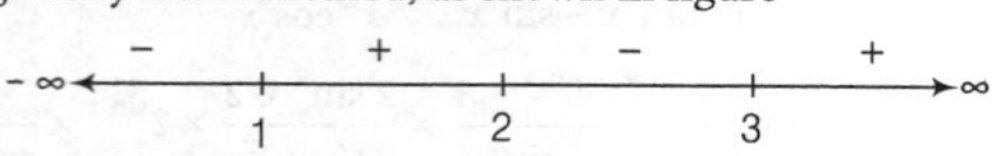

When $x>3$, then
$|(x-1)(x-2)(x-3)| = +(x-1)(x-2)(x-3)$
When $2<x<3$, then
$|(x-1)(x-2)(x-3)| = -(x-1)(x-2)(x-3)$
When $1<x<2$, then
$|(x-1)(x-2)(x-3)| = +(x-1)(x-2)(x-3)$
When $x<1$, then
$|(x-1)(x-2)(x-3)| = -(x-1)(x-2)(x-3)$

Thus, $$f(x)=\begin{cases}-1, & x<1\\ 1, & 1<x<2\\ -1, & 2<x<3\\ 1, & x>3\end{cases}$$

Shows limit exists at all points except at $x = 1, 2, 3$. Graphically, this is shown in given figure.

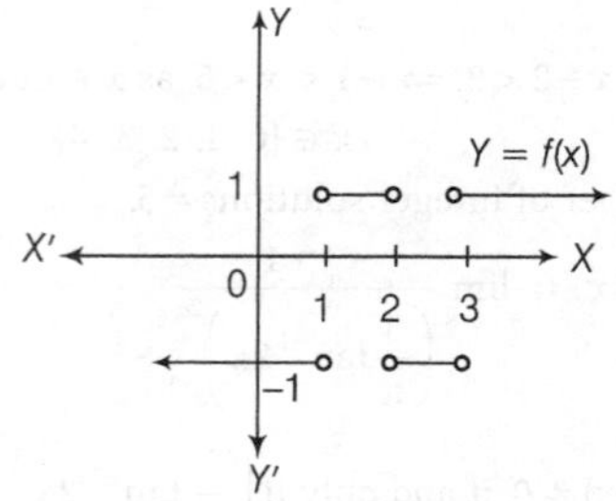

which shows limit doesn't exist at $x = 1, 2, 3$.

32. Here, $t_r = \dfrac{r}{1+r^2+r^4}$

$$= \frac{1}{2}\left(\frac{2r}{1+2r^2+r^4-r^2}\right) = \frac{1}{2}\left[\frac{2r}{(1+r^2)^2-(r^2)}\right]$$

$$= \frac{1}{2}\left[\frac{2r}{(1+r^2+r)(1+r^2-r)}\right]$$

$$= \frac{1}{2}\left[\frac{(r^2+r+1)-(r^2-r+1)}{(1+r^2+r)(1+r^2-r)}\right] \qquad [\because 2r = (r^2+r+1)-(r^2-r+1)]$$

$$= \frac{1}{2}\left[\frac{1}{1+r^2-r}-\frac{1}{1+r^2+r}\right] = \frac{1}{2}\left[\frac{1}{1+r(r-1)}-\frac{1}{1+r(r+1)}\right] \quad \ldots(\text{i})$$

Thus, $$t_r = \frac{1}{2}\left(\frac{1}{1+r(r-1)}-\frac{1}{1+r(r+1)}\right)$$

$$t_1 = \frac{1}{2}\left(\frac{1}{1+0}-\frac{1}{1+2}\right) = \frac{1}{2}\left(1-\frac{1}{3}\right)$$

$$t_2 = \frac{1}{2}\left(\frac{1}{1+2}-\frac{1}{1+2\cdot 3}\right) = \frac{1}{2}\left(\frac{1}{3}-\frac{1}{7}\right)$$

$$t_3 = \frac{1}{2}\left(\frac{1}{7}-\frac{1}{13}\right) = \frac{1}{2}\left(\frac{1}{7}-\frac{1}{13}\right)$$

....

....

$$t_n = \frac{1}{2}\left(\frac{1}{1+n(n-1)}-\frac{1}{1+n(n+1)}\right)$$

$$\therefore \quad \lim_{n\to\infty}\sum_{r=1}^{n} t_r = \frac{1}{2}\left(1-\frac{1}{1+n(n+1)}\right) \quad \ldots(\text{ii})$$

Hence, $$\lim_{n\to\infty}\sum_{r=1}^{n} t_r = \lim_{n\to\infty}\frac{1}{2}\left(1-\frac{1}{1+n(n+1)}\right) = \frac{1}{2}\left(1-\frac{1}{\infty}\right)$$

$$\therefore \quad \lim_{n\to\infty}\sum_{r=1}^{n} t_r = \frac{1}{2}$$

33. Here, $$\lim_{n\to\infty}\sum_{r=1}^{n}\cot^{-1}\left(\frac{r^3-r+\frac{1}{r}}{2}\right) = \lim_{n\to\infty}\sum_{r=1}^{n}\tan^{-1}\left(\frac{2r}{1-r^2+r^4}\right)$$

$$= \lim_{n\to\infty}\sum_{r=1}^{n}\tan^{-1}\left(\frac{2r}{1+(r^2-r)(r^2+r)}\right)$$

$$= \lim_{n\to\infty} \sum_{r=1}^{n} \tan^{-1}\left(\frac{(r^2+r)-(r^2-r)}{1+(r^2+r)(r^2-r)}\right)$$

$$= \lim_{n\to\infty} \sum_{r=1}^{n} \{\tan^{-1}(r^2+r) - \tan^{-1}(r^2-r)\}$$

$$= \lim_{n\to\infty} [(\tan^{-1}2 - \tan^{-1}0) + (\tan^{-1}6 - \tan^{-1}2) + (\tan^{-1}12 - \tan^{-1}6) + \ldots + \{(\tan^{-1}(n^2+n) - \tan^{-1}(n^2-n)\}]$$

$$= \lim_{n\to\infty} \{\tan^{-1}(n^2+n) - \tan^{-1}(0)\}$$

$$= \tan^{-1}(\infty) - \tan^{-1}(0) = \frac{\pi}{2}$$

$$\therefore \lim_{n\to\infty} \sum_{r=1}^{n} \cot^{-1}\left(\frac{r^3 - r + \frac{1}{r}}{2}\right) = \frac{\pi}{2}$$

34. Here, two straight lines $(\tan\alpha)\,x + (\sin\alpha)\,y = \alpha$ and $(\alpha\operatorname{cosec}\alpha)\,x + (\cos\alpha)\,y = 1$ have their point of intersection, as

$$x = \frac{\alpha\cos\alpha - \sin\alpha}{\sin\alpha - \alpha} \text{ and } y = \frac{\alpha - x\tan\alpha}{\sin\alpha}$$

$\therefore$ When $\alpha \to 0$, we obtain the point P.

i.e. $\lim\limits_{\alpha\to 0} x = \lim\limits_{\alpha\to 0} \dfrac{\alpha\cos\alpha - \sin\alpha}{\sin\alpha - \alpha}$ $\left[\dfrac{0}{0} \text{ form}\right]$

$$= \lim_{\alpha\to 0} \frac{-\alpha\sin\alpha + \cos\alpha - \cos\alpha}{\cos\alpha - 1} \quad \text{[applying L'Hospital's rule]}$$

$$= \lim_{\alpha\to 0} \frac{\alpha\left(2\sin\frac{\alpha}{2}\cos\frac{\alpha}{2}\right)}{2\sin^2\frac{\alpha}{2}} = \lim_{\alpha\to 0} \frac{\alpha}{\tan\frac{\alpha}{2}} = \lim_{\alpha\to 0} \frac{\frac{2\alpha}{2}}{\tan\frac{\alpha}{2}} = 2$$

Again, as $\lim\limits_{\alpha\to 0} y = \lim\limits_{\alpha\to 0} \dfrac{\alpha - x\tan\alpha}{\sin\alpha}$

$$= \lim_{\alpha\to 0}\left(\frac{\alpha}{\sin\alpha} - \frac{x}{\cos\alpha}\right) = \lim_{\alpha\to 0} \frac{\alpha}{\sin\alpha} - \lim_{\alpha\to 0} \frac{x}{\cos\alpha}$$

$$= 1 - 2 \quad \left[\text{as } \lim_{\alpha\to 0} x = 2\right]$$

$$\therefore \lim_{\alpha\to 0} y = -1$$

Hence, in limiting position $P(2, -1)$.

35. Here, $\lim\limits_{x\to 0}\left(1 + \dfrac{x^2 + f(x)}{x^2}\right)^{1/x}$

$$\Rightarrow \lim_{x\to 0}\left(1 + \frac{x^2 + f(x)}{x^2}\right)^{1/x} = L \text{ (say)}$$

exists only when

$$\lim_{x\to 0} \frac{x^2 + f(x)}{x^2} = 0 \quad [\text{it converts to } 1^\infty \text{ form}]$$

So, the least degree in $f(x)$ is of degree 2.

i.e. $f(x) = a_2x^2 + a_3x^3 + \ldots$

Now, $\lim\limits_{x\to 0}\left(1 + \dfrac{x^2 + f(x)}{x^2}\right)^{1/x} = e^2$

$$\Rightarrow e^{\lim\limits_{x\to 0}\left(\frac{x^2 + f(x)}{x^2}\right)\frac{1}{x}} = e^2 \Rightarrow e^{\lim\limits_{x\to 0} \frac{x^2 + f(x)}{x^3}} = e^2$$

$$\therefore \lim_{x\to 0} \frac{x^2 + f(x)}{x^3} = 2 \Rightarrow \lim_{x\to 0} \frac{x^2 + a_2x^2 + a_3x^3 + \ldots}{x^3} = 2$$

$\Rightarrow a_2 = -1$, $a_3 = 2$ and a_4, a_5 are any arbitrary constants. Since, we want polynomial of least degree.

Hence, $f(x) = -x^2 + 2x^3$

36. We know, $n \le [x] < n + 1 \Rightarrow [x] = n$

Here, $\dfrac{n\sin x}{x} \to n$, as $x \to 0$ but less than n.

Also, $\dfrac{n\tan x}{x} \to n$, as $x \to 0$ but more than n.

Thus, $n - 1 < \left[\dfrac{n\sin x}{x}\right] < n$, as $x \to 0$

$$\Rightarrow \left[\frac{n\sin x}{x}\right] = n - 1$$

Again, $n \le \left[\dfrac{n\tan x}{x}\right] < n + 1$, as $x \to 0$

$$\Rightarrow \left[\frac{n\tan x}{x}\right] = n$$

Thus, $\lim\limits_{x\to 0}\left(\left[\dfrac{n\sin x}{x}\right] + \left[\dfrac{n\tan x}{x}\right]\right) = (n-1) + (n) = (2n-1)$

$$\therefore \lim_{x\to 0}\left(\left[\frac{n\sin x}{x}\right] + \left[\frac{n\tan x}{x}\right]\right) = (2n-1)$$

37. Here, $\lim\limits_{x\to a} [\sqrt{2-x} + \sqrt{1+x}]$ and we know, we could only apply limit after defining greatest integral function.

Thus, finding range of $[\sqrt{2-x} + \sqrt{1+x}]$, when $x \in \left[0, \dfrac{1}{2}\right]$.

i.e. Let $f(x) = \sqrt{2-x} + \sqrt{1+x}$

For range $f'(x) = \dfrac{1}{2}\left(-\dfrac{1}{\sqrt{2-x}} + \dfrac{1}{\sqrt{1+x}}\right)$

$$= \frac{1}{2}\left(\frac{\sqrt{2-x} - \sqrt{1+x}}{\sqrt{2-x}\cdot\sqrt{1+x}}\right)$$

$\Rightarrow f'(x)$ will be the +ve for $\sqrt{2-x} > \sqrt{1+x}$.

$\Rightarrow f'(x)$ will be the +ve for $2 - x > 1 + x$.

$\Rightarrow f'(x)$ will be the +ve for $2x < 1$.

$$\Rightarrow x < \frac{1}{2}$$

$\therefore f(x)$ will be increasing for $x < \dfrac{1}{2}$.

$$\therefore f(0) = \sqrt{2} + 1$$

$$\Rightarrow f(1/2) = \sqrt{6}$$

which shows range of $f(x)$ is $[1 + \sqrt{2}, \sqrt{6}]$, when $x \in \left[0, \dfrac{1}{2}\right]$.

$$\therefore [\sqrt{2-x} + \sqrt{1+x}] = 2$$

$$\Rightarrow \lim_{x\to a} [\sqrt{2-x} + \sqrt{1+x}] = \lim_{x\to a} 2 = 2$$

38. Let $y=\sqrt{(\tan x-\sin x)+\sqrt{(\tan x-\sin x)+\sqrt{(\tan x-\sin x)+\ldots\infty}}}$

$\Rightarrow\quad y=\sqrt{(\tan x-\sin x)+y}$

$\Rightarrow\quad y^2-y-(\tan x-\sin x)=0$

$\Rightarrow\quad y=\dfrac{1+\sqrt{1+(\tan x-\sin x)\,4}}{2}$ [as $y>0$] ...(i)

Again, let $Z=\sqrt{x^3+\sqrt{x^3+\sqrt{x^3+\ldots\infty}}}$

$\Rightarrow\quad Z=\sqrt{x^3+Z}$

$\Rightarrow\quad Z^2-Z-x^3=0$

$\Rightarrow\quad Z=\dfrac{1+\sqrt{(1+4x^3)}}{2}$ [as $Z>0$] ...(ii)

$\therefore\quad \lim_{x\to0^+}\dfrac{-1+\sqrt{(\tan x-\sin x)+\sqrt{(\tan x-\sin x)+\ldots\infty}}}{-1+\sqrt{x^3+\sqrt{x^3+\sqrt{x^3+\ldots\infty}}}}$

$=\lim_{x\to0^+}\dfrac{-1+\left(\dfrac{1+\sqrt{1+4(\tan x-\sin x)}}{2}\right)}{-1+\left(\dfrac{1+\sqrt{1+4x^3}}{2}\right)}$ [from Eqs. (i) and (ii)]

$=\lim_{x\to0^+}\dfrac{-1+\sqrt{1+4(\tan x-\sin x)}}{-1+\sqrt{1+4x^3}}$

Rationalising numerator and denominator, we get

$\lim_{x\to0^+}\dfrac{4(\tan x-\sin x)(1+\sqrt{1+4x^3})}{4x^3(1+\sqrt{1+4(\tan x-\sin x)})}$

$=\lim_{x\to0^+}\dfrac{\left(\dfrac{\sin x}{\cos x}-\dfrac{\sin x}{1}\right)(1+\sqrt{1+4x^3})}{x^3(1+\sqrt{1+4(\tan x-\sin x)})}$

$=\lim_{x\to0^+}\dfrac{\sin x(1-\cos x)}{x^3\cos x}\cdot\dfrac{(1+\sqrt{1+4x^3})}{(1+\sqrt{1+4(\tan x-\sin x)})}$

$=\lim_{x\to0^+}\dfrac{\sin x}{x}\cdot\dfrac{2\sin^2\dfrac{x}{2}}{\dfrac{4x^2}{4}}\cdot\dfrac{1}{\cos x}\cdot\dfrac{(1+\sqrt{1+4x^3})}{(1+\sqrt{1+4(\tan x-\sin x)})}$

$=1\cdot\dfrac{1}{2}\cdot1\cdot\dfrac{(1+1)}{(1+1)}=\dfrac{1}{2}$

$\therefore\quad \lim_{x\to0^+}\dfrac{-1+\sqrt{(\tan x-\sin x)+\sqrt{(\tan x-\sin x)+\sqrt{(\tan x-\sin x)+\ldots\infty}}}}{-1+\sqrt{x^3+\sqrt{x^3+\sqrt{x^3+\ldots\infty}}}}=\dfrac{1}{2}$

39. Here, $\lim_{\theta\to0}\dfrac{\cos^2(1-\cos^2(1-\cos^2(1\ldots\cos^2\theta))\ldots)}{\sin\left(\dfrac{\pi(\sqrt{\theta+4}-2)}{\theta}\right)}$

$=\lim_{\theta\to0}\dfrac{\cos^2(\sin^2(\sin^2\ldots(\sin^2\theta))\ldots)}{\sin\left(\dfrac{\pi(\sqrt{\theta+4}-2)}{\theta}\right)}$

$=\lim_{\theta\to0}\dfrac{\cos^2(\sin^2(\sin^2\ldots(\sin^2\theta))\ldots)}{\sin\left(\pi\lim_{\theta\to0}\dfrac{\theta}{\theta(\sqrt{\theta+4}+2)}\right)}=\dfrac{\cos^2(0)}{\sin\left(\dfrac{\pi}{4}\right)}=\sqrt{2}$

40. Here, $\lim_{n\to\infty}a_n=\lim_{n\to\infty}a_{n+1}$.

So, let $l=\lim_{n\to\infty}a_n=\lim_{n\to\infty}a_{n+1}$

$\therefore\quad l=\lim_{n\to\infty}a_{n+1}=\lim_{n\to\infty}\sqrt{2+a_n}$

or $\quad l=\sqrt{2+l}$ $\left[\because \lim_{n\to\infty}a_n=l\right]$

$\Rightarrow\quad l^2=2+l\ \Rightarrow\ l^2-l-2=0$

$\Rightarrow\quad (l-2)(l+1)=0\ \Rightarrow\ l=2,-1$

$\therefore\quad \lim_{n\to\infty}a_n=2$ or -1

$\Rightarrow\quad \lim_{n\to\infty}a_n=2$ $\left[\text{neglecting } \lim_{n\to\infty}a_n=-1, \text{ as } a_n>0\right]$

41. $\lim_{x\to\infty}4x\dfrac{\tan^{-1}\left(\dfrac{1}{2x+3}\right)}{\left(\dfrac{1}{2x+3}\right)}\times\dfrac{1}{2x+3}=2$

$\Rightarrow\quad y^2+4y+5=2$

$\Rightarrow\quad y=-1,-3.$

42. $\lim_{x\to0}\dfrac{2\sin^2\left(\dfrac{x^2}{2}\right)}{2}\cdot\dfrac{\left(\dfrac{x^2}{2}\right)^2}{(4^x-1)}\cdot\dfrac{x}{x^4}=\dfrac{1}{2}\ln 4=\ln 2=1-\dfrac{1}{2}\ln\left(\dfrac{e^2}{4}\right)$

43. $f(x)=e^{[\cot x]}$, as $\cot x$ is negative in the II quadrant and

$\cot\dfrac{\pi}{2}=0\,[\cot x]=-1\ \Rightarrow\ \lim_{x\to\frac{\pi}{2}^+}f(x)=e^{-1}=\dfrac{1}{e}$

As $x\to\dfrac{\pi}{2}^-$, $\cot x$ is positive (being in I quadrant) and hence $[\cot x]=0\ \Rightarrow\ \lim_{x\to\frac{\pi}{2}^-}f(x)=e^0=1$

44. If $m<0$, then for values of x sufficiently close to 0.

$1+\dfrac{1}{m}<\dfrac{\sin x}{x}<1\ \Rightarrow\ m+1>m\dfrac{\sin x}{x}>m$

$\therefore\quad \left[m\dfrac{\sin x}{x}\right]=m\ \Rightarrow\ \lim_{x\to0}\left[m\dfrac{\sin x}{x}\right]=m$

If $m>0$, then for values of x sufficiently close to 0, we can have

$1-\dfrac{1}{m}<\dfrac{\sin x}{x}<1$

$\therefore\quad m-1<m\dfrac{\sin x}{x}<m$

$\therefore\quad \lim_{x\to0}\left[m\dfrac{\sin x}{x}\right]=m-1$

45. $\lim_{x \to 0} (1 + ax + bx^2)^{\frac{2}{x}} = e^3 \Rightarrow e^{\lim_{x \to \infty} \frac{2}{x}(ax + bx^2)} = e^3$

$\Rightarrow \quad e^{2a} = e^3$

Since, limit value does not involve b.

$\therefore$ b can have any value. Thus, $a = \frac{3}{2}$, $b \in R$

46. (a) $\lim_{x \to 0^+} f(x) = 1$

(b) $\lim_{x \to 1} f(x) = 2$, as RHL = LHL = 2 (at $x=1$)

(c) $\lim_{x \to 3^-} f(x) \to \infty$ and $\lim_{x \to 3^+} f(x) \to -\infty$

$\therefore$ $\lim_{x \to 3} f(x)$ does not exists.

(d) $\lim_{x \to 4} f(x) = 0$

47. RHL $= \lim_{x \to 0^+} \frac{\cot^{-1}(1/x)}{x} = \lim_{x \to 0^+} \frac{\tan^{-1}}{x} = 1$

LHL $= \lim_{x \to 0} \frac{\cot^{-1}(1/x)}{x} = \lim_{x \to 0} \frac{\pi + \tan^{-1} x}{x} = \infty$

$\therefore$ RHL exists and LHL does not exists.

48. As, $l = \lim_{x \to \infty} \left(\frac{x+1}{x-1}\right)^x = \lim_{x \to \infty} \left(1 + \frac{2}{x-1}\right)^x = e^{\lim_{x \to \infty} \frac{2x}{x-1}} = e^2$

$\therefore$ $[l] = 7$ and $\{l\} = e^2 - 7$.

49. (a) $\lim_{x \to 0^-} f(x) = \lim_{x \to 0^-} \frac{\sin[x]}{[x]} = \lim_{h \to 0} \frac{\sin(-1)}{(-1)} = \sin 1$

(b) $\lim_{x \to 0^+} f(x) = \lim_{h \to 0} [0+h] = 0$

$\therefore$ Limit does not exist at $x=0$.

50. (a) $\lim_{x \to 3} [[x]] - [2x-1]$

RHL $= \lim_{h \to 0} [[3+h]] - [6+2h-1] = \lim_{h \to 0} 3-5 = -2$

LHL $= \lim_{h \to 0} [[3-h]] - [6-2h-1] = \lim_{h \to 0} 2-4 = -2$

$\therefore$ Limit exists.

(b) $\lim_{x \to 1} [x] - x$

RHL $= \lim_{h \to 0} [1+h] - (1+h) = \lim_{h \to 0} (1-1-h) = 0$

LHL $= \lim_{h \to 0} [1-h] - (1+h) = \lim_{h \to 0} 0 - 1 + h = -1$

$\therefore$ Limit does not exist.

(c) $\lim_{x \to 0} \{x\}^2 - \{-x\}^2$

RHL $= \lim_{h \to 0} \{h\}^2 - \{-h\}^2 = \lim_{h \to 0} h^2 - (1-h)^2 = -1$

LHL $= \lim_{h \to 0} \{-h\}^2 - \{h\}^2 = \lim_{h \to 0} (1-h)^2 - h^2 = 1$

$\therefore$ Limit does not exist.

(d) $\lim_{x \to 0} \frac{\tan(\operatorname{sgn} x)}{\operatorname{sgn} x}$; RHL $= \lim_{h \to 0} \frac{\tan(1)}{1} = \tan 1$

LHL $= \lim_{h \to 0} \frac{\tan(-1)}{-1} = \tan 1$

$\therefore$ Limit exists.

51. (a) $\lim_{n \to \infty} \left[\frac{1}{2} + \frac{1}{2^2} + \ldots + \frac{1}{2^n}\right]$

$= \lim_{n \to \infty} \left[\frac{\frac{1}{2}\left(1 - \frac{1}{2^n}\right)}{1 - \frac{1}{2}}\right] = \lim_{n \to \infty} \left[1 - \frac{1}{2^n}\right] = 0$

(b) $\lim_{n \to \infty} (x-1)\{x\}$

RHL $= \lim_{h \to 0} (1+h-1)\{1+h\} = \lim_{h \to 0} h(h) = 0$

LHL $= \lim_{h \to 0} (1-h-1)\{1-h\} = \lim_{h \to 0} -h(1-h) = 0$

$\therefore \quad \lim_{x \to 1} (x-1)\{x\} = 0$

(c) As, $\frac{\tan x}{x} > 1$

$\therefore \quad \left[\frac{\tan x}{x}\right] \to 1$, as $x \to 0^+ \Rightarrow \lim_{x \to 0^+} \left[\frac{\tan x}{x}\right] = 1$

(d) $\left[\lim_{x \to 0^+} \frac{\tan x}{x}\right] = [1+h] = 1$

52. Put $x = \frac{\pi}{2} - h$

$l = \lim_{h \to 0} \frac{a^{\tan h} - a^{\sin h}}{\tan h - \sin h} = \lim_{h \to \infty} \frac{a^{\sin h}(a^{\tan h - \sin h} - 1)}{\tan h - \sin h} = \log_e a$

and $m = \lim_{x \to -\infty} (\sqrt{x^2 + ax} - \sqrt{x^2 - ax})$

$= \lim_{x \to -\infty} \frac{(x^2 + ax) - (x^2 - ax)}{\sqrt{x^2 + ax} + \sqrt{x^2 - ax}}$

$= \lim_{x \to -\infty} \frac{2ax}{|x|\left(\sqrt{1 + \frac{a}{x}} + \sqrt{1 - \frac{a}{x}}\right)} = -a$

53. $f(x) = \lim_{x \to \infty} \left[1 + \frac{ax+1}{bx+2} - 1\right]^x$

$= \lim_{x \to \infty} \left[1 + \frac{(a-b)x - 1}{bx+2}\right]^x = e^{\lim_{x \to \infty} \left[\frac{(a-b)x-1}{bx+2}\right]^x}$

When $a=b$, $f(x) = e^{\lim_{x \to \infty} \frac{-x}{bx+2}} = e^{-\frac{1}{b}}$ or $e^{-\frac{1}{a}}$.

When $a>b$, $f(x) = e^{\infty} = \infty$, does not exist.

When $a<b$, $f(x) = e^{-\infty} = 0$.

54. $\lim_{x \to 0} f(x) = \lim_{x \to 0} \frac{x \cdot (2^x - 1)}{2\sin^2 x/2} = \lim_{x \to 0} \frac{2 \cdot x^2 \frac{1}{4}}{\sin^2 \frac{x}{2}} \cdot \frac{2^x - 1}{x} = 2 \cdot \log_e 2 = \log_e 4$

and $\lim_{x \to \infty} 2^x \cdot \sin(\log 2/2^x) = \lim_{x \to \infty} \frac{\sin(\log 2/2^x)}{(\log 2/2^x)} \cdot \log_e 2 = \log_e 2$

55. As, $\lim_{x \to 3} \frac{x^3 + cx^2 + 5x + 12}{x^2 - 7x + 12} = l$. [finite]

$\therefore \quad 27 + 9c + 15 + 12 = 0$

$\Rightarrow \quad 9c + 54 = 0 \Rightarrow c = -6$

$\therefore \quad \lim_{x \to 3} \frac{x^3 + cx^2 + 5x + 12}{x^2 - 7x + 12} = \lim_{x \to 3} \frac{3x^2 + 2cx + 5}{2x - 7}$

$$=\frac{27+6c+5}{9-7}=\frac{27-36+5}{-1}$$

$\therefore \quad c=-6$ and $l=4$

56. $\displaystyle\lim_{x\to 0}\frac{\sin x}{x-\sin x}=\lim_{x\to 0}\frac{\frac{\sin x}{x}}{1-\frac{\sin x}{x}}=\infty$

Hence, $L=e^{\lim_{x\to 0}\left(\frac{\sin x}{x}-1\right)\frac{\sin x}{x-\sin x}}=e^{\lim_{x\to 0}\frac{-\sin x}{x}}=e^{-1}=\frac{1}{e}$

57. $\displaystyle\lim_{x\to 0}\frac{x-1+\cos x}{x}=\lim_{x\to 0}1-\frac{2\sin^2\frac{x}{2}}{x}=1$

Hence, $\quad L=e^{\lim_{x\to 0}\left(\frac{x-1+\cos x}{x}-1\right)\times\frac{1}{x}}$

$$=e^{\lim_{x\to 0}\left(\frac{-2\sin^2\frac{x}{2}}{x^2}\right)}=e^{-\frac{2}{4}}=e^{-1/2}$$

58. $L=e^{\lim_{x\to 0}\left(\frac{a^x+b^x+c^x}{3}-1\right)\frac{2}{x}}=e^{\lim_{x\to 0}\left(\frac{a^x-1}{x}+\frac{b^x-1}{x}+\frac{c^x-1}{x}\right)\frac{2}{3}}$

$$=e^{\frac{2}{3}(\log a+\log b+\log c)}=e^{\frac{2}{3}\log abc}=(abc)^{2/3}$$

Sol. (Q. Nos. 59 to 61)

$$f(x)=\lim_{n\to\infty}\left(\cos\sqrt{\frac{x}{n}}\right)^n=\lim_{n\to\infty}\left(1+\left(\cos\sqrt{\frac{x}{n}}-1\right)\right)^n$$

$$=e^{\lim_{n\to\infty}\left(\cos\sqrt{\frac{x}{n}}-1\right)n}=e^{-\lim_{n\to\infty}2\frac{\sin^2\left(\frac{1}{2}\sqrt{\frac{x}{n}}\right)\cdot n}{\left(\frac{1}{2}\sqrt{\frac{x}{n}}\right)^2}\times\left(\frac{1}{2}\sqrt{\frac{x}{n}}\right)^2}$$

$$=e^{-2\lim_{n\to\infty}\frac{\left(\frac{1}{2}\sqrt{\frac{x}{n}}\right)^2}{\frac{1}{n}}}=e^{-2\lim_{n\to\infty}\frac{1}{4}\frac{x/n}{1/n}}=e^{-\frac{x}{2}}$$

$$y=f(x)=e^{-x/2},\ x\ge 0,\ \text{range}=(0,1]$$

$\Rightarrow \quad g(x)=\lim_{n\to\infty}(1-x+x\sqrt[n]{e})^n$

$$=e^{\lim_{n\to\infty}x\frac{(e^{1/n}-1)}{1/n}}=e^x,\ \forall\ x\in R$$

$\Rightarrow \quad h(x)=\tan^{-1}(g^{-1}(f^{-1}(x))$

Let $\quad y^2=f(x)\Rightarrow y=e^{-x/2}$

$$-\frac{x}{2}=\ln y\ \Rightarrow x=2\ln\frac{1}{y}$$

$\Rightarrow f^{-1}(x)=2\ln\frac{1}{x}$, for $0<x\le 1$

$$y=g(x)=e^x\ \Rightarrow\ x=\ln y,\ g^{-1}(x)=\ln x$$

$\therefore\ g^{-1}\left(2\ln\frac{1}{x}\right)=\ln\left(2\ln\left(\frac{1}{x}\right)\right)$, for $0<x<1$

$\therefore \quad h(x)=\tan^{-1}\left(\ln\left(\ln\frac{1}{x^2}\right)\right)$, for $0<x<1$

59. $\displaystyle\lim_{x\to 0^+}\frac{\ln(f(x))}{\ln(g(x))}=\lim_{x\to 0}\frac{-x/2}{x}=-\frac{1}{2}$

60. Domain of $h(x)$ is $(0, 1)$.

61. $h(x)=\tan^{-1}(\ln(\ln 1/x^2))$, for $0<x<1$

$$1<\frac{1}{x^2}<\infty\quad\Rightarrow 0<\ln\frac{1}{x^2}<\infty$$

$\Rightarrow\ -\infty<\ln(\ln(1/x^2))<\infty$

$\therefore$ Range of $h(x)$ is $(-\pi/2, \pi/2)$.

Sol. (Q. Nos. 62 to 63)

Here, $K_1=(n+1)(n+2)(n+3)\dots(2n)$

$K_2=n\times n\times n\times\dots\times n$ times $=n^n$ and $K_3=(n)!$

62. $L=\displaystyle\lim_{n\to\infty}\left(\frac{K_1}{K_2}\right)^{1/n}=\lim_{n\to\infty}\left(\frac{(n+1)(n+2)\dots(2n)}{n\times n\times\dots\times n\text{ times}}\right)^{1/n}$

Taking log on both sides, we get

$$\log L=\lim_{n\to\infty}\frac{1}{n}\cdot\sum_{r=1}^{n}\log\left(1+\frac{r}{n}\right)=\int_0^1\log(1+x)\,dx=\log(4/e)$$

$\therefore\quad L=4/e$

63. Also, $M=\displaystyle\lim_{n\to\infty}\frac{K_1}{K_2+K_3}=\lim_{n\to\infty}\frac{(n+1)(n+2)\dots(n+n)}{n^n+n!}$

$$=\lim_{n\to\infty}\frac{n^n((1+1/n)(1+2/n)\dots(1+n/n))}{n^n\left[1+\frac{1\cdot 2\cdot 3\dots n}{n\cdot n\dots n}\right]}=1$$

Sol. (Q. Nos. 64 to 65)

Here, $f(1)=0\cdot 8$ and $g(1)=0\cdot 6\Rightarrow(f(1))^2+(g(1))^2=1$

$\therefore$ Let $f(1)=\cos\alpha$ and $g(1)=\sin\alpha$, then

$$f(2)=f(1)\cos(g(1))-g(1)\sin(g(1))$$

$$=\cos\alpha\cdot\cos(g(1))-\sin\alpha\cdot\sin(g(1))=\cos(\alpha+g(1))$$

Similarly, $f(n+1)=\cos(\alpha+g(1)+g(2)+\dots+g(n))$ ...(i)

Here, for all $x\in(0,\pi)$

$$\sin x<\pi-x\ \Rightarrow\ x+\sin x<\pi$$

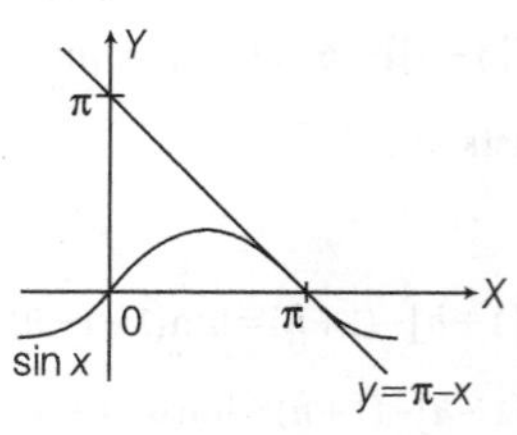

$\Rightarrow\quad \alpha+g(1)+g(2)+\dots+g(n)<\pi$

$\therefore$ As $n\to\infty$, $\alpha+g(1)+g(2)+\dots+g(n)\to\pi$ ...(ii)

64. $\displaystyle\lim_{n\to\infty}f(n)=\lim_{n\to\infty}\cos(\alpha+g(1)+g(2)+\dots+g(n))$

$$=\cos\pi=-1$$

65. Similarly, $\quad g(n)=\sin(\alpha+g(1)+g(2)+\dots+g(n-1))$

Again, $\quad g(2)=\sin(\alpha+g(1))$

As, $\quad\displaystyle\lim_{n\to\infty}g(n)=\lim_{n\to\infty}\sin(\alpha+g(1)+g(2)+\dots+g(n-1))$

$=\sin(\pi)=0$ [using Eq. (ii)]

Sol. (Q. Nos. 66 to 68)

Here, $\quad T(x)=\frac{1}{2}\left(\sin\left(\frac{x}{2}\right)\cdot\tan\left(\frac{x}{2}\right)\cdot\cos\left(\frac{\pi}{2}-\frac{x}{2}\right)\right)\times 2$

$$= \sin^2\left(\frac{x}{2}\right)\cdot\tan\left(\frac{x}{2}\right)$$

$\therefore \quad S(x) = \frac{1}{2}(1)^2\cdot(x-\sin x)$

66. $\lim_{x\to 0}\frac{T(x)}{x^3} = \lim_{x\to 0}\frac{\sin^2\frac{x}{2}\cdot\tan\frac{x}{2}}{x^3} = \frac{1}{8}$

67. $\lim_{x\to 0}\frac{S(x)}{x} = \lim_{x\to 0}\frac{1}{2}\frac{(x-\sin x)}{x} = 0$

68. $\lim_{x\to 0}\frac{T(x)}{S(x)} = \lim_{x\to 0}\frac{\sin^2(x/2)\cdot\tan(x/2)}{\frac{1}{2}(x-\sin x)} = \lim_{x\to 0}\frac{1}{4}\cdot\frac{x^3}{(x-\sin x)} = \frac{3}{2}$

69. (A) $\lim_{x\to\frac{\pi^+}{2}}\tan^{-1}(\tan x) = -\frac{\pi}{2}$, as $\lim_{x\to\frac{\pi^+}{2}}(\tan x) = -\infty$

(B) $\sum_{r=1}^{n}\frac{1}{2^r} = \left(1-\frac{1}{2^n}\right) < 1$, for all $n \in N$.Thus, $\lim_{n\to 0}\left[\sum_{r=1}^{n}\frac{1}{2^r}\right] = 0$

(C) As $x \to \infty$, $\frac{x}{x+1} \to 1^-$

and hence $\lim_{x\to\infty}\sec^{-1}\left(\frac{x}{x+1}\right)$ doesn't exist.

(D) Put $\frac{\pi}{2} - x = \theta$, then the given limit is

$$\lim_{\theta\to 0}\frac{\sin\theta}{(1-\cos\theta)^{2/3}} = \lim_{\theta\to 0}\frac{\sin\theta}{2^{2/3}\cdot\sin^{4/3}\frac{\theta}{2}}$$

$$= \lim_{\theta\to 0}\frac{2\sin\frac{\theta}{2}\cos\frac{\theta}{2}}{2^{2/3}\cdot\sin^{4/3}\frac{\theta}{2}} = 2^{1/2}\lim_{\theta\to 0}\frac{\cos\frac{\theta}{2}}{\sin^{1/3}\frac{\theta}{2}},$$

which doesn't exist as for $\lim_{\theta\to 0^+}$ limit is ∞ and for $\lim_{\theta\to 0^+}$ limit is $-\infty$.

70. Here, $y = \sec\theta$ and $x = \tan\theta$

(A) $\lim_{\theta\to\frac{\pi}{2}}(\sqrt{y}-\sqrt{x}) = \lim_{\theta\to\frac{\pi}{2}}(\sqrt{\sec\theta}-\sqrt{\tan\theta})$

$$= \lim_{\theta\to\frac{\pi}{2}}\frac{1-\sqrt{\sin\theta}}{\sqrt{\cos\theta}}$$

$$= \lim_{\theta\to\frac{\pi}{2}}\left(\frac{\cos\theta}{\sin\theta}\right)^{3/2} = 0 \quad \text{[using L'Hospital's rule]}$$

(B) $\lim_{\theta\to\frac{\pi}{2}}(y-x) = \lim_{\theta\to\frac{\pi}{2}}(\sec\theta-\tan\theta) = \lim_{\theta\to\frac{\pi}{2}}\frac{1-\sin\theta}{\cos\theta}$

$$= \lim_{\theta\to\frac{\pi}{2}}\frac{\cos\theta}{\sin\theta} = 0 \quad \text{[using L'Hospital's rule]}$$

(C) $\lim_{\theta\to\frac{\pi}{2}}(y^2-x^2) = \lim_{\theta\to\frac{\pi}{2}}(\sec^2\theta-\tan^2\theta) = 1$

(D) $\lim_{\theta\to\frac{\pi}{2}}(y^3-x^3) = \lim_{\theta\to\frac{\pi}{2}}(\sec^3\theta-\tan^3\theta)$

$$= \lim_{\theta\to\frac{\pi}{2}}\frac{1-\sin^3\theta}{\cos^3\theta}$$

$$= \lim_{\theta\to\frac{\pi}{2}}\frac{-3\sin^2\theta\cos\theta}{-3\cos^2\theta\sin\theta} = \frac{\sin\theta}{\cos\theta} \quad \text{[using L' Hospital's rule]}$$

$$= \lim_{\theta\to\frac{\pi}{2}}\tan\theta = \infty$$

71. (A) $l = \lim_{x\to 0}\frac{e^{x^2}-1+x-e^x+1}{2\frac{\sin^2 x}{x^2}\cdot x^2}$

$$= \frac{1}{2}\left[\lim_{x\to 0}\frac{e^{x^2}-1}{x^2} + \lim_{x\to 0}\frac{x-e^x+1}{x^2}\right]$$

$$= \frac{1}{2}\left[1-\lim_{x\to 0}\frac{e^x-x-1}{x^2}\right]$$

$$= \frac{1}{2}\left[1-\frac{1}{2}\right] = \frac{1}{4} \Rightarrow \frac{1}{l} = 4$$

(B) $l = \lim_{x\to 0}\left(\frac{3+x}{3-x}\right)^{1/x} = e^{\lim_{x\to 0}\frac{1}{x}\left(\frac{3+x}{3-x}-1\right)}$

$$= e^{\lim_{x\to 0}\frac{2x}{x(3-x)}} = e^{2/3} \Rightarrow 2+3 = 5$$

(C) $\lim_{x\to 0}\frac{(\tan^3 x - x^3)-(\tan x^3 - x^3)}{x^5}$

$$= \lim_{x\to 0}\frac{\tan^3 x - x^3}{x^5} - \underbrace{\lim_{x\to 0}\frac{\tan x^3 - x^3}{x^5}}_{\text{zero (by expansion)}}$$

$$= \lim_{x\to 0}\frac{(\tan x - x)}{x^3}\cdot\frac{(\tan^2 x + x\tan x + x^2)}{x^2} = \frac{1}{3}\times 3 = 1$$

(D) Rationalising gives

$$\lim_{x\to 0}\frac{(x+2\sin x)\left[\sqrt{(x^2+2\sin x+1)}+\sqrt{\sin^2 x - x + 1}\right]}{(x^2+2\sin x+1)-(\sin^2 x - x + 1)}$$

$$= 2\cdot\lim_{x\to 0}\frac{x+\sin 2x}{x^2-\sin^2 x+2\sin x+x}$$

$$= 2\cdot\lim_{x\to 0}\frac{1+\frac{\sin 2x}{x}}{x-\frac{\sin^2 x}{x}+2+1} = 2\left(\frac{1+2}{3}\right) = 2$$

72. (A) $l = \lim_{n\to\infty}\cos^2(\pi(\sqrt[3]{n^3+n^2+2n}-n))$

Consider $\lim_{n\to\infty}[(n^3+n^2+2n)^{1/3}-n]$

$$= \lim_{n\to\infty}\left[n\left(1+\left(\frac{1}{n}+\frac{2}{n^2}\right)\right)^{1/3}-n\right]$$

$$= \lim_{n\to\infty} n\left[\left\{1+\left(\frac{1}{n}+\frac{2}{n^2}\right)\right\}^{1/3}-1\right]$$

$$= \lim_{n \to \infty} n\left[1 + \frac{1}{3}\left(\frac{1}{n} + \frac{2}{n^2}\right) + \ldots - 1\right]$$

$$= \frac{1}{3} + \text{terms containing } \frac{1}{n}, \frac{1}{n^2}, \frac{1}{n^3}, \ldots = \frac{1}{3}$$

$$\therefore \quad l = \cos^2\left(\frac{\pi}{3}\right) = \frac{1}{4}$$

(B) $l = \lim_{n \to \infty} n \sin(2\pi\sqrt{1+n^2} - 2n\pi)$

$$= \lim_{n \to \infty} n \sin\left[\frac{2\pi(\sqrt{1+n^2} - n)}{(\sqrt{1+n^2}+n)}(\sqrt{1+n^2}+n)\right]$$

$$= \lim_{n \to \infty}\left[\frac{n \sin\left(\frac{2\pi}{\sqrt{1+n^2}+n}\right)}{\left(\frac{2\pi}{\sqrt{1+n^2}+n}\right)}\left(\frac{2\pi}{\sqrt{1+n^2}+n}\right)\right]$$

$$= \lim_{n \to \infty} \frac{2n\pi}{n\left(\sqrt{1+\frac{1}{n^2}}+1\right)} = \frac{2\pi}{2} = \pi$$

(C) $l = \lim_{n \to \infty} (-1)^n (-1)^{n-1} \sin\left(n\pi - \pi\sqrt{n^2 + \frac{n}{2} + 1}\right)$

$$= \lim_{n \to \infty} (-1)^{2n-1} \sin\pi\left[\frac{\left(n - \sqrt{n^2+\frac{n}{2}+1}\right)\left(n + \sqrt{n^2+\frac{n}{2}+1}\right)}{n + \sqrt{n^2+\frac{n}{2}+1}}\right]$$

$$= \lim_{n \to \infty} (-1)^{2n-1} \sin\pi\left(\frac{n^2 - n^2 - \frac{n}{2} - 1}{n\left(1 + \sqrt{1 + \frac{1}{2n} + \frac{1}{n^2}}\right)}\right)$$

$$= \lim_{n \to \infty} (-1)^{2n} \sin\pi\left(\frac{\frac{1}{2} + \frac{1}{n}}{1 + \sqrt{1 + \frac{1}{2n} + \frac{1}{n^2}}}\right)$$

$$= (1) \sin\frac{\pi}{4} = \frac{1}{\sqrt{2}}$$

Also, as $n \to \infty, \sin\frac{(n+1)\pi}{4n} \to \frac{1}{\sqrt{2}}$

$\therefore$ Final answer is $\frac{1}{2}$.

(D) $l = e^{\lim_{x \to \infty} x\left(\frac{x+a}{x-a} - 1\right)} = e^{\lim_{x \to \infty} x\left(\frac{2a}{x-a}\right)}$

$$= e^{\lim_{x \to \infty} x\left(\frac{2a}{1-(a/x)}\right)} = e^{2a}$$

$$\therefore \quad e^{2a} = e \Rightarrow a = 1/2$$

73. Here, $L = \lim_{x \to \infty} x \log x + 2x \cdot \log\sin\left(\frac{1}{\sqrt{x}}\right)$,

$$= \lim_{\theta \to 0} \frac{1}{\theta^2} \cdot \log\left(\frac{1}{\theta^2}\right) + \frac{2}{\theta^2} \cdot \log(\sin\theta) \qquad \left[\text{put } x = \frac{1}{\theta^2}\right]$$

$$= \lim_{\theta \to 0} \frac{2\log(\sin\theta) - 2\log\theta}{\theta^2}$$

$$= \lim_{\theta \to 0} \frac{2\left[\cot\theta - \frac{1}{\theta}\right]}{2\theta} \qquad \text{[using L' Hospital's rule]}$$

$$= \lim_{\theta \to 0} \frac{2[\theta\cot\theta - 1]}{2\theta^2} = \frac{\theta - \tan\theta}{\theta^2 \tan\theta}$$

$$= \lim_{\theta \to 0} \frac{\theta - \tan\theta}{\theta^3} \cdot \frac{\theta}{\tan\theta} = \lim_{\theta \to 0} \frac{\theta - \tan\theta}{\theta^3}$$

$$L = \lim_{\theta \to 0} \frac{\theta - (\theta + \theta^3/3 + \cdots)}{\theta^3} = -1/3$$

$$\therefore \quad \frac{-2}{L} = 6$$

74. We have, $\left(1 + \frac{1}{n}\right)^{n + x_n} = e$ ...(i)

On taking log both sides of Eq. (i), we get

$$(n + x_n)\log_e\left(1 + \frac{1}{n}\right) = 1$$

$$\Rightarrow n + x_n = \frac{1}{\log\left(1 + \frac{1}{n}\right)} \Rightarrow x_n = \frac{1}{\log\left(1 + \frac{1}{n}\right)} - n$$

Let $\frac{n+1}{n} = \mu \Rightarrow n\mu = n + 1$

$$\Rightarrow \quad n = \frac{1}{\mu - 1}$$

$$\therefore \quad \lim_{n \to \infty} x_n = \lim_{\mu \to 1}\left(\frac{1}{\log\mu} - \frac{1}{\mu - 1}\right)$$

$$= \lim_{\mu \to 1} \frac{(\mu - 1) - \log\mu}{(\mu - 1)\log\mu} \qquad \left[\frac{0}{0}\text{ form}\right]$$

$$= \lim_{\mu \to 1} \frac{1 - \frac{1}{\mu}}{(\mu - 1)\cdot\frac{1}{\mu} + \log\mu} \qquad \text{[using L' Hospital's rule]}$$

$$= \lim_{\mu \to 1} \frac{1/\mu^2}{1/\mu^2 + 1/\mu} = \frac{1}{2}$$

$$\therefore 2\lim_{n \to \infty} x_n = 1$$

75. Let $\left[\sqrt{n} + \frac{1}{2}\right] = K \Rightarrow K \le \sqrt{n} + \frac{1}{2} < (K + 1)$

$$\Rightarrow (K - 1/2)^2 \le n < (K + 1/2)^2$$

$$\Rightarrow K^2 - K + 1/4 \le n < K^2 + K + 1/4, K \in N, n \in N$$

$$\Rightarrow K^2 - K + 1 \le n \le K^2 + K$$

So, $S = \sum_{n=1}^{\infty} \frac{2^{f(n)} + 2^{-f(n)}}{2^n} = \sum_{n=1}^{\infty} \frac{2^K + 2^{-K}}{2^n}$

$$= \left(\frac{2 + 2^{-1}}{2} + \frac{2^1 + 2^{-1}}{2^2}\right) + \left(\frac{2^2 + 2^{-2}}{2^3} + \frac{2^2 + 2^{-2}}{2^4} + \frac{2^2 + 2^{-2}}{2^5} + \frac{2^2 + 2^{-2}}{2^6}\right)$$

As, $\begin{cases} K=1 \Rightarrow 1 \le n \le 2 \\ K=2 \Rightarrow 3 \le n \le 6 \\ K=3 \Rightarrow 7 \le n \le 12 \\ \vdots \quad \vdots \end{cases}$

$$\therefore S = \sum_{K=1}^{\infty} \sum_{n=K^2-K+1}^{K^2+K} \frac{2^K+2^{-K}}{2^n} = \sum_{K=1}^{\infty}(2^K+2^{-K}) \cdot \sum_{n=K^2-K+1}^{K^2+K} \frac{1}{2^n}$$

$$= \sum_{K=1}^{\infty}(2^K+2^{-K}) \cdot \left[\frac{1}{2^{K^2-K+1}} + \frac{1}{2^{K^2-K+2}} + \dots + \frac{1}{2^{K^2+K}}\right]$$

$$= \sum_{K=1}^{\infty}(2^K+2^{-K}) \frac{\frac{1}{2^{K^2-K+1}}\left\{1-\frac{1}{2^{2K}}\right\}}{1-1/2}$$

$$= \sum_{K=1}^{\infty}(2^K+2^{-K}) \cdot (2^{-K^2+K} - 2^{-K^2-K})$$

$$= \sum_{K=1}^{\infty}(2^{-K^2+2K} - 2^{-K^2} + 2^{-K^2} - 2^{-K^2-2K})$$

$$= \sum_{K=1}^{\infty}(2^{-K^2+2K} - 2^{-K^2-2K}) = \sum_{K=1}^{\infty}(2^{-K(K-2)} - 2^{-K(K+2)})$$

$$= (2^1-2^{-3}) + (2^0-2^{-8}) + (2^{-3}-2^{-15}) + (2^{-8}-2^{-24}) + \dots \infty$$

$$= 2+1 = 3$$

76. Arithmetic mean $= \dfrac{1\cdot(n-1)+2\cdot(n-2)+\dots+(n-1)\cdot 1}{(n-1)}$

$$S_n = \frac{\sum_{r=1}^{n-1} r(n-r)}{(n-1)} = \frac{n\cdot\sum_{r=1}^{n-1} r - \sum_{r=1}^{n-1} r^2}{(n-1)}$$

$$= \frac{n\cdot\frac{n(n-1)}{2!} - \frac{n(n-1)(2n-1)}{6}}{(n-1)} \Rightarrow S_n = \frac{n^2}{2} - \frac{n(2n-1)}{6}$$

$$\therefore \lim_{n\to\infty} \frac{n^2}{S_n} = \frac{n^2}{\frac{n^2}{2} - \frac{n(2n-1)}{6}} = \frac{1}{1/2 - 2/6} = 6$$

77. Here, $k = \sum_{n=1}^{\infty} \dfrac{6^n}{(3^n-2^n)(3^{n+1}-2^{n+1})}$

$$= \sum_{n=1}^{\infty} \frac{1}{\left[1-\left(\frac{2}{3}\right)^n\right]\left[3\cdot\left(\frac{3}{2}\right)^n - 2\right]}$$

$$= \sum_{n=1}^{\infty} \frac{1}{\left[1-\left(\frac{2}{3}\right)^n\right]\cdot\left(\frac{3}{2}\right)^n\cdot\left[3-2\cdot\left(\frac{2}{3}\right)^n\right]}$$

$$= \sum_{n=1}^{\infty} \frac{(2/3)^n\cdot(3-2)}{\left[1-\left(\frac{2}{3}\right)^n\right]\left[3-2\cdot\left(\frac{2}{3}\right)^n\right]}$$

$$= \sum_{n=1}^{\infty} \frac{\left[3-2\left(\frac{2}{3}\right)^n\right] - 3\left[1-\left(\frac{2}{3}\right)^n\right]}{\left[1-\left(\frac{2}{3}\right)^n\right]\left[3-2\cdot\left(\frac{2}{3}\right)^n\right]}$$

$$= \lim_{n\to\infty} \sum_{n=1}^{n} \frac{1}{1-\left(\frac{2}{3}\right)^n} - \frac{3}{3-2\cdot\left(\frac{2}{3}\right)^n}$$

$$= \lim_{n\to\infty}\left[\frac{1}{1-(2/3)} - \frac{3}{3-2\cdot(2/3)}\right] + \left[\frac{1}{1-(2/3)^2} - \frac{3}{3-2\cdot(2/3)^2}\right] + \dots + \left[\frac{1}{1-(2/3)^n} - \frac{3}{3-2\cdot(2/3)^n}\right]$$

$$= \lim_{n\to\infty}\left[\frac{1}{1-2/3} - \frac{1}{1-(2/3)^{n+1}}\right] = 3-1 = 2$$

78. We have, $\lim_{x\to\frac{\pi}{2}} \sqrt{\dfrac{\tan x - \sin\{\tan^{-1}(\tan x)\}}{\tan x + \cos^2(\tan x)}}$

Now, RHL at $x = \dfrac{\pi}{2}$,

$$= \lim_{x\to\frac{\pi}{2}^+} \sqrt{\frac{\tan x - \sin\{\tan^{-1}(\tan x)\}}{\tan x + \cos^2(\tan x)}}$$

$$= \lim_{x\to\frac{\pi}{2}^+} \sqrt{\frac{\tan x - \sin(x-\pi)}{\tan x + \cos^2(\tan x)}}$$

$$\left[\because \tan^{-1}(\tan x) = x - \pi, \text{ when } x > \frac{\pi}{2}\right]$$

$$= \lim_{x\to\frac{\pi}{2}^+} \sqrt{\frac{1+\frac{\sin x}{\tan x}}{1+\frac{\cos^2(\tan x)}{\tan x}}} = \sqrt{\frac{1+0}{1+0}} = 1$$

Again, LHL at $x = \dfrac{\pi}{2}$,

$$= \lim_{x\to\frac{\pi}{2}^-} \sqrt{\frac{\tan x - \sin\{\tan^{-1}(\tan x)\}}{\tan x + \cos^2(\tan x)}}$$

$$= \lim_{x\to\frac{\pi}{2}^-} \sqrt{\frac{\tan x - \sin(x)}{\tan x + \cos^2(\tan x)}} \quad \left[\because \tan^{-1}(\tan x) = x, \text{ when } x < \frac{\pi}{2}\right]$$

$$= \lim_{x\to\frac{\pi}{2}^-} \sqrt{\frac{1-\frac{\sin x}{\tan x}}{1+\frac{\cos^2(\tan x)}{\tan x}}} = \sqrt{\frac{1+0}{1+0}} = 1$$

$$\therefore \lim_{x\to\frac{\pi}{2}} \sqrt{\frac{\tan x - \sin\{\tan^{-1}(\tan x)\}}{\tan x + \cos^2(\tan x)}} = 1$$

79. We have, $A(t) = \int_0^t \sin x^2 \cdot dx$; $B(t) = \dfrac{t\sin t^2}{2}$.

$$\therefore \lim_{t\to 0} \frac{A(t)}{B(t)} = \lim_{t\to 0} \frac{2\int_0^t \sin x^2 \cdot dx}{t\sin t^2} = \lim_{t\to 0} \frac{2\int_0^t \sin x^2 dx}{t^3 \cdot \frac{\sin^2 t}{t^2}}$$

$$= \lim_{t\to 0} \frac{2\int_0^t \sin x^2 \cdot dx}{t^3} = \lim_{t\to 0} \frac{2\cdot\sin(t^2)}{3t^2}$$

[applying Newton-Leibnitz's rule followed by L' Hospital's rule]

$\therefore \quad \lim_{t \to 0} \frac{A(t)}{B(t)} = \frac{2}{3}$

Then, $\quad m + n = 2 + 3 = 5$

80. Here, $AB : \left[\int_0^{2t} \left(\frac{\sin x}{x} + 1\right) dx\right] x + y = 3t$

and $AC : 2tx + y = 0$

On solving, we get

$$AB : \left[\int_0^{2t} \left(\frac{\sin x}{x} + 1\right) dx - 2t\right] x = 3t \qquad [\because y = -2tx]$$

$$\therefore \quad x_A = \frac{3t}{\int_0^{2t} \left(\frac{\sin x}{x} + 1\right) dx - 2t}$$

$$= \frac{3}{\left(\frac{\sin 2t}{2t} + 1\right) 2 - 2} = \frac{3}{2 \cdot \frac{\sin 2t}{2t}}$$

Now, $\quad \lim_{t \to 0} \frac{3}{2 \cdot \left(\frac{\sin 2t}{2t}\right)} = \frac{3}{2} \quad$ [using L'Hospital's rule]

$\therefore \quad \frac{p}{q} = \frac{3}{2} \quad \Rightarrow p + q = 5$

81. Let $A_k = (2t, t^2)$

$\therefore$ Slope of $FA_k = \frac{t^2 - 1}{2t - 0}$

$$= \tan\left(\frac{\pi}{2} + \theta_k\right)$$

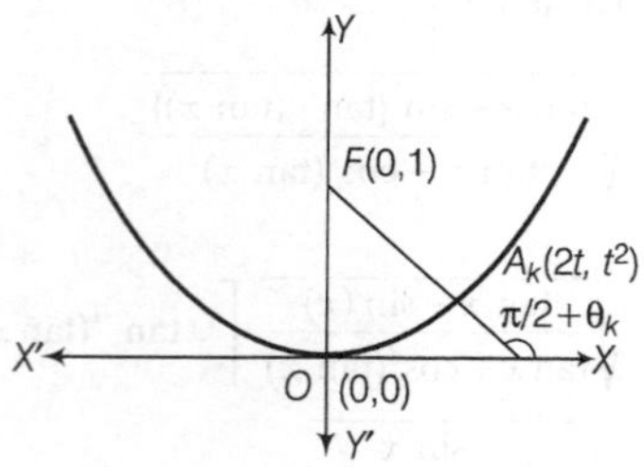

$$\tan(\theta_k) = \frac{2t}{1 - t^2} = \tan(2\phi), \text{ where } t = \tan\phi$$

$$\therefore \quad \phi = \frac{\theta_k}{2} = \frac{k\pi}{4n}, \text{ where } \tan\phi = t.$$

Also, $\quad FA_K = \sqrt{(t^2 - 1)^2 + (2t)^2}$

$$= t^2 + 1 = 1 + \tan^2\phi$$

$$= \sec^2\left(\frac{k\pi}{4n}\right)$$

$$\therefore \lim_{n \to \infty} \frac{1}{n} \cdot \sum_{k=1}^{n} FA_k = \lim_{n \to \infty} \frac{1}{n} \cdot \sum_{k=1}^{n} \sec^2\left(\frac{k\pi}{4n}\right)$$

$$= \int_0^1 \sec^2\left(\frac{\pi x}{4}\right) dx$$

$$= \frac{4}{\pi} \cdot \left[\tan\left(\frac{\pi x}{4}\right)\right]_0^1 = \frac{4}{\pi}$$

Hence, $m = 4$.

82. Here, $L = \lim_{h \to 0} \frac{\cos(\tan^{-1}(\tan(\pi/2 + h)))}{h}$

$$= \lim_{h \to 0} \frac{\cos(\tan^{-1}(-\cot h))}{h} = \lim_{h \to 0} \frac{\cos\left(-\frac{\pi}{2} + h\right)}{h} = \frac{\sin h}{h} = 1$$

$\therefore \quad \cos(2\pi L) = \cos(2\pi) = 1$

83. Here, $k = \lim_{n \to \infty} \prod_{r=2}^{n} \frac{(r-1)(r^2+r+1)}{(r+1)(r^2-r+1)}$

$$= \lim_{n \to \infty} \frac{2}{n(n+1)} \times \frac{n^2+n+1}{3} = \frac{2}{3}.$$

So, $\operatorname{cosec}\theta = \frac{2}{3} \Rightarrow$ Number of solution is zero.

84. Here, $\lim_{x \to \infty} \left(\frac{x+c}{x-c}\right)^x = 4 \Rightarrow e^{\lim_{x \to \infty} \left(\frac{2c}{x-c}\right) x} = 4$

$\Rightarrow \quad e^{2c} = 4 \Rightarrow e^c = 2 \qquad$ [only positive value]

$$\therefore \quad \frac{e^c}{2} = 1$$

85. $\lim_{x \to -\infty} \frac{(3x^4 + 2x^2) \cdot \sin\left(\frac{1}{x}\right) - x^3 + 5}{-x^3 + x^2 - x + 1} = k$

$$\Rightarrow \quad \lim_{x \to -\infty} \frac{\left(3 + \frac{2}{x^2}\right) \cdot \frac{\sin\left(\frac{1}{x}\right)}{\frac{1}{x}} - 1 + \frac{5}{x^3}}{-1 + \frac{1}{x} - \frac{1}{x^2} + \frac{1}{x^3}}$$

$$\Rightarrow \quad \frac{(3-1)}{-1} = k \Rightarrow \left|\frac{k}{2}\right| = 1$$

86. $f(x) = \lim_{t \to 0} \frac{2x}{\pi} \tan^{-1}\left(\frac{x}{t^2}\right)$

***Case* I** When $x > 0$,

$$f(x) = \lim_{t \to 0} \frac{2x}{\pi} \cdot \tan^{-1}\left(\frac{x}{t^2}\right) = \frac{2x}{\pi} \times \frac{\pi}{2} = x$$

***Case* II** When $x < 0$,

$$f(x) = \lim_{t \to 0} \frac{2x}{\pi} \cdot \tan^{-1}\left(\frac{x}{t^2}\right) = \frac{2x}{\pi} \cdot \left(\frac{-\pi}{2}\right) = -x$$

***Case* III** When $x = 0 \Rightarrow f(x) = 0$

$\therefore \quad f(x) = (x) \Rightarrow f(1) = 1$

87. Let $f(x) = \left(\frac{1}{\sqrt{n^2}} + \frac{1}{\sqrt{n^2+1}} + \frac{1}{\sqrt{n^2+2}} + \ldots + \frac{1}{\sqrt{n^2+2n}}\right)$

As, $\quad \frac{1}{\sqrt{n^2+2n}} < \frac{1}{\sqrt{n^2}} = \frac{1}{\sqrt{n^2}}$

$$\frac{1}{\sqrt{n^2+2n}} < \frac{1}{\sqrt{n^2+1}} < \frac{1}{\sqrt{n^2}}$$

$$\vdots$$

On adding, we get

$$\frac{2n+1}{\sqrt{n^2+2n}} < \frac{1}{\sqrt{n^2}} + \frac{1}{\sqrt{n^2+1}} + \ldots + \frac{1}{\sqrt{n^2+2n}} < \frac{2n+1}{\sqrt{n^2}}$$

On applying, limit $n \to \infty$

$$2 < \lim_{n\to\infty}\left(\frac{1}{\sqrt{n^2}}+\frac{1}{\sqrt{n^2+1}}+\frac{1}{\sqrt{n^2+2}}+\ldots+\frac{1}{\sqrt{n^2+2n}}\right) < 2$$

$$\therefore \quad \lim_{n\to\infty}\left(\frac{1}{\sqrt{n^2}}+\frac{1}{\sqrt{n^2+1}}+\ldots+\frac{1}{\sqrt{n^2+2n}}\right)=2$$

88. $l = \lim_{x\to 1^+} 2^{-2^{\frac{1}{1-x}}} = \lim_{h\to 0} 2^{-2^{-\frac{1}{h}}} = 2^0 = 1$

and $m = \lim_{h\to 0} \dfrac{(1+h)\cdot \sin h}{h} = 1$

$\therefore \quad l + m = 2$

89. As, $\dfrac{\sin x \tan x}{x^2} = \dfrac{4\cdot \tan^4 \dfrac{x}{2}}{x^2\cdot(1-\tan^4 x/2)}$

Hence, when $x \to 0$, $\dfrac{\sin x \cdot \tan x}{x^2} \to 1$. But $\dfrac{\sin x \cdot \tan x}{x^2} > 1$

$$\therefore \quad \lim_{x\to 0}\left[\frac{\sin x \tan x}{x^2}\right] = 1$$

90. Let $S_n = \dfrac{k}{2}\sum_{r=1}^{n} \dfrac{2r}{1\times 3\times 5\times \ldots \times(2r-1)\times(2r+1)}$

$$= \frac{k}{2}\sum_{r=1}^{n}\left[\frac{1}{1\times3\times5\times\ldots\times(2r-1)} - \frac{1}{1\times3\times5\times\ldots\times(2r+1)}\right]$$

$$= \frac{k}{2}\left[1 - \frac{1}{1\times3\times5\times\ldots\times(2n+1)}\right], \text{ as } n\to\infty.$$

$\lim_{n\to\infty} S_n = \dfrac{k}{2} = 1 \Rightarrow k^2 = 4$

91. Here, $\sin^4 x = \sin^2 x - \dfrac{1}{4}\cdot\sin^2 2x$

$$\therefore S_n = \left(\sin^2 x - \frac{1}{4}\cdot\sin^2 2x\right) + \left(\frac{1}{4}\cdot\sin^2 2x - \frac{1}{4^2}\cdot\sin^2 2^2 x\right)$$

$$+ \ldots + \left(\frac{1}{4^n}\cdot\sin^2 2^n x - \frac{1}{4^{n+1}}\cdot\sin^2 2^{n+1} x\right)$$

$$= \sin^2 x - \frac{1}{4^{n+1}}\cdot\sin^2 2^{n+1} x \Rightarrow f(x) = \sin^2 x,\ g(x) = \cos^2 x$$

$$\therefore \quad [\sqrt{f(x)} + \sqrt{g(x)}]^4 = 4$$

92. Here, $f(3) = 3f(1) = 1$

$f(4) = f(2+1+1) = f(2) + f(1) + f(1) = 2+1+1 = 4$ and so on.

In general, $f(x) = x$, for $x \in N$

$$\therefore \quad \lim_{n\to\infty}\sum_{r=1}^{n}\frac{(4r)f(3r)}{n^3} = \frac{12n(n+1)(2n+1)}{6n^3} = 4$$

93. As, $0 \le \dfrac{x-1}{3} < 1 \Rightarrow 1 \le x \le 4$

Then, $\lim_{n\to\infty}\dfrac{n}{n\cdot\left(\dfrac{x-1}{3}\right)^n + n\cdot 3 - 1} = \dfrac{1}{3}$, when $0 \le \dfrac{x-1}{3} < 1$

$\Rightarrow \quad x = 1, 2, 3$

$\therefore$ Number of integral values = 3.

94. As, $x^5 - y^5 = (x-y)(x^4 + x^3y + x^2y^2 + xy^3 + y^4)$

and the coefficient of x^n in

$(x-1)(x-2)(x+3)(x+10)(x+15) - x^5$

$$= -1 - 2 + 3 + 10 + 15 = 25$$

$$\therefore \lim_{x\to\infty}((x-1)(x-2)(x+3)(x+10)(x+15))^{1/3} - x)$$

$$= \frac{25}{5} = 5$$

95. Here, $k = \lim_{x\to\infty}(1 + [f(x)] + x^2 - 1)\{f(x)\}$

$$= e^{\lim_{x\to\infty}\frac{\left[\frac{\tan x}{x}\right] + x^2 - 1}{\left\{\frac{\tan x}{x}\right\}}} = e^{\lim_{x\to\infty}\frac{1 + x^2 - 1}{\frac{\tan x}{x} - \left[\frac{\tan x}{x}\right]}}$$

$$= e^{\lim_{x\to\infty}\frac{x^2}{\left(1 + \frac{x^2}{3} + \frac{x^4}{15} - 1\right)}} = e^3$$

$\therefore \quad [k/e] = [e^2] = 7$

96. Here, $(\sqrt{3}+1)^{2n} = I + f$ ...(i)

Let $\quad f = (\sqrt{3}-1)^{2n}$ ...(ii)

$\therefore \quad I + f + F =$ even integer, $0 < f,\ f < 1$

$\Rightarrow \quad f + F = 1 \Rightarrow 0 < f + F < 2$

$\Rightarrow \quad f + F = 1$

$\Rightarrow \quad \{(\sqrt{3}+1)\}^{2n} = f = 1 - f$

$\therefore \quad \lim_{n\to\infty} f = \lim_{n\to\infty} 1 - f = 1 - \lim_{n\to\infty}(\sqrt{3}-1)^{2n} = 1 - 0 = 1$

97. Here, $f(x) = a_0x^4 + a_1x^3 + a_2x^2 + a_3x + a_4$.

As, $\quad \lim_{x\to 0} 1 + \dfrac{f(x)}{x^2} = 2 \Rightarrow \lim_{x\to 0}\dfrac{f(x)}{x^2} = 1$

So, $\quad a_3 = a_4 = 0$ and $a_2 = 1$

$f(x) = a_0x^4 + a_1x^3 + x^2$

$\Rightarrow f'(x) = 4a_0x^3 + 3a_1x^2 + 2x$

As, $f(x)$ has extremum at $x = 1$ and $x = 2$.

$f'(1) = 4a_0 + 3a_1 + 2$

and $\quad f'(2) = 32a_0 + 12a_1 + 4$

So, $\quad a_0 = \dfrac{1}{4}$ and $a_1 = -1$

$\Rightarrow \quad f(2) = 0$

98. Here, $|x| = \log(\{x\})$ has no solution.

Hence, $\alpha = 0$

$$\therefore \lim_{x\to 0}\frac{x e^{ax} - b\sin x}{x^3} = \lim_{x\to 0}\frac{(1-b)x + ax^2 + x^3\left(\frac{a^2}{2} - \frac{b}{6}\right)}{x^3}$$

$\Rightarrow \quad a = 0,\ b = 1$

$\Rightarrow \quad a + b = 1$

99. Here, $\sin(x_{n+1} - x_n) + 2^{-(n+1)}\sin x_n \cdot \sin x_{n+1} = 0$

$\Rightarrow \quad \cot x_{n+1} - \cot x_n = 2^{-(n+1)}$

$\Rightarrow \quad \cot x_n - \cot x_{n-1} = 2^{-n}$

$\Rightarrow \quad \cot x_{n+1} - \cot x_1 = \dfrac{1}{2^2} + \dfrac{1}{2^3} + \ldots + \dfrac{1}{2^{n+1}}$

$\Rightarrow \quad \cot x_{n+1} = \dfrac{1}{2} + \dfrac{1}{2^2} + \ldots \dfrac{1}{2^{n+1}}$

$$\Rightarrow \lim_{n\to\infty} \cot x_{n+1} = 1$$

$$\therefore \lim_{n\to\infty} x_{n+1} = \frac{\pi}{4}$$

$$l = \frac{\pi}{4} \Rightarrow 4l = \pi$$

$$\therefore [4l] = 3$$

100. Here, $\lim_{x\to 0} \frac{x^2 \sin(\beta x)}{\alpha x - \sin x} = 1$

$$\Rightarrow \lim_{x\to 0} \frac{x^2\left(\beta x - \frac{(\beta x)^3}{3!} + \frac{(\beta x)^5}{5!} - \dots\right)}{\alpha x - \left(x - \frac{x^3}{3!} + \frac{x^5}{5!} - \dots\right)} = 1$$

$$\Rightarrow \lim_{x\to 0} \frac{x^3\left(\beta - \frac{\beta^3 x^2}{3!} + \frac{\beta^5 x^4}{5!} - \dots\right)}{(\alpha-1)x + \frac{x^3}{3!} + \frac{x^5}{5!} - \dots} = 1$$

Limit exists only, when $\alpha - 1 = 0$

$$\Rightarrow \alpha = 1 \quad \dots(i)$$

$$\therefore \lim_{x\to 0} \frac{x^3\left(\beta - \frac{\beta^3 x^2}{3!} + \frac{\beta^5 x^4}{5!} - \dots\right)}{x^3\left(\frac{1}{3!} - \frac{x^2}{5!} - \dots\right)} = 1$$

$$\Rightarrow 6\beta = 1 \quad \dots(ii)$$

From Eqs. (i) and (ii), we get

$$6(\alpha + \beta) = 6\alpha + 6\beta = 6 + 1 = 7$$

101. Given, $\lim_{\alpha\to 0}\left[\frac{e^{\cos(\alpha^n)} - e}{\alpha^m}\right] = -\frac{e}{2}$

$$\Rightarrow \lim_{\alpha\to 0} \frac{e\{e^{\cos(\alpha^n)-1} - 1\}}{\cos(\alpha^n) - 1} \cdot \frac{\cos(\alpha^n) - 1}{\alpha^m} = \frac{-e}{2}$$

$$\Rightarrow \lim_{\alpha\to 0} e\left\{\frac{e^{\cos(\alpha^n)-1} - 1}{\cos(\alpha^n) - 1}\right\} \cdot \lim_{\alpha\to 0} \frac{-2\sin^2 \frac{\alpha^n}{2}}{\alpha^m} = -\frac{e}{2}$$

$$\Rightarrow e \times 1 \times (-2) \lim_{\alpha\to 0} \frac{\sin^2\left(\frac{\alpha^n}{2}\right)}{\frac{\alpha^{2n}}{4}} \cdot \frac{\alpha^{2n}}{4\alpha^m} = \frac{-e}{2}$$

$$\Rightarrow e \times 1 \times -2 \times 1 \times \lim_{\alpha\to 0} \frac{\alpha^{2n-m}}{4} = \frac{-e}{2}$$

For this to be exists, $2n - m = 0 \Rightarrow \frac{m}{n} = 2$

102. Here, $\lim_{x\to 1} \frac{F(x)}{G(x)} = \frac{1}{14}$

$$\Rightarrow \lim_{x\to 1} \frac{F'(x)}{G'(x)} = \frac{1}{14} \quad \text{[using L'Hospital's rule]} \dots(i)$$

As, $F(x) = \int_{-1}^{x} f(t)\,dt$

$$\Rightarrow F'(x) = f(x) \quad \dots(ii)$$

and $G(x) = \int_{-1}^{x} t\,|f\{f(t)\}|\,dt$

$$\Rightarrow G'(x) = x\,|f\{f(x)\}| \quad \dots(iii)$$

$$\therefore \lim_{x\to 1} \frac{F(x)}{G(x)} = \lim_{x\to 1} \frac{F'(x)}{G'(x)} = \lim_{x\to 1} \frac{f(x)}{x\,|f\{f(x)\}|}$$

$$= \frac{f(1)}{1\,|f\{f(1)\}|} = \frac{\frac{1}{2}}{\left|f\left(\frac{1}{2}\right)\right|} \quad \dots(iv)$$

Given, $\lim_{x\to 1} \frac{F(x)}{G(x)} = \frac{1}{14}$

$$\therefore \frac{\frac{1}{2}}{\left|f\left(\frac{1}{2}\right)\right|} = \frac{1}{14} \Rightarrow \left|f\left(\frac{1}{2}\right)\right| = 7$$

103. Given, $\lim_{x\to 1}\left\{\frac{\sin(x-1) + a(1-x)}{(x-1) + \sin(x-1)}\right\}^{\frac{(1+\sqrt{x})(1-\sqrt{x})}{1-\sqrt{x}}} = \frac{1}{4}$

$$\lim_{x\to 1}\left\{\frac{\frac{\sin(x-1)}{(x-1)} - a}{1 + \frac{\sin(x-1)}{(x-1)}}\right\}^{1+\sqrt{x}} = \frac{1}{4}$$

$$\Rightarrow \left(\frac{1-a}{2}\right)^2 = \frac{1}{4}$$

$$\Rightarrow (a-1)^2 = 1$$

$$\Rightarrow a = 2 \text{ or } 0$$

Hence, the maximum value of a is 2.

104. $L = \lim_{x\to 0} \frac{a - \sqrt{a^2 - x^2} - \frac{x^2}{4}}{x^4}, a > 0$

$$= \lim_{x\to 0} \frac{a - a\cdot\left[1 - \frac{1}{2}\cdot\frac{x^2}{a^2} + \frac{\frac{1}{2}\left(\frac{1}{2} - 1\right)}{2}\cdot\frac{x^4}{a^4} - \dots\right] - \frac{x^2}{4}}{x^4}$$

$$= \lim_{x\to 0} \frac{\frac{x^2}{2a} + \frac{1}{8}\cdot\frac{x^4}{a^3} + \dots - \frac{x^2}{4}}{x^4}$$

Since, L is finite $\Rightarrow 2a = 4 \Rightarrow a = 2$

$$\therefore L = \lim_{x\to 0} \frac{1}{8\cdot a^3} = \frac{1}{64}$$

105. $\lim_{x\to\frac{\pi}{2}} \frac{\cot x\,(1 - \sin x)}{-8\left(x - \frac{\pi}{2}\right)^3}$

$$= \lim_{x\to\frac{\pi}{2}} \frac{\tan\left(\frac{\pi}{2} - x\right)}{8\left(\frac{\pi}{2} - x\right)} \cdot \frac{\left(1 - \cos\left(\frac{\pi}{2} - x\right)\right)}{\left(\frac{\pi}{2} - x\right)^2} = \frac{1}{8}\cdot 1 \cdot \frac{1}{2} = \frac{1}{16}$$

106. Given, $p = \lim_{x \to 0^+} (1 + \tan^2 \sqrt{x})^{\frac{1}{2x}}$ [1^∞ form]

$$= e^{\lim_{x \to 0^+} \frac{\tan^2 \sqrt{x}}{2x}} = e^{\frac{1}{2} \lim_{x \to 0^+} \left(\frac{\tan \sqrt{x}}{\sqrt{x}}\right)^2} = e^{\frac{1}{2}}$$

$\therefore \log p = \log e^{\frac{1}{2}} = \frac{1}{2}$

107. Let $l = \lim_{n \to \infty} \left[\frac{(n+1)\cdot(n+2)\dots(3n)}{n^{2n}}\right]^{\frac{1}{n}}$

$$= \lim_{n \to \infty} \left[\frac{(n+1)\cdot(n+2)\dots(n+2n)}{n^{2n}}\right]^{\frac{1}{n}}$$

$$= \lim_{n \to \infty} \left[\left(\frac{n+1}{n}\right)\left(\frac{n+2}{n}\right)\dots\left(\frac{n+2n}{n}\right)\right]^{\frac{1}{n}}$$

Taking log on both sides, we get

$$\log l = \lim_{n \to \infty} \frac{1}{n}\left[\log\left\{\left(1+\frac{1}{n}\right)\left(1+\frac{2}{n}\right)\dots\left(1+\frac{2n}{n}\right)\right\}\right]$$

$$\Rightarrow \log l = \lim_{n \to \infty} \frac{1}{n} \left[\log\left(1+\frac{1}{n}\right) + \log\left(1+\frac{2}{n}\right) + \dots + \log\left(1+\frac{2n}{n}\right)\right]$$

$$\Rightarrow \log l = \lim_{n \to \infty} \frac{1}{n} \sum_{r=1}^{2n} \log\left(1+\frac{r}{n}\right)$$

$$\Rightarrow \log l = \int_0^2 \log(1+x)\,dx$$

$$\Rightarrow \log l = \left[\log(1+x)\cdot x - \int \frac{1}{1+x}\cdot x\,dx\right]_0^2$$

$$\Rightarrow \log l = [\log(1+x)\cdot x]_0^2 - \int_0^2 \frac{x+1-1}{1+x}\,dx$$

$$\Rightarrow \log l = 2\cdot\log 3 - \int_0^2 \left(1 - \frac{1}{1+x}\right)dx$$

$$\Rightarrow \log l = 2\cdot\log 3 - [x - \log|1+x|]_0^2$$

$$\Rightarrow \log l = 2\cdot\log 3 - [2 - \log 3]$$

$$\Rightarrow \log l = 3\cdot\log 3 - 2$$

$$\Rightarrow \log l = \log 27 - 2$$

$$\therefore \quad l = e^{\log 27 - 2}$$

$$= 27\cdot e^{-2} = \frac{27}{e^2}$$

108. $\lim_{x\to 0}\frac{\sin(\pi\cos^2 x)}{x^2} = \lim_{x\to 0}\frac{\sin\pi(1-\sin^2 x)}{x^2}$

$$= \lim_{x\to 0}\frac{\sin(\pi - \pi\sin^2 x)}{x^2}$$

$$= \lim_{x\to 0}\frac{\sin(\pi\sin^2 x)}{x^2} \quad [\because \sin(\pi-\theta) = \sin\theta]$$

$$= \lim_{x\to 0}\frac{\sin\pi\sin^2 x}{\pi\sin^2 x}\times(\pi)\left(\frac{\sin^2 x}{x^2}\right) = \pi \quad \left[\because \lim_{\theta\to 0}\frac{\sin\theta}{\theta} = 1\right]$$

109. We have, $\lim_{x\to 0}\frac{(1-\cos 2x)(3+\cos x)}{x\tan 4x} = \lim_{x\to 0}\frac{2\sin^2 x\,(3+\cos x)}{x\times\frac{\tan 4x}{4x}\times 4x}$

$$= \lim_{x\to 0}\frac{2\sin^2 x}{x^2}\times\lim_{x\to 0}\frac{(3+\cos x)}{4}\times\frac{1}{\lim_{x\to 0}\frac{\tan 4x}{4x}}$$

$$= 2\times\frac{4}{4}\times 1 = 2 \quad \left[\because \lim_{\theta\to 0}\frac{\sin\theta}{\theta} = 1 \text{ and } \lim_{\theta\to 0}\frac{\tan\theta}{\theta} = 1\right]$$

110. Plan $\left(\frac{\infty}{\infty}\right)$ form

$$\lim_{x\to\infty}\frac{a_0x^n + a_1x^{n-1} + \dots + a_n}{b_0x^m + b_1x^{m-1} + \dots + b_m} = \begin{cases} 0, & \text{if } n<m \\ \frac{a_0}{b_0}, & \text{if } n=m \\ +\infty, & \text{if } n>m \text{ and } a_0b_0>0 \\ -\infty, & \text{if } n<m \text{ and } a_0b_0<0 \end{cases}$$

Description of Situation As to make degree of numerator equal to degree of denominator.

$$\therefore \quad \lim_{x\to\infty}\left(\frac{x^2+x+1}{x+1} - ax - b\right) = 4$$

$$\Rightarrow \quad \lim_{x\to\infty}\frac{x^2+x+1-ax^2-ax-bx-b}{x+1} = 4$$

$$\Rightarrow \quad \lim_{x\to\infty}\frac{x^2(1-a)+x(1-a-b)+(1-b)}{x+1} = 4$$

Here, we make degree of numerator = degree of denominator

$$\therefore \quad 1-a = 0$$

$$\Rightarrow \quad a = 1$$

and $\quad \lim_{x\to\infty}\frac{x(1-a-b)+(1-b)}{x+1} = 4$

$$\Rightarrow \quad 1-a-b = 4$$

$$\Rightarrow \quad b = -4 \quad [\because (1-a) = 0]$$

111. Plan To make the quadratic into simple form, we should eliminate radical sign.

Description of Situation As for given equation, when $a \to 0$ the equation reduces to identity in x.

i.e. $\quad ax^2 + bx + c = 0, \forall\, x \in R$ or $a = b = c \to 0$

Thus, first we should make above equation independent from coefficients as 0.

Let $a + 1 = t^6$. Thus, when $a \to 0, t \to 1$.

$$\therefore \quad (t^2-1)\,x^2 + (t^3-1)x + (t-1) = 0$$

$$\Rightarrow (t-1)\,\{(t+1)x^2 + (t^2+t+1)\,x + 1\} = 0, \text{ as } t \to 1$$

$$2x^2 + 3x + 1 = 0$$

$$\Rightarrow \quad 2x^2 + 2x + x + 1 = 0$$

$$\Rightarrow \quad (2x+1)(x+1) = 0$$

Thus, $\quad x = -1, -1/2$

or $\quad \lim_{a\to 0^+}\alpha(a) = -1/2$

and $\quad \lim_{a\to 0^+}\beta(a) = -1$

112. Here, $\lim_{x \to 0} \{1 + x \log(1+b^2)\}^{1/x}$ [1^∞ form]

$$= e^{\lim_{x \to 0} \{x \log(1+b^2)\} \cdot \frac{1}{x}}$$

$$= e^{\log(1+b^2)} = (1+b^2) \quad \text{...(i)}$$

Given, $\lim_{x \to 0} \{1 + x \log(1+b^2)\}^{1/x} = 2b \sin^2 \theta$

$$\Rightarrow \quad (1+b^2) = 2b \sin^2 \theta$$

$$\therefore \quad \sin^2 \theta = \frac{1+b^2}{2b} \quad \text{...(ii)}$$

By AM $\geq$ GM, $\dfrac{b + \frac{1}{b}}{2} \geq \left(b \cdot \frac{1}{b}\right)^{1/2}$

$$\Rightarrow \quad \frac{b^2+1}{2b} \geq 1 \quad \text{...(iii)}$$

From Eqs. (ii) and (iii), we get

$$\sin^2 \theta = 1$$

$$\Rightarrow \quad \theta = \pm \frac{\pi}{2}, \text{ as } \theta \in (-\pi, \pi]$$

113. Here, $\lim_{x \to 0} (\sin x)^{1/x} + \lim_{x \to 0} \left(\frac{1}{x}\right)^{\sin x}$

$$= 0 + \lim_{x \to 0} e^{\log\left(\frac{1}{x}\right)^{\sin x}} = e^{\lim_{x \to 0} \frac{\log(1/x)}{\operatorname{cosec} x}} \quad \left[\begin{array}{l} \lim_{x \to 0} (\sin x)^{1/x} \to 0 \\ \text{as, (decimal)}^\infty \to 0 \end{array}\right]$$

Applying L' Hospital's rule, we get

$$e^{\lim_{x \to 0} \frac{x\left(-\frac{1}{x^2}\right)}{-\operatorname{cosec} x \cot x}} = e^{\lim_{x \to 0} \frac{\sin x}{x} \tan x} = e^0 = 1$$

114. Here, $\lim_{h \to 0} \dfrac{f(2h + 2 + h^2) - f(2)}{f(h - h^2 + 1) - f(1)}$

[$\because f'(2) = 6$ and $f'(1) = 4$, given]

Applying L' Hospital's rule,

$$= \lim_{h \to 0} \frac{\{f'(2h + 2 + h^2)\} \cdot (2 + 2h) - 0}{\{f'(h - h^2 + 1)\} \cdot (1 - 2h) - 0}$$

$$= \frac{f'(2) \cdot 2}{f'(1) \cdot 1}$$

$$= \frac{6 \cdot 2}{4 \cdot 1} = 3 \quad \text{[using } f'(2)=6 \text{ and } f'(1) = 4]$$

115. Given, $\lim_{x \to 0} \dfrac{\{(a-n)nx - \tan x\} \sin nx}{x^2} = 0$

$$\Rightarrow \quad \lim_{x \to 0} \left\{(a-n)n - \frac{\tan x}{x}\right\} \frac{\sin nx}{nx} \times n = 0$$

$$\Rightarrow \quad \{(a-n)n - 1\} n = 0$$

$$\Rightarrow \quad (a-n)n = 1$$

$$\Rightarrow \quad a = n + \frac{1}{n}$$

116. $\lim_{x \to 0} \dfrac{(\cos x - 1)(\cos x - e^x)}{x^n}$

$$= \lim_{x \to 0} \frac{\left(-2\sin^2 \frac{x}{2}\right)\left\{\left(1 - \frac{x^2}{2!} + \frac{x^4}{4!} - \dots\right) - \left(1 + x + \frac{x^2}{2!} + \frac{x^3}{3!} + \dots\right)\right\}}{x^n}$$

$$= \lim_{x \to 0} \frac{\left(-2\sin^2 \frac{x}{2}\right)\left(-x - \frac{2x^2}{2!} - \frac{x^3}{3!} - \dots\right)}{4\left(\frac{x}{2}\right)^2 x^{n-2}}$$

$$= \lim_{x \to 0} \frac{\sin^2 \frac{x}{2}\left(1 + x + \frac{x^2}{3!} + \dots\right)}{2\left(\frac{x}{2}\right)^2 x^{n-3}}$$

Above limit is finite, if $n - 3 = 0$, i.e. $n = 3$.

117. Let $y = \left[\dfrac{f(1+x)}{f(1)}\right]^{1/x}$

$$\Rightarrow \log y = \frac{1}{x}[\log f(1+x) - \log f(1)]$$

$$\Rightarrow \quad \lim_{x \to 0} \log y = \lim_{x \to 0}\left[\frac{1}{f(1+x)} \cdot f'(1+x)\right]$$

[using L'Hospital's rule]

$$= \frac{f(1)}{f(1)} = \frac{6}{3}$$

$$\Rightarrow \log\left(\lim_{x \to 0} y\right) = 2 \Rightarrow \lim_{x \to 0} y = e^2$$

118. For $x \in R$, $\lim_{x \to \infty}\left(\dfrac{x-3}{x+2}\right)^x = \lim_{x \to \infty} \dfrac{(1 - 3/x)^x}{(1 + 2/x)^x} = \dfrac{e^{-3}}{e^2} = e^{-5}$

CHAPTER

06

Continuity and Differentiability

Learning Part

Session 1
- Continuous Function
- Continuity of a Function at a Point

Session 2
- Continuity in an Interval or Continuity at End Points

Session 3
- Discontinuity of a Function

Session 4
- Theorems Based on Continuity
- Continuity of Composite Function

Session 5
- Intermediate Value Theorem

Session 6
- Differentiability : Meaning of Derivative

Session 7
- Differentiability in an Interval

Practice Part
- JEE Type Examples
- Chapter Exercises

Arihant on Your Mobile !
Exercises with this symbol can be practised on your mobile. See title inside to activate for free.

Session 1

Continuous Function, Continuity of a Function at a Point

Introduction to Continuity

A real function is continuous at a fixed point if we can draw the graph of the function around that point without lifting the pen from the plane of the paper. In case one has to lift the pen at a point, the graph of the function is said to have a break or discontinuous at that point, say $x = a$.

Different types of situations, which may come up at $x = a$ along the graph, can be shown as below

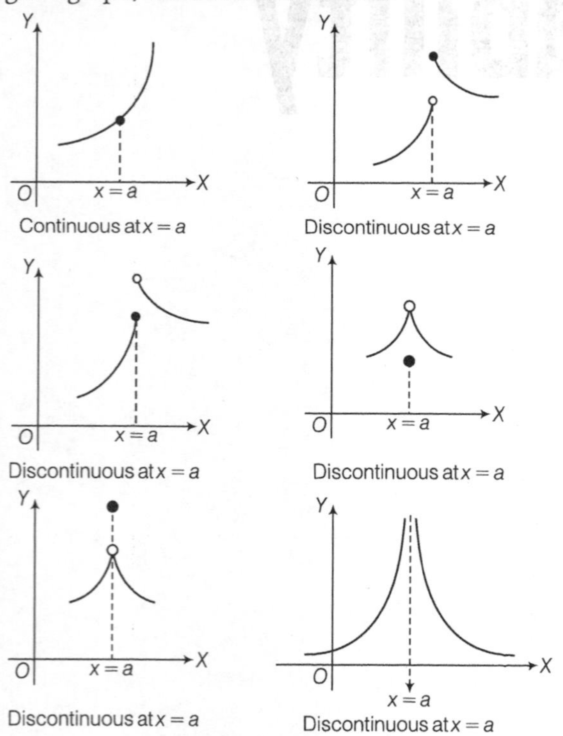

Figure 6.1

- There should be heading "continuity of a function at point"
- A function $f(x)$ is said to be continuous at $x = a$; where $a \in$ domain of $f(x)$, if $\lim_{x \to a^-} f(x) = \lim_{x \to a^+} f(x) = f(a)$

  i.e. LHL = RHL = Value of a function at $x = a$

  or $\quad \lim_{x \to a} f(x) = f(a)$
- If $f(x)$ is not continuous at $x = a$, we say that $f(x)$ is discontinuous at $x = a$, $f(x)$ will be discontinuous at $x = a$ in any of the following cases

(i) $\lim_{x \to a^-} f(x)$ and $\lim_{x \to a^+} f(x)$ exist but are not equal.

(ii) $\lim_{x \to a^-} f(x)$ and $\lim_{x \to a^+} f(x)$ exist and are equal but not equal to $f(a)$.

(iii) $f(a)$ is not defined.

(iv) Atleast one of the limit doesn't exist.

Now, we will discuss different conditions for a function to be discontinuous in detail

(i) $\lim_{x \to a^-} f(x)$ and $\lim_{x \to a^+} f(x)$ exists but are not equal.

Here, $\quad \lim_{x \to a^-} f(x) = l_1$

and $\quad \lim_{x \to a^+} f(x) = l_2$

$\therefore \lim_{x \to a^-} f(x)$ and $\lim_{x \to a^+} f(x)$ exist but are not equal.

Thus, $f(x)$ is discontinuous at $x = a$. It does not matter whether $f(a)$ exist or not.

Graphically This conditions can be illustrated as

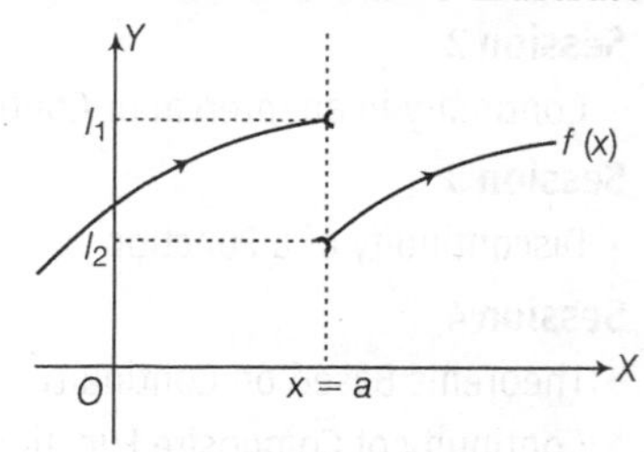

Figure 6.2

Example 1 If $f(x) = \dfrac{|x|}{x}$. Discuss the continuity at $x = 0$.

Sol. Here, $\quad f(x) = \dfrac{|x|}{x}$

$\therefore$ RHL at $x = 0$

Let $\quad x = 0 + h$

i.e. $\quad \lim_{x \to 0^+} f(x) = \lim_{h \to 0} f(0 + h) = \lim_{h \to 0} \dfrac{|0 + h|}{0 + h} = \lim_{h \to 0} \dfrac{h}{h} = 1$

$\Rightarrow \quad \lim_{x \to 0^+} f(x) = 1$

Again, LHL at $x = 0$

Let $\quad x = 0 - h$

i.e. $\lim_{x \to 0^-} f(x) = \lim_{h \to 0} f(0-h)$

$= \lim_{h \to 0} \frac{|0-h|}{0-h} = \lim_{h \to 0} \frac{h}{-h} = -1$

$\Rightarrow \lim_{x \to 0^-} f(x) = -1 \Rightarrow \lim_{x \to 0^+} f(x) \neq \lim_{x \to 0^-} f(x).$

Thus, $f(x)$ is discontinuous at $x = 0$.

Graphically Here,

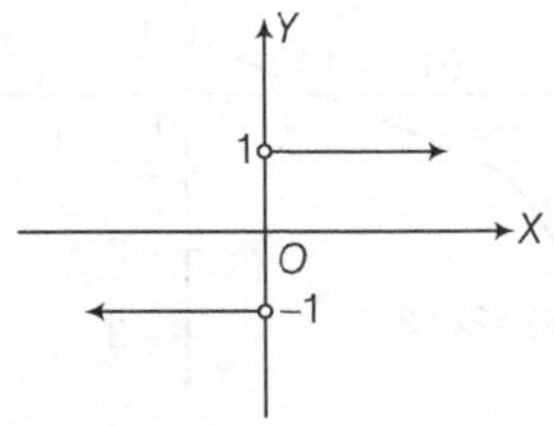

$$f(x) = \frac{|x|}{x} = \begin{cases} \frac{x}{x}, & x > 0 \\ -\frac{x}{x}, & x < 0 \end{cases} = \begin{cases} 1, & x > 0 \\ -1, & x < 0 \end{cases}$$

and $f(0) = \frac{0}{0}$ [indeterminate form]

$\Rightarrow$ It is not defined.

Which shows, the graph is broken at $x = 0$.

Where, $\lim_{x \to 0^-} f(x) = -1$ and $\lim_{x \to 0^+} f(x) = -1$

Thus, $\lim_{x \to 0} f(x)$ doesn't exist and hence function is discontinuous.

Remark

Here, $f(x)$ is not defined at $x = 0$, as $f(0) = \frac{0}{0}$. [indeterminate form]
So, we could say directly that the function is discontinuous at $x = 0$

(ii) $\lim_{x \to a^-} f(x)$ and $\lim_{x \to a^+} f(x)$ exist and are equal but not equal to $f(a)$.

Here, $\lim_{x \to a^-} f(x) = \lim_{x \to a^+} f(x) = L$

$f(a)$ is also defined but $f(a) \neq L$.

So, limit of $f(x)$ exists at $x = a$. But, it is not continuous at $x = a$.

Graphically This conditions can be illustrated as

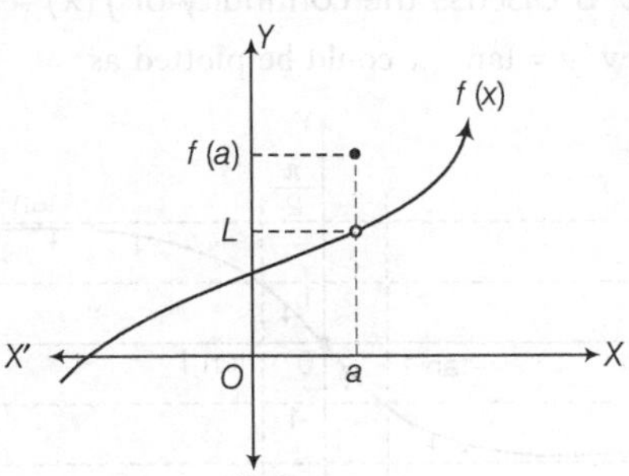

Figure 6.3

Example 2 If $f(x) = \begin{cases} 2x+3, & \text{when } x < 0 \\ 0, & \text{when } x = 0. \\ x^2+3, & \text{when } x > 0 \end{cases}$

Discuss the continuity at $x = 0$.

Sol. Here, $f(x) = \begin{cases} 2x+3, & \text{when } x < 0 \\ 0, & \text{when } x = 0 \\ x^2+3, & \text{when } x > 0 \end{cases}$

$\therefore$ RHL at $x = 0$, let $x = 0 + h$

i.e. $\lim_{x \to 0^+} f(x) = \lim_{h \to 0} f(0+h)$

$= \lim_{h \to 0} \{(0+h)^2 + 3\} = 3$

$\Rightarrow \lim_{x \to 0^+} f(x) = 3$

Again, LHL at $x = 0$, let $x = 0 - h$

i.e. $\lim_{x \to 0^-} f(x) = \lim_{h \to 0} f(0-h)$

$= \lim_{h \to 0} \{2(0-h) + 3\} = 3$

$\Rightarrow \lim_{x \to 0^-} f(x) = 3$

But $f(0) = 0$

Therefore, $\lim_{x \to 0^+} f(x) = \lim_{x \to 0^-} f(x) = 3 \neq f(0)$

Thus, $f(x)$ is discontinuous at $x = 0$.

Graphically

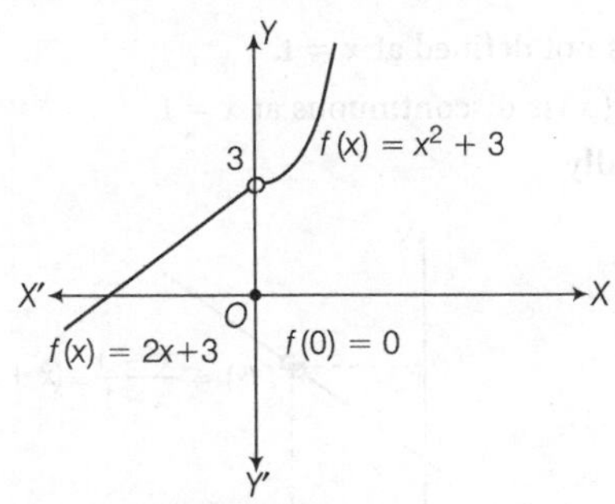

Here, $\lim_{x \to 0^-} f(x) = 3$

$\lim_{x \to 0^+} f(x) = 3$

$f(0) = 0$

Thus, $\lim_{x \to 0^-} f(x) = \lim_{x \to 0^+} f(x) = 3 \neq f(0)$

Hence, $f(x)$ is discontinuous at $x = 0$.

(iii) $f(a)$ is not defined.

Here, $\lim_{x \to a^+} f(x) = L$

and $\lim_{x \to a^-} f(x) = L$

But, $f(a)$ is not defined. So, $f(x)$ is discontinuous at $x = a$.

Graphically This conditions can be illustrated as

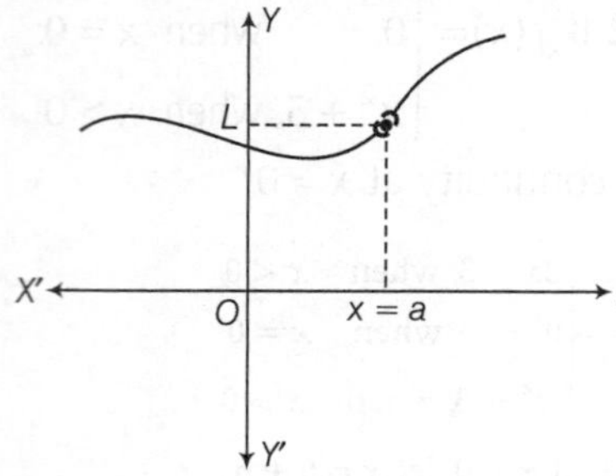

Figure 6.4

Example 3 If $f(x)=\dfrac{x^2-1}{x-1}$.

Discuss the continuity at $x=1$.

Sol. Here, $f(x)=\dfrac{x^2-1}{x-1}$

$$\lim_{x\to 1} f(x)=\lim_{x\to 1}\frac{x^2-1}{x-1}$$

$$=\lim_{x\to 1}\frac{(x-1)(x+1)}{(x-1)}$$

$$=\lim_{x\to 1}(x+1)=2$$

But $f(1)=\dfrac{0}{0}$ [indeterminate form]

$\therefore$ $f(1)$ is not defined at $x=1$.

Hence, $f(x)$ is discontinuous at $x=1$.

Graphically

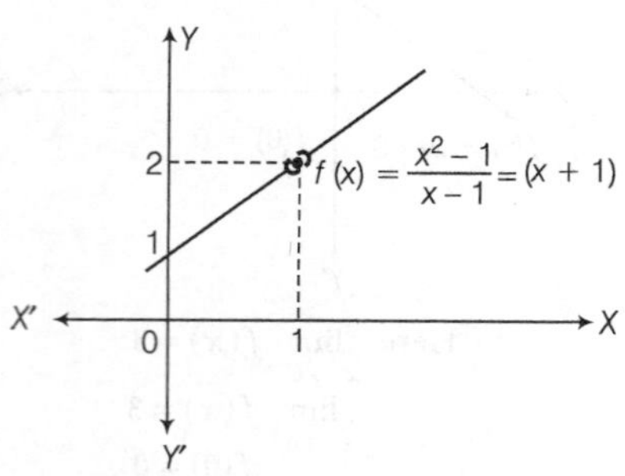

Here, $\lim_{x\to 1} f(x)=2$ but $f(1)$ is not defined.

So, $f(x)$ is discontinuous at $x=1$.

Example 4 Show that the function

$$f(x)=\begin{cases}2x+3, & -3\le x<-2\\ x+1, & -2\le x<0\\ x+2, & 0\le x\le 1\end{cases}$$

is discontinuous at $x=0$ and continuous at every other point in interval $[-3,1]$.

Sol. Graphically $f(x)=\begin{cases}2x+3, & -3\le x<-2\\ x+1, & -2\le x<0\\ x+2, & 0\le x\le 1\end{cases}$ is plotted as shown

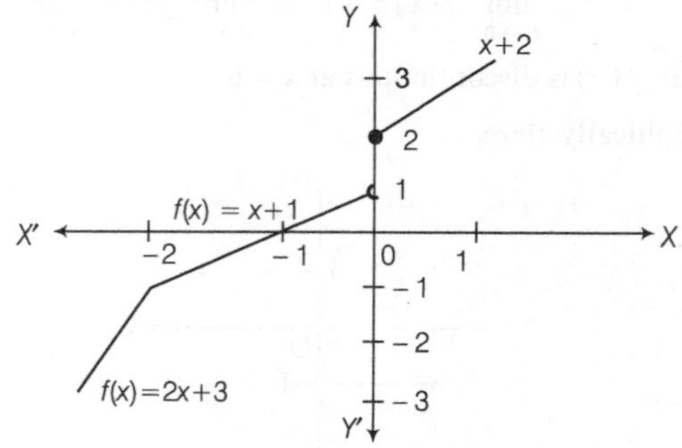

Here, if we observe the graph we could conclude that at $x=0$, $\lim_{x\to 0^-} f(x)=1$ and $\lim_{x\to 0^+} f(x)=2$, which shows that the function is discontinuous at $x=0$ and continuous at every other point in $[-3,1]$.

Example 5 Examine the function,

$$f(x)=\begin{cases}\dfrac{\cos x}{\pi/2-x}, & x\ne\pi/2\\ 1, & x=\pi/2\end{cases}$$ for continuity *at* $x=\pi/2$.

Sol. We have, $\lim_{x\to\pi/2} f(x)=\lim_{x\to\pi/2}\dfrac{\cos x}{\pi/2-x}$

[as $x\ne\pi/2$ but $x\to\pi/2$]

$$=\lim_{x\to\pi/2}\frac{\cos x}{\pi/2-x}\qquad\left[\frac{0}{0}\text{ form}\right]$$

$$=\lim_{x\to\pi/2}\frac{-\sin x}{0-1}\qquad\text{[applying L'Hospital's rule]}$$

$$\Rightarrow\quad \lim_{x\to\pi/2} f(x)=\sin\frac{\pi}{2}=1$$

Also, $f(\pi/2)=1$

$\therefore$ $\lim_{x\to\pi/2} f(x)=f(\pi/2)$

Thus, $f(x)$ is continuous at $x=\pi/2$.

Example 6 Discuss the continuity of $f(x)=[\tan^{-1}x]$.

Sol. We know, $y=\tan^{-1}x$ could be plotted as

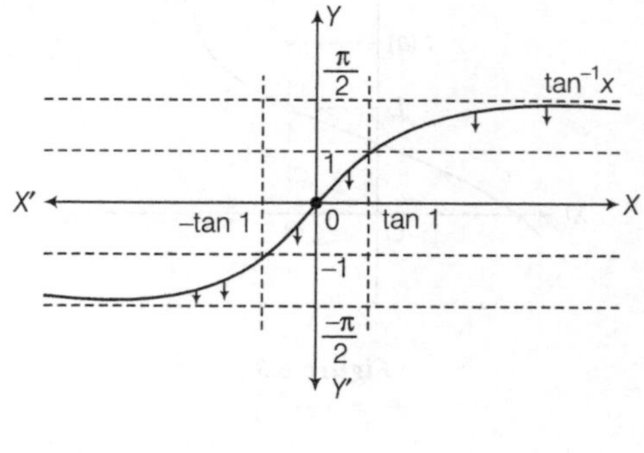

Thus, $f(x) = [\tan^{-1} x]$ could be plotted as

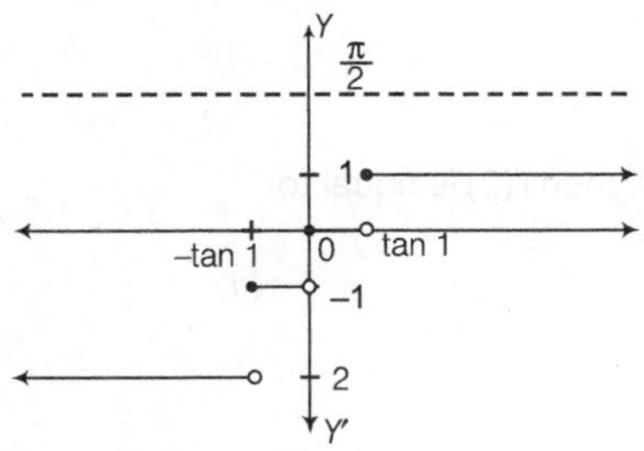

Which clearly shows the graph is broken at $\{-\tan 1, 0, \tan 1\}$.

$\therefore$ $f(x)$ is not continuous when $x \in \{-\tan 1, 0, \tan 1\}$.

Example 7 Let $y = f(x)$ be defined parametrically as $y = t^2 + t|t|$, $x = 2t - |t|, t \in R$. Then, at $x = 0$, find $f(x)$ and discuss continuity.

Sol. As, $\quad y = t^2 + t|t|$ and $x = 2t - |t|$

Thus, when $t \ge 0$

$\Rightarrow \quad x = 2t - t = t, y = t^2 + t^2 = 2t^2$

$\therefore \quad x = t$ and $y = 2t^2 \Rightarrow y = 2x^2, \forall\ x \ge 0$

Again, when $t < 0$

$\Rightarrow\ x = 2t + t = 3t$ and $y = t^2 - t^2 = 0 \Rightarrow\ y = 0, \forall\ x < 0$

Hence, $\quad f(x) = \begin{cases} 2x^2, & x \ge 0 \\ 0, & x < 0 \end{cases}$

which is clearly continuous for all x as shown graphically.

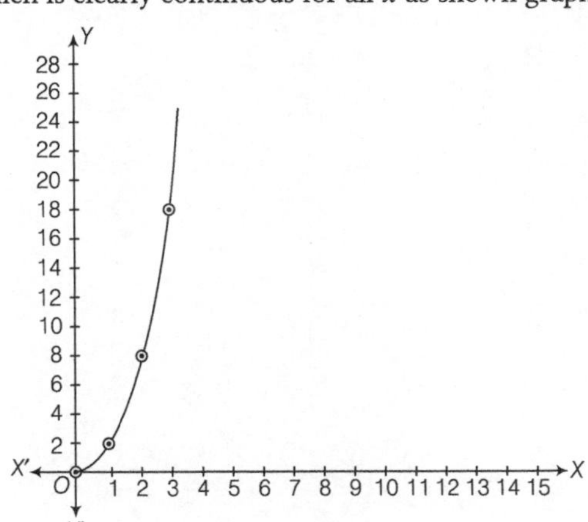

Example 8 Let

$$f(x) = \frac{e^{\tan x} - e^x + \log(\sec x + \tan x) - x}{\tan x - x}$$ be a continuous function at $x = 0$. The value of $f(0)$ equals

(a) $\frac{1}{2}$ (b) $\frac{2}{3}$

(c) $\frac{3}{2}$ (d) 2

Sol. For continuity at $x = 0$, we have

$$f(0) = \lim_{x \to 0} f(x)$$

$$= \lim_{x \to 0} \frac{e^{\tan x} - e^x}{\tan x - x} + \lim_{x \to 0} \frac{\log(\sec x + \tan x) - x}{\tan x - x}$$

$$= \lim_{x \to 0} e^x \frac{(e^{\tan x - x} - 1)}{(\tan x - x)} + \lim_{x \to 0} \frac{\sec x - 1}{\sec^2 x - 1}$$

[using L'Hospital's rule]

$$= 1 + \lim_{x \to 0} \frac{1}{\sec x + 1} = 1 + \frac{1}{2} = \frac{3}{2}$$

Hence, (c) is the correct answer.

Example 9 If $f(x) = \sqrt{\dfrac{1}{\tan^{-1}(x^2 - 4x + 3)}}$, then $f(x)$ is continuous for

(a) (1, 3) (b) $(-\infty, 0)$

(c) $(-\infty, 1) \cup (3, \infty)$ (d) None of these

Sol. Here,

$$f(x) = \sqrt{\frac{1}{\tan^{-1}(x^2 - 4x + 3)}}$$

For domain, $\quad \tan^{-1}(x^2 - 4x + 3) > 0$

$\Rightarrow \quad x^2 - 4x + 3 > 0$

$\Rightarrow \quad (x - 1)(x - 3) > 0$

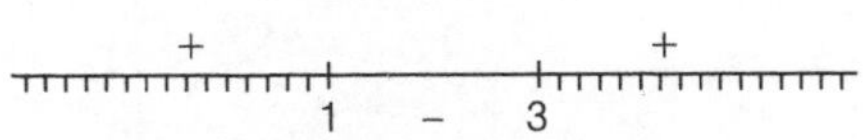

$\therefore \quad x \in (-\infty, 1) \cup (3, \infty)$

Since, every general function is continuous in its domain.

$\therefore f(x)$ is continuous for $x \in (-\infty, 1) \cup (3, \infty)$.

Hence, (c) is correct answer.

Exercise for Session 1

1. If function $f(x)=\dfrac{\sqrt{1+x}-\sqrt[3]{1+x}}{x}$ is continuous function at $x=0$, then $f(0)$ is equal to

(a) 2 (b) $\frac{1}{4}$ (c) $\frac{1}{6}$ (d) $\frac{1}{3}$

2. If $f(x)=\begin{cases}\dfrac{1}{e^{1/x}+1}, & x\neq 0\\ 0, & x=0\end{cases}$, then

(a) $\lim_{x\to 0^-} f(x)=0$ (b) $\lim_{x\to 0^+} f(x)=1$

(c) $f(x)$ is discontinuous at $x=0$ (d) $f(x)$ is continuous at $x=0$

3. If $f(x)=\begin{cases}\dfrac{x^2-(a+2)x+2a}{x-2}, & x\neq 2\\ 2, & x=2\end{cases}$ is continuous at $x=2$, then a is equal to

(a) 0 (b) 1 (c) -1 (d) 2

4. If $f(x)=\begin{cases}\dfrac{\log(1+2ax)-\log(1-bx)}{x}, & x\neq 0\\ k, & x=0\end{cases}$ is continuous at $x=0$, then k is equal to

(a) $2a+b$ (b) $2a-b$ (c) $b-2a$ (d) $a+b$

5. If $f(x)=\begin{cases}[x]+[-x], & x\neq 2\\ \lambda, & x=2\end{cases}$ and f is continuous at $x=2$, where $[\cdot]$ denotes greatest integer function, then λ is

(a) -1 (b) 0 (c) 1 (d) 2

Session 2

Continuity in an Interval or Continuity at End Points

Let a function $y = f(x)$ be defined on $[a, b]$. Then, the function $f(x)$ is said to be continuous at the left end $x = a$

If $\quad f(a) = \lim_{x \to a^+} f(x) \quad$ [need not check LHL]

The function $f(x)$ is said to be continuous at the right end $\quad x = b$

If $\quad f(b) = \lim_{x \to b^-} f(x) \quad$ [need not check RHL]

| Example 10 If $f(x) = [x]$, where $[\cdot]$ denotes greatest integral function. Then, check the continuity on $[1, 2]$.

Sol. **Graphically** $f(x) = [x]$ could be plotted as

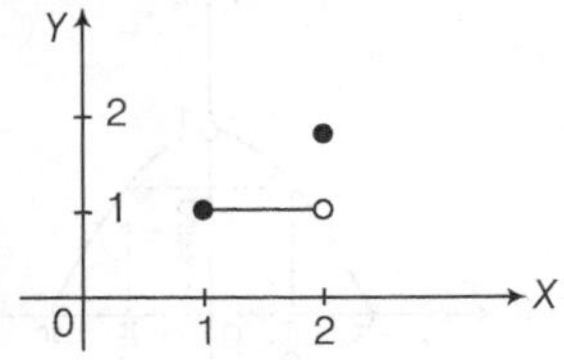

Here, in the graph $f(x)$ is continuous at all points, where $1 < x < 2$.

To check continuity at $x = 1$ and $x = 2$

(i) **To check continuity at $x = 1$**

Here, $\quad f(1) = 1$ and RHL at $x = 1$

$\Rightarrow \lim_{x \to 1^+} f(x) = 1$, i.e. $f(1) = \lim_{x \to 1^+} f(x)$

Thus, $f(x)$ is continuous at $x = 1$.

(ii) **To check continuity at $x = 2$**

Here, $\quad f(2) = 2$

But LHL at $x = 2 \Rightarrow \lim_{x \to 2^-} f(x) = 1$

which shows $f(2) \neq \lim_{x \to 2^-} f(x)$

Thus, $f(x)$ is discontinuous at $x = 2$.

From the above information, it becomes clear that $f(x)$ is continuous at all points on $[1, 2]$ except at $x = 2$,

i.e. $f(x)$ is continuous for $x \in [1, 2)$.

Exercise for Session 2

1. Let $f(x) = \begin{cases} -2\sin x & \text{for } -\pi \le x \le -\dfrac{\pi}{2} \\ a\sin x + b & \text{for } -\dfrac{\pi}{2} < x < \dfrac{\pi}{2} \\ \cos x & \text{for } \dfrac{\pi}{2} \le x \le \pi \end{cases}$. If f is continuous on $[-\pi, \pi)$, then find the values of a and b.

2. Draw the graph of the function $f(x) = x - |x - x^2|, -1 \le x \le 1$ and discuss the continuity or discontinuity of f in the interval $-1 \le x \le 1$.

3. Discuss the continuity of 'f' in $[0, 2]$, where $f(x) = \begin{cases} |4x - 5|\,[x] & \text{for } x > 1 \\ [\cos \pi x] & \text{for } x \le 1 \end{cases}$; where $[x]$ is greatest integer not greater than x.

4. Let $f(x) = \begin{cases} Ax - B, & x \le -1 \\ 2x^2 + 3Ax + B, & x \in (-1, 1] \\ 4, & x > 1 \end{cases}$

Statement I $f(x)$ is continuous at all x, if $A = \dfrac{3}{4}$ and $B = -\dfrac{1}{4}$.

Statement II Polynomial function is always continuous.

A. Both Statement I and Statement II are correct and Statement II is the correct explanation of Statement I

B. Both Statement I and Statement II are correct but Statement II is not the correct explanation of Statement I

C. Statement I is correct but Statement II is incorrect

D. Statement II is correct but Statement I is incorrect

(a) A $\qquad$ (b) B $\qquad$ (c) C $\qquad$ (d) D

Session 3

Discontinuity of a Function

A function f which is not continuous, said to be discontinuous functions, i.e. if there is a break in the function of any kind, then it is discontinuous functions.

There are two types of discontinuity

1. Removable discontinuity
2. Non-removable discontinuity

Removable Discontinuity

In this type of discontinuity, $\lim_{x \to a} f(x)$ necessarily exists, but is either not equal to $f(a)$ or $f(a)$ is not defined. Such function is said to have a removable discontinuity of the first kind.

In this case, it is possible to redefine the function in such a manner that $\lim_{x \to a} f(x) = f(a)$ and thus making function continuous.

These discontinuities can be further classified as

(i) Missing point discontinuity

(ii) Isolated point discontinuity

Examples of Missing Point Discontinuity

Here, $\lim_{x \to a} f(x)$ exists finitely but $f(a)$ is not defined. e.g.

(i) Let $f(x) = \dfrac{(x-1)(9-x^2)}{(x-1)}$,

clearly $f(1) \to \dfrac{0}{0}$ form

$\therefore$ $f(x)$ has missing point discontinuity. Shown as,

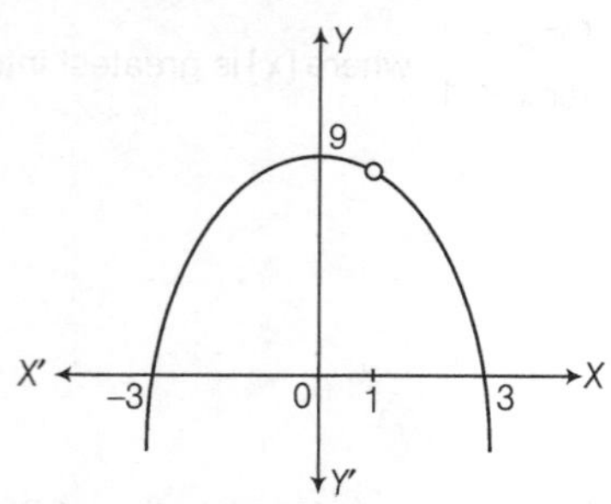

Figure 6.5

(ii) $f(x) = \dfrac{x^2-4}{x-2}$ has missing point discontinuity at $x = 2$.

Shown as,

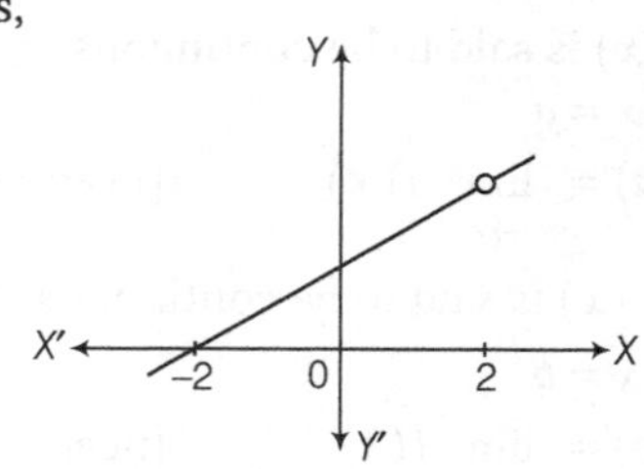

Figure 6.6

(iii) $f(x) = \dfrac{\sin x}{x}$ has missing point discontinuity at $x = 0$.

Shown as,

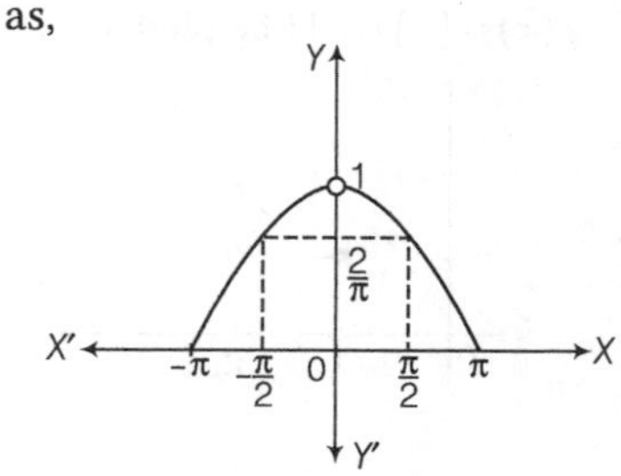

Figure 6.7

Examples of Isolated Point Discontinuity

Here, $\lim_{x \to a} f(x)$ exists and $f(a)$ also exists but $\lim_{x \to a} f(x) \neq f(a)$. e.g.

(i) Let $f(x) = [x] + [-x] \Rightarrow f(x) = \begin{cases} 0, & \text{if } x \in I \\ -1, & \text{if } x \notin I \end{cases}$

where x = Integer, has isolated point discontinuity, can be shown as

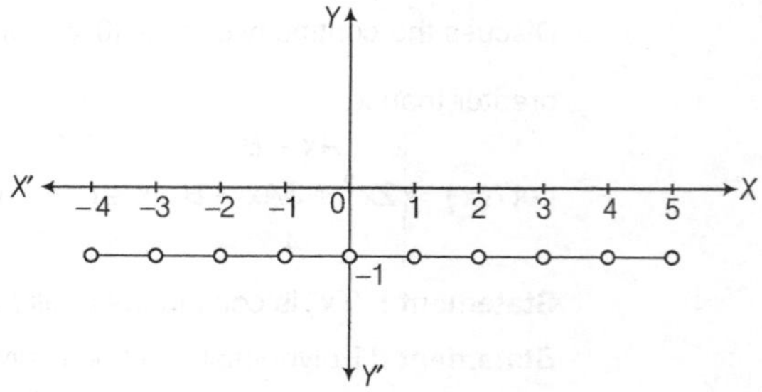

Figure 6.8

(ii) Let $f(x) = \text{sgn}(\cos 2x - 2\sin x + 3)$

$\Rightarrow f(x) = \text{sgn}(1 - 2\sin^2 x - 2\sin x + 3)$

$= \text{sgn}(2(2 + \sin x)(1 - \sin x))$

$$= \begin{cases} 0, \text{ if } x = 2n\pi + \dfrac{\pi}{2} \\ 1, \text{ if } x \neq 2n\pi + \dfrac{\pi}{2} \end{cases}$$

$\therefore f(x)$ has an isolated point discontinuity at $x = 2n\pi + \dfrac{\pi}{2}$.

Shown as,

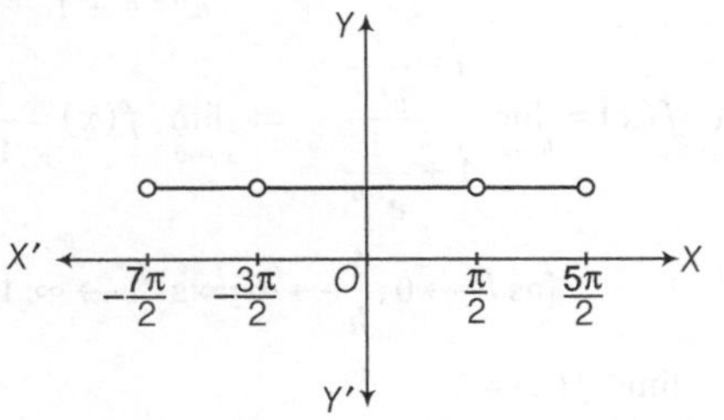

Figure 6.9

Non-removable Discontinuity

In this type of discontinuity $\lim_{x \to a} f(x)$ doesn't exist and therefore, it is not possible to redefine the function in any manner and make it continuous. Such function is said to have non-removable discontinuity or discontinuity of second kind.

Such discontinuities can further be classified into three types.

(i) Finite Discontinuity

If for a function $f(x)$: $\lim_{x \to a^-} f(x) = L_1$ and $\lim_{x \to a^+} f(x) = L_2$ and $L_1 \neq L_2$, then such function is said to have a finite discontinuity or a jump discontinuity.

In this case, the non-negative difference between the two limits is called the **Jump of Discontinuity**. A function having a finite number of jumps in a given interval is called a **Piecewise Continuous** or **Sectionally Continuous Function.**

Examples of Finite Discontinuity

(i) $f(x) \tan^{-1}\left(\dfrac{1}{x}\right)$ at $x = 0$

$$\left.\begin{aligned} &\text{RHL} \quad \text{i.e. } f(0^+) = \frac{\pi}{2} \\ &\text{LHL} \quad \text{i.e. } f(0^-) = -\frac{\pi}{2} \end{aligned}\right\} \text{jump} = \pi$$

(ii) $f(x) = \dfrac{|\sin x|}{x}$ at $x = 0$

$$\left.\begin{aligned} &\text{RHL} \quad \text{i.e. } f(0^+) = 1 \\ &\text{LHL} \quad \text{i.e. } f(0^-) = -1 \end{aligned}\right\} \text{jump} = 2$$

(ii) Infinite Discontinuity

If for a function $f(x)$: $\lim_{x \to a^-} f(x) = L_1$ and $\lim_{x \to a^+} f(x) = L_2$ and either L_1 or L_2 is infinity, then such function is said to have infinite discontinuity.

In other words, If $x = a$ is a vertical asymptote for the graph of $y = f(x)$, then f is said to have infinite discontinuity at a.

Examples of Infinite Discontinuity

(i) $f(x) = \dfrac{x}{1-x}$, at $x = 1$

RHL i.e. $f(1^+) = -\infty$

LHL i.e. $f(1^-) = \infty$

(ii) $f(x) = \dfrac{1}{x^2}$, at $x = 0$

RHL i.e. $f(0^+) = \infty$

LHL i.e. $f(0^-) = \infty$

(iii) Oscillatory Discontinuity

If for a function $f(x)$: $\lim_{x \to a} f(x)$ doesn't exist but oscillate between two finite quantities, then such function is said to have oscillatory discontinuity.

Examples of Oscillatory Discontinuity

(i) $f(x) = \sin\left(\dfrac{1}{x}\right)$

$\Rightarrow \lim_{x \to 0} f(x) =$ a value between -1 to 1.

$\therefore$ Limit doesn't exist, as it oscillates between -1 and 1 as $x \to 0$.

(ii) $f(x) = \cos\left(\dfrac{1}{x}\right)$

$\Rightarrow \lim_{x \to 0} \cos\left(\dfrac{1}{x}\right) =$ a value between -1 to 1.

$\therefore$ Limit doesn't exist, as it oscillates between -1 to 1 at $x \to 0$.

List of Continuous Functions

	Function $f(x)$	Interval in which $f(x)$ is continuous
1.	constant c	$(-\infty, \infty)$
2.	x^n, n is an integer ≥ 0	$(-\infty, \infty)$
3.	x^{-n}, n is a positive integer	$(-\infty, \infty)-\{0\}$
4.	$\lvert x-a \rvert$	$(-\infty, \infty)$
5.	$P(x) = a_0x^n + a_1x^{n-1} + ... + a_n$	$(-\infty, \infty)$
6.	$\dfrac{p(x)}{q(x)}$, where $p(x)$ and $q(x)$ are polynomial in x	$(-\infty, \infty)-\{x : q(x) = 0\}$
7.	$\sin x$	$(-\infty, \infty)$
8.	$\cos x$	$(-\infty, \infty)$
9.	$\tan x$	$(-\infty, \infty)-\left\{(2n+1)\dfrac{\pi}{2} : n \in I\right\}$
10.	$\cot x$	$(-\infty, \infty)-\{n\pi : n \in I\}$
11.	$\sec x$	$(-\infty, \infty)-\{(2n+1)\pi/2 : n \in I\}$
12.	$\operatorname{cosec} x$	$(-\infty, \infty)-\{n\pi : n \in I\}$
13.	e^x	$(-\infty, \infty)$
14.	$\log_e x$	$(0, \infty)$

Example 11 Examine the function, $f(x) = \begin{cases} x-1, & x<0 \\ 1/4, & x=0 \\ x^2-1, & x>0 \end{cases}$

Discuss the continuity and if discontinuous remove the discontinuity.

Sol. $\lim_{x \to 0^-} f(x) = \lim_{x \to 0^+} f(x) = -1$

But $f(0) = 1/4$. Thus, $f(x)$ has removable discontinuity and $f(x)$ could be made continuous by taking

$$f(0) = -1 \Rightarrow f(x) = \begin{cases} x-1, & x<0 \\ -1, & x=0 \\ x^2-1, & x>0 \end{cases}$$

Graphically $f(x)$ could be plotted as showing

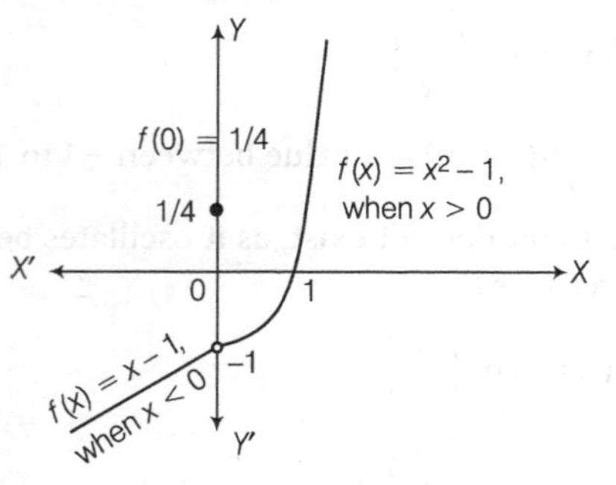

Example 12 Show the function, $f(x) = \begin{cases} \dfrac{e^{1/x}-1}{e^{1/x}+1}, & \text{when } x \neq 0 \\ 0, & \text{when } x = 0 \end{cases}$

has non-removable discontinuity at $x = 0$.

Sol. We have, $f(x) = \begin{cases} \dfrac{e^{1/x}-1}{e^{1/x}+1}, & \text{when } x \neq 0 \\ 0, & \text{when } x = 0 \end{cases}$

$\therefore$ RHL at $x = 0$, let $x = 0 + h$

$$\Rightarrow \lim_{x \to 0^+} f(x) = \lim_{h \to 0} f(0+h) = \lim_{h \to 0} \frac{e^{\frac{1}{0+h}}-1}{e^{\frac{1}{0+h}}+1} = \lim_{h \to 0} \frac{e^{\frac{1}{h}}-1}{e^{\frac{1}{h}}+1}$$

$$\Rightarrow \lim_{x \to 0^+} f(x) = \lim_{h \to 0} \frac{1-\frac{1}{e^{1/h}}}{1+\frac{1}{e^{1/h}}} \Rightarrow \lim_{x \to 0^+} f(x) = \frac{1-0}{1+0} = 1$$

$$\left[\text{as } h \to 0; \frac{1}{h} \to \infty \Rightarrow e^{1/h} \to \infty; 1/e^{1/h} \to 0\right]$$

$$\therefore \quad \lim_{x \to 0^+} f(x) = 1$$

Again, LHL at $x = 0$, let $x = 0 - h$

$$\Rightarrow \quad \lim_{x \to 0^-} f(x) = \lim_{h \to 0} f(0-h)$$

$$= \lim_{h \to 0} \frac{e^{-1/h}-1}{e^{-1/h}+1} = \frac{0-1}{0+1} = -1$$

$$[\text{as } h \to 0; e^{-1/h} \to 0]$$

$$\lim_{x \to 0^-} f(x) = -1$$

$$\Rightarrow \quad \lim_{x \to 0^+} f(x) \neq \lim_{x \to 0^-} f(x).$$

Thus, $f(x)$ has non-removable discontinuity.

Example 13 Show $f(x) = \dfrac{1}{|x|}$ has discontinuity of second kind at $x = 0$.

Sol. Here, $f(0) = \dfrac{1}{|0|}$, which shows function has discontinuity of second kind.

Graphically Here, the graph is broken at $x = 0$ as

$$\Rightarrow \quad \underset{x \to 0}{f(x) \to \infty}$$

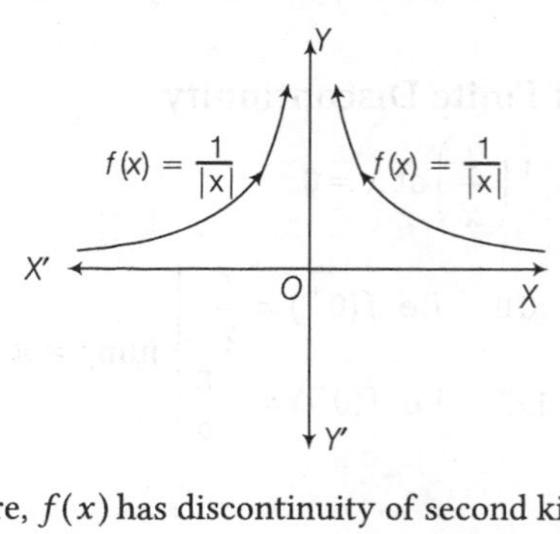

Therefore, $f(x)$ has discontinuity of second kind.

Exercise for Session 3

1. Which of the following function(s) has/have removable discontinuity at the origin?

(a) $f(x)=\dfrac{1}{1+2^{\cot x}}$ (b) $f(x)=\cos\left(\dfrac{|\sin x|}{x}\right)$

(c) $f(x)=x\sin\dfrac{\pi}{x}$ (d) $f(x)=\dfrac{1}{\ln|x|}$

2. Function whose jump (non-negative difference of LHL and RHL) of discontinuity is greater than or equal to one. is/are

(a) $f(x)=\begin{cases}\dfrac{(e^{1/x}+1)}{(e^{1/x}-1)}; & x<0\\ \dfrac{(1-\cos x)}{x}; & x>0\end{cases}$ (b) $g(x)=\begin{cases}\dfrac{(x^{1/3}-1)}{(x^{1/2}-1)}; & x>0\\ \dfrac{\ln x}{(x-1)}; & \dfrac{1}{2}<x<1\end{cases}$

(c) $u(x)=\begin{cases}\dfrac{\sin^{-1}2x}{\tan^{-1}3x}; & x\in\left(0,\dfrac{1}{2}\right]\\ \dfrac{|\sin x|}{x}; & x<0\end{cases}$ (d) $v(x)=\begin{cases}\log_3(x+2); & x>2\\ \log_{1/2}(x^2+5); & x<2\end{cases}$

3. Consider the piecewise defined function $f(x)=\begin{cases}\sqrt{-x}, & \text{if } x<0\\ 0, & \text{if } 0\le x\le 4\\ x-4, & \text{if } x>4\end{cases}$, choose the answer which best describes the continuity of this function.

(a) the function is unbounded and therefore cannot be continuous.
(b) the function is right continuous at $x=0$.
(c) the function has a removable discontinuity at 0 and 4, but is continuous on the rest of the real line.
(d) the function is continuous on the entire real line.

4. If $f(x)=\text{sgn}(\cos 2x-2\sin x+3)$, where sgn () is the signum function, then $f(x)$

(a) is continuous over its domain.
(b) has a missing point discontinuity.
(c) has isolated point discontinuity.
(d) has irremovable discontinuity

5. $f(x)=\dfrac{2\cos x-\sin 2x}{(\pi-2x)^2}$; $g(x)=\dfrac{e^{-\cos x}-1}{8x-4\pi}$

$h(x)=f(x)$ for $x<\pi/2=g(x)$ for $x>\pi/2$, then which of the followings does not hold?

(a) h is continuous at $x=\pi/2$
(b) h has an irremovable discontinuity at $x=\pi/2$
(c) h has a removable discontinuity at $x=\pi/2$
(d) $f\left(\dfrac{\pi^+}{2}\right)=g\left(\dfrac{\pi^-}{2}\right)$

Session 4

Theorems Based on Continuity; Continuity of Composite Function

Theorem 1 Sum, difference, product and quotient of two continuous functions is always a continuous function.

However, $h(x)=\dfrac{f(x)}{g(x)}$ is continuous at $x=a$ only if $g(a)\neq 0$.

Theorem 2 If $f(x)$ is continuous and $g(x)$ is discontinuous at $x=a$, then the product function $\phi(x)=f(x)\cdot g(x)$ is not necessarily be discontinuous at $x=a$.

e.g. (i) $f(x)=x$ and $g(x)=\begin{cases}\sin\dfrac{1}{x}, & x\neq 0\\ 0, & x=0\end{cases}$

where $f(x)$ is continuous and $g(x)$ is discontinuous at $x=0$.

But $\phi(x)=f(x)\cdot g(x)=\begin{cases}x\cdot\sin\dfrac{1}{x}, & x\neq 0\\ 0, & x=0\end{cases}$

is continuous at $x=0$.

(ii) $f(x)=\cos(2x-1)\dfrac{\pi}{2}$ is continuous at $x=1$ and $g(x)=[x]$ is discontinuous at $x=1$.

But $\phi(x)=f(x)\cdot g(x)=[x]\cos\left(\dfrac{2x-1}{2}\right)\pi$ is continuous at $x=1$.

Theorem 3 If $f(x)$ and $g(x)$ both are discontinuous at $x=a$, then the product function $\phi(x)=f(x)\cdot g(x)$ is not necessarily be discontinuous at $x=a$.

e.g. $f(x)=\begin{cases}1, & x\geq 0\\ -1, & x<0\end{cases}$ and $g(x)=\begin{cases}-1, & x\geq 0\\ 1, & x<0\end{cases}$

$\therefore\quad \phi(x)=f(x)\cdot g(x)=-1, \forall\, x\in R$

$\phi(x)$ is continuous, where $f(x)$ and $g(x)$ are discontinuous at $x=0$.

| Example 14 $f(x)=\begin{cases}\left(\tan\left(\dfrac{\pi}{4}+x\right)\right)^{1/x}, & x\neq 0\\ k, & x=0\end{cases}$.

For what value of k, $f(x)$ is continuous at $x=0$?

Sol. Here, $\lim\limits_{x\to 0} f(x)=\lim\limits_{x\to 0}\left\{\tan\left(\dfrac{\pi}{4}+x\right)\right\}^{1/x}$

$$\Rightarrow\quad \lim_{x\to 0} f(x)=\lim_{x\to 0}\left[\frac{1+\tan x}{1-\tan x}\right]^{1/x}\qquad [1^\infty \text{ form}]$$

$$\Rightarrow\quad \lim_{x\to 0} f(x)=\lim_{x\to 0}\left[1+\left(\frac{1+\tan x}{1-\tan x}-1\right)\right]^{1/x}$$

$$\Rightarrow\quad \lim_{x\to 0} f(x)=e^{\lim\limits_{x\to 0}\left(\frac{2\tan x}{1-\tan x}\right)\cdot\frac{1}{x}}$$

$$\Rightarrow\quad \lim_{x\to 0} f(x)=e^{2\lim\limits_{x\to 0}\frac{\tan x}{x(1-\tan x)}}=e^2$$

Here, $f(x)$ is continuous at $x=0$, when

$$\lim_{x\to 0} f(x)=f(0)$$

$$\Rightarrow\quad k=e^2$$

| Example 15 A function $f(x)$ is defined by,

$$f(x)=\begin{cases}\dfrac{[x^2]-1}{x^2-1}, & \text{for } x^2\neq 1\\ 0, & \text{for } x^2=1\end{cases}$$

Discuss the continuity of $f(x)$ at $x=1$.

Sol. We have, $f(x)=\begin{cases}\dfrac{[x^2]-1}{x^2-1}, & \text{for } x^2\neq 1\\ 0, & \text{for } x^2=1\end{cases}$

$$\Rightarrow\quad f(x)=\begin{cases}\dfrac{-1}{x^2-1}, & \text{for } 0<x^2<1\\ 0, & \text{for } x^2=1\\ \dfrac{1-1}{x^2-1}, & \text{for } 1<x^2<2\end{cases}$$

$$\Rightarrow\quad f(x)=\begin{cases}\dfrac{-1}{x^2-1}, & \text{for } 0<x^2<1\\ 0, & \text{for } x^2=1\\ 0, & \text{for } 1<x^2<2\end{cases}$$

$\therefore\quad$ RHL at $x=1$

$\Rightarrow \quad \lim_{x \to 1^+} f(x) = 0$

Also, LHL at $x^2 = 1$

$\Rightarrow \quad \lim_{x \to 1^-} f(x) = \lim_{h \to 0} f(1-h)$

$$= \lim_{h \to 0} \frac{-1}{(1-h)^2 - 1} = -\infty$$

$\therefore \lim_{x \to 1} f(x)$ doesn't exist. [as RHL $\neq$ LHL]

Hence, $f(x)$ is not continuous at $x = 1$.

Example 16 Discuss the continuity of the function

$$f(x) = \lim_{n \to \infty} \frac{\log(2+x) - x^{2n} \sin x}{1 + x^{2n}} \text{ at } x = 1.$$

Sol. We have, $f(1) = \lim_{n \to \infty} \frac{\log 3 - \sin 1}{2} = \frac{1}{2}(\log 3 - \sin 1)$ ...(i)

We know that, $\lim_{n \to \infty} x^{2n} = \begin{cases} 0, & \text{if } x^2 < 1 \\ \infty, & \text{if } x^2 > 1 \end{cases}$

$\therefore$ For $x^2 < 1$, we have

$$f(x) = \lim_{n \to \infty} \frac{\log(2+x) - x^{2n} \sin x}{1 + x^{2n}} = \log(2+x)$$

Again, for $x^2 > 1$, we have

$$f(x) = \lim_{n \to \infty} \frac{\frac{1}{x^{2n}} \log(2+x) - \sin x}{1 + \frac{1}{x^{2n}}} = -\sin(x)$$

Here, as $x \to 1$

$\lim_{x \to 1^-} f(x) = \log(3)$

and $\lim_{x \to 1^+} f(x) = -\sin(1)$

So, $\lim_{x \to 1^-} f(x) \neq \lim_{x \to 1^+} f(x)$

Therefore, $f(x)$ is not continuous at $x = 1$.

Example 17 Discuss the continuity of $f(x)$, where

$$f(x) = \lim_{n \to \infty} \left(\sin \frac{\pi x}{2}\right)^{2n}.$$

Sol. Since, $\lim_{n \to \infty} x^{2n} = \begin{cases} 0, & |x| < 1 \\ 1, & |x| = 1 \end{cases}$

$$\therefore \quad f(x) = \lim_{n \to \infty} \left(\sin\left(\frac{\pi x}{2}\right)\right)^{2n} = \begin{cases} 0, & \left|\sin \frac{\pi x}{2}\right| < 1 \\ 1, & \left|\sin \frac{\pi x}{2}\right| = 1 \end{cases}$$

Thus, $f(x)$ is continuous for all x, except for those values of x for which $\left|\sin \frac{\pi x}{2}\right| = 1$, i.e. x is an odd integer.

$\Rightarrow \quad x = (2n+1)$, where $n \in I$

Check continuity at $x = (2n+1)$

LHL $= \lim_{x \to (2n+1)^-} f(x) = 0$ and $f(2n+1) = 1$

Thus, LHL $\neq f(2n+1)$

$\Rightarrow f(x)$ **is discontinuous at $x = (2n+1)$**

[i.e. at odd integers]

Hence, $f(x)$ is discontinuous at $x = (2n+1)$; $n \in$ integer.

Example 18 Let

$$f(x) = \begin{cases} \{1 + |\sin x|\}^{a/|\sin x|}, & -\pi/6 < x < 0 \\ b, & x = 0 \\ e^{\tan 2x / \tan 3x}, & 0 < x < \pi/6 \end{cases}$$

Determine a and b such that $f(x)$ is continuous at $x = 0$. ***[IIT JEE 1994]***

Sol. Since, f is continuous at $x = 0$.

Therefore, RHL = LHL = $f(0)$

RHL at $x = 0$

$$\lim_{x \to 0^+} f(x) = \lim_{h \to 0} f(0+h) = \lim_{h \to 0} e^{\tan 2h / \tan 3h}$$

$$= \lim_{h \to 0} e^{\frac{\tan 2h}{2h} \cdot \frac{3h}{\tan 3h} \cdot \frac{2}{3}} = e^{2/3} \quad \text{...(i)}$$

Again, LHL at $x = 0$

$$\lim_{x \to 0^-} f(x) = \lim_{h \to 0} f(0-h)$$

$$= \lim_{h \to 0} \{1 + |\sin(0-h)|\}^{a/|\sin(0-h)|}$$

$$= \lim_{h \to 0} \{1 + |\sin h|\}^{\frac{a}{|\sin h|}}$$

$$= e^{\lim_{h \to 0} |\sin h| \cdot \frac{a}{|\sin h|}} = e^a \quad \text{...(ii)}$$

and $f(0) = b$ ...(iii)

Thus, $e^{2/3} = e^a = b \Rightarrow a = 2/3$ and $b = e^{2/3}$

Example 19 Fill in the blanks so that the resulting statement is correct.

Let $f(x) = [x] \sin\left(\frac{\pi}{[x+1]}\right)$, where $[\cdot]$ denotes greatest integral function. The domain of f is .... and the points of discontinuity of f in the domain are .... ***[IIT JEE 1996]***

Sol. Let $f(x) = [x] \sin\left(\frac{\pi}{[x+1]}\right)$

Domain of $f(x)$ is $x \in R$ excluding the points where $[x+1] = 0$ [$\because$ denominator can't be zero]

$\Rightarrow \quad 0 \le x + 1 < 1 \Rightarrow -1 \le x < 0$

i.e. for all $x \in [-1, 0)$, denominator is zero.

So, domain is $x \in R - [-1, 0)$.

$\Rightarrow$ Domain is $x \in (-\infty, -1) \cup [0, \infty)$

and the internal point discontinuity $\in [-1, 0)$.

Example 20 Let $f(x+y)=f(x)+f(y)$ for all x and y. If the function $f(x)$ is continuous at $x=0$, show that $f(x)$ is continuous for all x.

Sol. As, the function is continuous at $x = 0$, we have

$$\lim_{x\to 0^-} f(x) = \lim_{x\to 0^+} f(x) = f(0)$$

$$\Rightarrow \quad \lim_{h\to 0} f(0-h) = \lim_{h\to 0} f(0+h) = f(0)$$

$$\Rightarrow \quad \lim_{h\to 0} \{f(0)+f(-h)\} = \lim_{h\to 0} \{f(0)+f(h)\} = f(0)$$

[using $f(x+y)=f(x)+f(y)$]

$$\Rightarrow \quad \lim_{h\to 0} f(-h) = \lim_{h\to 0} f(h) = 0 \quad \text{...(i)}$$

Now, consider some arbitrary point $x = a$

$$\text{LHL} = \lim_{h\to 0} f(a-h) = \lim_{h\to 0} f(a)+f(-h)$$

[using $f(x+y)=f(x)+f(y)$]

$$= f(a) + \lim_{h\to 0} f(-h)$$

$$\text{LHL} = f(a)+0 = f(a)$$

[as $\lim_{h\to 0} f(-h) = 0$, using Eq. (i)]

$$\text{RHL} = \lim_{h\to 0} f(a+h) = \lim_{h\to 0} f(a)+f(h)$$

$$= f(a) + \lim_{h\to 0} f(h)$$

$$\text{RHL} = f(a)+0 = f(a)$$

[as $\lim_{h\to 0} f(h) = 0$, using Eq. (i)]

At any arbitrary point $(x = a)$,

$$\text{LHL} = \text{RHL} = f(a)$$

Therefore, function is continuous for all values of x, if it is continuous at 0.

Example 21 Let $f(x)$ be a continuous function defined for $1 \le x \le 3$. If $f(x)$ takes rational values for all x and $f(2)=10$, then find the value of $f(1.5)$. *[IIT JEE 1997]*

Sol. As, $f(x)$ is continuous in [1, 3], $f(x)$ will attain all values between $f(1)$ and $f(3)$. As, $f(x)$ takes rational values for all x and there are innumerable irrational values between $f(1)$ and $f(3)$ which implies that $f(x)$ can take rational values for all x, if $f(x)$ has a constant rational value at all points between $x = 1$ and $x = 3$.

So, $\quad f(2) = f(1.5) = 10$

Continuity of composite function

If the function $u = f(x)$ is continuous at the point $x = a$ and the function $y = g(u)$ is continuous at the point $u = f(a)$, then the composite function $y = (gof)(x) = g(f(x))$ is continuous at the point $x = a$.

Example 22 Discuss the continuity for

$$f(x) = \frac{1-u^2}{2+u^2}, \text{ where } u = \tan x.$$

Sol. Here, $u = \tan x$ is discontinuous at $n\pi \pm \frac{\pi}{2}, n \in I$

and $\quad f(x) = \dfrac{1-u^2}{2+u^2}$ is continuous at every $u \in R$.

Hence, $f(x)$ is continuous on; $x \in R - \left\{n\pi \pm \frac{\pi}{2}, n \in I\right\}$.

$$\text{Also,} \quad \lim_{x\to n\pi\pm\frac{\pi}{2}} f(x) = \lim_{u^2\to\infty} \frac{1-u^2}{2+u^2}$$

$$= \lim_{u\to\infty} \frac{\frac{1}{u^2}-1}{\frac{2}{u^2}+1} = -1$$

Hence, the points $n\pi \pm \frac{\pi}{2}, n \in I$ have removable discontinuity.

i.e. If $f(x)$ is defined as

$$f(x) = \begin{cases} \dfrac{1-u^2}{2+u^2}, & x \ne n\pi \pm \dfrac{\pi}{2} \text{ and } u = \tan x \\ -1, & x = n\pi \pm \dfrac{\pi}{2} \end{cases}, \text{ then}$$

$f(x)$ is continuous for all $x \in R$.

Example 23 Find the points of discontinuity of $y = \dfrac{1}{u^2+u-2}$, where $u = \dfrac{1}{x-1}$.

Sol. The function $u = f(x) = \dfrac{1}{x-1}$ is discontinuous at the point $x = 1$. ...(i)

The function $y = g(x) = \dfrac{1}{u^2+u-2} = \dfrac{1}{(u+2)(u-1)}$ is discontinuous at $u = -2$ and $u = 1$.

$$\text{when} \quad u = -2, \frac{1}{x-1} = u = -2$$

$$\Rightarrow \quad x - 1 = -\frac{1}{2}$$

$$\Rightarrow \quad x = 1/2 \quad \text{...(ii)}$$

$$\text{when} \quad u = 1, \frac{1}{x-1} = u = 1$$

$$\Rightarrow \quad x - 1 = 1$$

$$\Rightarrow \quad x = 2 \quad \text{...(iii)}$$

Hence, the composite function $y = g(f(x))$ is discontinuous at three points $x = \frac{1}{2}$, 1 and 2.

Exercise for Session 4

1. If $f(x)=\dfrac{1}{x^2-17x+66}$, then $f\left(\dfrac{2}{x-2}\right)$ is discontinuous at $x=$

(a) 2 (b) $\dfrac{7}{3}$ (c) $\dfrac{24}{11}$ (d) 6, 11

2. Let f be a continuous function on R.
If $f(1/4^n)=(\sin e^n)e^{-n^2}+\dfrac{n^2}{n^2+1}$, then $f(0)$ is

(a) not unique (b) 1
(c) data sufficient to find $f(0)$ (d) data insufficient to find $f(0)$

3. $f(x)$ is continuous at $x=0$, then which of the following are always true?

(a) $\lim_{x\to 0} f(x)=0$
(b) $f(x)$ is non continuous at $x=1$
(c) $g(x)=x^2 f(x)$ is continuous at $x=0$
(d) $\lim_{x\to 0^+}(f(x)-f(0))=0$

4. If $f(x)=\cos\left[\dfrac{\pi}{x}\right]\cos\left(\dfrac{\pi}{2}(x-1)\right)$; where $[x]$ is the greatest integer function of x, then $f(x)$ is continuous at

(a) $x=0$ (b) $x=1$
(c) $x=2$ (d) None of these

5. Let $f(x)=[x]$ and $g(x)=\begin{cases}0\ , & x\in Z\\ x^2, & x\in R-Z\end{cases}$, then (where [.] denotes greatest integer function)

(a) $\lim_{x\to 1} g(x)$ exists, but $g(x)$ is not continuous at $x=1$
(b) $\lim_{x\to 1} f(x)$ does not exist and $f(x)$ is not continuous at $x=1$
(c) gof is continuous for all x
(d) fog is continuous for all x

6. Let $f(x)=\begin{cases}a\sin^{2n}x, & \text{for } x\ge 0 \text{ and } n\to\infty\\ b\cos^{2m}x-1, & \text{for } x<0 \text{ and } m\to\infty\end{cases}$, then

(a) $f(0^-)\ne f(0^+)$ (b) $f(0^+)\ne f(0)$
(c) $f(0^-)=f(0)$ (d) f is continuous at $x=0$

7. Consider $f(x)=\lim_{n\to\infty}\dfrac{x^n-\sin x^n}{x^n+\sin x^n}$ for $x>0, x\ne 1, f(1)=0$, then

(a) f is continuous at $x=1$
(b) f has a finite discontinuity at $x=1$
(c) f has an infinite or oscillatory discontinuity at $x=1$
(d) f has a removal type of discontinuity at $x=1$

Session 5

Intermediate Value Theorem

If $f(x)$ is continuous in $[a, b]$ and $f(a) \neq f(b)$, then for any value $c \in (f(a), f(b))$, there is atleast one number x_0 in (a, b) for which $f(x_0) = c$.

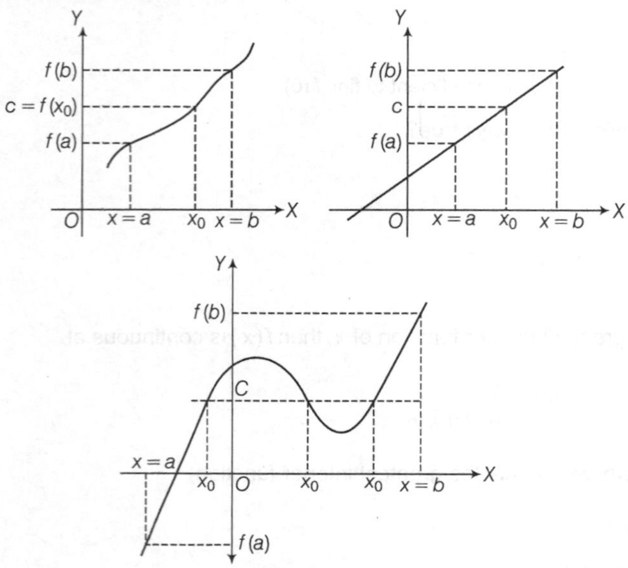

Figure 6.10

Example 24 Show that the function $f(x) = (x - a)^2 (x - b)^2 + x$ takes the value $\dfrac{a+b}{2}$ for some value of $x \in [a, b]$.

Sol. Here, $f(a) = a$ and $f(b) = b$

Also, $f(x)$ is continuous in $[a, b]$ and $\dfrac{a+b}{2} \in [a, b]$.

Therefore, using intermediate value theorem, there exists some $c \in [a, b]$, such that

$$f(c) = \frac{a+b}{2}$$

Example 25 Suppose that $f(x)$ is continuous in $[0, 1]$ and $f(0) = 0, f(1) = 0$. Prove that $f(c) = 1 - 2c^2$ for some $c \in (0, 1)$.

Sol. Let $g(x) = f(x) + 2x^2 - 1$ in $[0, 1]$, $g(0) = -1$, $g(1) = 1$

$\therefore$ By intermediate value theorem, there exists some

$$c \in (0, 1); g(c) = 0 \Rightarrow f(c) = 1 - 2c^2$$

Continuity for Rational and Irrational Function

Functions should be continuous only at one point and to be defined everywhere. [Single Point Continuity]

e.g.

(i) $f(x) = \begin{cases} x, & \text{if } x \in Q \\ 0, & \text{if } x \notin Q \end{cases}$, is continuous only at $x = 0$.

(ii) $f(x) = \begin{cases} x, & \text{if } x \in Q \\ -x, & \text{if } x \notin Q \end{cases}$, is continuous only at $x = 0$ and defined everywhere.

(iii) $f(x) = \begin{cases} x, & \text{if } x \in Q \\ 1 - x, & \text{if } x \notin Q \end{cases}$, is continuous only at $x = \dfrac{1}{2}$ and defined everywhere.

(iv) $f(x) = \begin{cases} x^2, & \text{if } x \in Q \\ 1, & \text{if } x \notin Q \end{cases}$, is continuous only at $x = 1$ or $x = -1$ and defined everywhere.

Exercise for Session 5

1. Examine the continuity at $x = 0$ of the sum function of the infinite series $\dfrac{x}{x+1} + \dfrac{x}{(x+1)(2x+1)} + \dfrac{x}{(2x+1)(3x+1)} + \dots \infty$

2. If $g : [a, b]$ onto $[a, b]$ is continuous, then show that there is some $c \in [a, b]$ such that $g(c) = c$.

3. Show that (a) a polynomial of an odd degree has atleast one real root.
   (b) a polynomial of an even degree has atleast two real roots, if it attains atleast one value opposite in sign to the coefficient of its highest-degree term.

4. Let $y_n(x) = x^2 + \dfrac{x^2}{1+x^2} + \dfrac{x^2}{(1+x^2)^2} + \dots + \dfrac{x^2}{(1+x^2)^{n-1}}$ and $y(x) = \lim_{n \to \infty} y_n(x)$. Discuss the continuity of $y_n(x)\,(n = 1, 2, 3, \dots n)$ and $y(x)$ at $x = 0$.

Session 6

Differentiability : Meaning of Derivative

The instantaneous rate of change of function with respect to the independent variable is called derivative. Let $f(x)$ be a given function of one variable and Δx denotes a number (positive or negative) to be added to the number x.

Let Δf denotes the corresponding change of f, then $\Delta f = f(x+\Delta x) - f(x)$.

$$\frac{\Delta f}{\Delta x} = \frac{f(x+\Delta x) - f(x)}{\Delta x}$$

If $\frac{\Delta f}{\Delta x}$ approaches a limit as Δx approaches to zero, this limit is the derivative of f at the point x. The derivative of a function f is denoted by symbols such as $f'(x)$, $\frac{df}{dx}$, $\frac{d}{dx} f(x)$ or $\frac{d\,f(x)}{dx}$.

$$\Rightarrow \quad \frac{df}{dx} = \lim_{\Delta x \to 0} \frac{\Delta f}{\Delta x} = \lim_{\Delta x \to 0} \frac{f(x+\Delta x) - f(x)}{\Delta x}$$

The derivative evaluated at a point a is written,

$$f'(a), \left(\frac{d\,f(x)}{dx}\right)_{x=a}, (f'(x))_{x=a} \text{ etc.}$$

Existence of Derivative at $x = a$

The derivative of a function $f(x)$ exists at $x = a$, if

$$\lim_{x \to a} \frac{f(x) - f(a)}{(x-a)} \text{ exists finitely.}$$

$$\text{or} \quad \lim_{x \to a^-} \frac{f(x) - f(a)}{(x-a)} = \lim_{x \to a^+} \frac{f(x) - f(a)}{(x-a)}$$

$$\text{or} \quad \lim_{h \to 0} \frac{f(a-h) - f(a)}{-h} = \lim_{h \to 0} \frac{f(a+h) - f(a)}{h}$$

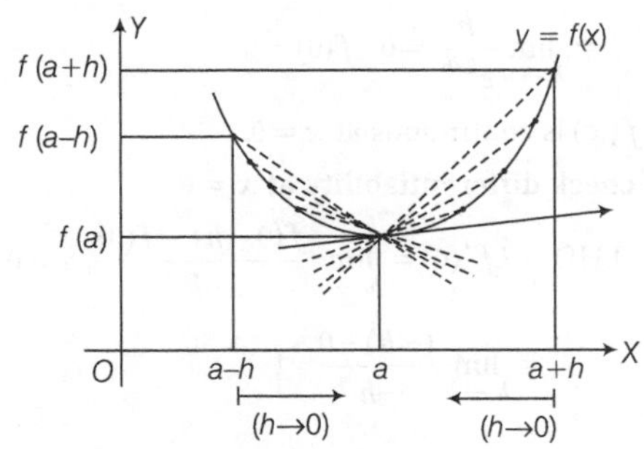

Figure 6.11

(a) **Right hand derivative** The right hand derivative of $f(x)$ at $x = a$ denoted by $f(a^+)$ is defined as

$$f'(a^+) = \lim_{h \to 0} \frac{f(a+h) - f(a)}{h}, \text{ provided the limit exists}$$

and is finite ($h > 0$).

(b) **Left hand derivative** The left hand derivative of $f(x)$ at $x = a$ denoted by $f(a^-)$ is defined as

$$f'(a^-) = \lim_{h \to 0} \frac{f(a-h) - f(a)}{-h}, \text{ provided the limit exists}$$

and is finite ($h > 0$).

Hence, $f(x)$ is said to be derivable or differentiable at $x = a$.

If $f'(a^+) = f'(a^-) =$ finite quantity and it is denoted by $f'(a)$; where $f'(a) = f'(a^-) = f'(a^+)$ and it is called derivative or differential coefficient of $f(x)$ at $x = a$.

Relation between Continuity and Differentiability

If a function is differentiable at a point, then it should be continuous at that point as well and a discontinuous function cannot be differentiable. This fact is proved in the following theorem.

Theorem If a function is differentiable at a point, it is necessarily continuous at that point. But the converse is not necessarily true.

Or

$f(x)$ is differentiable at $x = c \Rightarrow f(x)$ is continuous at $x = c$.

Proof Let a function $f(x)$ be differentiable at $x = c$.

Then, $\lim_{x \to c} \frac{f(x) - f(c)}{x - c}$ exists finitely.

Let $$\lim_{x \to c} \frac{f(x) - f(c)}{x - c} = f'(c) \quad \text{...(i)}$$

In order to prove that $f(x)$ is continuous at $x = c$, it is sufficient to show that $\lim_{x \to c} f(x) = f(c)$.

Now, $$\lim_{x \to c} f(x) = \lim_{x \to c} \left[\left(\frac{f(x) - f(c)}{x - c}\right)(x - c) + f(c)\right]$$

$$= \lim_{x \to c} \left[\left\{ \frac{f(x) - f(c)}{x - c} \right\} \cdot (x - c) \right] + f(c)$$

$$= \lim_{x \to c} \frac{f(x) - f(c)}{x - c} \cdot \lim_{x \to c} (x - c) + f(c)$$

$$= f'(c) \times 0 + f(c)$$

$$= f(c)$$

Hence, $f(x)$ is continuous at $x = c$.

Converse The converse of above theorem is not necessarily true, i.e. a function may be continuous at a point but may not be differentiable at that point.

e.g. The function $f(x) = |x|$ is continuous at $x = 0$ but it is not differentiable at $x = 0$, as shown by figure.

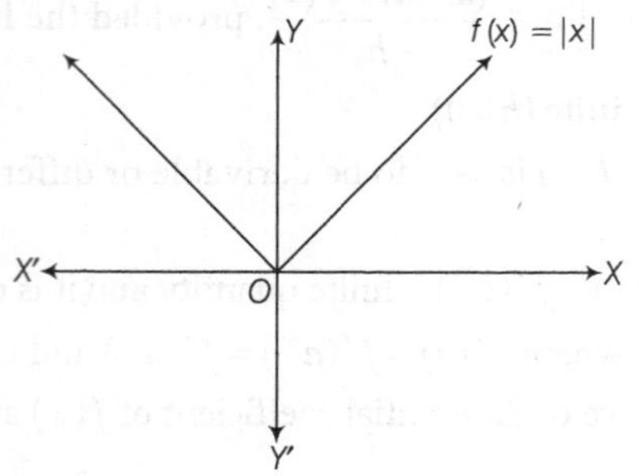

Figure 6.12

Which shows we have sharp edge at $x = 0$, hence not differentiable but continuous at $x = 0$.

| Example 26 Show that $f(x) = \begin{cases} x \sin \dfrac{1}{x}, & \text{when } x \neq 0 \\ 0, & \text{when } x = 0 \end{cases}$

is continuous but not differentiable at $x = 0$.

Sol. (a) **To check continuity at $x = 0$**

Here, $f(0) = 0$

Also, $\lim_{x \to 0} f(x) = \lim_{x \to 0} x \sin \dfrac{1}{x}$

$= 0 \times$ (A finite quantity that lies between -1 to $+1$)

$= 0$ [as $n \to 0$, $\sin \dfrac{1}{x} \to \sin \infty$, which is a finite quantity between -1 to $+1$]

$\therefore \lim_{x \to 0} f(x) = f(0)$ and hence, $f(x)$ is continuous.

Now,

(b) **To check differentiability at $x = 0$**

(LHD at $x = 0$)

$$= \lim_{x \to 0^-} \frac{f(x) - f(0)}{x - 0}$$

$$= \lim_{h \to 0} \frac{f(0 - h) - f(0)}{(0 - h) - (0)}$$

$$= \lim_{h \to 0} \frac{-h \sin(-1/h)}{(-h)}$$

$$= -\lim_{h \to 0} \sin\left(\frac{1}{h}\right)$$

[a number which oscillates between -1 and $+1$]

$\therefore$ (LHD at $x = 0$) doesn't exist.

Similarly, it could be shown that RHD at $x = 0$ doesn't exist.

Hence, $f(x)$ is continuous but not differentiable.

| Example 27 Let $f(x) = \begin{cases} x \exp\left[-\left(\dfrac{1}{|x|} + \dfrac{1}{x}\right)\right], & x \neq 0 \\ 0, & x = 0 \end{cases}$

Test whether (a) $f(x)$ is continuous at $x = 0$
(b) $f(x)$ is differentiable at $x = 0$.

[IIT JEE 1997]

Sol. Here, $f(x) = \begin{cases} x \exp\left[-\left(\dfrac{1}{|x|} + \dfrac{1}{x}\right)\right], & x \neq 0 \\ 0, & x = 0 \end{cases}$

$$\Rightarrow \quad f(x) = \begin{cases} xe^{-\left\{\frac{1}{x} + \frac{1}{x}\right\}}, & x > 0 \\ xe^{-\left\{\frac{1}{-x} + \frac{1}{x}\right\}}, & x < 0 \\ 0, & x = 0 \end{cases} \qquad \left[\because |x| = \begin{cases} x, & x \geq 0 \\ -x, & x < 0 \end{cases}\right]$$

$$\Rightarrow \quad f(x) = \begin{cases} xe^{-2/x}, & x > 0 \\ x, & x < 0 \\ 0, & x = 0 \end{cases} \qquad \text{...(i)}$$

(a) **To check continuity of $f(x)$ at $x = 0$**

$$\text{LHL} = \lim_{x \to 0^-} f(x) = \lim_{x \to 0^-} x$$

$$= \lim_{h \to 0} (0 - h) = 0$$

$$\text{RHL} = \lim_{x \to 0^+} f(x) = \lim_{x \to 0^+} xe^{-2/x}$$

$$= \lim_{h \to 0} \frac{h}{e^{2/h}} = 0, \; f(0) = 0$$

$\therefore f(x)$ is continuous at $x = 0$.

(b) **To check differentiability at $x = 0$**

$$\text{LHD} = Lf'(0) = \lim_{h \to 0} \frac{f(0 - h) - f(0)}{-h}, h > 0$$

$$= \lim_{h \to 0} \frac{(-h) - 0}{-h} = 1$$

$$\text{RHD} = Rf'(0) = \lim_{h \to 0} \frac{f(0+h) - f(0)}{h}$$

$$= \lim_{h \to 0} \frac{he^{\frac{-2}{h}} - 0}{h} = \lim_{h \to 0} e^{\frac{-2}{h}}$$

$$= e^{-\infty} = 0$$

$\therefore \quad Lf'(0) \neq Rf'(0)$

Therefore, $f(x)$ is not differentiable at $x = 0$.

Some Standard Results on Differentiability

Functions $f(x)$	Intervals in which $f(x)$ is differentiable
1. Polynomial	$(-\infty$ to $\infty)$
2. Exponential $(a^x, a > 0)$	$(-\infty$ to $\infty)$
3. Constant	$(-\infty$ to $\infty)$
4. Logarithmic	Each point in its domain
5. Trigonometric	Each point in its domain
6. Inverse trigonometric	Each point in its domain

Example 28 Let $f(x) = |x-1| + |x+1|$. Discuss the continuity and differentiability of the function.

Sol. Here, $f(x) = |x-1| + |x+1|$

$$\Rightarrow \quad f(x) = \begin{cases} (x-1) + (x+1), & \text{when} \quad x > 1 \\ -(x-1) + (x+1), & \text{when} \quad -1 \le x \le 1 \\ -(x-1) - (x+1), & \text{when} \quad x < -1 \end{cases}$$

$$\Rightarrow \quad f(x) = \begin{cases} 2x, & \text{when} \quad x > 1 \\ 2, & \text{when} \quad -1 \le x \le 1 \\ -2x, & \text{when} \quad x < -1 \end{cases}$$

Graphically The graph of the function is shown below.

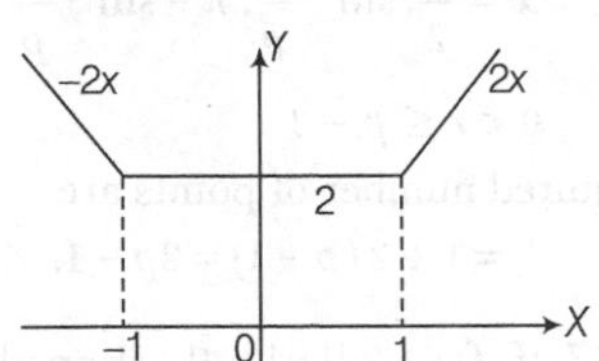

From the graph, it is clear that the function is continuous at all real x, also differentiable at all real x except at $x = \pm 1$. Since, it has sharp edges at $x = -1$ and $x = 1$.

At $x = 1$ we see that the slope from the right, i.e. RHD = 2, while slope from the left, i.e. LHD = 0.

Similarly, at $x = -1$ it is clear that RHD = 0, while LHD = −2

Aliter In this method, first of all we differentiate the function and from the derivative equality sign should be removed from doubtful points.

$$\text{Here, } f'(x) = \begin{cases} -2, & x < -1 \\ 0, & -1 < x < 1 \quad [\text{no equality on } -1 \text{ and } +1] \\ 2, & x > 1 \end{cases}$$

Now, at $x = 1$, $f'(1^+) = 2$, while $f'(1^-) = 0$ and

at $x = -1$, $f'(-1^+) = 0$, while $f'(-1^-) = -2$

Thus, $f(x)$ is not differentiable at $x = \pm 1$.

Remark

This method is not applicable when function is discontinuous.

Example 29 Discuss the continuity and differentiability for $f(x) = [\sin x]$ when $x \in [0, 2\pi]$; where $[\cdot]$ denotes the greatest integer function x.

Sol. In last chapter, we have discussed the plotting of curves, so $y = [\sin x]$ could be plotted as

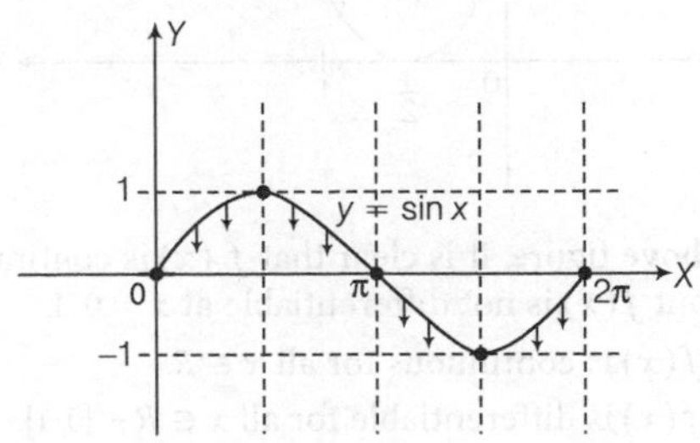

Or

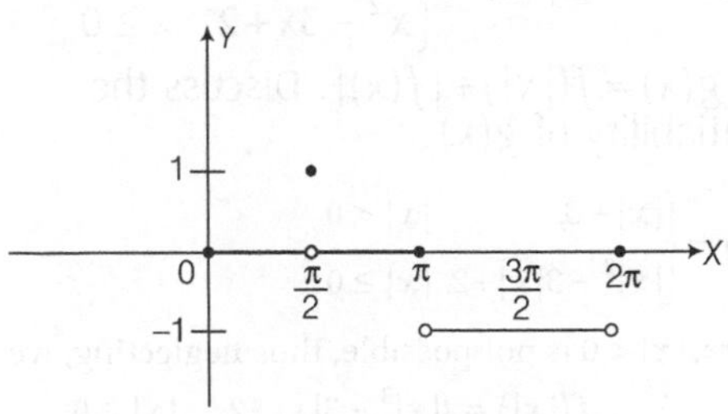

which shows $f(x) = [\sin x]$ is discontinuous for $x = \frac{\pi}{2}, \pi, 2\pi$, when $x \in [0, 2\pi]$ and $f(x)$ is not differentiable at $x = \frac{\pi}{2}, \pi, 2\pi$.

As we know, function is neither differentiable nor continuous at those points for which graph is broken.

Example 30 If $f(x)=\{|x|-|x-1|\}^2$, draw the graph of $f(x)$ and discuss its continuity and differentiability of $f(x)$.

Sol. We know, $|x|-|x-1|=\begin{cases}-x+x-1, & x<0\\ x+x-1, & 0\le x<1\\ x-(x-1), & 1\le x\end{cases}$

$$\Rightarrow \quad |x|-|x-1|=\begin{cases}-1, & x<0\\ 2x-1, & 0\le x<1\\ 1, & 1\le x\end{cases}$$

$$\therefore \quad f(x)=\{|x|-|x-1|\}^2$$

$$=\begin{cases}1, & \text{when } x<0 \text{ or } x\ge 1\\ (2x-1)^2, & \text{when } 0\le x<1\end{cases}$$

Graphically $f(x)$ could be shown as

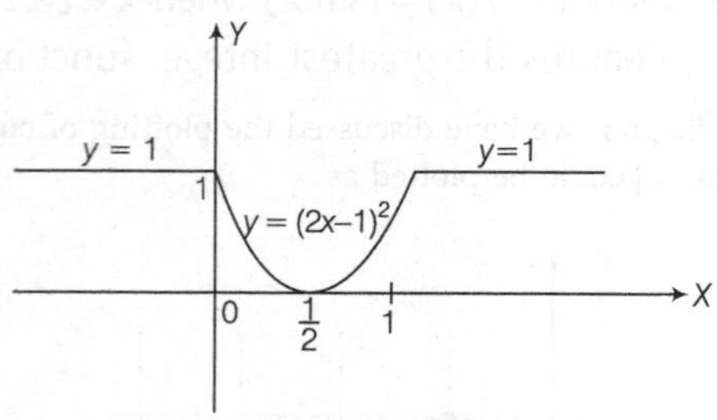

From above figure, it is clear that $f(x)$ is continuous for $x\in R$; but $f(x)$ is not differentiable at $x=0, 1$.

$\Rightarrow \quad f(x)$ is continuous for all $x\in R$.

$\Rightarrow \quad f(x)$ is differentiable for all $x\in R-\{0,1\}$.

Example 31 If $f(x)=\begin{cases}x-3, & x<0\\ x^2-3x+2, & x\ge 0\end{cases}$ and let $g(x)=f(|x|)+|f(x)|$. Discuss the differentiability of $g(x)$.

Sol. $f(|x|)=\begin{cases}|x|-3, & |x|<0\\ |x|^2-3|x|+2, & |x|\ge 0\end{cases}$

Where, $|x|<0$ is not possible, thus neglecting, we get

$$f(|x|)=\{|x|^2-3|x|+2, \quad |x|\ge 0$$

$$f(|x|)=\begin{cases}x^2+3x+2, & x<0\\ x^2-3x+2, & x\ge 0\end{cases} \quad \text{...(i)}$$

Again, $$|f(x)|=\begin{cases}|x-3|, & x<0\\ |x^2-3x+2|, & x\ge 0\end{cases}$$

$$|f(x)|=\begin{cases}(3-x), & x<0\\ (x^2-3x+2), & 0\le x<1\\ -(x^2-3x+2), & 1\le x<2\\ (x^2-3x+2), & 2\le x\end{cases} \quad \text{...(ii)}$$

Now, from Eqs. (i) and (ii), $g(x)=f(|x|)+|f(x)|$

$$g(x)=\begin{cases}x^2+2x+5, & x<0\\ 2x^2-6x+4, & 0\le x<1\\ 0, & 1\le x<2\\ 2x^2-6x+4, & x\ge 2\end{cases}$$

and $$g'(x)=\begin{cases}2x+2, & x<0\\ 4x-6, & 0<x<1\\ 0, & 1<x<2\\ 4x-6, & x>2\end{cases}$$

Therefore, $g(x)$ is continuous in $R-\{0\}$ and $g(x)$ is differentiable in $R-\{0,1,2\}$.

Example 32 Let $f(x)=[n+p\sin x], x\in(0,\pi), n\in Z$ and p is a prime number, where $[\cdot]$ denotes the greatest integer function. Then, find the number of points, where $f(x)$ is not differentiable.

Sol. Here, $f(x)=[n+p\sin x]$ is not differentiable at those points where $n+p\sin x$ is an integer.

As, p is a prime number.

$\Rightarrow \quad n+p\sin x$ is an integer if $\sin x=1,-1, r/p$

i.e. $$x=\frac{\pi}{2},-\frac{\pi}{2},\sin^{-1}\frac{r}{p},\pi-\sin^{-1}\frac{r}{p}$$

where $\quad 0\le r\le p-1$

But $\quad x\ne-\frac{\pi}{2},0$

$\therefore$ Function is not differentiable at

$$x=\frac{\pi}{2},\sin^{-1}\frac{r}{p},\pi-\sin^{-1}\frac{r}{p}$$

where $\quad 0<r\le p-1$

So, the required number of points are

$$=1+2(p-1)=2p-1.$$

Example 33 If $f(x)=||x|-1|$, then draw the graph of $f(x)$ and $fof(x)$ and also discuss their continuity and differentiability. Also, find derivative of $(fof)^2$ at $x=\frac{3}{2}$.

Sol. The graph of $f(x)$ is shown as

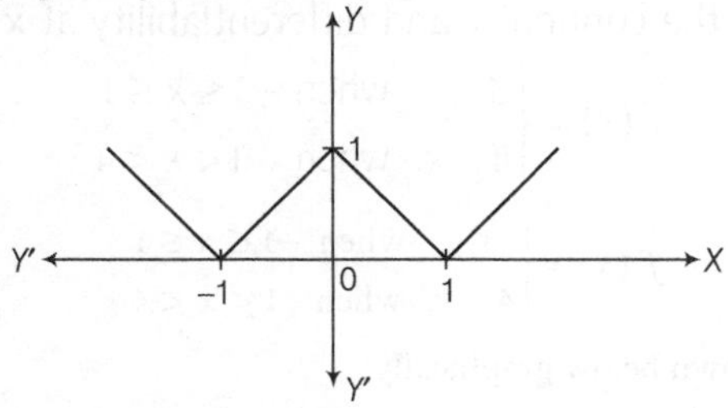

It is clear from the graph that, $f(x)$ is continuous for all x, but $f(x)$ is not differentiable at $x = \{-1, 0, 1\}$.

Now, $fof(x) = ||f(x)| - 1| = |f(x) - 1|$ [as $f(x) \geq 0$ for all x]

Now, if $f(x) \to f(x) - 1$, shift the graph one unit below the X-axis, i.e., as shown;

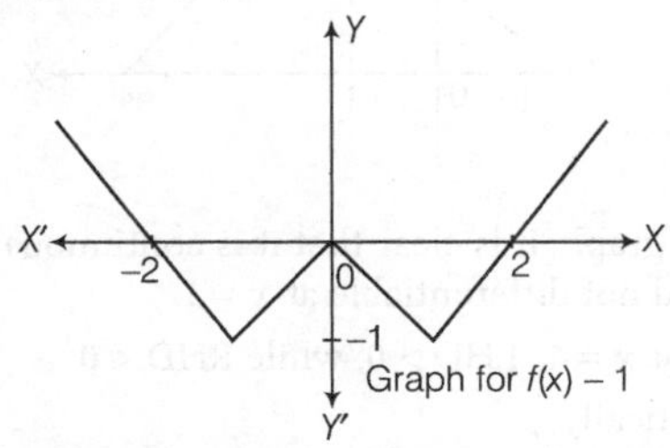

Graph for $f(x) - 1$

Thus, for graph of $fof(x) = |f(x) - 1|$ is taking image of the graph of $f(x) - 1$ below X-axis and leaving the portion above Y-axis as it is.

$\therefore$ Graph for $fof(x)$ is shown as

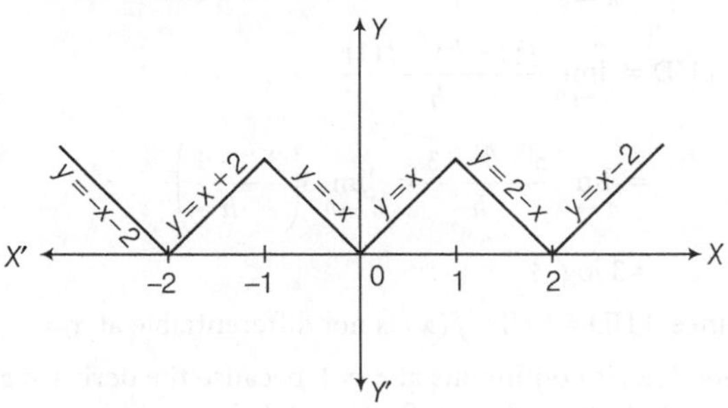

which is clearly continuous for all $x \in R$, but not differentiable at $x = \{-2, -1, 0, 1, 2\}$.

Also, $\quad fof(x) = 2 - x, 1 \leq x \leq 2$

$\therefore \quad (fof)^2 = (2 - x)^2$, when $1 \leq x \leq 2$

$\Rightarrow \quad \dfrac{d}{dx}(fof)^2 = 2(2 - x)(-1)$, when $1 \leq x \leq 2$

$\therefore \quad \dfrac{d}{dx}(fof)^2$ (when $x = 3/2$) $= -2(2 - 3/2) = -1$

$\Rightarrow \quad \dfrac{d}{dx}\{(fof)^2\}_{\text{at } x = 3/2} = -1$

Example 34 Draw the graph of the function $g(x) = f(x + l) + f(x - l)$, where

$$f(x) = \begin{cases} k\left\{1 - \dfrac{|x|}{l}\right\}, & \text{for } |x| \leq l \\ 0, & \text{for } |x| > l \end{cases}$$

Also, discuss the continuity and differentiability of the function $g(x)$.

Sol. We have, $f(x) = \begin{cases} k\left(1 - \dfrac{|x|}{l}\right), & |x| \leq l \\ 0, & |x| > l \end{cases}$

The graph of $f(x)$ is shown as

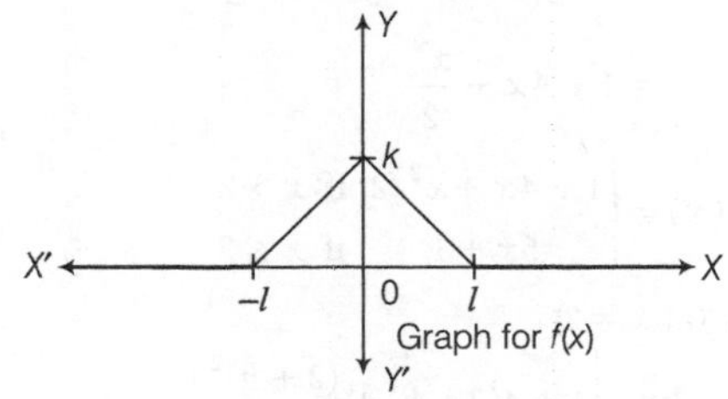

Graph for $f(x)$

Now, to plot $g(x) = f(x + l) + f(x - l)$, we shall first plot the graphs of $f(x + l)$ and $f(x - l)$ which are given as

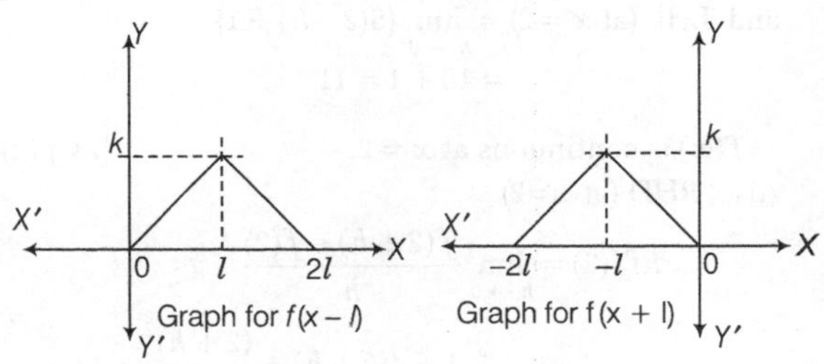

Graph for $f(x - l)$ Graph for $f(x + l)$

Thus, the graph of $g(x) = f(x + l) + f(x - l)$ is obtained by adding graph of $f(x + l)$ and $f(x - l)$ as

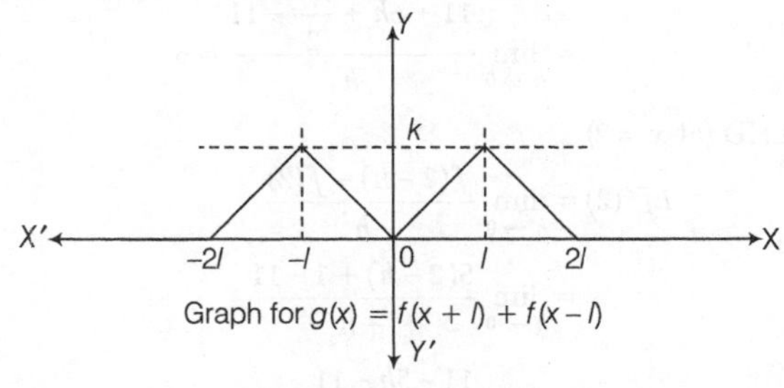

Graph for $g(x) = f(x + l) + f(x - l)$

From the above figure, $g(x)$ is continuous for all $x \in R$ and $g(x)$ is differentiable for all $x \in R - \{\pm 2l, \pm l, 0\}$.

Example 35 Let $f(x)=\begin{cases}\int_0^x\{5+|1-t|\}dt, & \text{if } x>2\\ 5x+1, & \text{if } x\le 2\end{cases}$.

Test $f(x)$ for continuity and differentiability for all real x.

Sol. Here, $f(x)$ for $x>2$.

$$f(x)=\int_0^x\{5+|1-t|\}dt$$

$$=\int_0^1(5+1-t)dt+\int_1^x(5+t-1)dt \quad [\text{since } x>2]$$

$$=\left(6t-\frac{t^2}{2}\right)_0^1+\left(4t+\frac{t^2}{2}\right)_1^x$$

$$=6-\frac{1}{2}+4x+\frac{x^2}{2}-4-\frac{1}{2}$$

$$=1+4x+\frac{x^2}{2}$$

$$\therefore \quad f(x)=\begin{cases}1+4x+x^2/2, & \text{if } x>2\\ 5x+1, & \text{if } x\le 2\end{cases}$$

$\therefore$ RHL (at $x=2$)

$$=\lim_{h\to 0}\left\{1+4(2+h)+\frac{(2+h)^2}{2}\right\}$$

$$=1+8+2=11$$

and LHL (at $x=2$) $=\lim_{h\to 0}\{5(2-h)+1\}$

$$=10+1=11$$

$\therefore f(x)$ is continuous at $x=2$. [as $f(2)=11$]

Also, RHD (at $x=2$)

$$Rf'(2)=\lim_{h\to 0}\frac{f(2+h)-f(2)}{h}$$

$$=\lim_{h\to 0}\frac{1+4(2+h)+\frac{(2+h)^2}{2}-11}{h}$$

$$=\lim_{h\to 0}\frac{11+6h+\frac{h^2}{2}-11}{h}=6$$

LHD (at $x=2$)

$$Lf'(2)=\lim_{h\to 0}\frac{f(2-h)-f(2)}{-h}$$

$$=\lim_{h\to 0}\frac{5(2-h)+1-11}{-h}$$

$$=\lim_{h\to 0}\frac{11-5h-11}{-h}=5$$

$\therefore f(x)$ is not differentiable at $x=2$.

Thus, $f(x)$ is continuous for all $x\in R$ and differentiable for all $x\in R-\{2\}$.

Example 36 Draw the graph of the function and discuss the continuity and differentiability at $x=1$ for,

$$f(x)=\begin{cases}3^x, & \text{when } -1\le x\le 1\\ 4-x, & \text{when } 1<x<4\end{cases}$$

Sol. Here, $f(x)=\begin{cases}3^x, & \text{when } -1\le x\le 1\\ 4-x, & \text{when } 1<x<4\end{cases}$

is shown below graphically.

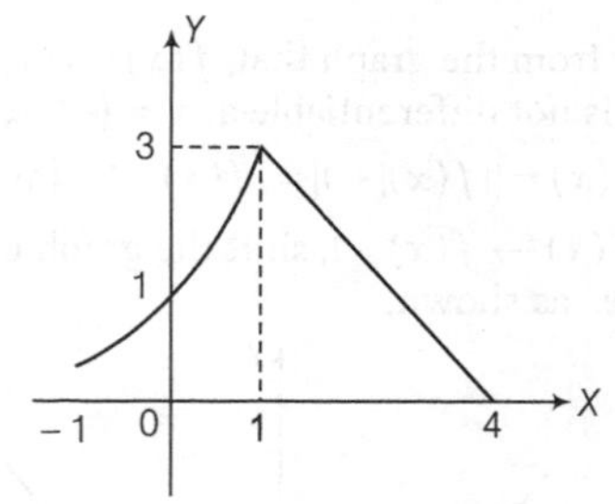

From the graph, it is clear that it is continuous for all x in $[-1, 4)$ and not differentiable at $x=1$.

Because at $x=1$, LHD >0, while RHD <0

Mathematically,

$$\text{RHD}=\lim_{h\to 0}\frac{f(1+h)-f(1)}{h}$$

$$=\lim_{h\to 0}\frac{4-(1+h)-3}{h}=-1$$

$$\text{LHD}=\lim_{h\to 0}\frac{f(1-h)-f(1)}{-h}$$

$$=\lim_{h\to 0}\frac{3^{(1-h)}-3}{-h}=\lim_{h\to 0}3\left(\frac{3^{-h}-1}{-h}\right)$$

$$=3\log_e 3$$

Since, LHD $\ne$ RHD, $f(x)$ is not differentiable at $x=1$.

But $f(x)$ is continuous at $x=1$, because the derivatives from both the sides are finite and definite.

Aliter

The given function is continuous.

Hence, $f'(x)=\begin{cases}3^x\log 3, & -1\le x<1\\ -1, & 1<x<4\end{cases}$

Here, $f'(1^+)=-1$

and $f'(1^-)=\lim_{x\to 1^-}3^x\log 3=3\log 3$

which shows, $f'(1^+)\ne f'(1^-)$

Therefore, $f(x)$ is not differentiable at $x=1$.

Example 37 Match the column I with column II.

	Column I		Column II
(i)	$\sin (\pi [x])$	(A)	differentiable everywhere
(ii)	$\sin \{(x-[x])\pi\}$	(B)	nowhere differentiable
		(C)	not differentiable at -1 and $+1$

where [ ] denotes greatest integral function. *[IIT JEE 1992]*

Sol. (i) We know, $[x] \in I, \forall\ x \in R$

$\therefore\ \sin(\pi[x]) = \sin(I\pi) = 0, \forall\ x \in R$

By theory, we know that every constant function is differentiable in its domain.

Thus, $\sin(\pi[x])$ is differentiable everywhere.

Hence, (i) $\leftrightarrow$ (A)

(ii) Again, $f(x) = \sin\{(x-[x])\pi\}$

Here, we know $x - [x] = \{x\}$,

then $\pi(x-[x]) = \pi\{x\}$

which is not differentiable at integral points.

$\therefore\ f(x) = \sin\{\pi(x-[x])\}$ is not differentiable at $x \in I$. [integers]

Hence, (ii) $\leftrightarrow$ (C)

Example 38 Fill in the blank, in the statement given below.

Let $f(x) = x|x|$. The set of points, where $f(x)$ is twice differentiable is .... *[IIT JEE 1992]*

Sol. The function $f(x) = x|x|$ can be written as,

$$f(x) = \begin{cases} x(x), & \text{if } x \ge 0 \\ x(-x), & \text{if } x < 0 \end{cases}$$

$$\Rightarrow \quad f(x) = \begin{cases} x^2, & \text{if } x \ge 0 \\ -(x)^2, & \text{if } x < 0 \end{cases}$$

$f(x)$ is differentiable, $\forall\ x \in R$.

$$\text{Again,} \quad f'(x) = \begin{cases} 2x, & \text{if } x > 0 \\ -2x, & \text{if } x < 0 \end{cases}$$

$$f''(x) = \begin{cases} 2, & \text{if } x > 0 \\ -2, & \text{if } x < 0 \end{cases}$$

Here, $f''(x)$ is discontinuous at $x = 0$

$\therefore\ f''(x)$ is not differentiable at $x = 0$.

Hence, $f(x)$ is twice differentiable, $\forall\ x \in R - \{0\}$.

Example 39 The function $f(x) = (x^2-1)|x^2-3x+2| + \cos(|x|)$ is not differentiable at

(a) -1 (b) 0 (c) 1 (d) 2

[IIT JEE 1992]

Sol. The function $f(x) = (x^2-1)|x^2-3x+2| + \cos(|x|)$ ...(i)

Here, $|x|$ is not differentiable at $x = 0$, but

$$\cos(|x|) = \begin{cases} \cos(-x), & x < 0 \\ \cos(x), & x \ge 0 \end{cases}$$

$$\Rightarrow \quad \cos(|x|) = \begin{cases} \cos(x), & x \ge 0 \\ \cos(x), & x < 0 \end{cases}$$

$\therefore\ \cos(|x|)$ is differentiable at $x = 0$ ...(ii)

Again, $|x^2 - 3x + 2| = |(x-1)(x-2)|$

$$= \begin{cases} (x-1)(x-2), & \text{if } x < 1 \\ -(x-1)(x-2), & \text{if } 1 \le x < 2 \\ (x-1)(x-2), & \text{if } x \ge 2 \end{cases} \quad \text{...(iii)}$$

$$\text{So,} \quad f(x) = \begin{cases} (x^2-1)(x-1)(x-2) + \cos x, & \text{if } -\infty < x < 1 \\ -(x^2-1)(x-1)(x-2) + \cos x, & \text{if } 1 \le x < 2 \\ (x^2-1)(x-1)(x-2) + \cos x, & \text{if } 2 \le x < \infty \end{cases}$$

Now, to check differentiability at $x = 1, 2$

[using shortcut method]

$$f'(x) = \begin{cases} (x^2-1)(2x-3) + (2x)(x^2-3x+2) - \sin x, & -\infty < x < 1 \\ -(x^2-1)(2x-3) - (2x)(x^2-3x+2) - \sin x, & 1 \le x < 2 \\ (x^2-1)(2x-3) + (2x)(x^2-3x+2) - \sin x, & 2 \le x < \infty \end{cases}$$

Thus, for $f'(1)$, we have

$$f'(1) = \begin{cases} -\sin 1, & x < 1 \\ -\sin 1, & x > 1 \end{cases}$$

Thus, $f(x)$ is differentiable at $x = 1$

$$\text{Also,} \quad f'(2) = \begin{cases} -3 - \sin 2, & x < 2 \\ 3 - \sin 2, & x > 2 \end{cases}$$

Thus, $f(x)$ is not differentiable at $x = 2$.

Hence, (d) is the correct answer.

Example 40 If $f(x) = \sum_{r=1}^{n} a_r |x|^r$, where a_i's are real constants, then $f(x)$ is

(a) continuous at $x = 0$, for all a_i
(b) differentiable at $x = 0$, for all $a_i \in R$
(c) differentiable at $x = 0$, for all $a_{2k+1} = 0$
(d) None of the above

Sol. We know that, $|x|^r, r = 0, 1, 2, \ldots$ are all continuous everywhere.

$\therefore\ f(x) = \sum_{r=1}^{n} a_r |x|^r$ is everywhere continuous.

Since, $|x|, |x|^3, |x|^5, \ldots$ are not differentiable at $x = 0$, whereas $|x|^2, |x|^4, \ldots$ are everywhere differentiable.

$\therefore\ f(x)=\sum_{r=1}^{n} a_r |x|^r$ is not differentiable at $x=0$, if anyone of $a_1, a_3, a_5, \ldots$ is non-zero.

Thus, for $f(x)$ to be differentiable at $x=0$, we must have

$a_1=a_3=a_5 \ldots =0$

i.e. $\quad a_{2k+1}=0$

Hence, (a) and (c) are the correct answers.

Example 41 Let f and g be differentiable functions satisfying $g'(a)=2$, $g(a)=b$ and $fog=I$ (Identity function). Then, $f'(b)$ is equal to

(a) 2 (b) $\frac{2}{3}$ (c) $\frac{1}{2}$ (d) None of these

Sol. We have, $fog=I$

$$\Rightarrow \quad f\{g(x)\}=x, \text{ for all } x\in R$$

$$\therefore \quad f'\{g(x)\}\cdot g'(x)=1, \text{ for all } x\in R$$

$$\Rightarrow \quad f'(g(a))=\frac{1}{g'(a)} \Rightarrow f'(b)=\frac{1}{2}$$

Hence, (c) is the correct answer.

Example 42 If $f(x)=\frac{x}{1+(\log x)(\log x)\ldots\infty}, \forall\ x\in[1,3]$ is non-differentiable at $x=k$. Then, the value of $[k^2]$, is (where $[\cdot]$ denotes greatest integer function).

(a) 5 (b) 6 (c) 7 (d) 8

Sol. Let $g(x)=(\log x)\cdot(\log x)\ldots\infty$

$$\therefore \quad g(x)=\begin{cases}1, & x=e\\ 0, & 1\le x<e\\ \infty, & x>e\end{cases}$$

$$\therefore \quad f(x)=\begin{cases}x, & 1\le x<e\\ x/2, & x=e\\ 0, & e<x\le 3\end{cases}$$

$\Rightarrow f(x)$ is neither continuous nor differentiable at $x=e=k$.

$$\therefore \quad [k^2]=7$$

Hence, (c) is the correct answer.

Example 43 If $f(x)=|1-x|$, then the points where $\sin^{-1}(f(|x|))$ is non-differentiable, are

(a) $\{0, 1\}$ (b) $\{0, -1\}$

(c) $\{0, 1, -1\}$ (d) None of these

Sol. Here, $\quad f(x)=|1-x|$

$$f(|x|)=|1-|x||=\begin{cases}x-1, & x>1\\ 1-x, & 0<x<1\\ 1+x, & -1\le x\le 0\\ -x-1, & x<-1\end{cases}$$

Here, for domain of $\sin^{-1}(f(|x|))$

$$-1\le f(|x|)\le 1$$

$$\Rightarrow \quad |1-|x||\le 1$$

$$\Rightarrow \quad -1\le 1-|x|\le 1$$

$$\Rightarrow \quad -2\le -|x|\le 0$$

$$\Rightarrow \quad 2\ge |x|\ge 0$$

$$\therefore \quad x\in[-2, 2]$$

Then, $\sin^{-1}(f(|x|))$, for $x\in[-2,2]$ is not differentiable at $x=\{0, 1, -1\}$.

Hence, (c) is the correct answer.

Exercise for Session 6

1. If a function $f(x)$ is defined as $f(x)=\begin{cases}-x, & x<0\\ x^2, & 0\le x\le 1\\ x^2-x+1, & x>1\end{cases}$, then

(a) $f(x)$ is differentiable at $x=0$ and $x=1$
(b) $f(x)$ is differentiable at $x=0$ but not at $x=1$
(c) $f(x)$ is not differentiable at $x=1$ but not at $x=0$
(d) $f(x)$ is not differentiable at $x=0$ and $x=1$

2. If $f(x)=x^3\operatorname{sgn}(x)$, then

(a) f is differentiable at $x=0$
(b) f is continuous but not differentiable at $x=0$
(c) $f'(0^-)=1$
(d) None of these

3. Which of the following is continuous everywhere in its domain but has atleast one point where it is not differentiable?

(a) $f(x)=x^{1/3}$
(b) $f(x)=\dfrac{|x|}{x}$
(c) $f(x)=e^{-x}$
(d) $f(x)=\tan x$

4. If $f(x)=\begin{cases}x+\{x\}+x\sin\{x\}, & \text{for } x\ne 0\\ 0, & \text{for } x=0\end{cases}$, where $\{x\}$ denotes the fractional part function, then

(a) f is continuous and differentiable at $x=0$
(b) f is continuous but not differentiable at $x=0$
(c) f is continuous and differentiable at $x=2$
(d) None of these

5. If $f(x)=\begin{cases}x\left(\dfrac{e^{1/x}-e^{-1/x}}{e^{1/x}+e^{1/x}}\right), & x\ne 0\\ 0, & x=0\end{cases}$, then at $x=0$, $f(x)$ is

(a) differentiable
(b) not differentiable
(c) $f'(0^+)=-1$
(d) $f'(0^-)=1$

Session 7

Differentiability in an Interval

(i) A function $f(x)$ defined in an open interval (a, b) is said to be differentiable or derivable in open interval (a, b), if it is differentiable at each point of (a, b).

(ii) A function $f(x)$ defined in a close interval $[a, b]$ is said to be differentiable or derivable at the end points a and b, if it is differentiable from the right at a and from the left at b. In other words, $\lim_{x \to a^+} \frac{f(x) - f(a)}{x - a}$ and $\lim_{x \to b^-} \frac{f(x) - f(b)}{x - b}$ both exist.

Example 44 Discuss the differentiability of $f(x) = \sin^{-1}\left(\frac{2x}{1+x^2}\right)$.

Sol. We have, $f(x) = \sin^{-1}\left(\frac{2x}{1+x^2}\right)$

$$\Rightarrow f'(x) = \frac{1}{\sqrt{1 - \left(\frac{2x}{1+x^2}\right)^2}} \times \frac{d}{dx}\left(\frac{2x}{1+x^2}\right)$$

$$= \frac{(1+x^2)}{\sqrt{(1+x^2)^2 - 4x^2}} \times \left[\frac{(1+x^2)(2) - 2x(2x)}{(1+x^2)^2}\right]$$

$$= \frac{(1+x^2)}{\sqrt{1+2x^2+x^4 - 4x^2}} \times \frac{(2+2x^2-4x^2)}{(1+x^2)^2}$$

$$= \frac{(1+x^2)}{\sqrt{1-2x^2+x^4}} \times \frac{(2-2x^2)}{(1+x^2)^2}$$

$$= \frac{(1+x^2)}{\sqrt{(1-x^2)^2}} \times \frac{(2-2x^2)}{(1+x^2)^2} = \frac{(1+x^2)}{|1-x^2|} \times \frac{2(1-x^2)}{(1+x^2)^2}$$

[since $1 + x^2 \neq 0$]

$$\Rightarrow f'(x) = \frac{1}{|(1-x^2)|} \times \frac{2(1-x^2)}{(1+x^2)} \quad \text{...(i)}$$

Here, in Eq. (i), $f'(x)$ exists only if, $|1-x^2| \neq 0$

$\Rightarrow \quad 1 - x^2 \neq 0$

$\Rightarrow \quad x^2 \neq 1 \Rightarrow x \neq \pm 1$

Thus, $f'(x)$ exists only, if $x \in R - \{-1, 1\}$.

$\therefore$ $f(x)$ is differentiable for all $x \in R - \{1, -1\}$.

Remark

The above example, can also be solved as follows

$$y = f(x) = \sin^{-1}\left(\frac{2x}{1+x^2}\right), \text{ let } x = \tan\theta$$

$$\therefore \quad y = \sin^{-1}\left(\frac{2\tan\theta}{1+\tan^2\theta}\right) \Rightarrow y = \sin^{-1}(\sin 2\theta)$$

$$\therefore \quad y = 2\theta \text{ or } y = 2\tan^{-1}x$$

$\frac{dy}{dx} = \frac{2}{1+x^2}$, which states $f'(x)$ exists for all $x \in R$. "Which is wrong as we have not checked the domain of $f(x)$." So, students are advised to solve these problems carefully, while applying this method.

Example 45 Let [ ] denotes the greatest integer function and $f(x) = [\tan^2 x]$, then ***[IIT JEE 1993]***

(a) $\lim_{x \to 0} f(x)$ doesn't exist (b) $f(x)$ is continuous at $x = 0$

(c) $f(x)$ is not differentiable at $x = 0$

(d) $f'(0) = 1$

Sol. Here, [ ] denotes the greatest integral function.

Thus, $\quad -45° < x < 45°$

$\Rightarrow \quad \tan(-45°) < \tan x < \tan(45°)$

$\Rightarrow \quad -1 < \tan x < 1 \Rightarrow 0 < \tan^2 x < 1$

Since, $\quad f(x) = [\tan^2 x] = 0$

Therefore, $f(x)$ is zero for all values of x from $(-45°)$ to $(45°)$. Thus, $f(x)$ exists when $x \to 0$ and also it is continuous at $x = 0$, $f(x)$ is differentiable at $x = 0$ and has a value 0. (i.e. $f(0) = 0$).

Hence, (b) is the correct answer.

Theorems of Differentiability

Theorem 1 If $f(x)$ and $g(x)$ are both derivable at $x = a$, $f(x) \pm g(x)$, $f(x) \cdot g(x)$ and $\frac{f(x)}{g(x)}$ will also be derivable at $x = a$ $\left\{\text{only if } g(a) \neq 0 \text{ for } \frac{f(x)}{g(x)}\right\}$.

Theorem 2 If $f(x)$ is derivable at $x = a$ and $g(x)$ is not differentiable at $x = a$, then $f(x) \pm g(x)$ will not be derivable at $x = a$.

e.g. $f(x) = \cos|x|$ is derivable at $x = 0$ and $g(x) = |x|$ is not derivable at $x = 0$.

Then, $\cos|x|+|x|$ is not derivable at $x=0$.

However, nothing can be said about the product function, as in this case

$$f(x)=x \text{ is derivable at } x=0$$
$$g(x)=|x| \text{ is not derivable at } x=0$$

But, $$f(x)\cdot g(x)=\begin{cases} x^2, & \text{if } x\ge 0 \\ -x^2, & \text{if } x<0 \end{cases}$$

which is derivable at $x=0$.

Theorem 3 If both $f(x)$ and $g(x)$ are non-derivable, then nothing can be said about the sum/difference/product function.

e.g. $f(x)=\sin|x|$, not derivable at $x=0$

$g(x)=|x|$, not derivable at $x=0$

Then, the function

$F(x)=\sin|x|+|x|$, not derivable at $x=0$

$G(x)=\sin|x|-|x|$, derivable at $x=0$

Theorem 4 If $f(x)$ is derivable at $x=a$ and $f(a)=0$ and $g(x)$ is continuous at $x=a$.

Then, the product function $F(x)=f(x)\cdot g(x)$ will be derivable at $x=a$.

Proof $$F'(a^+)=\lim_{h\to 0}\frac{f(a+h)\cdot g(a+h)-0}{h}=f'(a)\cdot g(a)$$

$$F'(a^-)=\lim_{h\to 0}\frac{f(a-h)\cdot g(a-h)-0}{-h}=f'(a)\cdot g(a)$$

$\therefore$ Derivable at $x=a$.

Theorem 5 Derivative of a continuous function need not be a continuous function.

e.g. $$f(x)=\begin{cases} x^2\cdot\sin\left(\dfrac{1}{x}\right), & \text{if } x\ne 0 \\ 0, & \text{if } x=0 \end{cases}$$

Here, $f(0^+)=0$ and $f(0^-)=0$

$\therefore$ Continuous at $x=0$.

and $$f'(x)=\begin{cases} 2x\cdot\sin\dfrac{1}{x}-x^2\cdot\cos\left(\dfrac{1}{x}\right)\cdot\dfrac{1}{x^2}, & x\ne 0 \\ 0, & x=0 \end{cases}$$

$\Rightarrow$ $f'(x)$ is not continuous at $x=0$.

$$\left[\text{as } \lim_{x\to 0} f'(x) \text{ doesn't exist}\right]$$

Remark

One must remember the formula which we can write as

$$\max\{f(x),g(x)\}=\frac{f(x)+g(x)}{2}+\left|\frac{f(x)-g(x)}{2}\right|$$

$$\min\{f(x),g(x)\}=\frac{f(x)+g(x)}{2}-\left|\frac{f(x)-g(x)}{2}\right|$$

Example 46 Let $h(x)=\min\{x,x^2\}$ for every real number of x. Then, *[IIT JEE 1998]*

(a) h is not continuous for all x

(b) h is differentiable for all x

(c) $h'(x)=1$ for all x

(d) h is not differentiable at two values of x

Sol. Here, $h(x)=\min\{x,x^2\}$ can be drawn on graph in two steps.

(a) Draw the graph of $y=x$ and $y=x^2$ also find their point of intersection.

Clearly, $x=x^2 \Rightarrow x=0,1$

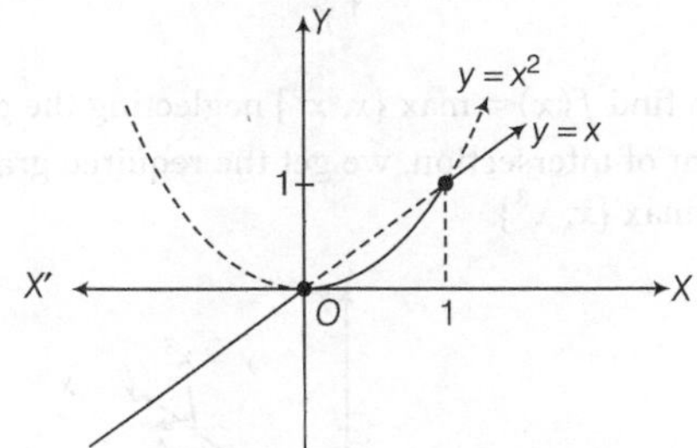

(b) To find $h(x)=\min\{x,x^2\}$ neglecting the graph above the point of intersection, we get

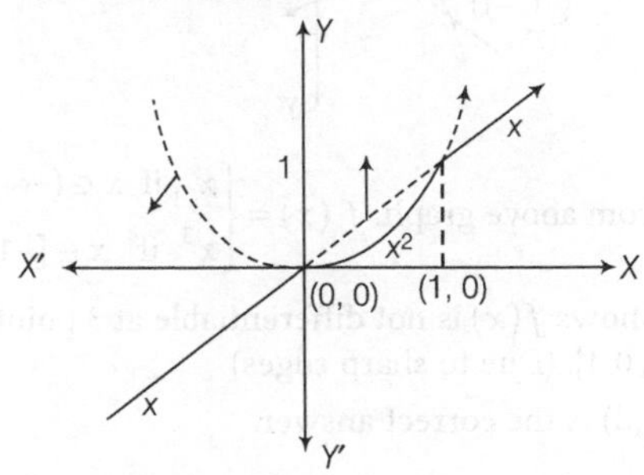

Thus, from the above graph,

$$h(x)=\begin{cases} x, & x\le 0 \text{ or } x\ge 1 \\ x^2, & 0\le x\le 1 \end{cases}$$

which shows $h(x)$ is continuous for all x. But not differentiable at $x=\{0,1\}$.

Thus, $h(x)$ is not differentiable at two values of x.

Hence, (d) is the correct answer.

Example 47 Let $f:R\to R$ be a function defined by $f(x)=\max\{x,x^3\}$. The set of all points where $f(x)$ is not differentiable, is *[IIT JEE 2001]*

(a) $\{-1,1\}$ (b) $\{-1,0\}$

(c) $\{0,1\}$ (d) $\{-1,0,1\}$

Sol. $f(x) = \max\{x, x^3\}$. Consider the graph separately of $y = x^3$ and $y = x$ and find their point of intersection;

Clearly, $\quad x^3 = x$

$\Rightarrow \quad x = 0, 1, -1$

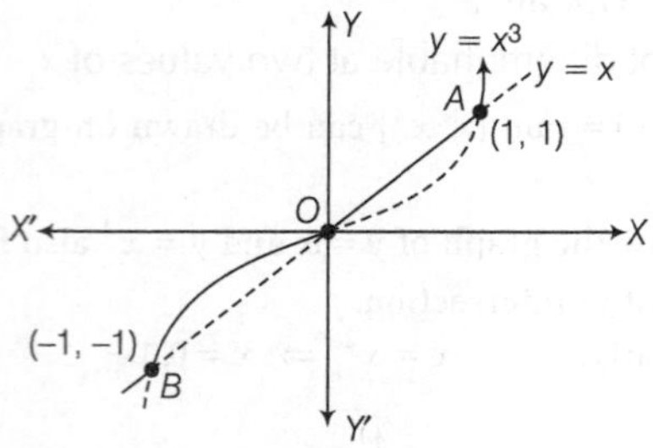

Now, to find $f(x) = \max\{x, x^3\}$ neglecting the graph below the point of intersection, we get the required graph of $f(x) = \max\{x, x^3\}$.

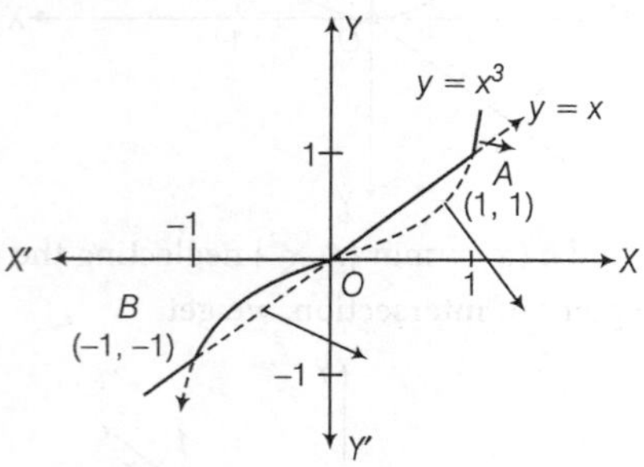

Thus, from above graph, $f(x) = \begin{cases} x, & \text{if } x \in (-\infty, -1] \cup [0, 1] \\ x^3, & \text{if } x \in [-1, 0] \cup [1, \infty) \end{cases}$

which shows $f(x)$ is not differentiable at 3 points, i.e. $x = \{-1, 0, 1\}$. (Due to sharp edges)

Hence, (d) is the correct answer.

Example 48 Let $f(x)$ be a continuous function, $\forall\ x \in R$, $f(0) = 1$ and $f(x) \neq x$ for any $x \in R$, then show $f(f(x)) > x$, $\forall\ x \in R^+$.

Sol. Let $g(x) = f(x) - x$

So, $g(x)$ is continuous and $g(0) = f(0) - 0$.

$\Rightarrow \quad g(0) = 1$

Now, it is given that $g(x) \neq 0$ for any $x \in R$

[as $f(x) \neq x$ for any $x \in R$]

So, $\quad g(x) > 0, \forall\ x \in R^+$

i.e. $\quad f(x) > x, \forall\ x \in R^+$

$\Rightarrow \quad f(f(x)) > f(x) > x, \ \forall\ x \in R^+$

or $\quad f(f(x)) > x, \forall\ x \in R^+$

Example 49 The total number of points of non-differentiability of

$f(x) = \max\left\{\sin^2 x, \cos^2 x, \dfrac{3}{4}\right\}$ in $[0, 10\pi]$, is

(a) 40
(b) 30
(c) 20
(d) 10

Sol. Here, $f(x) = \max\left\{\sin^2 x, \cos^2 x, \dfrac{3}{4}\right\}$

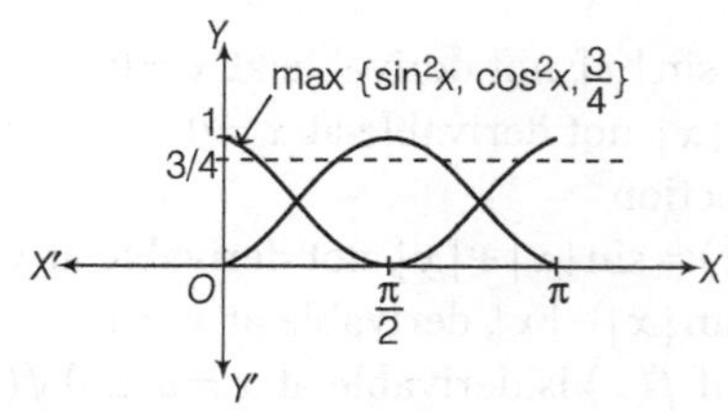

Since, $\sin^2 x$ and $\cos^2 x$ are periodic with period π and in $[0, \pi]$, there are four points of non-differentiability of $f(x)$.

$\therefore$ In $[0, 10\pi]$, there are 40 points of non-differentiability.

Hence, (a) is the correct answer.

Example 50 If $f(x) = |x + 1|\{|x| + |x - 1|\}$, then draw the graph of $f(x)$ in the interval $[-2, 2]$ and discuss the continuity and differentiability in $[-2, 2]$.

Sol. Here, $f(x) = |x + 1|\{|x| + |x - 1|\}$

$$f(x) = \begin{cases} (x+1)(2x-1), & -2 \le x < -1 \\ -(x+1)(2x-1), & -1 \le x < 0 \\ (x+1), & 0 \le x < 1 \\ (x+1)(2x-1), & 1 \le x \le 2 \end{cases}$$

Thus, the graph of $f(x)$ is

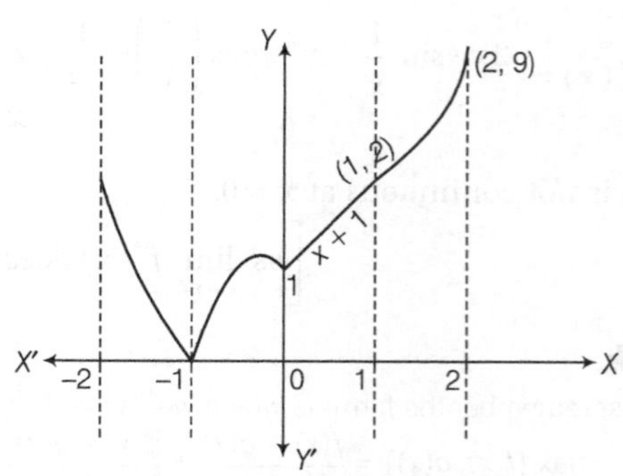

Clearly, continuous for $x \in R$ and has differentiability for $x \in R - \{-1, 0, 1\}$

Example 51 If the function

$$f(x)=\left[\frac{(x-2)^3}{a}\right]\sin(x-2)+a\cos(x-2),$$

(where [ ] denotes the greatest integer function) is continuous and differentiable in (4, 6), then

(a) $a\in[8, 64]$ (b) $a\in(0, 8]$

(c) $a\in[64, \infty)$ (d) None of these

Sol. We have, $x\in(4,6)\Rightarrow 2<x-2<4$

$$\Rightarrow \quad \frac{8}{a}<\frac{(x-2)^3}{a}<\frac{64}{a} \quad [\because a>0]$$

For $f(x)$ to be continuous and differentiable in $(4, 6)$, $\left[\frac{(x-2)^3}{a}\right]$ must attain a constant value for $x\in(4,6)$.

Clearly, this is possible only when $a\ge 64$.

In that case, we have

$f(x)=a\cos(x-2)$, which is continuous and differentiable

$\therefore \quad a\in[64,\infty)$

Hence, (c) is the correct answer.

Example 52 If $f(x)=x^2-2|x|$ and

$$g(x)=\begin{cases}\min\{f(t): -2\le t\le x, \ -2\le x\le 0\}\\ \max\{f(t): 0\le t\le x, \quad 0\le x\le 3\}\end{cases}$$

(i) Draw the graph of $f(x)$ and discuss its continuity and differentiability.

(ii) Find and draw the graph of $g(x)$. Also, discuss the continuity.

Sol. (i) Graph of $f(x)=\begin{cases}x^2-2x, & x\ge 0\\ x^2+2x, & x<0\end{cases}$ is shown as

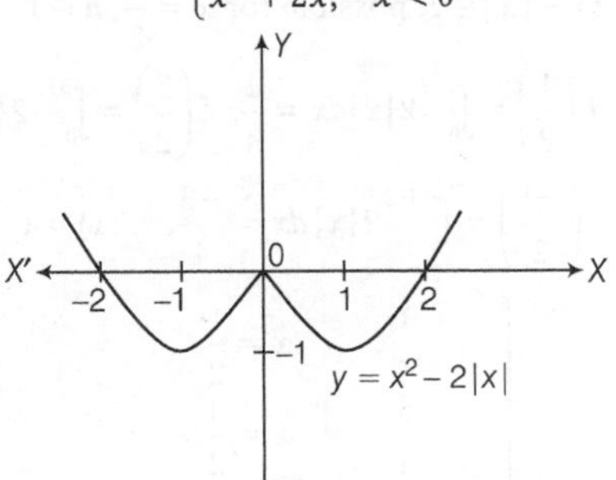

which shows $f(x)$ is continuous for all $x\in R$ and differentiable for all $x\in R-\{0\}$.

(ii) We know that,

If $f(x)$ is an increasing function on $[a, b]$, then

$\max\{f(t); a\le t\le x, \ a\le x\le b\}=f(x)$

$\min\{f(t); a\le t\le x, \ a\le x\le b\}=f(a)$

If $f(x)$ is decreasing function on $[a, b]$, then

$\max\{f(t); a\le t\le x, \ a\le x\le b\}=f(a)$

$\min\{f(t); a\le t\le x, \ a\le x\le b\}=f(x)$

From graph of $f(x)$,

$$g(x)=\begin{cases}f(x), & \text{for } -2\le x<-1\\ -1, & \text{for } -1\le x<0\\ 0, & \text{for } \quad 0\le x<1\\ f(x), & \text{for } \quad x\ge 1\end{cases}$$

$$\Rightarrow \quad g(x)=\begin{cases}x^2+2x, & \text{for } -2\le x<-1\\ -1, & \text{for } -1\le x<0\\ 0, & \text{for } \quad 0\le x<1\\ x^2-2x, & \text{for } \quad x\ge 1\end{cases}$$

Thus, graph of $g(x)$ is

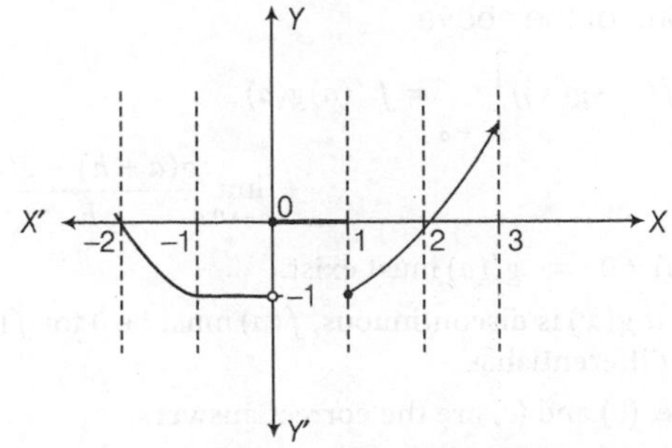

From above figure, it is clear that $g(x)$ is not continuous at $x=0, 1$.

Example 53 Let $f(x)=\phi(x)+\psi(x)$ and $\phi'(a), \psi'(a)$ are finite and definite. Then,

(a) $f(x)$ is continuous at $x=a$

(b) $f(x)$ is differentiable at $x=a$

(c) $f'(x)$ is continuous at $x=a$

(d) $f'(x)$ is differentiable at $x=a$

Sol. We know that the sum of two continuous (differentiable) functions is continuous (differentiable).

$\therefore$ $f(x)$ is continuous and differentiable at $x=a$.

Hence, (a) and (b) are the correct answers.

Example 54 If $f(x)=x+\tan x$ and $g(x)$ is the inverse of $f(x)$, then $g'(x)$ is equal to

(a) $\dfrac{1}{1+(g(x)-x)^2}$ (b) $\dfrac{1}{2+(g(x)+x)^2}$

(c) $\dfrac{1}{2+(g(x)-x)^2}$ (d) None of these

Sol. We have, $f(x)=x+\tan x$

$$\Rightarrow \quad f(f^{-1}(x))=f^{-1}(x)+\tan(f^{-1}(x))$$

$$\Rightarrow \quad x=g(x)+\tan(g(x)) \quad \text{...(i)}$$

$$[\because g(x)=f^{-1}(x)]$$

$$1=g'(x)+\sec^2(g(x))\cdot g'(x)$$

$$\Rightarrow \quad g'(x)=\frac{1}{1+\sec^2(g(x))}$$

$$\Rightarrow \quad g'(x) = \frac{1}{2+\tan^2(g(x))}$$

$$\Rightarrow \quad g'(x) = \frac{1}{2+(x-g(x))^2} \quad \text{[from Eq. (i)]}$$

Hence, (c) is the correct answer.

Example 55 If $f(x)$ is differentiable function and $(f(x)\cdot g(x))$ is differentiable at $x = a$, then

(a) g(x) must be differentiable at $x = a$
(b) g(x) is discontinuous, then $f(a) = 0$
(c) $f(a) \neq 0$, then g(x) must be differentiable
(d) None of the above

Sol. $$\left[\frac{d}{dx}(f(x)\cdot g(x))\right]_{x=a} = f'(a)g(a) + \lim_{h\to 0}\frac{g(a+h)-g(a)}{h}\cdot f(a)$$

If $f(a) \neq 0 \Rightarrow g'(a)$ must exist.

Also, if $g(x)$ is discontinuous, $f(a)$ must be 0 for $f(x)\cdot g(x)$ to be differentiable.

Hence, (b) and (c) are the correct answers.

Example 56 If $f(x) = [x^{-2}[x^2]]$, (where $[\cdot]$ denotes the greatest integer function) $x \neq 0$, then incorrect statement

(a) $f(x)$ is continuous everywhere
(b) $f(x)$ is discontinuous at $x = \sqrt{2}$
(c) $f(x)$ is non-differentiable at $x = 1$
(d) $f(x)$ is discontinuous at infinitely many points

Sol. Here, $\quad 0 \le [x^2] \le x^2$

$$\Rightarrow \quad 0 \le x^{-2}[x^2] \le 1 \Rightarrow [x^{-2}[x^2]] = 0 \text{ or } 1$$

$f(x)$ is discontinuous at $x^2 = n, n \in N \Rightarrow x = \sqrt{n}$

$\therefore f(x)$ is neither continuous nor differentiable at $x = \sqrt{n}, n \in N$.

Hence, (b), (c) and (d) are the correct answers.

Example 57 If $f(x) = \begin{cases} x^2(\operatorname{sgn}[x]) + \{x\}, & 0 \le x \le 2 \\ \sin x + |x-3|, & 2 \le x \le 4 \end{cases}$,

where [ ] and { } represents greatest integer and fractional part function respectively, then

(a) $f(x)$ is differentiable at $x = 1$
(b) $f(x)$ is continuous but non-differentiable at $x = 2$
(c) $f(x)$ is non-differentiable at $x = 2$
(d) $f(x)$ is discontinuous at $x = 2$

Sol. For continuity at $x = 1$

$$\lim_{x\to 1^+} f(x) = \lim_{x\to 1^+} x^2 \operatorname{sgn}[x] + \{x\} = 1 + 0 = 1$$

$$\lim_{x\to 1^-} f(x) = \lim_{x\to 1^-} x^2 \operatorname{sgn}[x] + \{x\} = 0 + 1 = 1$$

Also, $f(1) = 1$

$\therefore f(x)$ is Continuous at $x = 1$

$$f(x) = \begin{cases} x, & 0 \le x < 1 \\ x^2 + x - 1, & 1 \le x < 2 \end{cases}, \text{ non-differentiable at } x = 1$$

$$\lim_{x\to 2^-} f(x) = \lim_{x\to 2^-} x^2 \operatorname{sgn}[x] + \{x\} = 4\times 1 + 1 = 5$$

$$\lim_{x\to 2^+} f(x) = \lim_{x\to 2^+} \sin x + |x-3| = 1 + \sin 2$$

Thus, $f(x)$ is neither continuous nor differentiable at $x = 2$.

Hence, (c) and (d) are the correct answers.

Example 58 A real valued function $f(x)$ is given as

$$f(x) = \begin{cases} \int_0^x 2\{x\}dx, & x + \{x\} \in I \\ x^2 - x + \frac{1}{2}, & \frac{1}{2} < x < \frac{3}{2} \text{ and } x \neq 1, \\ x^2 - x + \frac{1}{6}, & \text{otherwise} \end{cases}$$

where [ ] denotes greatest integer less than or equals to x and { } denotes fractional part function of x. Then,

(a) $f(x)$ is continuous and differentiable in $x \in \left(-\frac{1}{2}, \frac{1}{2}\right)$
(b) $f(x)$ is continuous and differentiable in $x \in \left[-\frac{1}{2}, \frac{1}{2}\right]$
(c) $f(x)$ is continuous and differentiable in $x \in \left[\frac{1}{2}, \frac{3}{2}\right]$
(d) $f(x)$ is continuous but not differentiable in $x \in (0, 1)$

Sol. Here, $\quad x + \{x\} \in I \Rightarrow x + x - [x] \in I$

$$\Rightarrow \quad 2x - [x] \in I, \text{ possible for } x = \frac{n}{2}, n \in I$$

$$\therefore \quad f\left(\frac{1}{2}\right) = \int_0^{1/2} 2\{x\}dx = \frac{1}{4}, \; f\left(\frac{3}{2}\right) = \int_0^{3/2} 2\{x\}dx = \frac{5}{4}$$

$$\text{and} \quad f\left(\frac{-1}{2}\right) = \int_0^{-1/2} 2\{x\}dx = \frac{-3}{4}, \; f(1) = 1$$

$$\text{Then, } f(x) = \begin{cases} \frac{1}{4}, & x = \frac{1}{2} \\ \frac{5}{4}, & x = \frac{3}{2} \\ \frac{-3}{4}, & x = \frac{-1}{2} \\ 1, & x = 1 \\ x^2 - x + \frac{1}{2}, & \frac{1}{2} < x < \frac{3}{2} \text{ and } x \neq 1 \\ x^2 - x + \frac{1}{6}, & \text{otherwise} \end{cases}$$

Clearly, continuous for $x \in (0, 1)$ but not differentiable.

Hence, (d) is the correct answer.

Exercise for Session 7

1. If $f(x) = \sin(\pi (x - [x])), \forall x \in \left(-\frac{\pi}{2}, \frac{\pi}{2}\right)$, where $[\cdot]$ denotes the greatest integer function, then

(a) $f(x)$ is discontinuous at $x = \{-1, 0, 1\}$
(b) $f(x)$ is differentiable for $x \in \left(-\frac{\pi}{2}, \frac{\pi}{2}\right) - \{0\}$
(c) $f(x)$ is differentiable for $x \in \left(-\frac{\pi}{2}, \frac{\pi}{2}\right) - \{-1, 0, 1\}$
(d) None of these

2. Let $f(x) = \begin{cases} x - 1, & -1 \le x < 0 \\ x^2, & 0 \le x \le 1 \end{cases}$, $g(x) = \sin x$ and $h(x) = f(|g(x)|) + |f(g(x))|$. Then,

(a) $h(x)$ is continuous for $x \in [-1, 1]$
(b) $h(x)$ is differentiable for $x \in [-1, 1]$
(c) $h(x)$ is differentiable for $x \in [-1, 1] - \{0\}$
(d) $h(x)$ is differentiable for $x \in (-1, 1) - \{0\}$

3. If $f(x) = \begin{cases} |1 - 4x^2|, & 0 \le x < 1 \\ [x^2 - 2x], & 1 \le x < 2 \end{cases}$, where [ ] denotes the greatest integer function, then

(a) $f(x)$ is continuous for all $x \in [0, 2)$
(b) $f(x)$ is differentiable for all $x \in [0, 2) - \{1\}$
(c) $f(x)$ is differentiable for all $x \in [0, 2) - \left\{\frac{1}{2}, 1\right\}$
(d) None of these

4. Let $f(x) = \int_0^1 |x - t| t \, dt$, then

(a) $f(x)$ is continuous but not differentiable for all $x \in R$
(b) $f(x)$ is continuous and differentiable for all $x \in R$
(c) $f(x)$ is continuous for $x \in R - \left\{\frac{1}{2}\right\}$ and $f(x)$ is differentiable for $x \in R - \left\{\frac{1}{4}, \frac{1}{2}\right\}$
(d) None of these

5. Let $f(x)$ be a function such that $f(x + y) = f(x) + f(y)$ for all x and y and $f(x) = (2x^2 + 3x) \cdot g(x)$ for all x, where $g(x)$ is continuous and $g(0) = 3$. Then, $f'(x)$ is equal to

(a) 6 (b) 9 (c) 8 (d) None of these

6. If a function $g(x)$ which has derivatives $g'(x)$ for every real x and which satisfies the following equation $g(x + y) = e^y g(x) + e^x g(y)$ for all x and y and $g'(0) = 2$, then the value of $\{g'(x) - g(x)\}$ is equal to

(a) e^x (b) $\frac{2}{3}e^x$ (c) $\frac{1}{2}e^x$ (d) $2e^x$

7. Let $f: R \to R$ be a function satisfying $f\left(\frac{xy}{2}\right) = \frac{f(x) \cdot f(y)}{2}, \forall x, y \in R$ and $f(1) = f'(1) \ne 0$. Then, $f(x) + f(1 - x)$ is (for all non-zero real values of x)

(a) constant (b) can't be discussed (c) x (d) $\frac{1}{x}$

8. Let $f(x)$ be a derivable function at $x = 0$ and $f\left(\frac{x + y}{K}\right) = \frac{f(x) + f(y)}{K} (K \in R, K \ne 0, 2)$. Then, $f(x)$ is

(a) even function
(b) neither even nor odd function
(c) either zero or odd function
(d) either zero or even function

9. Let $f: R - (-\pi, \pi)$ be a differentiable function such that $f(x) + f(y) = f\left(\frac{x + y}{1 - xy}\right)$.

If $f(1) = \frac{\pi}{2}$ and $\lim_{x \to 0} \frac{f(x)}{x} = 2$. Then, $f(x)$ is equal to

(a) $2\tan^{-1} x$ (b) $\frac{1}{2}\tan^{-1} x$ (c) $\frac{\pi}{2}\tan^{-1} x$ (d) $2\pi \tan^{-1} x$

10. Let $f(x) = \sin x$ and $g(x) = \begin{cases} \max\{f(t), 0 \le t \le x\}, & \text{for } 0 \le x \le \pi \\ \frac{1 - \cos x}{2}, & \text{for } x > \pi \end{cases}$. Then, $g(x)$ is

(a) differentiable for all $x \in R$
(b) differentiable for all $x \in R - \{\pi\}$
(c) differentiable for all $x \in (0, \infty)$
(d) differentiable for all $x \in (0, \infty) - \{\pi\}$

JEE Type Solved Examples : Single Option Correct Type Questions

Ex. 1 *The values of a and b so that the function*

$$f(x)=\begin{cases} x+a\sqrt{2}\sin x, & 0\le x<\pi/4 \\ 2x\cot x+b, & \pi/4\le x\le \pi/2 \\ a\cos 2x-b\sin x, & \pi/2<x\le \pi \end{cases}$$ *is continuous for*

$x\in[0,\pi]$, *are*

(a) $a=\dfrac{\pi}{6}, b=-\dfrac{\pi}{6}$ (b) $a=-\dfrac{\pi}{6}, b=\dfrac{\pi}{12}$

(c) $a=\dfrac{\pi}{6}, b=-\dfrac{\pi}{12}$ (d) None of these

Sol. (c) Since, $f(x)$ is continuous for $x\in[0,\pi]$.

$\therefore$ $f(x)$ is continuous at $x=\pi/4$ and $x=\pi/2$ and hence to discuss continuity at $x=\pi/4$ and $x=\pi/2$.

Now, at $x=\pi/4$

Left hand limit (at $x=\pi/4$)

$$=\lim_{x\to\frac{\pi}{4}^-} f(x)=\lim_{x\to\frac{\pi}{4}^-}(x+a\sqrt{2}\sin x)$$

$$=\lim_{h\to 0}\{(\pi/4-h)+a\sqrt{2}\sin(\pi/4-h)\}$$

$$\therefore\quad \text{LHL}=\lim_{x\to\frac{\pi}{4}^-} f(x)=\frac{\pi}{4}+a$$

Right hand limit at $x=\pi/4$

$$=\lim_{x\to\frac{\pi}{4}^+} f(x)=\lim_{x\to\frac{\pi}{4}^+}(2x\cot x+b)$$

$$=\lim_{h\to 0} 2\left(\frac{\pi}{4}+h\right)\cot\left(\frac{\pi}{4}+h\right)+b$$

$$\Rightarrow\quad \text{RHL}=\lim_{x\to\frac{\pi}{4}^+} f(x)=\frac{\pi}{2}+b$$

Also, $f\left(\dfrac{\pi}{4}\right)=\dfrac{\pi}{2}+b$

For continuity, LHL (at $x=\pi/4$) = RHL (at $x=\pi/4$) = $f\left(\dfrac{\pi}{4}\right)$

$$\Rightarrow\quad \frac{\pi}{4}+a=\frac{\pi}{2}+b$$

$$\Rightarrow\quad a-b=\frac{\pi}{4} \qquad \text{...(i)}$$

At $x=\pi/2$

Left hand limit (at $x=\pi/2$)

$$=\lim_{x\to\frac{\pi}{2}^-}(2x\cot x+b)=\lim_{h\to 0}2\left(\frac{\pi}{2}-h\right)\cot\left(\frac{\pi}{2}-h\right)+b$$

$$\text{LHL}=b$$

Right hand limit (at $x=\pi/2$)

$$=\lim_{x\to\frac{\pi}{2}^+}(a\cos 2x-b\sin x)$$

$$=\lim_{h\to 0}\left[a\cos 2\left(\frac{\pi}{2}+h\right)-b\sin\left(\frac{\pi}{2}+h\right)\right]$$

$$\text{RHL}=-a-b$$

and $f(\pi/2)=2\dfrac{\pi}{2}\cot\dfrac{\pi}{2}+b=b$

For continuity, LHL (at $x=\pi/2$) = RHL (at $x=\pi/2$) = $f(\pi/2)$

$$\Rightarrow\quad b=-a-b\ \Rightarrow\ a+2b=0 \qquad \text{...(ii)}$$

On solving Eqs. (i) and (ii), we get

$$a=\pi/6,\ b=-\pi/12$$

Ex. 2 *Let f be an even function and $f'(0)$ exists, then $f'(0)$ is*

(a) 1 (b) 0

(c) -1 (d) -2

Sol. (b) Since, f is an even function.

So, $\quad f(-x)=f(x)$, for all x ...(i)

Also, $f'(0)$ exists.

So, $\quad Rf'(0)=Lf'(0)$

$$\Rightarrow\quad \lim_{h\to 0}\frac{f(0+h)-f(0)}{h}=\lim_{h\to 0}\frac{f(0-h)-f(0)}{-h}$$

$$\Rightarrow\quad \lim_{h\to 0}\frac{f(h)-f(0)}{h}=\lim_{h\to 0}\frac{f(-h)-f(0)}{-h}$$

$$\Rightarrow\quad \lim_{h\to 0}\frac{f(h)-f(0)}{h}=\lim_{h\to 0}\frac{f(h)-f(0)}{-h}$$

[using Eq. (i), $f(-h)=f(h)$]

$$\Rightarrow\quad \lim_{h\to 0}\frac{f(h)-f(0)}{h}=-\lim_{h\to 0}\frac{f(h)-f(0)}{h}$$

$$\Rightarrow\quad 2\lim_{h\to 0}\frac{f(h)-f(0)}{h}=0$$

$$\Rightarrow\quad 2f'(0)=0\Rightarrow f'(0)=0$$

Ex. 3 *The set of points where $x^2|x|$ is thrice differentiable, is*

(a) R (b) $R-\{0,1\}$

(c) $[0,\infty)$ (d) $R-\{0\}$

Sol. (d) Let $f(x)=x^2|x|$ which could be expressed as

$$f(x)=\begin{cases} -x^3, & x<0 \\ 0, & x=0 \\ x^3, & x>0 \end{cases}$$

This gives, $f'(x)=\begin{cases}-3x^2, & x<0\\ 0, & x=0\\ 3x^2, & x>0\end{cases}$

So, $f'(x)$ exists for all real x.

$$f''(x)=\begin{cases}-6x, & x<0\\ 0, & x=0\\ 6x, & x>0\end{cases}$$

So, $f''(x)$ exists for all real x.

$$f'''(x)=\begin{cases}-6, & x<0\\ 0, & x=0\\ 6, & x>0\end{cases}$$

However, $f'''(0)$ doesn't exist, since $f'''(0^-)=-6$ and $f'''(0^+)=6$ which are not equal.

Thus, the set of points where $f(x)$ is thrice differentiable is $R-\{0\}$.

Ex. 4 *The function* $f(x)=\dfrac{|x+2|}{\tan^{-1}(x+2)}$, *is continuous for*

(a) $x\in R$ (b) $x\in R-\{0\}$
(c) $x\in R-\{-2\}$ (d) None of these

Sol. (c) Clearly, f is continuous except possibility at $x=-2$

RHL (at $x=-2$) $=\lim\limits_{x\to-2^+} f(x)$

$$=\lim_{x\to-2^+}\frac{(x+2)}{\tan^{-1}(x+2)}=1$$

LHL (at $x=-2$) $=\lim\limits_{x\to-2^-} f(x)$

$$=\lim_{x\to-2^-}\frac{-(x+2)}{\tan^{-1}(x+2)}=-1$$

So, f is not continuous at $x=-2$.

Ex. 5 *If* $f(x)=\begin{cases}\dfrac{\sin[x^2]\pi}{x^2-3x-18}+ax^3+b, & \text{for } 0\le x\le 1\\ 2\cos\pi x+\tan^{-1}x, & \text{for } 1<x\le 2\end{cases}$ *differentiable function in* $[0, 2]$, *where* $[\cdot]$ *denotes the greatest integer function, then*

(a) $a=\frac{1}{6}, b=\frac{\pi}{4}-\frac{13}{6}$ (b) $a=-\frac{1}{6}, b=\frac{\pi}{4}$
(c) $a=-\frac{1}{6}, b=\frac{\pi}{4}-\frac{13}{6}$ (d) None of these

Sol. (a) Here, $[x^2]=0$, for all $0\le x<1$

and $[x^2]=1$, for $x=1$

$\therefore\ \sin[x^2]\pi=0$, for $0\le x\le 1$

Hence, $f(x)=\begin{cases}ax^3+b, & 0\le x\le 1\\ 2\cos\pi x+\tan^{-1}x, & 1<x\le 2\end{cases}$

As, $f(x)$ is differentiable in $[0, 2]$.

$\Rightarrow f(x)$ is continuous and differentiable at $x=1$

$$\Rightarrow \lim_{x\to1^-} f(x)=\lim_{x\to1^+} f(x)=f(1)$$

$$a+b=-2+\frac{\pi}{4}=a+b$$

$$\Rightarrow a+b=-2+\frac{\pi}{4} \quad \text{...(i)}$$

Again, since $f(x)$ is differentiable at $x=1$

(LHD at $x=1$) = (RHD at $x=1$)

$$\Rightarrow \lim_{x\to1^-}\frac{f(x)-f(1)}{x-1}=\lim_{x\to1^+}\frac{f(x)-f(1)}{x-1}$$

$$\Rightarrow \lim_{x\to1^-}\frac{(ax^3+b)-(a+b)}{x-1}$$

$$=\lim_{x\to1^+}\frac{(2\cos\pi x+\tan^{-1}x)-(a+b)}{x-1}$$

$$\Rightarrow 3a=\lim_{x\to1^+}\frac{-2\pi\sin\pi x+\frac{1}{1+x^2}}{1}$$

$$\Rightarrow 3a=\frac{1}{2} \text{ or } a=\frac{1}{6} \quad \text{...(ii)}$$

From Eqs. (i) and (ii), we get

$$a=\frac{1}{6} \text{ and } b=\frac{\pi}{4}-\frac{13}{6}$$

Ex. 6 *If* $g(x)=\lim\limits_{m\to\infty}\dfrac{x^m f(1)+h(x)+1}{2x^m+3x+3}$ *is continuous at* $x=1$ *and* $g(1)=\lim\limits_{x\to1}\{\log_e(ex)\}^{2/\log_e x}$, *then the value of* $2g(1)+2f(1)-h(1)$ *when* $f(x)$ *and* $h(x)$ *are continuous at* $x=1$, *is*

(a) 0 (b) 1 (c) 2 (d) 3

Sol. (b) Here, $g(1)=\lim\limits_{x\to1}\{\log e+\log_e x\}^{2/\log_e x}$

$$=\lim_{x\to1}\{1+\log_e x\}^{2/\log_e x}$$

$$=e^{\lim\limits_{x\to1}\log_e x\cdot\frac{2}{\log_e x}}$$

$$g(1)=e^2 \quad \text{...(i)}$$

Also, $\lim\limits_{x\to1^-} g(x)=\lim\limits_{x\to1^-}\lim\limits_{m\to\infty}\left\{\dfrac{x^m f(1)+h(x)+1}{2x^m+3x+3}\right\}$

$$=\lim_{m\to\infty}\left\{\lim_{x\to1^-}\frac{x^m f(1)+h(x)+1}{2x^m+3x+3}\right\}$$

$$=\frac{h(1)+1}{3+3} \quad \left[\text{as } x<1\Rightarrow \lim_{m\to\infty} x^m=0\right]$$

$$\lim_{x\to1^-} g(x)=\frac{h(1)+1}{6} \quad \text{...(ii)}$$

Again, $\lim_{x \to 1^+} g(x) = \lim_{x \to 1^+} \lim_{m \to \infty} \left\{ \frac{x^m f(1) + h(x) + 1}{2x^m + 3x + 3} \right\}$

$$= \lim_{m \to \infty} \left\{ \lim_{x \to 1^+} \frac{x^m f(1) + h(x) + 1}{2x^m + 3x + 3} \right\}$$

$$= \lim_{m \to \infty} \lim_{x \to 1^+} \frac{f(1) + h(x)/x^m + 1/x^m}{2 + 3/x^{m-1} + 3/x^m} = \frac{f(1)}{2}$$

$\therefore \quad \lim_{x \to 1^+} g(x) = \frac{f(1)}{2}$ ...(iii)

From Eqs. (i), (ii) and (iii), we get

$e^2 = \frac{h(1) + 1}{6} = \frac{f(1)}{2}$ [as $g(x)$ is continuous at $x = 1$]

$\Rightarrow \quad h(1) = 6e^2 - 1$ and $f(1) = 2e^2$

$\therefore \; 2g(1) + 2f(1) - h(1) = 2e^2 + 4e^2 - 6e^2 + 1 = 1$

$\Rightarrow 2g(1) + 2f(1) - h(1) = 1$

Ex. 7 *Let $g(x) = \log f(x)$, where $f(x)$ is a twice differentiable positive function on $(0, \infty)$ such that $f(x+1) = xf(x)$. Then, for $N = 1, 2, 3, \ldots, g''\left(N + \frac{1}{2}\right) - g''\left(\frac{1}{2}\right)$ is equal to*

[IIT JEE 2008]

(a) $-4\left\{1 + \frac{1}{9} + \frac{1}{25} + \ldots + \frac{1}{(2N-1)^2}\right\}$

(b) $4\left\{1 + \frac{1}{9} + \frac{1}{25} + \ldots + \frac{1}{(2N-1)^2}\right\}$

(c) $-4\left\{1 + \frac{1}{9} + \frac{1}{25} + \ldots + \frac{1}{(2N+1)^2}\right\}$

(d) $4\left\{1 + \frac{1}{9} + \frac{1}{25} + \ldots + \frac{1}{(2N+1)^2}\right\}$

Sol. (*a*) We have, $g(x) = \log f(x)$ ...(i)

$\Rightarrow \quad f(x) = e^{g(x)}$

$\Rightarrow \quad f(x+1) = e^{g(x+1)}$

$\Rightarrow \quad x f(x) = e^{g(x+1)}$ [given $f(x+1) = x f(x)$]

$\Rightarrow \log x + \log f(x) = \log e^{g(x+1)}$ [taking log both sides]

$\Rightarrow \quad \log x + g(x) = g(x+1)$ [from Eq. (i)]

i.e. $\quad g(x+1) - g(x) = \log x$ ...(ii)

On replacing x by $x - \frac{1}{2}$ in Eq. (ii), we get

$$g\left(x + \frac{1}{2}\right) - g\left(x - \frac{1}{2}\right) = \log\left(x - \frac{1}{2}\right)$$

$$= \log(2x - 1) - \log 2$$

$\therefore \quad g''\left(x + \frac{1}{2}\right) - g''\left(x - \frac{1}{2}\right) = \frac{-4}{(2x-1)^2}$ ...(iii)

On substituting $x = 1, 2, 3, \ldots, N$ in Eq. (iii) and adding

$$g''\left(N + \frac{1}{2}\right) - g''\left(\frac{1}{2}\right) = -4\left\{1 + \frac{1}{9} + \frac{1}{25} + \ldots + \frac{1}{(2N-1)^2}\right\}$$

Ex. 8 *Let $y = f(x)$ be a differentiable function, $\forall \; x \in R$ and satisfies;*

$$f(x) = x + \int_0^1 x^2 z \, f(z)\, dz + \int_0^1 x z^2 \, f(z)\, dz, \text{ then}$$

(a) $f(x) = \frac{20x}{119}(2 + 9x)$ (b) $f(x) = \frac{20x}{119}(4 + 9x)$

(c) $f(x) = \frac{10x}{119}(4 + 9x)$ (d) $f(x) = \frac{5x}{119}(4 + 9x)$

Sol. (*b*) We have, $f(x) = x + \int_0^1 x^2 z\, f(z)\, dz + \int_0^1 x z^2 f(z) dz$

or $\quad f(x) = x + x^2 \int_0^1 z\, f(z) dz + x \int_0^1 z^2 f(z) dz$

Let $\quad \lambda_1 = \int_0^1 z\, f(z) dz$ and $\lambda_2 = \int_0^1 z^2 f(z) dz$

$\therefore \quad f(x) = x + x^2 \lambda_1 + x\lambda_2$ ...(i)

Now, $\quad \lambda_1 = \int_0^1 z\, f(z) dz$

$$\lambda_1 = \int_0^1 z(z + z^2\lambda_1 + z\lambda_2) dz$$

[using Eq. (i), as $f(z) = z + z^2\lambda_1 + z\lambda_2$]

$\Rightarrow \quad \lambda_1 = (1 + \lambda_2) \int_0^1 z^2\, dz + \lambda_1 \int_0^1 z^3 dz$

$\Rightarrow \quad \lambda_1 = (1 + \lambda_2)\left(\frac{z^3}{3}\right)_0^1 + \lambda_1\left(\frac{z^4}{4}\right)_0^1$

$\Rightarrow \quad \lambda_1 = \frac{1 + \lambda_2}{3} + \frac{\lambda_1}{4}$

$\Rightarrow \; 9\lambda_1 - 4\lambda_2 = 4$ ...(ii)

Also, $\lambda_2 = \int_0^1 z^2 f(z)\, dz$ or $\lambda_2 = \int_0^1 z^2 \{z + z^2\lambda_1 + z\lambda_2\} dz$

[using Eq. (i), as $f(z) = z + z^2\lambda_1 + z\lambda_2$]

$\Rightarrow \quad \lambda_2 = \int_0^1 z^3 (1 + \lambda_2) dz + \lambda_1 \int_0^1 z^4\, dz$

$\Rightarrow \quad \lambda_2 = \frac{(1 + \lambda_2)}{4} + \frac{\lambda_1}{5}$

or $\quad 15\lambda_2 - 4\lambda_1 = 5$ ...(iii)

From Eqs. (ii) and (iii), we get

$$9\lambda_1 - 4\lambda_2 = 4 \text{ and } 15\lambda_2 - 4\lambda_1 = 5$$

$$\lambda_1 = \frac{80}{119}, \lambda_2 = \frac{61}{119} \quad \text{...(iv)}$$

Thus, Eq. (i) becomes

$$f(x) = x + \frac{80}{119}x^2 + \frac{61}{119}x \quad \text{[from Eq. (iv)]}$$

Hence, $\quad f(x) = \frac{20x}{119}(4 + 9x)$

Ex. 9 *A function $f: R \to R$ satisfies the equation $f(x+y) = f(x) \cdot f(y)$ for all, $f(x) \neq 0$. Suppose that the function is differentiable at $x = 0$ and $f'(0) = 2$. Then,*

(a) $f'(x) = 2f(x)$ (b) $f'(x) = f(x)$
(c) $f'(x) = f(x) + 2$ (d) $f'(x) = 2f(x) + x$

Sol. (a) We are given that

$$f(x+y) = f(x) \cdot f(y) \quad \text{...(i)}$$

and $$f'(0) = \lim_{h \to 0} \frac{f(h) - f(0)}{h} = 2 \quad \text{...(ii)}$$

Now, $$f'(x) = \lim_{h \to 0} \frac{f(x+h) - f(x)}{h}$$

$$\Rightarrow f'(x) = \lim_{h \to 0} \frac{f(x+h) - f(x+0)}{h}$$

$$\Rightarrow f'(x) = \lim_{h \to 0} \frac{f(x) \cdot f(h) - f(x) \cdot f(0)}{h} \quad \text{[using Eq. (i)]}$$

$$\Rightarrow f'(x) = f(x) \lim_{h \to 0} \frac{f(h) - f(0)}{h}$$

$$\Rightarrow f'(x) = 2 f(x) \quad \text{[using Eq. (ii)]}$$

Ex. 10 *Let f be a function such that $f(x + f(y)) = f(x) + y, \forall\ x, y \in R$, then find $f(0)$. If it is given that there exists a positive real δ, such that $f(h) = h$ for $0 < h < \delta$, then find $f'(x)$,*

(a) 0, 1 (b) −1, 0 (c) 2, 1 (d) −2, 0

Sol. (a) Let $x = 0, y = 0$ in $f(x + f(y)) = f(x) + y$

$$f(0 + f(0)) = f(0) + 0$$

$$\Rightarrow f(f(0)) = f(0)$$

$$\Rightarrow f(a) = a \text{ or } f(0) = 0$$

Given, $$f(h) = h$$

Then, $$f'(x) = \lim_{h \to 0} \frac{f(x+h) - f(x)}{h} \text{ for } 0 < h < \delta$$

$$\Rightarrow f'(x) = \lim_{h \to 0} \frac{f(x + f(h)) - f(x)}{h} \quad \text{[given } f(h) = h\text{]}$$

$$\Rightarrow f'(x) = \lim_{h \to 0} \frac{f(x) + h - f(x)}{h}$$

$$\Rightarrow f'(x) = \lim_{h \to 0} \frac{h}{h}$$

$$\Rightarrow f'(x) = 1$$

Ex. 11 *If the function of $f(x) = \left[\frac{(x-5)^2}{A}\right] \sin(x-5) + a\cos(x-2)$, where $[\cdot]$ denotes the greatest integer function, is continuous and differentiable in $(7, 9)$, then*

(a) $A \in [8, 64]$ (b) $A \in [0, 8)$
(c) $A \in [16, \infty)$ (d) $A \in [8, 16]$

Sol. (c) As we know, $[x]$ is neither continuous nor differentiable at integer values.

So, $f(x)$ is continuous and differentiable in (7, 9).

If $\left[\frac{(x-5)^2}{A}\right] = 0$

$$\Rightarrow A > (x-5)^2 \quad \text{...(i)}$$

As, $$x \in (7, 9) \Rightarrow x - 5 \in (2, 4)$$

$$\Rightarrow (x-5)^2 \in (4, 16) \quad \text{...(ii)}$$

From Eqs. (i) and (ii), we get

$$A \geq 16$$

Ex. 12 *If $f(x) = [2 + 5|n|\sin x]$, where $n \in I$ has exactly 9 points of non-derivability in $(0, \pi)$, then possible values of n are (where $[x]$ denotes greatest integer function)*

(a) ± 3 (b) ± 2
(c) ± 1 (d) None of these

Sol. (c) We have, $[2 + 5|n|\sin x] = 2 + [5|n|\sin x]$, let $y = 5|n|\sin x$

$\therefore$ Number of points of non-derivability

$$= 2(5|n| - 1) + 1 = 10|n| - 1$$

According to the question, $10|n| - 1 = 9$

$$\Rightarrow 10|n| = 10$$

$$\Rightarrow |n| = 1$$

Here, $n = \pm 1$

Ex. 13 *The number of points of discontinuity of $f(x) = [2x]^2 - \{2x\}^2$ (where $[\cdot]$ denotes the greatest integer function and $\{\cdot\}$ denotes the fractional part of x) in the interval $(-2, 2)$ are*

(a) 6 (b) 8
(c) 4 (d) 3

Sol. (a) Here, $f(x) = ([2x] + \{2x\})([2x] - \{2x\}) = 2x(2x - 2\{2x\})$

$$= 4x^2 - 4x\{2x\} \quad \text{...(i)}$$

We know, $\{2x\}$ is discontinuous at integers.

Here, $$-2 < x < 2$$

$$\Rightarrow -4 < 2x < 4$$

$\therefore$ Integer values of $2x = \{-3, -2, -1, 0, 1, 2, 3\}$, but $f(x) = 4x(x - \{2x\})$ is continuous at $x = 0$

$\therefore$ Total number of points of discontinuity are 6.

Ex. 14 *If $x \in R^+$ and $n \in N$, we can uniquely write $x = mn + r$, where $m \in W$ and $0 \leq r < n$. We define x mod $n = r$. e.g. 10.3 mod 3 = 1.3. The number of points of discontinuity of the function, $f(x) = (x \text{ mod } 2)^2 + (x \text{ mod } 4)$ in the interval $0 < x < 9$ is*

(a) 0 (b) 2
(c) 4 (d) None of these

Sol. (c) Here, $f(x)=(x \bmod 2)^2+(x \bmod 4)$

$$=\begin{cases} x^2+x, 0<x<2 \\ (x-2)^2+x, 2\le x<4 \\ (x-4)^2+(x-4), 4\le x<6 \\ (x-6)^2+(x-4), 6\le x<8 \\ (x-8)^2+(x-8), 8\le x<9 \end{cases}$$

Clearly, $f(2^-)=6, f(2^+)=2 \Rightarrow$ discontinuous at $x=2$.

$f(4^-)=8, f(4^+)=0 \Rightarrow$ discontinuous at $x=4$.

$f(6^-)=6, f(6^+)=2 \Rightarrow$ discontinuous at $x=6$.

$f(8^-)=8, f(8^+)=0 \Rightarrow$ discontinuous $x=8$.

$\therefore$ Number of points of discontinuity = 4.

Ex. 15 *Let $f:R\to R$ be a differentiable function at $x=0$ satisfying $f(0)=0$ and $f'(0)=1$, then the value of $\lim_{x\to 0}\frac{1}{x}\cdot\sum_{n=1}^{\infty}(-1)^n\cdot f\left(\frac{x}{n}\right)$, is*

(a) 0 (b) $-\log 2$
(c) 1 (d) e

Sol. (b) Let $L=\lim_{x\to 0}\frac{f(x/n)}{x}=\lim_{x\to 0}\frac{f(x/n)-f(0)}{n\cdot x/n}$, [using $f(0)=0$]

$$\therefore \quad L=\frac{1}{n}\lim_{x\to 0}\frac{f(x/n)-f(0)}{x/n}$$

$$\Rightarrow \quad L=\frac{1}{n}\cdot f'(0)=\frac{1}{n} \qquad \text{[as } f'(0)=1]$$

$$\therefore \quad \lim_{x\to 0}\frac{1}{x}\cdot\sum_{n=1}^{\infty}(-1)^n f\left(\frac{x}{n}\right)=\sum_{n=1}^{\infty}(-1)^n\cdot\frac{1}{n}$$

$$=-1+\frac{1}{2}-\frac{1}{3}+\frac{1}{4}-\frac{1}{5}+\frac{1}{6}-\ldots$$

$$=-\left(1-\frac{1}{2}+\frac{1}{3}-\frac{1}{4}+\frac{1}{5}-\frac{1}{6}+\ldots\right)=-\log 2$$

Ex. 16 *Let $f(x)$ is a function continuous for all $x\in R$ except at $x=0$ such that $f'(x)<0, \forall x\in(-\infty,0)$ and $f'(x)>0, \forall x\in(0,\infty)$. If $\lim_{x\to 0^+}f(x)=3$, $\lim_{x\to 0^-}f(x)=4$ and $f(0)=5$, then the image of the point $(0,1)$ about the line, $y\cdot\lim_{x\to 0}f(\cos^3 x-\cos^2 x)=x\cdot\lim_{x\to 0}f(\sin^2 x-\sin^3 x)$, is*

(a) $\left(\frac{12}{25},\frac{-9}{25}\right)$ (b) $\left(\frac{12}{25},\frac{9}{25}\right)$
(c) $\left(\frac{16}{25},\frac{-8}{25}\right)$ (d) $\left(\frac{24}{25},\frac{-7}{25}\right)$

Sol. (d) We have,

$$y\cdot\lim_{x\to 0}f(\cos^3 x-\cos^2 x)=x\cdot\lim_{x\to 0}f(\sin^2 x-\sin^3 x)$$

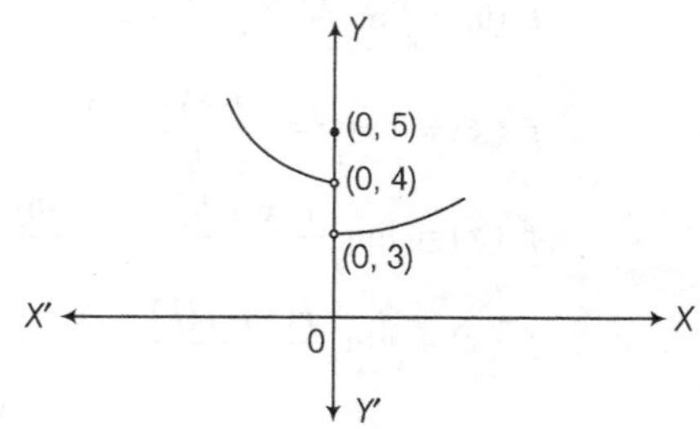

Here, $\cos^3 x-\cos x\to 0$ from LHS

$$\therefore \quad \lim_{x\to 0}f(\cos^3 x-\cos x)=4$$

$$\sin^2 x-\sin^3 x\to 0 \text{ from RHS}$$

$$\therefore \quad \lim_{x\to 0}f(\sin^2 x-\sin^3 x)\to 3$$

$\therefore$ Equation of straight line is $4y=3x$ ...(i)

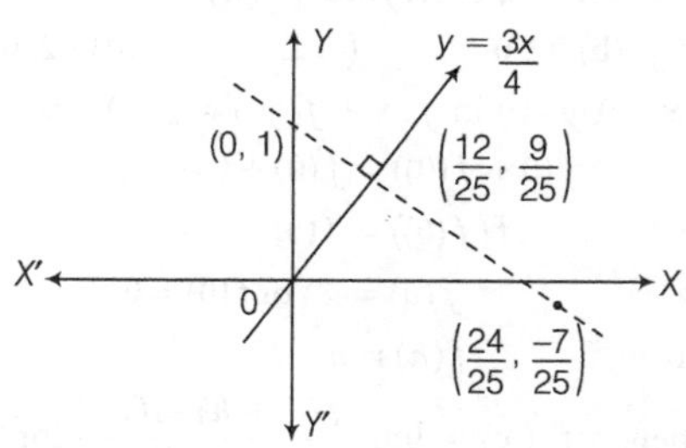

Equation of line perpendicular to $y=\frac{3}{4}x$ and passing through $(0,1)$, is $(y-1)=\frac{-4}{3}(x-0)$

$$\Rightarrow \quad y-1=\frac{-4}{3}x \qquad \text{...(ii)}$$

On solving Eqs. (i) and (ii), we get

$$x=\frac{12}{25}, y=\frac{9}{25}$$

Hence, image point is $\left(\frac{24}{25},\frac{-7}{25}\right)$.

JEE Type Solved Examples :
More than One Correct Option Type Questions

Ex. 17 *If $f(x)$ be such that $f(x) = \max(|3-x|, 3-x^3)$, then*

(a) $f(x)$ is continuous $\forall\ x \in R$
(b) $f(x)$ is differentiable $\forall\ x \in R$
(c) $f(x)$ is non-differentiable at three points only
(d) $f(x)$ is non-differentiable at four points only

Sol. (a,d) From the graph of $f(x)$, $f(x)$ is continuous, $\forall\ x \in R$ and $f(x)$ is not differentiable at $x = -1, 0, 1, 3$.

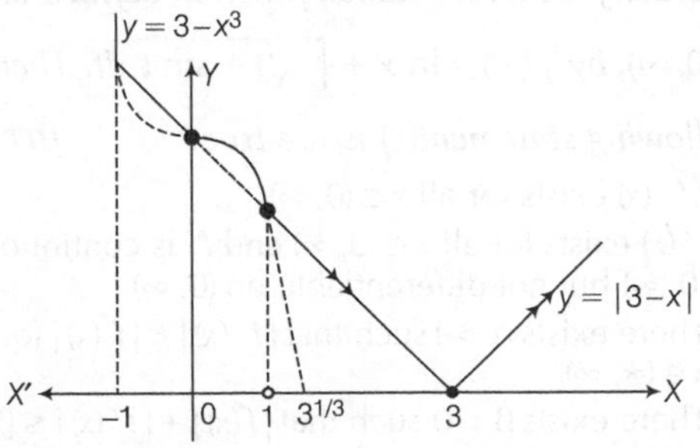

Ex. 18 *Let $f(x) = |x-1|([x]-[-x])$, then which of the following statement(s) is/are correct. (where $[\cdot]$ denotes greatest integer function.)*

(a) $f(x)$ is continuous at $x = 1$
(b) $f(x)$ is derivable at $x = 1$
(c) $f(x)$ is non-derivable at $x = 1$
(d) $f(x)$ is discontinuous at $x = 1$

Sol. (a, c) We have, $f(x) = |x-1|([x]-[-x])$

Here, $[x] = 0$, when $0 < x < 1$

$[-x] = -1$, when $0 < x < 1$

$[x] = 1$, when $1 < x < 2$

$[-x] = -2$, when $1 < x < 2$

$[x] = 1, [-x] = -1$, when $x = 1$

$$\therefore \quad f(x) = \begin{cases} (x-1)(1+2), & \text{when } 1 < x < 2 \\ 0, & \text{when } x = 1 \\ (1-x), & \text{when } 0 < x < 1 \end{cases}$$

$$\Rightarrow \quad f'(1^+) = \lim_{h \to 0} \frac{3(h) - 0}{h} = 3$$

$$\text{and } f'(1^-) = \lim_{h \to 0} \frac{(1-(1-h)) - 0}{-h} = -1$$

$\Rightarrow f(x)$ is continuous at $x = 1$ and non-derivable at $x = 1$.

Ex. 19 *If $y = f(x)$ defined parametrically by $x = 2t - |t-1|$ and $y = 2t^2 + t|t|$, then*

(a) $f(x)$ is continuous for all $x \in R$
(b) $f(x)$ is continuous for all $x \in R - \{2\}$
(c) $f(x)$ is differentiable for all $x \in R$
(d) $f(x)$ is differentiable for all $x \in R - \{2\}$

Sol. (a,d) Here, $x = 2t - |t-1|$ and $y = 2t^2 + t|t|$

Now, when $t < 0$,

$$x = 2t - \{-(t-1)\} = 3t - 1$$

$$\text{and} \quad y = 2t^2 - t^2 = t^2 \Rightarrow y = \frac{1}{9}(x+1)^2$$

When $0 \le t < 1$,

$$x = 2t - (-(t-1)) = 3t - 1$$

$$\text{and} \quad y = 2t^2 + t^2 = 3t^2 \Rightarrow y = \frac{1}{3}(x+1)^2$$

When $t \ge 1$,

$$x = 2t - (t-1) = t + 1$$

$$\text{and} \quad y = 2t^2 + t^2 = 3t^2 \Rightarrow y = 3(x-1)^2$$

$$\text{Thus,} \quad y = f(x) = \begin{cases} \frac{1}{9}(x+1)^2, & x < -1 \\ \frac{1}{3}(x+1)^2, & -1 \le x < 2 \\ 3(x-1)^2, & x \ge 2 \end{cases}$$

Now, to check continuity at $x = -1$ and 2.

Continuity at $x = -1$,

$$\text{LHL} = \lim_{h \to 0} f(-1-h) = \lim_{h \to 0} \frac{1}{9}(-1-h+1)^2 = 0$$

$$\text{RHL} = \lim_{h \to 0} f(-1+h) = \lim_{h \to 0} \frac{1}{3}(-1+h+1)^2 = 0$$

$$f(-1) = 0$$

$\therefore f(x)$ is continuous at $x = -1$.

Now, to check continuity at $x = 2$,

$$\text{LHL} = \lim_{h \to 0} f(2-h) = \lim_{h \to 0} \frac{1}{3}(2-h+1)^2 = 3$$

$$\text{RHL} = \lim_{h \to 0} f(2+h) = \lim_{h \to 0} 3(2+h-1)^2 = 3$$

$$f(2) = 3$$

Thus, $f(x)$ is continuous at $x = 2$.

Now, to check differentiability at $x = -1$ and 2.

Differentiability at $x = -1$,

$$\text{LHD} = Lf'(-1) = \lim_{h \to 0} \frac{f(-1-h) - f(-1)}{-h}$$

$$= \lim_{h \to 0} \frac{\frac{1}{9}(-1-h+1)^2 - 0}{-h} = 0$$

$$\text{RHD} = Rf'(-1) = \lim_{h \to 0} \frac{f(-1+h) - f(-1)}{h}$$

$$= \lim_{h \to 0} \frac{\frac{1}{3}(-1+h+1)^2 - 0}{h} = 0$$

Hence, $f(x)$ is differentiable at $x = -1$.

Differentiability at $x = 2$,

$$\text{LHD} = Lf'(2) = \lim_{h \to 0} \frac{f(2-h) - f(2)}{-h}$$

$$= \lim_{h \to 0} \frac{\frac{1}{3}(2-h+1)^2 - 3}{-h} = 2$$

$$\text{RHD} = Rf'(2) = \lim_{h \to 0} \frac{f(2+h) - f(2)}{h}$$

$$= \lim_{h \to 0} \frac{3(2+h-1)^2 - 3}{h} = 6$$

Hence, $f(x)$ is not differentiable at $x = 2$.

$\therefore f(x)$ is continuous for all x and differentiable for all x, except $x = 2$.

Ex. 20 *$f(x) = \sin^{-1}[e^x] + \sin^{-1}[e^{-x}]$, where $[\cdot]$ greatest integer function, then*

(a) domain of $f(x) = (-\ln 2, \ln 2)$

(b) range of $f(x) = \{\pi\}$

(c) $f(x)$ has removable discontinuity at $x = 0$

(d) $f(x) = \cos^{-1} x$ has only solution

Sol. (*a, c*) $0 < e^x < 2$ and $0 < e^{-x} < 2$

$\Rightarrow \quad -\infty < x < \log_e 2$ and $-\infty < -x < \log_e 2$

$\Rightarrow \quad (-\infty < x < \log_e 2)$

and $\quad (-\log_e 2 < x < \infty)$

$\Rightarrow \quad -\log_e 2 < x < \log_e 2$

$$\therefore \quad f(x) = \begin{cases} \pi, & x = 0 \\ \frac{\pi}{2}, & x \in (-\log_e 2, 0) \cup (0, \log_e 2) \end{cases}$$

Ex. 21 *$f : R \to R$ is one-one, onto and differentiable function and graph of $y = f(x)$ is symmetrical about the point (4, 0), then*

(a) $f^{-1}(2010) + f^{-1}(-2010) = 8$

(b) $\int_{-2010}^{2018} f(x)\,dx = 0$

(c) if $f'(-100) > 0$, then roots of $x^2 - f'(10)x - f'(10) = 0$ may be non-real

(d) if $f'(10) = 20$, then $f'(-2) = 20$

Sol. (*a, b, d*) Graph is symmetrical about (4, 0).

$\Rightarrow \quad f(4+x) = -f(4-x)$

$\Rightarrow \quad f(x) = -f(8-x) \quad$ ...(i)

Now, let $f(x) = 2010$, then $f(8-x) = -2010$

$\Rightarrow \quad f^{-1}(2010) + f^{-1}(-2010) = 8$

$\Rightarrow$ Option (a) is true.

and $\quad \int_{-2010}^{4} f(x)dx = -\int_{4}^{2018} f(x)\,dx$

$\Rightarrow$ Option (b) is true.

Also, $\quad D = (f'(10))^2 + 4f'(10) > 0$

As, $\quad f'(-100) > 0 \Rightarrow f'(10) \geq 0$

$\Rightarrow \quad x^2 - f'(10)x - f'(10) = 0$ has real roots.

$\Rightarrow$ Option (c) is false.

As, $\quad f'(4+x) = f'(4-x)$

$f'(x)$ is symmetric about $x = 4$

$\Rightarrow \quad f'(10) = f'(-2) = 20$

$\Rightarrow$ Option (d) is true.

Ex. 22 *Let f be a real valued function defined on the interval $(0, \infty)$, by $f(x) = \ln x + \int_0^x \sqrt{1+\sin t}\,dt$. Then, which of the following statement(s) is/are true?* **[IIT JEE 2010]**

(a) $f''(x)$ exists for all $x \in (0, \infty)$

(b) $f'(x)$ exists for all $x \in (0, \infty)$ and f' is continuous on $(0, \infty)$ but not differentiable on $(0, \infty)$

(c) There exists $\alpha > 1$ such that $|f'(x)| < |f(x)|$ for all $x \in (\infty, \infty)$

(d) There exists $\beta > 0$ such that $|f(x)| + |f'(x)| \leq \beta$ for all $x \in (0, \infty)$

Sol. (*b, c*) Here, $f'(x) = \frac{1}{x} + \sqrt{1+\sin x}$, $x > 0$ but $f(x)$ is not differentiable in $(0, \infty)$ as $\sin x$ may be -1 and then $f''(x) = -\frac{1}{x^2} + \frac{1}{2}\frac{\cos x}{\sqrt{1+\sin x}}$ will not exist.

$\Rightarrow f'(x)$ is continuous for all $x \in (0, \infty)$ but $f'(x)$ is not differentiable on $(0, \infty)$.

$\therefore$ Option (b) is true.

Also, $\quad f'(x) \leq 3$, if $x > 1$

and $\quad f(x) > 3$, if $x > e^3$

$\therefore$ Let $\quad \alpha = e^3$

$\Rightarrow$ Option (c) is true.

(d) is not possible, as $f(x) \to \infty$ when $x \to \infty$.

Ex. 23 *If $f(x) + f(y) = f\left(\frac{x+y}{1-xy}\right)$ for all $x, y \in R$ $(xy \neq 1)$ and $\lim_{x \to 0} \frac{f(x)}{x} = 2$, then*

(a) $f\left(\frac{1}{\sqrt{3}}\right) = \frac{\pi}{3}$

(b) $f\left(\frac{1}{\sqrt{3}}\right) = -\frac{\pi}{3}$

(c) $f'(1) = 1$

(d) $f'(1) = -1$

Sol. (*a, c*) $f(x) + f(y) = f\left(\frac{x+y}{1-xy}\right) \quad$ ...(i)

On putting $x = y = 0$, we get $f(0) = 0$

On putting $y = -x$, we get $f(x) + f(-x) = f(0)$

$\Rightarrow \quad f(-x) = -f(x) \quad$ ...(ii)

Also, $\lim_{x \to 0} \frac{f(x)}{x} = 2$

Now, $f'(x) = \lim_{h \to 0} \frac{f(x+h) - f(x)}{h}$ ...(iii)

$= \lim_{h \to 0} \frac{f(x+h) + f(-x)}{h}$

[using Eq. (ii), $-f(x) = f(-x)$]

$f'(x) = \lim_{h \to 0} \frac{f\left(\frac{x+h-x}{1-(x+h)(-x)}\right)}{h}$ [using Eq. (i)]

$\Rightarrow \quad f'(x) = \lim_{h \to 0} \left[\frac{f\left(\frac{h}{1+x(x+h)}\right)}{h}\right]$

$\Rightarrow \quad f'(x) = \lim_{h \to 0} \frac{f\left(\frac{h}{1+xh+x^2}\right)}{\left(\frac{h}{1+xh+x^2}\right)} \times \left(\frac{1}{1+xh+x^2}\right)$

$\Rightarrow \quad f'(x) = \lim_{h \to 0} \frac{f\left(\frac{h}{1+xh+x^2}\right)}{\left(\frac{h}{1+xh+x^2}\right)} \times \lim_{h \to 0} \frac{1}{1+xh+x^2}$

$\left[\text{using } \lim_{x \to 0} \frac{f(x)}{x} = 2\right]$

$\Rightarrow \quad f'(x) = 2 \times \frac{1}{1+x^2} \Rightarrow f'(x) = \frac{2}{1+x^2}$

On integrating both sides, we get

$f(x) = 2\tan^{-1}(x) + C$

Where $f(0) = 0 \Rightarrow C = 0$

Thus, $f(x) = 2\tan^{-1} x$

Hence, $f\left(\frac{1}{\sqrt{3}}\right) = 2\tan^{-1}\left(\frac{1}{\sqrt{3}}\right) = 2 \cdot \frac{\pi}{6} = \frac{\pi}{3}$

and $f'(1) = \frac{2}{1+1^2} = \frac{2}{2} = 1$

Ex. 24 *Let $f: R \to R$ is a function which satisfies condition $f(x+y^3) = f(x) + [f(y)]^3$ for all $x, y \in R$. If $f'(0) \geq 0$, then*

(a) $f(x) = 0$ only (b) $f(x) = x$ only

(c) $f(x) = 0$ or x only (d) $f(10) = 10$

Sol. *(c, d)* Given, $f(x+y^3) = f(x) + [f(y)]^3$ ...(i)

and $f'(0) \geq 0$ ...(ii)

On replacing x, y by 0,

$f(0) = f(0) + f(0)^3 \Rightarrow f(0) = 0$ ...(iii)

Also, $f'(0) = \lim_{h \to 0} \frac{f(0+h) - f(0)}{h} = \lim_{h \to 0} \frac{f(h)}{h}$ ...(iv)

Let $I = f'(0) = \lim_{h \to 0} \frac{f(0 + (h^{1/3})^3) - f(0)}{(h^{1/3})^3}$

$= \lim_{h \to 0} \frac{(f(h^{1/3}))^3}{(h^{1/3})^3} = \lim_{h \to 0} \left(\frac{f(h^{1/3})}{(h^{1/3})}\right)^3 = I^3$

$\Rightarrow \quad I = I^3$ or $I = 0, 1, -1$ as $f'(0) \geq 0$

$\therefore \quad f'(0) = 0, 1$ ...(v)

Thus, $f'(x) = \lim_{h \to 0} \frac{f(x+h) - f(x)}{h}$

$= \lim_{h \to 0} \frac{f(x + (h^{1/3})^3) - f(x)}{(h^{1/3})^3}$

$f'(x) = \lim_{h \to 0} \frac{f(x) + (f(h^{1/3}))^3 - f(x)}{(h^{1/3})^3}$ [using Eq. (i)]

$\Rightarrow \quad f'(x) = \lim_{h \to 0} \left(\frac{f(h^{1/3})}{(h^{1/3})}\right)^3 = (f'(0))^3$

$\Rightarrow \quad f'(x) = 0$ or 1 [as $f'(0) = 0$ or 1, using Eq. (v)]

On integrating both sides, we get

$f(x) = C$ or $x + C$ as $f(0) = 0$

$\Rightarrow \quad f(x) = 0$ or x

Thus, $f(10) = 0$ or 10

Ex. 25 *Let $f(x) = x^3 - x^2 + x + 1$ and $g(x) = \begin{cases} \max f(t), & 0 \leq t \leq x \text{ for } 0 \leq x \leq 1 \\ 3 - x, & 1 < x \leq 2 \end{cases}$*

Then, $g(x)$ in $[0, 2]$ is

(a) continuous for $x \in [0, 2] - \{1\}$

(b) continuous for $x \in [0, 2]$

(c) differentiable for all $x \in [0, 2]$

(d) differentiable for all $x \in [0, 2] - \{1\}$

Sol. *(b, d)* Here, $f(x) = x^3 - x^2 + x + 1$

$\Rightarrow f'(x) = 3x^2 - 2x + 1$, which is strictly increasing in $(0, 2)$.

$\therefore \quad g(x) = \begin{cases} f(x), & 0 \leq x \leq 1 \\ 3 - x, & 1 < x \leq 2 \end{cases}$

[as $f(x)$ is increasing, so $f(x)$ is maximum when $0 \leq t \leq x$]

So, $g(x) = \begin{cases} x^3 - x^2 + x + 1, & 0 \leq x \leq 1 \\ 3 - x, & 1 < x \leq 2 \end{cases}$

Also, $g'(x) = \begin{cases} 3x^2 - 2x + 1, & 0 \leq x \leq 1 \\ -1, & 1 < x \leq 2 \end{cases}$

Which clearly shows $g(x)$ is continuous for all $x \in [0, 2]$ but $g(x)$ is not differentiable at $x = 1$.

JEE Type Solved Examples : Passage Based Questions

Passage I

(Ex. Nos. 26 to 27)

In the given figure graph of
$y = P(x) = ax^5 + bx^4 + cx^3 + dx^2 + ex + f$, *is given.*

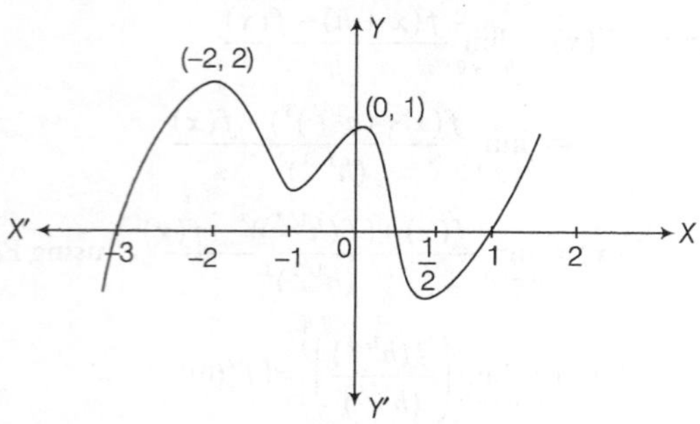

26. If $P''(x)$ has real roots α, β, γ, then $[\alpha]+[\beta]+[\gamma]$, is

(a) –2 (b) –3 (c) –1 (d) 0

27. The minimum number of real roots of equation $(P''(x))^2 + P'(x) \cdot P'''(x) = 0$, is

(a) 5 (b) 7 (c) 6 (d) 4

Sol. (Ex. Nos. 26 to 27) Here, $P'(x) = 0$ at $x = \left\{-2, -1, 0, \frac{1}{2}\right\}$

$\Rightarrow P''(x)$ will have real roots $\in (-2, -1), (-1, 0), \left(0, \frac{1}{2}\right)$.

$\Rightarrow [\alpha]+[\beta]+[\gamma] = -2-1+0 = -3$

Also, let $f(x) = (P''(x))^2 + P'(x) \cdot P'''(x)$

$\therefore \quad f(x) = \frac{d}{dx}(P'(x) \cdot P''(x))$

Since, $P'(x)$ has 4 real roots.

$\Rightarrow$ $P''(x)$ has 3 real roots.

$\Rightarrow$ $f(x)$ has 6 real roots.

26. (b) **27.** (c)

Passage II

(Ex. Nos. 28 to 30)

If α, β *(where* $\alpha < \beta$*) are the points of discontinuity of the function* $g(x) = f(f(f(x)))$, *where* $f(x) = \frac{1}{1-x}$ *and* $P(a, a^2)$ *is any point on XY-plane. Then,*

28. The points of discontinuity of $g(x)$ is

(a) $x = 0, -1$ (b) $x = 1$ only (c) $x = 0$ only (d) $x = 0, 1$

Sol. (*d*) Here, $f(x) = \frac{1}{1-x}$

$\Rightarrow \quad x \neq 1$

$$f(f(x)) = \frac{1}{1-f(x)} = \frac{1}{1-\frac{1}{1-x}} = \frac{x-1}{x}$$

$\Rightarrow \quad x \neq 0, 1$

and $f(f(f(x))) = \frac{f(x)-1}{f(x)} = \frac{\frac{1}{1-x}-1}{\frac{1}{x-1}} = x$, where $x \neq 0, 1$.

Here, point of discontinuity are $x = 0, 1$.

29. The domain of $f(g(x))$, is

(a) $x \in R$
(b) $x \in R - \{1\}$
(c) $x \in R - \{0, 1\}$
(d) $x \in R - \{0, 1, -1\}$

Sol. (*c*) Here, $g(x) = f(f(f(x))) = x, \ x \in R - \{0, 1\}$

$$f(g(x)) = f(x) = \frac{1}{1-x}$$

$\Rightarrow \quad x \neq 0, 1$

$\therefore$ Domain of $f(g(x)) \in R - \{0, 1\}$

30. If point $P(a, a^2)$ lies on the same side as that of (α, β) with respect to line $x + 2y - 3 = 0$, then

(a) $a \in \left(-\frac{3}{2}, 1\right)$ (b) $a \in R$

(c) $a \in \left(-\frac{3}{2}, 0\right)$ (d) $a \in (0, 1)$

Sol. (*a*) From graph, $a \in \left(-\frac{3}{2}, 1\right)$, as $(\alpha, \beta) = (0, 1)$.

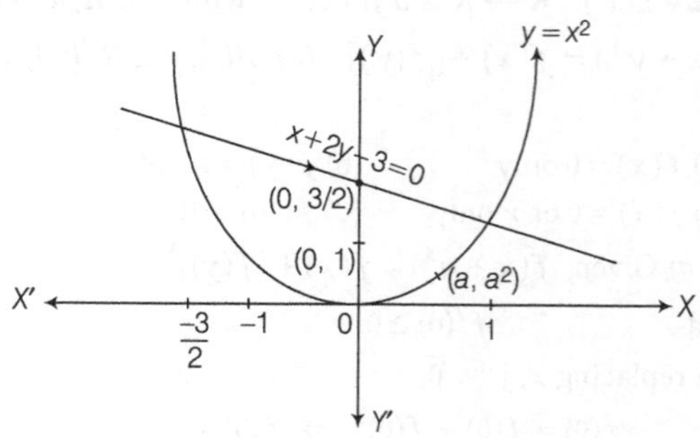

JEE Type Solved Examples :
Matching Type Questions

Ex. 31 *Let* $f(x)=\begin{cases}[x], & -2\le x<0\\ |x|, & 0\le x\le 2\end{cases}$

(where $[\cdot]$ *denotes the greatest integer function)* $g(x)=\sec x, x\in R-(2n+1)\pi/2$.
Match the following statements in Column I with their values in Column II in the interval $\left(-\frac{3\pi}{2},\frac{3\pi}{2}\right)$.

	Column I		Column II
(A)	Limit of *fog* exists at	(p)	-1
(B)	Limit of *gof* doesn't exist at	(q)	π
(C)	Points of discontinuity of *fog* is/are	(r)	$\frac{5\pi}{6}$
(D)	Points of differentiability of *fog* is/are	(s)	$-\pi$

Sol. (A) → (p, q, s), (B) → (p), (C) → (q, s), (D) → (p, r)

$$f(x)=\begin{cases}[x], & -2\le x<0\\ |x|, & 0\le x\le 2\end{cases}\Rightarrow f(x)=\begin{cases}-2, & -2\le x<-1\\ -1, & -1\le x<0\\ x, & 0\le x\le 2\end{cases}$$

$$g(x)=\sec x,\ x\in R-(2n+1)\frac{\pi}{2}$$

$$\Rightarrow\quad fog(x)=\begin{cases}-2, & -2\le \sec x<-1\\ -1, & -1\le \sec x<0\\ \sec x, & 0\le \sec x\le 2\end{cases}$$

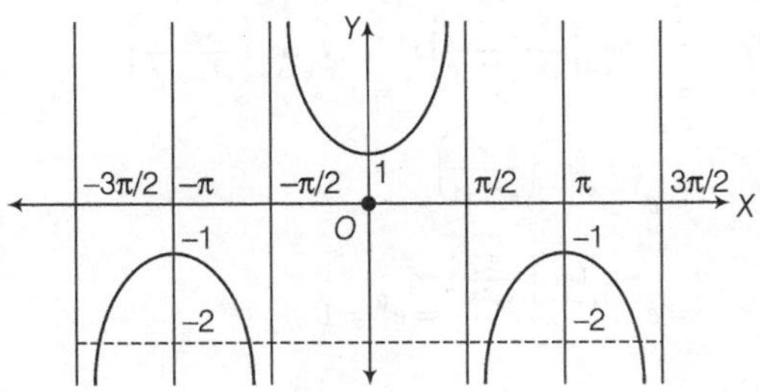

$$\therefore\quad fog=\begin{cases}-2, & x\in\left[-\frac{4\pi}{3},-\frac{2\pi}{3}\right]\cup\left[\frac{2\pi}{3},\frac{4\pi}{3}\right]-\{-\pi,\pi\}\\ -1, & x=-\pi,\pi\\ \sec x, & x\in\left[-\frac{\pi}{3},\frac{\pi}{3}\right]\end{cases}$$

Limit of *fog* exist at $x=-\pi,\pi,-1$ points of discontinuity of *fog* are $-\pi$, π and points of differentiability of *fog* are -1, $\frac{5\pi}{6}$.

$$gof=\begin{cases}\sec(-2), & x\in[-2,-1)-\left\{-\frac{\pi}{2}\right\}\\ \sec(-1), & x\in[-1,0)\\ \sec x, & x\in[0,2]-\left\{\frac{\pi}{2}\right\}\end{cases}$$

Limit of *gof* doesn't exist at $x=-1$.

Ex. 32 *Match the functions in Column I with the properties Column II.*

	Column I		Column II
(A)	$g:R\to Q$ (Rational number), $f:R\to Q$ (Rational number); f and g are continuous functions such that $\sqrt{3}\,f(x)+g(x)=3$, then $(1-f(x))^3+(g(x)-3)^3$ is	(p)	1
(B)	If $f(x)$, $g(x)$ and $h(x)$ are continuous and positive functions such that $f(x)+g(x)+h(x)=\sqrt{f(x)g(x)}+\sqrt{g(x)h(x)}+\sqrt{h(x)f(x)}$, then $f(x)+g(x)-2h(x)$ is	(q)	0
(C)	$y=f(x)$ satisfies the equation $y^3-2y^2(x+1)+4xy+(x^2-1)(y-2)=0$, then $y'(1)+y(1)$ would be equal to	(r)	2
(D)	If $y=f(x)$ satisfies $(xf(x))^{99}+(xf(x))^{98}+\ldots+(xf(x))+1=0$, then $(1+f(1))$ is	(s)	3
		(t)	-1

Sol. (A) → (p), (B) → (q), (C) → (p, r,s), (D) → (q)

(A) On comparing, $f(x)=0$, $g(x)=3$

$$\therefore\quad (1-f(x))^3+(g(x)-3)^3=1$$

(B) Here, $f(x)+g(x)+h(x)$

$$=\sqrt{f(x)\cdot g(x)}+\sqrt{g(x)\cdot h(x)}+\sqrt{h(x)\cdot f(x)}$$

$$\Rightarrow\quad \frac{1}{2}\{(\sqrt{f(x)}-\sqrt{g(x)})^2+(\sqrt{g(x)}-\sqrt{h(x)})^2+(\sqrt{h(x)}-\sqrt{f(x)})^2\}=0$$

$$\Rightarrow\quad f(x)=g(x)=h(x)$$

$$\therefore\quad f(x)+g(x)-2h(x)=0$$

(C) $y^3 - 2y^2(x+1) + 4xy + (x^2-1)(y-2) = 0$

Put $y = 2 \Rightarrow 8 - 8(x+1) + 8x = 0$

$\therefore$ $y = 2$ is solution.

Put $y = (x+1)$

$\Rightarrow (x+1)^3 - 2(x+1)^3 + 4x(x+1) + (x^2-1)(x-1)$

$= -(x+1)^3 + 4x(x+1) + (x+1)(x-1)^2$

$= -(x+1)^3 + (x+1)\{4x + (x-1)^2\}$

$= -(x+1)^3 + (x+1)^3 = 0$

$\therefore$ $y = (x+1)$ is the solution.

Similarly, $y = (x-1)$ is the solution

$\Rightarrow \quad y = 2, x+1, x-1$

$\therefore \quad \dfrac{dy}{dx} = 0, 1$

$$\Rightarrow \begin{cases} y'(1) + y(1) = 0 + 2 = 2, \text{ when } y = 2 \\ y'(1) + y(1) = 1 + 2 = 3, \text{ when } y = x+1 \\ y'(1) + y(1) = 1 + 0 = 1, \text{ when } y = x-1 \end{cases}$$

(D) $\dfrac{(xf(x))^{100} - 1}{xf(x) - 1} = 0 \Rightarrow xf(x) = -1$

As, $\quad xf(x) \neq 1$

$\therefore \quad xf(x) = -1$

$\Rightarrow \quad f(1) = -1$ or $1 + f(1) = 0$

Ex. 33 *Suppose a function $f(x)$ satisfies the following conditions*

$f(x+y) = \dfrac{f(x) + f(y)}{1 + f(x)f(y)}, \forall\, x, y$ *and* $f'(0) = 1$. *Also,*

$-1 < f(x) < 1, \forall\, x \in R$.

Match the entries of the following two columns.

	Column I		Column II
(A)	$f(x)$ is differentiable over the set	(p)	$R - (-1, 0, 1)$
(B)	$f(x)$ increases in the interval	(q)	R
(C)	Number of the solutions of $f(x) = 0$ is	(r)	0
(D)	The value of the limit $\lim\limits_{x\to\infty} [f(x)]^x$ is	(s)	1

Sol. (A) $\to$ (q), (B) $\to$ (q), (C) $\to$ (s), D $\to$ (s)

Put $\quad x = y = 0 \Rightarrow f(0) = 0$

Now, $f'(x) = \lim\limits_{h\to 0} \dfrac{f(x+h) - f(x)}{h}$

$$= \lim_{h\to 0} \frac{\dfrac{f(x) + f(h)}{1 + f(x)f(h)} - f(x)}{h}$$

$$= \lim_{h\to 0} \frac{f(h)[1 - \{f(x)\}^2]}{h[1 + f(x)f(h)]}$$

$$= \lim_{h\to 0} \left\{\frac{f(h) - f(0)}{h - 0}\right\}\left[\frac{1 - \{f(x)\}^2}{1 + f(x)f(h)}\right]$$

$$= f'(0)[1 - \{f(x)\}^2] = 1 - \{f^2(x)\}$$

$\therefore \quad f'(x) = 1 - \{f^2(x)\} \quad$ ...(i)

On integrating both sides, we get

$$\frac{1}{2}\ln\left[\frac{1 + f(x)}{1 - f(x)}\right] = x + c$$

or $\quad \dfrac{1 + f(x)}{1 - f(x)} = ke^{2x}$

Now, $\quad f(0) = 0 \Rightarrow k = 1$

$\therefore \quad f(x) = \dfrac{e^{2x} - 1}{e^{2x} + 1} = \dfrac{e^x - e^{-x}}{e^x + e^{-x}}$

Clearly, $f(x)$ is differentiable for all $x \in R$ and from Eq. (i), $f'(x) > 0$ for all $x \in R$. Again, $f(x)$ is an odd function, $f(x) = 0 \Rightarrow x = 0$.

Now, $\lim\limits_{x\to\infty} [f(x)]^x = \lim\limits_{x\to\infty} \left(\dfrac{e^x - e^{-x}}{e^x + e^{-x}}\right)^x$

$$= e^{\lim\limits_{x\to\infty}\left(\frac{e^x - e^{-x}}{e^x + e^{-x}} - 1\right)x} = e^{-2\lim\limits_{x\to\infty}\left(\frac{xe^{-x}}{e^x + e^{-x}}\right)}$$

$$= e^{-2\lim\limits_{x\to\infty}\left(\frac{x}{e^{2x} + 1}\right)}$$

$$= e^{-2\lim\limits_{x\to\infty}\left(\frac{1}{2e^{2x}}\right)} = e^0 = 1$$

JEE Type Solved Examples : Single Integer Answer Type Questions

Ex. 34 *Let* $f(x)=\begin{cases}\dfrac{1-\cos 4x}{x^2}, & x<0\\ a, & x=0\\ \dfrac{\sqrt{x}}{\sqrt{16+\sqrt{x}}-4}, & x>0\end{cases}$

Then, the value of a if possible, so that the function is continuous at $x=0$*, is ..... .*

Sol. (8) As, $f(x)$ is continuous at $x=0$.

$\therefore$ We must have,

$$\text{RHL (at } x=0) = \text{LHL (at } x=0) = f(0)$$

$$\text{RHL (at } x=0) = \lim_{x\to 0^+} f(x) = \lim_{x\to 0^+}\frac{\sqrt{x}}{\sqrt{16+\sqrt{x}}-4}$$

$$= \lim_{h\to 0}\frac{\sqrt{0+h}}{\sqrt{16+\sqrt{0+h}}-4}\times\frac{\sqrt{16+\sqrt{h}}+4}{\sqrt{16+\sqrt{h}}+4} \quad [\text{putting } x=0+h]$$

$$= \lim_{h\to 0}\frac{\sqrt{h}\{\sqrt{16+\sqrt{h}}+4\}}{16+\sqrt{h}-16}$$

$$= \lim_{h\to 0}\{\sqrt{16+\sqrt{h}}+4\} = 8$$

$$\text{Also, LHL (at } x=0) = \lim_{x\to 0^-} f(x) = \lim_{x\to 0^-}\frac{1-\cos 4x}{x^2}$$

$$= \lim_{h\to 0}\frac{1-\cos 4(0-h)}{(0-h)^2} \quad [\text{putting } x=0-h]$$

$$= \lim_{h\to 0}\frac{1-\cos 4h}{h^2}$$

$$= \lim_{h\to 0}\frac{2\sin^2 2h}{h^2}$$

$$= \lim_{h\to 0} 8\left(\frac{\sin 2h}{2h}\right)^2 = 8$$

And $f(0)=a$

Since, $f(x)$ is continuous at $x=0$

$\Rightarrow \quad f(0) = \text{RHL} = \text{LHL}$

or $\quad f(0)=8$ or $a=8$

Ex. 35 *Let* $f(x) = \text{maximum}\{4, 1+x^2, x^2-1\}, \forall x\in R$*. Then, the total number of points, where* $f(x)$ *is not differentiable, ...*

Sol. (2) We have, discussed in the last chapter for sketching maximum $\{4, 1+x^2, x^2-1\}$ as

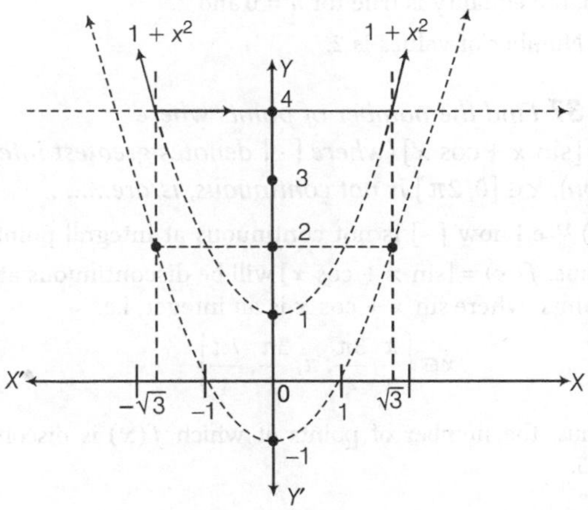

Or

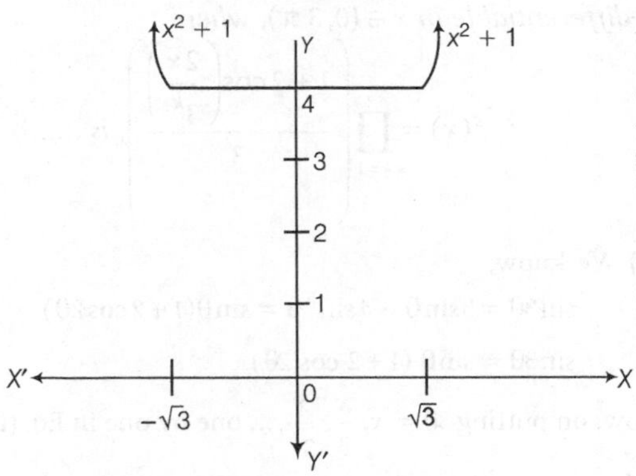

Thus, from above graph, we can simply say $f(x)$ is not differentiable at $x=\pm\sqrt{3}$.

$\therefore$ Not differentiable at 2 points.

Ex. 36 *Let* $f(x)=x^n$*, n being non-negative integer. Then, the number of value of n for which the equality* $f'(a+b)=f'(a)+f'(b)$ *is valid for all* $a, b>0$*, is*

Sol. (2) Since, $f(x)=x^n$, n being non-negative integer.

Then, $\quad f'(x)=nx^{n-1}$

$$f'(a)=na^{n-1},\ f'(b)=nb^{n-1},$$

$$f'(a+b)=n(a+b)^{n-1}$$

Now, the equality $f'(a+b)=f'(a)+f'(b)$ holds, if

$$n(a+b)^{n-1}=na^{n-1}+nb^{n-1}$$

or $$(a+b)^{n-1}=a^{n-1}+b^{n-1} \quad \text{...(i)}$$

Above statement is true, only if $(n-1)=1 \Rightarrow n=2$

i.e. $$(a+b)^{n-1}=a^{n-1}+b^{n-1} \quad [\text{if } n=2]$$

or $\quad (a+b)^1 = a^1 + b^1$

Also, when $n = 1$, then LHS < RHS

and when $n = 3, 4, 5, \ldots$, then LHS > RHS

Again, when $n = 0$, $f'(x) = 0$ for all x.

So, the equality is true for $n = 0$ and 2.

$\therefore$ Number of values is 2.

Ex. 37 *Find the number of points where $f(x) = [\sin x + \cos x]$ (where $[\cdot]$ denotes greatest integral function), $x \in [0, 2\pi]$ is not continuous, is/are…… .*

Sol. (5) We know $[\cdot]$ is not continuous at integral points.

Thus, $f(x) = [\sin x + \cos x]$ will be discontinuous at those points, where $\sin x + \cos x$ is an integer, i.e.

$$x \in \left\{\frac{\pi}{2}, \frac{3\pi}{4}, \pi, \frac{3\pi}{2}, \frac{7\pi}{4}\right\}$$

Thus, the number of points at which $f(x)$ is discontinuous is 5.

Ex. 38 *The number of points where $|xf(x)| + ||x-2|-1||$ is non-differentiable in $x \in (0, 3\pi)$, where*

$$f(x) = \prod_{k=1}^{\infty}\left(\frac{1+2\cos\left(\frac{2x}{3^k}\right)}{3}\right), \text{ is …… .}$$

Sol. (5) We know,

$$\sin 3\theta = 3\sin\theta - 4\sin^3\theta = \sin\theta(1+2\cos 2\theta)$$

$$\therefore \quad \sin 3\theta = \sin\theta\,(1+2\cos 2\theta) \quad \ldots\text{(i)}$$

Now, on putting $3\theta = x, \frac{x}{3}, \frac{x}{3^2}, \ldots$ one by one in Eq. (i), we get

$$\Rightarrow \quad \sin x = \sin\left(\frac{x}{3}\right)\cdot\left(1+2\cos\frac{2x}{3}\right)$$

and $$\sin\frac{x}{3} = \sin\left(\frac{x}{3^2}\right)\cdot\left(1+2\cos\frac{2x}{3^2}\right)$$

$$\therefore \quad \sin\frac{x}{3^2} = \sin\left(\frac{x}{3^3}\right)\cdot\left(1+2\cos\frac{2x}{3^3}\right) \ldots \text{ and so on}$$

$$\Rightarrow \quad \sin x = \sin\left(\frac{x}{3^n}\right)\cdot\prod_{k=1}^{n}\left(1+2\cos\frac{2x}{3^k}\right)$$

$$\Rightarrow \quad \frac{\sin x}{x} = \lim_{n\to\infty}\frac{\sin x/3^n}{x/3^n}\prod_{k=1}^{n}\frac{\left(1+2\cos\left(\frac{2x}{3^k}\right)\right)}{3^n}$$

$$\Rightarrow \quad \frac{\sin x}{x} = \prod_{k=1}^{\infty}\left(\frac{1+2\cos\left(\frac{2x}{3^k}\right)}{3}\right) = f(x) \quad \ldots\text{(ii)}$$

Thus, $|xf(x)| + ||x-2|-1| = \frac{|x||\sin x|}{|x|} + ||x-2|-1|$ is not differentiable at $\{\pi, 2\pi, 0, 1, 3\}$.

$\therefore$ Number of points are 5.

Ex. 39 *If $f\left(\frac{xy}{2}\right) = \frac{f(x)\cdot f(y)}{2}$; $x, y \in R$, $f(1) = f'(1)$.*

Then, $\frac{f(3)}{f'(3)}$ is …… .

Sol. (3) Here, $f'(x) = \lim_{h\to 0}\frac{f(x+h)-f(x)}{h}$

$$= \lim_{h\to 0}\frac{f\left(\frac{2x+2h}{2}\right)-f(x)}{h}$$

$$= \lim_{h\to 0}\frac{f\left(\frac{2x(1+h/x)}{2}\right)-f\left(\frac{2x\cdot 1}{2}\right)}{h}$$

$$= \lim_{h\to 0}\frac{\frac{f(2x)\cdot f(1+h/x)}{2}-\frac{f(2x)\cdot f(1)}{2}}{h}$$

$$= \frac{f(2x)}{2x}\cdot\lim_{h\to 0}\frac{f(1+h/x)-f(1)}{h/x}$$

$$= \frac{f(2x)}{2x}\cdot f'(1)$$

$$= \frac{f(2x)}{2x}\cdot f(1), \text{ given } f(1) = f'(1)$$

$$= \frac{f\left(\frac{2x\cdot 1}{2}\right)}{x}, \text{ using } f\left(\frac{xy}{2}\right) = \frac{f(x)\cdot f(y)}{2}$$

$$\therefore \quad f'(x) = \frac{f(x)}{x}$$

$$\Rightarrow \quad \frac{f(x)}{f'(x)} = x \Rightarrow \frac{f(3)}{f'(3)} = 3$$

Ex. 40 *Let $f: R \to R$ be a differentiable function satisfying $f(x) = f(y)\,f(x-y), \forall x, y \in R$ and $f'(0) = \int_0^4 \{2x\}\,dx$, where $\{\cdot\}$ denotes the fractional part function and $f'(-3) = \alpha e^{\beta}$. Then, $|\alpha+\beta|$ is equal to ……… .*

Sol. (4) Given, $f(x) = f(y)\,f(x-y)$

On replacing x by $(x+y)$,

$$f(x+y) = f(y)\cdot f(x)$$

$$\Rightarrow \quad f(x) = e^{kx} \Rightarrow f'(x) = ke^{kx}$$

But, $$f'(0) = \int_0^4 \{2x\}dx = 2$$

$$\Rightarrow \quad f'(0) = k = 2 \Rightarrow f'(x) = 2e^{2x}$$

$$\Rightarrow \quad f'(-3) = 2e^{-6} \Rightarrow |\alpha+\beta| = 4$$

Ex. 41 *Let $f(x)$ is a polynomial function and $(f(\alpha))^2+(f'(\alpha))^2=0$, then find $\lim_{x\to\alpha}\frac{f(x)}{f'(x)}\left[\frac{f'(x)}{f(x)}\right]$, where $[\cdot]$ denotes greatest integer function, is ....*

Sol. (1) Here, $(f(\alpha))^2+(f'(\alpha))^2=0 \Rightarrow f(\alpha)=f'(\alpha)=0$

$\therefore$ α is repeated root of $f(x)$.

Now, $\lim_{x\to\alpha}\frac{f(x)}{f'(x)}\left[\frac{f'(x)}{f(x)}\right]=\lim_{x\to\alpha}\frac{f(x)}{f'(x)}\left(\frac{f'(x)}{f(x)}-\left\{\frac{f'(x)}{f(x)}\right\}\right)$

$=\lim_{x\to\alpha}\left(1-\frac{f(x)}{f'(x)}\cdot\left\{\frac{f'(x)}{f(x)}\right\}\right)$

$\left[\text{as, } x=\alpha \text{ is the repeated root and } \left\{\frac{f(x)}{f'(x)}\right\} \text{ is bounded; } \therefore \lim_{x\to\alpha}\frac{f(x)}{f'(x)}=0\right]$

$=1-0=1$

Ex. 42 *Let $f:R\to R$ is a function satisfying $f(10-x)=f(x)$ and $f(2-x)=f(2+x), \forall\, x\in R$. If $f(0)=101$. Then, the minimum possible number of values of x satisfying $f(x)=101, x\in[0,25]$ is ....*

Sol. (9) Since, $f(10-x)=f(x)=f(4-x)$

$\Rightarrow \quad f(10-x)=f(4-x)$

Say, $\quad 4-x=t \Rightarrow f(6+t)=f(t)$

$\Rightarrow f(x)$ is periodic function with period 6.

So, for $x\in[0,25]$

$f(x)=101$ at $x=0, 6, 12, 18, 24$

Total numbers = 5

Since, $\quad f(2-x)=f(2+x)$

$\Rightarrow f(x)$ is symmetric about $x=2$ line.

Due to symmetry in one period length.

$f(x)=101$ has one solution at $x=4$ other than 0 and 6.

Now, $\quad f(x)=101$ at $x=4, 10, 16, 22$

Total numbers = 4

Hence, atleast minimum possible number of values of $x=9$.

Ex. 43 *If $f(x)$ is a differentiable function for all $x\in R$ such that $f(x)$ has fundamental period 2. $f(x)=0$ has exactly two solutions in $[0,2]$, also $f(0)\neq 0$. If minimum number of zeros of $h(x)=f'(x)\cos x-f(x)\sin x$ in $(0,99)$ is $120+k$, then k is ....*

Sol. (7) $h(x)=\frac{d}{dx}(f(x)\cdot\cos x)$

First find the minimum number of zeros of $(f(x)\cdot\cos x)=0$.

$f(x)=0$ has minimum 98 roots in $[0,99)$ and $\cos x=0$ has 31 roots is $[0,99)$.

Maximum common possible root is only 1.

Hence, minimum number of roots of $f(x)\cos x=0$ is 128.

Thus, $\frac{d}{dx}(f(x)\cos x)=0$ has minimum 127 roots.

$$127=120+7$$

Hence, value of k is 7.

Subjective Type Questions

Ex. 44 *Let $f(x)=\max\{2\sin x, 1-\cos x\}, \forall\, x\in(0,\pi)$, then discuss differentiability of $f(x)$ in $(0,\pi)$.*

Sol. We know that, $f(x)=$ maximum $\{2\sin x, 1-\cos x\}$ can be plotted as

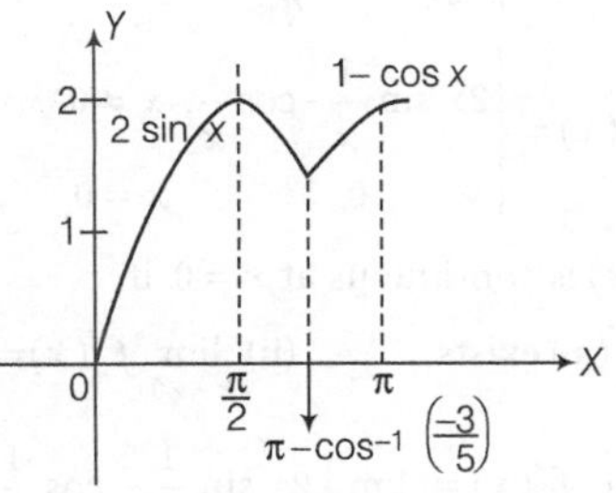

Thus, $f(x)=$ maximum $\{2\sin x, 1-\cos x\}$ is not differentiable.

When, $\quad 2\sin x=1-\cos x$

or $\quad 4\sin^2 x=(1-\cos x)^2$

or $\quad 4(1+\cos x)=(1-\cos x)$

or $\quad 4+4\cos x=1-\cos x$

or $\quad \cos x=-3/5$

$\Rightarrow \quad x=\cos^{-1}(-3/5)$

$\therefore f(x)$ is not differentiable at

$x=\pi-\cos^{-1}(3/5)$.

Ex. 45 *Discuss the continuity of the function $g(x)=[x]+[-x]$ at integral values of x.*

Sol. Here, x can assume two values

(a) integers (b) non-integers

(a) If x is an integer

$[x]=x$ and $[-x]=-x$

$\Rightarrow \quad g(x)=x-x=0$

(b) If x is not an integer

Let $x=n+f$, where n is an integer and $f\in(0,1)$

$\Rightarrow \quad [x] = [n + f] = n$

$\Rightarrow [-x] = [-n - f] = [(-n-1) + (1-f)] = (-n-1)$

$[\because 0 < f < 1 \Rightarrow (1-f) < 1]$

Hence, $g(x) = [x] + [-x] = n + (-n-1) = -1$

$\therefore \quad g(x) = \begin{cases} 0, & \text{if } x \text{ is an integer} \\ -1, & \text{if } x \text{ is not an integer} \end{cases}$

Let us discuss the continuity of $g(x)$ at a point $x = a$ [where $a \in$ integer]

$\text{LHL} = \lim_{x \to a^-} g(x) = -1$

$[\because$ as $x \to a^-$, x is not an integer$]$

$\text{RHL} = \lim_{x \to a^+} g(x) = -1$

$[\because$ as $x \to a^+$, x is not an integer$]$

But $g(a) = 0$ because a is an integer.
Hence, $g(x)$ has a removable discontinuity at integral values of x.

Ex. 46 *If $f(x) = \dfrac{\sin 2x + A\sin x + B\cos x}{x^3}$ is continuous at $x = 0$. Find the values of A and B. Also, find $f(0)$.*

Sol. As, $f(x)$ is continuous at $x = 0$.

$f(0) = \lim_{x \to 0} f(x)$ and both $f(0)$ and $\lim_{x \to 0} f(x)$ are finite.

$$\Rightarrow \quad f(0) = \lim_{x \to 0} \frac{\sin 2x + A\sin x + B\cos x}{x^3}$$

As, denominator $\to 0$ as $x \to 0$,
Numerator should also $\to 0$ as $x \to 0$
Which is possible only if (for $f(0)$ to be finite).

$\Rightarrow \quad \sin 2(0) + A\sin(0) + B\cos(0) = 0$

$\Rightarrow \quad B = 0$

$$\therefore \quad f(0) = \lim_{x \to 0} \frac{\sin 2x + A\sin x}{x^3}$$

$$\Rightarrow \quad f(0) = \lim_{x \to 0} \left(\frac{\sin x}{x}\right)\left(\frac{2\cos x + A}{x^2}\right)$$

$$= \lim_{x \to 0} \left(\frac{2\cos x + A}{x^2}\right)$$

Again, we can see that denominator $\to 0$ as $x \to 0$.

$\therefore$ Numerator should also approach to 0 as $x \to 0$

$[\because$ for $f(0)$ to be finite$]$

$\Rightarrow \quad 2 + A = 0$

$\Rightarrow \quad A = -2$

$$\Rightarrow \quad f(0) = \lim_{x \to 0} \left(\frac{2\cos x - 2}{x^2}\right) = \lim_{x \to 0} \left(\frac{-4\sin^2 x/2}{x^2}\right)$$

$$= \lim_{x \to 0} -\left(\frac{\sin^2 x/2}{x^2/4}\right) = -1$$

So, we get $\quad A = -2, B = 0$

and $\quad f(0) = -1$

Ex. 47 *Let $f : R \to R$ satisfying $|f(x)| \le x^2, \forall x \in R$, then show that $f(x)$ is differentiable at $x = 0$.*

Sol. Since, $\quad |f(x)| \le x^2, \forall x \in R$

$\therefore$ At $x = 0, |f(0)| \le 0$

$\Rightarrow \quad f(0) = 0 \quad$ ...(i)

$$\therefore \quad f'(0) = \lim_{h \to 0} \frac{f(h) - f(0)}{h} = \lim_{h \to 0} \frac{f(h)}{h} \quad \text{...(ii)}$$

$[f(0) = 0$, from Eq. (i)$]$

Now, $\quad \left|\dfrac{f(h)}{h}\right| \le |h|$

$\Rightarrow \quad -|h| \le \dfrac{f(h)}{h} \le |h|$

$\Rightarrow \quad \lim_{h \to 0} \dfrac{f(h)}{h} \to 0$

[using Cauchy-Squeeze theorem] ...(iii)

$\therefore$ From Eqs. (ii) and (iii), we get $f'(0) = 0$

i.e. $f(x)$ is differentiable at $x = 0$

Ex. 48 *Show that the function defined by*

$$f(x) = \begin{cases} x^2 \sin 1/x, & x \ne 0 \\ 0, & x = 0 \end{cases}$$

is differentiable for every value of x, but the derivative is not continuous for $x = 0$.

Sol. For $x \ne 0$,

$$f'(x) = 2x\sin(1/x) + x^2\left(-\frac{1}{x^2}\right)\cos\left(\frac{1}{x}\right)$$

$$f'(x) = 2x\sin\frac{1}{x} - \cos\frac{1}{x}$$

For $x = 0$,

$$f'(x) = \lim_{h \to 0} \frac{f(0+h) - f(0)}{h}$$

$$= \lim_{h \to 0} \frac{h^2 \sin\frac{1}{h} - 0}{h}$$

$$\Rightarrow \quad f'(x) = \lim_{h \to 0} h\sin\frac{1}{h} = 0$$

Thus, $\quad f'(x) = \begin{cases} 2x\sin\frac{1}{x} - \cos\frac{1}{x}, & x \ne 0 \\ 0, & x = 0 \end{cases}$

Now, $f'(x)$ is continuous at $x = 0$, if

(i) $\lim_{x \to 0} f'(x)$ exists $\quad$ (ii) $\lim_{x \to 0} f'(x) = f'(0)$

Again, $\lim_{x \to 0} f'(x) = \lim_{x \to 0}\left(2x\sin\frac{1}{x} - \cos\frac{1}{x}\right)$, doesn't exist

Since, $\lim_{x \to 0} \cos\frac{1}{x}$ doesn't exist.

Hence, $f'(x)$ is not continuous at $x = 0$.

Ex. 49 *If* $f(x)=\begin{cases} x-[x], & x\notin I \\ 1, & x\in I\end{cases}$

where I is an integer and $[\cdot]$ represents the greatest integer function and

$$g(x)=\lim_{n\to\infty}\frac{\{f(x)\}^{2n}-1}{\{f(x)\}^{2n}+1}, \text{ then}$$

(a) Draw graphs of $f(2x)$, $g(x)$ and $g\{g(x)\}$.

(b) Find the domain and range of these functions.

(c) Are these functions periodic? If yes, find their periods.

Sol. Here, $f(x)=\begin{cases} x-[x], & x\notin I \\ 1, & x\in I\end{cases}$

(a) **Graph of $f(2x)$**

As, $f(2x)=\begin{cases} 2x-[2x], & 2x\notin I \\ 1, & 2x\in I\end{cases}$

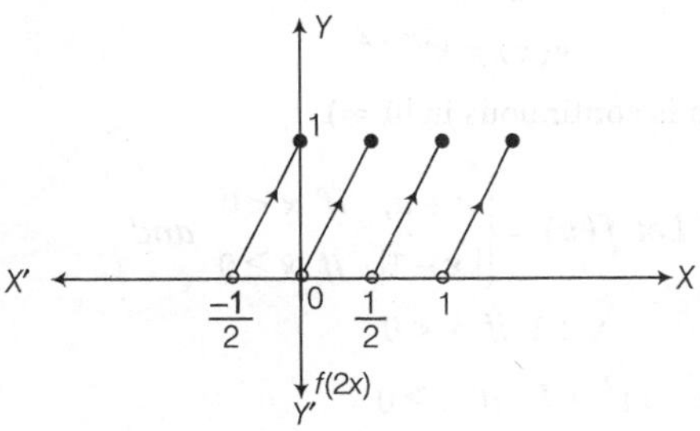

Graph of $g(x)$

Here, $g(x)=\lim_{n\to\infty}\frac{\{f(x)\}^{2n}-1}{\{f(x)\}^{2n}+1}$

$$g(x)=\begin{cases} 0, & x\in I \\ -1, & x\notin I, \text{ as } \lim_{n\to\infty}\{f(x)\}^{2n}=0\end{cases}$$

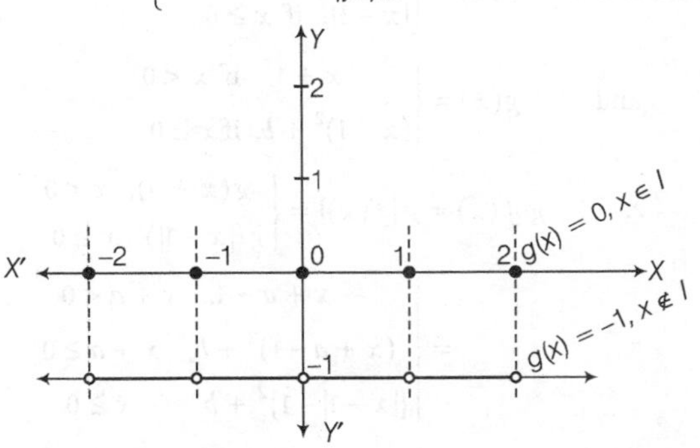

Graph of $g\{g(x)\}$

We have, $g(x)=\begin{cases} 0, & x\in I \\ -1, & x\notin I\end{cases}$

$\Rightarrow \quad g\{g(x)\}=\begin{cases} 0, & g(x)\in I \\ -1, & g(x)\notin I\end{cases}$

where $\quad g(x)\in\{0,-1\}$

and thus $g(x)\notin I$, should be neglected.

$\Rightarrow \quad g\{g(x)\}=0,\ x\in R$ and could be plotted as,

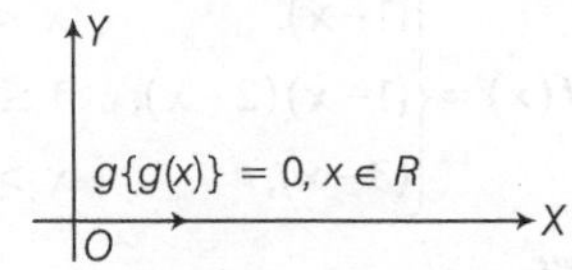

(b) From the above three graphs, we have

Domain of $f(2x)\in R$

and Range of $f(2x)\in(0,1]$

Domain of $g(x)\in R$

and Range of $g(x)\in\{0,-1\}$

Domain of $g(g(x))\in R$

and Range of $g(g(x))\in\{0\}$

(c) Here, $f(2x)$ and $g(x)$ are periodic with period 1/2 and 1. Also, $g\{g(x)\}$ is constant function. Therefore, periodic and its period is undetermined.

Ex. 50 *Prove that* $f(x)=[\tan x]+\sqrt{\tan x-[\tan x]}$. *(where $[\cdot]$ denotes greatest integer function) is continuous in* $\left[0,\frac{\pi}{2}\right)$.

Sol. Here, $\quad f(x)=[\tan x]+\sqrt{\tan x-[\tan x]}$

or $\quad g(x)=[x]+\sqrt{x-[x]}$, where $x=\tan x\ge 0$

Then, for $a\in N$. We discuss continuity of $f(x)$ as

LHL (at $x=a$)

$$=\lim_{x\to a^-}g(x)=\lim_{h\to 0}([a-h]+\sqrt{a-h-[a-h]})$$

$$=\lim_{h\to 0}((a-1)+\sqrt{a-h-a+1})=a-1+1=a$$

$$\left\{\begin{array}{l}\text{where, } a-1<a-h<a \\ \therefore [a-h]=a-1\end{array}\right.$$

Now, RHL (at $x=a$)

$$=\lim_{x\to a^+}g(x)=\lim_{h\to 0}([a+h]+\sqrt{a+h-[a+h]})$$

$$=\lim_{h\to 0}(a+\sqrt{a+h-a})=a$$

$$[\because a<a+h<a+1 \Rightarrow [a+h]=a]$$

and $\quad g(a)=[a]+\sqrt{a-[a]}=a$

as $\quad a\in N$

So, $g(x)$ is continuous, $\forall\ a\in N$, now $g(x)$ is clearly continuous in $(a-1,a)$, $\forall\ a\in N$.

Hence, $g(x)$ is continuous in $[0,\infty)$.

Now, let $\phi(x)=\tan x$ which is continuous in $[0,\pi/2]$.

So, $g\{\phi(x)\}$ is continuous in $[0,\pi/2)$.

Hence, $f(x)=[\tan x]+\sqrt{\tan x-[\tan x]}$ is continuous in $[0,\pi/2)$.

Ex. 51 *Determine the values of x for which the following functions fails to be continuous or differentiable*

$$f(x)=\begin{cases}(1-x), & x<1\\ (1-x)(2-x), & 1\le x\le 2\\ (3-x), & x>2\end{cases}$$

Justify your answer. **[IIT JEE 1995]**

Sol. By the given definition, it is clear that the function f is continuous and differentiable at all points except possibility at $x=1$ and $x=2$.

Continuity at $x=1$

$$\text{LHL}=\lim_{x\to 1^-}f(x)=\lim_{x\to 1^-}(1-x)=\lim_{h\to 0}(1-(1-h))=0$$

$$\text{RHL}=\lim_{x\to 1^+}f(x)=\lim_{x\to 1^+}(1-x)(2-x)$$

$$=\lim_{h\to 0}\{1-(1+h)\}\{2-(1+h)\}=0$$

Also, $f(1)=0$.

Hence, LHL = RHL = $f(1)$

Therefore, $f(x)$ is continuous at $x=1$.

Differentiability at $x=1$

$$Lf'(1)=\lim_{h\to 0}\frac{f(1-h)-f(1)}{-h}=\lim_{h\to 0}\frac{1-(1-h)-0}{-h}=-1$$

$$Rf'(1)=\lim_{h\to 0}\frac{f(1+h)-f(1)}{h}$$

$$=\lim_{h\to 0}\frac{\{1-(1+h)\}\{2-(1+h)\}-0}{h}=-1$$

Since, $L\{f'(1)\}=R\{f'(1)\}$ so, we get $f(x)$ is differentiable at $x=1$.

Continuity at $x=2$

$$\text{LHL}=\lim_{x\to 2^-}f(x)=\lim_{x\to 2^-}(1-x)(2-x)$$

$$=\lim_{h\to 0}(1-(2-h))(2-(2-h))=0$$

$$\text{RHL}=\lim_{x\to 2^+}f(x)=\lim_{x\to 2^+}(3-x)=\lim_{h\to 0}3-(2+h)=1$$

Since, RHL $\ne$ LHL

Therefore, $f(x)$ is not continuous at $x=2$.

As such $f(x)$ cannot be differentiable at $x=2$.

Hence, $f(x)$ is continuous and differentiable at all points except at $x=2$.

Ex. 52 *If $g(x)$ is continuous function in $[0,\infty)$ satisfying $g(1)=1$. If $\int_0^x 2x\cdot g^2(t)\,dt=\left(\int_0^x 2g(x-t)\,dt\right)^2$, find $g(x)$.*

Sol. Here, $\int_0^x g(t)\,dt=\int_0^x g(x-t)\,dt$, the given equation could be written as

$$x\int_0^x g^2(t)\,dt=2\left(\int_0^x g(t)\,dt\right)^2 \quad \text{...(i)}$$

On differentiating both sides w.r.t. 'x', we get

$$\int_0^x g^2(t)\,dt+xg^2(x)=4g(x)\left\{\int_0^x g(t)\,dt\right\}$$

or $$x\int_0^x g^2(t)\,dt+x^2g^2(x)=4xg(x)\int_0^x g(t)\,dt \quad \text{...(ii)}$$

Using Eq. (i),

$$2\left(\int_0^x g(t)\,dt\right)^2+x^2g^2(x)=4xg(x)\int_0^x g(t)\,dt$$

$$\Rightarrow 2\left(\int_0^x g(t)\,dt\right)^2-4xg(x)\int_0^x g(t)\,dt+x^2g^2(x)=0$$

which is quadratic in $\int_0^x g(t)\,dt$,

$$\therefore \int_0^x g(t)\,dt=\frac{2\pm\sqrt{2}}{2}xg(x)$$

On differentiating both sides w.r.t. x, we get

$$\frac{g'(x)}{g(x)}=\frac{1\pm\sqrt{2}}{x}$$

$$\Rightarrow g(x)=kx^{1\pm\sqrt{2}}, \text{ since } g(1)=1$$

$$\Rightarrow k=1$$

$$\therefore g(x)=x^{1\pm\sqrt{2}}$$

which is continuous in $[0,\infty)$.

Ex. 53 *Let $f(x)=\begin{cases}x+a, & \text{if } x<0\\ |x-1|, & \text{if } x\ge 0\end{cases}$ and $g(x)=\begin{cases}x+1, & \text{if } x<0\\ (x-1)^2+b, & \text{if } x\ge 0\end{cases}$ where a and b are non-negative real numbers. Determine the composite function gof. If $(gof)(x)$ is continuous for all real x. Determine the values of a and b. Further for these values of a and b, is gof differentiable at $x=0$. Justify your answer.* **[IIT JEE 2002]**

Sol. Here, $$f(x)=\begin{cases}x+a, & \text{if } x<0\\ |x-1|, & \text{if } x\ge 0\end{cases}$$

and $$g(x)=\begin{cases}x+1, & \text{if } x<0\\ (x-1)^2+b, & \text{if } x\ge 0\end{cases}$$

$$\therefore gof(x)=g\{f(x)\}=\begin{cases}g(x+a), & x<0\\ g(|x-1|), & x\ge 0\end{cases}$$

$$=\begin{cases}x+a+1, & x+a<0\\ (x+a-1)^2+b, & x+a\ge 0\\ \{|x-1|-1\}^2+b, & x\ge 0\end{cases}$$

$$=\begin{cases}x+a+1, & x<-a\\ (x+a-1)^2+b, & 0>x\ge -a\\ x^2+b, & 0\le x<1\\ (x-2)^2+b, & x\ge 1\end{cases}$$

$\Rightarrow gof(x)$ is continuous for all real x.

$\therefore gof(x)$ is continuous at $x=-a, 0, 1$.

Since, $gof(x)$ is continuous at $x=-a$.

$$\Rightarrow \lim_{x \to -a^-} gof(x) = \lim_{x \to -a^+} gof(x) = gof(-a)$$

$$\Rightarrow \lim_{x \to -a^-} (x + a + 1) = \lim_{x \to -a^+} (x + a - 1)^2 + b$$

$$= (-a + a - 1)^2 + b$$

$$\Rightarrow -a + a + 1 = (-a + a - 1)^2 + b = (-a + a - 1)^2 + b$$

$$\Rightarrow b = 0$$

and $gof(x)$ is continuous at $x = 0$.

$$\Rightarrow \lim_{x \to 0^-} gof(x) = \lim_{x \to 0^+} gof(x) = gof(0)$$

$$\Rightarrow (a-1)^2 + b = b \Rightarrow a = 1$$

Now, $(\text{LHD at } x = 0) = \frac{d}{dx}\{(x + a - 1)^2 + b\}_{\text{at } x = 0}$

$$= 2(a - 1) = 0 \qquad [\text{as } a = 1]$$

Again, $(\text{RHD at } x = 0) = \frac{d}{dx}\{(x^2 + b)\}_{\text{at } x = 0} = 0$

$\therefore$ $gof(x)$ is differentiable at $x = 0$.

Ex. 54 *If a function $f : [-2a, 2a] \to R$ is an odd function such that $f(x) = f(2a - x)$ for $x \in [a, 2a]$ and the left hand derivative at $x = a$ is 0, then find the left hand derivative at $x = -a$* **[IIT JEE 2003]**

Sol. It is given that, LHD (at $x = a$) = 0

$$\therefore \lim_{h \to 0} \frac{f(a-h) - f(a)}{-h} = 0$$

$$\Rightarrow \lim_{h \to 0} \frac{f(a-h) - f(a)}{h} = 0 \qquad \text{...(i)}$$

Now, LHD (at $x = -a$)

$$= \lim_{h \to 0} \frac{f(-a-h) - f(-a)}{-h}$$

$$= \lim_{h \to 0} \frac{-f(a+h) + f(a)}{-h} \qquad [\text{as } f(-x) = -f(x), \text{ given}]$$

$$= \lim_{h \to 0} \frac{f(a+h) - f(a)}{h} = \lim_{h \to 0} \frac{f[2a - (a-h)] - f(a)}{h}$$

$$= \lim_{h \to 0} \frac{f(a-h) - f(a)}{h} = 0 \qquad [\text{as } f(x) = f(2a - x)]$$

[using Eq. (i)]

$\therefore$ LHD (at $x = -a$) = 0

Ex. 55 *Discuss the continuity of $f(x)$ in $[0, 2]$, where*

$$f(x) = \begin{cases} [\cos \pi x], & x \le 1 \\ |2x - 3|[x - 2], & x > 1 \end{cases}$$

where $[\cdot]$ denotes the greatest integral function.

Sol. First we shall find the points where $f(x)$ may be discontinuous.

Consider $x \in [0, 1]$

$f(x) = [\cos \pi x]$ is discontinuous where $\cos \pi x \in I$.

In $[0, 1]$, $\cos \pi x$ is an integer at $x = 0, 1/2, 1$.

$\Rightarrow x = 0, x = \frac{1}{2}$ and $x = 1$ may be the points at which $f(x)$ may be discontinuous ...(i)

Now, consider $x \in (1, 2]$

$$f(x) = [x - 2]|2x - 3|$$

If $x \in (1, 2)$, then $[x - 2] = -1$ and if $x = 2$, then $[x - 2] = 0$

Also, $|2x - 3| = 0 \Rightarrow x = 3/2$

$\Rightarrow$ $x = 3/2$ and 2 may be the points at which $f(x)$ may be discontinuous ...(ii)

On combining Eqs. (i) and (ii), we have

$$x = 0, 1/2, 1, 3/2, 2$$

On dividing $f(x)$ by about the 5 critical points, we get

$$f(x) = \begin{cases} 1, \ x = 0 & \because \cos(\pi 0) = 1 \\ 0, \ 0 < x \le \frac{1}{2} & \because 0 \le \cos \pi x < 1 \Rightarrow [\cos \pi x] = 0 \\ -1, \frac{1}{2} < x \le 1 & \because -1 \le \cos \pi x < 0 \Rightarrow [\cos \pi x] = -1 \\ -(3 - 2x), 1 < x \le 3/2 & \because |2x - 3| = 3 - 2x \text{ and } [x - 2] = -1 \\ -(2x - 3), 3/2 < x < 2 & \because |2x - 3| = 2x - 3 \text{ and } [x - 2] = -1 \\ 0, \ x = 2 & \because [x - 2] = 0 \end{cases}$$

Checking continuity at $x = 0$

$$\text{RHL} = \lim_{x \to 0^+} f(x) = \lim_{x \to 0^+} 0 = 0$$

And $f(0) = 1$

As $\text{RHL} \ne f(0)$

Thus, $f(x)$ is discontinuous at $x = 0$

Checking continuity at $x = 1/2$

$$\text{LHL} = \lim_{x \to \frac{1}{2}^-} f(x) = \lim_{x \to \frac{1}{2}^-} 0 = 0$$

$$\text{RHL} = \lim_{x \to \frac{1}{2}^+} f(x) = \lim_{x \to \frac{1}{2}^+} (-1) = -1$$

As $\text{RHL} \ne \text{LHL}$

Therefore, $f(x)$ is discontinuous at $x = 1/2$

Checking continuity at $x = 1$

$$\text{LHL} = \lim_{x \to 1^-} f(x) = \lim_{x \to 1^-} (-1) = -1$$

$$\text{RHL} = \lim_{x \to 1^+} f(x) = \lim_{x \to 1^+} -(3 - 2x) = -1$$

And $f(1) = -1$

As $\text{RHL} = \text{LHL} = f(1)$

Therefore, $f(x)$ is continuous at $x = 1$

Checking continuity at $x = 3/2$

$$\text{LHL} = \lim_{x \to 3/2^-} (2x - 3) = 0$$

$$\text{RHL} = \lim_{x \to 3/2^+} (3 - 2x) = 0$$

And $f(3/2) = 0$

As $\text{RHL} = \text{LHL} = f(3/2)$

Hence, $f(x)$ is continuous at $x = 3/2$

Checking continuity at $x = 2$

$$\text{LHL} = \lim_{x \to 2^-} (3-2x) = -1$$

And $f(2) = 0$

As $\text{LHL} \neq f(2)$

Hence, $f(x)$ is discontinuous at $x = 2$

Thus, $f(x)$ is continuous when $x \in [0, 2] - \{0, 1/2, 2\}$.

Ex. 56 *Let f is a differentiable function such that $f(x) = x^2 + \int_0^x e^{-t} f(x-t)\, dt$. Find $f(x)$.*

Sol. We have, $f(x) = x^2 + \int_0^x e^{-t} f(x-t)\, dt$

$$\Rightarrow \quad f(x) = x^2 + \int_0^x e^{-(x-t)} f(x-(x-t))\, dt$$

$$\left[\text{using} \int_0^a f(x)dx = \int_0^a f(a-x)dx\right]$$

$$\Rightarrow \quad f(x) = x^2 + e^{-x}\int_0^x e^t f(t)\, dt \quad \text{...(i)}$$

On differentiating both the sides, we get

$$f'(x) = 2x + e^{-x}[e^x f(x)] - e^{-x}\int_0^x e^t f(t)\, dt \quad \text{...(ii)}$$

[using Leibnitz rule]

$$\Rightarrow \quad f(x) + f'(x) = x^2 + 2x + f(x)$$

$$\Rightarrow \quad f'(x) = x^2 + 2x \quad \text{...(iii)}$$

On integrating both the sides of Eq. (iii), we get

$$f(x) = \frac{x^3}{3} + x^2 + C$$

But $f(0) = 0 \Rightarrow C = 0$

Hence, $f(x) = \dfrac{x^3}{3} + x^2$

Ex. 57 *Let $f : R^+ \to R$ satisfies the functional equation $f(xy) = e^{xy-x-y}\{e^y f(x) + e^x f(y)\}, \forall\, x, y \in R^+$. If $f'(1) = e$, determine $f(x)$.*

Sol. Given that,

$$f(xy) = e^{xy-x-y}\{e^y f(x) + e^x f(y)\}, \forall\, x, y \in R^+ \quad \text{...(i)}$$

On putting $x = y = 1$, we get

$$f(1) = e^{-1}\{e^1 f(1) + e^1 f(1)\} \Rightarrow f(1) = 0 \quad \text{...(ii)}$$

Now, $f'(x) = \lim_{h \to 0} \dfrac{f(x+h) - f(x)}{h}$

$$= \lim_{h \to 0} \frac{f\left\{x\left(1+\frac{h}{x}\right)\right\} - f(x)}{h}$$

$$= \lim_{h \to 0} \frac{e^{x\left(1+\frac{h}{x}\right) - x - \left(1+\frac{h}{x}\right)}\left\{e^{1+\frac{h}{x}} f(x) + e^x f\left(1+\frac{h}{x}\right)\right\} - f(x)}{h}$$

$$= \lim_{h \to 0} \frac{e^h f(x) + e^{h-1-\frac{h}{x}+x} f\left(1+\frac{h}{x}\right) - f(x)}{h}$$

$$= \lim_{h \to 0} \frac{f(x)(e^h - 1) + e^{h-1-\frac{h}{x}+x}\left\{f\left(1+\frac{h}{x}\right) - f(1)\right\}}{h} \quad [\because f(1) = 0]$$

$$= \lim_{h \to 0} \frac{f(x)(e^h-1)}{h} + \lim_{h \to 0} \frac{e^{h-1-\frac{h}{x}+x}\left\{f\left(1+\frac{h}{x}\right) - f(1)\right\}}{\frac{h}{x} \cdot x}$$

$$= f(x) + \frac{e^{x-1} \cdot f'(1)}{x}$$

$$\left[\because \lim_{h \to 0} \frac{f\left(1+\frac{h}{x}\right) - f(1)}{\frac{h}{x}} = f'(1) \text{ and } \lim_{h \to 0} \frac{e^h - 1}{h} = 1\right]$$

$$= f(x) + \frac{e^x}{ex} \cdot f'(1)$$

$$\therefore\ f'(x) = f(x) + \frac{e^x}{x} \Rightarrow \frac{e^x}{x} = f'(x) - f(x) \quad [\because f'(1) = e]$$

$$\Rightarrow \quad \frac{1}{x} = \frac{e^x f'(x) - f(x) \cdot e^x}{e^{2x}}$$

$$\left[\text{as by quotient rule, we can write } \frac{e^x f'(x) - f(x) \cdot e^x}{(e^x)^2} = \frac{d}{dx}\left\{\frac{f(x)}{e^x}\right\}\right]$$

$$\therefore \quad \frac{1}{x} = \frac{d}{dx}\left\{\frac{f(x)}{e^x}\right\}$$

On integrating both the sides w.r.t. 'x', we get

$$\log|x| + C = \frac{f(x)}{e^x}$$

$$f(x) = e^x\{\log|x| + C\}$$

Since, $f(1) = 0 \Rightarrow C = 0$

Thus, $f(x) = e^x \log|x|$

Ex. 58 *Let f is a differentiable function such that $f'(x) = f(x) + \int_0^2 f(x)\, dx$, $f(0) = \dfrac{4-e^2}{3}$, find $f(x)$.*

Sol. It is given that

$$f'(x) = f(x) + \int_0^2 f(x)\, dx$$

$$\Rightarrow \quad f'(x) = f(x) + c \quad \left[\therefore c = \int_0^2 f(x)\, dx\right]$$

$$\Rightarrow \quad \frac{f'(x)}{f(x) + c} = 1$$

On integrating both the sides of above expression,

$$\log_e\{f(x) + c\} = x + c \Rightarrow f(x) + c = ke^x \quad \text{...(i)}$$

[k being constant of integration]

Since, $f(0)=\frac{4-e^2}{3}$ and $f(0)+c=k$

$$\therefore \quad k=\frac{4-e^2}{3}+c \Rightarrow c=k-\left(\frac{4-e^2}{3}\right) \quad \text{...(ii)}$$

From Eqs. (i) and (ii), we get

$$f(x)+k-\left(\frac{4-e^2}{3}\right)=k(e^x)$$

$$\Rightarrow \quad f(x)=k(e^x-1)+\left(\frac{4-e^2}{3}\right) \quad \text{...(iii)}$$

Now, to find constant of integration k.

On integrating both the sides from 0 to 2, we get

$$\int_0^2 f(x)\,dx=k\int_0^2(e^x-1)\,dx+\frac{4-e^2}{3}\int_0^2 dx$$

$$\Rightarrow \quad c=k(e^2-2-1)+\frac{4-e^2}{3}(2)$$

$$\Rightarrow \quad c=\frac{2}{3}(4-e^2)+k(e^2-3) \quad \text{...(iv)}$$

From Eqs. (ii) and (iv), we get

$$k-\left(\frac{4-e^2}{3}\right)=\frac{2}{3}(4-e^2)+k(e^2-3) \Rightarrow k=1$$

On putting $k=1$ in Eq. (iii), we get

$$f(x)=(e^x-1)+\frac{4-e^2}{3}$$

Hence, $\quad f(x)=e^x-\frac{(e^2-1)}{3}$

Ex. 59 *A function $f(x)$ satisfies the following property $f(x+y)=f(x)\cdot f(y)$.*
Show that the function is continuous for all values of x, if it is continuous at $x=1$.

Sol. As the function is continuous at $x=1$, we have

$$\lim_{x\to 1^-} f(x)=\lim_{x\to 1^+} f(x)=f(1)$$

$$\Rightarrow \quad \lim_{h\to 0} f(1-h)=\lim_{h\to 0} f(1+h)=f(1)$$

[using $f(x+y)=f(x)\cdot f(y)$]

$$\Rightarrow \quad \lim_{h\to 0} f(1)\,f(-h)=\lim_{h\to 0} f(1)\cdot f(h)=f(1)$$

$$\Rightarrow \quad \lim_{h\to 0} f(-h)=\lim_{h\to 0} f(h)=1 \quad \text{...(i)}$$

Now, consider some arbitrary points $x=a$.

Left hand limit

$$\Rightarrow \quad \lim_{h\to 0} f(a-h)=\lim_{h\to 0} f(a)\cdot f(-h)=f(a)\lim_{h\to 0} f(-h)$$

LHL $=f(a)$ [as $\lim_{h\to 0} f(-h)=1$, using Eq. (i)]

Right hand limit

$$\Rightarrow \quad \lim_{h\to 0} f(a+h)=\lim_{h\to 0} f(a)\cdot f(h)=f(a)\lim_{h\to 0} f(h)$$

RHL $=f(a)$ [as $\lim_{h\to 0} f(h)=1$, using Eq. (i)]

Hence, at any arbitrary point $(x=a)$

$$\text{LHL}=\text{RHL}=f(a)$$

Therefore, function is continuous for all values of x, if it is continuous at 1.

Ex. 60 *Let $f\left(\frac{x+y}{2}\right)=\frac{f(x)+f(y)}{2}$ for all real x and y. If $f'(0)$ exists and equals to -1 and $f(0)=1$, then find $f'(x)$.* **[IIT JEE 1995]**

Sol. Given equation is $f\left(\frac{x+y}{2}\right)=\frac{f(x)+f(y)}{2}$ ...(i)

On putting $y=0$ and $f(0)=1$ in Eq. (i), we have

$$f\left(\frac{x}{2}\right)=\frac{1}{2}[f(x)+1] \quad [\because f(0)=1]$$

$$\Rightarrow \quad f(x)=2f\left(\frac{x}{2}\right)-1 \quad \text{...(ii)}$$

Now, $f'(x)=\lim_{h\to 0}\frac{f(x+h)-f(x)}{h}$

$$=\lim_{h\to 0}\frac{f\left(\frac{2x+2h}{2}\right)-f(x)}{h}$$

$$=\lim_{h\to 0}\frac{\frac{f(2x)+f(2h)}{2}-f(x)}{h} \quad \text{[using Eq. (i)]}$$

$$=\lim_{h\to 0}\frac{f(2x)+f(2h)-2f(x)}{2h}$$

$$=\lim_{h\to 0}\frac{2f(x)-1+f(2h)-2f(x)}{2h} \quad \text{[using Eq. (ii)]}$$

$$\lim_{h\to 0}\frac{f(2h)-1}{2h}=f'(0)$$

Therefore, $\quad f'(x)=-1$

Ex. 61 *Let* $f(x)=1+4x-x^2, \forall\ x\in R$

$g(x)=\max\{f(t);\ x\le t\le(x+1);\ 0\le x<3\}$

$\quad=\min\{(x+3);\ 3\le x\le 5\}$

Verify continuity of $g(x)$, for all $x\in[0,5]$.

Sol. Here, $\quad f(t)=1+4t-t^2$

$$f'(t)=4-2t$$

When $f'(t)=0$, then $t=2$

At $t=2$, $f(t)$ has a maxima.

Since, $g(x)=\max\{f(t)$ for $t\in[x,x+1], 0\le x<3\}$

$$\therefore \quad g(x)=\begin{cases} f(x+1), & \text{if } [x, x+1]<2 \\ f(2), & \text{if } x\le 2\le x+1 \\ f(x), & \text{if } 2<[x, x+1]\end{cases}$$

$$g(x)=\begin{cases}4+2x-x^2, & \text{if } 0\le x<1\\ 5, & \text{if } 1\le x\le 2\\ 1+4x-x^2, & \text{if } 2<x<3\\ 6, & \text{if } 3\le x\le 5\end{cases}$$

Which is clearly continuous for all $x \in [0, 5]$ except $x = 3$.

$\therefore$ $g(x)$ is continuous for $x = [0, 3) \cup (3, 5]$

Ex. 62 *Let $f(x)=x^4-8x^3+22x^2-24x$ and*

$$g(x)=\begin{cases}\min f(x); & x\le t\le x+1, -1\le x\le 1\\ x-10; & x>1\end{cases}$$

Discuss the continuity and differentiability of $g(x)$ in $[-1, \infty)$.

Sol. Here, $f(x)=x^4-8x^3+22x^2-24x$

$\Rightarrow$ $f'(x)=4x^3-24x^2+44x-24$

or $f'(x)=4(x-1)(x-2)(x-3)$

Which shows $f(x)$ is increasing in $[1, 2] \cup [3, \infty)$ and decreasing in $(-\infty, 1] \cup [2, 3]$.

Thus, minimum $f(x); x \le t \le x+1, -1 \le x \le 1$

$$\Rightarrow \quad \text{Minimum } f(x)=\begin{cases}f(x+1), & -1\le x\le 0\\ f(1), & 0<x\le 1\end{cases}$$

$$\text{Thus,}\quad g(x)=\begin{cases}f(x+1), & -1\le x\le 0\\ f(1), & 0<x\le 1\\ x-10, & x>1\end{cases}$$

$$=\begin{cases}(x+1)^4-8(x+1)^3+22(x+1)^2-24(x+1), & -1\le x\le 0\\ 1-8+22-24, & 0<x\le 1\\ x-10, & x>1\end{cases}$$

$$g(x)=\begin{cases}x^4-4x^3+4x^2-9, & -1\le x\le 0\\ -9, & 0<x\le 1\\ x-10, & x>1\end{cases}$$

$$\text{Also,}\quad g'(x)=\begin{cases}4x^3-12x^2+8x, & -1\le x\le 0\\ 0, & 0<x\le 1\\ 1, & x>1\end{cases}$$

Which clearly shows $g(x)$ is continuous in $[-1, \infty)$ but not differentiable at $x = 1$.

Ex. 63 *Let $g(x)=\int_0^x f(t)\,dt$, where f is such that $1/2 \le f(t) \le 1$ for $t \in [0, 1]$ and $0 \le f(t) \le 1/2$ for $t \in [1, 2]$. Then, find the interval in which $g(2)$ lies.* **[IIT JEE 2000]**

Sol. Here, $g(x)=\int_0^x f(t)dt \Rightarrow g(2)=\int_0^2 f(t)dt$

$$\Rightarrow \quad g(2)=\int_0^1 f(t)dt+\int_1^2 f(t)dt \qquad \text{...(i)}$$

Now, $1/2 \le f(t) \le 1$ for $t \in [0, 1]$

We get, $\int_0^1 \frac{1}{2}dt \le \int_0^1 f(t)dt \le \int_0^1 1\,dt$

$$\Rightarrow \quad \frac{1}{2} \le \int_0^1 f(t)dt \le 1 \qquad \text{...(ii)}$$

Again, $0 \le f(t) \le \frac{1}{2}$, for $t \in [1, 2]$

$\Rightarrow$ $\int_1^2 0\,dt \le \int_1^2 f(t)dt \le \int_1^2 \frac{1}{2}dt$

$$\Rightarrow \quad 0 \le \int_1^2 f(t)dt \le \frac{1}{2} \qquad \text{...(iii)}$$

On adding Eq. (ii) and (iii), we get

$$0+\frac{1}{2} \le \int_0^1 f(t)dt+\int_1^2 f(t)dt \le 1+\frac{1}{2}$$

or $\frac{1}{2} \le g(2) \le \frac{3}{2} \Rightarrow g(2) \in \left[\frac{1}{2}, \frac{3}{2}\right]$

Ex. 64 *Let f be a one-one function such that*

$$f(x)\cdot f(y)+2=f(x)+f(y)+f(xy), \forall\ x, y \in R-\{0\}$$

and $f(0)=1$, $f'(1)=2$. Prove that $3(\int f(x)\,dx)-x(f(x)+2)$ is constant.

Sol. We have, $f(x)\cdot f(y)+2=f(x)+f(y)+f(xy)$ ...(i)

On replacing $x, y \to 1$, we get

$$(f(1))^2+2=3f(1)$$

$\Rightarrow$ $f^2(1)-3f(1)+2=0$

$\Rightarrow$ $f(1)=2$ or 1

But $f(1)$ cannot be equal to one as $f(0)=1$

$$\Rightarrow \quad f(1)=2 \qquad \text{...(ii)}$$

On replacing y by $1/x$ in Eq. (i), we get

$f(x)\cdot f(1/x)+2=f(x)+f(1/x)+f(1)$

$\Rightarrow$ $f(x)\cdot f(1/x)+2=f(x)+f(1/x)+2$ [using Eq. (ii)]

$\Rightarrow$ $f(x)\cdot f(1/x)=f(x)+f(1/x)$

$\Rightarrow$ $f(x)=\dfrac{f(1/x)}{f(1/x)-1}$

$$\text{and} \quad f(1/x)=\frac{f(x)}{f(x)-1} \qquad \text{...(iii)}$$

Now, $f'(x)=\lim_{h\to 0}\dfrac{f(x+h)-f(x)}{h}$

$$=\lim_{h\to 0}\frac{f(x+h)+\dfrac{f(1/x)}{1-f(1/x)}}{h}$$

$$=\lim_{h\to 0}\frac{f(x+h)-f(x+h)\cdot f(1/x)+f(1/x)}{h\{1-f(1/x)\}}$$

$$=\lim_{h\to 0}\frac{f(x+h)-f(x+h)-f(1/x)-f\left(\dfrac{x+h}{x}\right)+2+f(1/x)}{h\{1-f(1/x)\}}$$

[using Eq. (i)]

$$= \lim_{h \to 0} \frac{f\left(1+\frac{h}{x}\right)-2}{h\{f(1/x)-1\}} = \lim_{h \to 0} \frac{f(1+h/x)-f(1)}{\frac{h}{x}\cdot x\,\{f(1/x)-1\}}$$

[using Eq. (ii)]

$$= \frac{f'(1)}{x\,\{f(1/x)-1\}}$$

$$\left[\because \text{from Eq. (iii), } f(x)\cdot f\left(\frac{1}{x}\right) = \frac{f(x)f\left(\frac{1}{x}\right)}{\left\{f\left(\frac{1}{x}\right)-1\right\}\{f(x)-1\}}\right]$$

$$f'(x) = \frac{2\{f(x)-1\}}{x} \Rightarrow xf'(x) = 2\{f(x)-1\}$$

On integrating both the sides of the above expression, we get

$$x\,f(x) - \int f(x)\,dx = 2\int f(x)\,dx - 2x + \lambda,$$

where λ is the constant of integral

$$\Rightarrow \quad 3\int f(x)\,dx = x\,\{2+f(x)\} - \lambda$$

Hence, $3\int f(x)\,dx - x\,\{2+f(x)\} = \lambda$ (constant)

Ex. 65 *Let $f: R \to R$, such that $f'(0) = 1$ and $f(x+2y) = f(x) + f(2y) + e^{x+2y}(x+2y) - x\cdot e^x - 2y\cdot e^{2y} + 4xy,\ \forall\ x, y \in R$. Find $f(x)$.*

Sol. We have,

$$f(x+2y) = f(x) + f(2y) + e^{x+2y}(x+2y) - x\cdot e^x - 2y\cdot e^{2y} + 4xy$$

On replacing $x, y \to 0$, we get

$$f(0) = f(0) + f(0) + 0 - 0 - 0 + 0 \Rightarrow f(0) = 0$$

On replacing $2y \to -x$, we get

$$f(0) = f(x) + f(-x) - x\cdot e^x + xe^{-x} - 2x^2$$

$$\Rightarrow \quad -f(x) = f(-x) - x\cdot e^x + xe^{-x} - 2x^2 \quad \text{...(i)}$$

Now, $f'(x) = \lim_{h \to 0} \frac{f(x+h)-f(x)}{h}$

$$= \lim_{h \to 0} \frac{f(x+h) + f(-x) - xe^x + x\cdot e^{-x} - 2x^2}{h}$$

$$\Rightarrow f'(x) = \lim_{h \to 0} \frac{\left[\begin{array}{c} f(h) - e^h\cdot h + (x+h)\,e^{(x+h)} - xe^{-x} + \\ 2(x+h)\,x - xe^x + xe^{-x} - 2x^2 \end{array}\right]}{h}$$

$$= \lim_{h \to 0} \frac{f(h) - e^h\cdot h + x\cdot e^x\cdot e^h + h\cdot e^x\cdot e^h + 2hx - xe^x}{h}$$

$$= \lim_{h \to 0} \frac{f(h)-f(0)}{h} + \frac{h\,(e^x-1)}{h}e^h + \frac{xe^x\,(e^h-1)}{h} + \frac{2hx}{h}$$

$$= f'(0) + (e^x - 1) + x\cdot e^x + 2x$$

$$\therefore \quad f'(x) = 1 + e^x - 1 + xe^x + 2x$$

$$\text{or} \quad f'(x) = e^x\,(x+1) + 2x \quad \text{...(ii)}$$

On integrating Eq. (ii) both sides, we get

$$f(x) = \int e^x\,(x+1)\,dx + 2\int x\,dx$$

$$= (x+1)e^x - \int 1\cdot e^x dx + 2\frac{x^2}{2} + C$$

$$= (x+1)\,e^x - e^x + x^2 + C$$

$$\Rightarrow \quad f(x) = x^2 + xe^x + C$$

But $f(0) = 0 \Rightarrow C = 0$

So, $f(x) = x^2 + x\cdot e^x$

Ex. 66 *Let f be a function such that $f(xy) = f(x)\cdot f(y), \forall\ y \in R$ and $f(1+x) = 1 + x(1+g(x))$, where $\lim_{x \to 0} g(x) = 0$. Find the value of $\int_1^2 \frac{f(x)}{f'(x)}\cdot\frac{1}{1+x^2}dx$.*

Sol. We know, $f'(x) = \lim_{h \to 0} \frac{f(x+h)-f(x)}{h}$

$$= \lim_{h \to 0} \frac{f\left(x\left(1+\frac{h}{x}\right)\right) - f(x)}{h}$$

$$\Rightarrow \quad f'(x) = \lim_{h \to 0} \frac{f(x)\cdot f\left(1+\frac{h}{x}\right) - f(x)}{h}$$

[given $f(xy) = f(x)\cdot f(y)$]

$$\Rightarrow \quad f'(x) = \lim_{h \to 0} \frac{f(x)\cdot\left\{1 + \frac{h}{x}(1+g\,(h/x))\right\} - f(x)}{h}$$

[given $f(1+x) = 1 + x\,(1+g\,(x))$]

$$\Rightarrow \quad f'(x) = \lim_{h \to 0} \frac{f(x)\cdot\left\{1 + \frac{h}{x}(1+g\,(h/x)) - 1\right\}}{h}$$

$$\therefore \quad f'(x) = \frac{f(x)}{x} \quad [\text{as } \lim_{x \to 0} g(x) = 0] \quad \text{...(i)}$$

To find, $\int_1^2 \frac{f(x)}{f'(x)}\cdot\frac{1}{1+x^2}dx = \int_1^2 \frac{x}{1+x^2}dx$ [using Eq. (i)]

$$= \frac{1}{2}[\log|1+x^2|]_1^2 = \frac{1}{2}[\log(5/2)]$$

Ex. 67 *If $f(x) = ax^2 + bx + c$ is such that $|f(0)| \le 1$, $|f(1)| \le 1$ and $|f(-1)| \le 1$, prove that $|f(x)| \le 5/4, \forall\ x \in [-1, 1]$.*

Sol. We have, $\quad f(x) = ax^2 + bx + c \quad$ ...(i)

$$\Rightarrow \quad f(-1) = a - b + c \quad \text{...(ii)}$$

$$\Rightarrow \quad f(0) = c \quad \text{...(iii)}$$

$$\Rightarrow \quad f(1) = a + b + c \quad \text{...(iv)}$$

From Eqs. (ii), (iii) and (iv), we get

$$a = \frac{1}{2}[f(-1) + f(1) - 2f(0)],$$

$$b = \frac{1}{2}[f(1) - f(-1)] \text{ and } c = f(0)$$

On substituting the values of a, b and c in Eq. (i), we get

$$\Rightarrow \quad f(x) = \frac{1}{2}[f(-1) + f(1) - 2f(0)]x^2 + \frac{1}{2}[f(1) - f(-1)]x + f(0)$$

$$f(x) = \frac{x(x-1)}{2} f(-1) - (x+1)(x-1) f(0) + \frac{(x+1)x}{2} f(1) \quad \text{...(v)}$$

Since, $|f(-1)|, |f(0)|$ and $|f(1)|$ are ≤ 1, we have

$$2|f(x)| \leq |x(x-1)| + 2|x^2 - 1| + |x(x+1)| \quad \text{...(vi)}$$

In the interval, $x \in [-1, 1]$

$$0 \leq 1 + x \leq 2, \; 0 \leq 1 - x \leq 2 \text{ and } 0 \leq 2(1 - x^2) \leq 2$$

$$\Rightarrow \quad 2|f(x)| \leq |x|(1 - x + 1 + x) + 2(1 - x^2)$$

$$2|f(x)| \leq 2(|x| + 1 - x^2)$$

Therefore, $\quad |f(x)| \leq \left(|x| - \frac{1}{2}\right)^2 + 5/4 \leq 5/4$

$$\Rightarrow \quad |f(x)| \leq \frac{5}{4}, \forall \; x \in [-1, 1]$$

Ex. 68 *Let $\alpha + \beta = 1$, $2\alpha^2 + 2\beta^2 = 1$ and $f(x)$ be a continuous function such that $f(2 + x) + f(x) = 2$ for all $x \in [0, 2]$ and $p = \int_0^4 f(x)\,dx - 4$, $q = \frac{\alpha}{\beta}$. Then, find the least positive integral value of 'a' for which the equation $ax^2 - bx + c = 0$ has both roots lying between p and q, where $a, b, c \in N$.*

Sol. Given, $\quad \alpha + \beta = 1 \quad$ ...(i)

$$2\alpha^2 + 2\beta^2 = 1 \quad \text{...(ii)}$$

On solving Eqs. (i) and (ii), we get

$$\alpha = \beta = \frac{1}{2} \Rightarrow q = \frac{\alpha}{\beta} = 1 \quad \text{...(iii)}$$

and given $f(2 + x) + f(x) = 2, \forall \; x \in [0, 2] \quad$ ...(iv)

Now, $\quad p = \int_0^4 f(x)\,dx - 4 = \int_0^2 f(x)\,dx + \int_2^4 f(x)\,dx - 4$

$$= \int_0^2 f(x)\,dx + \int_0^2 f(t+2)\,dt - 4$$

[let $x = t + 2$ for second integration]

$$= \int_0^2 f(x)\,dx + \int_0^2 \{2 - f(x)\}\,dx - 4$$

[given, $f(2 + x) + f(x) = 2$]

$$\Rightarrow \quad p = \int_0^2 f(x)\,dx + 2\int_0^2 dx - \int_0^2 f(x)\,dx - 4 = 0 \quad \text{...(v)}$$

Then, from Eqs. (v) and (iii), $p = 0, q = 1$

Let the roots of equation $ax^2 - bx + c = 0$ be α and β.

$$\therefore \quad f(x) = ax^2 - bx + c = a(x - \alpha)(x - \beta) \quad \text{...(vi)}$$

Since, equation $f(x) = 0$ has both roots lying between 0 and 1.

$$\therefore \quad f(0) \cdot f(1) > 0 \quad \text{...(vii)}$$

But $\quad f(0) \cdot f(1) = c(a - b + c) = \text{an integer} \quad$ ...(viii)

$\therefore$ Least value of $f(0) \cdot f(1) = 1 \quad$ ...(ix)

Now, from Eq. (vi),

$$f(0) \cdot f(1) = a\,\alpha\beta\, a(1 - \alpha)(1 - \beta)$$

$$= a^2\, \alpha\beta(1 - \alpha)(1 - \beta) \quad \text{...(x)}$$

As we know,

$\alpha(1 - \alpha)$ has greatest value $\frac{1}{4}$ at $\alpha = \frac{1}{2}$ and $\beta(1 - \beta)$ has greatest value $\frac{1}{4}$ at $\beta = \frac{1}{2}$. But $\alpha \neq \beta$

Thus, from Eq. (x) greatest value of $f(0) \cdot f(1) < \frac{a^2}{16}$ ...(xi)

$\therefore$ From Eqs. (ix) and (xi), we get $1 < \frac{a^2}{16}$

$$\Rightarrow \quad a^2 - 16 > 0$$

$$\Rightarrow \quad a < -4 \text{ or } a > 4 \quad [\because a \in N]$$

$\Rightarrow$ Least value of $a = 5 \quad$ [as $a \in$ natural number]

Ex. 69 *Prove that the function $f(x) = a\sqrt{x - 1} + b\sqrt{2x - 1} - \sqrt{2x^2 - 3x + 1}$, where $a + 2b = 2$ and $a, b \in R$ always has a root in $(1, 5)$, $\forall \; b \in R$.*

Sol. Let $b > 0$, then $f(1) = b > 0$

and $\quad f(5) = 2a + 3b - 6 = 2(a + 2b) - b - 6$

$$= 4 - b - 6 = -(2 + b) < 0$$

Hence, by IVT, $\exists$ some $c \in (1, 5)$ such that

$$\Rightarrow \quad f(c) = 0$$

If $b = 0$, then $a = 2$

$$f(x) = 2\sqrt{x - 1} - \sqrt{2x^2 - 3x + 1} = 0$$

$$\Rightarrow \quad 4(x - 1) = 2x^2 - 3x + 1 = (2x - 1)(x - 1)$$

$$\Rightarrow \quad (x - 1)(2x - 5) = 0$$

$$\Rightarrow \quad x = \frac{5}{2}$$

Hence, $f(x) = 0$, if $x = \frac{5}{2}$, which lies in $(1, 5)$.

If $b < 0$, $f(1) = b < 0$ and $f(2) = a + b\sqrt{3} - \sqrt{3}$

$$= (a + 2b) + (\sqrt{3} - 2)b - \sqrt{3}$$

$$= (2 - \sqrt{3}) - (2 - \sqrt{3})b$$

$$= (2 - \sqrt{3})(1 - b) > 0 \quad [\text{as } b < 0]$$

Hence, $f(1)$ and $f(2)$ have opposite signs

$\exists$ some $c \in (1, 2) \subset (1, 5)$ for which $f(c) = 0$.

Ex. 70 *Let $\alpha \in R$. Prove that a function $f : R \to R$ is differentiable at α, if and only if there is a function $g : R \to R$ which is continuous at $x = \alpha$ and satisfies $f(x) - f(\alpha) = g(x)(x - \alpha)$ for all $x \in R$.* ***[IIT JEE 2001]***

Sol. If $g(x)$ is continuous at $x = \alpha$.

$$\Rightarrow \lim_{x \to \alpha} g(x) = g(\alpha) \quad \text{...(i)}$$

and $f(x) - f(\alpha) = g(x)(x - \alpha), \forall x \in R$

$$\Rightarrow \frac{f(x) - f(\alpha)}{(x - \alpha)} = g(x)$$

$$\Rightarrow \lim_{x \to \alpha} \frac{f(x) - f(\alpha)}{(x - \alpha)} = \lim_{x \to \alpha} g(x) \Rightarrow f'(\alpha) = g(\alpha)$$

Therefore, $f(x)$ is differentiable at $x = \alpha$ ...(ii)

Conversely, f is differentiable at $x = \alpha$, then $\lim_{x \to \alpha} \frac{f(x) - f(\alpha)}{(x - \alpha)}$ exists finitely.

Let
$$g(x) = \begin{cases} \frac{f(x) - f(\alpha)}{x - \alpha}, & x \neq \alpha \\ f'(\alpha), & x = \alpha \end{cases}$$

Clearly, $\lim_{x \to \alpha} g(x) = f'(\alpha)$

Thus, $g(x)$ is continuous at $x = \alpha$.

Hence, $f(x)$ is differentiable at $x = \alpha$, iff $g(x)$ is continuous at $x = \alpha$.

Continuity and Differentiability Exercise 1 :
Single Option Correct Type Questions

1. If $f(x)=\begin{cases}\sin\dfrac{\pi x}{2}, & x<1\\ [x], & x\geq 1\end{cases}$, where $[x]$ denotes the greatest integer function, then

(a) $f(x)$ is continuous at $x=1$
(b) $f(x)$ is discontinuous at $x=1$
(c) $f(1^{+})=0$
(d) $f(1^{-})=-1$

2. Consider $f(x)=\begin{cases}\dfrac{8^x-4^x-2^x+1}{x^2}, & x>0\\ e^x\sin x+\pi x+k\log 4, & x<0\end{cases}$

Then, $f(0)$ so that $f(x)$ is continuous at $x=0$, is

(a) log 4 (b) log 2
(c) (log 4)(log 2) (d) None of these

3. Let $f(x)=\begin{cases}\dfrac{a(1-x\sin x)+b\cos x+5}{x^2}, & x<0\\ 3, & x=0\\ \left[1+\left(\dfrac{cx+dx^3}{x^2}\right)\right]^{1/x}, & x>0\end{cases}$

If f is continuous at $x=0$, then $(a+b+c+d)$ is

(a) 5 (b) -5 (c) $\log_e 3-5$ (d) $5-\log_e 3$

4. If $f(x)=\begin{cases}\cos^{-1}\{\cot x\}, & x<\pi/2\\ \pi[x]-1, & x\geq\pi/2\end{cases}$, then the jump of discontinuity, (where [·] denotes greatest integer and {} denotes fractional part function) is

(a) 1 (b) $\pi/2$ (c) $\dfrac{\pi}{2}-1$ (d) 2

5. Let $f:[0,1]\xrightarrow{\text{onto}}[0,1]$ be a continuous function, then $f(x)=x$ holds for

(a) at least one $x\in[0,1]$ (b) at least one $x\in[1,2]$
(c) at least one $x\in[-1,0]$ (d) can't be discussed

6. If $f(x)=\dfrac{x+1}{x-1}$ and $g(x)=\dfrac{1}{x-2}$, then $(fog)(x)$ is discontinuous at

(a) $x=3$ only (b) $x=2$ only
(c) $x=2$ and 3 (d) $x=1$ only

7. Let $y_n(x)=x^2+\dfrac{x^2}{(1+x^2)}+\dfrac{x^2}{(1+x^2)^2}+\ldots+\dfrac{x^2}{(1+x^2)^{n-1}}$ and $y(x)=\lim\limits_{n\to\infty}y_n(x)$, then $y_n(x)$, $n=1, 2, 3, \ldots, n$ and $y(x)$ is

(a) continuous for $x\in R$ (b) continuous for $x\in R-\{0\}$
(c) continuous for $x\in R-\{1\}$ (d) data unsufficient

8. If $g(x)=\begin{cases}\dfrac{1+a^x\cdot x\cdot\log a-a^x}{a^x\cdot x^2}, & \text{for } x\leq 0\\ \dfrac{2^x\cdot a^x-x\log 2-x\log a-1}{x^2}, & \text{for } x>0\end{cases}$

where $a>0$, then a for which $g(x)$ is continuous, is

(a) $-\dfrac{1}{\sqrt{2}}$ (b) $\dfrac{1}{\sqrt{2}}$ (c) 2 (d) -2

9. Let $f(x)=\begin{cases}\dfrac{\left(\dfrac{\pi}{2}-\sin^{-1}(1-\{x\}^2)\right)\cdot\sin^{-1}(1-\{x\})}{\sqrt{2}(\{x\}-\{x\}^3)}, & x\neq 0,\\ \dfrac{\pi}{2}, & x=0\end{cases}$ where {·} is fractional part of x, then

(a) $f(0^{+})=-\dfrac{\pi}{2}$
(b) $f(0^{-})=\dfrac{\pi}{4\sqrt{2}}$
(c) $f(x)$ is continuous at $x=0$
(d) None of the above

10. Let $f(x)=\begin{cases}\operatorname{sgn}(x)+x, & -\infty<x<0\\ -1+\sin x, & 0\leq x\leq\pi/2\\ \cos x, & \pi/2\leq x<\infty\end{cases}$, then the number of points, where $f(x)$ is not differentiable, is/are

(a) 0 (b) 1
(c) 2 (d) 3

11. Let $f(x)=\begin{cases}\dfrac{1}{|x|}, & |x|\geq 1\\ ax^2+b, & |x|<1\end{cases}$ be continuous and differentiable for all x. Then, a and b are

(a) $-\dfrac{1}{2},\dfrac{3}{2}$ (b) $\dfrac{1}{2},-\dfrac{3}{2}$
(c) $\dfrac{1}{2},\dfrac{3}{2}$ (d) None of these

12. If $f(x)=\begin{cases}A+Bx^2, & x<1\\ 3Ax-B+2, & x\geq 1\end{cases}$, then A and B, so that $f(x)$ is differentiable at $x=1$, are

(a) $-2, 3$ (b) $2, -3$
(c) 2, 3 (d) $-2, -3$

13. If $f(x)=\begin{cases}|x-1|([x]-x), & x\neq 1\\ 0, & x=1\end{cases}$, then

(a) $f'(1^{+})=0$ (b) $f'(1^{-})=0$
(c) $f'(1^{-})=-1$ (d) $f(x)$ is differentiable at $x=1$

14. If $f(x)=\begin{cases}[\cos \pi x], & x \le 1\\ 2\{x\}-1, & x>1\end{cases}$, where $[\cdot]$ and $\{\cdot\}$ denote greatest integer and fractional part of x, then

(a) $f'(1^-)=2$ (b) $f'(1^+)=2$

(c) $f'(1^-)=-2$ (d) $f'(1^+)=0$

15. If $f(x)=\begin{cases}x-3, & x<0\\ x^2-3x+2, & x\ge 0\end{cases}$, then $g(x)=f(|x|)$ is

(a) $g'(0^+)=-3$

(b) $g'(0^-)=-3$

(c) $g'(0^+)=g'(0^-)$

(d) $g(x)$ is not continuous at $x=0$

16. If $f(x)=\begin{cases}\left\{x+\frac{1}{3}\right\}[\sin \pi x], & 0\le x<1\\ [2x]\operatorname{sgn}\left(x-\frac{4}{3}\right), & 1\le x\le 2\end{cases}$, where $[\cdot]$ and $\{\cdot\}$ denote greatest integer and fractional part of x respectively, then the number of points of non-differentiability, is

(a) 3 (b) 4 (c) 5 (d) 6

17. Let f be differentiable function satisfying $f\left(\frac{x}{y}\right)=f(x)-f(y)$ for all $x, y>0$. If $f'(1)=1$, then $f(x)$ is

(a) $2\log_e x$ (b) $3\log_e x$ (c) $\log_e x$ (d) $\frac{1}{2}\log_e x$

18. Let $f(x+y)=f(x)+f(y)-2xy-1$ for all x and y. If $f'(0)$ exists and $f'(0)=-\sin\alpha$, then $f\{f'(0)\}$ is

(a) -1 (b) 0

(c) 1 (d) 2

19. A derivable function $f: R^+ \to R$ satisfies the condition $f(x)-f(y)\ge \log\left(\frac{x}{y}\right)+x-y, \forall\, x, y \in R^+$.

If g denotes the derivative of f, then the value of the sum $\sum_{n=1}^{100} g\left(\frac{1}{n}\right)$ is

(a) 5050 (b) 5510 (c) 5150 (d) 1550

20. If $\frac{d(f(x))}{dx}=e^{-x}f(x)+e^{x}f(-x)$, then $f(x)$ is, (given $f(0)=0$)

(a) an even function

(b) an odd function

(c) neither even nor odd function

(d) can't say

21. Let $f:(0,\infty)\to R$ be a continuous function such that $f(x)=\int_0^x t\, f(t)\,dt$.

If $f(x^2)=x^4+x^5$, then $\sum_{r=1}^{12} f(r^2)$ is equal to

(a) 216 (b) 219

(c) 222 (d) 225

22. For $x>0$, let $h(x)=\begin{cases}\frac{1}{q}, & \text{if } x=\frac{p}{q}\\ 0, & \text{if } x \text{ is irrational}\end{cases}$, where p and $q>0$ are relatively prime integers, then which one of the following does not hold good?

(a) $h(x)$ is discontinuous for all x in $(0,\infty)$

(b) $h(x)$ is continuous for each irrational in $(0,\infty)$

(c) $h(x)$ is discontinuous for each rational in $(0,\infty)$

(d) $h(x)$ is not derivable for all x in $(0,\infty)$

23. Let $f(x)=\frac{g(x)}{h(x)}$, where g and h are continuous functions on the open interval (a, b). Which of the following statements is true for $a<x<b$?

(a) f is continuous at all x for which $x\ne 0$

(b) f is continuous at all x for which $g(x)=0$

(c) f is continuous at all x for which $g(x)\ne 0$

(d) f is continuous at all x for which $h(x)\ne 0$

24. If $f(x)=\frac{2\cos x-\sin 2x}{(\pi-2x)^2}$, $g(x)=\frac{e^{-\cos x}-1}{8x-4\pi}$ and

$h(x)=\begin{cases}f(x), & \text{for } x<\pi/2\\ g(x), & \text{for } x>\pi/2\end{cases}$,

then which of the following holds?

(a) h is continuous at $x=\pi/2$

(b) h has an irremovable discontinuity at $x=\pi/2$

(c) h has a removable discontinuity at $x=\pi/2$

(d) $f\left(\frac{\pi^+}{2}\right)=g\left(\frac{\pi^-}{2}\right)$

25. If $f(x)=\frac{x-e^x+\cos 2x}{x^2}$, $x\ne 0$ is continuous at $x=0$, then

(a) $f(0)=\frac{5}{2}$ (b) $[f(0)]=-2$

(c) $\{f(0)\}=-0.5$ (d) $[f(0)]\cdot\{f(0)\}=-1.5$

where, $[x]$ and $\{x\}$ denote the greatest integer and fractional part function.

26. Consider the function $f(x)=\begin{cases}x\{x\}+1, & \text{if } 0\le x<1\\ 2-\{x\}, & \text{if } 1\le x\le 2\end{cases}$, where $\{x\}$ denotes the fractional part function. Which one of the following statements is not correct?

(a) $\lim_{x\to 1} f(x)$ exists

(b) $f(0)\ne f(2)$

(c) $f(x)$ is continuous in $[0, 2]$

(d) Rolle's theorem is not applicable to $f(x)$ in $[0, 2]$

27. Let $f(x)=\begin{cases}\dfrac{2^x+2^{3-x}-6}{\sqrt{2^{-x}-2^{1-x}}}, & \text{if } x>2\\[2mm] \dfrac{x^2-4}{x-\sqrt{3x-2}}, & \text{if } x<2\end{cases}$, then

(a) $f(2)=8 \Rightarrow f$ is continuous at $x=2$
(b) $f(2)=16 \Rightarrow f$ is continuous at $x=2$
(c) $f(2^-)\neq f(2^+) \Rightarrow f$ is discontinuous
(d) f has a removable discontinuity at $x=2$

28. Let $[x]$ denotes the integral part of $g(x)=x-[x], x\in R$. Let $f(x)$ be any continuous function with $f(0)=f(1)$, then the function $h(x)=f(g(x))$
(a) has finitely many discontinuities
(b) is discontinuous at some $x=c$
(c) is continuous on R
(d) is a constant function

29. Let f be a differentiable function on the open interval (a, b).
I. f is continuous on the closed interval $[a, b]$.
II. f is bounded on the open interval (a, b).
III. If $a<a_1<b_1<b$ and $f(a_1)<0<f(b_1)$, then there exists a number c such that $a_1<c<b_1$ and $f(c)=0$.
Which of the above statements must be true?
(a) I and II
(b) I and III
(c) II and III
(d) Only III

30. Number of points, where the function $f(x)=(x^2-1)|x^2-x-2|+\sin(|x|)$ is not differentiable, is
(a) 0 (b) 1
(c) 2 (d) 3

31. Consider function $f: R-\{-1, 1\}\to R.\ f(x)=\dfrac{x}{1-|x|}$.
Then, which of the statements is incorrect?
(a) It is continuous at the origin
(b) It is not derivable at the origin
(c) The range of the function is R
(d) f is continuous and derivable in its domain

32. If the functions $f: R\to R$ and $g: R\to R$ are such that $f(x)$ is continuous at $x=\alpha$ and $f(\alpha)=a$ and $g(x)$ is discontinuous at $x=a$ but $g(f(x))$ is continuous at $x=\alpha$, where $f(x)$ and $g(x)$ are non-constant functions.
(a) $x=\alpha$ is a extremum of $f(x)$ and $x=a$ is an extremum of $g(x)$
(b) $x=\alpha$ may not be an extremum of $f(x)$ and $x=a$ is an extremum of $g(x)$
(c) $x=\alpha$ is an extremum of $f(x)$ and $x=a$ may not be an extremum of $g(x)$
(d) None of the above

33. The total number of points of non-differentiability of $f(x)=\min\left[|\sin x|, |\cos x|, \dfrac{1}{4}\right]$ in $(0, 2\pi)$ is
(a) 8 (b) 9 (c) 10 (d) 11

34. The function $f(x)=[x]^2-[x^2]$, where $[y]$ is the greatest integer less than or equal to y, is discontinuous at
(a) all integers
(b) all integers except 0 and 1
(c) all integers except 0
(d) all integers except 1

35. The function $f(x)=(x^2-1)|x^2-6x+5|+\cos|x|$ is not differentiable at
(a) -1 (b) 0 (c) 1 (d) 5

36. Let $f(x)=\begin{cases}e^{\frac{-1}{x^2}}, & \text{if } x\neq 0\\ 0, & \text{if } x=0\end{cases}$

Then, $f'(0)$ is equal to
(a) 0 (b) 1
(c) -1 (d) doesn't exist

37. Given $f(x)=\dfrac{e^x-\cos 2x-x}{x^2}$, for $x\in R-\{0\}$

$$g(x)=\begin{cases} f(\{x\}), & \text{for } n<x<n+\frac{1}{2}\\ f(1-\{x\}), & \text{for } n+\frac{1}{2}\le x<n+1, n\in I\\ \frac{5}{2}, & \text{otherwise}\end{cases}$$

$\left\{\begin{array}{l}\text{where } \{x\} \text{ denotes}\\ \text{fractional part function,}\end{array}\right.$ then $g(x)$ is

(a) discontinuous at all integral values of x only
(b) continuous everywhere except for $x=0$
(c) discontinuous at $x=n+\dfrac{1}{2}, n\in I$ and at some $x\in I$
(d) continuous everywhere

38. The function $g(x)=\begin{cases}x+b, & x<0\\ \cos x, & x\ge 0\end{cases}$ cannot be made differentiable at $x=0$,
(a) if b is equal to zero (b) if b is not equal to zero
(c) if b takes any real value (d) for no value of b

39. The graph of function f contains the point $P(1, 2)$ and $Q(s, r)$. The equation of the secant line through P and Q is $y=\left(\dfrac{s^2+2s-3}{s-1}\right)x-1-s$. The value of $f'(1)$ is
(a) 2 (b) 3
(c) 4 (d) non-existent

40. Consider

$$f(x)=\left[\frac{2(\sin x-\sin^3 x)+|\sin x-\sin^3 x|}{2(\sin x-\sin^3 x)-|\sin x-\sin^3 x|}\right], x\neq\frac{\pi}{2} \text{ for}$$

$x\in(0,\pi)$, $f(\pi/2)=3$, where [ ] denotes the greatest integer function, then

(a) f is continuous and differentiable at $x=\pi/2$
(b) f is continuous but not differentiable at $x=\pi/2$
(c) f is neither continuous nor differentiable at $x=\pi/2$
(d) None of the above

41. If $f(x+y)=f(x)+f(y)+|x|y+xy^2, \forall\ x, y\in R$ and $f'(0)=0$, then

(a) f need not be differentiable at every non-zero x
(b) f is differentiable for all $x\in R$
(c) f is twice differentiable at $x=0$
(d) None of the above

42. Let $f(x)=\max\{|x^2-2|x||, |x|\}$ and $g(x)=\min\{|x^2-2|x||, |x|\}$, then

(a) both $f(x)$ and $g(x)$ are non-differentiable at 5 points
(b) $f(x)$ is not differentiable at 5 points whether $g(x)$ is non-differentiable at 7 points
(c) number of points of non-differentiability for $f(x)$ and $g(x)$ are 7 and 5 points, respectively
(d) both $f(x)$ and $g(x)$ are non-differentiable at 3 and 5 points, respectively

43. Let $g(x)=\begin{cases}3x^2-4\sqrt{x}+1, & \text{for } x<1\\ ax+b, & \text{for } x\geq 1\end{cases}$. If $g(x)$ is the continuous and differentiable for all numbers in its domain, then

(a) $a=b=4$ (b) $a=b=-4$
(c) $a=4$ and $b=-4$ (d) $a=-4$ and $b=4$

44. Let $f(x)$ be continuous and differentiable function for all reals and $f(x+y)=f(x)-3xy+f(y)$.

If $\lim_{h\to 0}\frac{f(h)}{h}=7$, then the value of $f'(x)$ is

(a) $-3x$ (b) 7
(c) $-3x+7$ (d) $2f(x)+7$

45. Let $[x]$ be the greatest integer function and

$$f(x)=\frac{\sin\frac{1}{4}\pi[x]}{[x]}.$$ Then, which one of the following does not hold good?

(a) Not continuous at any point
(b) Continuous at 3/2
(c) Discontinuous at 2
(d) Differentiable at 4/3

46. If $f(x)=\begin{cases}b([x]^2+[x])+1, & \text{for } x\geq -1\\ \sin(\pi(x+a)), & \text{for } x<-1\end{cases}$, where $[x]$ denotes the integral part of x, then for what values of a, b the function is continuous at $x=-1$?

(a) $a=2n+(3/2);\ b\in R;\ n\in I$
(b) $a=4n+2;\ b\in R;\ n\in I$
(c) $a=4n+(3/2);\ b\in R^+;\ n\in I$
(d) $a=4n+1;\ b\in R^+;\ n\in I$

47. If both $f(x)$ and $g(x)$ are differentiable functions at $x=x_0$, then the function defined as, $h(x)=\max\{f(x), g(x)\}$

(a) is always differentiable at $x=x_0$
(b) is never differentiable at $x=x_0$
(c) is differentiable at $x=x_0$ when $f(x_0)\neq g(x_0)$
(d) cannot be differentiable at $x=x_0$, if $f(x_0)=g(x_0)$

48. Number of points of non-differentiability of the function

$$g(x)=[x^2]\{\cos^2 4x\}+\{x^2\}[\cos^2 4x]+x^2\sin^2 4x +[x^2][\cos^2 4x]+\{x^2\}\{\cos^2 4x\} \text{ in } (-50, 50),$$

where $[\cdot]$ and $\{\cdot\}$ are greatest integer function and fractional part of x, is equal to

(a) 98
(b) 99
(c) 100
(d) 0

49. If $f(x)=\frac{\{x\}g(x)}{\{x\}g(x)}$ is a periodic function with period $\frac{1}{4}$, where $g(x)$ is a differentiable function, then (where $\{\cdot\}$ denotes fractional part of x).

(a) $g'(x)$ has exactly three roots in $\left(\frac{1}{4},\frac{5}{4}\right)$
(b) $g(x)=0$ at $x=\frac{k}{4}$, where $k\in I$
(c) $g(x)$ must be non-zero function
(d) $g(x)$ must be periodic function

50. If $f\left(\frac{x}{y}\right)=\frac{f(x)}{f(y)}$ for all $x, y\in R$, $y\neq 0$ and $f'(x)$ exists for all x, $f(2)=4$. Then, $f(5)$ is

(a) 3 (b) 5
(c) 25 (d) None of these

Continuity and Differentiability Exercise 2 : More than One Option Correct Type Questions

51. Indicate the correct alternative, if $f(x) = \frac{x}{2} - 1$, then on the interval $[0, \pi]$,

(a) $\tan(f(x))$ and $\frac{1}{f(x)}$ are both continuous

(b) $\tan(f(x))$ and $\frac{1}{f(x)}$ are both discontinuous

(c) $\tan(f(x))$ and $f^{-1}(x)$ are both continuous

(d) $\tan(f(x))$ is continuous but $\frac{1}{f(x)}$ is not continuous

52. On the interval $I = [-2, 2]$, if the function

$$f(x) = \begin{cases} (x+1)\, e^{-\left(\frac{1}{|x|} + \frac{1}{x}\right)}, & x \neq 0 \\ 0, & x = 0 \end{cases}$$

then which of the following hold good?

(a) $f(x)$ is continuous for all values of $x \in I$

(b) $f(x)$ is continuous for $x \in I - \{0\}$

(c) $f(x)$ assumes all intermediate values from $f(-2)$ to $f(2)$

(d) $f(x)$ has a maximum value equal to $3/e$

53. If $f(x) = \begin{cases} 3 - \left[\cot^{-1}\left(\dfrac{2x^3 - 3}{x^2}\right)\right], & \text{for } x > 0 \\ \{x^2\} \cos(e^{1/x}), & \text{for } x < 0 \end{cases}$, where $\{x\}$ and $[x]$ denote fractional part and the greatest integer function respectively, then which of the following statements does not hold good?

(a) $f(0^-) = 0$

(b) $f(0^+) = 0$

(c) $f(0) = 0 \Rightarrow$ Continuous at $x = 0$

(d) Irremovable discontinuity at $x = 0$

54. If $f(x) = \begin{cases} b([x]^2 + [x]) + 1, & \text{for } x > -1 \\ \sin(\pi(x+a)), & \text{for } x < -1 \end{cases}$, where $[x]$ denotes the integral part of x, then for what values of a and b, the function is continuous at $x = -1$?

(a) $a = 2n + \frac{3}{2}; b \in R, n \in I$

(b) $a = 4n + 2; b \in R, n \in I$

(c) $a = 4n + \frac{3}{2}; b \in R^+, n \in I$

(d) $a = 4n + 1; b \in R^+, n \in I$

55. Let $[x]$ be the greatest integer function, then

$$f(x) = \frac{\sin \frac{1}{4} \pi [x]}{[x]} \text{ is}$$

(a) not continuous at any point

(b) continuous at $x = \frac{3}{2}$

(c) discontinuous at $x = 2$

(d) differentiable at $x = \frac{4}{3}$

56. If $f(x) = \begin{cases} (\sin^{-1} x)^2 \cos\left(\dfrac{1}{x}\right), & x \neq 0 \\ 0, & x = 0 \end{cases}$, then $f(x)$ is

(a) continuous nowhere in $-1 \le x \le 1$

(b) continuous everywhere in $-1 \le x \le 1$

(c) differentiable nowhere in $-1 \le x \le 1$

(d) differentiable everywhere in $-1 \le x \le 1$

57. Let $f(x) = \cos x$ and

$$H(x) = \begin{cases} \min(f(t) : 0 \le t < x), & \text{for } 0 \le x \le \pi/2 \\ \dfrac{\pi}{2} - x, & \text{for } \dfrac{\pi}{2} < x \le 3 \end{cases}, \text{ then}$$

(a) $H(x)$ is continuous and derivable in $[0, 3]$

(b) $H(x)$ is continuous but not derivable at $x = \frac{\pi}{2}$

(c) $H(x)$ is neither continuous nor derivable at $x = \frac{\pi}{2}$

(d) maximum value of $H(x)$ in $[0, 3]$ is 1

58. If $f(x) = 3(2x+3)^{2/3} + 2x + 3$, then

(a) $f(x)$ is continuous but not differentiable at $x = -\frac{3}{2}$

(b) $f(x)$ is differentiable at $x = 0$

(c) $f(x)$ is continuous at $x = 0$

(d) $f(x)$ is differentiable but not continuous at $x = -\frac{3}{2}$

59. If $f(x) = \begin{cases} \dfrac{x \cdot \ln(\cos x)}{\ln(1 + x^2)}, & x \neq 0 \\ 0, & x = 0 \end{cases}$, then

(a) f is continuous at $x = 0$

(b) f is continuous at $x = 0$ but not differentiable at $x = 0$

(c) f is differentiable at $x = 0$

(d) f is not continuous at $x = 0$

60. Let $[x]$ denotes the greatest integer less than or equal to x. If $f(x)=[x \sin \pi x]$, then $f(x)$ is
(a) continuous at $x=0$
(b) continuous in $(-1, 0)$
(c) differentiable at $x=1$
(d) differentiable in $(-1, 1)$

61. The function $f(x)=[|x|]-|[x]|$, where $[x]$ denotes greatest integer function
(a) is continuous for all positive integers
(b) is discontinuous for all non-positive integers
(c) has finite number of elements in its range
(d) is such that its graph does not lie above the X-axis

62. The function $f(x)=\sqrt{1-\sqrt{1-x^2}}$
(a) has its domain $-1 \le x \le 1$
(b) has finite one sided derivates at the point $x=0$
(c) is continuous and differentiable at $x=0$
(d) is continuous but not differentiable at $x=0$

63. Consider the function $f(x)=|x^3+1|$. Then,
(a) domain of $f(x) \in R$
(b) range of f is R^+
(c) f has no inverse
(d) f is continuous and differentiable for every $x \in R$

64. f is a continuous function in $[a, b]$, g is a continuous function in $[b, c]$. A function $h(x)$ is defined as

$$h(x)=\begin{cases} f(x) \text{ for } x \in [a, b) \\ g(x) \text{ for } x \in (b, c] \end{cases}$$ If $f(b)=g(b)$, then

(a) $h(x)$ has a removable discontinuity at $x=b$
(b) $h(x)$ may or may not be continuous in $[a, c]$
(c) $h(b^-)=g(b^+)$ and $h(b^+)=f(b^-)$
(d) $h(b^+)=g(b^-)$ and $h(b^-)=f(b^+)$

65. Which of the following function(s) has/have the same range?
(a) $f(x)=\dfrac{1}{1+x}$
(b) $f(x)=\dfrac{1}{1+x^2}$
(c) $f(x)=\dfrac{1}{1+\sqrt{x}}$
(d) $f(x)=\dfrac{1}{\sqrt{3-x}}$

66. If $f(x)=\sec 2x+\operatorname{cosec} 2x$, then $f(x)$ is discontinuous at all points in
(a) $\{n\pi, n \in N\}$
(b) $\left\{(2n \pm 1)\dfrac{\pi}{4}, n \in I\right\}$
(c) $\left\{\dfrac{n\pi}{4}, n \in I\right\}$
(d) $\left\{(2n \pm 1)\dfrac{\pi}{8}, n \in I\right\}$

67. Let $f(x)=\begin{cases} x^n \sin\left(\dfrac{1}{x^2}\right), & x \ne 0 \\ 0, & x=0 \end{cases}$, $(n \in I)$, then
(a) $\lim_{x \to 0} f(x)$ exists for every $n>1$
(b) f is continuous at $x=0$ for $n>1$
(c) f is differentiable at $x=0$ for every $n>1$
(d) None of the above

68. A function is defined as $f(x)=\begin{cases} e^x, & x \le 0 \\ |x-1|, & x>0 \end{cases}$, then $f(x)$ is
(a) continuous at $x=0$
(b) continuous at $x=1$
(c) differentiable at $x=0$
(d) differentiable at $x=1$

69. Let $f(x)=\int_{-2}^{x} |t+1|\, dt$, then
(a) $f(x)$ is continuous in $[-1, 1]$
(b) $f(x)$ is differentiable in $[-1, 1]$
(c) $f'(x)$ is continuous in $[-1, 1]$
(d) $f'(x)$ is differentiable in $[-1, 1]$

70. A function $f(x)$ satisfies the relation $f(x+y)=f(x)+f(y)+xy(x+y), \forall\, x, y \in R$. If $f'(0)=-1$, then
(a) $f(x)$ is a polynomial function
(b) $f(x)$ is an exponential function
(c) $f(x)$ is twice differentiable for all $x \in R$
(d) $f'(3)=8$

71. If $f(x)=\begin{cases} 3x^2+12x-1, & -1 \le x \le 2 \\ 37-x, & 2<x \le 3 \end{cases}$, then
(a) $f(x)$ is increasing on $[-1, 2]$
(b) $f(x)$ is continuous on $[-1, 3]$
(c) $f'(2)$ doesn't exist
(d) $f(x)$ has the maximum value at $x=2$

72. If $f(x)=0$ for $x<0$ and $f(x)$ is differentiable at $x=0$, then for $x>0$, $f(x)$ may be
(a) x^2
(b) x
(c) $-x$
(d) $-x^{3/2}$

Continuity and Differentiability Exercise 3 : Statements I and II Type Questions

- **Directions** (Q. Nos. 73 to 82) *For the following questions, choose the correct answer from the codes* (a), (b), (c) *and* (d) *defined as follows :*
  - (a) Statement I is correct, Statement II is also correct; Statement II is the correct explanation of Statement I
  - (b) Statement I is correct, Statement II is also correct; Statement II is not the correct explanation of Statement I
  - (c) Statement I is correct, Statement II is incorrect
  - (d) Statement I is incorrect, Statement II is correct

73. **Statement I** $f(x) = \sin x + [x]$ is discontinuous at $x = 0$.

Statement II If $g(x)$ is continuous and $f(x)$ is discontinuous, then $g(x) + f(x)$ will necessarily be discontinuous at $x = a$.

74. Consider $f(x) = \begin{cases} 2\sin(a\cos^{-1} x), & \text{if } x \in (0, 1) \\ \sqrt{3}, & \text{if } x = 0 \\ ax + b, & \text{if } x < 0 \end{cases}$

Statement I If $b = \sqrt{3}$ and $a = \frac{2}{3}$, then $f(x)$ is continuous in $(-\infty, 1)$.

Statement II If a function is defined on an interval I and limit exists at every point of interval I, then function is continuous in I.

75. Let $f(x) = \begin{cases} \dfrac{\cos x - e^{x^2/2}}{x^3}, & x \neq 0 \\ 0, & x = 0 \end{cases}$, then

Statement I $f(x)$ is continuous at $x = 0$.

Statement II $\lim_{x \to 0} \dfrac{\cos x - e^{-x^2/2}}{x^3} = -\dfrac{1}{12}$

76. **Statement I** The equation $\dfrac{x^3}{4} - \sin \pi x + \dfrac{2}{3} = 0$ has atleast one solution in $[-2, 2]$.

Statement II Let $f : [a, b] \to R$ be a function and c be a number such that $f(a) < c < f(b)$, then there is at least one number $n \in (a, b)$ such that $f(n) = c$.

77. **Statement I** Range of $f(x) = x\left(\dfrac{e^{2x} - e^{-2x}}{e^{2x} + e^{-2x}}\right) + x^2 + x^4$ is not R.

Statement II Range of a continuous even function cannot be R.

78. Let $h(x) = f_1(x) + f_2(x) + f_3(x) + \dots + f_n(x)$, where $f_1(x), f_2(x), f_3(x), \dots, f_n(x)$ are real valued functions of x.

Statement I $f(x) = |\cos|x|| + \cos^{-1}(\text{sgn } x) + |\ln x|$ is not differentiable at 3 points in $(0, 2\pi)$.

Statement II Exactly one function, is $f_i(x), i = 1, 2, \dots, n$ is not differentiable and the rest of the function is differentiable at $x = a$ makes $h(x)$ not differentiable at $x = a$.

79. **Statement I** $f(x) = |x| \sin x$ is differentiable at $x = 0$.

Statement II If $g(x)$ is not differentiable at $x = a$ and $h(x)$ is differentiable at $x = a$, then $g(x) \cdot h(x)$ cannot be differentiable at $x = a$.

80. **Statement I** $f(x) = |\cos x|$ is not derivable at $x = \dfrac{\pi}{2}$.

Statement II If $g(x)$ is differentiable at $x = a$ and $g(a) = 0$, then $|g(x)|$ is non-derivable at $x = a$.

81. Let $f(x) = x - x^2$ and $g(x) = \{x\}, \forall x \in R$, where $\{\}$ denotes fractional part function.

Statement I $f(g(x))$ will be continuous, $\forall x \in R$.

Statement II $f(0) = f(1)$ and $g(x)$ is periodic with period 1.

82. Let $f(x) = \begin{cases} -ax^2 - b|x| - c, & -\alpha \le x < 0 \\ ax^2 + b|x| + c, & 0 \le x \le \alpha \end{cases}$, where a, b, c are positive and $\alpha > 0$, then

Statement I The equation $f(x) = 0$ has at least one real root for $x \in [-\alpha, \alpha]$.

Statement II Values of $f(-\alpha)$ and $f(\alpha)$ are opposite in sign.

Continuity and Differentiability Exercise 4 : Passage Based Questions

Passage I (Q. Nos. 83 to 85)

A man leaves his home early in the morning to have a walk. He arrives at a junction of roads A and B as shown in figure. He takes the following steps in later journies :

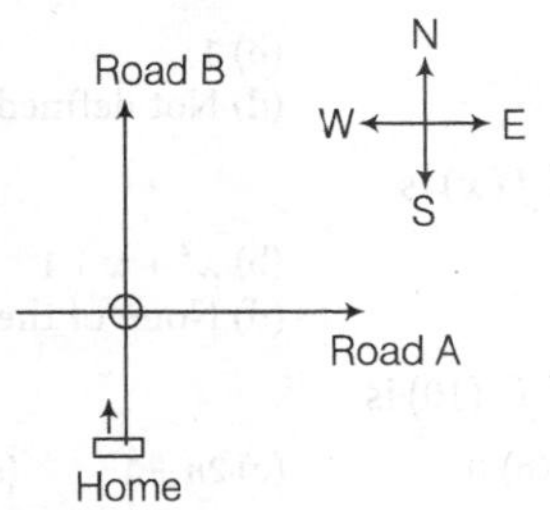

(i) *1 km in North direction.*

(ii) *Changes direction and moves in North-East direction for $2\sqrt{2}$ km.*

(iii) *Changes direction and moves Southwards for distance of 2 km.*

(iv) *Finally, he changes the direction and moves in South-East direction to reach road A again.*

Visible/invisible path *The path traced by the man in the direction parallel to road A and road B is called invisible path, the remaining path is called visible.*

Visible points *The point about which the man changes direction are called visible points, except the point from where he changes direction last time.*

Now, if roads A and B are taken as X-axis and Y-axis, then visible point representing the graph of $y = f(x)$.

83 The value of x at which the function is discontinuous, is

(a) 2 (b) 0 (c) 1 (d) 3

84. The value of x for which *fof* is discontinuous, is

(a) 0 (b) 1 (c) 2 (d) 3

85. If $f(x)$ is periodic with period 3, then $f(19)$ is

(a) 2 (b) 3
(c) 19 (d) None of these

Passage II (Q. Nos. 86 to 89)

Let f be a function that is differentiable everywhere and that has the following properties :

(i) $f(x) > 0$ (ii) $f'(0) = -1$

(iii) $f(-x) = \dfrac{1}{f(x)}$ and $f(x+h) = f(x) \cdot f(h)$

A standard result is $\dfrac{f'(x)}{f(x)} dx = \log|f(x)| + C$.

86. Range of $f(x)$ is

(a) R (b) $R - \{0\}$
(c) R^+ (d) $(0, e)$

87. The range of the function $\Delta = f(|x|)$ is

(a) $[0, 1]$ (b) $[0, 1)$
(c) $(0, 1]$ (d) None of these

88. The function $y = f(x)$ is

(a) odd (b) even
(c) increasing (d) decreasing

89. If $h(x) = f'(x)$, then $h(x)$ is given by

(a) $-f(x)$ (b) $\dfrac{1}{f(x)}$ (c) $f(x)$ (d) $e^{f(x)}$

Passage III (Q. Nos. 90 to 92)

Let $y = f(x)$ be defined in $[a, b]$, then

(i) *Test of continuity at $x = c, a < c < b$*

(ii) *Test of continuity at $x = a$*

(iii) *Test of continuity at $x = b$*

Case I *Test of continuity at $x = c, a < c < b$*

If $y = f(x)$ be defined at $x = c$ and its value $f(c)$ be equal to limit of $f(x)$ as $x \to c$, i.e. $f(c) = \lim_{x \to c} f(x)$

or $\lim_{x \to c^-} f(x) = f(c) = \lim_{x \to c^+} f(x)$

or LHL $= f(c) =$ RHL

Then, $y = f(x)$ is continuous at $x = c$.

Case II *Test of continuity at $x = a$*

If RHL $= f(a)$

Then, $f(x)$ is said to be continuous at the end point $x = a$.

Case III *Test of continuity at $x = b$, if LHL $= f(b)$*

Then, $f(x)$ is continuous at right end $x = b$.

90. If $f(x) = \begin{cases} \sin x, & x \le 0 \\ \tan x, & 0 < x < 2\pi \\ \cos x, & 2\pi \le x < 3\pi \\ 3\pi, & x = 3\pi \end{cases}$, then $f(x)$ is discontinuous at

(a) $\dfrac{\pi}{2}, \dfrac{3\pi}{2}, 2\pi$ (b) $0, \dfrac{\pi}{2}, \pi, \dfrac{3\pi}{2}, 3\pi$

(c) $\dfrac{\pi}{2}, 2\pi$ (d) None of these

91. Number of points of discontinuity of $[2x^3 - 5]$ in $[1, 2)$ is (where $[\cdot]$ denotes the greatest integral function)

(a) 14 (b) 13
(c) 10 (d) None of these

92. Max $([x], |x|)$ is discontinuous at

(a) $x = 0$
(b) ϕ
(c) $x = n, n \in I$
(d) None of the above

Passage IV (Q. Nos. 93 to 95)

$f(x) = \cos x$ *and* $H_1(x) = \min\{f(t), 0 \le t < x\};$

$$0 \le x \le \frac{\pi}{2} = \frac{\pi}{2} - x; \frac{\pi}{2} < x \le \pi$$

$f(x) = \cos x$ *and* $H_2(x) = \max\{f(t), 0 \le t \le x\};$

$$0 \le x \le \frac{\pi}{2} = \frac{\pi}{2} - x; \frac{\pi}{2} < x \le \pi$$

$g(x) = \sin x$ *and* $H_3(x) = \min\{g(t), 0 \le t \le x\};$

$$0 \le x \le \frac{\pi}{2} = \frac{\pi}{2} - x; \frac{\pi}{2} < x \le \pi$$

$g(x) = \sin x$ *and* $H_4(x) = \max\{g(t), 0 \le t \le x\};$

$$0 \le x \le \frac{\pi}{2} = \frac{\pi}{2} - x; \frac{\pi}{2} < x \le \pi$$

93. Which of the following is true for $H_2(x)$?

(a) Continuous and derivable in $[0, \pi]$

(b) Continuous but not derivable at $x = \frac{\pi}{2}$

(c) Neither continuous nor derivable at $x = \frac{\pi}{2}$

(d) None of the above

94. Which of the following is true for $H_3(x)$?

(a) Continuous and derivable in $[0, \pi]$

(b) Continuous but not derivable at $x = \frac{\pi}{2}$

(c) Neither continuous nor derivable at $x = \frac{\pi}{2}$

(d) None of the above

95. Which of the following is true for $H_4(x)$?

(a) Continuous and derivable in $[0, \pi]$

(b) Continuous but not derivable at $x = \frac{\pi}{2}$

(c) Neither continuous nor derivable at $x = \frac{\pi}{2}$

(d) None of the above

Passage V (Q. Nos. 96 to 99)

Let $f(x)$ be a real valued function not identically zero, which satisfied the following conditions

I. $f(x + y^{2n+1}) = f(x) + (f(y))^{2n+1}, n \in N, x, y$ *are any real numbers.*

II. $f'(0) \ge 0$

96. The value of $f(1)$ is

(a) 0 (b) 1
(c) 2 (d) Not defined

97. The value of $f(x)$ is

(a) $2x$ (b) $x^2 + x + 1$
(c) x (d) None of these

98. The value of $f'(10)$ is

(a) 10 (b) 0 (c) $2n + 1$ (d) 1

99. The function $f(x)$ is

(a) odd
(b) even
(c) neither even nor odd
(d) both even as well as odd

Passage VI (Q. Nos. 100 to 101)

If $f: R \to (0, \infty)$ is a differentiable function $f(x)$ satisfying $f(x + y) - f(x - y) = f(x) \cdot \{f(y) - f(-y)\}, \forall\, x, y \in R$, *$(f(y) \ne f(-y)$ for all $y \in R)$ and $f'(0) = 2010$. Now, answer the following questions*

100. Which of the following is true for $f(x)$?

(a) $f(x)$ is one-one and into

(b) $\{f(x)\}$ is non-periodic, where $\{\cdot\}$ denotes fractional part of x

(c) $f(x) = 4$ has only two solutions

(d) $f(x) = f'(x)$ has only one solution

101. The value of $\frac{f'(x)}{f(x)}$ is

(a) 2016 (b) 2014 (c) 2012 (d) 2010

Continuity and Differentiability Exercise 5: Matching Type Questions

102. Match the column.

Column I	Column II
(A) If $f(x)=\begin{cases}\sin\{x\}; & x<1\\ \cos x+a; & x\geq 1\end{cases}$, where $\{x\}$ denotes the fractional part function, such that $f(x)$ is continuous at $x=1$. If $\lvert K\rvert=\dfrac{a}{\sqrt{2}\sin\dfrac{(4-\pi)}{4}}$, then K is	(p) 1
(B) If the function $f(x)=\dfrac{1-\cos(\sin x)}{x^2}$ is continuous at $x=0$, then $f(0)$ is	(q) 0
(C) If $f(x)=\begin{cases}x, & x\in\theta\\ 1-x, & x\notin\theta\end{cases}$, then the values of x at which $f(x)$ is continuous, is	(r) -1
(D) If $f(x)=x+\{-x\}+[x]$, where $[x]$ and $\{x\}$ represent greatest integer and fractional part of x, then the values of x at which $f(x)$ is discontinuous, is	(s) $\dfrac{1}{2}$

103. Match the column.

Column I	Column II
(A) Number of points where the function $f(x)=\begin{cases}1+\left[\cos\dfrac{\pi x}{2}\right], & 1<x\leq 2\\ 1-\{x\}, & 0\leq x<1 \text{ and}\\ \lvert\sin \pi x\rvert, & -1\leq x<0\end{cases}$ $f(1)=0$ is continuous but non-differentiable, where $[\cdot]$ denotes greatest function and $\{\cdot\}$ denotes fractional part of x, is	(p) 0
(B) If $f(x)=\begin{cases}x^2e^{1/x}, & x\neq 0\\ 0, & x=0\end{cases}$, then $f(0^-)$ is	(q) 1
(C) The number of points at which $f(x)=\dfrac{1}{1+2/f(x)}$ is not differentiable, where $f(x)=\dfrac{1}{1+1/x}$, is	(r) 2
(D) Number of points where tangent does not exist for the curve $y=\text{sgn}(x^2-1)$, is	(s) 3

104. Match the column.

Column I	Column II
(A) The number of values of x in $(0, 2\pi)$, where the function $f(x)=\dfrac{\tan x+\cot x}{2}-\left\lvert\dfrac{\tan x-\cot x}{2}\right\rvert$ is continuous but not differentiable, is	(p) 2
(B) The number of points where the function $f(x)=\min\{1, 1+x^3, x^2-3x+3\}$ is non-derivable, is	(q) 0
(C) The number of points where $f(x)=(x+4)^{1/3}$ is non-differentiable, is	(r) 4
(D) Consider $f(x)=\begin{cases}\dfrac{-\pi}{2}\log\left(\dfrac{2x}{\pi}\right)+\dfrac{\pi}{2}, & 0<x\leq\dfrac{\pi}{2}\\ \sin^{-1}(\sin x), & \dfrac{\pi}{2}<x\leq\dfrac{3\pi}{2}\end{cases}$ Number of points in $\left(0, \dfrac{3\pi}{2}\right)$, where $f(x)$ is non-differentiable, is	(s) 1

105. Match the entries of the following two columns.

Column I	Column II
(A) $f(x)=\begin{cases}x+1, & \text{if } x<0\\ \cos x, & \text{if } x\geq 0\end{cases}$ at $x=0$ is	(p) continuous
(B) For every $x\in R$, the function $g(x)=\dfrac{\sin(\pi[x-\pi])}{1+[x]^2}$, where $[x]$ denotes the greatest integer function, is	(q) differentiability
(C) $h(x)=\sqrt{\{x\}^2}$, where $\{x\}$ denotes fractional part function for all $x\in I$, is	(r) discontinuous
(D) $k(x)=\begin{cases}x^{\frac{1}{\ln x}}, & \text{if } x\neq 1\\ e, & \text{if } x=1\end{cases}$ at $x=1$ is	(s) non-derivable

106. Match the entries of the following two columns.

Column I	Column II
(A) $\lim\limits_{x\to\infty}\left(e^{\sqrt{x^4+1}}-e^{(x^2+1)}\right)$ is	(p) e
(B) For $a>0$, let $f(x)=\begin{cases}\dfrac{a^x+a^{-x}-2}{x^2}, & \text{if } x>0\\ 3\ln(a-x)-2, & \text{if } x\leq 0\end{cases}$. If f is continuous at $x=0$, then a equals	(q) e^2
(C) Let $L=\lim\limits_{x\to a}\dfrac{x^x-a^a}{x-a}$ and $M=\lim\limits_{x\to a}\dfrac{x^a-a^x}{x-a}$ $(a>0)$. If $L=2M$, then the value of a is equal to	(r) non-existent

Continuity and Differentiability Exercise 6 : Single Integer Answer Type Questions

107. Number of points of discontinuity of $f(x) = \tan^2 x - \sec^2 x$ in $(0, 2\pi)$ is ....

108. Number of point(s) of discontinuity of the function $f(x) = [x^{1/x}], x > 0$, (where [ ] denotes the greatest integral function) is ....

109. Let $f(x) = x + \cos x + 2$ and $g(x)$ be the inverse function of $f(x)$, then $g'(3)$ equals to ....

110. Let $f(x) = x \tan^{-1}(x^2) + x^4$. Let $f^k(x)$ denotes kth derivative of $f(x)$ w.r.t. x, $k \in N$.

If $f^{2m}(0) \neq 0$, $m \in N$, then m equals to ....

111. Let $f_1(x)$ and $f_2(x)$ be twice differentiable functions, where $F(x) = f_1(x) + f_2(x)$ and $G(x) = f_1(x) - f_2(x), \forall x \in R, f_1(0) = 2$ and $f_2(0) = 1$. If $f_1'(x) = f_2(x)$ and $f_2'(x) = f_1(x), \forall x \in R$, then the number of solutions of the equation $(F(x))^2 = \dfrac{9x^4}{G(x)}$ is ....

112. Suppose the function $f(x) - f(2x)$ has the derivative 5 at $x = 1$ and derivative 7 at $x = 2$. The derivative of the function $f(x) - f(4x) - 10x$ at $x = 1$ is equal to ....

113. In a ΔABC, angles A, B, C are in AP.

If $f(x) = \lim_{A \to C} \dfrac{\sqrt{3 - 4\sin A \sin C}}{|A - C|}$, then $f'(x)$ is equal to ....

114. Let $f(x) = \begin{cases} x\dfrac{\left(\frac{3}{4}\right)^{1/x} - \left(\frac{3}{4}\right)^{-1/x}}{\left(\frac{3}{4}\right)^{1/x} + \left(\frac{3}{4}\right)^{-1/x}}, & x \neq 0 \\ 0, & x = 0 \end{cases}$

If $P = f'(0^-) - f'(0^+)$, then

$$4\left(\lim_{x \to P-} \frac{(\exp((x+2)\log 4))^{\frac{[x+1]}{4}} - 16}{4^x - 16}\right)$$ is ....

(where $[x]$ denotes greatest integer function.)

115. Let $f(x) = -x^3 + x^2 - x + 1$ and

$g(x) = \begin{cases} \min(f(t)), 0 \le t \le x \text{ and } 0 \le x \le 1 \\ x - 1, \quad 1 < x \le 2 \end{cases}$

Then, the value of $\lim_{x \to 1} g(g(x))$ is...... .

116. The number of points at which the function $f(x) = (x - |x|)^2 (1 - x + |x|)^2$ is not differentiable in the interval $(-3, 4)$ is ....

117. If $f(x) = \begin{cases} \dfrac{\frac{\pi}{2} - \sin^{-1}(1 - \{x\}^2)\sin^{-1}(1 - \{x\})}{\sqrt{2}(\{x\} - \{x\}^3)}, & x > 0 \\ k, & x = 0 \\ \dfrac{A\sin^{-1}(1 - \{x\})\cos^{-1}(1 - \{x\})}{\sqrt{2\{x\}}(1 - \{x\})}, & x < 0 \end{cases}$

is continuous at $x = 0$, then the value of $\sin^2 k + \cos^2\left(\dfrac{A\pi}{\sqrt{2}}\right)$, is...... (where $\{\cdot\}$ denotes fractional part of x).

Continuity and Differentiability Exercise 7 : Subjective Type Questions

118. Discuss the continuity of the function $f(x) = [[x]] - [x - 1]$, where [ ] denotes the greatest integral function.

119. Examine the continuity or discontinuity of the following :

(i) $f(x) = [x] + [-x]$ (ii) $g(x) = \lim_{n \to \infty} \dfrac{x^{2n} - 1}{x^{2n} + 1}$

120. Examine the continuity and differentiability at points $x = 1$ and $x = 2$.

The function f defined by

$f(x) = \begin{cases} x[x], & 0 \le x < 2 \\ (x - 1)[x], & 2 \le x \le 3 \end{cases}$ (where $[\cdot]$ denotes the greatest integral function less than or equal to x).

121. Let f be twice differentiable function, such that $f''(x)=-f(x)$ and $f'(x)=g(x)$, $h(x)=[f(x)]^2+[g(x)]^2$. Find $h(10)$, if $h(5)=11$.

122. A function $f:R\to R$ satisfies the equation $f(x+y)=f(x)\cdot f(y)$ for all x, y in R and $f(x)\neq 0$ for any x in R. Let the function be differentiable at $x=0$ and $f'(0)=2$. Show that $f'(x)=2f(x)$ for all x in R. Hence, determine $f(x)$.

123. A function $f:R\to R$, where R is a set of real numbers, satisfying the equation $f\left(\frac{x+y}{3}\right)=\frac{f(x)+f(y)+f(0)}{3}$ for all x, y in R. If the function is differentiable at $x=0$, then show that it is differentiable for all x in R.

124. Let $f(x+y)=f(x)+f(y)+2xy-1$ for all real x, y and $f(x)$ be differentiable functions. If $f'(0)=\cos\alpha$, then prove that $f(x)>0, \forall x\in R$.

Continuity and Differentiability Exercise 8 : Questions Asked in Previous 10 Years' Exams

(i) JEE Advanced & IIT-JEE

125. For every pair of continuous function $f, g:[0,1]\to R$ such that $\max\{f(x): x\in[0,1]\}=\max\{g(x): x\in[0,1]\}$. The correct statement(s) is/are **[More than One Correct Option, 2014]**

(a) $[f(c)]^2+3f(c)=[g(c)]^2+3g(c)$ for some $c\in[0,1]$

(b) $[f(c)]^2+f(c)=[g(c)]^2+3g(c)$ for some $c\in[0,1]$

(c) $[f(c)]^2+3f(c)=[g(c)]^2+g(c)$ for some $c\in[0,1]$

(d) $[f(c)]^2=[g(c)]^2$ for some $c\in[0,1]$

126. Let $f:R\to R$ and $g:R\to R$ be respectively given by $f(x)=|x|+1$ and $g(x)=x^2+1$. Define $h:R\to R$ by

$$h(x)=\begin{cases}\max\{f(x), g(x)\}, & \text{if } x\le 0\\ \min\{f(x), g(x)\}, & \text{if } x>0\end{cases}.$$

The number of points at which $h(x)$ is not differentiable is **[Integer Answer Type, 2014]**

127. Let $f(x)=\begin{cases}x^2\left|\cos\frac{\pi}{x}\right|, & x\neq 0\ x\in R,\\ 0, & x=0\end{cases}$, then f is **[One Correct Option, 2012]**

(a) differentiable both at $x=0$ and at $x=2$
(b) differentiable at $x=0$ but not differentiable at $x=2$
(c) not differentiable at $x=0$ but differentiable at $x=2$
(d) differentiable neither at $x=0$ nor at $x=2$

128. For every integer n, let a_n and b_n be real numbers. Let function $f:R\to R$ be given by

$$f(x)=\begin{cases}a_n+\sin\pi x, & \text{for } x\in[2n, 2n+1]\\ b_n+\cos\pi x, & \text{for } x\in(2n-1, 2n)\end{cases},$$

for all integers n.
If f is continuous, then which of the following hold(s) for all n? **[More than One Correct Option, 2012]**

(a) $a_{n-1}-b_{n-1}=0$ (b) $a_n-b_n=1$
(c) $a_n-b_{n+1}=1$ (d) $a_{n-1}-b_n=-1$

129. Let $f:R\to R$ be a function such that $f(x+y)=f(x)+f(y), \forall x, y\in R$. If $f(x)$ is differentiable at $x=0$, then **[More than One Correct Option, 2011]**

(a) $f(x)$ is differentiable only in a finite interval containing zero
(b) $f(x)$ is continuous for all $x\in R$
(c) $f'(x)$ is constant for all $x\in R$
(d) $f(x)$ is differentiable except at finitely many points

130. If $f(x)=\begin{cases}-x-\frac{\pi}{2}, & x\le-\frac{\pi}{2}\\ -\cos x, & -\frac{\pi}{2}<x\le 0\\ x-1, & 0<x\le 1\\ \ln x, & x>1\end{cases}$, then **[More than One Correct Option, 2011]**

(a) $f(x)$ is continuous at $x=-\frac{\pi}{2}$
(b) $f(x)$ is not differentiable at $x=0$
(c) $f(x)$ is differentiable at $x=1$
(d) $f(x)$ is differentiable at $x=-\frac{3}{2}$

131. For the function $f(x)=x\cos\frac{1}{x}, x\ge 1$, **[More than One Correct Option, 2009]**

(a) for at least one x in the interval $[1,\infty)$, $f(x+2)-f(x)<2$
(b) $\lim_{x\to\infty} f'(x)=1$
(c) for all x in the interval $[1,\infty)$, $f(x+2)-f(x)>2$
(d) $f'(x)$ is strictly decreasing in the interval $[1,\infty)$

132. Let $g(x)=\frac{(x-1)^n}{\log\cos^m(x-1)}; 0<x<2$, m and n are integers, $m\neq 0$, $n>0$ and let p be the left hand derivative of $|x-1|$ at $x=1$. If $\lim_{x\to 1^+} g(x)=p$, then **[One Correct Option, 2008]**

(a) $n=1, m=1$ (b) $n=1, m=-1$
(c) $n=2, m=2$ (d) $n>2, m=n$

133. Let f and g be real valued functions defined on interval $(-1, 1)$ such that $g''(x)$ is continuous, $g(0) \neq 0$, $g'(0) = 0$, $g''(0) \neq 0$ and $f(x) = g(x) \sin x$.

Statement I $\lim_{x \to 0} [g(x) \cos x - g(0)] \operatorname{cosec} x = f''(0)$.

Statement II $f'(0) = g(0)$.

For the above question, choose the correct answer from the codes (a), (b), (c) and (d) defined as follows. **[Assertion and Reason, 2008]**

(a) Statement I is true, Statement II is also true; Statement II is the correct explanation of Statement I
(b) Statement I is true, Statement II is also true; Statement II is not the correct explanation of Statement I
(c) Statement I is true; Statement II is false
(d) Statement I is false; Statement II is true

134. In the following, $[x]$ denotes the greatest integer less than or equal to x. **[Matching type Question, 2007]**

	Column I		Column II
A.	$x\|x\|$	p.	continuous in $(-1, 1)$
B.	$\sqrt{\|x\|}$	q.	differentiable in $(-1, 1)$
C.	$x+[x]$	r.	strictly increasing in $(-1, 1)$
D.	$\|x-1\|+\|x+1\|$	s.	not differentiable atleast at one point in $(-1, 1)$

135. If $f(x) = \min\{1, x^2, x^3\}$, then **[More than One Correct, 2006]**

(a) $f(x)$ is continuous everywhere
(b) $f(x)$ is continuous and differentiable everywhere
(c) $f(x)$ is not differentiable at two points
(d) $f(x)$ is not differentiable at one point

136. Let $f(x) = ||x| - 1|$, then points where, $f(x)$ is not differentiable is/are **[One Correct Option, 2005]**

(a) $0, \pm 1$ (b) ± 1
(c) 0 (d) 1

137. If f is a differentiable function satisfying $f\left(\frac{1}{n}\right) = 0, \forall$ $n \geq 1, n \in I$, then **[One Correct Option, 2005]**

(a) $f(x) = 0, x \in (0, 1]$
(b) $f'(0) = 0 = f(0)$
(c) $f(0) = 0$ but $f'(0)$ not necessarily zero
(d) $|f(x)| \leq 1, x \in (0, 1]$

138. The domain of the derivative of the functions

$$f(x) = \begin{cases} \tan^{-1} x, & \text{if } |x| \leq 1 \\ \frac{1}{2}(|x| - 1), & \text{if } |x| > 1 \end{cases} \text{ is}$$

[One Correct Option, 2002]

(a) $R - \{0\}$ (b) $R - \{1\}$
(c) $R - \{-1\}$ (d) $R - \{-1, 1\}$

139. The left hand derivative of $f(x) = [x] \sin(\pi x)$ at $x = k$, k is an integer, is **[One Correct Option, 2001]**

(a) $(-1)^k (k-1)\pi$
(b) $(-1)^{k-1} (k-1)\pi$
(c) $(-1)^k k\pi$
(d) $(-1)^{k-1} k\pi$

140. Which of the following functions is differentiable at $x = 0$? **[One Correct Option, 2001]**

(a) $\cos(|x|) + |x|$ (b) $\cos(|x|) - |x|$
(c) $\sin(|x|) + |x|$ (d) $\sin(|x|) - |x|$

(ii) JEE Main & AIEEE

141. For $x \in R$, $f(x) = |\log 2 - \sin x|$ and $g(x) = f(f(x))$, then

(a) g is not differentiable at $x = 0$ **[2016, JEE Main]**
(b) $g'(0) = \cos(\log 2)$
(c) $g'(0) = -\cos(\log 2)$
(d) g is differentiable at $x = 0$ and $g'(0) = -\sin(\log 2)$

142. If the function $g(x) = \begin{cases} k\sqrt{x+1}, & 0 \leq x \leq 3 \\ mx + 2, & 3 < x \leq 5 \end{cases}$ is differentiable, then the value of $k + m$ is **[2015, JEE Main]**

(a) 2 (b) $\frac{16}{5}$
(c) $\frac{10}{3}$ (d) 4

143. If f and g are differentiable functions in (0, 1) satisfying $f(0) = 2 = g(1)$, $g(0) = 0$ and $f(1) = 6$, then for some $c \in]0, 1[$ **[2014, JEE Main]**

(a) $2f'(c) = g'(c)$ (b) $2f'(c) = 3g'(c)$
(c) $f'(c) = g'(c)$ (d) $f'(c) = 2g'(c)$

144. If $f : R \to R$ is a function defined by $f(x) = [x] \cos\left(\frac{2x-1}{2}\right)\pi$, where $[x]$ denotes the greatest integer function, then f is **[2012 AIEEE]**

(a) continuous for every real x
(b) discontinuous only at $x = 0$
(c) discontinuous only at non-zero integral values of x
(d) continuous only at $x = 0$

Answers

Exercise for Session 1

1. (c) 2. (c) 3. (a) 4. (a) 5. (a)

Exercise for Session 2

1. $a = -1, b = 1$ 2. f is continuous in $-1 \le x \le 1$.
3. $f(x)$ is continuous everywhere in $[0, 2]$ except for $x = \frac{1}{2}$, 1 and 2.
4. (d)

Exercise for Session 3

1. (b,c,d) 2. (a,c,d) 3. (d) 4. (c) 5. (a,c,d)

Exercise for Session 4

1. (a,b,c) 2. (b,c) 3. (c,d) 4. (b,c) 5. (a,b,c) 6. (a) 7. (b)

Exercise for Session 5

1. discontinuity at $x = 0$
4. $y_n(x)$ is continuous at $x = 0$ for all n and $y(x)$ is discontinuous at $x = 0$.

Exercise for Session 6

1. (d) 2. (a) 3. (a) 4. (d) 5. (b)

Exercise for Session 7

1. (c) 2. (c) 3. (c) 4. (b) 5. (b) 6. (d)
7. (a) 8. (c) 9. (a) 10. (c)

Chapter Exercises

1. (a) 2. (c) 3. (c) 4. (c) 5. (a) 6. (c)
7. (b) 8. (b) 9. (b) 10. (b) 11. (a) 12. (c)
13. (a) 14. (b) 15. (a) 16. (c) 17. (c) 18. (c)
19. (c) 20. (b) 21. (b) 22. (b) 23. (d) 24. (b)
25. (d) 26. (c) 27. (c) 28. (c) 29. (d) 30. (c)
31. (b) 32. (c) 33. (d) 34. (d) 35. (d) 36. (a)
37. (d) 38. (d) 39. (c) 40. (a) 41. (b) 42. (b)
43. (c) 44. (c) 45. (c) 46. (a) 47. (c) 48. (d)
49. (b) 50. (c) 51. (c,d) 52. (b,c,d) 53. (a,b,c) 54. (a,c)
55. (b,c,d) 56. (b,d) 57. (a,d) 58. (a,b,c) 59. (a,c) 60. (a, b, c, d)
61. (a, b, c, d) 62. (a, b, d) 63. (a, c) 64. (a,c)
65. (b, c) 66. (a, b, c) 67. (a, b, c)
68. (a, b) 69. (a, b, c, d) 70. (a, c, d)
71. (a, b, c, d) 72. (a,d) 73. (a) 74. (c)
75. (a) 76. (c) 77. (a) 78. (a) 79. (c) 80. (c)
81. (a) 82. (d) 83. (a) 84. (b,c) 85. (a) 86. (c)
87. (a) 88. (d) 89. (a) 90. (a) 91. (b) 92. (b)
93. (c) 94. (b) 95. (c) 96. (b) 97. (c) 98. (d)
99. (a) 100. (b) 101. (d)
102. (A) $\rightarrow$ (p,r); (B) $\rightarrow$ (s); (C) $\rightarrow$ (s); (D) $\rightarrow$ (p,q,r)
103. (A) $\rightarrow$ (q); (B) $\rightarrow$ (p); (C) $\rightarrow$ (s); (D) $\rightarrow$ (p)
104. (A) $\rightarrow$ (r); (B) $\rightarrow$ (p); (C) $\rightarrow$ (s); (D) $\rightarrow$ (q)
105. (A) $\rightarrow$ (p, s); (B) $\rightarrow$ (p, q); (C) $\rightarrow$ (r, s); (D) $\rightarrow$ (p, q)
106. (A) $\rightarrow$ (r); (B) $\rightarrow$ (p, q); (C) $\rightarrow$ (p)
107. (2) 108. (1) 109. (1) 110. (2) 111. (2)
112. (9) 113. (0) 114. (2) 115. (1) 116. (0) 117. (2)
125. (a,d) 126. (3) 127. (b) 128. (b, d) 129. (b,c) 130. (a, b, c, d)
131. (b, c, d) 132. (c) 133. (b)
134. (A) $\rightarrow$ (p, q, r); (B) $\rightarrow$ (p, s); (C) $\rightarrow$ (r, s); (D) $\rightarrow$ (p, q)
135. (a,d) 136. (a) 137. (b) 138. (d) 139. (a) 140. (d)
141. (b) 142. (a) 143. (d) 144. (a)

Solutions

1. $f(x)=\begin{cases}\sin\dfrac{\pi x}{2}, & x<1\\ [x], & x\ge 1\end{cases}$

$f(1)=1,\ f(1^+)=\lim_{h\to 0}[1+h]=1$

$f(1^-)=\lim_{h\to 0}\sin\dfrac{\pi}{2}(1-h)=\lim_{h\to 0}\sin\left(\dfrac{\pi}{2}-\dfrac{\pi h}{2}\right)=\lim_{h\to 0}\cos\dfrac{\pi h}{2}=1$

$\therefore\ f(x)$ is continuous at $x=1$.

2. $f(0^+)=\lim_{h\to 0}\dfrac{8^h-4^h-2^h+1}{h^2}=\lim_{h\to 0}\dfrac{(4^h-1)}{h}\dfrac{(2^h-1)}{h}$

$=(\log 4)(\log 2)$

$f(0^-)=\lim_{h\to 0}(e^{-h}\sin(-h)+\pi(-h)+k\log 4)=k\log 4$

Since, $f(x)$ is continuous at $x=0$

$\Rightarrow\quad k\log 4=(\log 4)(\log 2)=f(0).$

$\therefore\quad f(0)=(\log 4)(\log 2)$, when $k=\log 2$

3. $f(0^-)=\lim_{x\to 0^-}\dfrac{a(1-x\sin x)+b\cos x+5}{x^2}=f(0)$

$=\lim_{x\to 0^-}\dfrac{(a+b+5)+\left(-a-\dfrac{b}{2}\right)x^2+\ldots}{x^2}=3$

$\Rightarrow\quad a+b+5=0$ and $-a-\dfrac{b}{2}=3$

$\Rightarrow\quad a=-1,\ b=-4$

$\Rightarrow\quad f(0^+)=\lim_{x\to 0^+}\left[1+\left(\dfrac{cx+dx^3}{x^2}\right)\right]^{1/x}$ exists.

$\Rightarrow\quad \lim_{x\to 0^+}\dfrac{cx+dx^3}{x^2}=0\Rightarrow c=0$

Now, $\lim_{x\to 0^+}(1+dx)^{1/x}=\lim_{x\to 0^+}((1+dx)^{1/dx})^d=e^d$

So, $\quad e^d=3\Rightarrow d=\log_e 3$

$\therefore\quad a+b+c+d=\log_e 3-5$

4. $f\left(\dfrac{\pi^+}{2}\right)=\lim_{h\to 0}\pi\left[\dfrac{\pi}{2}+h\right]-1=(\pi-1)$

$f\left(\dfrac{\pi^-}{2}\right)=\lim_{h\to 0}\cos^{-1}\left\{\cot\left(\dfrac{\pi}{2}-h\right)\right\}=\lim_{h\to 0}\cos^{-1}\{\tan h\}=\dfrac{\pi}{2}$

$\therefore$ Jump of discontinuity $=(\pi-1)-\dfrac{\pi}{2}=\dfrac{\pi}{2}-1$

5. Consider $\quad g(x)=f(x)-x$

$g(0)=f(0)-0=f(0)\ge 0\qquad [\because 0\le f(x)\le 1]$

$g(1)=f(1)-1\le 0$

$\Rightarrow\quad g(0)\cdot g(1)\le 0$

$\Rightarrow\quad g(x)=0$ has at least one root in $[0, 1]$.

$\Rightarrow\quad f(x)=x$ for at least one root in $[0, 1]$.

6. Here, $f(x)=\dfrac{x+1}{x-1}$, discontinuous at $x=1$

$g(x)=\dfrac{1}{x-2}$, discontinuous at $x=2$

$f(g(x))=\dfrac{g(x)+1}{g(x)-1}$, discontinuous at $g(x)=1$

$\Rightarrow\quad \dfrac{1}{x-2}=1$

$\Rightarrow\quad x=3$

$\therefore\ (fog)(x)$ is discontinuous at $x=2$ and 3.

7. Here, $y_n(x)=\begin{cases}\dfrac{x^2\left[\dfrac{1}{(1+x^2)^n}-1\right]}{\dfrac{1}{1+x^2}-1}=(1+x^2)\cdot\left[1-\dfrac{1}{(1+x^2)^n}\right], & \text{when } x\ne 0,\ n\in N\\ 0, & \text{when } x=0, n\in N\end{cases}$

$\therefore\ y(x)=\lim_{n\to\infty}y_n(x)=\begin{cases}(1+x^2), & x\ne 0\\ 0, & x=0\end{cases}$

$\therefore\ y(x)$ is discontinuous at $x=0$.

8. $g(0^-)=\lim_{h\to 0}\dfrac{a^h-1-h\log a}{h^2}$

$=\lim_{h\to 0}\dfrac{\left[1+h\log a+\dfrac{h^2}{2!}(\log a)^2+\ldots\right]-1-h\log a}{h^2}$

$=\lim_{h\to 0}\left[\dfrac{(\log a)^2}{2!}\cdot 1+\dfrac{(\log a)^3}{3!}\cdot h\ldots\right]=\dfrac{(\log a)^2}{2}$

$g(0^+)=\lim_{h\to 0}\dfrac{2^h a^h-h\log 2-h\log a-1}{h^2}$

$=\lim_{h\to 0}\dfrac{\left[1+h\log(2a)+\dfrac{h^2}{2!}(\log(2a))^2\ldots\right]-h\log 2-h\log a-1}{h^2}$

$=\dfrac{(\log(2a))^2}{2}$

Since, $g(x)$ is continuous.

$\Rightarrow\quad \dfrac{(\log a)^2}{2}=\dfrac{(\log(2a))^2}{2}\Rightarrow(\log a)^2=(\log 2+\log a)^2$

$\Rightarrow\quad \log a=-\dfrac{1}{2}\log 2=\log 2^{-1/2}$

$\Rightarrow\quad a=\dfrac{1}{\sqrt{2}}$

9. $f(0^+)=\lim_{h\to 0}\dfrac{\left(\dfrac{\pi}{2}-\sin^{-1}(1-h^2)\right)\cdot\sin^{-1}(1-h)}{\sqrt{2}(h-h^3)}$

$=\lim_{h\to 0}\dfrac{(\cos^{-1}(1-h^2))\cdot\sin^{-1}(1-h)}{\sqrt{2}(1-h^2)\cdot h}=\dfrac{\pi}{2}$

$$f(0^-)=\lim_{h\to 0}\frac{\left(\frac{\pi}{2}-\sin^{-1}(1-(1-h^2))\right)\cdot\sin^{-1}(1-(1-h))}{\sqrt{2}((1-h)-(1-h)^3)}$$

$$=\lim_{h\to 0}\frac{\frac{\pi}{2}\sin^{-1}h}{\sqrt{2}(1-h)(2-h)h}=\frac{\pi}{4\sqrt{2}}$$

10. $f(x)=\begin{cases}-1+x, & -\infty<x<0\\ -1+\sin x, & 0\le x<\pi/2\\ \cos x, & \pi/2\le x<\infty\end{cases}$

$$f'(x)=\begin{cases}1, & -\infty<x<0\\ \cos x, & 0<x<\pi/2\\ -\sin x, & \pi/2<x<\infty\end{cases}$$

$$f'(0^-)=1,\ f'(0^+)=1$$

$\Rightarrow f(x)$ is differentiable at $x=0$

$$f'\left(\frac{\pi^-}{2}\right)=0,\ f'(\pi/2^+)=-1$$

$\Rightarrow f(x)$ is not differentiable at $x=\frac{\pi}{2}$.

$\therefore$ Number of points of non-differentiability is 1.

11. $f(x)=\begin{cases}-\frac{1}{x}, & x\le -1\\ ax^2+b, & -1<x<1\\ \frac{1}{x}, & x\ge 1\end{cases}$

Since, function is continuous everywhere.

$\therefore$ LHL = RHL at $x=-1$

$\Rightarrow$ $f(-1^-)=1,\ f(-1^+)=a+b$

$\Rightarrow$ $a+b=1$ ...(i)

$f(x)$ is differentiable at $x=-1$

$$f'(x)=\begin{cases}\frac{1}{x^2}, & x<-1\\ 2ax, & -1<x<1\\ -\frac{1}{x^2}, & x>1\end{cases}$$

$f'(-1^-)=1,\ f'(-1^+)=-2a\Rightarrow -2a=1,\ a=-1/2$ ...(ii)

From Eqs. (i) and (ii), we get

$$a=-\frac{1}{2},\ b=\frac{3}{2}$$

12. $f(1^-)=A+B,\ f(1^+)=3A-B+2$

Since, continuous $A+B=3A-B+2$

$\Rightarrow$ $2A-2B=-2\Rightarrow A-B=-1$ ...(i)

$$f'(x)=\begin{cases}2Bx, & x<1\\ 3A, & x>1\end{cases}$$

$$f'(1^+)=3A,\ f'(1^-)=2B$$

$\Rightarrow$ $3A=2B$ ...(ii)

On solving Eqs. (i) and (ii), we get

$$A=2,\ B=3$$

13. $f'(1^+)=\lim_{h\to 0}\frac{f(1+h)-f(1)}{h}$

$$=\lim_{h\to 0}\frac{|1+h-1|([1+h]-(1+h))}{h}=0=\lim_{h\to 0}\frac{h(1-1-h)}{h}=0$$

$$f'(1^-)=\lim_{h\to 0}\frac{f(1-h)-f(1)}{-h}=\lim_{h\to 0}\frac{|-h|([1-h]-(1-h))}{-h}$$

$$=\lim_{h\to 0}\frac{h(0-1+h)}{-h}=1$$

$\Rightarrow$ $f'(1^+)=0,\ f'(1^-)=1$

$\therefore f(x)$ is not differentiable at $x=1$.

14. $f'(1^-)=\lim_{h\to 0}\frac{f(1-h)-f(1)}{-h}=\lim_{h\to 0}\frac{[\cos\pi(1-h)]+1}{-h}$

$$=\lim_{h\to 0}\frac{-1+1}{-h}=0$$

$$f'(1^+)=\lim_{h\to 0}\frac{f(1+h)-f(1)}{h}$$

$$=\lim_{h\to 0}\frac{2\{1+h\}-1+1}{h}$$

$$=\lim_{h\to 0}\frac{2h}{h}=2$$

$\therefore$ $f'(1^-)=0,\ f'(1^+)=2$

15. $f(x)=\begin{cases}x-3, & x<0\\ x^2-3x+2, & x\ge 0\end{cases}$

$$f(|x|)=\begin{cases}|x|-3, & |x|<0 \text{ not possible}\\ |x|^2-3|x|+2, & |x|\ge 0\end{cases}$$

$\Rightarrow$ $g(x)=\begin{cases}x^2+3x+2, & x\le 0\\ x^2-3x+2, & x\ge 0\end{cases}$

$\Rightarrow g(x)$ is continuous at $x=0$.

$$g'(x)=\begin{cases}2x+3, & x\le 0\\ 2x-3, & x\ge 0\end{cases}$$

$$g'(0^+)=-3,\ g'(0^-)=3$$

16. $f(x)=\begin{cases}0, & 0\le x<1/2\\ 5/6, & x=1/2\\ 0, & 1/2<x<1\\ -2, & 1\le x<4/3\\ 0, & x=4/3\\ 2, & 4/3<x<3/2\\ 3, & 3/2\le x<2\\ 4, & x=2\end{cases}$

Hence, $f(x)$ is neither continuous nor differentiable at $x=\frac{1}{2},1,\frac{4}{3},\frac{3}{2},2$.

$\therefore$ Number of points is 5.

17. Put $x=y=1$ in given rule

$\Rightarrow$ $f(1)=f(1)-f(1)=0$

$$\therefore\ f'(x)=\lim_{h\to 0}\frac{f(x+h)-f(x)}{h}=\lim_{h\to 0}\frac{f\left(\frac{x+h}{x}\right)}{h}$$

$$= \lim_{h \to 0} \left(\frac{f\left(1 + \frac{h}{x}\right) - f(1)}{\frac{h}{x}} \right) \frac{1}{x} = \frac{f'(1)}{x} = \frac{1}{x}$$

[given, $f'(1) = 1$]

On integrating both sides, $f(x) = \log x + C$

On putting $x = 1$, we get $C = 0 \Rightarrow f(x) = \log_e x$

18. $f'(x) = \lim_{h \to 0} \frac{f(x+h) - f(x)}{h}$

$$= \lim_{h \to 0} \frac{f(x) + f(h) - 2xh - 1 - f(x)}{h}$$

$$f'(x) = \lim_{h \to 0} \left(-2x + \frac{f(h) - 1}{h} \right) = -2x + f'(0)$$

On integrating both sides, we get

$\Rightarrow \quad f(x) = -x^2 + f'(0)x + c$

$\Rightarrow \quad f(x) = -x^2 - (\sin \alpha)x + c \quad$ [given $f'(0) = -\sin \alpha$]

$\Rightarrow \quad f(0) = c \Rightarrow c = 1$

So, $f\{f'(0)\} = f\{-\sin \alpha\} = -\sin^2 \alpha + \sin^2 \alpha + 1$

$= 1$

19. $f'(x) = \lim_{h \to 0} \frac{f(x+h) - f(x)}{h} \geq \lim_{h \to 0} \frac{\log\left(\frac{x+h}{x}\right) + x + h - x}{h}$

$$= \lim_{h \to 0} \frac{\log\left(1 + \frac{h}{x}\right)}{\frac{h}{x}} \frac{1}{x} + 1 \geq \frac{1}{x} + 1 \quad \text{...(i)}$$

$$f'(x) = \lim_{h \to 0} \frac{f(x-h) - f(x)}{-h} \leq \lim_{h \to 0} \frac{\log\left(\frac{x-h}{x}\right) + x - h - x}{-h}$$

$$= \lim_{h \to 0} \frac{\log\left(1 - \frac{h}{x}\right)}{-h} + 1 \leq \frac{1}{x} + 1 \quad \text{...(ii)}$$

$\therefore$ From Eqs. (i) and (ii), we get $f'(x) = \frac{1}{x} + 1 = g(x)$

$$\Rightarrow \sum_{n=1}^{100} g\left(\frac{1}{n}\right) = g\left(\frac{1}{1}\right) + g\left(\frac{1}{2}\right) + \dots + g\left(\frac{1}{100}\right)$$

$= (1 + 1) + (2 + 1) + (3 + 1) + \dots + (100 + 1)$

$= (1 + 2 + \dots + 100) + 100$

$= 5050 + 100 = 5150$

20. Here, $\frac{d(f(x))}{dx} = e^{-x} f(x) + e^x f(-x)$

Let $\quad \frac{d(f(x))}{dx} = g(x)$, where $g(x) = e^{-x} f(x) + e^x f(-x)$

$\therefore \quad g(-x) = e^x f(-x) + e^{-x} f(x) = g(x)$

$\Rightarrow$ $g(x)$ is even function.

Hence, $f(x)$ should be an odd function as $f(0) = 0$.

21. We have, $f(x^2) = \int_0^{x^2} t f(t)\,dt = x^4 + x^5 \quad$ ...(i)

On differentiating both the sides w.r.t. x, we get

$$x^2 f(x^2) \cdot 2x = 4x^3 + 5x^4$$

$$\Rightarrow \quad f(x^2) = 2 + \frac{5}{2} x \quad \text{...(ii)}$$

$$\therefore \sum_{r=1}^{12} f(r^2) = \sum_{r=1}^{12} \left(2 + \frac{5}{2} r \right) = 24 + \frac{5}{2} \cdot \frac{(12)(13)}{2} = 24 + 195 = 219$$

Hence, $\quad \sum_{r=1}^{12} f(r^2) = 219$

22. Let $x = \frac{2}{3}$, which is rational.

$$\Rightarrow \quad h\left(\frac{2}{3}\right) = \frac{1}{3}$$

$\lim_{t \to 0} h\left(\frac{2}{3} + t\right) = 0 \Rightarrow$ Discontinuous at $x \in Q$

Let $\quad x = \sqrt{2} \notin Q$

$h(\sqrt{2}) = 0$, consider $\sqrt{2} = 1.4142135624$

$$h(\sqrt{2}) = h\left(\frac{14142135624}{10^{10}}\right) = \frac{1}{10^{10}} \to 0$$

Hence, h is continuous for all irrational.

23. By theorem, if g and h are continuous functions on the open interval (a, b), then g/h is also continuous at all x in the open interval (a, b), where $h(x) \neq 0$.

24. $h(x) = \begin{cases} \frac{2\cos x - \sin 2x}{(\pi - 2x)^2}, & x < \frac{\pi}{2} \\ \frac{e^{-\cos x} - 1}{8x - 4\pi}, & x > \frac{\pi}{2} \end{cases}$

LHL at $x = \pi/2$

$$\text{LHL} = \lim_{h \to 0} \frac{2\sin h - \sin 2h}{4h^2} = \lim_{h \to 0} \frac{2\sin h\,(1 - \cos h)}{4h^2} = 0$$

$$\text{RHL} = \lim_{h \to 0} \frac{e^{\sin h} - 1}{8((\pi/2) + h) - 4\pi}$$

$$= \lim_{h \to 0} \frac{e^{\sin h} - 1}{8h} \cdot \frac{\sin h}{\sin h} = \frac{1}{8}$$

$\Rightarrow h(x)$ is discontinuous at $x = \pi/2$.

Irremovable discontinuity at $x = \pi/2$.

$$f\left(\frac{\pi^+}{2}\right) = 0 \text{ and } g\left(\frac{\pi^-}{2}\right) = \frac{1}{8}$$

$$\Rightarrow \quad f\left(\frac{\pi^+}{2}\right) \neq g\left(\frac{\pi^-}{2}\right)$$

25. $\lim_{x \to 0} \frac{x - e^x + 1 - (1 - \cos 2x)}{x^2} = -\frac{1}{2} - 2 = -\frac{5}{2}$

Hence, for continuity, $f(0) = -\frac{5}{2}$

$\therefore [f(0)] = -3$; $\{f(0)\} = \left\{-\frac{5}{2}\right\} = \frac{1}{2}$

Hence, $\quad [f(0)] \cdot \{f(0)\} = -\frac{3}{2} = -1.5$

26. $f(1^+) = f(1^-) = f(1) = 2$

$f(0) = 1, \quad f(2) = 2$

$f(2^-) = 1,$

$\Rightarrow$ f is not continuous at $x = 2$.

27. $f(2^+) = 8,\ f(2^-) = 16$

28. $g(x) = x - [x] = \{x\}$

f is continuous with $f(0) = f(1)$

$h(x) = f(g(x)) = f(\{x\})$

Let the graph of f is as shown in the figure satisfying

$$f(0) = f(1)$$

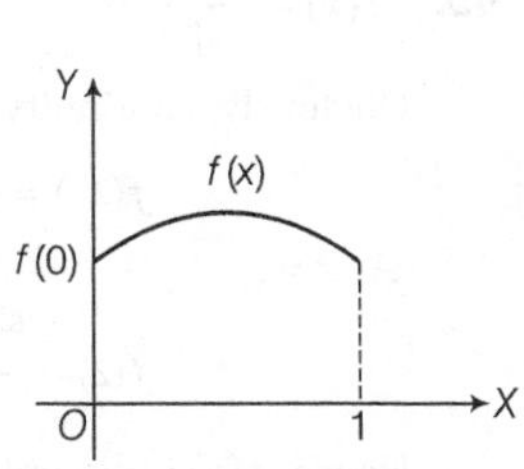

Now,

$h(0) = f(\{0\}) = f(0) = f(1)$

$h(0.2) = f(\{0.2\}) = f(0.2)$

$h(1.5) = f(\{1.5\}) = f(0.5)$ etc.

Hence, the graph of $h(x)$ will be a periodic graph as shown

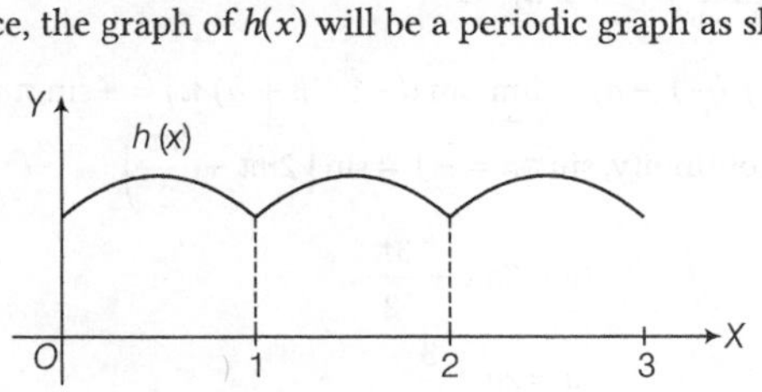

$\Rightarrow h$ is continuous in R.

29. Statements I and II are false. The function $f(x) = 1/x,\ 0 < x < 1$ is a counter example.

Statement III is true. Apply the intermediate value theorem to f on the closed interval $[a_1, b_1]$.

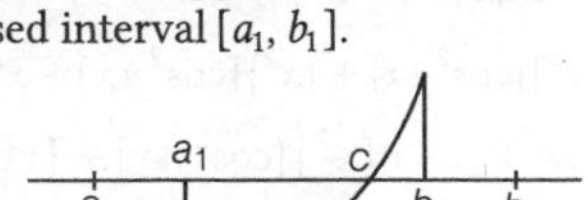

30. Since, the given function is not differentiable, at $x = 0$ and 2.

Hence, the number of points is 2.

31. $f(x) = \begin{cases} \dfrac{x}{1-x}, & \text{if } x \ge 0,\ x \ne 1 \\ \dfrac{x}{1+x}, & \text{if } x < 0,\ x \ne -1 \end{cases}$

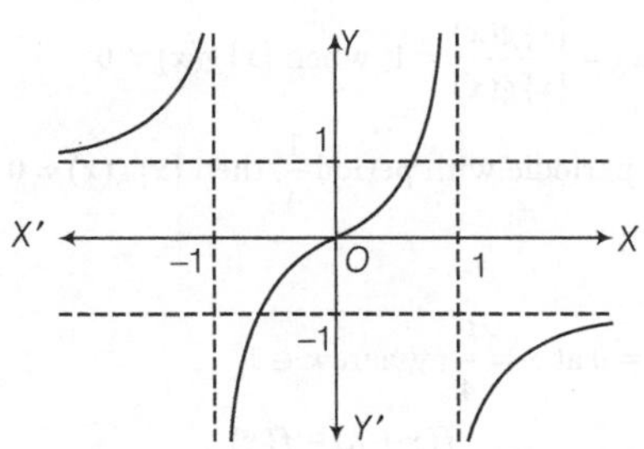

and $f'(x) = \begin{cases} \dfrac{1}{(1-x)^2}, & \text{if } x > 0, x \ne 1 \\ \dfrac{1}{(1+x)^2}, & \text{if } x < 0,\ x \ne -1 \end{cases}$

32. $g[f(x)]$ is continuous at $x = \alpha$

$\Rightarrow \quad g[f(\alpha)] = g(a)$

Also, $\quad \lim_{x \to \alpha} g(f(x)) = g(a)$

$\Rightarrow \quad g[f(\alpha^-)] = g[f(\alpha^+)] = g(a)$

$\Rightarrow g(x)$ takes same limiting values at $f(\alpha^-),\ f(\alpha^+)$ and $f(\alpha)$.

$\Rightarrow f(\alpha^-) = f(\alpha^+) \Rightarrow x = \alpha$ is an extremum of $f(x)$ and $x = a$ may not be an extremum of $g(x)$.

33. From the graph, number of points of non-differentiability = 11

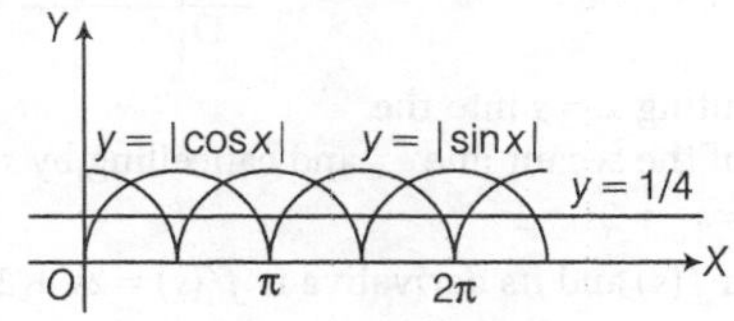

34. Let n be any integer other than 1.

$$\lim_{x \to n^-} f(x) = \lim_{h \to 0} ([n-h]^2 - [(n-h)^2])$$

$$= (n-1)^2 - (n^2 - 1) = 2 - 2n$$

$$\lim_{x \to n^+} f(x) = \lim_{h \to 0} [n+h]^2 - [(n+h)^2] = n^2 - n^2 = 0$$

$\therefore$ LHL $\ne$ RHL unless $n = 1$.

Hence, $f(x)$ is discontinuous at all integral values except 1.

35. At $x = 5$, $f'(x) = \lim_{x \to 5} \dfrac{\{(x-1)^2 (x+1)|x-5| + \cos|x|\} - \cos 5}{x-5}$

$$= \lim_{x \to 5} \frac{96|x-5|}{x-5} = +96, \text{ if } x > 5 \text{ and } -96, \text{ if } x < 5$$

Hence, $f'(5)$ doesn't exist.

This ambiguity doesn't occur at other points.

$\therefore f(x)$ is not differentiable at $x = 5$.

36. $f'(0^+) = \lim_{x \to 0^+} \dfrac{f(x) - f(0)}{x} = \lim_{x \to 0^+} \dfrac{e^{\frac{-1}{x^2}} - 0}{x} = \lim_{x \to 0^+} \dfrac{1/x}{e^{1/x^2}}$

$$= \lim_{x \to 0^+} \frac{-1/x^2}{e^{1/x^2} \cdot \left(-\dfrac{2}{x^3}\right)} = \lim_{x \to 0^+} \frac{x}{2e^{1/x^2}} = 0$$

As, f is even, so $f'(0^-) = f'(0^+) = 0$. Thus, $f'(0) = 0$

37. $\lim_{h \to 0} g(n+h) = \lim_{h \to 0} \dfrac{e^h - \cos 2h - h}{h^2}$

$$= \lim_{h \to 0} \frac{e^h - h - 1}{h^2} + \lim_{h \to 0} \frac{(1 - \cos 2h)}{4h^2} \cdot 4 = \frac{1}{2} + 2 = \frac{5}{2}$$

$$\lim_{h \to 0} g(n-h) = \frac{e^{1-\{n-h\}} - \cos 2(1 - \{n-h\}) - (1 - \{n-h\})}{(1 - \{n-h\})^2}$$

$$= \lim_{h \to 0} \frac{e^h - \cos 2h - h}{h^2} [\{n-h\} = \{-h\} = 1 - h] = \frac{5}{2}$$

$g(n) = \dfrac{5}{2}$. Hence, $g(x)$ is continuous, $\forall\ x \in I$.

Hence, $g(x)$ is continuous, $\forall\ x \in R$.

38. $g'(0^+) = \lim_{h \to 0} \dfrac{\cos h - 1}{h} = 0$

$g'(0^-) = \lim_{h \to 0} \dfrac{-h + b - 1}{-h}$ for existence of limit $b = 1$

Thus, $g'(0^-) = 1$

Hence, g cannot be made differentiable for no value of b.

39. By definition $f'(1)$ is the limit of the slope of the secant line when $s \to 1$.

Thus, $f'(1) = \lim\limits_{s \to 1} \dfrac{s^2 + 2s - 3}{s - 1}$

$= \lim\limits_{s \to 1} \dfrac{(s-1)(s+3)}{s-1}$

$= \lim\limits_{s \to 1} (s+3) = 4$

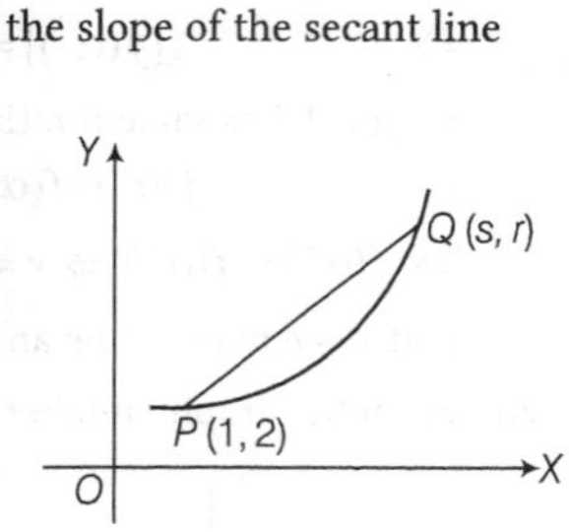

Aliter

By substituting $x = s$ into the equation of the secant line and cancelling by $s - 1$. Again, we get $y = s^2 + 2s - 1$.

This is $f(s)$ and its derivative is $f'(s) = 2s + 2$, so $f'(1) = 4$.

40. In the immediate neighbourhood of $x = \pi/2$, $\sin x > \sin^3 x \Rightarrow |\sin x - \sin^3 x| = \sin x - \sin^3 x$

Hence, for $x \neq \pi/2$,

$$f(x) = \left[\frac{2(\sin x - \sin^3 x) + \sin x - \sin^3 x}{2(\sin x - \sin^3 x) - \sin x + \sin^3 x}\right]$$

$$= \frac{3\sin x - 3\sin^3 x}{\sin x - \sin^3 x} = 3$$

Hence, f is continuous and differentiable at $x = \pi/2$.

41. $f'(x) = \lim\limits_{h \to 0} \dfrac{f(x+h) - f(x)}{h} = \lim\limits_{h \to 0} \dfrac{f(h) + |x|h + xh^2}{h}$

$\because \quad f(0) = 0$

Hence, $\quad f'(x) = \lim\limits_{h \to 0} \left[\dfrac{f(h) - f(0)}{h} + |x| + xh\right]$

$\Rightarrow \quad f'(x) = f'(0) + |x| = |x|$

42. $f(x)$ is non-differentiable at $x = \alpha, \beta, 0, \gamma, \delta$

and $g(x)$ is non-differentiable at $x = \alpha, -2, \beta, 0, \gamma, 2, \delta$.

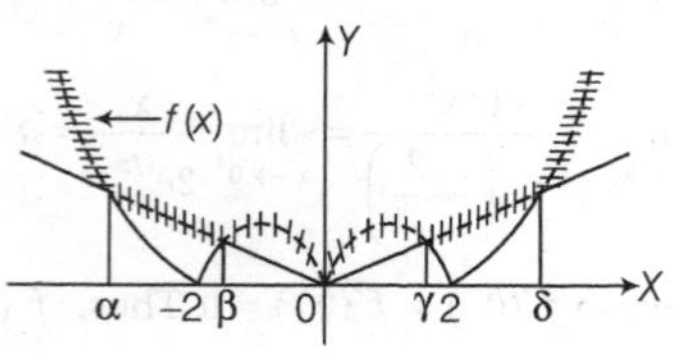

43. We have, $g(x) = \begin{cases} 3x^2 - 4\sqrt{x} + 1, & \text{for } x < 1 \\ ax + b, & \text{for } x \geq 1 \end{cases}$

For differentiability at $x = 1$, $g'(1^+) = g'(1^-)$

$$a = 6 - \frac{4}{2\sqrt{1}}$$

$\Rightarrow \quad a = 6 - 2 = 4$

For continuity at $x = 1$, $g(1^+) = g(1^-)$

$$a + b = 3 - 4 + 1 \Rightarrow a + b = 0$$

$\Rightarrow \quad b = -4$

Hence, $\quad a = 4$ and $b = -4$

44. $f'(x) = \lim\limits_{h \to 0} \dfrac{f(x+h) - f(x)}{h} = \lim\limits_{h \to 0} \dfrac{-3xh + f(h)}{h}$

$= \lim\limits_{h \to 0} \left\{-3x + \dfrac{f(h)}{h}\right\} = -3x + \lim\limits_{h \to 0} \dfrac{f(h)}{h}$

$= -3x + 7$

45. $f(x) = \dfrac{\sin \dfrac{\pi[x]}{4}}{[x]}$

Obviously, continuity at $x = 3/2$

$$f(2^-) = \sin\frac{\pi}{4} = \frac{1}{\sqrt{2}}$$

At $x = 2$,

$$f(2) = \frac{\sin\dfrac{\pi}{2}}{2} = \frac{1}{2}$$

Hence, $f(x)$ is discontinuous at $x = 2$.

46. $f(-1) = b(1-1) + 1 = 1$

and $\lim\limits_{h \to 0} f(-1+h) = 1$

$\lim\limits_{h \to 0} f(-1-h) = \lim\limits_{h \to 0} \sin((-1-h+a)\pi) = -\sin \pi a$

For continuity, $\sin \pi a = -1 = \sin\left(2n\pi + \dfrac{3\pi}{2}\right)$

$\Rightarrow \quad \pi a = 2n\pi + \dfrac{3\pi}{2}$

$\Rightarrow \quad a = 2n + \dfrac{3}{2}$

Hence, $\quad a = 2n + \dfrac{3}{2}$, $n \in I$ and $b \in R$

47. Consider the graph of $h(x) = \max(x, x^2)$ at $x = 0$ and $x = 1$

For (d): $h(x) = \max(x^2, -x^2)$

48. Here, $g(x) = [x^2]\{\cos^2 4x\} + \{x^2\}[\cos^2 4x] + x^2 \sin^2 4x + [x^2][\cos^2 4x] + \{x^2\}\{\cos^2 4x\}$

$= [x^2](\{\cos^2 4x\} + [\cos^2 4x]) + \{x^2\}([\cos^2 4x] + \{\cos^2 4x\}) + x^2 \sin^2 4x$

$= ([x^2] + \{x^2\})(\{\cos^2 4x\} + [\cos^2 4x]) + x^2 \sin^2 4x$

$= x^2 \cos^2 4x + x^2 \sin^2 4x$

$= x^2$

Clearly, $g(x)$ is always differentiable.

$\therefore$ Number of points of non-differentiability is 0.

49. Here, $f(x) = \dfrac{\{x\}g(x)}{\{x\}g(x)} = 1$, when $\{x\}g(x) \neq 0$

If $f(x)$ is periodic with period $\dfrac{1}{4}$, then $\{x\}g(x) \neq 0$ with period $\dfrac{1}{4}$.

$\Rightarrow g(x) = 0$ at $x = \dfrac{k}{4}$, where $k \in I$.

50. Here, $\quad f'(x) = \lim\limits_{h \to 0} \dfrac{f(x+h) - f(x)}{h}$

$$= \lim_{h \to 0} \frac{f\left(x\left(1 + \dfrac{h}{x}\right)\right) - f\left(\dfrac{1}{1/x}\right)}{h}$$

$$= \lim_{h \to 0} \frac{f\left(\dfrac{1 + h/x}{1/x}\right) - f\left(\dfrac{1}{1/x}\right)}{h}$$

$$= \lim_{h\to 0} \frac{\dfrac{f(1+h/x)}{f(1/x)} - \dfrac{f(1)}{f(1/x)}}{h}$$

$$= \lim_{h\to 0} \frac{f(1+h/x)-f(1)}{h/x} \cdot \frac{1}{xf\left(\frac{1}{x}\right)} = f'(1)\cdot\frac{1}{x}\cdot\frac{f(x)}{f(1)}$$

$\therefore \quad \dfrac{dy}{dx} = k\cdot\dfrac{y}{x}$, where $\dfrac{f'(1)}{f(1)} = k,\ f(x) = y.$

$\Rightarrow \quad \int \dfrac{dy}{y} = \int k\dfrac{dx}{x}$

$\Rightarrow \quad \log y = k\log x + \log C$

$\Rightarrow \quad \log y = \log(x^k \cdot C)$

$\Rightarrow \quad y = C\cdot x^k$

$\therefore \quad f(x) = C\cdot x^k$

Put $x = 2, y = 1$ in $f\left(\dfrac{x}{y}\right) = \dfrac{f(x)}{f(y)}$

$\Rightarrow \quad f(2) = \dfrac{f(2)}{f(1)} \Rightarrow f(1) = 1$

Also, $f(x) = C\cdot x^k$, put $x = 1$

$\Rightarrow \quad f(1) = C \Rightarrow C = 1$

$\therefore \quad f(x) = x^k$, now $f(2) = 4$

$\Rightarrow \quad f(2) = 2^k$

$4 = 2^k \Rightarrow k = 2$

$\therefore \quad f(x) = x^2 \Rightarrow f(5) = 25$

51. $\tan(f(x)) = \tan\left(\dfrac{x}{2} - 1\right),\quad x\in[0,\pi]$

$$0 \le x \le \pi \Rightarrow -1 \le \frac{x}{2} - 1 \le \frac{\pi}{2} - 1$$

$\therefore \tan(f(x))$ is continuous in $[0,\pi]$.

$\dfrac{1}{f(x)} = \dfrac{2}{x-2}$ is not defined at $x = 2 \in [0,\pi]$.

$y = \dfrac{x-2}{2} \Rightarrow f^{-1}(x) = 2x + 2$ is continuous in R.

52. $\lim\limits_{x\to 0^+} (x+1)e^{-\frac{2}{x}} = \lim\limits_{x\to 0^+} \dfrac{(x+1)}{e^{2/x}} = 0$

$\lim\limits_{x\to 0^-} (x+1)e^0 = 1$

Hence, continuous for $x \in I - \{0\}$, assumes all intermediate values from $f(-2)$ to $f(2)$ and maximum value $\dfrac{3}{e}$ at $x = 2$.

53. RHL $= \lim\limits_{x\to 0^+}\left(3 - \left[\cot^{-1}\left(\dfrac{2x^3-3}{x^2}\right)\right]\right)$

$= 3 - [\cot^{-1}(-\infty)] = 3 - 3 = 0$

LHL $= \lim\limits_{x\to 0^-} \{x^2\}\cos\left(e^{-\frac{1}{x}}\right)$

$= \lim\limits_{h\to 0} h^2 \cos\left(e^{\frac{1}{h}}\right) = 0$

and $f(0) = 0$

Hence, $f(x)$ is continuous at $x = 0$.

54. $\lim\limits_{x\to -1^+} f(x) = \lim\limits_{x\to -1^+} (b([x]^2 + [x]) + 1)$

$= \lim\limits_{h\to 0} (b([-1+h]^2 + [-1+h]) + 1)$

$= \lim\limits_{h\to 0} (b((-1)^2 - 1) + 1) = 1$

$\Rightarrow \quad b \in R$

$\lim\limits_{x\to -1^-} f(x) = \lim\limits_{x\to -1^-} \sin(\pi(x+a))$

$= \lim\limits_{h\to 0} \sin(\pi((-1-h)+a))$

$\therefore \quad \sin\pi a = -1$

$\Rightarrow \quad \pi a = 2n\pi + \dfrac{3\pi}{2} \Rightarrow a = 2n + \dfrac{3}{2}$

55. $f(x) = \begin{cases} \dfrac{\sin\frac{\pi}{4}}{1} = \dfrac{1}{\sqrt{2}}, & 1 \le x < 2 \\ \dfrac{\sin\frac{\pi}{2}}{1} = \dfrac{1}{2}, & 2 \le x < 3 \end{cases}$

Hence, $f(x)$ is continuous at $\dfrac{3}{2}$, differentiable at $\dfrac{4}{3}$ and discontinuous at 2.

56. Since, $\sin^{-1}x$ and $\cos(1/x)$ are continuous and differentiable in $x \in [-1,1] - \{0\}$.

Now, at $x = 0$

$$f'(0^-) = \lim_{h\to 0} \frac{f(-h) - f(0)}{-h} = 0$$

$$= \lim_{h\to 0} \frac{(\sin^{-1}(-h))^2 \cos\left(-\frac{1}{h}\right) - 0}{-h} = 0$$

$$f'(0^+) = \lim_{h\to 0} \frac{(\sin^{-1}h)^2 \cos\left(\frac{1}{h}\right) - 0}{h} = 0$$

Hence, LHD = RHD. So, $f(x)$ is continuous and differentiable everywhere in $-1 \le x \le 1$.

57. $H(x) = \begin{cases} \cos x, & 0 \le x < \dfrac{\pi}{2} \\ \dfrac{\pi}{2} - x, & \dfrac{\pi}{2} < x \le 3 \end{cases}$

$$H'\left(\frac{\pi}{2}^-\right) = -\sin x = -1$$

$$H'\left(\frac{\pi}{2}^+\right) = -1$$

Hence, $H(x)$ is continuous and derivable in $[0, 3]$ and has maximum value 1 in $[0, 3]$.

58. Here, $f(x) = 3(2x+3)^{2/3} + 2x + 3$

$$\Rightarrow \quad f'(x) = \frac{4}{(2x+3)^{1/3}} + 2$$

Now, $\quad 2x + 3 \neq 0 \Rightarrow x \neq -\frac{3}{2}$

Hence, $f(x)$ is continuous but not differential at $x = -3/2$.

Also, $f(x)$ is differentiable and continuous at $x = 0$.

59. $$f'(0^+) = \lim_{h \to 0} \frac{h \ln(\cos h)}{h \ln(1+h^2)} = \lim_{h \to 0} \frac{\ln(\cos h)^{1/h^2}}{\frac{\ln(1+h^2)}{h^2}}$$

$$= \lim_{h \to 0} \frac{1}{h^2}(\cos h - 1) = -\frac{1}{2}; \text{ similarly } f'(0^-) = -\frac{1}{2}$$

Hence, f is continuous and derivable at $x = 0$.

60. $$f(x) = \begin{cases} 0, & 0 < x < 1 \\ 0, & x = 0 \text{ or } 1 \text{ or } -1 \\ 0, & -1 < x < 0 \end{cases} \Rightarrow f(x) = 0 \text{ for all in } [-1, 1]$$

61. $$[|x|] - |[x]| = \begin{cases} 0, & x = -1 \\ -1, & -1 < x < 0 \\ 0, & 0 \le x \le 1 \\ 0, & 1 < x \le 2 \end{cases}$$

$\Rightarrow$ Range is $\{0, -1\}$. The graph is

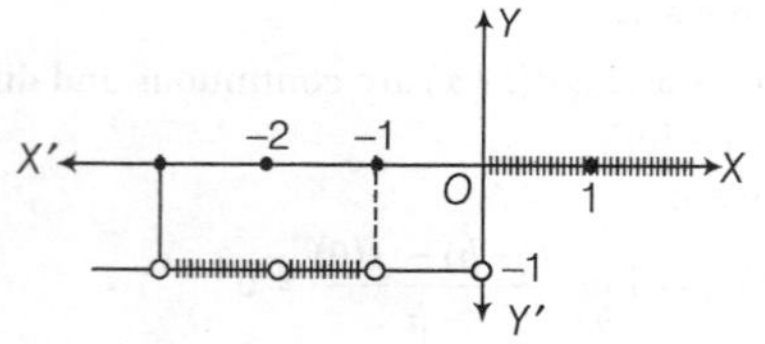

62. $f'(0^+) = \frac{1}{\sqrt{2}}, f'(0^-) = -\frac{1}{\sqrt{2}}$;

$$f(x) = \frac{\sqrt{x^2}}{\sqrt{1+\sqrt{1-x^2}}} = \frac{|x|}{\sqrt{1+\sqrt{1-x^2}}}$$

63. Range is $R^+ \cup \{0\} \Rightarrow$ Option (b) is not correct.

f is not differentiable at $x = -1$

As, $$f(x) = \begin{cases} x^3 + 1, & \text{if } x \ge -1 \\ -(x^3+1), & \text{if } x < -1 \end{cases}$$

$$\Rightarrow \quad f'(x) = \begin{cases} 3x^2, & \text{if } x > -1 \\ -3x^2, & \text{if } x < -1 \end{cases}$$

$f'(-1^+) = 3, f'(-1^-) = -3 \Rightarrow f$ is not differentiable at $x = -1$.

Also, f is not bijective, hence it has no inverse.

64. Given, f is continuous in $[a, b]$ ...(i)

$\Rightarrow$ g is continuous in $[b, c]$ ...(ii)

$\Rightarrow$ $f(b) = g(b)$ ...(iii)

$$\Rightarrow \quad \left.\begin{aligned} h(x) &= f(x) \text{ for } x \in [a, b) \\ &= g(x) \text{ for } x \in (b, c] \end{aligned}\right\} \quad \text{...(iv)}$$

$\Rightarrow h(x)$ is continuous in $[a, b) \cup (b, c]$ [using Eqs. (i) and (ii)]

Also, $\quad f(b^-) = f(b), g(b^+) = g(b)$

[using Eqs. (i) and (ii)] ...(iv)

$\therefore \quad h(b^-) = f(b^-) = f(b) = g(b) = g(b^+) = h(b^+)$

[using Eqs. (iv) and (v)]

Now, verify each alternative. Of course! $g(b^-)$ and $f(b^+)$ are undefined.

$$h(b^-) = f(b^-) = f(b) = g(b) = g(b^+)$$

and $\quad h(b^+) = g(b^+) = g(b) = f(b) = f(b^-)$

Hence, $\quad h(b^-) = h(b^+) = f(b) = g(b)$

and $h(b)$ is not defined.

65. (a) Domain is $R - \{-1\}$; Range $= R - \{0\}$

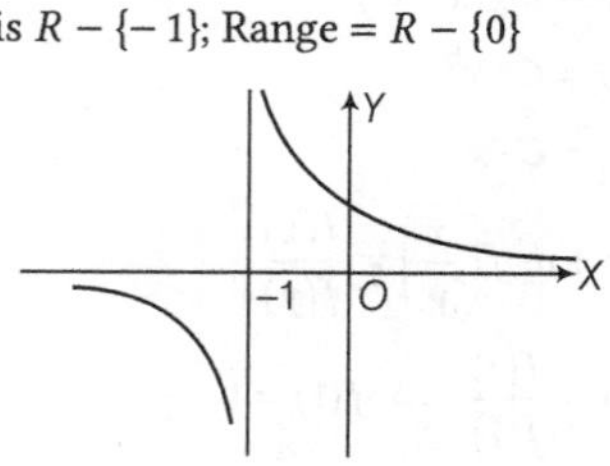

(b) Domain is $x \in R$; Range $= (0, 1]$

(c) Domain is $[0, \infty)$; Range $= (0, 1]$

(d) Domain is $(-\infty, 3)$; Range $= (0, \infty)$

66. $$f(x) = \sec 2x + \operatorname{cosec} 2x = \frac{2(\sin 2x + \cos 2x)}{2\cos 2x \sin 2x}$$

$$= \frac{2(\sin 2x + \cos 2x)}{\sin 4x}$$

is discontinuous, where $4x = n\pi, n \in I$ or $x = \frac{n\pi}{4}$.

Options (a) and (b) also satisfy the condition, since they are subsets of option (c).

67. $$\lim_{x \to 0} f(x) = \lim_{x \to 0} x^n \cdot \sin\left(\frac{1}{x^2}\right) = 0, \text{ if } n > 0$$

and hence true for $n > 1$.

Since, $f(0) = 0$, $f(x)$ is continuous at $x = 0$.

Now, $$\lim_{x \to 0} \frac{f(x) - f(0)}{x - 0} = \lim_{x \to 0} \frac{x^n \sin\frac{1}{x^2}}{x}$$

$$= \lim_{x \to 0} x^{n-1} \sin\left(\frac{1}{x^2}\right) = 0, \text{ if } n > 1.$$

Hence, $f(x)$ is differentiable at $x = 0$, if $n > 1$.

68. $$f(x) = \begin{cases} e^x, & x \le 0 \\ 1 - x, & 0 < x < 1 \\ x - 1, & x \ge 1 \end{cases}$$

$\Rightarrow \quad \lim_{x \to 0^-} f(x) = 1, \lim_{x \to 0^+} f(x) = 1$

and $\quad \lim_{x \to 1^-} f(x) = 0, \lim_{x \to 1^+} f(x) = 0$

Hence, $f(x)$ is continuous at $x = 0$ and 1.

$$f'(0^-) = 1 \text{ and } f'(0^+) = -1.$$

Hence, $f(x)$ is not differentiable at $x = 1$.

69. $f(x)=\int_{-2}^{x}|t+1|\,dt=-\int_{-2}^{-1}(t+1)\,dt+\int_{-1}^{x}(t+1)\,dt$

$$=\frac{1}{2}+\left(\frac{t^2}{2}+t\right)_{-1}^{x}=\frac{x^2}{2}+x+1,\text{ for } x\geq -1$$

$f(x)$ is a quadratic polynomial.

$\therefore$ $f(x)$ is continuous as well as differentiable in $[-1, 1]$.

Also, $f'(x)$ is continuous as well as differentiable in $[-1, 1]$.

70. We have, $f(x+y)=f(x)+f(y)+xy\,(x+y)$

$$f(0)=0$$

$$\therefore \quad \lim_{h\to 0}\frac{f(h)}{h}=-1$$

$$\therefore \quad \lim_{h\to 0}\frac{f(x+h)-f(x)}{h}$$

$$=\lim_{h\to 0}\frac{f(x)+f(h)+xh\,(x+h)-f(x)}{h}$$

$$=\lim_{h\to 0}\frac{f(h)}{h}+\lim_{h\to 0}x\,(x+h)=-1+x^2$$

$$\therefore \quad f'(x)=-1+x^2$$

$$\therefore \quad f(x)=\frac{x^3}{3}-x+c$$

$\therefore$ $f(x)$ is a polynomial function,

$f(x)$ is twice differentiable for all $x\in R$ and

$f'(3)=3^2-1=8$.

71. We have, $f'(x)=6x+12$

For $f(x)$ is increasing, $f'(x)\geq 0 \Rightarrow x\geq -2$

Hence, $f(x)$ is increasing in $[-1, 2]$

$\lim_{x\to 2^+} f(x)=35,\ \lim_{x\to 2^-} f(x)=35$ and $f(2)=35$

Hence, $f(x)$ is continuous on $[-1, 3]$, $f'(2^-)=24$ and $f'(2^+)=-1$.

Hence, $f'(2)$ doesn't exist for maximum, $f(2)=35$

$$f(-1)=-10, \qquad f(3)=34$$

Hence, $f(x)$ has maximum value at $x=2$.

72. Since, $f(x)=0$, $x<0$ and differentiable at $x=0$, LHD $=0$ (function is on X-axis for $x<0$). If $f(x)$,

(a) $x^2, x>0$

RHD at $x=0$,

$f'(0)=2\times 0=0$ (possible)

(d) $-x^{3/2}, x>0$

RHD at $x=0$,

$f'(0)=-3/2\,x^{1/2}=-3/2\times 0=0$ (possible)

73. We know that, $\sin x$ is periodic function in $[0, 2\pi]$.

$\therefore$ $\sin x$ is continuous at $x=0$

Now, $\lim_{x\to 0}[x]$

RHL $=\lim_{x\to 0^+}[x]=\lim_{h\to 0}[x+h]=x=0 \qquad [\because [x+h]=x]$

LHL $=\lim_{x\to 0^-}[x]=\lim_{h\to 0}[x-h]=(x-1) \qquad [\because [x-h]=(x-1)]$

$=0-1=-1$

$\therefore$ RHL $\neq$ LHL

So, $[x]$ is discontinuous at $x=0$.

We know that, sum of continuous and discontinuous functions is discontinuous.

74. We have,

$$f(x)=\begin{cases} 2\sin(a\cos^{-1}x), & \text{if } x\in(0,1)\\ \sqrt{3}, & \text{if } x=0\\ ax+b, & \text{if } x<0\end{cases}$$

Continuity at $x=0$

(LHL at $x=0$) $=\lim_{x\to 0^-} ax+b$

$=\lim_{h\to 0} a(0-h)+b=\lim_{h\to 0} -ah+b$

$=b$

(RHL at $x=0$) $=\lim_{x\to 0^+} 2\sin(a\cos^{-1}x)$

$=\lim_{h\to 0} 2\sin(a\cos^{-1}(0+h))$

$=\lim_{h\to 0} 2\sin a\cos^{-1}h$

$=2\sin a\cos^{-1}h=2\sin a\cos^{-1}0$

$=2\sin\frac{a\pi}{2}$

$f(0)=\sqrt{3}$

For $f(x)$ to be continuous at $x=0$,

LHL = RHL = $f(0)$

$\therefore \quad b=2\sin\frac{a\pi}{a}=\sqrt{3}$

$\therefore \quad b=\sqrt{3}$ and $\sin\frac{a\pi}{2}=\frac{\sqrt{3}}{2}$

$\Rightarrow b=\sqrt{3}$ and $\frac{a\pi}{2}=\frac{\pi}{3}$

$\Rightarrow b=\sqrt{3}$ and $a=\frac{2}{3}$

So, Statement I is correct.

Since, for $x<0$, $f(x)=ax+b$

which is a polynomial function and will be continuous for $(-\infty, 0)$.

Again, for $x\in(0,1)$,

$f(x)=2\sin(a\cos^{-1}x)$, which is trigonometric function will be continuous for $x\in(0,1)$.

$\because$ $f(x)$ is continuous in $(-\infty, 1)$.

$\therefore$ Statement II is also correct.

75. We have, $f(x)=\begin{cases}\dfrac{\cos x-e^{-\frac{x^2}{2}}}{x^3}, & x\neq 0\\ 0, & x=0\end{cases}$

Clearly, $f(0)=0$

Now consider, $\lim_{x\to 0} f(x)=\lim_{x\to 0}\dfrac{\cos x-e^{-\frac{x^2}{2}}}{x^3} \quad \left[\dfrac{0}{0}\text{ form}\right]$

$=\lim_{x\to 0}\dfrac{-\sin x+e^{-\frac{x^2}{2}}\cdot x}{3x^2} \quad \left[\dfrac{0}{0}\text{ form}\right]$

$$= \lim_{x \to 0} \frac{-\cos x + e^{-\frac{x^2}{2}} - e^{-\frac{x^2}{2}} \cdot x^2}{6x} \qquad \left[\frac{0}{0} \text{ form}\right]$$

$$= \lim_{x \to 0} \frac{\sin x - e^{-\frac{x^2}{2}} \cdot x - 2xe^{-\frac{x^2}{2}} + e^{-\frac{x^2}{2}}}{6}$$

$$= 0$$

Thus, $\lim_{x \to 0} f(x) = f(0)$

$\Rightarrow f$ is continuous at $x = 0$

Hence, option (c) is correct.

76. Let $f(x) = \frac{x^3}{4} - \sin \pi x + \frac{2}{3}$. Then, $f(x)$ will be continuous function. ($\because$ Sum and difference of two continuous function is continuous)

Here, $f(2) = \frac{8}{3}$ and $f(-2) = -\frac{4}{3}$ $\qquad [\because \sin n\pi = 0, \forall\, n \in Z]$

Now, as $f(-2) < 0 < f(2)$, therefore by intermediate value theorem we can say that there exists atleast one point $n \in [-2, 2]$. Such that $f(n) = 0$

Hence, $f(x) = 0$, i.e. $\frac{x^3}{4} - \sin \pi x + \frac{2}{3} = 0$ has atleast one solution in $[-2, 2]$.

Clearly, Statement II is wrong. Because for this to be true, $f(x)$ should be a continuous function (by intermediate value theorem).

Hence, option (c) is correct.

77. We have,

$$f(x) = x\left(\frac{e^{2x} - e^{-2x}}{e^{2x} + e^{-2x}}\right) + x^2 + x^4$$

$$\therefore f(-x) = (-x)\left(\frac{e^{2(-x)} - e^{-2(-x)}}{e^{2(-x)} + e^{-2(-x)}}\right) + (-x)^2 + (-x)^4$$

$$= -x\left(\frac{e^{-2x} - e^{+2x}}{e^{-2x} + e^{+2x}}\right) + x^2 + x^4$$

$$= -x\left[-\left(\frac{e^{2x} - e^{-2x}}{e^{2x} + e^{-2x}}\right)\right] + x^2 + x^4$$

$$= x\left(\frac{e^{2x} - e^{-2x}}{e^{2x} + e^{-2x}}\right) + x^2 + x^4$$

$$= f(x)$$

$\therefore f(x)$ is even function and even function can't has range equal to R.

78. $y = |\ln x|$ not differentiable at $x = 1$.

$y = |\cos |x||$ is not differentiable at $x = \frac{\pi}{2}, \frac{3\pi}{2}$.

$y = \cos^{-1}(\text{sgn } x) = \cos^{-1}(1) = 0$ differentiable, $\forall\, x \in (0, 2\pi)$.

79. $f'(0^+) = \lim_{h \to 0} \frac{h \sin h - 0}{h} = 0$

$f'(0^-) = \lim_{h \to 0} \frac{h \sin(-h) - 0}{-h} = 0$

$f(x)$ is differentiable at $x = 0$.

e.g. $x|x|$ is derivable at $x = 0$.

80. Consider $g(x) = x^3$ at $x = 0$, $g(0) = 0$

$|g(x)|$ is derivable at $x = 0$.

Actually nothing definite can be said. Also, for $g(x) = x - 1$ with $g(1) = 0$.

Then, $|g(x)|$ is not derivable at $x = 1$.

81.

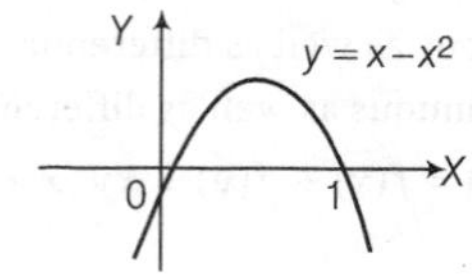

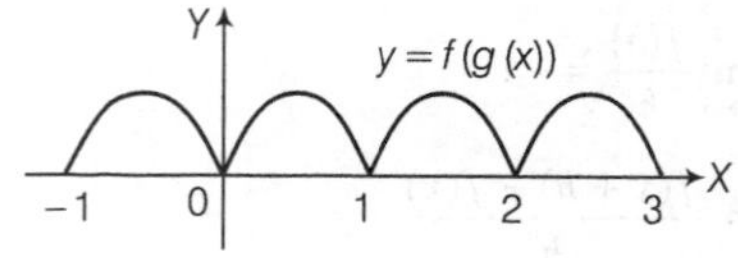

82. $f(x)$ is discontinuous at $x = 0$ and $f(x) < 0, \forall\, x \in [-\alpha, 0)$ and $f(x) > 0, \forall\, x \in [0, \alpha]$.

Sol. (Q. Nos. 83 to 85)

$$f(x) = \begin{cases} x + 1, & 0 \le x \le 2 \\ -x + 3, & 2 < x < 3 \end{cases}$$

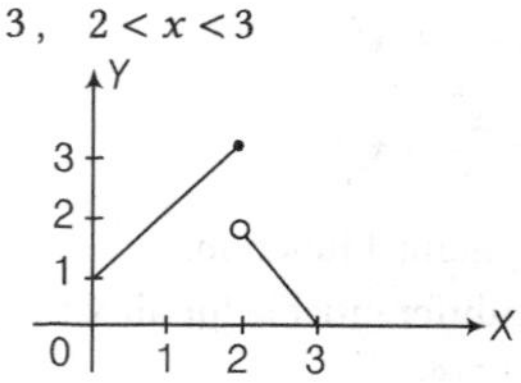

$\therefore f(x)$ is discontinuous at $x = 2$.

$$(fof)(x) = \begin{cases} x + 2, & 0 \le x \le 1 \\ -x + 2, & 1 < x \le 2 \\ -x + 4, & 2 < x < 3 \end{cases}$$

$\Rightarrow (fof)(x)$ is discontinuous at $x = 1, 2$

and $\qquad f(19) = f(3 \times 6 + 1) = f(1) = 2$

83. (a) **84.** (b, c) **85.** (a)

Sol. (Q. Nos. 86 to 89)

Since, $f(-x) = \frac{1}{f(x)}$

$\therefore$ At $x = 0$

$\Rightarrow f(0) = \frac{1}{f(0)} \Rightarrow f^2(0) = 1 \Rightarrow f(0) = +1$, as $f(x) > 0$.

$$\Rightarrow \qquad f'(x) = \lim_{h \to 0} \frac{f(x + h) - f(x)}{h}$$

$$f'(x) = \lim_{h \to 0} \frac{f(x) \cdot f(h) - f(x)}{h}$$

$$\therefore \qquad f'(x) = f(x) \cdot \left(\lim_{h \to 0} \frac{f(h) - 1}{h}\right) = f(x) \cdot f'(0)$$

$$\Rightarrow \qquad \int \frac{f'(x)}{f(x)} = \int -1\, dx \Rightarrow \log f(x) = -x + c$$

$$f(x) = e^{-x} \cdot \lambda \text{ at } x = 0,\ \lambda = 1 \Rightarrow f(x) = e^{-x}$$

$\Rightarrow$ Range of $f(x) \in R^+$. $\Rightarrow$ Range of $f(|x|)$ is $[0, 1]$.

$\Rightarrow f(x)$ is decreasing function

and $\qquad f'(x) = -e^{-x} = -f(x)$.

86. (c) **87.** (a) **88.** (d) **89.** (a)

90. $f(x) = \begin{cases} \sin x, & x \le 0 \\ \tan x, & 0 < x < 2\pi \\ \cos x, & 2\pi \le x < 3\pi \\ 3\pi, & x = 3\pi \end{cases}$

$f(x)$ is discontinuous at $\frac{\pi}{2}, \frac{3\pi}{2}, 2\pi$.

91. $f(x) = [2x^3 - 5]$

Since $1 \le x < 2 \Rightarrow 1 \le x^3 < 8$

$-3 \le 2x^3 - 5 < 11$

Now, $2x^3 - 5$ is discontinuous at integer points.

$\therefore$ $-2, -1, 0, 1, 2, 3, 4, 5, 6, 7, 8, 9, 10.$

Hence, number of points of discontinuity = 13

92. Max $([x], |x|)$, hence discontinuity at $x = \phi$.

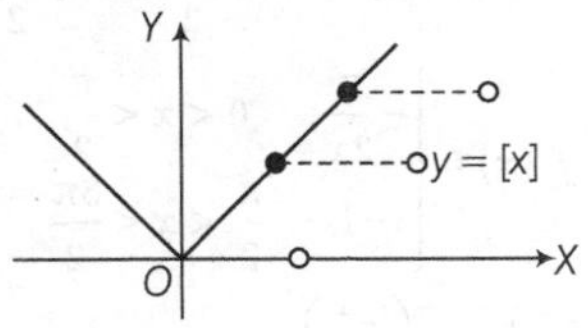

93. From figure, option (c) is correct.

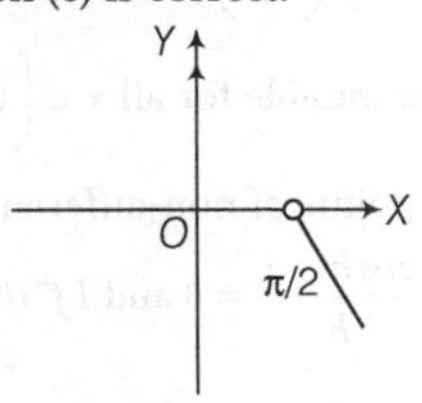

94. From figure, option (b) is correct.

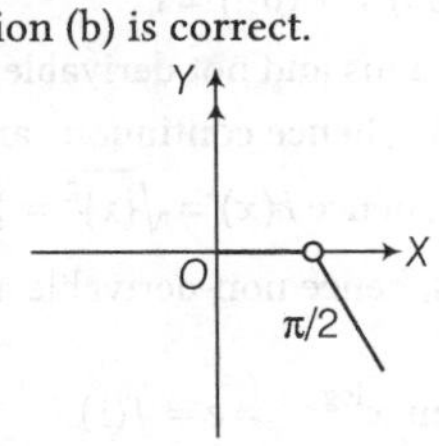

95. From figure, option (c) is correct.

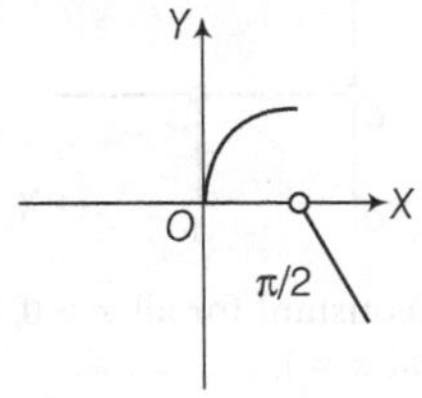

Sol. (Q. Nos. 96 to 99)

Here, $f(x + y^{2n+1}) = f(x) + (f(y))^{2n+1}$

On differentiating, we get

$$f'(x + y^{2n+1}) = f'(x)$$

$$\left[\because x \text{ and } y \text{ are independent, so } \frac{dy}{dx} = 0\right]$$

$\Rightarrow$ $f'(x)$ is constant, say $f'(x) = k$.

On integrating, we get $f(x) = kx + c$

Now, $f(0) = 0 \Rightarrow c = 0$ and $f(1) = 1 \Rightarrow k = 1$

$\therefore$ $f(x) = x \Rightarrow f'(x) = 1$, for all $x \in R$

$\therefore$ $f(x) = x$ is an odd function.

96. (b) **97.** (c) **98.** (d) **99.** (a)

We know, $f'(x) = \lim_{h \to 0} \frac{f(x+h) - f(x)}{h}$

and $f'(x) = \lim_{h \to 0} \frac{f(x-h) - f(x)}{-h}$

On adding, we get

$$2f'(x) = \lim_{h \to 0} \frac{f(x+h) - f(x)}{h} + \lim_{h \to 0} \frac{f(x-h) - f(x)}{-h} \quad \text{...(i)}$$

$$\Rightarrow \quad 2f'(x) = \lim_{h \to 0} \frac{f(x+h) - f(x-h)}{h}$$

$$\Rightarrow \quad 2f'(x) = \lim_{h \to 0} f(x) \cdot \frac{\{f(h) - f(-h)\}}{h}$$

[using $f(x+y) - f(x-y) = f(x)\{f(y) - f(-y)\}$]...(ii)

Also, $2f'(0) = \lim_{h \to 0} \left(\frac{f(h) - f(0)}{h} + \frac{f(-h) - f(0)}{-h} \right)$ [using Eq. (i)]

$$= \lim_{h \to 0} \frac{f(h) - f(-h)}{h} \quad \text{...(iii)}$$

From Eqs. (ii) and (iii), we get

$$2f'(x) = f(x) \cdot 2f'(0) \Rightarrow \frac{f'(x)}{f(x)} = f'(0)$$

$$\therefore \quad \frac{f'(x)}{f(x)} = 2010 \quad \text{...(iv)}$$

On integrating both sides, we get

$\log(f(x)) = 2010x + c$, as $f(0) = 1$

$\therefore$ $c = 0 \Rightarrow f(x) = e^{2010x}$

Thus, $\{f(x)\}$ is non-periodic. ...(v)

100. (b) **101.** (d)

102. (A) $\lim_{h \to 0} \sin\{1 - h\} = \cos 1 + a$

$\Rightarrow$ $\lim_{h \to 0} \sin(1 - h) - \cos 1 = a$

$\Rightarrow$ $a = \sin 1 - \cos 1$

Now, $|K| = \dfrac{\sin 1 - \cos 1}{\sqrt{2}\left(\sin 1 \cdot \frac{1}{\sqrt{2}} - \cos 1 \cdot \frac{1}{\sqrt{2}}\right)} = 1 \Rightarrow K = \pm 1$

(B) $f(0) = \lim_{x \to 0} \dfrac{2\sin^2\left(\frac{\sin x}{2}\right)}{x^2 \cdot \left(\frac{\sin x}{2}\right)^2} \cdot \left(\frac{\sin x}{2}\right)^2 = \frac{1}{2}$

(C) Function should have same rule for θ and $\theta' \Rightarrow x = 1 - x$

$\Rightarrow$ $x = \frac{1}{2}$

(D) $f(x) = x + \{-x\} + [x]$

x is continuous at $x \in R$.

Check at $x = I$ (where I is integer)

$f(I^+) = 2I + 1$ or $f(I^-) = 2I - 1$

So, $f(x)$ is discontinuous at every integer, i.e. $\{1, 0, -1\}$.

103. (A) $f(x)\begin{cases} 1-1, & 1<x\le 2 \\ 0, & x=1 \\ 1-x, & 0\le x<1 \\ -\sin\pi x, & -1\le x<0 \end{cases}$

At $x=0$, $f(x)$ is not continuous and not differentiable. At $x=1$, $f(x)$ is continuous and non-differentiable. At $x=2$ and -1, $f(x)$ is continuous and differentiable.

(B) $f(0^-)=\lim\limits_{h\to 0} h^2 e^{-\frac{1}{h}}=\lim\limits_{h\to 0}\dfrac{h^2}{e^{1/h}}=0$

(C) $f(x)=\dfrac{x}{x+1}$, not defined at $x=-1$.

$$g(x)=\frac{f(x)}{f(x)+2}$$

$g(x)$ is not defined at $f(x)=-2$

$$\Rightarrow \quad \frac{x}{x+1}=-2\Rightarrow x=-\frac{2}{3}$$

Also, $x=0$ is not in domain.

$\therefore$ $f(x)$ is not differentiable at 3 points.

(D) $y=\text{sgn}(x^2-1)=\begin{cases} 1, & x^2-1>0 \\ 0, & x^2-1=0 \\ -1, & x^2-1<0 \end{cases}=\begin{cases} 1, & |x|>1 \\ 0, & |x|=1 \\ -1, & |x|<1 \end{cases}$

$\therefore$ Tangent exists for all x.

$\therefore$ Number of points where tangent does not exists is 0.

104. (A) $f(x)=\dfrac{\tan x+\cot x}{2}-\left|\dfrac{\tan x-\cot x}{2}\right|$

$$=\begin{cases} \cot x, & \tan x\ge \cot x \\ \tan x, & \cot x>\tan x \end{cases}$$

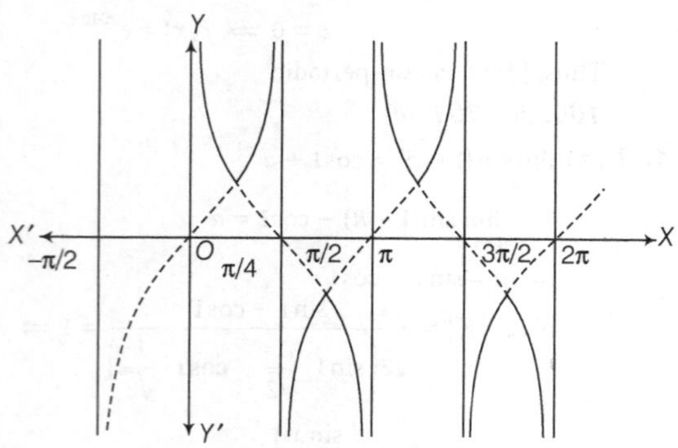

There are 4 points where the function is continuous but not differentiable in $(0, 2\pi)$.

(B) $f(x)=\min\{1, 1+x^3, x^2-3x+3\}$ can be shown as

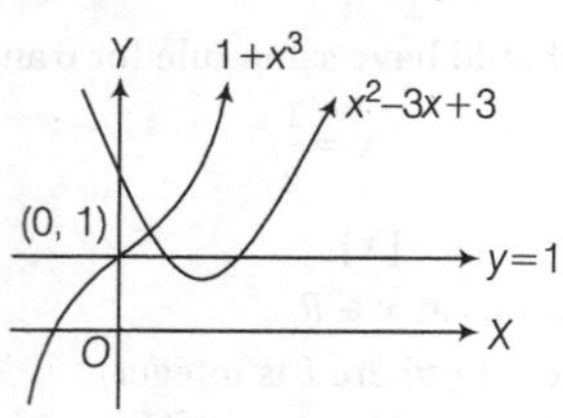

$$\therefore \quad f(x)=\begin{cases} 1+x^3, & x\le 0 \\ 1, & 0\le x\le 1 \text{ or } x\ge 2 \\ x^2-3x+3, & 1\le x\le 2 \end{cases}$$

Clearly, $f(x)$ is not differentiable at 2 points.

(C) $f(x)=(x+4)^{\frac{1}{3}}$

$\Rightarrow f'(x)=\dfrac{1}{3(x+4)^{\frac{2}{3}}}$

$\therefore$ Functions non-derivable at $x=-4$, i.e., at one point.

(D) $f(x)=\begin{cases} -\dfrac{\pi}{2}\cdot\log\left(\dfrac{2x}{\pi}\right)+\dfrac{\pi}{2}, & 0<x\le\dfrac{\pi}{2} \\ \pi-x, & \dfrac{\pi}{2}<x<\dfrac{3\pi}{2} \end{cases}$

$$f'(x)=\begin{cases} -\dfrac{\pi}{2x}, & 0<x<\dfrac{\pi}{2} \\ -1, & \dfrac{\pi}{2}<x<\dfrac{3\pi}{2} \end{cases}$$

$$f'\left(\frac{\pi^-}{2}\right)=f'\left(\frac{\pi^+}{2}\right)=-1$$

$\therefore$ $f(x)$ is differentiable for all $x\in\left(0, \dfrac{3\pi}{2}\right)$.

$\therefore$ Number of points of non-differentiable is 0.

105. (A) $Rf'(0)=\lim\limits_{h\to 0}\dfrac{\cos h-1}{h}=0$ and $Lf'(0)=\lim\limits_{h\to 0}\dfrac{-h+1-1}{-h}=-1$

Obviously,

$$f(0)=f(0^-)=f(0^+)=1$$

Hence, continuous and not derivable.

(B) $g(x)=0$ for all x, hence continuous and derivable.

(C) As $0\le\{x\}<1$, hence $h(x)=\sqrt{\{x\}^2}=\{x\}$ which is discontinuous, hence non-derivable all $x\in I$.

(D) $\lim\limits_{x\to 1} x^{\frac{1}{\ln x}}=\lim\limits_{x\to 1} x^{\log_x e}=e=f(1)$

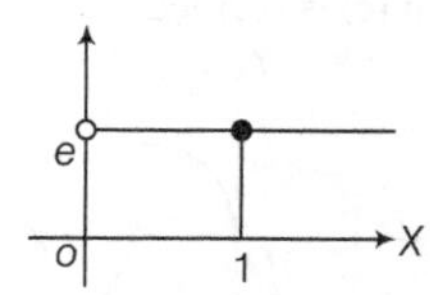

Hence, $k(x)$ is constant for all $x>0$, hence continuous and differentiable at $x=1$.

106. (A) $l=\lim\limits_{x\to\infty} e^{x^2+1}\left[e^{\sqrt{x^4+1}-(x^2+1)}-1\right]$

Consider, $\lim\limits_{x\to\infty}\left[\sqrt{x^4+1}-(x^2+1)\right]$

$$=\lim_{x\to\infty}\frac{x^4+1-(x^4+1+2x^2)}{\sqrt{x^4+1}+(x^2+1)}$$

$$=\lim_{x\to\infty}\frac{-2x^2}{x^2\left(\sqrt{1+(1/x^4)}+1+1/x^2\right)}=-1$$

Now, as $x \to \infty$, $\sqrt{x^4+1} - (x^2+1) \to -1$

$\infty \times \left(\frac{1}{e} - 1\right) \to -\infty$ and hence limit doesn't exist.

(B) $f(0^+) = \lim\limits_{h \to 0} \dfrac{a^{2h} - 2a^h + 1}{h^2} = \lim\limits_{h \to 0} \left(\dfrac{a^h - 1}{h}\right)^2 = \ln^2 a$

$f(0^-) = \lim\limits_{h \to 0} 3 \ln (a+h) - 2 = 3 \ln a - 2 = f(0)$

For continuous

$\ln^2 a = 3 \ln a - 2 \quad \ln^2 a - 3 \ln a + 2 = 0$

$\Rightarrow \quad (\ln a - 2)(\ln a - 1) = 0 ; \; a = e^2$ or $a = e$

(C) $L = a^a \ln ae$

$\Rightarrow \quad M = a^a$

$\therefore \quad a^a \ln ae = 2a^a$

$\therefore \; \ln ae = 2 \Rightarrow ae = e^2 \Rightarrow a = e$

107. $\tan^2 x$ is discontinuous at $x = \frac{\pi}{2}, \frac{3\pi}{2}$

$\sec^2 x$ is discontinuous at $x = \frac{\pi}{2}, \frac{3\pi}{2}$

$\Rightarrow$ Number of discontinuities = 2

108. Let $g(x) = x^{1/x}, g'(x) = x^{1/x} \dfrac{1 - \ln x}{x^2}$

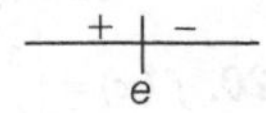

$\Rightarrow \quad g_{\max} = e^{1/e} \in (1, 2)$

$\Rightarrow \quad \lim\limits_{x \to 0} x^{1/x} = 0, \; \lim\limits_{x \to \infty} x^{1/x} = 1$

So, $\quad f(x) = \begin{cases} 0, & 0 < x < 1 \\ 1, & 1 \le x < \infty \end{cases}$

109. $f(x) = x + \cos x + 2, f(0) = 3 \Rightarrow g(3) = 0$

$g(f(x)) = x \Rightarrow g'(f(x)) \cdot f'(x) = 1$, putting $x = 0$

$g'(3) \cdot f'(0) = 1$

Now, $\quad f'(x) = 1 - \sin x \Rightarrow f'(0) = 1 \Rightarrow \; g'(3) = 1$

110. Let $g(x) = x \tan^{-1}(x^2)$. It is an odd function.

So, $\quad g^{2m}(0) = 0.$

Let $\quad h(x) = x^4$

So, $\quad f(x) = g(x) + h(x)$

$\Rightarrow \quad f^{2m}(0) = g^{2m}(0) + h^{2m}(0)$

$= h^{2m}(0) \neq 0$

It happens when $2m = 4 \Rightarrow m = 2$

111. $F(x) = 3e^x$ and $G(x) = e^{-x}$

The equation $9x^4 = (F(x))^2 G(x)$ becomes $x^4 = e^x$

Hence, number of solutions = 2

112. $y' = f'(x) - 2f'(2x)$

$y'(1) = f'(1) - 2f'(2) = 5 \quad$ ...(i)

and $\quad y'(2) = f'(2) - 2f'(4) = 7 \quad$ ...(ii)

Now, let $\quad y = f(x) - f(4x) - 10x$

$y' = f'(x) - 4f'(4x) - 10$

$y'(1) = f'(1) - 4f'(4) - 10 \quad$ ...(iii)

On substituting the value of $f'(2) = 7 + 2f'(4)$ in Eq. (i), we get

$f'(1) - 2[7 + 2f'(4)] = 5$

$f'(1) - 4f'(4) = 19$

$\Rightarrow \quad f'(1) - 4f'(4) - 10 = 9$

113. A, B, C are in AP.

$\therefore \quad 2B = A + C$ and $A + B + C = 180°$

$\therefore \quad B = 60°$

$\therefore \quad \cos B = \dfrac{1}{2} = \dfrac{a^2 + c^2 - b^2}{2ac}$

$\Rightarrow \quad a^2 + c^2 = b^2 + ac$

$\Rightarrow \quad (a - c)^2 = b^2 - ac$

or $\quad |\sin A - \sin C| = \sqrt{\sin^2 B - \sin A \sin C}$

$\Rightarrow 2\cos\left(\dfrac{A+C}{2}\right)\left|\sin\left(\dfrac{A-C}{2}\right)\right| = \sqrt{\dfrac{3}{4} - \sin A \sin C}$

$\Rightarrow \quad 2\left|\sin\left(\dfrac{A-C}{2}\right)\right| = \sqrt{3 - 4\sin A \sin C}$

$\therefore \quad \lim\limits_{A \to C} \dfrac{\sqrt{3 - 4\sin A \sin C}}{|A - C|} = \lim\limits_{A \to C} \dfrac{2\left|\sin\left(\dfrac{A-C}{2}\right)\right|}{|A - C|}$

$= \lim\limits_{A \to C} \left| \dfrac{\sin\left(\dfrac{A-C}{2}\right)}{\left(\dfrac{A-C}{2}\right)} \right|$

$= |1| = 1 \Rightarrow f(x) = 1$

$\therefore \quad f'(x) = 0$

114. $\text{RHD} = \lim\limits_{h \to 0} h \dfrac{\left(\left(\frac{3}{4}\right)^{1/h} - \left(\frac{3}{4}\right)^{-1/h}\right) - 0}{\left(\left(\frac{3}{4}\right)^{1/h} + \left(\frac{3}{4}\right)^{-1/h}\right) \cdot h} = -1$

$\text{LHD} = \lim\limits_{h \to 0} (-h) \dfrac{\left(\left(\frac{3}{4}\right)^{-1/h} - \left(\frac{3}{4}\right)^{1/h}\right) - 0}{\left(\left(\frac{3}{4}\right)^{-1/h} + \left(\frac{3}{4}\right)^{1/h}\right) \cdot (-h)} = 1$

$\therefore \quad P = f'(0^-) - f'(0^+) = 1 - (-1) = 2$

Now, $4 \cdot \left(\lim\limits_{x \to 2^-} \dfrac{(\exp((x+2)\log 4))^{\frac{[x+1]}{4}} - 16}{4^x - 16} \right)$

$= 4 \cdot \left[\lim\limits_{x \to 2^-} \dfrac{(4^{x+2})^{\frac{[x+1]}{4}} - 16}{4^x - 16} \right] = 4 \times \dfrac{1}{2} = 2$

115. Here, $f(x) = -x^3 + x^2 - x + 1 \Rightarrow f'(x) = -3x^2 + 2x - 1$

$\Rightarrow \quad D = 4 - 12 = -8 < 0$

$\therefore \; f(x)$ is decreasing.

Thus, min $f(t) = f(x)$, as $f(x)$ is decreasing and $0 \le t \le x$.

$$\therefore \quad g(x)=\begin{cases}-x^3+x^2-x+1, & 0<x\le 1\\ x-1, & 1<x\le 2\end{cases}$$

$$\Rightarrow \quad \lim_{x\to 1^+} g(g(x)) = \lim_{x\to 1^+} g(0^+)=1$$

$$\Rightarrow \quad \lim_{x\to 1^-} g(g(x)) = \lim_{x\to 1^-} g(0^+)=1$$

$$\therefore \quad \lim_{x\to 1} g(g(x))=1$$

116. We have, $f(x)=(x-|x|)^2(1-x+|x|)^2$

$$\Rightarrow \quad f(x)=\begin{cases}(2x)^2(1-2x)^2, & x<0\\ 0, & x\ge 0\end{cases}$$

$$=\begin{cases}16x^4-16x^3+4x^2, & x<0\\ 0, & x\ge 0\end{cases}$$

Clearly, $f(x)$ is continuous as well as derivable for all $x\in R$.

$\therefore$ Number of points of non-differentiable = 0.

117. $$\text{RHL} = \lim_{h\to 0}\frac{\left(\frac{\pi}{2}-\sin^{-1}(1-h^2)\right)\sin^{-1}(1-h)}{\sqrt{2}(h-h^3)}$$

$$=\lim_{h\to 0}\frac{\cos^{-1}(1-h^2)\sin^{-1}(1-h)}{\sqrt{2}h(1-h^2)}$$

$$=\lim_{h\to 0}\frac{\sin^{-1}\sqrt{2h^2-h^4}\cdot\sin^{-1}(1-h)}{\sqrt{2}h(1+h)(1-h)}$$

$$=\lim_{h\to 0}\frac{\sin^{-1}(h\sqrt{2-h^2})\sin^{-1}(1-h)}{\sqrt{2}h(1+h)(1-h)}$$

$$=\lim_{h\to 0}\frac{1}{\sqrt{2}}\cdot\frac{\sin^{-1}(h\sqrt{2-h^2})}{h\cdot\sqrt{2-h^2}}\cdot\frac{\sqrt{2-h^2}}{1+h}\cdot\frac{\sin^{-1}(1-h)}{(1-h)}$$

$$=\frac{1}{\sqrt{2}}\cdot\sqrt{2}\cdot\frac{\sin^{-1}(1)}{1}=\frac{\pi}{2}\Rightarrow k=\frac{\pi}{2} \qquad \text{...(i)}$$

$$\text{LHL}=\lim_{h\to 0}\frac{A\sin^{-1}(1-(1-h))\cdot\cos^{-1}(1-(1-h))}{\sqrt{2(1-h)}\cdot(1-(1-h))}$$

$$=\lim_{h\to 0}\frac{A\cdot\sin^{-1}(h)\cdot\cos^{-1}(h)}{\sqrt{2(1-h)}\cdot h}=\frac{A\cdot\pi/2}{\sqrt{2}}=\frac{A\pi}{2\sqrt{2}}$$

$$\Rightarrow \quad \frac{A\pi}{2\sqrt{2}}=\frac{\pi}{2}\Rightarrow A=\sqrt{2}$$

Hence, $\sin^2 k+\cos^2\left(\frac{A\pi}{\sqrt{2}}\right)=\sin^2\left(\frac{\pi}{2}\right)+\cos^2(\pi)=1+1=2.$

118. $f(x)=[[x]]-[x-1]$

or $f(x)=[x]-[x]+1$

or $f(x)=1$, which is constant function and which is continuous for all real numbers.

119. (i) $f(x)=[x]+[-x]=\begin{cases}x-x, & x\in \text{integers}\\ [x]-1-[x], & x\notin\text{integers}\end{cases}$

$$\therefore\ f(x)=\begin{cases}0, & x\in\text{integers}\\ -1, & x\notin\text{integers}\end{cases}$$

which shows the graph of $f(x)$ as shown in figure.

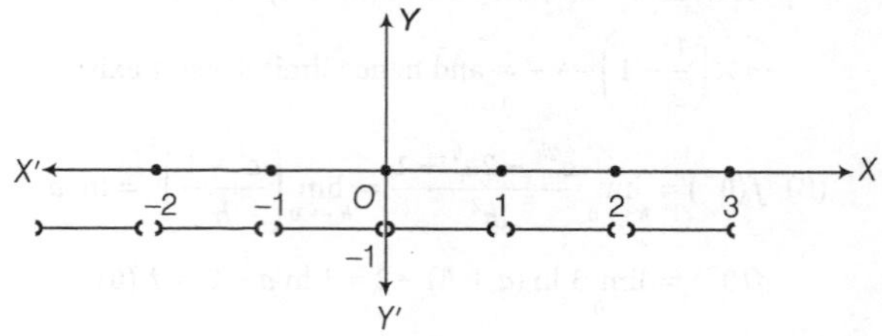

Thus, $f(x)$ is discontinuous at $x\in$ integers.

(ii) $$g(x)=\lim_{n\to\infty}\frac{x^{2n}-1}{x^{2n}+1}=\begin{cases}-1, & |x|<1\\ 0, & |x|=1\\ 1, & |x|>1\end{cases}$$

which can be shown as in the figure.

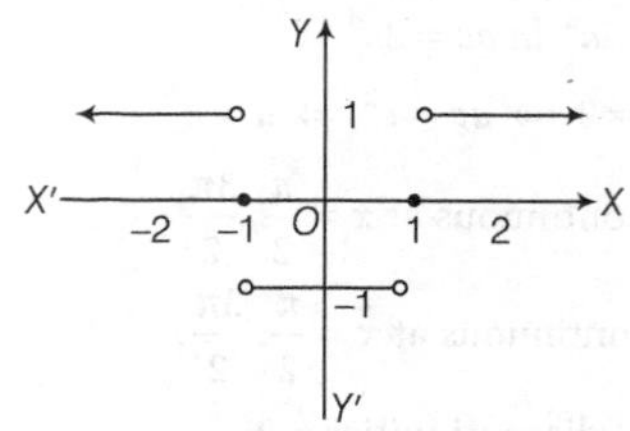

Thus, $g(x)$ is discontinuous at $x=\pm 1$.

120. $f(x)=\begin{cases}x[x], & 0\le x<2\\ (x-1)[x], & 2\le x\le 3\end{cases}$

To check continuity at $x=1$

RHL (at $x=1$) $=\lim_{h\to 0}(1+h)[1+h]=1$

LHL (at $x=1$) $=\lim_{h\to 0}(1-h)[1-h]=0$

Hence, the function is discontinuous at $x=1$.

To check continuity at $x=2$

RHL (at $x=2$) $=\lim_{h\to 0}(2+h-1)[2+h]=2$

LHL (at $x=2$) $=\lim_{h\to 0}(2-h)[2-h]=2\Rightarrow f(2)=(2-1)[2]=2$

Hence, the function is continuous at $x=2$.

To check differentiability at $x=2$

$$\text{RHD (at } x=2)=\lim_{h\to 0}\frac{f(2+h)-f(2)}{h}$$

$$=\lim_{h\to 0}\frac{(2+h-1)[2+h]-2}{h}$$

$$=\lim_{h\to 0}\frac{(1+h)2-2}{h}=2$$

$$\text{LHD (at } x=2)=\lim_{h\to 0}\frac{(2-h)[2-h]-2}{-h}$$

$$\Rightarrow \lim_{h\to 0}\frac{2-h-2}{-h}=1$$

which shows $f(x)$ is not differentiable at $x=2$.

Also, $f(x)$ is not differentiable at $x=1$, as $f(x)$ at $x=1$ is not continuous.

121. Differentiating both the sides, we get

$$h'(x) = 2f(x)\cdot f'(x) + 2g(x)\cdot g'(x)$$

$$h'(x) = 2f(x)\cdot g(x) + 2g(x)\cdot f''(x)$$

[as $g(x) = f'(x)$, $g'(x) = f''(x)$]

$$h'(x) = 2f(x)\cdot g(x) - 2g(x)\cdot f(x) \quad [\because f''(x) = -f(x)]$$

$$h'(x) = 0$$

$\therefore$ $h(x)$ must be constant function.

Given, $h(5) = 11$

Hence, $h(10) = 11$

122. Given that,

$$f(x+y) = f(x)\cdot f(y) \text{ for all } x \in R \quad \text{...(i)}$$

On putting $x = y = 0$ in Eq. (i), we get

$$f(0)\{f(0)-1\} = 0$$

$\Rightarrow$ $f(0) = 0$ or $f(0) = 1$

If $f(0) = 0$, then

$$f(x) = f(x+0) = f(x)\cdot f(0)$$

$\Rightarrow$ $f(x) = 0$ for all $x \in R$

which is not true, so $f(0) = 1$

$$\therefore\ f'(x) = \lim_{h\to 0} \frac{f(x+h)-f(x)}{h}$$

$$= \lim_{h\to 0} \frac{f(x)\cdot f(h) - f(x)}{h} \quad \text{[from Eq. (i)]}$$

$$= \lim_{h\to 0} f(x)\left\{\frac{f(h)-1}{h}\right\}$$

$$= f(x)\cdot f'(0)$$

$$= 2f(x) \quad \text{[given } f'(0) = 2\text{]}$$

or $\dfrac{f'(x)}{f(x)} = 2$

On integrating both the sides, we get $\log_e f(x) = 2x + C$

At $x = 0$, $f(0) = 1$

Hence, $\log f(1) = 2(0) + C$

$\Rightarrow$ $\log 1 = 0 + C$

$\Rightarrow$ $C = 0$

$\Rightarrow$ $\log_e f(x) = 2x + 0$

$\therefore$ $f(x) = e^{2x}$

123. Since, $f(x)$ is differentiable at $x = 0$.

$$\Rightarrow \lim_{h\to 0} \frac{f(0+h)-f(0)}{h} = p \text{ (say)} \quad \text{...(i)}$$

Then, $f'(x) = \lim_{h\to 0} \dfrac{f(x+h)-f(x)}{h}$

or $f'(x) = \lim_{h\to 0} \dfrac{f\left\{\dfrac{3x+3h}{3}\right\} - f\left(\dfrac{3x+3\cdot 0}{3}\right)}{h}$

or $f'(x) = \lim_{h\to 0} \dfrac{f(3x)+f(3h)+f(0)-f(3x)-f(0)-f(0)}{3h}$

or $f'(x) = \lim_{h\to 0} \dfrac{f(3h)-f(0)}{3h}$

$$\left[\text{using } f\left(\frac{x+y}{3}\right) = \frac{f(x)+f(y)+f(0)}{3}\right]$$

or $f'(x) = f'(0)$ or $f'(x) = p$ (let) [from Eq. (i)]

$\therefore$ $f(x) = px + q$

which shows $f(x)$ is differentiable for all x in R.

124. We have,

$$f'(x) = \lim_{h\to 0} \frac{f(x+h)-f(x)}{h}$$

$$= \lim_{h\to 0} \frac{f(x)+f(h)+2hx-1-f(x)}{h}$$

[using given definition]

$$= \lim_{h\to 0}\left\{2x + \frac{f(h)-1}{h}\right\}$$

Now, substituting $x = y = 0$ in the given functional relation, we get

$$f(0) = f(0) + f(0) + 0 - 1$$

$\Rightarrow$ $f(0) = 1$

$$\therefore\ f'(x) = 2x + \lim_{h\to 0} \frac{f(h)-f(0)}{h}$$

$$= 2x + f'(0)$$

$\Rightarrow$ $f'(x) = 2x + \cos\alpha$

On integrating, $f(x) = x^2 + x\cos\alpha + C$

Here, at $x = 0$, $f(0) = 1$

$\therefore$ $1 = C \Rightarrow f(x) = x^2 + x\cos\alpha + 1$

It is a quadratic in x with discriminant,

$$D = \cos^2\alpha - 4 < 0$$

and coefficient of $x^2 = 1 > 0$

$\therefore$ $f(x) > 0$, $\forall\, x \in R$

125. **Plan** If a continuous function has values of opposite sign inside an interval, then it has a root in that interval.

$$f, g : [0, 1] \to R$$

We take two cases.

***Case* I** Let f and g attain their common maximum value at p.

$\Rightarrow$ $f(p) = g(p)$,

where $p \in [0,1]$

***Case* II** Let f and g attain their common maximum value at different points.

$\Rightarrow$ $f(a) = M$ and $g(b) = M$

$\Rightarrow$ $f(a) - g(a) > 0$ and $f(b) - g(b) < 0$

$\Rightarrow$ $f(c) - g(c) = 0$ for some $c \in [0,1]$ as f and g are continuous functions.

$\Rightarrow$ $f(c) - g(c) = 0$ for some $c \in [0,1]$ for all cases. ...(i)

Option (a) $\Rightarrow f^2(c) - g^2(c) + 3[f(c) - g(c)] = 0$

which is true from Eq. (i).

Option (d) $\Rightarrow f^2(c) - g^2(c) = 0$, which is true from Eq. (i).

Now, if we take $f(x) = 1$ and $g(x) = 1$, $\forall x \in [0,1]$

Options (b) and (c) does not hold.

Hence, options (a) and (d) are correct.

126. **Plan**

(i) In these type of questions, we draw the graph of the function.

(ii) The points at which the curve taken a sharp turn, are the points of non-differentiability. Curve of $f(x)$ and $g(x)$ are

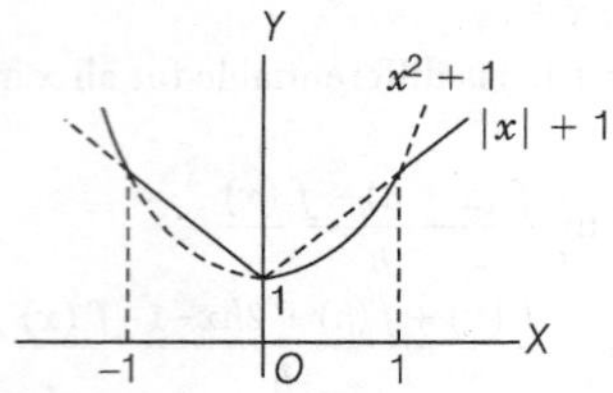

$h(x)$ is not differentiable at $x=\pm 1$ and 0.

As, $h(x)$ take sharp turns at $x=\pm 1$ and 0.

Hence, number of points of non-differentiability of $h(x)$ is 3.

127. Plan To check differentiability at a point we use RHD and LHD at a point and if RHD = LHD, then $f(x)$ is differentiable at the point.

Description of Situation

As, $$R\{f'(x)\}=\lim_{h\to 0}\frac{f(x+h)-f(x)}{h}$$

and $$L\{f'(x)\}=\lim_{h\to 0}\frac{f(x-h)-f(x)}{-h}$$

To check differentiable at $x=0$,

$$R\{f'(0)\}=\lim_{h\to 0}\frac{f(0+h)-f(0)}{h}$$

$$=\lim_{h\to 0}\frac{h^2\left|\cos\frac{\pi}{h}\right|-0}{h}=\lim_{h\to 0}h\cdot\left|\cos\frac{\pi}{h}\right|=0$$

$$L\{f'(0)\}=\lim_{h\to 0}\frac{f(0-h)-f(0)}{-h}=\lim_{h\to 0}\frac{h^2\left|\cos\left(-\frac{\pi}{h}\right)\right|-0}{-h}=0$$

So, $f(x)$ is differentiable at $x=0$.

To check differentiability at $x=2$,

$$R\{f'(2)\}=\lim_{h\to 0}\frac{f(2+h)-f(2)}{h}$$

$$=\lim_{h\to 0}\frac{(2+h)^2\left|\cos\left(\frac{\pi}{2+h}\right)\right|-0}{h}=\lim_{h\to 0}\frac{(2+h)^2\cdot\cos\left(\frac{\pi}{2+h}\right)}{h}$$

$$=\lim_{h\to 0}\frac{(2+h)^2\cdot\sin\left(\frac{\pi}{2}-\frac{\pi}{2+h}\right)}{h}$$

$$=\lim_{h\to 0}\frac{(2+h)^2\cdot\sin\left(\frac{\pi h}{2(2+h)}\right)}{h\cdot\frac{\pi}{2(2+h)}}\cdot\frac{\pi}{2(2+h)}=\pi$$

and $$L\{f'(2)\}=\lim_{h\to 0}\frac{f(2-h)-f(2)}{-h}$$

$$=\lim_{h\to 0}\frac{(2-h)^2\cdot\left|\cos\frac{\pi}{2-h}\right|-2^2\cdot\left|\cos\frac{\pi}{2}\right|}{-h}$$

$$=\lim_{h\to 0}\frac{(2-h)^2\left(-\cos\frac{\pi}{2-h}\right)-0}{-h}$$

$$=\lim_{h\to 0}\frac{-(2-h)^2\cdot\sin\left(\frac{\pi}{2}-\frac{\pi}{2-h}\right)}{h}$$

$$=\lim_{h\to 0}\frac{(2-h)^2\cdot\sin\left(-\frac{\pi h}{2(2-h)}\right)}{h\times\frac{-\pi}{2(2-h)}}\times\frac{-\pi}{2(2-h)}=-\pi$$

Thus, $f(x)$ is differentiable at $x=0$ but not differentiable at $x=2$.

128. $f(2n)=a_n,\quad f(2n^+)=a_n$

$f(2n^-)=b_n+1\Rightarrow a_n-b_n=1$

$f(2n+1)=a_n\quad\Rightarrow f\{(2n+1)^-\}=a_n$

$f\{(2n+1)^+\}=b_{n+1}-1$

$\Rightarrow\quad a_n=b_{n+1}-1$ or $a_n-b_{n+1}=-1$

or $\quad a_{n-1}-b_n=-1$

129. $f(x+y)=f(x)+f(y)$, as $f(x)$ is differentiable at $x=0$.

$\Rightarrow\quad f'(0)=k$ ...(i)

Now, $$f'(x)=\lim_{h\to 0}\frac{f(x+h)-f(x)}{h}$$

$$=\lim_{h\to 0}\frac{f(x)+f(h)-f(x)}{h}=\lim_{h\to 0}\frac{f(h)}{h}\quad\left[\frac{0}{0}\text{ form}\right]$$

[given, $f(x+y)=f(x)+f(y),\ \forall\ x,y$

$\therefore\quad f(0)=f(0)+f(0)$,

when $x=y=0\Rightarrow f(0)=0$]

Using L'Hospital's rule,

$$\lim_{h\to 0}\frac{f'(h)}{1}=f'(0)=k\quad\text{...(ii)}$$

$\Rightarrow\quad f'(x)=k$, integrating both sides, we get

$f(x)=kx+C$, as $f(0)=0\Rightarrow C=0$

$\therefore\quad f(x)=kx$

$\therefore f(x)$ is continuous for all $x\in R$ and $f'(x)=k$, i.e. constant for all $x\in R$. Hence, (b) and (c) are correct.

130. We have, $$f(x)=\begin{cases}-x-\frac{\pi}{2}, & x\le-\frac{\pi}{2}\\ -\cos x, & -\frac{\pi}{2}<x\le 0\\ x-1, & 0<x\le 1\\ \log x, & x>1\end{cases}$$

Continuity at $x=-\frac{\pi}{2}$, $f\left(-\frac{\pi}{2}\right)=-\left(-\frac{\pi}{2}\right)-\frac{\pi}{2}=0$

$$\text{RHL}=\lim_{h\to 0}-\cos\left(-\frac{\pi}{2}+h\right)=0$$

$\therefore$ Continuous at $x=\left(-\frac{\pi}{2}\right)$.

Continuity at $x=0\Rightarrow f(0)=-1$

$$\text{RHL}=\lim_{h\to 0}(0+h)-1=-1$$

$\therefore$ Continuous at $x=0$. Continuity at $x=1$, $f(1)=0$

$$\text{RHL}=\lim_{h\to 0}\log(1+h)=0$$

$\therefore$ Continuous at $x=1$

$$f'(x)=\begin{cases}-1, & x\le-\frac{\pi}{2}\\ \sin x, & -\frac{\pi}{2}<x\le 0\\ 1, & 0<x\le 1\\ \frac{1}{x}, & x>1\end{cases}$$

Differentiable at $x = 0$, LHD = 0, RHD = 1

$\therefore$ Not differentiable at $x = 0$

Differentiable at $x = 1$, LHD = 1, RHD = 1

$\therefore$ Differentiable at $x = 1$.

Also, for $x = -\frac{3}{2} \Rightarrow f(x) = -\cos x$

$\therefore$ Differentiable at $x = -\frac{3}{2}$

131. Given, $f(x) = x\cos\frac{1}{x}$, $x \geq 1$

$\Rightarrow \quad f'(x) = \frac{1}{x}\sin\frac{1}{x} + \cos\frac{1}{x}$

$\Rightarrow \quad f''(x) = -\frac{1}{x^3}\cos\left(\frac{1}{x}\right)$

Now $\lim_{x\to\infty} f'(x) = 0 + 1 = 1 \Rightarrow$ Option (b) is correct.

Now, $x \in [1, \infty) \Rightarrow \frac{1}{x} \in (0, 1] \Rightarrow f''(x) < 0$

Option (d) is correct.

As, $f'(1) = \sin 1 + \cos 1 > 1$

$f'(x)$ is strictly decreasing and $\lim_{x\to\infty} f'(x) = 1$

So, graph of $f'(x)$ is shown as below

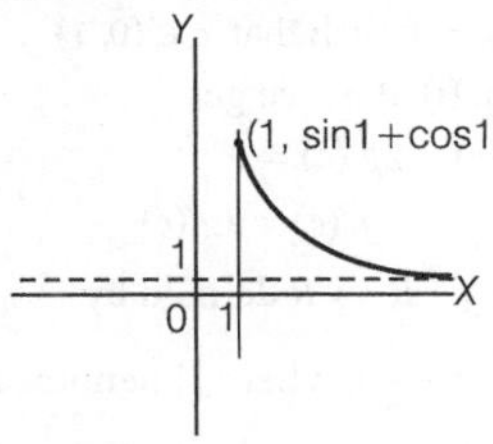

Now, in $[x, x+2]$, $x \in [1, \infty)$, $f(x)$ is continuous and differentiable so by LMVT, $f'(x) = \frac{f(x+2) - f(x)}{2}$

As, $\quad f'(x) > 1$

For all $x \in [1, \infty)$

$\Rightarrow \quad \frac{f(x+2) - f(x)}{2} > 1 \Rightarrow f(x+2) - f(x) > 2$

For all $x \in [1, \infty)$.

132. Given, $g(x) = \frac{(x-1)^n}{\log\cos^m(x-1)}$; $0 < x < 2$, $m \neq 0$, n are integers

and $|x - 1| = \begin{cases} x - 1, & x \geq 1 \\ 1 - x, & x < 1 \end{cases}$

The left hand derivative of $|x - 1|$ at $x = 1$ is $p = -1$.

Also, $\lim_{x\to 1^+} g(x) = p = -1$

$\Rightarrow \quad \lim_{h\to 0} \frac{(1 + h - 1)^n}{\log\cos^m(1 + h - 1)} = -1$

$\Rightarrow \quad \lim_{h\to 0} \frac{h^n}{\log\cos^m h} = -1 \Rightarrow \lim_{h\to 0} \frac{h^n}{m\log\cos h} = -1$

$\Rightarrow \quad \lim_{h\to 0} \frac{n\cdot h^{n-1}}{m\frac{1}{\cos h}(-\sin h)} = -1$ [using L'Hospital's rule]

$\Rightarrow \quad \lim_{h\to 0}\left(-\frac{n}{m}\right)\cdot\frac{h^{n-2}}{\left(\frac{\tan h}{h}\right)} = -1 \Rightarrow \left(\frac{n}{m}\right)\lim_{h\to 0}\frac{h^{n-2}}{\left(\frac{\tan h}{h}\right)} = 1$

$\Rightarrow \quad n = 2$ and $\frac{n}{m} = 1$

$\therefore \quad m = n = 2$

133. We have, $\lim_{x\to 0}\frac{g(x)\cos x - g(0)}{\sin x}$ $\left[\frac{0}{0}\text{ form}\right]$

$= \lim_{x\to 0}\frac{g'(x)\cos x - g(x)\sin x}{\cos x} = 0$

Statement I

Since, $f(x) = g(x)\sin x$

$f'(x) = g'(x)\sin x + g(x)\cos x$

and $f''(x) = g''(x)\sin x + 2g'(x)\cos x - g(x)\sin x$

$\Rightarrow \quad f''(0) = 0$

Thus, $\lim_{x\to 0}[g(x)\cos x - g(0)]\operatorname{cosec} x = 0 = f''(0)$

$\Rightarrow$ Statement I is true.

Statement II $f'(x) = g'(x)\sin x + g(x)\cos x \Rightarrow f'(0) = g(0)$

Statement II is not a correct explanation of Statement I.

134. Match the Columns

A. $x|x|$ is continuous, differentiable and strictly increasing in $(-1, 1)$.

B. $\sqrt{|x|}$ is continuous in $(-1, 1)$ and not differentiable at $x = 0$.

C. $x + [x]$ is strictly increasing in $(-1, 1)$ and discontinuous at $x = 0$

$\Rightarrow$ not differentiable at $x = 0$.

D. $|x - 1| + |x + 1| = 2$ in $(-1, 1)$

$\Rightarrow$ The function is continuous and differentiable in $(-1, 1)$.

135. Here, $f(x) = \min\{1, x^2, x^3\}$ which could be graphically shown as

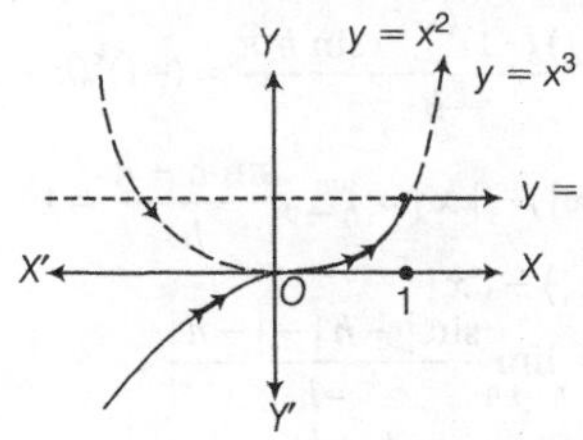

$\Rightarrow f(x)$ is continuous for $x \in R$ and not differentiable at $x = 1$ due to sharp edge. Hence, (a) and (d) are correct answers.

136. Using graphical transformation. As, we know that, the function is not differentiable at sharp edges.

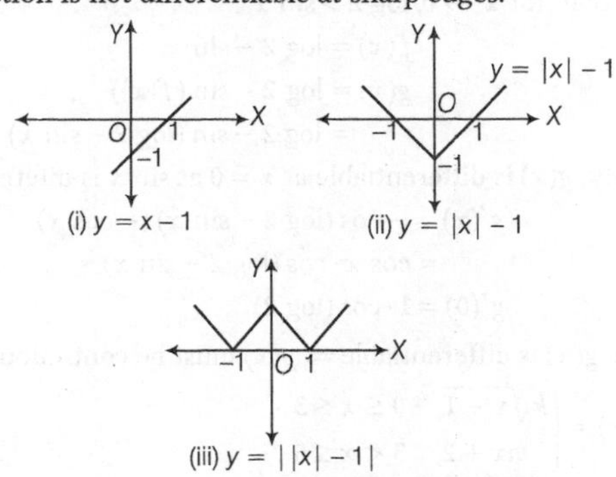

In function, $y = ||x| - 1|$, we have 3 sharp edges at $x = -1, 0, 1$.

Hence, $f(x)$ is not differentiable at $\{0, \pm 1\}$.

137. Given, $f(1) = f\left(\frac{1}{2}\right) = f\left(\frac{1}{3}\right) = \ldots = \lim_{n \to \infty} f\left(\frac{1}{n}\right) = 0$

as $f\left(\frac{1}{n}\right) = 0$; $n \in$ integers and $n \geq 1$.

$\Rightarrow \quad \lim_{n \to \infty} f\left(\frac{1}{n}\right) = 0$

$\Rightarrow \quad f(0) = 0$

Since, there are infinitely many points in neighbourhood of $x = 0$.

$\therefore \quad f(x) = 0$

$\Rightarrow \quad f'(x) = 0 \Rightarrow f'(0) = 0$

Hence, $\quad f(0) = f'(0) = 0$

138. Given, $f(x) = \begin{cases} \frac{1}{2}(-x-1), & \text{if } x < -1 \\ \tan^{-1} x, & \text{if } -1 \leq x \leq 1 \\ \frac{1}{2}(x-1), & \text{if } x > 1 \end{cases}$

$f(x)$ is discontinuous at $x = -1$ and $x = 1$.

$\therefore$ Domain of $f'(x) \in R - \{-1, 1\}$

139. Given, $f(x) = [x] \sin \pi x$

If x is just less than k, $[x] = k - 1$

$\Rightarrow \quad f(x) = (k-1) \sin \pi x.$

LHD of $f(x) = \lim_{x \to k} \frac{(k-1) \sin \pi x - k \sin \pi k}{x - k}$

$= \lim_{x \to k} \frac{(k-1) \sin \pi x}{x - k},$

$= \lim_{h \to 0} \frac{(k-1) \sin \pi (k-h)}{-h}$ [where $x = k - h$]

$= \lim_{h \to 0} \frac{(k-1)(-1)^{k-1} \cdot \sin h\pi}{-h} = (-1)^k (k-1)\pi$

140. RHD of $\sin(|x|) - |x| = \lim_{h \to 0} \frac{\sin h - h}{h} = 1 - 1 = 0$ [$\because f(0) = 0$]

LHD of $\sin(|x|) - |x|$

$= \lim_{h \to 0} \frac{\sin|-h| - |-h|}{-h}$

$= \lim_{h \to 0} \frac{\sin h - h}{-h} = 0$

Therefore, (d) is the answer.

141. We have, $f(x) = |\log 2 - \sin x|$ and $g(x) = f(f(x))$, $x \in R$

Note that, for $x \to 0$, $\log 2 > \sin x$

$\therefore \quad f(x) = \log 2 - \sin x$

$\Rightarrow \quad g(x) = \log 2 - \sin(f(x))$

$= \log 2 - \sin(\log 2 - \sin x)$

Clearly, $g(x)$ is differentiable at $x = 0$ as $\sin x$ is diferentiable.

Now, $\quad g'(x) = -\cos(\log 2 - \sin x)(-\cos x)$

$= \cos x \cdot \cos(\log 2 - \sin x)$

$\Rightarrow \quad g'(0) = 1 \cdot \cos(\log 2)$

142. Since, $g(x)$ is differentiable $\Rightarrow g(x)$ must be continuous.

$\therefore g(x) = \begin{cases} k\sqrt{x+1}, & 0 \leq x \leq 3 \\ mx + 2, & 3 < x \leq 5 \end{cases}$

At $x = 3$, $\quad$ RHL $= 3m + 2$

and at $x = 3$, $\quad$ LHL $= 2k$

$\therefore \quad 2k = 3m + 2 \quad$ ...(i)

Also, $\quad g'(x) = \begin{cases} \frac{k}{2\sqrt{x+1}}, & 0 \leq x < 3 \\ m, & 3 < x \leq 5 \end{cases}$

$\therefore \quad L\{g'(3)\} = \frac{k}{4}$ and $R\{g'(3)\} = m$

$\Rightarrow \quad \frac{k}{4} = m$, i.e. $k = 4m \quad$ ...(ii)

On solving Eqs. (i) and (ii), we get

$k = \frac{8}{5}, m = \frac{2}{5}$

$\Rightarrow \quad k + m = 2$

143. Given, $f(0) = 2 = g(1)$, $g(0) = 0$ and $f(1) = 6$

f and g are differentiable in (0, 1).

Let $\quad h(x) = f(x) - 2g(x) \quad$ ...(i)

$h(0) = f(0) - 2g(0) = 2 - 0 = 2$

and $\quad h(1) = f(1) - 2g(1) = 6 - 2(2) = 2$

$\Rightarrow \quad h(0) = h(1) = 2$

Hence, using Rolle's theorem,

$h'(c) = 0$, such that $c \in (0, 1)$

Differentiating Eq. (i) at c, we get

$\Rightarrow \quad f'(c) - 2g'(c) = 0$

$\Rightarrow \quad f'(c) = 2g'(c)$

144. Given A function $f: R \to R$ defined by

$f(x) = [x] \cos \pi \left(x - \frac{1}{2}\right)$, where [] denotes the greatest integer function.

To discuss The continuity of function f.

Now, $\cos x$ is continuous, $\forall x \in R$.

$\Rightarrow \cos \pi \left(x - \frac{1}{2}\right)$ is also continuous, $\forall x \in R$.

Hence, the continuity of f depends upon the continuity of $[x]$.

Since, $[x]$ is discontinuous, $\forall x \in I$.

So, we should check the conitnuity of f at $x = n, \forall n \in I$.

LHL at $x = n$ is given by

$f(n^-) = \lim_{x \to n^-} f(x)$

$= \lim_{x \to n^-} [x] \cos \pi \left(x - \frac{1}{2}\right) = (n-1) \cos \frac{(2n-1)\pi}{2} = 0$

RHL at $x = n$ is given by

$f(n^+) = \lim_{x \to n^+} f(x) = \lim_{x \to n^+} [x] \cos \pi \left(x - \frac{1}{2}\right)$

$= (n) \cos \frac{(2n-1)\pi}{2} = 0$

Also, value of the function at $x = n$ is

$f(n) = [n] \cos \pi \left(n - \frac{1}{2}\right) = (n) \cos \frac{(2n-1)\pi}{2} = 0$

$\therefore \quad f(n^+) = f(n^-) = f(n)$

Hence, f is continuous at $x = n, \forall n \in I$.

CHAPTER

07

dy/dx as a Rate Measurer and Tangents, Normals

Learning Part

Session 1

- Derivative as the Rate of Change
- Velocity and Acceleration in Rectilinear Motion

Session 2

- Differential and Approximation
- Geometrical Meaning of Δx, Δy, dx and dy

Session 3

- Slope of Tangent and Normal
- Equation of Tangent
- Equation of Normal

Session 4

- Angle of Intersection of Two Curves
- Length of Tangent, Subtangent, Normal and Subnormal

Session 5

- Rolle's Theorem
- Lagrange's Mean Value Theorem

Session 6

- Application of Cubic Functions

Practice Part

- JEE Type Examples
- Chapter Exercises

Session 1

Derivative as the Rate of Change, Velocity and Acceleration in Rectilinear Motion

Derivative as the Rate of Change

If a variable quantity y is some function of time t, i.e. $y = f(t)$, then small change in time Δt have a corresponding change Δy in y.

Thus, the average rate of change $= \dfrac{\Delta y}{\Delta t}$

When limit $\Delta t \to 0$ is applied, the rate of change becomes instantaneous and we get the rate of change of y with respect to the instant t.

i.e. $$\lim_{\Delta t \to 0} \frac{\Delta y}{\Delta t} = \frac{dy}{dt}$$

Hence, it is clear that **the rate of change of any variable with respect to some other variable is the derivative of first variable with respect to other variable.**

The differential coefficient of y with respect to x i.e. $\dfrac{dy}{dx}$ is nothing but the rate of increase of y relative to x.

Example 1 If the radius of a circle is increasing at a uniform rate of 2 cm/s, then find the rate of increase of area of circle, at the instant when the radius is 20 cm.

Sol. Given, $\dfrac{dr}{dt} = 2$ cm/s [where, r is radius and t is time]

Now, area of circle is given by $A = \pi r^2$

Differentiating it with respect to time t, we get

$$\frac{dA}{dt} = 2\pi r \frac{dr}{dt} \Rightarrow \frac{dA}{dt} = 2\pi \cdot 20 \cdot 2 \text{ cm}^2/\text{s}$$

$$\Rightarrow \frac{dA}{dt} = 80\pi \text{ cm}^2/\text{s}$$

Thus, the rate of change of area of circle with respect to time is 80π cm^2/s.

Example 2 On the curve $x^3 = 12y$, find the interval at which the abscissa changes at a faster rate than the ordinate.

Sol. Given, $$x^3 = 12y$$

Differentiating it with respect to y, we get

$$3x^2 \frac{dx}{dy} = 12 \Rightarrow \frac{dx}{dy} = \frac{12}{3x^2}$$

In the interval, at which the abscissa changes at a faster rate than the ordinate, we must have

$$\left|\frac{dx}{dy}\right| > 1 \text{ or } \left|\frac{12}{3x^2}\right| > 1$$

or $$\frac{4}{x^2} > 1 \Rightarrow \frac{4 - x^2}{x^2} > 0, \text{ when } x \neq 0$$

$$\Rightarrow 4 - x^2 > 0;\ x \neq 0 \Rightarrow x^2 - 4 < 0;\ x \neq 0$$

$$\Rightarrow (x-2)(x+2) < 0;\ x \neq 0$$

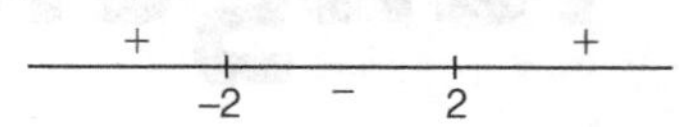

Using number line rule, we have $-2 < x < 2; x \neq 0$

Thus, $x \in (-2, 2) - \{0\}$ is the required interval at which abscissa changes at a faster rate than the ordinate.

Velocity and Acceleration in Rectilinear Motion

The velocity of a moving particle is defined as the rate of change of its displacement with respect to time and the acceleration is defined as the rate of change of its velocity with respect to time.

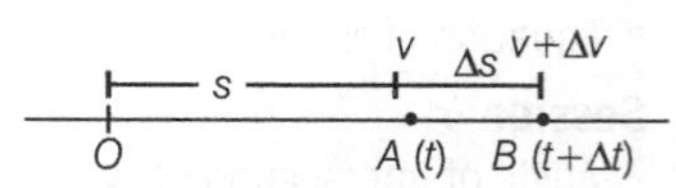

Figure 7.1

Let a particle A move rectilinearly as shown in figure. Let s be the displacement from a fixed point O along the path at time t; s is considered to be positive on right of the point O and negative on the left of it.

Also, Δs is positive when s increases, i.e. when the particle moves towards right.

Thus, if Δs be the increment in s in time Δt, then average velocity in this interval is

$$\text{Average velocity} = \frac{\Delta s}{\Delta t}$$

and the instantaneous velocity, i.e. velocity at time t is

$$v = \lim_{\Delta t \to 0} \frac{\Delta s}{\Delta t} = \frac{ds}{dt}$$

If the velocity varies, then there is change of velocity Δv in time Δt. Hence, the acceleration at time t is

$$\text{Acceleration } (a) = \lim_{\Delta t \to 0} \frac{\Delta v}{\Delta t} = \frac{dv}{dt}$$

Example 3 If the displacement of a particle is given by $s = \left(\frac{1}{2}t^2 + 4\sqrt{t}\right)$ m. Find the velocity and acceleration at $t = 4$ seconds.

Sol. We have, the displacement of particle is given by,

$$s = \left(\frac{1}{2}t^2 + 4\sqrt{t}\right) \text{ m}$$

As, we know velocity $v = \frac{ds}{dt} = \left(t + \frac{2}{\sqrt{t}}\right)$ m/s ...(i)

and acceleration $a = \frac{d^2s}{dt^2} = \frac{dv}{dt} = \left(1 - \frac{1}{t^{3/2}}\right)$ m/s ...(ii)

Now, velocity when $t = 4$, is $v = \left(4 + \frac{2}{\sqrt{4}}\right)$ m/s [from Eq. (i)]

$\Rightarrow \quad v = 5$ m/s

and acceleration when $t = 4$, is $a = \left(1 - \frac{1}{4^{3/2}}\right)$ m/s^2 [from Eq. (i)]

$\Rightarrow \quad a = \left(1 - \frac{1}{8}\right)$ m/s^2 $\Rightarrow a = \frac{7}{8}$ m/s^2

Example 4 If $s = \frac{1}{2}t^3 - 6t$, then find the acceleration at time when the velocity tends to zero.

Sol. We have the displacement s is given by, $s = \frac{1}{2}t^3 - 6t$

Now, velocity $\quad v = \frac{ds}{dt} = \left(\frac{3t^2}{2} - 6\right)$ ...(i)

and acceleration, $a = \frac{dv}{dt} = \frac{d^2s}{dt^2} = (3t)$ ...(ii)

To find acceleration, when velocity $\to 0$

Now, when velocit y $\to 0$, then $\frac{3t^2}{2} - 6 = 0$ [from Eq. (i)]

$\Rightarrow \quad t^2 = 4 \Rightarrow t = 2$

Thus, acceleration when velocity tends to zero, is

$a = 3t = 6$ [from Eq. (ii)]

Example 5 If r is the radius, S the surface area and V the volume of a spherical bubble, prove that

(i) $\frac{dV}{dt} = 4\pi r^2 \frac{dr}{dt}$ (ii) $\frac{dV}{dS} \propto r$.

Sol. (i) Since, $V = \frac{4}{3}\pi r^3$

$\therefore \quad \frac{dV}{dt} = \frac{4}{3}\pi \cdot 3r^2 \frac{dr}{dt} = 4\pi r^2 \frac{dr}{dt}$ ...(i)

(ii) We know, $S = 4\pi r^2 \Rightarrow \frac{dS}{dt} = 8\pi r \frac{dr}{dt}$ ...(ii)

Thus, $\frac{dV}{dS} = \frac{dV/dt}{dS/dt} = \frac{1}{2} \cdot r$ [from Eqs. (i) and (ii)]

$\Rightarrow \quad \frac{dV}{dS} = \frac{1}{2}r$ or $\frac{dV}{dS} \propto r$

Example 6 A man who is 1.6 m tall walks away from a lamp which is 4 m above ground at the rate of 30 m/min. How fast is the man's shadow lengthening?

Sol. Let $PQ = 4$ m be the height of lamp and $AB = 1.6$ m be height of man. Let the end of shadow is R and it is at a distance of l from A when the man is at a distance of x from PQ, at some instance.

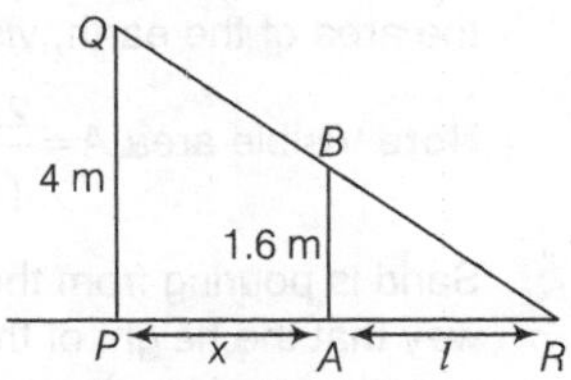

Since, ΔPQR and ΔABR are similar, we have

$$\frac{PQ}{AB} = \frac{PR}{AR} \Rightarrow \frac{4}{1.6} = \frac{x + l}{l}$$

$\Rightarrow \quad 2x = 3l \quad \Rightarrow 2\frac{dx}{dt} = 3\frac{dl}{dt} \quad \left[\text{given } \frac{dx}{dt} = 30 \text{ m/min}\right]$

$\therefore \quad \frac{dl}{dt} = \frac{2}{3} \cdot 30$ m/min $= 20$ m/min (lengthening)

Example 7 If x and y are the sides of two squares such that $y = x - x^2$, find the rate of change of the area of the second square with respect to the first square.

Sol. Given, x and y are sides of two squares, thus the area of two squares are x^2 and y^2.

We have to obtain $\frac{d(y^2)}{d(x^2)} = \frac{2y\frac{dy}{dx}}{2x} = \frac{y}{x} \cdot \frac{dy}{dx}$ ...(i)

The given curve is $\quad y = x - x^2$

$\Rightarrow \quad \frac{dy}{dx} = 1 - 2x$ ...(ii)

Thus, $\quad \frac{d(y^2)}{d(x^2)} = \frac{y}{x}(1 - 2x)$ [from Eqs. (i) and (ii)]

or $\quad \frac{d(y^2)}{d(x^2)} = \frac{(x - x^2)(1 - 2x)}{x}$

$\Rightarrow \quad \frac{d(y^2)}{d(x^2)} = 2x^2 - 3x + 1$

Exercise for Session 1

1. The surface area of a spherical bubble is increasing at the rate of 2 cm^2/s. Find the rate at which the volume of the bubble is increasing at the instant, if its radius is 6 cm.
2. A particle moves along the curve $y = \frac{2}{3}x^3 + 1$. Find the point on the curve at which y-coordinate is changing twice as fast as x-coordinate.
3. The area of an expanding rectangle is increasing at the rate of 48 cm^2/s. The length of rectangle is always equal to square of the breadth. At which rate the length is increasing at the instant when the breadth is 4 cm?
4. An edge of a variable cube is increasing at the rate of 10 cm/s. How fast the volume of the cube is increasing when the edge is 5 cm long?
5. An air-force plane is ascending vertically at the rate of 100 km/h. If the radius of the earth is r km, how fast is the area of the earth, visible from the plane, increasing at 3 min after it started ascending?

   Note Visible area $A = \frac{2\pi r^2 h}{r+h}$, where h is the height of the plane above the earth.
6. Sand is pouring from the pipe at the rate of 12 cm^3/s. The falling sand forms a cone on the ground in such a way that the height of the cone is always one-sixth of the radius of the base. How fast is the height of the sand cone increasing when the height is 4 cm?
7. Water is dripping out from a conical funnel, at the uniform rate of 2 cm^3/s, through a tiny hole at the vertex of the bottom. When the slant height of the water is 4 cm, find the rate of decrease of the slant height of the water, if the vertical angle of funnel is 120°.
8. From a cylindrical drum containing oil and kept vertical, the oil leaking at the rate of 10 cm^3/s. If the radius of the drum is 10 cm and height is 50 cm, then find the rate at which level of oil is changing when oil level is 20 cm.
9. A kite is 120 m high and 130 m of string is out. If the kite is moving away horizontally at the rate of 52 m/s, find the rate at which the string is being paid out.
10. A ladder 13 m long leans against a wall. The foot of the ladder is pulled along the ground away from the wall, at the rate of 1.5 m/s. How fast is the angle θ between the ladder and the ground changing when the foot of the ladder is 12 m away from the wall?
11. Water is running into a conical vessel, 15 cm deep and 5 cm in radius, at the rate of 0.1 cm^3/s. When the water is 6 cm deep, find at what rate is
    (a) the water level rising?
    (b) the water surface area increasing?
    (c) the wetted surface of the vessel increasing?
12. Height of a tank in the form of an inverted cone is 10 m and radius of its circular base is 2 m. The tank contains water and it is leaking through a hole at its vertex at the rate of 0.02 m^3/s. Find the rate at which the water level changes and the rate at which the radius of water surface changes when height of water level is 5 m.
13. At what points of the ellipse $16x^2 + 9y^2 = 400$ does the ordinate decreases at the same rate at which the abscissa increases?
14. A satellite travels in a circular orbit of radius R. If its x-coordinate decreases at the rate of 2 units/s at the point (a, b), how fast is the y-coordinate changing?
15. The ends of a rod AB which is 5 m long moves along two grooves OX, OY which are at right angles. If A moves at a constant speed of $\frac{1}{2}$ m/s, what is the speed of B, when it is 4 m from O?

Session 2

Differential and Approximation, Geometrical Meaning of Δx, Δy, dx and dy

Differential and Approximation

We are familiar with the symbol dy/dx representing derivative of y w.r.t. x.

Now, we shall give meaning to the symbols dx and dy such that the symbol dy/dx represents the **quotient of dy and dx.**

Let $y = f(x)$ be a function of x.

Let Δx denotes small change in x and let the corresponding change in y be Δy.

Then, $$\lim_{\Delta x \to 0} \frac{\Delta y}{\Delta x} = \frac{dy}{dx} \Rightarrow \frac{\Delta y}{\Delta x} = \frac{dy}{dx} \text{ (approx)}$$

$$\Rightarrow \quad \frac{\Delta y}{\Delta x} = \frac{dy}{dx} + \varepsilon, \text{ where } \varepsilon \to 0 \text{ as } \Delta x \to 0$$

$$\Rightarrow \quad \Delta y = \frac{dy}{dx} \cdot \Delta x + \varepsilon \cdot \Delta x \quad \text{...(i)}$$

Now, Δy consists of two parts. The part $\varepsilon \cdot \Delta x$ is very small and hence negligible. The part $\frac{dy}{dx} \cdot \Delta x$ is called **principal part of Δy** and is denoted by **dy** and is called **differential of y.**

Thus, $$dy = \frac{dy}{dx} \cdot \Delta x \quad \text{...(ii)}$$

$$\Rightarrow \quad dx = \frac{dy}{dy} \cdot \Delta x$$

$$= (1)\Delta x = \Delta x \quad \text{...(iii)}$$

From Eqs. (ii) and (iii), we get

$$dy = \frac{dy}{dx} \cdot dx$$

$\Rightarrow$ (Differential of y) = (Differential coefficient of y w.r.t. x) $\cdot$ (Differential of x)

$$\Rightarrow \quad \frac{\text{Differential of } y}{\text{Differential of } x} = (\text{Differential coefficient of } y \text{ w.r.t. } x)$$

From Eq. (i), $$\Delta y = \frac{dy}{dx} \cdot \Delta x \text{ (approx)}$$

$$\Rightarrow \quad \Delta y = dy \text{ (approx)}$$

Hence, approximately

$$dy = \Delta y = (y + \Delta y) - y = f(x + \Delta x) - f(x)$$

$$\Rightarrow \quad \frac{dy}{dx} \cdot dx = f(x + \Delta x) - f(x) \text{ (approx)}$$

$$\boxed{f(x + \Delta x) - f(x) = f'(x) \cdot \Delta x \text{ (approx)}}$$

Errors

Definition of Absolute Error

Δx or dx is called absolute error in x.

Definition of Relative Error

$\frac{\Delta x}{x}$ or $\frac{dx}{x}$ is called relative error in x.

Definition of Percentage Error

$\left(\frac{\Delta x}{x}\right) \cdot 100$ or $\left(\frac{dx}{x}\right) \cdot 100$ is called percentage error in x.

Geometrical Meaning of $\Delta x, \Delta y, dx$ and dy

Let us take a point $A(x, y)$ on the curve $y = f(x)$, where $f(x)$ is differentiable real function.

Let $B(x + \Delta x, y + \Delta y)$ be a neighbouring point on the curve, where Δx denotes a small change in x and Δy is the corresponding change in y.

From the figure, it is clear that if Δx and Δy are sufficiently small quantities, then

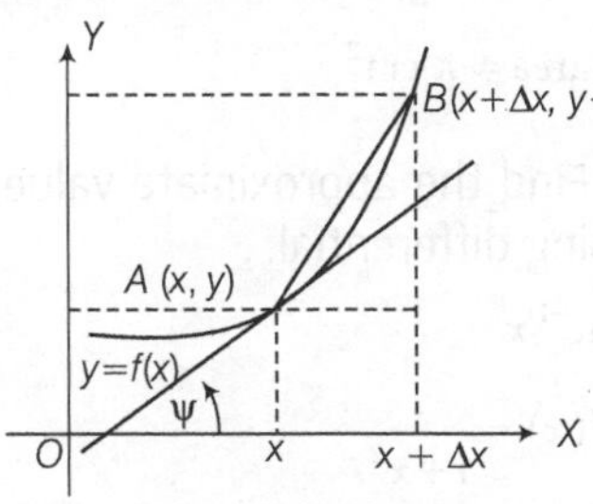

Figure 7.2

$$\frac{\Delta y}{\Delta x} = \tan \psi \cong \frac{dy}{dx} = f'(x)$$

Hence, approximate change in the value of y, called its differential, is given by $\Delta y = f'(x) \cdot \Delta x$

Example 8 Use differential to approximate $\sqrt{101}$.

Sol. Let $f(x) = \sqrt{x}$, then $f'(x) = \dfrac{1}{2\sqrt{x}}$.

Now, let $x = 100$ and $\Delta x = 1$, so that $x + \Delta x = 101$

Then, by definition $f(x+\Delta x) - f(x) = f'(x)\cdot \Delta x$, we get

$$\sqrt{101} - \sqrt{100} = f'(\sqrt{100})\cdot 1$$

$$\sqrt{101} = \sqrt{100} + \frac{1}{2\sqrt{100}}\cdot 1$$

$$= 10 + \frac{1}{20} = 10.05$$

Example 9 Use differential to approximate $(66)^{1/3}$.

Sol. Let $\quad f(x) = (x)^{1/3}$

Then, $\quad f'(x) = \dfrac{1}{3x^{2/3}}$

Now, let $x = 64$, $\Delta x = 2$, so that $x + \Delta x = 66$

Then, by definition $f(x + \Delta x) - f(x) = f'(x)\cdot \Delta x$

$\Rightarrow \quad f(66) - f(64) = f'(64)\cdot(2)$

$\Rightarrow \quad (66)^{1/3} - (64)^{1/3} = \dfrac{1}{3(64)^{2/3}} \times 2$

$\Rightarrow \quad (66)^{1/3} = 4 + \dfrac{2}{48} = \dfrac{97}{24} = 4.0416$

Example 10 If the radius of a circle increases from 5 cm to 5.1 cm, find the increase in area.

Sol. Let r be the radius of the circle. Then, $r = 5$ cm and $r + dr = 5.1$ cm, so that $dr = 0.1$ cm.

Let A denotes the areas of the circle.

Then, $\quad A = \pi r^2 \quad \therefore \dfrac{dA}{dr} = 2\pi r$

$\therefore$ At $r = 5$, $\dfrac{dA}{dr} = 10\pi$

Now, $\quad dA = \dfrac{dA}{dr}\cdot dr = 10\pi \times (0.1) = \pi \text{ cm}^2$

$\therefore$ Increase in area $= \pi \text{ cm}^2$

Example 11 Find the approximate value of $\tan^{-1}(0.999)$ using differential.

Sol. Let $f(x) = \tan^{-1} x$

Then, $\quad f'(x) = \dfrac{1}{1+x^2}$

Now, let $x = 1$ and $\Delta x = -0.001$. So that $x + \Delta x = 0.999$

Then, by definition,

$f(x + \Delta x) - f(x) = f'(x)\cdot \Delta x$, we get

$f(0.999) - f(1) = f'(1)\cdot(-0.001)$

$\Rightarrow \quad \tan^{-1}(0.999) = \tan^{-1}(1) - (0.001)\cdot\dfrac{1}{1+1^2}$

$$= \frac{\pi}{4} - 0.0005 = \frac{1}{4}\times\frac{22}{7} - 0.0005 = 0.7852$$

Example 12 The time T of oscillation of a simple pendulum of length l is given by $T = 2\pi\cdot\sqrt{\dfrac{l}{g}}$.

Find percentage error in T corresponding to

(i) an increase of 2% in the value of l.

(ii) decrease of 2% in the value of l.

Sol. (i) We have, $T = 2\pi\cdot\sqrt{\dfrac{l}{g}}$

$\Rightarrow \quad \log T = \log 2 + \log\pi + \dfrac{1}{2}\log l - \dfrac{1}{2}\log g$

Differentiating w.r.t. l, we get

$$\frac{1}{T}\frac{dT}{dl} = 0 + 0 + \frac{1}{2l} - 0 \Rightarrow \frac{dT}{T} = \frac{1}{2}\frac{dl}{l}$$

$$\therefore \quad \frac{dT}{T}\times 100 = \frac{1}{2}\left(\frac{dl}{l}\times 100\right) = \frac{1}{2}(2)\% = 1\%$$

$$\left[\because \frac{dl}{l}\times 100 = 2\%\right]$$

Percentage error in $T = 1\%$.

(ii) We have, $\dfrac{dT}{T} = \dfrac{1}{2}\dfrac{dl}{l}$

$$\Rightarrow \frac{dT}{T}\times 100 = \frac{1}{2}\left(\frac{dl}{l}\times 100\right) = \frac{1}{2}(-2)\% = -1\%$$

$[\because$ decrease in $l = 2\%]$

Percentage error in $T = -1\%$.

Example 13 In an acute ΔABC, if sides a and b are constants and the base angles A and B vary, show that $\dfrac{dA}{\sqrt{a^2 - b^2\sin^2 A}} = \dfrac{dB}{\sqrt{b^2 - a^2\sin^2 B}}$.

Sol. $\dfrac{a}{\sin A} = \dfrac{b}{\sin B}$ or $b\sin A = a\sin B$

$\Rightarrow \quad b\cos A\, dA = a\cos B\, dB$

$$\Rightarrow \quad \frac{dA}{a\cos B} = \frac{dB}{b\cos A}$$

$$\Rightarrow \quad \frac{dA}{a\sqrt{1-\sin^2 B}} = \frac{dB}{b\sqrt{1-\sin^2 A}}$$

$$\Rightarrow \quad \frac{dA}{a\sqrt{1-\dfrac{b^2\sin^2 A}{a^2}}} = \frac{dB}{b\sqrt{1-\dfrac{a^2\sin^2 B}{b^2}}}$$

$$\Rightarrow \quad \frac{dA}{\sqrt{a^2 - b^2\sin^2 A}} = \frac{dB}{\sqrt{b^2 - a^2\sin^2 B}}$$

Exercise for Session 2

1. Use differential to approximate $\sqrt{51}$.
2. Use differential to approximate $\log(9.01)$. (Given, $\log 3 = 1.0986$)
3. If the error committed in measuring the radius of a circle is 0.01%, find the corresponding error in calculating the area.
4. The pressure p and the volume V of a gas are connected by the relation $pV^{1/4} = a$ (constant). Find the percentage increase in pressure corresponding to a decrease of $\left(\frac{1}{2}\right)$% in volume.
5. If in a ΔABC, the side c and the angle C remain constant, while the remaining elements are changed slightly. Using differential, show that $\frac{da}{\cos A} + \frac{db}{\cos B} = 0$.
6. If a ΔABC, inscribed in a fixed circle, is slightly varied in such a way as to have its vertices always on the circle, show that $\frac{da}{\cos A} + \frac{db}{\cos B} + \frac{dc}{\cos C} = 0$.

Session 3

Slope of Tangent and Normal, Equation of Tangent, Equation of Normal

Slope of Tangent and Normal

Slope of Tangent

Let $y = f(x)$ be a continuous curve and let $P(x_1, y_1)$ be a point on it.

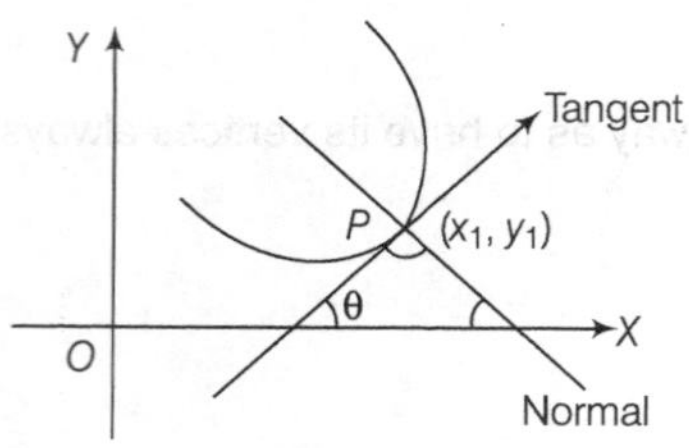

Figure 7.3

Then, $\left(\frac{dy}{dx}\right)_{(x_1, y_1)}$ is the slope of tangent to the curve $y = f(x)$ at a point $P(x_1, y_1)$.

$$\Rightarrow \quad \left(\frac{dy}{dx}\right)_P = \tan\theta = \text{Slope of tangent at } P$$

where, θ is the angle which the tangent at $P(x_1, y_1)$ forms with the positive direction of X-axis as shown in the figure.

Remarks

(i) Horizontal tangent If tangent is parallel to X-axis, then

$$\theta = 0° \Rightarrow \tan\theta = 0$$

$$\therefore \quad \left(\frac{dy}{dx}\right)_{(x_1, y_1)} = 0$$

(ii) Vertical tangent If tangent is perpendicular to X-axis or parallel to Y-axis, then

$$\theta = 90° \Rightarrow \tan\theta = \infty \quad \text{or} \quad \cot\theta = 0$$

$$\therefore \quad \left(\frac{dx}{dy}\right)_{(x_1, y_1)} = 0$$

Slope of Normal

We know that the normal to the curve at $P(x_1, y_1)$ is a line perpendicular to tangent at $P(x_1, y_1)$ and passes through P.

∴ Slope of the normal at

$$P = -\frac{1}{\text{Slope of the tangent at } P}$$

$$\Rightarrow \text{Slope of normal at } P(x_1, y_1) = -\frac{1}{\left(\frac{dy}{dx}\right)_{(x_1, y_1)}}$$

$$\text{or} \quad \text{Slope of normal at } P(x_1, y_1) = -\left(\frac{dx}{dy}\right)_{(x_1, y_1)}$$

Remarks

(i) Horizontal normal If normal is parallel to X-axis, then

$$-\left(\frac{dx}{dy}\right)_{(x_1, y_1)} = 0 \text{ or } \left(\frac{dx}{dy}\right)_{(x_1, y_1)} = 0$$

(ii) Vertical normal If normal is perpendicular to X-axis or parallel to Y-axis, then $-\left(\frac{dy}{dx}\right)_{(x_1, y_1)} = 0$

Example 14 Find the slopes of the tangent and normal to the curve $x^3 + 3xy + y^3 = 2$ at (1, 1).

Sol. Given equation of curve is $x^3 + 3xy + y^3 = 2$.

Differentiating it w.r.t. x, we get

$$3x^2 + 3x\frac{dy}{dx} + 3y + 3y^2\frac{dy}{dx} = 0$$

$$\Rightarrow \quad \frac{dy}{dx} = -\frac{(3x^2 + 3y)}{(3x + 3y^2)} \Rightarrow \frac{dy}{dx} = -\frac{(x^2 + y)}{(x + y^2)}$$

$$\Rightarrow \quad \left(\frac{dy}{dx}\right)_{(1,1)} = -\left(\frac{2}{2}\right) = -1$$

$$\therefore \quad \text{Slope of tangent at } (1,1) = \left(\frac{dy}{dx}\right)_{(1,1)} = -1$$

$$\text{and slope of normal at } (1,1) = -\frac{1}{\left(\frac{dy}{dx}\right)_{(1,1)}} = \frac{-1}{-1} = 1$$

Example 15 Find the point on the curve $y = x^3 - 3x$ at which tangent is parallel to X-axis.

Sol. Let the point at which tangent is parallel to X-axis be $P(x_1, y_1)$.

Then, it must lie on curve.

Therefore, we have $y_1 = x_1^3 - 3x_1$ ...(i)

Differentiating $y = x^3 - 3x$ w.r.t. x, we get

$$\frac{dy}{dx} = 3x^2 - 3 \Rightarrow \left(\frac{dy}{dx}\right)_{(x_1, y_1)} = 3x_1^2 - 3$$

Since, the tangent is parallel to X-axis.

$$\therefore \quad \left(\frac{dy}{dx}\right)_{(x_1,\, y_1)} = 0 \Rightarrow 3x_1^2 - 3 = 0$$

$$\Rightarrow \quad x_1 = \pm 1 \quad \text{...(ii)}$$

From Eqs. (i) and (ii), we get

When $x_1 = 1$, then $y_1 = 1 - 3 = -2$

When $x_1 = -1$, then $y_1 = -1 + 3 = 2$

$\therefore$ Points at which tangent is parallel to X-axis are $(1, -2)$ and $(-1, 2)$.

Example 16 Find the point on the curve $y = x^3 - 2x^2 - x$ at which the tangent line is parallel to the line $y = 3x - 2$.

Sol. Let $P(x_1, y_1)$ be the required point.

Then, we have $\quad y_1 = x_1^3 - 2x_1^2 - x_1 \quad$ ...(i)

Differentiating the curve $y = x^3 - 2x^2 - x$ w.r.t. x, we get

$$\frac{dy}{dx} = 3x^2 - 4x - 1 \Rightarrow \left(\frac{dy}{dx}\right)_{(x_1,\, y_1)} = 3x_1^2 - 4x_1 - 1$$

Since, tangent at (x_1, y_1) is parallel to the line $y = 3x - 2$.

$\therefore$ Slope of the tangent at $P(x_1, y_1)$ = Slope of the line

$$y = 3x - 2 \Rightarrow \left(\frac{dy}{dx}\right)_{(x_1,\, y_1)} = 3$$

$$\Rightarrow \quad 3x_1^2 - 4x_1 - 1 = 3 \Rightarrow 3x_1^2 - 4x_1 - 4 = 0$$

$$\Rightarrow \quad (x_1 - 2)(3x_1 + 2) = 0 \Rightarrow x_1 = 2, -2/3 \quad \text{...(ii)}$$

From Eqs. (i) and (ii), we get

When $x_1 = 2$, then

$$y_1 = 8 - 8 - 2 \Rightarrow y_1 = -2$$

When $\quad x_1 = -2/3$, then $y_1 = x_1^3 - 2x_1^2 - x_1$

$$\Rightarrow \quad y_1 = \frac{-8}{27} - \frac{8}{9} + \frac{2}{3} \Rightarrow y_1 = \frac{-14}{27}$$

Thus, the point at which tangent is parallel to $y = 3x - 2$ are $(2, -2)$ and $\left(-\frac{2}{3}, -\frac{14}{27}\right)$.

Example 17 In which of the following cases, the function $f(x)$ has a vertical tangent at $x = 0$?

(i) $f(x) = x^{1/3}$ $\quad$ (ii) $f(x) = \text{sgn}\, x$

(iii) $f(x) = x^{2/3}$ $\quad$ (iv) $f(x) = \sqrt{|x|}$

(v) $f(x) = \begin{cases} 0, \text{if } x < 0 \\ 1, \text{if } x \ge 0 \end{cases}$

Sol. **Vertical Tangent**

Concept $y = f(x)$ has a vertical tangent at the points $x = x_0$, if

$$\lim_{h \to 0} \frac{f(x_0 + h) - f(x_0)}{h} = \infty \text{ or } -\infty \text{ but not both.}$$

Here, the functions $f(x) = x^{1/3}$ and $f(x) = \text{sgn}\, x$ both have a vertical tangent at $x = 0$, but $f(x) = x^{2/3}, f(x) = \sqrt{|x|}$ and $f(x) = \begin{cases} 0, \text{ if } x < 0 \\ 1, \text{ if } x \ge 0 \end{cases}$ have no vertical tangent.

Explanation

(i) $f(x) = x^{1/3}$

$$\left.\begin{aligned} f'(0^+) &= \lim_{h \to 0} \frac{h^{1/3}}{h} = \frac{1}{h^{2/3}} \to \infty \\ f'(0^-) &= \lim_{h \to 0} \frac{(-h)^{1/3}}{-h} = \frac{1}{(-h)^{2/3}} = \frac{1}{h^{2/3}} \to \infty \end{aligned}\right\}$$

$\Rightarrow f(x)$ has a vertical tangent at $x = 0$.

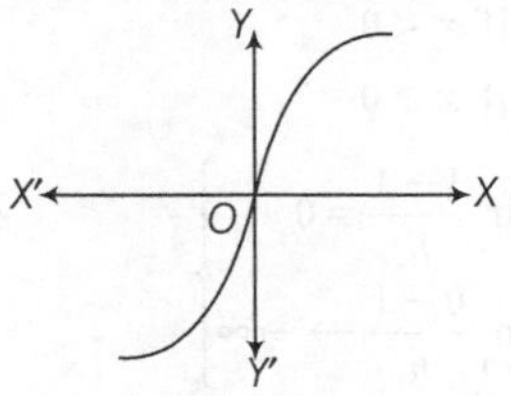

(ii) $f(x) = \text{sgn}\, x = \begin{cases} 1, \text{ if } x > 0 \\ 0, \text{ if } x = 0 \\ -1, \text{ if } x < 0 \end{cases}$

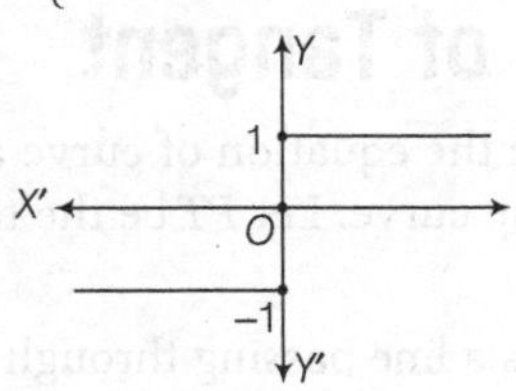

$$\left.\begin{aligned} f'(0^+) &= \lim_{h \to 0} \frac{1 - 0}{h} \to \infty \\ f'(0^-) &= \lim_{h \to 0} \frac{-1}{-h} \to \infty \end{aligned}\right\}$$

$\Rightarrow f(x)$ has a vertical tangent at $x = 0$.

(iii) $f(x) = x^{2/3}$

$$\left.\begin{aligned} f'(0^+) &= \lim_{h \to 0} \frac{h^{2/3}}{h} \to \infty \\ f'(0^-) &= \lim_{h \to 0} \frac{(-h)^{2/3}}{-h} = -\frac{1}{h^{1/3}} \to -\infty \end{aligned}\right\}$$

$\Rightarrow$ No vertical tangent at $x = 0$.

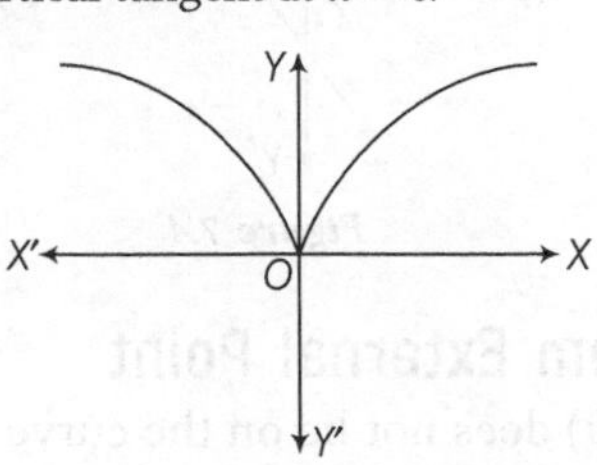

(iv) $f(x)=\sqrt{|x|}=\begin{cases}\sqrt{x}, \text{ if } x\ge 0\\ \sqrt{-x}, \text{ if } x<0\end{cases}$

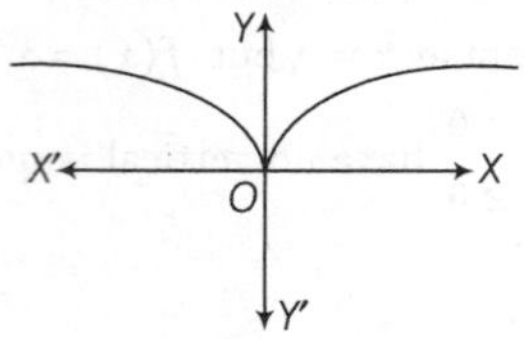

$$\left.\begin{aligned} f'(0^+) &= \lim_{h\to 0}\frac{\sqrt{h}-0}{h}\to\infty \\ f'(0^-) &= \lim_{h\to 0}\frac{\sqrt{h}}{-h}\to-\infty \end{aligned}\right\}$$

$\Rightarrow$ No vertical tangent at $x=0$.

(v) $f(x)=\begin{cases}0, \text{ if } x<0\\ 1, \text{ if } x\ge 0\end{cases}$

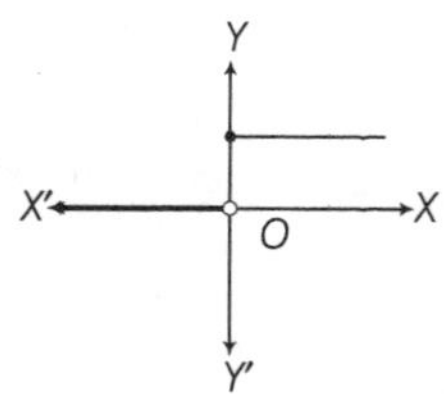

$$\left.\begin{aligned} f'(0^+) &= \lim_{h\to 0}\frac{1-1}{h}=0 \\ f'(0^-) &= \lim_{h\to 0}\frac{0-1}{h}\to-\infty \end{aligned}\right\}$$

$\Rightarrow$ No vertical tangent at $x=0$.

Equation of Tangent

Let $y=f(x)$ be the equation of curve and point (x_1, y_1) be any point on the curve. Let PT be the tangent at point (x_1, y_1).

Since, tangent is a line passing through the point $P(x_1, y_1)$ and having slope $m=\left(\frac{dy}{dx}\right)_{(x_1,y_1)}$, therefore by coordinate geometry, the equation of tangent is

$$\boldsymbol{y-y_1=m(x-x_1)\Rightarrow y-y_1=\left(\frac{dy}{dx}\right)_{(x_1,y_1)}\cdot(x-x_1).}$$

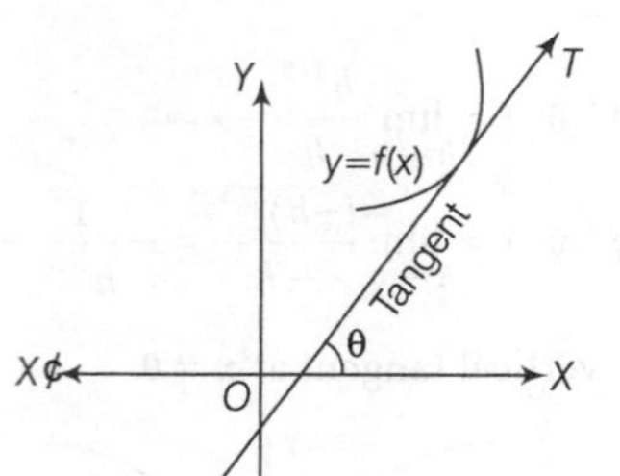

Figure 7.4

Tangent from External Point

If a point $P(a, b)$ does not lie on the curve $y=f(x)$, then the equation of all possible tangents to the curve $y=f(x)$, passing through (a, b), can be found by solving for the point of contact Q.

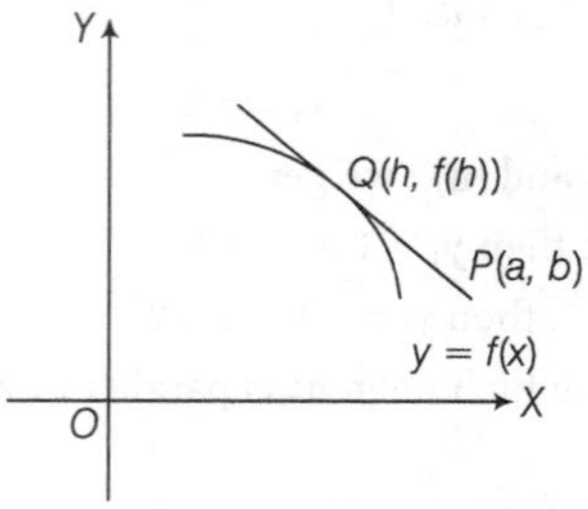

Figure 7.5

Then, $\quad f'(h)=\dfrac{f(h)-b}{h-a}$

and equation of tangent is

$$\boldsymbol{y-b=\frac{f(h)-b}{h-a}\cdot(x-a)}$$

Example 18 Find value of c such that line joining points (0, 3) and (5, – 2) becomes tangent to $y=\dfrac{c}{x+1}$.

Sol. Equation of line joining $A(0, 3)$ and $B(5, -2)$ is $x+y=3$.

Solving the line and curve, we get

$$3-x=\frac{c}{x+1}$$

$$\Rightarrow \qquad x^2-2x+(c-3)=0 \qquad \text{...(i)}$$

For tangency, roots of this equation must be equal.

Hence, discriminant of quadratic equation = 0.

$$\Rightarrow \qquad 4=4(c-3)\Rightarrow c=4$$

Hence, required value of c is 4.

Equation of Normal

We know that normal to curve at any point is a straight line passes through that point and is perpendicular to the tangent to the curve at that point.

Since, the slope of tangent at $P(x_1, y_1)=\left(\frac{dy}{dx}\right)_{(x_1,y_1)}$

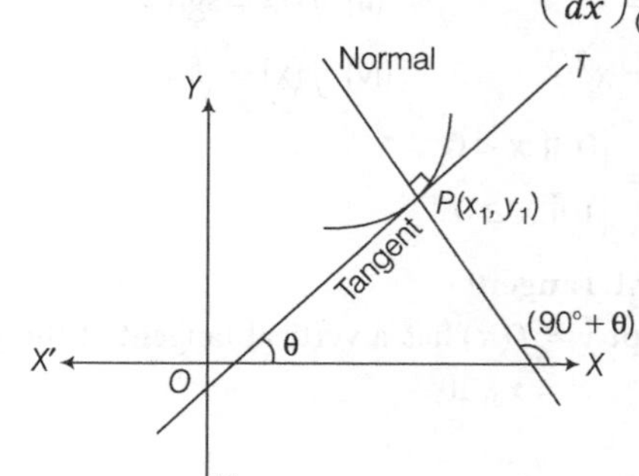

Figure 7.6

Now, as normal is perpendicular to the tangent.

$\therefore$ Slope of normal at $P(x_1, y_1) = \dfrac{-1}{\left(\dfrac{dy}{dx}\right)_{(x_1, y_1)}}$

Hence, from coordinate geometry equation of normal is

$$y - y_1 = -\frac{1}{\left(\dfrac{dy}{dx}\right)_{(x_1, y_1)}}(x - x_1)$$

or $$y - y_1 = -\left(\frac{dx}{dy}\right)_{(x_1, y_1)}(x - x_1)$$

Example 19 Find the equation of tangent and normal to the curve $2y = 3 - x^2$ at (1, 1).

Sol. The equation of given curve is,

$$2y = 3 - x^2 \quad \text{...(i)}$$

Differentiating Eq. (i) w.r.t. x, we get

$$2\left(\frac{dy}{dx}\right) = -2x \Rightarrow \frac{dy}{dx} = -x$$

$$\Rightarrow \left(\frac{dy}{dx}\right)_{(1,1)} = -1 \quad \text{...(ii)}$$

Now, the equation of tangent at (1, 1) is,

$$\Rightarrow \frac{y-1}{x-1} = \left(\frac{dy}{dx}\right)_{(1,1)} \Rightarrow \frac{y-1}{x-1} = -1$$

$$\Rightarrow y - 1 = -x + 1$$

$$\Rightarrow y + x = 2 \quad \text{[required equation of tangent]}$$

and the equation of the normal at (1, 1) is,

$$\frac{y-1}{x-1} = -\frac{1}{(dy/dx)_{(1,1)}} = 1$$

$$\therefore \quad y - x = 0 \quad \text{[required equation of normal]}$$

Example 20 Find the equation of tangent and normal to the parabola $y^2 = 4ax$ at the point $(at^2, 2at)$.

Sol. The equation of given curve is,

$$y^2 = 4ax \quad \text{...(i)}$$

Differentiating Eq. (i) w.r.t. x, we get $2y\dfrac{dy}{dx} = 4a$

$$\Rightarrow \left(\frac{dy}{dx}\right)_{(at^2, 2at)} = \frac{4a}{4at} = \frac{1}{t} \quad \text{...(ii)}$$

Now, the equation of tangent at $(at^2, 2at)$ is

$$\frac{y - 2at}{x - at^2} = \left(\frac{dy}{dx}\right)_{(at^2, 2at)} = \frac{1}{t} \quad \text{[using Eq. (ii)]}$$

$$\Rightarrow (y - 2at)t = x - at^2$$

$$\Rightarrow yt - 2at^2 = x - at^2 \Rightarrow yt = x + at^2$$

[required equation of tangent]

and the equation of normal at $(at^2, 2at)$ is,

$$\frac{y - 2at}{x - at^2} = -\frac{1}{\left(\dfrac{dy}{dx}\right)_{(at^2, 2at)}} = -t \quad \text{[using Eq. (ii)]}$$

$$\Rightarrow y - 2at = -xt + at^3$$

$$\Rightarrow y + xt = 2at + at^3 \quad \text{[required equation of normal]}$$

Example 21 Find the point on the curve $y - e^{xy} + x = 0$ at which we have vertical tangent.

Sol. The equation of given curve is,

$$y - e^{xy} + x = 0 \quad \text{...(i)}$$

Differentiating Eq. (i) w.r.t. x, we get

$$\frac{dy}{dx} - e^{xy}\left\{1 \cdot y + x \cdot \frac{dy}{dx}\right\} + 1 = 0$$

$$\Rightarrow \frac{dy}{dx}(1 - xe^{xy}) = -1 + y \cdot e^{xy}$$

$$\Rightarrow \frac{dy}{dx} = \frac{-1 + y \cdot e^{xy}}{1 - xe^{xy}} \quad \text{...(ii)}$$

Let at point (x_1, y_1) on the curve, we have a vertical tangent (i.e. a tangent parallel to Y-axis). Then,

$$\left(\frac{dy}{dx}\right)_{(x_1, y_1)} = \infty \text{ or } \left(\frac{dx}{dy}\right)_{(x_1, y_1)} = 0$$

$$\Rightarrow \frac{1 - x_1e^{x_1y_1}}{-1 + y_1e^{x_1y_1}} = 0 \Rightarrow 1 - x_1e^{x_1y_1} = 0$$

$$\Rightarrow x_1e^{x_1y_1} = 1,$$

which is possible only if $x_1 = 1$ and $y_1 = 0$.

Thus, the required point is (1, 0).

Remarks

For standard curves students are advised to use direct method of finding equation of tangent.

For the curve of the form

$ax^2 + 2hxy + by^2 + 2gx + 2fy + c = 0$, replace

x^2 by xx_1; $2x$ by $x + x_1$;

y^2 by yy_1; $2y$ by $y + y_1$

and xy by $\dfrac{xy_1 + x_1y}{2}$

Then, equation of tangent is,

$axx_1 + h(xy_1 + yx_1) + byy_1 + g(x + x_1) + f(y + y_1) + c = 0$

e.g.

(i) Find equation of tangent to the curve $2y = x^2 + 3$ at (x_1, y_1).

Sol. On replacing $2y$ by $y + y_1$ and x^2 by xx_1, we get

$(y + y_1) = xx_1 + 3$, which is the required equation of tangent.

(ii) Find equation of tangent to the curve $y^2 = 4ax$ at $(at^2, 2at)$.

Sol. Clearly, the equation of curve $y^2 = 4ax$ is a standard equation of curve, therefore on replacing y^2 by yy_1 and $2x$ by $x + x_1$, we can get the equation of tangent. Thus, the required equation is given by

$$yy_1 = 2a(x + x_1)$$

where $(x_1, y_1) = (at^2, 2at)$

Hence, the required equation of tangent is

$$y(2at) = 2a(x + at^2) \Rightarrow yt = x + at^2$$

(iii) Find the equation of tangent, at point $P(x_1, y_1)$, to the curve

$$\frac{x^2}{a^2} + \frac{y^2}{b^2} = 1$$

Sol. Clearly, the equation of curve $\frac{x^2}{a^2} + \frac{y^2}{b^2} = 1$ is a standard equation of curve, therefore equation of the tangent can be obtained by replacing x^2 *by* xx_1 and y^2 *by* yy_1.

i.e. $$\frac{xx_1}{a^2} + \frac{yy_1}{b^2} = 1$$

Some Important Points Regarding Tangent and Normal

1. If a curve passes through the origin, then the equation of the tangent at the origin can be directly written by equating the lowest degree terms appearing in the equation of the curve to zero.

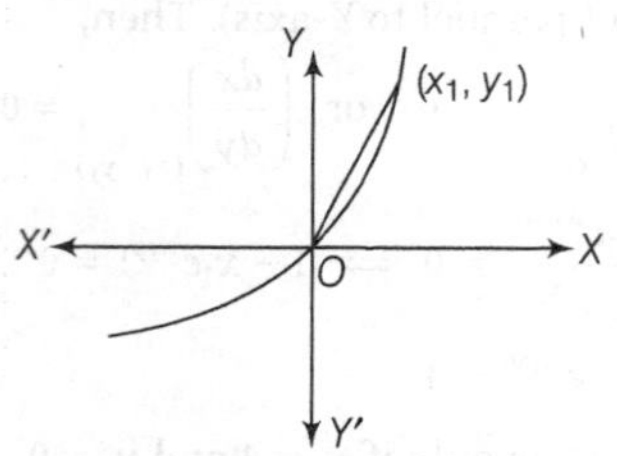

Figure 7.7

Proof Let the equation of the curve be

$$a_1x + b_1y + a_2x^2 + b_2xy + c_2y^2 = 0 \quad \text{...(i)}$$

Tangent $y - 0 = \lim\limits_{\substack{x_1 \to 0 \\ y_1 \to 0}} \frac{y_1}{x_1}(x - 0)$

Now, Eq. (i) becomes

$$a_1 + b_1\frac{y_1}{x_1} + a_2x_1 + b_2\frac{y_1}{x_1} \cdot x_1 + c_2\frac{y_1}{x_1} \cdot y_1 = 0 \quad \text{...(ii)}$$

at $x_1 \to 0$ and $y_1 \to 0, \frac{y_1}{x_1} \to m$

From Eq. (ii), $a_1 + b_1 m = 0$ $\qquad \left[\because m = -\frac{a_1}{b_1}\right]$

Hence, tangent is $y = -\frac{a_1}{b_1}x$

$$\Rightarrow \quad a_1x + b_1y = 0$$

e.g.

(i) Equation of tangent at origin, to the curve $x^2 + y^2 + 2gx + 2fy = 0$ is $gx + fy = 0$.

(ii) Equation of tangent at origin to the curve $x^3 + y^3 - 3x^2y + 3xy^2 + x^2 - y^2 = 0$ is $x^2 - y^2 = 0$.

(iii) Equation of tangent at origin, to the curve $x^3 + y^3 - 3xy = 0$ is $xy = 0$.

2. If the curve is $x^4 + y^4 = x^2 + y^2$, then the equation of the tangent would be $x^2 + y^2 = 0$ which would indicate that the origin is an isolated point on the graph.

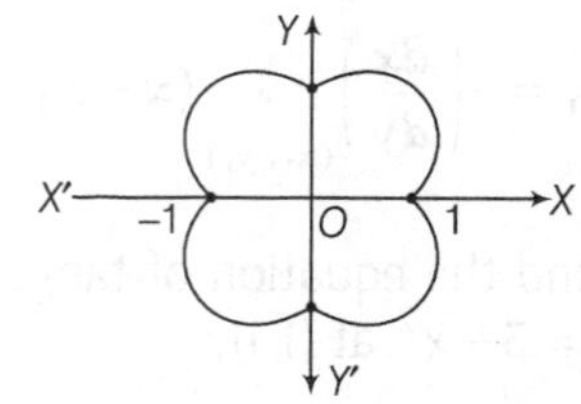

Figure 7.8

3. Same line could be the tangent as well as normal to a given curve at a given point.

e.g. In $x^3 + y^3 - 3xy = 0$ [folium of descartes]

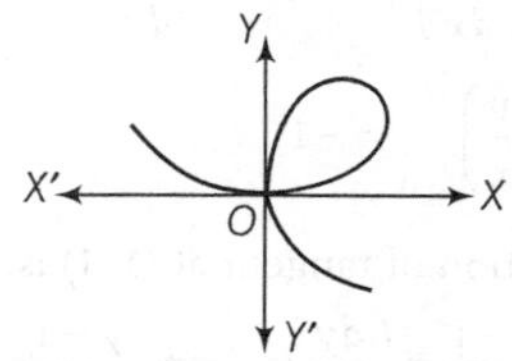

Figure 7.9

The line pair $xy = 0$ is both the tangent as well as normal at $x = 0$.

4. Some common parametric coordinates on a curve

(i) For $x^{2/3} + y^{2/3} = a^{2/3}$, take parametric coordinate $x = a\cos^3\theta$ and $y = a\sin^3\theta$.

(ii) For $\sqrt{x} + \sqrt{y} = \sqrt{a}$, take $x = a\cos^4\theta$ and $y = a\sin^4\theta$.

(iii) For $\frac{x^n}{a^n} + \frac{y^n}{b^n} = 1$, take $x = a(\cos\theta)^{2/n}$ and $y = b(\sin\theta)^{2/n}$.

(iv) For $c^2(x^2 + y^2) = x^2y^2$, take $x = c\sec\theta$ and $y = c\,\text{cosec}\,\theta$.

(v) For $y^2 = x^3$, take $x = t^2$ and $y = t^3$.

Example 22 Find the sum of the intercepts on the axes of coordinates by any tangent to the curve $\sqrt{x} + \sqrt{y} = 2$.

Sol. Here, equation of curve is $\sqrt{x} + \sqrt{y} = 2$.

Whose parametric coordinates are given by,

$$\sqrt{x} = 2\cos^2\theta$$

and $\quad \sqrt{y} = 2\sin^2\theta$

i.e. $\quad x = 4\cos^4\theta$

and $\quad y = 4\sin^4\theta$

$$\therefore \quad \frac{dy}{dx} = \frac{4\times 4\sin^3\theta\cdot\cos\theta}{4\times 4\cos^3\theta(-\sin\theta)} = -\tan^2\theta$$

Now, equation of tangent is $\dfrac{y-4\sin^4\theta}{x-4\cos^4\theta} = -\tan^2\theta$

$$\therefore \quad x\text{-intercept}, \left|4\cos^4\theta + \frac{4\sin^4\theta}{\tan^2\theta}\right| = 4\cos^2\theta$$

$$\text{and } y\text{-intercept}, \left|4\sin^4\theta + \frac{4\sin^4\theta}{\tan^2\theta}\right| = 4\sin^2\theta$$

Hence, the sum of intercepts made on the axes of coordinates is,

$$4\cos^2\theta + 4\sin^2\theta = 4$$

Example 23 The tangent, represented by the graph of the function $y = f(x)$, at the point with abscissa $x = 1$ form an angle of $\pi/6$, at the point $x = 2$ form an angle of $\pi/3$ and at the point $x = 3$ form an angle of $\pi/4$. Then, find the value of,

$$\int_1^3 f'(x)f''(x)\,dx + \int_2^3 f''(x)\,dx.$$

Sol. Given, at $x = 1$, $\dfrac{dy}{dx} = \tan\pi/6 = 1/\sqrt{3}$

or at $\quad x = 1,\ f'(1) = \tan\pi/6 = 1/\sqrt{3}$

Also, at $\quad x = 2,\ f'(2) = \tan\pi/3 = \sqrt{3}$

and at $\quad x = 3,\ f'(3) = \tan\pi/4 = 1$

Then, $\quad \displaystyle\int_1^3 f'(x)f''(x)\,dx + \int_2^3 f''(x)\,dx$

$$= \int_{f'(1)}^{f'(3)} t\,dt + (f'(x))_2^3$$

[putting $f'(x) = t \Rightarrow f''(x)\,dx = dt$]

$$= \frac{1}{2}(t^2)_{f'(1)}^{f'(3)} + \{f'(3) - f'(2)\}$$

$$= \frac{1}{2}\{(f'(3))^2 - (f'(1))^2\} + \{f'(3) - f'(2)\}$$

$$= \frac{1}{2}\left\{(1)^2 - \left(\frac{1}{\sqrt{3}}\right)^2\right\} + \{1-\sqrt{3}\} = \frac{1}{2}\left(1-\frac{1}{3}\right) + (1-\sqrt{3})$$

$$= \frac{4}{3} - \sqrt{3} = \frac{4-3\sqrt{3}}{3}$$

Example 24 Find the equation for family of curves for which the length of normal is equal to the radius vector.

Sol. Let $P(x, y)$ be the point on the curve.

$$OP = \text{radius vector} = \sqrt{x^2+y^2}$$

$$PN = \text{length of normal}$$

Now, $\quad \tan\phi = -\dfrac{1}{\left(\dfrac{dy}{dx}\right)}$ and $PN = \dfrac{y}{\sin\phi}$

It is given $\quad OP = PN$

$$\Rightarrow \quad \sqrt{x^2+y^2} = y\sqrt{1+\left(\frac{dy}{dx}\right)^2}$$

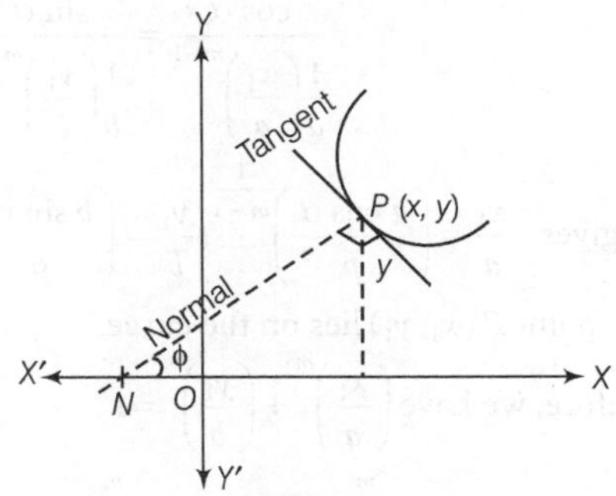

Figure 7.10

$$\Rightarrow \quad x^2 + y^2 = y^2\left[1+\left(\frac{dy}{dx}\right)^2\right]$$

$$\Rightarrow \quad x^2 = y^2\left(\frac{dy}{dx}\right)^2 \Rightarrow \frac{dy}{dx} = \pm\frac{x}{y}$$

or $\quad y\,dy = \pm x\,dx$

Integrating both the sides, we get

$y^2 = \pm x^2 + C$, which is the required family of curves.

Example 25 Find the condition that the line $x\cos\alpha + y\sin\alpha = p$ may touch the curve $\left(\dfrac{x}{a}\right)^m + \left(\dfrac{y}{b}\right)^m = 1$.

Sol. Given equation of curve is $\left(\dfrac{x}{a}\right)^m + \left(\dfrac{y}{b}\right)^m = 1$

Differentiating the equation of curve w.r.t. x, we get

$$m\left(\frac{x}{a}\right)^{m-1}\cdot\frac{1}{a} + m\left(\frac{y}{b}\right)^{m-1}\cdot\frac{1}{b}\cdot\frac{dy}{dx} = 0$$

On simplifying, we get $\dfrac{dy}{dx} = \dfrac{-b^m x^{m-1}}{a^m y^{m-1}}$

Now, at any point $P(x_1, y_1)$ on the curve,

$$\text{slope of tangent} = \left(\frac{dy}{dx}\right)_{(x_1, y_1)} = \frac{-b^m x_1^{m-1}}{a^m y_1^{m-1}}$$

$\therefore$ Equation of tangent at P is, $y - y_1 = \dfrac{-b^m x_1^{m-1}}{a^m y_1^{m-1}}(x - x_1)$

$$\Rightarrow \quad \frac{yy_1^{m-1}}{b^m} - \frac{y_1^m}{b^m} = -\frac{xx_1^{m-1}}{a^m} + \frac{x_1^m}{a^m}$$

$$\text{i.e. } \frac{x}{a}\left(\frac{x_1}{a}\right)^{m-1} + \frac{y}{b}\left(\frac{y_1}{b}\right)^{m-1} = \left(\frac{x_1}{a}\right)^m + \left(\frac{y_1}{b}\right)^m = 1$$

Since, P lies on the curve, therefore the equation of tangent at $P(x_1, y_1)$ on the curve is,

$$\frac{x}{a}\left(\frac{x_1}{a}\right)^{m-1}+\frac{y}{b}\left(\frac{y_1}{b}\right)^{m-1}=1 \quad \text{...(i)}$$

Also, we have $x\cos\alpha + y\sin\alpha = p$ ...(ii)

If Eq. (ii) is the equation of tangent, then coefficients of Eqs. (i) and (ii) must be proportional for point (x_1, y_1).

i.e. $$\frac{\cos\alpha}{\frac{1}{a}\left(\frac{x_1}{a}\right)^{m-1}}=\frac{\sin\alpha}{\frac{1}{b}\left(\frac{y_1}{b}\right)^{m-1}}=\frac{p}{1}$$

This gives $\frac{x_1}{a}=\left(\frac{a\cos\alpha}{p}\right)^{\frac{1}{m-1}}, \frac{y_1}{b}=\left(\frac{b\sin\alpha}{p}\right)^{\frac{1}{m-1}}$

Since, point $P(x_1, y_1)$ lies on the curve.

Therefore, we have $\left(\frac{x_1}{a}\right)^m+\left(\frac{y_1}{b}\right)^m=1$

i.e. $$\left(\frac{a\cos\alpha}{p}\right)^{\frac{m}{m-1}}+\left(\frac{b\sin\alpha}{p}\right)^{\frac{m}{m-1}}=1$$

i.e. $$(a\cos\alpha)^{\frac{m}{m-1}}+(b\sin\alpha)^{\frac{m}{m-1}}=p^{\frac{m}{m-1}},$$

which is the required condition.

Example 26 If tangent and normal to the curve $y=2\sin x+\sin 2x$ are drawn at $P\left(x=\frac{\pi}{3}\right)$, then area of the quadrilateral formed by the tangent, the normal at P and the coordinate axes is

(a) $\frac{\pi}{3}$ (b) 3π

(c) $\frac{\pi\sqrt{3}}{2}$ (d) None of these

Sol. Here, $\frac{dy}{dx}=0$ at $\left(x=\frac{\pi}{3}, y=\frac{3\sqrt{3}}{2}\right)$

$\Rightarrow$ Tangent at $x=\frac{\pi}{3}$ is parallel to X-axis.

$\Rightarrow$ Equation of tangent is, $y=\frac{3\sqrt{3}}{2}$

Also, equation of normal is, $x=\frac{\pi}{3}$

Now, area of quadrilateral $=\frac{\pi}{3}\cdot\frac{3\sqrt{3}}{2}=\frac{\pi\sqrt{3}}{2}$ sq unit

Hence, (c) is the correct answer.

Example 27 The maximum value of the sum of the intercepts made by any tangent to the curve $(a\sin^2\theta, 2a\sin\theta)$ with the axes is

(a) $2a$ (b) $a/4$ (c) $a/2$ (d) a

Sol. The curve is $(a\sin^2\theta, 2a\sin\theta) \Rightarrow x=a\sin^2\theta, y=2a\sin\theta$

Now, the equation of any tangent to the curve is,

$$\frac{y-2a\sin\theta}{x-a\sin^2\theta}=\frac{1}{\sin\theta} \Rightarrow y\sin\theta = x + a\sin^2\theta$$

$$\Rightarrow \frac{x}{-a\sin^2\theta}+\frac{y}{a\sin\theta}=1$$

$\Rightarrow$ Sum of intercepts $= a(\sin^2\theta+\sin\theta)$

$$=a\left\{\left(\sin\theta+\frac{1}{2}\right)^2-\frac{1}{4}\right\}$$

Which is maximum, when $\sin\theta = 1$

i.e. (Sum of intercept)$_{max} = 2a$

Hence, (a) is the correct answer.

Example 28 If $g(x)$ is a curve which is obtained by the reflection of $f(x)=\frac{e^x-e^{-x}}{2}$ by the line $y=x$, then

(a) $g(x)$ has more than one tangent parallel to X-axis

(b) $g(x)$ has more than one tangent parallel to Y-axis

(c) $y=-x$ is a tangent to $g(x)$ at $(0, 0)$

(d) $g(x)$ has no extremum

Sol. As $g(x)$ is a curve, obtained by the reflection of

$$f(x)=\frac{e^x-e^{-x}}{2} \text{ on } y=x.$$

so $g(x)$ is inverse of $f(x)$

$$\therefore \quad g(x)=\log(x+\sqrt{1+x^2})=f^{-1}(x)$$

$$\Rightarrow \quad g'(x)=\frac{1}{x+\sqrt{1+x^2}}\cdot\left(1+\frac{2x}{2\sqrt{1+x^2}}\right)$$

$$=\frac{1}{\sqrt{1+x^2}}\neq 0, \forall x\in R$$

$\Rightarrow g(x)$ has no tangent parallel to X-axis. Also, $g'(x)$ is always defined, $\forall x \in R$.

$\Rightarrow g(x)$ has no tangent parallel to Y-axis.

Since $g'(x)>0 \ \forall x\in R$,

therefore $g(x)$ doesn't have any extremum.

Hence, (d) is the correct answer.

Exercise for Session 3

1. If the line $ax + by + c = 0$ is normal to the $xy + 5 = 0$, then a and b have
   (a) same sign
   (b) opposite sign
   (c) cannot be discussed
   (d) None of these

2. The equation of tangent drawn to the curve $y^2 - 2x^3 - 4y + 8 = 0$ from the point (1, 2) is given by
   (a) $y - 2(1 \pm \sqrt{2}) = \pm 2\sqrt{3}\,(x - 2)$
   (b) $y - 2(1 \pm \sqrt{3}) = \pm 2\sqrt{2}\,(x - 2)$
   (c) $y - 2\,(1 \pm \sqrt{3}) = \pm 2\sqrt{3}\,(x - 2)$
   (d) None of these

3. The equation of the tangents to the curve $(1 + x^2)y = 1$ at the points of its intersection with the curve $(x + 1)y = 1$, is given by
   (a) $x + y = 1,\ y = 1$
   (b) $x + 2y = 2,\ y = 1$
   (c) $x - y = 1,\ y = 1$
   (d) None of these

4. The tangent lines for the curve $y = \int_0^x 2\,|\,t\,|\,dt$ which are parallel to the bisector of the first coordinate angle, is given by
   (a) $y = x + \frac{3}{4},\ y = x - \frac{1}{4}$
   (b) $y = -x + \frac{1}{4},\ y = -x + \frac{3}{4}$
   (c) $x + y = 2,\ x - y = 1$
   (d) None of these

5. The equation of normal to $x + y = x^y$, where it intersects X-axis, is given by
   (a) $x + y = 1$
   (b) $x - y - 1 = 0$
   (c) $x - y + 1 = 0$
   (d) None of these

6. The equation of normal at any point θ to the curve $x = a\cos\theta + a\theta\sin\theta,\ y = a\sin\theta - a\theta\cos\theta$ is always at a distance of
   (a) $2a$ unit from origin
   (b) a unit from origin
   (c) $\frac{1}{2}a$ unit from origin
   (d) None of these

7. If the tangent at (x_0, y_0) to the curve $x^3 + y^3 = a^3$ meets the curve again at (x_1, y_1), then $\frac{x_1}{x_0} + \frac{y_1}{y_0}$ is equal to
   (a) a
   (b) $2a$
   (c) 1
   (d) None of these

8. The area bounded by the axes of reference and the normal to $y = \log_e x$ at (1, 0), is
   (a) 1 sq unit
   (b) 2 sq units
   (c) $\frac{1}{2}$ sq unit
   (d) None of these

9. If $\frac{x}{a} + \frac{y}{b} = 2$ touches the curve $\frac{x^n}{a^n} + \frac{y^n}{b^n} = 2$ at the point (α, β), then
   (a) $\alpha = a^2, \beta = b^2$
   (b) $\alpha = a, \beta = b$
   (c) $\alpha = -2a, \beta = 4b$
   (d) $\alpha = 3a, \beta = -2b$

10. The equation of tangents to the curve $y = \cos(x + y),\ -2\pi \le x \le 2\pi$ that are parallel to the line $x + 2y = 0$, is
    (a) $x + 2y = \frac{\pi}{2}$ and $x + 2y = -\frac{3\pi}{2}$
    (b) $x + 2y = \frac{\pi}{2}$ and $x + 2y = \frac{3\pi}{2}$
    (c) $x + 2y = 0$ and $x + 2y = \pi$
    (d) None of these

Session 4

Angle of Intersection of Two Curves, Length of Tangent, Subtangent, Normal and Subnormal

Angle of Intersection of Two Curves

The angle of intersection of two curves is defined as the angle between the tangents to the two curves at their point of intersection.

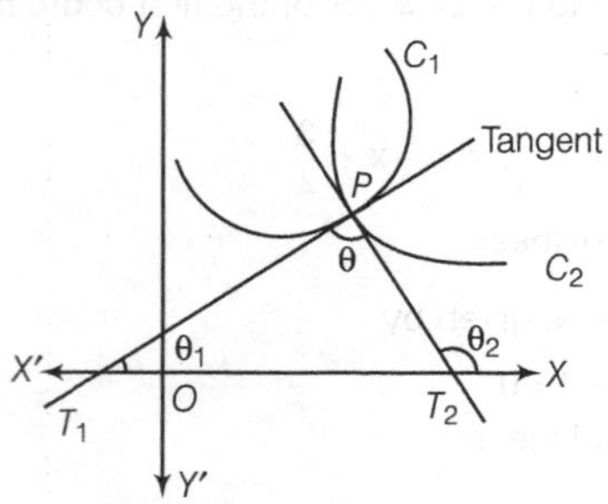

Figure 7.11

Let C_1 and C_2 be two curves having equations $y = f(x)$ and $y = g(x)$, respectively.

Let PT_1 and PT_2 be tangents to the curves C_1 and C_2 at their point of intersection.

Let θ be the angle between the two tangents PT_1 and PT_2 and θ_1 and θ_2 are the angles made by tangents with the positive direction of X-axis in anti-clockwise sense.

Then, $$m_1 = \tan\theta_1 = \left(\frac{dy}{dx}\right)_{C_1}$$

$$m_2 = \tan\theta_2 = \left(\frac{dy}{dx}\right)_{C_2}$$

From the figure it follows, $\theta = \theta_2 - \theta_1$

$$\Rightarrow \quad \tan\theta = \tan(\theta_2 - \theta_1) = \frac{\tan\theta_2 - \tan\theta_1}{1 + \tan\theta_2 \tan\theta_1}$$

$$\Rightarrow \quad \tan\theta = \left|\frac{\left(\frac{dy}{dx}\right)_{C_1} - \left(\frac{dy}{dx}\right)_{C_2}}{1 + \left(\frac{dy}{dx}\right)_{C_1}\left(\frac{dy}{dx}\right)_{C_2}}\right|$$

Angle of intersection of these curves is defined as acute angle between the tangents.

Orthogonal Curves If the angle of intersection of two curves is a right angle, then the two curves are said to be orthogonal and the curves are called orthogonal curves.

∴ If the curves are orthogonal, then $\theta = \pi/2$

$$\Rightarrow \quad 1 + \left(\frac{dy}{dx}\right)_{C_1}\left(\frac{dy}{dx}\right)_{C_2} = 0$$

$$\Rightarrow \quad \left(\frac{dy}{dx}\right)_{C_1}\left(\frac{dy}{dx}\right)_{C_2} = -1$$

Condition for Two Curves to Touch

If two curves touch each other, then

$$\theta = 0$$

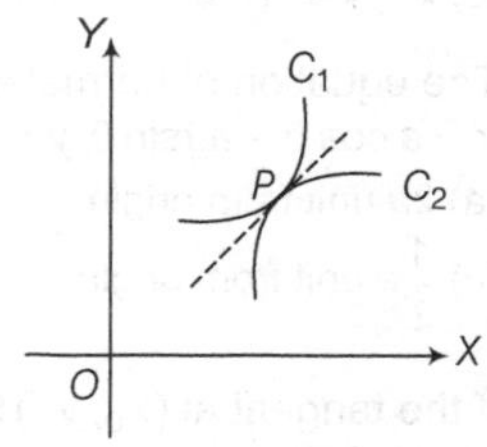

Figure 7.12

$$\Rightarrow \quad \tan\theta = 0$$

$$\Rightarrow \quad \left(\frac{dy}{dx}\right)_{C_1} - \left(\frac{dy}{dx}\right)_{C_2} = 0$$

$$\Rightarrow \quad \left(\frac{dy}{dx}\right)_{C_1} = \left(\frac{dy}{dx}\right)_{C_2}$$

| Example 29 Find the angle of intersection of the curves $y = x^2$ and $y = 4 - x^2$.

Sol. For the intersection points of the given curves, consider

$$x^2 = 4 - x^2$$

$$\Rightarrow \quad x = \pm\sqrt{2}$$

Now, at $x = \sqrt{2}$, $\frac{dy}{dx}$ for first curve $= 2x|_{\text{at } x=\sqrt{2}} = 2\sqrt{2}$

While at $x = \sqrt{2}$,

$\frac{dy}{dx}$ for second curve $= -2x|_{\text{at } x=\sqrt{2}} = -2\sqrt{2}$

Hence, if θ_1 is the acute angle of intersection of the curves, then

$$\tan\theta_1 = \left|\frac{2\sqrt{2}-(-2\sqrt{2})}{1+2\sqrt{2}(-2\sqrt{2})}\right| = \left|\frac{4\sqrt{2}}{-7}\right|$$

$$\therefore \quad \theta_1 = \tan^{-1}\left(\frac{4\sqrt{2}}{7}\right) \quad \text{...(i)}$$

Again, at $x = -\sqrt{2}$,

$$\left(\frac{dy}{dx}\right) \text{ for first curve} = -2\sqrt{2}$$

and $$\left(\frac{dy}{dx}\right) \text{ for second curve} = 2\sqrt{2}$$

Hence, if θ_2 is the acute angle of intersection of the curves, then

$$\tan\theta_2 = \left|\frac{-2\sqrt{2}-2\sqrt{2}}{1+(-2\sqrt{2})(2\sqrt{2})}\right|$$

$$\therefore \quad \theta_2 = \tan^{-1}\left(\frac{4\sqrt{2}}{7}\right) \quad \text{...(ii)}$$

From Eqs. (i) and (ii) the two acute angles are equal.

Example 30 Find the acute angle between the curves $y = |x^2 - 1|$ and $y = |x^2 - 3|$ at their points of intersection when $x > 0$.

Sol. For the intersection of the given curves,

$$|x^2 - 1| = |x^2 - 3| \Rightarrow (x^2 - 1)^2 = (x^2 - 3)^2$$

$$\Rightarrow \quad (x^2 - 1)^2 - (x^2 - 3)^2 = 0$$

$$\Rightarrow \quad [(x^2 - 1) - (x^2 - 3)][(x^2 - 1) + (x^2 - 3)] = 0$$

$$\Rightarrow \quad 2[2x^2 - 4] = 0 \Rightarrow 2x^2 = 4 \Rightarrow x = \pm\sqrt{2}$$

Neglecting $x = -\sqrt{2}$, as $x > 0$

Now, we have, point of intersection as $x = \sqrt{2}$.

Here, $y = |x^2 - 1| = (x^2 - 1)$ in the neighbourhood of $x = \sqrt{2}$ and $y = -(x^2 - 3)$ in the neighbourhood of $x = \sqrt{2}$.

Now, at $x = \sqrt{2}$, $\frac{dy}{dx}$ for first curve

$$\Rightarrow \quad \left(\frac{dy}{dx}\right)_{C_1} = 2x = 2\sqrt{2}$$

and $\frac{dy}{dx}$ for second curve $\left(\frac{dy}{dx}\right)_{C_2} = -2x = -2\sqrt{2}$

Hence, if θ is angle between them, then

$$\tan\theta = \left|\frac{2\sqrt{2}-(-2\sqrt{2})}{1+2\sqrt{2}(-2\sqrt{2})}\right| = \left|\frac{4\sqrt{2}}{-7}\right| = \left(\frac{4\sqrt{2}}{7}\right)$$

$$\therefore \quad \theta = \tan^{-1}\left(\frac{4\sqrt{2}}{7}\right)$$

Example 31 Find the angle of intersection of curves, $y = [|\sin x| + |\cos x|]$ and $x^2 + y^2 = 5$, where $[\cdot]$ denotes the greatest integral function.

Sol. We know that, $\quad 1 \le |\sin x| + |\cos x| \le \sqrt{2}$

$$\therefore \quad y = [|\sin x| + |\cos x|] = 1$$

Let P and Q be the points of intersection of given curves.

Clearly, the given curves meet at points where $y = 1$, so we get $x^2 + 1 = 5$

$$\Rightarrow \quad x = \pm 2$$

Now, we have $P(2, 1)$ and $Q(-2, 1)$

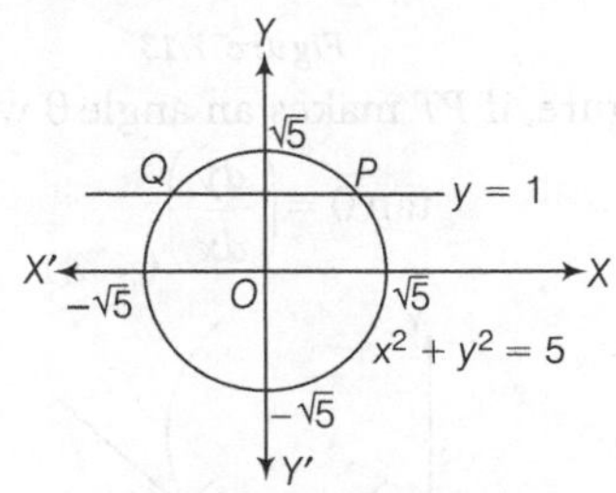

Differentiating $x^2 + y^2 = 5$ w.r.t. x, we get

$$2x + 2y\frac{dy}{dx} = 0$$

$$\Rightarrow \quad \frac{dy}{dx} = -\frac{x}{y}$$

$$\Rightarrow \quad \left(\frac{dy}{dx}\right)_{(2,\,1)} = -2$$

and $$\left(\frac{dy}{dx}\right)_{(-2,\,1)} = 2$$

Clearly, the slope of line $y = 1$ is zero and the slope of the tangents at P and Q are (-2) and (2), respectively.

Thus, the angle of intersection is $\tan^{-1}(2)$.

Length of Tangent, Subtangent, Normal and Subnormal

Length of Tangent The length of the segment PT, i.e. the portion of the tangent intercepted between the point of contact and X-axis is called the length of tangent.

Subtangent and its Length The projection of the segment PT along X-axis is called the subtangent. Here, ST is the length of subtangent.

Length of Normal The length of segment PN, i.e. the portion of the normal intercepted between the point on the curve and X-axis is called the length of normal.

Subnormal and its Length

The projection of the segment PN along X-axis is called the subnormal. Here, SN is the length of sub normal.

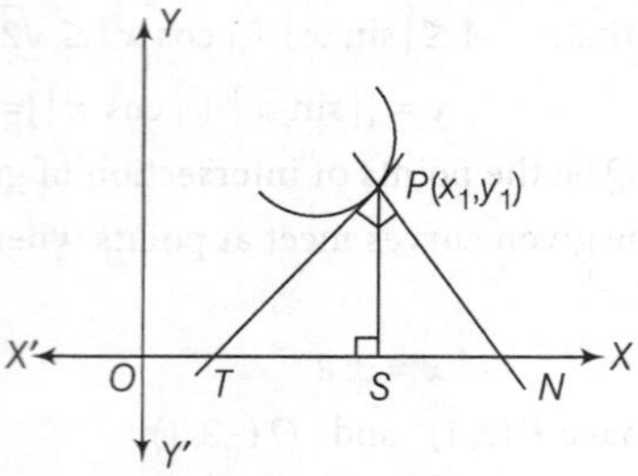

Figure 7.13

From the figure, if PT makes an angle θ with X-axis, then

$$\tan\theta = \left(\frac{dy}{dx}\right)_{(x_1, y_1)}$$

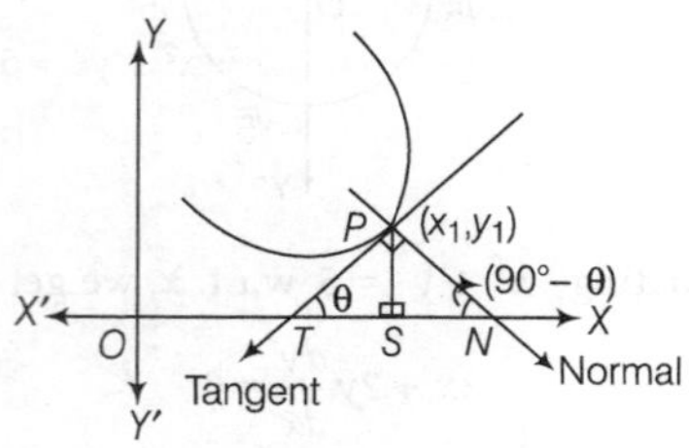

Figure 7.14

Also, we have $\dfrac{ST}{PS} = \cot\theta \Rightarrow ST = PS\cot\theta$

or **subtangent** $= ST = \left| y_1 \left(\dfrac{dx}{dy}\right)_{(x_1, y_1)} \right|$

Similarly, $\dfrac{SN}{PS} = \cot(90° - \theta)$

$\Rightarrow$ Subnormal $= SN = PS\tan\theta$

$\Rightarrow$ **Subnormal** $= SN = \left| y_1 \left(\dfrac{dy}{dx}\right)_{(x_1, y_1)} \right|$

Now, **length of tangent**

$$= PT = \left| \sqrt{y_1^2 + y_1^2 \left(\frac{dx}{dy}\right)^2_{(x_1, y_1)}} \right|$$

$$\Rightarrow \quad PT = \left| y_1 \sqrt{1 + \left(\frac{dx}{dy}\right)^2_{(x_1, y_1)}} \right|$$

and **length of normal** $= PN = \left| \sqrt{y_1^2 + y_1^2 \left(\dfrac{dy}{dx}\right)^2_{(x_1, y_1)}} \right|$

$$\Rightarrow \quad PN = \left| y_1 \sqrt{1 + \left(\frac{dy}{dx}\right)^2_{(x_1, y_1)}} \right|$$

Example 32 Show that for the curve $y = be^{x/a}$, the subtangent is of constant length and the subnormal varies as the square of ordinate.

Sol. The equation of given curve is $y = be^{x/a}$

Let us consider a point (x_1, y_1) on the curve.

Then, we have $\qquad y_1 = be^{x_1/a} \qquad$ ...(i)

Differentiating the curve $y = be^{x/a}$ w.r.t. x, we get

$$\frac{dy}{dx} = be^{x/a} \cdot \frac{1}{a}$$

$$\therefore \quad \left(\frac{dy}{dx}\right)_{(x_1, y_1)} = \frac{b}{a} e^{x_1/a} \qquad \text{...(ii)}$$

Thus, the length of subtangent $= \left| y_1 \left(\dfrac{dx}{dy}\right)_{(x_1, y_1)} \right|$

$$= \left| y_1 \cdot \frac{a}{be^{x_1/a}} \right| = \left| be^{x_1/a} \cdot \frac{a}{be^{x_1/a}} \right|$$

$= a$ (constant) $\qquad$ [using Eqs. (i) and (ii)]

$\Rightarrow$ Subtangent is of constant length a.

Again, length of subnormal $= \left| y_1 \left(\dfrac{dy}{dx}\right)_{(x_1, y_1)} \right|$

$$= \left| be^{x_1/a} \cdot \frac{be^{x_1/a}}{a} \right| = \frac{1}{a}(be^{x_1/a})^2 = \frac{1}{a}y^2 \qquad \text{[using Eq. (i)]}$$

Therefore, subnormal varies as the square of ordinate.

Example 33 Find the length of tangent, subtangent, normal and subnormal to $y^2 = 4ax$ at $(at^2, 2at)$.

Sol. The equation of given curve is

$$y^2 = 4ax \qquad \text{...(i)}$$

Differentiating Eq. (i) w.r.t. x, we get

$$2y\frac{dy}{dx} = 4a$$

$$\Rightarrow \quad \left[\frac{dy}{dx}\right]_{(at^2, 2at)} = \frac{4a}{4at} = \frac{1}{t} \qquad \text{...(ii)}$$

Now, the length of tangent at $(at^2, 2at)$ is

$$= y_1 \sqrt{1 + \left(\frac{dx}{dy}\right)^2_{(x_1, y_1)}} = 2at\sqrt{1 + t^2} \qquad \text{[using Eq. (ii)]}$$

length of normal at $(at^2, 2at)$ is

$$= y_1 \sqrt{1 + \left(\frac{dy}{dx}\right)^2_{(x_1, y_1)}} = 2at\sqrt{1 + 1/t^2} = 2a\sqrt{t^2 + 1}$$

length of subtangent is

$$\frac{y_1}{\left[\dfrac{dy}{dx}\right]_{(x_1, y_1)}} = \frac{2at}{1/t} = 2at^2$$

and length of subnormal is

$$y_1\left[\frac{dy}{dx}\right]_{(x_1,\,y_1)} = 2at\cdot\frac{1}{t} = 2a$$

Example 34 Find the equation of tangent and normal, the length of subtangent and subnormal of the circle $x^2+y^2=a^2$ at the point (x_1, y_1).

Sol. The equation of given curve is,

$$x^2+y^2=a^2 \qquad \text{...(i)}$$

Differentiating Eq. (i) w.r.t. x, we get

$$2x+2y\frac{dy}{dx}=0$$

$$\Rightarrow \left(\frac{dy}{dx}\right)_{(x_1,\,y_1)} = -\frac{x_1}{y_1} \qquad \text{...(ii)}$$

Thus, the equation of tangent is,

$$y-y_1=\left[\frac{dy}{dx}\right]_{(x_1,\,y_1)}(x-x_1)$$

$$\Rightarrow y-y_1=-\frac{x_1}{y_1}(x-x_1) \qquad \text{[from Eq. (ii)]}$$

$$\Rightarrow yy_1-y_1^2=-xx_1+x_1^2$$

$$\Rightarrow xx_1+yy_1=x_1^2+y_1^2$$

$$\Rightarrow xx_1+yy_1=a^2 \qquad [\text{using Eq. (i) as } (x_1,y_1) \text{ lies on } x^2+y^2=a^2 \Rightarrow x_1^2+y_1^2=a^2]$$

While the equation of normal is,

$$y-y_1=\frac{y_1}{x_1}(x-x_1)$$

$$\Rightarrow x_1y-x_1y_1=xy_1-x_1y_1 \Rightarrow xy_1-x_1y=0$$

$$\text{The length of subtangent} = \left|y_1\cdot\left(\frac{dx}{dy}\right)_{(x_1,\,y_1)}\right|$$

$$=\left|y_1\cdot\left(\frac{-y_1}{x_1}\right)\right| \qquad \text{[using Eq. (ii)]}$$

$$\Rightarrow \text{The length of subtangent} = \left|\frac{y_1^2}{x_1}\right|$$

$$\text{While the length of subnormal} = \left|y_1\cdot\left(\frac{dy}{dx}\right)_{(x_1,\,y_1)}\right|$$

$$=\left|y_1\left(-\frac{x_1}{y_1}\right)\right| = |x_1|$$

Example 35 If the relation between subnormal SN and subtangent ST on the curve, $by^2=(x+a)^3$ is $p(SN)=q(ST)^2$, then find the value of $\frac{p}{q}$.

Sol. The equation of given curve is $by^2=(x+a)^3$

Differentiating both the sides w.r.t. x, we get

$$2by\frac{dy}{dx}=3(x+a)^2\cdot 1$$

$$\Rightarrow \frac{dy}{dx}=\frac{3}{2}\frac{(x+a)^2}{by}$$

$\therefore$ Length of subnormal

$$\Rightarrow SN = y\frac{dy}{dx}=\frac{3}{2}\cdot\frac{(x+a)^2}{b} \qquad \text{...(i)}$$

and Length of subtangent

$$\Rightarrow ST = y\frac{dx}{dy}=\frac{2by^2}{3(x+a)^2} \qquad \text{...(ii)}$$

$$\because \frac{p}{q}=\frac{(ST)^2}{(SN)} \qquad \text{[given]}$$

$$\therefore \frac{p}{q}=\frac{(2by^2)^2\cdot 2b}{\{3(x+a)^2\}^2\cdot 3(x+a)^2} \qquad \text{[using Eqs. (i) and (ii)]}$$

$$=\frac{8b}{27}\cdot\frac{\{(x+a)^3\}^2}{(x+a)^6}=\frac{8b}{27} \qquad [\text{using, } by^2=(x+a)^3]$$

$$\therefore \frac{p}{q}=\frac{8b}{27}$$

Example 36 If the length of subnormal is equal to length of subtangent at any point (3,4) on the curve $y=f(x)$ and the tangent at (3, 4) to $y=f(x)$ meets the coordinate axes at A and B, then maximum area of the ΔOAB where O is origin, is

(a) $\frac{45}{2}$ (b) $\frac{49}{2}$ (c) $\frac{25}{2}$ (d) $\frac{81}{2}$

Sol. Length of subnormal = Length of subtangent

$$\Rightarrow \frac{dy}{dx}=\pm 1$$

If $\frac{dy}{dx}=1$, then equation of tangent is,

$$y-4=x-3 \Rightarrow y-x=1$$

$$\therefore \text{Area of } \Delta OAB=\frac{1}{2}\times 1\times 1=\frac{1}{2} \qquad \text{...(i)}$$

If $\frac{dy}{dx}=-1$, then equation of tangent is,

$$y-4=-x+3 \Rightarrow x+y=7$$

$$\therefore \text{Area of } \Delta OAB=\frac{1}{2}\times 7\times 7=\frac{49}{2} \qquad \text{...(ii)}$$

Clearly, maximum area $=\frac{49}{2}$

Hence, (b) is the correct answer.

Exercise for Session 4

1. The angle of intersection of $y = a^x$ and $y = b^x$, is given by

(a) $\tan\theta = \left|\dfrac{\log(ab)}{1-\log(ab)}\right|$ (b) $\tan\theta = \left|\dfrac{\log(a/b)}{1+\log a \log b}\right|$

(c) $\tan\theta = \left|\dfrac{\log(a/b)}{1-\log(a/b)}\right|$ (d) None of these

2. The angle between the curves $x^2 + 4y^2 = 32$ and $x^2 - y^2 = 12$, is

(a) $\dfrac{\pi}{3}$ (b) $\dfrac{\pi}{4}$ (c) $\dfrac{\pi}{6}$ (d) $\dfrac{\pi}{2}$

3. If $ax^2 + by^2 = 1$ cuts $a'x^2 + b'y^2 = 1$ orthogonally, then

(a) $\dfrac{1}{a} - \dfrac{1}{a'} = \dfrac{1}{b} - \dfrac{1}{b'}$ (b) $\dfrac{1}{a} + \dfrac{1}{a'} = \dfrac{1}{b} + \dfrac{1}{b'}$

(c) $\dfrac{1}{a} + \dfrac{1}{b} = \dfrac{1}{a'} + \dfrac{1}{b'}$ (d) None of these

4. The length of subtangent to the curve, $y = e^{x/a}$, is

(a) $2a$ (b) a (c) $a/2$ (d) $a/4$

5. The length of normal to the curve $x = a(\theta + \sin\theta)$, $y = a(1-\cos\theta)$, at $\theta = \dfrac{\pi}{2}$ is

(a) $2a$ (b) a (c) $\sqrt{2}a$ (d) $2\sqrt{2}a$

6. The length of tangent to the curve $y = x^3 + 3x^2 + 4x - 1$ at a point $x = 0$, is

(a) $\dfrac{1}{4}$ (b) $\dfrac{\sqrt{17}}{4}$ (c) $\dfrac{3}{4}$ (d) $\dfrac{\sqrt{15}}{4}$

7. The possible values of p such that the equation $px^2 = \log x$ has exactly one solution, are

(a) R (b) R^+ (c) $(-\infty, 0] \cup \left\{\dfrac{1}{2e}\right\}$ (d) $[0, \infty)$

8. If curve $y = 1 - ax^2$ and $y = x^2$ intersect orthogonally then a is

(a) $\dfrac{1}{2}$ (b) $\dfrac{1}{3}$ (c) 2 (d) 3

9. The length of tangent to the curve $x = a\left(\cos t + \log\tan\dfrac{t}{2}\right)$, $y = a(\sin t)$, is

(a) ax (b) ay (c) a (d) xy

10. The lines tangent to the curves $y^3 - x^2y + 5y - 2x = 0$ and $x^4 - x^3y^2 + 5x + 2y = 0$ at the origin intersect at an angle θ equal to

(a) $\dfrac{\pi}{6}$ (b) $\dfrac{\pi}{4}$ (c) $\dfrac{\pi}{3}$ (d) $\dfrac{\pi}{2}$

Session 5

Rolle's Theorem, Lagrange's Mean Value Theorem

Rolle's Theorem

The theorem was named after the French Mathematician Michel Rolle (1652–1719). Who first gave it in his book *Methode pour resoudre lesegalites* (1691).

Statement

Let f be a real-valued function defined on the closed interval $[a, b]$ such that

(i) $f(x)$ is continuous in the closed interval $[a, b]$

(ii) $f(x)$ is differentiable in the open interval $]a, b[$ and

(iii) $f(a) = f(b)$

Then, there is atleast one value c of x in open interval $]a, b[$ for which $f'(c) = 0$.

Analytical Proof

Now, Rolle's theorem is valid for a function such that

(i) $f(x)$ is continuous in the closed interval $[a, b]$

(ii) $f(x)$ is differentiable in open interval $]a, b[$ and

(iii) $f(a) = f(b)$

So, generally two cases arise in such circumstances.

***Case* I** $f(x)$ is constant in the interval $[a, b]$, then $f'(x) = 0$ for all $x \in [a, b]$.

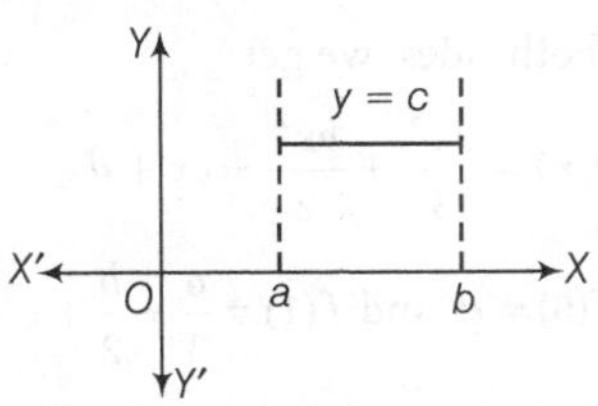

Figure 7.15

Hence, Rolle's theorem follows and we can say, $f'(c) = 0$, where $a < c < b$

***Case* II** $f(x)$ is not constant in the interval $[a, b]$ and since $f(a) = f(b)$.

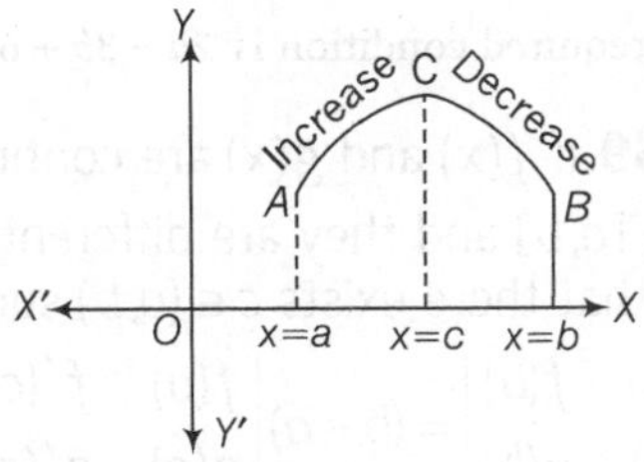

Figure 7.16

The function should either increase or decrease when x assumes values slightly greater than a.

Now, let $f(x)$ increases for $x > a$.

Since, $f(a) = f(b)$, hence the function increase at some value $x = c$ and decrease upto $x = b$.

Clearly, at $x = c$ function has maximum value.

Now, let h be a small positive quantity, then from definition of maximum value of the function,

$$f(c+h) - f(c) < 0$$

and $$f(c-h) - f(c) < 0$$

$\therefore$ $$\frac{f(c+h) - f(c)}{h} < 0$$

and $$\frac{f(c-h) - f(c)}{-h} > 0$$

So, $$\lim_{h \to 0} \frac{f(c+h) - f(c)}{h} \le 0$$

and $$\lim_{h \to 0} \frac{f(c-h) - f(c)}{-h} \ge 0 \quad \text{...(i)}$$

But, if $$\lim_{h \to 0} \frac{f(c+h) - f(c)}{h} \ne \lim_{h \to 0} \frac{f(c-h) - f(c)}{-h}$$

Then, Rolle's theorem cannot be applied because in such case,

$$\text{RHD at } x = c \ne \text{LHD at } x = c.$$

Hence, $f(x)$ is not differentiable at $x = c$, which contradicts the condition of Rolle's theorem.

$\therefore$ Only one possible solution arises, when

$$\lim_{h \to 0} \frac{f(c+h) - f(c)}{h} = \lim_{h \to 0} \frac{f(c-h) - f(c)}{-h} = 0$$

which implies that, $f'(c) = 0$ where $a < c < b$.

Hence, Rolle's theorem is proved.

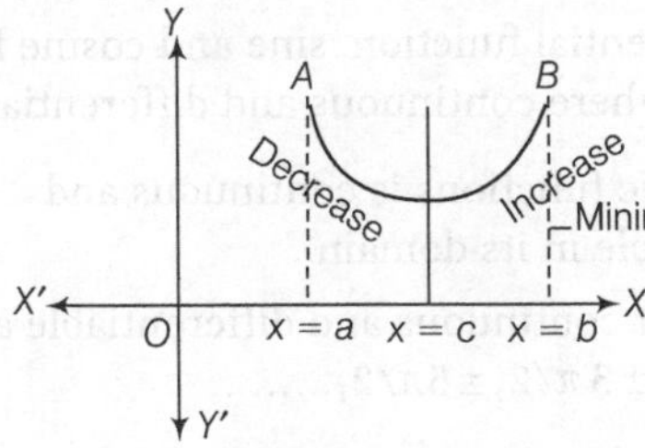

Figure 7.17

Similarly, the case where $f(x)$ decreases in the interval $a < x < c$ and then increases in the interval $c < x < b$, $f'(c) = 0$, but when $x = c$, the minimum value of $f(x)$ exists in the interval $[a, b]$.

Geometrical Proof

Consider the portion AB of the curve $y = f(x)$, lying between $x = a$ and $x = b$, such that

(i) it goes continuously from A to B

(ii) it has tangent at every point between A and B

(iii) ordinate of A = ordinate of B

From the figure, it is clear that $f(x)$ **increases in the interval** AC_1, which implies that $f'(x) > 0$ in this region and decreases in the interval C_1B which implies $f'(x) < 0$ in this region. Now, **since there is unique tangent to be drawn on the curve lying in between A and B and since each of them has a unique slope, i.e. unique value of** $f'(x)$.

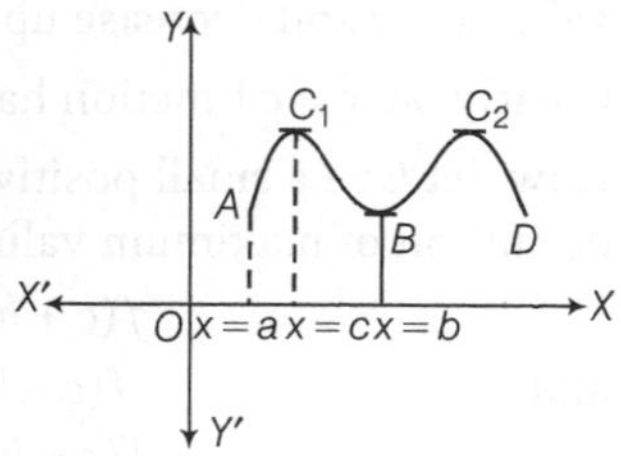

Figure 7.18

So, due to continuity and differentiability of the function $f(x)$ in the region A to B, there is a point $x = c$, where $f'(c)$ should be zero.

Hence, $f'(c) = 0$, where $a < c < b$

Thus, Rolle's theorem is proved.

Remarks

Generally, two types of problems are formulated on Rolle's theorem.

(i) To check the applicability of Rolle's theorem to a given function on a given interval

(ii) To verify Rolle's theorem for a given function in a given interval.

In both the types of problems, first we check whether $f(x)$ satisfies the conditions of Rolle's theorem or not.

Some Important Results Regarding Rolle's Theorem

(i) A polynomial function is everywhere continuous and differentiable.

(ii) The exponential function, sine and cosine functions are everywhere continuous and differentiable.

(iii) Logarithmic functions is continuous and differentiable in its domain.

(iv) $\tan x$ is not continuous and differentiable at $x = \pm\pi/2, \pm 3\pi/2, \pm 5\pi/2, \ldots\ldots\ldots$

(v) $|x|$ is not differentiable at $x = 0$.

(vi) If $f'(x)$ tends to $\pm\infty$ as $x \to k$, then $f(x)$ is not differentiable at $x = k$.

e.g. If $f(x) = (2x-1)^{1/2}$, then $f'(x) = \dfrac{1}{\sqrt{2x-1}}$ is such that as

$$x \to \left(\frac{1}{2}\right)^+ \Rightarrow f'(x) \to \infty$$

So, $f(x)$ is not differentiable at $x = 1/2$.

Example 37 Verify Rolle's theorem for the function $f(x) = x^3 - 3x^2 + 2x$ in the interval $[0, 2]$.

Sol. Here, we observe that

(a) $f(x)$ is polynomial and since polynomials are always continuous, so $f(x)$ is continuous in the interval $[0, 2]$.

(b) $f'(x) = 3x^2 - 6x + 2$, which exists for all $x \in (0, 2)$. So, $f(x)$ is differentiable for all $x \in (0, 2)$ and

(c) $f(0) = 0$, $f(2) = 2^3 - 3\cdot(2)^2 + 2(2) = 0$

$\therefore \quad f(0) = f(2)$

Thus, all the conditions of Rolle's theorem are satisfied.

So, there must exists some $c \in (0, 2)$ such that $f'(c) = 0$

$\Rightarrow \quad f'(c) = 3c^2 - 6c + 2 = 0 \Rightarrow c = 1 \pm \dfrac{1}{\sqrt{3}}$.

Where, $c = 1 \pm \dfrac{1}{\sqrt{3}} \in (0, 2)$, thus Rolle's theorem is verified.

Example 38 If $ax^2 + bx + c = 0$, $a, b, c \in R$, then find the condition that this equation would have at least one root in $(0, 1)$.

Sol. Let $f'(x) = ax^2 + bx + c$

Integrating both sides, we get

$$f(x) = \frac{ax^3}{3} + \frac{bx^2}{2} + cx + d$$

$$\Rightarrow \quad f(0) = d \text{ and } f(1) = \frac{a}{3} + \frac{b}{2} + c + d$$

Now, for Rolle's theorem to be applicable, we should have

$$f(0) = f(1)$$

$$\Rightarrow \quad d = \frac{a}{3} + \frac{b}{2} + c + d$$

$$\Rightarrow \quad 2a + 3b + 6c = 0$$

Here, the required condition is $2a + 3b + 6c = 0$.

Example 39 If $f(x)$ and $g(x)$ are continuous functions in $[a, b]$ and they are differentiable in (a, b), then prove that there exists $c \in (a, b)$ such that

$$\begin{vmatrix} f(a) & f(b) \\ g(a) & g(b) \end{vmatrix} = (b-a)\begin{vmatrix} f(a) & f'(c) \\ g(a) & g'(c) \end{vmatrix}$$

Sol. Consider a function, $\phi(x) = f(a)\,g(x) - f(x)\,g(a)$ for all $x \in [a, b]$.

As, $\phi(x)$ is continuous and differentiable on (a, b).

$$\therefore \quad \phi'(c) = \frac{\phi(b) - \phi(a)}{b - a} \text{ for some } c \in (a, b). \quad \text{...(i)}$$

Now, as $\phi(x) = f(a)\,g(x) - f(x)\,g(a)$

$$\therefore \quad \phi'(x) = f(a)\,g'(x) - f'(x)\,g(a)$$

$$\Rightarrow \quad \phi'(c) = f(a)\,g'(c) - f'(c)\,g(a)$$

$$\Rightarrow \quad \phi'(c) = \begin{vmatrix} f(a) & f'(c) \\ g(a) & g'(c) \end{vmatrix}$$

Also, $\phi(b) = f(a)\,g(b) - f(b)\,g(a) = \begin{vmatrix} f(a) & f(b) \\ g(a) & g(b) \end{vmatrix}$

and $\phi(a) = f(a)\,g(a) - f(a)\,g(a) = 0$

$\therefore$ Eq. (i) reduces to;

$$\begin{vmatrix} f(a) & f'(c) \\ g(a) & g'(c) \end{vmatrix} = \frac{1}{(b-a)} \begin{vmatrix} f(a) & f(b) \\ g(a) & g(b) \end{vmatrix}$$

or

$$\begin{vmatrix} f(a) & f(b) \\ g(a) & g(b) \end{vmatrix} = (b-a) \begin{vmatrix} f(a) & f'(c) \\ g(a) & g'(c) \end{vmatrix}$$

| Example 40 Use Rolle's theorem to find the condition for the polynomial equation $f(x) = 0$ to have a repeated real roots. Hence, or otherwise prove that the equation;

$$1 + \frac{x}{1!} + \frac{x^2}{2!} + \ldots + \frac{x^n}{n!} = 0, \text{ cannot have repeated roots.}$$

Sol. By Rolle's theorem, we can say that between any two roots of a polynomial there is always a root of its derivative. Thus, if α is a repeated root of a polynomial $f(x)$, then there must be a root of $f'(x)$ in the interval.

$\Rightarrow \quad f'(\alpha) = 0$

i.e. $f(\alpha) = f'(\alpha) = 0$, for α to be a repeated root.

Let $\phi(x) = 1 + \frac{x}{1!} + \frac{x^2}{2!} + \ldots + \frac{x^n}{n!}$ has a repeated root α.

$$\Rightarrow \quad \phi(\alpha) = 0$$

and

$$\phi'(\alpha) = 0$$

$$\Rightarrow \quad 1 + \frac{\alpha}{1!} + \frac{\alpha^2}{2!} + \ldots + \frac{\alpha^n}{n!} = 0$$

and

$$1 + \alpha + \frac{\alpha^2}{2!} + \ldots + \frac{\alpha^{n-1}}{(n-1)!} = 0$$

Solving above equations, we get

$$\frac{\alpha^n}{n!} = 0$$

or $\alpha = 0$, thus 0 is the repeated root of $\phi(x) = 0$.

But, 0 doesn't satisfy $\phi(x)$.

$\therefore$ There is no repeated root of $\phi(x) = 0$.

| Example 41 Let $f(x)$ and $g(x)$ be differentiable functions such that $f'(x)g(x) \neq f(x)g'(x)$ for any real x. Show that between any two real solutions of $f(x) = 0$, there is atleast one real solution of $g(x) = 0$.

Sol. Let a, b be the solutions of $f(x) = 0$.

Suppose $g(x)$ is not equal to zero for any x belonging to $[a, b]$.

Now, consider $\quad h(x) = f(x)/g(x)$

Since, $g(x)$ is not equal to zero, therefore $h(x)$ is differentiable and continuous in $[a, b]$.

Also, $\quad h(a) = h(b) = 0$

[as $f(a) = 0$ and $f(b) = 0$ but $g(a)$ and $g(b) \neq 0$]

Applying Rolle's theorem for $h(x)$ in $[a, b]$, we get

$h'(c) = 0$ for some c belonging to (a, b)

$\Rightarrow \quad f(c)g'(c) = f'(c)g(c)$, for some $c \in (a, b)$.

This gives the contradiction. **Hence proved.**

| Example 42 Consider the function

$$f(x) = \begin{cases} x \sin \dfrac{\pi}{x}, & \text{for } x > 0 \\ 0, & \text{for } x = 0 \end{cases}, \text{ then the number of points}$$

in (0, 1), where the derivative $f'(x)$ tends to zero is

(a) 0 (b) 1 (c) 2 (d) infinite *[IIT 2010]*

Sol. $f(x)$ tends to zero at points, where $\sin \frac{\pi}{x} = 0$

i.e. $\frac{\pi}{x} = k\pi,\ k = 1, 2, 3, 4, \ldots$

Hence, $x = \frac{1}{k}$.

Also, $\quad f'(x) = \sin\frac{\pi}{x} - \frac{\pi}{x}\cos\frac{\pi}{x}, \quad \text{if } x \neq 0$

Since, the function has a derivative at any interior point of the interval (0, 1), also continuous in [0, 1] and $f(0) = f(1)$, therefore Rolle's theorem is applicable to any one of the intervals $\left[\frac{1}{2}, 1\right], \left[\frac{1}{3}, \frac{1}{2}\right], \ldots, \left[\frac{1}{k+1}, \frac{1}{k}\right]$.

Hence, there exists some c in each of these intervals, where $f'(c) = 0 \Rightarrow$ Infinite points.

Hence, (d) is the correct answer.

Lagrange's Mean Value Theorem

First Form

If a function $f(x)$,

(i) is continuous in the closed interval $[a, b]$ and

(ii) is differentiable in the open interval $]a, b[$

Then, there is at least one value $c \in (a, b)$, such that

$$f'(c) = \frac{f(b) - f(a)}{b - a}$$

Proof Consider the function,

$$\phi(x) = f(x) - \frac{f(b)-f(a)}{b-a}x$$

Since, $f(x)$ is continuous in $[a, b]$.

$\therefore$ $\phi(x)$ is also continuous in $[a, b]$.

Since, $f'(x)$ exists in (a, b), therefore $\phi'(x)$ also exists in (a, b) and

$$\phi'(x) = f'(x) - \frac{f(b)-f(a)}{b-a} \quad \text{...(i)}$$

Also, we have $\phi(a) = \phi(b)$.

Thus, $\phi(x)$ satisfies all the conditions of Rolle's theorem.

$\therefore$ There is atleast one value of c of x between a and b, such that $\phi'(c) = 0$

On substituting $x = c$ in Eq. (i), we get

$$f'(c) = \frac{f(b)-f(a)}{b-a}, \text{ which proves the theorem.}$$

Second Form

If we write $b = a + h$, then $c = a + \theta h$, where $0 < \theta < 1$. $[\because a < c < b]$

Thus, the mean value theorem can be stated as follows:

If (i) $f(x)$ is continuous in closed interval $[a, a+h]$.

(ii) $f'(x)$ exists in the open interval $]a, a+h[$, then there exists at least one number θ $(0 < \theta < 1)$, such that $f(a+h) = f(a) + hf'(a + \theta h)$

Geometrical Interpretation of Lagrange's Theorem

Let A, B be the points on the curve $y = f(x)$ corresponding to $x = a$ and $x = b$, so that $A = [a, f(a)]$ and $B = [b, f(b)]$

Now, slope of chord $AB = \dfrac{f(b)-f(a)}{b-a}$

The slope of the chord $AB = f'(c)$, the slope of the tangent to the curve at $x = c$.

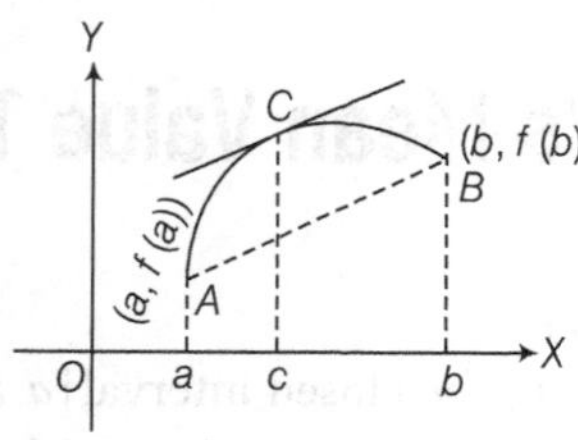

Figure 7.19

Hence, the Lagrange's mean value theorem asserts that if a curve has a tangent at each of its points, then there is a point c on this curve in between A and B, the tangent at which is parallel to the chord AB.

Example 43 Find c of the Lagrange's mean value theorem for which $f(x) = \sqrt{25 - x^2}$ in $[1, 5]$.

Sol. It is clear that $f(x)$ has a definite and unique value for each $x \in [1, 5]$.

Thus, for every point in the interval $[1, 5]$ the value of $f(x)$ is equal to the limit of $f(x)$.

So, $f(x)$ is continuous in the interval $[1, 5]$.

Also, $f'(x) = \dfrac{-x}{\sqrt{25 - x^2}}$, which clearly exists for all x in open interval $(1, 5)$

Hence, $f(x)$ is differentiable in $(1, 5)$.

So, there must be a value $c \in (1, 5)$ such that,

$$f'(c) = \frac{f(5) - f(1)}{5-1} = \frac{0 - \sqrt{24}}{4} = \frac{-\sqrt{6}}{2}$$

But $$f'(c) = \frac{-c}{\sqrt{25 - c^2}}$$

$$\therefore \quad \frac{-c}{\sqrt{25-c^2}} = \frac{-\sqrt{6}}{2} \Rightarrow 4c^2 = 6(25 - c^2)$$

$$\Rightarrow \quad c = \pm\sqrt{15}$$

Clearly, $c = \sqrt{15} \in (1, 5)$. So, for $c = \sqrt{15}$ Lagrange's theorem is satisfied.

Example 44 Let $f(x)$ and $g(x)$ be differentiable for $0 \le x \le 2$ such that $f(0) = 2, g(0) = 1$ and $f(2) = 8$. Let there exists a real number c in $[0, 2]$ such that $f'(c) = 3g'(c)$, then the value of $g(2)$ must be

(a) 2 (b) 3 (c) 4 (d) 5

Sol. As $f(x)$ and $g(x)$ are continuous and differentiable in $[0, 2]$, therefore there exists atleast one value 'c' such that

$$\frac{f'(c)}{g'(c)} = \frac{f(2)-f(0)}{g(2)-g(0)} \Rightarrow \frac{8-2}{g(2)-1} = 3$$

$$\Rightarrow \quad g(2) - 1 = 2 \Rightarrow g(2) = 3$$

Hence, (b) is the correct answer.

Example 45 If $f(x) = \log_e x$, $g(x) = x^2$ and $c \in (4, 5)$, then $c \log\left(\dfrac{4^{25}}{5^{16}}\right)$ is equal to

(a) $c\log_e 5 - 8$ (b) $2(c^2 \log_e 4 - 8)$

(c) $2(c^2 \log_e 5 - 8)$ (d) $c\log_e 4 - 8$

Sol. Let $\phi(x) = x^2 (\log_e 4) - 16\log_e x$, which is continuous in [4, 5] and differentiable in (4, 5).

Then, by Lagrange's theorem, we get

$$\frac{\phi(5) - \phi(4)}{5-4} = \phi'(c), \text{ for some } c \in (4, 5)$$

Now, $\phi(5) - \phi(4) = \log_e\left(\frac{4^{25}}{5^{16}}\right)$

Also, $\phi'(c) = \frac{2}{c}(c^2 \log_e 4 - 8)$

$\therefore \quad \phi'(c) = \frac{\phi(5) - \phi(4)}{5-4}$

$\Rightarrow \quad \frac{2}{c}(c^2 \log_e 4 - 8) = \log_e\left(\frac{4^{25}}{5^{16}}\right)$

or $\quad c \log\left(\frac{4^{25}}{5^{16}}\right) = 2(c^2 \log 4 - 8)$

Hence, (b) is the correct answer.

Example 46 If $0 < a < b < \frac{\pi}{2}$ and $f(a,b) = \frac{\tan b - \tan a}{b - a}$, then

(a) $f(a, b) \geq 2$ (b) $f(a, b) > 1$

(c) $f(a, b) \leq 1$ (d) None of these

Sol. Consider the function $f(x) = \tan x$, defined on $[a, b]$ such that $a, b \in \left(0, \frac{\pi}{2}\right)$.

Applying Lagrange's mean value theorem, we have

$$f'(c) = \frac{f(b) - f(a)}{b - a} \text{ for some } c \in (a, b)$$

$\Rightarrow \quad \sec^2 c = \frac{\tan b - \tan a}{b - a}$

$\Rightarrow \quad f(a, b) = \sec^2 c$

$\Rightarrow \quad f(a, b) > 1 \quad [\because \sec^2 c > 1 \text{ as } c \in (0, \pi/2)]$

Hence, (b) is the correct answer.

Example 47 In [0, 1] Lagrange's mean value theorem is not applicable to *[IIT JEE 2003]*

(a) $f(x) = \begin{cases} \frac{1}{2} - x, & x < \frac{1}{2} \\ \left(\frac{1}{2} - x\right)^2, & x \geq \frac{1}{2} \end{cases}$

(b) $f(x) = \begin{cases} \frac{\sin x}{x}, & x \neq 0 \\ 1, & x = 0 \end{cases}$

(c) $f(x) = x|x|$

(d) $f(x) = |x|$

Sol. For the function $f(x)$ given in option (a), we have (LHD at $x = 1/2$) $= -1$ and (RHD at $x = 1/2$) $= 0$.

So, it is not differentiable at $x = 1/2 \in (0, 1)$.

$\therefore$ Lagrange's mean value theorem is not applicable.

Hence, (a) is the correct answer.

Example 48 Let $f(x)$ satisfy the requirements of Lagrange's mean value theorem in [0, 2]. If $f(0) = 0$ and $|f'(x)| \leq 1/2$ for all $x \in [0, 2]$, then

(a) $f(x) \leq 2$ (b) $|f(x)| \leq 2x$

(c) $|f(x)| \leq 1$ (d) $f(x) = 3$,

for atleast one $x \in [0, 2]$

Sol. Let $x \in (0, 2)$. Since, $f(x)$ satisfies the requirements of Lagrange's mean value theorem in [0, 2]. So, it also satisfies in [0, x]. Consequently, there exists $c \in (0, x)$ such that

$$f'(c) = \frac{f(x) - f(0)}{x - 0} \Rightarrow f'(c) = \frac{f(x)}{x}$$

$\Rightarrow \quad \left|\frac{f(x)}{x}\right| = |f'(c)| \leq 1/2 \quad [\because |f'(x)| \leq 1/2]$

$\Rightarrow \quad |f(x)| \leq \frac{x}{2}$

$\Rightarrow \quad |f(x)| \leq 1 \quad [\because x \in (0, 2), \therefore x \leq 2]$

Hence, (c) is the correct answer.

Example 49 Let $f : [2, 7] \to [0, \infty)$ be a continuous and differentiable function. Then, the value of $(f(7) - f(2)) \frac{(f(7))^2 + (f(2))^2 + f(2) \cdot f(7)}{3}$ is (where $c \in (2, 7)$)

(a) $3f^2(c) f'(c)$ (b) $5f^2(c) \cdot f(c)$

(c) $5f^2(c) \cdot f'(c)$ (d) None of these

Sol. Let $g(x) = f^3(x) \Rightarrow g'(x) = 3f^2(x) \cdot f'(x)$

$\because \quad f : [2, 7] \to [0, \infty) \quad \therefore \quad g : [2, 7] \to [0, \infty)$

Using Lagrange's mean value theorem on $g(x)$, we get

$$g'(c) = \frac{g(7) - g(2)}{5}, c \in (2, 7)$$

$$\Rightarrow 3f^2(c) f'(c) = \frac{f^3(7) - f^3(2)}{5}$$

$$\Rightarrow 5f^2(c) \cdot f'(c) = \frac{(f(7) - f(2))(f^2(7) + f^2(2) + f(7) \cdot f(2))}{3}$$

Hence, (c) is the correct answer.

Example 50 The equation $\sin x + x \cos x = 0$ has atleast one root in the interval .

(a) $\left(-\frac{\pi}{2}, 0\right)$ (b) $(0, \pi)$ (c) $\left(-\frac{\pi}{2}, \frac{3\pi}{2}\right)$ (d) None of these

Sol. Consider the function given by,

$$f(x) = \int (\sin x + x\cos x)\,dx = x\sin x$$

we observe that $f(0) = f(\pi) = 0$

$\therefore$ 0 and π are two roots of $f(x) = 0$.

Consequently, $f'(x) = 0$, i.e. $\sin x + x\cos x = 0$ has at least one root in $(0, \pi)$.

Hence, (b) is the correct answer.

Example 51 Between any two real roots of the equation $e^x \sin x - 1 = 0$, the equation $e^x \cos x + 1 = 0$ has

(a) at least one root (b) at most one root
(c) exactly one root (d) no root

Sol. Let $f(x) = e^{-x} - \sin x$ and α and β be two roots of the equation $e^x \sin x - 1 = 0$ such that $\alpha < \beta$.

Then, $e^{\alpha}\sin\alpha = 1$ and $e^{\beta}\sin\beta = 1$

$\Rightarrow$ $e^{-\alpha} - \sin\alpha = 0$ and $e^{-\beta} - \sin\beta = 0$ ...(i)

Clearly, $f(x)$ is continuous on $[\alpha, \beta]$ and differentiable on (α, β). Also, $f(\alpha) = f(\beta) = 0$ [using Eq. (i)]

$\therefore$ By Rolle's theorem there exists $c \in (\alpha, \beta)$ such that

$$f'(c) = 0$$

$\Rightarrow$ $-e^{-c} - \cos c = 0$

$\Rightarrow$ $e^c \cos c + 1 = 0$

$\Rightarrow$ $x = c$ is root of $e^x \cos x + 1 = 0$; where $c \in (\alpha, \beta)$

Hence, (a) is the correct answer.

Exercise for Session 5

1. If $f(x) = x^{\alpha}\log x$ and $f(0) = 0$, then the value of 'α' for which Rolle's theorem can be applied in [0, 1], is

(a) -2 (b) -1
(c) 0 (d) $\frac{1}{2}$

2. If a, b, c are non-zero real numbers such that $\int_0^1 (1 + \cos^8 x)(ax^2 + bx + c)\,dx$ $= \int_0^2 (1 + \cos^8 x)(ax^2 + bx + c)\,dx = 0$, then the equation $ax^2 + bx + c = 0$ will have

(a) one root between 0 and 1 and another between 1 and 2
(b) both the roots between 0 and 1
(c) both the roots between 1 and 2
(d) None of the above

3. Let $n \in N$, if the value of c prescribed in Rolle's theorem for the function $f(x) = 2x\,(x - 3)^n$ on [0, 3] is $\frac{3}{4}$, then n is equal to

(a) 1 (b) 3
(c) 5 (d) 7

4. If $f(x) > x; \forall\, x \in R$. Then, the equation $f(f(x)) - x = 0$, has *[IIT]*

(a) atleast one real root
(b) more than one real root
(c) no real root if $f(x)$ is a polynomial and one real root if $f(x)$ is not a polynomial
(d) no real root

5. If $f(x)$ is twice differentiable function such that $f(a) = 0, f(b) = 2, f(c) = -1, f(d) = 2, f(e) = 0$, where $a < b < c < d < e$, then the minimum number of zeroes of $g(x) = (f'(x))^2 + f''(x)\cdot f(x)$ in the interval $[a, e]$, is *[IIT]*

(a) 4 (b) 5
(c) 6 (d) 7

Session 6

Application of Cubic Functions

Application of Cubic Functions

The cubic function $y = ax^3 + bx^2 + cx + d$ will have

1. One real and two imaginary roots
(always monotonic), $\forall\ x \in R$

Condition $f'(x) \geq 0$ or $f'(x) \leq 0$ together with either $f'(x) = 0$ has no root (i.e. $D < 0$) or $f'(x) = 0$ has a root $x = \alpha$, then $f(\alpha) = 0$.

e.g. $\quad y = x^3 - 2x^2 + 5x + 4$

$$y' = 3x^2 - 4x + 5 \qquad [D < 0]$$

$$y = (x-2)^3 \Rightarrow y' = 3(x-2)^2 = 0$$

$\Rightarrow \quad x = 2$, also $f(2) = 0$ gives $x = 2, y(2) = 0$

In this case, if $f'(x) = 0$ has a root $x = \alpha$ and $f(\alpha) = 0$ this would mean $f(x) = 0$ has repeated roots, which is dealt with separately.

2. Exactly one root and non-monotonic

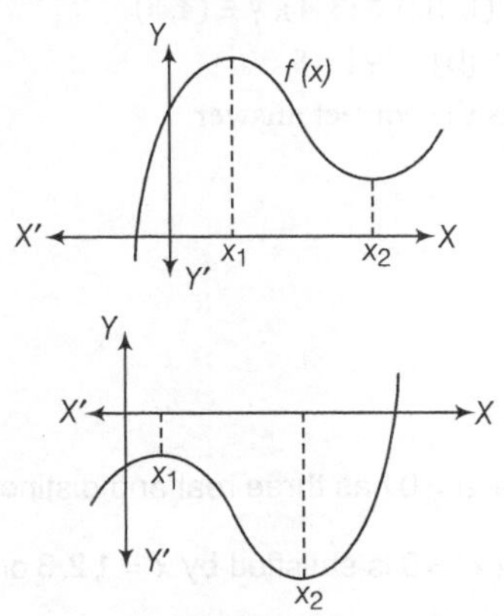

Figure 7.20

$$f(x_1) \cdot f(x_2) > 0$$

where x_1 and x_2 are the roots of $f'(x) = 0$

3. Two coincident and One different $f(x_1) \cdot f(x_2) = 0$

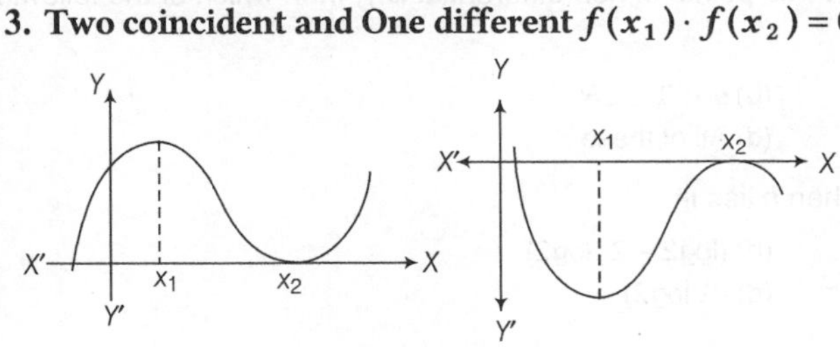

Figure 7.21

where x_1 and x_2 are the roots of $f'(x) = 0$

4. All three distinct real roots

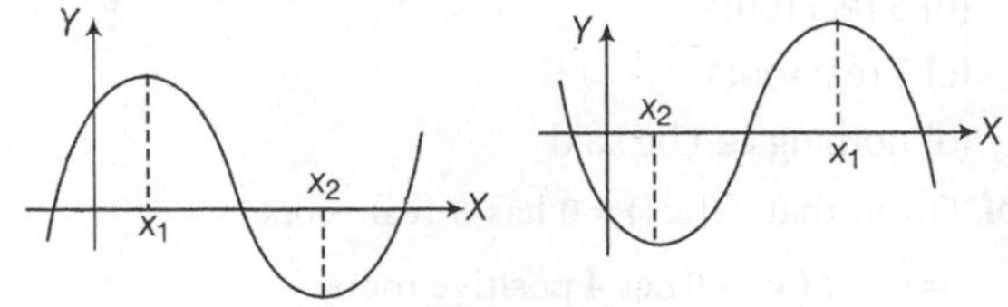

Figure 7.22

$f(x_1) \cdot f(x_2) < 0$

where x_1 and x_2 are the roots of $f'(x) = 0$

5. All the three roots are coincident

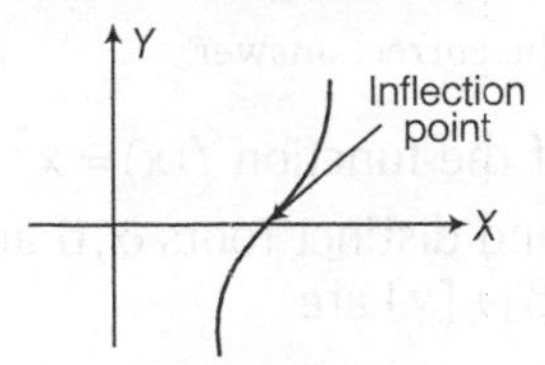

Figure 7.23

$$f'(x) \geq 0 \quad \text{or} \quad f'(x) \leq 0$$

and $\quad f(\alpha) = 0$

where α is a root of $f'(x) = 0$

e.g. $\quad y = (x-1)^3$

Remarks

(i) Graph of every cubic polynomial must have exactly one point of inflection.

(ii) In case (4), if $f(a)$, $f(b)$, $f(c)$ and $f(d)$ alternatively change sign.

Example 52 If the cubic function $f(x) = x^3 + px + q$ has 3 distinct real roots, then prove that $4p^3 + 27q^2 < 0$.

Sol. $f'(x) = 3x^2 + p = 3x^2 + 0 \cdot x + p$

Now, we have $x_1 + x_2 = 0$ and $x_1x_2 = \frac{p}{3}$, where x_1, x_2 are roots of $f'(x) = 0$

According to given condition, we have

$(x_1^3 + px_1 + q)(x_2^3 + px_2 + q) < 0$

$$\Rightarrow \quad x_1^3 \cdot x_2^3 + px_1^3x_2 + qx_1^3 + p^2x_1x_2 + px_1x_2^3 + qx_2^3 + pqx_1 + q^2 + pqx_2 < 0$$

$$\Rightarrow (x_1x_2)^3 + px_1x_2(x_1^2 + x_2^2) + q(x_1^3 + x_2^3) + pq(x_1 + x_2) + p^2x_1x_2 + q^2 < 0$$

$$\Rightarrow (x_1x_2)^3 + px_1x_2\{(x_1 + x_2)^2 - 2x_1x_2\} + q\{(x_1 + x_2)^3 - 3(x_1x_2)(x_1 + x_2)\} + pq(x_1 + x_2) + p^2x_1x_2 + q^2 < 0$$

$$\Rightarrow \quad \frac{p^3}{27} + \frac{p^2}{3}\left\{-\frac{2p}{3}\right\} + p^2\left\{\frac{p}{3}\right\} + q^2 < 0$$

$$\Rightarrow \quad 4p^3 + 27q^2 < 0$$

Example 53 If $f(x)$ is a polynomial of degree 5 with real coefficients such that $f(|x|) = 0$ has 8 real roots, then $f(x) = 0$ has

(a) 4 real roots
(b) 5 real roots
(c) 3 real roots
(d) nothing can be said

Sol. Given that $f(|x|) = 0$ has 8 real roots.

$\Rightarrow \quad f(x) = 0$ has 4 positive roots.

Since, $f(x)$ is a polynomial of degree 5, $f(x)$ cannot have even number of real roots.

$\Rightarrow f(x)$ has all the five roots real, in which four positive and one root is negative.

Hence, (b) is the correct answer.

Example 54 If the function $f(x) = x^3 - 9x^2 + 24x + c$ has three real and distinct roots α, β and γ, then the value of $[\alpha]+[\beta]+[\gamma]$ are

(a) 5, 6
(b) 6, 7
(c) 7, 8
(d) None of the above

Sol. Take $y = x^3 - 9x^2 + 24x$

$$\Rightarrow \quad \frac{dy}{dx} = 3x^2 - 18x + 24 = 3(x^2 - 6x + 8)$$

$$= 3(x-2)(x-4)$$

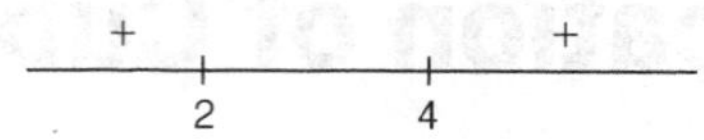

for three real roots of

$f(x) = x^3 - 9x^2 + 24x + c, f(2)f(4) < 0$

$\therefore (c+20)(c+16) < 0$

$c \in (-20, -16)$

Now, if $c \in (-20, -18)$, $\alpha \in (1, 2)$, $\beta \in (2, 3)$, $\gamma \in (4, 5)$

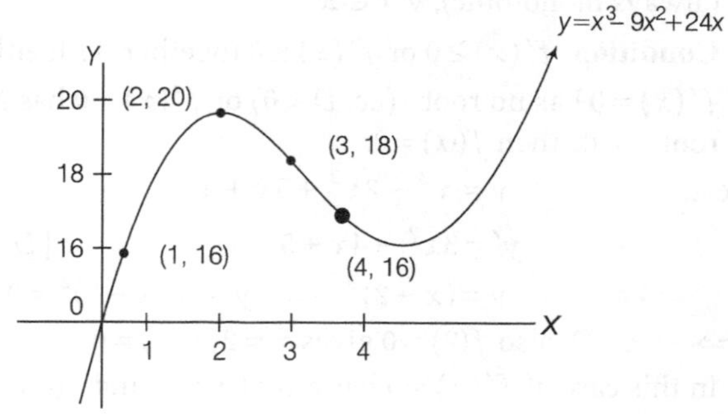

$\Rightarrow \quad [\alpha]+[\beta]+[\gamma]=7$

If $\quad c \in (-18, -16)$

$\Rightarrow \quad \alpha \in (1, 2), \beta \in (3, 4), \gamma \in (4, 5)$

$\Rightarrow \quad [\alpha]+[\beta]+[\gamma] = 8$

Hence, (c) is the correct answer.

Exercise for Session 6

1. Find all the possible values of the parameter a, so that $x^3 - 3x + a = 0$ has three real and distinct roots.

2. $f(x)$ is a polynomial of degree 4 with real coefficients such that $f(x) = 0$ is satisfied by $x = 1, 2, 3$ only, then $f'(1) \cdot f'(2) \cdot f'(3)$ is equal to

(a) 0 (b) 2
(c) –1 (d) None of these

3. If the function $f(x) = |x^2 + a|x| + b|$ has exactly three points of non-differentiability, then which of the following can be true?

(a) $b = 0, a < 0$ (b) $b < 0, a \in R$
(c) $b > 0, a \in R$ (d) All of these

4. If the equation $e^{||x|-2|+b} = 2$ has four solutions, then b lies in

(a) $(\log 2, -\log 2)$ (b) $(\log 2 - 2, \log 2)$
(c) $(-2, \log 2)$ (d) $(0, \log 2)$

JEE Type Solved Examples :
Single Option Correct Type Questions

• **Ex. 1** *Number of integral value(s) of k for which the equation $4x^2 - 16x + k = 0$ has one root lie between 1 and 2 and other root lies between 2 and 3, is*

(a) 1 (b) 2 (c) 3 (d) 4

Sol. (c) Here, $f(1)\cdot f(2) < 0$

$\Rightarrow \quad (k-12)(k-16) < 0$

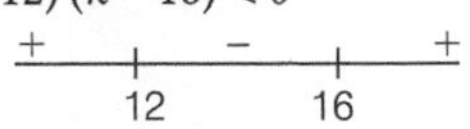

$\Rightarrow \quad 12 < k < 16 \quad$ ...(i)

Also, $\quad f(2)\cdot f(3) < 0$

$\Rightarrow \quad (k-16)(k-12) < 0 \Rightarrow 12 < k < 16 \quad$ ...(ii)

From Eqs. (i) and (ii), we get

$k = \{13, 14, 15\}$

$\therefore$ Number of integral values of $k = 3$

• **Ex. 2** *Let $a, b, c \in R$ such that two of them are equal and satisfy* $\begin{vmatrix} 2a & b & c \\ b & c & 2a \\ c & 2a & b \end{vmatrix} = 0$, *then equation $24ax^2 + 4bx + c = 0$ has*

(a) at least one root in $(0, 1/2)$
(b) at least one root in $(-1/2, 1/2)$
(c) at least one root in $(-1, 0)$
(d) at least two roots in $(0, 2)$

Sol. (a) Given determinant is

$$2a(bc - 4a^2) - b(b^2 - 2ac) + c(2ab - c^2) = 0$$

$$\Rightarrow \quad 6abc - 8a^3 - b^3 - c^3 = 0$$

or $\quad (2a + b + c)[(2a-b)^2 + (b-c)^2 + (c-2a)^2] = 0$

$\Rightarrow \quad 2a + b + c = 0 \quad$ ...(i)

$[\because b \neq c]$

Let $f(x) = 8ax^3 + 2bx^2 + cx \Rightarrow f'(x) = 24ax^2 + 4bx + c$

Here, $f(0) = 0$ and $f(1/2) = a + b/2 + c/2 = \dfrac{2a + b + c}{2}$

$\therefore \quad f(1/2) = 0 \quad$ [using Eq. (i)]

So, $f(x)$ satisfy Rolle's theorem and hence $f'(x) = 0$ must have at least one root in $(0, 1/2)$.

• **Ex. 3** *The set of values of 'a' for which the equation $\log_e(a \log_e x) = \log_e x$ has more than one solution is*

(a) $(1, \infty)$ (b) (e, ∞) (c) $(0, e)$ (d) $(1, e)$

Sol. (b) Here, $\log_e(a \log_e x) = \log_e x$

$\Rightarrow \quad a \log_e x = e^{\log_e x} \Rightarrow a \log_e x = x$

If we put $a = e$, then $\log_e x = x/e$ will have only one solution at $x = e$.

$[\because$ the line $y = \dfrac{x}{e}$ is tangent to the curve $y = \log_e x$ at $x = e]$

$\Rightarrow \log_e x = \dfrac{x}{a}$ has more than one solution for $a \in (e, \infty)$.

• **Ex. 4** *The tangent to the hyperbola $y = \dfrac{x+9}{x+5}$ passing through the origin is*

(a) $x + 25y = 0$ (b) $5x + y = 0$
(c) $5x - y = 0$ (d) $x - 25y = 0$

Sol. (a) Here, $y = 1 + \dfrac{4}{x+5} \Rightarrow (dy/dx)$ at $(x_1, y_1) = -\dfrac{4}{(x_1+5)^2}$

Now, equation of tangent, is

$$y - \left(1 + \frac{4}{x_1+5}\right) = -\frac{4}{(x_1+5)^2}(x - x_1)$$

Since, it passes through (0, 0), therefore

$$-1 - \frac{4}{x_1+5} = \frac{4x_1}{(x_1+5)^2}$$

$\Rightarrow \quad -(x_1+5)^2 - 4(x_1+5) = 4x_1$

$\Rightarrow \quad (x_1+5)^2 + 4(x_1+5) + 4x_1 = 0$

$\Rightarrow \quad x_1^2 + 18x_1 + 45 = 0$

$\Rightarrow \quad (x_1+15)(x_1+3) = 0 \Rightarrow x_1 = -15$ or -3

So, equation of tangents are

$$x + 25y = 0 \text{ or } x + y = 0$$

• **Ex. 5** *The tangent to the curve $y = e^x$ drawn at the point (c, e^c) intersects the line joining the points $(c-1, e^{c-1})$ and $(c+1, e^{c+1})$* ***[IIT JEE 2007]***

(a) on the left of $x = c$ (b) on the right of $x = c$
(c) at no point (d) at all points

Sol. (a) Slope of the line joining the points $(c-1, e^{c-1})$ and $(c+1, e^{c+1})$ is equal to $\dfrac{e^{c+1} - e^{c-1}}{2} > e^c$

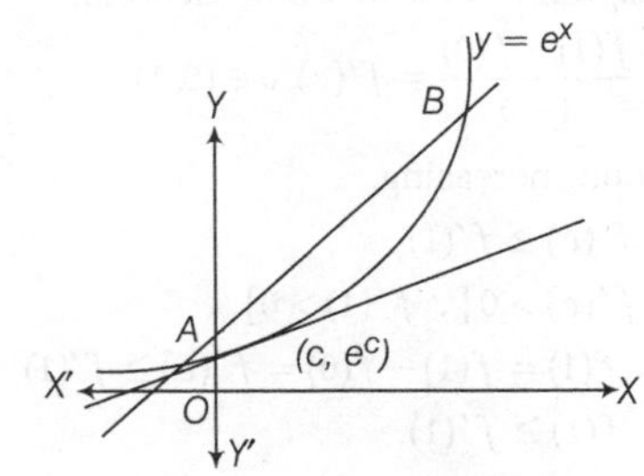

$\Rightarrow$ Tangent to the curve $y = e^x$ will intersect the given line to the left of the line $x = c$.

Aliter

The equation of the tangent to the curve $y = e^x$ at (c, e^c) is

$$y - e^c = e^c (x - c) \quad \text{...(i)}$$

Equation of the line joining the given points is

$$y - e^{c-1} = \frac{e^c (e - e^{-1})}{2} [x - (c - 1)] \quad \text{...(ii)}$$

Eliminating y from Eqs. (i) and (ii), we get

$$[x - (c - 1)] [2 - (e - e^{-1})] = 2e^{-1}$$

or $$x - c = \frac{e + e^{-1} - 2}{2 - (e - e^{-1})} < 0$$

$\Rightarrow$ $$x < c$$

JEE Type Solved Examples : More than One Correct Option Type Questions

Ex. 6 *The coordinate of the point(s) on the graph of the function, $f(x) = \frac{x^3}{3} - \frac{5x^2}{2} + 7x - 4$ where the tangent drawn cuts-off intercepts from the coordinate axes which are equal in magnitude but opposite in sign, is*

(a) $\left(2, \frac{8}{3}\right)$ (b) $\left(3, \frac{7}{2}\right)$

(c) $\left(1, \frac{5}{6}\right)$ (d) None of these

Sol. (*a*, *b*) Since, intercepts are equal in magnitude but opposite in sign, therefore

$$\left[\frac{dy}{dx}\right]_P = 1$$

Now, $\frac{dy}{dx} = x^2 - 5x + 7 = 1$

$\Rightarrow$ $x^2 - 5x + 6 = 0$

$\therefore$ $x = 2$ or 3

Ex. 7 *Let $f : [0, 1] \to R$ be a differentiable function with non-increasing derivative such that $f(0) = 0$, $f'(1) > 0$, then*

(a) $f(1) \geq f'(1)$

(b) $f'(c) \neq 0$ for any $c \in (0, 1)$

(c) $f(1/2) > f(1)$

(d) None of the above

Sol. (*a*, *b*) By Lagrange's mean value theorem,

$$\frac{f(1) - f(0)}{1 - 0} = f'(c), c \in (0, 1)$$

$\because f'(x)$ is non-increasing.

$\therefore$ $f'(c) \geq f'(1)$

$\Rightarrow$ $f'(c) > 0 \, [\because f'(1) > 0]$

Also, $f(1) = f(1) - f(0) = f'(c) \geq f'(1)$

$\therefore$ $f(1) \geq f'(1)$.

Ex. 8 *If $f(x)$ is continuous and derivable, $\forall x \in R$ and $f'(c) = 0$ for exactly 2 real values of 'c', then the number of real and distinct values of 'd' for which $f(d) = 0$ can be*

(a) 1 (b) 2 (c) 3 (d) 4

Sol. (*a*, *b*, *c*) If $f'(x) = 0$ has n real roots, then

$f(x) = 0$ has atmost $(n + 1)$ real roots.

Now, if $f'(c) = 0$ for exactly 2 real values of c.

Then, following cases may arise

(a)

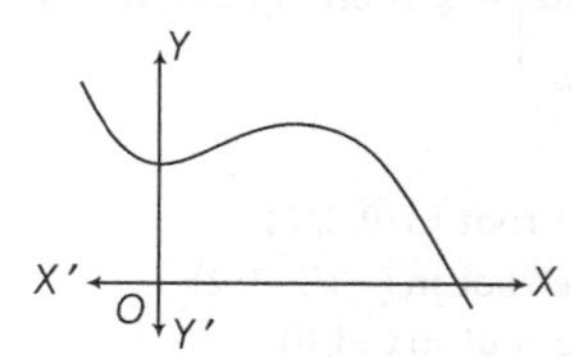

$\Rightarrow$ $f(x) = 0$ have 1 real root.

(b)

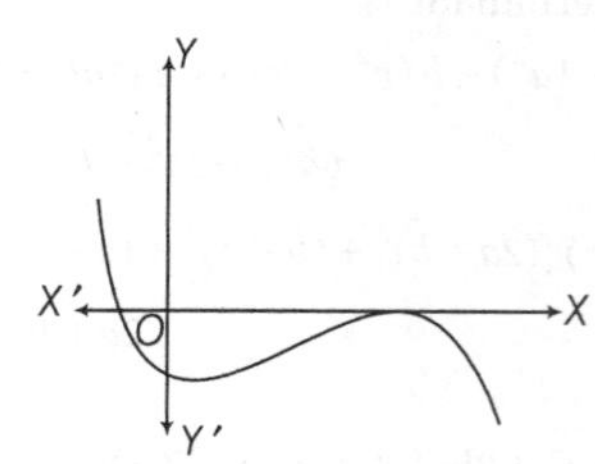

$\Rightarrow$ $f(x) = 0$ has 2 real roots.

(c)

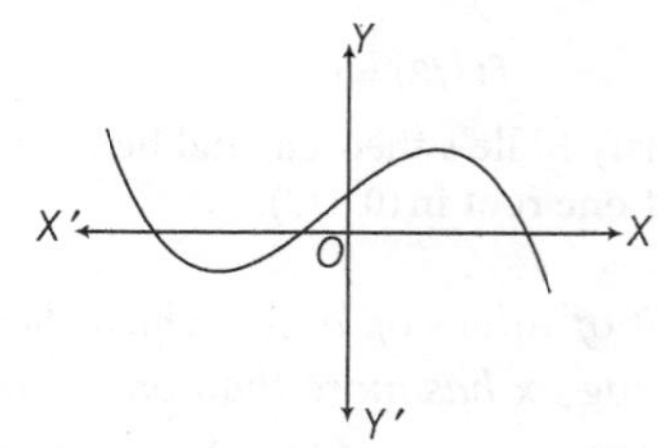

$\Rightarrow$ $f(x) = 0$ has 3 real roots.

Ex. 9 *Which of the following statements are true, where $\phi(x)$ is a polynomial?*

(a) Between any two roots of $\phi(x)=0$ there exists at least one root of $\phi'(x)+\lambda\phi(x)=0$

(b) Between any two roots of $\phi(x)=0$ there exists at least one root of $x\phi'(x)+\lambda\phi(x)=0$

(c) Between any two roots of $\phi(x)=0$ there exists at least one root of $(x^2+1)\phi'(x)+\phi(x)=0$

(d) Between any two roots of $\phi(x)=0$ there exists at least one root of $\phi'(x)+x\phi(x)=0$

Sol. (*c*, *d*) If α and β are the consecutive roots of $\phi(x)=0$, then $\phi'(\alpha)\cdot\phi'(\beta)\le 0$

Let $\quad \phi(x)=(x-1)(x-2)$

$$\phi'(x)=2x-3$$

$$\therefore \quad \phi'(x)+\lambda\phi(x)=0$$

$$\Rightarrow \quad (2x-3)+\lambda(x-1)(x-2)=0$$

Must have at least one root $\in(\alpha,\beta)$.

Similarly, $(x^2+1)\phi'(x)+\phi(x)=0$

and $\phi'(x)+x\phi(x)=0$ has at least one root $\in(\alpha,\beta)$.

Ex. 10 *Let $f(x)$ be twice differentiable function such that $f''(x)<0$ in $[0, 2]$. Then,*

(a) $f(0)+f(2)=2f(c), 0<c<2$

(b) $f(0)+f(2)=2f(1)$

(c) $f(0)+f(2)>2f(1)$

(d) $f(0)+f(2)<2f(1)$

Sol. (*a*, *d*) By intermediate mean value theorem, we get

$$\frac{f(0)+f(2)}{2}=f(c), \quad 0<c<2 \quad \text{...(i)}$$

By Lagrange's mean value theorem, we get

$$f(1)-f(0)=f'(c_1), \quad 0<c_1<1 \quad \text{...(ii)}$$

$$f(2)-f(1)=f'(c_2), \quad 1<c_2<2 \quad \text{...(iii)}$$

On subtracting Eq. (ii) from Eq. (iii), we get

$$f(2)+f(0)-2f(1)=f'(c_2)-f'(c_1) \quad \text{...(iv)}$$

Again, by Lagrange's mean value theorem, we get

$$f''(c_3)=\frac{f'(c_2)-f'(c_1)}{c_2-c_1}, \text{ for some } c_3\in(c_1, c_2)$$

$$\Rightarrow \quad f'(c_2)-f'(c_1)=(c_2-c_1)f''(c_3)<0 \quad [f''(x)<0] \text{ ...(v)}$$

From Eqs. (iv) and (v), we get

$$f(2)+f(0)-2f(1)<0 \Rightarrow f(0)+f(2)<2f(1)$$

JEE Type Solved Examples : Statements I and II Type Questions

- **Directions** (Q. Nos. 11 to 13) *For the following questions, choose the correct answers from the option (a), (b), (c) and (d) defined as follows*

(a) Statement I is true, Statement II is also true; Statement II is the correct explanation of Statement I

(b) Statement I is true, Statement II is also true; Statement II is not the correct explanation of Statement I

(c) Statement I is true, Statement II is false

(d) Statement I is false, Statement II is true

Ex. 11 **Statement I** *The tangent at $x=1$ to the curve $y=x^3-x^2-x+2$ again meets the curve at $x=-2$.*

Statement II *When an equation of a tangent solved with the curve, repeated roots are obtained at the point of tangency.*

Sol. (*d*) When $x=1$, then $y=1$. Also, $y'=3x^2-2x-1$

$$\Rightarrow \quad y'|_{x=1}=0 \Rightarrow \text{equation of tangent is } y=1.$$

Solving $x^3-x^2-x+2=1$

or $\quad x^3-x^2-x+1=0$, we get $x=-1, 1$

$\therefore$ The tangent meets the curve again at $x=-1$.

$\therefore$ Statement I is false and Statement II is true.

Ex. 12 **Statement I** *The ratio of length of tangent to length of normal is inversely proportional to the ordinate of the point of tengency at the curve $y^2=4ax$.*

Statement II *Length of normal and tangent to a curve*

$$y=f(x) \text{ is } \left|y\sqrt{1+m^2}\right| \text{ and } \left|\frac{y\sqrt{1+m^2}}{m}\right|,$$

where $m=\dfrac{dy}{dx}$.

Sol. (*a*) Let the slope of the tangent be denoted by $\tan\psi$ and the length of tangent $=y\,\text{cosec}\,\psi$.

Then, length of normal $=y\sec\psi$

$$\therefore \quad \frac{\text{Length of tangent}}{\text{Length of normal}}=\cot\psi\propto\frac{1}{y}$$

$\therefore$ Statement I is true.

$$\text{Length of normal}=\left|y\sqrt{1+\left(\frac{dy}{dx}\right)^2}\right|$$

$$= \left| y\sqrt{1+m^2} \right|$$

$$\text{Length of tangent} = \left| y\sqrt{1+\left(\frac{dx}{dy}\right)^2} \right| = \left| \frac{y\sqrt{1+m^2}}{m} \right|$$

$\therefore$ Statement II is true and explain Statement I.

Ex. 13 Statement I *Tangent drawn at the point* (0, 1) *to the curve* $y = x^3 - 3x + 1$ *meets the curve thrice at one point only.*

Statement II *Tangent drawn at the point* (1, − 1) *to the curve* $y = x^3 - 3x + 1$ *meets the curve at one point only.*

Sol. (c) $\frac{dy}{dx} = 3x^2 - 3$

Statement I $\left(\frac{dy}{dx}\right)_{\text{at }(0,\,1)} = -3$

$\therefore$ Equation of the tangent is $y - 1 = -3(x - 0)$

i.e. $y = -3x + 1$

Now, $-3x + 1 = x^3 - 3x + 1 \Rightarrow x = 0$

$\therefore$ The tangent meets the curve at one point only.

$\therefore$ Statement I is true.

Statement II $\left(\frac{dy}{dx}\right)_{\text{at }(1,\,-1)} = 0$

$\therefore$ Equation of the tangent is $y + 1 = 0\,(x - 1)$

i.e. $y = -1$

Now, $-1 = x^3 - 3x + 1$

$\Rightarrow x^3 - 3x + 2 = 0$

$\Rightarrow (x - 1)(x^2 + x - 2) = 0$

$\Rightarrow$ The tangent meets the curve at two points.

$\therefore$ Statement II is false.

JEE Type Solved Examples : Passage Based Questions

Passage I (Ex. Nos. 14 to 18)

Let $f(x) = x^3 + ax^2 + bx + c$ *be the given cubic polynomial and* $f(x) = 0$ *be the corresponding cubic equation, where* $a, b, c \in R$. *Now,* $f'(x) = 3x^2 + 2ax + b$

Let $D = 4a^2 - 12b = 4(a^2 - 3b)$ *be the discriminant of the equation* $f'(x) = 0$.

Ex. 14 *If* $D = 4(a^2 - 3b) < 0$, *then*

(a) $f(x)$ has all real roots
(b) $f(x)$ has one real and two imaginary roots
(c) $f(x)$ has repeated roots
(d) None of the above

Ex. 15 *If* $D = 4(a^2 - 3b) > 0$ *and* $f(x_1) \cdot f(x_2) > 0$, *where* x_1, x_2 *are the roots of* $f(x)$, *then*

(a) $f(x)$ has all real and distinct roots
(b) $f(x)$ has three real roots but one of the roots would be repeated
(c) $f(x)$ would have just one real root
(d) None of the above

Ex. 16 *If* $D = 4(a^2 - 3b) > 0$ *and* $f(x_1) \cdot f(x_2) < 0$, *where* x_1, x_2 *are the roots of* $f(x)$, *then*

(a) $f(x)$ has all real and distinct roots
(b) $f(x)$ has three real roots but one of the roots would be repeated
(c) $f(x)$ would have just one real root
(d) None of the above

Ex. 17 *If* $D = 4(a^2 - 3b) > 0$ *and* $f(x_1) \cdot f(x_2) = 0$, *where* x_1, x_2 *are the roots of* $f(x)$, *then*

(a) $f(x)$ has all real and distinct roots
(b) $f(x)$ has three real roots but one of the roots would be repeated
(c) $f(x)$ would have just one real root
(d) $f(x)$ has three real roots but all are same

Ex. 18 *If* $D = 4(a^2 - 3b) = 0$, *then*

(a) $f(x)$ has all real and distinct roots
(b) $f(x)$ has three real roots but one of the roots would be repeated
(c) $f(x)$ would have just one real root
(d) None of the above

Sol. (Ex. Nos. 14 to 18)

***Case* I** If $D < 0 \Rightarrow f'(x) > 0, \forall\, x \in R$

That means $f(x)$ would be an increasing function of x.

Also, $\lim_{x \to \infty} f(-x) = -\infty$ and $\lim_{x \to \infty} f(x) = \infty$

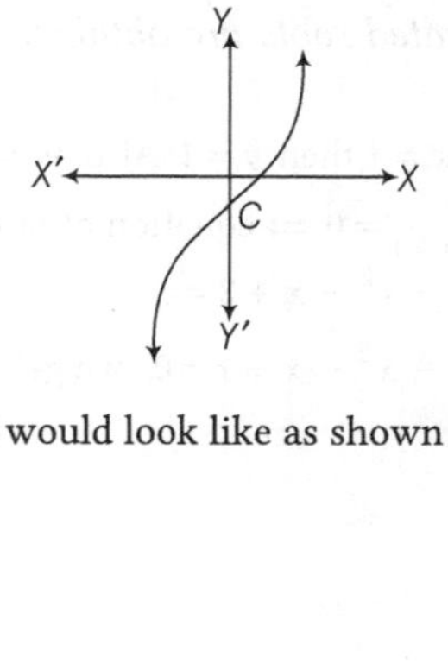

The graph of $f(x)$ would look like as shown in this figure.

It is clear that the graph of $f(x)$ would intersect the X-axis only once.

That means we would have just one real root (say x_0).

***Case* II** Clearly, $x_0 > 0$, if $c < 0$ and $x_0 < 0$, if $c > 0$.

If $D > 0$, $f'(x) = 0$ would have two real roots (say x_1 and x_2, let $x_1 < x_2$).

$\Rightarrow \quad f'(x) = 3(x - x_1)(x - x_2)$

Clearly, $\quad f'(x) < 0, x \in (x_1, x_2)$

$f'(x) > 0, x \in (-\infty, x_1)\ (x_2, \infty)$

That means $f(x)$ would increase in $(-\infty, x_1)$ and (x_2, ∞) and would decrease in (x_1, x_2). Hence, $x = x_1$ would be a point of local maxima and $x = x_2$ would be a point of local minima. Thus, the graph of $y = f(x)$ could have these five possibilities.

Thus, the following results are obtained from the above graphs

(a) $f(x_1) \cdot f(x_2) > 0$, $f(x) = 0$ would have just one real root.

(b) $f(x_1) \cdot f(x_2) < 0$, $f(x) = 0$ would have three real and distinct roots.

(c) $f(x_1) \cdot f(x_2) = 0$, $f(x) = 0$ would have three real roots about one of the roots would be repeated.

***Case* III** If $D = 0$, $f'(x) = 3(x - x_1)^2$, where x_1 is root of $f'(x) = 0$

$\Rightarrow \quad f(x) = (x - x_1)^3 + k$

$\therefore \quad f(x) = 0$ has three real roots, if $k = 0$

$f(x) = 0$ have one real root, if $k \neq 0$.

14. (*b*) Here, $f(x) = 0$ and $D < 0$

$\Rightarrow f'(x) > 0, \ \forall x \in R$

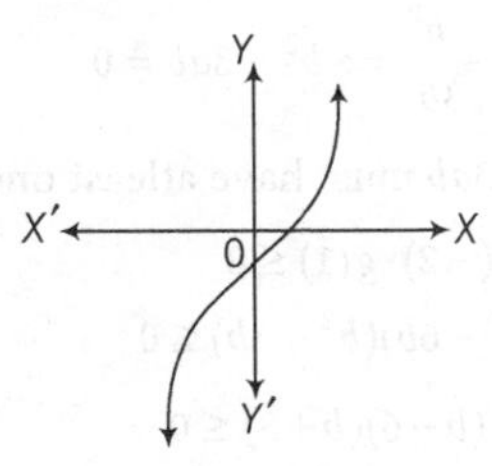

15. (*c*) Here, $f(x) = 0$ with one real root $x = \alpha$ and other two imaginary roots.

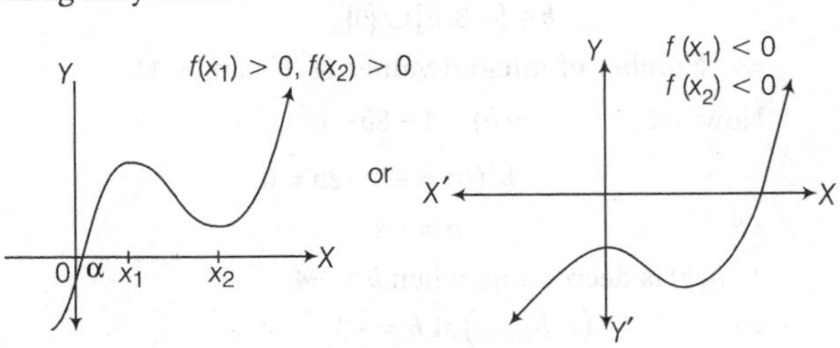

From both graph, $f(x_1)\, f(x_2) > 0$

$\therefore$ $f(x)$ would have one real roots.

16. (*a*) Here, $f(x) = 0$ with three distinct root α, β, γ.

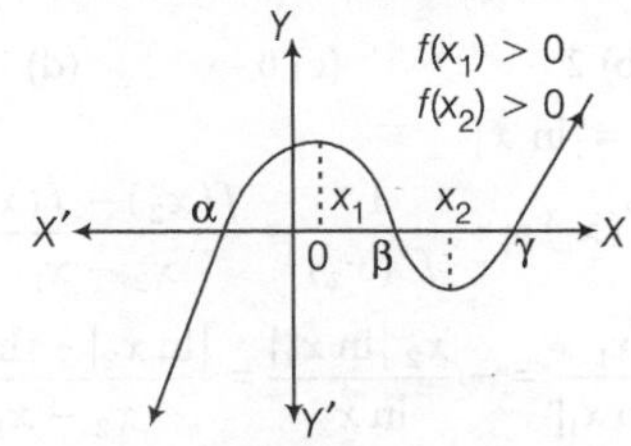

From graph, $f(x_1)\, f(x_2) < 0$

$\therefore$ $f(x)$ has real and distinct roots.

17. (*b*) Here, $f(x) = 0$ with three roots $x = \alpha, x_2$ (x_2 being repeated root).

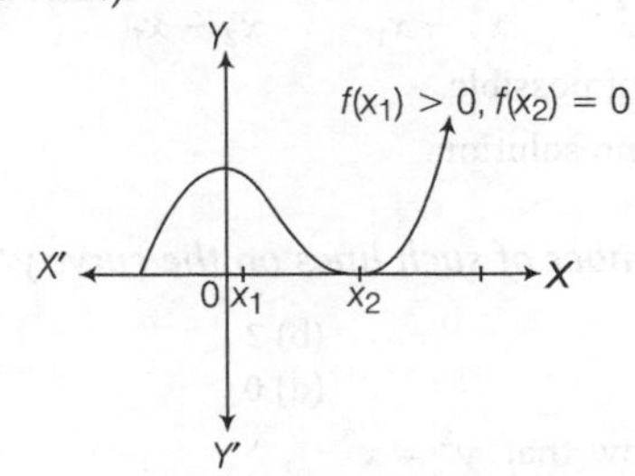

18. (*b*) Here, $f(x) = 0$ with three real roots $x = x_1, \alpha$ (x_1 being repeated root).

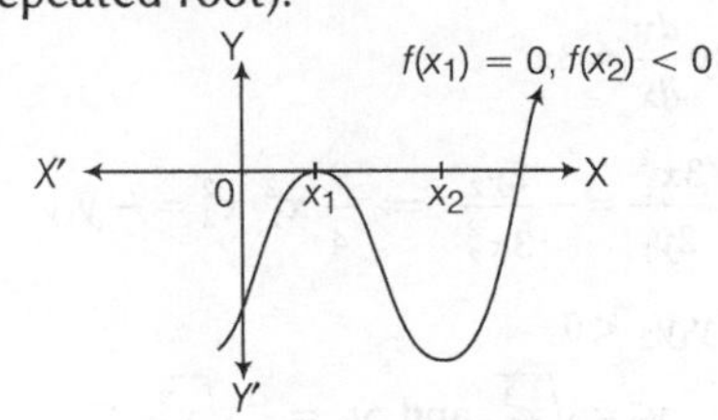

Passage II (Ex. Nos. 19 to 21)

- *If $y = f(x)$ is a curve and if there exists two points $A(x_1, f(x_1))$ and $B(x_2, f(x_2))$ on it such that $f'(x_1) = -\dfrac{1}{f'(x_2)} = \dfrac{f(x_2) - f(x_1)}{x_2 - x_1}$, then the tangent at x_1 is normal at x_2 for that curve. Now, answer the following questions.*

Ex. 19 *Number of such lines on the curve $y = \sin x$ is*

(a) 1 (b) 0 (c) 2 (d) infinite

Sol. (*b*) Let $f(x) = y = \sin x$

Then, $\quad f'(x) = \dfrac{dy}{dx} = \cos x$

$$\therefore \quad \cos x_1 = -\frac{1}{\cos x_2} = \frac{\sin x_2 - \sin x_1}{x_2 - x_1}$$

i.e. $\quad \cos x_1 \cos x_2 = -1$

$\therefore \quad \sin x_1 = \sin x_2 = 0$

$\therefore$ There is no solution.

Ex. 20 *Number of such lines on the curve $y = |\ln x|$ is*

(a) 1 (b) 2 (c) 0 (d) infinite

Sol. (c) $f(x) = y = |\ln x|$

$$\therefore \quad f'(x_1) = -\frac{1}{f'(x_2)} = \frac{f(x_2) - f(x_1)}{x_2 - x_1}$$

$$\Rightarrow \quad \frac{\ln x_1}{x_1 |\ln x_1|} = -\frac{x_2 |\ln x_2|}{\ln x_2} = \frac{|\ln x_2| - |\ln x_1|}{x_2 - x_1} \quad \text{...(i)}$$

$\Rightarrow \ln x_1 \cdot \ln x_2 < 0$

Let $0 < x_1 < 1$, then $1 < x_2$ and $x_1 \cdot x_2 = 1$

From Eq. (i), we get

$$-\frac{1}{x_1} = -x_2 = \frac{\ln x_2 + \ln x_1}{x_2 - x_1} = \frac{\ln x_1 x_2}{x_2 - x_1} = 0,$$

which is not possible.

$\therefore$ There is no solution.

Ex. 21 *Number of such lines on the curve $y^2 = x^3$, is*

(a) 1 (b) 2
(c) 3 (d) 0

Sol. (b) We know that, $y^2 = x^3$

$$\therefore \quad y = \sqrt{x^3} \text{ or } -\sqrt{x^3}$$

$$2y\frac{dy}{dx} = 3x^2$$

$$\frac{3x_1^2}{2y_1} = -\frac{2y_2}{3x_2^2} \Rightarrow \frac{9}{4}x_1^2 x_2^2 = -y_1 y_2$$

$$\therefore \quad y_1 y_2 < 0$$

Let $\quad y_1 = \sqrt{x_1^3}$ and $y_2 = -\sqrt{x_2^3}$

Thus, $$\frac{3x_1^2}{2\sqrt{x_1^3}} = -\frac{2\sqrt{x_2^3}}{3x_2^2} = \frac{-\sqrt{x_2}^3 - \sqrt{x_1^3}}{x_2 - x_1}$$

$$\Rightarrow \quad \frac{3\sqrt{x_1}}{2} = \frac{2}{3\sqrt{x_2}} = \frac{\sqrt{x_2^3} + \sqrt{x_1^3}}{x_1 - x_2}$$

$$\Rightarrow \quad \sqrt{x_1 x_2} = \frac{4}{9}$$

and $3x_1\sqrt{x_1} - 3\sqrt{x_1}\, x_2 = 2\sqrt{x_2^3} + 2\sqrt{x_1^3}$

$$\Rightarrow \quad 3(\sqrt{x_1})^3 - \frac{3 \times 16}{81\sqrt{x_1}} = 2 \cdot \frac{64}{729\sqrt{x_1^3}} + 2\sqrt{x_1^3}$$

$$\Rightarrow \quad 3x_1^3 - \frac{16}{27}x_1 = \frac{128}{729} + 2x_1^3$$

$$\Rightarrow \quad x_1^3 - \frac{16}{27}x_1 = \frac{128}{729}$$

$$\Rightarrow \quad 739x_1^3 - 432x_1 - 12 = 0$$

Consider, $\quad h(t) = 729x^3 - 432t - 128$

$$h'(t) = 3 \times 729t^2 - 432 = 0$$

Gives $\quad t = \pm\frac{4}{9}$

$$h\left(-\frac{4}{9}\right) = 0$$

$\therefore$ There are two distinct solutions of

$$729x_1^3 - 432x_1 - 128 = 0.$$

Passage III (Ex. Nos. 22 and 23)

Let $f(x) = \int_0^x (|t-1| - |t+2| + t - 2)\, dt$, such that $f''(a) \neq 1$. If vectors $a\hat{\mathbf{i}} - b^2\hat{\mathbf{j}}$ and $\hat{\mathbf{i}} + 3b\hat{\mathbf{j}}$ are parallel for at least one a, then

Ex. 22 *Number of integral values of 'b' can be*

(a) 5 (b) 10
(c) 11 (d) 13

Ex. 23 *Maximum value of $(1 - 8b - b^2)$ is*

(a) 4 (b) 8
(c) 12 (d) 16

Sol. (Ex. Nos. 22 and 23) Here, $f(x) = \int_0^x (|t-1| - |t+2| + t - 2)\, dt$

$$\Rightarrow \quad f'(x) = |x-1| - |x+2| + x - 2$$

or $$f'(x) = \begin{cases} x+1, & x < -2 \\ -x-3, & -2 \le x \le 1 \\ x-5, & x > 1 \end{cases}$$

Clearly, $f''(a) \neq 1$ only when $a \in [-2, 1]$ ...(i)

Since, vectors are parallel.

$$\therefore \quad \frac{a}{1} = \frac{-b^2}{3b} \Rightarrow b^2 + 3ab = 0 \quad \text{...(ii)}$$

Let $g(a) = b^2 + 3ab$ must have atleast one root in $[-2, 1]$.

$$\therefore \quad g(-2) \cdot g(1) \le 0$$

$$\Rightarrow \quad (b^2 - 6b)(b^2 + 3b) \le 0$$

$$\Rightarrow \quad b^2(b-6)(b+3) \le 0$$

$+$ $\quad -3 \quad$ $-$ $\quad 6 \quad$ $+$

$$\therefore \quad b \in [-3, 6] \cup \{0\}$$

$\Rightarrow$ Number of integral values of 'b' can be 11.

Now, let $\quad h(b) = 1 - 8b - b^2$

$$h'(b) = -8 - 2b = 0$$

$$\Rightarrow \quad b = -4$$

$\therefore$ $h(b)$ is decreasing, when $b > -4$

$$\Rightarrow \quad (h(b)_{\max}) \text{ at } b = -3$$

$$\therefore \quad (h(-3))_{\max} = 16$$

22. (c)
23. (d)

JEE Type Solved Examples :
Matching Type Questions

- **Ex. 24** *Match the statements of Column I with values of Column II.*

	Column I		Column II
(A)	The sides of a triangle vary slightly in such a way that its circumradius remains constant, if $\frac{da}{\cos A}+\frac{db}{\cos B}+\frac{dc}{\cos C}+1=\lvert m\rvert$, then the value of m is	(p)	1
(B)	If the length of subtangent to the curve $x^2y^2=16$ at the point $(-2, 2)$ is $\lvert k\rvert$, then the value of k is	(q)	-1
(C)	If the curve $y=2e^{2x}$ intersects the Y-axis at an angle $\cot^{-1}\lvert(8n-4)/3\rvert$, then the value of n is	(r)	2
(D)	If the area of a triangle formed by normal at the point $(1, 0)$ on the curve $x=e^{\sin y}$ with axes is $\lvert 2t+1\rvert/6$ sq units, then the value of t is	(s)	-2

Sol. (A) → (p, q); (B) → (r, s); (C) → (r, q); (D) → (p, s)

(A) We know that in any ΔABC,

$$\frac{a}{\sin A}=\frac{b}{\sin B}=\frac{c}{\sin C}=2R$$

$$\Rightarrow \quad a=2R\sin A,\, b=2R\sin B,\, c=2R\sin C$$

$$\Rightarrow \quad \frac{da}{dA}=2R\cos A,\frac{db}{dB}=2R\cos B,\frac{dc}{dC}=2R\cos C$$

$$\Rightarrow \quad \frac{da}{\cos A}+\frac{db}{\cos B}+\frac{dc}{\cos C}=2R(dA+dB+dC)$$

Also, $A+B+C=\pi$

$\therefore \quad dA+dB+dC=0$

Hence, $$\frac{da}{\cos A}+\frac{db}{\cos B}+\frac{dc}{\cos C}=0$$

or $$\frac{da}{\cos A}+\frac{db}{\cos B}+\frac{dc}{\cos C}+1=1$$

$$\Rightarrow \quad \lvert m\rvert=1\Rightarrow m=\pm1$$

(B) We have, $x^2y^2=16$

$$\Rightarrow \quad 2x^2y\frac{dy}{dx}+2xy^2=0 \quad\Rightarrow\quad \frac{dy}{dx}=-y/x$$

$$\text{Length of subtangent}=\left|y\frac{dx}{dy}\right|=\left|-y\frac{x}{y}\right|=\lvert x\rvert$$

$\Rightarrow$ Length of subtangent = 2

$\therefore \quad k=\pm 2$

(C) We have $y=2e^{2x}$, which intersect Y-axis at $(0, 2)$

Now, $$\frac{dy}{dx}=4e^{2x}$$

$$\therefore \quad \left(\frac{dy}{dx}\right)_{(0,2)}=4$$

$\therefore$ Angle of intersection with Y-axis

$$=\frac{\pi}{2}-\tan^{-1}4=\cot^{-1}4$$

Hence $$\left|\frac{8n-4}{3}\right|=4,\ n=2 \text{ or } -1$$

(D) We have, $x=e^{\sin y}$

$$\Rightarrow \quad \frac{dx}{dy}=\cos y\, e^{\sin y}$$

Slope of normal at $(1, 0)=-1$

Equation of normal is $x+y=1$

Area of triangle $=\frac{1}{2}$

$$\therefore \quad \frac{\lvert 2t+1\rvert}{6}=\frac{1}{2}$$

$$\Rightarrow \quad t=1,-2$$

- **Ex. 25** *Match the statements of Column I with values of Column II.*

	Column I		Column II
(A)	Rolle's theorem in $[-1, 1]$ does not hold, if	(p)	$f(x)=\begin{cases}\left(\frac{1}{2}-x\right)^2, & x>\frac{1}{2}\\ \left(\frac{1}{2}-x\right), & x\le\frac{1}{2}\end{cases}$
(B)	LMVT in $[0, 1]$ does not hold, if	(q)	$f(x)=\lvert x\rvert$
(C)	Rolle's theorem in $[0, 1]$ does not hold, if	(r)	$f(x)=\begin{cases}1, & 0\le x<\frac{1}{2}\\ 2, & \frac{1}{2}\le x\le 1\end{cases}$
		(s)	$f(x)=\log\lvert x\rvert$
		(t)	$f(x)=x^2+x+1$

Sol. (A) → (p,q,r,s,t); (B) → (p, r, s); (C) → (p,r,s,t)

(p) $$f(x)=\begin{cases}\left(\dfrac{1}{2}-x\right)^2, & x>\dfrac{1}{2}\\ \left(\dfrac{1}{2}-x\right), & x\le\dfrac{1}{2}\end{cases}$$

$$\therefore\quad f'(x)=\begin{cases}-2\left(\dfrac{1}{2}-x\right), & x>\dfrac{1}{2}\\ -1, & x\le\dfrac{1}{2}\end{cases}$$

$$f'\left(\frac{1}{2}^+\right)\ne f'\left(\frac{1}{2}^-\right)$$

$\Rightarrow f(x)$ is not differentiable at $x=\dfrac{1}{2}$.

(q) $f(x)=|x|$ is not differential at $x=0$.

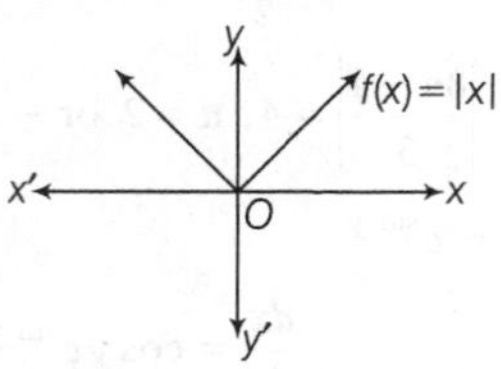

(r) $$f(x)=\begin{cases}1, & 0\le x<\dfrac{1}{2}\\ 2, & \dfrac{1}{2}\le x<1\end{cases}$$

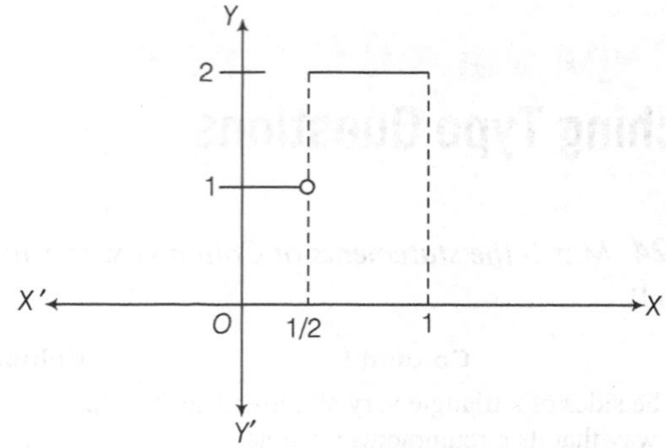

$\Rightarrow f(x)$ is discontinuous at $x=\dfrac{1}{2}$.

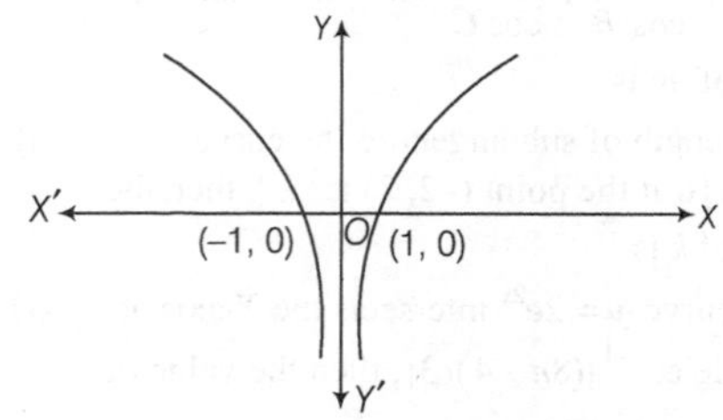

$f(x)=\log|x|$ is discontinuous at $x=0$.

(t) $f(x)=x^2+x+1$

$\Rightarrow\quad f(-1)=1$

and $\quad f(1)=3$

$\Rightarrow\quad f(-1)\ne f(1)$

JEE Type Solved Examples : Single Integer Answer Type Questions

Ex. 26 *Suppose that $f(0)=-3$ and $f'(x)\le 5$ for all values of x. Then, the largest value which $f(2)$ can assume is ....*

Sol. (7) Using LMVT in [0, 2]

$$\frac{f(2)-f(0)}{2-0}=f'(c)\qquad [\text{where } c\in(0,2)]$$

$$\Rightarrow\quad \frac{f(2)+3}{2}\le 5$$

$$\Rightarrow\quad f(2)\le 7$$

Ex. 27 *Let C be the curve $y=x^3$ (where x assumes all real values). The tangent at A meets the curve again at B. If the gradient at B is k times the gradient at A, then k is equal to ....*

Sol. (4) $$\frac{dy}{dx}=3x^2=3t^2 \text{ at 'A'}$$

$$\therefore\quad 3t^2=\frac{T^3-t^3}{T-t}=T^2+Tt+t^2$$

$$\Rightarrow\quad T^2+Tt-2t^2=0$$

$$\Rightarrow\quad (T-t)(T+2t)=0$$

$$\Rightarrow\quad T=t \text{ or } T=-2t\qquad [T=t \text{ is not possible}]$$

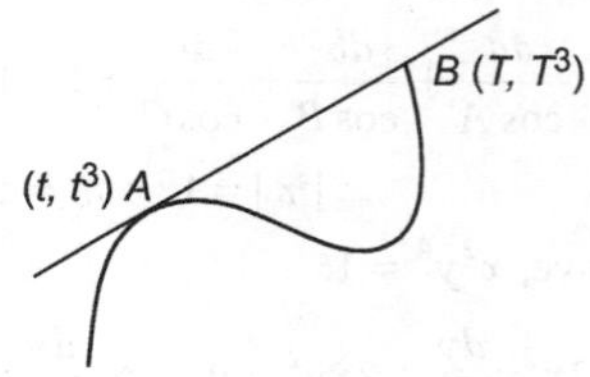

Now, $$m_A=\frac{t^3}{t}=t^2;\ m_B=T^2$$

$$\Rightarrow\quad \frac{m_B}{m_A}=\frac{T^2}{t^2}=\frac{4t^2}{t^2}\qquad [\text{using } T=-2t]$$

$$\therefore\quad k=4$$

Ex. 28 *Consider the two graphs $y = 2x$ and $x^2 - xy + 2y^2 = 28$. The absolute value of the tangent of the angle between the two curves at the points where they meet, is ....*

Sol. (2) $y = 2x,\ x^2 - xy + 2y^2 = 28$

Solving the point of intersection are (2, 4) and (– 2, – 4).

For 1st curve, $\frac{dy}{dx} = 2 = m_1$ ...(i)

For 2nd curve, $\frac{dy}{dx} = \frac{y - 2x}{4y - x} \Rightarrow m_2 = 0$ ...(ii)

$\therefore \quad \tan\theta = 2$

Ex. 29 *At the point $P(a, a^n)$ on the graph of $y = x^n$ $(n \in N)$ in the first quadrant a normal is drawn. The normal intersects the Y-axis at the point (0, b). If $\lim_{a \to 0} b = \frac{1}{2}$, then n equals......... .*

Sol. (2) $y = x^n \Rightarrow \frac{dy}{dx} = n\,x^{n-1} = na^{n-1}$ at $p(a, a^n)$

Slope of normal $= -\frac{1}{na^{n-1}}$

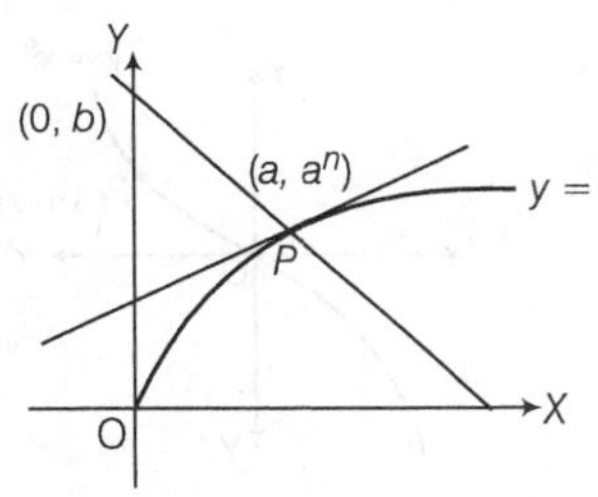

Equation of normal is $y - a^n = -\frac{1}{na^{n-1}}(x - a)$

Put $x = 0$ to get y-intercept, then $y = a^n + \frac{1}{na^{n-2}}$

Hence, $b = a^n + \frac{1}{na^{n-2}}$

Now, $\lim_{a \to 0} b = \begin{cases} 0, & \text{if } n < 2 \\ \frac{1}{2}, & \text{if } n = 2 \\ \infty, & \text{if } n > 2 \end{cases} \Rightarrow \lim_{a \to 0} b = \frac{1}{2}$, if $n = 2$

Ex. 30 *The maximum possible integral value of $\frac{\beta - \alpha}{\tan^{-1}\beta - \tan^{-1}\alpha}$, where $0 < \alpha < \beta < \sqrt{3}$ is*

Sol. (3) Let $f(x) = \tan^{-1} x,\ 0 < \alpha < \beta < \sqrt{3}$.

Then, by using LMVT, $\frac{\tan^{-1}\beta - \tan^{-1}\alpha}{\beta - \alpha} = \frac{1}{1 + c^2}$,

where $0 < \alpha < c < \beta < \sqrt{3}$

So, $\frac{1}{4} < \frac{1}{1 + c^2} < \frac{1}{1}$

$\Rightarrow \quad \frac{1}{4} < \frac{\tan^{-1}\beta - \tan^{-1}\alpha}{\beta - \alpha} < 1$

or $\quad 1 < \frac{\beta - \alpha}{\tan^{-1}\beta - \tan^{-1}\alpha} < 4$

$\therefore$ Maximum possible integral value is 3.

Ex. 31 *Let P be a point on the curve $c_1 : y = \sqrt{2 - x^2}$ and Q be a point on the curve $c_2 : xy = 9$, both P and Q be in the first quadrant. If d denotes the minimum distance between P and Q, then d^2 is, ...*

Sol. (8) Here, c_1 is a semi-circle and c_2 is a rectangular hyperbola. PQ will be minimum if the normal at P on semi-circle is also a normal at Q on $xy = 9$.

Let the normal at P be $y = mx\ (m > 0)$. ...(i)

Solving with $xy = 9$

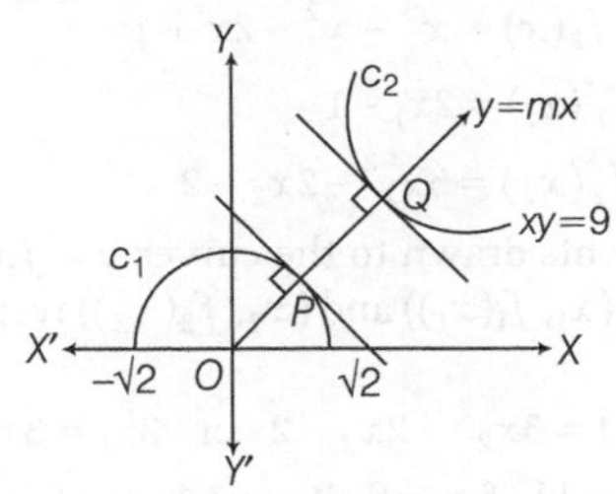

$mx^2 = 9 \Rightarrow x = 3/\sqrt{m}$

$\Rightarrow \quad y = 9\sqrt{m}/3$

$\therefore \quad Q(3/\sqrt{m}, 3\sqrt{m})$

Differentiating $xy = 9 \Rightarrow x\,dy/dx + y = 0$

$\Rightarrow \quad dy/dx = -y/x$

$\therefore \quad |dy/dx|_Q = -\frac{3\sqrt{m}\sqrt{m}}{3} = -m$

Now, tangent at P and Q must be parallel.

$\therefore \quad -m = -1/m \Rightarrow m^2 = 1 \Rightarrow m = 1$

$\therefore$ Normal at P and Q is $y = x$

Now, we get $P(1, 1)$ and $Q(3, 3)$.

$\therefore \quad (PQ)^2 = d^2 = 4 + 4 = 8$

Subjective Type Questions

● **Ex. 32** *If the line $x \cos\alpha + y \sin\alpha = P$ is the normal to the curve $(x+a)y = c^2$, then show that*

$$\alpha \in \left(2n\pi + \frac{\pi}{2}, (2n+1)\pi\right) \cup \left(2n\pi + \frac{3\pi}{2}, (2n+2)\pi\right).$$

Sol. Here, $y = \dfrac{c^2}{x+a} \Rightarrow \dfrac{dy}{dx} = -\dfrac{c^2}{(x+a)^2}$

Slope of normal is $\dfrac{(x+a)^2}{c^2} > 0$ for all x.

$\because$ $x \cos\alpha + y \sin\alpha = P$ is normal

$\therefore$ $-\dfrac{\cos\alpha}{\sin\alpha} > 0 \Rightarrow \cot\alpha < 0$

i.e. α lies in II or IV quadrant.

So, $\alpha \in \left(2n\pi + \dfrac{\pi}{2}, (2n+1)\pi\right) \cup \left(2n\pi + \dfrac{3\pi}{2}, (2n+2)\pi\right)$

● **Ex. 33** *Find the total number of parallel tangents of $f_1(x) = x^2 - x + 1$ and $f_2(x) = x^3 - x^2 - 2x + 1$.*

Sol. Here, $f_1(x) = x^2 - x + 1$

and $f_2(x) = x^3 - x^2 - 2x + 1$

$\Rightarrow$ $f_1'(x_1) = 2x_1 - 1$

and $f_2'(x_2) = 3x_2^{\ 2} - 2x_2 - 2$

Let the tangents drawn to the curves $y = f_1(x)$ and $y = f_2(x)$ at $(x_1, f_1(x_1))$ and $(x_2, f_2(x_2))$ respectively are parallel.

$\therefore$ $2x_1 - 1 = 3x_2^{\ 2} - 2x_2 - 2$ or $2x_1 = 3x_2^{\ 2} - 2x_2 - 1$

Which is possible for infinite numbers of ordered pairs.

$\therefore$ It will have infinite number of parallel tangents.

● **Ex. 34** *Find the point on the curve $3x^2 - 4y^2 = 72$, which is nearest to the line $3x + 2y + 1 = 0$.*

Sol. Slope of the given line $3x + 2y + 1 = 0$ is $(-3/2)$.

Let us locate the point (x_1, y_1) on the curve at which the tangent is parallel to the given line.

Differentiating the curve both sides with respect to x, we get

$$6x - 8y\frac{dy}{dx} = 0$$

$\Rightarrow$ $\left(\dfrac{dy}{dx}\right)_{(x_1, y_1)} = \dfrac{3x_1}{4y_1} = \dfrac{-3}{2}$ [$\because$ parallel to $3x + 2y = 1$]

Also, the point (x_1, y_1) lies on $3x^2 - 4y^2 = 72$

$\therefore$ $3x_1^2 - 4y_1^2 = 72$

$\Rightarrow$ $3\dfrac{x_1^2}{y_1^2} - 4 = \dfrac{72}{y_1^2} \Rightarrow 3(4) - 4 = \dfrac{72}{y_1^2}$ $\left[\text{as } \dfrac{x_1}{y_1} = -2\right]$

$\Rightarrow$ $y_1^2 = 9 \Rightarrow y_1 = \pm 3$

Required points are $(-6, 3)$ and $(6, -3)$.

Distance of $(-6, 3)$ from the given line

$$= \frac{|-18 + 6 + 1|}{\sqrt{13}} = \frac{11}{\sqrt{13}}$$

and distance of $(6, -3)$ from the given line

$$= \frac{|18 - 6 + 1|}{\sqrt{13}} = \frac{13}{\sqrt{13}} = \sqrt{13}$$

Thus, $(-6, 3)$ is the required point.

● **Ex. 35** *Tangent at a point P_1 (other than $(0, 0)$) on the curve $y = x^3$ meets the curve again at P_2. The tangent at P_2 meets the curve at P_3 and so on, show that the abscissa of $P_1, P_2, P_3, \ldots, P_n$, form a GP. Also, find the ratio of $\dfrac{ar(\Delta P_1P_2P_3)}{ar(\Delta P_2P_3P_4)}$.*

Sol. Let $P_1(x_1, y_1)$ be a point on the curve.

$$y = x^3 \quad \text{...(i)}$$

Then, we have $y_1 = x_1^3$ ...(ii)

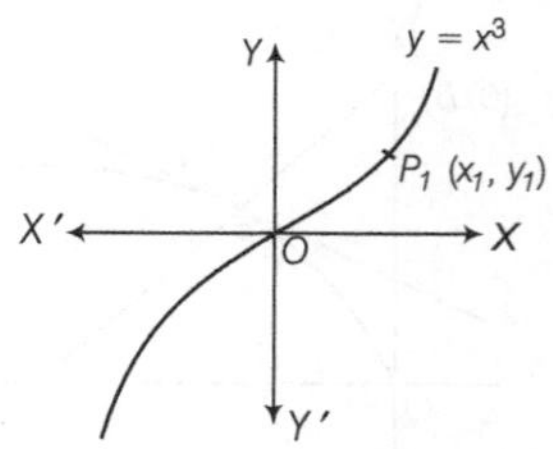

Now, $\dfrac{dy}{dx} = 3x^2$

$\therefore$ Slope of the tangent at $P_1 = m_1 = 3x_1^2$

$\therefore$ Equation of the tangent at $P_1(x_1, y_1)$ is

$$y - x_1^3 = 3x_1^2(x - x_1) \text{ or } y = 3x_1^2 x - 2x_1^3 \quad \text{...(iii)}$$

Solving Eqs. (i) and (iii), we get

$$x^3 - 3x_1^2 x + 2x_1^3 = 0$$

i.e. $(x - x_1)\cdot(x^2 + xx_1 - 2x_1^2) = 0$

i.e. $(x - x_1)(x - x_1)(x + 2x_1) = 0$

$\therefore$ $x = x_1$ (neglecting) or $x = -2x_1$

$\therefore$ $x_2 = -2x_1, y_2 = x_2^3 = -8x_1^3$

$\therefore$ $P_2(x_2, y_2) = (-2x_1, -8x_1^3)$

Now, we find P_3, the point where the tangent, at P_2 meets the curve.

Slope of the tangent at $P_2 = \left(\dfrac{dy}{dx}\right)_{(x_2, y_2)}$

$$= 3x_2^2 = 3\cdot 4x_1^2 = 12x_1^2$$

$\therefore$ Equation of tangent at P_2 is,

$$y - x_2^3 = 3x_2^2(x - x_2) \quad \text{...(iv)}$$

To get, $P_3 = (x_3, y_3)$, solve Eqs. (i) and (iv), we get

$$P_3 = (x_3, y_3) = (-2x_2, -8x_2^3) = (4x_1, 64x_1^3) \text{ and so on.}$$

$\therefore$ Abscissa of $P_1, P_2, P_3, \ldots$ are given by $x_1, -2x_1, 4x_1, -8x_1, \ldots$, which forms an GP with common ratio $= -2$.

Now, $$\text{a}(\Delta P_1P_2P_3) = \frac{1}{2}\begin{vmatrix} x_1 & y_1 & 1 \\ x_2 & y_2 & 1 \\ x_3 & y_3 & 1 \end{vmatrix}$$

$$\Rightarrow \text{ar}(\Delta P_1P_2P_3) = \frac{1}{2}\begin{vmatrix} x_1 & x_1^3 & 1 \\ -2x_1 & -8x_1^3 & 1 \\ 4x_1 & 64x_1^3 & 1 \end{vmatrix}$$

$$\Rightarrow \text{ar}(\Delta P_1P_2P_3) = \frac{x_1^4}{2}\begin{vmatrix} 1 & 1 & 1 \\ -2 & -8 & 1 \\ 4 & 64 & 1 \end{vmatrix}$$

Similarly, $$\text{ar}(\Delta P_2P_3P_4) = \frac{1}{2}\begin{vmatrix} x_2 & y_2 & 1 \\ x_3 & y_3 & 1 \\ x_4 & y_4 & 1 \end{vmatrix}$$

$$\Rightarrow \text{ar}(\Delta P_2P_3P_4) = \frac{1}{2}\begin{vmatrix} -2x_1 & -8x_1^3 & 1 \\ 4x_1 & 64x_1^3 & 1 \\ -8x_1 & -512x_1^3 & 1 \end{vmatrix}$$

$$\Rightarrow \text{ar}(\Delta P_2P_3P_4) = 8x_1^4\begin{vmatrix} 1 & 1 & 1 \\ -2 & -8 & 1 \\ 4 & 64 & 1 \end{vmatrix}$$

$$\therefore \frac{\text{ar}(\Delta P_1P_2P_3)}{\text{ar}(\Delta P_2P_3P_4)} = \frac{1}{16}$$

Ex. 36 *Determine all polynomial $P(x)$ with rational coefficient so that for all x with $|x| \le 1$;*

$$P(x) = P\left(\frac{-x + \sqrt{3 - 3x^2}}{2}\right).$$

Sol. Here, $P(x) = P\left(\frac{-x + \sqrt{3 - 3x^2}}{2}\right)$ for $|x| \le 1$ ...(i)

Let $x = 0 \Rightarrow P(0) = P(\sqrt{3}/2)$

Which shows, $P(x) - P(0)$ is divisible by $x\left(x - \frac{\sqrt{3}}{2}\right)$.

Since, $P(x) - P(0)$ has rational coefficients and $\frac{\sqrt{3}}{2}$ is one of the roots.

$\therefore$ $-\frac{\sqrt{3}}{2}$ is also a root of $P(x) - P(0)$.

Thus, $x\left(x - \frac{\sqrt{3}}{2}\right)\left(x + \frac{\sqrt{3}}{2}\right) = x^3 - \frac{3}{4}x = \frac{4x^3 - 3x}{4}$ is factor of $P(x) - P(0)$.

$$\therefore \quad P(x) = P(0) + (3x - 4x^3)P_1(x) \quad \text{for } |x| \le 1 \quad \text{...(ii)}$$

As, $$P(x) = P\left(\frac{-x + \sqrt{3 - 3x^2}}{2}\right)$$

$$\Rightarrow 3\left(\frac{-x + \sqrt{3 - 3x^2}}{2}\right) - 4\left(\frac{-x + \sqrt{3 - 3x^2}}{2}\right) = 3x - 4x^3$$

$$\therefore \quad P_1(x) = P_1\left(\frac{-x + \sqrt{3 - 3x^2}}{2}\right)$$

$$\therefore \quad P_1(x) = P_1(0) + (3x - 4x^3)P_2(x) \quad \text{[using Eq. (ii)]}$$

$$\Rightarrow P(x) = P(0) + (3x - 4x^3)(P_1(0)) + (3x - 4x^3)P_2(x))$$

$$= P(0) + (3x - 4x^3)P_1(0) + (3x - 4x^3)^2 P_2(x)$$

Thus, in general,

$$P(x) = a_0 + a_1(3x - 4x^3) + a_2(3x - 4x^3)^2 + \ldots + (3x - 4x^3)^k \cdot k(x)$$

Where $k(x)$ is a polynomial with rational coefficient.

Ex. 37 *Let $f(x) = (x - a)(x - b)(x - c), a < b < c$. Show that $f'(x) = 0$ has two roots one belonging to (a, b) and other belonging to (b, c).*

Sol. Here, $f(x)$ being a polynomial, is continuous and differentiable for all real values of x.

We also have, $f(a) = f(b) = f(c)$

If we apply Rolle's theorem to $f(x)$ in $[a, b]$ and $[b, c]$ we observe that $f'(x) = 0$ would have atleast one root in (a, b) and atleast one root in (b, c).

But $f'(x) = 0$ is a polynomial of degree two, hence $f'(x) = 0$ cannot have more than two roots.

It implies that exactly one root of $f'(x) = 0$ would lie in (a, b) and exactly one root of $f'(x) = 0$ would lie in (b, c).

Ex. 38 *If at each point of the curve $y = x^3 - ax^2 + x + 1$, the tangents is inclined at an acute angle with the positive direction of the X-axis, then find the interval of a.*

Sol. As, $y = x^3 - ax^2 + x + 1$ and the tangent is inclined at an acute angle with the positive direction of X-axis.

$$\therefore \quad \frac{dy}{dx} \ge 0 \Rightarrow 3x^2 - 2ax + 1 \ge 0, \text{ for all } x \in R$$

[we know, $ax^2 + bx + c \ge 0$ for all $x \in R$ $\Rightarrow a > 0$ and $D \le 0$]

$$\Rightarrow (2a)^2 - 4(3)(1) \le 0 \Rightarrow 4(a^2 - 3) \le 0$$

$$\Rightarrow (a - \sqrt{3})(a + \sqrt{3}) \le 0 \quad \therefore -\sqrt{3} \le a \le \sqrt{3}$$

Ex. 39 *Show that there is no cubic curve for which the tangent lines at two distinct points coincide.*

Sol. Suppose $y \equiv ax^3 + bx^2 + cx + d = 0\ (a \ne 0)$ be a cubic curve.

We assume that (x_1, y_1) and $(x_2, y_2), (x_1 < x_2)$ are two distinct points on the curve at which tangents coincide.

Then, by mean value theorem, there exists $x_3\ (x_1 < x_3 < x_2)$ such that

$$\frac{y_2 - y_1}{x_2 - x_1} = y'(x_3)$$

Since, tangent x_1, x_2, x_3 are solutions of equation

$$3ax^2 + 2bx + c = M$$

But, it is a quadratic and thus cannot have more than two roots. Therefore, no such cubic is possible.

Ex. 40 *The tangent at a point P to the rectangular hyperbola $xy = c^2$ meets the lines $x - y = 0, x + y = 0$ at Q and R, Δ_1 is the area of the Δ OQR, where O is the origin. The normal at P meets the X-axis at M and Y-axis at N. If Δ_2 is the area of the ΔOMN, show that Δ_2 varies inversely as the square of Δ_1.*

Sol. Tangent at $P(x_1, y_1)$ is $xy_1 + yx_1 = 2c^2$

The point of intersection of this line with $x - y = 0$ is given by,

$$x(x_1 + y_1) = 2c^2, \text{ i.e. } x = 2c^2/(x_1 + y_1)$$

$$\therefore \quad Q \text{ is } \left(\frac{2c^2}{x_1 + y_1}, \frac{2c^2}{x_1 + y_1}\right)$$

The point of intersection of the tangent with $x + y = 0$ is given by,

$$x(y_1 - x_1) = 2c^2, \quad x = \frac{2c^2}{y_1 - x_1}$$

$$\therefore R \text{ is } \left(\frac{2c^2}{y_1 - x_1}, \frac{2c^2}{x_1 - y_1}\right)$$

Now, $\quad \Delta_1 = \dfrac{1}{2}\{a_1b_2 - a_2b_1\} = \dfrac{1}{2}\{-a_1a_2 - a_2a_1\} = -a_1a_2$

$$\Delta_1 = \frac{4c^4}{x_1^2 - y_1^2}$$

The normal at (x_1, y_1) is $y - y_1 = \dfrac{x_1}{y_1}(x - x_1)$

Intersection with $y = 0$ is given by, $x - x_1 = -\dfrac{y_1^2}{x_1}$

$$\Rightarrow \quad x = \frac{x_1^2 - y_1^2}{x_1} \quad \therefore \text{ M is } \left(\frac{x_1^2 - y_2^2}{x_1}, 0\right)$$

intersection with $x = 0$ is given by $\quad y - y_1 = -\dfrac{x_1^2}{y_1}$

$$\Rightarrow \quad y = \frac{y_1^2 - x_1^2}{y_1}. \quad \therefore N \text{ is } \left(0, \frac{y_1^2 - x_1^2}{y_1}\right)$$

Now, $\quad \Delta_2 = \dfrac{1}{2}\dfrac{(x_1^2 - y_1^2)}{x_1}\cdot\dfrac{(y_1^2 - x_1^2)}{y_1}$

$$= -\frac{1}{2}\frac{(x_1^2 - y_1^2)^2}{x_1y_1} = -\frac{1}{2}\frac{(x_1^2 - y_1^2)^2}{c^2}$$

$$\therefore \Delta_1^2\,\Delta_2 = \frac{16c^8}{(x_1^2 - y_1^2)^2}\left(-\frac{1}{2}\right)\frac{(x_1^2 - y_1^2)^2}{c^2} = -8c^6 = \text{constant}$$

$$\Rightarrow \quad \Delta_2 \propto \frac{1}{\Delta_1^2}$$

or Δ_2 varies inversely as the square of Δ_1.

Ex. 41 *If the function of $f : [0, 4] \to R$ is differentiable, then show that*

(i) $(f(4))^2 - (f(0))^2 = 8f'(a)f(b)$ for some $a, b \in (0, 4)$.

(ii) $\int_0^4 f(t)\,dt = 2\{\alpha f(\alpha^2) + \beta f(\beta^2)\}$ for some $0 < \alpha, \beta < 2$.

[IIT JEE 2003]

Sol. (i) Here, f is differentiable $\Rightarrow f$ is also continuous.

Now, by Lagrange's mean value theorem, there exist $a \in (0, 4)$ such that

$$f'(a) = \frac{f(4) - f(0)}{4 - 0} = \frac{f(4) - f(0)}{4} \quad \text{...(i)}$$

Also, by intermediate mean value theorem, there exists $b \in (0, 4)$ such that

$$f(b) = \frac{f(4) + f(0)}{2} \quad \text{...(ii)}$$

From Eqs. (i) and (ii), we get

$$f'(a)\,f(b) = \frac{f(4) - f(0)}{4} \times \frac{f(4) + f(0)}{2}$$

$$f'(a)\,f(b) = \frac{(f(4))^2 - (f(0))^2}{8}$$

$$\Rightarrow (f(4))^2 - (f(0))^2 = 8f'(a)f(b) \text{ for some, } a, b \in (0, 4)$$

(ii) Putting $t = z^2$, we have

$$\int_0^4 f(t)\,dt = \int_0^2 2z\,f(z^2)\,dz$$

Consider the function $\phi(t)$ given by,

$$\phi(t) = \int_0^t 2z\,f(z^2)\,dz; t \in [0, 2]$$

Clearly, $\phi(t)$ being an integral function of a continuous function, is continuous and differentiable on [0, 2].

$\therefore$ By Lagrange's mean value theorem, there exists $c \in (0, 2)$ such that $\dfrac{\phi(2) - \phi(0)}{2 - 0} = \phi'(c)$

$$\Rightarrow \quad \frac{\int_0^2 2z\,f(z^2)\,dz - \int_0^0 2z\,f(z^2)\,dz}{2 - 0} = 2cf(c^2)$$

[using $\phi'(t) = 2t\,f(t^2)$]

$$\Rightarrow \quad \int_0^2 2z\,f(z^2)\,dz = 4cf(c^2) \quad \text{...(i)}$$

Also, by intermediate mean value theorem for $c \in (0, 2)$, there exist $\alpha, \beta \in (0, 2)$ such that

$$\frac{\phi'(\alpha) + \phi'(\beta)}{2} = \phi'(c), \text{ where } 0 < \alpha < c < \beta < 2$$

$$\Rightarrow \quad 2\alpha f(\alpha^2) + 2\beta\,f(\beta^2) = 2\{2c\,f(c^2)\} \quad \text{...(ii)}$$

From Eqs. (i) and (ii), we get

$$\int_0^2 2z\,f(z^2)\,dz = 2\alpha\,f(\alpha^2) + 2\beta\,f(\beta^2),$$

for some $0 < \alpha, \beta < 2$

$$\Rightarrow \quad \int_0^4 f(t)\,dt = 2(\alpha\,f(\alpha^2) + \beta\,f(\beta^2)),$$

for some $0 < \alpha, \beta < 2$

dy/dx as Rate Measurer and Tangents, Normals Exercise 1 : Single Option Correct Type Questions

1. Consider the cubic equation $f(x) = x^3 - nx + 1 = 0$ where $n \geq 3, n \in N$, then $f(x) = 0$ has

(a) at least one root in (0, 1) (b) at least one root in (1, 2)
(c) at least one root in (−1, 0) (d) data insufficient

2. If the normal to $y = f(x)$ at (0, 0) is given by $y - x = 0$, then $\lim_{x \to 0} \dfrac{x^2}{f(x^2) - 20f(9x^2) + 2f(99x^2)}$ is

(a) 1/19 (b) −1/19 (c) 1/2 (d) does not exist

3. Tangent to a curve intersects the Y-axis at a point. A line perpendicular to this tangent through P passes through another point (1, 0). The differential equation of the curve is

(a) $y\dfrac{dy}{dx} - x\left(\dfrac{dy}{dx}\right)^2 = 1$ (b) $x\dfrac{d^2y}{dx^2} + \left(\dfrac{dy}{dx}\right)^2 = 1$
(c) $y\dfrac{dy}{dx} + x = 1$ (d) None of these

4. The number of points in the rectangle $\{(x, y): |x| \leq 9, |y| \leq 3\}$ which lie on the curve $y^2 = x + \sin x$ and at which the tangent to the curve is parallel to X-axis, is

(a) 3 (b) 2 (c) 4 (d) None of these

5. If $a, b, c, d \in R$ such that $\dfrac{a + 2c}{b + 3d} = -\dfrac{4}{3}$, then the equation $ax^3 + bx^2 + cx + d = 0$ has

(a) at least one root in (−1, 0) (b) at least one root in (0, 1)
(c) no root in (−1, 1) (d) no root in (0, 2)

6. If $3(a + 2c) = 4(b + 3d) \neq 0$, then the equation $ax^3 + bx^2 + cx + d = 0$ will have

(a) no real solution
(b) at least one real root in (− 1, 0)
(c) at least one real root in (0, 1)
(d) None of the above

7. Let $f(x)$ be a differentiable function in the interval (0, 2), then the value of $\int_0^2 f(x)\,dx$

(a) $f(c)$ where $c \in (0, 2)$ (b) $2f(c)$, where $c \in (0, 2)$
(c) $f'(c)$, where $c \in (0, 2)$ (d) None of these

8. Let $f(x)$ be a fourth differentiable function such that $f(2x^2 - 1) = 2xf(x), \forall x \in R$, then $f^{iv}(0)$ is equal to (where $f^{iv}(0)$ represents fourth derivative of $f(x)$ at $x = 0$)

(a) 0 (b) 1
(c) − 1 (d) data insufficient

9. The curve $x + y - \ln(x + y) = 2x + 5$ has a vertical tangent at the point (α, β). Then, $\alpha + \beta$ is equal to

(a) − 1 (b) 1 (c) 2 (d) − 2

10. Let $y = f(x)$, $f: R \to R$ be an odd differentiable function such that $f'''(x) > 0$ and
$g(\alpha, \beta) = \sin^8 \alpha + \cos^8 \beta + 2 - 4 \sin^2 \alpha \cos^2 \beta$.
If $f''(g(\alpha, \beta)) = 0$, then $\sin^2 \alpha + \sin^2 \beta$ is equal to

(a) 0 (b) 1 (c) 2 (d) 3

11. A polynomial of 6th degree $f(x)$ satisfies $f(x) = f(2 - x), \forall x \in R$, if $f(x) = 0$ has 4 distinct and 2 equal roots, then sum of the roots of $f(x) = 0$ is

(a) 4 (b) 5 (c) 6 (d) 7

12. Let a curve $y = f(x), f(x) \geq 0, \forall x \in R$ has property that for every point P on the curve length of subnormal is equal to abscissa of P. If $f(1) = 3$, then $f(4)$ is equal to

(a) $-2\sqrt{6}$ (b) $2\sqrt{6}$ (c) $3\sqrt{5}$ (d) None of these

13. If a variable tangent to the curve $x^2y = c^3$ makes intercepts, a, b on X and Y-axes respectively, then the value of a^2b is

(a) $27c^3$ (b) $\dfrac{4}{27}c^3$
(c) $\dfrac{27}{4}c^3$ (d) $\dfrac{4}{9}c^3$

14. Let $f(x) = \begin{vmatrix} 1 & 1 & 1 \\ 3 - x & 5 - 3x^2 & 3x^3 - 1 \\ 2x^2 - 1 & 3x^5 - 1 & 7x^8 - 1 \end{vmatrix}$. Then, the equation $f(x) = 0$ has

(a) no real root
(b) atmost one real root
(c) at least two real roots
(d) exactly one real root in (0, 1) and no other real root

15. The graphs $y = 2x^3 - 4x + 2$ and $y = x^3 + 2x - 1$ intersect at exactly 3 distinct points. The slope of the line passing through two of these points is

(a) equal to 4 (b) equal to 6
(c) equal to 8 (d) not unique

16. In which of the following functions Rolle's theorem is applicable?

(a) $f(x) = \begin{cases} x, & 0 \leq x < 1 \\ 0, & x = 1 \end{cases}$ on [0, 1]

(b) $f(x) = \begin{cases} \dfrac{\sin x}{x}, & -\pi \leq x < 0 \\ 0, & x = 0 \end{cases}$ on $[-\pi, 0]$

(c) $f(x)=\dfrac{x^2-x-6}{x-1}$ on $[-2, 3]$

(d) $f(x)=\begin{cases}\dfrac{x^3-2x^2-5x+6}{x-1}, & \text{if } x \neq 1 \\ -6, & \text{if } x=1\end{cases}$ on $[-2, 3]$

17. The figure shows a right triangle with its hypotenuse OB along the Y-axis and its vertex A on the parabola $y=x^2$.

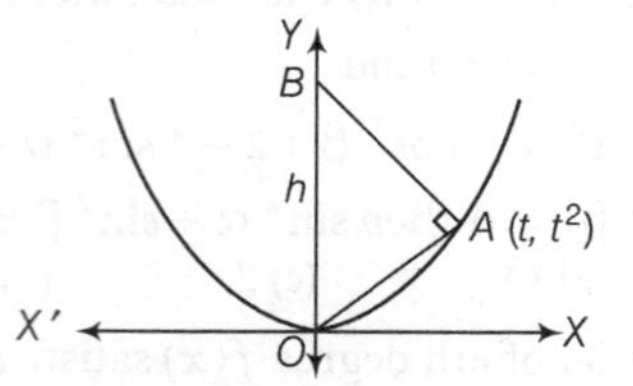

Let h represents the length of the hypotenuse which depends on the x-coordinate of the point A. The value of $\lim\limits_{t \to 0}(h)$ is equal to

(a) 0 (b) 1/2 (c) 1 (d) 2

18. Number of positive integral value(s) of 'a' for which the curve $y=a^x$ intersects the line $y=x$ is

(a) 0 (b) 1 (c) 2 (d) more than 2

19. If $f(x)=4-\left(\dfrac{1}{2}-x\right)^{2/3}$, $g(x)=\begin{cases}\dfrac{\tan[x]}{x}, & x \neq 0 \\ 1, & x=0\end{cases}$

$h(x)=\{x\}$, $k(x)=5^{\log_2(x+3)}$, where $[x]$ and $\{x\}$ denote the greatest integer and fraction part function, then in $[0, 1]$, Lagrange's mean value theorem is not applicable to

(a) f, g, h (b) h, k (c) f, g (d) g, h, k

20. If the function $f(x)=x^4+bx^2+8x+1$ has a horizontal tangent and a point of inflection for the same value of x, then the value of b is equal to

(a) -1 (b) 1 (c) 6 (d) -6

21. Coffee is coming out from a conical filter, with height and diameter both are 15 cm into a cylindrical coffee pot with a diameter 15 cm. The rate at which coffee comes out from the filter into the pot is 100 cu cm/min.

The rate (in cm/min) at which the level in the pot is rising at the instance when the coffee in the pot is 10 cm, is

(a) $\dfrac{9}{16\pi}$ (b) $\dfrac{25}{9\pi}$ (c) $\dfrac{5}{3\pi}$ (d) $\dfrac{16}{9\pi}$

22. A horse runs along a circle with a speed of 20 km/h. A lantern is at the centre of the circle. A fence is there along the tangent to the circle at the point at which the horse starts. The speed with which the shadow of the horse moves along the fence at the moment when it covers 1/8 of the circle (in km/h) is

(a) 20 (b) 40 (c) 30 (d) 60

23. Water runs into an inverted conical tent at the rate of 20 cu ft/min and leaks out at the rate of 5 cu ft/min. The height of the water is three times the radius of the water's surface. The radius of the water surface is increasing when the radius is 5 ft, is

(a) $\dfrac{1}{5\pi}$ ft/min (b) $\dfrac{1}{10\pi}$ ft/min (c) $\dfrac{1}{15\pi}$ ft/min (d) None of these

24. Let $f(x)=x^3-3x^2+2x$. If the equation $f(x)=k$ has exactly one positive and one negative solution, then the value of k equals

(a) $-\dfrac{2\sqrt{3}}{9}$ (b) $-\dfrac{2}{9}$

(c) $\dfrac{2}{3\sqrt{3}}$ (d) $\dfrac{1}{3\sqrt{3}}$

25. The x-intercept of the tangent at any arbitrary point of the curve $\dfrac{a}{x^2}+\dfrac{b}{y^2}=1$ is proportional to

(a) square of the abscissa of the point of tangency

(b) square root of the abscissa of the point of tangency

(c) cube of the abscissa of the point of tangency

(d) cube root of the abscissa of the point of tangency

26. If $f(x)$ is continuous and differentiable over $[-2, 5]$ and $-4 \le f'(x) \le 3$ for all x in $(-2, 5)$, then the greatest possible value of $f(5)-f(-2)$ is

(a) 7 (b) 9

(c) 15 (d) 21

27. A curve is represented parametrically by the equations $x=t+e^{at}$ and $y=-t+e^{at}$, where $t \in R$ and $a>0$. If the curve touches the axis of x at the point A, then the coordinates of the point A are

(a) $(1, 0)$ (b) $(1/e, 0)$

(c) $(e, 0)$ (d) $(2e, 0)$

28. At any two points of the curve represented parametrically by $x=a(2\cos t-\cos 2t)$, $y=a(2\sin t-\sin 2t)$, the tangents are parallel to the axis of x corresponding to the values of the parameter t differing from each other by

(a) $2\pi/3$ (b) $3\pi/4$ (c) $\pi/2$ (d) $\pi/3$

29. Let $F(x)=\displaystyle\int_{\sin x}^{\cos x} e^{(1+\sin^{-1}(t))^2}\,dt$ on $\left[0, \dfrac{\pi}{2}\right]$, then

(a) $F''(c)=0$ for all $c \in \left(0, \dfrac{\pi}{2}\right)$

(b) $F''(c)=0$ for some $c \in \left(0, \dfrac{\pi}{2}\right)$

(c) $F'(c)=0$ for some $c \in \left(0, \dfrac{\pi}{2}\right)$

(d) $F(c) \neq 0$ for all $c \in \left(0, \dfrac{\pi}{2}\right)$

30. If $f'(1)=1$ and $\frac{d}{dx}(f(2x))=f'(x), \forall\ x>0$. If $f'(x)$ is differentiable, then there exists a number $c\in(2,4)$ such that $f''(c)$ is equal to

(a) $-1/4$
(b) $-1/8$
(c) $1/4$
(d) $1/8$

31. Let $f(x)$ and $g(x)$ be two functions which are defined and differentiable for all $x\ge x_0$. If $f(x_0)=g(x_0)$ and $f'(x)>g'(x)$ for all $x>x_0$, then

(a) $f(x)<g(x)$ for some $x>x_0$
(b) $f(x)=g(x)$ for some $x>x_0$
(c) $f(x)>g(x)$ only for some $x>x_0$
(d) $f(x)>g(x)$ for all $x>x_0$

32. The range of values of m for which the line $y=mx$ and the curve $y=\frac{x}{x^2+1}$ enclose a region, is

(a) $(-1,1)$ (b) $(0,1)$
(c) $[0,1]$ (d) $(1,\infty)$

33. Let S be a square with sides of length x. If we approximate the change in size of the area of S by $h\cdot\left.\frac{dA}{dx}\right|_{x=x_0}$, when the sides are changed from x_0 to x_0+h, then the absolute value of the error in our approximation, is

(a) h^2 (b) $2hx_0$ (c) x_0^2 (d) h

34. Consider $f(x)=\int_1^x\left(t+\frac{1}{t}\right)dt$ and $g(x)=f'(x)$ for $x\in\left[\frac{1}{2},3\right]$. If P is a point on the curve $y=g(x)$ such that the tangent to this curve at P is parallel to a chord joining the points $\left(\frac{1}{2},g\left(\frac{1}{2}\right)\right)$ and $(3,g(3))$ of the curve, then the coordinates of the point P are

(a) can't be found out (b) $\left(\frac{7}{4},\frac{65}{28}\right)$
(c) $(1,2)$ (d) $\left(\sqrt{\frac{3}{2}},\frac{5}{\sqrt{6}}\right)$

dy/dx as Rate Measurer & Tangents, Normals Exercise 2 : More than One Option Correct Type Questions

35. For the curve represented parametrically by the equation, $x=2\log(\cot t)+1$ and $y=\tan t+\cot t$, then

(a) tangent at $t=\frac{\pi}{4}$ is parallel to X-axis
(b) normal at $t=\frac{\pi}{4}$ is parallel to Y-axis
(c) tangent at $t=\frac{\pi}{4}$ is parallel to $y=x$
(d) normal at $t=\frac{\pi}{4}$ is parallel to $y=x$

36. Consider the curve $f(x)=x^{1/3}$, then

(a) the equation of tangent at $(0,0)$ is $x=0$
(b) the equation of normal at $(0,0)$ is $y=0$
(c) normal to the curve does not exist at $(0,0)$
(d) $f(x)$ and its inverse meet at exactly 3 points

37. The angle at which the curve $y=ke^{kx}$ intersects Y-axis is

(a) $\tan^{-1}(k^2)$ (b) $\cot^{-1}(k^2)$
(c) $\sin^{-1}\left(\frac{1}{\sqrt{1+k^4}}\right)$ (d) $\sec^{-1}(\sqrt{1+k^4})$

38. Let $f(x)=8x^3-6x^2-2x+1$, then

(a) $f(x)=0$ has no root in $(0,1)$
(b) $f(x)=0$ has at least one root in $(0,1)$
(c) $f'(c)$ vanishes for some $c\in(0,1)$
(d) None of the above

39. If $f(0)=f(1)=f(2)=0$ and function $f(x)$ is twice differentiable in $(0,2)$ and continuous in $[0,2]$, then which of the following is/are definitely true?

(a) $f''(c)=0;\ \forall\ c\in(0,2)$
(b) $f'(c)=0$; for at least two $c\in(0,2)$
(c) $f'(c)=0$; for exactly one $c\in(0,2)$
(d) $f''(c)=0$; for at least one $c\in(0,2)$

40. Equation $\frac{1}{(x+1)^3}-3x+\sin x=0$ has

(a) no real root
(b) two real and distinct roots
(c) exactly one negative root
(d) exactly one root between -1 and ∞

41. If f is an odd continuous function in $[-1,1]$ and differentiable in $(-1,1)$, then

(a) $f'(A)=f(1)$ for some $A\in(-1,0)$
(b) $f'(B)=f(1)$ for some $B\in(0,1)$
(c) $n(f(A))^{n-1}f'(A)=(f(1))^n$ for some $A\in(-1,0)$, $n\in N$
(d) $n(f(B))^{n-1}f'(B)=(f(1))^n$ for some $B\in(0,1)$, $n\in N$

42. The parabola $y=x^2+px+q$ intersects the straight line $y=2x-3$ at a point with abscissa 1. If the distance between the vertex of the parabola and the X-axis is least, then

(a) $p=0$ and $q=-2$
(b) $p=-2$ and $q=0$

(c) least distance between the vertex of the parabola and X-axis is 2
(d) least distance between the vertex of the parabola and X-axis is 1

43. The abscissa of the point on the curve $\sqrt{xy} = a + x$, the tangent at which cuts off equal intersects from the coordinate axes, is $(a > 0)$
(a) $\frac{a}{\sqrt{2}}$ (b) $-\frac{a}{\sqrt{2}}$ (c) $a\sqrt{2}$ (d) $-a\sqrt{2}$

44. If the side of a triangle vary slightly in such a way that its circumradius remains constant, then $\frac{da}{\cos A} + \frac{db}{\cos B} + \frac{dc}{\cos C}$ is equal to
(a) $6R$ (b) $2R$
(c) 0 (d) $2R\,(dA + dB + dC)$

45. Let $f(x)$ satisfy the requirements of Lagrange's mean value theorem in $[0, 1]$, $f(0) = 0$ and $f'(x) \le 1 - x, \forall\ x \in (0, 1)$, then
(a) $f(x) \ge x$ (b) $|f(x)| \ge 1$
(c) $f(x) \le x\,(1 - x)$ (d) $f(x) \le 1/4$

46. For function $f(x) = \frac{\ln x}{x}$, which of the following statements are true?
(a) $f(x)$ has horizontal tangent at $x = e$
(b) $f(x)$ cuts the X-axis only at one point
(c) $f(x)$ is many-one function
(d) $f(x)$ has one vertical tangent

47. Equation of a line which is tangent to both the curves $y = x^2 + 1$ and $y = -x^2$ is
(a) $y = \sqrt{2}\,x + \frac{1}{2}$ (b) $y = \sqrt{2}\,x - \frac{1}{2}$
(c) $y = -\sqrt{2}\,x + \frac{1}{2}$ (d) $y = -\sqrt{2}\,x - \frac{1}{2}$

48. Let $F(x) = (f(x))^2 + (f'(x))^2$, $F(0) = 6$ where $f(x)$ is thrice differentiable function such that $|f(x)| \le 1$ for all $x \in [-1, 1]$, then choose the correct statement(s).
(a) There is at least one point in each of the intervals $(-1, 0)$ and $(0, 1)$ where $|f'(x)| \le 2$
(b) There is at least one point in each of the intervals $(-1, 0)$ and $(0, 1)$ where $F(x) \le 5$
(c) There is no point of local maxima of $F(x)$ in $(-1, 1)$
(d) For some $c \in (-1, 1)$, $F(c) \ge 6$, $F'(c) = 0$ and $F''(c) \le 0$

49. If the Rolle's theorem is applicable to the function f defined by

$$f(x) = \begin{cases} ax^2 + b, & |x| \le 1 \\ 1, & |x| = 1 \\ \frac{c}{|x|}, & |x| > 1 \end{cases}$$

in the interval $[-3, 3]$, then which of the following alternative(s) is/are correct?
(a) $a + b + c = 2$ (b) $|a| + |b| + |c| = 3$
(c) $2a + 4b + 3c = 8$ (d) $4a^2 + 4b^2 + 5c^2 = 15$

dy/dx as Rate Measurer & Tangents, Normals Exercise 3 : Statements I and II Type Questions

- **Directions** (Q. Nos. 50 to 56) *For the following questions, choose the correct answer from the options (a), (b), (c) and (d) defined as follows*
  (a) Statement I is true, Statement II is also true; Statement II is the correct explanation of Statement I
  (b) Statement I is true, Statement II is also true; Statement II is not the correct explanation of Statement I
  (c) Statement I is true, Statement II is false
  (d) Statement I is false, Statement II is true

50. Statement I If $g(x)$ is a differentiable function $g(1) \ne 0$, $g(-1) \ne 0$ and Rolle's theorem is not applicable to $f(x) = \frac{x^2 - 1}{g(x)}$ in $[-1, 1]$, then $g(x)$ has atleast one root in $(-1, 1)$.

Statement II If $f(a) = f(b)$, then Rolle's theorem is applicable for $x \in (a, b)$.

51. Statement I Shortest distance between $|x| + |y| = 2$ and $x^2 + y^2 = 16$ is $4 - \sqrt{2}$.

Statement II Shortest distance between the two smooth curves lies along the common normal.

52. Statement I If $\alpha_1, \alpha_2, \alpha_3, \ldots, \alpha_n$ are the n real roots of a polynomial equation of nth degree with real coefficients such that sum of the roots taken $r\,(1 \le r \le n)$ at a time is positive, then all the roots are positive.

Statement II The number of times sign of coefficients change while going left to right of a polynomial equation is the number of maximum positive roots.

53. Statement I Tangents at two distinct points of a cubic polynomial cannot coincide.

Statement II If $P(x)$ is a polynomial of degree $n\,(n \ge 2)$, then $P'(x) + k$ cannot hold for n or more distinct values of x.

54. **Statement I** For $f(x)=\begin{cases}\frac{1}{2}-x, & x<\frac{1}{2}\\ \left(\frac{1}{2}-x\right)^2, & x\geq\frac{1}{2}\end{cases}$, mean value theorem is applicable in the interval [0, 1].

Statement II For application of mean value theorem, $f(x)$ must be continuous in [0, 1] and differentiable in (0, 1).

55. Let $f(x)=\ln(2+x)-\frac{2x+2}{x+3}$.

Statement I The function $f(x)=0$ has a unique solution in the domain of $f(x)$.

Statement II If $f(x)$ is continuous in $[a, b]$ and is strictly monotonic in (a, b), then f has a unique root in (a, b).

56. Consider the polynomial function

$$f(x)=\frac{x^7}{7}-\frac{x^6}{6}+\frac{x^5}{5}-\frac{x^4}{4}+\frac{x^3}{3}-\frac{x^2}{2}+x$$

Statement I The equation $f(x)=0$ cannot have two or more roots.

Statement II Rolle's theorem is not applicable for $y=f(x)$ on any interval $[a, b]$, where $a, b\in R$

dy/dx as Rate Measurer & Tangents, Normals Exercise 4 : Passage Based Questions

Passage I (Q. Nos. 57 to 59)

We say an equation $f(x)=g(x)$ is consistent, if the curves $y=f(x)$ and $y=g(x)$ touch or intersect at at least one point. If the curves $y=f(x)$ and $y=g(x)$ do not intersect or touch, then the equation $f(x)=g(x)$ is said to be inconsistent, i.e. has no solution.

57. The equation $\cos x+\cos^{-1}x=\sin x+\sin^{-1}x$ is

(a) consistent and has infinite number of solutions
(b) consistent and has finite number of solutions
(c) inconsistent
(d) None of the above

58. The equation $\sin x=x^2+x+1$ is

(a) consistent and has infinite number of solutions
(b) consistent and has finite number of solutions
(c) inconsistent
(d) consistent and has unique solution

59. Among the following equations, which is consistent in $(0, \pi/2)$?

(a) $\sin x+x^2=0$ (b) $\cos x=x$
(c) $\tan x=x$ (d) All of these

Passage II (Q. Nos. 60 to 62)

To find the point of contact $P\equiv(x_1, y_1)$ of a tangent to the graph of $y=f(x)$ passing through origin O, we equate the slope of tangent to $y=f(x)$ at P to the slope of OP. Hence, we solve the equation $f'(x_1)=\frac{f(x_1)}{x_1}$ to get x_1 and y_1.

60. The equation $|\ln mx|=px$, where m is a positive constant, has a single root for

(a) $0<p<\frac{m}{e}$ (b) $p<\frac{e}{m}$
(c) $0<p<\frac{e}{m}$ (d) $p>\frac{m}{e}$

61. The equation $|\ln mx|=px$, where m is a positive constant, has exactly two roots for

(a) $p=\frac{m}{e}$ (b) $p=\frac{e}{m}$
(c) $0<p\leq\frac{e}{m}$ (d) $0<p\leq\frac{m}{e}$

62. The equation $|\ln mx|=px$, where m is a positive constant, has exactly three roots for

(a) $p<\frac{m}{e}$ (b) $0<p<\frac{m}{e}$
(c) $0<p<\frac{e}{m}$ (d) $p<\frac{e}{m}$

Passage III (Q. Nos. 63 to 64)

Consider the family of circles: $x^2+y^2-3x-4y-c_i=0, c_i\in N$ $(i=1, 2, 3, \ldots, n)$.
Also, let all circles intersects X-axis at integral points only and $c_1<c_2<c_3<c_4\ldots<c_n$. A point (x, y) is said to be integral point, if both coordinates x and y are integers.

63. If circle $x^2+y^2-3x-4y-(c_2-c_1)=0$ and circle $x^2+y^2=r^2$ have only one common tangent, then

(a) $r=1/2$
(b) tangent passes through (10, 0)
(c) (3, 4) lies outside the circle $x^2+y^2=r^2$
(d) $c_2=2r+c_1$

64. The ellipse $4x^2+9y^2=36$ and hyperbola $a^2x^2-y^2=4$ intersect orthogonally, then the equation of circle through the points of intersection of two conics is

(a) $x^2+y^2=(c_5)^2$ (b) $x^2+y^2=\frac{c_4}{7}$
(c) $x^2+y^2=c_3-2c_1$ (d) $x^2+y^2=c_7/14$

dy/dx as Rate Measurer & Tangents, Normals Exercise 5 : Matching Type Questions

65. Match the statements of Column I with values of Column II.

	Column I		Column II
(A)	The equation $x \log x = 3 - x$ has at least one root in	(p)	(0, 1)
(B)	If $27a + 9b + 3c + d = 0$, then the equation $4ax^3 + 3bx^2 + 2cx + d = 0$ has at least one root in	(q)	(1, 3)
(C)	If $c = \sqrt{3}$ and $f(x) = x + \frac{1}{x}$, then interval of x in which LMVT is applicable for $f(x)$ is	(r)	(0, 3)
(D)	If $c = \frac{1}{2}$ and $f(x) = 2x - x^2$, then the interval of x in which LMVT is applicable for $f(x)$, is	(s)	(– 1, 1)

66. Match the statements of Column I with values of Column II.

	Column I		Column II
(A)	A circular plate is expanded by heat from radius 5 cm to 5.06 cm. Approximate increase in area is	(p)	4
(B)	If an edge of a cube increases by 1%, then percentage increase in volume is	(q)	0.6π
(C)	If the rate of decrease of $\frac{x^2}{2} - 2x + 5$ is twice the rate of decrease of x, then x is equal to (rate of decrease is non-zero)	(r)	3
(D)	Rate of increase in area of equilateral triangle of side 15 cm, when each side is increasing at the rate of 0.1 cm/s, is	(s)	$\frac{3\sqrt{3}}{4}$

dy/dx as Rate Measurer & Tangents, Normals Exercise 6 : Single Integer Answer Type Questions

67. A point is moving along the curve $y^3 = 27x$. The interval in which the abscissa changes at slower rate than ordinate, is (a, b). Then, $(a + b)$ is ....

68. The slope of the curve $2y^2 = ax^2 + b$ at $(1, -1)$ is -1. Then, $(a - b)$ is ....

69. Let $f(1) = -2$, $f'(x) \geq 4.2$ for $1 \leq x \leq 6$. The smallest possible value of $f(6) - 16$ is ....

70. Let $f(x) = \begin{cases} -x^2, & \text{for } x < 0 \\ x^2 + 8, & \text{for } x \geq 0 \end{cases}$. Then, the absolute value of x-intercept of the line that is tangent to the graph of $f(x)$, is ....

71. The tangent to the graph of the function $y = f(x)$ at that point with abscissa, $x = a$ forms with the X- axis an angle of $\frac{\pi}{3}$ and the point with abscissa at $x = b$ at an angle $\frac{\pi}{4}$, then the value of $\left| \int_a^b f'(x) \cdot f''(x)\, dx \right|$ is ....

72. Two curves $C_1 : y = x^2 - 3$ and $C_2 : y = kx^2$, $k \in R$ intersect each other at two different points. The tangent drawn to C_2 at one of the points of intersection $A(a, y_1)$, $(a > 0)$ meets C_1 again at $B(1, y_2)$.
The value of a is ....

73. Consider the function $f(x) = 8x^2 - 7x + 5$ on the interval $[-6, 6]$. The value of c that satisfies the conclusion of the mean value theorem, is ....

74. Suppose that f is differentiable for all x and that $f'(x) \leq 2$ for all x. If $f(1) = 2$ and $f(4) = 8$, then $f(2)$ has the value equal to ....

75. Suppose a, b and c are positive integers with $a < b < c$ such that $1/a + 1/b + 1/c = 1$. The value of $(a + b + c - 5)$ is ....

Subjective Type Questions

76. Show that a tangent to an ellipse whose segment intercepted by the axes is the shortest, is divided at the point of tangency into two parts respectively, is equal to the semi-axes of the ellipse.

77. Tangents are drawn from the origin to the curve $y = \sin x$. Prove that points of contact lie on $y^2 = \frac{x^2}{1 + x^2}$.

78. If f is a continuous function with $\int_0^x f(t)\,dt \to \infty$ as $|x| \to \infty$, then show that every line $y = mx$ intersects the curve $y^2 + \int_0^x f(t)\,dt = 2$.

79. Find the equation of the straight line which is a tangent at one point and normal at another point to the curve $y = 8t^3 - 1$, $x = 4t^2 + 3$.

80. Let a curve $y = f(x)$ passes through (1, 1), at any point P on the curve tangent and normal are drawn to intersect the X-axis at Q and R, respectively. If $QR = 2$, find the equation of all such possible curves.

81. Show that the angle between the tangent at any point P and the line joining P to the origin 'O' is the same at all points of the curve $\log(x^2 + y^2) = c\tan^{-1}(y/x)$, where c is constant.

82. If the equation of two curves are $y^2 = 4ax$ and $x^2 = 4ay$

(i) Find the angle of intersection of two curves.

(ii) Find the equation of common tangents to these curves.

83. A straight line intersects the three concentric circles at A, B, C. If the distance of the line from the centre of the circles is 'P', prove that the area of the triangle formed by the tangents to the circle at A, B, C is $\left(\frac{1}{2P} \cdot AB \cdot BC \cdot CA\right)$.

84. Find the equation of all possible curves such that length of intercept made by any tangent on X-axis is equal to the square of x-coordinate of the point of tangency. Given that the curve passes through (2, 1).

85. The tangent to the curve $y = x - x^3$ at a point P meets the curve again at Q. Prove that one point of trisection of PQ lies on the Y-axis. Find the locus of the other points of trisection.

86. Determine all the curves for which the ratio of the length of the segment intercepted by any tangent on the Y-axis to the length of the radius vector is constant.

87. If t be a real number satisfying $2t^3 - 9t^2 + 30 - a = 0$, then the values of the parameter a for which the equation $x + \frac{1}{x} = t$ gives six real and distinct values of x.

dy/dx as Rate Measurer & Tangents, Normals Exercise 8 : Questions Asked in Previous 10 Years' Exams

(i) JEE Advanced & IIT-JEE

88. The slope of the tangent to the curve $(y - x^5)^2 = x(1 + x^2)^2$ at the point (1, 3) is **[Integer Type Question 2014]**

89. Let $f(x) = 2 + \cos x$, for all real x.

Statement I For each real t, there exists a point c in $[t, t + \pi]$, such that $f'(c) = 0$. Because

Statement II $f(t) = f(t + 2\pi)$ for each real t. **[Assertion and Reason 2007]**

(a) Statement I is correct, Statement II is also correct; Statement II is the correct explanation of Statement I

(b) Statement I is correct, Statement II is also correct; Statement II is not the correct explanation of Statement I

(c) Statement I is correct; Statement II is incorrect

(d) Statement I is incorrect; Statement II is correct

90. If $|f(x_1) - f(x_2)| \le (x_1 - x_2)^2, \forall x_1, x_2 \in R$. Find the equation of tangent to the curve $y = f(x)$ at the point (1, 2). **[Analytical Descriptive 2005]**

91. The point(s) on the curve $y^3 + 3x^2 = 12y$, where the tangent is vertical, is (are) **[One Correct Option 2002]**

(a) $\left(\pm\frac{4}{\sqrt{3}}, -2\right)$ (b) $\left(\pm\sqrt{\frac{11}{3}}, 0\right)$ (c) $(0, 0)$ (d) $\left(\pm\frac{4}{\sqrt{3}}, 2\right)$

92. If the normal to the curve $y = f(x)$ at the point (3, 4) makes an angle $\frac{3\pi}{4}$ with the positive X-axis, then $f'(3)$ is equal to

[One Correct Option 2000]

(a) −1 (b) −3/4 (c) 4/3 (d) 1

(ii) JEE Main & AIEEE

93. The normal to the curve $y(x-2)(x-3) = x + 6$ at the point where the curve intersects the Y-axis passes through the point **[2017 JEE Main]**

(a) $\left(\frac{1}{2}, \frac{1}{3}\right)$ (b) $\left(-\frac{1}{2}, -\frac{1}{2}\right)$ (c) $\left(\frac{1}{2}, \frac{1}{2}\right)$ (d) $\left(\frac{1}{2}, -\frac{1}{3}\right)$

94. The normal to the curve $x^2 + 2xy - 3y^2 = 0$ at (1,1) **[2015 JEE Main]**

(a) does not meet the curve again
(b) meets in the curve again the second quadrant
(c) meets the curve again in the third quadrant.
(d) meets the curve again in the fourth quadrant

Answers

Exercise for Session 1

1. 6 cm^3/s 2. (1, 5/3) and (−1, 1/3)
3. 8 cm/s 4. 750 cm^3/s 5. $\frac{200\pi r^3}{(r+5)^2}$ km^2/h
6. $\frac{1}{48\pi}$ cm/s 7. $-\frac{1}{3\pi}$ cm/s 8. $\frac{1}{10\pi}$ cm/s
9. 20 m/s 10. $\frac{-3}{10}$ rad/s 11. (a) $\frac{1}{40\pi}$ cm/s (b) $\frac{1}{30}$ cm^2/s (c) $\frac{\sqrt{10}}{30}$ cm^2/s
12. $\frac{0.004}{\pi}$ m/s and $\frac{0.02}{\pi}$ m/s 13. $\left(3, \frac{16}{3}\right), \left(-3, \frac{-16}{3}\right)$
14. $\left(\frac{2a}{b}\right)$ units/s 15. $-\frac{3}{8}$ m/s

Exercise for Session 2

1. $\frac{50}{7}$ 2. 2.1983 3. 0.02%
4. $\left(\frac{1}{8}\right)$%

Exercise for Session 3

1. (a) 2. (c) 3. (b) 4. (a) 5. (b)
6. (b) 7. (d) 8. (c) 9. (b) 10. (a)

Exercise for Session 4

1. (b) 2. (d) 3. (a) 4. (b) 5. (c)
6. (b) 7. (c) 8. (b) 9. (c) 10. (d)

Exercise for Session 5

1. (d) 2. (a) 3. (b) 4. (d) 5. (c)

Exercise for Session 6

1. Real and distinct roots if $a \in (-2, 2)$.
2. (a) 3. (a) 4. (b)

Chapter Exercises

1. (a) 2. (b) 3. (a) 4. (b) 5. (b)
6. (b) 7. (b) 8. (a) 9. (b) 10. (b)
11. (c) 12. (b) 13. (c) 14. (c) 15. (c)
16. (d) 17. (c) 18. (b) 19. (a) 20. (d)
21. (d) 22. (b) 23. (a) 24. (a) 25. (c)
26. (d) 27. (d) 28. (a) 29. (b) 30. (b)
31. (d) 32. (b) 33. (a) 34. (d)
35. (a, b) 36. (a, b, d) 37. (b, c) 38. (b, c)
39. (b, d) 40. (b, c, d)
41. (a, b, d) 42. (b, d) 43. (a, b) 44. (c, d)
45. (c, d) 46. (a, b, c) 47. (a, c)
48. (a, b, d) 49. (a, b, c, d) 50. (c)
51. (d) 52. (a) 53. (d) 54. (d) 55. (c)
56. (a) 57. (b) 58. (c) 59. (b) 60. (d) 61. (a)
62. (b) 63. (b) 64. (d)
65. (A) → (q); (B) → (r); (C) → (q); (D) → (p)
66. (A) → (q); (B) → (r); (C) → (p); (D) → (s)
67. (0) 68. (2) 69. (3) 70. (1) 71. (1)
72. (3) 73. (0) 74. (4) 75. (6)
79. $\pm\sqrt{2}(27x - 105)$
80. $\log y - x = \pm\left(\log - \left|\frac{1-\sqrt{1-y^2}}{y}\right| + \sqrt{1-y^2}\right) - 1$
82. (i) $Q = \tan^{-1}\left(\frac{3}{4}\right)$ (ii) $x + y = a = 0$
84. Possible curve are $y = \frac{x}{2(x-1)}$ or $y = \frac{3x}{2(1+x)}$.
85. $y = x - 5x^3$ 86. $(y + \sqrt{x^2 + y^2})\, x^{k-1} = c_1$
87. No real value 88. 8 89. (b) 90. $y - 2 = 0$
91. (d) 92. (d) 93. (c) 94. (d)

Solutions

1. Here, $f(x) = x^3 - nx + 1$

$f(0) = 1$ and $f(1) = 1 - n + 1 = 2 - n$

$\because \quad n \geq 3$

$\therefore \quad f(1) < 0$ and also we have, $f(0) > 0$

$\therefore$ $f(x)$ must have at least one real root in $(0, 1)$.

2. Given, slope of normal to $y = f(x)$ is 1.

$$\Rightarrow \left(-\frac{1}{f'(x)}\right)_{(0,\,0)} = 1$$

$$\Rightarrow \quad f'(0) = -1 \quad \text{...(i)}$$

Now, $\displaystyle \lim_{x \to 0} \frac{x^2}{f(x^2) - 20f(9x^2) + 2f(99x^2)}$

$$= \lim_{x \to 0} \frac{2x}{f'(x^2)\cdot 2x - 20f'(9x^2)\cdot 18x + 2f'(99x^2)\cdot(198x)}$$

[using L' Hospital's rule]

$$= \lim_{x \to 0} \frac{1}{f'(x^2) - 180f'(9x^2) + f'(99x^2)\cdot(198)}$$

$$= \frac{1}{f'(0) - 180\cdot f'(0) + 198\cdot f'(0)}$$

$$= \frac{1}{-1 + 180 - 198}$$

$$= -\frac{1}{19}$$

3. The equation of the tangent at the point $R(x, f(x))$ is

$$Y - f(x) = f'(x)(X - x) \quad \text{...(i)}$$

The point of intersection on Y-axis, is $P(0, f(x) - x\,f'(x))$.

The slope of line perpendicular to tangent at R, is

$$m_{PQ} = \frac{(f(x) - xf'(x)) - 0}{0 - 1} \quad \text{...(ii)}$$

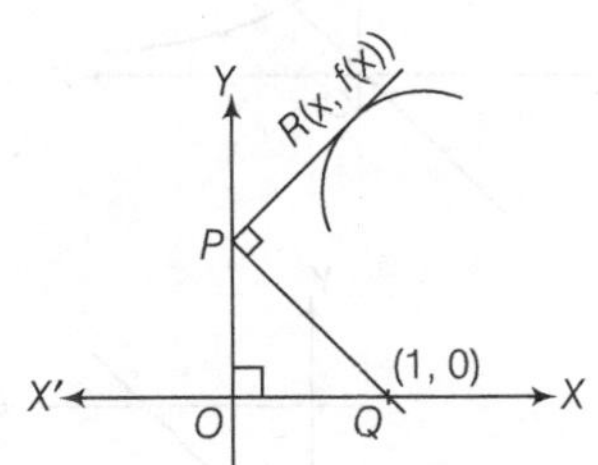

$\therefore \quad f'(x)\cdot m_{PQ} = -1$

$$\Rightarrow \quad f'(x)\cdot\frac{(f(x) - xf'(x))}{-1} = -1$$

$$\Rightarrow \quad f(x)f'(x) - x(f'(x))^2 = 1$$

4. Here, $y^2 = x + \sin x$

$\Rightarrow \quad 2y\,dy/dx = 1 + \cos x$

For horizontal tangent, $dy/dx = 0$

$\therefore \quad \cos x = -1 \Rightarrow x = (2n + 1)\pi$

Since, $\quad y^2 = x + \sin x$ and $|y| \leq 3$

$\Rightarrow \quad 0 \leq x \leq 9$

$\Rightarrow \quad 0 \leq (2n + 1)\pi \leq 9$

$\therefore \quad n = 0 \Rightarrow x = \pi$

$\Rightarrow \quad y^2 = \pi$ or $y = \pm\sqrt{\pi}$

Thus, points are $(\pi, \sqrt{\pi})$ and $(\pi, -\sqrt{\pi})$.

$\therefore$ Number of points is 2.

5. Here, $\dfrac{a + 2c}{b + 3d} = -\dfrac{4}{3}$

$\Rightarrow \quad 3a + 6c = -4b - 12d$

$\Rightarrow \quad 3a + 4b + 6c + 12d = 0$

or $\quad \dfrac{a}{4} + \dfrac{b}{3} + \dfrac{c}{2} + d = 0 \quad$ ...(i)

Consider, $\quad f(x) = \dfrac{ax^4}{4} + \dfrac{bx^3}{3} + \dfrac{cx^2}{2} + dx$

Then, $\quad f(0) = 0 = f(1) \quad$ [using Eq. (i)] ...(ii)

$\therefore$ $f(x)$ satisfy the condition of Rolle's theorem in $[0, 1]$.

Hence, $f'(x) = 0$ has at least one solution in $(0, 1)$.

6. Let $f(x) = \dfrac{ax^4}{4} + \dfrac{bx^3}{3} + \dfrac{cx^2}{2} + dx,$

which is continuous and differentiable.

Also, $\quad f(0) = 0,\ f(-1) = \dfrac{a}{4} - \dfrac{b}{3} + \dfrac{c}{2} - d$

$$= \frac{1}{4}(a + 2c) - \frac{1}{3}(b + 3d) = 0$$

So, according to Rolle's theorem, there exists at least one root of $f'(x) = 0$ in $(-1, 0)$.

7. Let us consider $g(t) = \displaystyle\int_0^t f(x)\,dx$

Applying LMVT in $(0, 2)$, we get

$$\frac{g(2) - g(0)}{2 - 0} = g'(c), \text{ where } c \in (0, 2)$$

$$\Rightarrow \quad \int_0^2 f(x)\,dx = 2f(c), \text{ where } c \in (0, 2)$$

8. Replace x by $-x$

$\Rightarrow x[f(x) + f(-x)] = 0 \Rightarrow f(x)$ is an odd function.

$\Rightarrow f^{iv}(x)$ is also odd $\Rightarrow f^{iv}(0) = 0$

9. Given, $x + y - \ln(x + y) = 2x + 5$

$$\Rightarrow 1 + \frac{dy}{dx} - \frac{1}{x + y}\left(1 + \frac{dy}{dx}\right) = 2$$

$$\Rightarrow \quad \frac{dy}{dx} = \frac{x + y + 1}{x + y - 1}$$

$$\Rightarrow \quad \left(\frac{dy}{dx}\right)_{(\alpha,\,\beta)} = \frac{\alpha + \beta + 1}{\alpha + \beta - 1} = \infty, \text{ when } \alpha + \beta = 1$$

10. Since, $f''(x)$ is odd function.

$\therefore \quad g(\alpha, \beta) = 0$

$\Rightarrow \quad (\sin^4 \alpha - 1)^2 + (\cos^4 \beta - 1)^2 + 2(\sin^2 \alpha - \cos^2 \beta)^2 = 0$

$\Rightarrow \quad \sin^2 \alpha + \sin^2 \beta = 1$

11. Let α be the root of $f(x) = 0$.

$\Rightarrow \quad f(\alpha) = f(2 - \alpha) = 0$

$\therefore$ $f(x)$ has 4 distinct and 2 equal roots.

$\therefore$ Sum of the roots = 6

12. Given, $y\dfrac{dy}{dx} = x$

$\Rightarrow \quad y\,dy = x\,dx \Rightarrow y^2 = x^2 + c$

$\because \quad f(1) = 3$

$\therefore \quad 9 = 1 + c \Rightarrow c = 8 \Rightarrow y^2 = x^2 + 8$

$\Rightarrow \quad f(x) = \sqrt{x^2 + 8}$

$\Rightarrow \quad f(4) = \sqrt{16 + 8} = 2\sqrt{6}$

13. Given, $x^2 y = c^3$

$\Rightarrow \quad x^2 \dfrac{dy}{dx} + 2xy = 0$

$\Rightarrow \quad \dfrac{dy}{dx} = -\dfrac{2y}{x}$

Equation of tangent at (x, y), is $Y - y = -\dfrac{2y}{x}(X - x)$

$Y = 0$, gives $X = \dfrac{3x}{2} = a$

and $\quad X = 0$, gives $Y = 3y = b$

Now, $\quad a^2 b = \dfrac{9x^2}{4} \cdot 3y = \dfrac{27}{4} x^2 y = \dfrac{27}{4} c^3$

14. Clearly, $f(0) = f(1) = 0$ and $f(x)$ is a polynomial of degree 10.

Therefore, by LMVT, we must have at least one root in (0, 1).

Since, the degree of $f(x)$ is even.

$\therefore$ It has at least two real roots.

15. Let (x_1, y_1) and (x_2, y_2) be two intersect points.

Given, $\quad y = x^3 + 2x - 1$ and $y = 2x^3 - 4x + 2$

$\therefore \quad y_1 = 2x_1^3 - 4x_1 + 2 \quad$ ...(i)

and $\quad 2y_1 = 2x_1^3 + 4x_1 - 2 \quad$ ...(ii)

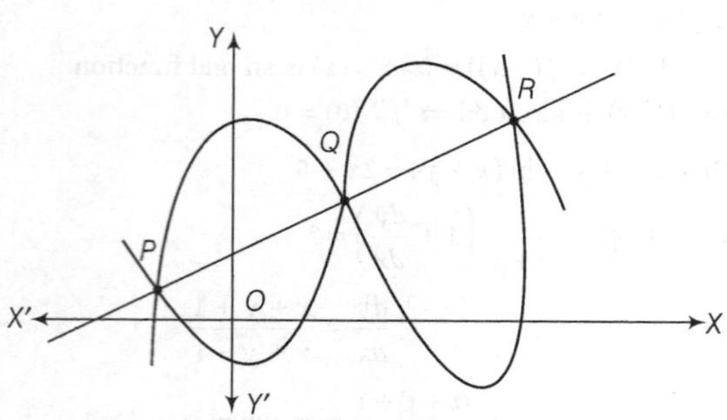

Subtracting Eq. (i) from Eq. (ii), we get

$y_1 = 8x_1 - 4 \quad$ ...(iii)

Similarly, $\quad y_2 = 8x_2 - 4 \quad$ ...(iv)

Now, from Eqs. (iii) and (iv), we get $y_2 - y_1 = 8(x_2 - x_1)$

$\therefore \quad \dfrac{y_2 - y_1}{x_2 - x_1} = 8$

16. (a) Discontinuous at $x = 1$, therefore not applicable.

(b) Discontinuous at $x = 0$, therefore not applicable.

(c) Discontinuity at $x = 1 \Rightarrow$ not applicable.

(d) Note that $x^3 - 2x^2 - 5x + 6 = (x - 1)(x^2 - x - 6)$.

Hence, $f(x) = x^2 - x - 6$, if $x \neq 1$ and $f(1) = -6$

$\Rightarrow$ f is continuous at $x = 1$.

So, $f(x) = x^2 - x - 6$ is continuous throughout the interval $[-2, 3]$.

Also, note that $f(-2) = f(3) = 0$. Hence, Rolle's theorem is applicable.

17. Let $A = (t, t^2)$. Then, $m_{OA} = t, m_{AB} = -\dfrac{1}{t}$

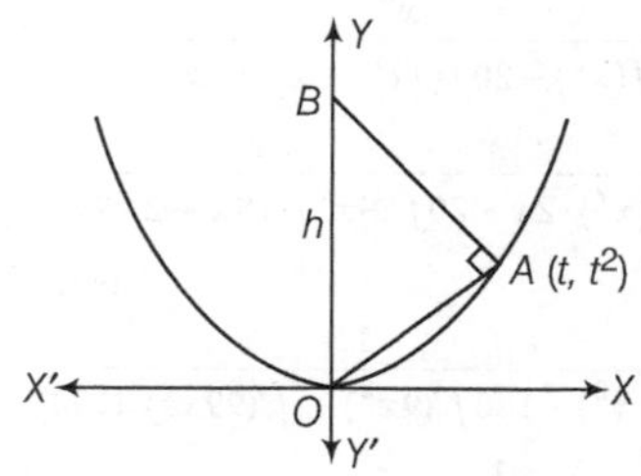

Equation of AB, $\quad y - t^2 = -\dfrac{1}{t}(x - t)$

Putting $\quad x = 0$

$\Rightarrow \quad h = t^2 + 1$

Now, $\quad \lim_{t \to 0} (h) = \lim_{t \to 0} (1 + t^2) = 1$

18. For $0 < a \leq 1$, the line always intersects $y = a^x$

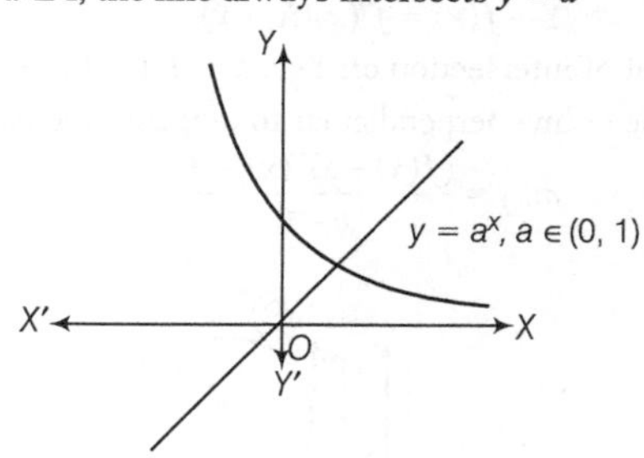

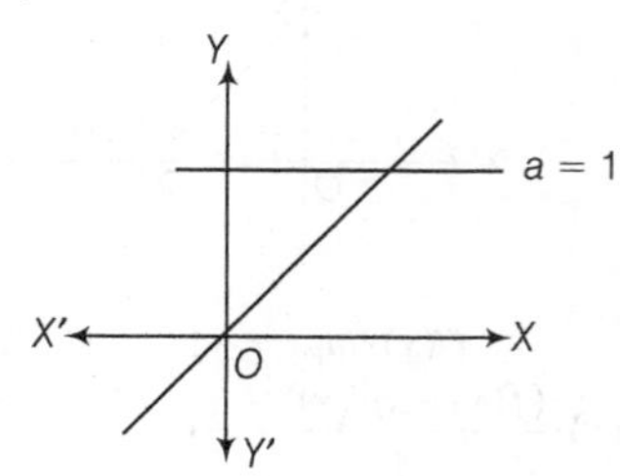

For $a > 1$, say $a = e$. Consider $f(x) = e^x - x$, $f'(x) = e^x - 1$

$f'(x) > 0$ for $x > 0$ and $f'(x) < 0$ for $x < 0$

$\therefore$ $f(x)$ is increasing ($\uparrow$) for $x > 0$
and decreasing ($\downarrow$) for $x < 0$

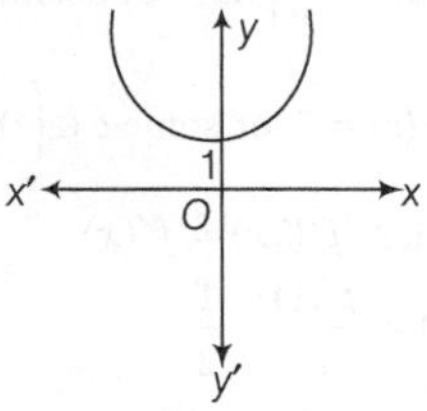

$y = e^x$ always lies above $y = x$, i.e. $e^x - x \geq 1$ for $a > 1$
Hence, the line $y = x$ intersect when $a \in (0, 1]$.

19. $\because$ f is not differentiable at $x = \frac{1}{2}$,
g is not continuous in $[0, 1]$ at $x = 0$ and 1,
and h is not continuous in $[0, 1]$ at $x = 1$

$$k(x) = (x+3)^{\ln_2 5} = (x+3)^p, \text{ where } 2 < p < 3$$

$\therefore$ None of these, f, g, h follows Lagrange's mean value theorem.

20. $f'(x) = 0$ and $f''(x) = 0$ for the same $x = x_1$ [say]

Now, $\quad f'(x) = 4x^3 + 2bx + 8$

$$f'(x_1) = 2[2x_1^3 + bx_1 + 4] = 0 \quad \text{...(i)}$$

$$f''(x_1) = 2[6x_1^2 + b] = 0 \quad \text{...(ii)}$$

From Eq. (ii), $\quad b = -6x_1^2$

Substituting this value of b in Eq. (i), we get

$$2x_1^3 + (-6x_1^3) + 4 = 0$$

$$\Rightarrow \quad 4x_1^3 = 4$$

$$\Rightarrow \quad x_1 = 1$$

Hence, $\quad b = -6$

21. For a cylindrical pot, $V = \pi r^2 h$

$$\Rightarrow \quad \frac{dV}{dt} = \pi\left[r^2\frac{dh}{dt} + h \cdot 2r\frac{dr}{dt}\right] \quad \left[r = \text{constant}, \frac{dr}{dt} = 0\right]$$

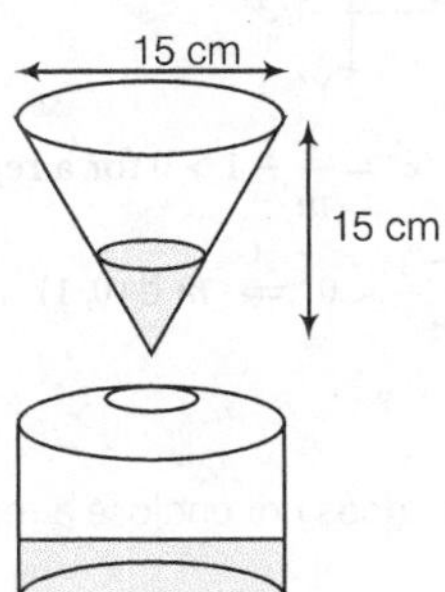

$$\Rightarrow \quad 100 = \pi r^2\frac{dh}{dt} \Rightarrow 100 = \pi \cdot \frac{225}{4} \cdot \frac{dh}{dt} \quad \left[\because r = \frac{15}{2} \text{ cm}\right]$$

$$\Rightarrow \quad \frac{dh}{dt} = \frac{400}{225\pi} = \frac{16}{9\pi} \text{ cm/min}$$

22. $\tan\theta = x/r \Rightarrow x = r\tan\theta$

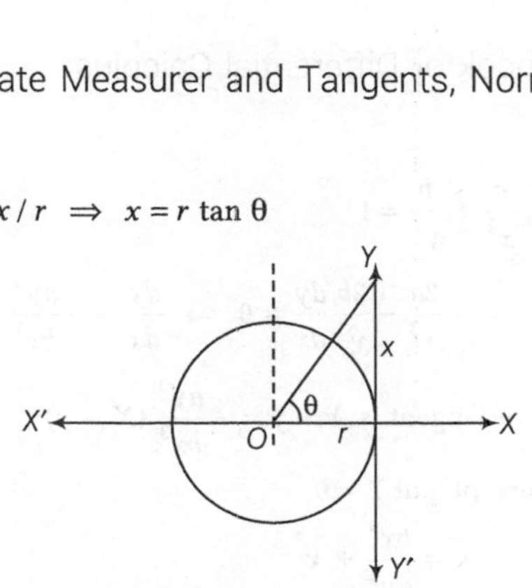

$$\Rightarrow \quad dx/dt = r\sec^2\theta\,(d\theta/dt) = r\,\omega\sec^2\theta = v\sec^2\theta$$

when, $\quad \theta = 2\pi/8,\ dx/dt = v\sec^2(\pi/4) = 2v = 40$ km/h.

23. $\frac{dV}{dt} = 15,\ h = 3r,\ V = \frac{1}{3}\pi x^2 y,\ \frac{dx}{dt} = ?$, when $x = 5$

$$\frac{x}{y} = \frac{r}{h} = \frac{1}{3} \Rightarrow V = \frac{2}{3}\pi r^2 3x = \pi x^3$$

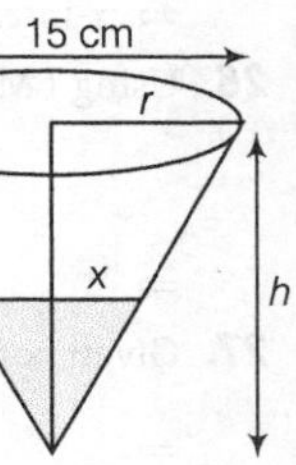

$$\frac{dV}{dt} = 3\pi x^2\frac{dx}{dt}$$

$$\Rightarrow \quad 15 = 3\pi \cdot 25\frac{dx}{dt}$$

$$\Rightarrow \quad \frac{dx}{dt} = \frac{1}{5\pi}$$

24. Given, $f(x) = x(x^2 - 3x + 2)$
$\Rightarrow f(x) = x(x-2)(x-1)$
Graph of $y = f(x)$ is shown as

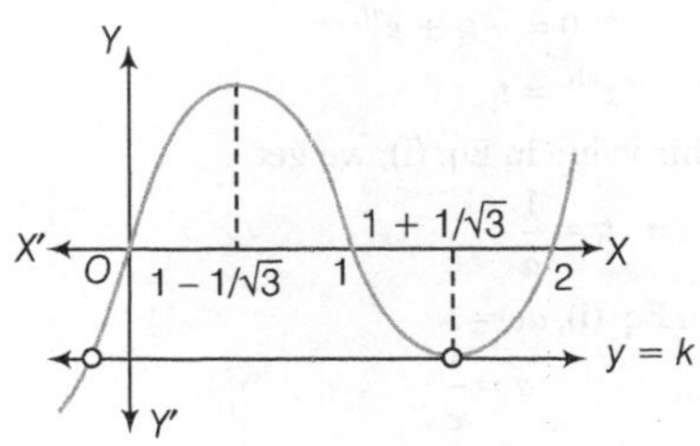

Now, for exactly one positive and one negative solution of the equation $f(x) = K$

We should have, $\quad k = f\left(1 + \frac{1}{\sqrt{3}}\right)$

$$\left[\because -\frac{1}{\sqrt{3}} \text{ and } 1 + \frac{1}{\sqrt{3}} \text{ are the roots of } f'(x) = 0\right]$$

$$\therefore \quad k = \underbrace{\left(1 + \frac{1}{\sqrt{3}}\right)}_{x}\underbrace{\left(\frac{1}{\sqrt{3}} - 1\right)}_{x-2}\underbrace{\left(\frac{1}{\sqrt{3}}\right)}_{x-1}$$

$$= \left(\frac{1}{3} - 1\right)\left(\frac{1}{\sqrt{3}}\right) = -\frac{2}{3\sqrt{3}}$$

$$= -\frac{2\sqrt{3}}{3}$$

25. We have, $\dfrac{a}{x^2}+\dfrac{b}{y^2}=1$

$$\Rightarrow \quad -\frac{2a}{x^3}-\frac{2b}{y^3}\frac{dy}{dx}=0 \Rightarrow \frac{dy}{dx}=-\frac{ay^3}{bx^3} \quad \text{...(i)}$$

Equation of tangent is $Y-y=-\dfrac{ay^3}{bx^3}(X-x)$

For x-intercept, put $Y=0$

$$\therefore \quad X=\frac{bx^3}{ay^2}+x$$

$$\Rightarrow \quad X=x\left[\frac{bx^2+ay^2}{ay^2}\right]=x\left[\frac{x^2y^2}{ay^2}\right]=\frac{x^3}{a}$$

[using Eq. (i)]

$\Rightarrow$ x-intercept is proportional to cube of abscissa.

26. Using LMVT in $[-2, 5]$, we get

$$-4\le\frac{f(5)-f(-2)}{7}\le 3$$

$$\Rightarrow \quad -28\le f(5)-f(-2)\le 21$$

27. Given, $x=t+e^{at}$, $y=-t+e^{at}$

$$\Rightarrow \quad \frac{dx}{dt}=1+ae^{at}, \frac{dy}{dt}=-1+ae^{at}, \frac{dy}{dx}=\frac{-1+ae^{at}}{1+ae^{at}}$$

At the point A, $y=0$ and $\dfrac{dy}{dx}=0$ for some $t=t_1$

$$\therefore \quad ae^{at_1}=1 \quad \text{...(i)}$$

Also, $\quad 0=-t_1+e^{at_1}$

$$\therefore \quad e^{at_1}=t_1 \quad \text{...(ii)}$$

Putting this value in Eq. (i), we get

$$at_1=1 \Rightarrow t_1=\frac{1}{a}$$

Now from Eq. (i), $ae=1$

$$\Rightarrow \quad a=\frac{1}{e}$$

Hence, $x_A=t_1+e^{at_1}=e+e=2e$

$\Rightarrow$ $A\equiv(2e, 0)$

28. Given, $x=a(2\cos t-\cos 2t)$, $y=a(2\sin t-\sin 2t)$

$$\therefore \quad \frac{dy}{dx}=\frac{\cos t-\cos 2t}{\sin 2t-\sin t}=0$$

$$\Rightarrow \quad \cos 2t=\cos t$$

$$\Rightarrow \quad \cos 2t=\cos(2\pi-t)$$

$$\Rightarrow \quad t=2\pi/3$$

29. $F'(x)=e^{(1+\sin^{-1}(\cos x))^2}\cdot(-\sin x)-e^{(1+x)^2}\cdot\cos x$

$$\therefore \quad F'(0)=0-e=-e$$

and $\quad F'\left(\dfrac{\pi}{2}\right)=-e-0=-e$

Hence, Rolle's theorem is applicable for the function $F'(x)$.

So, there exists c in $\left(0, \dfrac{\pi}{2}\right)$ for which $F''(c)=0$ as

Rolle's theorem is applicable for $F'(x)$ in $\left[0, \dfrac{\pi}{2}\right]$.

Also, $\quad F(0)=\int_0^1 f(t)\,dt$ and $F\left(\dfrac{\pi}{2}\right)=\int_1^0 f(t)\,dt$

Hence, $F(0)$ and $F\left(\dfrac{\pi}{2}\right)$ have opposite signs.

$$\Rightarrow \quad F(c)=0 \text{ for some } c\in\left(0, \frac{\pi}{2}\right)$$

30. Given, $f'(1)=1$, $2\cdot f'(2x)=f'(x)$

Put $x=1$, $f'(2)=\dfrac{f'(1)}{2}=\dfrac{1}{2}$

and $\quad f'(4)=\dfrac{1}{2}f'(2)=\dfrac{1}{4}$

Applying LMVT for $y=f'(x)$ in $[2, 4]$, we get

$$f''(c)=\frac{f'(4)-f'(2)}{2}=\frac{\frac{1}{4}-\frac{1}{2}}{2}=-\frac{1}{8}$$

31. Consider $\phi(x)=f(x)-g(x)\Rightarrow\phi'(x)=f'(x)-g'(x)>0$.

Clearly, $\phi(x)$ is also continuous and derivable in $[x_0, x]$.

Using LMVT for $\phi(x)$ in $[x_0, x]$, we get

$$\phi'(c)=\frac{\phi(x)-\phi(x_0)}{x-x_0}$$

Since, $\quad \phi'(x)=f'(x)-g'(x)>0$ for all $x>x_0$

$$\therefore \quad \phi'(c)>0$$

Hence, $\quad \phi(x)-\phi(x_0)>0$

$$\Rightarrow \quad \phi(x)>\phi(x_0)$$

$$\Rightarrow \quad \phi(x)>0 \quad [\because \phi(x_0)=f(x_0)-g(x_0)=0]$$

$$\Rightarrow \quad f(x)-g(x)>0$$

32. Solving, $mx=\dfrac{x}{x^2+1} \Rightarrow x^2+1=\dfrac{1}{m}$ or $x=0$

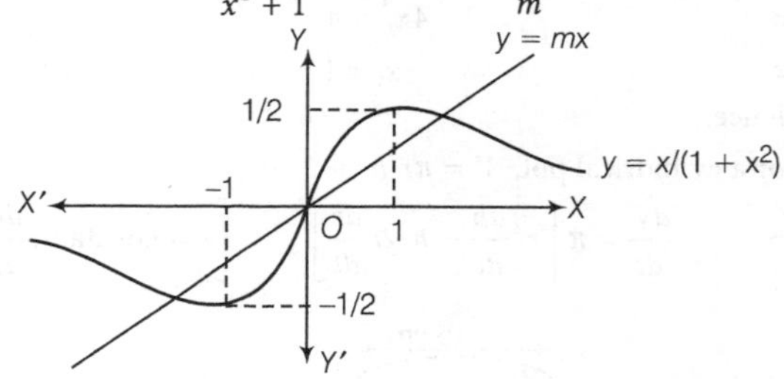

$$\Rightarrow \quad x^2=\frac{1}{m}-1>0 \text{ for a region}$$

$$\Rightarrow \quad \frac{m-1}{m}<0 \Rightarrow m\in(0, 1)$$

Remark

For $m=0$ or 1, the line does not enclose a region.

33. $\because$ Side $=x$

$\therefore$ Area $A=x^2 \Rightarrow \dfrac{dA}{dx}=2x$. So, $\left\{\left(\dfrac{dA}{dx}\right)_{x=x_0}\times h\right\}=2x_0h$

The exact change in the area of S when x is changed from x_0 to x_0+h, is

$$(x_0+h)^2-x_0^2=x_0^2+2x_0h+h^2-x_0^2=2x_0h+h^2$$

The difference between the exact change and the approximate change is $2x_0h+h^2-2x_0h=h^2$

34. $f(x) = \int_1^x \left(t + \frac{1}{t}\right) dt \Rightarrow f'(x) = x + \frac{1}{x}$

$\therefore \quad g(x) = x + \frac{1}{x}$

For $x \in \left[\frac{1}{2}, 3\right]$,

$g\left(\frac{1}{2}\right) = 2 + \frac{1}{2} = \frac{5}{2}, \ g(3) = 3 + \frac{1}{3} = \frac{10}{3}$

Let $P \equiv (c, g(c)), c \in \left[\frac{1}{2}, 3\right]$

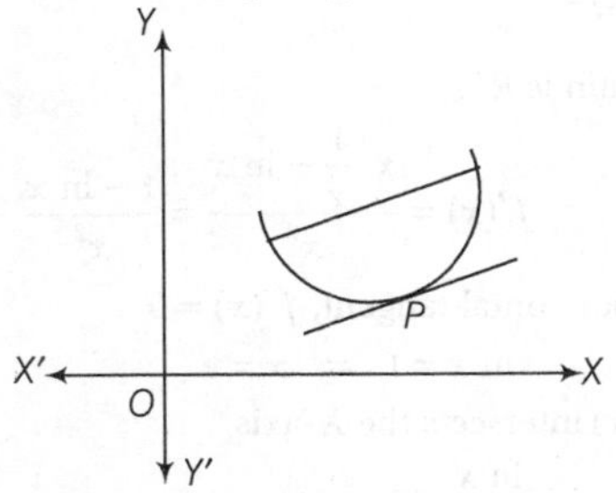

By LMVT, $g'(c) = \dfrac{g(3) - g\left(\frac{1}{2}\right)}{3 - \frac{1}{2}}$

$\therefore \quad 1 - \frac{1}{c^2} = \dfrac{\frac{10}{3} - \frac{5}{2}}{3 - \frac{1}{2}}$

$\Rightarrow \quad c^2 = \frac{3}{2}$

$\Rightarrow \quad c = \sqrt{\frac{3}{2}}$

$\therefore \quad g(c) = \sqrt{\frac{3}{2}} + \dfrac{1}{\sqrt{\frac{3}{2}}} = \frac{5}{\sqrt{6}}$

Thus, $P \equiv \left(\sqrt{\frac{3}{2}}, \frac{5}{\sqrt{6}}\right)$

35. Here, $\dfrac{dx}{dt} = \dfrac{2(-\operatorname{cosec}^2 t)}{\cot t}$

$\Rightarrow \quad \left.\dfrac{dx}{dt}\right|_{\text{at } t = \frac{\pi}{4}} = -4$

and $\dfrac{dy}{dt} = \sec^2 t - \operatorname{cosec}^2 t$

$\Rightarrow \quad \left.\dfrac{dy}{dt}\right|_{\text{at } t = \frac{\pi}{4}} = 0$

$\therefore \quad \dfrac{dy}{dx} = 0$

Hence, tangent is parallel to X-axis and its normal is parallel to Y-axis.

36. Here, $f'(x) = \dfrac{1}{3x^{2/3}}$

$\Rightarrow f'(0) \to \infty$ and tangent is vertical at $x = 0$.

Equation of tangent at (0, 0) is $x = 0$.

Equation of normal is $y = 0$.

Now, $f(x) = f^{-1}(x) \Rightarrow x^{1/3} = x^3$

$\Rightarrow \quad x^9 = x \Rightarrow x = 0, 1, -1$

37. Here, $\dfrac{dy}{dx} = k^2 e^{kx}$

$\Rightarrow \left(\dfrac{dy}{dx}\right)_{x=0} = k^2 = \tan\theta$, where θ is angle made by X- axis.

Let ϕ be the angle made by Y-axis.

$\therefore \quad \tan\theta = \tan\left(\frac{3\pi}{2} - \phi\right) = \cot\phi$

$\Rightarrow \quad \cot\phi = k^2 \Rightarrow \phi = \cot^{-1} k^2$

$\Rightarrow \quad \phi = \sin^{-1}\left(\dfrac{1}{\sqrt{1 + k^4}}\right)$

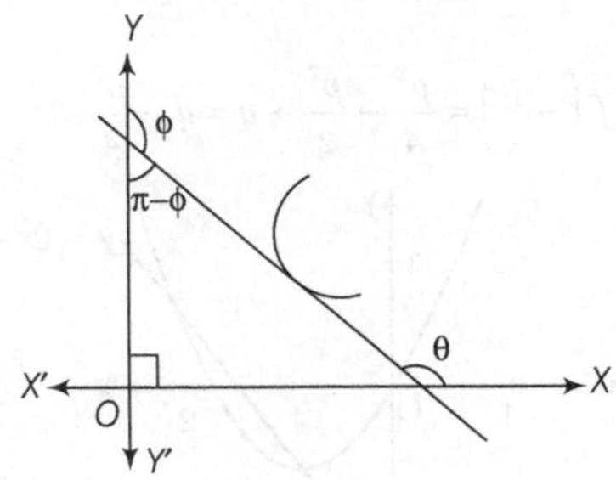

38. As, $f(x)$ is continuous in [0, 1] and differentiable in (0, 1) and $f(0) = f(1) = 1$.

$\therefore$ By Rolle's theorem, there must exist at least one $x = c \in (0, 1)$ such that $f'(c) = 0$

$\therefore f'(c)$ vanishes for some $c \in (0, 1)$.

Now, $f(0) = 1$, $f(1/2) = -\frac{1}{2}$ and $f(1) = 1$

$\therefore$ By intermediate value theorem, $f(x)$ must have one root belongs to $\left(0, \frac{1}{2}\right)$ and other in the interval $\left(\frac{1}{2}, 1\right)$.

39. Here, $f(0) = f(1)$ and f is continuous in [0, 1] and derivable in (0, 1).

$\therefore f'(c_1) = 0$ for at least one $c_1 \in (0, 1)$

Similarly, as $f(1) = f(2)$

$\therefore \ f'(c_2) = 0$ for at least one $c_2 \in (1, 2)$. $\Rightarrow f'(c_1) = f'(c_2)$

$\Rightarrow \ f''(c) = 0$ for at least one $c \in (c_1, c_2)$.

40. Let $f(x) = \dfrac{1}{(x+1)^3} - 3x + \sin x$

Domain of f is $(-\infty, -1) \cup (-1, \infty)$.

$$f'(x) = -3\left[\frac{1}{(x+1)^4} + 1\right] + \cos x \Rightarrow f'(x) < 0$$

$\Rightarrow f$ is decreasing.

Also, $\lim_{x \to -1^+} f(x) \to \infty, \ \lim_{x \to -1^-} f(x) \to -\infty$

and $\lim_{x \to \infty} f(x) \to -\infty, \ \lim_{x \to -\infty} f(x) \to \infty$

$\Rightarrow f(x) = 0$ has exactly two roots.

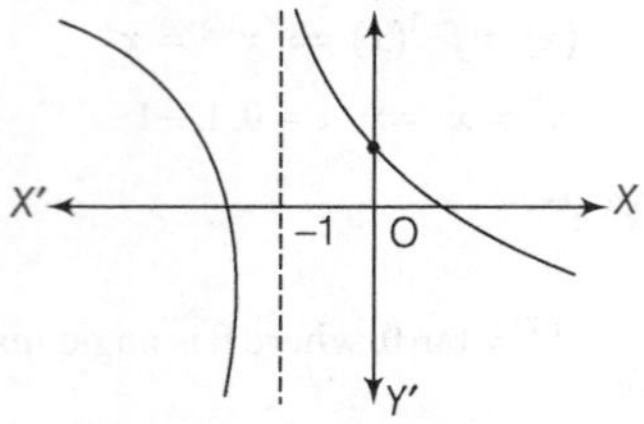

41. Clearly, $f(-1) = -f(1)$ and $f(0) = 0$. For (a) and (b) apply LMVT for the function $f(x)$ in $(-1, 0)$ and $(0, 1)$, respectively. For (d) apply LMVT for $(f(x))^n$ in $(0, 1)$.

42. When $x = 1, y = -1$ [from the line]

Thus, it must lie on the parabola $y = x^2 + px + q$

$\Rightarrow -1 = 1 + p + q \Rightarrow p + q = -2$

$\therefore$ Now, distance of the vertex of the parabola from the X-axis is

$$d = f\left(-\frac{p}{2}\right) = \frac{p^2}{4} - \frac{p^2}{2} + q = q - \frac{p^2}{4}$$

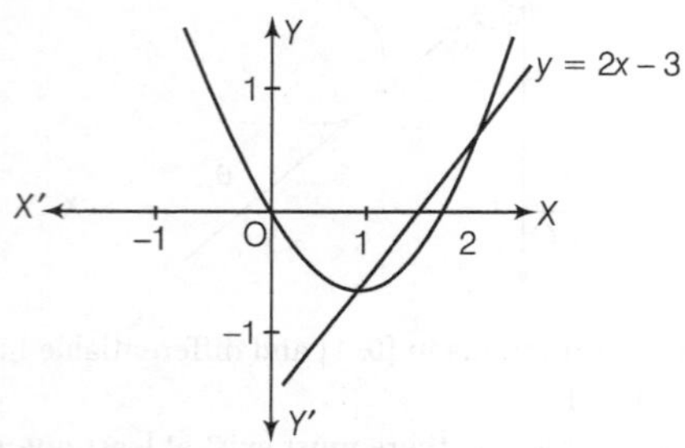

Substituting $q = -2 - p$, we get

$$d = -2 - p - \frac{p^2}{4}$$

Now, take $g(p) = -2 - p - \dfrac{p^2}{4}$

So, $$g'(p) = -1 - \frac{p}{2} = 0 \Rightarrow p = -2$$

Hence, $q = 0$

Note that least distance of the vertex from X-axis is 1.

43. Given, $\sqrt{xy} = a + x \Rightarrow xy = a^2 + x^2 + 2ax$

$$\Rightarrow \quad y = \frac{a^2}{x} + x + 2a$$

$$\Rightarrow \quad \frac{dy}{dx} = -\frac{a^2}{x^2} + 1 = -1$$

$$\Rightarrow \quad 2x^2 = a^2$$

$$\Rightarrow \quad x = \pm \frac{a}{\sqrt{2}}$$

44. Given, $$\frac{a}{\sin A} = \frac{b}{\sin B} = \frac{c}{\sin C} = 2R \quad \text{[say]}$$

$\therefore$ $da = 2R \cos A\, dA$, $db = 2R \cos B\, dB$, $dc = 2R \cos C\, dC$

$$\therefore \quad \frac{da}{\cos A} + \frac{db}{\cos B} + \frac{dc}{\cos C} = 2R\,(dA + dB + dC) \quad \text{...(i)}$$

Also, $A + B + C = \pi$

So, $dA + dB + dC = 0$ ...(ii)

From Eqs. (i) and (ii), we get

$$\frac{da}{\cos A} + \frac{db}{\cos B} + \frac{dc}{\cos C} = 0$$

45. $\dfrac{f(x) - f(0)}{x - 0} = f'(c) \le 1 - x$ for some $c \in (0, 1)$.

$\Rightarrow f(x) \le x(1 - x) \le 1/4$

46. Here, $f(x) = \dfrac{\ln x}{x}$ ...(i)

$\therefore$ Domain is R^+.

$$\because \quad f'(x) = \frac{x \cdot \frac{1}{x} - \ln x}{x^2} = \frac{1 - \ln x}{x^2}$$

(a) For horizontal tangent, $f'(x) = 0$

$\Rightarrow \ln x = 1 \Rightarrow x = e$ [true]

(b) If $f(x)$ intersects the X-axis.

$\dfrac{\ln x}{x} = 0 \Rightarrow x = 1$ [true]

(c) $f'(x)$ is positive, if $x \in (0, e)$ and $f'(x)$ is negative, if $x \in (e, \infty)$

$\therefore$ $f(x)$ is not monotonic.

$\therefore$ $f(x)$ is many-one.

(d) For vertical tangent, $f'(x) = \infty$ [true]

$$\Rightarrow \frac{1 - \ln x}{x^2} = \infty \Rightarrow \frac{x^2}{1 - \ln x} = 0$$

$\Rightarrow x = 0$ which is not in the domain of $f(x)$. [false]

47. Let the tangent line be $y = ax + b$

The equation for its intersection with the upper parabola is

$$x^2 + 1 = ax + b$$

$$\Rightarrow \quad x^2 - ax + (1 - b) = 0$$

This has a double root when $a^2 - 4(1 - b) = 0$

or $$a^2 + 4b = 4$$

For the lower parabola, $ax + b = -x^2$

$$\Rightarrow \quad x^2 + ax + b = 0$$

This has a double root when $a^2 - 4b = 0$

On subtracting these two equations, we get $8b = 4$ or $b = \dfrac{1}{2}$

On adding these equations, we get $2a^2 = 4$ or $a = \pm\sqrt{2}$

The tangent lines are $y = \sqrt{2}\,x + \dfrac{1}{2}$ and $y = -\sqrt{2}\,x + \dfrac{1}{2}$

48. For some $\alpha \in (0, 1)$, $|f'(\alpha)| = \left|\dfrac{f(1) - f(0)}{1 - 0}\right| \le |f(1)| + |f(0)|$

$\Rightarrow |f'(\alpha)| \le 1 + 1 = 2$

Similarly, for some $\beta \in (-1, 0)$, $|f'(\beta)| \le 2$

Also, $F(x) = (f(x))^2 + (f'(x))^2$

$\Rightarrow F(\alpha) = (f(\alpha))^2 + (f'(\alpha))^2 \le 1 + 4 \le 5$

Similarly, $F(\beta) \le 5$ for some $\beta \in (-1, 0)$

As, $F(0) = 6$, so there must be a point of local maxima for $F(x)$ in $(-1, 1)$ and at the point of maxima, say $x = c$,

$$F(c) \geq 6 \Rightarrow F'(c) = 0 \text{ and } F''(c) \leq 0$$

49. As, Rolle's theorem is applicable, the function should be continuous and differentiable in $[-3, 3]$.

So, at $x = 1$ it is continuous

$$\Rightarrow \lim_{x \to 1^-} f(x) = \lim_{x \to 1^+} f(x) = f(1)$$

$$\Rightarrow a + b = c = 1 \quad \text{...(i)}$$

Since, differentiable at $x = 1$, therefore

$$f'(1^+) = f'(1^-)$$

$$\Rightarrow \lim_{h \to 0} \frac{\frac{c}{1+h} - 1}{h} = 2a$$

$$\Rightarrow 2a = \lim_{h \to 0} \frac{c - 1 - h}{h(h+1)} \text{ exists only when } c = 1$$

$$\Rightarrow 2a = \lim_{h \to 0} \frac{-1}{(1+h)} = -1$$

$$\therefore a = -1/2 \text{ and } c = 1 \quad \text{...(ii)}$$

From Eqs. (i) and (ii), we get

$$b = 3/2 \quad \text{...(iii)}$$

$$\therefore a + b + c = 2 \quad \text{...(iv)}$$

$$|a| + |b| + |c| = 3 \quad \text{...(v)}$$

$$2a + 4b + 3c = 8 \quad \text{...(vi)}$$

$$4a^2 + 4b^2 + 5c^2 = 15 \quad \text{...(vii)}$$

50. **Statement I** As, $f(-1) = f(1)$ and Rolle's theorem is not applicable, then it implies that $f(x)$ is either discontinuous or $f'(x)$ does not exist at at least one point in $(-1, 1)$.

$\Rightarrow g(x) = 0$ for at at least one value of x in $(-1, 1)$.

Statement II is false. Consider the example in Statement I.

51. Common normal is $y = x$

Solving, $y = x$ with $x + y = 2$, we get $A(1, 1)$

and with $x^2 + y^2 = 16$, we get $B(2\sqrt{2}, 2\sqrt{2})$

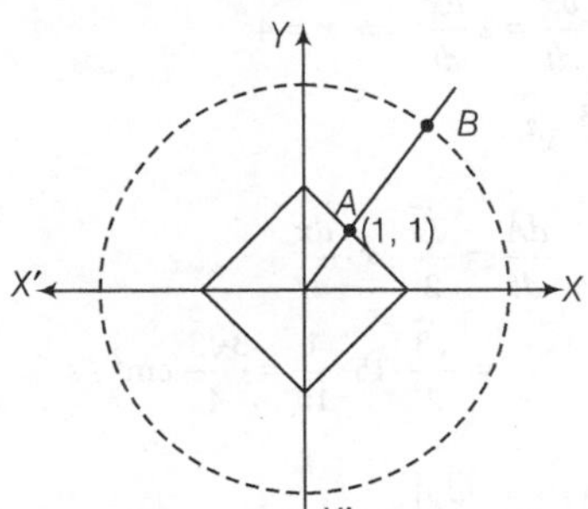

The shortest distance between the given curves is $AB = 4 - \sqrt{2}$. But as the curves are not smooth, check at slope points. The coordinates in 1st quadrant are (2, 0) and (4, 0) and here distance = 2.

$\therefore$ $4 - \sqrt{2}$ is not the shortest.

52. If $P(x) = 0$ is a polynomial equation, then $P(-x) = 0$ has no positive root.

$\Rightarrow P(x) = 0$ cannot have negative roots.

53. Let $A(a, P(a))$, $B(b, P(b))$, then slope of $AB = P'(a) = P'(b)$ from LMVT there exists $c \in (a, b)$, where $P'(c) =$ slope of AB.

54.
$$f(x) = \begin{cases} \frac{1}{2} - x, & x < \frac{1}{2} \\ \left(\frac{1}{2} - x\right)^2, & x \geq \frac{1}{2} \end{cases}$$

$$\Rightarrow f'(x) = \begin{cases} -1, & x < \frac{1}{2} \\ 2\left(\frac{1}{2} - x\right)(-1), & x > \frac{1}{2} \end{cases}$$

Left hand derivative at $x = 1/2$ is (-1) and right hand derivative at $x = 1/2$ is 0, so the function is not differentiable at $x = 1/2$.

55. $f(x) = \ln(2 + x) - \frac{2x + 2}{x + 3}$ is continuous in $[-2, \infty)$.

$$f'(x) = \frac{1}{x + 2} - \frac{4}{(x + 3)^2} = \frac{(x + 3)^2 - 4(x + 2)}{(x + 2)(x + 3)^2}$$

$$= \frac{x^2 + 2x + 1}{(x + 2)(x + 3)^2} = \frac{(x + 1)^2}{(x + 2)(x + 3)^2} > 0$$

$[f'(x) = 0 \text{ at } x = -1]$

$\Rightarrow f$ is increasing in $(-2, \infty)$.

Also, $\lim_{x \to -2^+} f(x) \to -\infty$

and $\lim_{x \to \infty} f(x) \to \infty \Rightarrow$ unique root.

56. Let $f(x) = 0$ has two roots say $x = r_1$ and $x = r_2$, where $r_1, r_2 \in [a, b]$.

$$\Rightarrow f(r_1) = f(r_2)$$

Hence, there must exist some $c \in (r_1, r_2)$, where $f'(c) = 0$

But $f'(x) = x^6 - x^5 + x^4 - x^3 + x^2 - x + 1$

for $x \geq 1$, $f'(x) = (x^6 - x^5) + (x^4 - x^3) + (x^2 - x) + 1 > 0$

for $x \leq 1$, $f'(x) = (1 - x) + (x^2 - x^3) + (x^4 - x^5) + x^6 > 0$

Hence, $f'(x) > 0$ for all x.

$\therefore$ Rolle's theorem fails.

$\Rightarrow f(x) = 0$ cannot have two or more roots.

57. Let $f(x) = \cos x - \sin x + \cos^{-1} x - \sin^{-1} x$, $x \in [-1, 1]$

$\because f(-1) f(1) < 0$

$\therefore$ There exists at least one $c \in (-1, 1)$ such that $f(c) = 0$.

Hence, the curves $y = \cos x + \cos^{-1} x$ and $y = \sin x + \sin^{-1} x$ intersect each other at at least one point.

58. Given, $\sin x = x^2 + x + 1 = \left(x + \frac{1}{2}\right)^2 + \frac{3}{4}$

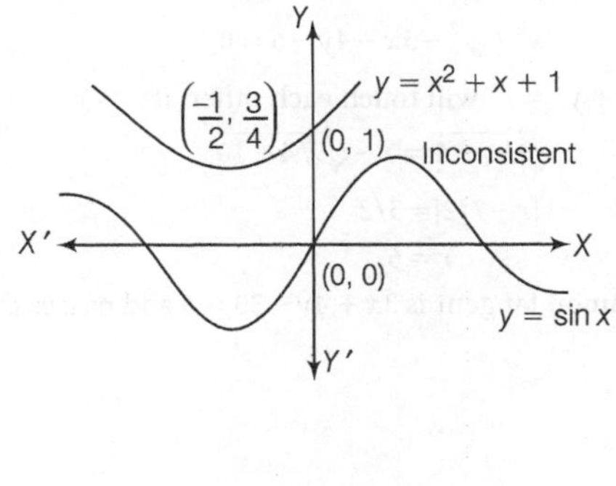

59.

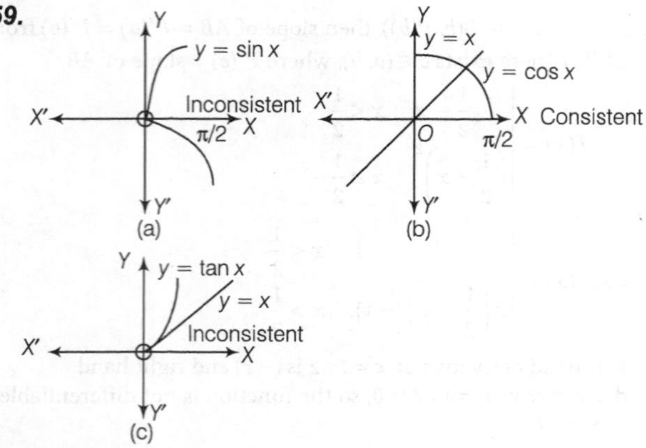

Sol. (Q. Nos. 60 to 62)

Slope of tangent at $P =$ Slope of OP

$$\Rightarrow \quad \frac{1}{t} = \frac{\ln mt}{t}$$

$$\Rightarrow \quad t = \frac{e}{m}$$

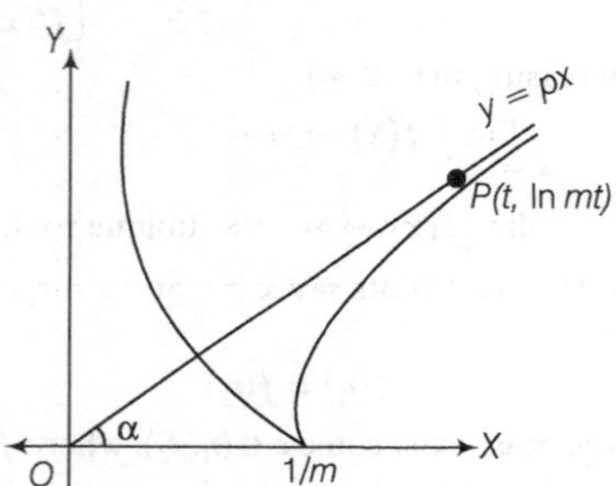

$$\Rightarrow \quad P \equiv \left(\frac{e}{m}, 1\right)$$

$$\Rightarrow \quad \tan \alpha = p = \frac{m}{e}$$

60. $p > m/e$ **61.** $p = m/e$

62. $0 < p < m/e$

63. Putting $y = 0$, we get $x^2 - 3x - c_i = 0$

As roots are integers and $D = 9 + 4c_i$ must be perfect square, therefore $9 + 4c_i = (2\lambda + 1)^2, \lambda \in I$.

$$\Rightarrow \quad c_i = \lambda^2 + \lambda - 2$$

$$\Rightarrow \quad c_k = k(k+3), \quad k = 1, 2, 3, \ldots$$

$$\therefore \quad c_1 = 4, c_2 = 10, \ldots$$

$$\Rightarrow \quad x^2 + y^2 - 3x - 4y - (c_2 - c_1) = 0$$

or $\quad x^2 + y^2 - 3x - 4y - 6 = 0$

and $x^2 + y^2 = r^2$ will touch each other, if

$$\sqrt{9/4 + 4} = |r - \sqrt{9/4 + 10}|$$

$$\Rightarrow \quad |r - 7/2| = 5/2$$

$$\Rightarrow \quad r = 6$$

$\therefore$ Common tangent is $3x + 4y - 30 = 0$ and passes through (10, 0).

64. An ellipse and hyperbola intersect orthogonally.

They must be confocal $\Rightarrow a = 2$

Let point $P(\alpha, \beta)$ lies on both the curves, then

$$4\alpha^2 + 9\beta^2 = 36 \quad \ldots\text{(i)}$$

and $\quad 4\alpha^2 - \beta^2 = 4 \quad \ldots$(ii)

On adding Eqs. (i) and (ii), we get

$$8\alpha^2 + 8\beta^2 = 40$$

$$\Rightarrow \quad \alpha^2 + \beta^2 = 5$$

or $\quad x^2 + y^2 = \frac{c_7}{14}$, as $c_7 = 70$.

65. (A) $f'(x) = \log x - \frac{3}{x} + 1$

$$\Rightarrow \quad f(x) = (x - 3)\log x + c$$

$$\Rightarrow \quad f(1) = f(3)$$

(B) $f'(x) = 4ax^3 + 3bx^2 + 2cx + d$

$$\Rightarrow \quad f(x) = ax^4 + bx^3 + cx^2 + dx + e$$

$$\Rightarrow \quad f(0) = f(3) \quad [\because 27a + 9b + 3c + d = 0]$$

(C) $\frac{f(b) - f(a)}{b - a} = f'(\sqrt{3}) = \frac{2}{3}$

$$\Rightarrow \quad \frac{ab - 1}{ab} = \frac{2}{3}$$

(D) $\frac{f(b) - f(a)}{b - a} = f'\left(\frac{1}{2}\right) = 1$

$$\Rightarrow \quad a + b = 1$$

66. (A) Here, $r = 5$ cm, $\Delta r = 0.06$

$$\because \quad A = \pi r^2,$$

$$\therefore \quad dA = 2\pi r\, dr = 2\pi r \cdot \Delta r = 10\pi \times 0.06 = 0.6\,\pi$$

(B) $v = x^3, dv = 3x^2\, dx$

$$\frac{dv}{v} \times 100 = 3\frac{dx}{x} \times 100 = 3 \times 1 = 3$$

(C) $(x - 2)\frac{dx}{dt} = 2\frac{dx}{dt} \Rightarrow x = 4$

(D) $A = \frac{\sqrt{3}}{4}x^2$

$$\frac{dA}{dt} = \frac{\sqrt{3}}{2} \cdot x \cdot \frac{dx}{dt}$$

$$= \frac{\sqrt{3}}{2} \cdot 15 \cdot \frac{1}{10} = \frac{3\sqrt{3}}{4}\ \text{cm}^2/\text{s}$$

67. $\because \left|\frac{dx}{dt}\right| < \left|\frac{dy}{dt}\right| \Rightarrow \left|\frac{dy}{dx}\right| > 1$

and $\quad 3y^2 \cdot \frac{dy}{dx} = 27$ or $\frac{dy}{dx} = \frac{9}{y^2}$

$$\therefore \quad \frac{9}{y^2} > 1 \Rightarrow y^2 < 9$$

$$\Rightarrow \quad -3 < y < 3 \Rightarrow -27 < y^3 < 27$$

$$\Rightarrow \quad -27 < 27x < 27 \Rightarrow -1 < x < 1$$

$$\therefore \quad x \in (-1, 1) \Rightarrow a + b = 0$$

68. Here, $4y\dfrac{dy}{dx} = 2ax$

$\therefore \quad \left.\dfrac{dy}{dx}\right|_{(1,-1)} = \dfrac{-a}{2} = -1$

$\Rightarrow \quad a = 2$

Also, $2y^2 = ax^2 + b$ at $(1, -1)$ is

$2 = a + b$

$\Rightarrow \quad b = 0$

$\therefore \quad a - b = 2$

69. Using LMVT, for some $c \in (1, 6)$, we get

$$f'(c) = \frac{f(6) - f(1)}{5}$$

$$= \frac{f(6) + 2}{5} \geq 4.2$$

$\Rightarrow \quad f(6) \geq 19$

$\Rightarrow \quad f(6) - 16 \geq 3$

$\therefore$ Least value of $f(6) - 16 = 3$

70. Let $y = mx + c$ be a tangent to $f(x)$.

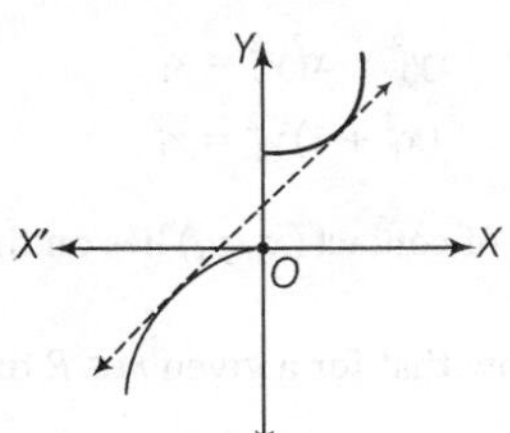

For, $x \geq 0$, intersection point is given by

$mx + c = x^2 + 8 \quad [\because y = x^2 + 8, \text{ for } x \geq 0]$

$\Rightarrow \quad x^2 - mx + (8 - c) = 0$

For line to be tangent, $D = 0$

$\therefore \quad m^2 = 4(8 - c) \quad \ldots\text{(i)}$

Again, for $x < 0$

$mx + c = -x^2 \quad [\because y = -x^2, \text{ for } x < 0]$

$\Rightarrow \quad x^2 + mx + c = 0$

Now, $\quad D = 0$

$\Rightarrow \quad m^2 = 4c \quad \ldots\text{(ii)}$

From Eqs. (i) and (ii), we get

$c = 4, m = 4$

$\therefore$ Tangent is $y = 4x + 4$

Putting $\quad y = 0$, we get

$x = -1$

$\therefore$ Absolute value of x-intercept is 1.

71. Here, $f'(a) = \sqrt{3}$ and $f'(b) = 1$

$$\therefore \quad \left|\int_a^b f'(x) \cdot f''(x)\, dx\right| = \left|\left(\frac{(f'(x))^2}{2}\right)_a^b\right| = \frac{1}{2}\left|(f'(b))^2 - (f'(a))^2\right|$$

$$= \frac{1}{2}|1 - 3| = 1$$

72. Point $A(a, y_1)$ lies on C_1 and C_2.

$\therefore \quad y_1 = a^2 - 3$ and $y_1 = ka^2$.

$\Rightarrow \quad a^2 - 3 = ka^2 \quad \ldots\text{(i)}$

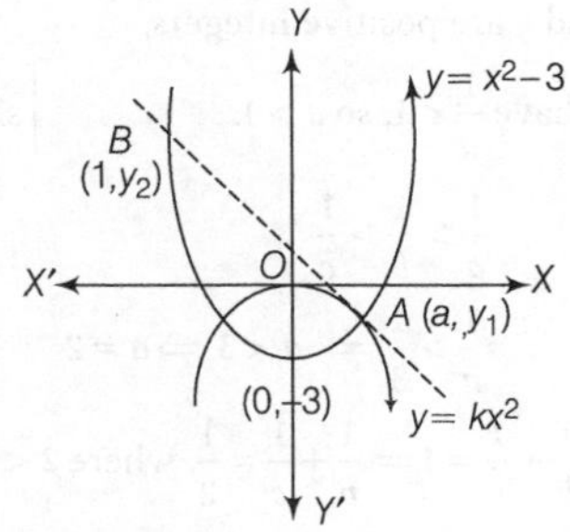

Now, $\quad y = kx^2$

$\Rightarrow \quad \dfrac{dy}{dx} = 2kx$

$\therefore \quad \left(\dfrac{dy}{dx}\right)_{(a, y_1)} = 2ka = \dfrac{y_2 - y_1}{1 - a}$

But, $\quad y_2 = 1 - 3 = -2$

$\therefore \quad 2ka = \dfrac{-2 - (a^2 - 3)}{1 - a}$

$\Rightarrow \quad 2ka = \dfrac{1 - a^2}{1 - a} = 1 + a$

$\Rightarrow \quad 2ak = 1 + a \quad \ldots\text{(ii)}$

Substituting $k = \dfrac{a^2 - 3}{a^2}$ from Eq. (i) in Eq. (ii), we get

$$\frac{2a(a^2 - 3)}{a^2} = 1 + a$$

$\Rightarrow \quad 2a^2 - 6 = a + a^2$

$\Rightarrow \quad a^2 - a - 6 = 0$

$\Rightarrow \quad a = 3, -2$

$\therefore \quad a = 3 \quad [\because -2 \text{ is rejected as } a > 0]$

73. $f'(c) = 16c - 7 = \dfrac{f(6) - f(-6)}{12}$

$$= \frac{(8 \cdot 36 - 7 \cdot 6 + 5) - (8 \cdot 36 + 7 \cdot 6 + 5)}{12}$$

$$= -\frac{2 \cdot 7 \cdot 6}{12} = -7$$

$\Rightarrow 16c = 0 \Rightarrow c = 0$

74. Using LMVT for f in $[1, 2]$, we get, for some $c \in (1, 2)$

$$\frac{f(2) - f(1)}{2 - 1} = f'(c) \leq 2$$

$$f(2) - f(1) \leq 2$$

$\Rightarrow \quad f(2) \leq 4 \quad \ldots\text{(i)}$

Again, using LMVT in $[2, 4]$, we get, for some $d \in (2, 4)$

$\Rightarrow \quad \dfrac{f(4) - f(2)}{4 - 2} = f'(d) \leq 2$

$\therefore\quad f(4)-f(2)\le 4,\quad 8-f(2)\le 4,\quad 4\le f(2)$

$\Rightarrow\quad f(2)\ge 4$ ...(ii)

From Eqs. (i) and (ii), we get $f(2)=4$

75. As, a, b and c are positive integers.

We must have $\frac{1}{a}<1$, so $a>1$. $\left[\text{similarly, } \frac{1}{b}<1, \frac{1}{c}<1\right]$

Since, $\frac{1}{a}>\frac{1}{b}>\frac{1}{c}$

$\Rightarrow\quad \frac{1}{a}>\frac{1}{3}\Rightarrow a<3\Rightarrow a=2$

$\therefore\quad \frac{1}{a}+\frac{1}{b}+\frac{1}{c}=1\Rightarrow\frac{1}{b}+\frac{1}{c}=\frac{1}{2}$, where $2<b<c$.

Similarly, $\frac{1}{b}>\frac{1}{4}$, so $b<4\Rightarrow b=3$

Now, $\frac{1}{a}+\frac{1}{b}+\frac{1}{c}=1$

$\Rightarrow\quad \frac{1}{2}+\frac{1}{3}+\frac{1}{c}=1$

$\Rightarrow\quad c=6$

$\therefore\quad (a+b+c-5)=2+3+6-5=6$

76. Equation of ellipse is $\frac{x^2}{a^2}+\frac{y^2}{b^2}=1$

Equation of tangent at $(a\cos\theta, b\sin\theta)$ is

$$\frac{x}{a}\cos\theta+\frac{y}{b}\sin\theta=1$$

Intercept on the X-axis $=(a\sec\theta)$

Intercept on the Y-axis $=(b\operatorname{cosec}\theta)$

Length of intercept of the tangent by the axes

$$=\sqrt{a^2\sec^2\theta+b^2\operatorname{cosec}^2\theta}$$

Let $\quad l=a^2\sec^2\theta+b^2\operatorname{cosec}^2\theta$

$\Rightarrow\quad \frac{dl}{d\theta}=2a^2\sec^2\theta\tan\theta-2b^2\operatorname{cosec}^2\theta\cot\theta$

Now, $\quad \frac{dl}{d\theta}=0$

$\Rightarrow\quad a^2\sin^4\theta=b^2\cos^4\theta\Rightarrow\frac{a}{b}=\cot^2\theta$

$\Rightarrow\quad \sin^2\theta=\frac{b}{a+b},\quad \cos^2\theta=\frac{a}{a+b}$

Distance between $(a\sec\theta, 0)$ and point of tangency $(a\cos\theta, b\sin\theta)$ is

$$=\sqrt{a^2(\sec\theta-\cos\theta)^2+b^2\sin^2\theta}$$

$$=\sqrt{a^2\cos^2\theta(\sec^2\theta-1)^2+b^2\sin^2\theta}$$

$$=\sqrt{a^2\cos^2\theta\left(\frac{a+b}{a}-1\right)^2+b^2\sin^2\theta}$$

$$=\sqrt{a^2\cos^2\theta\frac{b^2}{a^2}+b^2\sin^2\theta}=b$$

Similarly, distance between $(0, b\operatorname{cosec}\theta)=a$

77. Let (x_1, y_1) be a point of contact of tangents from the origin (0, 0) to the curve $y=\sin x$.

Here, $\quad y=\sin x$

$\therefore\quad \frac{dy}{dx}=\cos x$

$\Rightarrow\quad \left(\frac{dy}{dx}\right)_{(x_1, y_1)}=\cos x_1$

Now, equation of tangent at (x_1, y_1) is

$$y-y_1=\left(\frac{dy}{dx}\right)_{(x_1, y_1)}(x-x_1)$$

$\Rightarrow\quad y-y_1=(\cos x_1)(x-x_1)$

$\because$ It passes through (0, 0).

$\therefore\quad -y_1=(\cos x_1)(-x_1)$ ...(i)

Also, (x_1, y_1) lies on the curve.

So, $\quad y_1=\sin x_1$ ...(ii)

Squaring and adding Eqs. (i) and (ii), we get

$$\left(\frac{y_1}{x_1}\right)^2+y_1^2=\cos^2 x_1+\sin^2 x_1=1$$

or $\quad y_1^2+x_1^2y_1^2=x_1^2$

i.e. $\quad (x_1^2+1)y_1^2=x_1^2$

$\therefore$ The point of contact (x_1, y_1) lies on the curve $y^2=\frac{x^2}{x^2+1}$.

78. We must show that for a given $m\in R$ these exists $x\in R$ such that,

$$m^2x^2+\int_0^x f(t)\,dt=2$$

Let $\quad f(x)=\int_0^x[2m^2t+f(t)]\,dt-2,\ x\in R$

Since, $f(x)$ is continuous and $2m^2t^2$ is continuous, therefore

$\int_0^x[2m^2t+f(t)]\,dt$ continuous on R,

$\therefore\quad f$ is continuous on R, also

$$f(0)=\int_0^0[2m^2t+f(t)]\,dt-2=0-2=-2$$

and $\quad f(x)=\int_0^x[2m^2t+f(t)]\,dt-2$

where, $\quad f(x)=m^2x^2+\int_0^x f(t)\,dt-2\to\infty$

$\left[\because \int_0^x f(t)\,dt\to\infty\right]$

As, $\quad |x|\to\infty$

Thus, these exists some $a\in R$ such that;

$$f(x)>1,\ \text{for } |x|>a$$

Note that f is a continuous on $[0, a+1]$ and $f(0)\,f(a+1)<0$. By the intermediate value theorem of continuous functions, we have that there exists some $b\in(0, a+1)$ such that $f(b)=0$, i.e. there exists a real β which satisfies the equation

$$m^2x^2+\int_0^x f(t)\,dt=2$$

79. Tangent at any point $P(t_1)$, i.e. $(4t_1^2+3, 8t_1^3-1)$ be normal to the curve at $Q(t_2)$, i.e. $(4t_2^2+3, 8t_2^3-1)$.

The equation of the tangent at t_1 is

$$y-(8t_1^3-1)=\left(\frac{dy}{dx}\right)_{t_1}\cdot\{x-(4t_1^2+3)\}$$

or $$y-(8t_1^3-1)=\left(\frac{dy/dt}{dx/dt}\right)_{t_1}\cdot\{x-(4t_1^2+3)\}$$

or $$y-(8t_1^3-1)=\frac{24t_1^2}{8t_1}\cdot\{x-(4t_1^2+3)\}$$

or $$y-(8t_1^3-1)=3t_1\{x-(4t_1^2+3)\} \quad \text{...(i)}$$

Clearly, slope of tangent at t_1 = slope of tangent at t_2.

$$\therefore \left(\frac{dy}{dx}\right)_{t_1}=\frac{-1}{\left(\frac{dy}{dx}\right)_{t_2}} \quad i.e.\ 3t_1=\frac{-1}{3t_2} \quad \text{...(ii)}$$

$\Rightarrow$ Equation of normal at t_2 is $y-(8t_2^{\,3}-1)=3t_1\{x-(4t_2^{\,3}+3)\}$ ...(iii)

On subtracting Eq. (iii) from Eq. (i), we get

$$(8t_2^3-1)-(8t_1^3-1)=3t_1\{(4t_2^2+3)-(4t_1^2+3)\}$$

$$\Rightarrow 2t_2^2=t_1t_2+t_1^2$$

$$\Rightarrow 2\cdot\left(\frac{-1}{9t_1}\right)^2=-\frac{1}{9}+t_1^2 \quad \text{[using Eq. (ii)]}$$

$$\Rightarrow 2=-9t_1^2+81t_1^4$$

$$\therefore 81t_1^4-9t_1^2-2=0$$

$$\Rightarrow t_1=\pm\frac{\sqrt{2}}{3}$$

Putting in Eq. (i), the equation is

$$27(y+1)\mp16\sqrt{2}=\pm\sqrt{2}(27x-105)$$

80. Equation of tangent at (x, y) is

$$Y-y=\frac{dx}{dy}(X-x)$$

$$\therefore Q=\left(x-y\frac{dx}{dy},0\right)$$

Equation of normal at (x, y) is

$$Y-y=-\frac{dx}{dy}(X-x)$$

$$\therefore R=\left(x+y\frac{dy}{dx},0\right)$$

Given, $QR=2$

$$\Rightarrow y\frac{dy}{dx}+y\frac{dx}{dy}=2$$

$$\Rightarrow y\left(\frac{dy}{dx}\right)^2-2\left(\frac{dy}{dx}\right)+y=0$$

$$\Rightarrow \frac{dy}{dx}=\frac{2\pm\sqrt{4-4y^2}}{2y}=\frac{1\pm\sqrt{1-y^2}}{y}$$

$$\Rightarrow \frac{y\,dy}{1\pm\sqrt{1-y^2}}=dx \text{ or } \frac{1\mp\sqrt{1-y^2}}{y}dy=dx$$

On integrating both the sides, we get

$$\Rightarrow \log y\mp\left(\log\left|\frac{1-\sqrt{1-y^2}}{y}\right|+\sqrt{1-y^2}\right)=x+c$$

The curve passes through (1,1), so $c=-1$

Hence, the possible curves

$$\log y-x=\pm\left(\log\left|\frac{1-\sqrt{1-y^2}}{y}\right|+\sqrt{1-y^2}\right)-1$$

81. Let the point $P(x, y)$ be on the curve,

$$\log(x^2+y^2)=c\tan^{-1}\left(\frac{y}{x}\right)$$

Differentiating both the sides w.r.t. 'x', we get

$$\frac{2x+2yy'}{(x^2+y^2)}=\frac{c(xy'-y)}{(x^2+y^2)}$$

$$\Rightarrow y'=\frac{2x+cy}{cx-2y}=m_1 \text{ (say)}$$

Slope of $OP=\frac{y}{x}=m_2$ (say)

Let the angle between the tangent at P and OP be θ.

Then, $$\tan\theta=\left|\frac{m_1-m_2}{1+m_1m_2}\right|=\left|\frac{\frac{2x+cy}{cx-2y}-\frac{y}{x}}{1+\frac{2xy+cy^2}{x(cx-2y)}}\right|=\frac{2}{c}$$

$$\therefore \theta=\tan^{-1}\left(\frac{2}{c}\right)=\text{constant.}$$

Hence, the angle between the tangent at any point P and the line joining P to the origin O is the same.

82. (i) The given curves are

$$y^2=4ax \quad \text{...(i)}$$

and $$x^2=4ay \quad \text{...(ii)}$$

Point of intersection of Eqs. (i) and (ii) are $(0, 0)$ and $(4a, 4a)$.

From Eq. (i), $$\frac{dy}{dx}=\frac{2a}{y}=m_1 \text{ (say)}$$

From Eq. (ii), $$\frac{dy}{dx}=\frac{x}{2a}=m_2 \text{ (say)}$$

Let the angle of intersection of two curves is θ, then

$$\tan\theta=\left|\frac{m_1-m_2}{1+m_1m_2}\right|=\left|\frac{\frac{2a}{y}-\frac{x}{2a}}{1+\frac{x}{y}}\right|=\left|\frac{4a^2-xy}{2a(x+y)}\right|$$

$$\therefore (\tan\theta)_{(0,0)}=\infty \text{ or } \theta=90^\circ$$

and $$(\tan\theta)_{(4a,4a)}=\left|\frac{\frac{1}{2}-2}{1+1}\right|=\left|-\frac{3}{4}\right|=\frac{3}{4}$$

Hence, $$\theta=\tan^{-1}\left(\frac{3}{4}\right)$$

(ii) The given curves are $y^2 = 4ax$ ...(i)

and $x^2 = 4ay$ ...(ii)

Tangents of Eq. (i) in terms of slope m is

$$y = mx + \frac{a}{m} \quad \text{...(iii)}$$

Now, Eq. (iii) is also tangent of Eq. (ii).

Eliminating y from Eqs. (ii) and (iii), we have

$$x^2 = 4a\left(mx + \frac{a}{m}\right)$$

$$\Rightarrow \quad x^2 - 4amx - \frac{4a^2}{m} = 0 \quad \text{...(iv)}$$

$$\Rightarrow \quad B^2 - 4AC = 0$$

$$\Rightarrow \quad 16a^2m^2 - 4\left(-\frac{4a^2}{m}\right) = 0$$

$$\Rightarrow \quad m^3 = -1 \text{ or } m = -1$$

From Eq. (iii) common tangent is

$$y = -x - a \text{ or } y + x + a = 0$$

Hence, the common tangent is $x + y + a = 0$.

83. Let the coordinate system be chosen such that the given straight line is $x = p$ and the equations of the circles are $x^2 + y^2 = a^2, x^2 + y^2 = b^2, x^2 + y^2 = c^2$.

The line $x = p$ cuts these circles at A, B and C, respectively.

The coordinates of these points are $A(p, \sqrt{a^2 - p^2})$, $B(p, \sqrt{b^2 - p^2})$ and $C(p, \sqrt{c^2 - p^2})$.

Equations of the tangents at these points are

$$px + \sqrt{a^2 - p^2}y = a^2, px + \sqrt{b^2 - p^2}\, y = b^2$$

and $px + \sqrt{c^2 - p^2}\, y = c^2$.

These tangents intersect at

$$\left[\frac{p^2 - \sqrt{a^2 - p^2}\sqrt{b^2 - p^2}}{p}, \sqrt{a^2 - p^2} + \sqrt{b^2 - p^2}\right],$$

$$\left[\frac{p^2 - \sqrt{b^2 - p^2}\sqrt{c^2 - p^2}}{p}, \sqrt{b^2 - p^2} + \sqrt{c^2 - p^2}\right],$$

$$\left[\frac{p^2 - \sqrt{c^2 - p^2}\sqrt{c^2 - p^2}}{p}, \sqrt{c^2 - p^2} + \sqrt{a^2 - p^2}\right].$$

Area of Δ formed by the tangents at A, B, C is

$$\Delta = \frac{1}{2}\begin{vmatrix} \frac{p^2 - \sqrt{a^2 - p^2}\sqrt{b^2 - p^2}}{p} & \sqrt{a^2 - p^2} + \sqrt{b^2 - p^2} & 1 \\ \frac{p^2 - \sqrt{b^2 - p^2}\sqrt{c^2 - p^2}}{p} & \sqrt{b^2 - p^2} + \sqrt{c^2 - p^2} & 1 \\ \frac{p^2 - \sqrt{c^2 - p^2}\sqrt{c^2 - p^2}}{p} & \sqrt{c^2 - p^2} + \sqrt{a^2 - p^2} & 1 \end{vmatrix}$$

$$= \frac{(\sqrt{a^2 - p^2} - \sqrt{c^2 - p^2})(\sqrt{c^2 - p^2} - \sqrt{b^2 - p^2})(\sqrt{b^2 - p^2} - \sqrt{a^2 - p^2})}{2p}$$

$$= \frac{CA \cdot BC \cdot AB}{2p}$$

84. Let the curve be $y = f(x)$ and tangent drawn at $P(x, y)$ meets the X-axis at T.

We have, $OT = x^2$

Equation of tangent at $P(x, y)$;

$$Y - y = f'(x)(X - x)$$

$$\Rightarrow \quad T \equiv \left(\left|x - \frac{y}{f'(x)}\right|, 0\right)$$

$$\Rightarrow \quad \left|x - \frac{y}{f'(x)}\right| = x^2$$

$$\Rightarrow \quad x - \frac{f(x)}{f'(x)} = \pm x^2$$

$$\Rightarrow \quad \frac{x f'(x) - f(x)}{x^2} = \pm f'(x)$$

$$\Rightarrow \quad \frac{d}{dx}\left(\frac{f(x)}{x}\right) = \pm f'(x)$$

On integrating both the sides, we get $\frac{f(x)}{x} = \pm f(x) + c$

Since, the curve passes through (2, 1).

$$\therefore \quad \frac{1}{2} = \pm 1 + c$$

$$\Rightarrow \quad c = -\frac{1}{2}, \frac{3}{2}$$

$$\Rightarrow \quad f(x) = \frac{x}{2(x-1)} \text{ or } f(x) = \frac{3x}{2(1+x)}$$

Hence, possible curves are $y = \frac{x}{2(x-1)}$ and $y = \frac{3x}{2(1+x)}$.

85. For $y = x - x^3, \frac{dy}{dx} = 1 - 3x^2$

Therefore, the equation of the tangent at the point $P(x_1, y_1)$ is

$$y - y_1 = (1 - 3x_1^2)(x - x_1)$$

It meets the curve again at $Q(x_2, y_2)$.

Hence, $\quad x_2 - x_2^3 - (x_1 - x_1^3) = (1 - 3x_1^2)(x_2 - x_1)$

$$\Rightarrow (x_2 - x_1)[1 - (x_2^2 + x_1x_2 + x_1^2)] = (x_2 - x_1)(1 - 3x_1^2)$$

$$\Rightarrow \quad 1 - x_2^2 - x_1x_2 - x_1^2 = 1 - 3x_1^2$$

$$\Rightarrow \quad x_2^2 + x_1x_2 - 2x_1^2 = 0$$

$$\Rightarrow \quad \left(x_2 + \frac{x_1}{2}\right)^2 = \frac{9x_1^2}{4} \Rightarrow x_2 + \frac{x_1}{2} = \pm\frac{3x_1}{2}$$

Since, $x_1 \neq x_2$, we have $x_2 = -2x_1$

$\Rightarrow$ Q is $(-2x_1, -2x_1 + 8x_1^3)$.

If $L_1(\alpha, \beta)$ is the point of trisection of PQ, then $\alpha = \frac{2x_1 - 2x_1}{3} = 0$. Hence, L_1 lies on the Y-axis. If $L_2(h, k)$ is the other point of trisection, then $h = \frac{x_1 - 4x_1}{3} = -x_1$ and

$$k = \frac{y_1 - 4x_1 + 16x_1^3}{3}$$

i.e. $$k = \frac{x_1 - x_1^3 - 4x_1 + 16x_1^3}{3} = -x_1 + 5x_1^3$$

$$\Rightarrow \quad k = h - 5h^3$$

$\therefore$ Locus of (h, k) is $y = x - 5x^3$.

86. Let the curve be $y = f(x)$

The equation of the tangent at any point (x, y) is

$$Y - y = \frac{dy}{dx}(X - x)$$

Its intercept on the y-axis is given by $(X = 0)$

$$Y = y - x\frac{dy}{dx} = k\sqrt{x^2 + y^2}$$

So, $x\frac{dy}{dx} - y + k\sqrt{x^2 + y^2} = 0$ is the differential equation governing the curve. This can be written as

$$\frac{dy}{dx} = \frac{y - k\sqrt{x^2 + y^2}}{x}$$

Let $y = vx$, so that $\frac{dy}{dx} = v + x\frac{dv}{dx}$

The differential equation becomes

$$v + x\frac{dv}{dx} = v - k\sqrt{1 + v^2}$$

$$\Rightarrow \quad \frac{dv}{\sqrt{1 + v^2}} + k\frac{dx}{x} = 0$$

$$\Rightarrow \quad \log|v + \sqrt{1 + v^2}| + k\log|x| = c \quad \text{(on integrating)}$$

$$\Rightarrow \quad \log\left|\frac{y}{x} + \sqrt{1 + \frac{y^2}{x^2}}\right| + k\log|x| = c$$

$$\Rightarrow \quad \log|y + \sqrt{x^2 + y^2}| + (k-1)\log|x| = c$$

$$\Rightarrow \quad (y + \sqrt{x^2 + y^2})\, x^{k-1} = c_1$$

87. We have, $2t^3 - 9t^2 + 30 - a = 0$

Any real root t_0 of this equation gives two real and distinct values of x if $|t_0| > 2$.

Thus, we need to find the condition for the equation in t to have three real and distinct roots none of which lies in $[-2, 2]$.

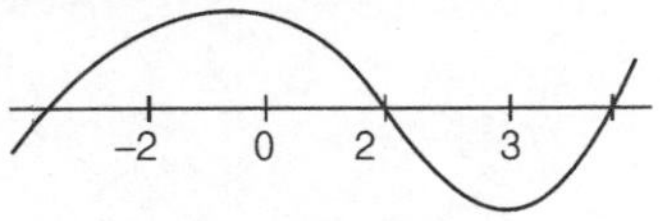

Let $\quad f(t) = 2t^3 - 9t^2 + 30 - a$

$\Rightarrow \quad f'(t) = 6t^2 - 8t = 0 \quad \Rightarrow \quad t = 0, 3$

So, the equation $f(t) = 0$ has three real and distinct roots, if $f(0) \cdot f(3) < 0$

$\Rightarrow \quad (30 - a)(54 - 81 + 30 - a) < 0$

$\Rightarrow \quad (30 - a)(3 - a) < 0$

$\Rightarrow \quad (a - 3)(a - 30) < 0$, using number line rule i.e.,

$\Rightarrow \quad a \in (3, 30)$ ...(i)

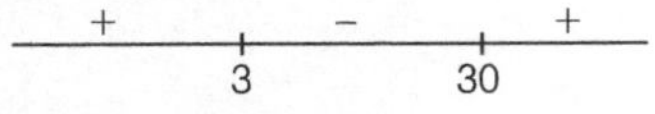

Also, none of the roots lie in $[-2, 2]$

if $\quad f(-2) > 0 \quad$ and $\quad f(2) > 0$

$\Rightarrow \quad (-16 - 36 + 30 - a) > 0 \quad$ and $\quad (16 - 36 + 30 - a) > 0$

$\Rightarrow \quad a + 22 < 0 \quad$ and $\quad a - 10 < 0$

$\Rightarrow \quad a < -22$ and $\quad a < 10$

$\Rightarrow \quad a < -22$ ...(ii)

From Eqs. (i) and (ii) no real value of a exists.

88. Slope of tangent at the point (x_1, y_1) is given by $\left(\frac{dy}{dx}\right)_{(x_1, y_1)}$.

Given curve, $\quad (y - x^5)^2 = x(1 + x^2)^2$

$$\Rightarrow \quad 2(y - x^5)\left(\frac{dy}{dx} - 5x^4\right) = (1 + x^2)^2 + 2x(1 + x^2) \cdot 2x$$

Put $x = 1$ and $y = 3$, then

$$dy/dx = 8$$

89. Given, $f(x) = 2 + \cos x, \forall x \in R$

Statement I There exists a point $\in [t, t + \pi]$, where $f'(c) = 0$

Hence, Statement I is true.

Statement II $f(t) = f(t + 2\pi)$ is true. But Statement II is not correct explanation for Statement I.

90. As, $\quad |f(x_1) - f(x_2)| \le (x_1 - x_2)^2, \forall x_1, x_2 \in R$

$\Rightarrow \quad |f(x_1) - f(x_2)| \le |x_1 - x_2|^2 \quad$ [as $x^2 = |x|^2$]

$$\therefore \quad \left|\frac{f(x_1) - f(x_2)}{x_1 - x_2}\right| \le |x_1 - x_2|$$

$$\Rightarrow \quad \lim_{x_1 \to x_2}\left|\frac{f(x_1) - f(x_2)}{x_1 - x_2}\right| \le \lim_{x_1 \to x_2}|x_1 - x_2|$$

$$\Rightarrow \quad |f'(x_1)| \le 0, \forall x_1 \in R$$

$\therefore \quad |f'(x)| \le 0$, which shows $|f'(x)| = 0$

[as modulus is non-negative or $|f'(x)| \ge 0$]

$\therefore \quad f'(x) = 0$ or $f(x)$ is constant function.

$\Rightarrow$ Equation of tangent at (1, 2) is

$$\frac{y - 2}{x - 1} = f'(x)$$

or $\quad y - 2 = 0 \quad [\because f'(x) = 0]$

$\therefore \quad y - 2 = 0$ is required equation of tangent.

91. Given, $\quad y^3 + 3x^2 = 12y$ ...(i)

On differentiating w.r.t. x, we get

$$3y^2\frac{dy}{dx} + 6x = 12\frac{dy}{dx}$$

$$\Rightarrow \quad \frac{dy}{dx} = \frac{6x}{12 - 3y^2}$$

$$\Rightarrow \quad \frac{dx}{dy} = \frac{12 - 3y^2}{6x}$$

For vertical tangent,

$$\frac{dx}{dy} = 0$$

$$\Rightarrow \quad 12 - 3y^2 = 0$$

$$\Rightarrow \quad y = \pm 2$$

On putting, $y = 2$ in Eq. (i), we get $x = \pm\frac{4}{\sqrt{3}}$ and again putting $y = -2$ in Eq. (i), we get $3x^2 = -16$, no real solution.

So, the required point is $\left(\pm\frac{4}{\sqrt{3}}, 2\right)$.

92. Slope of tangent to the curve, $y = f(x)$ is

$$\left.\frac{dy}{dx}\right|_{(3, 4)} = f'(x)_{(3, 4)}$$

Therefore, slope of normal $= -\dfrac{1}{f'(x)_{(3,4)}} = -\dfrac{1}{f'(3)}$

But $$-\frac{1}{f'(3)} = \tan\left(\frac{3\pi}{4}\right) \quad \text{[given]}$$

$$\Rightarrow \quad \frac{-1}{f'(3)} = \tan\left(\frac{\pi}{2}+\frac{\pi}{4}\right) = -1$$

$$f'(3) = 1$$

93. We have, $y = \dfrac{x+6}{(x-2)(x-3)}$

Point of intersection with Y-axis (0, 1)

$$y' = \frac{(x^2-5x+6)(1)-(x+6)(2x-5)}{(x^2-5x+6)^2}$$

$\Rightarrow y' = 1$ at point (0, 1).

$\therefore$ Slope of normal is -1.

Hence, equation of normal is $x + y = 1$.

$\therefore \left(\dfrac{1}{2}, \dfrac{1}{2}\right)$ satisfy it.

94. Given equation of curve is

$$x^2 + 2xy - 3y^2 = 0 \quad \text{...(i)}$$

On differentiating w.r.t. x, we get

$$2x + 2xy' + 2y - 6yy' = 0 \Rightarrow y' = \frac{x+y}{3y-x}$$

At $x = 1, y = 1, y' = 1$ i.e. $\left(\dfrac{dy}{dx}\right)_{(1,1)} = 1$

Equation of normal at (1, 1) is

$$y - 1 = -\frac{1}{1}(x-1) \Rightarrow y - 1 = -(x-1)$$

$$\Rightarrow \quad x + y = 2 \quad \text{...(ii)}$$

On solving Eqs. (i) and (ii) simultaneously, we get

$$\Rightarrow \quad x^2 + 2x(2-x) - 3(2-x)^2 = 0$$

$$\Rightarrow \quad x^2 + 4x - 2x^2 - 3(4 + x^2 - 4x) = 0$$

$$\Rightarrow \quad -x^2 + 4x - 12 - 3x^2 + 12x = 0$$

$$\Rightarrow \quad -4x^2 + 16x - 12 = 0$$

$$\Rightarrow \quad 4x^2 - 16x + 12 = 0$$

$$\Rightarrow \quad x^2 - 4x + 3 = 0$$

$$\Rightarrow \quad (x-1)(x-3) = 0$$

$$\therefore \quad x = 1, 3$$

Now, when $x = 1$, then $y = 1$

and when $x = 3$, then $y = -1$.

$\therefore \quad P = (1, 1)$ and $Q = (3, -1)$

Hence, normal meets the curve again at (3, −1) in fourth quadrant.

Aliter

Given, $$x^2 + 2xy - 3y^2 = 0$$

$$\Rightarrow \quad (x-y)(x+3y) = 0$$

$$\Rightarrow \quad x - y = 0 \text{ or } x + 3y = 0$$

Equation of normal at (1, 1) is

$$y - 1 = -1(x-1)$$

$$\Rightarrow \quad x + y - 2 = 0$$

It intersects $x + 3y = 0$ at (3, −1) and hence normal meets the curve again in fourth quadrant.

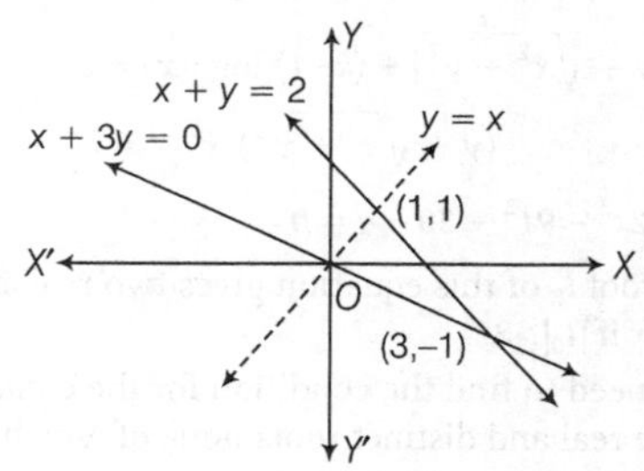

CHAPTER

08

Monotonicity, Maxima and Minima

Learning Part

Session 1
- Monotonicity

Session 2
- Critical Points

Session 3
- Comparison of Functions Using Calculus

Session 4
- Introduction to Maxima and Minima
- Methods of Finding Extrema of Continuous Functions
- Convexity/Concavity and Point of Inflection
- Concept of Global Maximum/Minimum

Session 5
- Maxima and Minima of Discontinuous Functions

Practice Part

- JEE Type Examples
- Chapter Exercises

Arihant on Your Mobile !
Exercises with this symbol can be practised on your mobile. See title inside to activate for free.

Session 1

Monotonicity

Monotonicity

A function is said to be monotonic if it is either increasing or decreasing in its domain.

Strictly Increasing

Consider a function represented by the following graph.

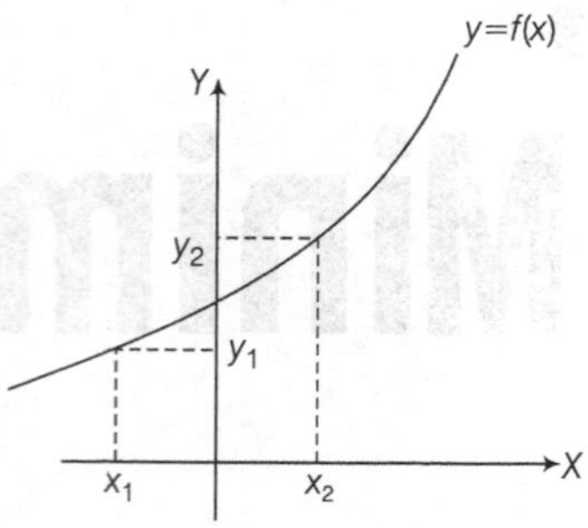

Figure 8.1

i.e. $x_1 < x_2$

implies $f(x_1) < f(x_2)$

Such a function is called a strictly increasing function or a monotonically increasing function.

For two different input arguments x_1 and x_2, where $x_1 < x_2$, $y_1 = f(x_1)$ will always be less than $y_2 = f(x_2)$.

The word 'monotonically' apparently has its origin in the word monotonous.

e.g. A monotonous routine is one in which one follows the same routine repeatedly or continuously.

Similarly, a monotonically increasing function is one that increases continuously.

e.g. consider $f(x) = [x]$.

For this function $x_1 < x_2$ does not always imply $f(x_1) < f(x_2)$.

However, $x_1 < x_2$ does imply $f(x_1) \le f(x_2)$.

In other words, $f(x) = [x]$ is not strictly (or monotonically) increases. It will nevertheless be termed **increasing**.

Strictly Decreasing

Now, we consider a function represented by the following graph.

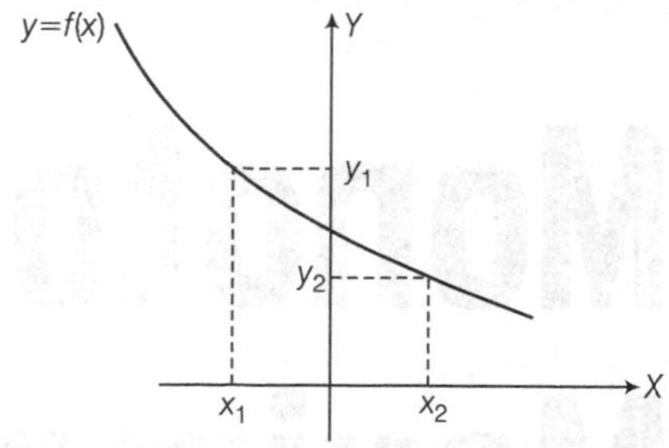

Figure 8.2

For two different input arguments x_1 and x_2, where $x_1 < x_2$, $y_1 = f(x_1)$ will always be greater than $y_2 = f(x_2)$.

i.e. $x_1 < x_2 \Rightarrow f(x_1) > f(x_2)$

Such a function is called a strictly decreasing function or a monotonically decreasing function. e.g. consider $f(x) = -[x]$.

For this function $x_1 < x_2$ does not imply $f(x_1) > f(x_2)$.

However, $x_1 < x_2 \Rightarrow f(x_1) \ge f(x_2)$

In other words, $f(x) = -[x]$ is not strictly decreasing. It would only be termed **decreasing**. The following table list down a few Examples of functions and their behaviour in different intervals.

Function	Behaviour
$f(x) = x$	Strictly increasing on R
$f(x) = x^2$	Strictly decreasing on $(-\infty, 0]$ Strictly increasing on $[0, \infty)$
$f(x) = \sqrt{x}$	Strictly increasing on $[0, \infty)$
$f(x) = x^3$	Strictly increasing on R
$f(x) = \|x\|$	Strictly decreasing on $(-\infty, 0]$ Strictly increasing on $[0, \infty)$
$f(x) = \frac{1}{x}$	Neither decreasing nor increasing on R. Strictly decreasing on $(-\infty, 0)$, Strictly decreasing on $(0, \infty)$
$f(x) = [x]$	Increasing on R
$f(x) = \{x\}$	Neither increasing nor decreasing on R. However, strictly increasing on $[n, n+1)$, where $n \in Z$
$f(x) = \sin x$	Neither increasing nor decreasing on R. Strictly increasing on $\left[\left(2n - \frac{1}{2}\right)\pi, \left(2n + \frac{1}{2}\right)\pi\right]; n \in Z$ Strictly decreasing on $\left[\left(2n + \frac{1}{2}\right)\pi, \left(2n + \frac{3}{2}\right)\pi\right]; n \in Z$

Function	Behaviour
$f(x)=\cos x$	Neither increasing nor decreasing on R. Strictly increasing on $[(2n-1)\pi, 2n\pi]; n \in Z$ Strictly decreasing on $[2n\pi, (2n+1)\pi]; n \in Z$
$f(x)=\tan x$	Neither increasing nor decreasing on R. Strictly increasing on $\left[\left(n-\frac{1}{2}\right)\pi, \left(n+\frac{1}{2}\right)\pi\right]; n \in Z$
$f(x)=e^x$	Strictly increasing on R
$f(x)=e^{-x}$	Strictly decreasing on R
$f(x)=\ln x$	Strictly increasing on $(0, \infty)$

Monotonicity with the Help of Derivative

Let us now deduce the condition(s) on the derivative of a function $f(x)$ which determines whether $f(x)$ is increasing/decreasing on a given interval. We are assuming that $f(x)$ is everywhere differentiable.

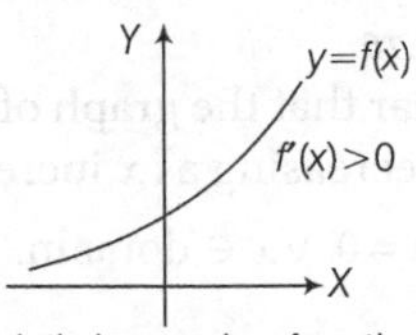

Strictly increasing function

Figure 8.3

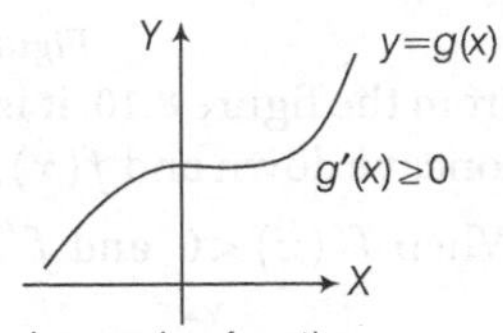

Increasing function

Figure 8.4

In figure 8.3, the function $y=f(x)$ is a strictly increasing function. Notice that the slope of the tangent drawn at any point on this curve is always positive.

Hence, a sufficient condition for $f(x)$ to be **strictly increasing** on a given domain D is $\boldsymbol{f'(x)>0\ \forall\ x \in D}$ (Later on, we will see that this is not a necessary condition for a function to be strictly increasing).

In figure 8.4, the function $y=g(x)$ is not strictly increasing though it is increasing. Notice that $g'(x)>0$ or $g'(x)=0, \forall x$. $g'(x)$ is never negative.

Hence, a sufficient condition for $g(x)$ to be **increasing** on a given domain D is $\boldsymbol{g'(x) \geq 0,\ \forall\ x \in D}$. Now, consider $f(x)$ and $g(x)$ in given figure.

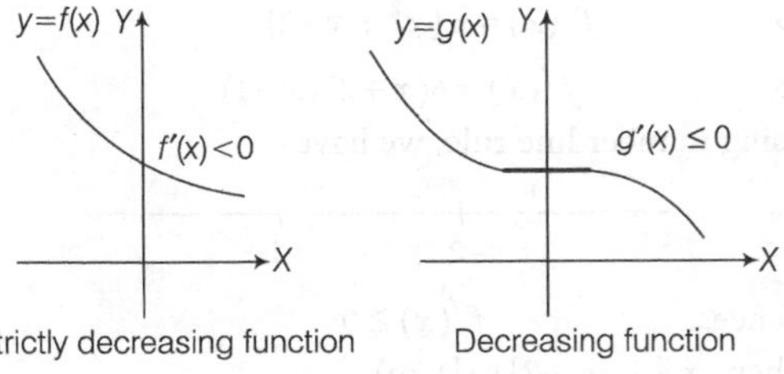

Strictly decreasing function   Decreasing function

Figure 8.5

Extending as above, we get the conditions for a **strictly decreasing and decreasing** functions

Strictly decreasing : $f'(x)<0\ \forall\ x \in D$

Decreasing : $g'(x) \leq 0\ \ \forall\ x \in D$

Note

(i) Above conditions on the derivatives to be applied, the function must be differentiable in the given domain. However, these conditions will hold good even if the function is non differentiable, at a finite number (or infinitely countable number) of points. e.g. $f(x)=[x]+\sqrt{\{x\}}$ is strictly increasing on R. However, $f(x)$ is non-differentiable at all integers (a countable set).

(ii) A function must be continuous for above conditions to be applied. Consider $y=\{x\}$. This is non-differentiable (due to discontinuities) at all integers. At all other points, $y'=1>0$. However, we know that $y=\{x\}$ is not strictly increasing. Similarly, $y=\frac{1}{x}$ is non-differentiable (and non-continuous) at $x=0$. At all other points, $y'=\frac{-1}{x^2}<0$, so y should be strictly decreasing on $R\setminus\{0\}$. However, it is not strictly decreasing on $R\setminus\{0\}$ although it is strictly decreasing in the separate intervals $(-\infty,0)$ and $(0,\infty)$.
Therefore, we see that discontinuous functions cannot be subjected to the derivative condition. Even though they may be discontinuous only at finite (on infinitely countable) number of points.

Properties

(i) If $f(x)$ is strictly increasing, then $f^{-1}(x)$ is also strictly increasing. Similarly, if $f(x)$ is strictly decreasing, then $f^{-1}(x)$ is also strictly decreasing.

(ii) If $f(x)$ and $g(x)$ have the same monotonicity (both increasing or decreasing) on $[a,b]$, then $f(g(x))$ and $g(f(x))$ are monotonically increasing on $[a,b]$.

(iii) If $f(x)$ and $g(x)$ have opposite monotonicity on $[a,b]$, then $f(g(x))$ and $g(f(x))$ are strictly decreasing on $[a,b]$.

(iv) The inverse of a continuous function is continuous

(v) If $f'(x)>0, \forall\ x \in (a,b)$ except for a finite (or an infinitely countable) number of points, where $f'(x)=0$, $f(x)$ is still strictly increasing on (a,b). That is why, we said earlier that $f'(x)>0, \forall\ x \in D$ is not necessary condition for strict increase. e.g. In a later Example, we will consider the graph of the function $f(x)=x+\cos x$. We will see that $f'(x)=1-\sin x$ is not always positive $\left(\text{at } x=2n\pi+\frac{\pi}{2}, n \in Z, f'(x)=0\right)$; even then, $f(x)$ increases strictly because the points at which $f'(x)=0$ are countable.

(vi) Similarly if $f'(x)<0, \forall\ x \in (a,b)$ except for a finite (or an infinitely countable) number of points, where $f'(x)=0$, $f(x)$ is still strictly decreasing on (a,b).

Classification of Strictly Increasing Functions

Increasing functions can be classified as

(i) **Concave up** When $f'(x) > 0$ and $f''(x) > 0$, $\forall x \in$ domain.

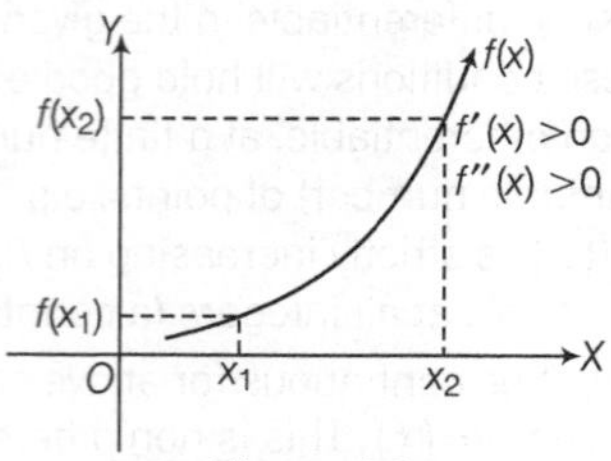

Figure 8.6

From the figure 8.6, it is clear that the graph of $f(x)$ is concave up and increasing as x increases.

(ii) **Concave down** When $f'(x) > 0$ and $f''(x) < 0$, $\forall x \in$ domain.

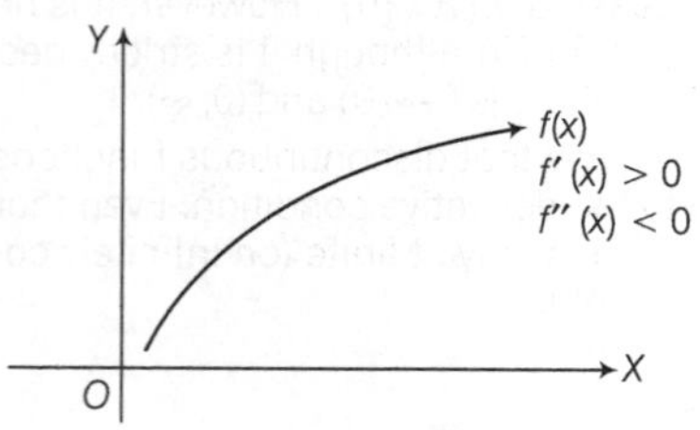

Figure 8.7

From the figure 8.7 it is clear that the graph of $f(x)$ is concave down and increasing as x increases.

(iii) When $f'(x) > 0$ and $f''(x) = 0, \forall x \in$ domain.

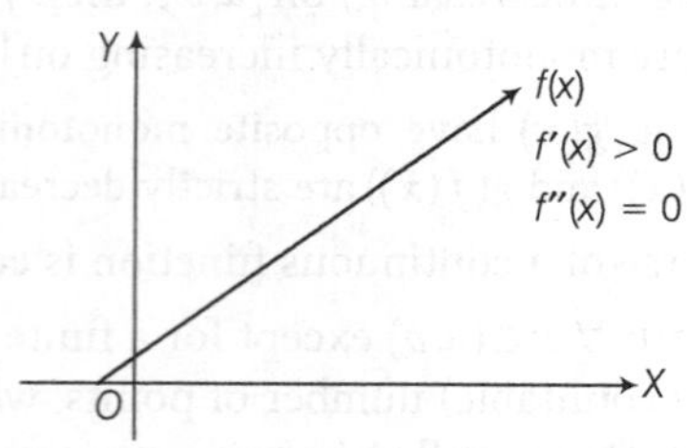

Figure 8.8

From the figure 8.8, it is clear that the graph of $f(x)$ is neither concave up nor concave down but still increasing as x increases.

Classification of Strictly Decreasing Functions

Decreasing functions can be classified as follows :

(i) **Concave up** When $f'(x) < 0$ and $f''(x) > 0, \forall x \in$ domain.

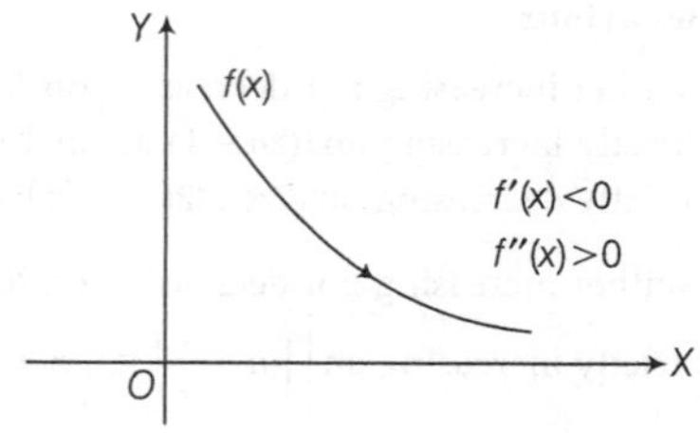

Figure 8.9

From the figure 8.9, it is clear that the graph of the function $f(x)$ is concave up and decreasing as x increases.

(ii) **Concave down** When $f'(x) < 0$ and $f''(x) < 0$, $\forall x \in$ domain.

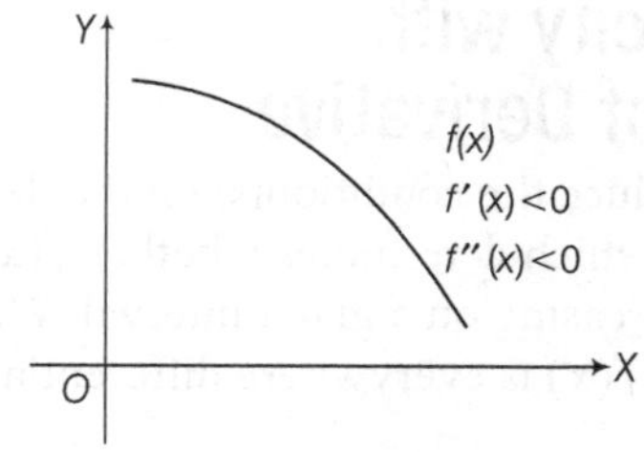

Figure 8.10

From the figure 8.10, it is clear that the graph of $f(x)$ is concave down and $f(x)$ is decreasing as x increases.

(iii) When $f'(x) < 0$ and $f''(x) = 0, \forall x \in$ domain.

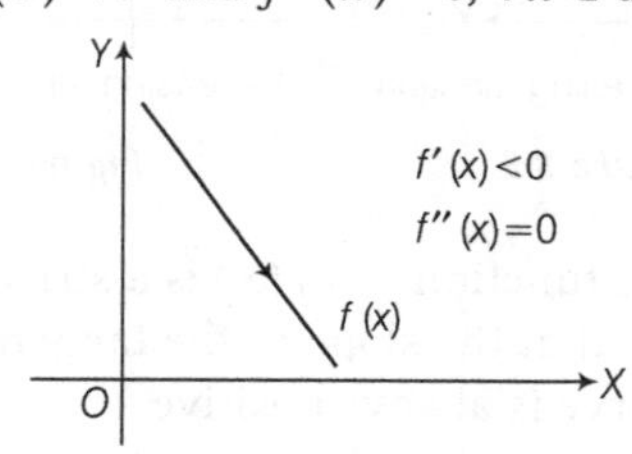

Figure 8.11

From the figure 8.11, it is clear that the graph of $f(x)$ is neither concave up nor concave down but $f(x)$ is decreasing as x increases.

Example 1 Find the interval in which $f(x) = 2x^3 + 3x^2 - 12x + 1$ is increasing.

Sol. Given, $f(x) = 2x^3 + 3x^2 - 12x + 1$

Differentiating both the sides, we have

$$f'(x) = 6x^2 + 6x - 12$$

$$\Rightarrow \quad f'(x) = 6(x^2 + x - 2)$$

$$\Rightarrow \quad f'(x) = 6(x+2)(x-1)$$

Using number line rule, we have

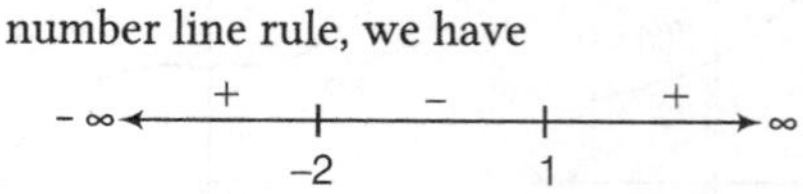

Hence, $\quad f'(x) \geq 0$

when $x \in (-\infty, -2] \cup [1, \infty)$

$\Rightarrow f(x)$ is increasing when $x \in (-\infty, -2] \cup [1, \infty)$

Example 2 Find the interval in which $f(x) = x^3 - 3x^2 - 9x + 20$ is strictly increasing or strictly decreasing.

Sol. Given, $f(x) = x^3 - 3x^2 - 9x + 20$

$$f'(x) = 3x^2 - 6x - 9$$

$$\Rightarrow \quad f'(x) = 3(x^2 - 2x - 3)$$

$$\Rightarrow \quad f'(x) = 3(x-3)(x+1)$$

Using number line method as shown in figure,

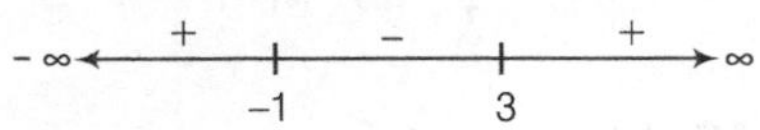

For $f'(x) > 0$
$x \in (-\infty, -1) \cup (3, \infty)$ and $f'(x) < 0$,

For $x \in (-1, 3)$

Thus, $f(x)$ is strictly increasing for $x \in (-\infty, -1) \cup (3, \infty)$ and strictly decreasing for $x \in (-1, 3)$.

Example 3 Show that the function $f(x) = x^2$ is a strictly increasing function on $(0, \infty)$.

Sol. Given, $f(x) = x^2 \Rightarrow f'(x) = 2x$

$\Rightarrow \quad f'(x) > 0$ for $x \in (0, \infty) \qquad [\because x \in (0, \infty) \Rightarrow 2x > 0]$

Thus, $f(x)$ is strictly increasing for $x \in (0, \infty)$.

Example 4 Find the interval of increasing or decreasing of the $f(x) = \int_{-1}^{x} (t^2 + 2t)(t^2 - 1)\, dt$.

Sol. Given, $f(x) = \int_{-1}^{x} (t^2 + 2t)(t^2 - 1)\, dt$

On differentiating both the sides, we have

$$f'(x) = (x^2 + 2x)(x^2 - 1)\left\{\frac{d}{dx}(x)\right\} - (1-2)(1-1)\left\{\frac{d}{dx}(-1)\right\}$$

[using Leibnitz rule]

$$\Rightarrow \quad f'(x) = (x^2 + 2x)(x^2 - 1)$$

$$\Rightarrow \quad f'(x) = x(x+2)(x+1)(x-1)$$

Using number line rule as shown in figure,

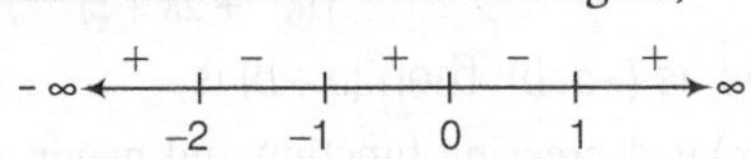

Clearly, $f'(x) \geq 0$ when $x \in (-\infty, -2] \cup [-1, 0] \cup [1, \infty)$

and $f'(x) \leq 0$ when $x \in [-2, -1] \cup [0, 1]$.

Hence, $f(x)$ is increasing, when $x \in (-\infty, -2] \cup [-1, 0] \cup [1, \infty)$ and $f(x)$ is decreasing, when $x \in [-2, -1] \cup [0, 1]$.

Remark

In above example, Leibnitz rule is used which is stated as

$$\frac{d}{dx}\left[\int_{\phi(x)}^{\psi(x)} f(t)\, dt\right] = f(\psi(x))\left\{\frac{d}{dx}\psi(x)\right\} - f(\phi(x))\left\{\frac{d}{dx}\phi(x)\right\}.$$

Example 5 The function $f(x) = \sin^4 x + \cos^4 x$ increasing, if

(a) $0 < x < \pi/8$ (b) $\pi/4 < x < 3\pi/8$
(c) $3\pi/8 < x < 5\pi/8$ (d) $5\pi/8 < x < 3\pi/4$

[IIT JEE 1999]

Sol. Here, $f(x) = \sin^4 x + \cos^4 x$

$$\Rightarrow \quad f'(x) = 4\sin^3 x \cdot \cos x + 4\cos^3 x(-\sin x)$$

$$f'(x) = 4\sin x \cos x(\sin^2 x - \cos^2 x)$$

$$f'(x) = 2(\sin 2x)(-\cos 2x) \Rightarrow f'(x) = -\sin 4x$$

Now, $f'(x) \geq 0$, if $\sin 4x \leq 0 \Rightarrow \pi \leq 4x \leq 2\pi$

$\Rightarrow \quad \pi/4 \leq x \leq \pi/2$

Here, $\left(\frac{\pi}{4}, \frac{3\pi}{8}\right)$ is only subset of $\left[\frac{\pi}{4}, \frac{\pi}{2}\right]$.

Hence, (b) is the correct answer.

Example 6 Let $f(x) = \int_0^x e^t (t-1)(t-2)\, dt$. Then, f decreases in the interval

[IIT JEE 2000]

(a) $(-\infty, -2)$ (b) $(-2, -1)$
(c) $[1, 2]$ (d) $(2, \infty)$

Sol. Here, $f(x) = \int_0^x e^t (t-1)(t-2)\, dt$

$$f'(x) = e^x (x-1)(x-2),$$

[using Leibnitz rule]

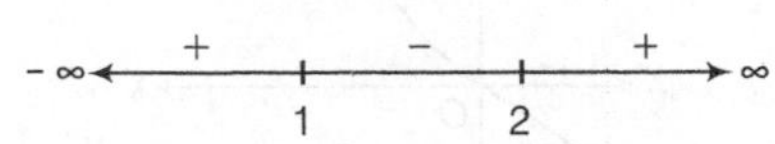

Using number line rule for $f'(x)$, we get

$f'(x) \leq 0$ when $1 \leq x \leq 2$, as e^x is always positive.

$\therefore$ f decreases when $1 \leq x \leq 2$

Hence, (c) is the correct answer.

Example 7 If $f(x) = x \cdot e^{x(1-x)}$, then $f(x)$ is

[IIT JEE 2000]

(a) increasing on $\left[-\frac{1}{2}, 1\right]$ (b) decreasing on R

(c) increasing on R (d) decreasing on $\left[-\frac{1}{2}, 1\right]$

Sol. Here, $f'(x) = x \cdot e^{x(1-x)} \cdot (1-2x) + 1 \cdot e^{x(1-x)}$

$$f'(x) = e^{x(1-x)}[x - 2x^2 + 1]$$

$$f'(x) = -e^{x(1-x)}(x-1)(2x+1)$$

Using number line rule for $f'(x)$ we get, $f'(x) \geq 0$, when $x \in \left[-\frac{1}{2}, 1\right]$ as shown in figure.

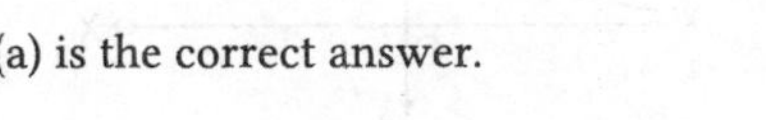

Hence, (a) is the correct answer.

Example 8 Find the interval for which $f(x) = x - \sin x$ is increasing or decreasing.

Sol. Given, $f(x) = x - \sin x$,

Differentiating both the sides w.r.t. x, we have

$$f'(x) = 1 - \cos x$$

We know, $-1 \le \cos x \le 1$

or $\cos x \le 1$

$\Rightarrow$ $1 - \cos x \ge 0$

Therefore, $f'(x) \ge 0, \forall x \in R$

Which shows $f(x)$ is increasing for the entire number scale.

i.e. all real numbers.

Example 9 Discuss the nature of following functions graphically.

(i) $f(x) = x^3$ (ii) $f(x) = \dfrac{1}{|x|}$

(iii) $f(x) = e^x$ (iv) $f(x) = [x]$

Sol. (i) $f(x) = x^3$ can be graphically plotted as shown in the following figure, which shows $f(x) = x^3$ is strictly increasing in R.

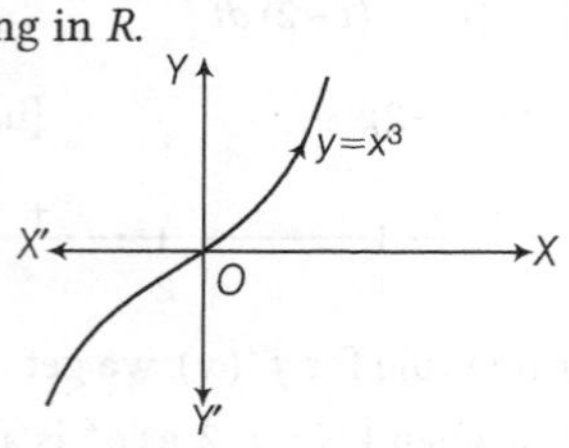

(ii) $f(x) = \dfrac{1}{|x|}$ can be graphically plotted as shown in the following figure, which shows $f(x) = \dfrac{1}{|x|}$ is strictly increasing in $]-\infty, 0[$ and strictly decreasing in $]0, \infty[$.

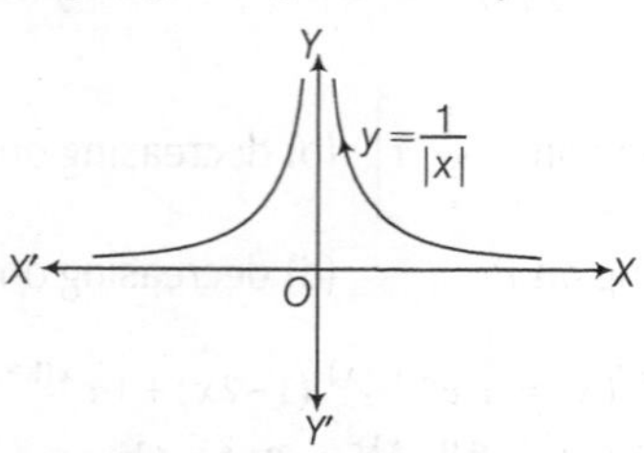

(iii) $f(x) = e^x$ can be graphically plotted as shown in the following figure, which shows $f(x) = e^x$ is strictly increasing in R.

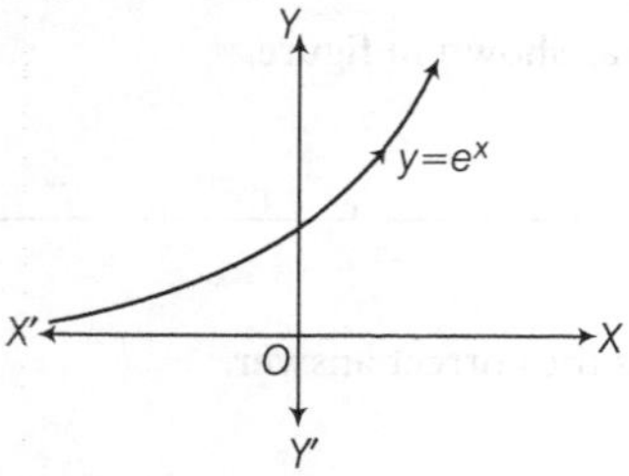

(iv) $f(x) = [x]$ can be graphically plotted as shown in the following figure, which shows $f(x) = [x]$ is increasing but not strictly increasing i.e. non-decreasing in R.

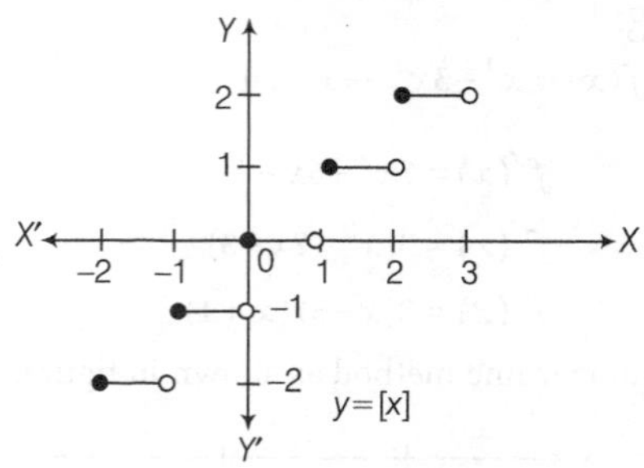

Example 10 If $H(x_0) = 0$ for some $x = x_0$ and $\dfrac{d}{dx} H(x) > 2cx\, H(x)$ for all $x \ge x_0$, where $c > 0$, then prove that $H(x)$ cannot be zero for any $x > x_0$.

Sol. Given that, $\dfrac{d}{dx} H(x) > 2\, cx\, H(x)$

$\Rightarrow$ $\dfrac{d}{dx} H(x) - 2\, cx\, H(x) > 0$

$\Rightarrow$ $\dfrac{d}{dx} \{H(x)\} e^{-cx^2} - 2\, cx e^{-cx^2} \cdot H(x) > 0$

$\Rightarrow$ $\left\{\dfrac{d}{dx} H(x)\right\} e^{-cx^2} + H(x) \left\{\dfrac{d}{dx} e^{-cx^2}\right\} > 0$

$\Rightarrow$ $\left\{\dfrac{d}{dx} H(x) \cdot e^{-cx^2}\right\} > 0$

$\therefore$ $H(x)\, e^{-cx^2}$ is an strictly increasing function.

But $H(x_0) = 0$ and e^{-cx^2} is always positive.

$\Rightarrow$ $H(x_0) > 0$ for all $x > x_0$

$\Rightarrow$ $H(x)$ cannot be zero for any $x > x_0$.

Example 11 If $f(x)$ is a decreasing function and attain positive values, then the set of value of 'k', for which the major axis of the ellipse $\dfrac{x^2}{f(k^2 + 2k + 5)} + \dfrac{y^2}{f(k + 11)} = 1$, is the X-axis, is $(-a, b)$, then $(a + b)$ is

Sol. Here, $f(x)$ is decreasing function and major axis is X-axis

$\Rightarrow$ $f(k^2 + 2k + 5) > f(k + 11)$

[$\because$ for the ellipse $\dfrac{x^2}{a^2} + \dfrac{y^2}{b^2} = 1$ major axis lies along X-axis, then $a^2 > b^2$]

As, $f(x)$ is decreasing, therefore

$$k^2 + 2k + 5 < k + 11$$

$\Rightarrow$ $k^2 + k - 6 < 0$

$\Rightarrow$ $(k + 3)(k - 2) < 0$

$\Rightarrow$ $k \in (-3, 2) = (-a, b)$

$\therefore$ $a + b = 5$

Example 12 Let $f(x) = 3x + 5$, then show that $f(x)$ is strictly increasing and $f^{-1}(x)$ exists and is strictly increasing for $x \in R$.

Sol. Here, $f(x) = 3x + 5$

$f'(x) = 3 > 0$, which is strictly increasing for $x \in R$.

Now, finding $f^{-1}(x)$, let $f(x) = y$, $y = 3x + 5$

or $$x = \frac{y-5}{3}$$

or $$f^{-1}(y) = \frac{y-5}{3} \quad [\text{as } f(x) = y \Rightarrow x = f^{-1}(y)]$$

$$f^{-1}(x) = \frac{x-5}{3}$$

Which shows $f^{-1}(x) = \frac{x-5}{3}$ exists for all $x \in R$ and is strictly increasing as $\frac{d}{dx}(f^{-1}(x)) = \frac{1}{3} > 0$ for all $x \in R$.

Example 13 Let $\phi(x) = \sin(\cos x)$, then check whether it is increasing or decreasing in $[0, \pi/2]$.

Sol. Given, $\phi(x) = \sin(\cos x)$

$\Rightarrow \quad \phi'(x) = \cos(\cos x) \cdot (-\sin x)$

$\Rightarrow \quad \phi'(x) = -\cos(\cos x) \cdot \sin x$

$$\left[\because \cos(\cos x) > 0 \text{ and } \sin x > 0, \forall x \in \left(0, \frac{\pi}{2}\right)\right]$$

Therefore, it is clearly decreasing for $x \in [0, \pi/2]$ as $\phi'(x) \le 0$.

Aliter

Here, $f(x) = \sin x$ and $g(x) = \cos x$ are increasing and decreasing in $[0, \pi/2]$.

$\Rightarrow (fog)(x) = \phi(x) = \sin(\cos x)$ is decreasing.

Example 14 Let $\phi(x) = \cos(\cos x)$, then check whether it is increasing or decreasing in $[0, \pi/2]$.

Sol. Given, $\phi(x) = \cos(\cos x)$

$\Rightarrow \quad \phi'(x) = -\sin(\cos x) \cdot (-\sin x)$

$\Rightarrow \quad \phi'(x) = \sin x \sin(\cos x)$

$$\left[\because \sin(\cos x) > 0 \text{ and } \sin x > 0, \forall x \in \left(0, \frac{\pi}{2}\right)\right]$$

Therefore, it is clearly increasing for $x \in [0, \pi/2]$ as $\phi'(x) \ge 0$.

Aliter

Here, $f(x) = \cos x$ and $g(x) = \cos x$ are decreasing in $[0, \pi/2]$.

$\Rightarrow (fog)(x) = \cos(\cos x) = \phi(x)$ is increasing.

Example 15 Let $f(x) = \begin{cases} xe^{ax}; & x \le 0 \\ x + ax^2 - x^3; & x > 0 \end{cases}$ where a is positive constant. Find the interval in which $f'(x)$ is increasing. ***[IIT JEE 1996]***

Sol. Given, $f(x) = \begin{cases} xe^{ax}; & x \le 0 \\ x + ax^2 - x^3; & x > 0 \end{cases}$

Differentiating both the sides, we have

$$f'(x) = \begin{cases} axe^{ax} + e^{ax}; & x \le 0 \\ 1 + 2ax - 3x^2; & x > 0 \end{cases}$$

Again, differentiating both sides, we have

$$f''(x) = \begin{cases} 2ae^{ax} + a^2xe^{ax}; & x \le 0 \\ 2a - 6x; & x > 0 \end{cases}$$

Now, $f''(x) = 0$, then in the interval $x \le 0$ the root is $x = -\frac{2}{a}$ and in interval $x > 0$ root is $x = \frac{a}{3}$

Using sign scheme or number line rule as shown in figure, we get

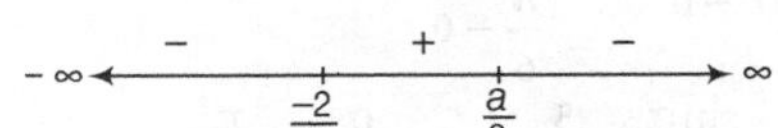

$f'(x)$ decreases on $\left(-\infty, -\frac{2}{a}\right) \cup \left[\frac{a}{3}, \infty\right)$ and increases on $\left[-\frac{2}{a}, \frac{a}{3}\right]$.

Example 16 If $a < 0$ and $f(x) = e^{ax} + e^{-ax}$ is monotonically decreasing. Find the interval to which x belongs.

Sol. Given, $a < 0$ and $f(x) = e^{ax} + e^{-ax}$ is monotonically decreasing. ...(i)

$\Rightarrow \quad f'(x) < 0$

$\Rightarrow \quad ae^{ax} - ae^{-ax} < 0$

$\Rightarrow \quad a\left(\frac{e^{2ax} - 1}{e^{ax}}\right) < 0$ ...(ii)

As from Eq. (i), $a < 0$

$\Rightarrow \quad (e^{2ax} - 1) > 0$

$\Rightarrow \quad e^{2ax} > 1$

$\Rightarrow \quad 2ax > 0$

$\Rightarrow \quad ax > 0$

$\Rightarrow \quad x < 0$ [as $a < 0$]

Thus, $f(x)$ is monotonically decreasing, if $x \in (-\infty, 0)$.

Example 17 If $0 < \alpha < \frac{\pi}{6}$, then the value of $(\alpha \operatorname{cosec} \alpha)$ is

(a) less than $\frac{\pi}{3}$ (b) more than $\frac{\pi}{3}$

(c) less than $\frac{\pi}{6}$ (d) more than $\frac{\pi}{6}$

Sol. From the figure, we can say

Slope of $OA >$ Slope of OB

where $A = (\alpha, \sin\alpha)$ and $B\left(\frac{\pi}{6}, \sin\frac{\pi}{6}\right)$.

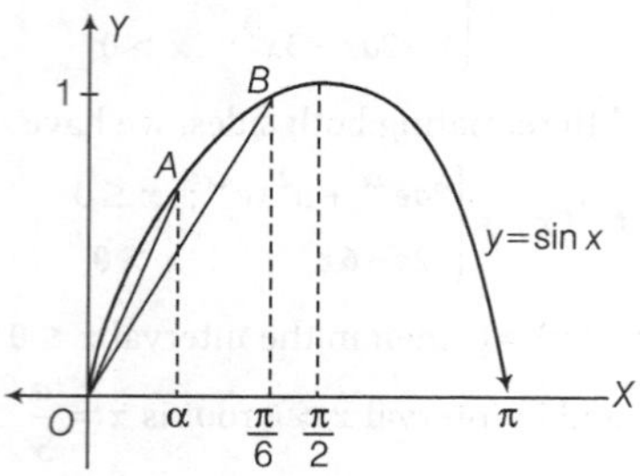

$$\Rightarrow \quad \frac{\sin\alpha - 0}{\alpha - 0} > \frac{\sin\frac{\pi}{6} - 0}{\frac{\pi}{6} - 0}$$

$$\Rightarrow \quad \frac{\sin\alpha}{\alpha} > \frac{3}{\pi} \text{ or } \frac{\alpha}{\sin\alpha} < \frac{\pi}{3} \Rightarrow \alpha \operatorname{cosec}\alpha < \frac{\pi}{3}$$

Hence, (a) is the correct answer.

Example 18 If $f(x) = ax^3 + bx^2 + cx + d$, where a, b, c and d are real numbers and $3b^2 < c^2$, is an increasing cubic function and $g(x) = af'(x) + bf''(x) + c^2$, then

(a) $\int_a^x g(t)\,dt$ is a decreasing function

(b) $\int_a^x g(t)\,dt$ is an increasing function

(c) $\int_a^x g(t)\,dt$ is a neither increasing nor a decreasing function

(d) None of the above

Sol. $f'(x) = 3ax^2 + 2bx + c > 0$ [since, $f(x)$ is increasing]

$$\Rightarrow \quad a > 0 \text{ and } b^2 - 3ac < 0 \Rightarrow a > 0 \text{ and } b^2 < 3ac$$

Also, $g(x) = af'(x) + bf''(x) + c^2$

$$g(x) = 3a^2x^2 + 2abx + ac + 6abx + 2b^2 + c^2$$

$$g(x) = 3a^2x^2 + 8abx + (2b^2 + c^2 + ac)$$

where $D = 64a^2b^2 - 4 \cdot 3a^2 \cdot (2b^2 + c^2 + ac)$

$$= 4a^2(16b^2 - 6b^2 - 3c^2 - 3ac)$$

$$= 4a^2(10b^2 - 3c^2 - 3ac) < 4a^2(10b^2 - 3c^2 - b^2)$$

[as $3ac > b^2 \Rightarrow -3ac < -b^2$]

$$= 4a^2(9b^2 - 3c^2) = 12a^2(3b^2 - c^2) \text{ [given } 3b^2 < c^2]$$

$\therefore \quad D < 0 \Rightarrow g(x) > 0, \forall x \in R$

$\therefore \int_a^x g(t)\,dt$ is an increasing function.

$$\left[\because \frac{d}{dx}\left(\int_a^x g(t)dt\right) = g(x) > 0\right]$$

Hence, (b) is the correct answer.

Example 19 If $f: R \to R$, $f(x)$ is a differentiable bijective function, then which of the following is true?

(a) $(f(x) - x)\, f''(x) < 0, \forall x \in R$

(b) $(f(x) - x)\, f''(x) > 0, \forall x \in R$

(c) If $(f(x) - x)\, f''(x) > 0$, then $f(x) = f^{-1}(x)$ has no solution

(d) If $(f(x) - x)\, f''(x) > 0$, then $f(x) = f^{-1}(x)$ has at least one real solution

Sol. As, $(f(x) - x)\, f''(x) < 0, \forall x \in R$

$$\Rightarrow \quad (f(x) - x) > 0 \text{ and } f''(x) < 0)$$

or $\quad (f(x) - x) < 0$ and $f''(x) > 0)$

Can't be true as $f(x) - x > 0$ and $f'(x)$ are decreasing. Then, $f(x)$ has to intersect the line $y = x$.

Similarly, $f(x) - x < 0$ and $f'(x)$ is increasing, is not possible.

Also, $\quad f(x) - x \neq 0$

$$\Rightarrow \quad f(x) = f^{-1}(x) \text{ has no solution.}$$

Hence, (c) is the correct answer.

Example 20 If $f(x)$ and $g(x)$ are two positive and increasing functions, then

(a) $(f(x))^{g(x)}$ is always increasing

(b) if $(f(x))^{g(x)}$ is decreasing, then $f(x) < 1$

(c) if $(f(x))^{g(x)}$ is increasing, then $f(x) > 1$

(d) if $f(x) > 1$, then $(f(x))^{g(x)}$ is increasing

Sol. Let $h(x) = (f(x))^{g(x)}$

$$\Rightarrow \quad \log(h(x)) = g(x)\{\log f(x)\}$$

$$\Rightarrow \quad \frac{1}{h(x)} \cdot h'(x) = \frac{g(x)}{f(x)} \cdot f'(x) + \{\log f(x)\}\, g'(x)$$

and $h(x)$ is increasing, if $\log(f(x)) > 0 \Rightarrow f(x) > 1$

Hence, (d) is the correct answer.

Example 21 If the function $y = \sin(f(x))$ is monotonic for all values of x [where $f(x)$ is continuous], then the maximum value of the difference between the maximum and the minimum value of $f(x)$, is

(a) π (b) 2π (c) $\frac{\pi}{2}$ (d) None of these

Sol. As, $y = \sin(f(x))$ is monotonic for

$$f(x) \in \left[2n\pi - \frac{\pi}{2}, 2n\pi + \frac{\pi}{2}\right]$$

or

$$\left[2n\pi + \frac{\pi}{2}, 2n\pi + \frac{3\pi}{2}\right]$$

$\therefore$ The maximum value of difference is π.

Hence, (a) is the correct answer.

Example 22 If $f''(x) > 0$ and $f'(1) = 0$ such that $g(x) = f(\cot^2 x + 2\cot x + 2)$, where $0 < x < \pi$, then the interval in which $g(x)$ is decreasing is

(a) $(0, \pi)$ (b) $\left(\frac{\pi}{2}, \pi\right)$

(c) $\left(\frac{3\pi}{4}, \pi\right)$ (d) $\left(0, \frac{3\pi}{4}\right)$

Sol. Here, $g(x) = f(\cot^2 x + 2\cot x + 2)$

$$\Rightarrow \quad g'(x) = f'(\cot^2 x + 2\cot x + 2) \cdot \{-2\cot x \operatorname{cosec}^2 x - 2\operatorname{cosec}^2 x\}$$

for $g(x)$ to be decreasing, $g'(x) < 0$

$$\Rightarrow \quad f'\{(\cot x + 1)^2 + 1\} \cdot (-2\operatorname{cosec}^2 x)(\cot x + 1) < 0$$

$$\Rightarrow \quad f'\{(\cot x + 1)^2 + 1\} \cdot (\cot x + 1) > 0 \quad \text{...(i)}$$

As $f''(x) > 0 \Rightarrow f'(x)$ is increasing, then

$$f'\{(\cot x + 1)^2 + 1\} > f'(1) = 0, \ \forall \ x \in \left(0, \frac{3\pi}{4}\right) \cup \left(\frac{3\pi}{4}, \pi\right)$$

Thus, Eq. (i) holds, if $\cot x + 1 > 0$

$$\Rightarrow \quad \cot x > -1, \forall \ x \in \left(0, \frac{3\pi}{4}\right)$$

Hence, (d) is the correct answer.

Exercise for Session 1

1. The curve $y = f(x)$ which satisfies the condition $f'(x) > 0$ and $f''(x) < 0$ for all real x, is

(a) 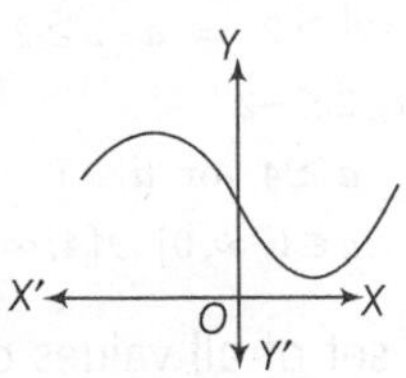

(b) 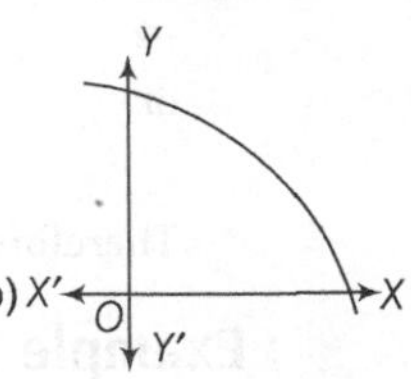

(c) 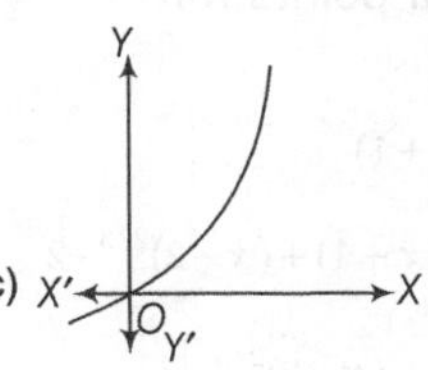

(d) 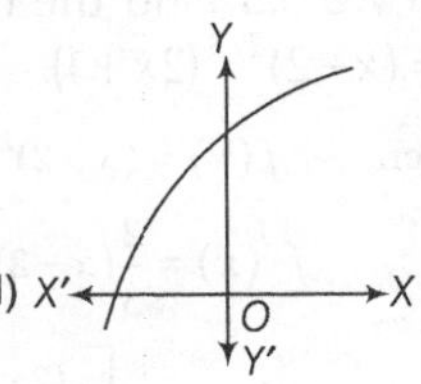

2. The interval in which $f(x) = \cot^{-1} x + x$ increases, is

(a) R (b) $(0, \infty)$ (c) $R - \{n\pi\}$ (d) None of these

3. The interval in which $f(x) = 3\cos^4 x + 10\cos^3 x + 6\cos^2 x - 3$ increase or decrease in $(0, \pi)$

(a) decreases on $\left(\frac{\pi}{2}, \frac{2\pi}{3}\right)$ and increases on $\left(0, \frac{\pi}{2}\right) \cup \left(\frac{2\pi}{3}, \pi\right)$ (b) decreases on $\left(\frac{\pi}{2}, \pi\right)$ and increases on $\left(0, \frac{\pi}{2}\right)$

(c) decreases on $\left(0, \frac{\pi}{2}\right) \cup \left(\frac{2\pi}{3}, \pi\right)$ and increases on $\left(\frac{\pi}{2}, \frac{2\pi}{3}\right)$ (d) decreases on $\left(0, \frac{\pi}{2}\right)$ and increases on $\left(\frac{\pi}{2}, \pi\right)$

4. The interval in which $f(x) = \int_0^x \{(t+1)(e^t - 1)(t-2)(t+4)\}\, dt$ increases and decreases

(a) increases on $(-\infty, -4) \cup (-1, 0) \cup (2, \infty)$ and decreases on $(-4, -1) \cup (0, 2)$

(b) increases on $(-\infty, -4) \cup (-1, 2)$ and decreases on $(-4, -1) \cup (2, \infty)$

(c) increases on $(-\infty, -4) \cup (2, \infty)$ and decreases on $(-4, 2)$

(d) increases on $(-4, -1) \cup (0, 2)$ and decreases on $(-\infty, -4) \cup (-1, 0) \cup (2, \infty)$

5. The interval of monotonicity of the function $f(x) = \frac{x}{\log_e x}$, is

(a) increases when $x \in (e, \infty)$ and decreases when $x \in (0, e)$

(b) increases when $x \in (e, \infty)$ and decreases when $x \in (0, e) - \{1\}$

(c) increases when $x \in (0, e)$ and decreases when $x \in (e, \infty)$

(d) increases when $x \in (0, e) - \{1\}$ and decreases when $x \in (e, \infty)$

6. Let $f(x) = x^3 + ax^2 + bx + 5\sin^2 x$ be an increasing function on the set R. Then,

(a) $a^2 - 3b + 15 > 0$ (b) $a^2 - 3b + 5 < 0$ (c) $a^2 - 3b + 15 < 0$ (d) $a^2 - 3b + 5 > 0$

7. Let $g(x) = f(x) + f(1-x)$ and $f''(x) > 0, \forall \ x \in (0, 1)$. Then, $g(x)$ is

(a) increasing on $\left(0, \frac{1}{2}\right)$ and decreasing on $\left(\frac{1}{2}, 1\right)$ (b) increasing on $\left(\frac{1}{2}, 1\right)$ and decreasing on $\left(0, \frac{1}{2}\right)$

(c) increasing on $(0, 1)$ (d) decreasing on $(0, 1)$

Session 2

Critical Points

Critical Points

It is a collection of points for which,

(i) $f(x)$ does not exist (ii) $f'(x)$ does not exist

(iii) $f'(x)=0$.

All the values of x obtained from above conditions are said to be the critical points.

It should be noted that critical points are the interior points of an interval.

Example 23 Find the critical points for $f(x)=(x-2)^{2/3}(2x+1)$.

Sol. Given, $f(x)=(x-2)^{2/3}(2x+1)$

$$\Rightarrow \quad f'(x)=\frac{2}{3}(x-2)^{-1/3}(2x+1)+(x-2)^{2/3}\cdot 2$$

or $$f'(x)=2\left[\frac{(2x+1)}{3(x-2)^{1/3}}+\frac{(x-2)^{2/3}}{1}\right]$$

Clearly, $f'(x)$ is not defined at $x=2$, so $x=2$ is a critical point.

Another critical point is given by

$$f'(x)=0$$

i.e. $$2\left[\frac{(2x+1)+3(x-2)}{(x-2)^{1/3}}\right]=0$$

$$\Rightarrow \quad 5x-5=0 \Rightarrow x=1$$

Hence, $x=1$ and $x=2$ are two critical points of $f(x)$.

Example 24 Find all the values of a for which the function

$f(x)=(a^2-3a+2)\cos\left(\frac{x}{2}\right)+(a-1)x$, possess critical points.

Sol. Given, $f(x)=(a^2-3a+2)\cos\left(\frac{x}{2}\right)+(a-1)x$

$$\Rightarrow \quad f'(x)=-\frac{1}{2}(a-1)(a-2)\sin\left(\frac{x}{2}\right)+(a-1)$$

$$\Rightarrow \quad f'(x)=(a-1)\left[1-\frac{1}{2}(a-2)\sin\left(\frac{x}{2}\right)\right]$$

If $f(x)$ possess critical points, then

$$f'(x)=0$$

$$\Rightarrow \quad (a-1)\left[1-\left(\frac{a-2}{2}\right)\sin\frac{x}{2}\right]=0$$

$$\Rightarrow \quad a=1 \text{ and } 1-\left(\frac{a-2}{2}\right)\sin\frac{x}{2}=0,$$

but at $a=1 \Rightarrow f(x)=0$

Thus, $$\sin\left(\frac{x}{2}\right)=\frac{2}{a-2}$$

$$\Rightarrow \quad \left|\frac{2}{a-2}\right|\le 1$$

$$\Rightarrow \quad |a-2|\ge 2 \Rightarrow a-2\ge 2$$

or $$a-2\le -2$$

$$a\ge 4 \text{ or } a\le 0$$

Therefore, $$a\in(-\infty,0]\cup[4,\infty)$$

Example 25 The set of all values of 'b' for which the function

$f(x)=(b^2-3b+2)(\cos^2 x-\sin^2 x)+(b-1)x+\sin 2$

does not possess stationary points is

(a) $[1,\infty)$ (b) $(0,1)\cup(1,4)$

(c) $\left(\frac{3}{2},\frac{5}{2}\right)$ (d) None of these

Sol. Here, $f(x)=(b^2-3b+2)(\cos^2 x-\sin^2 x)+(b-1)x+\sin 2$

$$=(b^2-3b+2)\cos 2x+(b-1)x+\sin 2$$

$$\Rightarrow \quad f'(x)=(b^2-3b+2)\cdot(-2\sin 2x)+(b-1).$$

As, $f(x)$ does not possess stationary points.

$$\Rightarrow \quad f'(x)\ne 0$$

$$\Rightarrow \quad (b-1)(b-2)(-2\sin 2x)+(b-1)\ne 0,$$

for any $x\in R$

$$\Rightarrow \quad (b-1)\{1-2(b-2)\sin 2x\}\ne 0$$

$$\Rightarrow \quad \left|\frac{1}{2(b-2)}\right|>1 \text{ and } b\ne 1$$

$$\Rightarrow \quad -\frac{1}{2}<b-2<\frac{1}{2} \text{ and } b\ne 1$$

$$\Rightarrow \quad \frac{3}{2}<b<\frac{5}{2} \Rightarrow b\in\left(\frac{3}{2},\frac{5}{2}\right)$$

Hence, (c) is the correct answer.

Example 26 Find the set of critical points of the function

$$f(x)=x-\log x+\int_2^x\left(\frac{1}{z}-2-2\cos 4z\right)dz.$$

Sol. Here, $f(x)=x-\log x+\int_2^x\left(\frac{1}{z}-2-2\cos 4z\right)dz$

$\Rightarrow \quad f'(x)=1-\frac{1}{x}+\left(\frac{1}{x}-2-2\cos 4x\right)(1)-0$

$= -1-2\cos 4x$ [using Leibnitz rule]

Put $f'(x)=0$,

$\Rightarrow \quad \cos 4x=-\frac{1}{2}$ or $\cos 4x=\cos\left(\pi-\frac{\pi}{3}\right)$

or $\quad 4x=2n\pi+\frac{2\pi}{3}, n\in$ Integer

$\Rightarrow \quad x=\frac{n\pi}{2}\pm\frac{\pi}{6}, n\in I$

But $\log x$ is defined for $x>0$

$\therefore$ For $n=0$, $x=\pm\frac{\pi}{6}$

$\Rightarrow \quad x=\frac{\pi}{6}$ [neglecting $x=-\pi/6$]

$\therefore$ Set of critical points $=\left\{\frac{\pi}{6},\frac{n\pi}{2}\pm\frac{\pi}{6}\right\}$, where $n\in N$.

Exercise for Session 2

1. Determine all the critical points for the function $f(x)=6x^5+33x^4-30x^3+100$.
2. Find the critical points of $f(x)=x^{2/3}(2x-1)$.
3. Determine all the critical points for the function $f(x)=xe^{x^2}$.
4. The number of critical points of $f(x)=\max\{\sin x,\cos x\}, \forall x\in(-2\pi,2\pi)$ is
   (a) 5 (b) 6
   (c) 7 (d) 8

Session 3

Comparison of Functions Using Calculus

Comparison of Functions Using Calculus

If we want to compare $\phi(x)$ and $g(x)$ then we consider a function $f(x) = \phi(x) - g(x)$ or $f(x) = g(x) - \phi(x)$ and check whether $f(x)$ is increasing or decreasing in given domain of $\phi(x)$ and $g(x)$. The procedure is illustrated by following examples :

Example 27 Using calculus, find the order relation between x and $\tan^{-1} x$ when $x \in [0, \infty)$.

Sol. Let $f(x) = x - \tan^{-1}(x)$

$$\Rightarrow \quad f'(x) = 1 - \frac{1}{1+x^2}$$

$$\Rightarrow \quad f'(x) = \frac{x^2}{1+x^2} \geq 0, \forall x \in [0, \infty)$$

Thus, $f(x)$ is an increasing function.

As, we know $x_1 \leq x_2 \Rightarrow f(x_1) \leq f(x_2)$ for increasing function

$$\therefore \quad x \geq 0, \forall x \in [0, \infty)$$

$$\Rightarrow \quad f(x) \geq f(0), \forall x \in [0, \infty)$$

$$\Rightarrow \quad x - \tan^{-1} x \geq 0 - \tan^{-1}(0), \forall x \in [0, \infty)$$

$$\Rightarrow \quad x \geq \tan^{-1} x, \forall x \in [0, \infty)$$

Thus, the above relation is the order relation between x and $\tan^{-1} x$.

Example 28 Using calculus, find the order relation between x and $\tan^{-1} x$ when $x \in (-\infty, 0]$.

Sol. Let $f(x) = (x) - \tan^{-1}(x)$

$$\Rightarrow \quad f'(x) = 1 - \frac{1}{1+x^2}$$

$$\Rightarrow \quad f'(x) = \frac{x^2}{1+x^2} \geq 0, \forall x \in (-\infty, 0]$$

Thus, $f(x)$ is increasing function.

As, we know $x_1 \leq x_2 \Rightarrow f(x_1) \leq f(x_2)$ for increasing function.

$$\therefore \quad x \leq 0, \forall x \in (-\infty, 0]$$

$$\Rightarrow \quad f(x) \leq f(0), \forall x \in (-\infty, 0]$$

$$\Rightarrow \quad x - \tan^{-1} x \leq 0, \forall x \in (-\infty, 0]$$

$$\Rightarrow \quad x \leq \tan^{-1} x, \forall x \in (-\infty, 0]$$

Thus, the above relation is the order relation between x and $\tan^{-1} x$.

Example 29 For all $x \in (0, 1)$ *[IIT JEE 2000]*

(a) $e^x < 1 + x$ (b) $\log_e(1 + x) < x$

(c) $\sin x > x$ (d) $\log_e x > x$

Sol. (a) Let $f(x) = e^x - 1 - x$

$\Rightarrow f'(x) = e^x - 1 > 0, \forall x \in (0, 1)$

So, $f(x)$ is increasing, when $0 < x < 1$

$\Rightarrow \quad f(x) > f(0)$ or $e^x - 1 - x > 0$

$\Rightarrow \quad e^x > 1 + x$

Hence, (a) is false.

(b) Let $g(x) = \log(1 + x) - x$

$\Rightarrow g'(x) = \dfrac{1}{1+x} - 1 = \dfrac{-x}{1+x} < 0, \forall x \in (0, 1)$

So, $g(x)$ is decreasing, when $0 < x < 1$

$\Rightarrow \quad g(0) > g(x) \Rightarrow \log(1 + x) < x$

Hence, (b) is true.

(c) Let $h(x) = \sin x - x \Rightarrow h'(x) = \cos x - 1 < 0, \forall x \in (0, 1)$

So, $h(x)$ is decreasing, when $0 < x < 1 \Rightarrow h(x) < h(0)$

$\Rightarrow \quad \sin x < x$

Hence, (c) is false.

(d) Let $g(x) = \log x - x \quad \Rightarrow g'(x) = \dfrac{1}{x} - 1$

$\therefore g'(x) > 0, \forall x \in (0, 1)$ or $g(x) < g(1)$

$\Rightarrow \quad \log x - x < -1$

$\Rightarrow x - 1 > \log x$ or $x > \log x$

Hence, (d) is false.

Thus, (b) is the correct answer.

Example 30 Prove that

$$\left(\tan^{-1}\frac{1}{e}\right)^2 + \frac{2e}{\sqrt{e^2+1}} < (\tan^{-1} e)^2 + \frac{2}{\sqrt{e^2+1}}.$$

Sol. Let us consider a function $f(x)$,

i.e. $\quad f(x) = (\tan^{-1} x)^2 + \dfrac{2}{\sqrt{x^2+1}}$ for all $x \in R$

$$\therefore \quad f'(x) = \frac{2\tan^{-1} x}{1+x^2} - \frac{2x}{(x^2+1)^{3/2}}$$

$$= \frac{2}{1+x^2}\left[\tan^{-1} x - \frac{x}{\sqrt{x^2+1}}\right]$$

$$\Rightarrow \quad f'(x) = \frac{2}{1+x^2} g(x),$$

where $g(x) = \tan^{-1} x - \frac{x}{\sqrt{x^2+1}}$ ...(i)

$$\therefore \quad g'(x) = \frac{1}{1+x^2} - \frac{1}{\sqrt{x^2+1}} + \frac{x^2}{(x^2+1)^{3/2}}$$

$$= \frac{1}{1+x^2} - \frac{1}{(x^2+1)^{3/2}}$$

$$= \frac{1}{1+x^2}\left(1 - \frac{1}{\sqrt{(x^2+1)}}\right) > 0 \text{ for all } x \in R$$

$\Rightarrow$ $g(x)$ is increasing for $x \in R$

$\Rightarrow$ $g(x) > g(0)$ for all $x \in R$

$$\Rightarrow \quad \tan^{-1} x - \frac{x}{\sqrt{x^2+1}} > 0 \quad \text{...(ii)}$$

$\therefore$ From Eqs. (i) and (ii), $f'(x) > 0$ for all $x > 0$

$\therefore$ $f(x)$ is increasing for all $x > 0$

$$\Rightarrow \quad f(1/e) < f(e)$$

$$\Rightarrow \quad \left(\tan^{-1}\frac{1}{e}\right)^2 + \frac{2e}{\sqrt{e^2+1}} < (\tan^{-1} e)^2 + \frac{2}{\sqrt{e^2+1}}$$

Exercise for Session 3

1. Show that $\sin x < x < \tan x$ for $0 < x < \pi/2$.

2. Show that $\frac{x}{(1+x)} < \log(1+x) < x$ for $x > 0$.

3. Show that $x - \frac{x^3}{6} < \sin x$ for $0 < x < \frac{\pi}{2}$.

4. If $ax^2 + \frac{b}{x} \ge c$ for all positive x, where $a, b > 0$, then

   (a) $27ab^2 \ge 4c^3$ (b) $27ab^2 < 4c^3$

   (c) $4ab^2 \ge 27c^3$ (d) None of these

5. If $ax + \frac{b}{x} \ge c$ for all positive x, where $a, b > 0$, then

   (a) $ab < \frac{c^2}{4}$ (b) $ab \ge \frac{c^2}{4}$

   (c) $ab \ge \frac{c}{4}$ (d) None of these

Session 4

Introduction to Maxima and Minima, Methods of Finding Extrema of Continuous Functions, Convexity/Concavity & Point of Inflection, Concept of Global Maximum/Minimum

Introduction to Maxima and Minima

By the maximum/minimum value of function $f(x)$ we mean local or regional maximum/minimum value and not the greatest/least value attainable by the function. It is also possible in a function that local maximum at one point is smaller than local minimum at another point. Sometimes, we use the word extrema for maxima and minima.

Definition A function $f(x)$ is said to have a maximum at $x = a$, if $f(a)$ is greatest of all values in the suitably small neighbourhood of a, where $x = a$ is an interior point in the domain of $f(x)$.

Analytically, this means $\boldsymbol{f(a) \geq f(a+h)}$ and $\boldsymbol{f(a) \geq f(a-h)}$, where $h \geq 0$ (very small quantity).

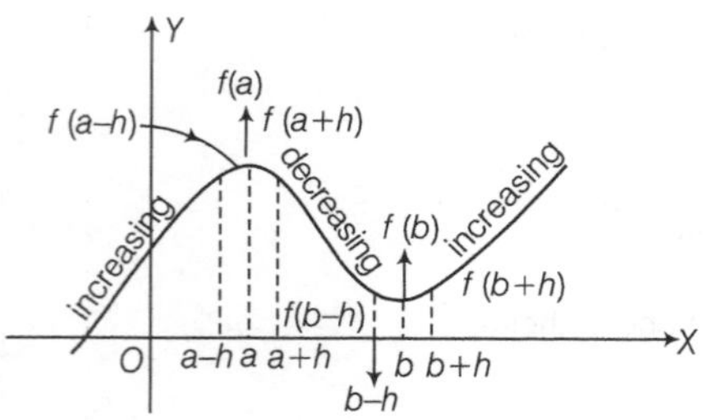

Figure 8.12

Similarly, a function $y = f(x)$ is said to have a minimum at $x = b$, if $f(b)$ is smallest of all values in the suitably small neighbourhood of b, where $x = b$ is an interior point in the domain of $f(x)$.

Analytically, $\boldsymbol{f(b) \leq f(b+h)}$ and $\boldsymbol{f(b) \leq f(b-h)}$, where $h \geq 0$ (very small quantity).

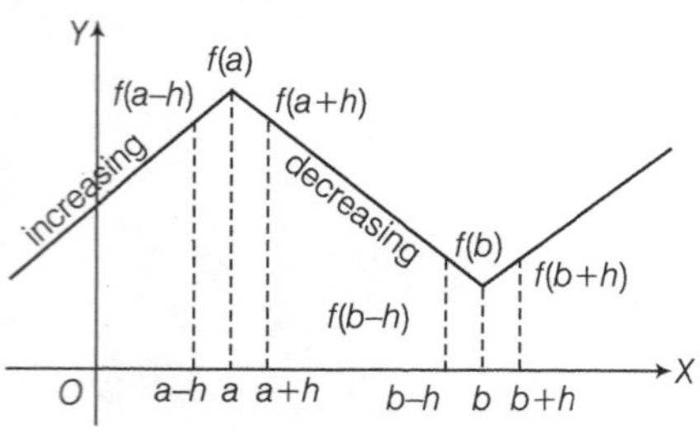

Figure 8.13

Methods of Finding Extrema of Continuous Functions

The following tests apply to a continuous function in order to get the extrema

First Derivative Test

As we know the function attains maximum, when it has assumed its maximum value and attains minimum, when it has assumed its minimum value which could be shown as

(i) At a critical point $x = a$

(a) **When $f(x)$ attains maximum at $(x = a)$**

Consider the following graph

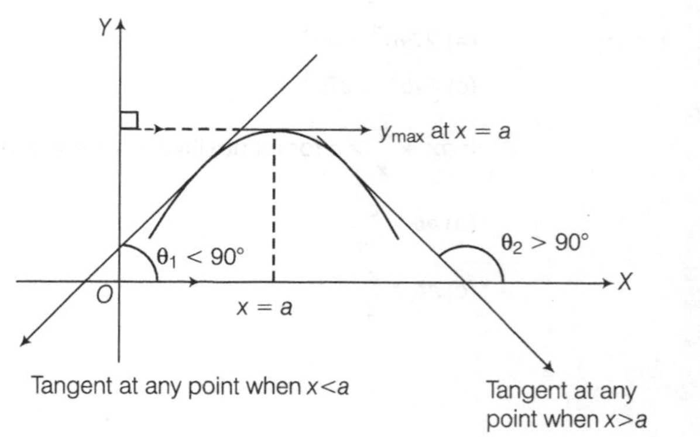

Figure 8.14

From above graph, we see that

$$\begin{cases} \text{for } x < a, \theta_1 < 90^\circ \Rightarrow \tan\theta_1 > 0 \text{ or increasing for } x < a \\ \text{for } x = a, \tan\theta = 0 \Rightarrow \text{neither increasing nor decreasing for } x = a \\ \text{for } x > a, \theta_2 > 90^\circ \Rightarrow \tan\theta_2 < 0 \text{ or decreasing for } x > a \end{cases}$$

Thus, we can say,

$f(x)$ is maximum at some point $(x = a)$.

$$\Rightarrow \begin{cases} f(x) \text{ is increasing for } x < a \\ f(x) \text{ is decreasing for } x > a \end{cases}$$

(b) **When $f(x)$ attains minimum at $(x = a)$**
Consider the following graph

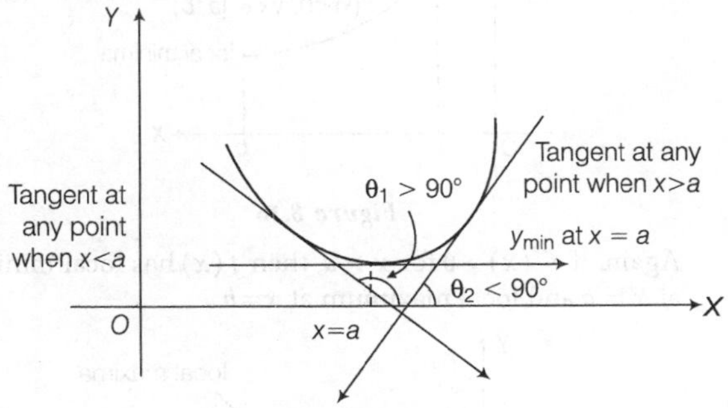

Figure 8.15

From the above graph, we see that

$$\begin{cases} \text{for } x < a, \theta_1 > 90^\circ \Rightarrow \tan\theta_1 < 0 \text{ or decreasing when } x < a \\ \text{for } x = a, \tan\theta = 0 \Rightarrow \text{neither increasing nor decreasing for } x = a \\ \text{for } x > a, \theta_2 < 90^\circ \Rightarrow \tan\theta_2 > 0 \text{ or increasing when } x > a \end{cases}$$

Thus, we can say,
$f(x)$ is minimum at some points '$x = a$',

$$\Rightarrow \begin{cases} f(x) \text{ is decreasing for } x < a \\ f(x) \text{ is increasing for } x > a \end{cases}$$

Here, some of the examples are given to make it more clear.

***Remark** (Ex. Nos. 31-33)*
Statement of example Number 31, 32, 33 can be used directly as result.

Example 31 If $f'(x)$ changes from positive to negative at x_0 while moving from left to right,
i.e. $f'(x) > 0, \quad x < x_0$
$f'(x) < 0, \quad x > x_0$, *then $f(x)$ has local maximum value at $x = x_0$.*

Sol. Consider the following graph

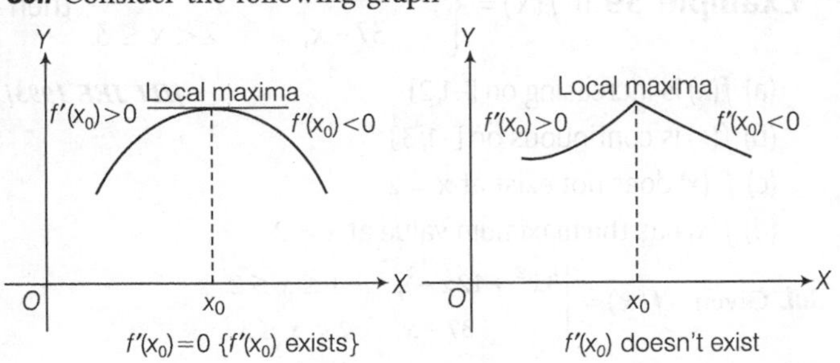

In both the graph we see that $f'(x)$ changes its sign from positive to negative when we move through x_0 and the function has local maxima at $x = x_0$.

Example 32 If $f'(x)$ changes from negative to positive at x_0 while moving from left to right,
i.e. $f'(x) < 0, x < x_0$
$f'(x) > 0, x > x_0$,
then $f(x)$ has local minimum value at $x = x_0$

Sol. Consider the following graph

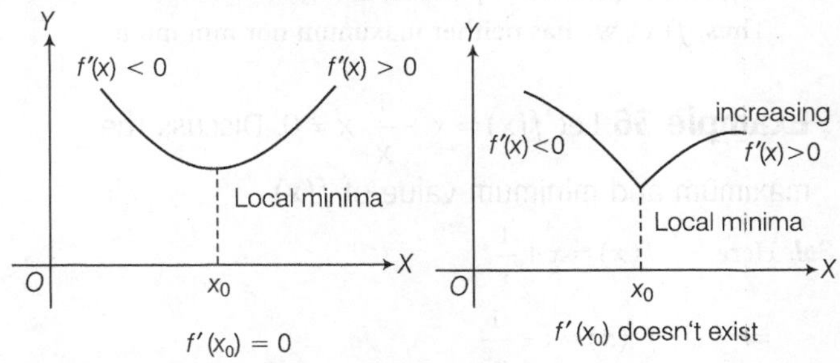

In both the graph we see that $f'(x)$ changes its sign from negative to positive when we move through x_0 and the curve has local minima at $x = x_0$.

Example 33 If sign of $f'(x)$ doesn't change at x_0, while moving from left to right, then $f(x)$ has neither a maximum nor a minimum at x_0.

Sol. Consider the following graph

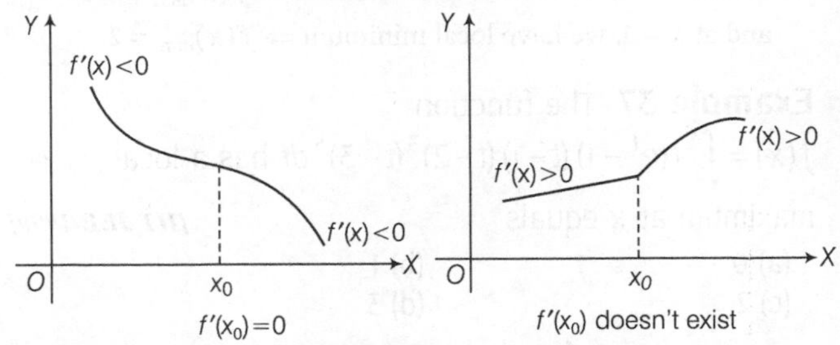

In both the graph we see that $f'(x)$ has same sign when we move through x_0, therefore $f(x)$ has neither maxima nor minima at $x = x_0$.

Example 34 Let $f(x) = x^3 - 3x^2 + 6$ find the point at which $f(x)$ assumes local maximum and local minimum.

Sol. Given, $\quad f(x) = x^3 - 3x^2 + 6$

$\Rightarrow \quad f'(x) = 3x^2 - 6x = 3x(x-2)$

Using number line rule,
$f'(x)$ changes sign from +ve to –ve at $x = 0$ and $f'(x)$ changes sign from –ve to +ve at $x = 2$.

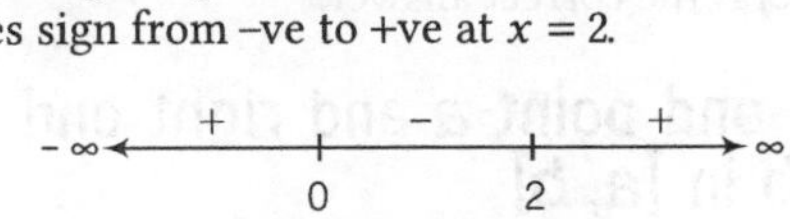

Therefore, at $x = 0$, we have local maxima and at $x = 2$, we have local minima.

Example 35 Let $f(x)=x^3$ find the point at which $f(x)$ assumes local maximum and local minimum.

Sol. Given, $f(x)=x^3$

$\Rightarrow$ $f'(x)=3x^2$

Here, $f'(x)\geq 0$ for all x, it does not change sign from negative to positive or positive to negative.

Thus, $f(x)$, we has neither maximum nor minimum.

Example 36 Let $f(x)=x+\frac{1}{x}$, $x\neq 0$. Discuss the maximum and minimum value of $f(x)$.

Sol. Here, $f(x)=x+\frac{1}{x}$

$\Rightarrow$ $f'(x)=1-\frac{1}{x^2}$

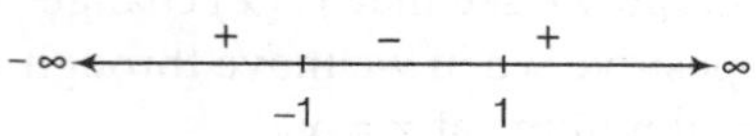

$$f'(x)=\frac{x^2-1}{x^2}=\frac{(x-1)(x+1)}{x^2}$$

Using number line rule, we have maximum at $x=-1$ and minimum at $x=1$

$\therefore$ At $x=-1$, we have local maximum $\Rightarrow f(x)_{max}=-2$

and at $x=1$, we have local minimum $\Rightarrow f(x)_{min}=2$

Example 37 The function $f(x)=\int_{-1}^{x} t(e^t-1)(t-1)(t-2)^3(t-3)^5\,dt$ has a local maximum at x equals **[IIT JEE 1999]**

(a) 0 (b) 1

(c) 2 (d) 3

Sol. Given, $f(x)=\int_{-1}^{x} t(e^t-1)(t-1)(t-2)^3(t-3)^5\,dt$

$$f'(x)=x(e^x-1)(x-1)(x-2)^3(x-3)^5$$

[using Leibnitz rule]

Using number line rule for $f'(x)$, we get the following figure which shows local maxima at $x=2$ as $f'(x)$ changes from (+ve) to (–ve) and local minima at $x=1$ and $x=3$ as $f'(x)$ changes from (–ve) to (+ve).

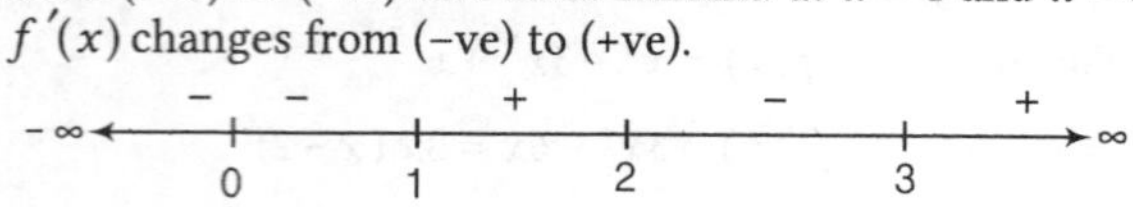

$\therefore$ Local minima at $x=1,3$ and local maxima at $x=2$.

Hence, (c) is the correct answer.

(ii) At left end point *a* and right end point *b* in [*a*, *b*]

Let $f(x)$ be defined on $[a,b]$ and if $f'(x)<0$ for $x>a$, then $f(x)$ has local maximum at $x=a$ and local minimum at $x=b$.

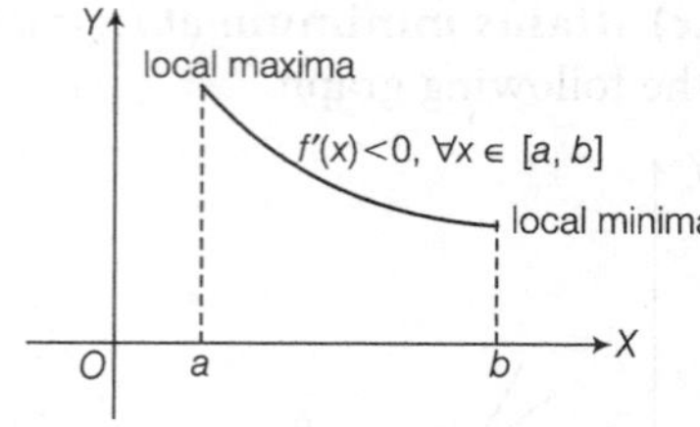

Figure 8.16

Again, if $f'(x)>0$ for $x>a$, then $f(x)$ has local minimum at $x=a$ and local maximum at $x=b$.

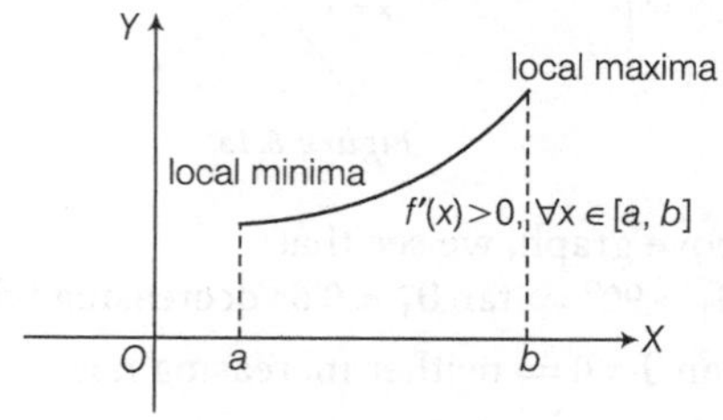

Figure 8.17

We can summarise above result as

(a) If a function is strictly increasing in $[a,b]$, then $\begin{cases} f(a) \text{ is local minimum} \\ f(b) \text{ is local maximum} \end{cases}$

(b) If a function is strictly decreasing in $[a,b]$, then $\begin{cases} f(a) \text{ is local maximum} \\ f(b) \text{ is local minimum} \end{cases}$

Example 38 Find the local maximum and local minimum of $f(x)=x^3+3x$ in $[-2,4]$.

Sol. Given, $f(x)=x^3+3x$

$\Rightarrow f'(x)=3x^2+3$ which is strictly increasing for all $x\in R$ and thus, increasing for $[-2,4]$.

Hence, local minimum is $f(-2)=(-2)^3+3(-2)=-14$

and local maximum is $f(4)=(4)^3+3(4)=76$

Example 39 If $f(x)=\begin{cases} 3x^2+12x-1, & -1\leq x\leq 2 \\ 37-x, & 2<x\leq 3 \end{cases}$, then **[IIT JEE 1993]**

(a) $f(x)$ is increasing on $[-1,2]$

(b) $f(x)$ is continuous on $[-1,3]$

(c) $f'(x)$ does not exist at $x=2$

(d) $f(x)$ has the maximum value at $x=2$

Sol. Given, $f(x)=\begin{cases} 3x^2+12x-1, & -1\leq x\leq 2 \\ 37-x, & 2<x\leq 3 \end{cases}$

$$\Rightarrow \quad f'(x)=\begin{cases} 6x+12, & -1\leq x<2 \\ -1, & 2<x\leq 3 \end{cases}$$

(a) Which shows $f'(x)>0$ for $x\in[-1,2]$

So, $f(x)$ is increasing on $[-1,2]$. Hence, (a) is correct.

(b) For continuity of $f(x)$. [check at $x = 2$]

$\because \quad f(x) = 3x^2 + 12x - 1, -1 \le x \le 2$

It is a polynomial, therefore continuous in $[-1, 2]$.

Also, for $(2,3]$, $f(x) = 37 - x$ which is again a polynomial, therefore continuous in $(2,3]$.

Now, we have to check continuity of $f(x)$ at $x = 2$

$$\text{LHL} = \lim_{h\to 0} f(2-h) = \lim_{h\to 0} 3(2-h)^2 + 12(2-h) - 1 = 35$$

$$\text{RHL} = \lim_{h\to 0} f(2+h) = \lim_{h\to 0} 37 - (2+h) = 35$$

RHL = 35, LHL = 35 and $f(2) = 35$

So, (b) is correct.

(c) As discussed in previous chapter,

$$Rf'(2) = \lim_{h\to 0} \frac{37 - (2+h) - 35}{h} = -1$$

and $$Lf'(2) = \lim_{h\to 0} \frac{3(2-h)^2 + 12(2-h) - 1 - 35}{-h} = 24$$

So, not differentiable at $x = 2$

Hence, (c) is correct.

(d) We know $f(x)$ is increasing on $[-1, 2]$ and decreasing on $(2, 3]$. Thus, maximum at $x = 2$

Hence, (d) is correct.

$\therefore$ Hence, (a), (b), (c) and (d) all are correct answers.

Example 40 Let $f(x) = \sin x - x$ on $[0, \pi/2]$, find local maximum and local minimum.

Sol. Given, $f(x) = \sin x - x$

$$f'(x) = \cos x - 1, \ \forall x \in [0, \pi/2]$$

$(\cos x - 1) \le 0, \forall x \in [0, \pi/2]$, as $\cos x \le 1$

$\therefore \quad f'(x) \le 0, \forall x \in [0, \pi/2]$

$\therefore$ $f(x)$ is decreasing for $x \in [0, \pi/2]$.

Hence, maximum value of $f(x)$ is at $x = 0$

i.e. $f_{\max}(0) = \sin 0 - 0 = 0$

and minimum value of $f(x)$ is at $x = \pi/2$

i.e. $f_{\min}\left(\frac{\pi}{2}\right) = \sin\frac{\pi}{2} - \frac{\pi}{2} = 1 - \frac{\pi}{2}$

"n^{th} Derivative Test"

First we find the root of $f'(x) = 0$. Suppose $x = a$ is one of the roots of $f'(x) = 0$.

Now, find $f''(x)$ at $x = a$

(i) If $f''(a)$ = negative, then $f(x)$ is maximum at $x = a$

(ii) If $f''(a)$ = positive, then $f(x)$ is minimum at $x = a$

(iii) If $f''(a)$ = zero

Then, we find $f'''(x)$ at $x = a$

If $f'''(a) \ne 0$, then $f(x)$ has neither maximum nor minimum (inflexion point), at $x = a$.

But, if $f'''(a) = 0$, then find $f^{iv}(a)$

If $f^{iv}(a)$ = positive, then $f(x)$ is minimum at $x = a$.

If $f^{iv}(a)$ = negative, then $f(x)$ is maximum at $x = a$.

and so on, process is repeated till point is discussed.

Convexity/Concavity and Point of Inflection

Observe the two graphs sketched in the figure below. What is the difference between them? Although they are both increasing, the first graph's rate of increase in itself increasing whereas the rate of increase is decreasing in case of the second graph.

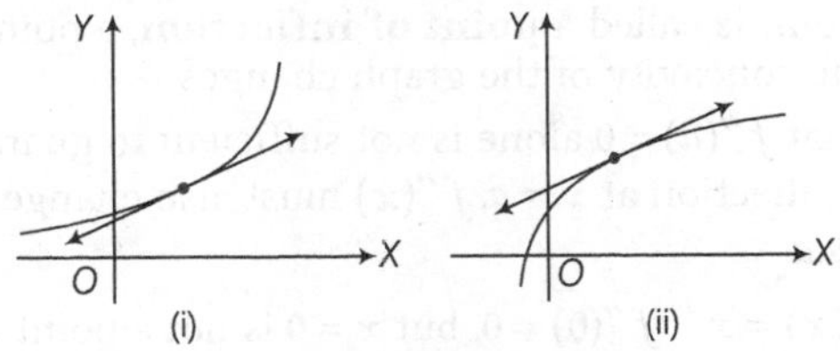

Figure 8.18

On graph (i), if you draw a tangent any where, the entire curve will lie above this tangent. Such a curve is called a **concave upwards curve.**

For graph (ii), the entire curve will lie below any tangent drawn to itself. Such a curve is called a **concave downwards curve.**

The concavity's nature can of course be restricted to particular intervals. e.g. A graph might be concave upwards in some interval while concave downwards in another shown in fig. 8.19.

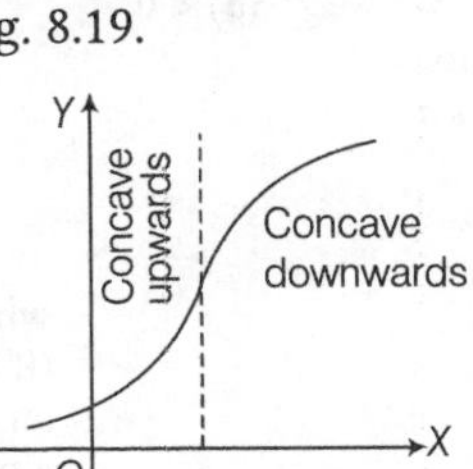

Figure 8.19

Relation of Concavity with the Derivative

Let us again consider a graph in figure. This is a concave upwards curve. We see that the rate of increase of the graph itself increases with increasing x, i.e. rate of increase of slope is positive.

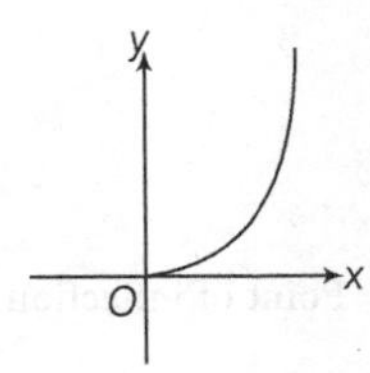

Figure 8.20

$$\frac{d}{dx}\left(\frac{dy}{dx}\right)>0 \Rightarrow \frac{d^2y}{dx^2}>0$$

Similarly, for **a concave downwards curve,** $\frac{d^2y}{dx^2}<0$

The nature of concavity is simply related to the sign of the second derivative.

We now finally come to what we mean by point of inflection.

Consider $f(x)=x^3$ again $f'(0)=f''(0)=0$

$f''(x)<0$ for $x<0$

and >0 for $x>0$

$\Rightarrow f''(x)$ changes sign as x crosses 0.

$\Rightarrow f(x)$ changes the nature of its concavity as x crosses 0.

Such a point is called a **point of inflection**, a point at which the concavity of the graph changes.

Notice that $f''(a)=0$ alone is not sufficient to guarantee a point of inflection at $x=a$. $f''(x)$ must also change sign as x crosses a.

e.g. In $f(x)=x^4$, $f''(0)=0$, but $x=0$ is not a point of inflection since $f''(x)$ does not change its sign as x crosses 0. From the higher order derivative test, we know that $x=0$ is a local minimum for $f(x)=x^4$.

A summary of results for maxima, minima and point of Inflection

	First order derivative test	Second order derivative test	Higher order derivative test
Max	$f'(a)=0$ $f'(x)$ changes sign from +ve to −ve as x crosses a	$f'(a)=0$ $f''(a)<0$	$f'(a)=0$ $f''(a)=0$ ⋮ $f^{n-1}(a)=0$ $f^n(a)<0$ where n is even (If n is odd, $x=a$ is not an extremum point; it is a point of inflextion)
Min.	$f'(a)=0$ $f'(x)$ changes sign from −ve to +ve as x crosses a	$f'(a)=0$ $f''(a)>0$	$f'(a)=0$ $f''(a)=0$ ⋮ $f^{n-1}(a)=0$ $f^n(a)>0$ where n is even (If n is odd, $x=a$ is not an extremum point; it is a point of inflextion)
Point of inflection			$f''(x)$ change sign at $x=a$

Example 41 Let $f(x)=x(x-1)^2$, find the point at which $f(x)$ assumes maximum and minimum.

Sol. Given,
$$f(x)=x(x-1)^2$$
$$f'(x)=2x(x-1)+(x-1)^2$$
$$f'(x)=(x-1)[2x+x-1]$$
$$f'(x)=(x-1)(3x-1)$$

Using number line rule for $f'(x)$, we have the following figure which shows $f'(x)$ changes sign from + ve to − ve at $x=1/3$. Hence, at $x=1/3$, we have maximum and $f'(x)$ changes sign from − ve to + ve at $x=1$.

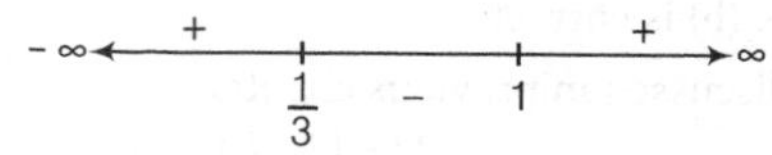

Hence, $f(x)$ is minimum at $x=1$.

Aliter

We have, $f'(x)=(x-1)(3x-1)$ and $f''(x)=6x-4$

Let $f'(x)=0 \Rightarrow x=1,1/3$ [critical points]

$\therefore$ $f''(1)=2>0$, i.e. minimum at $x=1$

and $f''(1/3)=-2<0$, i.e. maximum at $x=1/3$

Example 42 Let $f(x)=(x-1)^4$ discuss the point at which $f(x)$ assumes maximum or minimum value.

Sol. Given, $f(x)=(x-1)^4$

$f'(x)=4(x-1)^3$

Using number line rule.

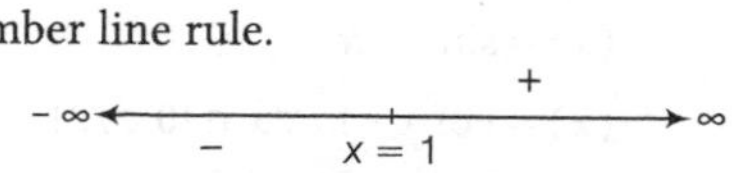

Shows $f'(x)$ changes sign from −ve to + ve.

Therefore, $f(x)$ assumes minimum at $x=1$.

Aliter

Given, $f(x)=(x-1)^4 \Rightarrow f'(x)=4(x-1)^3$

Let $f'(x)=0 \Rightarrow x=1$

Now, $f''(x)=12(x-1)^2$ which is zero at $x=1$

i.e. $f''(1)=0$

Thus, finding $f'''(x)=24(x-1)$ which is again zero at $x=1$

i.e. $f'''(1)=0$

Again, finding $f^{iv}(x)=24$ which is positive at $x=1$,

i.e. $f^{iv}(1)>0$

Therefore, minimum at $x=1$.

Example 43 Discuss the function $f(x)=x^6-3x^4+3x^2-5$, and plot the graph.

Sol. The domain is obviously R.

$f(x)$ is an even function.

Since $f(x)$ is a polynomial function, it is continuous and differentiable on R.

$f'(x) = 6x^5 - 12x^3 + 6x = 6x(x^4 - 2x^2 + 1) = 6x(x^2 - 1)^2$

$f'(x) = 0 \Rightarrow 0, \pm 1$

$$\begin{aligned} f''(x) &= 6(x^2 - 1)^2 + 24x^2(x^2 - 1) \\ &= (x^2 - 1)\{6(x^2 - 1) + 24x^2\} \\ &= (x^2 - 1)(30x^2 - 6) \quad ...(i) \\ &= 6(5x^4 - 6x^2 + 1) \end{aligned}$$

$f''(0) = 6, f''(\pm 1) = 0$

$\Rightarrow x = 0$ is a point of local minimum and $x = \pm 1$ are points of inflection

(verify that $f''(x)$ does not change sign as x crosses ± 1).

Now, $f'(x) > 0$ if $x > 0$ and $f'(x) < 0$ if $x < 0$.

Therefore, $f(x)$ decreases on $(-\infty, 0)$ and increases on $(0, -\infty)$.

There is one more important fact we must take into account. $f''(x)$ has roots ± 1 and additionally, $\pm \frac{1}{\sqrt{5}}$.

[from Eq. (i)]

Therefore, at these four points the convexity of the graph changes

$\Rightarrow f''(x) > 0 \, \forall \; x \in (-\infty, -1) \cup \left(\frac{-1}{\sqrt{5}}, \frac{1}{\sqrt{5}}\right) \cup (1, \infty)$ so that $f(x)$ is concave upwards in these intervals.

$\Rightarrow f''(x) < 0 \; \forall \; x \in \left(-1, \frac{-1}{\sqrt{5}}\right) \cup \left(\frac{1}{\sqrt{5}}, 1\right)$ so that $f(x)$ is concave downwards in these intervals.

$f(0) = -5, f(\pm 1) = -4, f(\pm 2) = 23$

Therefore one root each of $f(x)$ lies in $(-2, -1)$ and $(1, 2)$.

This information is sufficient to accurately draw the graph of the given function.

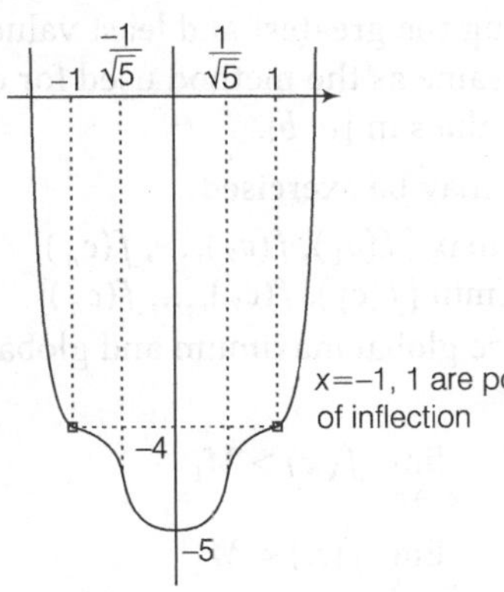

Example 44 Discuss the function

$f(x) = \frac{1}{2}\sin 2x + \cos x$., and plot its graph.

Sol. The domain of $f(x)$ is R.

$f(x)$ is periodic with period 2π and therefore we need to analyse it only in $[0, 2\pi]$.

$f(x)$ is continuous and differentiable on R.

$$\begin{aligned} f'(x) &= \cos 2x - \sin x \\ &= 1 - 2\sin^2 x - \sin x \\ &= (1 + \sin x)(1 - 2\sin x) \end{aligned}$$

This is 0 in $[0, 2\pi]$ when $x = \frac{\pi}{6}, \frac{5\pi}{6}, \frac{3\pi}{2}$

$$f''(x) = -2\sin 2x - \cos x$$

Now, $f''\left(\frac{\pi}{6}\right) < 0, f''\left(\frac{5\pi}{6}\right) > 0$ and $f''\left(\frac{3\pi}{2}\right) = 0$

$\Rightarrow x = \frac{\pi}{6}$ is a local maximum for $f(x)$; $f\left(\frac{\pi}{6}\right) = \frac{3\sqrt{3}}{4}$

$x = \frac{5\pi}{6}$ is a local minimum for $f(x)$; $f\left(\frac{5\pi}{6}\right) = \frac{-3\sqrt{3}}{4}$

$x = \frac{3\pi}{2}$ is a point of inflection; $f\left(\frac{3\pi}{2}\right) = 0$

We now need to analyse the sign of $f''(x)$.

$$\begin{aligned} f''(x) &= -2\sin 2x - \cos x \\ &= -4\sin x \cos x - \cos x \\ &= -\cos x(1 + 4\sin x) \end{aligned}$$

This is 0 in $[0, 2\pi]$ when

$$x = \frac{\pi}{2}, \pi + \sin^{-1}\frac{1}{4}, \frac{3\pi}{2}, 2\pi - \sin^{-1}\frac{1}{4}$$

We see that $f(x)$ will change its convexity at four different points.

$\Rightarrow f''(x) > 0 \; \forall \; x \in \left(\frac{\pi}{2}, \pi + \sin^{-1}\frac{1}{4}\right) \cup \left(\frac{3\pi}{2}, 2\pi - \sin^{-1}\frac{1}{4}\right)$

So that $f(x)$ is concave upwards in these intervals

$$\Rightarrow \quad f''(x) < 0 \; \forall \; x \in \left(0, \frac{\pi}{2}\right) \cup \left(\pi + \sin^{-1}\frac{1}{4}, \frac{3\pi}{2}\right) \cup \left(2\pi - \sin^{-1}\frac{1}{4}, 2\pi\right)$$

So that, $f(x)$ is concave downwards in these intervals.

$$f(0) = 1, f\left(\frac{\pi}{2}\right) = 0, f(2\pi) = 1$$

The graph has been plotted below for $[0, 2\pi]$

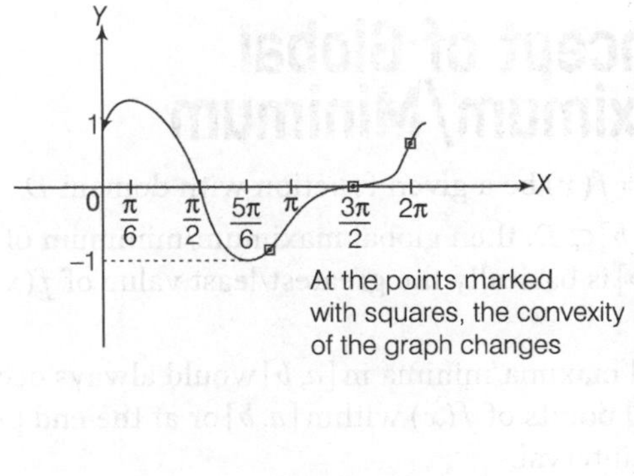

Example 45 Discuss the function $y = x + \ln(x^2 - 1)$ and plot its graph.

Sol. The domain is given by $x^2 - 1 > 0$

$$\Rightarrow \quad D = R \setminus [-1, 1]$$

$f(x)$ is continuous and differentiable on D.

$$\therefore \quad \lim_{x \to 1^+} y = -\infty; \ \lim_{x \to 1^-} y = -\infty$$

$\Rightarrow x = \pm 1$ are vertical asymptotes to the curve.

Verify that the graph has no other asymptotes.

$$y' = 1 + \frac{2x}{x^2 - 1}$$

$$\Rightarrow \quad y' = 0 \text{ when } x^2 + 2x - 1 = 0$$

$$\Rightarrow \quad x = -1 \pm \sqrt{2}$$

$x = -1 + \sqrt{2}$ does not belong to D.

$x = -1 - \sqrt{2}$ is an extremum point.

$$\Rightarrow \quad y'' = -\frac{2(x^2 + 1)}{(x^2 - 1)^2} < 0, \forall\ x$$

$\Rightarrow$ The curve is always concave downwards so that $x = -1 - \sqrt{2}$ is a point of local maximum.

$$\lim_{x \to +\infty} y = \infty$$

$$\lim_{x \to -\infty} y = -\infty$$

$\therefore$ Based on this data, the graph can be plotted as shown below.

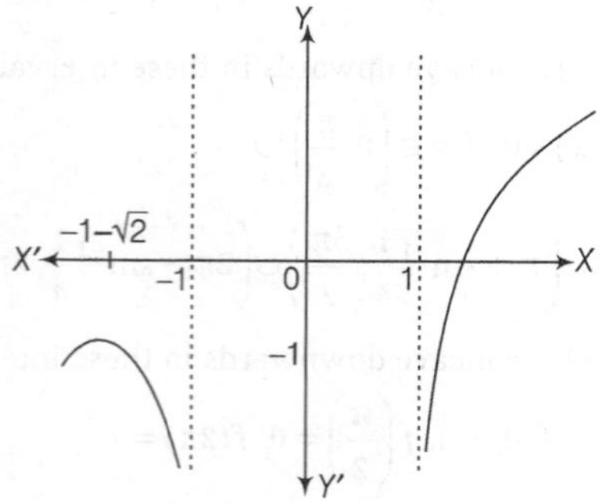

Concept of Global Maximum/Minimum

Let $y = f(x)$ be a given function with domain D.

Let $[a, b] \subseteq D$, then global maximum/minimum of $f(x)$ in $[a, b]$ is basically the greatest/least value of $f(x)$ in $[a, b]$.

Global maxima/minima in $[a, b]$ would always occur at critical points of $f(x)$ within $[a, b]$ or at the end points of the interval.

Global Maximum/Minimum in $[a, b]$

In order to find the global maximum and minimum of $f(x)$ in $[a, b]$, find out all critical points of $f(x)$ in $[a, b]$ (i.e. all points at which $f'(x) = 0$).

Let $c_1, c_2, c_3, \ldots, c_n$ be the points at which $f'(x) = 0$ and let $f(c_1), f(c_2), \ldots, f(c_n)$ be the values of the function at these points,

$$\max \{f(a), f(c_1), f(c_2), \ldots, f(c_n), f(b)\} = M_1 \quad \text{(say)}$$

$$\min \{f(a), f(c_1), f(c_2), \ldots, f(c_n), f(b)\} = M_2 \quad \text{(say)}$$

Then, M_1 is the greatest value or global maxima in $[a, b]$ and M_2 is the least value or global minima in $[a, b]$.

Example 46 Let $f(x) = 2x^3 - 9x^2 + 12x + 6$. Discuss the global maxima and minima of $f(x)$ in $[0, 2]$.

Sol. Given, $\quad f(x) = 2x^3 - 9x^2 + 12x + 6$

$$\Rightarrow \quad f'(x) = 6x^2 - 18x + 12$$

$$\Rightarrow \quad f'(x) = 6(x^2 - 3x + 2)$$

$$\Rightarrow \quad f'(x) = 6(x - 1)(x - 2)$$

Put $\quad f'(x) = 0$

$$\therefore \quad x = 1, 2 \quad [\text{say } c_1 \text{ and } c_2]$$

Then, for global maximum or global minimum.

We have, $\quad f(0) = 6,\ f(1) = 11,\ f(2) = 10,$

$\therefore$ Global maximum $\Rightarrow M_1 = \max \{6, 10, 11\} = 11$

and global minimum $\Rightarrow M_2 = \min \{6, 10, 11\} = 6$

$\therefore f(1) = 11$ global maximum and $f(0) = 6$ global minimum.

Global Maximum/Minimum in (a, b)

Method for obtaining the greatest and least values of $f(x)$ in (a, b) is almost the same as the method used for obtaining the greatest and least values in $[a, b]$.

However a caution may be exercised

let $\quad M_1 = \max \{f(c_1), f(c_2), \ldots, f(c_n)\}$

and $\quad M_2 = \min \{f(c_1), f(c_2), \ldots, f(c_n)\}$

Now, M_1 and M_2 are global maximum and global minimum, respectively.

But, if $\quad \lim_{x \to a^+} f(x) > M_1$

or $\quad \lim_{x \to b^-} f(x) < M_2$

$\Rightarrow f(x)$ would not possess global maximum or global minimum in (a, b).

This means that if **limiting values at the end points are greater than M_1 or less than M_2, then global maximum or global minimum does not exist in (a, b).**

Example 47 Let $f(x)=2x^3-9x^2+12x+6$. Discuss the global maxima and global minima of $f(x)$ in (1, 3).

Sol. Given, $f(x)=2x^3-9x^2+12x+6$

$\Rightarrow \quad f'(x)=6x^2-18x+12 \quad \Rightarrow f'(x)=6(x-1)(x-2)$

Put $f'(x)=0 \Rightarrow x=1, 2$

$\therefore \quad f(1)=11$ and $f(2)=10$ ...(i)

Let us consider the open interval (1, 3).

Clearly, $x=2$ is the only point in (1, 3).

And $f(2)=10$ [from Eq. (i)]

Now, $\lim_{x\to 1^+} f(x)=\lim_{h\to 0} f(1+h)=\lim_{h\to 0} 2(1+h)^3-9(1+h)^2+12(1+h)+6=11$

and $\lim_{x\to 3^-} f(x)=\lim_{h\to 0} f(3-h)=\lim_{h\to 0} 2(3-h)^3-9(3-h)^2+12(3-h)+6=15$

Thus, $x=2$ is the point of global minima in (1, 3) and global maxima does not exist in (1, 3).

Remark

Based on the above discussion, we can summarize things in a single graph as given below.

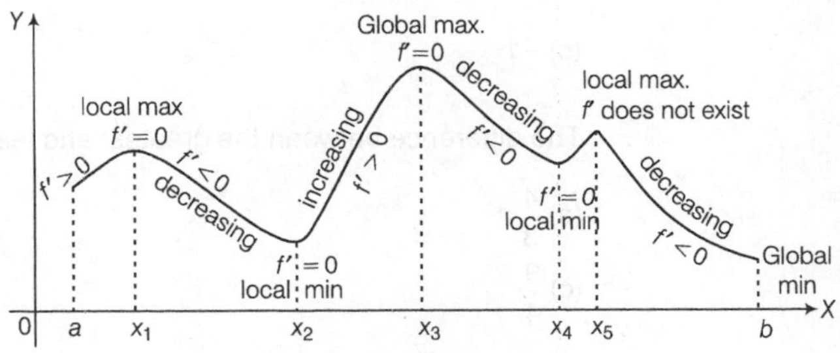

Figure 8.21

Exercise for Session 4

1. The minimum value of x^x is attained when x is equal to
   (a) e (b) e^{-1} (c) 1 (d) e^2

2. The function f is defined by $f(x)=x^p(1-x)^q$ for all $x\in R$, where p,q are positive integers, has a maximum value, for x is equal to
   (a) $\frac{pq}{p+q}$ (b) 1 (c) 0 (d) $\frac{p}{p+q}$

3. The least area of the circle circumscribing any right triangle of area S is
   (a) πS (b) $2\pi S$ (c) $\sqrt{2}\pi S$ (d) $4\pi S$

4. The coordinates of the point on the curve $x^2=4y$, which is atleast distance from the line $y=x-4$ is
   (a) (2, 1) (b) (– 2, 1)
   (c) (– 2, –1) (d) none of these

5. The largest area of a rectangle which has one side on the x-axis and the two vertices on the curve $y=e^{-x^2}$ is
   (a) $\sqrt{2}\,e^{-1/2}$ (b) $2\,e^{-1/2}$
   (c) $e^{-1/2}$ (d) none of these

6. Let $f(x)=\ln(2x-x^2)+\sin\frac{\pi x}{2}$. Then,
   (a) graph of f is symmetrical about the line $x=1$ (b) graph of f is symmetrical about the line $x=2$
   (c) maximum value of f is 1 (d) minimum value of f does not exist

7. The sum of the legs of a right triangle is 9 cm. When the triangle rotates about one of the the legs, a cone results which has the maximum volume. Then,
   (a) slant height of such a cone is $3\sqrt{5}$ (b) maximum volume of the cone is 32π
   (c) curved surface of the cone is $18\sqrt{5}\pi$ (d) semi vertical angle of cone is $\tan^{-1}\sqrt{2}$

8. Least value of the function $f(x)=2^{x^2}-1+\frac{2}{2^{x^2}+1}$ is
   (a) 0 (b) $\frac{3}{2}$ (c) $\frac{2}{3}$ (d) 1

9. The greatest and the least value of the function, $f(x)=\sqrt{1-2x+x^2}-\sqrt{1+2x+x^2}$, $x\in(-\infty,\infty)$ are

(a) 2, −2
(b) 2, −1
(c) 2, 0
(d) none

10. The minimum value of the polynomial $x(x+1)(x+2)(x+3)$ is

(a) 0
(b) $\frac{9}{16}$
(c) −1
(d) $-\frac{3}{2}$

11. The difference between the greatest and least values of the functions, $f(x)=\cos x+\frac{1}{2}\cos 2x-\frac{1}{3}\cos 3x$ is

(a) $\frac{4}{3}$
(b) 1
(c) $\frac{9}{4}$
(d) $\frac{1}{6}$

12. The point at which the slope of the tangent of the function, $f(x)=e^x\cdot\cos x$ attains minima, when $x\in[0,2\pi]$ is

(a) $\frac{\pi}{4}$
(b) $\frac{\pi}{2}$
(c) $\frac{3\pi}{4}$
(d) π

13. If λ,μ are real numbers such that, $x^3-\lambda x^2+\mu x-6=0$ has its real roots and positive, then the minimum value of μ is

(a) $3(6)^{1/3}$
(b) $3(6)^{2/3}$
(c) $(6)^{1/3}$
(d) $(6)^{2/3}$

14. The points for which the function $f(x)=\int_1^x\{2(t-1)(t-2)^3+3(t-1)^2(t-2)^2\}\,dt$ attains maxima and minima, is

(a) maximum when $x=\frac{7}{5}$ and minimum when $x=1$
(b) maximum when $x=1$ and minimum when $x=0$
(c) maximum when $x=1$ and minimum when $x=2$
(d) maximum when $x=1$ and minimum when $x=\frac{7}{5}$

15. The set of values of 'a' for which the function $f(x)=\frac{ax^3}{3}+(a+2)x^2+(a-1)x+2$ possess a negative point of inflection.

(a) $(-\infty,-2)\cup(0,\infty)$
(b) $\{-4/5\}$
(c) $(-2,0)$
(d) empty set

Session 5

Maxima and Minima of Discontinuous Functions

Maxima and Minima of Discontinuous Functions

In discontinuous functions we don't apply any of the methods discussed earlier but we observe certain conditions and their graphs which would give you more clear picture.

Minimum of Discontinuous Functions

Let $f(x)$ be a function discontinuous or not differentiable at $x = a$, then the following four cases arise for minimum at $x = a$

Case (i) From figure 8.22

$$f(a) < f(a+h) \qquad f(a) < f(a-h)$$

Case (ii) From figure 8.23

$$f(a) < f(a+h) \qquad f(a) < f(a-h)$$

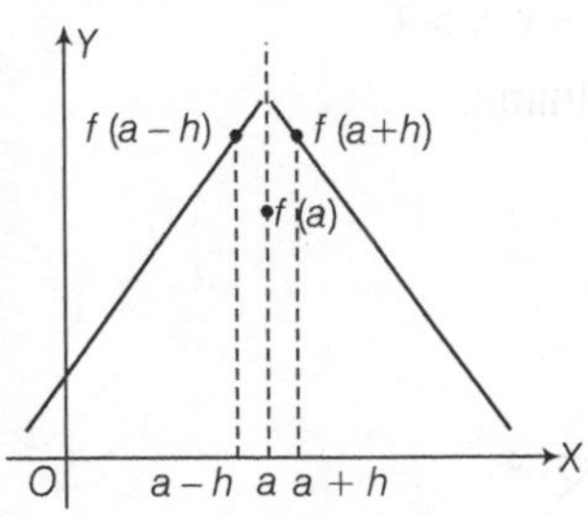

Figure 8.22

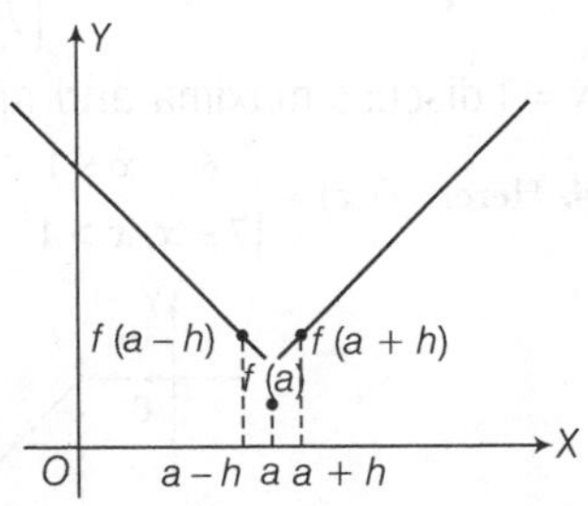

Figure 8.23

Case (iii) From figure 8.24

$$f(a) < f(a+h) \qquad f(a) \le f(a-h)$$

Case (iv) From figure 8.25

$$f(a) \le f(a+h) \qquad f(a) < f(a-h)$$

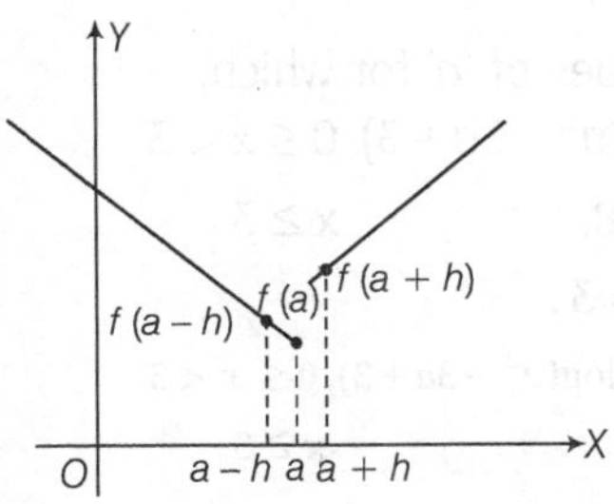

Figure 8.24

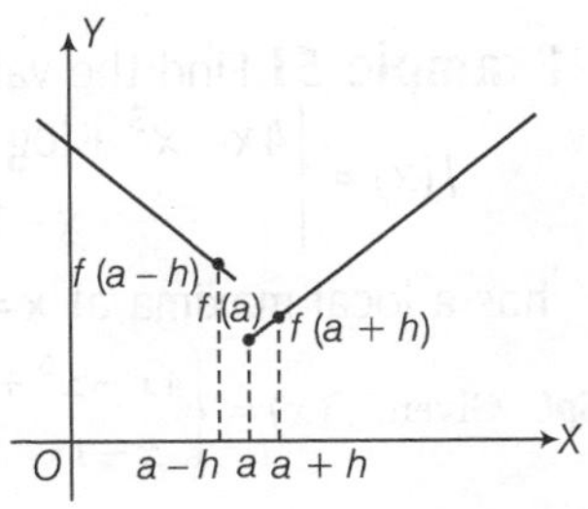

Figure 8.25

from all the four above mentioned cases for minimum of discontinuous functions, we have

$$f(a) \le f(a+h) \text{ and } f(a) \le f(a-h)$$

Example 48 Discuss the minima of $f(x) = \{x\}$, (where {.} denotes the fractional part of x) for $x = 6$.

Sol. As we have discussed for discontinuous functions, minimum at $x = a$ is attained when,

$f(a) \le f(a+h)$ and $f(a) \le f(a-h) \Rightarrow f(x)$ is minimum at $x = a$

Here, $f(x) = \{x\}$ is discontinuous function at $x = 6$

where, $f(6) = 0$

$$f(6+h) > 0 \text{ and } f(6-h) > 0$$

So, $\quad f(6) < f(6+h)$ and $f(6) < f(6-h)$

$\Rightarrow \quad f(x)$ is minimum at $x = 6$.

$\therefore \quad f(x)$ attains local minima for $x = 6$.

Example 49 Let $f(x) = \begin{cases} |x-2| + a^2 - 9a - 9, & \text{if } x < 2 \\ 2x - 3, & \text{if } x \ge 2 \end{cases}$.

Then, find the value of 'a' for which $f(x)$ has local minimum at $x = 2$.

Sol. We have, $f(x) = \begin{cases} |x-2| + a^2 - 9a - 9, & \text{if } x < 2 \\ 2x - 3, & \text{if } x \ge 2 \end{cases}$

$f(x)$ has local minima at $x = 2$.

Since, $f(x) = 2x - 3$ for $x \ge 2$ [is strictly increasing]

$$\therefore \quad \lim_{x \to 2^-} f(x) \ge f(2) \text{ or } \lim_{h \to 0} f(2-h) \ge f(2)$$

$$\Rightarrow \lim_{h \to 0} \{|2-h-2| + a^2 - 9a - 9\} \ge 1 \quad [\because f(2) = 2 \times 2 - 3 = 1]$$

$$\Rightarrow \quad a^2 - 9a - 10 \ge 0$$

$$\Rightarrow \quad (a+1)(a-10) \ge 0$$

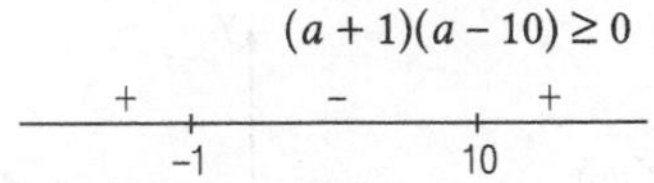

$$\Rightarrow \quad a \le -1 \text{ or } a \ge 10$$

Maximum of Discontinuous Functions

Let $f(x)$ be a function discontinuous or not differentiable at $x = a$, then the following four cases arise for maximum at $x = a$.

(i) From figure 8.26

$$f(a) > f(a+h)$$
$$f(a) > f(a-h)$$

(ii) From figure 8.27

$$f(a) > f(a+h)$$
$$f(a) > f(a-h)$$

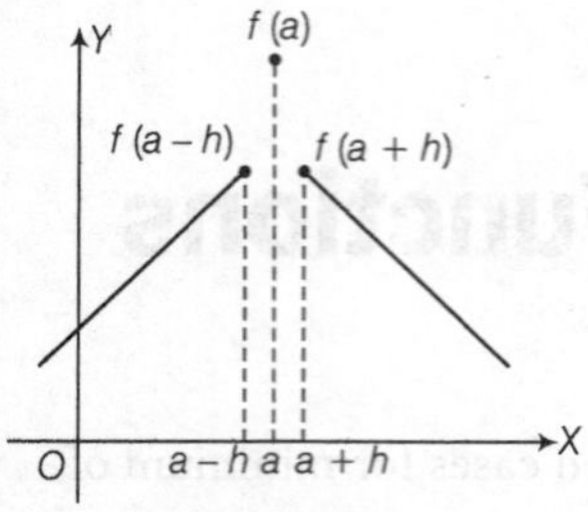

Figure 8.26

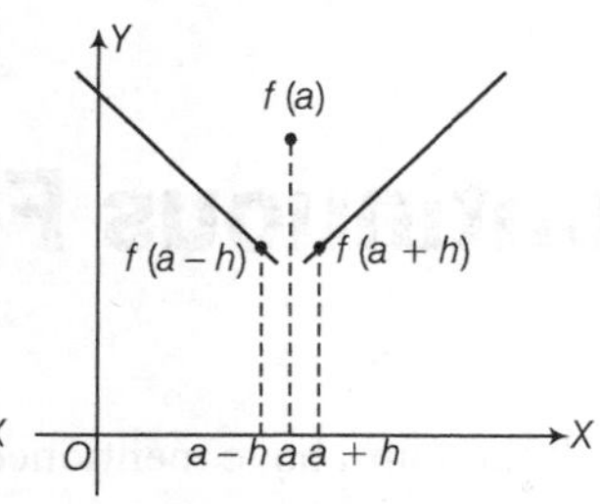

Figure 8.27

(iii) From figure 8.28

$f(a) > f(a+h)$

$f(a) \geq f(a-h)$

(iv) From figure 8.29

$f(a) \geq f(a+h)$

$f(a) > f(a-h)$

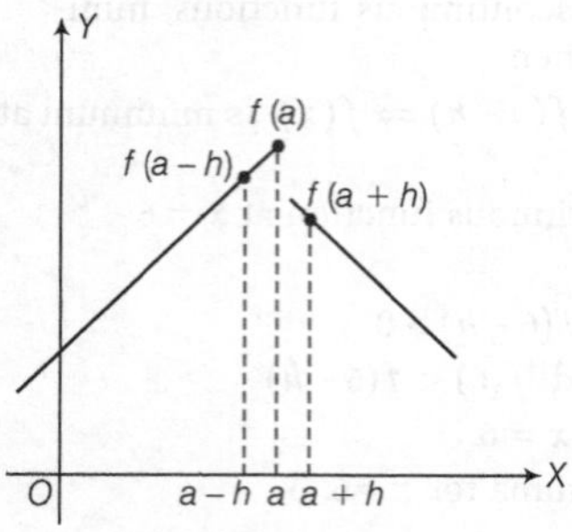

Figure 8.28

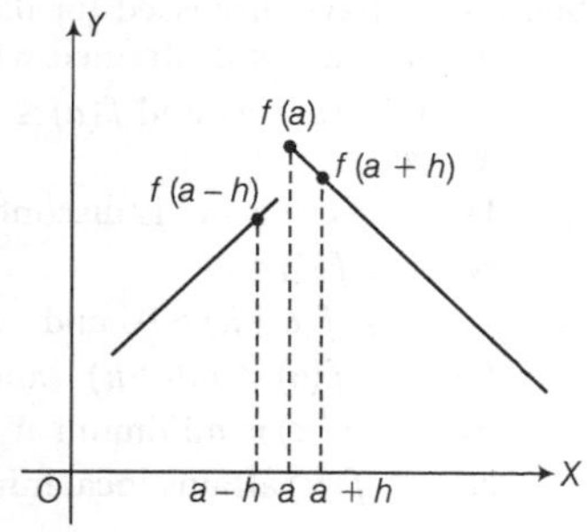

Figure 8.29

From all the above four mentioned cases for maximum of discontinuous functions, we have

$$f(a) \geq f(a+h), f(a) \geq f(a-h)$$

Neither Maximum Nor Minimum for Discontinuous Functions

Let $f(x)$ be a function discontinuous or not differentiable at $x = a$, then the following cases arise for neither maximum nor minimum at $x = a$

(i) $f(a) < f(a-h)$

$f(a) \geq f(a+h)$

(ii) $f(a) \leq f(a+h)$

$f(a) > f(a-h)$

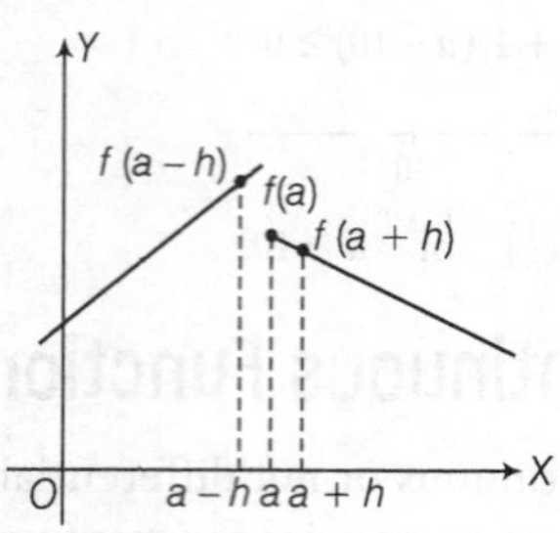

Figure 8.30

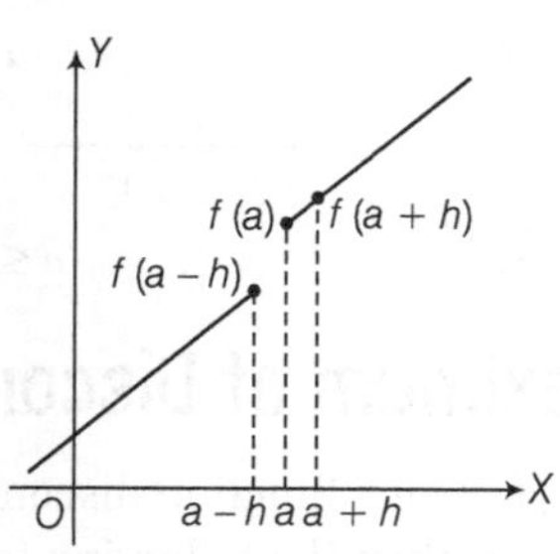

Figure 8.31

(iii) $f(a) < f(a+h)$

$f(a) \geq f(a-h)$

(iv) $f(a) \leq f(a-h)$

$f(a) > f(a+h)$

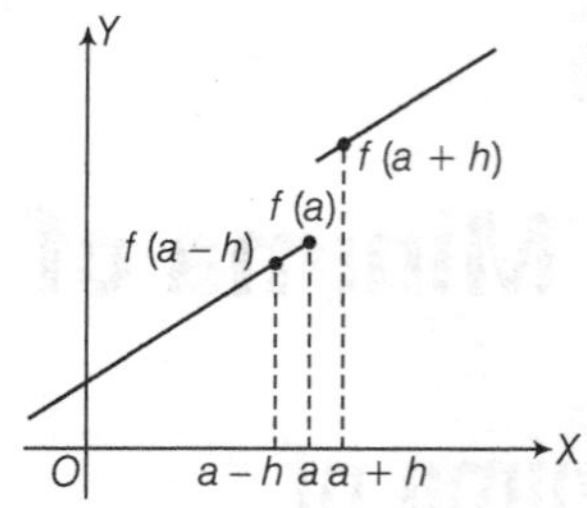

Figure 8.32

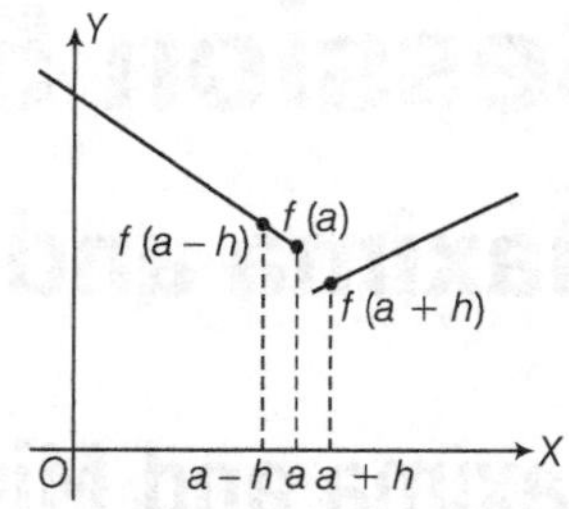

Figure 8.33

(v) $f(a) < f(a-h)$

$f(a) > f(a+h)$

(vi) $f(a) < f(a-h)$

$f(a) \geq f(a+h)$

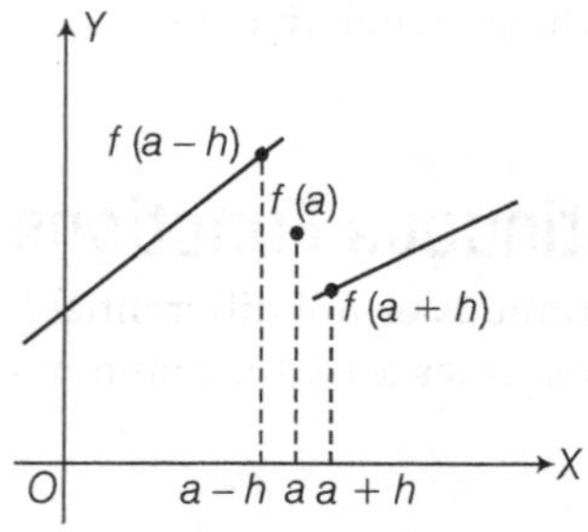

Figure 8.34

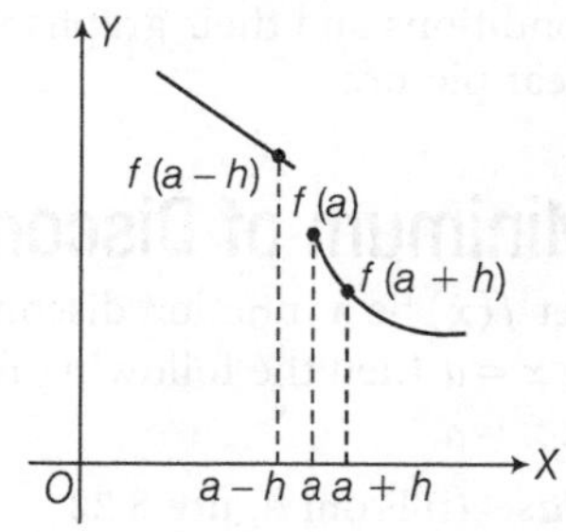

Figure 8.35

In all above cases no extremum exist.

Example 50 Let $f(x) = \begin{cases} 6, & x \leq 1 \\ 7-x, & x > 1 \end{cases}$, then for $f(x)$ at $x = 1$ discuss maxima and minima.

Sol. Here, $f(x) = \begin{cases} 6, & x \leq 1 \\ 7-x, & x > 1 \end{cases}$

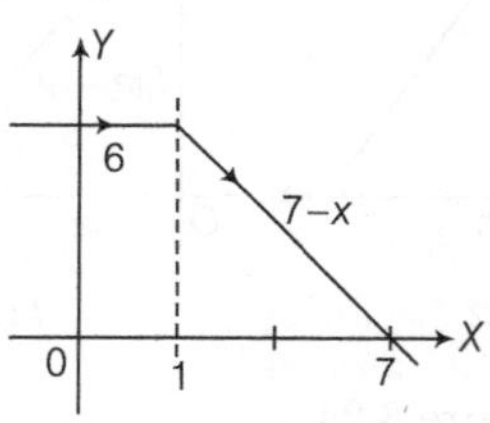

Clearly, $f(x)$ is not differentiable at $x = 1$

$\Rightarrow \quad f(1) = 6 \Rightarrow f(1-h) = 6$

and $\quad f(1+h) = 7-(1+h) = 6-h < 6$

Thus, $x = 1$ is a point of maxima.

Example 51 Find the values of 'a' for which,

$$f(x) = \begin{cases} 4x - x^3 + \log(a^2 - 3a + 3), & 0 \leq x < 3 \\ x - 18, & x \geq 3 \end{cases}$$

has a local maxima at $x = 3$.

Sol. Given, $f(x) = \begin{cases} 4x - x^3 + \log(a^2 - 3a + 3), & 0 \leq x < 3 \\ x - 18, & x \geq 3 \end{cases}$

Since, function attains maxima at $x = 3$

$\Rightarrow \quad f(3) \geq f(3-0)$

$\Rightarrow \quad -15 \geq 12 - 27 + \log(a^2 - 3a + 3)$

$$\Rightarrow \qquad \log(a^2 - 3a + 3) \le 0$$

Where for 'log' to exists,

$$a^2 - 3a + 3 > 0 \text{ and } \log(a^2 - 3a + 3) \le 0$$

$$\Rightarrow \qquad 0 < a^2 - 3a + 3 \le 1$$

i.e. $\qquad (a-2)(a-1) \le 0$

Using number line rule, we have

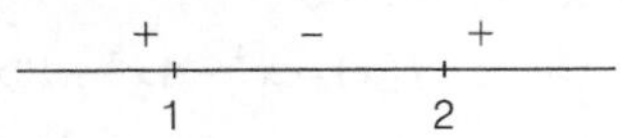

i.e. $\qquad 1 \le a \le 2$

$\therefore$ $f(x)$ attains local maxima at $x = 3$, when $a \in [1, 2]$.

Nature of Roots of Cubic Polynomials

Let $f(x) = ax^3 + bx^2 + cx + d$ be the given cubic polynomial and $f(x) = 0$ be the corresponding cubic equations, where $a, b, c, d \in R$ and $a > 0$.

Now, $\quad f'(x) = 3ax^2 + 2bx + c$

Let $D = 4b^2 - 12ac = 4(b^2 - 3ac)$ be the discriminant of the equations $f'(x) = 0$.

Now, we have the following three cases.

***Case* I** If $D < 0 \Rightarrow f'(x) > 0, \forall x \in R$, i.e. $f(x)$ would be an increasing function of x. Also, $\lim_{x \to \infty} f(x) = \infty$ and $\lim_{x \to -\infty} f(x) = -\infty$.

Since, $f(x)$ is an increasing function, so it will intersect the X-axis at only once.

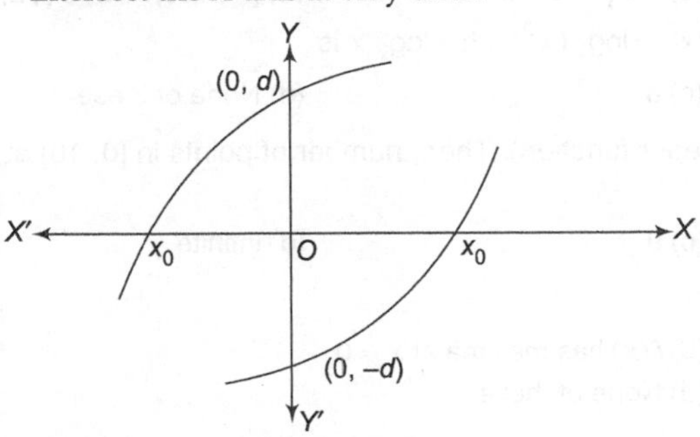

Figure 8.36

Clearly, $x_0 > 0$ if $d < 0$ and $x_0 < 0$, if $d > 0$

***Case* II** If $D > 0, f'(x) = 0$ would have two real roots (say) α and β (where $\alpha < \beta$)

$$\Rightarrow \qquad f'(x) = 3a(x - \alpha)(x - \beta)$$

$$\Rightarrow \qquad \begin{cases} f'(x) < 0, x \in (\alpha, \beta) \\ f'(x) > 0, x \in (-\infty, \alpha) \cup (\beta, \infty) \\ f'(x) = 0, x \in \{\alpha, \beta\} \end{cases}$$

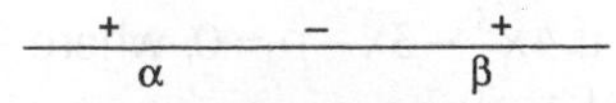

Here, $x = \alpha$, will be point of local maxima and $x = \beta$ will be point of local minima.

***Case* III** If $D = 0$, $f'(x) = 3a(x - \alpha)^2$, where α is a root of $f'(x) = 0$, then $f(x) = a(x - \alpha)^3 + K$. If $K = 0$, then $f(x) = 0$ has three equal real roots. if $K \ne 0$, then $f(x) = 0$ has one real root.

Graphical Representation of Roots of Cubic Polynomials

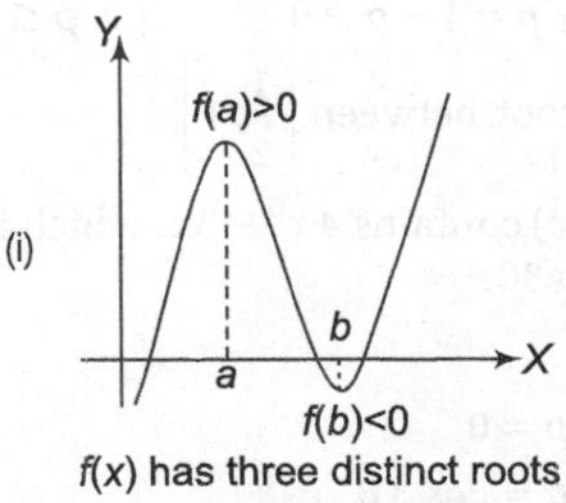

Figure 8.37

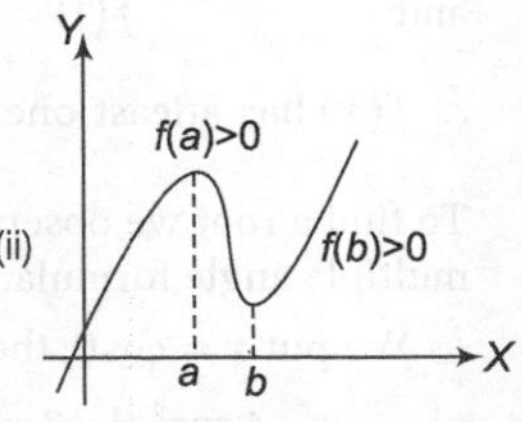

Figure 8.38

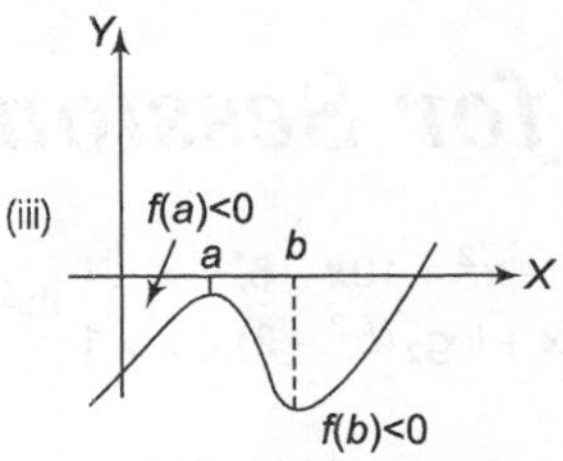

Figure 8.39

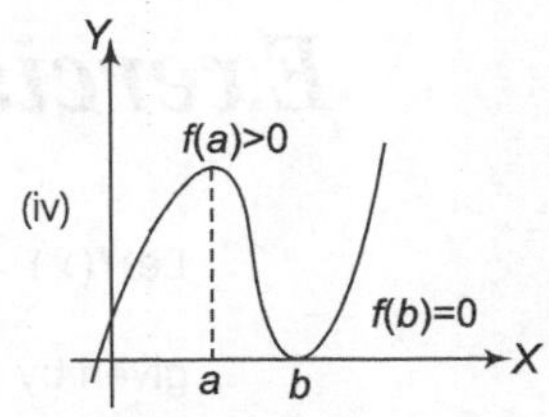

Figure 8.40

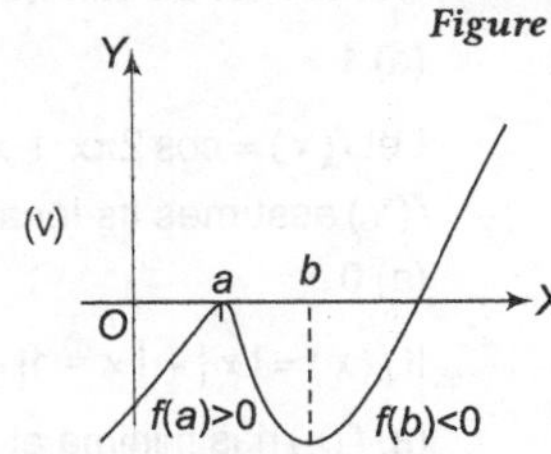

Figure 8.41

Conclusions

(i) If $f(a) \cdot f(b) < 0$, then $f(x)$ has three distinct roots (see figure 8.37).

(ii) If $f(a) \cdot f(b) > 0$, then $f(x)$ has one real and two imaginary roots (see figure 8.38 and 8.39).

(iii) If $f(a) \cdot f(b) = 0$, then $f(x)$ has three roots but two roots are identical (see figure 8.40 and 8.41).

Example 52 If $4x^3-3x-p=0$, where $-1\le p\le 1$ has unique root in $\left[\frac{1}{2},1\right]$, then the root is **[IIT JEE 2005]**

(a) $\frac{\cos^{-1}p}{3}$ (b) $\cos\left(\frac{1}{3}\cos^{-1}p\right)$

(c) $\cos(\cos^{-1}p)$ (d) None of these

Sol. Given that, $-1\le p\le 1$

Let $f(x)=4x^3-3x-p$

Now, $f\left(\frac{1}{2}\right)=\frac{1}{2}-\frac{3}{2}-p=-1-p\le 0$ $[\because p\ge -1]$

and $f(1)=4-3-p=1-p\ge 0$ $[\because p\le 1]$

$\therefore$ $f(x)$ has atleast one real root between $\left[\frac{1}{2},1\right]$.

To find a root we observe $f(x)$ contains $4x^3-3x$, which is multiple angle formula for $\cos 3\theta$.

$\therefore$ We put $x=\cos\theta$, then

$$4\cos^3\theta-3\cos\theta-p=0$$

$$\Rightarrow \quad p=\cos 3\theta$$

$$\Rightarrow \quad \theta=\frac{1}{3}\cos^{-1}(p)$$

$\therefore$ Root is $\cos\left(\frac{1}{3}\cos^{-1}p\right)$

Hence, (b) is the correct answer.

Example 53 The number of distinct real roots of $x^4-4x^3+12x^2+x-1=0$ **[IIT JEE 2011]**

Sol. Given, $f(x)=x^4-4x^3+12x^2+x-1$

$$\Rightarrow \quad f'(x)=4x^3-12x^2+24x+1$$

$$\Rightarrow \quad f''(x)=12x^2-24x+24$$

$$\Rightarrow \quad f''(x)=12(x^2-2x+2)$$

$$\Rightarrow \quad f''(x)=12((x-1)^2+1)\Rightarrow f''(x)>0$$

$\Rightarrow$ $f'(x)$ is an increasing cubic function.

$\Rightarrow$ $f'(x)$ has only one real root and two imaginary root.

$\therefore f(x)$ cannot have all distinct roots, atmost two real roots.

Now, $f(-1)=15, f(0)=-1, f(1)=9$

$\therefore f(x)$ must have one root in (−1, 0) and other in (0,1).

Hence, there are 2 real roots.

Exercise for Session 5

1. Let $f(x)=\begin{cases} x^3-x^2+10x-5, & x\le 1 \\ -2x+\log_2(b^2-2), & x>1 \end{cases}$ the set of values of b for which $f(x)$ have greatest value at $x=1$ is given by

   (a) $1<b\le 2$ (b) $b=\{1,2\}$ (c) $b\in(-\infty,-1)$ (d) $[-\sqrt{130},-\sqrt{2})\cup(\sqrt{2},\sqrt{130}]$

2. Number of solution(s) satisfying the equations, $3x^2-2x^3=\log_2(x^2+1)-\log_2 x$ is

   (a) 1 (b) 2 (c) 3 (d) None of these

3. Let $f(x)=\cos 2\pi x+x-[x]$ ([.] denotes the greatest integer function). Then, number of points in [0, 10] at which $f(x)$ assumes its local maximum value, is

   (a) 0 (b) 10 (c) 9 (d) infinite

4. If $f(x)=|x|+|x-1|+|x-2|$, then

   (a) $f(x)$ has minima at $x=1$ (b) $f(x)$ has maxima at $x=0$

   (c) $f(x)$ has neither maxima nor minima at $x=3$ (d) None of these

5. The function $f(x)=1+[\cos x]x$, in $0<x\le\frac{\pi}{2}$

   (a) has a minimum value 0 (b) has a maximum value 2

   (c) is continuous in $\left[0,\frac{\pi}{2}\right]$ (d) is not differentiable at $x=\frac{\pi}{2}$

6. If $\lim_{x\to a} f(x)=\lim_{x\to a}[f(x)]$ ([.] denotes the greater integer function) and $f(x)$ is non-constant continuous function, then

   (a) $\lim_{x\to a} f(x)$ is irrational (b) $\lim_{x\to a} f(x)$ is non-integer

   (c) $f(x)$ has local maxima at $x=a$ (d) $f(x)$ has local minima at $x=a$

7. Find the value of a if $x^3-3x+a=0$ has three distinct real roots.

8. Prove that there exist exactly three non- similar isosceles ΔABC such that $\tan A+\tan B+\tan C=100$.

JEE Type Solved Examples :

Single Option Correct Type Questions

Ex. 1 *The values of 'K' for which the point of minimum of the function $f(x) = 1 + K^2x - x^3$ satisfy the inequality $\frac{(x^2 + x + 2)}{(x^2 + 5x + 6)} < 0$, belongs*

(a) $(-3\sqrt{3}, \infty)$

(b) $(-3\sqrt{3}, -2\sqrt{3}) \cup (0, \infty)$

(c) $(-3\sqrt{3}, -2\sqrt{3}) \cup (2\sqrt{3}, 3\sqrt{3})$

(d) $(0, \infty)$

Sol. Here, $\frac{x^2 + x + 2}{x^2 + 5x + 6} < 0 \Rightarrow \frac{\left(x + \frac{1}{2}\right)^2 + \frac{7}{4}}{(x+2)(x+3)} < 0$

+   −3   −   −2   +

where $\left(x + \frac{1}{2}\right)^2 + \frac{7}{4}$ is always positive.

$\therefore$ Using number line rule for $(x + 2)(x + 3)$ as shown above, we get

$$\frac{x^2 + x + 2}{x^2 + 5x + 6} < 0 \Rightarrow -3 < x < -2 \quad \text{...(i)}$$

Now, consider $f(x) = 1 + K^2x - x^3$

$$f'(x) = K^2 - 3x^2, \; f''(x) = -6x$$

For maximum/minimum let $f'(x) = 0$,

$$\Rightarrow \quad x = \pm \frac{|K|}{\sqrt{3}}$$

Let $\quad x_1 = \frac{|K|}{\sqrt{3}}$ and $x_2 = -\frac{|K|}{\sqrt{3}}$

$\therefore \quad f''(x_1) < 0$ and $f''(x_2) > 0$

$\Rightarrow f(x)$ is maximum at $x = x_1$ and $f(x)$ is minimum at $x = x_2$.

$\therefore \quad -3 < x_2 < -2$

$\Rightarrow \quad -3 < \frac{-|K|}{\sqrt{3}} < -2 \quad$ [from Eq. (i)]

$\Rightarrow \quad 3\sqrt{3} > |K| > 2\sqrt{3}$

$\Rightarrow \quad K \in (-3\sqrt{3}, -2\sqrt{3}) \cup (2\sqrt{3}, 3\sqrt{3})$

Hence, (c) is the correct answer.

Ex. 2 *The values of a and b for which all the extrema of the function, $f(x) = a^2x^3 - 0.5\,ax^2 - 2x - b$, is positive and the minimum is at the point $x_0 = \frac{1}{3}$, are*

(a) when $a = -2 \Rightarrow b < -\frac{11}{27}$ and when $a = 3 \Rightarrow b < -\frac{1}{2}$

(b) when $a = 3 \Rightarrow b < -\frac{11}{27}$ and when $a = 2 \Rightarrow b < -\frac{1}{2}$

(c) when $a = -2 \Rightarrow b < -\frac{1}{2}$ and when $a = 3 \Rightarrow b < -\frac{11}{27}$

(d) None of the above

Sol. Given, $\quad f(x) = a^2x^3 - 0.5ax^2 - 2x - b$

$\Rightarrow \quad f'(x) = 3a^2x^2 - ax - 2$

For extrema, $f'(x) = 0$

$\Rightarrow \quad 3a^2x^2 - ax - 2 = 0$ at $x = \frac{1}{3}$

$\left[\text{as at } x = \frac{1}{3} \text{ function is minimum}\right]$

$\therefore \quad 3a^2\left(\frac{1}{3}\right)^2 - a\left(\frac{1}{3}\right) - 2 = 0 \Rightarrow \frac{a^2}{3} - \frac{a}{3} - 2 = 0$

$\Rightarrow \quad a^2 - a - 6 = 0$ or $a = -2, 3$

So, their arise two cases as

***Case* I** At $a = 3$, if function attains minimum and is positive.

Then, $\quad 9\left(\frac{1}{3}\right)^3 - (0.5)(3)\left(\frac{1}{3}\right)^2 - 2\left(\frac{1}{3}\right) - b > 0$

$\left[\because \text{minimum at } x = \frac{1}{3} \text{ and } a = 3 \Rightarrow f\left(\frac{1}{3}\right) > 0 \text{ when } a = 3\right]$

$\Rightarrow \quad b < \frac{1}{3} - \frac{1.5}{9} - \frac{2}{3} \Rightarrow b < -\frac{1}{2}$

***Case* II** At $a = -2$, if function attains minimum and is positive.

Then, $\quad (-2)^2\left(\frac{1}{3}\right)^3 - (0.5)(-2)\left(\frac{1}{3}\right)^2 - 2\left(\frac{1}{3}\right) - b > 0$

$\left[\begin{array}{l}\text{since minimum at } x = \frac{1}{3}, \text{ when } a = -2 \\ \Rightarrow f\left(\frac{1}{3}\right) > 0, \text{ when } a = -2\end{array}\right]$

$\Rightarrow \quad b < \frac{4}{27} + \frac{1}{9} - \frac{2}{3}$ or $b < -\frac{11}{27}$

$\therefore$ When $a = 3 \Rightarrow b < -\frac{1}{2}$ and when $a = -2 \Rightarrow b < -\frac{11}{27}$

Hence, (a) is the correct answer.

Ex. 3 *If $f''(x) + f'(x) + f^2(x) = x^2$ is the differential equation of a curve and let P be the point of maxima, then number of tangents which can be drawn from P to $x^2 - y^2 = a^2$ is/are*

(a) 2   (b) 1

(c) 0   (d) either 1 or 2

Sol. At point of maxima $f'(x)=0$ and $f''(x)<0$

$$\Rightarrow \quad f''(x)=x^2-f^2(x)\le 0$$

(Since, the curve is $x^2-y^2=a^2$ and $x^2-f^2(x)\le 0$

$\therefore x_1^2-y_1^2<a^2 \Rightarrow$ point lies outside the hyperbola)

$\Rightarrow$ Point $P(x, f(x))$ lies outside $x^2-y^2=a^2$

$\therefore$ Two tangents can be drawn.

Hence, (a) is the correct answer.

Ex. 4 *Let* $f(x)=\int_0^x \cos\left(\frac{t^2+2t+1}{5}\right)dt,\ 0<x<2$, *then* $f(x)$

(a) increases monotonically
(b) decreasing monotonically
(c) has one point of local maximum
(d) has one point of local minima

Sol. We have, $f'(x)=\cos\left(\frac{x^2+2x+1}{5}\right)\times 1=\cos\left(\frac{(x+1)^2}{5}\right)$

Since, $\quad 0<x<2 \Rightarrow 1/5<\frac{(x+1)^2}{5}>9/5$

Now, $\quad \cos\left(\frac{(x+1)^2}{5}\right)=0$

$$\Rightarrow \quad \frac{(x+1)^2}{5}=\frac{\pi}{2}$$

$$f'(x)>0, \text{ if } 0<x<(\sqrt{5\pi/2}-1)$$

and $\quad f'(x)<0, \text{ if } (\sqrt{5\pi/2}-1)<x<2$

$\therefore \quad x=(\sqrt{5\pi/2}-1)$ is a point of local maxima

Hence, (c) is the correct answer.

Ex. 5 *As 'x' ranges over the interval* $(0,\infty)$, *the function* $f(x)=\sqrt{9x^2+173x+900}-\sqrt{9x^2+77x+900}$, *ranges over*

(a) (0, 4) (b) (0, 8)
(c) (0, 12) (d) (0, 16)

Sol. $\because \sqrt{9x^2+173x+900}>\sqrt{9x^2+77x+900},\ \forall\ x\in(0,\infty)$

$$\Rightarrow \quad f(x)>0,\ \forall\ x\in(0,\infty)$$

Now, $\quad f(x)=\dfrac{96x}{\sqrt{9x^2+173x+900}+\sqrt{9x^2+77x+900}}$ [rationalise]

$$\therefore \lim_{x\to\infty} f(x)=\lim_{x\to\infty}\frac{96}{\sqrt{9+\frac{173}{x}+\frac{900}{x^2}}+\sqrt{9+\frac{77}{x}+\frac{900}{x^2}}}$$

$$=\frac{96}{6}=16$$

$\therefore$ Range of $f(x)\in(0,16)$

Hence, (d) is the correct answer.

Ex. 6 *Let* $g:[1,6]\to[0,\infty)$ *be a real valued differentiable function satisfying* $g'(x)=\frac{2}{x+g(x)}$ *and* $g(1)=0$, *then the maximum value of g cannot exceed*

(a) $\log 2$ (b) $\log 6$ (c) $6\log 2$ (d) $2\log 6$

Sol. Here, $g'(x)=\frac{2}{x+g(x)}>0$, for all $x\in[1,6]$

$\Rightarrow g(x)$ is increasing function on $[1,6]$

$$\therefore \quad \int_1^6 g'(x)\,dx\le\int_1^6 \frac{2}{x}dx$$

$$\Rightarrow \quad (g(x))_1^6\le 2(\log_e x)_1^6$$

$$\Rightarrow \quad g(6)-g(1)\le 2\log 6$$

$$\therefore \quad g(6)\le 2\log 6, \text{ as } g(1)=0$$

$\Rightarrow g$ cannot exceed $2\log 6$.

Hence, (d) is the correct answer.

Ex. 7 *The minimum value of the function,* $f(x)=x^{3/2}+x^{-3/2}-4\left(x+\frac{1}{x}\right)$. *for all permissible real values of x, is*

(a) −10 (b) −6 (c) −7 (d) −8

Sol. Here, $f(x)=x^{3/2}+x^{-3/2}-4\left(x+\frac{1}{x}\right)$

$$f(x)=\left(\sqrt{x}+\frac{1}{\sqrt{x}}\right)^3-3\left(\sqrt{x}+\frac{1}{\sqrt{x}}\right)-4\left(\left(\sqrt{x}+\frac{1}{\sqrt{x}}\right)^2-2\right)$$

Let $\quad \sqrt{x}+\frac{1}{\sqrt{x}}=t,$ $\quad [x>0]$

$$\therefore \quad g(t)=t^3-3t-4(t^2-2)$$

Now, $\quad g(t)=t^3-4t^2-3t+8$, where $t\in[2,\infty)$

$$g'(t)=3t^2-8t-3=(t-3)(3t+1)$$

$$g'(t)=0 \Rightarrow t=3\ (t\ne -1/3)$$

$$g''(t)=6t-8$$

$$\Rightarrow \quad g''(3)=10>0$$

$\therefore$ Minimum value is

$$g(3)=27-36-9+8=-10$$

Hence, (a) is the correct answer.

Ex. 8 *The tangent to the curve* $y=1-x^2$ *at* $x=\alpha$, *where* $0<\alpha<1$, *meets the axes at P and Q. If* α *varies and the minimum value of the area of the* ΔOPQ *is k times the area bounded by the axis and the part of the curve for* $0<x<1$, *then k is equal to*

(a) $2/\sqrt{3}$ (b) 75/16 (c) 25/18 (d) 2/3

Sol. Here, equation of tangent

$$y+2\alpha x=\alpha^2+1 \quad \text{...(i)}$$

From Eq. (i) intersect the axes at

$$P\left(\frac{\alpha^2+1}{2\alpha}, 0\right) \text{ and } Q(0, \alpha^2+1)$$

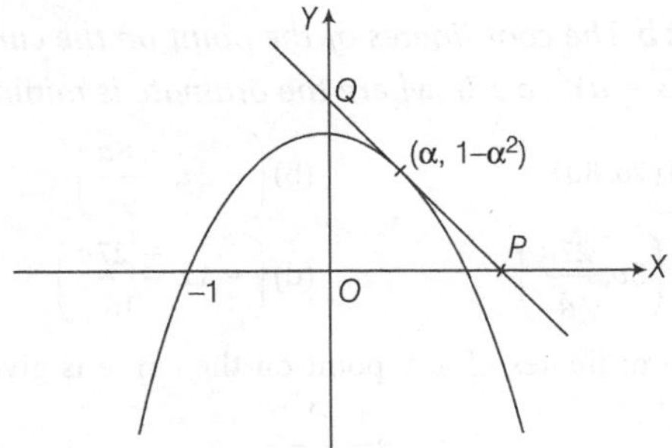

$$\therefore \text{ Area of } \Delta OPQ = \frac{1}{2}\cdot\frac{(\alpha^2+1)^2}{2\alpha} = \frac{(\alpha^2+1)^2}{4\alpha}$$

$$f(\alpha) = \frac{1}{4}(\alpha^3 + 2\alpha + 1/\alpha)$$

$$f'(\alpha) = \frac{1}{4}(3\alpha^2 + 2 - 1/\alpha^2) = \frac{1}{4}\frac{(3\alpha^4 + 2\alpha^2 - 1)}{\alpha^2}$$

$$= \frac{1}{4}\cdot\frac{(\alpha^2+1)(3\alpha^2-1)}{\alpha^2}$$

$$= \frac{1}{4}\frac{(\alpha^2+1)(\sqrt{3}\alpha - 1)(\sqrt{3}\alpha+1)}{\alpha^2}$$

Using number line rule, we have

$+ \quad -1/\sqrt{3} \quad - \quad 1/\sqrt{3} \quad +$

$\therefore$ Minimum when $\alpha = 1/\sqrt{3}$

Thus, minimum area of triangle $= 4/3\sqrt{3}$

$$\text{Given, } \frac{4}{3\sqrt{3}} = k\int_0^1 (1-x^2)\,dx = k\left(x - \frac{x^3}{3}\right)_0^1 = \frac{2k}{3}$$

$$\Rightarrow \qquad k = 2/\sqrt{3}$$

● **Ex. 9** *Least natural number a for which* $x + ax^{-2} > 2, \forall\, x \in [0, \infty)$ *is*

(a) 1 (b) 2 (c) 5 (d) None of these

Sol. Here, $x + ax^{-2} > 2$

$$\Rightarrow \qquad x^3 - 2x^2 + a > 0$$

Let $\quad f(x) = x^3 - 2x^2 + a$

Since, $f(x) > 0, \forall\, x \in [0, \infty), f(0) > 0$

and $\quad \min f(x) > 0$

$\Rightarrow \qquad a > 0$ and for minimum $f(x)$

$- \quad 4/3 \quad +$

$$f'(x) = 3x^2 - 4x = 0$$

$$\Rightarrow \qquad x = 0, \frac{4}{3}$$

$\therefore\ f(x)$ is minimum at $x = \frac{4}{3}$

$$f\left(\frac{4}{3}\right) > 0 \Rightarrow a > \frac{32}{27}$$

Hence, (b) is the correct answer.

● **Ex. 10** *If* $k\sin^2 x + \frac{1}{k}\operatorname{cosec}^2 x = 2, x \in \left(0, \frac{\pi}{2}\right)$, *then* $\cos^2 x + 5\sin x\cos x + 6\sin^2 x$ *is equal to*

(a) $\frac{k^2+5k+6}{k^2}$ (b) $\frac{k^2-5k+6}{k^2}$

(c) 6 (d) None of these

Sol. Given, $\qquad k\sin^2 x + \frac{1}{k\sin^2 x} = 2$

$$\Rightarrow \qquad \left(\sqrt{k}\sin x - \frac{1}{\sqrt{k}\sin x}\right)^2 = 0$$

$$\Rightarrow \qquad \sin^2 x = \frac{1}{k}$$

So, $\cos^2 x + 5\sin x\cos x + 6\sin^2 x$

$$= \frac{k-1}{k} + \frac{5\sqrt{k-1}}{k} + \frac{6}{k}$$

$$= \frac{k+5+5\sqrt{k-1}}{k}$$

Hence, (d) is the correct answer.

● **Ex. 11** *The least value of the expression* $x^2 + 4y^2 + 3z^2 - 2x - 12y - 6z + 14$ *is*

(a) 0

(b) 1

(c) no least value

(d) None of the above

Sol. Let $f(x, y, z) = x^2 + 4y^2 + 3z^2 - 2x - 12y - 6z + 14$

$$= (x-1)^2 + (2y-3)^2 + 3(z-1)^2 + 1$$

For least value of $f(x, y, z)$

$$x - 1 = 0, 2y - 3 = 0 \text{ and } z - 1 = 0$$

$$\therefore \qquad x = 1,\ y = \frac{3}{2},\ z = 1$$

Hence, least value of $f(x, y, z)$ is $f\left(1, \frac{3}{2}, 1\right) = 1$

Hence, (b) is the correct answer.

Ex. 12 *On the interval* $\left[\frac{5\pi}{4}, \frac{4\pi}{3}\right]$, *the least value of the function* $f(x) = \int_{5\pi/4}^{x} (3\sin t + 4\cos t)\, dt$ *is*

(a) $\frac{3}{2} + \frac{1}{\sqrt{2}} - 2\sqrt{3}$ (b) $\frac{3}{2} - \frac{1}{\sqrt{2}} + 2\sqrt{3}$

(c) $\frac{3}{2} - \frac{1}{\sqrt{2}} - 2\sqrt{3}$ (d) None of these

Sol. Given, $f(x) = \int_{5\pi/4}^{x} (3\sin t + 4\cos t)\, dt,\ x \in \left[\frac{5\pi}{4}, \frac{4\pi}{3}\right]$.

$$f'(x) = 3\sin x + 4\cos x$$

$f'(x) < 0$ as $\sin x, \cos x$ are negative for $x \in \left[\frac{5\pi}{4}, \frac{4\pi}{3}\right]$.

$$\Rightarrow \quad f(x)|_{\min} = f\left(\frac{4\pi}{3}\right) = \int_{5\pi/4}^{4\pi/3} (3\sin t + 4\cos t)\, dt$$

$$= \frac{3}{2} + \frac{1}{\sqrt{2}} - 2\sqrt{3}$$

Hence, (a) is the correct answer.

Ex. 13 *For any real* θ, *the maximum value of* $\cos^2(\cos\theta) + \sin^2(\sin\theta)$ *is*

(a) 1 (b) $1 + \sin^2 1$ (c) $1 + \cos^2 1$ (d) does not exist

Sol. The maximum value of $\cos^2(\cos\theta)$ is 1 and that of $\sin^2(\sin\theta)$ is $\sin^2 1$, both exists for $\theta = \frac{\pi}{2}$.

So, maximum value is $1 + \sin^2 1$.

Hence, (b) is the correct answer.

Ex. 14 *If* $\sin\theta + \cos\theta = 1$, *then the minimum value of* $(1 + \operatorname{cosec}\theta)(1 + \sec\theta)$ *is*

(a) 3 (b) 4 (c) 6 (d) 9

Sol. We know that, AM $\geq$ GM

$$\frac{\sin\theta + \cos\theta}{2} \geq \sqrt{\sin\theta\cos\theta} \quad \Rightarrow \quad \sin\theta\cos\theta \leq \frac{1}{4}$$

Now, let $\sin\theta = x, \cos\theta = y$

and $(1 + \operatorname{cosec}\theta)(1 + \sec\theta) \geq p$

$$\Rightarrow \quad \left(1 + \frac{1}{x}\right)\left(1 + \frac{1}{y}\right) \geq p$$

$$\Rightarrow \quad \left(\frac{1+x}{x}\right)\left(\frac{1+y}{y}\right) \geq p$$

$$\Rightarrow \quad xy + x + y + 1 \geq pxy$$

$$\Rightarrow \quad x + y + 1 \geq (p-1)\, xy$$

$$\Rightarrow \quad 2 \geq (p-1)\, xy$$

[since, $x + y = 1$]

$$\Rightarrow \quad xy \leq \frac{2}{p-1}$$

$$\Rightarrow \frac{2}{p-1} = \frac{1}{4} \Rightarrow p - 1 = 8 \Rightarrow p = 9$$

Hence, (d) is the correct answer.

Ex. 15 *The coordinates of the point on the curve* $x^3 = y(x-a)^2, a > 0$, *where the ordinate is minimum*

(a) $(2a, 8a)$ (b) $\left(-2a, \frac{-8a}{9}\right)$

(c) $\left(3a, \frac{27a}{4}\right)$ (d) $\left(-3a, \frac{-27a}{16}\right)$

Sol. The ordinates of any point on the curve is given by

$$y = \frac{x^3}{(x-a)^2}$$

$$\frac{dy}{dx} = \frac{x^2(x-3a)}{(x-a)^3}$$

Now, $\frac{dy}{dx} = 0$

$$\Rightarrow \quad x = 0 \text{ or } x = 3a$$

$$\left[\frac{d^2y}{dx^2}\right]_{x=0} = 0 \text{ and } \left[\frac{d^2y}{dx^2}\right]_{x=3a} = \frac{72a^5}{(2a)^6} > 0$$

So, y is minimum at $x = 3a$ and is equal to $\frac{27a}{4}$.

Hence, (c) is the correct answer.

Ex. 16 *If* $x^2 + y^2 + z^2 = 1$ *for* $x, y, z \in R$, *then the maximum value of* $x^3 + y^3 + z^3 - 3xyz$ *is*

(a) $\frac{1}{2}$ (b) 1 (c) 2 (d) 3

Sol. Let $t = xy + yz + zx$, so $-\frac{1}{2} \leq t \leq 1$

$$\therefore \quad x^3 + y^3 + z^3 - 3xyz$$

$$= (x + y + z)(x^2 + y^2 + z^2 - xy - yz - zx)$$

$$= \sqrt{(1+2t)}\,(1-t)$$

Let $f(t) = (1 + 2t)(1-t)^2$

$$f'(t) = 6t(t-1) = 0$$

+   −   +
0   1

Clearly, $t_{\max} = f(0) = 1$

Hence, (b) is the correct answer.

Ex. 17 *If* $a, b \in R$ *distinct numbers satisfying* $|a-1| + |b-1| = |a| + |b| = |a+1| + |b+1|$, *then the minimum value of* $|a-b|$ *is*

(a) 3 (b) 0

(c) 1 (d) 2

Sol. Let $a < b$ and $f(x) = |x - a| + |x - b|, \forall\ x \in R$

So, $f(x)$ is decreasing in $(-\infty, a]$ constant in $[a, b]$ and increasing in $[b, \infty)$, we have

$$f(0) = f(1) = f(-1)$$

$\Rightarrow \quad \{-1, 0, 1\} \in [a, b]$

$\therefore \quad |a - b|_{\min} = 2$

Hence, (d) is the correct answer.

Ex. 18 *Let $f(x) = x^4 + ax^3 + 3x^2 + bx + 1$, $a, b \in R$. If $f(x) \geq 0, \forall\ x \in R$, then the maximum value of $a^2 + b^2$ is equal to*

(a) 10 (b) 12
(c) 16 (d) None of these

Sol. Given, $f(x) = x^4 + ax^3 + 3x^2 + bx + 1$ and $f(x) \geq 0$

$$\Rightarrow \left(x^2 + \frac{a}{2}x\right)^2 + \left(3 - \frac{a^2 + b^2}{4}\right)x^2 + \left(\frac{b}{2}x + 1\right)^2 \geq 0$$

which holds only when $3 - \dfrac{a^2 + b^2}{4} \geq 0$

$$\Rightarrow \quad a^2 + b^2 \leq 12$$

Hence, (b) is the correct answer.

Ex. 19 *The maximum value of $(\sqrt{-3 + 4x - x^2} + 4)^2 + (x - 5)^2$ (where $1 \leq x \leq 3$) is*

(a) 34 (b) 36
(c) 32 (d) 20

Sol. Here, $(\sqrt{-3 + 4x - x^2} + 4)^2 + (x - 5)^2$, represents the square of the distance between circle $y = \sqrt{-3 + 4x - x^2}$ and point $(5, -4)$.

Clearly, the centre and radius of the circle respectively are $(2, 0)$ and 1.

i.e. Maximum distance between $x^2 + y^2 - 4x + 3 = 0$ and $(5, -4)$,

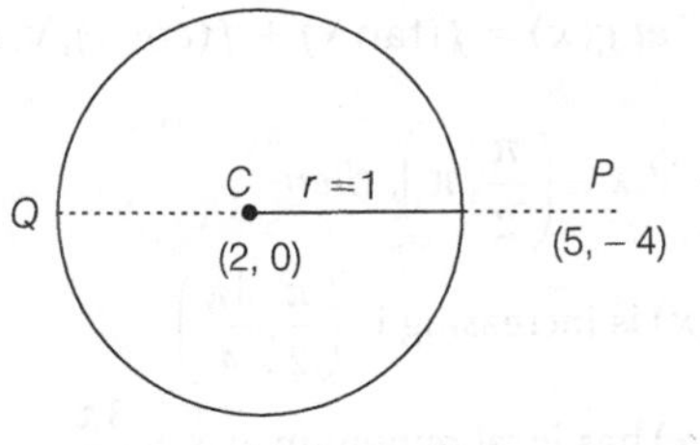

$$\Rightarrow \quad PQ^2 = (PC + \text{radius})^2$$

$$= (\sqrt{(5-2)^2 + (4-0)^2} + 1)^2$$

$$= 6^2 = 36$$

Hence, (b) is the correct answer.

Ex. 20 *If $a > b > 0$ and $f(\theta) = \dfrac{(a^2 - b^2)\cos\theta}{a - b\sin\theta}$, then the maximum value of $f(\theta)$, is*

(a) $2\sqrt{a^2 + b^2}$ (b) $\sqrt{a^2 + b^2}$
(c) $\sqrt{a^2 - b^2}$ (d) $\sqrt{b^2 - a^2}$

Sol. Here, $f(\theta) = \dfrac{(a^2 - b^2)\cos\theta}{a - b\sin\theta} = \dfrac{(a^2 - b^2)}{a\sec\theta - b\tan\theta}$

or $\quad f(\theta) = \dfrac{a^2 - b^2}{h(\theta)}$, where $h(\theta) = a\sec\theta - b\tan\theta$

$\therefore f(\theta)$ is maximum and minimum as $h(\theta)$ is minimum and maximum respectively.

$\Rightarrow \quad h(\theta) = a\sec\theta - b\tan\theta$

$\Rightarrow \quad h'(\theta) = \sec\theta\,(a\tan\theta - b\sec\theta)$

For maximum and minimum put $h'(\theta) = 0$

$\Rightarrow \quad \sin\theta = \dfrac{b}{a} \qquad [\sec\theta \neq 0] \ldots \text{(i)}$

Also, $\quad h''(\theta) = a\sec^3\theta + a\sec\theta \cdot \tan^2\theta - 2b\sec^2\theta\tan\theta$

or $\quad h''(\theta) = \dfrac{a + a\sin^2\theta - 2b\sin\theta}{\cos^3\theta}$

When $\sin\theta = \dfrac{b}{a}$

$$(h''(\theta)) = \frac{a + a \cdot \dfrac{b^2}{a^2} - 2b \cdot \dfrac{b}{a}}{\left(1 - \dfrac{b^2}{a^2}\right)^{3/2}} = \frac{a^2 - b^2}{a\left(\dfrac{a^2 - b^2}{a^2}\right)^{3/2}} > 0$$

$[\text{as } a > b]$

$\Rightarrow h(\theta)$ is minimum, when $\sin\theta = \dfrac{b}{a}$

$\therefore \quad f(\theta)$ is maximum, when

$$\sin\theta = \frac{b}{a} \Rightarrow f_{\max}(\theta) = \sqrt{a^2 - b^2}$$

Hence, (c) is the correct answer.

Ex. 21 *A solid cylinder of height H has a conical portion of same height and radius 1/3rd of height removed from it. Rain water is accumulating in it, at the rate equal to π times the instantaneous radius of the water surface inside the hole, the time after which hole will filled with water is*

(a) $\dfrac{H^2}{3}$ (b) H^2 (c) $\dfrac{H^2}{6}$ (d) $\dfrac{H^2}{4}$

Sol. Here radius, $r = \dfrac{H}{3}$, $\dfrac{x}{r} = \dfrac{y}{H} \Rightarrow 3x = y \Rightarrow \dfrac{dv}{dt} = \pi x$

$$\Rightarrow \quad \frac{d}{dt}\left(\frac{1}{3}\pi x^2 y\right) = \pi x \Rightarrow 3\int_0^r x\,dx = \int_0^t dt$$

$$\Rightarrow \quad 3\frac{r^2}{2} = t \Rightarrow \frac{3}{2} \cdot \frac{H^2}{9} = t \Rightarrow t = \frac{H^2}{6}$$

Hence, (c) is the correct answer.

Ex. 22 *If composite function $f_1(f_2(f_3(....(f_n(x)))))$ n times, is an increasing function and if r of f_i's are decreasing function while rest are increasing, then maximum value of function is*

(a) $\frac{n^2-1}{4}$ when n is an even number

(b) $\frac{n^2}{4}$ when n is an odd number

(c) $\frac{n^2-1}{4}$ when n is an odd number

(d) None of the above

Sol. r must be an even integer because two decreasing are required to make it increasing function.

Let $\quad y = r(n-r)$,

When n is odd

$$r = \frac{n-1}{2} \text{ or } \frac{n+1}{2} \text{ for maximum value of } y$$

When n is even

$$r = \frac{n}{2} \text{ for maximum value of } y$$

$\therefore$ Maximum $(y) = \frac{n^2-1}{4}$, when n is odd and $\frac{n^2}{4}$

When n is even.

Hence, (c) is the correct answer.

JEE Type Solved Examples :
More than One Correct Option Type Questions

Ex. 23 *Let $f(x) = \sin x + ax + b$. Then, $f(x) = 0$ has*

(a) only one real root which is positive, if $a > 1, b < 0$

(b) only one real root which is negative, if $a > 1, b > 0$

(c) only one real root which is negative, if $a < -1, b < 0$

(d) None of the above

Sol. Given, $f(x) = \sin x + ax + b$

$\Rightarrow f'(x) = -\cos x + a$, if $a > 1$, then $f(x)$ is entirely increasing. So, $f(x) = 0$ has only one real root, which is positive, if $f(x) < 0$ and negative, if $f(0) > 0$. Similarly, when $a < -1$ then, $f(x)$ entirely decreasing. Therefore, $f(x)$ has only one real root which is positive, if $f(0) < 0$ and negative, if $f(0) > 0$.

Hence, (a), (b) and (c) are the correct answers.

Ex. 24 *If $a > 0, b > 0, c > 0$ and $a + b + c = abc$, then atleast one of the numbers a, b, c exceeds*

(a) $\frac{3}{2}$ (b) $\frac{17}{10}$

(c) 2 (d) $\frac{13}{10}$

Sol. We may suppose $a \geq b \geq c$

$$abc = a + b + c \geq 3c$$

So, $\quad ab \geq 3, a \geq b \text{ and } a \geq \sqrt{3} > \frac{17}{10}$

Aliter $(a+b+c)^3 \geq 27\, abc \geq 27\,(a+b+c)$

$\therefore \quad a + b + c \geq 3\sqrt{3}$

$\Rightarrow$ Atleast one of them $\geq \sqrt{3}$

Hence, (a), (b) and (d) are the correct answers.

Ex. 25 *Let $f(x, y) = x^2 + 2xy + 3y^2 - 6x - 2y$, where $x, y \in R$, then*

(a) $f(x, y) \geq -11$ (b) $f(x, y) \geq -10$

(c) $f(x, y) > -11$ (d) $f(x, y) > -12$

Sol. Let $z = x^2 + 2xy + 3y^2 - 6x - 2y$

$\Rightarrow \quad x^2 + 2xy + 3y^2 - 6x - 2y - z = 0$ as $x \in R$

$\therefore \quad D \geq 0$

$\Rightarrow \quad 4(y-3)^2 - 4(3y^2 - 2y - z) \geq 0$

$\Rightarrow \quad y^2 + 9 - 6y - 3y^2 + 2y + z \geq 0$

$\Rightarrow \quad -2y^2 - 4y + 9 + z \geq 0$

$\Rightarrow \quad z \geq 2(y^2 + 2y + 1) - 11 = 2(y+1)^2 - 11$

$\Rightarrow \quad z \geq -11$

Hence, (a), (c) and (d) are the correct answers.

Ex. 26 *Let $g(x) = f(\tan x) + f(\cot x), \forall\, x \in \left(\frac{\pi}{2}, \pi\right)$. If $f''(x) < 0, \forall\, x \in \left(\frac{\pi}{2}, \pi\right)$, then*

(a) $g(x)$ is increasing in $\left(\frac{\pi}{2}, \frac{3\pi}{4}\right)$

(b) $g(x)$ has local minimum at $x = \frac{3\pi}{4}$

(c) $g(x)$ is decreasing in $\left(\frac{3\pi}{4}, \pi\right)$

(d) $g(x)$ has local maximum at $x = \frac{3\pi}{4}$

Sol. Given, $\quad g(x) = f(\tan x) + f(\cot x)$

$$g'(x) = f'(\tan x)\sec^2 x - f'(\cot x)\,\text{cosec}^2\, x$$

For increasing $g'(x) > 0$

$$f'(\tan x) > f'(\cot x)$$

$$[\because f''(x) < 0 \text{ and } \tan x < \cot x, \forall\, x \left(\frac{\pi}{4}, \frac{\pi}{2}\right)]$$

Also, $\sec^2 x > \operatorname{cosec}^2 x,\ x \in \left(\frac{\pi}{2}, \frac{3\pi}{4}\right)$

$$g'(x) > 0 \Rightarrow g(x) \text{ is increasing in } \left(\frac{\pi}{2}, \frac{3\pi}{4}\right)$$

Similarly, $g(x)$ is decreasing in $\left(\frac{3\pi}{4}, \pi\right)$

Also, $g(x)$ has local maximum at $x = \frac{3\pi}{4}$. Hence, (a), (c) and (d) are the correct answers.

Ex. 27 *The function $f(x) = \int_0^x \sqrt{1 - t^4}\, dt$ is such that*

(a) it is defined on the interval $[-1, 1]$
(b) it is an increasing function
(c) it is an odd function
(d) the point (0, 0) is the point of inflection

Sol. $f'(x) = \sqrt{1 - x^4} > 0$ in $(-1, 1) \Rightarrow f$ is increasing

Now, $f(x) + f(-x) = \int_0^x \sqrt{1 - t^4}\, dt + \int_0^{-x} \sqrt{1 - t^4}\, dt$

$$\Rightarrow \int_0^x \sqrt{1 - t^4}\, dt + \left(-\int_0^x \sqrt{1 - y^4}\, dy\right) (t = -y) = 0$$

$\Rightarrow f(x)$ is odd

Again, $f''(x) = \frac{-4x^3}{2\sqrt{1 - x^4}}$ which vanished at $x = 0$

and changes sign $\Rightarrow (0, 0)$ is point of inflection since, f is well defined in $[-1, 1]$.

Hence, (a), (b), (c) and (d) are the correct answers.

Ex. 28 *The function $\frac{\sin(x + a)}{\sin(x + b)}$ has no maxima or minima, if*

(a) $b - a = n\pi, n \in I$ (b) $b - a = (2n + 1)\pi, n \in I$
(c) $b - a = 2n\pi, n \in I$ (d) None of these

Sol. Given, $f(x) = \frac{\sin(x + a)}{\sin(x + b)}$

$$f'(x) = \frac{\sin(x + b) \times \cos(x + a) - \sin(x + a)\cos(x + b)}{\sin^2(x + b)}$$

$$= \frac{\sin(b - a)}{\sin^2(x + b)}$$

If $\sin(b - a) = 0$, then $f'(x) = 0$

$\Rightarrow f(x)$ will be constant.

i.e. $b - a = n\pi$ or $b - a = (2n + 1)\pi$

or $b - a = 2n\pi$,

then $f(x)$ has no minima.

Hence, (a), (b) and (c) are the correct answers.

Ex. 29 *Let $F(x) = 1 + f(x) + (f(x))^2 + (f(x))^3$, where $f(x)$ is an increasing differentiable function and $F(x) = 0$ has a positive root, then*

(a) $F(x)$ is an increasing function (b) $F(0) \le 0$
(c) $f(0) \le -1$ (d) $F'(0) > 0$

Sol. Given, $F(x) = 1 + f(x) + (f(x))^2 + (f(x))^3$

$F'(x) = (1 + 2f(x) + 3(f(x))^2)\, f'(x) > 0$, so $F(x)$ is increasing.

So, $F(0) < 0 \Rightarrow (1 + f(0))(1 + f(0))^2) < 0 \Rightarrow f(0) < -1$

Hence, (a), (b), (c) and (d) are the correct answers.

Ex. 30 *The extremum values of the function*

$f(x) = \frac{1}{\sin x + 4} - \frac{1}{\cos x - 4}$, *where $x \in R$ is*

(a) $\frac{4}{8 - \sqrt{2}}$ (b) $\frac{2\sqrt{2}}{8 - \sqrt{2}}$ (c) $\frac{2\sqrt{2}}{4\sqrt{2} + 1}$ (d) $\frac{4\sqrt{2}}{8 + \sqrt{2}}$

Sol. Given, $f(x) = \frac{1}{\sin x + 4} - \frac{1}{\cos x - 4}$

$$\Rightarrow f'(x) = \frac{-\cos x}{(\sin x + 4)^2} + \frac{\sin x}{(\cos x - 4)^2}$$

$$\Rightarrow f'(x) = 0 \Rightarrow (\sin x + \cos x)\text{ (non-zero quantity)} = 0 \Rightarrow \tan x = -1$$

$$\Rightarrow x = \frac{3\pi}{4} \text{ or } \frac{7\pi}{4}$$

Global minimum $= x = 2n\pi + \left(\frac{3\pi}{4}\right)$

Global maximum $= x = 2n\pi + \left(\frac{7\pi}{4}\right)$

$$M = \frac{4}{8 - \sqrt{2}},\ m = \frac{4}{8 + \sqrt{2}} = \frac{2\sqrt{2}}{4\sqrt{2} + 1}$$

Hence, (a) and (c) are the correct answers.

Ex. 31 *The function $f(x) = x^{1/3}(x - 1)$*

(a) has 2 inflection points
(b) is strictly increasing for $x > \frac{1}{4}$ and strictly decreasing for $x < \frac{1}{4}$
(c) is concave down in $\left(-\frac{1}{2}, 0\right)$
(d) area enclosed by the curve lying in the fourth quadrant is $\frac{9}{28}$

Sol. $y = x^{1/3}(x - 1)$

$$\frac{dy}{dx} = \frac{4}{3} x^{1/3} - \frac{1}{3} \cdot \frac{1}{x^{2/3}} = \frac{1}{3x^{2/3}}[4x - 1]$$

$x^{2/3}$ is always positive and $x = \frac{1}{4}$ has a local minima.

Hence, f is increasing for $x > \frac{1}{4}$

and f is decreasing for $x < \frac{1}{4}$

Now, $f'(x) = \frac{4}{3} x^{1/3} - \frac{1}{3} \cdot x^{-2/3}$

[non-existence at $x = 0$, vertical tangent]

$$f''(x) = \frac{4}{9} \cdot \frac{1}{x^{2/3}} + \frac{1}{3} \cdot \frac{2}{3} \cdot \frac{1}{x^{5/3}} = \frac{2}{9x^{2/3}} \left[2 + \frac{1}{x} \right]$$

$$= \frac{2}{9x^{2/3}} \left[\frac{2x+1}{x} \right]$$

$\therefore \quad f''(x) = 0$ at $x = -\frac{1}{2}$ [inflection point]

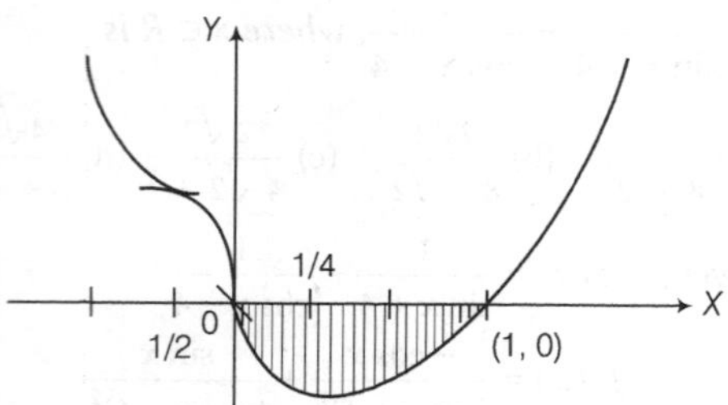

Graph of $f(x)$ is as $A = \int_0^1 (x^{4/3} - x^{1/3})\, dx$

$$= \left[\frac{3}{7} x^{3/7} - \frac{3}{4} x^{4/3} \right]_0^1 = \left| \frac{3}{7} - \frac{3}{4} \right| = 3 \left| \frac{4-7}{28} \right| = \frac{9}{28}$$

Hence, (a), (b), (c) and (d) are the correct answers.

Ex. 32 *Assume that inverse of the function f is denoted by g, then which of the following statement hold good?*

(a) If f is increasing, then g is also increasing
(b) If f is decreasing, then g is increasing
(c) The function f is injective
(d) The function g is onto

Sol. If f and g are inverse, then $(fog)(x) = x$

$$f'[g(x)]g'(x) = 1$$

If f is increasing $\Rightarrow f' > 0 \Rightarrow$ Sign of g' is also positive.
Therefore, option (a) is correct.
If f is decreasing $\Rightarrow f' < 0 \Rightarrow$ Sign of g' is negative.
Therefore, option (b) is false.
Since, f has an inverse.
$\Rightarrow$ f is bijective $\Rightarrow f$ is injective
Therefore, option (c) is correct.
Inverse of a bijective mapping is bijective.
$\Rightarrow g$ is also bijective $\Rightarrow g$ is onto
Therefore, option (d) is correct.
Hence, (a), (c) and (d) are the correct answers

JEE Type Solved Examples : Statements I and II Type Questions

Directions (Ex. Nos. 33 to 39) *For the following questions, choose the correct answers from the codes (a), (b), (c) and (d) defined as follows*

(a) Statement I is true, Statement II is also true, Statement II is the correct explanation of Statement I.
(b) Statement I is true, Statement II is also true, Statement II is not the correct explanation of Statement I
(c) Statement I is true, Statement II is false
(d) Statement I is false, Statement II is true

Ex. 33 Statement I *Among all the rectangles of the given perimeter, the square has the largest area. Also, among all the rectangles of given area, the square has the least perimeter.*

Statement II *For $x > 0, y > 0$, if $x + y$ = constant, then xy will be maximum for $y = x$ and if xy = constant, then $x + y$ will be minimum for $y = x$.*

Sol. Statement II $x + y = k$

then $xy = x(k - x) = f(x), f'(x) = k - 2x = 0$

$\Rightarrow \quad x = \frac{k}{2}, y = \frac{k}{2}$

$\Rightarrow \quad y = x$

$$x + y = x + \frac{k}{x} = f(x), f'(x) = 1 - \frac{k}{x^2} = 0$$

$\Rightarrow \quad x = \sqrt{k}, y = \sqrt{k}$

So, Statement II is true and it explains Statement I.
Hence, (a) is the correct answer.

Ex. 34 Statement I *The function $f(x) = (x^3 + 3x - 4)(x^2 + 4x - 5)$ has local extremum at $x = 1$.*

Statement II *$f(x)$ is continuous and differentiable and $f'(1) = 0$.*

Sol. Statement I is correct because $f(1)^- > f(1) < f(1)^+$.

Statement II is correct as $f(x)$ has a repeated root at $x = 1$.

Statement II is not the correct explanation of Statement I as $f'(c) = 0$ doesn't imply that f has an extrema at $x = c$.

Hence, (b) is the correct answer.

Ex. 35 **Statement I** *If $f(x)$ is increasing function with upward concavity, then concavity of $f^{-1}(x)$ is also upwards.*

Statement II *If $f(x)$ is decreasing function with upwards concavity, then concavity of $f^{-1}(x)$ is also upwards.*

Sol. Let $g(x)$ be the inverse function of $f(x)$.

Then, $f(g(x)) = x$

$\therefore \quad f'(g(x)) \cdot g'(x) = 1$

i.e. $$g'(x) = \frac{1}{f'(g(x))}$$

$$\therefore \quad g'' = -\frac{1}{(f'(g(x)))^2} \cdot f''(g(x)) \cdot g'(x)$$

In Statement I $f''(g(x)) > 0$

and $g'(x) > 0$

$\Rightarrow \quad g''(x) < 0$

$\Rightarrow$ Concavity of $f^{-1}(x)$ is downwards.

$\therefore$ Statement I is false.

In Statement II $f''(g(x)) > 0$

and $g'(x) < 0$

$\Rightarrow \quad g''(x) > 0$

$\Rightarrow$ Concavity of $f^{-1}(x)$ is upwards.

$\therefore$ Statement II is true.

Hence, (d) is the correct answer.

Ex. 36 **Statement I** *The minimum distance of the fixed point $(0, y_0)$, where $0 \le y_0 \le \frac{1}{2}$, from the curve $y = x^2$ is y_0.*

Statement II *Maxima and minima of a function is always a root of the equation $f'(x) = 0$.*

Sol. Let the point on the parabola be (t, t^2).

Let d be the distance between (t, t^2) and $(0, y_0)$,

then $$d^2 = t^2 + (t^2 - y_0)^2 = t^4 + (1 - 2y_0)t^2 + y_0^2$$

$$= z^2 + (1 - 2y_0)z + y_0^2, z \ge 0$$

Its vertex is at $x = y_0 - \frac{1}{2} < 0$

$\therefore$ The minimum value of d^2 is at $z = 0$, i.e. $t^2 = 0$

$\therefore \quad d = y_0$

Statement I is true. Statement II is false because extremum can occur at a point where $f'(x)$ does not exist.

Hence, (c) is the correct answer.

Ex. 37 *Let $f: R \to R$ is differentiable and strictly increasing function throughout its domain.*

Statement I *If $|f(x)|$ is also strictly increasing function, then $f(x) = 0$ has no real roots.*

Statement II *At ∞ or $-\infty$, $f(x)$ may approach to 0, but cannot be equal to zero.*

Sol. Suppose $f(x) = 0$ has a real root say $x = a$, then $f(x) < 0$ for $x < a$. Thus, $|f(x)|$ becomes strictly decreasing on $(-\infty, a)$, which is a contradiction.

Hence, (a) is the correct answer.

Ex. 38 **Statement I** *$f(x) = x + \cos x$ is strictly increasing.*

Statement II *If $f(x)$ is strictly increasing, then $f'(x)$ may tend to zero at some finite number of points.*

Sol. Given, $f(x) = x + \cos x$

$\therefore \quad f'(x) = 1 - \sin x > 0, \ \forall \ x \in R,$

except at $x = 2n\pi + \frac{\pi}{2}$

and $f'(x) = 0$ at $x = 2n\pi + \frac{\pi}{2}$

$\therefore$ $f(x)$ is strictly increasing.

Statement II is true but does not explain Statement I.

Statement II gives $f'(x)$ may tend to zero at finite number of points but in Statement I $f'(x)$ tend to zero at infinite number of points.

Hence, (b) is the correct answer.

Ex. 39 **Statement I** *The largest term in the sequence*

$$a_n = \frac{n^2}{n^3 + 200}, n \in N \text{ is } \frac{(400)^{2/3}}{600}$$

Statement II *$f(x) = \frac{x^2}{x^3 + 200}$, $x > 0$, then at $x = (400)^{1/3}$, $f(x)$ is maximum.*

Sol. Statement II Given, $f(x) = \frac{x^2}{x^3 + 200}$

$$f'(x) = \frac{(x^3 + 200)2x - 3x^2x^2}{(x^3 + 200)^2} = \frac{-x^4 + 400x}{(x^3 + 200)^2}$$

$$x \to 0^+ \ f(x) = 0^+ \ \Rightarrow \ x = 400^{1/3} \ f(x) = \frac{400^{2/3}}{600}$$

$$x \to \infty \ f(x) \to 0$$

So, Statement II is true. But Statement I is false as $x \in N$.

Hence, (d) is the correct answer.

JEE Type Solved Examples : Passage Based Questions

Passage I

(Ex. Nos. 40 to 42)

Let x_1, x_2, x_3, x_4 be the roots (real or complex) of the equation $x^4 + ax^3 + bx^2 + cx + d = 0$. If $x_1 + x_2 = x_3 + x_4$ and $a, b, c, d \in R$, then

Ex. 40 *If $a = 2$, then the value of $b - c$ is*

(a) -1 (b) 1 (c) -2 (d) 2

Ex. 41 *If $b < 0$, then how many different values of a, we may have*

(a) 3 (b) 2 (c) 1 (d) 0

Ex. 42 *If $b + c = 1$ and $a \neq -2$, then for real values of $a, c \in$*

(a) $\left(-\infty, \frac{1}{4}\right)$ (b) $(-\infty, 3)$

(c) $(-\infty, 1)$ (d) $(-\infty, 4)$

Sol. (Ex. Nos. 40 to 42)

Let $x^4 + ax^3 + bx^2 + cx + d$

$$= (x - x_1)(x - x_2)(x - x_3)(x - x_4)$$

Let $(x - x_1)(x - x_2) = x^2 + px + q$

and $(x - x_3)(x - x_4) = x^2 + px + r$

$\therefore$ $q = x_1x_2$ and $r = x_3x_4$

$x^4 + ax^3 + bx^2 + cx + d$

$$= x^4 + 2px^3 + (p^2 + q + r)x^2 + p(q + r)x + qr$$

$\therefore$ $a = 2p, b = p^2 + q + r, c = p(q + r), d = qr$

Clearly, $a^3 - 4ab + 8c = 0$ ...(i)

40. If $a = 2 \Rightarrow b - c = 1$

Hence, (b) is the correct answer.

41. Investigating the nature of the cubic equation of a.

Let $f(a) = a^3 - 4ab + 8c$

$f'(a) = 3a^2 - 4b$

If $b < 0 \Rightarrow f'(a) > 0$

$\therefore$ The equation $a^3 - 4ab + 8c = 0$, has only one real root.

Hence, (c) is the correct answer.

42. Substituting $c = 1 - b$ in Eq. (i), we have

$$(a + 2)[(a - 1)^2 + 3 - 4b] = 0$$

$\Rightarrow$ $4b - 3 > 0$

$\Rightarrow$ $b > \frac{3}{4} \Rightarrow c < \frac{1}{4}$.

Hence, (a) is the correct answer.

Passage II

(Ex. Nos. 43 to 45)

Consider a ΔOAB formed by the point $O(0, 0)$, $A(2, 0)$, $B(1, \sqrt{3})$. $P(x, y)$ is an arbitrary interior point of triangle moving in such a way that $d(P, OA) + d(P, AB) + d(P, OB) = \sqrt{3}$, where $d(P, OA)$, $d(P, AB)$, $d(P, OB)$ represent the distance of P from the sides OA, AB and OB, respectively.

Ex. 43 *Area of region representing all possible position of point P is equal to*

(a) $2\sqrt{3}$ (b) $\sqrt{6}$ (c) $\sqrt{3}$ (d) None of these

Ex. 44 *If the point P moves in such a way that $d(P, OA) \leq \min(d(P, OB), d(P, AB))$, then area of region representing all possible position of point P is equal to*

(a) $\sqrt{3}$ (b) $\sqrt{6}$ (c) $\frac{1}{\sqrt{3}}$ (d) $\frac{1}{\sqrt{6}}$

Ex. 45 *If the point P moves in such a way that $d(P, OA) \geq \min(d(P, OB), d(P, AB))$, then area of region representing all possible position of point P is equal to*

(a) $\sqrt{3}$ (b) $\sqrt{6}$ (c) $1/\sqrt{3}$ (d) $\frac{1}{\sqrt{6}}$

Sol. (Ex. Nos. 43 to 45)

43. ΔOAB is clearly equilateral

$$\Delta\ OAB = \Delta\ OPA + \Delta\ APB + \Delta\ OPB = \frac{\sqrt{3}}{4} \times 4 = \frac{1}{2} \cdot 2$$

$$(d(P, OA) + (P, AB) + (P, OB))$$

$$d(P, OA) + d(P, AB) + d(P, OB) = \frac{4}{\sqrt{3}}$$

Hence, (c) is the correct answer.

44. We must have, $d(P, OA) \leq d(P, OB)$ as well as $d(P, OA) \leq d(P, AB)$, then P lies either on or below the angle bisector of $\angle BOA$ and $\angle BAO$ area

$$= \frac{1}{3} \Delta\ OAB = \frac{1}{\sqrt{3}}.$$

Hence, (c) is the correct asnwer.

45. We must have $d(P, OA) \geq d(P, OB)$ as well as $d(P, OA) \geq d(P, AB)$, then P must be above bisector of $\angle BOA$ and $\angle BAO$.

Area of triangle $= \frac{1}{3} \Delta\ OAB = \frac{1}{\sqrt{3}}$

Hence, (c) is the correct answer.

Passage III

(Ex. Nos. 46 to 48)

Let $f(x) = ax^2 + bx + c, a, b, c \in R$. It is given $|f(x)| \le 1, |x| \le 1$

Ex. 46 *The possible value of $|a + c|$, if $\frac{8}{3}a^2 + 2b^2$ is maximum, is given by*

(a) 1 (b) 0 (c) 2 (d) 3

Ex. 47 *The possible value of $|a + b|$, if $\frac{8}{3}a^2 + 2b^2$ is maximum, is given by*

(a) 1 (b) 0 (c) 2 (d) 3

Ex. 48 *The possible maximum value of $\frac{8}{3}a^2 + 2b^2$ is given by*

(a) 32 (b) $\frac{32}{3}$

(c) $\frac{2}{3}$ (d) $\frac{16}{3}$

Sol. (Ex. Nos. 46 to 48)

We know that for $|u| \le 1; |v| \le 1$, then $|u - v| \le 2$

Now, $|f(1) - f(0)| \le 2 \Rightarrow |a + b| \le 2$

$\Rightarrow (a + b)^2 \le 4$ ...(i)

Also, $|f(-1) - f(0)| \le 2 \Rightarrow |a - b| \le 2$

$\Rightarrow (a - b)^2 \le 4$ ...(ii)

Now, $4a^2 + 3b^2 = 2(a + b)^2 + 2(a - b)^2 - b^2 \le 16$

$\Rightarrow (4a^2 + 3b^2)_{max} = 16$ when $b = 0$

$\Rightarrow |a + b| = |a - b| = |a| = 2$

Also, $|f(1) - f(0)| = |(a + c) - c| = |a| = 2$

$\Rightarrow |a + c| = |c| = 1$

The possible ordered triplet (a, b, c) are $(2, 0, -1)$ or $(-2, 0, 1)$.

Also, $\frac{8}{3}a^2 + 2b^2 = \frac{2}{3}(4a^2 + 3b^2) \le \frac{2}{3} \times 16$

Ans. 46. (a) **47.** (c) **48.** (b)

Passage IV

(Ex. Nos. 49 to 53)

The absolute maximum and minimum values of functions can be found by their monotonic and asymptotic behaviour provided they exist. We may agree that finite limiting values may be regarded as absolute maximum or minimum. For instance the absolute maximum value of $\frac{1}{1+x^2}$ is unity. It is attained at $x = 0$ while absolute minimum value of the same function is zero which is a limiting value of ($x \to \infty$ or $x \to -\infty$)

Ex. 49 *The function $x^4 - 4x + 1$ will have*

(a) absolute maximum value
(b) absolute minimum value
(c) both absolute maximum and minimum values
(d) None of the above

Ex. 50 *The absolute minimum value of the function $\frac{x - 2}{\sqrt{x^2 + 1}}$ is*

(a) -1 (b) $\frac{1}{2}$ (c) $-\sqrt{5}$ (d) None of these

Ex. 51 *The absolute minimum and maximum values of the function $\frac{x^2 - x + 1}{x^2 + x + 1}$ is*

(a) 1 and 3 (b) $\frac{1}{2}$ and 3

(c) $\frac{1}{3}$ and 3 (d) None of these

Ex. 52 *The function $f(x) = \frac{4}{x - 1} - \frac{9}{x + 1}$ will*

(a) have absolute maximum value $-\frac{1}{2}$

(b) have absolute minimum value $-\frac{25}{2}$

(c) not lie between $-\frac{25}{2}$ and $-\frac{1}{2}$

(d) always be negative

Ex. 53 *Which of the following functions will have absolute minimum value?*

(a) cot (sin x) (b) tan (log x)

(c) $x^{2005} - x^{1947} + 1$ (d) $x^{2006} + x^{1947} + 1$

Sol. (Ex. Nos. 49 to 53)

49. Since, $f'(x) = 4x^3 - 4 = 0 \Rightarrow x = 1$

$\Rightarrow$ There is only one extrema which is minima.

$\Rightarrow$ 1 is a point of absolute minima.

Hence, (b) is the correct answer.

50. Since, $f'(x) = \frac{1 + 2x}{(x^2 + 1)^{3/2}}$

$\Rightarrow f(x)$ is increasing in $\left(-\frac{1}{2}, \infty\right)$ and decreasing in $\left(-\infty, -\frac{1}{2}\right)$

$\Rightarrow$ Absolute minimum occurs at $x = -\frac{1}{2}$, we have $= -\sqrt{5}$

Hence, (c) is the correct answer.

51. $yx^2 + yx + y = x^2 - x + 1$

$x^2(1-y) - x(1+y) + (1-y) = 0 \because D \geq 0$

$\Rightarrow (1+y)^2 - 4(1-y)^2 \geq 0 \Rightarrow -(3y-1)(y-3) \geq 0$

$\Rightarrow \quad \frac{1}{3} \leq y \leq 3$

Hence, (c) is the correct answer.

52. Should be the correct choice (we can prove by using monotonically that f cannot lie between $-\frac{25}{2}$ and $-\frac{1}{2}$. This is an Ex. of a function whose maximum (local) value is smaller than minimum value).

Hence, (c) is the correct answer.

53. As even degree polynomial will have absolute minimum essentially.

Hence, (d) is the correct answer.

Passage V

(Ex. Nos. 54 to 56)

We are given the curves $y = \int_{-\infty}^{x} f(t)\,dt$ *through the point* $\left(0, \frac{1}{2}\right)$ *and* $y = f(x)$, *where* $f(x) > 0$ *and* $f(x)$ *is differentiable,* $\forall\, x \in R$ *through* $(0, 1)$. *If tangents drawn to both the curves at the points with equal abscissae intersect on the same point on the X-axis, then*

Ex. 54 *Number of solutions* $f(x) = 2ex$ *is equal to*

(a) 0 (b) 1
(c) 2 (d) None of these

Ex. 55 $\lim_{x \to \infty} (f(x))^{f(-x)}$ *is*

(a) 3 (b) 6
(c) 1 (d) None of these

Ex. 56 *The function* $f(x)$ *is*

(a) increasing for all x (b) non-monotonic
(c) decreasing for all x (d) None of these

Sol. (Ex. Nos. 54 to 56)

We have, the equations of the tangents to the curves $y = \int_{-\infty}^{x} f(t)\,dt$ and $y = f(x)$ at arbitrary points on them are

$$Y - \int_{-\infty}^{x} f(t)\,dt = f(x)(X - x) \quad \text{...(i)}$$

and

$$Y - f(x) = f'(x)(X - X) \quad \text{...(ii)}$$

As Eqs. (i) and (ii), intersect at the same point on X-axis.

On putting $Y = 0$ and equating x-coordinates, we have

$$x - \frac{f(x)}{f'(x)} = x - \frac{\int_{-\infty}^{x} f(t)\,dt}{f(x)} \Rightarrow \frac{f(x)}{\int_{-\infty}^{x} f(t)\,dt} = \frac{f'(x)}{f(x)}$$

$$\Rightarrow \quad \int_{-\infty}^{x} f(t) = cf(x) \quad \text{...(iii)}$$

As, $f(0) = 1 \Rightarrow \int_{-\infty}^{0} f(t)\,dt = \frac{1}{2} = c \times 1 \Rightarrow c = \frac{1}{2}$

$\Rightarrow \int_{-\infty}^{x} f(t)\,dt = \frac{1}{2} f(x)$, differentiating both the sides and on integrating and using boundary condition, we get $f(x) = e^{2x}$, $y = 2ex$ is tangent to $y = e^{2x}$

$\Rightarrow$ Number of solutions = 1

Clearly, $f(x)$ is increasing for all x.

$$\lim_{x \to \infty} (e^{2x})^{e^{-2x}} = 1 \quad [\infty^0 \text{ form}]$$

Ans. 54. (b) **55.** (c) **56.** (a)

Passage VI

(Ex. Nos. 57 to 59)

Let $f(x) = \left(1 + \frac{1}{x}\right)^x \ (x > 0)$ *and*

$$g(x) = \begin{cases} x \ln(1 + (1/x)), & \text{if } 0 < x \leq 1 \\ 0, & \text{if } x = 0 \end{cases}$$

Ex. 57 $\lim_{x \to 0^+} g(x)$

(a) is equal to 0 (b) is equal to 1
(c) is equal to e (d) is non-existent

Ex. 58 *The function* f

(a) has a maxima but not minima
(b) has a minima but not maxima
(c) has both of maxima and minima
(d) is a monotonic

Ex. 59 $\lim_{n \to \infty} \left\{ f\left(\frac{1}{n}\right) \cdot f\left(\frac{2}{n}\right) \cdot f\left(\frac{3}{n}\right) \cdots f\left(\frac{n}{n}\right) \right\}^{1/n}$ *equals*

(a) $\sqrt{2}\,e$ (b) $\sqrt{2e}$
(c) $2\sqrt{e}$ (d) $\sqrt{e}$

Sol. (Ex. Nos. 57 to 59)

57. $$\lim_{x \to 0^+} x \ln\left(1 + \frac{1}{x}\right) = \lim_{x \to 0^+} \frac{\ln\left(\frac{x+1}{x}\right)}{\frac{1}{x}} \quad \left[\frac{\infty}{\infty} \text{ form}\right]$$

Using L'Hospital's rule,

$$l = \lim_{x \to 0} -\left(\frac{1}{x+1} - \frac{1}{x}\right) x^2 = \lim_{x \to 0} \left(\frac{1}{x} - \frac{1}{x+1}\right) \cdot x^2$$

$$= \lim_{x \to 0} \frac{1}{x(x+1)} \cdot x^2 = \lim_{x \to 0} \frac{x}{(x+1)} = 0$$

Hence, (a) is the correct answer.

58. $\lim_{x \to 0} f(x) = 1$ (can be verified)

$\lim_{x \to \infty} f(x) = e$

Also, f is increasing for all

$x > 0$

$\Rightarrow$ (d) (can be verified)

Hence, (d) is the correct answer.

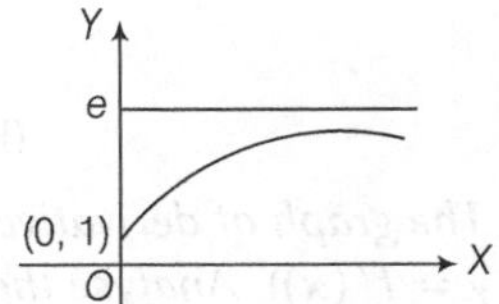

59. Given, $\lim_{n \to \infty} \left(f\left(\frac{1}{n}\right) f\left(\frac{2}{n}\right) \dots f\left(\frac{n}{n}\right) \right)^{1/n}$

$$l = \left(\prod_{k=1}^{n} \left(1 + \frac{n}{k}\right)^{k/n} \right)^{1/n}$$

[given $f(x) = (1 + 1/x)^x$ and $f(k/n) = \left(1 + \frac{n}{k}\right)^{k/n}$]

Taking log,

$$\ln l = \lim_{n \to \infty} \frac{1}{n} \cdot \sum_{k=1}^{n} \ln \left(1 + \frac{n}{k}\right)^{k/n}$$

$$= \lim_{n \to \infty} \frac{1}{n} \sum_{k=1}^{n} \frac{k}{n} \ln \left(1 + \frac{1}{k/n}\right)$$

$$= \int_0^1 \underbrace{x}_{\text{II}} \underbrace{\ln \left(1 + \frac{1}{x}\right)}_{\text{I}} dx = \left[\ln \left(1 + \frac{1}{x}\right) \cdot \frac{x^2}{2} \right]_0^1$$

$$+ \int_0^1 \left(\frac{1}{x} - \frac{1}{x+1} \right) \cdot \frac{x^2}{2} dx$$

$$= \left(\frac{1}{2} \ln 2 - 0 \right) + \frac{1}{2} \int_0^1 \frac{x + 1 - 1}{x + 1} dx$$

$$= \frac{1}{2} \ln 2 + \frac{1}{2} [x - \ln (x+1)]_0^1$$

$$= \frac{1}{2} \ln 2 + \frac{1}{2} [(1 - \ln 2) - 0] = \frac{1}{2}; \ l = \sqrt{e}$$

Hence, (d) is the correct answer.

Passage VII

(Ex. Nos. 60 to 62)

Consider the cubic $f(x) = 8x^3 + 4ax^2 + 2bx + a$, where $a, b \in R$.

Ex. 60 *For $a = 1$, if $y = f(x)$ is strictly increasing, $\forall\ x \in R$, then maximum range of the values of b is*

(a) $\left(-\infty, \frac{1}{3}\right]$ (b) $\left(\frac{1}{3}, \infty\right)$

(c) $\left[\frac{1}{3}, \infty\right)$ (d) $(-\infty, \infty)$

Ex. 61 *For $b = 1$, if $y = f(x)$ is non-monotonic, then the sum of the integral values of $a \in [1, 100]$, is*

(a) 4950 (b) 5049

(c) 5050 (d) 5047

Ex. 62 *If the sum of the base 2 logarithms of the roots of the cubic $f(x) = 0$ is 5, then the value of a is*

(a) -64 (b) -8 (c) -128 (d) -256

Sol. (Ex. Nos. 60 to 62)

60. $a = 1$

$$f(x) = 8x^3 + 4x^2 + 2bx + 1$$

$$f'(x) = 24x^2 + 8x + 2b = 2(12x^2 + 4x + b)$$

For increasing function $f'(x) \geq 0, \forall\ x \in R$

$\therefore \quad D \leq 0$

$\Rightarrow \quad 16 - 48b \leq 0 \Rightarrow b \geq \frac{1}{3}$

Hence, (c) is the correct answer.

61. If $b = 1$

$$f(x) = 8x^3 + 4ax^2 + 2x + a$$

$$f'(x) = 24x^2 + 8ax + 2 \text{ or } 2(12x^2 + 4ax + 1)$$

For non-monotonic $f'(x) = 0$ must have distinct roots hence,

$D > 0$, i.e. $16a^2 - 48 > 0$

$\Rightarrow \quad a^2 > 3$

$\therefore \quad a > \sqrt{3}$ or $a < -\sqrt{3}$

$\therefore \quad a \in 2, 3, 4, \dots,$

Sum $= 5050 - 1 = 5049$

Hence, (b) is the correct answer.

62. If x_1, x_2 and x_3 are the roots, then

$$\log_2 x_1 + \log_2 x_2 + \log_2 x_3 = 5$$

$$\log_2 (x_1 x_2 x_3) = 5$$

$$x_1 x_2 x_3 = 32$$

$$-\frac{a}{8} = 32 \Rightarrow a = -256$$

Hence, (d) is the correct answer.

Passage VIII

(Ex. Nos. 63 to 65)

Let $f(x) = \begin{cases} \max \ \{t^3 - t^2 + t + 1, 0 \leq t \leq x\}, 0 \leq x \leq 1 \\ \min \ \{3 - t, 1 < t \leq x\}, 1 < x \leq 2 \end{cases}$ *and*

$$g(x) = \begin{cases} \max \ \{3/8\,t^4 + 1/2\,t^3 - 3/2\,t^2 + 1, 0 \leq t \leq x\}, 0 \leq x \leq 1 \\ \min \ \{3/8\,t + 1/32 \sin^2 \pi t + 5/8, 1 \leq t \leq x\}, 1 \leq x \leq 2 \end{cases}$$

Define $f(x)$ and $g(x)$ expilicitly and then answer the following questions.

Ex. 63 *The function $f(x), \forall\ x \in [0, 2]$ is*

(a) continuous and differentiable

(b) continuous but not differentiable

(c) discontinuous and not differentiable

(d) None of the above

Ex. 64 *Which of the following is true.*

(a) $\lim_{x\to 1^-}(fog)(x) > \lim_{x\to 1^+}(gof)(x)$

(b) $\lim_{x\to 1^-}(fog)(x) < \lim_{x\to 1^+}(gof)(x)$

(c) $\lim_{x\to 1^-}(fog)(x) = \lim_{x\to 1^+}(gof)(x)$

(d) None of the above

Ex. 65 *Let* $z(x) = \frac{d}{dx} f(x)^{g(x)}$ *and* $y(x) = \frac{d}{dx} g(x)^{f(x)}$, *then* $z(x)$ *and* $y(x)$ *vanish simultaneously at*

(a) $x = -1/3$ (b) $x = 0$
(c) $x = 1$ (d) No real value of x

Sol. (Ex. Nos. 63 to 65)

Consider $f(x)$ in [0, 1]

$$f'(t) = 3t^2 - 2t + 1 > 0, \forall\, t \in (0, 1)$$

$\Rightarrow$ $f(t)$ is increasing on (0, 1)

Maximum occurs at $t = x$

and $$f(x) = \begin{cases} x^3 - x^2 + x + 1, & 0 \le x \le 1 \\ 3 - x, & 1 < x \le 2 \end{cases}$$

Again, consider $g(x)$ in (0, 1)

$$g(t) = 3/8t^4 + 1/2t^3 - 3/2t^2 + 1$$

$$g'(t) = 3/2t^3 + 3/2t^2 - 3t = 3/2t(t^2 + t - 2)$$

$$= 3/2t(t-1)(t+2)$$

$g(t)$ decreases in [0, 1) $\Rightarrow$ maximum occurs when $t = 0$ and $g(0) = 1$, again consider $g(x)$ function in [1, 2].

$$g(t) = 3/8t + 1/32\sin^2 \pi t + 5/8$$

$$g'(t) = 3/8 + \pi/32\sin(2\pi t) > 0, \forall\, t \in R$$

$\therefore$ $g(t)$ is an increasing function in [1, 2].

$\Rightarrow$ Minimum occurs when $t = 1$ and $g(1) = 1$

Hence, $$g(x) = \begin{cases} 1, & 0 \le x \le 1 \\ 1, & 1 < x \le 2 \end{cases} = 1$$

$\Rightarrow$ $$g(x) = 1, \forall\, x \in [0, 2]$$

63. $f(x)$ is continuous but not differentiable at $x = 1$

64. $\lim_{x\to 1^-} fog(x) = f(1)$ and $\lim_{x\to 1^+} gof(x) = 1$, also $f(1) > 1$

65. $z(x) = \frac{d}{dx} f(x)^{g(x)} = \frac{d}{dx}(f(x))^1, \forall\, x \in [0, 1) \cup (1, 2]$

$= f'(x), \forall\, x \in [0, 1) \cup (1, 2]$

and $$y(x) = \frac{d}{dx}(g(x))^{f(x)}$$

$$= \frac{d}{dx}(1)^{f(x)} = 0,$$

$\forall\, x \in [0, 1) \cup (1, 2]$

Hence, the functions $y(x)$ and $z(x)$ can vanish simultaneously at $f'(x) = 0$, which is not possible for any real x.

Passage IX

(Ex. Nos. 66 to 69)

The graph of derivative of a function $f(x)$ *is given* (i.e. $y = f'(x)$). *Analyse the graph in the given domain and answer the following questions, if it is given that* $f(0) = 0$

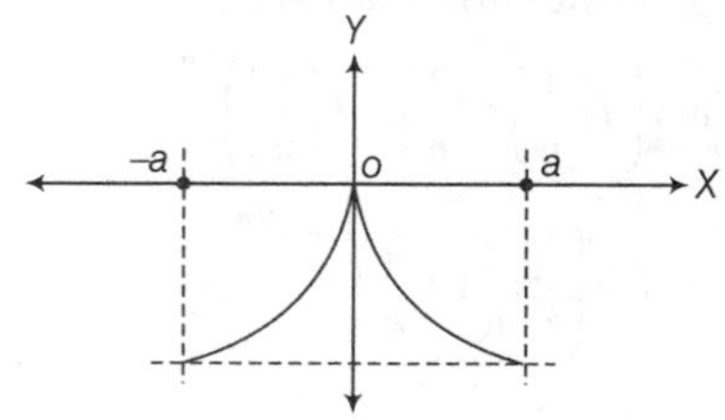

Ex. 66 *The function* $f'(x)$ *is*

(a) even function (b) odd function
(c) neither even nor odd (d) indefinite

Ex. 67 *The function* $f(x)$ *is*

(a) even function (b) odd function
(c) neither even nor odd (d) indefinite

Ex. 68 *The graph of* $y = f(x)$ *has*

(a) no inflexion point (b) one point of inflexion
(c) one extreme point (d) two extreme points

Ex. 69 *The function* $f(x)$ *for* $-a \le x \le a$, *is*

(a) always decreasing
(b) always increasing
(c) increasing for $(-a, 0)$ and decreasing for $(0, a)$
(d) increasing for $(0, a)$ and decreasing for $(-a, 0)$

Sol. (Ex. Nos 66 to 69)

66. The graph of $y = f'(x)$ is symmetrical about Y-axis, so $f(x)$ is an even function.

67. $f(0) = 0$, so $f(x)$ is an odd function (i.e. derivative of an odd function is an even function)

68. Here, $f''(x) > 0$, as $x < 0$ and $f''(x) < 0$, as $x > 0$
$\therefore$ $x = 0$, is the point of inflexion

69. $f'(x) \le 0, \forall\, x$
So, $f(x)$ is always decreasing.

Passage X

(Ex. Nos. 70 to 72)

If a function (continuous and twice differentiable) is always concave upward in an interval, then its graph lies always below the segment joining extremities of the graph in that interval and vice-versa.

Ex. 70 *If* $\sin x + x \ge |k| x^2, \forall\, x \in [0, \pi/2]$, *then the greatest value of* k, *is*

(a) $\frac{-2(2+\pi)}{\pi^2}$ (b) $\frac{2(2+\pi)}{\pi^2}$

(c) can't be determined finitely

(d) zero

Ex. 71 *Let $f(x)$, $f'(x)$ and $f''(x)$ are all positive, $\forall x \in [0, 7]$. If $f^{-1}(x)$ exists, then $3f^{-1}(4) - f^{-1}(2) - 2f^{-1}(5)$, is*

(a) always positive (b) always negative
(c) non-negative (d) non-positive

Ex. 72 *Let $f: R^+ \to R^+$ is such that $f''(x) \geq 0$, $\forall x \in [a, b]$, then value of $\int_a^b f(x)dx$, cannot exceed*

(a) $\frac{(f(a)+f(b))(b-a)}{3}$ (b) $\frac{(f(b)-f(a))(b-a)}{2}$
(c) $\frac{(f(b)+f(a))(b-a)}{2}$ (d) None of these

Sol. (Ex. Nos. 70 to 72)

70. $f(x) = \sin x + x \Rightarrow f'(x) = \cos x + 1$

$\Rightarrow f''(x) = -\sin x$

f is concave downward for $x \in [0, \pi/2]$

Irrespective of k, $g(x) = |k|x^2$ is concave upward

So, if $g(\pi/2) \leq f(\pi/2)$, then $f(x) \geq g(x)$

$\Rightarrow \quad 1 + \pi/2 \geq |k|\frac{\pi^2}{4} \Rightarrow k \in \left[\frac{(-2\pi+4)}{\pi^2}, \frac{(2\pi+4)}{\pi^2}\right]$

71. $f''(x) > 0$

$\Rightarrow$ f is concave upwards $\Rightarrow f^{-1}$ is concave downwards

Consider point dividing the join of this segment in ratio 2:1 is given as $\left(4, \frac{2f^{-1}(5)+f^{-1}(2)}{3}\right)$ when upon a point $(4, f^{-1}(4))$ on graph of $f^{-1}(x)$ is always above it.

$\Rightarrow \quad 3f^{-1}(4) - f^{-1}(2) - 2f^{-1}(5) > 0$

72. $\int_a^b f(x)dx$, is area bounded by curve in first quadrant.

$\frac{(f(a)+f(b))(b-a)}{2}$ is area of trapezium $ABCD$, which is always more than or equal to $\int_a^b f(x)dx$.

Passage XI

(Ex. Nos. 73 to 74)

Consider a twice differentiable function $f(x)$ of degree four symmetrical to line $x = 1$, defined as $f: R \to R$ and $f''(2) = 0$, then

Ex. 73 *The sum of roots of the cubic, $f'(x) = 0$, is*

(a) 0 (b) 1 (c) 2 (d) 9/5

Ex. 74 *If $f(1) = 0$, $f(2) = 1$, then the value of $f(3)$, is*

(a) 6/7 (b) 7/5 (c) 8/5 (d) 9/5

Sol. (Ex. Nos. 73 to 74)

Since, $f(x)$ is symmetric about $x = 1$ and it is twice differentiable.

So, $f'(x)$ must have one root $x = 1$

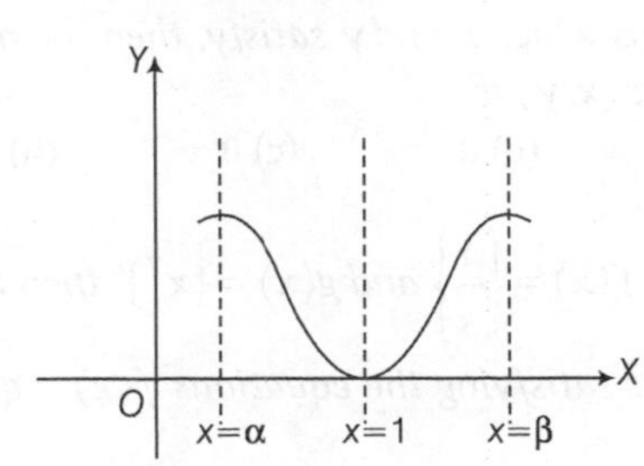

$\therefore \quad f'(x) = a(x-1)(x-\alpha)(x-\beta)$

$= a(x-1)[x^2 - (\alpha+\beta)x + \alpha\beta]$

$= a(x-1)[x^2 - 2x + \alpha\beta]$, as $\frac{\alpha+\beta}{2} = 1$

$\therefore \quad f'(x) = a(x^3 - 2x^2 + \alpha\beta x - x^2 + 2x - \alpha\beta)$

$f''(x) = a(3x^2 - 6x + \alpha\beta + 2)$

$f''(2) = a(12 - 12 + \alpha\beta + 2) = a(\alpha\beta + 2) = 0$

$\therefore \quad \alpha\beta = -2$ and $\alpha + \beta = 2$

$\Rightarrow \quad \alpha - \frac{2}{\alpha} = 2 \Rightarrow \alpha^2 - 2\alpha - 2 = 0$

$\therefore \quad \alpha = 1 - \sqrt{3}, \beta = 1 + \sqrt{3}$

So, sum $= 1 + \alpha + \beta = 3$

$f'(x) = a(x-1)(x^2 - 2x - 2) = a(x-1)[(x-1)^2 - 3]$

On integrating, we get

$f(x) = a\left[\frac{(x-1)^4}{4} - \frac{3(x-1)^2}{2}\right] + C$

$\{\because f(1) = 0$, so $C = 0$ and $f(2) = 1$, so $a = -4/5\}$

$\Rightarrow \quad f(x) = -4/5\left[\frac{(x-1)^4}{4} - \frac{3(x-1)^2}{2}\right]$

$f(3) = -\frac{4}{5}\left[\frac{16}{4} - \frac{3}{2}(2)^2\right] = 8/5$

Passage XII

(Ex. Nos. 75 to 76)

Let $y = f(x)$ and $y = g(x)$ be the two functions, then the number of solutions of these two functions means the number of values of x for which these two function gives same of y as it is shown below.

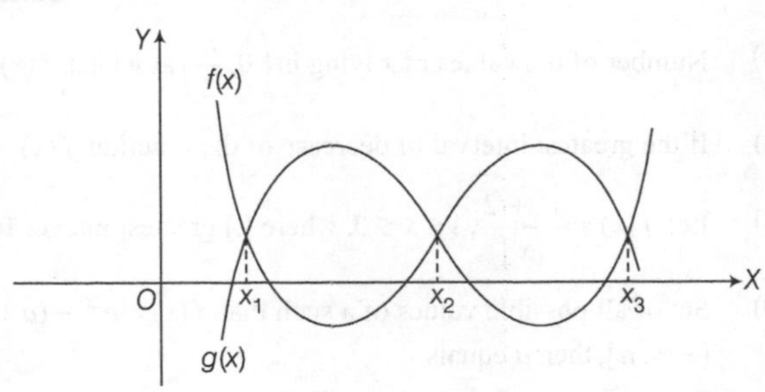

Ex. 75 *Let* $\max\{|x+y|, |x-y|\} = 1$ *and* $y = x - [x]$ *be two equations which x and y satisfy, then the number of ordered pairs* (x, y) *is*

(a) 6 (b) 8 (c) 0 (d) ∞

Ex. 76 *If* $f(x) = \left\{\dfrac{1}{x}\right\}$ *and* $g(x) = \{x^2\}$, *then the number of positive roots satisfying the equations* $f(x) = g(x)$ *such that* $2 < x^2 < 3$

(a) 1 (b) 0 (c) 3 (d) 2

Sol. (Ex. Nos. 75 to 76)

75. Here, $y = x - [x] = \{x\}$ and we know graph of fractional part of x.

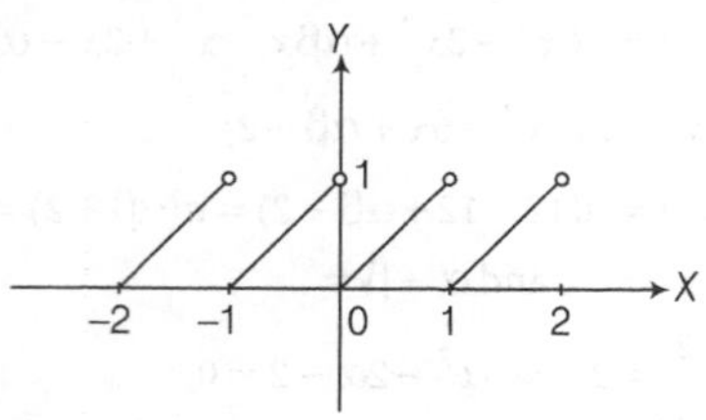

Also, we have $\max\{|x+y|, |x-y|\} = 1$

If $|x+y| \ge |x-y|$, then $|x+y| = 1$

On squaring above inequality, we get

$$x^2 + y^2 + 2xy \ge x^2 + y^2 - 2xy$$

$\Rightarrow$ $4xy \ge 0$, then $|x+y| = 1$, i.e. if function lies in I or III quadrant,

$$|x+y| = 1 \Rightarrow x + y = \pm 1 \qquad \text{...(i)}$$

and of $|x+y| \le |x-y|$, then $|x-y| = 1$

On squaring both sides $4xy \le 0$ means x and y lies in II or IV quadrant. So, if curves lies in II or IV quadrant, then

$$|x-y| = 1 \Rightarrow x - y = \pm 1 \qquad \text{...(ii)}$$

So, graph is,

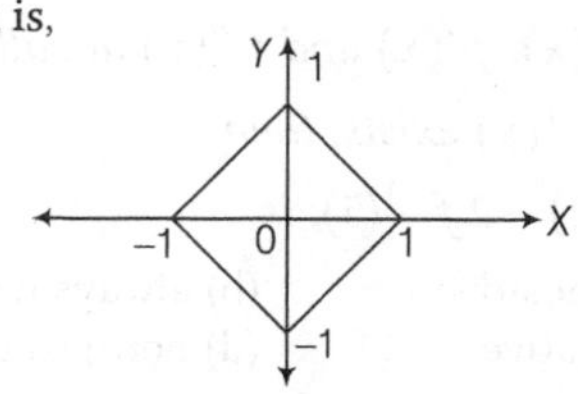

Combining both graph,
when $-1 \le x < 0$, then both graph coincide.

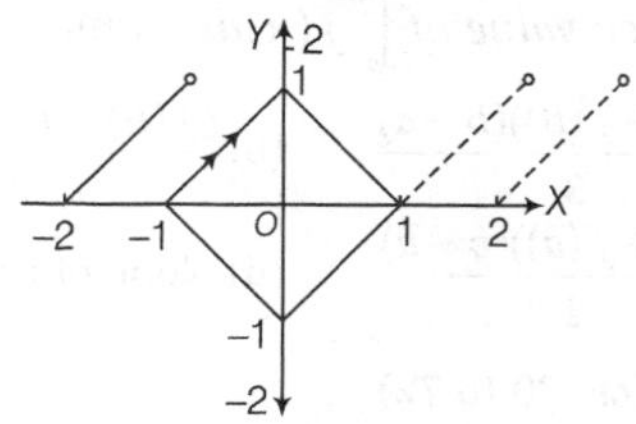

So, number of solutions is infinite.

76. If $2 < x^2 < 3 \Rightarrow \sqrt{2} < x < \sqrt{3}$ [as $x > 0$]

So, if $\sqrt{2} < x < \sqrt{3}$, then $\dfrac{1}{\sqrt{3}} < \dfrac{1}{x} < \dfrac{1}{\sqrt{2}}$

$$\Rightarrow \quad \left\{\frac{1}{x}\right\} = \frac{1}{x} \qquad \text{...(i)}$$

$$\text{and} \quad \{x^2\} = x^2 - 2 \qquad \text{...(ii)}$$

So, from Eqs. (i) and (ii), we get

$$\left\{\frac{1}{x}\right\} = \{x^2\}$$

$$\therefore \quad \frac{1}{x} = x^2 - 2 \Rightarrow x^3 - 2x - 1 = 0$$

$$\Rightarrow (x+1)(x^2 - x - 1) = 0$$

So, $\quad x = -1, \dfrac{1 \pm \sqrt{5}}{2}$, but as $x > 0$

$$\therefore \quad x = \frac{1+\sqrt{5}}{2} \quad \left(\frac{1-\sqrt{5}}{2} \text{ and } -1 \text{ are neglected}\right)$$

JEE Type Solved Examples :
Matching Type Questions

Ex. 77 *Match the statements of Column I with values of Column II.*

	Column I		Column II
(A)	Number of the values of x lying in $\left(0, \dfrac{\pi}{2}\right)$ at which $f(x) = \ln(\sin x)$ is not monotonic, is	(p)	0
(B)	If the greatest interval of decrease of the function $f(x) = x^3 - 3x + 2$ is $[a, b]$, then $a + b$ equals	(q)	2
(C)	Let $f(x) = \dfrac{x^2 + 2}{[x]}$, $1 \le x \le 3$, where [.] greatest integer function, then least value of $f(x)$ is	(r)	-3
(D)	Set of all possible values of a such that $f(x) = e^{2x} - (a+1)e^x + 2x$ is monotonically increasing for all $x \in R$ is $(-\infty, a]$, then a equals	(s)	3

Sol. (A) → (p), (B) → (p), (C) → (s), (D) → (s)

(A) $f(x) = \ln(\sin x) \Rightarrow f'(x) = \dfrac{\cos x}{\sin x} > 0$

∴ Required number of values of x is 0.

(B) $f'(x) = 3x^2 - 3 \le 0$, if $-1 \le x \le 1$

$\therefore \quad a = -1, b = 1 \;\; \therefore \;\; a + b = 0$

(C) $f(x) = \begin{cases} x^2 + 2, & 1 \le x < 2 \\ \dfrac{x^2+2}{2}, & 2 \le x < 3 \\ \dfrac{x^2+2}{3}, & x = 3 \end{cases} \Rightarrow f'(x) = \begin{cases} 2x, & 1 < x < 2 \\ x, & 2 < x < 3 \end{cases}$

∴ Least value of $f(x)$ in [1, 2] is 3.

Least value of $f(x)$ in [2, 3) is 3.

$$f(3) = \frac{11}{3}$$

∴ Least value of $f(x)$ is 3.

(D) $f(x) = e^{2x} - (a+1)e^x + 2x$

$f'(x) = 2e^{2x} - (a+1)e^x + 2$

Now, $2e^{2x} - (a+1)e^x + 2 \ge 0$, for all $x \in R$

$\Rightarrow \quad 2\left(e^x + \dfrac{1}{e^x}\right) - (a+1) \ge 0$, for all $x \in R$

$\Rightarrow \quad 4 - (a+1) \ge 0 \Rightarrow a \le 3 \;\; \therefore \; a = 3$

• ***Ex. 78** Match the statements of Column I with values of Column II.*

Column I		Column II
(A) The dimensions of the rectangle of perimeter 36 cm, which sweeps out the largest volume when revolved about one of its sides, are	(p)	6
(B) Let $A(-1, 2)$ and $B(2, 3)$ be two fixed points, A point P lying on $y = x$ such that perimeter of ΔPAB is minimum, then sum of the abscissa and ordinate of point P, is	(q)	12
(C) If x_1 and x_2 are abscissae of two points on the curve $f(x) = x - x^2$ in the interval [0, 1], then maximum value of expression $(x_1 + x_2) - (x_1^2 + x_2^2)$ is	(r)	4
(D) The number of non-zero integral values of a for which the function $f(x) = x^4 + ax^3 + \dfrac{3x^2}{2} + 1$ is concave upward along the entire real line is	(s)	$\dfrac{1}{2}$

Sol. (A) → (p, q), (B) → (r), (C) → (s), (D) → (r)

(A) Perimeter of the rectangle is 36 cm.

If one side is x, then the other side $= 18 - x$

If the rectangle is revolved around the side x, then volume swept out $V = \pi x(18 - x)^2$

$$\frac{dV}{dx} = \pi[(18 - x)^2 - 2x(18 - x)]$$

$$= \pi(18 - x)(18 - x - 2x)$$

$\therefore \qquad x = 6$ and $y = 12$

(B) $A(-1, 2)$, $B(2, 3)$ and P in a point on $y = x$. Perimeter of ΔPAB is minimum when $PA + PB$ is minimum image of $A(-1, 2)$ in the line $y = x$ is $A'(2, -1)$. Equation of $A'B$ is $x = 2$, hence P is (2, 2).

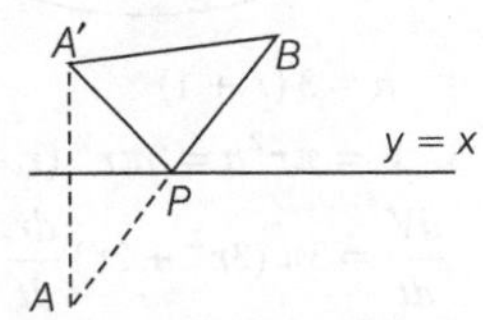

(C) Let (x_1, y_1) and (x_2, y_2) are two points.

$\therefore \quad y_1 + y_2 = (x_1 + x_2) - (x_1^2 + x_2^2)$

Now, $\quad y = x - x^2 = -\left(x - \dfrac{1}{2}\right)^2 + \dfrac{1}{4}$

$\therefore \; (y_1 + y_2)_{max} = 2 \times \dfrac{1}{4} = \dfrac{1}{2}$

(D) $f''(x) = 12x^2 + 6ax + 3 \ge 0, \forall \; x \in R \Rightarrow a \in [-2, 2]$

⇒ Number of non-zero integer values of a is 4.

JEE Type Solved Examples :
Single Integer Answer Type Questions

• ***Ex. 79** The function $S(x) = \int_0^x \sin\left(\dfrac{\pi t^2}{2}\right) dt$ has two critical points in the interval [1, 2.4]. One of the critical points is a local minimum and the other is a local maximum. The local minimum occurs at x equals ....*

Sol. Given, $S(x) = \int_0^x \sin\left(\dfrac{\pi t^2}{2}\right) dt$, $S'(x) = \sin\left(\dfrac{\pi x^2}{2}\right) = 0$

$\Rightarrow \quad \dfrac{\pi x^2}{2} = n\pi \Rightarrow x^2 = 2n \;\; [1 \le x^2 \le 5.76$ as is given]

Hence, $\quad n = 1$ or 2

$x = \sqrt{2}$ or $x = 2$, $S''(x) = \cos\left(\dfrac{\pi x^2}{2}\right) \cdot \pi x$

$S''(\sqrt{2}) < 0$ and $S''(2) > 0$

$\Rightarrow \quad$ minimum at $x = 2$

Ex. 80 *The radius of a right circular cylinder increases at a constant rate. Its altitude is a linear function of the radius and increases three times as fast as radius. When the radius is 1 cm the altitude is 6 cm. When the radius is 6 cm, the volume is increasing at the rate of 1 cu cm/s. When the radius is 36 cm, the volume is increasing at a rate of n cu cm/s. The value of n/11 is equal to ....*

Sol. Here, $\frac{dr}{dt} = c$ and $h = ar + b$

Also, $\frac{dh}{dt} = 3\frac{dr}{dt}$ [given]

$\therefore \quad a\frac{dr}{dt} = 3\frac{dr}{dt} \Rightarrow a = 3$

Hence, $h = 3r + b$

when $r = 1, h = 6 \Rightarrow 6 = 3 + b \Rightarrow b = 3$

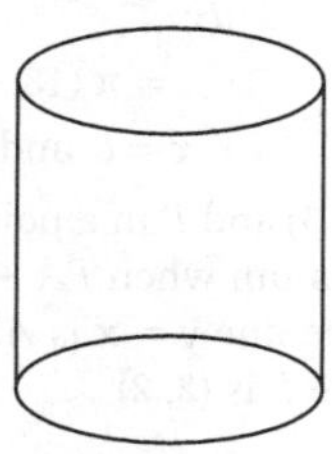

$\therefore \quad h = 3(r+1)$

$$V = \pi r^2 h = 3\pi r^2 (r+1) = 3\pi (r^3 + r^2)$$

$$\frac{dV}{dt} = 3\pi (3r^2 + 2r)\frac{dr}{dt}$$

where $r = 6, \frac{dV}{dt} = 1$ cc/s

$\therefore \quad 1 = 3\pi (108 + 12)\frac{dr}{dt} \Rightarrow 360\pi \frac{dr}{dt} = 1$

Again when $r = 23, \frac{dV}{dt} = n$

$$n = 3\pi ((3.36)^2 + 2.36)\frac{dr}{dt}$$

$$n = 3\pi \cdot 36 (110) \cdot \frac{1}{360\pi} \Rightarrow n = 33$$

$\Rightarrow \quad \frac{n}{11} = \frac{33}{11} = 3$

Ex. 81 *The set of all points where $f(x)$ is increasing is $(a, b) \cup (c, \infty)$, then find $[a + b + c]$ {where [.] denotes GIF}. Given that*

$$f(x) = 2f\left(\frac{x^2}{2}\right) + f(6 - x^2), \forall x \in R$$

and $f''(x) > 0, \forall x \in R$.

Sol. Here, $f(x) = 2f\left(\frac{x^2}{2}\right) + f(6 - x^2)$

$$f'(x) = 2f'\left(\frac{x^2}{2}\right) \cdot x - 2x f'(6 - x^2)$$

$$f'(x) = 2x\left(f'\left(\frac{x^2}{2}\right) - f'(6 - x^2)\right)$$

$$f'\left(\frac{x^2}{2}\right) > f'(6 - x)^2, \text{ if } \frac{x^2}{2} > 6 - x^2$$

[$\because f'(x)$ is increasing]

$$\frac{x^2}{2} > 6 - x^2 \Rightarrow x^2 > 4$$

$\Rightarrow \quad f'\left(\frac{x^2}{2}\right) - f'(6 - x^2) > 0$

when $x < -2$ or $x > 2 \Rightarrow f'(x) > 0$

when $x \; (-2, 0) \cup (2, \infty)$

$\therefore \quad a + b + c = 0$

$\Rightarrow \quad [a + b + c] = 0$

Ex. 82 *The graphs $y = 2x^3 - 4x + 2$ and $y = x^3 + 2x - 1$ intersect at exactly 3 distinct points. The slope of the line passing through two of these points is equal to......... .*

Sol. Let (x_1, y_1) and (x_2, y_2) be two of these points. Given, $y = x^3 + 2x - 1$ and $y = 2x^3 - 4x + 2$

$\therefore \quad y_1 = 2x_1^3 - 4x_1 + 2$ ...(i)

and $2y_1 = 2x_1^3 + 4x_1 - 2$ ...(ii)

Subtracting Eq. (i) from Eq. (ii), we get

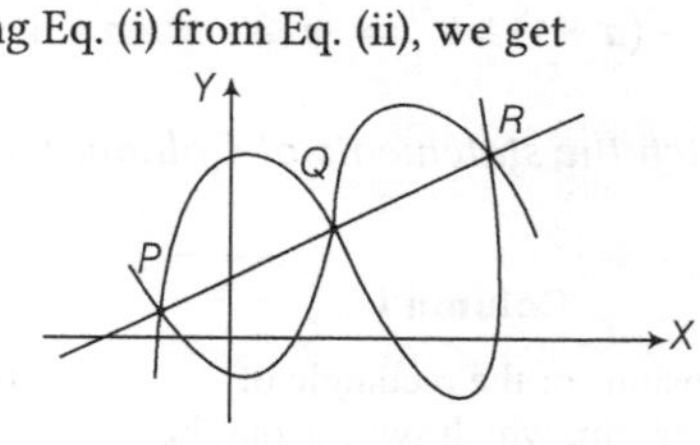

$y_1 = 8x_1 - 4$ ...(iii)

Similarly, $y_2 = 8x_2 - 4$ ...(iv)

$$y_2 - y_1 = 8(x_2 - x_1)$$

$$\frac{y_2 - y_1}{x_2 - x_1} = 8$$

Ex. 83 *The length of the shortest path that begins at the point (2, 5), touches the X-axis and then ends at a point on the circle $x^2 + y^2 + 12x - 20y + 120 = 0$, is......... .*

Sol. Circles with centre $(-6, 10)$ and radius

$$= \sqrt{36 + 100 - 120} = 4$$

Now, let $(a, 0)$ be a point on the X-axis. If y is the distance from A to P and P to M

$$y = \sqrt{(a-2)^2 + 25} + \sqrt{(a+6)^2 + 100} - 4$$

$$\frac{dy}{dx} = \frac{2(a-2)}{2\sqrt{(a-2)^2 + 25}} + \frac{2(a+6)}{2\sqrt{(a+6)^2 + 100}}$$

$\frac{dy}{da}$ can be zero only if $a - 2 > 0$ and $a + 6 < 0$ not possible or $a - 2 < 0$ and $a + 6 > 0$, hence $a \in (-6, 2)$

Solving $\frac{dy}{da} = 0$, gives $a = 10$ (rejected) or $a = -\frac{2}{3}$

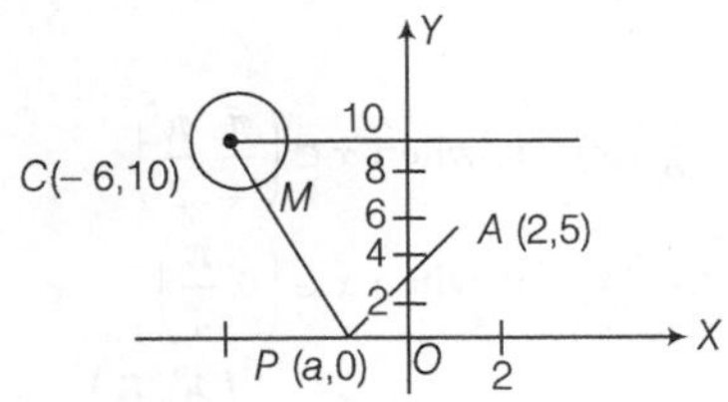

Hence, $y_{\min} = \sqrt{\frac{64}{9} + 25} + \sqrt{\frac{256}{9} + 100} - 4$

$= \frac{17}{3} + \frac{\sqrt{1156}}{3} - 4 = \frac{17}{3} + \frac{34}{3} - 4 = 17 - 4 = 13$

Ex. 84 *If $f:[1,\infty) \to R$: $f(x)$ is monotonic and differentiable function and $f(1) = 1$, then number of solutions of the equation $f(f(x)) = \frac{1}{x^2 - 2x + 2}$ is/are ....*

Sol. Let $g(x) = f(f(x))$, given $f(x)$ is monotonic and differentiable

$\therefore$ $g'(x) = f'(f(x)) \cdot f'(x)$, since $f(x)$ is monotonic

$\Rightarrow$ $f'(x) \cdot f'(f(x)) > 0$ for all $x \ge 1$

Here, $g(1) = f(f(1))$

$\therefore$ $g(1) = f(1) = 1$ [given, $f(1) = 1$] ...(i)

Since, $g(x)$ is increasing for $x \ge 1$.

$\therefore$ $g(x) \ge g(1) \Rightarrow \frac{1}{x^2 - 2x + 2} \ge 1$

$\Rightarrow$ $\frac{1}{(x-1)^2 + 1} \ge 1$, i.e. only possible when $x = 1$.

$\therefore$ Number of solutions of $f(f(x))$ is 1.

Ex. 85 *The set of values of 'a' for which the equation $x^4 + 4x^3 + ax^2 + 4x + 1 = 0$ has all its roots real is given by $(a_1, a_2] \cup \{a_3\}$. Then, $|a_3 + a_2|$ is*

Sol. Here, $x^4 + 4x^3 + ax^2 + 4x + 1 = 0$

dividing by x^2 $\left(x^2 + \frac{1}{x^2}\right) + 4\left(x + \frac{1}{x}\right) + a = 0$

$\Rightarrow$ $\left(x + \frac{1}{x}\right)^2 + 4\left(x + \frac{1}{x}\right) + (a-2) = 0$

Put $x + \frac{1}{x} = y$, ($y \ge 2$ or $y \le -2$)

$\therefore$ $y^2 + 4y + (a-2) = 0$

For y to be real $D \ge 0$.

$16 - 4(a-2) \ge 0 \Rightarrow a \le 6$ ...(i)

Also, if $f(y) = y^2 + 4y + (a-2) \Rightarrow f(-2) \cdot f(2) \ge 0$

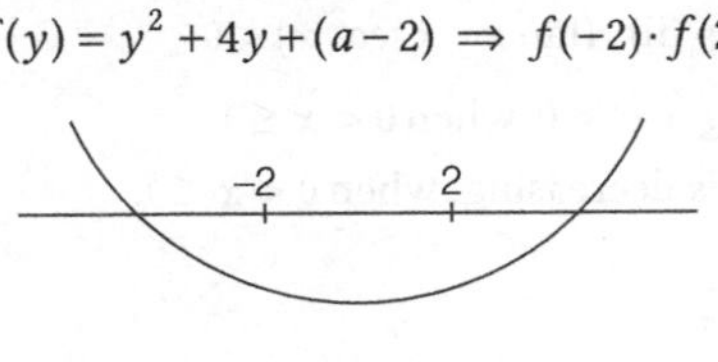

$\Rightarrow$ $(4 + 8 + a - 2)(4 - 8 + a - 2) \ge 0$

$\Rightarrow$ $(a + 10)(a - 6) \ge 0$ ...(ii)

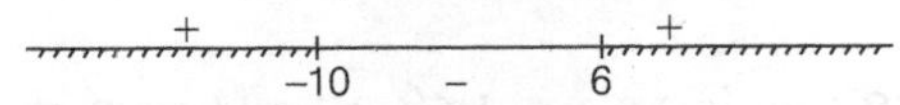

From Eqs. (i) and (ii), we get $a \in (-\infty, -10] \cup \{6\}$

when $a = 6$, $y^2 + 4y + 4 = 0 \Rightarrow y = -2$

$\therefore$ $a \in (-\infty, -10] \cup \{6\}$

$a_2 = -10, a_3 = 6 \Rightarrow |a_2 + a_3| = 4$

Ex. 86 *Let $f(x)$ be a cubic polynomial defined by $f(x) = \frac{x^3}{3} + (a-3)x^2 + x - 13$. Then, the sum of all possible value (s) of 'a' for which $f(x)$ has negative point of local minimum in the interval $[1, 5]$, is*

Sol. We have, $f(x) = \frac{x^3}{3} + (a-3)x^2 + x - 13$

$\therefore$ For $f(x)$ have negative point of local minimum, the equation $f'(x) = 0$ must have two distinct negative roots.

Now, $f'(x) = x^2 + 2(a-3)x + 1$

$\therefore$ Following condition(s) must be satisfied simultaneously

(i) $D > 0$ (ii) $\alpha + \beta < 0$ (iii) $\alpha\beta > 0$

Now, $D > 0 \Rightarrow 4(a-3)^2 - 4 > 0 \Rightarrow (a-2)(a-4) > 0$

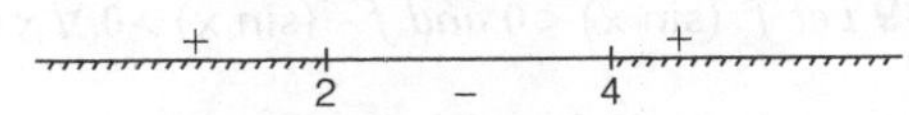

$\therefore$ $a \in (-\infty, 2) \cup (4, \infty)$ ...(i)

Also, $-2(a-3) < 0 \Rightarrow a - 3 > 0 \Rightarrow a > 3$ ...(ii)

and product of root(s) $= 1 > 0, \forall a \in R$

$\therefore$ (i) $\cap$ (ii) $\cap$ (iii) $\Rightarrow a \in (4, \infty)$ ...(iii)

$\therefore$ $a = 5$, as $a \in [1, 5]$

Thus, sum of all possible values is 5

Ex. 87 *Consider a polynomial $P(x)$ of the least degree that has a maximum equal to 6 at $x = 1$ and a minimum equal to 2 at $x = 3$. Then, the value of $P(2) + P'(0) - 7$, is*

Sol. The polynomial is an everywhere differentiable function. Therefore, the points of extremum can only be roots of derivative. Further more, the derivative of a polynomial is a polynomial.

The polynomial of the least degree with roots $x_1 = 1$ and $x_2 = 3$ has the form $a(x-1)(x-3)$.

Hence, $P'(x) = a(x-1)(x-3) = a(x^2 - 4x + 3)$

Since, at $x = 1$, there must be $P(1) = 6$, we have

$P(x) = \int_1^x P'(x)dx + 6 = a\int_1^x (x^2 - 4x + 3)dx + 6$

$= a\left(\frac{x^3}{3} - 2x^2 + 3x - \frac{4}{3}\right) + 6$

Also, $P(3) = 2 \Rightarrow a = 3$

Hence, $P(x) = x^3 - 6x^2 + 9x + 2$

Now, $P(2) = 8 - 24 + 18 + 2 = 4$, also

$P'(x) = 3(x^2 - 4x + 3) \Rightarrow P'(0) = 9$

$\therefore$ $P(2) + P'(0) - 7 = 6$

Subjective Type Questions

Ex. 88 *Let $g'(x) > 0$ and $f'(x) < 0, \forall x \in R$, then show*
(i) $g(f(x+1)) < g(f(x-1))$ (ii) $f(g(x+1)) < f(g(x-1))$

Sol. Here, $g'(x) > 0$ and $f'(x) < 0, \forall x \in R$

i.e. $g(x)$ is increasing [or if $x_1 > x_2 \Rightarrow g(x_1) > g(x_2)$]

and $f(x)$ is decreasing
[or if $x_1 > x_2 \Rightarrow f(x_1) < f(x_2), \forall x \in R$]

$\therefore \quad f(x+1) < f(x-1) \quad$ ...(i)

and $\quad g(x+1) > g(x-1)$
[as $(x+1) > (x-1)$] ...(ii)

Case I As, $g(x)$ is increasing (so greater input gives greater output)

$\Rightarrow \quad g(f(x-1)) > g(f(x+1)) \quad$ [using Eq. (i)]

or $\quad g(f(x+1)) < g(f(x-1))$

Case II $f(x)$ is decreasing (so greater input gives smaller output)

$\Rightarrow \quad f(g(x+1)) < f(g(x-1)) \quad$ [using Eq. (ii)]

Ex. 89 *Let $f'(\sin x) < 0$ and $f''(\sin x) > 0, \forall x \in \left(0, \frac{\pi}{2}\right)$ and $g(x) = f(\sin x) + f(\cos x)$, then find the interval in which $g(x)$ is increasing and decreasing.*

Sol. Here, $f'(\sin x) < 0$ and $f''(\sin x) > 0, \forall x \in \left(0, \frac{\pi}{2}\right)$ ...(i)

and $\quad g(x) = f(\sin x) + f(\cos x)$

$\Rightarrow \quad g'(x) = f'(\sin x) \cdot \cos x + f'(\cos x)(-\sin x)$

$\Rightarrow \quad g''(x) = \{-f'(\sin x) \cdot \sin x + f''(\sin x) \cdot \cos^2 x\}$
$\quad - \{f'(\cos x) \cdot \cos x - f''(\cos x) \cdot \sin^2 x\}$ ...(ii)

As, $f'(\sin x) < 0, f''(\sin x) > 0$,

$\sin x > 0, \cos x > 0, \forall x \in \left(0, \frac{\pi}{2}\right)$ ...(iii)

$\therefore$ From Eqs. (ii) and (iii), we can say

$$g''(x) = \underbrace{\{-f'(\sin x) \cdot \cos x\}}_{+\text{ve}} + \underbrace{\{f''(\sin x) \cdot \cos^2 x\}}_{+\text{ve}} + \underbrace{\{f''(\cos x) \cdot \sin^2 x\}}_{+\text{ve}} + \underbrace{\{-f'(\cos x) \cdot \cos x\}}_{+\text{ve}}$$

$\Rightarrow \quad g''(x) > 0, \forall x \in \left(0, \frac{\pi}{2}\right)$ ...(iv)

$\Rightarrow \quad g'(x)$ is increasing in $\left(0, \frac{\pi}{2}\right)$ ...(v)

Now, putting $g'(x) = 0$,

$g'(x) = f'(\sin x) \cdot \cos x - f'(\cos x) \cdot \sin x = 0$

$\Rightarrow \quad x = \frac{\pi}{4}$

and $\quad g'(x) > 0$, when $x \in \left(\frac{\pi}{4}, \frac{\pi}{2}\right)$

$g'(x) < 0$, when $x \in \left(0, \frac{\pi}{4}\right)$

$\therefore$ $g(x)$ is increasing, when $x \in \left(\frac{\pi}{4}, \frac{\pi}{2}\right)$

$g(x)$ is decreasing, when $x \in \left(0, \frac{\pi}{4}\right)$.

Ex. 90 *If $f(x) = \frac{x}{\sin x}$ and $g(x) = \frac{x}{\tan x}$, where $0 < x \le 1$, then in this interval find that $f(x)$ and $g(x)$ are increasing or decreasing.*

Sol. Here, $\quad f(x) = \frac{x}{\sin x}$

$\Rightarrow \quad f'(x) = \frac{\sin x \cdot 1 - x \cdot \cos x}{\sin^2 x}$ ...(i)

where $\sin^2 x$ is always positive when $0 < x \le 1$.

But to check numerator we again let, $h(x) = \sin x - x \cos x$

$\Rightarrow \quad h'(x) = \cos x - 1 \cdot \cos x + x \sin x = x \sin x$, which is +ve for $0 < x \le 1$

$\therefore \quad h'(x) > 0$

$\Rightarrow h(x)$ is increasing, when $0 < x \le 1$

$\Rightarrow \quad h(0) < h(x)$

$\Rightarrow \quad 0 < \sin x - x \cos x$

$\therefore$ In Eq. (i), $\quad (\sin x - x \cos x) > 0$,

$\Rightarrow \quad f'(x) > 0$, when $0 < x \le 1$

Hence, $f(x)$ is increasing, when $0 < x \le 1$

Again, $\quad g(x) = \frac{x}{\tan x} \quad$ [given]

$\Rightarrow \quad g'(x) = \frac{\tan x \cdot 1 - x \cdot \sec^2 x}{\tan^2 x}$ ...(ii)

where $\tan^2 x > 0$

we let $\quad \phi(x) = \tan x - x \sec^2 x$

$\Rightarrow \quad \phi'(x) = \sec^2 x - \sec^2 x - x\,(2\sec x) \cdot (\sec x \tan x)$

$\phi'(x) = -2x \sec^2 x \tan x$

As $\quad \phi'(x) < 0, \forall 0 < x \le 1$

[$\because x \in (0,1], \sec^2 x > 0, \tan x > 0$]

$\therefore \quad \phi(x)$ is decreasing, when $0 < x \le 1$.

$\Rightarrow \phi(0) > \phi(x)$ or $0 > \tan x - x \sec^2 x$

$\therefore$ In Eq. (ii), $(\tan x - x \sec^2 x) < 0$

$\Rightarrow \quad g'(x) < 0$, when $0 < x \le 1$

$\therefore$ $g(x)$ is decreasing, when $0 < x \le 1$.

Ex. 91 *Let $f:[0, \infty) \to [0, \infty)$ and $g:[0, \infty) \to [0, \infty)$ be non-increasing and non-decreasing function and $h(x) = g(f(x))$ and if f and g are differentiable for all points in their respective domains and $h(0) = 0$. Then, show $h(x)$ is always identically zero.*

Sol. Here, $h(x) = g(f(x))$, since, $g(x) \in [0, \infty)$

$h(x) \ge 0, \forall\ x \in$ domain

Also, $h'(x) = g'(f(x)) \cdot f'(x) \le 0$, as $g'(x) \ge 0$

and $h(x) \le 0, \forall\ x \in$domain as $h(0) = 0$

Hence, $h(x) = 0, \forall\ x \in$ domain

Ex. 92 *A cubic function $f(x)$ tends to zero at $x = -2$ and has relative maximum/ minimum at $x = -1$ and $x = \frac{1}{3}$. If $\int_{-1}^{1} f(x)\, dx = \frac{14}{3}$. Find the cubic function $f(x)$.* **[IIT JEE 1992]**

Sol. $f(x)$ is a cubic polynomial. Therefore, $f'(x)$ is a quadratic polynomial and $f(x)$ has relative maximum and minimum at $x = \frac{1}{3}$ and $x = -1$, respectively.

$\therefore$ -1 and $\frac{1}{3}$ are roots of $f'(x) = 0$

$$\Rightarrow \quad f'(x) = a(x+1)\left(x - \frac{1}{3}\right) = a\left(x^2 + \frac{2}{3}x - \frac{1}{3}\right)$$

Now, integrating w.r.t. x, we get

$$f(x) = a\left[\frac{x^3}{3} + \frac{x^2}{3} - \frac{x}{3}\right] + c, \text{ c is constant of integration.}$$

Again, $f(-2) = 0$ [given]

$$\therefore \quad f(-2) = a\left(\frac{-8}{3} + \frac{4}{3} + \frac{2}{3}\right) + c = 0$$

$$\Rightarrow \quad \frac{-2a}{3} + c = 0 \quad \text{or} \quad c = \frac{2a}{3}$$

$$\Rightarrow \quad f(x) = a\left(\frac{x^3}{3} + \frac{x^2}{3} - \frac{x}{3}\right) + \frac{2a}{3}$$

$$\Rightarrow \quad f(x) = a\left(\frac{x^3}{3} + \frac{x^2}{3} - \frac{x}{3} + \frac{2}{3}\right)$$

Again, $\int_{-1}^{1} f(x)\, dx = \frac{14}{3}$ [given]

$$\Rightarrow \quad \frac{a}{3}\int_{-1}^{1}(x^3 + x^2 - x + 2)\, dx = \frac{14}{3}$$

Because, $y = x^3$ and $y = -x$ are odd functions.

So, $\int_{-1}^{1} x^3 dx = \int_{-1}^{1} x\, dx = 0$

$$\therefore \quad \frac{a}{3}\left[\frac{2x^3}{3} + 4x\right]_0^1 = \frac{14}{3} \Rightarrow a = 3$$

$$\therefore \quad f(x) = x^3 + x^2 - x + 2$$

Ex. 93 *Given that, $S = |\sqrt{x^2 + 4x + 5} - \sqrt{x^2 + 2x + 5}|$ for all real x, then find the maximum value of S^4.*

Sol. Here, $S = \left|\sqrt{(x+2)^2 + 1} - \sqrt{(x+1)^2 + 4}\right|$

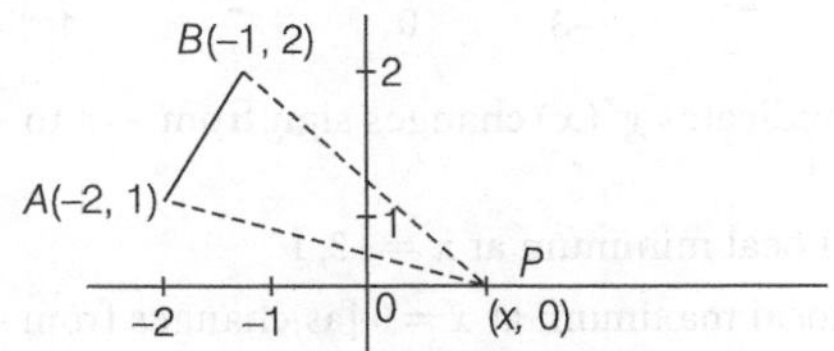

$$\therefore \quad S = \left|\sqrt{(x-(-2))^2 + (1-0)^2} - \sqrt{(x-(-1))^2 + (2-0)^2}\right|$$

$$= |PA - PB| \quad \text{...(i)}$$

Since, $|PA - PB| \le AB$, using triangle law.

$$\therefore \quad |PA - PB|_{\max} = \sqrt{1+1} = \sqrt{2}$$

$$[\because AB = \sqrt{-1 + 2^2 + (2-1)^2} = \sqrt{2}] \quad \text{...(ii)}$$

$$\therefore \quad S_{\max} = \sqrt{2}$$

$$\Rightarrow \quad S^4 = 4$$

Ex. 94 *If $f(x) = \max|2 \sin y - x|$, (where $y \in R$), then find the minimum value of $f(x)$.*

Sol. Here, $f(x) = \max|2\sin y - x| = \begin{cases} 2 - x, & x \le 0 \\ 2 + x, & x > 0 \end{cases}$

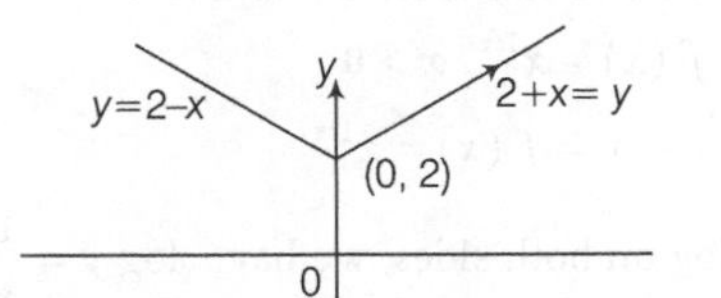

$\therefore$ Minimum value of $f(x) = 2$

Ex. 95 *Find the maximum and minimum value of $f(x) = \dfrac{40}{3x^4 + 8x^3 - 18x^2 + 60}$.*

Sol. Given, $f(x) = \dfrac{40}{3x^4 + 8x^3 - 18x^2 + 60}$ is maximum or minimum according by $3x^4 + 8x^3 - 18x^2 + 60$ is minimum or maximum.

Then, $f_{\max}$, if $3x^4 + 8x^3 - 18x^2 + 60$ is minimum.

$f_{\min}$, if $3x^4 + 8x^3 - 18x^2 + 60$ is maximum.

Let $g(x) = 3x^4 + 8x^3 - 18x^2 + 60$

$\Rightarrow \quad g'(x) = 12x(x^2 + 2x - 3)$

$\Rightarrow \quad g'(x) = 12x(x+3)(x-1)$

Using number line rule,

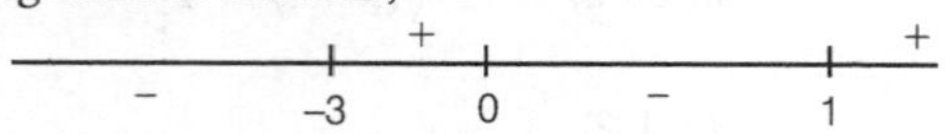

which indicates $g'(x)$ changes sign from −ve to +ve at $x = -3, 1$.

$\therefore$ Local minimum at $x = -3, 1$

and local maximum at $x = 0$ [as changes from +ve to −ve]

$\Rightarrow$ $g(x)$ is maximum at $x = 0$

i.e. $g_{\max}(0) = 60$

and for $g(x)$ to be minimum,

$$g_{\min}(-3) = 243 - 216 - 162 + 60 = -75$$

$$g_{\min}(1) = 3 + 8 - 18 + 60 = 53$$

Substituting these values in Eq. (i), we get

$f_{\max}(x)$ when $g_{\min}$, i.e. $f(x) = \dfrac{40}{-75}$ and $\dfrac{40}{53}$

$$\text{Maximum value} = \frac{40}{53}, \frac{-8}{15}$$

$f_{\min}(x)$ when $g_{\max}$ i.e. $f(x) = \dfrac{40}{60} = \dfrac{2}{3}$

$$\text{Minimum value} = \frac{2}{3}$$

Ex. 96 *Use the function $f(x) = x^{1/x}, x > 0$ and determine the bigger of the two numbers e^{π} and π^{e}.*

Sol. Given, $f(x) = x^{1/x}, x > 0$

Let $y = f(x) = x^{1/x}$

Taking log on both sides, we have $\log y = \dfrac{1}{x} \log x$

Differentiating both the sides, we have

$$\frac{1}{y} \cdot \frac{dy}{dx} = \frac{1}{x^2} + \log x \left(-\frac{1}{x^2}\right)$$

or $$\frac{dy}{dx} = y\left[\frac{1 - \log x}{x^2}\right]$$

$\therefore$ $$f'(x) = \frac{x^{1/x}}{x^2}[1 - \log x]$$

Let $f'(x) = 0$,

$\Rightarrow \quad \log x = 1$ or $x = e$

Again, $$f''(x) = \frac{x^{1/x}}{x^2}\left[0 - \frac{1}{x}\right] + (1 - \log x)\frac{d}{dx}\left(\frac{x^{1/x}}{x^2}\right)$$

$\therefore$ $$f''(e) = \frac{e^{1/e}}{e^2}\left(-\frac{1}{e}\right) + 0$$

$\Rightarrow \quad f''(e) < 0$

$\therefore$ 'f' has a maximum at $x = e$

But $x = e$ is the only extreme value.

$\therefore$ f has the greatest value at $x = e$

$\Rightarrow \quad f(e) > f(\pi)$, for all $x > 0$

$\Rightarrow \quad (e)^{\frac{1}{e}} > (\pi)^{\frac{1}{\pi}}$

$\Rightarrow \quad e^{\pi} > \pi^{e}$

Ex. 97 *Using the relation $2(1 - \cos x) < x^2, x \neq 0$ or otherwise prove that $\sin(\tan x) \geq x$ for all $x \in \left[0, \frac{\pi}{4}\right]$.* **[IIT JEE 2003]**

Sol. Let $f(x) = \sin(\tan x) - x$

Then, $f'(x) = \cos(\tan x) \cdot \sec^2 x - 1$

$\Rightarrow \quad f'(x) = \cos(\tan x)\{1 + \tan^2 x\} - 1$

$\Rightarrow \quad f'(x) = \tan^2 x \cos(\tan x) - \{1 - \cos(\tan x)\}$

[using $2(1 - \cos x) < x^2$]

$$f'(x) > \tan^2 x \cos(\tan x) - \frac{1}{2}\tan^2 x$$

$\Rightarrow \quad f'(x) > \dfrac{1}{2}\tan^2 x\{2\cos(\tan x) - 1\}$

[again, using $2(1 - \cos x) < x^2$]

$$f'(x) > \frac{1}{2}\tan^2 x(1 - \tan^2 x)$$

$\Rightarrow \quad f'(x) \geq 0, \forall \; x \in \left[0, \frac{\pi}{4}\right]$

$\Rightarrow$ $f(x)$ is increasing function for all $x \in \left[0, \frac{\pi}{4}\right]$.

$\therefore \quad f(x) \geq f(0)$, for all $x \in \left[0, \frac{\pi}{4}\right]$

$\Rightarrow \quad \sin(\tan x) - x > \sin(\tan 0) - 0$

$\Rightarrow \quad \sin(\tan x) \geq x$, for all $x \in \left[0, \frac{\pi}{4}\right]$

Ex. 98 *If $P(1) = 0$ and $\frac{d}{dx}\{P(x)\} > P(x)$ for all $x \geq 1$, then prove that $P(x) > 0$ for all $x > 1$.* **[IIT JEE 2003]**

Sol. Here, $\dfrac{d}{dx}\{P(x)\} > P(x)$, for all $x \geq 1$

or $\quad \dfrac{d}{dx}\{P(x)\}e^{-x} > P(x)e^{-x}$, for all $x \geq 1$

$\Rightarrow \quad \dfrac{d}{dx}\{P(x)\}e^{-x} - P(x)e^{-x} > 0$, for all $x \geq 1$

$\Rightarrow \quad \dfrac{d}{dx}\{e^{-x}P(x)\} > 0$, for all $x \geq 1$

$\Rightarrow \quad P(x)e^{-x}$ is an increasing function for all $x \geq 1$

$\Rightarrow \quad P(x)\,e^{-x} > P(1)\,e^{-1}$, for all $x > 1$

$\Rightarrow \quad P(x)\,e^{-x} > 0$, for all $x > 1$ [as $P(1) = 0$, given]

Thus, $\quad P(x) > 0$, for all $x > 1$ [as $e^{-x} > 0$, for all x]

Ex. 99 *In the graph of the function $y = \dfrac{3}{\sqrt{2}} x \log x$, where $x \in [e^{-1.5}, \infty[$; find the point $P(x, y)$ such that the segment of the tangent to the graph of the function at the point, intercepted between the point P and Y-axis, is shortest.*

Sol. Given, $\quad y = \dfrac{3}{\sqrt{2}} x \log x$

Differentiating w.r.t. x, the given function, we have

$$\frac{dy}{dx} = \frac{3}{\sqrt{2}}\left(x \cdot \frac{1}{x} + \log_e x\right) = \frac{3}{\sqrt{2}}(1 + \log_e x)$$

Let the point $P(x, y) \Rightarrow P\left(x, \dfrac{3}{\sqrt{2}} x \log_e x\right)$

Hence, the equation of tangent to the curve at the point $P\left(x, \dfrac{3}{\sqrt{2}} x \log_e x\right)$ is

$$Y - \frac{3}{\sqrt{2}} x \log_e x = \frac{3}{\sqrt{2}}(1 + \log_e x)(X - x)$$

When it cuts Y-axis, $\quad X = 0$

Thus, $\quad y = \dfrac{3}{\sqrt{2}} x \log_e x - \dfrac{3}{\sqrt{2}}(1 + \log_e x)\, x$

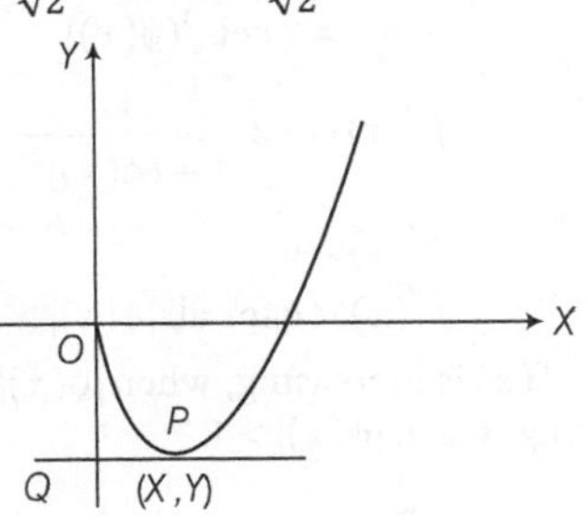

So, $\quad y = -\dfrac{3}{\sqrt{2}} x$

Hence, the given tangent intersects axis at

$$Q\left(0, -\frac{3}{\sqrt{2}} x\right)$$

Now, $\quad PQ^2 = x^2 + \left(y + \dfrac{3}{\sqrt{2}} x\right)^2$

$$= x^2 + \left(\frac{3}{\sqrt{2}} x(1 + \log_e x)\right)^2$$

$$PQ^2 = x^2\left[1 + \frac{9}{2}(1 + \log_e x)^2\right]$$

Differentiating w.r.t. x, we have

$$\frac{d(PQ^2)}{dx} = x^2\left[\frac{9}{2} \cdot 2(1 + \log_e x) \cdot \frac{1}{x}\right] + \left[1 + \frac{9}{2}(1 + \log_e x)^2\right] 2x$$

$$= x\,[9(1 + \log_e x) + 2 + 9(1 + \log_e x)^2]$$

For extremum $\quad \dfrac{d(PQ^2)}{dx} = 0$

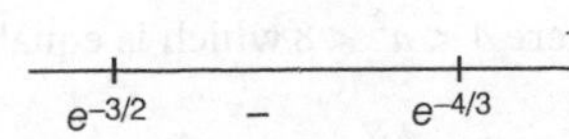

Since, $x \neq 0$, $9(1 + \log x)^2 + 9(1 + \log x) + 2 = 0$

$\Rightarrow \quad (1 + \log x) = -1/3, -2/3$

$\Rightarrow \quad \log x = -4/3, -5/3$

$\Rightarrow \quad x = e^{-4/3}, e^{-5/3}$

$x = e^{-5/3}$ lie outside the interval $(e^{-1.5}, \infty)$.

The sign scheme for $\dfrac{d(PQ^2)}{dx}$ is shown in figure, which shows that PQ^2 is minimum.

Therefore, PQ is minimum when $x = e^{-4/3}$.

Ex. 100 *John has x children from his first wife. Mary has $(x+1)$ children from her first husband. They marry and have children of their own. The whole family has 24 children. Assuming that the two children of same parents do not fight, then find the maximum possible number of fights that can take place.*

Sol. Since, the whole family has 24 children, those of John and Mary are

$$24 - x - (x + 1)$$

i.e. $\quad (23 - 2x)$

Now, F = Total number of fights.

= (number of fights when a John's child fights a Mary's child) + (number of fights when a John child fights a John-Mary's child) + (number of fights when a Mary's child fights a John-Mary's child)

$= x(x+1) + x(23 - 2x) + (x + 1)(23 - 2x)$

$= 23 + 45x - 3x^2$

For maximum, $\quad \dfrac{dF}{dx} = 0$

$\Rightarrow \quad 45 - 6x = 0$ or $x = 7.5$

$\Rightarrow \quad \dfrac{d^2F}{dx^2} = 45 > 0$

$\therefore f(x)$ is minimum when $x = 7.5$

But in this case fractional value is not possible. The nearest integral values are $x = 7$ and $x = 8$.

In either case the total number of fights $= 23 + 45 \times 7 - 3(7)^2$
$= 191$

Ex. 101 *Find the point on the curve $4x^2 + a^2y^2 = 4a^2$; $4 < a^2 < 8$ that is farthest from the point $(0, -2)$.*

Sol. The equation of given curve can be expressed as shown in the figure

$$\frac{x^2}{a^2} + \frac{y^2}{4} = 1, \text{ where } 4 < a^2 < 8 \text{ which is equation of ellipse.}$$

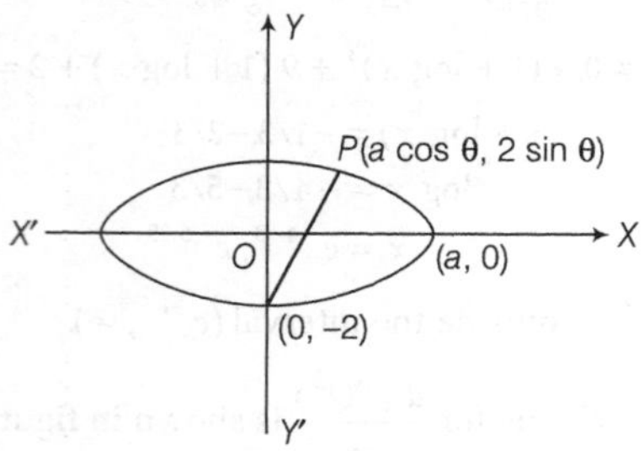

Hence, let us consider a point $P(a \cos \theta, 2 \sin \theta)$ on the ellipse.

Let the distance of $P(a \cos \theta, 2 \sin \theta)$ from $(0, -2)$ be L.

Then, $L^2 = (a \cos \theta - 0)^2 + (2 \sin \theta + 2)^2$

Differentiating w.r.t. θ, we have

$$\frac{d(L^2)}{d\theta} = a^2 \cdot 2 \cos \theta (-\sin \theta) + 4 \cdot 2(\sin \theta + 1) \cdot \cos \theta$$
$$= \cos \theta [-2a^2 \sin \theta + 8 \sin \theta + 8] = 0$$

$\Rightarrow$ Either $\cos \theta = 0$ or $(8 - 2a^2) \sin \theta + 8 = 0$

i.e. $\theta = \pi/2$ or $\sin \theta = \dfrac{4}{a^2 - 4}$

Since, $a^2 < 8$

$\Rightarrow$ $a^2 - 4 < 4$

$\Rightarrow$ $\dfrac{4}{a^2 - 4} > 1 \Rightarrow \sin \theta > 1$, which is not possible.

Further, $\dfrac{d^2(L^2)}{d\theta^2} = \cos \theta [-2a^2 \cos \theta + 8 \cos \theta]$
$+ (-\sin \theta)[-2a^2 \sin \theta + 8 \sin \theta + 8]$

At $\theta = \pi/2$, $\dfrac{d^2(L^2)}{d\theta^2} = 0 - [16 - 2a^2] = 2(a^2 - 8) < 0$ [as $a^2 < 8$]

Hence, L is maximum at $\theta = \pi/2$ and the farthest point is $(0, 2)$.

Ex. 102 *Let $f(x) = \sin^{-1}\left(\dfrac{2\phi(x)}{1 + \phi^2(x)}\right)$, then find the interval in which $f(x)$ is increasing or decreasing.*

Sol. Here, $f(x) = \sin^{-1}\left(\dfrac{2\phi(x)}{1 + \phi^2(x)}\right)$

Case I $|\phi(x)| < 1$

Let $\phi(x) = \tan \theta$

$\therefore$ $f(x) = \sin^{-1}\left(\dfrac{2 \tan \theta}{1 + \tan^2 \theta}\right) = \sin^{-1}(\sin 2\theta) = 2\theta$

$\Rightarrow$ $f(x) = 2 \tan^{-1}\{\phi(x)\}$

$\Rightarrow$ $f'(x) = \dfrac{2}{1 + \{\phi(x)^2\}} \cdot \phi'(x)$

where $\phi'(x) > 0 \Rightarrow f'(x) > 0 \Rightarrow f(x)$ is increasing [since $\phi(x) = \tan \theta$, which is increasing]

Case II When $|\phi(x)| > 1$ or $\left|\dfrac{1}{\phi(x)}\right| < 1$

Now, put $\dfrac{1}{\phi(x)} = \tan \theta$

$\therefore$ $f(x) = \sin^{-1}\left(\dfrac{2 \cdot \dfrac{1}{\phi(x)}}{1 + \left(\dfrac{1}{\phi(x)}\right)^2}\right)$

$\Rightarrow$ $f(x) = \sin^{-1}\left(\dfrac{2 \tan \theta}{1 + \tan^2 \theta}\right)$

$f(x) = \sin^{-1}(\sin 2\theta) = 2\theta = 2 \tan^{-1}\left(\dfrac{1}{\phi(x)}\right)$

$= 2 \cot^{-1}(\phi(x))$

$\Rightarrow$ $f'(x) = -2 \cdot \dfrac{1}{1 + (\phi(x))^2} \cdot \phi'(x)$

where $\phi'(x) > 0$

$\Rightarrow$ $f'(x) < 0$ for all $|\phi(x)| > 1$

Hence, $f(x)$ is increasing, when $|\phi(x)| < 1$ and $f(x)$ is decreasing, when $|\phi(x)| > 1$.

Ex. 103 *Find the minimum value of*
$$f(x) = ||x + 2| - 2|x - 2|| + |x|.$$

Sol. Here, $f(x) = ||x + 2| - 2|x - 2|| + |x|$, which gives rise to four cases as

Case I $x < -2$

$f(x) = |-(x + 2) + 2(x - 2)| - x$
$= |-x - 2 + 2x - 4| - x$
$= |x - 6| - x = -(x - 6) - x$

$\Rightarrow$ $f(x) = -2x + 6$ ...(i)

Case II $-2 \le x < 0$

$f(x) = |(x + 2) + 2(x - 2)| - x$
$= |3x - 2| - x = -(3x - 2) - x$

$\Rightarrow$ $f(x) = -4x + 2$ ...(ii)

***Case* III** $0 \le x < 2$

$$f(x) = |x + 2 + 2(x - 2)| + x$$
$$= |3x - 2| + x$$
$$= \begin{cases} -(3x - 2) + x, & \text{for } 0 \le x < \dfrac{2}{3} \\ (3x-2)+x, & \text{for } \dfrac{2}{3} \le x < 2 \end{cases}$$

$$\Rightarrow \quad f(x) = \begin{cases} -2x + 2, & 0 \le x < 2/3 \\ 4x - 2, & 2/3 \le x < 2 \end{cases} \quad \text{...(iii)}$$

***Case* IV** $x \ge 2$

$$f(x) = |x + 2 - 2(x - 2)| + x = |-x + 6| + x = |x - 6| + x$$
$$= \begin{cases} -(x-6)+x, & 2 \le x < 6 \\ (x - 6) + x, & x \ge 6 \end{cases}$$

$$f(x) = \begin{cases} 6, & 2 \le x < 6 \\ 2x - 6, & x \ge 6 \end{cases} \quad \text{...(iv)}$$

From Eqs. (i), (ii), (iii) and (iv), we have the following figure.

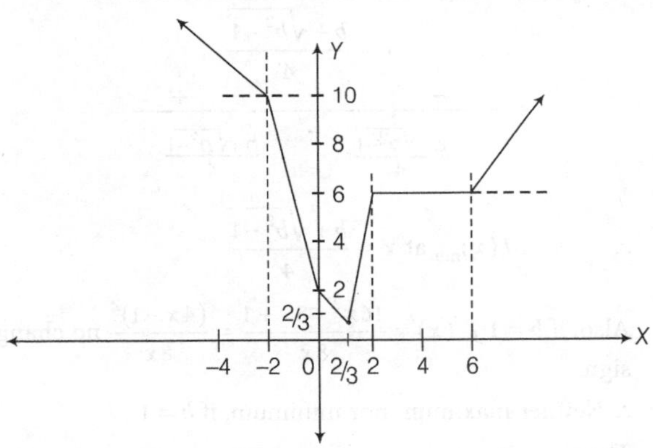

From the graph, minimum value of $f(x) = \dfrac{2}{3}$.

Ex. 104 *Which normal to the curve $y = x^2$ forms the shortest chord?* ***[IIT JEE 1992]***

Sol. Let (t, t^2) be any point on the parabola $y = x^2$

Now, $\dfrac{dy}{dx} = 2x \Rightarrow \left(\dfrac{dy}{dx}\right)_{(t,\, t^2)} = 2t$, which is the slope of tangent.

So, the slope of the normal to $y = x^2$ at (t, t^2) is $\left(-\dfrac{1}{2t}\right)$.

$\therefore$ The equation of the normal to $y = x^2$ at (t, t^2) is

$$y - t^2 = \left(-\frac{1}{2t}\right)(x - t) \quad \text{...(i)}$$

Suppose Eq. (i) meets the curve again at $B\,(t_1, t_1^2)$, then

$$t_1^2 - t^2 = -\frac{1}{2t}(t_1 - t) \Rightarrow t_1 + t = -\frac{1}{2t}$$

$$\Rightarrow \quad t_1 = -t - \frac{1}{2t} \quad \text{...(ii)}$$

Let L be the length of the chord AB (as normal)

$$L = AB^2 = (t - t_1)^2 + (t^2 - t_1^2)^2$$
$$= (t - t_1)^2\,[1 + (t + t_1)^2]$$
$$= \left(t + t + \frac{1}{2t}\right)^2 \left[1 + \left(t - t - \frac{1}{2t}\right)^2\right] \quad \text{[using Eq. (ii)]}$$
$$= \left(2t + \frac{1}{2t}\right)^2 \left(1 + \frac{1}{4t^2}\right)$$
$$L = 4t^2\left(1 + \frac{1}{4t^2}\right)^3$$

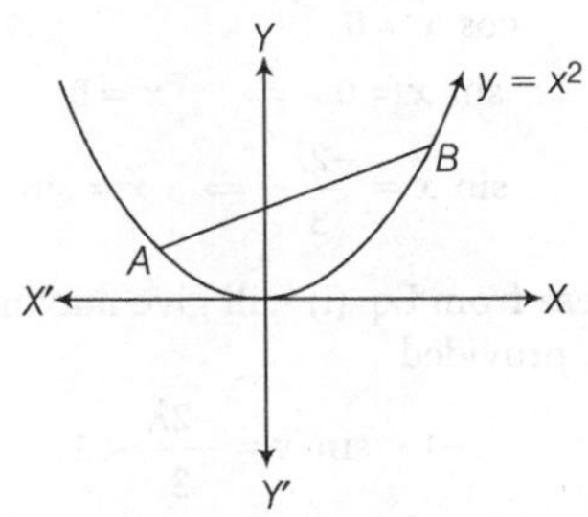

$$\Rightarrow \quad \frac{dL}{dt} = 8t\left(1 + \frac{1}{4t^2}\right)^3 + 12t^2\left(1 + \frac{1}{4t^2}\right)^2 \cdot \left(-\frac{2}{4t^3}\right)$$

$$\Rightarrow \quad \frac{dL}{dt} = 2\left(1 + \frac{1}{4t^2}\right)^2 \left[4t\left(1 + \frac{1}{4t^2}\right) - \frac{3}{t}\right]$$

$$\Rightarrow \quad \frac{dL}{dt} = 2\left(1 + \frac{1}{4t^2}\right)^2 \left(4t - \frac{2}{t}\right) = 4\left(1 + \frac{1}{4t^2}\right)^2 \left(2t - \frac{1}{t}\right)$$

For extremum, let $\dfrac{dL}{dt} = 0$

$$\Rightarrow \quad t = \pm\frac{1}{\sqrt{2}}$$

Again,

$$\frac{d^2L}{dt^2} = 8\left(1 + \frac{1}{4t^2}\right)\left(-\frac{1}{2t^2}\right)\left(2t - \frac{1}{t}\right) + 4\left(1 + \frac{1}{4t^2}\right)^2\left(2 + \frac{1}{t^2}\right)$$

$$\Rightarrow \quad \left(\frac{d^2L}{dt^2}\right)_{t = \pm\frac{1}{\sqrt{2}}} > 0$$

$\therefore$ Minimum when $t = \pm\dfrac{1}{\sqrt{2}}$

Thus, points are $A\left(\pm\frac{1}{\sqrt{2}}, \frac{1}{2}\right)$ and $B(\mp\sqrt{2}, 2)$.

$\Rightarrow$ Equation of normal AB is $\sqrt{2}x + 2y - 2 = 0$ and $\sqrt{2}x - 2y + 2 = 0$.

Ex. 105 *Let $f(x) = \sin^3 x + \lambda \sin^2 x$ where $-\pi/2 < x < \pi/2$. Find the intervals in which λ should lie in order that $f(x)$ has exactly one minimum.*

Sol. Given, $f(x) = \sin^3 x + \lambda \sin^2 x$

$$\therefore \quad f'(x) = 3\sin^2 x(\cos x) + \lambda \cdot 2\sin x(\cos x)$$

$$= \sin x \cos x(3\sin x + 2\lambda)$$

For extremum, let $f'(x) = 0$

$$\therefore \quad \sin x = 0, \quad \cos x = 0,$$

$$\sin x = -2\lambda/3$$

Since, $\quad -\pi/2 < x < \pi/2$

$$\therefore \quad \cos x \neq 0$$

$$\Rightarrow \quad \sin x = 0 \Rightarrow x = 0$$

and $\quad \sin x = \frac{-2\lambda}{3} \Rightarrow x = \sin^{-1}\left(\frac{-2\lambda}{3}\right) \quad$ ...(i)

One of these from Eq. (i) will give maximum and one minimum, provided

$$-1 < \sin x = \frac{-2\lambda}{3} < 1$$

i.e. $\quad -1 < \frac{-2\lambda}{3} < 1$

$$\Rightarrow \quad -3 < -2\lambda < 3 \Rightarrow -3 < 2\lambda < 3$$

i.e. $\quad -3/2 < \lambda < 3/2$

But, if $\lambda = 0$, then $\sin x = 0$ has only one solution.

$$\therefore \quad \lambda \in (-3/2, 3/2) - \{0\}$$

$$\Rightarrow \quad \lambda \in (-3/2, 0) \cup (0, 3/2)$$

For this value of λ there are two distinct solutions.

Since, $f(x)$ is continuous, these solutions give one maximum and one minimum because for a continuous function, between two maxima there must lie one minima and *vice-versa*.

Ex. 106 *Determine the points of maxima and minima of the function, $f(x) = \frac{1}{8}\log x - bx + x^2$, $x > 0$ when $b \geq 0$ is a constant.* *[IIT JEE 1996]*

Sol. Here, $f(x) = \frac{1}{8}\log x - bx + x^2$ is defined and continuous for all $x > 0$.

Then, $\quad f'(x) = \frac{1}{8x} - b + 2x$

or $\quad f'(x) = \frac{16x^2 - 8bx + 1}{8x}$

For extrema let $f'(x) = 0$

$$\Rightarrow \quad 16x^2 - 8bx + 1 = 0$$

So, $\quad x = \frac{8b \pm \sqrt{64(b^2 - 1)}}{2 \times 16}$

or $\quad x = \frac{b \pm \sqrt{b^2 - 1}}{4}$

Obviously the roots are real, if $b^2 - 1 \geq 0$

$$\Rightarrow \quad b > 1 \quad [\text{as } b > 0]$$

Hence, when $b > 1$, then using number line rule for $f'(x)$ as shown in given figure.

We know $f'(x)$ changes sign from +ve to −ve at

$$x = \frac{b - \sqrt{b^2 - 1}}{4}$$

$$\therefore \quad f(x)_{max} \text{ at } x = \frac{b - \sqrt{b^2 - 1}}{4}$$

and $f'(x)$ changes sign from −ve to +ve at

$$x = \frac{b + \sqrt{b^2 - 1}}{4}$$

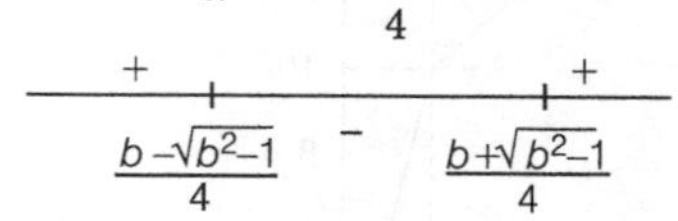

$$\therefore \quad f(x)_{min} \text{ at } x = \frac{b + \sqrt{b^2 - 1}}{4}$$

Also, if $b = 1$ $f'(x) = \frac{16x^2 - 8x + 1}{8x} = \frac{(4x - 1)^2}{8x}$ no change in sign.

$\therefore$ Neither maximum nor minimum, if $b = 1$

Thus,

$$f(x) = \begin{cases} f(x)0_{max} \text{ when } x = \frac{b - \sqrt{b^2 - 1}}{4} \text{ and } b > 1 \\ f(x)_{min} \text{ when } x = \frac{b + \sqrt{b^2 - 1}}{4} \text{ and } b > 1 \\ f(x) \text{ neither maximum nor minimum when } b = 1 \end{cases}$$

Ex. 107 *Find the points on the curve $ax^2 + 2bxy + ay^2 = c$, $0 < a < b < c$, whose distance from the origin is minimum.*

Sol. Let $P(x, y)$ be a point on the curve $ax^2 + 2bxy + ay^2 = c$, whose distance from the origin is r.

$\therefore \quad x = r\cos\theta$ and $y = r\sin\theta$

As, $r\cos\theta$ and $r\sin\theta$ lies on $ax^2 + 2bxy + ay^2 = c$

$$\Rightarrow ar^2\cos^2\theta + 2br^2\sin\theta\cos\theta + ar^2\sin^2\theta = c$$

$$\Rightarrow \quad (a + b\sin 2\theta)r^2 = c$$

$$\Rightarrow \quad r^2 = \frac{c}{a + b\sin 2\theta} \quad \text{...(i)}$$

From Eq. (i) r is minimum when $(a + b \sin 2\theta)$ is maximum, i.e. $\sin 2\theta$ is maximum.

i.e. $$2\theta = \frac{\pi}{2} \text{ or } \frac{5\pi}{2} \Rightarrow \theta = \frac{\pi}{4}, \frac{5\pi}{4}$$

For θ, maximum value of $(a + b \sin 2\theta) = a + b$

$$\therefore \quad r_{\min} = \sqrt{\frac{c}{a+b}}$$

Also, when $\theta = \frac{\pi}{4}$, $P\left(\frac{r}{\sqrt{2}}, \frac{r}{\sqrt{2}}\right) = \left(\sqrt{\frac{c}{2(a+b)}}, \sqrt{\frac{c}{2(a+b)}}\right)$

Again, when $\theta = \frac{5\pi}{4}$,

$$P\left(\frac{-r}{\sqrt{2}}, \frac{-r}{\sqrt{2}}\right) = \left(-\sqrt{\frac{c}{2(a+b)}}, -\sqrt{\frac{c}{2(a+b)}}\right)$$

Thus, the required points are $\pm\left(\sqrt{\frac{c}{2(a+b)}}, \sqrt{\frac{c}{2(a+b)}}\right)$.

Ex. 108 *Find the value of n, for which $f(x) = (x^2 - 4)^n (x^2 - x + 1), n \in N$ assumes a local minima at $x = 2$.*

Sol. Here, $f(x) = (x^2 - 4)^n (x^2 - x + 1)$ assumes local minima at $x = 2$

$\Rightarrow \quad f(2) < f(2-h)$ and $f(2) < f(2+h)$, where $h > 0$

where $\quad f(2) = 0$

$\Rightarrow \quad f(2-h) > 0$ and $f(2+h) > 0, \forall\, h > 0$

$\Rightarrow \quad (-h)^n (4-h)^n \cdot \{h^2 - 3h + 1\} > 0$

and $\quad h^n (4+h)^n (h^2 + 5h + 1) > 0$

i.e. $\quad (-h)^n > 0$

$[\because (4-h) > 0, h^2 - 3h + 1 > 0, 4 + h > 0, h^2 + 5h + 1 > 0, \forall\, h > 0]$

$\Rightarrow n \in$ even number.

Ex. 109 *The interval to which b may belong so that the function,*

$$f(x) = \left(1 - \frac{\sqrt{21 - 4b - b^2}}{b+1}\right) x^3 + 5x + \sqrt{16},$$

increases for all x.

Sol. If $f(x) = \left(1 - \frac{\sqrt{21-4b-b^2}}{b+1}\right) x^3 + 5x + \sqrt{16}$, increases we must have $f'(x) > 0, \forall\, x \in$ real number.

Then, $\quad f'(x) = \left(1 - \frac{\sqrt{21-4b-b^2}}{b+1}\right) 3x^2 + 5 > 0, \forall\, x \in R$

[as we know $ax^2 + bx + c > 0, \forall\, x \in R$, we must have $a > 0$ and $D < 0$]

$$\therefore \quad 1 - \frac{\sqrt{21-4b-b^2}}{b+1} > 0$$

and $\quad (0)^2 - 4 \times 3\left(1 - \frac{\sqrt{21-4b-b^2}}{b+1}\right) 5 < 0$

$$\Rightarrow \quad 1 - \frac{\sqrt{21-4b-b^2}}{b+1} > 0$$

The above inequality holds, when, (i) $b + 1 < 0$ and (ii) $21 - 4b - b^2 > 0$

$\therefore \quad b < -1$ and $b^2 + 4b - 21 < 0$

$\Rightarrow \quad b < -1$ and $(b+7)(b-3) < 0$

$\Rightarrow \quad b < -1$ and $-7 < b < 3$ [using number line rule]

$\therefore \quad b \in (-7, -1)$ ...(i)

Again, when $b + 1 > 0$, $f(x)$ will be increasing for all x, if

$$21 - 4b - b^2 > 0 \text{ and } 1 > \frac{\sqrt{21-4b-b^2}}{b+1}$$

$\Rightarrow \quad b^2 + 4b - 21 < 0$

and $\quad (b+1)^2 > (21 - 4b - b^2)$ [as $b + 1 > 0$]

$\Rightarrow \quad (b+7)(b-3) < 0$ and $b^2 + 3b - 10 > 0$

$\Rightarrow \quad (-7 < b < 3)$ and $(b < -5$ or $b > 2)$

$\Rightarrow \quad 2 < b < 3$ ...(ii)

From Eqs. (i) and (ii), we have concluded that.

$$b \in (-7, -1) \cup (2, 3)$$

Ex. 110 *Find the set of all values of 'a' for which $f(x) = \left(\frac{\sqrt{a+4}}{1-a} - 1\right) x^5 - 3x + \log 5$ monotonically decreases for all x.*

Sol. Given, $f(x) = \left(\frac{\sqrt{a+4}}{1-a} - 1\right) x^5 - 3x + \log 5$ decreases for all x.

Then, $\quad f'(x) = \left(\frac{\sqrt{a+4}}{1-a} - 1\right) 5x^4 - 3 < 0, \forall\, x \in R$

i.e. $\quad 5\left(\frac{\sqrt{a+4}}{1-a} - 1\right) x^4 - 3 < 0, \forall\, x \in R$

$$\Rightarrow \quad \left(\frac{\sqrt{a+4}}{1-a} - 1\right) < 0, \forall\, x \in R$$

[as $ax^2 + bx + c < 0 \Rightarrow a < 0$ and $D < 0$] ...(i)

Now, two cases arise

***Case* I** If $\quad 1 - a < 0 \Rightarrow a > 1$,

then $\quad \sqrt{a+4} > (1-a)$ and $a + 4 > 0$

Which is always true as LHS > 0 and RHS < 0

$\therefore$ Above inequality is true for all $a > 1$ ...(ii)

Case II If $1-a>0 \Rightarrow a<1$

$$\Rightarrow \sqrt{a+4} < 1-a \text{ and } a > -4$$

$$\Rightarrow a^2-3a-3>0 \text{ and } a>-4$$

$$\Rightarrow a < \frac{3-\sqrt{21}}{2}$$

or $$a > \frac{3+\sqrt{21}}{2} \text{ and } a > -4$$

$$\Rightarrow -4 < a < \frac{3-\sqrt{21}}{2} \quad \text{...(iii)}$$

From Eqs. (ii) and (iii), we conclude

$$a \in \left(-4, \frac{3-\sqrt{21}}{2}\right) \cup (1, \infty)$$

Ex. 111 *Let $a+b=4$, where $a<2$ and let $g(x)$ be a differentiable function. If $\frac{dg}{dx}>0, \forall\ x$ prove that $\int_0^b g(x)\,dx + \int_0^a g(x)\,dx$ increases as $(b-a)$ increases.*

[IIT JEE 1997]

Sol. Let $(b-a)=t$ and since $a+b=4$, we have $a=\frac{4-t}{2}$

and $$b=\frac{t+4}{2} \quad \text{...(i)}$$

where $t>0$ (as $a<2$ and $b>2$)

Let $$\int_0^a g(x)\,dx + \int_0^b g(x)\,dx = \phi(t)$$

$$\Rightarrow \phi(t) = \int_0^{\frac{4-t}{2}} g(x)\,dx + \int_0^{\frac{4+t}{2}} g(x)\,dx$$

$$\therefore \phi'(t) = g\left(\frac{4-t}{2}\right)\cdot\left(\frac{-1}{2}\right) + g\left(\frac{4+t}{2}\right)\left(\frac{1}{2}\right)$$

[using Leibnitz rule]

$$\Rightarrow \phi'(t) = \frac{1}{2}\left[g\left(\frac{4+t}{2}\right) - g\left(\frac{4-t}{2}\right)\right] \quad \text{...(ii)}$$

Since, $g(x)$ is increasing and we know

$$x_1 > x_2$$

$$\Rightarrow g(x_1) > g(x_2) \quad \left[\because \frac{dg}{dx} > 0\right]$$

Here, $\frac{4+t}{2} > \frac{4-t}{2}$ and $g(x)$ is increasing.

$$\therefore g\left(\frac{4+t}{2}\right) > g\left(\frac{4-t}{2}\right) \quad \text{...(iii)}$$

Then, $\phi'(t) = \frac{1}{2}\left[g\left(\frac{4+t}{2}\right) - g\left(\frac{4-t}{2}\right)\right] > 0$ [using Eq. (iii)]

$$\Rightarrow \phi'(t) > 0$$

Hence, $\phi(t)$ increases as t increases.

or $\int_0^a g(x)\,dx + \int_0^b g(x)\,dx$ is increasing as $(b-a)$ increases.

Ex. 112 *Let $g(x) = 2f\left(\frac{x}{2}\right) + f(2-x)$ and $f''(x)<0$, $\forall\ x \in (0,2)$. Find the intervals of increase and decrease of $g(x)$.*

Sol. We have, $$g(x) = 2f\left(\frac{x}{2}\right) + f(2-x)$$

$$\Rightarrow g'(x) = 2f'\left(\frac{x}{2}\right)\cdot\left(\frac{1}{2}\right) + f'(2-x)\cdot(-1)$$

$$\Rightarrow g'(x) = f'\left(\frac{x}{2}\right) - f'(2-x) \quad \text{...(i)}$$

We are given that $f''(x)<0, \forall\ x \in (0,2)$

It means that $f'(x)$ would be decreasing on $(0, 2)$.

Now, two cases arise

Case I $\frac{x}{2} > (2-x)$ and $f'(x)$ is decreasing.

$$\Rightarrow f'\left(\frac{x}{2}\right) < f'(2-x), \forall\ x > \frac{4}{3}$$

$$\left[\text{As } \frac{x}{2} > 2-x \Rightarrow x > \frac{4}{3}\right]$$

or $$g'(x) = f'\left(\frac{x}{2}\right) - f'(2-x) < 0, \forall\ \frac{4}{3} < x < 2$$

$\therefore$ $g(x)$ is decreasing in $\left(\frac{4}{3}, 2\right)$. ...(ii)

Case II $\frac{x}{2} < (2-x)$ and $f'(x)$ is decreasing.

$$\Rightarrow f\left(\frac{x}{2}\right) > f(2-x), \forall\ x < \frac{4}{3} \quad \left[\text{As } \frac{x}{2} < 2-x \Rightarrow x < \frac{4}{3}\right]$$

or $$g'(x) = f\left(\frac{x}{2}\right) - f(2-x) > 0, \forall\ 0 < x < \frac{4}{3}$$

$\therefore$ $g(x)$ is increasing in $\left(0, \frac{4}{3}\right)$. ...(iii)

From Eqs. (ii) and (iii), we conclude $g(x)$ is increasing in $\left(0, \frac{4}{3}\right)$ and decreasing in $\left(\frac{4}{3}, 2\right)$.

Ex. 113 *Let $f'(x) > 0$ and $f''(x) > 0$ where $x_1 < x_2$. Then, show $f\left(\frac{x_1 + x_2}{2}\right) < \frac{f(x_1) + f(x_2)}{2}$.*

Sol. As we have discussed in theory, if $f'(x) > 0$ and $f''(x) > 0$, then graphically it can be expressed as shown in the following figure

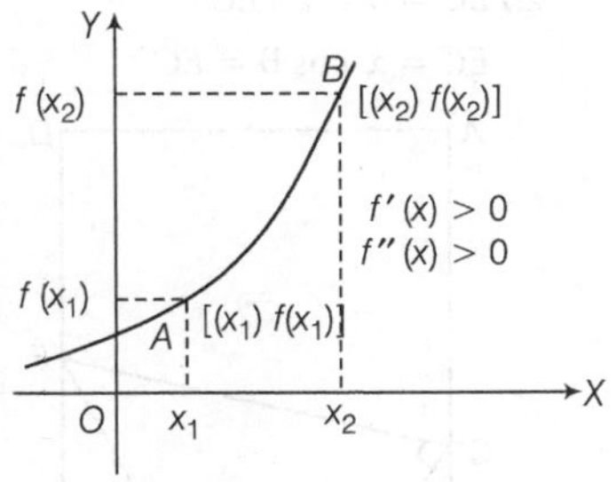

We know, $\quad x_1 < \frac{x_1 + x_2}{2} < x_2$

and $\left(\frac{x_1 + x_2}{2}, \frac{f(x_1) + f(x_2)}{2}\right)$ is mid-point of the chord joining A and B.

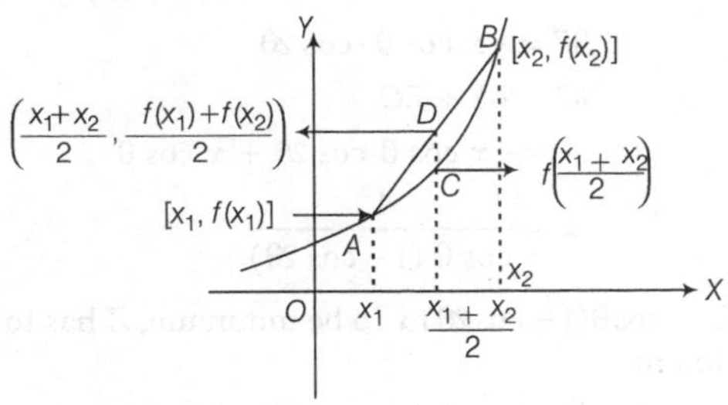

Thus, it can be expressed, from the figure as

$$f\left(\frac{x_1 + x_2}{2}\right) < \frac{f(x_1) + f(x_2)}{2}$$

Ex. 114 *Let $f'(x) > 0$ and $f''(x) < 0$ where $x_1 < x_2$. Then, show $f\left(\frac{x_1 + x_2}{2}\right) > \frac{f(x_1) + f(x_2)}{2}$.*

Sol. As we know, if $f'(x) > 0$ and $f''(x) < 0$. Then, it could be expressed graphically as shown in the following figure

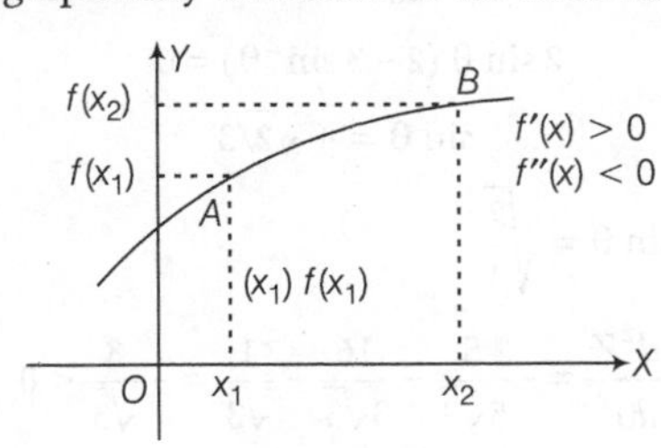

We know, $\quad x_1 < \frac{x_1 + x_2}{2} < x_2$

and $\left(\frac{x_1 + x_2}{2}, \frac{f(x_1) + f(x_2)}{2}\right)$ is the mid-point of chord joining A and B.

Thus, it can be expressed as shown in figure

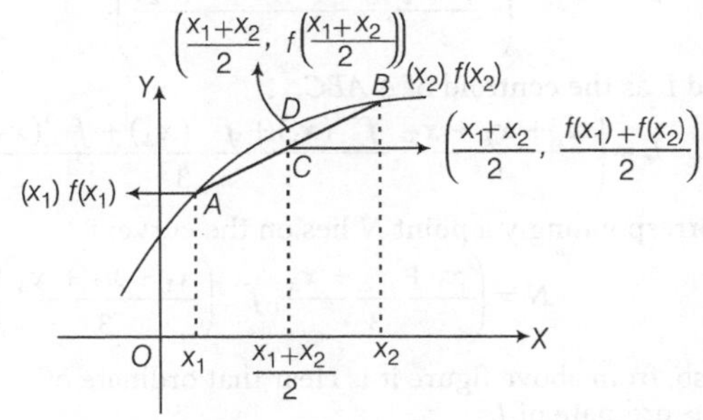

From the adjacent figure, $f\left(\frac{x_1 + x_2}{2}\right) > \frac{f(x_1) + f(x_2)}{2}$.

Ex. 115 *If $f(x)$ is monotonically increasing function for all $x \in R$, such that $f''(x) > 0$ and $f^{-1}(x)$ exists, then prove that $\frac{f^{-1}(x_1) + f^{-1}(x_2) + f^{-1}(x_3)}{3} < f^{-1}\left(\frac{x_1 + x_2 + x_3}{3}\right)$*

Sol. Let $g(x) = f^{-1}(x)$

Since, g is the inverse of f.

$\Rightarrow \quad fog(x) = gof(x) = x \Rightarrow g'(x) = \frac{1}{f'(g(x))}$

$\Rightarrow \quad g'(x) > 0 \quad$ [as $f(x)$ is increasing] ...(i)

$\Rightarrow$ $g(x)$ is increasing for all $x \in R$.

$\Rightarrow$ $f^{-1}(x)$ is increasing for all $x \in R$.

Again, $\quad g'(x) = \frac{1}{f'(g(x))}$

$\Rightarrow \quad g''(x) = -\frac{1}{(f'(g(x)))^2} f''(g(x))\, g'(x)$, for all $x \in R$

$\Rightarrow \quad g''(x) < 0,$

{as $g'(x) > 0$, from relation

(i) $f''(x) > 0$, given

$\Rightarrow$ $g'(x)$ is decreasing for all $x \in R$.

$\Rightarrow$ $f^{-1}(x)$ is increasing and $\frac{d}{dx}\{f^{-1}(x)\}$ is decreasing.

Thus, the graph for $f^{-1}(x)$ could be plotted as

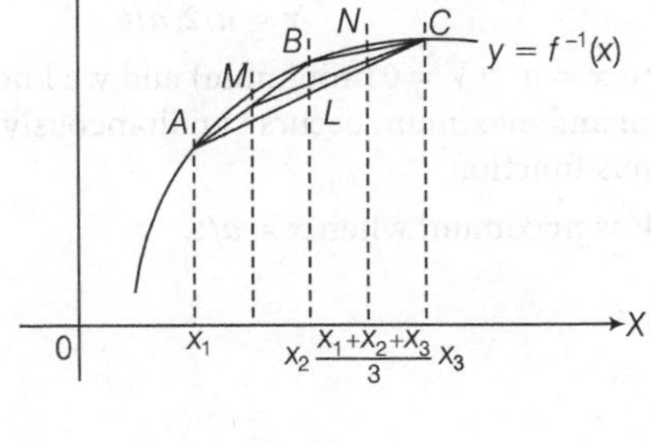

In above figure, we have taken three points A, B, C as; $A(x_1, f^{-1}(x_1)), B(x_2, f^{-1}(x_2)), C(x_3, f^{-1}(x_3))$.

Also, M is the mid-point of AB as

$$\left(\frac{x_1+x_2}{2}, \frac{f^{-1}(x_1)+f^{-1}(x_2)}{2}\right)$$

and L as the centroid of ΔABC,

i.e. $$L = \left(\frac{x_1+x_2+x_3}{3}, \frac{f^{-1}(x_1)+f^{-1}(x_2)+f^{-1}(x_3)}{3}\right)$$

Correspondingly a point N lies on the curve;

$$N = \left(\frac{x_1+x_2+x_3}{3}, f^{-1}\left(\frac{x_1+x_2+x_3}{3}\right)\right)$$

Also, from above figure it is clear that ordinate of $N >$ ordinate of L

$$\Rightarrow f^{-1}\left(\frac{x_1+x_2+x_3}{3}\right) > \frac{f^{-1}(x_1)+f^{-1}(x_2)+f^{-1}(x_3)}{3}$$

Ex. 116 *A box of maximum volume with top open is to be made by cutting out four equal squares from four corners of a square tin sheet of side length a feet and then folding up the flaps. Find the side of the square cut-off.*

Sol. Volume of the box is, $V = (a-2x)^2 \cdot x$ i.e. squares of side x are cut out, then we will get a box with a square base of side $(a-2x)$ and height x.

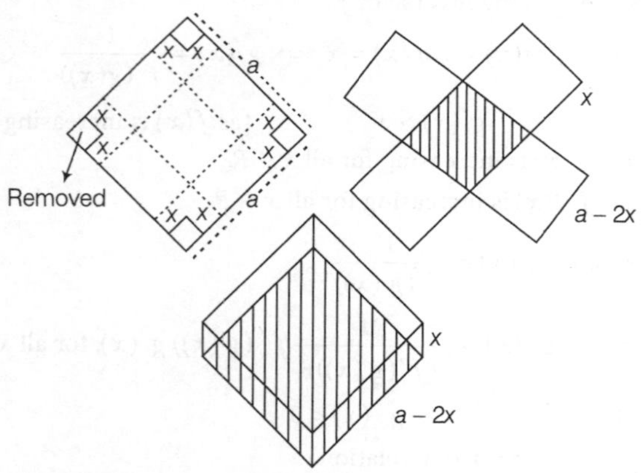

Volume of box $(V) = (2a-x)^2 \cdot x$

$$\therefore \quad \frac{dV}{dx} = (a-2x)^2 + x \cdot 2(a-2x) \cdot (-2) = (a-2x)(a-6x)$$

For V to be extremum, $\frac{dV}{dx} = 0$

$$\Rightarrow \quad x = a/2, a/6$$

But when $x = a/2$; $V = 0$ (minimum) and we know minimum and maximum occurs simultaneously in a continuous function.

Hence, V is maximum when $x = a/6$.

Ex. 117 *One corner of a long rectangular sheet of paper of width 1 unit is folded over so as to reach the opposite edge of the sheet. Find the minimum length of the crease.*

Sol. Let $ABCD$ be the rectangular sheet whose corner C is folded over along EF so as to reach the edge AB at C'.

Let $\quad EF = x$

$$\angle FEC = \theta = \angle FEC'$$

$$\therefore \quad EC = x\cos\theta = EC'$$

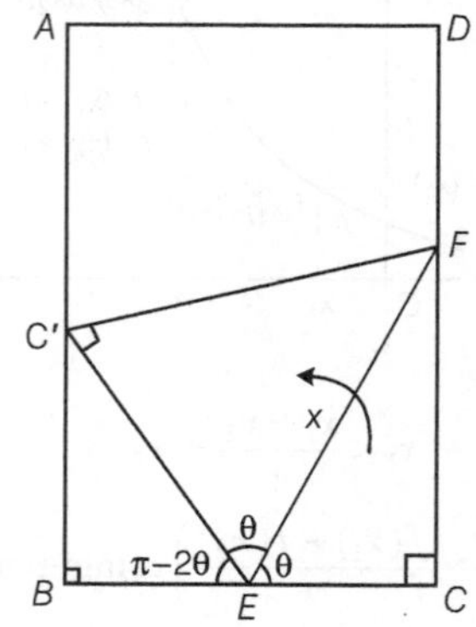

From $\Delta BEC'$, we have $BE = C'E\cos(\pi - 2\theta)$

$$\Rightarrow \quad BE = -x\cos\theta \cdot \cos 2\theta$$

$$\therefore \quad BC = BE + EC$$

$$1 = -x\cos\theta\cos 2\theta + x\cos\theta$$

$$\Rightarrow \quad x = \frac{1}{\cos\theta(1-\cos 2\theta)} \quad \text{...(i)}$$

Let $Z = \cos\theta(1-\cos 2\theta)$ x to be minimum, Z has to be maximum.

$$Z = \cos\theta(1-\cos 2\theta) \quad \text{...(ii)}$$

Differentiating Eq. (ii) w.r.t. θ, we get

$$\frac{dZ}{d\theta} = \cos\theta(0 + 2\sin 2\theta) - \sin\theta(1-\cos 2\theta)$$

and $$\frac{d^2Z}{d\theta^2} = \cos\theta(4\cos 4\theta) - 2\sin 2\theta \cdot \sin\theta - \sin\theta(2\sin 2\theta) - \cos\theta(1-\cos 2\theta)$$

For maximum/minimum,

$$\frac{dZ}{d\theta} = 0$$

$$\Rightarrow \quad 2\sin\theta(2 - 3\sin^2\theta) = 0$$

$$\therefore \quad \sin\theta = +\sqrt{2/3} \quad [\because \sin\theta \neq 0]$$

When $\sin\theta = \sqrt{\frac{2}{3}}$

$$\Rightarrow \quad \frac{d^2Z}{d\theta^2} = -\frac{5}{5\sqrt{3}} - \frac{16}{3\sqrt{3}} - \frac{1}{\sqrt{3}} = -\frac{8}{\sqrt{3}} < 0$$

Hence, Z is maximum.

$\Rightarrow \quad x = \dfrac{1}{Z}$ is minimum [from Eq. (i)]

$\therefore$ x is minimum.

$$\Rightarrow \quad x_{\min} \Rightarrow \frac{1}{Z} = \frac{1}{(1/\sqrt{3})\left(1+\frac{1}{3}\right)} = \frac{3\sqrt{3}}{4} \text{ unit}$$

Ex. 118 *Find the volume of the greatest right circular cone that can be described by the revolution about a side of a right angled triangle of hypotenuse 1 ft.*

Sol. Let ABC be right angled triangle.

Let the cone be revolved about AB.

$$AC = 1 \text{ ft} \quad \text{[given]}$$

Let $AB = a$ = height of cone

$$\therefore \quad BC = \sqrt{1-a^2} = \text{radius of cone}$$

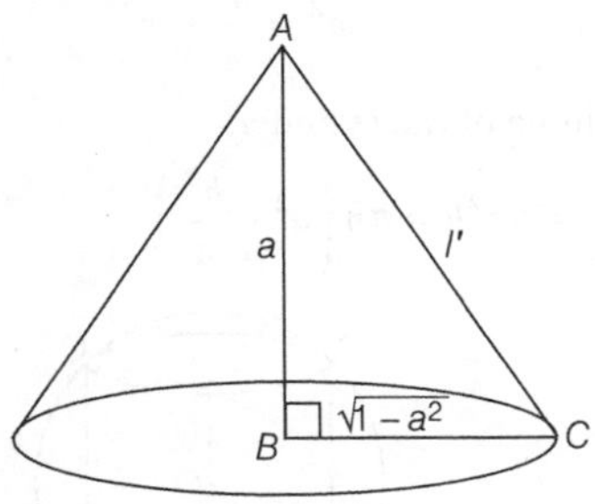

$$\text{Volume of cone} = \frac{1}{3}\pi(1-a^2)a$$

$$V = \frac{1}{3}\pi(a-a^3)$$

$$\Rightarrow \quad \frac{dV}{da} = \frac{1}{3}\pi(1-3a^2)$$

and $$\frac{d^2V}{da^2} = \frac{1}{3}\pi(-6a) < 0$$

$\therefore$ Maximum volume when $\dfrac{dV}{da} = 0$

i.e. when $a = \dfrac{1}{\sqrt{3}}$

Putting $a = \dfrac{1}{\sqrt{3}}$, we get $V_{\max} = \dfrac{2\sqrt{3}}{27}\pi$ cu ft

Ex. 119 *A window of fixed perimeter (including the base of the arc) is in the form of a rectangle surmounted by a semi-circle. The semi-circular portion is fitted with coloured glass while the rectangular portion is fitted with clear glass. The clear glass transmits three times as much light per square metre as the coloured glass does. What is the ratio of the sides of the rectangle so that the window transmits the maximum light?* **[IIT JEE 1991]**

Sol. Let $2b$ be the diameter of the circular portion and a be the lengths of the other side of the rectangle.

Total perimeter $= 2a + 4b + \pi b = K$ [say]...(i)

Now, let the light transmission rate (per square metre) of the coloured glass be L and Q be the total amount of transmitted light.

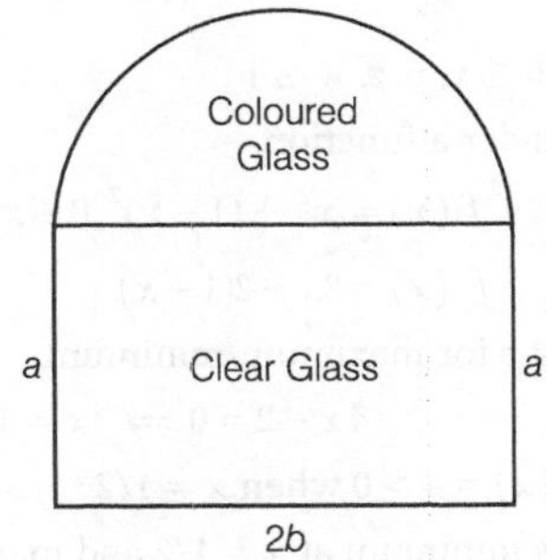

Then, $$Q = 2ab\cdot(3L) + \frac{1}{2}\pi b^2\cdot(L)$$

$$Q = \frac{L}{2}[\pi b^2 + 12ab]$$

$$Q = \frac{L}{2}[\pi b^2 + 6b(K - 4b - \pi b)]$$

$$Q = \frac{L}{2}[6Kb - 24b^2 - 5\pi b^2]$$

$$\therefore \quad \frac{dQ}{db} = \frac{L}{2}[6K - 48b - 10\pi b] = 0$$

$$b = \frac{6K}{48 + 10\pi} \quad \text{...(ii)}$$

and $$\frac{d^2Q}{db^2} = \frac{L}{2}[-48 - 10\pi] < 0$$

Thus, Q is maximum and from Eqs. (i) and (ii)

$$(48 + 10\pi)b = 6K = 6\{2a + 4b + \pi b\}$$

Thus, $$\text{the ratio} = \frac{2b}{a} = \frac{6}{6+\pi}$$

Ex. 120 *Let S be a square of unit area. Consider any quadrilateral which has one vertex on each side of S. If a, b, c and d denote the lengths of the sides of the quadrilateral, prove that* $2 \le a^2 + b^2 + c^2 + d^2 \le 4$ **[IIT JEE 1997]**

Sol. Let S be the square of unit area and $ABCD$ be the quadrilateral of sides a, b, c and d.

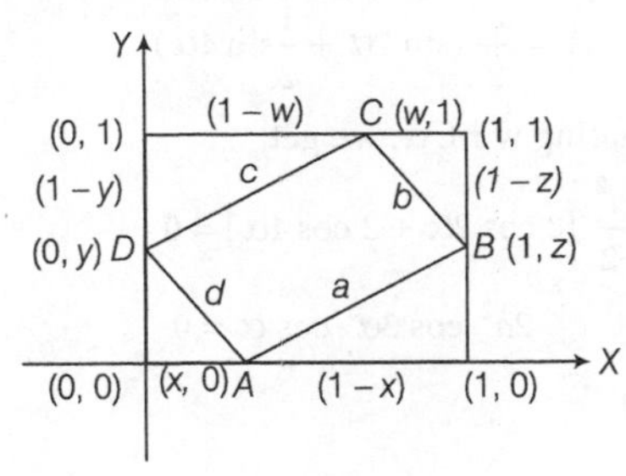

Here, $a^2 = (1 - x^2) + z^2,\ b^2 = w^2 + (1 - z)^2$

$c^2 = (1 - w^2) + (1 - y)^2, d^2 = x^2 + y^2$

Adding all the above, $a^2 + b^2 + c^2 + d^2$

$$= \{x^2 + (1 - x)^2\} + \{y^2 + (1 - y)^2\} + \{z^2 + (1 - z)^2\} + \{w^2 + (1 - w)^2\}$$

where $0 \le x, y, z, w \le 1$

Let us consider a function,

$$f(x) = x^2 + (1 - x)^2, 0 \le x \le 1$$

Then, $f'(x) = 2x - 2(1 - x)$

Let $f'(x) = 0$ for maximum/minimum.

$\Rightarrow \quad 4x - 2 = 0 \Rightarrow x = 1/2$

Again, $f''(x) = 4 > 0$ when $x = 1/2$

$\therefore$ $f(x)$ is minimum at $x = 1/2$ and maximum at $x = 1$

$\Rightarrow \quad 2 \le a^2 + b^2 + c^2 + d^2 \le 4$

Ex. 121 *Show that a triangle of maximum area that can be inscribed in a circle of radius a is an equilateral triangle.*

Sol. Let BC be one of the sides of the triangle and the third vertex A should be in a position that the altitude AD is maximum (for area of the triangle to be maximum).

For that the Δ ABC must be symmetric about AD.

i.e. D should be the mid-point of BC.

Let $\angle A = 2\alpha$

$\therefore \quad \angle BOD = \angle COD = 2\alpha$

Thus, $AD = AO + OD = a + a\cos 2\alpha$

and $CD = a \sin 2\alpha$

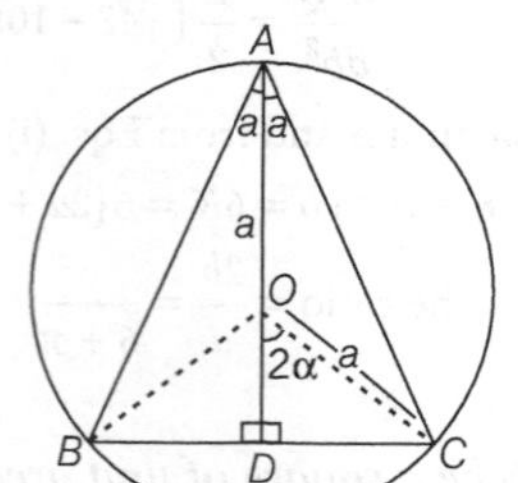

Hence, area of Δ $ABC = A = \frac{1}{2} \cdot AD \cdot BC$

Area, $A = \frac{1}{2} \cdot a(1 + \cos 2\alpha) \cdot a \sin 2\alpha$

$$A = \frac{a^2}{2}\left(\sin 2\alpha + \frac{1}{2}\sin 4\alpha\right)$$

Differentiating w.r.t. α, we get

$$\frac{dA}{d\alpha} = \frac{a^2}{2}[2\cos 2\alpha + 2\cos 4\alpha] = 0$$

$\therefore \quad 2a^2 \cos 3\alpha \cdot \cos \alpha = 0$

$\Rightarrow$ Either $\cos 3\alpha = 0$ or $\cos \alpha = 0$

$\Rightarrow \quad \alpha = \pi/6, \pi/2$

But $\alpha = \pi/2$ is not possible.

Now, $\frac{d^2A}{d\alpha^2} = 2a^2[-\sin 2\alpha - 4\sin 4\alpha] = -\text{ve at } \alpha = \pi/6$

$\therefore$ A is maximum when $\alpha = \pi/6$.

Also, $\angle A = \pi/3$ and triangle is isosceles.

Hence, $\Delta\, ABC$ must be equilateral.

Ex. 122 *Show that the height of the cylinder of maximum volume that can be inscribed in a sphere of radius a is $2a/\sqrt{3}$.*

Sol. Let h be the height and r be the radius of the cylinder. Let O be the centre of the sphere.

From the figure,

$$OA^2 = OP^2 + PA^2$$

i.e. $$a^2 = \frac{h^2}{4} + r^2 \quad \text{...(i)}$$

Now, volume of the cylinder,

$$V = \pi r^2 h = \pi h\left(a^2 - \frac{h^2}{4}\right) \quad \text{[using Eq. (i)]}$$

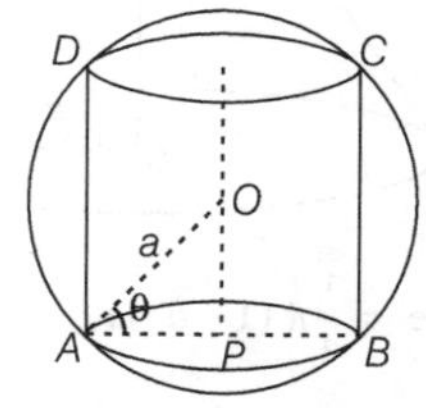

Differentiating w.r.t. h, we have

$$\frac{dV}{dh} = \pi\left[a^2 - \frac{3h^2}{4}\right] = 0 \text{ for extremum}$$

$\Rightarrow \quad h = \frac{2a}{\sqrt{3}}$

Also, $\frac{d^2V}{dh^2} = -\frac{3\pi}{2}h < 0$

at $h = \frac{2a}{\sqrt{3}}$

Thus, volume is maximum when $h = \frac{2a}{\sqrt{3}}$.

Ex. 123 *Let $A(p^2, p)$, $B(q^2, -q)$ and $C(r^2, -r)$ be the vertices of the $\Delta\, ABC$. A parallelogram AFDE is drawn with vertices D, E, F on the line segments BC, CA and AB, respectively. Show that maximum area of such a parallelogram is $\frac{1}{4}(p - q)(q - r)(r - p)$.*

Sol. Let $AF = x = DE$ and $AE = y = DF$

Since, Δ's CAB and CED are similar.

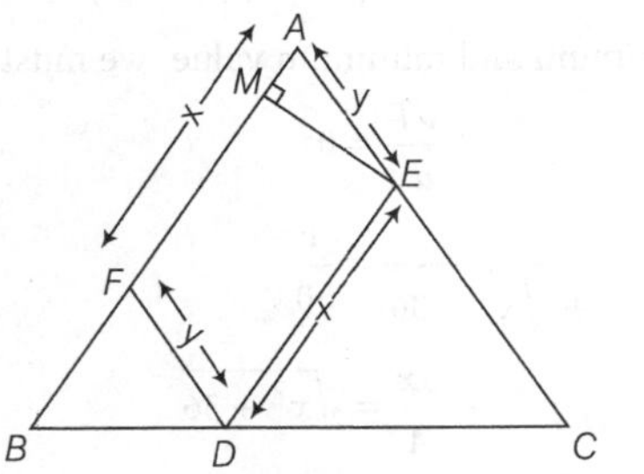

We have, $\dfrac{CE}{CA} = \dfrac{DE}{AB}$ [as shown in figure]

$\Rightarrow \dfrac{b-y}{b} = \dfrac{x}{c}$

[here, $BC = a$, $AC = b$ and $AB = c$] ...(i)

Now, area of parallelogram,

$$S = AF \cdot EM = xy \sin A$$

$$S = x \cdot b\left(1 - \frac{x}{c}\right)\sin A \quad \text{[from Eq. (i)] ...(ii)}$$

Differentiating w.r.t. x, we have

$$\frac{dS}{dx} = \frac{b}{c}(c - 2x)\sin A \quad \text{[where } \sin A \text{ is constant]}$$

For extremum, $\dfrac{dS}{dx} = 0$

$\Rightarrow x = \dfrac{c}{2}$

Also, $\dfrac{d^2S}{dx^2} = -\dfrac{2b}{c} < 0$

at $x = \dfrac{c}{2}$

Hence, S is maximum when $x = \dfrac{c}{2}$

Now, $S_{max} = \dfrac{1}{4} bc \sin A$ [from Eq. (ii)]

$$S_{max} = \frac{1}{2}\left(\frac{1}{2} bc \sin A\right)$$

$$S_{max} = \frac{1}{2}(\text{area of } \Delta\ ABC)$$

$$S_{max} = \frac{1}{4}\begin{vmatrix} p^2 & -p & 1 \\ q^2 & -q & 1 \\ r^2 & -r & 1 \end{vmatrix}$$

$$S_{max} = \frac{1}{4}(p-q)(q-r)(r-p)$$

Ex. 124 *LL' be latusrectum of the parabola $y^2 = 4ax$ and PP' is a double ordinate between the vertex and the latusrectum. Show that the area of trapezium $PP'L'L$ is maximum when the distance of PP' from the vertex is $\frac{a}{9}$.*

Sol. Let the double ordinate PP' be drawn at a distance $x = at^2$ from the origin. (vertex).

Thus, the coordinate of P is $(at^2, 2at)$ as shown in figure

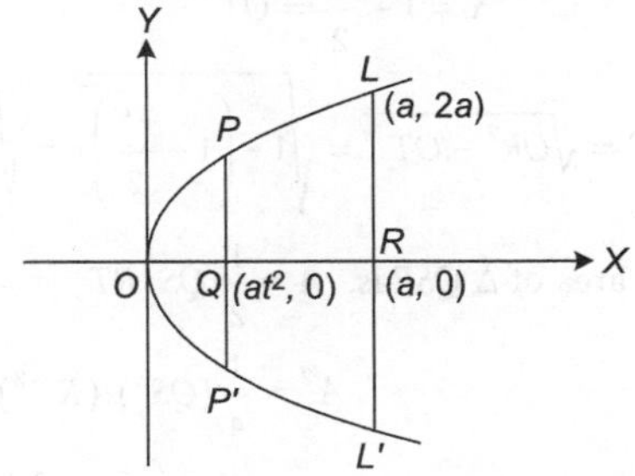

Hence, the area of trapezium is,

$$A = \frac{1}{2}(PP' + LL')\cdot QR$$

$$= \frac{1}{2}(4at + 4a)\cdot(a - at^2)$$

$$A = 2a^2 (t+1)(1-t^2)$$

$\therefore \dfrac{dA}{dt} = 2a^2(t+1)(1-3t) = 0$ [for extremum]

$\Rightarrow t = -1, 1/3$

Also, $\dfrac{d^2A}{dt^2} = 2a^2(-6t - 2)$

Thus, A is maximum when $t = 1/3$ as $\dfrac{d^2A}{dt^2} < 0$

Hence, $x = at^2 = \dfrac{a}{9}$ is the point at which area of trapezium is maximum.

Ex. 125 *The circle $x^2 + y^2 = 1$ intersects the X-axis at P and Q. Another circle with centre at Q and variable radius intersects the first circle at R, above the X-axis and the line segment PQ at S. Find the maximum area of the ΔQSR.*

[IIT JEE 1994]

Sol. The centre of the circle

$$x^2 + y^2 = 1 \quad \text{...(i)}$$

is (0, 0) and radius $OP = 1 = OQ$

So, coordinates of Q are (1,0).

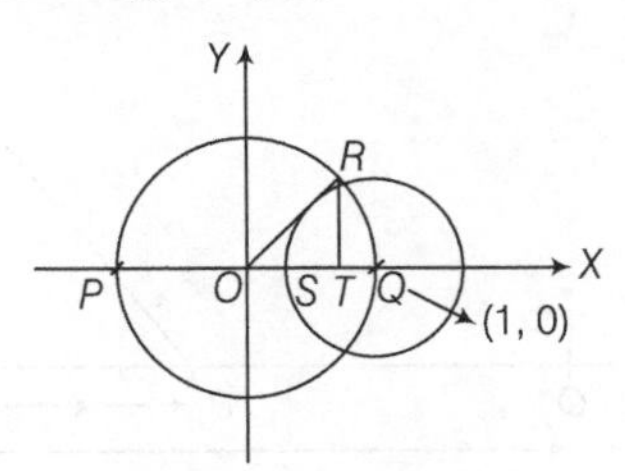

Let the radius of the variable circle be r.

Hence, its equation is

$$(x-1)^2 + (y)^2 = r^2 \quad \text{...(ii)}$$

Subtracting Eq. (ii) from Eq. (i), we get

$$2x - 1 = 1 - r^2$$

$$x = 1 - \frac{r^2}{2} = OT \qquad \text{...(iii)}$$

Now, $RT = \sqrt{OR^2 - OT^2} = \sqrt{1 - \left(1 - \frac{r^2}{2}\right)^2} = \sqrt{r^2 - \frac{r^4}{4}}$ ...(iv)

Now, the area of $\Delta\, QSR$ is, $A = \frac{1}{2} \cdot QS \cdot RT$

$\therefore \quad A^2 = \frac{1}{4}(QS^2) \cdot (RT^2)$

$\Rightarrow \quad A^2 = \frac{1}{4} r^2 \left(r^2 - \frac{r^4}{4}\right)$ [using Eqs. (ii) and (iv)]

$\Rightarrow \quad A^2 = \frac{1}{16}(4r^4 - r^6) = f(r)$ (say)

Thus, $\frac{df(r)}{dr} = \frac{1}{16}(16r^3 - 6r^5) = 0$ [for extremum]

$\Rightarrow \quad r = 2\sqrt{\frac{2}{3}}$

Also, $\frac{d^2 f(r)}{dr^2} = \frac{1}{16}(48r^2 - 30r^4) = -\frac{16}{3} < 0$

where $r = 2\sqrt{\frac{2}{3}}$

Hence, area is maximum at $r = 2\sqrt{\frac{2}{3}}$ and $A_{max} = \frac{4}{3\sqrt{3}}$ sq unit.

Ex. 126 *A despatch rider is in open country at a distance of 6 km from the nearest point P of a straight road. He wishes to proceed as quickly as possible to a point Q on the road 20 km from P. If his maximum speed across country is 40 km/h, 50 km/h on road, then at what distance from P, he should touch the road.*

Sol. Let A be the initial position of rider.

Then, $PB = x \Rightarrow QB = 20 - x$

and $AB = \sqrt{AP^2 + BP^2} = \sqrt{6^2 + x^2}$ km

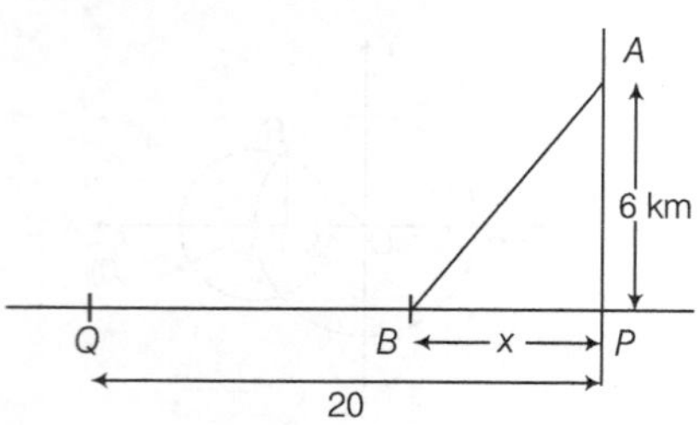

$\therefore$ Total time T,

$$T = \frac{\sqrt{x^2 + 36}}{40} + \frac{20 - x}{50}$$

$\Rightarrow \quad \frac{dT}{dx} = \frac{1}{40} \cdot \frac{(2x)}{2\sqrt{x^2 + 36}} - \frac{1}{50}$

For maximum and minimum value, we must have

$$\frac{dT}{dx} = 0$$

$\Rightarrow \quad \frac{x}{40\sqrt{x^2 + 36}} = \frac{1}{50}$

or $\quad \frac{5x}{4} = \sqrt{x^2 + 36}$

or $\quad \frac{25x^2}{16} - x^2 = 36$

$\Rightarrow \quad x^2 = \frac{36 \times 16}{9}$

$\Rightarrow \quad x = \frac{6 \times 4}{3} = 8$ km

Ex. 127 *From point A located on a highway a boy has to get his bus to his school B located in the field at a distance l from the highway in the least possible time. At what distance from D should the bus leave the highway when the bus moves n times slower in the field than on the highway?*

Sol. Let $AD = s$ and $CD = x$, where C is a point where the bus leaves the highway.

Also, let the speed of the bus is 'v' on the highway.

$\therefore$ Total time taken,

$$t = \frac{AC}{v} + \frac{BC}{(v/n)}$$

$\Rightarrow \quad t = \frac{s - x}{v} + \frac{n\sqrt{l^2 + x^2}}{v}$

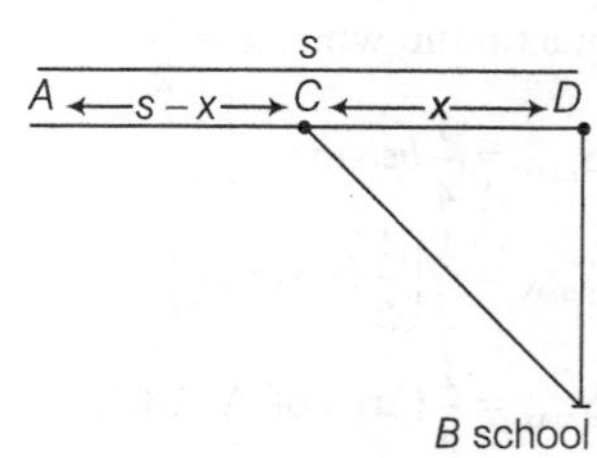

Differentiating w.r.t. x, we have

$$\frac{dt}{dx} = \frac{1}{v}\left[-1 + \frac{n}{2\sqrt{l^2 + x^2}}(+2x)\right] = 0 \quad \text{[for extremum]}$$

$\Rightarrow \quad n^2x^2 = l^2 + x^2$

$\Rightarrow \quad x = \frac{l}{\sqrt{n^2 - 1}}$

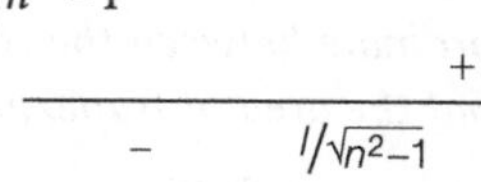

Thus, 't' is minimum when $x = \frac{l}{\sqrt{n^2 - 1}}$, as shown in figure.

Ex. 128 *Two men are walking on a path $x^3 + y^3 = a^3$ when the first man arrives at a point (x_1, y_1), he finds the second man in the direction of his own instantaneous motion. If the coordinates of the second man are (x_2, y_2), then show that*

$$\frac{x_2}{x_1} + \frac{y_2}{y_1} + 1 = 0.$$

Sol. Since, (x_1, y_1) and (x_2, y_2) lies on the curve.

$$\therefore \quad x_1^3 + y_1^3 = a^3 \quad \text{...(i)}$$

$$\text{and} \quad x_2^3 + y_2^3 = a^3 \quad \text{...(ii)}$$

Subtracting Eqs. (i) from (ii), we get

$$(x_2^3 - x_1^3) + (y_2^3 - y_1^3) = 0$$

$$\text{or} \quad x_2^3 - x_1^3 = -(y_2^3 - y_1^3) \quad \text{...(iii)}$$

Now, differentiating both sides of $x^3 + y^3 = a^3$ w.r.t. x, we get

$$3x^2 + 3y^2 \frac{dy}{dx} = 0$$

Slope of tangent at $(x_1, y_1) = -\dfrac{x_1^2}{y_1^2}$

$\therefore$ The equation of tangent at (x_1, y_1)

$$y - y_1 = -\frac{x_1^2}{y_1^2}(x - x_1)$$

It passes through (x_2, y_2).

$$\therefore \quad y_2 - y_1 = \frac{-x_1^2}{y_1^2}(x_2 - x_1)$$

$$\text{or} \quad x_1^2(x_2 - x_1) = -y_1^2(y_2 - y_1) \quad \text{...(iv)}$$

Dividing Eq. (iii) by Eq. (iv), we get

$$\frac{x_2^3 - x_1^3}{x_1^2(x_2 - x_1)} = \frac{-(y_2^3 - y_1^3)}{-y_1^2(y_2 - y_1)}$$

$$\Rightarrow \quad \frac{x_2^2 + x_1^2 + x_1x_2}{x_1^2} = \frac{y_2^2 + y_1^2 + y_1y_2}{y_1^2}$$

$$\Rightarrow \quad \left(\frac{x_2}{x_1}\right)^2 + \left(\frac{x_2}{x_1}\right) + 1 = \left(\frac{y_2}{y_1}\right)^2 + \left(\frac{y_2}{y_1}\right) + 1$$

$$\Rightarrow \quad \left(\frac{x_2}{x_1}\right)^2 - \left(\frac{y_2}{y_1}\right)^2 = \left(\frac{y_2}{y_1}\right) - \left(\frac{x_2}{x_1}\right)$$

$$\Rightarrow \quad \left(\frac{x_2}{x_1} - \frac{y_2}{y_1}\right) \cdot \left(\frac{x_2}{x_1} + \frac{y_2}{y_1}\right) + \left(\frac{x_2}{x_1} - \frac{y_2}{y_1}\right) = 0$$

$$\Rightarrow \quad \left(\frac{x_2}{x_1} - \frac{y_2}{y_1}\right) \cdot \left(\frac{x_2}{x_1} + \frac{y_2}{y_1} + 1\right) = 0$$

$$\text{i.e.} \quad \text{either } \frac{x_2}{x_1} = \frac{y_2}{y_1}$$

$$\text{or} \quad \frac{x_2}{x_1} + \frac{y_2}{y_1} + 1 = 0$$

$$\text{But,} \quad \frac{x_2}{x_1} \neq \frac{y_2}{y_1}$$

$$\therefore \quad \frac{x_2}{x_1} + \frac{y_2}{y_1} + 1 = 0$$

Ex. 129 *In still water a boat moves with a velocity which is K times less than velocity the river has current. At what angle to the stream direction must the boat move to minimize drifting.*

Sol. Let the flow velocity of river be u and the velocity of boat in still water be v.

$$\text{Thus,} \quad v = u/K$$

Also, let the boat moves at an angle θ with direction of stream.

Now, the velocity of boat in the river is vector resultant of the velocity of boat and velocity of following water or water current, which can be written as,

$$\bar{v}_B = (u - v\cos\alpha)\hat{\mathbf{i}} + (u\sin\alpha)\hat{\mathbf{j}} = (u + v\cos\theta)\hat{\mathbf{i}} + (u\sin\theta)\hat{\mathbf{j}}$$

Hence, the time taken to cross the river

$$= \frac{d}{u\sin\theta} \quad \text{[where, } d = \text{width of river]}$$

Thus, the drift $s = (u + v\cos\theta)\cdot d$

$$\Rightarrow \quad s = d\left(\operatorname{cosec}\theta + \frac{v}{u}\cot\theta\right)$$

$$\Rightarrow \quad \frac{ds}{d\theta} = d\left(-\operatorname{cosec}\theta\cot\theta - \frac{v}{u}\operatorname{cosec}^2\theta\right) = 0$$

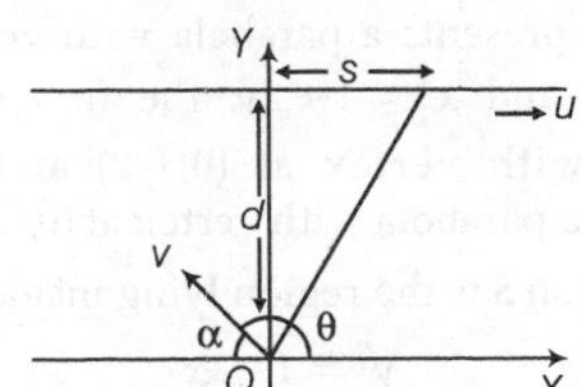

$$\Rightarrow \quad -\frac{v}{u}\operatorname{cosec}^2\theta = \operatorname{cosec}\theta\cot\theta$$

$$\Rightarrow \quad \operatorname{cosec}\theta = -1/K$$

$$\Rightarrow \quad \theta = \cos^{-1}(-1/K)$$

Ex. 130 *Consider a square with vertices at (1, 1) (−1, 1) (−1, −1) and (1, −1). Let S be the region consisting of all points inside the square which are nearer to the origin than to any edge. Sketch the region S and find its area.* **[IIT JEE 1995]**

Sol. Let the square $ABCD$ and the equations of the sides of the square are as follows

$AB : y = 1, \quad BC : x = -1, \quad CD : y = -1, \quad DA : x = 1$

Let the region be S and (x, y) is any point inside it. Then, according to given conditions,

$$\sqrt{x^2+y^2} < |1-x|, |1+x|, |1-y|, |1+y|$$

$$\Rightarrow (x^2+y^2) < x^2-2x+1,\ x^2+2x+1,\ y^2-2y+1,\ y^2+2y+1$$

$$\Rightarrow y^2 < 1-2x,\ y^2 < 1+2x,\ x^2 < 1-2y \text{ and } x^2 < 2y+1$$

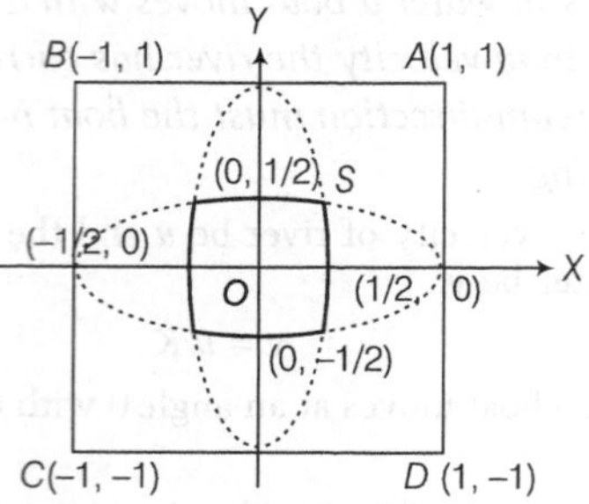

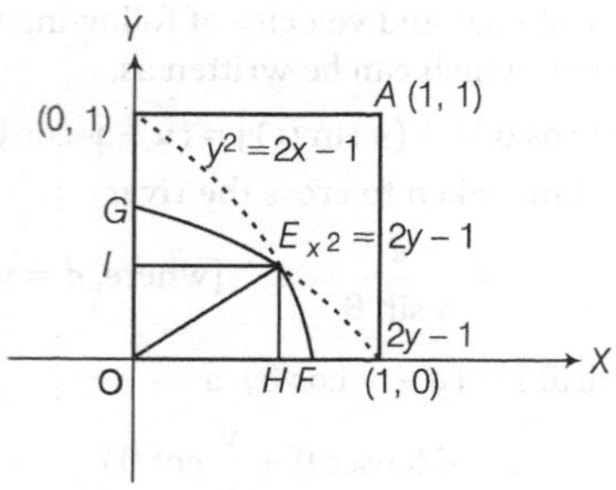

Now, in $y^2 = 1-2x$ and $y^2 = 2x+1$, the first equation represents a parabola with vertex $(1/2, 0)$ and second equation represents a parabola with vertex $(-1/2, 0)$ and in $x^2 = 1-2y$ and $x^2 = 1+2y$, the first equation represents parabola with vertex at $(0, 1/2)$ and second equation represents a parabola with vertex at $(0, -1/2)$.

So, the region S is the region lying inside the four parabolas.

$$y^2 = 1-2x$$
$$y^2 = 1+2x$$
$$x^2 = 1-2y$$
$$x^2 = 1+2y$$

Now, S is symmetrical in all four quadrants.

$\therefore$ $S = 4 \times$ area lying in first quadrant.

Now, $y^2 = 1-2x$

and $x^2 = 1-2y$ intersect on $y = x$

The point of intersection is $E(\sqrt{2}-1, \sqrt{2}-1)$.

$\therefore$ Area of region $OEFO$ = Area of $\Delta\ OEH$ + Area of region $HEFH$

$$= \frac{1}{2}(\sqrt{2}-1)^2 + \int_{\sqrt{2}-1}^{1/2} \sqrt{1-2x}\, dx$$

$$= \frac{1}{2}(2+1-2\sqrt{2}) + \frac{2}{3}[(1-2x)^{3/2}]_{\sqrt{2}-1}^{1/2}$$

$$= \frac{1}{2}(3-2\sqrt{2}) + \frac{1}{3}(3-2\sqrt{2})^{3/2}$$

$$= \frac{1}{2}(3-2\sqrt{2}) + \frac{1}{3}(\sqrt{2}-1)^3$$

$$= \frac{1}{2}(3-2\sqrt{2}) + \frac{1}{3}(5\sqrt{2}-7) = \frac{1}{6}[4\sqrt{2}-5]$$

Similarly, area of region $OEGO = \frac{1}{6}(4\sqrt{2}-5)$

So, area of S lying in first quadrant $= \frac{2}{6}(4\sqrt{2}-5)$

Hence, $S = \frac{4}{3}(4\sqrt{2}-5)$

Ex. 131 *If x is increasing at the rate of 2 cm/s at the instant when $x = 3$ cm and $y = 1$ cm, at what rate y must be changing in order that the quantity $(2xy - 3x^2y)$ shall be neither increasing nor decreasing.*

If $S = \{(a, b) \in R \times R : x = a, y = b, 2xy - 3x^2y = constant \Rightarrow \frac{dy}{dx} > 0\}$

$S' = \{(x, y) \in A \times B : -1 \le A \le 1$ and $-1 \le B \le 1\}$, then find the area $S \cap S'$.

Sol. Here, $x = 3$ cm, $y = 1$ cm and $\frac{dx}{dt} = 2$ cm/s

Let $f = 2xy - 3x^2y$

$$\Rightarrow \frac{df}{dt} = 2y\frac{dx}{dt} + 2x\frac{dy}{dt} - 3x^2\frac{dy}{dt} - 6xy\frac{dx}{dt}$$

$$= (2x - 3x^2)\frac{dy}{dx} + (2y - 6xy)\frac{dx}{dt} \quad \text{...(i)}$$

$\because f$ is neither increasing nor decreasing.

$$\therefore \left(\frac{df}{dt}\right)_{(3,1)} = 0$$

$$\Rightarrow (2\times 3 - 27)\frac{dy}{dt} + (2-18)2 = 0$$

$$\Rightarrow dy/dt = -32/21$$

Now, for S, f = constant

$$\Rightarrow 0 = (2x - 3x^2)\frac{dy}{dt} + (2y - 6xy)\frac{dx}{dt}$$

$$\Rightarrow \frac{dy}{dx} = \frac{6xy - 2y}{2x - 3x^2}$$

$$\Rightarrow \frac{dy}{dx} > 0 \Rightarrow \frac{6xy - 2y}{2x - 3x^2} > 0$$

$$\Rightarrow \frac{y(3x-1)}{x(3x-2)} < 0,$$

Thus, there arise two cases

Case I $y(3x-1) > 0$ and $x(3x-2) < 0$

$\Rightarrow$ $(y > 0$ and $x > 1/3)$ or $(y < 0$ and $x < 1/3)$

and $\left(0 < x < \frac{2}{3}\right)$

$$\Rightarrow \left(y > 0 \text{ and } \frac{1}{3} < x < \frac{2}{3}\right) \text{ or } \left(y < 0 \text{ and } 0 < x < \frac{1}{3}\right)$$

***Case* II** $y(3x-1)<0$ and $x(3x-2)>0$

$\Rightarrow \quad (y>0 \text{ and } x<1/3)$

or $\quad (y<0 \text{ and } x>1/3)$

and $\quad (x<0 \text{ or } x>2/3)$

$\Rightarrow \quad (y>0 \text{ and } x<0)$

or $\quad (y<0 \text{ and } x>2/3)$

Thus, S is shaded portion shown in figure (i).

Also, S' represents the portion shown in figure (ii).

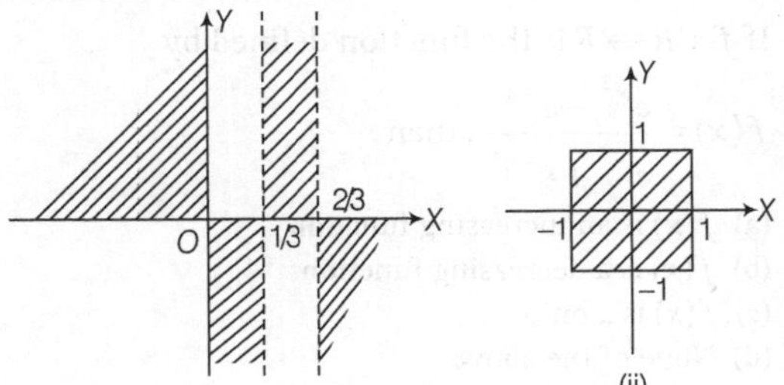

(i) (ii)

Thus, area for $S \cap S'$ is shown in figure (iii).

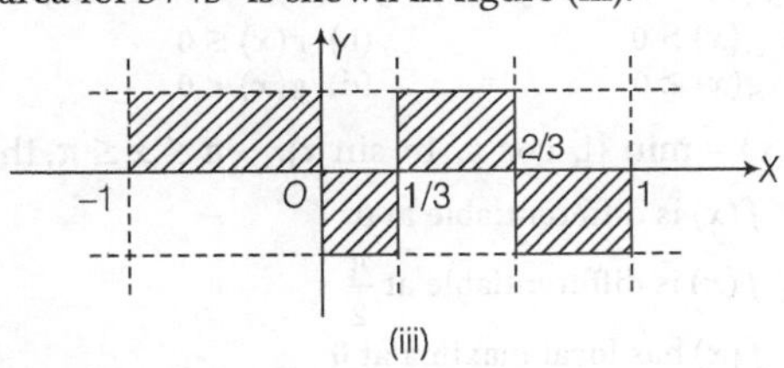

(iii)

Thus, the area of $S \cap S' = 2$

- **Ex. 132** *In how many parts an integer $N \geq 5$ should be divide so that the product of the parts is maximized?*

Sol. Let $x_1+x_2+x_3+\ldots+x_n=N$

$\therefore$ To maximize the value of $x_1x_2x_3\ldots x_n$

Using AM > GM

$$\frac{x_1+x_2+\ldots+x_n}{n} > (x_1x_2\ldots x_n)^{1/n}$$

$$\Rightarrow \quad x_1x_2\ldots x_n \leq \left(\frac{x_1+x_2+\ldots+x_n}{n}\right)^n$$

which shows maximum value of $x_1x_2\ldots x_n$ is obtained when $x_1=x_2=x_3=\ldots=x_n$

Now, function to be maximized is $\left(\frac{x_1+x_2+\ldots+x_n}{n}\right)^n$

which is discrete function of n. In order to arrive at some possible neighbourhood first we make it continuous. Thus, by changing the variable n to x.

We write $\quad f(x)=\left(\frac{N}{x}\right)^x$

For maxima $f'(x)=0$

i.e. $\quad \left(\frac{N}{x}\right)^x\left\{\log\frac{N}{x}-1\right\}=0 \Rightarrow \frac{N}{x}=e$

or $x=\frac{N}{e}$, now as $x \in$ integer

$\Rightarrow$ The nearest integer is $\left[\frac{N}{e}\right]$ or $\left[\frac{N}{e}\right]+1$.

Monotonicity, Maxima and Minima Exercise 1 :
Single Option Correct Type Questions

1. If $f : [1, 10] \rightarrow [1, 10]$ is a non-decreasing function and $g : [1, 10] \rightarrow [1, 10]$ is a non-increasing function. Let $h(x) = f(g(x))$ with $h(1) = 1$. Then, $h(2)$

(a) lies in (1, 2) (b) is more than two
(c) is equal to one (d) is not defined

2. P is a variable point on the curve $y = f(x)$ and A is a fixed point in the plane not lying on the curve. If PA^2 is minimum, then the angle between PA and the tangent at P is

(a) $\frac{\pi}{4}$ (b) $\frac{\pi}{3}$
(c) $\frac{\pi}{2}$ (d) None of these

3. Let $f(x) = \begin{cases} 1 + \sin x, & x < 0 \\ x^2 - x + 1, & x \geq 0 \end{cases}$. Then,

(a) f has a local maximum at $x = 0$
(b) f has a local minimum at $x = 0$
(c) f is increasing everywhere
(d) f is decreasing everywhere

4. If m and n are positive integers and $f(x) = \int_1^x (t-a)^{2n} (t-b)^{2m+1} dt$, $a \neq b$, then

(a) $x = b$ is a point of local minimum
(b) $x = b$ is a point of local maximum
(c) $x = a$ is a point of local minimum
(d) $x = a$ is a point of local maximum

5. Let $f(x) = x^{n+1} + a \cdot x^n$, where a is a positive real number. Then, $x = 0$ is a point of

(a) local minimum for any integer n
(b) local maximum for any integer n
(c) local minimum if n is an even integer
(d) local minimum if n is an odd integer

6. If $f(x) = \begin{cases} 2x^2 + \frac{2}{x^2}, & 0 < |x| \leq 2 \\ 3, & x > 2 \end{cases}$, then

(a) $x = 1, -1$ are the points of global minima
(b) $x = 1, -1$ are the points of local minima
(c) $x = 0$ is the points of local minima
(d) None of the above

7. Given a function $y = x^x$, $x > 0$ and $0 < x < 1$. The values of x for which the function attain values exceeding the values of its inverse are

(a) $0 < x < 1$ (b) $1 < x < \infty$
(c) $0 < x < 2$ (d) None of these

8. $\sin x + \cos x = y^2 - y + a$ has no value of x for any y, if a belongs

(a) $(0, \sqrt{3})$ (b) $(-\sqrt{3}, 0)$
(c) $(-\infty, -\sqrt{3})$ (d) $(\sqrt{3}, \infty)$

9. If $f : R \rightarrow R$ is the function defined by $f(x) = \frac{e^{x^2} - e^{-x^2}}{e^{x^2} + e^{-x^2}}$, then

(a) $f(x)$ is an increasing function
(b) $f(x)$ is a decreasing function
(c) $f(x)$ is a onto
(d) None of the above

10. Let $f(x)$ be a quadratic expression positive for all real x. If $g(x) = f(x) - f'(x) + f''(x)$, then for any real x

(a) $g(x) > 0$ (b) $g(x) \leq 0$
(c) $g(x) \geq 0$ (d) $g(x) < 0$

11. $f(x) = \min \{1, \cos x, 1 - \sin x\}$, $-\pi \leq x \leq \pi$, then

(a) $f(x)$ is differentiable at 0
(b) $f(x)$ is differentiable at $\frac{\pi}{2}$
(c) $f(x)$ has local maxima at 0
(d) None of the above

12. $f(x) = \begin{cases} 2 - |x^2 + 5x + 6|, & x \neq -2 \\ a^2 + 1, & x = -2 \end{cases}$, then the range of a, so that $f(x)$ has maxima at $x = -2$ is

(a) $|a| \geq 1$ (b) $|a| < 1$ (c) $a > 1$ (d) $a < 1$

13. Maximum number of real solution for the equation $ax^n + x^2 + bx + c = 0$, where $a, b, c \in R$ and n is an even positive number, is

(a) 2 (b) 3 (c) 4 (d) infinite

14. Maximum area of rectangle whose two sides are $x = x_0$, $x = \pi - x_0$ and which is inscribed in a region bounded by $y = \sin x$ and X-axis is obtained when $x_0 \in$

(a) $\left(\frac{\pi}{4}, \frac{\pi}{3}\right)$ (b) $\left(\frac{\pi}{6}, \frac{\pi}{4}\right)$
(c) $\left(0, \frac{\pi}{6}\right)$ (d) None of these

15. $f(x) = -1 + kx + k$ neither touches nor intersect the curve $f(x) = \ln x$, then minimum value of $k \in$

(a) $\left(\frac{1}{e}, \frac{1}{\sqrt{e}}\right)$ (b) (e, e^2)
(c) $\left(\frac{1}{\sqrt{e}}, e\right)$ (d) None of these

16. Let $f(x)$ be a polynomial with real coefficients satisfies $f(x) = f'(x) \times f'''(x)$. If $f(x) = 0$ satisfies $x = 1, 2, 3$ only, then the value of $f'(1) \times f'(2) \times f'(3)$ is equal to
(a) positive (b) negative
(c) 0 (d) inadequate data

17. A curve whose concavity is directly proportional to the logarithm of its x-coordinates at any of the curve, is given by
(a) $c_1 \cdot x^2(2\log x - 3) + c_2 x + c_3$
(b) $c_1 x^2(2\log x + 3) + c_2 x + c_3$
(c) $c_1 x^2(2\log x) + c_2$
(d) None of the above

18. $f(x) = 4\tan x - \tan^2 x + \tan^3 x, \forall x \neq n\pi + \frac{\pi}{2}, \forall n \in I$, then
(a) $f(x)$ is increasing for all $x \in R$
(b) $f(x)$ is decreasing for all $x \in R$
(c) $f(x)$ is increasing in its domain
(d) None of the above

19. If $f(x) = \begin{cases} 3 + |x - k|, & \text{for } x \le k \\ a^2 - 2 + \dfrac{\sin(x-k)}{x-k}, & \text{for } x > k \end{cases}$ has minimum at $x = k$, then
(a) $a \in R$ (b) $|a| < 2$
(c) $|a| > 2$ (d) $1 < |a| < 2$

20. Let f be a linear function with properties $f(1) \le f(2)$, $f(3) \ge f(4)$ and $f(5) = 5$, then which of the following is true?
(a) $f(0) < 0$ (b) $f(0) = 0$
(c) $f(1) < f(0) < f(-1)$ (d) $f(0) = 5$

21. If $P(x)$ is polynomial satisfying $P(x^2) = x^2 P(x)$ and $P(0) = -2$, $P'(3/2) = 0$ and $P(1) = 0$.
The maximum value of $P(x)$ is
(a) $-\frac{1}{3}$ (b) $-\frac{1}{4}$
(c) $-\frac{1}{2}$ (d) None of these

22. If the curve $x^2 = -4(y - a)$ does not intersect the curve $y = [x^2 - x + 1]$ (where [.] denotes the greatest integer function) in $\left[0, \frac{1+\sqrt{5}}{2}\right]$, then
(a) $\frac{1}{3} < a < 1$ (b) $-1 < a < 1$
(c) $\frac{1}{4} < a < 1$ (d) None of these

23. Analyze the following graph of $f'(x)$

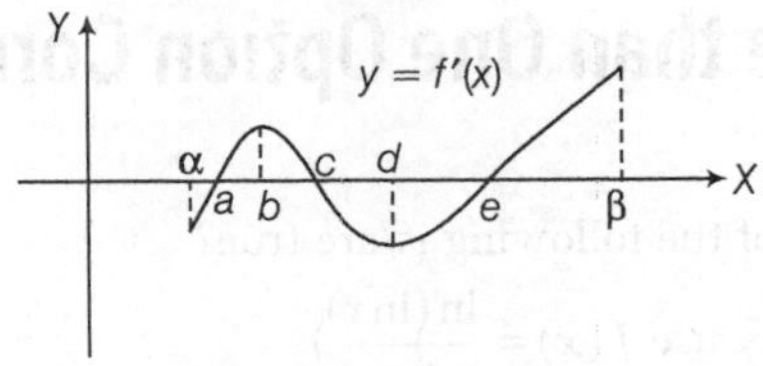

which is incorrect about $f(x)$ for $\alpha < x < \beta$?
(a) only three extreme points
(b) two inflexion points
(c) $f'''(x) > 0$ for $d < x < e$
(d) $x = e$ is the point of local maxima

24. Let $f(x) = x^2 - 2x$ and $g(x) = f(f(x) - 1) + f(5 - f(x))$, then
(a) $g(x) < 0, \forall x \in R$
(b) $g(x) < 0$, for some $x \in R$
(c) $g(x) \ge 0$, for some $x \in R$
(d) $g(x) \ge 0, \forall x \in R$

25. Let $f: N \to N$ be such that $f(n+1) > f(f(n))$ for all $n \in N$, then
(a) $f(x) = x^2 - x + 1$
(b) $f(x) = x - 1$
(c) $f(x) = x^2 + 1$
(d) None of the above

26. The equation $|2ax - 3| + |ax + 1| + |5 - ax| = \frac{1}{2}$ possesses
(a) infinite number of real solutions for some $a \in R$
(b) finitely many real solutions for some $a \in R$
(c) no real solutions for some $a \in R$
(d) no real solutions for all $a \in R$

27. If $\int_0^x 2x f^2(t)\,dt = \left(\int_0^x 2f(x-t)\,dt\right)^2$ for $f(1) = 1$ and $f(x)$ is continuous function for $x > 0$ and $\{a_n\}$ is a sequence such that $a_{n+1} = a_n + \sqrt{1 + a_n^2}$ for $a = 0$; if $f(x)$ is an increasing function, then $\lim\limits_{n \to \infty} \frac{a_k}{2^{n-1}} =$
(where $k = f(n^{\sqrt{2}-1})$) is
(a) $\pi/4$ (b) $4/\pi$ (c) π (d) $\pi/2$

28. A function f is defined by $f(x) = |x|^m |x - 1|^n, \forall x \in R$. The maximum value of the function is ($m, n \in N$)
(a) 1 (b) $m^n \cdot n^m$
(c) $\frac{m^m \cdot n^n}{(m+n)^{m+n}}$ (d) $\frac{(mn)^{mn}}{(m+n)^{m+n}}$

Monotonicity, Maxima and Minima Exercise 2 : More than One Option Correct Type Questions

29. Which of the following is/are true?

(you may use $f(x)=\dfrac{\ln(\ln x)}{\ln x}$)

(a) $(\ln 2.1)^{\ln 2.2} > (\ln 2.2)^{\ln 2.1}$

(b) $(\ln 4)^{\ln 5} < (\ln 5)^{\ln 4}$

(c) $(\ln 30)^{\ln 31} > (\ln 31)^{\ln 30}$

(d) $(\ln 28)^{30} < (\ln 30)^{\ln 28}$

30. If $\lim\limits_{x\to a} f(x) = \lim\limits_{x\to a} [f(x)]$ (where [·] denotes the greatest integer function) and $f(x)$ is non-constant continuous function, then

(a) $\lim\limits_{x\to a} f(x)$ is an integer

(b) $\lim\limits_{x\to a} f(x)$ is non-integer

(c) $f(x)$ has local maximum at $x = a$

(d) $f(x)$ has local minimum at $x = a$

31. Let S be the set of real values of parameter λ for which the equation $f(x)=2x^3-3(2+\lambda)x^2+12\lambda x$ has exactly one local maximum and exactly one local minimum. Then, S is a subset of

(a) $(-4, \infty)$ (b) $(-3, 3)$

(c) $(3, \infty)$ (d) $(-\infty, 0)$

32. $h(x)=3f\left(\dfrac{x^2}{3}\right)+f(3-x^2)$, $\forall\ x\in(-3, 4)$, where $f''(x)>0$, $\forall\ x\in(-3, 4)$, then $h(x)$ is

(a) increasing in $\left(\dfrac{3}{2}, 4\right)$

(b) increasing in $\left(-\dfrac{3}{2}, 0\right)$

(c) decreasing in $\left(-3, -\dfrac{3}{2}\right)$

(d) decreasing in $\left(0, \dfrac{3}{2}\right)$

33. $f(x)=\tan^{-1}(\sin x+\cos x)$, then $f(x)$ is increasing in

(a) $\left(-\dfrac{\pi}{2}, \dfrac{\pi}{4}\right)$ (b) $\left(-\dfrac{\pi}{4}, \dfrac{\pi}{4}\right)$

(c) $\left(\dfrac{5\pi}{4}, \dfrac{3\pi}{2}\right)$ (d) $\left(-2\pi, -\dfrac{7\pi}{4}\right)$

34. If maximum and minimum values of the determinant

$$\begin{vmatrix} 1+\sin^2 x & \cos^2 x & \sin 2x \\ \sin^2 x & 1+\cos^2 x & \sin 2x \\ \sin^2 x & \cos^2 x & 1+\sin 2x \end{vmatrix}$$

are α and β, then

(a) $\alpha+\beta^{99}=4$

(b) $\alpha^3-\beta^{17}=26$

(c) $(\alpha^{2n}-\beta^{2n})$ is always an even integer for $n\in N$

(d) a triangle can be drawn having it's sides as α, β and $\alpha-\beta$

35. Let $f(x)=\begin{cases} x^2+4x, & -3\le x\le 0 \\ -\sin x, & 0<x\le\dfrac{\pi}{2} \\ -\cos x-1, & \dfrac{\pi}{2}<x\le\pi \end{cases}$. Then,

(a) $x=-2$ is the point of global minima

(b) $x=\pi$ is the point of global maxima

(c) $f(x)$ is non-differentiable at $x=\dfrac{\pi}{2}$

(d) $f(x)$ is discontinuous at $x=0$

36. Let $f(x)=ab\sin x+b\sqrt{1-a^2}\cos x+c$, where $|a|<1$, $b>0$, then

(a) maximum value of $f(x)$ is b, if $c=0$

(b) difference of maximum and minimum value of $f(x)$ is $2b$

(c) $f(x)=c$, if $x=-\cos^{-1}a$

(d) $f(x)=c$, if $x=\cos^{-1}a$

37. If $f(x)=\displaystyle\int_{x^m}^{x^n}\frac{dt}{\ln t}$, $x>0$ and $n>m$, then

(a) $f'(x)=\dfrac{x^{m-1}(x-1)}{\ln x}$

(b) $f(x)$ is decreasing for $x>1$

(c) $f(x)$ is increasing in $(0, 1)$

(d) $f(x)$ is increasing for $x>1$

38. $f(x)=\sqrt{x-1}+\sqrt{2-x}$ and $g(x)=x^2+bx+c$ are two given functions such that $f(x)$ and $g(x)$ attain their maximum and minimum values respectively for same value of x, then

(a) $f(x)_{\max}$ at $x=\dfrac{1}{2}$ (b) $f(x)_{\max}$ at $x=\dfrac{3}{2}$

(c) $b=3$ (d) $b=-3$

39. If $f(x) = a^{\{a^{|x|} \operatorname{sgn} x\}}$; $g(x) = a^{[a^{|x|} \operatorname{sgn} x]}$ for $a > 0, a \neq 1$ and $x \in R$, where {} and [·] denote the fractional part and integral part functions respectively, then which of the following statements can hold good for the function $h(x)$? where $(\ln a)\, h(x) = (\ln f(x) + \ln g(x))$

(a) h is even and increasing
(b) h is odd and decreasing
(c) h is even and decreasing
(d) h is odd and increasing

40. For the function $f(x) = \ln(1 - \ln x)$, which of the following do not hold good?

(a) increasing in (0, 1) and decreasing in (1, e)
(b) decreasing in (0, 1) and increasing in (1, e)
(c) $x = 1$ is the critical number for $f(x)$
(d) f has two asymptotes

41. The function $f(x) = \begin{cases} x+2, & \text{if } x < -1 \\ x^2, & \text{if } -1 \le x < 1 \\ (x-2)^2, & \text{if } x \ge 1 \end{cases}$

(a) is continuous for all $x \in R$
(b) is continuous but not differentiable, $\forall\, x \in R$
(c) is such that $f'(x)$ change its sign exactly twice
(d) has two local maxima and two local minima

42. A function f is defined by $f(x) = \int_0^{\pi} \cos t \cos(x - t)\, dt$, $0 \le x \le 2\pi$, then which of the following hold(s) good?

(a) $f(x)$ is continuous but not differentiable in $(0, 2\pi)$
(b) Maximum value of f is π
(c) There exists atleast one $c \in (0, 2\pi)$ if $f'(c) = 0$
(d) Minimum value of f is $-\dfrac{\pi}{2}$

43. Let $f(x) = \dfrac{x-1}{x^2}$, then which of the following is correct?

(a) $f(x)$ has minima but no maxima
(b) $f(x)$ increases in the interval (0, 2) and decreases in the interval $(-\infty, 0) \cup (2, \infty)$
(c) $f(x)$ can come down in $(-\infty, 0) \cup (2, 3)$
(d) $x = 3$ is the point of inflection

44. Let $f(x)$ be differentiable function on the interval $(-\infty, \infty)$ such that $f(1) = 5$ and $\lim\limits_{a \to x} \dfrac{af(x) - xf(a)}{a - x} = 2$, for all $x \in R$. Then, which of the following alternative(s) is/are correct?

(a) $f(x)$ has an inflection point
(b) $f'(x) = 3, \forall\, x \in R$
(c) $\int_0^2 f(x)\, dx = 10$
(d) Area bounded by $f(x)$ with coordinate axes is (2/3)

45. Let $f : (0, \infty) \to (0, \infty)$ be a derivable function and $F(x)$ is the primitive of $f(x)$ such that $2(F(x) - f(x)) = f^2(x)$ for any real positive x

(a) f is strictly increasing (b) $\lim\limits_{x \to \infty} \dfrac{f(x)}{x} = 1$
(c) f is strictly decreasing (d) f is non-monotonic

Monotonicity, Maxima and Minima Exercise 3 : Statements I and II Type Questions

Directions (Q. Nos. 46 to 54) *For the following questions, choose the correct answers from the codes (a), (b), (c) and (d) defined as follows.*

(a) Statement I is true, Statement II is also true, Statement II is the correct explanation of Statement I
(b) Statement I is true, Statement II is also true, Statement II is not the correct explanation of Statement I
(c) Statement I is true, Statement II is false
(d) Statement I is false, Statement II is true

46. Statement I The equation $3x^2 + 4ax + b = 0$ has atleast one root in (0, 1), if $3 + 4a = 0$.

Statement II $f(x) = 3x^2 + 4x + b$ is continuous and differentiable in (0, 1).

47. Statement I For the function $f(x) = \begin{cases} 15 - x, & x < 2 \\ 2x - 3, & x \ge 2 \end{cases}$ $x = 2$ has neither a maximum nor a minimum point.

Statement II $f'(x)$ does not exist at $x = 2$.

48. Statement I $\phi(x) = \int_0^x (3\sin t + 4\cos t)\, dt$, $x \in \left[\dfrac{\pi}{6}, \dfrac{\pi}{3}\right]$ $\phi(x)$ attains its maximum value at $x = \dfrac{\pi}{3}$.

Statement II $\phi(x) = \int_0^x (3\sin t + 4\cos t)\, dt$, $\phi(x)$ is increasing function in $\left[\dfrac{\pi}{6}, \dfrac{\pi}{3}\right]$.

49. Let $f(x)$ be a twice differentiable function in $[a, b]$, given that $f(x)$ and $f''(x)$ has same sign in $[a, b]$.

Statement I $f'(x)=0$ has at the most one real root in $[a, b]$.

Statement II An increasing function can intersect the X-axis at the most once.

50. Let $u=\sqrt{c+1}-\sqrt{c}$ and $v=\sqrt{c}-\sqrt{c-1}$, $c>1$ and let $f(x)=\ln(1+x), \forall x \in(-1, \infty)$.

Statement I $f(u)>f(v), \forall c>1$ because

Statement II $f(x)$ is increasing function, hence for $u>v, f(u)>f(v)$.

51. Let $f(0)=0, f\left(\frac{\pi}{2}\right)=1, f\left(\frac{3\pi}{2}\right)=-1$ be a continuous and twice differentiable function.

Statement I $|f''(x)| \leq 1$ for atleast one $x \in\left(0, \frac{3\pi}{2}\right)$ because

Statement II According to Rolle's theorem, if $y=g(x)$ is continuous and differentiable, $\forall x \in[a, b]$ and $g(a)=g(b)$, then there exists atleast one c such that $g'(c)=0$.

52. **Statement I** For any ΔABC

$$\sin\left(\frac{A+B+C}{3}\right) \geq \frac{\sin A+\sin B+\sin C}{3}$$

Statement II $y=\sin x$ is concave downward for $x \in(0, \pi]$.

53. **Statement I** If $f(x)=[x](\sin x+\cos x-1)$ (where $[\cdot]$ denotes the greatest integer function), then $f'(x)=[x](\cos x-\sin x)$ for any $x \in$ integer.

Statement II $f'(x)$ does not exist for any $x \in$ integer.

54. $f(x)$ is a polynomial of degree 3 passing through origin having local extrema at $x=\pm a$.

Statement I Ratio of areas in which $f(x)$ cuts the circle $x^2+y^2=36$ is 1 : 1.

Statement II Both $y=f(x)$ and the circle are symmetric about origin.

Monotonicity, Maxima and Minima Exercise 4 : Passage Based Questions

Passage I

(Q. Nos. 55 to 57)

Let $f(x)=\frac{1}{1+x^2}$, let m be the slope, a be the x-intercept and b be the y-intercept of a tangent to $y=f(x)$.

55. Abscissa of the point of contact of the tangent for which m is greatest, is

(a) $\frac{1}{\sqrt{3}}$ (b) 1

(c) -1 (d) $-\frac{1}{\sqrt{3}}$

56. Value of b for the tangent drawn to the curve $y=f(x)$ whose slope is greatest, is

(a) $\frac{9}{8}$ (b) $\frac{3}{8}$

(c) $\frac{1}{8}$ (d) $\frac{5}{8}$

57. Value of a for the tangent drawn to the curve $y=f(x)$ whose slope is greatest, is

(a) $-\sqrt{3}$

(b) 1

(c) -1

(d) $\sqrt{3}$

Passage II

(Q. Nos. 58 to 60)

Consider the function $f(x)=\max\{x^2,(1-x)^2, 2x(1-x)\}, x \in[0,1]$

58. The interval in which $f(x)$ is increasing, is

(a) $\left(\frac{1}{3}, \frac{2}{3}\right)$ (b) $\left(\frac{1}{3}, \frac{1}{2}\right)$

(c) $\left(\frac{1}{3}, \frac{1}{2}\right) \cup\left(\frac{1}{2}, \frac{2}{3}\right)$ (d) $\left(\frac{1}{3}, \frac{1}{2}\right) \cup\left(\frac{2}{3}, 1\right)$

59. Let RMVT is applicable for $f(x)$ on (a, b), then $a+b+c$ is (where c is the point such that $f'(c)=0$)

(a) $\frac{2}{3}$ (b) $\frac{1}{3}$

(c) $\frac{1}{2}$ (d) $\frac{3}{2}$

60. The interval in which $f(x)$ is decreasing, is

(a) $\left(\frac{1}{3}, \frac{2}{3}\right)$ (b) $\left(\frac{1}{3}, \frac{1}{2}\right)$

(c) $\left(0, \frac{1}{3}\right) \cup\left(\frac{1}{2}, \frac{2}{3}\right)$ (d) $\left(\frac{1}{3}, \frac{1}{2}\right) \cup\left(\frac{2}{3}, 1\right)$

Passage III

(Q. Nos. 61 to 63)

$f(x), g(x), h(x)$ all are continuous and differentiable functions in $[a, b]$ also $a < c < b$ and $f(a) = g(a) = h(a)$. Point of intersection of the tangent at $x = c$ with chord joining $x = a$ and $x = b$ is on the left of c in $y = f(x)$ and on the right in $y = h(x)$. And tangent at $x = c$ is parallel to the chord in case of $y = g(x)$.
Now, answer the following questions.

61. If $f'(x) > g'(x) > h'(x)$, then

(a) $f(b) < g(b) < h(b)$ (b) $f(b) > g(b) > h(b)$
(c) $f(b) \le g(b) \le h(b)$ (d) $f(b) \ge g(b) \ge h(b)$

62. If $f(b) = g(b) = h(b)$, then

(a) $f'(c) = g'(c) = h'(c)$ (b) $f'(c) > g'(c) > h'(c)$
(c) $f'(c) < g'(c) < h'(c)$ (d) None of these

63. If $c = \dfrac{a+b}{2}$ for each b, then

(a) $g(x) = Ax^2 + Bx + C$
(b) $g(x) = \log x$
(c) $g(x) = \sin x$
(d) $g(x) = e^x$

Passage IV

(Q. Nos. 64 to 66)

In the non-decreasing sequence of odd integers $(a_1, a_2, a_3, \ldots) = \{1, 3, 3, 3, 5, 5, 5, 5, 5, \ldots\}$ each positive odd integer k appears k times. It is a fact that there are integers b, c and d such that for all positive integer n, $a_n = b[\sqrt{n+c}] + d$ (where [.] denotes greatest integer function).

64. The possible value of $b + c + d$ is

(a) 0 (b) 1 (c) 2 (d) 4

65. The possible value of $\dfrac{b-2d}{8}$ is

(a) 0 (b) 1
(c) 2 (d) 4

66. The possible value of $\dfrac{c+d}{2b}$ is

(a) 0 (b) 1
(c) 2 (d) 4

Passage V

(Q. Nos. 67 to 69)

Let $g(x) = a_0 + a_1x + a_2x^2 + a_3x^3$ and $f(x) = \sqrt{g(x)}$, $f(x)$ have its non-zero local minimum and maximum values at -3 and 3, respectively. If $a_3 \in$ the domain of the function

$$h(x) = \sin^{-1}\left(\frac{1+x^2}{2x}\right)$$

67. The value of $a_1 + a_2$ is equal to

(a) 30 (b) -30
(c) 27 (d) -27

68. The value of a_0 is

(a) equal to 50
(b) greater than 54
(c) less than 54
(d) less than 50

69. $f(-10)$ is defined for

(a) $a_0 > 830$
(b) $a_0 < 830$
(c) $a_0 = 830$
(d) None of the above

Passage VI

(Q. Nos. 70 to 74)

If $f : D \to R$, $f(x) = \dfrac{x^2 + bx + c}{x^2 + b_1x + c_1}$, where α, β are the roots of the equation $x^2 + bx + c = 0$ and α_1, β_1 are the roots of $x^2 + b_1x + c_1 = 0$. Now, answer the following questions for $f(x)$. A combination of graphical and analytical approach may be helpful in solving these problems. (If α_1 and β_1 are real, then $f(x)$ has vertical asymptote at $x = \alpha_1, \beta_1$). Then,

70. If $\alpha_1 < \alpha < \beta_1 < \beta$, then

(a) $f(x)$ is increasing in (α_1, β_1)
(b) $f(x)$ is decreasing in (α, β)
(c) $f(x)$ is decreasing in (β_1, β)
(d) $f(x)$ is decreasing in $(-\infty, \alpha)$

71. If $\alpha_1 < \beta_1 < \alpha < \beta$, then

(a) $f(x)$ has a maxima in $[\alpha_1, \beta_1]$ and a minima is $[\alpha, \beta]$
(b) $f(x)$ has a minima in (α_1, β_1) and a maxima in (α, β)
(c) $f'(x) > 0$ where ever defined
(d) $f'(x) < 0$ where ever defined

72. If the equations $x^2 + bx + c = 0$ and $x^2 + b_1x + c_1 = 0$ do not have real roots, then

(a) $f'(x) = 0$ has real and distinct roots
(b) $f'(x) = 0$ has real and equal roots
(c) $f'(x) = 0$ has imaginary roots
(d) nothing can be said

73. In the above problem, $\lim_{x\to\infty} [f(x)] \cdot \lim_{x\to\infty} [f(x)]$ (where [.] denotes the greatest integer function) is equal to

(a) 1
(b) 0
(c) -1
(d) does not exist

74. In the last problem, if $b > b_1$, then

(a) x-coordinate of point of minima is greater than the x-coordinate of point of maxima
(b) x-coordinate of point of minima is less than x-coordinate of point of maxima
(c) it also depends upon c and c_1
(d) nothing can be said

Passage VII

(Q. Nos. 75 to 77)

Consider the function $f(x) = \dfrac{x^2}{x^2 - 1}$

75. The interval in which f is increasing is

(a) $(-1, 1)$
(b) $(-\infty, -1) \cup (-1, 0)$
(c) $(-\infty, \infty) - \{-1, 1\}$
(d) $(0, 1) \cup (1, \infty)$

76. If f is defined from $R - \{-1, 1\} \rightarrow R$, then f is

(a) injective but not surjective
(b) surjective but not inective
(c) injective as well as surjective
(d) neither injective nor surjective

77. f has

(a) local maxima but not local minima
(b) local minima but not local maxima
(c) both local maxima and local minima
(d) neither local maxima nor local minima

Passage VIII

(Q. Nos. 78 to 80)

Suppose you do not know the function $f(x)$, however some information about $f(x)$ is listed below.
Read the following carefully before attempting the questions

(i) *$f(x)$ is continuous and defined for all real numbers*
(ii) *$f'(-5) = 0$, $f'(2)$ is not defined and $f'(4) = 0$*
(iii) *$(-5, 12)$ is a point which lies on the graph of $f(x)$*
(iv) *$f''(2)$ is undefined, but $f''(x)$ is negative everywhere else*
(v) *The signs of $f'(x)$ is given below*

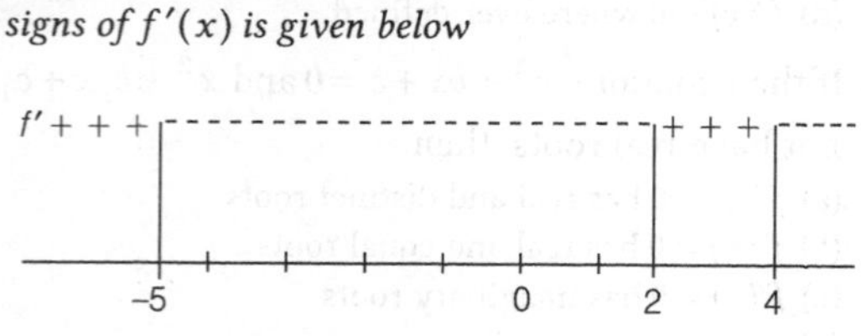

78. On the possible graph of $y = f(x)$, we have

(a) $x = -5$ is a point of relative minima
(b) $x = 2$ is a point of relative maxima
(c) $x = 4$ is a point of relative minima
(d) graph of $y = f(x)$ must have a geometrically sharp corner

79. From the possible graph of $y = f(x)$, we can say that

(a) there is exactly one point of inflection on the curve
(b) $f(x)$ increases on $-5 < x < 2$ and $x > 4$ and decreases on $-\infty < x < -5$ and $2 < x < 4$
(c) the curve is always concave down
(d) curve always concave up

80. Possible graph of $y = f(x)$ is

(a)

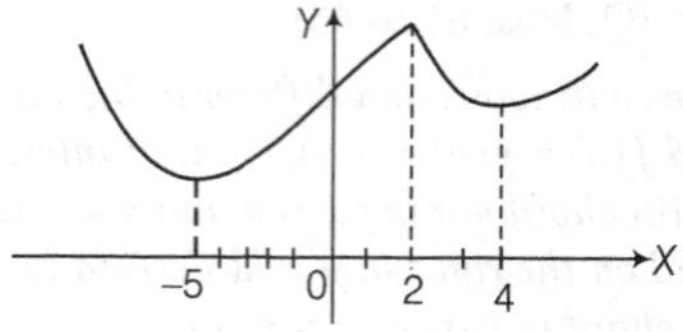

(b)

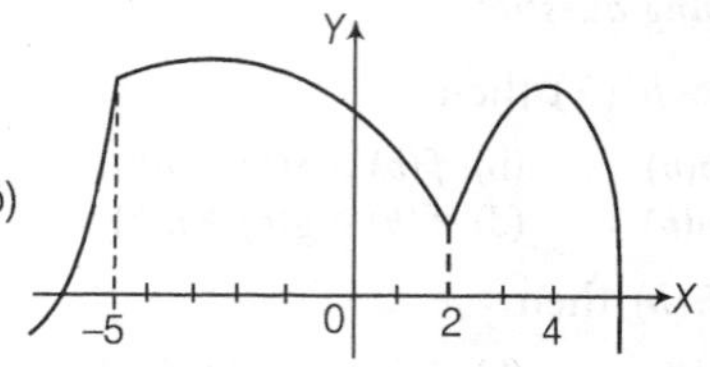

(c)

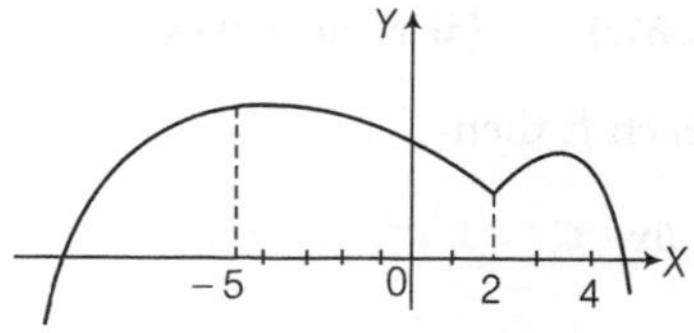

(d)

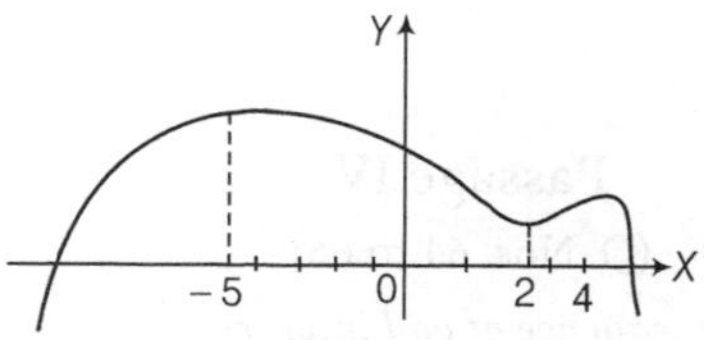

Passage IX

(Q. Nos. 81 to 83)

Let $f(x) = e^{(p+1)x} - e^x$ *for real number* $p > 0$, *then*

81. The value of $x = s_p$ for which $f(x)$ is minimum, is

(a) $\dfrac{-\log_e(p+1)}{p}$ (b) $-\log_e(p+1)$
(c) $-\log_e p$ (d) $\log_e\left(\dfrac{p+1}{p}\right)$

82. Let $g(t) = \int_t^{t+1} f(x) e^{t-x} dx$. The value of $t = t_p$, for which $g(t)$ is minimum, is

(a) $-\log_e \dfrac{(e^p - 1)}{p}$ (b) $-\dfrac{1}{p}\log_e\left(\dfrac{e^p - 1}{p}\right)$
(c) $-\dfrac{1}{p}\log_e\left(\dfrac{(p+1)(e^p - 1)}{p}\right)$ (d) $-\log_e((p+1)(e^p - 1))$

83. Use the fact that $1 + \dfrac{p}{2} \le \dfrac{e^p - 1}{p} \le 1 + \dfrac{p}{2} + p^2$ $(0 < p \le 1)$, the value of $\lim\limits_{p \to 0^+} (s_p - t_p)$ is

(a) 0 (b) $\dfrac{1}{2}$ (c) 1 (d) non-existent

Passage X

(Q. Nos. 84 to 86)

Consider f, g and h be three real valued function defined on R.
Let $f(x) = \sin 3x + \cos x$, $g(x) = \cos 3x + \sin x$ *and* $h(x) = f^2(x) + g^2(x)$.

84. The length of a longest interval in which the function $g = h(x)$ is increasing, is

(a) $\pi/8$ (b) $\pi/4$ (c) $\pi/6$ (d) $\pi/2$

85. The general solution of the equation $h(x) = 4$, is

(a) $(4n+1)\pi/8$ (b) $(8n+1)\pi/8$
(c) $(2n+1)\pi/4$ (d) $(7n+1)\pi/4$

86. Number of point(s) where the graphs of the two function, $y = f(x)$ and $y = g(x)$ intersects in $[0, \pi]$, is

(a) 2 (b) 3
(c) 4 (d) 5

Passage XI

(Q. Nos. 87 to 89)

Consider f, g and h be three real valued functions defined on R.

Let $f(x) = \begin{cases} -1, & x < 0 \\ 0, & x = 0 \\ 1, & x > 0 \end{cases}$, $g(x) = x(1 - x^2)$ *and* $h(x)$ *be such that*

$h''(x) = 6x - 4$.

Also, $h(x)$ *has local minimum value* 5 *at* $x = 1$.

87. The equation of tangent at $m(2, 7)$ to the curve $y = h(x)$, is

(a) $5x + y = 17$ (b) $x + 5y = 37$
(c) $x - 5y + 33 = 0$ (d) $5x - y = 3$

88. The area bounded by $y = h(x)$, $y = g(f(x))$ between $x = 0$ and $x = 2$ equals

(a) $23/2$ (b) $20/3$ (c) $32/3$ (d) $40/3$

89. Range of function $\sin^{-1}\sqrt{(fog(x))}$ is

(a) $(0, \pi/2)$ (b) $\{0, \pi/2\}$
(c) $\{-\pi/2, 0, \pi/2\}$ (d) $\{\pi/2\}$

Passage XII

(Q. Nos. 90 to 92)

Consider f, g and h be three real valued differentiable functions defined on R.

Let $g(x) = x^3 + g''(1)x^2 + (3g'(1) - g''(1) - 1)x + 3g'(1)$
$f(x) = x\,g(x) - 12x + 1$
and $f(x) = (h(x))^2$, *where* $h(0) = 1$.

90. The function $y = f(x)$ has

(a) Exactly one local minima and no local maxima
(b) Exactly one local maxima and no local minima
(c) Exactly one local maxima and two local minima
(d) Exactly two local maxima and one local minima

91. Which of the following is/are true for the function $y = g(x)$?

(a) $g(x)$ monotonically decreases inEREDEDDD $\left(-\infty, 2 - \frac{1}{\sqrt{3}}\right) \cup \left(2 + \frac{1}{\sqrt{3}}, \infty\right)$
(b) $g(x)$ monotonically increases in $\left[2 - \frac{1}{\sqrt{3}}, 2 + \frac{1}{\sqrt{3}}\right)$
(c) There exists exactly one tangent to $y = g(x)$ which is parallel to the chord joining the points $(1, g(1))$ and $(3, g(3))$
(d) There exists exactly two distinct lagrange's mean value in $(0, 4)$ for the function $y = g(x)$

92. Which one of the following does not hold good for $y = h(x)$?

(a) Exactly one critical point
(b) No point of inflexion
(c) Exactly one real zero in $(0, 3)$
(d) Exactly one tangent parallel to X-axis

Monotonicity, Maxima and Minima Exercise 5 : Matching Type Questions

93. Match the following :

	Column I		Column II
(A)	The maximum value attained by $y = 10 - \lvert x - 10 \rvert, -1 \le x \le 3$ is	(p)	3
(B)	If $P(t^2, 2t)$, $t \in [0, 2]$ is an arbitrary point on parabola $y^2 = 4x$, Q is foot of perpendicular from focus S on the tangent at P, then maximum area of ΔPQS is	(q)	$\frac{1}{3}$
(C)	If $a + b = 1$, $a, b > 0$, then maximum value of $\sqrt{\left(1 + \frac{1}{a}\right)\left(1 + \frac{1}{b}\right)}$ is	(r)	5
(D)	For real values of x, the greatest and least value of expression $\frac{x+2}{2x^2 + 3x + 6}$ is	(s)	$-\frac{1}{13}$

94. Match the entries of the following two columns.

Column I	Column II
(A) The least value of the function $f(x) = 2\cdot 3^{3x} - 3^{2x}\cdot 4 + 2\cdot 3^x$ in $[-1, 1]$ is	(p) 5
(B) The minimum value of the polynomial $f(x) = (x-1)x(x+1)$ is	(q) -1
(C) The value of the polynWmial $\int_{-1}^{3}(\lvert x-2\rvert-[x])dx$ (where $[\cdot]$ denotes the greatest integer function) is	(r) 3
(D) If period of the function $f(x) = \sin 36x \tan 42x$ is p, then $\frac{18p}{\pi}$ equals	(s) 0

95. Match the following :

Column I	Column II
(A) $f(x) = \int_0^{2x^2}(t-2)(t+1)^3(t-3)^2\,dt$ is/has in $(-1, 1)$	(p) local maxima
(B) $f(x) = \begin{cases}\sin\left(\frac{\pi x}{4}\right), & x \le 2\\ 9-4x, & x > 2\end{cases}$ is/has in (0,2)	(q) local minima
(C) $f(x) = \{2x\}$ denotes fractional part of x is/has in (0, 1)	(r) continuous
(D) $f(x) = \begin{cases}\lvert\lvert x-2\rvert-2\rvert, & x < 2\\ [x], & x \ge 2\end{cases}$ (where $[\cdot]$ denotes the greatest integer function), then $f(x)$ is/has in $(-1, 4)$	(s) non-differentiable

96. Match the statements of Column I with the values of Column II.

Column I	Column II
(A) $f(x) = \cos \pi x + 10x + 3x^2 + x^3$, $-2 \le x \le 3$. The absolute minimum value of $f(x)$ is	(p) 3/4
(B) If $x \in [-1, 1]$, then the minimum value of $f(x) = x^2 + x + 1$ is	(q) 2
(C) Let $f(x) = \frac{4}{3}x^3 - 4x$, $0 \le x \le 2$. Then, the global minimum value of the function is	(r) -15
(D) Let $f(x) = 6 - 12x + 9x^2 - 2x^3$, $1 \le x \le 4$. Then, the absolute maximum value of $f(x)$ in the interval is	(s) $-8/3$

Monotonicity, Maxima and Minima Exercise 6 :
Single Integer Answer Type Questions

97. A particular substance is being cooled by a streem of cold air (temperature of the air is constant and is 5°C) where rate of cooling is directly proportional to square of difference of temperature of the substance and the air. If the substance is cooled from 40°C to 30°C in 15 min and temperature after 1 hour is T°C, then find the value of $[T]/2$, where $[.]$ represents the greatest integer function.

98. The minimum value of $\dfrac{\tan\left(x + \frac{\pi}{6}\right)}{\tan x}$ is ....

99. The figure shows a right triangle with its hypotenuse OB along the Y-axis and its vertex A on the parabola $y = x^2$. Let h represents the length of the hypotenuse which depends on the x-coordinate of the point A.
The value of $\lim_{x\to 0}(h)$ equals ....

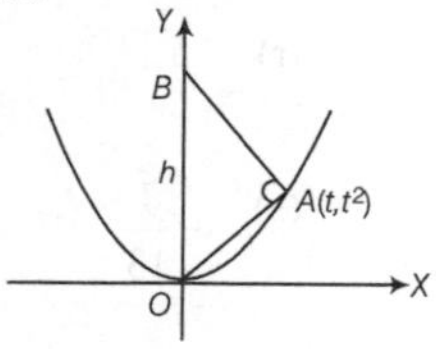

100. Number of positive integral values of a for which the curve $y = a^x$ intersects the line $y = x$ is ....

101. The least value of a for which the equation, $\frac{4}{\sin x} + \frac{1}{1-\sin x} = a$ has atleast one solution in the interval $(0, \pi/2)$ is ....

102. Let $f(x) = \begin{cases} x^{3/5} & \text{if } x \le 1\\ -(x-2)^3 & \text{if } x > 1\end{cases}$, then the number of critical points on the graph of the function are ....

103. The graph of $y = f''(x)$ for a function f is shown. Number of points of inflection for $y = f(x)$ is ....

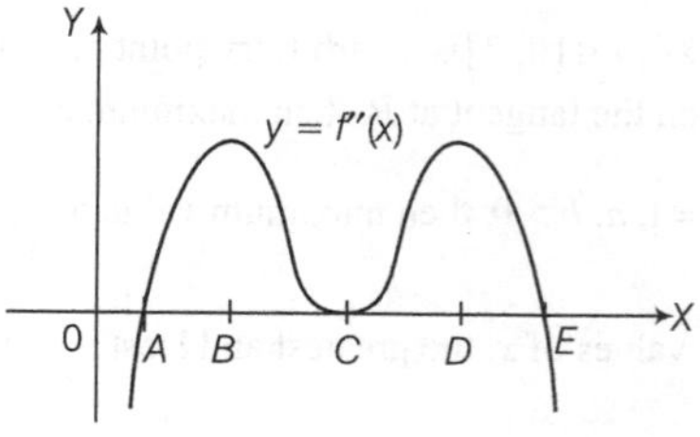

104. Number of critical points of the function, $f(x)=\frac{2}{3}\sqrt{x^3}-\frac{x}{2}+\int_1^x\left(\frac{1}{2}+\frac{1}{2}\cos 2t-\sqrt{t}\right)dt$ which lie in the interval $[-2\pi, 2\pi]$ is ....

105. Let $f(x)$ and $g(x)$ be two continuous functions defined from $R\to R$, such that $f(x_1)>f(x_2)$ and $g(x_1)<g(x_2)$, $\forall\ x_1>x_2$, then the least integral value of α for which $f(g(\alpha^2-2\alpha))>f(g(3\alpha-4))$ is ....

106. If the function $f(x)=\frac{t+3x-x^2}{x-4}$, where t is a parameter, has a minimum and a maximum, then the greatest value of t is ....

107. If $f(x)=(x-a)(x-b)$ for $a, b\in R$, then minimum number of roots of equation $\pi(f'(x))^2\cos(\pi f(x))+\sin(\pi f(x))\cdot f''(x)=0$ in (α,β), where $f(\alpha)=3=f(\beta)$ is .... (here, $\alpha<a<b<\beta$)

108. If absolute maximum value of $f(x)=\frac{1}{|x-4|+1}+\frac{1}{|x+8|+1}$ is $\frac{p}{q}$, (p, q are coprime) the $(p-q)$ is ....

Monotonicity, Maxima and Minima Exercise 7 : Questions Asked in Previous 10 Years' Exams

(i) JEE Advanced & IIT-JEE

109. The least value of $\alpha\in R$ for which $4\alpha x^2+\frac{1}{x}\geq 1$, for all $x>0$, is **[One Correct Option 2016 Adv.]**

(a) $\frac{1}{64}$ (b) $\frac{1}{32}$

(c) $\frac{1}{27}$ (d) $\frac{1}{25}$

110. The number of points in $(-\infty,\infty)$ for which $x^2-x\sin x-\cos x=0$, is **[One Correct Option, 2013 Adv.]**

(a) 6 (b) 4

(c) 2 (d) 0

111. Let $f:R\to(0,\infty)$ and $g:R\to R$ be twice differentiable functions such that f'' and g'' are continuous funcitons of R. Suppose $f'(2)=g(2)=0, f''(2)\neq 0$ and $g'(2)\neq 0$. If $\lim_{x\to 2}\frac{f(x)g(x)}{f'(x)g'(x)}=1$, then **[More than One Correct option, 2016 Adv.]**

(a) f has a local minimum at $x=2$

(b) f has a local maximum at $x=2$

(c) $f''(2)>f(2)$

(d) $f(x)-f''(x)=0$ for atleast one $x\in R$

112. If $f:(0,\infty)\to R$ be given by $f(x)=\int_{1/x}^{x}e^{-\left(t+\frac{1}{t}\right)}\frac{dt}{t}$. Then, **[More than One Correct Option, 2014 Adv.]**

(a) $f(x)$ is monotonically increasing on $[1,\infty)$

(b) $f(x)$ is monotonically decreasing on $[0,1)$

(c) $f(x)+f\left(\frac{1}{x}\right)=0, \forall x\in(0,\infty)$

(d) $f(2^x)$ is an odd function of x on R

113. The function $f(x)=2|x|+|x+2|-||x+2|-2|x||$ has a local minimum or a local maximum at x is equal to **[More than One Correct Option, 2013 Adv.]**

(a) -2 (b) $\frac{-2}{3}$

(c) 2 (d) 2/3

114. A rectangular sheet of fixed perimeter with sides having their lengths in the ratio 8 : 15 is converted into an open rectangular box by folding after removing squares of equal area from all four corners. If the total area of removed squares is 100, the resulting box has maximum volume. The lengths of the sides of the rectangular sheet are **[More than One Correct Option, 2013 Adv.]**

(a) 24 (b) 32 (c) 45 (d) 60

115. A vertical line passing through the point $(h, 0)$ intersects the ellipse $\frac{x^2}{4}+\frac{y^2}{3}=1$ at the points P and Q. If the tangents to the ellipse at P and Q meet at the point R. If $\Delta(h)=$ area of the ΔPQR, $\Delta_1=\max_{1/2\leq h\leq 1}\Delta(h)$ and $\Delta_2=\min_{1/2\leq h\leq 1}\Delta(h)$, then $\frac{8}{\sqrt{5}}\Delta_1-8\Delta_2$ is equal to **[Integer Type Question, 2013 Adv.]**

116. Let f, g and h be real-valued functions defined on the interval [0, 1] by $f(x)=e^{x^2}+e^{-x^2}$, $g(x)=xe^{x^2}+e^{-x^2}$ and $h(x)=x^2e^{x^2}+e^{-x^2}$. If a, b and c denote respectively, the absolute maximum of f, g and h on [0, 1], then **[One Correct Option, 2010]**

(a) $a=b$ and $c\neq b$ (b) $a=c$ and $a\neq b$

(c) $a\neq b$ and $c\neq b$ (d) $a=b=c$

117. The total number of local maxima and local minima of the function $f(x)=\begin{cases}(2+x)^3, & -3<x\le -1\\ x^{\frac{2}{3}}, & -1<x<2\end{cases}$ is **[One Correct Option, 2008]**

(a) 0 (b) 1 (c) 2 (d) 3

118. If the function $g:(-\infty,\infty)\to\left(-\frac{\pi}{2},\frac{\pi}{2}\right)$ is given by $g(u)=2\tan^{-1}(e^u)-\frac{\pi}{2}$. Then, g is **[One Correct Option, 2008]**

(a) even and is strictly increasing in $(0,\infty)$
(b) odd and is strictly decreasing in $(-\infty,\infty)$
(c) odd and is strictly increasing in $(-\infty,\infty)$
(d) neither even nor odd but is strictly increasing in $(-\infty,\infty)$

119. The second degree polynomial $f(x)$, satisfying $f(0)=0$, $f(1)=1$, $f'(x)>0\ \forall x\in(0,1)$ **[One Correct Option, 2005]**

(a) $f(x)=\phi$
(b) $f(x)=ax+(1-a)x^2$, $\forall\ a\in(0,\infty)$
(c) $f(x)=ax+(1-a)x^2$, $a\in(0,2)$
(d) No such polynomial

120. If $f(x)=x^3+bx^2+cx+d$ and $0<b^2<c$, then in $(-\infty,\infty)$ **[One Correct Option, 2004]**

(a) $f(x)$ is strictly increasing function
(b) $f(x)$ has a local maxima
(c) $f(x)$ is strictly decreasing function
(d) $f(x)$ is bounded

121. If $f(x)=x^2+2bx+2c^2$ and $g(x)=-x^2-2cx+b^2$, such that $\min f(x)>\max g(x)$, then the relation between b and c, is **[One Correct Option, 2003]**

(a) No real value of b and c (b) $0<c<b\sqrt{2}$
(c) $|c|<|b|\sqrt{2}$ (d) $|c|>|b|\sqrt{2}$

122. The length of a longest interval in which the function $3\sin x-4\sin^3 x$ is increasing, is **[One Correct Option, 2002]**

(a) $\frac{\pi}{3}$ (b) $\frac{\pi}{2}$ (c) $\frac{3\pi}{2}$ (d) π

123. The maximum value of $(\cos\alpha_1)\cdot(\cos\alpha_2)\cdot\ldots\cdot(\cos\alpha_n)$, under the restrictions $0\le\alpha_1,\alpha_2,\ldots,\alpha_n\le\frac{\pi}{2}$ and $(\cot\alpha_1)\cdot(\cot\alpha_2)\cdot\ldots\cdot(\cot\alpha_n)=1$ is **[One Correct Option, 2001]**

(a) $\frac{1}{2^{n/2}}$ (b) $\frac{1}{2^n}$ (c) $\frac{1}{2n}$ (d) 1

124. If $f(x)=\begin{cases}e^x, & 0\le x\le 1\\ 2-e^{x-1}, & 1<x\le 2\\ x-e, & 2<x\le 3\end{cases}$ and $g(x)=\int_0^x f(t)\,dt$, $x\in[1,3]$, then **[More than One Correct Option, 2006]**

(a) $g(x)$ has local maxima at $x=1+\log_e 2$ and local minima at $x=e$
(b) $f(x)$ has local maxima at $x=1$ and local minima at $x=2$
(c) $g(x)$ has no local minima
(d) $f(x)$ has no local maxima

125. If $f(x)$ is a cubic polynomial which has local maximum at $x=-1$. If $f(2)=18$, $f(1)=-1$ and $f'(x)$ has local minimum at $x=0$, then **[More than One Correct Option, 2006]**

(a) the distance between $(-1,2)$ and $(a,f(a))$, where $x=a$ is the point of local minima, is $2\sqrt{5}$
(b) $f(x)$ is increasing for $x\in[1,2\sqrt{5}]$
(c) $f(x)$ has local minima at $x=1$
(d) the value of $f(0)=5$

126. Consider the function $f:(-\infty,\infty)\to(-\infty,\infty)$ defined by $f(x)=\frac{x^2-ax+1}{x^2+ax+1}$; $0<a<2$. Which of the following is true? **[Passage Based Question, 2008]**

(a) $(2+a)^2 f''(1)+(2-a)^2 f''(-1)=0$
(b) $(2-a)^2 f''(1)-(2+a)^2 f''(-1)=0$
(c) $f'(1)\,f'(-1)=(2-a)^2$
(d) $f'(1)\,f'(-1)=-(2+a)^2$

127. Which of the following is true? **[Passage Based Question, 2008]**

(a) $f(x)$ is decreasing on $(-1,1)$ and has a local minimum at $x=1$
(b) $f(x)$ is increasing on $(-1,1)$ and has a local maximum at $x=1$
(c) $f(x)$ is increasing on $(-1,1)$ but has neither a local maximum nor a local minimum at $x=1$
(d) $f(x)$ is decreasing on $(-1,1)$ but has neither a local maximum nor a local minimum at $x=1$

128. Let $g(x)=\int_0^{e^x}\frac{f'(t)}{1+t^2}\,dt$. Which of the following is true? **[Passage Based Question, 2008]**

(a) $g'(x)$ is positive on $(-\infty,0)$ and negative on $(0,\infty)$
(b) $g'(x)$ is negative on $(-\infty,0)$ and positive on $(0,\infty)$
(c) $g'(x)$ changes sign on both $(-\infty,0)$ and $(0,\infty)$
(d) $g'(x)$ does not change sign $(-\infty,\infty)$

129. For the circle $x^2+y^2=r^2$, find the value of r for which the area enclosed by the tangents drawn from the point $P(6,8)$ to the circle and the chord of contact is maximum. **[Subjective Type Question, 2003]**

130. Find a point on the curve $x^2+2y^2=6$ whose distance from the line $x+y=7$, is minimum. **[Subjective Type Question, 2003]**

131. Let $f:R\to R$ be defined as $f(x)=|x|+|x^2-1|$. Total number of points at which f attains either a local maximum or a local minimum is **[Integer Type Question, 2012]**

132. Let $p(x)$ be a real polynomial of least degree which has a local maximum at $x=1$ and a local minimum at $x=3$. If $p(1)=6$ and $p(3)=2$, then $p'(0)$ is equal to **[Integer Type Question, 2012]**

133. Let f be a function defined on R (the set of all real numbers) such that $f'(x) = 2010(x-2009)(x-2010)^2 (x-2011)^3 (x-2012)^4, \forall x \in R$. If g is a function defined on R with values in the interval $(0, \infty)$ such that $f(x) = \ln(g(x)), \forall x \in R$, then the number of points in R at which g has a local maximum is ....

[Integer Type Question, 2010]

134. The maximum value of the expression $\dfrac{1}{\sin^2 \theta + 3\sin\theta \cos\theta + 5\cos^2 \theta}$ is .... **[Integer Type Question, 2010]**

135. The maximum value of the function $f(x) = 2x^3 - 15x^2 + 36x - 48$ on the set $A = \{x \mid x^2 + 20 \le 9x\}$ is ... . **[Integer Type Question, 2009]**

(ii) JEE Main & AIEEE

136. A wire of length 2 units is cut into two parts which are bent respectively to form a square of side = x units and a circle of radius = r units. If the sum of the areas of the square and the circle so formed is minimum, then

[2016 JEE Main]

(a) $2x = (\pi + 4)r$ (b) $(4 - \pi)x = \pi r$
(c) $x = 2r$ (d) $2x = r$

137. If $x = -1$ and $x = 2$ are extreme points of $f(x) = \alpha \log|x| + \beta x^2 + x$, then **[2014 JEE Main]**

(a) $\alpha = -6, \beta = \frac{1}{2}$ (b) $\alpha = -6, \beta = -\frac{1}{2}$
(c) $\alpha = 2, \beta = -\frac{1}{2}$ (d) $\alpha = 2, \beta = \frac{1}{2}$

138. Let $a, b \in R$ be such that the function f given by $f(x) = \log|x| + bx^2 + ax, x \ne 0$ has extreme values at $x = -1$ and $x = 2$.

Statement I f has local maximum at $x = -1$ and at $x = 2$.

Statement II $a = \frac{1}{2}$ and $b = \frac{-1}{4}$ **[2012 AIEEE]**

(a) Statement I is false, Statement II is true
(b) Statement I is true, Statement II is true; Statement II is a correct explanation of Statement I
(c) Statement I is true, Statement II is true; Statement II is not a correct explanation of Statement I
(d) Statement I is true, Statement II is false

Answers

Exercise for Session 1

1. (d) 2. (a) 3. (c) 4. (a) 5. (b) 6. (c) 7. (b)

Exercise for Session 2

1. $-5, 0, \frac{3}{5}$ 2. $0, \frac{1}{5}$ 3. No critical points 4. (c)

Exercise for Session 3

4. (a) 5. (b)

Exercise for Session 4

1. (b) 2. (d) 3. (a) 4. (a) 5. (a)
6. (c, d) 7. (a,c) 8. (d) 9. (a) 10. (c)
11. (c) 12. (d) 13. (b) 14. (d) 15. (a)

Exercise for Session 5

1. (d) 2. (a) 3. (b) 4. (a, c) 5. (c) 6. (d)
7. $-2 < a < 2$

Chapter Exercises

1. (c) 2. (c) 3. (a) 4. (a) 5. (c)
6. (b) 7. (a) 8. (d) 9. (d)
10. (a) 11. (c) 12. (a) 13. (d) 14. (b)
15. (a) 16. (c) 17. (a) 18. (c) 19. (c)
20. (d) 21. (b) 22. (c) 23. (c) 24. (d)
25. (d) 26. (d) 27. (b) 28. (c) 29. (b,c)
30. (a,d) 31. (c,d) 32. (a,b,c,d)
33. (a,b,c,d) 34. (a,b,c) 35. (a,b,c)
36. (a,b,c) 37. (c,d) 38. (b,d)
39. (b,d) 40. (a,b,c) 41. (a,b,d)
42. (a,b) 43. (b,c,d) 44. (b,c,d)
45. (a,b) 46. (d) 47. (d) 48. (a) 49. (a)
50. (d) 51. (a) 52. (b) 53. (b) 54. (a)
55. (d) 56. (a) 57. (a) 58. (d) 59. (d)
60. (c) 61. (b) 62. (c) 63. (a) 64. (c)
65. (a) 66. (a) 67. (c) 68. (b) 69. (a)
70. (a) 71. (a) 72. (a) 73. (b) 74. (b)
75. (b) 76. (d) 77. (a) 78. (d) 79. (c)
80. (c) 81. (a) 82. (c) 83. (b) 84. (b)
85. (a) 86. (c) 87. (d) 88. (c) 89. (b)
90. (c) 91. (d) 92. (c)
93. (A) $\to$ (p), (B) $\to$ (r), (C) $\to$ (p), (D) $\to$ (q, s)
94. (A) $\to$ (s), (B) $\to$ (q), (C) $\to$ (r), (D) $\to$ (r)
95. (A) $\to$ (p, r), (B) $\to$ (p, r, s), (C) $\to$ (p, s), (D) $\to$ (q, s)
96. (A) $\to$ (r), (B) $\to$ (p), (C) $\to$ (s), (D) $\to$ (q)
97. (9) 98. (3) 99. (1) 100. (1) 101. (9)
102. (3) 103. (2) 104. (4) 105. (2) 106. (3)
107. (4) 108. (1) 109. (c) 110. (c) 111. (a,d) 112. (a, c, d)
113. (a,b) 114. (a,c) 115. (9) 116. (d) 117. (a) 118. (c)
119. (c) 120. (a) 121. (d) 122. (a) 123. (a) 124. (a,b)
125. (b,c) 126. (a) 127. (d) 128. (b) 129. (5) 130. (2,1)
131. (5) 132. (9) 133. (1) 134. (2) 135. (7) 136. (c)
137. (c) 138. (c)

Solutions

1. Since, f is non-decreasing and g is non-increasing, so h is a non-increasing function. Now, $h(1) = 1$

$\Rightarrow h(x)$ is a constant function

$\Rightarrow \quad h(2) = 1$

2. Since, maximum or minimum distance between two curve always measure along common normal. AP is perpendicular on the tangent drawn to the curve.

3. f is continuous at 0 and $f'(0-) > 0$ and $f'(0+) < 0$. Thus, f has a local maximum at 0.

4. Here, $\quad f(x) = \int_1^x (t-a)^{2n}(t-b)^{2m+1} dt, a \neq b$

$f'(x) = (x-a)^{2n}(x-b)^{2m+1}.$

Obviously, $f'(a-), f'(a+) > 0$

while $\quad f'(b-) < 0$ and $f'(b+) > 0$

Hence, $x = b$ is a point of local minima.

5. $f'(x) = x^{n-1}[(n+1)x + an]$.

If n is even then, $f'(0-) < 0$ and $f'(0+) > 0$

Hence, 0 is a point of minimum when n is even.

6. $$f'(x) = 4x - \frac{4}{x^3} = \frac{4}{x^3}(x^4 - 1) = \frac{4}{x^3}(x-1)(x+1)(x^2+1)$$

$\Rightarrow f'(x) = 0$ at $x = 1$ and -1

$f''(x) = 4 + \frac{12}{x^4} > 0$ for $x = \pm 1$

$\Rightarrow \quad x = 1$ and -1 are points of local minima.

7. Since, $y_{\min} = \frac{1}{e^{1/e}} > \frac{1}{e}$. So, graph always lie above the line $y = x$.

Hence, $0 < x < 1$ is correct answer.

8. $y^2 - y + a = \left(y - \frac{1}{2}\right)^2 + a - \frac{1}{4}$

Since, $-\sqrt{2} \leq \sin x + \cos x \leq \sqrt{2}$, given equation will have no real value of x for any y, if $a - \frac{1}{4} > \sqrt{2}$

i.e. $a \in \left(\sqrt{2} + \frac{1}{4}, \infty\right)$

$\Rightarrow a \in (\sqrt{3}, \infty) \qquad \left[\text{as } \sqrt{2} + \frac{1}{4} < \sqrt{3}\right]$

9. $f(x) = 1 - \frac{2e^{-x^2}}{e^{x^2} + e^{-x^2}} = 1 - \frac{2}{e^{2x^2} + 1}$

As $x \to +\infty$, $f(x) \to 1$

As $x \to -\infty$, $f(x) \to 1$

$\Rightarrow f(x)$ is increasing as well as decreasing in some intervals. Since, the range of $f(x)$ is $[0, 1)$ which does not coincide with the co-domain R and hence f is not an onto function.

10. Let $f(x) = ax^2 + bx + c > 0, \forall\, x \in R$

$\Rightarrow \quad b^2 - 4ac < 0$ and $a > 0 \qquad$ ... (i)

Now, $\quad g(x) = f(x) - f'(x) + f''(x)$

$= (ax^2 + bx + c) - (2ax + b) + 2a$

$= ax^2 + (b - 2a)x + (2a - b + c)$

$\Rightarrow \quad \Delta = (b-2a)^2 - 4a(2a - b + c)$

$= (b^2 - 4ac) - 4a^2 < 0 \qquad$ [using Eq. (i)]

$\Rightarrow \quad g(x) > 0, \forall\, x \in R$

11. We have, $f(x) = \min\{1, \cos x, 1 - \sin x\}$

$$f(x) = \begin{cases} \cos x, & -\pi \leq x \leq 0 \\ 1 - \sin x, & 0 < x \leq \frac{\pi}{2} \\ \cos x, & \frac{\pi}{2} < x \leq \pi \end{cases} \Rightarrow f'(x) = \begin{cases} -\sin x, & -\pi \leq x \leq 0 \\ -\cos x, & 0 < x \leq \frac{\pi}{2} \\ -\sin x, & \frac{\pi}{2} < x \leq \pi \end{cases}$$

Therefore, $f'(0) = 0$

Hence, $f(x)$ has local maximum at 0 and $f(x)$ is not differentiable at $x = 0$ and $\frac{\pi}{2}$

Aliter

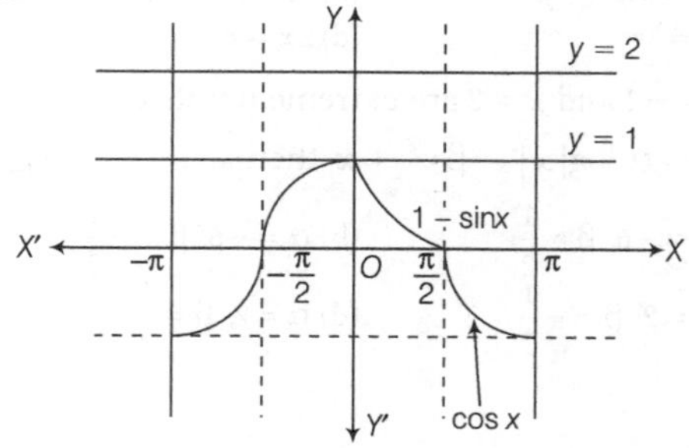

From the graph it is clear that $f(x)$ is not differentiable at $x = 0, \frac{\pi}{2}$ and $f(x)$ has occurs local maximum at $x = 0$.

12. $f(x)$ will have maxima only, if $a^2 + 1 \geq 2 \Rightarrow |a| \geq 1$

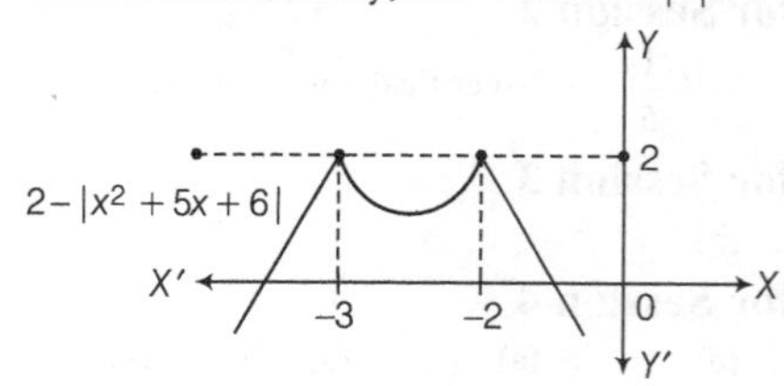

13. $ax^n = -x^2 - bx - c$. For $a = -1$ and $b = c = 0$ and $n = 2$, it will have infinite solutions.

14. $A = \text{Area} = \sin x(\pi - 2x)$ (as $l = \pi - 2x$, $b = \sin x$)

$\frac{dA}{dx} = (\pi - 2x)\cos x - 2\sin x = 0 \quad \Rightarrow \tan x = \frac{\pi}{2} - x$

Let $f(x) = \tan x + x - \frac{\pi}{2}$

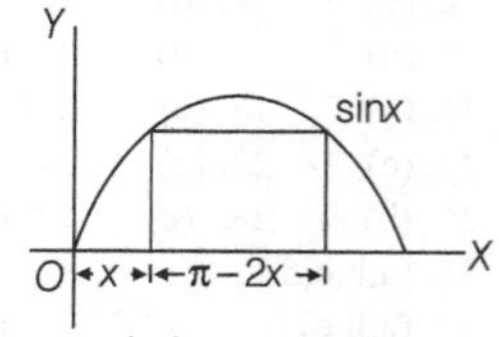

$f\left(\frac{\pi}{6}\right)$ is negative, $f\left(\frac{\pi}{4}\right)$ is positive.

So, one root lies between $\left(\frac{\pi}{6}, \frac{\pi}{4}\right)$.

15. $f(x)+1=k(x+1)$ always passes through $(-1,-1)$. Clearly, its maximum slope can go upto ∞. For minimum slope this line should touch $y=\ln x$.

$\frac{dy}{dx}=\frac{1}{x}=k$. So, $\left(\frac{1}{k},-\ln k\right)$ is point of tangency.

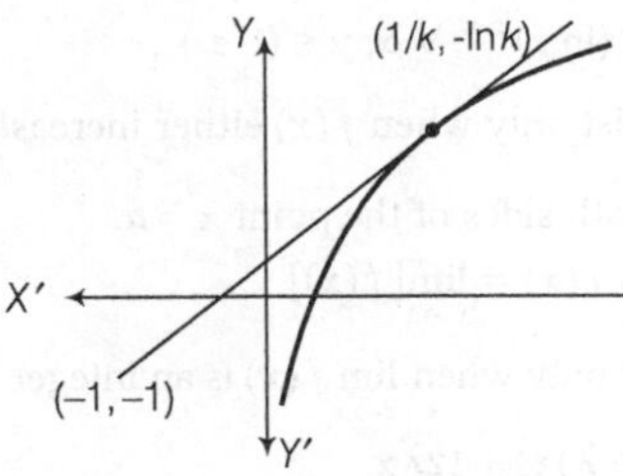

Now, $\frac{-\ln k+1}{\frac{1}{k}+1}=k \Rightarrow -\ln k+1=k+1$

$\Rightarrow -\ln k=k$

Let $f(x)=k+\ln k$

$f\left(\frac{1}{e}\right)=\frac{1}{e}-1$ [negative]

$f\left(\frac{1}{\sqrt{e}}\right)=\frac{1}{\sqrt{e}}-\frac{1}{2}$ [positive]

So, one root must lie between $\frac{1}{e}$ and $\frac{1}{\sqrt{e}}$.

16. $f(x)=f'(x)\times f'''(x)$ is satisfied only by the polynomial of degree 4. Since, $f(x)=0$ satisfies $x=1, 2, 3$ only. It is clear that one of the roots is repeated twice.

$\Rightarrow f'(1)f'(2)f'(3)=0$

17. $\frac{d^2y}{dx^2}=k\log x \Rightarrow \frac{dy}{dx}=k(x\log x-x)+A$

$=kx(\log x-1)+A$

$=k\left[(\log x-1)\frac{x^2}{2}-\frac{x^2}{4}\right]+Ax+B$

$=k\frac{x^2}{4}(2\log x-3)+Ax+B$

$\Rightarrow y=c_1(2\log x-3)x^2+c_2x+c_3$

18. $f'(x)=3\sec^2\left[\left(\tan x-\frac{1}{3}\right)^2+\frac{11}{9}\right]>0$ for all x in its domain.

19. $\lim_{x\to k^-} f(x)=\lim_{h\to 0} 3+h=3 \Rightarrow f(k)=3$

and $\lim_{x\to k^+} f(x)=\lim_{h\to 0} a^2-2+\frac{\sin h}{h}=a^2-2+1$

Since, $f(x)$ has minimum at $x=k$

$f(k^-)>f(k)$

and $f(k^+)>f(k) \Rightarrow \alpha^2-2+1>3$

$|a|>2$

20. Let $f(x)=mx+b$

$\because f(1)\le f(2) \Rightarrow m\ge 0$, similarly

$f(3)\ge f(4) \Rightarrow m\le 0 \Rightarrow m=0$

$\therefore f(0)=f(5)=5$

21. Let $P(x)$ is of degree n, then

$\Rightarrow 2n=n+2$ [as $P(x^2)=x^2P(x)$]

$\Rightarrow n=2$

$P(x)=ax^2+bx+c$ form

$P(0)=-2$

$\Rightarrow c=-2$ ...(i)

$P(1)=0$

$\Rightarrow a+b+c=0$ ...(ii)

$P'\left(\frac{3}{2}\right)=0 \Rightarrow 3a+b=0$

$\Rightarrow \frac{a}{b}=\frac{-1}{3} \Rightarrow P(x)=-x^2+3x-2$

Hence, maximum $P(x)=-\frac{1}{4}$

22. In [0, 1], $[x^2-x+1]=0$

$\ln\left[1,\frac{1+\sqrt{5}}{2}\right]$, $[x^2-x+1]=1 \Rightarrow x^2=-4(y-a)$

Put $y=0$, $x^2=4(a-1) \Rightarrow 4a>1$

Put $y=1$, $x^2=4(a-1) \Rightarrow a<1$

$\Rightarrow \frac{1}{4}<a<1.$

23. Clearly, $f'(a)=0$, $f'(c)=0$, $f'(e)=0$, $x=b$ and $x=d$ two points of inflexion $f'''(x)>0$, $d<x<e$.
$x=e$ is point of local minima.

24. $g(x)=f(x^2-2x-1)+f(5-x^2+2x)$

$=2x^4-8x^3-4x^2+24x+18$

$g'(x)=8x^3-24x^2-8x+24$

$g'(x)=0 \Rightarrow x=-1, 1, 3$

We observe that $g(x)\ge \min\{g(-1), g(1), g(3)\}=0$

$\therefore g(x)\ge 0, \forall x\in R$

25. Using the fact that every set of natural numbers have the smallest element.

$\therefore f(f(1))$ is the smallest element in $\{f(f(1)), f(2), f(f(2)), \ldots\}$

Same argument implies that $f(1)=1$

Repeating the argument for $f\{n\ge 2\}\to\{n\ge 2\}$, we get $f(2)=2$

$\therefore$ Clearly, $f(x)=x$

26. $\frac{1}{2}=|2ax-3|+|ax+1|+|5-ax|\ge|2ax-3-(ax+1)+5-ax|=1$

which is impossible.

27. Given, $\int_0^x 2xf^2(t)dt=\left(\int_0^x 2f(x-t)dt\right)^2$

$\Rightarrow x\cdot\int_0^x f^2(t)dt=2\left(\int_0^x f(t)dt\right)^2$, using $\int_0^a f(x)dx$

$=\int_0^a f(a-x)dx$

Differentiating, we get

$x\cdot f^2(x)+\int_0^x f^2(t)dt=4\int_0^x f(t)dt\cdot f(x)$

[using Leibnitz rule]

$\Rightarrow x^2f^2(x)+x\cdot\int_0^x f^2(t)dt=4x\cdot f(x)\cdot\int_0^x f(t)dt$

$$\Rightarrow \quad x^2 f^2(x) + 2\left(\int_0^x f(t)\,dt\right)^2 = 4x f(x)\cdot \int_0^x f(t)\,dt$$

$$\Rightarrow \quad 2\left(\int_0^x f(t)\,dt\right)^2 - 4x\cdot f(x)\cdot \int_0^x f(t)\,dt + x^2 f^2(x) = 0$$

$$\int_0^x f(t)\,dt = \frac{4x\, f(x) \pm \sqrt{16x^2 f^2(x) - 8x^2 f^2(x)}}{4} \qquad \left(\because ax^2 + bx + c = 0 \Rightarrow x = \frac{-b \pm \sqrt{d}}{2a}\right)$$

$$\Rightarrow \quad \sqrt{2}\int_0^x f(t)\,dt = (\sqrt{2} \pm 1)\, x f(x)$$

Again, differentiating, we get

$$\sqrt{2}\, f(x) = (\sqrt{2} \pm 1)(f(x) + x f'(x))$$

$$\Rightarrow \quad \sqrt{2}\, f(x) = (\sqrt{2} \pm 1)\, f(x) + (\sqrt{2} \pm 1)\, x\cdot f'(x)$$

$$\Rightarrow \quad \pm f(x) = (\sqrt{2} \pm 1)\cdot x f'(x)$$

$$\Rightarrow \quad \int \pm \frac{1}{\sqrt{2} \pm 1}\cdot \frac{1}{x}\,dx = \int \frac{f'(x)}{f(x)}\,dx$$

$$\Rightarrow \quad \pm(\sqrt{2} \pm 1)\cdot \log x = \log|f(x)| + C, \text{ as } f(1) = 1 \Rightarrow C = 0$$

$$\Rightarrow \quad f(x) = x^{\sqrt{2}\pm 1}$$

$$\therefore \quad f(x) = x^{\sqrt{2}+1} \qquad [\text{as } f'(x) > 0]$$

$$\therefore \quad f(n^{\sqrt{2}-1}) = n \qquad \text{...(i)}$$

Now, let $a_n = \cot \theta_n$

$$\Rightarrow \quad a_{n+1} = \cot\left(\frac{\theta_n}{2}\right)$$

$$\Rightarrow \quad \theta_{n+1} = \frac{1}{2}\theta_n$$

$$\because \quad \theta_0 = \pi/2$$

$$\Rightarrow \quad \theta_1 = \frac{1}{2}\left(\frac{\pi}{2}\right), \theta_2 = \frac{1}{2^2}\left(\frac{\pi}{2}\right)$$

$$\Rightarrow \quad \theta_n = \frac{1}{2^n}\times\frac{\pi}{2} \qquad \therefore \; a_n = \cot\left(\frac{\pi}{2^{n+1}}\right)$$

$$\Rightarrow \quad \lim_{n\to\infty}\frac{a_n}{2^{n-1}} = \frac{1}{2^{-1}\cdot \pi/2} = 4/\pi$$

28. We have, $f(x) = |x|^m |x-1|^n$

$$f(x) = \begin{cases} (-1)^{m+n} x^m \cdot (x-1)^n, & \text{if } \quad x < 0 \\ (-1)^n x^m (x-1)^n, & \text{if } \quad 0 \le x < 1 \\ x^m (x-1)^n, & \text{if } \quad x \ge 1 \end{cases}$$

Let, $g(x) = x^m (x-1)^n$, then

$$g'(x) = mx^{m-1}\cdot (x-1)^n + n x^m \cdot (x-1)^{n-1}$$

$$= x^{m-1}\cdot (x-1)^{n-1}\cdot (m(x-1) + n\cdot x) = 0$$

Now, $f'(x) = 0$

$$\Rightarrow \quad g'(x) = 0$$

$$\Rightarrow \quad x = 0 \text{ or } \frac{m}{m+n} \text{ or } 1.$$

$$f(0) = 0,\; f(1) = 0$$

$$f\left(\frac{m}{m+n}\right) = (-1)^n \cdot \frac{m^m \cdot n^n \cdot (-1)^n}{(m+n)^{m+n}} \qquad \left[\text{as, } 0 < \frac{m}{m+1} < 1\right]$$

$$\therefore \text{ Maximum value} = \frac{m^m \cdot n^n}{(m+n)^{m+n}}$$

29. Let $f(x) = \dfrac{\ln(\ln x)}{\ln x} \Rightarrow f'(x) = \dfrac{1 - \ln(\ln x)}{x(\ln x)^2} < 0, \forall\, x > e^e$

So, $f(x)$ is increasing in $(1, e^e)$ and decreasing in (e^e, ∞).

Therefore, $x > y$

$$\Rightarrow \quad (\ln x)^{\ln y} < (\ln y)^{\ln x}, \forall\, x, y \in (e^e, \infty) \text{ and } x < y$$

$$\Rightarrow \quad (\ln x)^{\ln y} < (\ln y)^{\ln x}, \forall\, x, y \in (1, e^e)$$

30. $\lim\limits_{x\to a}[f(x)]$ can exist only when $f(x)$ either increases or decreases at both sides of the point $x = a$.

Since, $\lim\limits_{x\to a} f(x) = \lim\limits_{x\to a}[f(x)]$

So, this can occur only when $\lim\limits_{x\to a} f(x)$ is an integer.

31. $f(x) = 2x^3 - 3(2+\lambda)x^2 + 12\lambda x$

$$\Rightarrow \quad f'(x) = 6x^2 - 6(2+\lambda)x + 12\lambda$$

$$\Rightarrow \quad f'(x) = 0 \Rightarrow x = 2, \lambda$$

If $f(x)$ has exactly one local maximum and exactly one local minimum, then $\lambda \ne 2.1$

32. $h(x) = 3f\left(\dfrac{x^2}{3}\right) + f(3 - x^2) \Rightarrow h'(x) = 2x\left(f'\left(\dfrac{x^2}{3}\right) - f'(3 - x^2)\right)$

$$\Rightarrow \quad f'\left(\frac{x^2}{3}\right) > f'(3 - x^2), \forall\, x \text{ such that } \frac{x^2}{3} > 3 - x^2 \Rightarrow x^2 > \frac{9}{4}$$

$$\Rightarrow \quad f'\left(\frac{x^2}{3}\right) < f'(3 - x^2), \forall\, x \text{ such that } x^2 < \frac{9}{4}$$

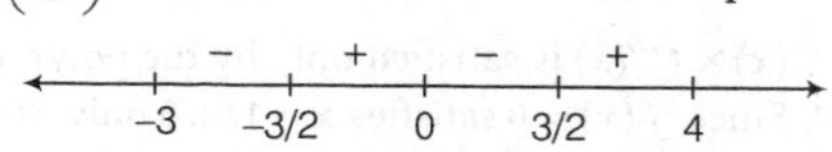

using number line rule

$$\Rightarrow \quad h(x) \text{ increases in } \left(-\frac{3}{2}, 0\right) \cup \left(\frac{3}{2}, 4\right)$$

and $h(x)$ decreases in $\left(-3, -\dfrac{3}{2}\right) \cup \left(0, \dfrac{3}{2}\right)$

33. $f'(x) = \dfrac{\cos x - \sin x}{1 + (\sin x + \cos x)^2} \ge 0 \Rightarrow \cos x \ge \sin x$

$$\Rightarrow \quad x \in \left[-2\pi, -\frac{7\pi}{4}\right] \cup \left[-\frac{3\pi}{4}, \frac{\pi}{4}\right] \cup \left[\frac{5\pi}{4}, 2\pi\right]$$

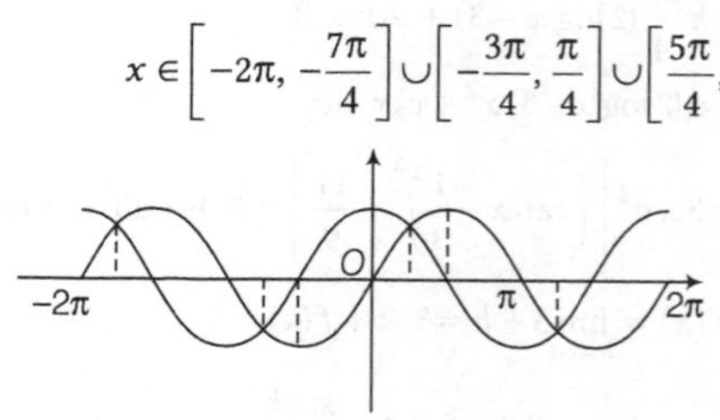

34. Applying $C_1 \to C_1 + C_2$, we get $\begin{vmatrix} 2 & \cos^2 x & \sin 2x \\ 2 & 1 + \cos^2 x & \sin 2x \\ 1 & \cos^2 x & 1 + \sin 2x \end{vmatrix}$

Applying $R_2 \to R_2 - R_1$ and $R_3 \to R_3 - R_1$, we get

$$\begin{vmatrix} 2 & \cos^2 x & \sin 2x \\ 0 & 1 & 0 \\ -1 & 0 & 1 \end{vmatrix} = 2 + \sin 2x$$

Since, the maximum value of $\sin 2x$ is 1 and minimum value of $\sin 2x$ is (-1). Therefore, $\alpha = 3$, $\beta = 1$.

35.

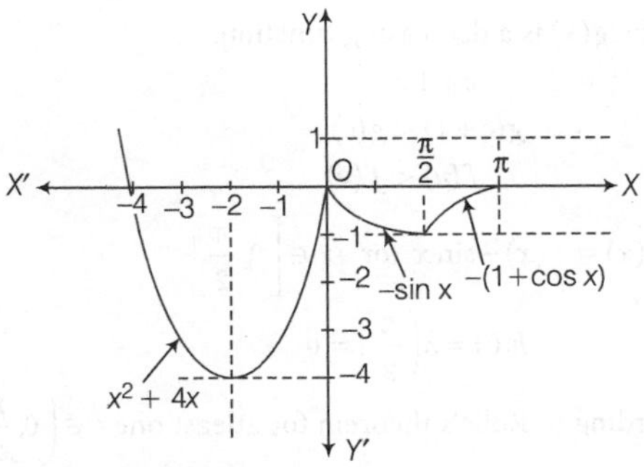

From above figure clearly options (a), (b) and (c) are correct.

36. $f(x) = \sqrt{a^2b^2 + b^2 - b^2a^2}\sin(x + \alpha) + c$

$= b\sin(x + \alpha) + c$, where $\tan\alpha = \dfrac{\sqrt{1-a^2}}{a}$

$f_{\max} = b + c$

$f_{\min} = -b + c$

$(f(x))_{\max} - (f(x))_{\min} = 2b$

Also, at $x = -\cos^{-1} a$, $f(x) = c$.

37. $f'(x) = \dfrac{n \cdot x^{n-1}}{(\ln x^n)} - \dfrac{1 \cdot mx^{m-1}}{(\ln x^m)}$,

Clearly, (c) and (d) are the answers.

38. $f(x)$ occurs maximum at $x = \dfrac{3}{2}$

$g(x)$ occurs maximum at $x = \dfrac{-b}{2}$

$\therefore \quad \dfrac{3}{2} = \dfrac{-b}{2} \Rightarrow b = -3$

39. $h(x) = \dfrac{\ln(f(x) \cdot g(x))}{\ln a} = \dfrac{\ln a^{\{a^{|x|} \text{ sgn } x\} + [a^{|x|} \cdot \text{ sgn } x]}}{\ln a}$

$= \{a^{|x|} \cdot \text{sgn } x\} + [a^{|x|} \cdot \text{sgn } x] = a^{|x|} \text{ sgn } x$

$= \begin{cases} a^x \text{ for } x > 0 \\ 0 \text{ for } x = 0 \\ -a^{-x} \text{ for } x < 0 \end{cases} \qquad [\because \{y\} + [y] = y]$

$\Rightarrow h(x)$ is an odd function.

40. $f(x) = \ln(1 - \ln x)$, Domain $(0, e)$

$$f'(x) = -\frac{1}{(1 - \ln x)} \cdot \frac{1}{x} < 0$$

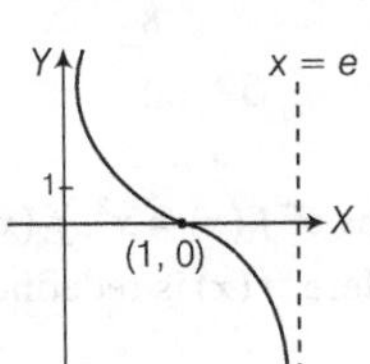

$\Rightarrow$ Decreasing, $\forall$ x in its domain

$\Rightarrow$ (a) and (b) are incorrect.

$f'(1) = -1 \Rightarrow$ (c) is also incorrect.

Also, $\quad f(1) = 0, \lim\limits_{x \to e} \to -\infty, \lim\limits_{x \to 0^+} f(x) \to \infty$

$$f''(x) = \frac{-\ln x}{x^2 (1 - \ln x)^2}$$

$f''(1) = 0$ which is a point of inflection as shown in graph Y-axis and $x = e$ are two asymptotes.

41. f is obvious continuous, $\forall$ $x \in R$ and not derivable at -1 and 1.

$f'(x)$ changes sign 4 times at $-1, 0, 1, 2$

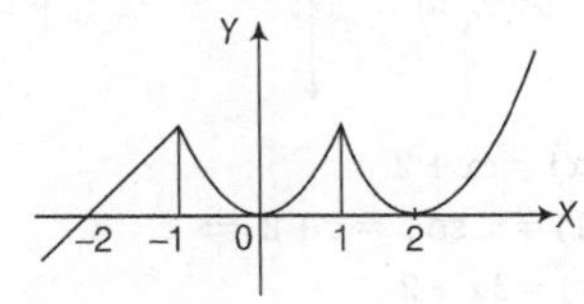

Local maxima at 1 and -1.

Local minima at $x = 0$ and 2.

42. $\quad f(x) = \int_0^{\pi} \cos t \cos(x - t)\, dt \qquad$ ...(i)

$= \int_0^{\pi} -\cos t \cdot \cos(x - \pi + t)\, dt$

$f(x) = \int_0^{\pi} -\cos t \cdot \cos(x + t)\, dt \qquad$ ...(ii)

On adding Eqs. (i) and (ii), we get

$2f(x) = \int_0^{\pi} \cos t\,(2 \cos x \cdot \cos t)\, dt$

$2f(x) = 2 \cos x \int_0^{\pi/2} \cos^2 t\, dt$

$f(x) = \dfrac{\pi \cos x}{2}$ Now, verify.

Only (a) and (b) are correct.

Aliter Convert the integer and into sum of two cosine functions.

43. $f'(x) = \dfrac{2 - x}{x^3}$ and $f''(x) = \dfrac{x - 3}{x^4}$ Now, interpret

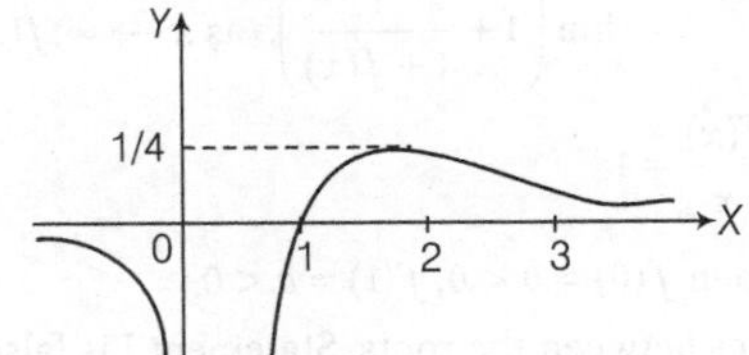

44. We have, $\lim\limits_{a \to x} \dfrac{a f(x) - x f(a)}{a - x} \qquad \left[\dfrac{0}{0}\text{ form}\right]$

$\Rightarrow \quad \lim\limits_{a \to x} \dfrac{f(x) - x f'(a)}{1} = 2$

$\Rightarrow \quad f(x) - x f'(x) = 2$

$\Rightarrow \quad \dfrac{f'(x)}{f(x) - 2} = \dfrac{1}{x}$

On integrating both sides w.r.t. x, we get

$$\int \frac{f'(x)}{f(x) - 2} dx = \int \frac{1}{x} dx$$

$\Rightarrow \quad \log(f(x)-2) = \log x + \log c$

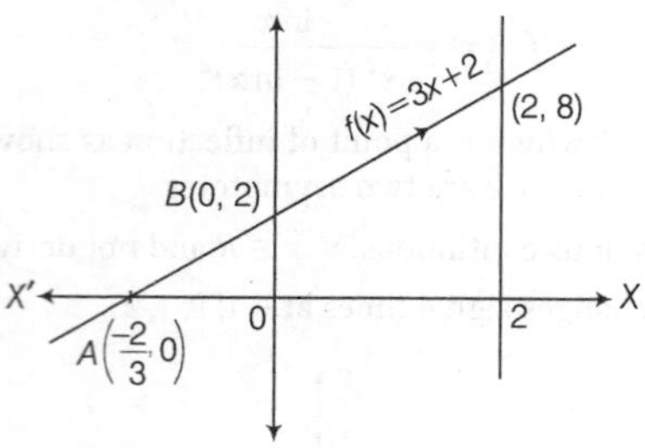

$\Rightarrow \quad f(x) = cx + 2$

As, $\quad f(1) = 5$, so $5 = c + 2 \Rightarrow c = 3$

Hence, $\quad f(x) = 3x + 2$

Clearly, area $(\Delta OAB) = \frac{1}{2}\left(\frac{2}{3}\right)(2) = \frac{2}{3}$

Also, $\quad \int_0^2 (3x+2)\,dx = \left(\frac{3x^2}{2} + 2x\right)_0^2$

$= 6 + 4 = 10$

45. Given, $2(F(x) - f(x)) = f^2(x)$ and $\frac{d\,F(x)}{dx} = f(x)$ ...(i)

$\therefore \quad F(x) = \frac{f^2(x)}{2} + f(x)$

$\Rightarrow \quad F'(x) = f(x)\cdot f'(x) + f'(x)$

$\therefore \quad f(x) = f(x)\cdot f'(x) + f'(x)$, using $F'(x) = f(x)$

$\Rightarrow \quad f'(x) = \frac{f(x)}{1 + f(x)}$, as $f(0,\infty) \to (0,\infty)$

$\therefore \quad f'(x) > 0$

Hence, f is strictly increasing and $\lim_{x\to\infty} \frac{f(x)}{x} = \lim_{x\to\infty} \frac{f'(x)}{1}$

[using L'Hospital's rule]

$= \lim_{x\to\infty}\left(1 + \frac{1}{1+f(x)}\right)$, as $x \to \infty; f(x) \to \infty$

$\therefore \quad \lim_{x\to 0} \frac{f(x)}{x} = 1$

46. If $b < 0$, then $f(0) = b < 0, f(1) = b < 0$

$\therefore$ 0, 1 lies between the roots, Statement I is false.

47. $x = 2$ is a point of local minima.

48. $\phi'(x) = 3\sin x + 4\cos x > 0$

$\phi(x)$ is increasing in $\left[\frac{\pi}{6}, \frac{\pi}{3}\right]$.

$\therefore f(x)$ attain maximum value at $x = \frac{\pi}{3}$.

49. Let $g(x) = f(x)\cdot f'(x)$

$\Rightarrow g'(x) > 0$ in $[a, b]$.

50. Let $g(x) = \sqrt{x} - \sqrt{x-1},\ x > 1$

$\Rightarrow g'(x) = \frac{1}{2\sqrt{x}} - \frac{1}{2\sqrt{x-1}} = \frac{1}{2}\left(\frac{\sqrt{x-1}-\sqrt{x}}{\sqrt{x(x-1)}}\right)$

$= \frac{-1}{2\sqrt{x}\sqrt{x-1}(\sqrt{x}+\sqrt{x-1})} < 0, \forall\ x > 1$

Hence, $g(x)$ is a decreasing function.

$\Rightarrow \quad c + 1 > c$

$g(c+1) < g(c)$

$\Rightarrow \quad f(u) < f(v)$

51. Let $h(x) = f(x) - \sin x$ or $x \in \left[0, \frac{\pi}{2}\right]$

$\Rightarrow \quad h(0) = h\left(\frac{\pi}{2}\right) = 0$

According to Rolle's theorem for atleast one $c \in \left(0, \frac{\pi}{2}\right)$.

$h'(c) = f'(c) - \sin c = 0$

for $\quad x \in \left[\frac{\pi}{2}, \frac{3\pi}{2}\right], h\left(\frac{\pi}{2}\right) = h\left(\frac{5\pi}{2}\right) = 0$

According to Rolle's theorem for atleast one $d \in \left(\frac{\pi}{2}, \frac{3\pi}{2}\right)$.

$h'(d) = f'(d) - \cos d = 0$

for $\ x \in [c, d], h'(c) = h'(d) = 0$

According to Rolle's theorem for atleast one $x \in (c, d)$.

$h'(x) = f''(x) + \sin x = 0$

$\Rightarrow |f''(x)| \le 1$ for atleast one $x \in \left(0, \frac{3\pi}{2}\right)$.

52. Statements I and II both are true but Statement II does not explain Statement I.

53. Statement II is true as $f'(a)^+ = f'(a)^-, \forall\ a \in I$.

Statement I is true and is obtained from differentiable rule.

54. Statement II is correct as $y = f(x)$ is odd and hence Statement I is correct.

Sol. **(Q. Nos. 55 to 57)**

Here, we have $f'(x) = \frac{-2x}{(1+x^2)^2}$ and $f''(x) = \frac{6x^2-2}{(1+x^2)^3}$

$\therefore\ f'(x)$ is maximum at $x = -\frac{1}{\sqrt{3}}$

If m is greatest, then $m = \frac{3\sqrt{3}}{8}$

y-coordinate of the point of contact is $\frac{3}{4}$.

$\therefore$ Equation of the tangent is $y - \frac{3}{4} = \frac{3\sqrt{3}}{8}\left(x + \frac{1}{\sqrt{3}}\right)$

$\therefore \quad a = -\sqrt{3}$ and $b = \frac{9}{8}$

55. (d) **56.** (a) **57.** (a)

Sol. **(Q. Nos. 58 to 60)**

We draw the graphs of $f_1(x) = x^2$, $f_2(x) = (1-x)^2$ and $f_3(x) = 2x(1-x)$. Here, $f(x)$ is redefined as

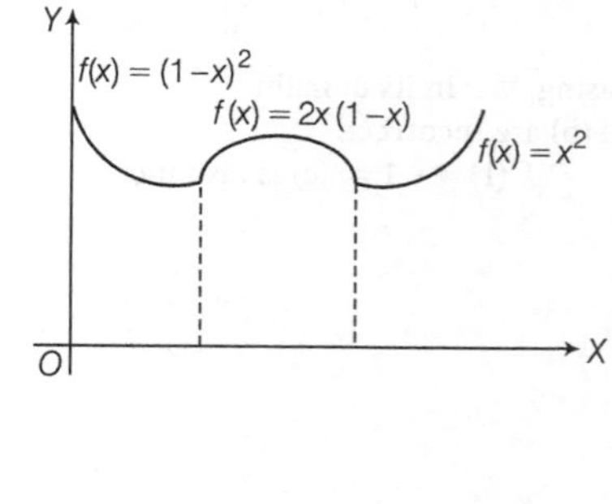

$$f(x)=\begin{cases}(1-x)^2, & 0\le x<\dfrac{1}{3}\\ 2x(1-x), & \dfrac{1}{3}\le x\le\dfrac{2}{3}\\ x^2, & \dfrac{2}{3}<x\le 1\end{cases}$$

Interval of increase of $f(x)$ is $\left(\frac{1}{3},\frac{1}{2}\right)\cup\left(\frac{2}{3},1\right)$.

Interval of decrease of $f(x)$ is $\left(0,\frac{1}{3}\right)\cup\left(\frac{1}{2},\frac{2}{3}\right)$.

Clearly, Rolle's theorem is applicable on $\left[\frac{1}{3},\frac{2}{3}\right]$, where $f(x)=2x(1-x)$.

$$\Rightarrow \quad f'(c)=2-4c=0 \Rightarrow c=\frac{1}{2}$$

$$\Rightarrow \quad a+b+c=\frac{1}{3}+\frac{2}{3}+\frac{1}{2}=\frac{3}{2}$$

58. (d) **59.** (d) **60.** (c)

Sol. (Q. Nos. 61 to 63)

According to paragraph

$$\frac{f(b)-f(a)}{b-a}>f'(c),\ \frac{g(b)-g(a)}{b-a}=g'(c) \text{ and } \frac{h(b)-h(a)}{b-a}<h'(c)$$

As, $f'(x)>g'(x)>h'(x)$

$$\Rightarrow \quad \frac{f(b)-f(a)}{b-a}>\frac{g(b)-g(a)}{b-a}>\frac{h(b)-h(a)}{b-a}$$

If $g(x)=Ax^2+Bx+C$

$$\Rightarrow \quad \frac{g(b)-g(a)}{b-a}=\frac{A(b^2-a^2)+B(b-a)}{b-a}$$

$$\Rightarrow \quad 2A\frac{(b+a)}{2}+B=g'\left(\frac{b+a}{2}\right)$$

61. (b) **62.** (c) **63.** (a)

Sol. (Q. Nos. 64 to 66)

$a_n>a_{n-1}$ iff $n+c$ is a perfect square, since $a_2>a_1$ and $a_5>a_4$.

$\Rightarrow$ $2+c$ and $5+c$ are perfect squares

$$\Rightarrow \quad c=-1 \quad \text{...(i)}$$

Now, $\quad a_2=3=b\,[\sqrt{2+c}]+d$

$$\therefore \quad b+d=2 \quad \text{...(ii)}$$

Also, $\quad a_{10}>a_9$

$$\Rightarrow \quad a_{10}=7=3b+d \quad \text{...(iii)}$$

On solving Eqs. (ii) and (iii), we get $b=2,\ d=1$

64. (c) **65.** (a) **66.** (a)

Sol. (Q. Nos 67 to 69)

$D_h=\{-1,1\}$, as minimum occurs before maxima

$\therefore \quad a_3=-1$

Now, $\quad g(x)=a_0+a_1x+a_2x^2-x^3$

$$g'(x)=a_1+2a_2x-3x^2=-3(x-3)(x+3)=-3x^2=27$$

$\therefore \quad a_1=27,\ a_2=0$

$\therefore \quad a_1+a_2=27$

Also, $\quad g(-3)>0$ and $g(3)>0 \Rightarrow a_0>54$ and $a_0<-54$

$\therefore \quad a_0>54$

Now, $\quad g(x)=a_0+27x-x^3 \Rightarrow f(x)=\sqrt{a_0+27x-x^3}$

$$f(-10)=\sqrt{a_0+270-1000}$$

Clearly, $f(-10)$ is defined for $a_0>830$

67. (c) **68.** (b) **69.** (a)

Sol. (Q. Nos. 70 to 74)

70. Graph of $f(x)$ is shown.

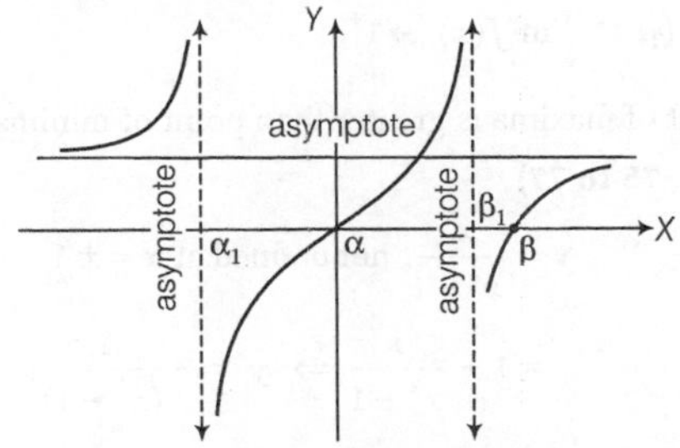

Clearly, $f(x)$ is increasing in (α_1,β_1).

71. Clearly, $f(x)$ has a maximum in $[\alpha_1,\beta_1]$ and a minima in $[\alpha,\beta]$, shown as.

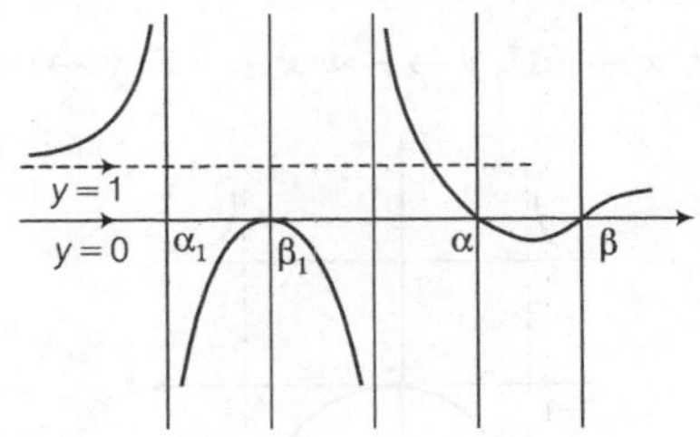

72. $f(x)$ has one of the two graphs.

$\Rightarrow$ $f'(x)=0$ has real and distinct roots.

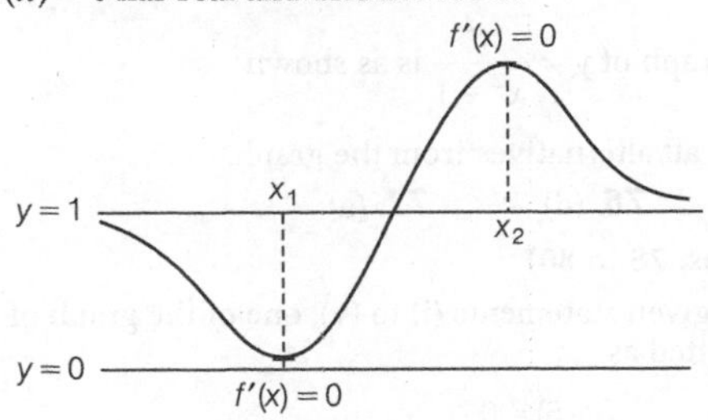

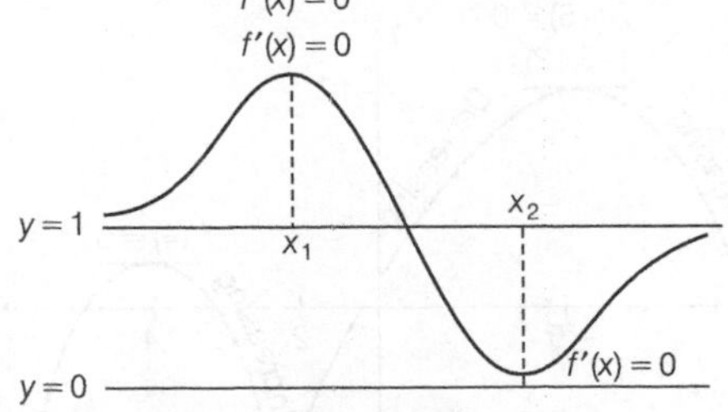

73. Clearly, $\lim\limits_{x\to\infty}[f(x)]=0$ and $\lim\limits_{x\to\infty}[f(x)]=1$

$$\therefore \quad \lim_{x\to\infty}[f(x)]\lim_{x\to\infty}[f(x)]=0$$

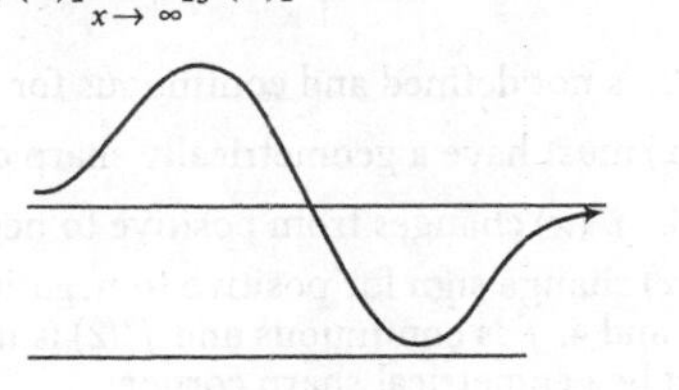

74. $f(x) = \dfrac{x^2+bx+c}{x^2+b_1x+c_1} = 1 + \dfrac{(b-b_1)x+(c-c_1)}{x^2+b_1x+c_1}$

$$= 1 + \frac{\dfrac{(b-b_1)}{x} + \dfrac{(c-c_1)}{x^2}}{1 + \dfrac{b_1}{x} + \dfrac{c_1}{x^2}}$$

For $b > b_1$, $\quad \lim\limits_{x\to\infty} f(x) \to 1^+$

$\Rightarrow$ Point of maxima is greater than point of minima.

Sol. **(Q. Nos. 75 to 77)**

$$y = \frac{x^2}{x^2-1}, \text{ not defined at } x = \pm 1$$

$$= 1 + \frac{1}{x^2-1} \Rightarrow y' = -\frac{2x}{(x^2-1)^2}$$

$$\frac{dy}{dx} = 0$$

$\Rightarrow \qquad x = 0 \qquad$ [point of maxima]

as $x \to 1^+, y \to \infty, \; x \to 1^-, y \to -\infty$

Similarly, $x \to -1^+, y \to -\infty, \; x \to -1^-, y \to \infty$

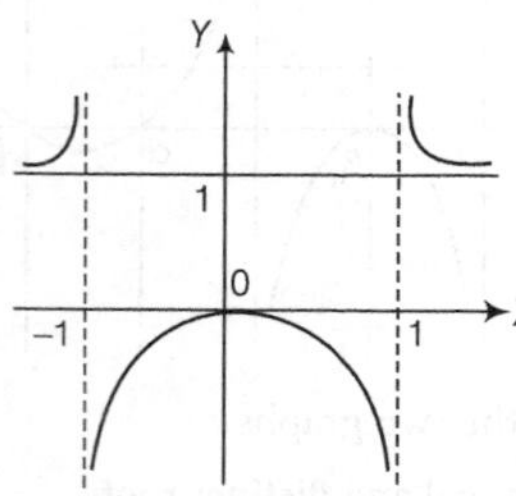

The graph of $y = \dfrac{x^2}{x^2-1}$ is as shown

verify all alternatives from the graph.

75. (b) **76.** (d) **77.** (a)

Sol. **(Q. Nos. 78 to 80)**

From given statements (i) to (v), one of the graph of $f(x)$ can be plotted as

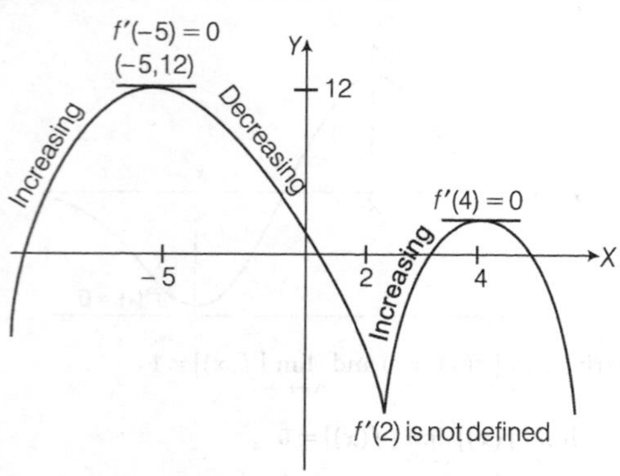

78. Since, $f'(2)$ is not defined and continuous for $x \in R$.

$\Rightarrow y = f(x)$ must have a geometrically sharp corner at $x = 2$.

79. At $x = -5$, $f'(x)$ changes from positive to negative and at $x = 4$, $f'(x)$ change sign for positive to negative, hence maxima at $x = -5$ and 4. f is continuous and $f'(2)$ is not defined, hence $x = 2$ must be geometrical sharp corner.

80. Clearly, one of the graph of $f(x)$ is

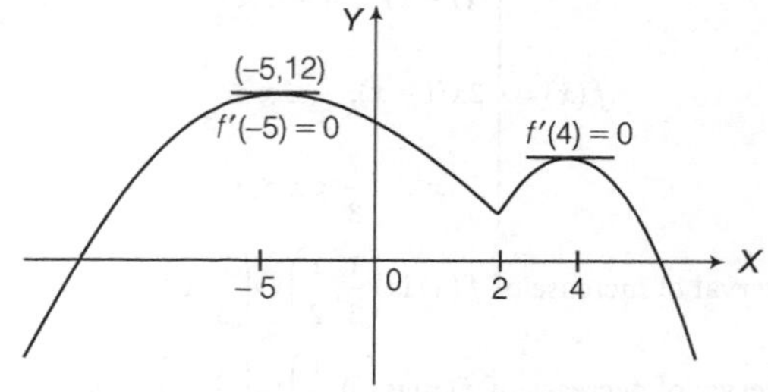

Sol. **(Q. Nos. 81 to 83)**

$$f(x) = e^{(p+1)x} - e^x = e^x(e^{px}-1)$$

$$f'(x) = e^x(e^{px}-1) + e^x \cdot pe^{px}$$

$$= e^x[(p+1)e^{px} - 1] = 0 \qquad [e^x \neq 0]$$

$$\Rightarrow \qquad e^{px} = \frac{1}{p+1} \Rightarrow px = -\log_e(p+1)$$

$$x = -\frac{\log_e(p+1)}{p} \qquad \text{...(i)}$$

Now, $\qquad f''(x) = (p+1)^2 e^{(p+1)x} - e^x$

$$= e^x[(p+1)^2 e^{px} - 1]$$

$$\therefore \quad f''\left(-\frac{\log(p+1)}{p}\right) = e^x[(p+1)-1] > 0$$

Hence, $x = s_p = -\dfrac{\log_e(p+1)}{p}$

$$g(t) = \int_t^{t+1} (e^{px}\cdot e^x - e^x)e^t \cdot e^{-x}\,dx = e^t \int_t^{t+1}(e^{px}-1)\,dx$$

On integrating, we get

$$g(t) = e^t \cdot \left(\frac{e^{px}}{p} - x\right)_t^{t+1} = e^t\left[\frac{e^{pt}(e^p-1)}{p} - 1\right]$$

$$\therefore \quad g(t) = \frac{(e^p-1)e^{(p+1)t}}{p} - e^t$$

$$g'(t) = (p+1)\frac{(e^p-1)}{p}e^{(p+1)t} - e^t = 0$$

$$\Rightarrow \quad \frac{(p+1)(e^p-1)}{p}e^{pt} = 1 \Rightarrow e^{pt} = \frac{p}{(p+1)(e^p-1)}$$

$$pt = \log_e\left(\frac{p}{(p+1)(e^p-1)}\right)$$

$$\therefore \quad t = -\frac{1}{p}\log\left(\frac{(p+1)(e^p-1)}{p}\right) = t_p$$

Also, $s_p - t_p = \dfrac{1}{p}\log\left(\dfrac{(p+1)(e^p-1)}{p}\right) - \dfrac{\log(p+1)}{p}$

$$= \frac{1}{p}\log\left(\frac{e^p-1}{p}\right)$$

Hence, $\lim\limits_{p\to 0^+}(s_p - t_p) = \lim\limits_{p\to 0^+}\dfrac{1}{p}\log\left(\dfrac{e^p-1}{p}\right)$

$$= \lim_{p\to 0}\frac{\log\left(1+\dfrac{e^p-p-1}{p}\right)}{\left(\dfrac{e^p-p-1}{p}\right)\cdot p}\cdot\left(\frac{e^p-p-1}{p}\right) = \lim_{p\to 0^+}\left(\frac{e^p-p-1}{p^2}\right) = \frac{1}{2}$$

81. (c) **82.** (c) **83.** (b)

Sol. (Q. Nos. 84 to 86)

84. Here, $h(x) = f^2(x) + g^2(x)$

$$= (\sin 3x + \cos x)^2 + (\cos 3x + \sin x)^2 = 2 + 2\sin 4x$$

Graph of $y = \sin 4x$

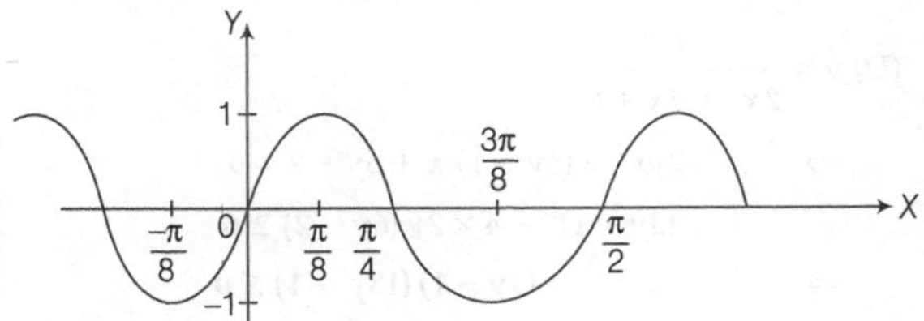

Clearly, $h(x)$ is periodic function with period $\pi/2$ and from above graph, the length of a longest interval in which the function $y = h(x)$ is increasing $= \pi/8 - (-\pi/8) = \pi/4$

85. We have, $h(x) = 4$

$\Rightarrow \quad 2 + 2\sin 4x = 4 \Rightarrow \sin 4x = 1$

$\therefore \quad 4x = 2n\pi + \pi/2 = (4n+1)\pi/2$

$\Rightarrow \quad x = (4n+1)\pi/8, n \in I$

86. We have, $f(x) = g(x)$

$\Rightarrow \quad \sin 3x + \cos x = \cos 3x + \sin x$

$\Rightarrow \quad \sin 3x - \sin x = \cos 3x - \cos x$

$\Rightarrow \quad 2\sin x(\cos 2x + \sin 2x) = 0$

$\Rightarrow \quad$ either $\sin x = 0$ or $\tan 2x = -1$

$\Rightarrow \quad x = 0, \pi, 3\pi/8, 7\pi/8$

$\therefore$ Number of solution = 4

Sol. (Q. Nos. 87 to 89)

87. We have, $h''(x) = 6x - 4$

$\Rightarrow \quad h'(x) = 3x^2 - 4x + c$

As, $h'(1) = 0 \Rightarrow 0 = -1 + c$ or $c = 1$

So, $h'(x) = 3x^2 - 4x + 1$

$\Rightarrow \quad h(x) = x^3 - 2x^2 + x + k$

Also, $h(1) = 5 \Rightarrow k = 5$

$\therefore \quad h(x) = x^3 - 2x^2 + x + 5$

Now, $h'(2) = 5$

$\therefore$ The equation of tangent at $m(2, 7)$ to $y = h(x)$, is

$(g - 7) = 5(x - 2)$

$\Rightarrow \quad 5x - y = 3$

88. Also, $g(f(x)) = 0, \forall x \in R$

$$\therefore \text{ Required area} = \int_0^2 h(x)\,dx = \left(\frac{x^4}{4} - \frac{2x^3}{3} + \frac{x^2}{2} + 5x\right)_0^2$$

$$= 4 - \frac{16}{3} + 2 + 10 = 32/3$$

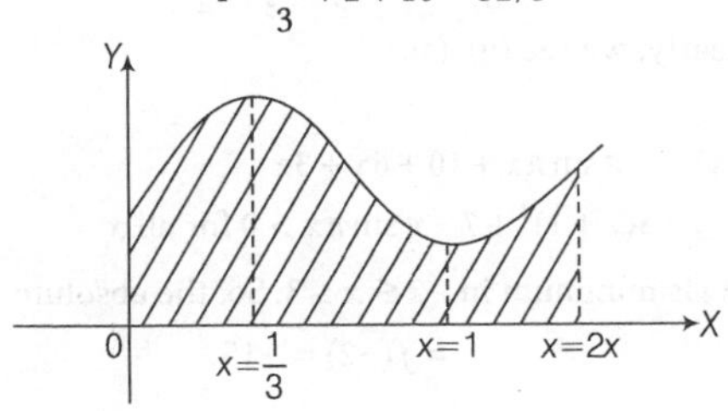

89. Also, range of $f(g(x)) = \{-1, 0, 1\}$

$\therefore \quad \sqrt{f(g(x))} \in \{0, 1\}$

Hence, range of $\sin^{-1}\sqrt{(fog)(x)} \in \{0, \pi/2\}$

Sol. (Q. Nos. 90 to 92)

We have, $g(x) = x^3 + g''(1)x^2 + \{3g'(1) - g''(1) - 1\}x + 3g'(1)$

Let $g'(1) = a, g''(1) = b$, then

$g(x) = x^3 + bx^2 + (3a - b - 1)x + 3a$

Differentiating both sides w.r.t. x, we get

$g'(x) = 3x^2 + 2bx + (3a - b - 1)$

Put $x = 1$,

$\Rightarrow \quad g'(1) = 3 + 2b + 3a - b - 1 \quad [\because g'(1) = a]$

$\Rightarrow \quad a = b + 3a + 2$ or $2a + b = -2 \quad$ ...(i)

Again differentiating, we get

$g''(x) = 6x + 2b$

Put $x = 1$,

$\Rightarrow \quad g''(1) = 6 + 2b \quad [\because g''(1) = b]$

$\Rightarrow \quad b = 6 + 2b$

$\Rightarrow \quad b = -6 \quad$ ...(ii)

From Eq. (i), $a = 2$

$\therefore \quad g(x) = x^3 - 6x^2 + 11x + 6$

Given, $f(x) = x\,g(x) - 12x + 1$

$= x^4 - 6x^3 + 11x^2 + 6x - 12x + 1$

$= x^4 - 6x^3 + 11x^2 - 6x + 1$

$= (x^2+1)^2 - 2x^2 + 11x^2 - 6x^3 - 6x$

$= (x^2+1)^2 - 6x(x^2+1) + (3x)^2$

$= (x^2 + 1 - 3x)^2 = \{h(x)\}^2$, given

$\because \quad h(0) = 1$

$\therefore \quad h(x) = x^2 - 3x + 1$

90. $f(x) = (x^2 - 3x + 1)^2$

$\because \quad f'(x) = 2(x^2 - 3x + 1)(2x - 3) = 0$

$\Rightarrow \quad x = \dfrac{3}{2}, \dfrac{3 \pm \sqrt{5}}{2}$

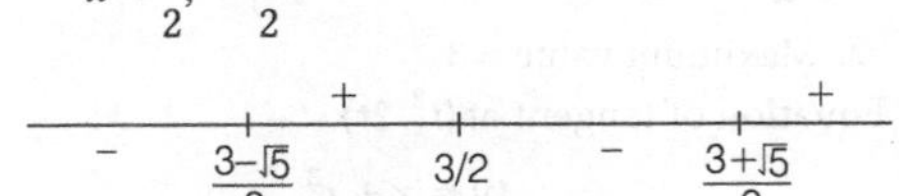

Clearly, $f(x)$ has local maxima at $x = 3/2$ and local minima at $x = \dfrac{3 \pm \sqrt{5}}{2}$

$\therefore$ $f(x)$ has exactly one local maxima and two local minima

91. We have,

$g(x) = x^3 - 6x^2 + 11x + 6$

$g'(x) = 3x^2 - 12x + 11$

$= 3(x-2)^2 - 1 = 3[(x-2)^2 - 1/3]$

$$\therefore \quad g'(x) > 0 \Rightarrow x \in \left(-\infty, 2 - \frac{1}{\sqrt{3}}\right) \cup \left(2 + \frac{1}{\sqrt{3}}, \infty\right)$$

and $$g'(x) < 0 \Rightarrow x \in \left(2 - \frac{1}{\sqrt{3}}, 2 + \frac{1}{\sqrt{3}}\right)$$

$\therefore$ $g(x)$ monotonically increases for $x \in \left(-\infty, 2-\frac{1}{\sqrt{3}}\right) \cup \left(2+\frac{1}{\sqrt{3}}, \infty\right)$ and monotonically decreases for $x \in \left(2-\frac{1}{\sqrt{3}}, 2+\frac{1}{\sqrt{3}}\right)$

For $x \in [1, 3]$

$$g(x) = (x-1)(x-2)(x-3) + 12$$

$\Rightarrow$ $g(1) = 12$ and $g(3) = 12$

$\therefore$ By Rolle's theorem in [1, 3], we have $g'(c) = 0$

$\Rightarrow$ $c = 2 \pm \frac{1}{\sqrt{3}}$ [both $\in (1, 3)$]

$\therefore$ There exists two distinct tangents to the curve $y = g(x)$ which are parallel to the chord joining $(1, g(1))$ and $(3, g(3))$.

For $x \in [0, 4]$

$$g(0) = 6 \text{ and } g(4) = 18$$

$\therefore$ By LMVT

$$g'(c) = \frac{g(4) - g(0)}{4 - 0} = \frac{18-6}{4} = 3$$

$\therefore$ $3c^2 - 12c + 11 = 3$

$\Rightarrow$ $3c^2 - 12c + 8 = 0$

$\Rightarrow$ $c = 2 \pm 2/\sqrt{3}$ [both $\in (0, 4)$]

$\therefore$ There exists exactly two distinct lagrange's mean value in $(0, 4)$ for $y = g(x)$.

92. We have, $h(x) = x^2 - 3x + 1 = \left(x - \frac{3}{2}\right)^2 - \frac{5}{4}$

The curve $y = h(x)$ is an upward parabola, intersecting X-axis at two distinct points.

$\therefore$ $h(x)$ has exactly one critical point (i.e. the vertex) and no point of inflexion.

Also, $h(x) = 0$

$\Rightarrow$ $x = \frac{3 \pm \sqrt{5}}{2}$ [both $\in (0, 3)$]

$\therefore$ $h(x) = 0$ has exactly two distinct zeroes in $(0, 3)$.

93. (A) $y = 10 - (10 - x), \ -1 \le x \le 3$

$= x$

$\therefore$ Maximum value = 3

(B) Equation of tangent at $(t^2, 2t)$

$$ty = x + t^2$$

this tangent means Y-axis at $Q(0, t)$

Here, $QS \times PQ = t \times \left(\frac{-1}{t}\right) = -1$

Area of triangle PQS, $\Delta = \frac{1}{2}\begin{vmatrix} t^2 & 2t & 1 \\ 0 & t & 1 \\ 1 & 0 & 1 \end{vmatrix}$

$\Rightarrow$ $\Delta = \frac{1}{2}[3t^2 + t]$

for $\in [0, 2]$, $\frac{d\Delta}{dt} \ge 0$

Δ is maximum at $t = 2$

$\therefore$ $\max \Delta = \frac{1}{2}[2^3 + 2] = 5$

(C) Let $y = \left(1 + \frac{1}{a}\right)\left(1 + \frac{1}{b}\right) = \left(1 + \frac{1}{1-b}\right)\left(1 + \frac{1}{b}\right)$

$\Rightarrow b^2(y-1) + b(1-y) + 2 = 0$

Then, $(1-y)^2 - 8(y-1) \le 0$

$\Rightarrow$ $1 \le \sqrt{y} \le 3$

(D) $y = \frac{x+2}{2x^2 + 3x + 6}$

$\Rightarrow$ $2yx^2 + (3y-1)x + 6y - 2 = 0$

$(3y-1)^2 - 4 \times 2y(6y-2) \ge 0$

$\Rightarrow$ $(3y-1)(13y+1) \le 0$

$\Rightarrow$ $\frac{-1}{13} \le y \le \frac{1}{3}$

94. (A) $3^x = t \Rightarrow Q(t) = 2t^3 - 4t^2 + 2t$ in $t \in \left[\frac{1}{3}, 3\right]$

$\Rightarrow Q'(t) < 0$ in $\left(\frac{1}{3}, 1\right)$ and $Q'(t) > 0$ in $(1, 3) \Rightarrow Q(t)_{\min} = Q(1) = 0$

(B) Take $x^2 + x = t \Rightarrow Q(t) = t(t-2) \Rightarrow Q(t)_{\min} = -1$

(C) $\int_{-1}^{3} |x-2|\, dx = 5$ and $\int_{-1}^{3} [x] dx = 2 \Rightarrow I = 3$

(D) Period of $\sin 36x = \frac{\pi}{18}$

Period of $\tan 42x = \frac{\pi}{42} \Rightarrow p = \frac{\pi}{6}$

95. (A) $f'(x) = 8x(x-1)(x+1)$ [positive factor]

$\Rightarrow$ Local maxima at $x = 0$

(B) Clearly, we see (p), (r), (s)

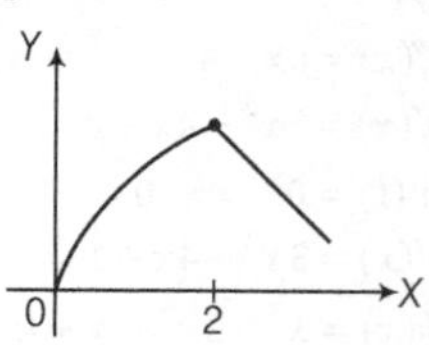

(C) $f\left(\frac{1}{2}\right)^{-} > f\left(\frac{1}{2}\right) < f\left(\frac{1}{2}\right)^{+}$

Also, $f(x)$ is non-differentiable at $x = \frac{1}{2}$

$\Rightarrow$ Clearly, we see (p), (s).

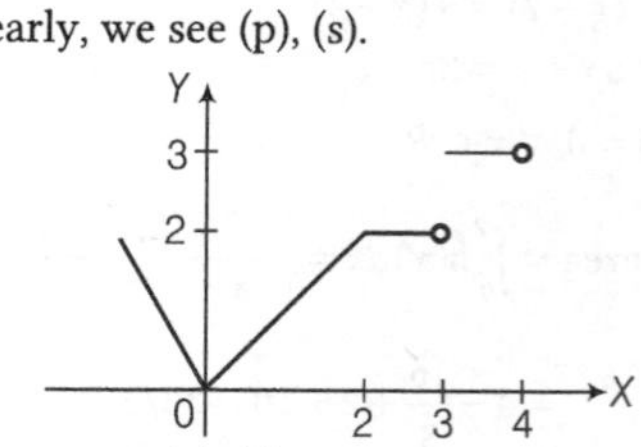

(D) Clearly, we see (q), (s).

96.(A) $f'(x) = -\pi \sin \pi x + 10 + 6x + 3x^2$

$= 3(x+1)^2 + 7 - \pi \sin \pi x > 0$ for all x

$\therefore$ $f(x)$ is minimum in $-2 \le x \le 3$. So, the absolute minimum $= f(-2) = -15$

(B) $f'(x) = 2x + 1$. Therefore, for $-1 \le x < -\frac{1}{2}$, we get

$f'(x) < 0$ and for $-\frac{1}{2} < x \le 1$, we get $f'(x) > 0$

$\therefore$ $f(x)$ is minimum decrease in $f\left[-1, -\frac{1}{2}\right]$.

and minimum decrease in $\left(-\frac{1}{2}, 1\right]$.

$$\therefore \ \min f(x) = f\left(-\frac{1}{2}\right) = \left(-\frac{1}{2}\right)^2 + \left(-\frac{1}{2}\right) + 1 = \frac{3}{4}$$

(C) $f'(x) = 4(x^2 - 1)$. So, for $0 \le x < 1$, we get $f'(x) < 0$, i.e. $f(x)$ is monotonically decreasing and for $1 < x \le 2$, we get $f'(x) > 0$, $f(x)$ is monotonically increasing.

$$\therefore \quad \min f(x) = f(1) = \frac{4}{3} - 4 = -\frac{8}{3}.$$

(D) $f'(x) = -12 + 18x - 6x^2$

$$= -6(x^2 - 3x + 2) = -6(x-1)(x-2)$$

$\therefore$ $f'(x) > 0$, if $1 < x < 2$

and $f'(x) < 0$, if $2 < x \le 4$

$\therefore$ $f(x)$ is monotonically increasing in $1 < x < 2$ and monotonically decreasing in $2 < x \le 4$.

$\therefore$ Absolute maximum = The greatest among $\{f(1), f(2)\}$

= The greatest among $\{1, 2\} = 2$

97. $\frac{dT}{dt} = k(T-5)^2$

$$\frac{dT}{(T-5)^2} = k\,dt$$

$$\Rightarrow \quad \frac{(T-5)^{-1}}{-1} = kt + c$$

$$\therefore \quad -(40-5)^{-1} = c \quad \text{i.e.} \quad c = -\frac{1}{35}$$

$$\therefore \quad -\frac{1}{T-5} = kt - \frac{1}{35}$$

After 15 min

$$-\frac{1}{30-5} = 15k - \frac{1}{35}$$

$$\therefore \quad k = \frac{1}{15}\left(\frac{1}{35} - \frac{1}{25}\right) = \frac{-10}{15 \times 35 \times 25} = \frac{-2}{75 \times 35}$$

Temperature after 60 min is given by

$$-\frac{1}{T-5} = 60\left(\frac{-2}{75 \times 35}\right) - \frac{1}{35}$$

$$\Rightarrow \quad \frac{1}{T-5} = \frac{120 + 75}{75 \times 35} + 5 = \frac{195}{75 \times 35}$$

$$\Rightarrow \quad T = \frac{75 \times 35}{195} + 5 = \frac{5 \times 35}{13} + 5 = \frac{175}{13} + 5$$

$$\Rightarrow \quad [T] = 18$$

$$\therefore \quad [T]/2 = 18/2 = 9$$

98. $y = \dfrac{2\sin\left(x + \frac{\pi}{6}\right)\cos x}{2\sin x \cos\left(x + \frac{\pi}{6}\right)} = \dfrac{\sin\left(2x + \frac{\pi}{6}\right) + \sin\frac{\pi}{6}}{\sin\left(2x + \frac{\pi}{6}\right) - \sin\frac{\pi}{6}}$

$$= 1 + \frac{1}{\sin\left(2x + \frac{\pi}{6}\right) - \sin\frac{\pi}{6}}$$

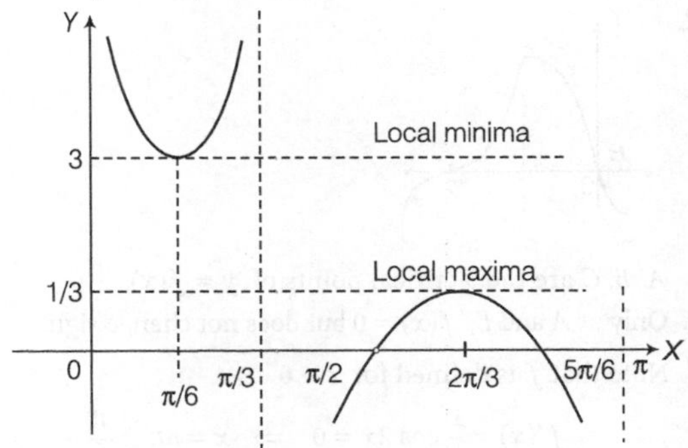

y is minimum, if $2x + \frac{\pi}{6} = \frac{\pi}{2} \Rightarrow x = \frac{\pi}{6}$

$$\Rightarrow \quad y_{\min} = 1 + 2 = 3$$

99. Let $A = (t, t^2)$, $m_{OA} = t$, $m_{AB} = -\frac{1}{t}$

Equation of AB, $y - t^2 = -\frac{1}{t}(x - t)$

Put $x = 0$, $h = t^2 + 1$

Now, $\lim_{x \to 0}(h) = \lim_{t \to 0}(1 + t^2) = 1$ [as $x \to 0$, then $t \to 0$]

100. For $0 < a \le 1$ the line always intersects $y = a^x$

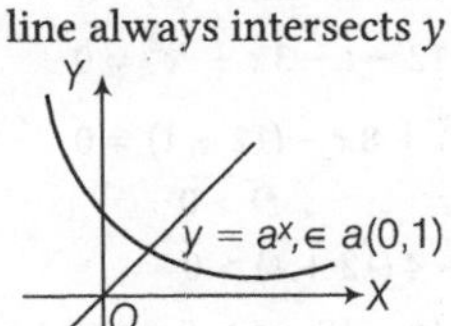

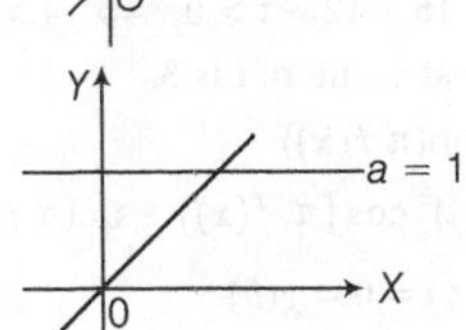

For $a > 1$ say $a = e$ consider $f(x) = e^x - x$

$$f'(x) = e^x - 1$$

$f'(x) > 0$ for $x > 0$ and $f'(x) < 0$ for $x < 0$

$\therefore f(x)$ is increasing ($\uparrow$) for $x > 0$ and decreasing ($\downarrow$) for $x < 0$

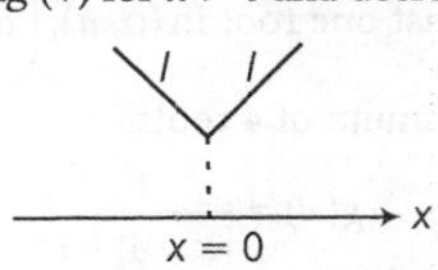

$y = e^x$ always lies above $y = x$, i.e. $e^x - x \ge 1$ for $a > 1$, hence never intersects $\Rightarrow a \in (0, 1]$

101. $\frac{dy}{dx} = -\frac{4\cos x}{\sin^2 x} + \frac{\cos x}{(1-\sin x)^2} = 0$ gives $\sin x = \frac{2}{3}$

note that $f(x) \to \infty$ as $x \to 0^+$ or $x \to \frac{\pi^-}{2}$ and between two maxima, we have a minima.

So, $$a = \frac{4}{2/3} + \frac{1}{1-2/3} = 9$$

102.

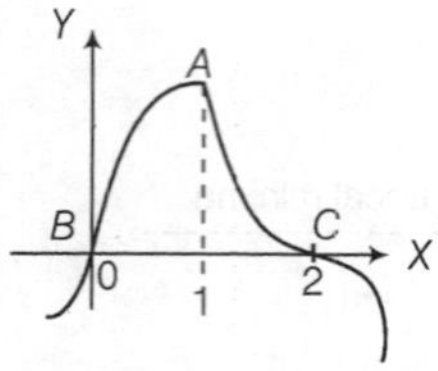

A, B, C are the 3 critical points of $y = f(x)$.

103. Only at A and E, $f(x) = 0$ but does not change sign.

104. Note that f is defined for $x > 0$

$$f'(x) = \frac{1}{2}\cos 2x = 0 \Rightarrow x = n\pi \pm \frac{\pi}{4}$$

105. Obviously, f is increasing and g is decreasing in (x_1, x_2), hence $f(g(\alpha^2 - 2\alpha)) > f(g(3\alpha - 4))$ as f is increasing

$\Rightarrow$ $$g(\alpha^2 - 2\alpha) > g(3\alpha - 4)$$

$\therefore$ $$\alpha^2 - 2\alpha < 3\alpha - 4 \text{ as } g \text{ is decreasing.}$$

$$\alpha^2 - 5\alpha + 4 < 0$$

$\Rightarrow$ $$(\alpha - 1)(\alpha - 4) < 0 \Rightarrow \alpha \in (1, 4)$$

106. $f(x) = \frac{t + 3x - x^2}{x - 4}$, $f'(x) = \frac{(x-4)(3-2x) - (t + 3x - x^2)}{(x-4)^2}$

For maximum or minimum, $f'(x) = 0$

$$-2x^2 + 11x - 12 - t - 3x + x^2 = 0$$

$$-x^2 + 8x - (12 + t) = 0$$

For one M and m, $D > 0$

$$64 - 4(12 + t) > 0$$

$$16 - 12 - t > 0 \Rightarrow 4 > t \text{ or } t < 4$$

Hence, the greatest value of t is 3.

107. Let $g(x) = f'(x)\sin(\pi f(x))$ ...(i)

$\therefore$ $g'(x) = \pi(f'(x))^2 \cos[\pi f(x)] + \sin(\pi f(x)) \cdot f''(x)$ ...(ii)

Here, $g(a) = 0 = g(b)$

and $g(\alpha) = g(\beta) = 0$

$$g\left(\frac{\alpha+\beta}{2}\right) = 0$$

$\therefore$ According to Rolle's theorem,

$g'(x) = 0$ has atleast one root in (α, a), $\left(a, \frac{a+b}{2}\right)$, $\left(\frac{a+b}{2}, b\right)$ and (b, β), i.e. minimum of 4 roots.

108. Let $$g(x) = \frac{1}{|x-4|+1}$$

$\Rightarrow$ $$g(x+12) = \frac{1}{|x+8|+1}$$

When, $x < -8$ both $g(x)$ and $g(x+12)$ are increasing, hence maximum value can't occur in this interval.

Similarly, for $x \in (4, \infty)$ both $g(x)$ and $g(x+12)$ are decreasing, hence maximum value can't occur in this interval.

So, now for all values of $x \in (-8, 4)$

$$f(x) = \frac{1}{x+9} + \frac{1}{5-x}$$

$\Rightarrow$ $$f'(x) = -\frac{1}{(x+9)^2} + \frac{1}{(5-x)^2} = \frac{(x+9)^2 - (5-x)^2}{(x+9)^2(5-x)^2}$$

$\therefore$ $$f'(x) = \frac{14(2x+4)}{(x+9)^2(5-x)^2}$$

$$\overset{-}{}\ \ \overset{+}{}$$
$$-2 = x$$

$\therefore$ Minimum at $x = -2$, and maximum occurs at $x = +4$ or $x = -8$.

Here, $$f(4) = f(-8) = \frac{14}{13} = \frac{p}{q}$$

$\therefore$ $$(p - q) = 1$$

109. Here, to find the least value of $\alpha \in R$, for which $4\alpha x^2 + \frac{1}{x} \geq 1$, for all $x > 0$.

i.e. to find the minimum value of α when $y = 4\alpha x^2 + \frac{1}{x}$; $x > 0$ attains minimum value of α.

$\therefore$ $$\frac{dy}{dx} = 8\alpha x - \frac{1}{x^2} \quad \text{...(i)}$$

Now, $$\frac{d^2y}{dx^2} = 8\alpha + \frac{2}{x^3} \quad \text{...(ii)}$$

When $\frac{dy}{dx} = 0$, then $8x^3\alpha = 1$

At $x = \left(\frac{1}{8\alpha}\right)^{1/3}$, $\frac{d^2y}{dx^2} = 8\alpha + 16\alpha = 24\alpha$, Thus, y attains minimum when

$$x = \left(\frac{1}{8\alpha}\right)^{1/3}; \alpha > 0.$$

$\therefore$ y attains minimum when $x = \left(\frac{1}{8\alpha}\right)^{1/3}$.

i.e. $$4\alpha\left(\frac{1}{8\alpha}\right)^{2/3} + (8\alpha)^{1/3} \geq 1$$

$\Rightarrow$ $$\alpha^{1/3} + 2\alpha^{1/3} \geq 1$$

$\Rightarrow$ $$3\alpha^{1/3} \geq 1 \Rightarrow \alpha \geq \frac{1}{27}$$

Hence, the least value of α is $\frac{1}{27}$.

110. **Plan** The given equation contains algebraic and trigonometric functions called transcendental equation. To solve transcendental equations we should always plot the graph for LHS and RHS.

Here, $x^2 = x\sin x + \cos x$

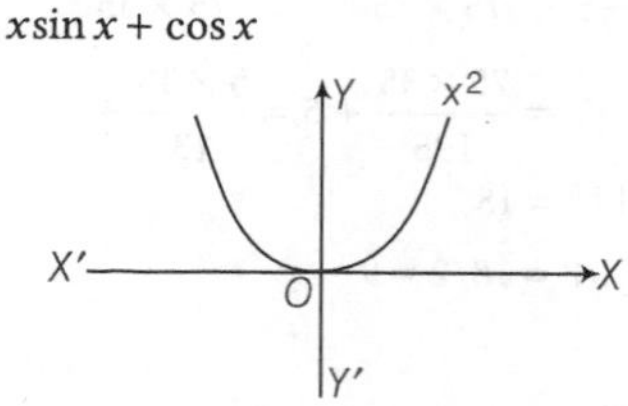

Let $f(x) = x^2$ and $g(x) = x\sin x + \cos x$

We know that, the graph for $f(x) = x^2$

To plot, $g(x) = x\sin x + \cos x$

$$g'(x) = x\cos x + \sin x - \sin x$$

$$g'(x) = x\cos x \quad \text{...(i)}$$

$$g''(x) = -x\sin x + \cos x \quad \text{...(ii)}$$

Put $g'(x) = 0 \Rightarrow x\cos x = 0$

$\therefore$ $x = 0, \dfrac{\pi}{2}, \dfrac{3\pi}{2}, \dfrac{5\pi}{2}, \dfrac{7\pi}{2}$

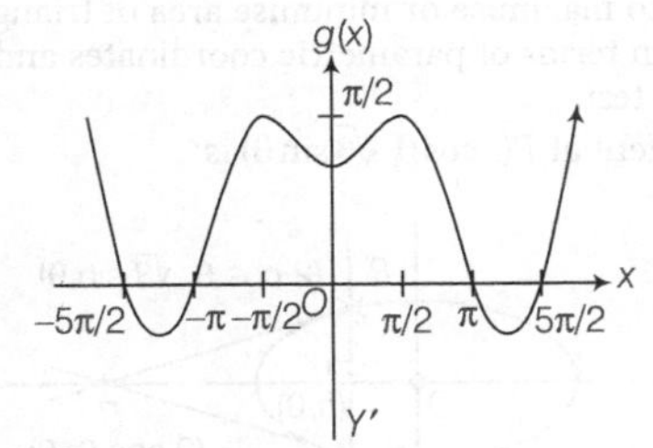

At $x = 0, \dfrac{3\pi}{2}, \dfrac{7\pi}{2}, \ldots, f''(x) > 0$, so minimum

At $x = \dfrac{\pi}{2}, \dfrac{5\pi}{2}, \dfrac{9\pi}{2}, \ldots, f'(x) < 0$, so maximum

So, graph of $f(x)$ and $g(x)$ are shown as

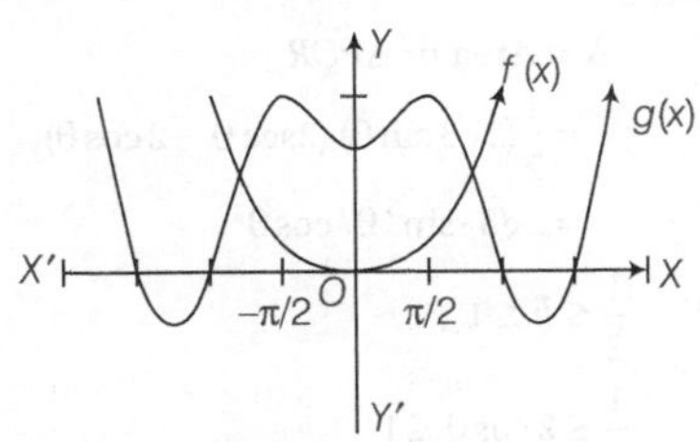

So, number of solutions are 2.

111. Here, $\displaystyle\lim_{x\to 2} \frac{f(x)\cdot g(x)}{f'(x)\cdot g'(x)} = 1$

$\Rightarrow$ $\displaystyle\lim_{x\to 2} \frac{f(x)g'(x) + f'(x)g(x)}{f''(x)g'(x) + f'(x)g''(x)} = 1$ [using L'Hospital's rule]

$\Rightarrow$ $\dfrac{f(2)g'(2) + f'(2)g(2)}{f''(2)g'(2) + f'(2)g''(2)} = 1$

$\Rightarrow$ $\dfrac{f(2)g'(2)}{f''(2)g'(2)} = 1$ $[\because f'(2) = g(2) = 0]$

$\Rightarrow$ $f(2) = f''(2)$ ...(i)

$\therefore$ $f(x) - f''(x) = 0$ for atleast one $x \in R$.

$\Rightarrow$ Option (d) is correct.

Also, $f: R \to (0, \infty)$

$\Rightarrow$ $f(2) > 0$

$\therefore$ $f''(2) = f(2) > 0$ [from Eq. (i)]

Since, $f'(2) = 0$ and $f''(2) > 0$

$\therefore f(x)$ attains local minimum at $x = 2$.

$\Rightarrow$ Option (a) is correct.

112. Given, $f(x) = \displaystyle\int_{\frac{1}{x}}^{x} \frac{e^{-\left(t+\frac{1}{t}\right)}}{t}\,dt$

$$f'(x) = 1\cdot\frac{e^{-\left(x+\frac{1}{x}\right)}}{x} - \left(\frac{-1}{x^2}\right)\frac{e^{-\left(\frac{1}{x}+x\right)}}{1/x}$$

$$= \frac{e^{-\left(x+\frac{1}{x}\right)}}{x} + \frac{e^{-\left(x+\frac{1}{x}\right)}}{x} = \frac{2e^{-\left(x+\frac{1}{x}\right)}}{x}$$

As $f'(x) > 0, \forall x \in (0, \infty)$

$\therefore$ $f(x)$ is monotonically increasing on $(0, \infty)$.

$\Rightarrow$ Option (a) is correct and option (b) is wrong.

Now, $f(x) + f\left(\dfrac{1}{x}\right) = \displaystyle\int_{1/x}^{x} \frac{e^{-\left(t+\frac{1}{t}\right)}}{t}\,dt + \int_{x}^{1/x} \frac{e^{-\left(t+\frac{1}{t}\right)}}{t}\,dt$

$= 0, \forall x \in (0, \infty)$

Now, let $g(x) = f(2^x) = \displaystyle\int_{2^{-x}}^{2^x} \frac{e^{-\left(t+\frac{1}{t}\right)}}{t}\,dt$

$g(-x) = f(2^{-x}) = \displaystyle\int_{2^x}^{2^{-x}} \frac{e^{-\left(t+\frac{1}{t}\right)}}{t}\,dt = -g(x)$

$\therefore$ $f(2^x)$ is an odd function of x on R.

113. **Plan**

We know that, $|x| = \begin{cases} x, & \text{if } x \ge 0 \\ -x, & \text{if } x < 0 \end{cases}$

$\Rightarrow$ $|x - a| = \begin{cases} x - a, & \text{if } x \ge a \\ -(x - a), & \text{if } x < a \end{cases}$

and for non-differentiable continuous function, the maximum or minimum can be checked with graph as

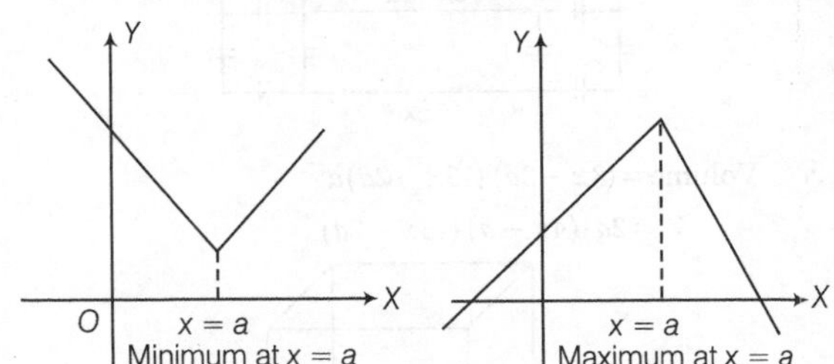

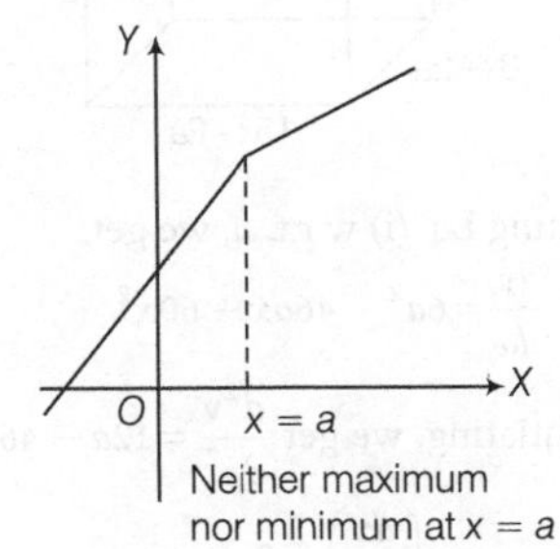

Here, $f(x) = 2|x| + |x + 2| - ||x + 2| - 2|x||$

$$= \begin{cases} -2x - (x + 2) + (x - 2), & \text{if when } x \le -2 \\ -2x + x + 2 + 3x + 2, & \text{if when } -2 < x \le -2/3 \\ -4x, & \text{if when } -\frac{2}{3} < x \le 0 \\ 4x, & \text{if when } 0 < x \le 2 \\ 2x + 4, & \text{if when } x > 2 \end{cases}$$

$$= \begin{cases} -2x-4, & \text{if} \quad x \le -2 \\ 2x+4, & \text{if} \quad -2 < x \le -2/3 \\ -4x, & \text{if} \quad -\frac{2}{3} < x \le 0 \\ 4x, & \text{if} \quad 0 < x \le 2 \\ 2x+4, & \text{if} \quad x > 2 \end{cases}$$

Graph for $y = f(x)$ is shown as

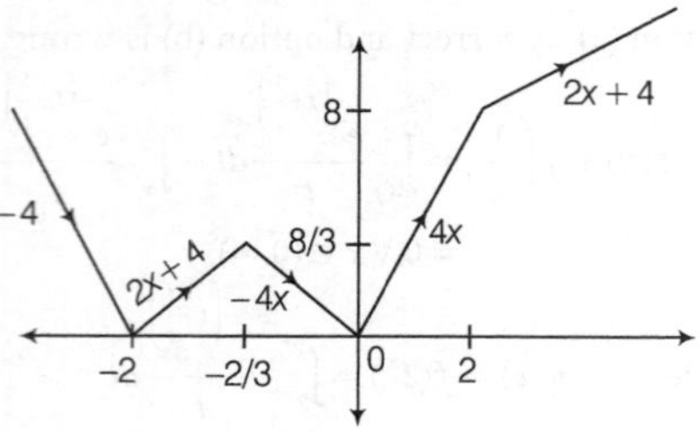

114. Plan The problem is based on the concept to maximise volume of cuboid, i.e. to form a function of volume, say $f(x)$ find $f'(x)$ and $f''(x)$.

Put $f'(x) = 0$ and check $f''(x)$ to be +ve or −ve for minimum and maximum, respectively.

Here, $\quad l = 15x - 2a,\ b = 8x - 2a$ and $h = a$

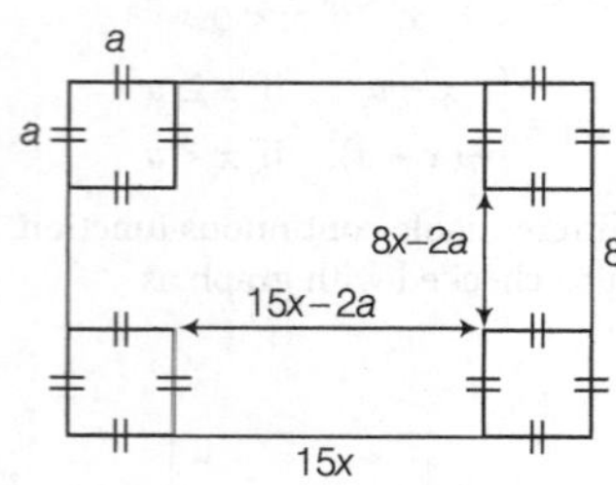

$\therefore \quad$ Volume $= (8x - 2a)(15x - 2a)a$

$$V = 2a \cdot (4x - a)(15x - 2a) \qquad \text{...(i)}$$

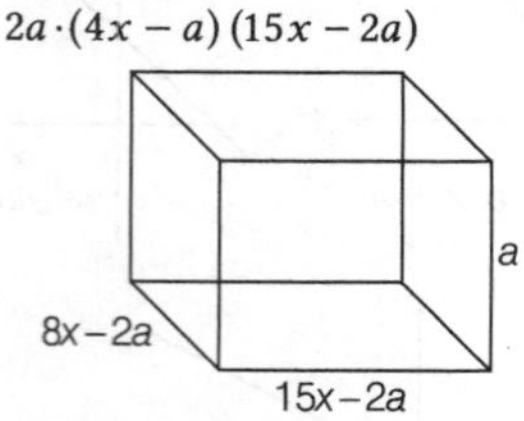

On differentiating Eq. (i) w.r.t. a, we get

$$\frac{dv}{da} = 6a^2 - 46ax + 60x^2$$

Again, differentiating, we get $\dfrac{d^2v}{da^2} = 12a - 46x$

Here, $\quad \left(\dfrac{dv}{da}\right) = 0$

$\Rightarrow \quad 6x^2 - 23x + 15 = 0$

At $a = 5,\ x = 3, \dfrac{5}{6}$

$\Rightarrow \quad \left(\dfrac{d^2v}{da^2}\right) = 2(30 - 23x)$

At $x = 3, \left(\dfrac{d^2v}{da^2}\right) = 2(30 - 69) < 0$

$\therefore$ Maximum when $x = 3$, also at $x = \dfrac{5}{6}$

$\Rightarrow \quad \left(\dfrac{d^2v}{da^2}\right) > 0$

$\therefore$ At $x = 5/6$, volume is minimum.

Thus, sides are $8x = 24$ and $15x = 45$

115. Plan As to maximise or minimise area of triangle, we should find area in terms of parametric coordinates and use second derivative test.

Here, tangent at $P(2\cos\theta, \sqrt{3}\sin\theta)$ is

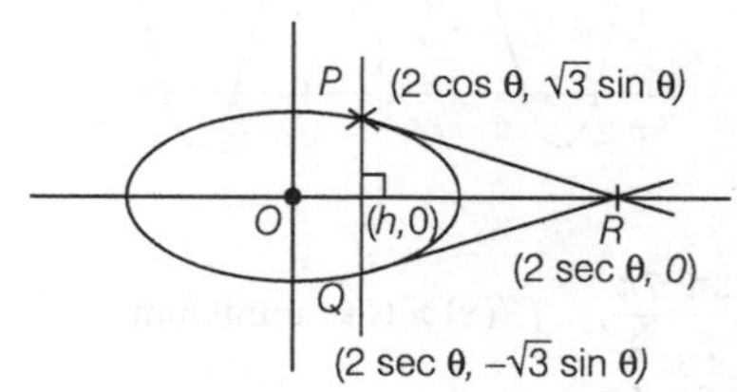

$$\frac{x}{2}\cos\theta + \frac{y}{\sqrt{3}}\sin\theta = 1$$

$\therefore \quad R(2\sec\theta, 0)$

$\Rightarrow \quad \Delta =$ Area of ΔPQR

$$= \frac{1}{2}(2\sqrt{3}\sin\theta)(2\sec\theta - 2\cos\theta)$$

$$= 2\sqrt{3} \cdot \sin^3\theta/\cos\theta \qquad \text{...(i)}$$

Since, $\quad \dfrac{1}{2} \le h \le 1$

$\therefore \quad \dfrac{1}{2} \le 2\cos\theta \le 1$

$\Rightarrow \quad \dfrac{1}{4} \le \cos\theta \le \dfrac{1}{2} \qquad \text{...(ii)}$

$$\therefore \frac{d\Delta}{d\theta} = \frac{2\sqrt{3}\{\cos\theta \cdot 3\sin^2\theta\cos\theta - \sin^3\theta(-\sin\theta)\}}{\cos^2\theta}$$

$$= \frac{2\sqrt{3}\cdot\sin^2\theta}{\cos^2\theta}[3\cos^2\theta + \sin^2\theta] = \frac{2\sqrt{3}\sin^2\theta}{\cos^2\theta}\cdot[2\cos^2\theta + 1]$$

$$= 2\sqrt{3}\tan^2\theta(2\cos^2\theta + 1) > 0$$

When $\quad \dfrac{1}{4} \le \cos\theta \le \dfrac{1}{2}$,

$\therefore \quad \Delta_1 = \Delta_{max}$ occurs at $\cos\theta = \dfrac{1}{4} = \left(\dfrac{2\sqrt{3}\cdot\sin^3\theta}{\cos\theta}\right)$

When $\cos\theta = \dfrac{1}{4} = \dfrac{45\sqrt{5}}{8}$

$\quad \Delta_2 = \Delta_{min}$ occurs at $\cos\theta = \dfrac{1}{2} = \left(\dfrac{2\sqrt{3}\sin^3\theta}{\cos\theta}\right)$

When $\cos\theta = \dfrac{1}{2} = \dfrac{9}{2}$

$\therefore \quad \dfrac{8}{\sqrt{5}}\Delta_1 - 8\Delta_2 = 45 - 36 = 9$

116. Given function, $f(x) = e^{x^2} + e^{-x^2}$, $g(x) = xe^{x^2} + e^{-x^2}$ and $h(x) = x^2e^{x^2} + e^{-x^2}$ are strictly increasing on [0, 1].
Hence, at $x = 1$, the given function attains absolute maximum all equal to $e + 1/e$.

$\Rightarrow \quad a = b = c$

117. Given, $f(x) = \begin{cases} (2+x)^3, & \text{if } -3 < x \le -1 \\ x^{2/3}, & \text{if } -1 < x < 2 \end{cases}$

$\Rightarrow \quad f'(x) = \begin{cases} 3(x+2)^2, & \text{if } -3 < x < -1 \\ \frac{2}{3}x^{-\frac{1}{3}}, & \text{if } -1 < x < 2 \end{cases}$

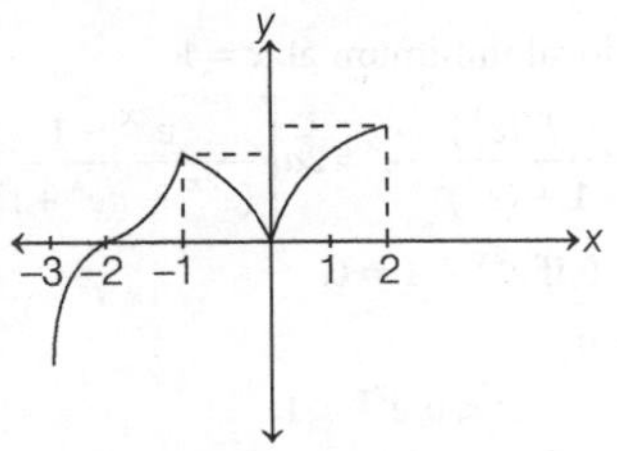

Clearly, $f'(x)$ changes its sign at $x = -1$ from +ve to –ve and so $f(x)$ has local maxima at $x = -1$.
Also, $f'(0)$ does not exist but $f'(0^-) < 0$ and $f'(0^+) > 0$. It can only be inferred that $f(x)$ has a possibility of a minima at $x = 0$. Hence, the given function has one local maxima at $x = -1$ and one local minima at $x = 0$.

118. Given, $g(u) = 2\tan^{-1}(e^u) - \frac{\pi}{2}$ for $u \in (-\infty, \infty)$

$$g(-u) = 2\tan^{-1}(e^{-u}) - \frac{\pi}{2} = 2(\cot^{-1}(e^u)) - \frac{\pi}{2}$$

$$= 2\left(\frac{\pi}{2} - \tan^{-1}(e^u)\right) - \frac{\pi}{2}$$

$$= \pi/2 - 2\tan^{-1}(e^u) = -g(u)$$

$\therefore \quad g(-u) = -g(u)$

$\Rightarrow$ $g(u)$ is an odd function.

We have, $g(u) = 2\tan^{-1}(e^u) - \pi/2$

$$g'(u) = \frac{2e^u}{1 + e^{2u}}$$

$g'(u) > 0, \forall x \in R$ $\quad [\because e^u > 0]$

So, $g'(u)$ is increasing for all $x \in R$.

119. Let $f(x) = bx^2 + ax + c$

$\because \quad f(0) = 0 \Rightarrow c = 0$

and $\quad f(1) = 1 \Rightarrow a + b = 1$

$\therefore \quad f(x) = ax + (1 - a)x^2$

Also, $f'(x) > 0$, for $x \in (0, 1)$

$\Rightarrow \quad a + 2(1 - a)x > 0$

$\Rightarrow \quad a(1 - 2x) + 2x > 0$

$\Rightarrow \quad a > \frac{2x}{2x - 1} \Rightarrow 0 < a < 2$, since $x \in (0, 1)$

$\therefore \quad f(x) = ax + (1 - a)x^2$, $0 < a < 2$

120. Given, $\quad f(x) = x^3 + bx^2 + cx + d$

$\Rightarrow \quad f'(x) = 3x^2 + 2bx + c$

As we know that, if $ax^2 + bx + c > 0, \forall x$, then $a > 0$ and $D < 0$.

Now, $\quad D = 4b^2 - 12c = 4(b^2 - c) - 8c$

[where, $b^2 - c < 0$ and $c > 0$]

$\therefore \quad D = (-\text{ve})$ or $D < 0$

$\Rightarrow f'(x) = 3x^2 + 2bx + c > 0 \, \forall x \in (-\infty, \infty)$ [as $D < 0$ and $a > 0$]

Hence, $f(x)$ is strictly increasing function.

121. Given $f(x) = x^2 + 2bx + 2c^2$ and $g(x) = -x^2 - 2cx + b^2$

Then, $f(x)$ is minimum and $g(x)$ is maximum at $\left(x = \frac{-b}{4a} \text{ and } f(x) = \frac{-D}{4a}\right)$, respectively.

$$\therefore \quad \min f(x) = \frac{-(4b^2 - 8c^2)}{4} = (2c^2 - b^2)$$

$$\text{and} \quad \max g(x) = -\frac{(4c^2 + 4b^2)}{4(-1)} = (b^2 + c^2)$$

Now, $\min f(x) > \max g(x)$

$\Rightarrow \quad 2c^2 - b^2 > b^2 + c^2 \Rightarrow c^2 > 2b^2 \Rightarrow |c| > \sqrt{2}|b|$

122. Let $f(x) = 3\sin x - 4\sin^3 x = \sin 3x$

The longest interval in which $\sin x$ is increasing is of length π.
So, the length of largest interval in which $f(x) = \sin 3x$ is increasing is $\frac{\pi}{3}$.

123. Given, $\cot \alpha_1 \cdot \cot \alpha_2 \cdot \ldots \cdot \cot \alpha_n = 1$

$$\Rightarrow \quad \frac{\cos \alpha_1}{\sin \alpha_1} \cdot \frac{\cos \alpha_2}{\sin \alpha_2} \cdot \frac{\cos \alpha_3}{\sin \alpha_3} \cdot \ldots \cdot \frac{\cos \alpha_n}{\sin \alpha_n} = 1$$

Let $\cos \alpha_1 \cdot \cos \alpha_2 \cdot \cos \alpha_3 \cdot \ldots \cdot \cos \alpha_n = k$ ...(i)

and $\quad \sin \alpha_1 \cdot \sin \alpha_2 \cdot \sin \alpha_3 \cdot \ldots \cdot \sin \alpha_n = k$ ...(ii)

Again, on multiplying Eqs. (i) and (ii), we get

$(\cos \alpha_1 \cdot \cos \alpha_2 \cdot \cos \alpha_3 \cdot \ldots \cdot \cos \alpha_n)$
$\times (\sin \alpha_1 \cdot \sin \alpha_2 \cdot \sin \alpha_3 \cdot \ldots \cdot \sin \alpha_n) = k^2$

$$k^2 = \frac{1}{2 \times 2 \times \ldots n \text{ times}}(2\sin \alpha_1 \cos \alpha_1)$$
$$(2\sin \alpha_2 \cos \alpha_2) \ldots (2\sin \alpha_n \cos \alpha_n)$$

$$\Rightarrow \quad k^2 = \frac{1}{2^n}(\sin 2\alpha_1)(\sin 2\alpha_2) \ldots (\sin 2\alpha_n)$$

$$\le \frac{1}{2^n} \sin 2\alpha_i \le 1, \forall 1 \le i < n \Rightarrow k \le \frac{1}{2^{n/2}}$$

124. Given, $\quad f(x) = \begin{cases} e^x, & \text{if } 0 \le x \le 1 \\ 2 - e^{x-1}, & \text{if } 1 < x \le 2 \\ x - e, & \text{if } 2 < x \le 3 \end{cases}$

and $\quad g(x) = \int_0^x f(t)\, dt$

$\Rightarrow \quad g'(x) = f(x)$

Put $\quad g'(x) = 0 \Rightarrow x = 1 + \log_e 2$ and $x = e$.

Also, $\quad g''(x) = \begin{cases} e^x, & \text{if } 0 \le x \le 1 \\ -e^{x-1}, & \text{if } 1 < x \le 2 \\ 1, & \text{if } 2 < x \le 3 \end{cases}$

At $x = 1 + \log_e 2$, $g''(1 + \log_e 2) = -e^{\log_e 2} < 0$, $g(x)$ has a local maximum.

Also, at $x = e$,

$g''(e) = 1 > 0$, $g(x)$ has a local minima.

$\because$ $f(x)$ is discontinuous at $x = 1$, then we get local maxima at $x = 1$ and local minima at $x = 2$.

Hence, (a) and (b) are correct answers.

125. Since, $f(x)$ has local maxima at $x = -1$ and $f'(x)$ has local minima at $x = 0$.

$\therefore \quad f''(x) = \lambda x$

On integrating, we get

$$f'(x) = \lambda \frac{x^2}{2} + c \quad [\because f'(-1) = 0]$$

$$\Rightarrow \quad \frac{\lambda}{2} + c = 0 \Rightarrow \lambda = -2c \quad \text{...(i)}$$

Again, integrating on both sides, we get

$$f(x) = \lambda \frac{x^3}{6} + cx + d$$

$$\Rightarrow \quad f(2) = \lambda\left(\frac{8}{6}\right) + 2c + d = 18 \quad \text{...(ii)}$$

and $$f(1) = \frac{\lambda}{6} + c + d = -1 \quad \text{...(iii)}$$

From Eqs. (i), (ii) and (iii), we get

$$f(x) = \frac{1}{4}(19x^3 - 57x + 34)$$

$$\therefore \quad f'(x) = \frac{1}{4}(57x^2 - 57) = \frac{57}{4}(x-1)(x+1)$$

For maxima or minima, put $f'(x) = 0$

$\Rightarrow \quad x = 1, -1$

Now $\quad f''(x) = \frac{1}{4}(114x)$

At $x = 1$, $f''(x) > 0$, minima

At $x = -1$, $f''(x) < 0$, maxima

$\therefore$ $f(x)$ is increasing for $[1, 2\sqrt{5}]$.

$\therefore$ $f(x)$ has local maxima at $x = -1$ and $f(x)$ has local minima at $x = 1$.

Also, $\quad f(0) = 34/4$

Hence, (b) and (c) are the correct answers.

126. $$f(x) = \frac{(x^2 + ax + 1) - 2ax}{x^2 + ax + 1} = 1 - \frac{2ax}{x^2 + ax + 1}$$

$$f'(x) = -\left[\frac{(x^2 + ax + 1)\cdot 2a - 2ax(2x + a)}{(x^2 + ax + 1)^2}\right]$$

$$= -\left[\frac{-2ax^2 + 2a}{(x^2 + ax + a)^2}\right] = 2a\left[\frac{(x^2 - 1)}{(x^2 + ax + 1)^2}\right] \quad \text{...(i)}$$

$$f''(x) = 2a\left[\frac{(x^2 + ax + 1)^2(2x) - 2(x^2 - 1)(x^2 + ax + 1)(2x + a)}{(x^2 + ax + 1)^4}\right]$$

$$= 2a\left[\frac{2x(x^2 + ax + 1) - 2(x^2 - 1)(2x + a)}{(x^2 + ax + 1)^3}\right] \quad \text{...(ii)}$$

Now, $$f''(1) = \frac{4a(a+2)}{(a+2)^3} = \frac{4a}{(a+2)^2}$$

and $$f''(-1) = \frac{4a(a-2)}{(2-a)^3} = \frac{-4a}{(a-2)^2}$$

$$\therefore \quad (2 + a)^2 f''(1) + (2 - a)^2 f''(-1) = 4a - 4a = 0$$

127. When $x \in (-1, 1)$,

$x^2 < 1$

$\Rightarrow x^2 - 1 < 0$

$\therefore$ $f'(x) < 0$, $f(x)$ is decreasing.

Also, at $\quad x = 1$, $f''(1) = \frac{4a}{(a+2)^2} > 0 \quad [\because 0 < a < 2]$

So, $f(x)$ has local minimum at $x = 1$.

128. $$g'(x) = \frac{f'(e^x)}{1 + (e^x)^2} \cdot e^x = 2a\left[\frac{e^{2x} - 1}{(e^{2x} + ae^x + 1)^2}\right]\left(\frac{e^x}{1 + e^{2x}}\right)$$

$g'(x) = 0$, if $e^{2x} - 1 = 0$,

i.e. $\quad x = 0$

If $\quad x < 0, e^{2x} < 1$

$\Rightarrow \quad g'(x) < 0$

129. To maximize area of $\Delta\, APB$, we know that, $OP = 10$ and $\sin\theta = r/10$, where $\theta \in (0, \pi/2)$ ... (i)

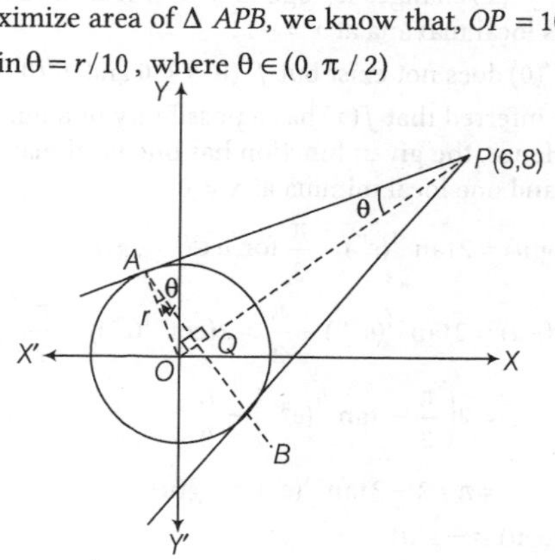

$\therefore \quad$ Area $= \frac{1}{2}(2AQ)(PQ)$

$= AQ \cdot PQ = (r\cos\theta)(10 - OQ)$

$= (r\cos\theta)(10 - r\sin\theta)$

$= 10\sin\theta\cos\theta(10 - 10\sin^2\theta)$ [from Eq. (i)]

$\Rightarrow \quad A = 100\cos^3\theta\sin\theta$

$\Rightarrow \quad \frac{dA}{d\theta} = 100\cos^4\theta - 300\cos^2\theta\cdot\sin^2\theta$

Put $\quad \frac{dA}{d\theta} = 0$

$\Rightarrow \quad \cos^2\theta = 3\sin^2\theta$

$\Rightarrow \quad \tan\theta = 1/\sqrt{3}$

$\Rightarrow \quad \theta = \pi/6$

At which $\frac{dA}{d\theta} < 0$, thus when $\theta = \pi/6$, area is maximum

From Eq. (i), $r = 10\sin\frac{\pi}{6} = 5$ units

130. Let us take a point $P(\sqrt{6}\cos\theta, \sqrt{3}\sin\theta)$ on $\frac{x^2}{6}+\frac{y^2}{3}=1$.

Now, to minimise the distance from P to given straight line $x+y=7$, shortest distance exists along the common normal.

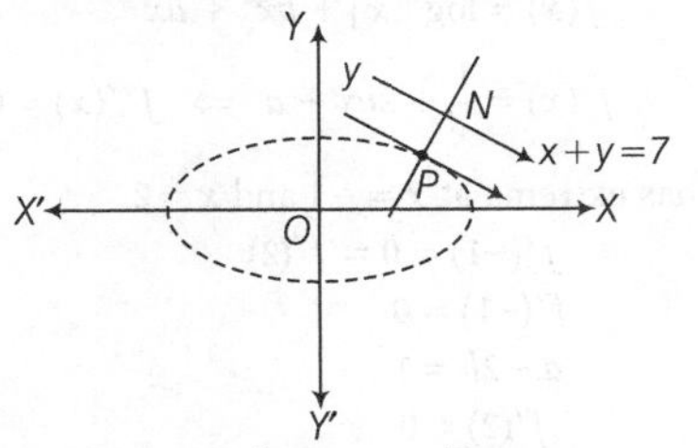

Slope of normal at $P=\frac{a^2/x_1}{b^2/y_1}=\frac{\sqrt{6}\sec\theta}{\sqrt{6}\operatorname{cosec}\theta}=\sqrt{2}\tan\theta=1$

So, $\cos\theta=\sqrt{\frac{2}{3}}$ and $\sin\theta=\frac{1}{\sqrt{3}}$

Hence, required point is $P(2, 1)$.

131. **Plan**

(i) Local maximum and local minimum are those points at which $f'(x)=0$, when defined for all real numbers.

(ii) Local maximum and local minimum for piecewise functions are also been checked at sharp edges.

Description of Situation $y=|x|=\begin{cases} x, & \text{if } x\geq 0 \\ -x, & \text{if } x<0 \end{cases}$

Also, $y=|x^2-1|=\begin{cases} (x^2-1), & \text{if } x\leq -1 \text{ or } x\geq 1 \\ (1-x^2), & \text{if } -1\leq x\leq 1 \end{cases}$

$$y=|x|+|x^2-1|=\begin{cases} -x+1-x^2, & \text{if } x\leq -1 \\ -x+1-x^2, & \text{if } -1\leq x\leq 0 \\ x+1-x^2, & \text{if } 0\leq x\leq 1 \\ x+x^2-1, & \text{if } x\geq 1 \end{cases}$$

$$=\begin{cases} -x^2-x+1, & \text{if } x\leq -1 \\ -x^2-x+1, & \text{if } -1\leq x\leq 0 \\ -x^2+x+1, & \text{if } 0\leq x\leq 1 \\ x^2+x-1, & \text{if } x\geq 1 \end{cases}$$

which could be graphically shown as

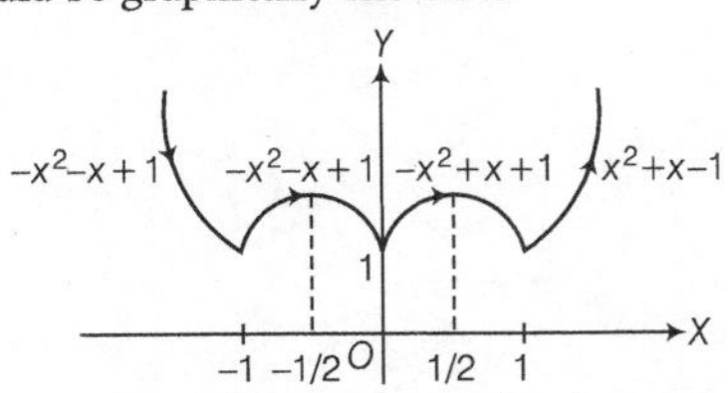

Thus, $f(x)$ attains maximum at $x=\frac{1}{2},\frac{-1}{2}$ and $f(x)$ attains minimum at $x=-1,0,1$.

$\Rightarrow$ Total number of points = 5

132. **Plan** If $f(x)$ is least degree polynomial having local maximum and local minimum at α and β.

Then, $f'(x)=\lambda(x-\alpha)(x-\beta)$

Here, $p'(x)=\lambda(x-1)(x-3)=\lambda(x^2-4x+3)$

On integrating both sides between 1 to 3, we get

$$\int_1^3 p'(x)\,dx=\int_1^3 \lambda(x^2-4x+3)\,dx$$

$$\Rightarrow \quad (p(x))_1^3=\lambda\left(\frac{x^3}{3}-2x^2+3x\right)_1^3$$

$$\Rightarrow \quad p(3)-p(1)=\lambda\left((9-18+9)-\left(\frac{1}{3}-2+3\right)\right)$$

$$\Rightarrow \quad 2-6=\lambda\left\{\frac{-4}{3}\right\} \Rightarrow \lambda=3$$

$$\Rightarrow \quad p'(x)=3(x-1)(x-3)$$

$$\therefore \quad p'(0)=9$$

133. Let $g(x)=e^{f(x)}, \forall x\in R \Rightarrow g'(x)=e^{f(x)}\cdot f'(x)$

$\Rightarrow f'(x)$ changes its sign from positive to negative in the neighbourhood of $x=2009$

$\Rightarrow f(x)$ has local maxima at $x=2009$.

So, the number of local maximum is one.

134. Let $f(\theta)=\frac{1}{\sin^2\theta+3\sin\theta\cos\theta+5\cos^2\theta}$

Again let, $g(\theta)=\sin^2\theta+3\sin\theta\cos\theta+5\cos^2\theta$

$$=\frac{1-\cos 2\theta}{2}+5\left(\frac{1+\cos 2\theta}{2}\right)+\frac{3}{2}\sin 2\theta$$

$$=3+2\cos 2\theta+\frac{3}{2}\sin 2\theta$$

$$\therefore \quad g(\theta)_{\min}=3-\sqrt{4+\frac{9}{4}}=3-\frac{5}{2}=\frac{1}{2}$$

$\therefore$ Maximum value of $f(\theta)=\frac{1}{1/2}=2$

135. Given, $A=\{x\,|\,x^2+20\leq 9x\}=\{x\,|\,x\in[4,5]\}$

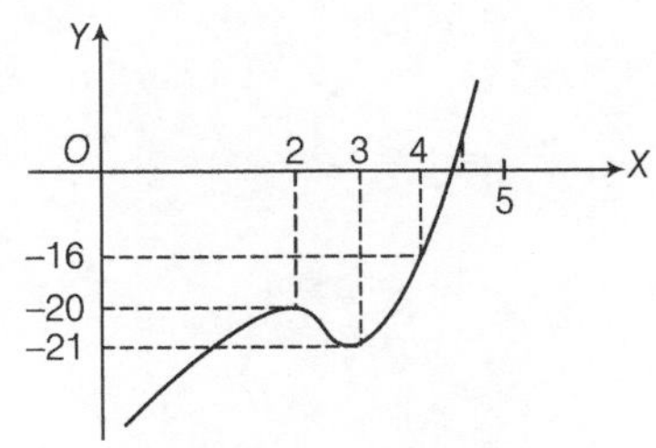

Now, $f'(x)=6(x^2-5x+6)$

Put $f'(x)=0 \Rightarrow x=2, 3$

$f(2)=-20, f(3)=-21, f(4)=-16, f(5)=7$

From graph, maximum value of $f(x)$ on set A is $f(5)=7$.

136. According to given information, we have

Perimeter of square + Perimeter of circle = 2 units

$$\Rightarrow \quad 4x+2\pi r=2$$

$$\Rightarrow \quad r=\frac{1-2x}{\pi} \quad \text{...(i)}$$

Now, let A be the sum of the areas of the square and the circle.

Then, $A=x^2+\pi r^2=x^2+\pi\frac{(1-2x)^2}{\pi^2}$

$$\Rightarrow \quad A(x)=x^2+\frac{(1-2x)^2}{\pi}$$

Now, for minimum value of $A(x)$, $\frac{dA}{dx} = 0$

$$\Rightarrow \quad 2x + \frac{2(1-2x)}{\pi}\cdot(-2) = 0 \Rightarrow x = \frac{2-4x}{\pi}$$

$$\Rightarrow \quad \pi x + 4x = 2$$

$$\Rightarrow \quad x = \frac{2}{\pi + 4} \quad \text{...(ii)}$$

Now, from Eq. (i), we get

$$r = \frac{1 - 2\cdot\frac{2}{\pi+4}}{\pi} = \frac{\pi + 4 - 4}{\pi(\pi+4)} = \frac{1}{\pi + 4} \quad \text{...(iii)}$$

From Eqs. (ii) and (iii), we get $x = 2r$

137. Here, $x = -1$ and $x = 2$ are extreme points of $f(x) = \alpha \log|x| + \beta x^2 + x$, then

$$f'(x) = \frac{\alpha}{x} + 2\beta x + 1$$

$$\therefore \quad f'(-1) = -\alpha - 2\beta + 1 = 0 \quad \text{...(i)}$$

[at extreme point, $f'(x) = 0$]

$$\text{and} \quad f'(-2) = \frac{\alpha}{2} + 4\beta + 1 = 0 \quad \text{...(ii)}$$

On solving Eqs. (i) and (ii), we get

$$\alpha = 2, \beta = -\frac{1}{2}$$

138. Given

(i) A function f, such that

$$f(x) = \log|x| + bx^2 + ax, x \neq 0$$

(ii) The function f has extrema at $x = -1$ and $x = 2$, *i.e.,* $f'(-1) = f'(2) = 0$ and $f''(-1) \neq 0 \neq f''(2)$.

Now, given function f is given by

$$f(x) = \log|x| + bx^2 + ax$$

$$\Rightarrow \quad f'(x) = \frac{1}{x} + 2bx + a \Rightarrow f''(x) = \frac{-1}{x^2} + 2b$$

Since, f has extrema at $x = -1$ and $x = 2$.

Hence, $f'(-1) = 0 = f'(2)$

$$f'(-1) = 0$$

$$\Rightarrow \quad a - 2b = 1 \quad \text{...(i)}$$

$$\text{and} \quad f'(2) = 0$$

$$\Rightarrow \quad a + 4b = \frac{-1}{2} \quad \text{...(ii)}$$

On solving Eqs. (i) and (ii), we get

$$a = \frac{1}{2} \text{ and } b = \frac{-1}{4}$$

$$\Rightarrow \quad f''(x) = \frac{-1}{x^2} + \frac{-1}{2} = -\left(\frac{x^2+2}{2x^2}\right)$$

$\Rightarrow f''(-1) < 0$ and $f''(2) < 0$

Hence, f has local maxima at both $x = -1$ and $x = 2$.

Hence, Statement I is correct.

Also, while solving for Statement I, we found the values of a and b, which justify that Statement II is also correct.

However, Statement II does not explain Statement I in any way.